The

DICTIONARY
of the BIOLOGICAL
SCIENCES

The author is particularly indebted to the
following scholars who have
reviewed this work in manuscript:

DR. LEE F. BRAITHWAITE
    Brigham Young University
DR. ROBERT H. BURRIS
    University of Wisconsin
DR. EARL L. CORE
    University of West Virginia
DR. KENNETH W. CUMMINS
    University of Pittsburgh
DR. MARJORIE A. DARKEN
    Lederle Laboratories
DR. DAVID E. DAVIS
    Pennsylvania State University
DR. RICHARD M. FOX
    Carnegie Museum
DR. FREDERICK J. GOTTLIEB
    University of Pittsburgh
DR. HORTON H. HOBBS
    The Smithsonian Institution
DR. MALCOLM JOLLIE
    Northern Illinois University
DR. L. L. LANGLEY
    National Institutes of Health
DR. CLARENCE J. McCOY, JR.
    Carnegie Museum
DR. BURT L. MONROE, JR.
    University of Louisville
DR. LEONARD P. SCHULTZ
    The Smithsonian Institution
DR. NELSON T. SPRATT
    University of Minnesota
DR. IAN SUSSEX
    Yale University
DR. GEORGE WALLACE
    Carnegie Museum
DR. G. B. WILSON
    Michigan State University

The

DICTIONARY
of the **BIOLOGICAL SCIENCES**

PETER GRAY

Andrey Avinoff Professor of Biology
University of Pittsburgh

 VAN NOSTRAND REINHOLD COMPANY

NEW YORK   CINCINNATI   ATLANTA   DALLAS   SAN FRANCISCO
LONDON   TORONTO   MELBOURNE

Van Nostrand Reinhold Company Regional Offices:
New York   Cincinnati   Chicago   Millbrae   Dallas

Van Nostrand Reinhold Company International Offices:
London   Toronto   Melbourne

Copyright © 1967 by Litton Educational Publishing, Inc.
Library of Congress Catalog Card Number: 67-24690
ISBN: 0-442-15590-5

Published by Van Nostrand Reinhold Company
450 West 33rd Street, New York, N.Y. 10001

Published simultaneously in Canada by
Van Nostrand Reinhold Ltd.

16   15   14   13   12   11   10

To

KENNETH T. RICHARDSON, M.D.

without whose

generous, skillful, and continued help

this work

could never have been completed

PREFACE

*"The preface of a book affords the author an opportunity of speaking to his reader in a comparatively direct and personal manner, and of acquainting the prospective user of the book with the considerations which impelled the author to write it."**

I little imagined, when I acquired my first degree nearly 40 years ago, that I would ultimately become both an encyclopedist and a lexicographer; yet the one derives directly from the other. It was the infeasibility of indexing the "Encyclopedia of the Biological Sciences" in a manner that would permit enough individual words to be found that led me to the conviction that a separate dictionary was a necessity.

The task of the lexicographer is in many ways more enjoyable than that of the encyclopedist. It is true that he does a great deal more work, but he is also far more free to be opinionated, stubborn, prejudiced and, indeed, to exhibit all those human traits that the editor of an encyclopedia must suppress in himself while encouraging in others. He may even, secure in the knowledge that this sin against contemporary scholarship will be forever lost in a mass of verbiage, permit himself an occasional moment of humor.

My colleagues frequently asked "How do you compile a dictionary?" The answer is perfectly simple. You get together a large library and then go through, page by page, several hundred selected volumes, entering words, and their apparent meanings, on index cards that are then filed in alphabetical order. After three years—the first year is quite enjoyable—you find yourself with about 60,000 cards which are the raw material from which a dictionary might be developed. The next stage is to laboriously go through these cards, eliminating duplications and resolving conflicting meanings by reference to the original literature and to other books not previously consulted. At this stage of the game, it became apparent

that a thesauric method of entry, grouping together words derived from identical roots or developed from the same noun, would be as useful to the reader of the published work as it would be to me. It is, for example, just as infuriating when one wishes to check whether "cribiform" is indeed synonymous with "madreporiform," or to find whether "Stanley crane" is just another name for "blue crane," to have to pass from one end of a bank of filing cabinets to the other, as it is to turn 400 pages of a dictionary. The cards were accordingly rearranged in this manner, the advantages of which are more fully explained in the Introduction and briefly indicated in the Explanatory Note.

You next have to go through the indices of many books, and through numerous specialized dictionaries and glossaries, seeking errors of omission and comission. At the end you will have about 45,000 cards, a completely frayed nervous system, and a deep doubt as to whether it is worth continuing which conflicts with a feeling that you are too deeply enmeshed to be able to escape. Moreover, your troubles, as distinct from your labors, are only just starting.

Neither I, nor my publishers, at any time suffered from the delusion that I had a specialized knowledge of every branch of biology. We were in complete agreement that many specialists should review those entries in the manuscript that pertained to their fields. The difficulty was to find a method of doing this that would combine efficiency with economy. We ended by typing the first draft manuscript, with marginal key numbers indicating subject relevance, on stencils and duplicating an edition of thirty copies.

*"The Bookman's Glossary," 3rd ed., New York, Bowker, [c. 1951]. Quoted by permission of the R. R. Bowker Co.

It was next necessary to find nearly a score of reviewers. I very speedily became convinced of the continued truth of the statement I made in the preface to the "Encyclopedia of the Biological Sciences" that "biologists the world over are the most scholarly, the most courteous, and the most cooperative group with which any person could be priviliged to associate." This is not to say that they always agreed with me or with each other, or that they were invariably correct. I was not infrequently embarrassed to find the generic and specific names by which I had defined an organism replaced, on various sheets, by two other generic and three other specific names. The disagreements in English were as frequent as in Latin. One reviewer, though not a mammalogist, irately deleted the word "numbat" and substituted "wombat," indicating his annoyance at my stupidity in a searing marginal comment. Reference to the present work, or for that matter to Webster, will show how embarrassing the retention of this correction would have been for both of us. All this should emphasize both the care taken by the reviewers and the fact that any residual inaccuracies in this volume are wholly my fault.

Since this, as the first edition of any dictionary, will contain numerous misspellings, it is of interest to examine the reasons, apart from fatigue (to which lexicographers are unduly subject) and stupidity (of which I do not think they have more than their fair share), that account for these errors; the more so as modern improvements in technology and education have markedly increased both the production and the retention of faulty spelling. When the first edition of Jackson's "Glossary of Botanic Terms" was published in 1900, it passed directly from manuscript to the type of the compositor, whose inability to make out the author's handwriting was the chief cause of a not infrequent confusion, for example, between "tt," "th," tl," "pl" and "pr." Remnants of these misunderstandings remain in the printing of 1953, the most recent that I have seen. Many of the entries in the present work were, however, voice recorded on magnetic tape, and it will surprise etymologists less than biologists that this produced numerous exchanges between "b" and "v," between "t" and "p," and between "f" and "s"; also, of course, my handwritten cards were as difficult for my typist as Jackson's manuscript was for his printer.

One of the reasons why all these errors were not immediately seen and corrected by me, my proofreader, the printer's proofreader or the publisher is the method of reading taught to our generation. Those of us who do much reading employ the method of sentence, rather than word, perception. Thus, embedded in a sentence of obvious meaning, the words "diminuitive" and "interstinal" look just as good as the real thing, particularly when seen on a photo-set proof in 8-point type. I must, however, agree with Lord Palmerston's remark about another matter, "This, Sir, may be ample explanation but it is insufficient excuse." I would be most grateful if those who detect errors of spelling or, what I believe to be much rarer, errors of fact would, in their charity, send me a postcard so that the second edition will have fewer blemishes than I fear may be found in the first.

I would be equally grateful if specialists who become annoyed by my ignorance of their speciality would be equally charitable. I offer as a shining example a remark of one of the most ruthless of the reviewers who, cheerfully wielding the blunt pencil which is his trademark, said "You know, considering how little you know about the subject, I think you've done remarkably well." I could not wish for a better scholarly epitaph.

PETER GRAY

Pittsburgh, Pa.
April, 1967

INTRODUCTION

The need for a "Dictionary of the Biological Sciences" became apparent to me when I was preparing the "Encyclopedia of the Biological Sciences" (New York, Reinhold Publishing Corp., 1961). There are numerous taxa and technical terms that are not important enough to warrant an entry in an encyclopedia but that require a brief dictionary definition, and there are also many hundreds of descriptive terms which should be defined but for which no justifiable place could be found in the necessarily restricted index to an encyclopedia.

SCOPE OF THIS WORK

This work iş not, as its name might indicate, a complete account of all words used by the English-speaking peoples in the description of organisms, their anatomy, function and mutual interactions.

In the first place, many words used by biologists are in general use in the English language, or their meaning is self-evident. For example, space is not wasted in defining such terms as "pre-anal" and "post-anal" since these quite obviously mean, respectively, "in front of" and "behind" the anus.

There are, however, many cases in which ordinary English words are used in a sense very different from that generally understood. A glance at the definitions assigned to the words "aesthete," "cognate," "connivent," "innate," "literate," "neophyte" (under-phyte) and "voluble" will make this distinction clear.

At the other end of the scale are many technical terms, mostly classic or pseudo-classic compounds, which are so highly specialized as to be of no interest to the general biologist. There is, for example, a vocabulary of more than 200 words devoted entirely to the anatomy of the larvae of carabid beetles, while Tuxen 1956* lists some 1500 words used exclusively to de-

*Full bibliographical references to this, and other works mentioned in the Introduction, will be found in the "List of Works Consulted" on page 599.

scribe the genitalia of insects. I have selected from these specialized terms those in general usage, but I have omitted those that are of interest only to a specialist working in a restricted field who will, in any case, have derived his vocabulary from the original literature.

A major difficulty has been to determine who are the "English-speaking peoples" and what are "English" words. It may be pointed out that, in addition to the United Kingdom and the greater part of North America, English is spoken, or at least used as a language of scholarly communication, in considerable areas of Africa and Asia, in most of the Australasian zone, and in many parts of the West Indies. It is obvious that the word "kangaroo" is as English as "mongoose" even though both these terms are gross corruptions from other languages. It is not so generally realized that "kauri" is as commonly used in New Zealand as is "oak" in Great Britain or North America and must therefore be accepted as English. The most difficult questions come from such more or less bilingual localities as Quebec, New Mexico and South Africa. I have been guided, in my selection of words from these areas, by a consideration of whether two forms are currently in use. Thus, in Quebec, "*oiseau de pluie,*" as the pileated woodpecker is locally known, is obviously a French term and cannot be included in the present work. In South Africa, however, the word "bok" is scarcely ever used interchangeably with its English equivalent "buck," and I have therefore included such terms as "gemsbok" since this is the word that an English-speaking individual would use to describe the animal in question.

The position with regard to Latin is more complex. The word "cranium" is indubitably now a part of the English language, but the word "aedeagus" is not. I have been forced, however, to include the latter since it is a technical term for which there is no exact English equivalent. In addition to these general considerations, a few words must be said of the nature of the terms included.

Taxa

This work contains the definitions or, where necessary, the synonymy of nearly all taxa of ordinal rank and above, and the great majority of families. These are defined in the customary biological terms and, where possible, reference is made to the English or vernacular names of the animals or plants involved. Genera and species are not the subject of main entries, save in the case of a few horticultural misnomers (*e.g.*, "Nasturtium") in which the Latin name of one genus is taken to be the English name of another.

Vernacular Names of Organisms

There appears no limit to the names that the English-speaking peoples have bestowed on plants and animals native to the countries in which they live. Many of these names occur only in local dialects and cannot be included in a general dictionary. I have been guided in my selection of names by the usage in contemporary books native to the countries involved, though I have included a few obsolete terms.

The confusion between vernacular names in different parts of the world is almost incredible, as witness the multitude of birds to which the name "robin" has been applied. In this section, I have had constantly in mind the needs of the unfortunate foreigner endeavoring to cope with these terms as well as the needs of native speakers, and I have attempted to give some precision by assigning Latin binomials rather than vague terms as definitions. The need for this is well exemplified by the fact that any standard French-English dictionary will inform the reader that the English word for "*bec figue*" is "fig-eater." In spite of the fact that the "*bec figue*" is a bird, most contemporary American dictionaries define "fig-eater" as a "scarabaeid beetle." As an example of the necessity for obsolete terms, I might add that I once sought the English name of a fish in a Russian-English dictionary and was rewarded with the information that the animal in question was called, in English, a "dorse."

There is a strong feeling, on the part of many taxonomists, that vernacular names should have the same degree of precision established for Latin binomials. This can never be, since vernacular names are established by local custom and not by international rules. There is some justification for the publication of lists of "correct" vernacular names when the organisms in question are the object of commercial transactions. Thus, Kelsey and Dayton (1942) provides a means of communication between nurseryman and gardener, and Lafoon (1960) lists principally insects of economic importance and is thus of value to manufacturers of pesticides in establishing the accuracy of their claims. It is difficult to take so kindly a view of the American Fisheries Society's 1960 list since only a very small fraction of the "correct" vernacular names apply to organisms of interest to either commercial or recreational fisherman. It is, for example, difficult to believe that the correct English names of all 91 American darters are commonly used by the local inhabitants of the areas in which they occur. Moreover, no amount of academic prodding will persuade a local fisherman that he should not call a wall-eyed pike a pickerel, if the latter is the name by which it is locally known. Indeed, in this list, commerce may be too well served since the Society's official endorsement of the practice of calling a Pacific pleuronectid the "Dover Sole" permits unscrupulous restaurateurs to serve under that name a fish both zoologically and gastronomically very different from the European epicure's *Solea solea*.

There can be no possible objection to the use of anglicized forms of Latin names as English words, provided that they are not capitalized and that the simplified spelling often inherent in the transfer is not used in the original. Thus paramecium and ameba are widely used in a far less restrictive sense (*e.g.*, "the amebas living in man") than that which must be retained for Paramoecium and Amoeba. An endeavor to simplify the spelling of Latin words, while retaining them in that language, results in nothing but confusion.

Descriptive Terms

I have endeavored to confine the descriptive terms in this dictionary to those which either are peculiar to biology or have a peculiar meaning in that science. Such terms as "giant" and "dwarf" are not separately defined although they not infrequently occur as parts of compound names of organisms. In most cases, I have not given both the substantive and the adjectival form of the word, though I have often indicated, when defining the root, how these forms are constructed. In most cases, alternate adjectival forms, usually ending in "-ous" or "-ic" are synonymous and definitions are given under one with a reference from the other. Adjectival forms are, however, given when compounds are pendent from them, and I have given both adjectival forms (*e.g.*, metabolic, metabolous) when a distinction in meaning is generally accepted.

Roots

Several thousand of the more common roots from which biological terms are constructed have been included in this dictionary, though the confusion between "c" and "k," and between the aspirated and unaspirated Greek "iota," is as infuriating as it is universal. An individual will cheerfully refer to a cinematographic record of a kinetic phenomenon, apparently without realizing that he is violating every rule of logic and common sense. The second of the two confusions cited makes it difficult to realize that "histology" and "meristem" derive from the same root which, in point of fact, means a web-like structure, as sections of cells appeared when first seen under the microscope.

Personal Names

Names and dates of individuals are given only when they have produced an eponymous adjective, the origin of which may be of interest.

Latin and Greek

It has already been pointed out that it is difficult to determine when a word of classic origin becomes English, when it is a classic word used as an English word, or when it is a classic word unnecessarily used to replace an English word. As this is an English dictionary, I have omitted the third of the three classes mentioned. I have also disregarded such long, descriptive Latin terms as "*extensor pollicis longus et indicus*"; the proper place to seek the meaning of this is an anatomical dictionary.

It is unfortunate that there are many biologists who suffer from the delusion that a classic or pseudo-classic word is less ambiguous than its English equivalent. Actually there are as many ambiguities in Latin and Greek as in English, so that the casual use of the classic tongues by those ill-acquainted with them has led to much confusion. For example, the word *retinaculum* (*pl. retinaculi*) has a completely different meaning from the word *retinacula* (*pl. retinaculae*), yet even so normally accurate a volume as Webster's Third New International Dictionary gives the plural of *retinaculum* as *retinacula* and proceeds to group under this one nonexistent word all those technical terms properly described by the two.

Moreover, the invention of pseudo-classic monstrosities frequently obscures a meaning that was plain in English. There are, for example, some plants that are fertilized by carrion beetles. This statement is perfectly clear, particularly if made in England, where the term "carrion beetle" has a restricted meaning. Unfortunately, the clarity of the English has not prevented the appearance of the word "necroentomophilous." To be a "lover-of-insecteaters-of-the-dead"—which is apparently what the woolly minded jargoneer who invented this word supposed it to mean—is a very different thing from being fertilized by a carrion beetle.

Jargon

Jargon is defined, in the volume by Webster just cited, as "the technical terminology ...of specialists...in a particular...area of knowledge. Often a pretentious or unnecessarily obscure and esoteric terminology." This is an admirable definition of a great deal of contemporary biological writing. Indeed, much biological jargon, as that of small boys at school and of thieves in their haunts, is apparently designed to conceal from outsiders the meaning of the subject under discussion. This is achieved either by inventing a new word to replace one already well known or by using an existing word in a completely different meaning. The former case is well illustrated by the inclusion of the word "geotome" in Hanson (1962), particularly as the definition there given is so loose as to leave one in some doubt as to whether a "spade" or a "trowel" is intended. The second type of obfuscation is exemplified by the definition of "jetsam" in Carpenter (1962), where this word is said to mean "the area of the beach where miscellaneous floating material is washed up and deposited . . . = flotsam." This sort of thing not only has Eleanor of Aquitaine a'birling in her grave but must also cause foreigners well acquainted with the language to have some doubt as to the degree of scholarship exhibited by contemporary English-speaking biologists. The fact that I have drawn two examples from ecology should not be taken to mean that I regard them as worse than any other branch of biologists in this regard. The customary use of "sacrifice," when "kill" is intended, is a pretentious stupidity for which no excuse exists.

Plurals

Plurals have not been recorded in this work unless they are irregular (*scutum, scuti*), both irregular and confusing (*retinaculum, retinaculi*; *retinacula, retinaculae*), or of a form (*mos, mores*; *salpinx, salpinges*) likely to occasion difficulty to one untrained in the classic tongues. I cannot, however, pass this opportunity to express a strong personal hope that Latin words terminating in -us, taken into English, should be regarded as collective terms, and thus both singular and plural in their own right. "A field of narcissus" is as grateful to the ear as it is to the eye; "a field of narcissuses" is a hideously sibilant cacophony. I have a colleague who insists on referring to the "peripatusses," by which he means the Onycophora, or velvet worms, under the honest impression that this multiplication of "s's" confers legitimate citizenship. I cannot help wondering whether the English word "pal" would become a citizen of France were one to pretend that its plural was "paux."

A curious custom, the origin of which I do not know, is the insistence of American ichthyologists that several of one species are fish, but that several of more than one species are fishes. This distinction is not noted in any major dictionary but is observed in this volume.

Hyphenation and Compounding

The hyphenation and compounding of English words is more the result of custom than of rule. I have, however, followed the principles laid down by Ball (1951) where these were not at variance with an expressed preference of any of the expert reviewers.

ARRANGEMENT OF MATERIAL IN THIS BOOK

Early in the compilation of this volume, it became apparent to me that there was as much need for a biological thesaurus, in which words were arranged according to their meaning, as there was for a biological dictionary in which words must be arranged according to the letters of which they are composed. There is certainly not room for two volumes, so that the present work endeavors to resolve this problem by arranging, in dictionary manner, words that are likely to be sought for their own sake as words, but reverting to a thesauric style for those that will be sought primarily for their meaning. An individual, for example, seeking a taxon will be most readily led to it by a strict form of alphabetization. The same reasoning applies to a mutant or gene symbol. If, however, it is desired to find out the meaning of a word ending in "-anthous" it seems to me far more logical to have all the words so ending placed together. If, for example, one has forgotten the antithesis of "homeothermic," one can be reasonably sure that it also ends in "-thermic," but there is no possible reason to seek under the "P's" where the correct word would be found were it alphabetized by its initial. These points are well illustrated by the several pages of entries following the definition of -gen-. The main classes of entry are therefore alphabetized as follows:

Taxa

All taxa start with an initial capital and are arranged in alphabetical order.

Mutants and Gene Symbols

Gene symbols, conventionally capitalized where necessary to indicate dominance, are arranged in alphabetical order. Names of mutants are placed in brackets to distinguish them from definitions of the same word, not used as the name of a mutant (*e.g.*, anthocyanin, [anthocyanin]), and are also arranged in alphabetical order.

Vernacular Names of Organisms

These are alphabetized according to the principal part of the word. For example, the "brittle star," the "sea star" and the "sun star" are all found together, since this seems a far more logical arrangement than to place the "sun star" with the "sun flower" and the "sea star" with the "sea weed." Even when the principal part of the name is compound (*e.g.*, "hornbill") the principal part of the compound (*i.e.*, "-bill") is used for alphabetizing, so that words of similar meaning, or names derived from the same structures, come in the same place. Numerous cross-references have been made at points of ambiguity. Similarly all the "ants," "trees," "weeds," etc., are placed each with their own kind.

Chemical Terms

The chemical and biochemical terms given in this dictionary are alphabetized according to the same system as that used for the vernacular names of organisms. That is, all the "acids" are together, as are also all the "aminases" and the like, for the reason that an individual is far more likely to wish to enquire whether a certain type of "aminase" is known than to seek an unknown "aminase" under a prefix with which he may not be acquainted.

Descriptive Terms

All descriptive terms in this dictionary are entered in thesauric style according to the operative portion of the word. Thus, for example, all of the multitudinous words ending in "-form" are to be found under "-form" and not scattered all over the dictionary. This seems to require little apology since it not only serves to bring together all words of basically similar meaning, but may, hopefully, prevent the continued invention of new words for those which already exist. The reader is reminded that, in the case of the few compound Latin terms given, the adjectival, or secondary, part of the word usually follows the operative part, so that, for example, "foramen magnum" is given under "foramen" and not "magnum." Words presenting an ambiguity, particularly to those not well acquainted with roots, have been cross-referenced. For example, the word "pollac-anthous" might easily be sought under "-acanthous" from which it is referred to "-anthous," the correct root.

"Without qualification"

The terms "without qualification" or "usually without qualification" are frequently used in definitions of vernacular names of organisms. They mean only that, unless the name itself is qualified by a further adjective, the Latin binomial indicates the form intended. This is necessitated by the fact that in various parts of the world the same vernacular name is used for two different organisms so that "*Brit.*, without qualification" is by no means always the same thing as "*U.S.*, without qualification." The entry "robin" should render clear the necessity for this procedure.

ABBREVIATIONS

commonly used in this dictionary

adj.	adjective, adjectival
Afr.	Africa, African
anat.	anatomy, anatomical
Austral.	Australian, Australia
biol.	biology, biological
bot.	botany, botanical
Brit.	Britain, British
Can.	Canada, Canadian
cent.	central
cf.	*confer* (compare)
circ.	*circa* (approximately)
comb. form	combining form
contemp.	contemporary
dial.	dialect
ecol.	ecology, ecological
e.g.	*exempli gratia* (for example)
E.I.	East Indies
embryol.	embryology, embryological
entomol.	entomology, entomological
epon. adj.	eponymous adjective
Eur.	Europe, European
E.U.S.	Eastern United States
Fr.	French
Ger.	German
Gr.	Greek
hort.	horticulture, horticultural
i.e.	*id est* (that is)
I.P.	Indian Peninsula
jarg.	jargon
L.	Latin
loc. cit.	*loco citato* (in the place cited)
N.Z.	New Zealand
obs.	obsolete
pl.	plural
q.v.	*quod vide* (which see)
S.Afr.	South Africa, South African
sing.	singular
sp.	spelling
Span.	Spanish
subst.	substantive
S.U.S.	Southern United States
syn.	synonym, synonymous
U.S.	United States
W.Afr.	West Africa
W.I.	West Indies
W.U.S.	Western United States
zool.	zoology, zoological

ACKNOWLEDGMENTS

My greatest debt, apart from that acknowledged in the dedication, is to the University of Pittsburgh which, by providing me with both adequate leisure and admirable facilities with which to utilize this leisure effectively, has made possible the production of this work in the brief space of five years.

I am particularly indebted to the following scholars who have reviewed this work in manuscript:

DR. LEE F. BRAITHWAITE,
  Brigham Young University
DR. ROBERT H. BURRIS,
  University of Wisconsin
DR. EARL L. CORE,
  University of West Virginia
DR. KENNETH W. CUMMINS,
  University of Pittsburgh
DR. MARJORIE A. DARKEN,
  Lederle Laboratories
DR. DAVID E. DAVIS,
  Pennsylvania State University
DR. RICHARD M. FOX,
  Carnegie Museum
DR. FREDERICK J. GOTTLIEB,
  University of Pittsburgh
DR. HORTON H. HOBBS,
  The Smithsonian Institution

DR. MALCOLM JOLLIE,
  Northern Illinois University
DR. L. L. LANGLEY,
  National Institutes of Health
DR. CLARENCE J. McCOY, JR.,
  Carnegie Museum
DR. BURT L. MONROE, JR.,
  University of Louisville
DR. LEONARD P. SCHULTZ,
  The Smithsonian Institution
DR. NELSON T. SPRATT,
  University of Minnesota
DR. IAN SUSSEX,
  Yale University
DR. GEORGE WALLACE,
  Carnegie Museum
DR. G. B. WILSON,
  Michigan State University

Though each of these experts was asked only to review entries in the field of his speciality, evidence on the copy returned showed that almost all had read every one of the 1800 pages sent to him. Indeed, I think it is safe to say that there is no word in this dictionary that has not been scrutinized by at least a dozen professional biologists. Such errors that remain are entirely my responsibility.

Among the many others who have helped are three successive secretaries. Mrs. Janet Natowitz was responsible for typing almost all of the 60,000 index cards from which this dictionary was ultimately organized. Mrs. Danelle McDonald Cinowalt finished the cards and transcribed the edited cards to 1600 stencils, and Miss Bernadette N. Brown finished the stencils and then kept track of everything as the book went through press. Miss Penny Frederickson helped in the monumental task of alphabetizing and filing cards, Miss Carol Maloney took charge of sorting and filing the 32,000 sheets of corrected copy sent back by the reviewers mentioned above, and Mr. Alan R. Christopher helped me to read both galley and page proof. Mrs. Catherine Linsenmayer correlated the reviewer-corrected entries with the master copy of the manuscript and also checked the proofs against this to make sure that the corrections and insertions have been made. The Reinhold Publishing Corporation, as always, provided wise counsel and active help. At the executive level, Mr. James B. Ross, Mr. Charles Hutchinson, and Mr.

Martin B. Berke did all that they could to smooth what was inevitably a troubled path, and my debt to Mrs. Alberta Gordon and her assistants is that of all authors to all competent, helpful copyeditors.

CONTENTS

EXPLANATORY NOTE
How to use this dictionary

1. Latin names of taxa (e.g. Chephalopoda, Monachinae) and mutants (e.g. [adenine 1], [prickle]), the latter bracketed to distinguish them from identical words in a nonmutant sense, are entered in alphabetical order.

2. Organisms (e.g. bug, finch, fern, grass), organs (e.g. artery, bone, gland, leaf) enzymes (e.g. carboxylase, hydrogenase), structures (e.g. shell, tail) and other substantive forms (e.g. law, theory) are alphabetized under the name of the organism, organ, enzyme, structure or other substantive without regard to qualifying adjectives. Thus "Greater Antillean bullfinch" is under **finch**, "common carotid artery" is under **artery**, "methyl crotonoyl -CoA carboxylase" is under **carboxylase**, the "sculptured nutshell" is under **shell** and "Wagner's separation theory" is under **theory**.

3. Compounded technical terms are alphabetized under the principal, usually terminal, root. The following roots in particular have numerous sub-entries:

-androus	-ferous	-morphic	-sperm
-angium	-florescent	-morphism	-sphere
-anthous	-form	-morphosis	-spore
-biont	-gamete	-phagous	-sporous
-biosis	-gamy	-philous	-stele
-blast	-gen	-phore	-stome
-carp	-genesis	-phyll	-styly
-carpous	-genous	-phyte	-taxis
-chrome	-gone	-plasm	-theca
-chyma	-gonidium	-plast	-thelium
-coel	-gynous	-ploid	-tome
-colous	-istem	-pod	-tropic
-cyst	-kinesis	-podium	-trophy
-cyte	-logy	-sociation	-tropism
-derm	-more	-some	-type
-dont	-morph	-species	-zooid

A *see* [Agouti]

-a- *comb. form* meaning "without". An -n is sometimes appended for the sake of euphony (e.g. anaerobic) or, more rarely, a -p (e.g. aphydrotaxis)

a *see* [arc], [non-agouti], and [anthocyanin absent]. This symbol is also used as a yeast genetic marker for mating type

a_1, a_2 *see* [anthocyanin]

A/a *see* [Sex]

a/a a yeast genetic marker for heterozygote mating type

aa *see* [anarista]

aardvark the tubulidentate mammal *Orycteropus capensis* (= cape anteater)

aasvogel *S.Afr.* a term applied to Old World vultures in general

ab see [abrupt]

abaca the musaceous tree *Musa textilis* and the fibers derived from it

abacate = avocado

abadavine *Brit. obs.* = siskin

abalone *U.S.* popular name of gastropod mollusks of the family Haliotidae, (*cf.* ormer) and particularly the genus Haliotis

abb *see* [abbreviated]

[abbreviated] a mutant (gene symbol *abb*) mapped at 105.5 on chromosome II of *Drosophila melanogaster*. The phenotypic expression is a reduction in bristle size

abd *see* [abdominal]

abdomen the posterior region of the animal body; in lower vertebates it terminates anteriorly in the heart region but in mammals its cavity is separated from the thorax by the diaphragm; in most arthropods it is clearly demarcated from the thorax or cephalothorax

petiolated abdomen an insect abdomen that, as in many Hymenoptera, is attached to the thorax by a stalk or petiole

sessile abdomen an insect abdomen that is not petiolate nor appreciably narrowed at its attachment with the thorax

[abdomen rotatum] a mutant (gene symbol *ar*) mapped at between 0 and 0.2 on chromosome IV of *Drosophila melanogaster*. The phenotypic abdomen is twisted clockwise

abdominal pertaining to the abdomen. Specifically an enlarged transverse scale on the abdomen of snakes.

[abdominal] a mutant (gene symbol *abd*) mapped at 27.0± on chromosome III of *Drosophila melanogaster*. The phenotypic expression is the breaking of the abdominal bands

Abdominalia a suborder of cirripede Crustacea containing the single family Alcippidae. There is a voluminous mantle but no shell

abduction drawing apart or away from

abele *U.S.* the salicaceous plant *Populus alba* (= white poplar)

[abero] a mutant (gene symbol *abr*) mapped at 83.0± on chromosome II of *Drosophila melanogaster*. The phenotype has a regular abdominal band and frayed wings

abietic pertaining to firs

Abietineae a plant taxon erected to contain the families Pinaceae, Taxodiaceae, and Cupressaceae

abiu the sapotaceous tree *Lucuma caimito*

ablactation inarching

ablastin an antibody that inhibits the division of microorganisms

ablation removal or cutting away

abomasum the fourth division of the artiodactyl stomach (*cf.* psalterium, rumen, reticulum) sometimes identified as the "true" stomach

abr *see* [abero]

Abrocomidae a family of hystricomorph rodents commonly called the rat-chinchillas; they differ from the Chinchillidae in having a naked tail

[abrupt] a mutant (gene symbol *ab*) mapped at 44.0 on chromosome II of *Drosophila melanogaster*. The phenotypic expression is a shortened L5 vein

[Abruptex] a mutant (gene symbol *Ax*) mapped at 3.0± on the X chromosome of *Drosophila melanogaster*. The phenotypic expression includes short wings with incomplete veins

abscissous cut square at the edges

absinth 1 (*see also* absinth 2) the compositaceous herb *Artemisia absinthum* (= wormwood)

absinth 2 (*see also* absinth 1) *U.S.* the geometrid lepidopteran insect *Tephroclystis absinthiata*

-abyss- *comb. form* meaning "bottomless" but commonly used for "very deep"

abyssal pertaining to great depths

-ac- *comb. form.* meaning a sharp point

ac *see* [achaete] and [aciform]

ac-1, 5-7, 12, 14–18, 21, 22, 28, 29, 31, 51, 55, 76, 115, 141, 157, 209 *see* "acetate-1", etc.

Ac-1 *see* [Acetate-1]

acr a yeast genetic marker for growth inhibition by actidione

acacia any of many trees, mostly of the leguminous genus Acacia, but applied to many other genera of similar appearance

ant acacia any of several spps. of the acacia which, being hollow, harbor ants

1

bastard acacia the leguminous tree *Robinia pseudoacacia* (= locust)

false acacia the leguminous shrub *Robinia pseudoacacia* (= black locust)

rose acacia the leguminous shrub *Robinia hispida*

three-thorned acacia the leguminous tree *Gleditsia tricantha*

Acalyptratae an assemblage of those myodarian schizophorous cyclorrhaphous brachyceran dipteran insects that lack a calypter

acanaceous prickly

-acanth- *comb. form* meaning "thistle head"

Acanthaceae that family of tubuliflorous dicotyledons which contains the well-known Acanthus. They may be distinguished from the closely related Scrophulariaceae by the two-celled ovary with two to four ovules

acanthaceous spiny

Acanthineae a suborder of tubifloral angiosperms containing the single family Acanthaceae

Acanthisittidae a small family of passeriform birds containing the New Zealand wrens. They strongly resemble the Trogodytidae in habit and appearance but differ from them in having long slender legs on which the outer and middle toes are joined basally

Acanthizidae a family of passeriform birds usually included with the Sylviidae

Acanthobdellida an order of hirudinian annelids erected to contain the single genus Acanthobdella, which, though clearly a leech, has chaetae on the second and sixth segment

Acanthocephala a phylum of pseudocoelomate bilateral animals distinguished by an anterior eversible proboscis armed with hooks. Sometimes called the spiny-headed worms

Acanthodii a class of fossil gnathostomatous craniate Chordata commonly called the spiny sharks, distinguished by true bone in the skeleton and a hyoidean gill slit

acanthoid a whip-like seta found on the legs of some arachnids

Acanthomeridae = Pantophthalmidae

Acanthopterygii an enormous order of actinopterygians commonly called the "spiny-rayed fishes" because of the stiff spines in the fins. The Acanthopterygii are sometimes regarded as a superorder in which case the suborders, noted elsewhere in this dictionary, are regarded as orders

acanthous having spines or thorns

anacanthous lacking spines or thorns

oxyacanthous sharp-spined

pyracanthous with flame colored thorns

Acanthuridae a large family of acanthopterygian fishes known as surgeon-fish by reason of the hinged "knife" carried in a groove on the dorsal surface

Acanthuroidea a suborder of acanthopterygian fishes containing the families Zanclidae and Acanthuridae. They are distinguished from the Percoidea by the articulation of the skull to the vertebra and by the acronurous larvae (*q.v. under* larva)

acara aquarists' name for cichlid fishes of the genus Aequidens

Acarina that order of arachnid arthropods which contains the animals commonly called mites and ticks. They are readily distinguished from all other arthropods by the fact that the mouth parts are more or less distinctly set off from the rest of the body on a "false head" or gnathasoma. They strongly resemble the Phalangida (*q.v.*) and Araneida (*q.v.*),

but differ from the former in lacking a segmented abdomen and from the latter because there is a distinct division between the cephalothorax and abdomen

acc *see* [acclinal wing]

acceleration the development of a genetically controlled character more rapidly than occurs in the parent or ancestor

accentor properly, any bird that "cheeps" but sometimes confined to those of the family Prunellidae or the genus Prunella

alpine accentor *P. collaris*

mountain accentor *P. montanella*

robin accentor *P. rubeculoides*

Accentoridae = Prunellidae

Accipitrinae a subfamily of Accipitridae containing the tains, *inter alia*, the hawks, eagles, kites, harriers, and Old World vultures. The family differs from the Falconidae in skeletal details and in many details of their soft anatomy such as the formula of the leg muscles

Accipitrinae a subfamily of Accipitridae containing the buzzards, eagles and short-winged hawks that are distinguished from the buzzard hawks by their shorter wings, long tail and direct flapping flight

acclimatization the process by which an organism becomes adapted to, or tolerant of, a new environment

[acclinal wing] a mutant (gene symbol *acc*) mapped at 54.5 on the X chromosome of *Drosophila melanogaster*. The upheld wings of the phenotype slope backward

acclivous with a gentle upward slope

accrescent said of an organ, or organism, that increases in size with age. In *bot.* designates particularly those which continue to grow after flowering

accrete to grow together

accumbent said of an organ that lies against another

ace *I.P.* any of several species of hesperiid lepidopteran insects of the genus Halpe and its near allies

straw ace *I.P.* the hesperiid lepidopteran insect *Pithauria stramineipennis*

-aceae suffix indicating familial rank in plant taxonomy. The *zool.* equivalent is -idae

-aceous *comb. suffix* meaning "to be full of", or "to possess abundantly"

Aceraceae that family of sapindale dicotyledons which contains the maples. The peculiar fruit (*see* samara) distinguishes this family from the closely allied Sapindoseae

acerate needle-shaped

acerose needle-shaped

acervate piled in a heap

coacervate in *bot.* clustered. Used as a noun, the word usually refers to aggregates of colloidal particles in suspension

acervulous having the appearance of small clusters or heaps

-acetabul- *comb. form* meaning "saucer"

acetabulum 1 (*see also* acetabulum 2 and 3) the deep socket in the innominate bone at the junction of its three components and into which the head of the femur fits

acetabulum 2 (*see also* acetabulum 1 and 3) a sucking disc with a raised rim and a central depression, most commonly used of platyhelminth suckers but also applied to similar structures in many other groups acetabulum 3 (*see also* acetabulum 1 and 2) in insects, any cavity into which a joint articulates, particularly the cavity in the coxa; also, a conical cavity at the front end of some larvae

acetase a general term for enzymes acting on acetate linkages

  fumarylacetoacetase an enzyme that catalyzes 4-fumarylacetoacetate to acetoacetate and fumarate

  oxaloacetase an enzyme that catalyzes the hydrolysis of oxaloacetate to oxalate and acetate

[**Acetate-1**] *either* a mutant (gene symbol *ac-1*) in linkage group II of *Neurospora crassa*. The phenotypic expression is a requirement for both acetate and ethanol. *Or* a mutant (gene symbol *ac-1*) in linkage group VII of *Chlamydomonas reinhardi*. The phenotypic expression is a requirement for acetate and a white colony

[**Acetate-5**] a mutant (gene symbol *ac-5*) in linkage group VII of *Chlamydomonas reinhardi*. The phenotypic expression is a requirement for acetate, slow growth and a pale yellow colony

[**Acetate-6**] a mutant (gene symbol *ac-6*) in linkage group VII of *Chlamydomonas reinhardi*. The phenotypic expression is a requirement for acetate

[**Acetate-7**] a mutant (gene symbol *ac-7*) in linkage group XI of *Chlamydomonas reinhardi*. The phenotypic expression is a requirement for acetate and a yellow-green colony

[**Acetate-12**] a mutant (gene symbol *ac-12*) in linkage group II of *Chlamydomonas reinhardi*. The phenotypic expression is a requirement for acetate

[**Acetate-14**] a mutant (gene symbol *ac-14*) in linkage group I of *Chlamydomonas reinhardi*. The phenotypic expression is a requirement for acetate

[**Acetate-15**] a mutant (gene symbol *ac-15*) in linkage group IX of *Chlamydomonas reinhardi*. The phenotypic expression is a requirement for acetate

[**Acetate-16**] a mutant (gene symbol *ac-16*) in linkage group X of *Chlamydomonas reinhardi*. The phenotypic expression is a requirement for acetate and an inability to fix CO_2

[**Acetate-17**] a mutant (gene symbol *ac-17*) in linkage group III of *Chlamydomonas reinhardi*. The phenotypic expression is a requirement for acetate

[**Acetate-18**] a mutant (gene symbol *ac-18*) in linkage group V of *Chlamydomonas reinhardi*. The phenotypic expression is a requirement for acetate

[**Acetate-21**] a mutant (gene symbol *ac-21*) in linkage group XI of *Chlamydomonas reinhardi*. The phenotypic expression is a requirement for acetate and an inability to fix CO_2

[**Acetate-22**] a mutant (gene symbol *ac-22*) in linkage group VII of *Chlamydomonas reinhardi*. The phenotypic expression is a requirement for acetate and a pale green colony with slow growth

[**Acetate-26**] a mutant (gene symbol *ac-26*) in linkage group III of *Chlamydomonas reinhardi*. The phenotypic expression is a requirement for acetate

[**Acetate-28**] a mutant (gene symbol *ac-28*) in linkage group III of *Chlamydomonas rienhardi*. The phenotypic expression is a requirement for acetate

[**Acetate-29**] a mutant (gene symbol *ac-29*) in linkage group VI of *Chlamydomonas reinhardi*. The phenotypic expression is a requirement for acetate and a yellow color

[**Acetate-31**] a mutant (gene symbol *ac-31*) in linkage group V of *Chlamydomonas reinhardi*. The phenotypic expression is a requirement for acetate and a yellow color

[**Acetate-51**] a mutant (gene symbol-*ac-51*) in linkage group IX of *Chlamydomonas reinhardi*. The phenotypic expression is a requirement for acetate

[**Acetate-55**] a mutant (gene symbol *ac-55*) in linkage group IV of *Chlamydomonas reinhardi*. The phenotypic expression is a requirement for acetate and a yellow color

[**Acetate-76**] a mutant (gene symbol *ac-76*) in linkage group I of *Chlamydomonas reinhardi*. The phenotypic expression is a requirement for acetate

[**Acetate-115**] a mutant (gene symbol *ac-115*) in linkage group I of *Chlamydomonas reinhardi*. The phenotypic expression is a requirement for acetate and an inability to fix CO_2

[**Acetate-141**] a mutant (gene symbol *ac-141*) in linkage group III of *Chlamydomonas reinhardi*. The phenotypic expression is a requirement for acetate and an inability to fix CO_2

[**Acetate-157**] a mutant (gene symbol *ac-157*) in linkage group VIII of *Chlamydomonas reinhardi*. The phenotypic expression is a requirement for acetate

[**Acetate-209**] a mutant (gene symbol *ac-209*) in linkage group I of *Chlamydomonas reinhardi*. The phenotypic expression is a requirement for acetate

[achaete] a mutant (gene symbol *ac*) mapped O+ on the X-chromosome of *Drosophila melanogaster*. The phenotype is characterized by the absence of postdorsocentral chaetae

Achatocarpaceae a small family of centrospermous dicotyledons

achene a dry, indehiscent, one-seeded fruit in which the thin pericarp is fused with the testa in one localized area (*cf.* caryopsis)

  diachene = cremocarp

  epiachene = hypoachene

  hypoachene an achene derived from an inferior ovary

  periachene any achene derived from a partially superior ovary

  polachene a cremocarp derived from five carpels

  triachene a seed-like fruit composed of three one-seeded carpels invested by an epigynous calyx (*cf.* cremocarp)

Achreioptera an order proposed in place of the coleopteran family Platypsyllidae (*q.v.*)

Achromatiaceae a family of beggiatoale schizomycetes occuring as large (up to 100 microns) individual, spherical, or ovoid cells, that move over a substrate with the slow rolling movement typical of the Beggiatoales. Some contain large included masses of calcium carbonate. Practically nothing is known of this group

Achromobacteraceae a family of gram negative, rod shaped eubacteriale schizomycetes, most of which are found in water and soil but some of which, (particularly of the genus Alcaligenes) are intestinal pathogens that have been confused with Vibrio (*cf.* Spirillaceae)

achyr- *comb. form* meaning "chaff"

-aci- *see* -ac-

acia a cuticular plate on the insect mandible

acicular needle-shaped

acid a compound capable of combining with a base. Very few acids are noted in this dictionary and those mostly for purposes of cross reference

  amino acid an organic acid with an amino group at one end and therefore possessing both acidic and basic linkages. Many, but not all, of the 39 amino acids are known to take part in protein synthesis

  anthranilic acid = o-aminobenzoic acid

ascorbic acid *see* vitamin C

p-aminobenzoic acid $C_7H_7NO_2$. Also called PABA and, at one time, vitamin B_x

α-aminobutyric acid $Ch_3CH_2CH(NH_2)$-COOH. A rare amino acid

aminosuccinic acid $HOOCCH_2CH(NH_2)COOH$. A common amino acid not essential in rat nutrition

aspartic acid = aminosuccinic acid

cysteic acid, α-amino-β-sulfopropionic acid. $HOOCCH(NH_2)CH_2SO_3H$ an amino acid isolated principally from hair after hydrolysis and oxidation

folic acid, pteroyl glutamic acid, N-[4-{[2-amino-4-hydroxy-6-pteridyl)-methyl]-amino}-benzoyl]-

L-glutamic acid a widely distributed amino acid, 2-aminopentanedioic acid, $HO_2CCH_2CH_2CH \cdot (NH_2)CO_2H$. Its monosodium salt is derived in very large quantities by fermentation and sold in crystalline form under various names as a food seasoning, or in solution as soy sauce.

pteroylglutamic acid = folic acid

nicotinic acid = niacin

pantothenicacid D(d)-N-(α, γ-dihydroxy-β,β-demethyl-butyryl 0-β' -alanine. A water soluble B complex vitamin. Definitely required by birds as an antidermatitis agent. The requirements for other forms are not clear

shikimic acid $C_6H_6(OH)_3COOH$, a precursor of aromatic amino acids, first isolated from the Japanese star anise (*q.v.*)

[aciform] a mutant (gene symbol *ac*) in linkage group I of *Habrobracon juglandis*. The name derives from the needle-like appearance of the terminal half of the phenotypic antenna

-acin- *comb. form* meaning a grape seed

-acinac- *comb. form* meaning "a scimitar"

acinaceous said of a fruit containing many seeds, or a nut containing many kernels

acine a single member of a compound fruit. Occasionally used for individual fruits in a bunch, or for terminal lobes of a gland

acinose = botryoidal

acinus has the same meaning as acine (*q.v.*) but is preferred by histologists for the terminal excretory lobes of alveolar glands

Acipenseridae a family of Chondrostei containing the sturgeons. They are readily distinguished by the heavy plate-like scales along the side

Acipenseriformes = Chondrostei

-acm- *comb. form* meaning "a point" in the sense of maximum or minimum of a cyclic curve

Acmaeidae a family of gastropod mollusks commonly called limpets. They are distinguished by lacking a spiral shell at any stage and by the absence of an opening at the apex of the shell

acme in *biol.* compounds of this word, also sometimes spelled acmy, are used to represent phases or periods, rather than sharp peaks in, or points on, a curve

epacme that phase in the development of a population of organisms when the vigor of the population is steadily increasing

heteracme = dichogamy

paracme that phase in the development of a race when vigor is declining

synacme the condition of having the stamens and pistils mature synchnously

-acmic *adj. suffix* from acme (*q.v.*)

diacmic said of plankton which has two maxima in one season

monacmic said of a plankton which blooms once a year only

Acnidaria a term at one time used to distinguish the Ctenophora, which lack nematocysts, as a subphylum of the Coelenterata. The Ctenophora are now usually regarded as a separate phylum

Acoela an order of turbellarian platyhelminths distinguished by having a mouth but no intestine

Acoelomata a term coined to define those phyla of the animal kingdom which lack a true coelom. This taxon therefore embraces the Platyhelminthes and the Nemertea

aconitase = aconitate hydratase

aconite any of numerous herbs of the ranunculaceous genus Aconitum

winter aconite any of several herbs of the ranunculaceous genus Eranthis

Acorduleceridae = Pergidae

acorn (*see also* weevil, worm) the nut of the oak, or any structure or organism of similar shape

sea acorn any of numerous species of sessile barnacles, mostly of the genus Balanus

acr symbol for a bacterial mutant showing resistance to acridine

Acraeidae a family of nymphaloid lepidopteran insects, mostly African though there are a few representatives in Australia, Asia, and Latin America. They have sparsely scaled long forewings and rounded hindwings

Acragidae = Dalceridae

acrandrous said of organisms that bear antheridia on the apex of a shoot

Acraniata a taxonomic division containing those chordate animals that lack a skull

Acrasieae a class of Myxomycophyta of doubtful affinities. They differ from all other Myxomycophyta in lacking flagellated swarm cells

Acrididae a family of orthopteran insects containing the short-horned grasshoppers. The short antennae are distinctive

-acro- *comb. form* meaning "summit", "peak", "point" or, particularly when used as a prefix, "apical"

Acroceratidae = Acroceridae

Acroceridae a small family of orthoraphous brachyceran dipteran insects with a hump-backed thorax and a minute head from which derives their popular name of small-headed fly. The larvae are internal parasites of spiders

Acrochordidae a small family of aglyph colubrid snakes commonly called oriental water snakes. Frequently listed as a subfamily of Colubridae. They are adapted to marine, or estuarine, life by a laterally compressed tail

Acrolepiidae a family of lepidopteran insects usually included in the Plutellidae

Acrolophidae a family of tineoid lepidopteran insects the larvae of which are commonly called burrowing webworms. The adults are distinguished by the unusually hairy eyes

acron the most anterior segment of a larva. Occasionally used as synonymous with prostome

Acrotonae a division of the orchidaceous subfamily Monandrae distinguished by the fact that the caudicle and basidia arise from the apices of the pollinia

-act- *comb. form*, better spelled -akt- meaning "rocky coast"

actad a plant or plant community growing on a rocky shore

ACTH = adrenocorticotropin (*see under* -tropin)

actic pertaining to rocky coasts though occasionally used as synonymous with "littoral" and often confined rigidly to the area between tide marks

-actin- *comb. form* meaning "ray", more properly spelled -aktin- (*cf.*-actino)

actinal said of those skeletal elements of an echinoderm, that are developed around the left coelom

  abactinal said of those skeletal elements of an echinoderm which are developed around the right coelom

actine a rayed sponge spicule

  hexactine a sponge spicule consisting of three axes crossing at right angles to produce six rays extending from a single point

  stauractine a hexactine sponge spicule that has lost one complete axis and therefore appears to belong to a tetractine type

  tetractine a sponge spicule with two axes crossing at right angles

Actiniaria an order of Zoantharia commonly called sea anemones. They are distinguished from the Antipatharia and Madreporaria by the adsence of a skeleton, from the Ceriantharia and Zoanthidia by the presence of a pedal disc

Actinidraceae a family of parietale dicotyledonous shrubs or woody vines

Actinistia an order of osteichthyes thought to be extinct until the discovery in 1938 of the living *Latimeria chalumnae* near East London, South, Africa. Since then numerous specimens of this species have been found off the African coast

-actino- *comb. form* from the same root as -actin-(*q.v.*) but usually taken to mean a "ray of light", and frequently misused in taxonomic nomenclature in the sense of "star"

Actinomycetaceae a family of actinomycetale schizomycetes producing a nonseptate mycelium in the early stages that may either remain in this form and produce spores on aerial hyphae or break up into rod-shaped segments. Some cause lesions in the lungs of man and domestic animals while others infect lymph nodes and ducts

Actinomycetales an order of schizomycetes that develop a branching mycelium-like structure, some even bearing conidia, and thus strongly resembling phychomycetes. A great majority are found in soil

Actinomycetes = Actinomycetales

Actinomyxidia an order of cnidosporidean Sporozoa distinguished by possessing three rayed spores with three polar capsules

Actinoplanaceae a family of actinomycetale schizomycetes distinguished by the production of sporangia. The vegetative mycelium is commonly formed under water on decomposing plant materials

Actinopoda a subclass of sarcodinan Protozoa. Distinguished from the Rhizopoda (*q.v.*) by the fact that the pseudopodia are stiff and radiating

Actinopterygii a subclass of bony fish distinguished by the possession of rayed fins

Actinozoa = Anthozoa

activator an essential inorganic component of many enzyme reactions

-acule- *comb. form* meaning a "prickle"

aculea a lepidopteran hair-like or spine-like alar seta. In other arthropods a small seta, particularly one which is immovable

aculeate prickly

aculei = microtrichia

acumen literally a sharp, or stinging, point but used in *biol.* for the terminal spine of a crustacean rostrum

acuminate tapering to a point

  abruptly acuminate terminating in a short point on a broad base

acus the needle-like leaf of a pine

-acut- *comb. form* meaning "sharp"

-ad 1 (*see also* -ad 2) *comb. suffix* used in formation of collective nouns particularly in *biol.* for ecological groupings of plants (*e.g.* lophad, a plant of crests or ridges.) These forms are entered under the substantive portion of the compound (*e.g.* loph-)

-ad 2 (*see also* -ad 1) *adverb. suffix* indicating direction (*e.g.* ventrad, in the direction of the belly or downward). These words are entered under the substantive part of the compound (*e.g.* ventr-)

ad 1 a yeast genetic marker indicating a requirement for adenine

ad 2 *see* [arcoid]

ad -1 to ad -8 *see* [Adenine -1] to [Adenine -8]

Adam-and-Eve *either* the orchid *Aplectrum hymenale* or the crassulaceous herb *Sempervivum tectoreum*

adanale a small sclerite at the base of an insect wing

-adant- *comb. form* meaning "attached"

adaptation the ability of an organism to cope successfully with its environment

  mutual adaptation an adaptation resulting in a symbiosis harmful to neither partner

  pre-adaptation the possession of a mutation not advantageous in the organism's present environment but which may adopt it to a new environment

adder 1 (*see also* adder 2) a term loosely applied to any of many snakes either venomous or thought to be venomous. *Brit.* without qualification, the viperine *Vipera berus*

  banded adder = krait

  death adder *Austral.* the elapine *Acanthophis antarcticus*

  hissing adder *U.S.* = puff adder *U.S.*

  milk adder *U.S.* the colubrid *Lampropeltis triangulum*

  night adder any of several species of African viperid of the genus Causus

  puff adder any of many species of viperid of the *Afr.* genus Bitis. Usually, without qualification, *B. lachesis. U.S.* = puffing adder

  puffing adder *U.S.* any of several species of colubrid of the genus Heterodon

  spreading adder *U.S.* = puff adder *U.S.*

adder 2 (*see also* adder 1, fern, mouth, tongue) any of many organisms thought to have the appearance, or habit, of adders or other snakes

  deaf-adder *Brit.* the anguid saurian *Anguis fragilis* (= slow-worm)

  flying adder *Brit.* = dragonfly

  sea adder *Brit.* any marine stickleback. *Brit.* and *U.S.* any of several pipefishes

  stone adder *Austral.* the geckonid saurian reptile *Diplodactylus vittatus* (= stone gecko)

addra a Saharan gazelline antelopid (*Addra dama*), the legs and neck of which are very long but not so disproportionately so as those of the gerenuk

adduction a moving together

ade a symbol for a bacterial mutant showing a requirement for adenine

Adelchorda = Hemichordata

Adelidae a family of incurvarioid lepidopteran insects distinguished by their small size and extremely long and delicate antennae. The larvae resemble, but are less well known than, the larvae of the coleophorids in their ability to build portable cases. These forms are sometimes called fairy moths

-adelo- *comb. form* meaning "not obvious", "concealed" or, rarely, "unknown"

Adelogamicae a no longer acceptable botanical taxon at one time containing the fungi and the lichens

-adelph- *comb. form* meaning "brother" but put to a wide variety of uses in biol

-adelphia names of taxa having this suffix are listed alphabetically

adelphous pertaining to brothers, or siblings, or to almost any replicated structures. In *bot.* usually applied with reference to the androecium

isoadelphous having similar parts, particularly floral parts

monadelphous having only one of a structure of which there are usually several. Any other *Gr.* numeral form may be combined with -adelphous to indicate the number of parts. In *bot* usually refers to an androecium with stamen filaments connate into a single tube

polyadelphous having many replicated parts, particularly glands or stamens

aden- *comb. prefix* meaning "gland"

adenase an enzyme hydrolyzing adenine to hypoxanthine and ammonia

adenine a 6-amino purine base deriving its name from the original source from which it was isolated

[adenine-i] a mutant (gene symbol *ad-1*) in linkage group VI of *Neurospora crassa.* The phenotypic expression is a requirement for adenine

[adenine-2] a mutant (gene symbol *ad-2*) found in linkage group III of *Neurospora crassa.* The phenotypic expression is a requirement for either adenine or hypoxanthine

[adenine-3] a mutant (gene symbol *ad-3*) on linkage group I in *Neurospora crassa.* The phenotypic expression is a requirement for adenine

[adenine-4] a mutant (gene symbol *ad-4*) on linkage group III of *Neurospora crassa.* The phenotypic expression is a requirement for adenine and an inability to use hypoxanthine

[adenine-5] a mutant (gene symbol *ad-5*) in linkage group I of *Neurospora crassa.* The phenotypic expression is a requirement for adenine

[adenine-6] a mutant (gene symbol *ad-6*) in linkage group IV of *Neurospora crassa.* The phenotypic expression is a requirement for adenine

[adenine-7] a mutant (gene symbol *ad-7*) in linkage group V of *Neurospora crassa.* The phenotypic expression is a requirement for adenine

[adenine-8] a mutant (gene symbol *ad-8*) linkage group VI of *Neurospora crassa.* The phenotypic expression is a requirement for adenine

adenoid gland-like. Particularly gland-like masses of lymphoid tissue, and specifically such a mass at the back of the pharynx

adenosine a nucleoside the phosphate derivatives of which are a primary energy transfer system in living materials

adenosine diphosphate (ADP) formed in biokinetic systems from ATP

adenosine monophosphate (AMP) formed in biokinetic systems from ADP

adenosine triphosphate (ATP) a major energy donor in biokinetic systems

Adephaga a suborder of coleopteran insects distinguished from all other Coleoptera by the fact that the hind coxae are immoveably fused to the metasternum

ADH vasopressin

Adimeridae = Monoedidae

Adineda a suborder containing the Dinoflagellata that lack both grooves and a cellulose test

adipose fatty

adjutant either of two ciconiid birds of the genus Leptoptilos, usually without qualification *L. dubius* (= **greater adjutant**)

lesser adjutant *L. javanensis*

adl symbol for a bacterial mutant capable of utilizing adonitol

adminculum one of many lines, or rows, or patches, of external teeth on a pupal case

admiral any of numerous nymphalid lepidopteran insects thought to resemble, by their gaudy colors or haughty manner, high ranking naval officers of past empires. In the *I.P.* there are many admirals of the genus Limenitis (*cf.* commodore, commander) and in the *U.S.* the term is sometimes applied to *Basilarchia archippus,* more usually called the viceroy. Some specifically designated admirals are the following

Australian admiral *Pyramesis itea*

blue admiral *I.P. Vanessa canace*

red admiral *Brit.* and *U.S. Vanessa atalanta, I.P. Vanessa indica*

long tail admiral *S. Afr.* the nymphalid *Antanartia hippomene*

short tail admiral *S. Afr.* the nymphalid *Antanartia schoeneia*

white admiral *U.S. Limenitis arthemis* (= banded purple), *Brit. Limenitis sibylla, I.P. Limenitis trivena*

adnate said of those parts of an organism which are fused to each other (*cf.* adnexed, connate)

coadnate coherent

adnation the condition resulting from the congenital growing together of unlike parts

admexa a part which supplements, but is not part of, a specific structure

adnexed said of those parts of an organism, usually a plant, which are pressed against, but not fused with, each other (*cf.* adnate)

adolescariae a term once employed for the cercariae and metacercariae larva (*cf.* parthenitatae and meritae)

ADP = adenosine diphosphate

adrenaline = epinephrine

noradrenaline = norepinephrine

adrenine = epinephrine

adth symbol for a bacterial mutant showing a multiple requirement for adenine plus thiamine

-adunc- *comb. form* meaning "hooked"

aduncate smoothly curved, or, rarely, hook shaped

adustous dusky

adventitia the connective tissue sheath of an organ

ae *see* [amylase negative]

aecidium a cup-shaped sporocarp that produces chains of spores from its inner surface

aedeagus the chitinous intromittent organ of male insects (*cf.* penis)

Aegeriidae a family of yponomeutoid lepidopteran insects commonly called clear-winged moths for the reason that the great majority lack scales on both the fore- and hindwings. Many strongly resemble hymenopterous insects and are called hornet moths or even hornets. The larvae are stem borers and many pests belong to this family

Aegialitidae = Eurystethidae

Aegidae a family of cymothoid isopod crustacea distinguished by their broad flattened body and the fact that they are parasites of fish

Aegithinidae = Irenidae

Aegothelidae a small family of caprimulgiform birds containing the owlet-frogmouth. They are distinguished by the large erect oral bristles with hair like barbs

aequilate of uniform width throughout the length

Aegyreidae = Nymphalidae

-aeithal- *comb. form* meaning an "evergreen thicket"

-ael-, aell- *comb. form* meaning a "high wind" or "storm"

aeneous brassy

Aeolosomatidae a family of microscopic fresh-water oliogochaete annelids having a ventrally ciliated prostomium and with the chaetae in four bundles on each segment

Aeolothripidae a family of terebrantian thysanopteran insects distinguished by the upwardly curved ovipositor. They are commonly called broad winged, or banded, thrips

-aequi- *see* -equi-

-aer- *comb. form* meaning "air"

aereous bronzy

aerobe an organism using air. The adjective or its derivatives is more frequently used

   aerobic (*see also* anoxybiontic) said of an organism or life process that utilizes, or can only exist in, the presence of oxygen

   anaerobic said of an organism or life process that does not utilize, or cannot exist in, the presence of oxygen

   kathaerobic said of an organism, frequently a protozoan, capable of living in well oxygenated water free of organic matter

Aerocharidae = Eurycerotidae

aeroplane *Austral.* any of several species of nymphalid lepidopteran insects

   black-and-white aeroplane either of two species of the genus Neptis

   Cape York aeroplane *Acca venilia*

   common aeroplane *Phaedyma shepherdi*

   orange aeroplane *Rahinda consimilis*

aeruginose with the color of verdigris

Aesculaceae = Hippocastanaceae

Aeshnidae a family of anisopteran odonatan insects distinguished by the fact that the eyes are partially fused

aesthetask = asthete

aesthete any invertebrate sense organ; mostly used of sensory nerve endings, but also applied to sensory hairs and bristles of arthropods

   megalaesthete a giant sensory nerve ending in Mollusca, particularly chitons, thought to be photosensitive

   microaesthete a sensory nerve ending in Mollusca, particularly chitons, thought to be tactile

-aesthetism sometimes used in *bot.* in the sense of -tropism (*q.v.*)

aestivation 1 (*see also* aestivation 2) a form of dormancy, not unlike hibernation, in which some organisms pass the summer months

aestivation 2 (*see also* aestivation 1) the arrangement of parts in an unfolding flower

coaetaneous = coetaneous

-aethal- *comb. form* meaning "soot"

aethalium the mass formed by the fusion of sporangia of Myxomycetes

-aetio- *comb. form* meaning "cause", usually, but not always, transliterated as -etio-

Aetoxiaceae a small family of sapindalous dicotyledons

Aflagellatae a class of phycomycete fungi distinguished by the fact that none produce flagellated reproductive bodies

ag *see* [agitans] and [Andrus' green stem]

-aga- *comb. form* meaning "beach" (*cf*-psamath-)

agad a beach plant

Agamae (*obs.*) = Cryptogamae

Agamidae a family of Old World lacertilian reptiles differing from the similar New World Iguanidae in having acrodont teeth

Agamophyta = Protophyta

Agaontidae a family of small chalcicoid apocritan hymenopteran insects commonly called fig insects, or fig wasps, because they are uniquely necessary for the cross pollination of figs

agar-agar a complex polysaccharide derived from several red algae particularly *Gelidium spp.*

agaric properly an agaricaceous fungus often applied to members of other families

   fly agaric the poisonous amanitaceous fungus *Amanita muscaria*. (*cf.* destroying angel)

   punk agaric any of several species of the polyporaceous genus Fomes

Agaricaceae a family of agaricale basidiomycete fungi readily distinguished from the closely allied Amanitaceae by the absence of a volva. There is a well-marked annulus

Agaricales that class of basidiomycete fungi which contains all the forms having a fleshy cap with gills on the underside of the cap. Commonly known as toadstools and mushrooms

Agaristidae a family of noctuoid lepidopteran insects commonly called forester moths. They characteristically possess two spots in each wing and clavate antennae

Agavaceae a family of liliiflorous monocotyledonous angiosperms containing fibrous leaved genera included by some within the Amaryllidaceae

agave (*see also* cactus) a large genus of amaryllidaceous herbs, known by this name in both English and Latin, mostly found in S. and S.W.U.S. and S. Amer. The so-called century plant is the best known

Agdistidae a family of lepidopteran insects now usually included in the Pterophoridae

age the period of time during which an organism or phenomenon has existed

   ecological age age in relation to the ability to breed and therefore divided into pre-, post- and actual

aging the continuous process of development in an adult, primarily degenerative in character

Agelenidae a family of araneid arthropods commonly called funnel-web spiders. They are relatively large spiders distinguished not only by the funnel-like web but also by the large chelicerae and unusually elongate hind spinnerets

aggedula the sporangium of a moss

-aggeio- *comb. form* meaning "a vessel" now invariably transliterated -angio-

agglutinated properly "stuck together", but used in *entomol.* to describe a larva with an unusually heavy chitinous sheath

agglutination the process of sticking or being stuck together, particularly applied to erythrocytes

agglutinin an antibody causing clumping of cells

   autohemagglutinin an antibody causing clumping of red cells in the organism that produces it

aggregation the clumping of individuals in a biotic community

aggression the actions of an organism seeking to dominate another organism

[**agitans**] a mutant (gene symbol *ag*) in linkage group III of the mouse. Phenotypic expression is a pronounced tremor commencing at birth and leading to death in three to four weeks

Aglossa a group of aquatic anuran amphibia distinguished by the absence of a tongue (= Pipidae)

Aglypha an assemblage of colubrid snakes distinguished by the fact that the teeth are solid and not grooved (*cf.* Opisthoglypha and Proteroglypha)

agnate related solely through the male line (*cf.* cognate)

Agnatha a class of craniate chordates containing the extinct ostracoderms and the living cyclostomes or Marsipobranchii. They are distinguished, as the name indicates, by the absence of a hinged mandible

agnation the condition of spear (*q.v.*) descent

Agonidae a small marine family of percomorph fishes having elongated bodies covered with bony plates the edges of which are frequently saw toothed. Commonly called sea poachers and alligator fish

agonisis = certation

agouti 1 (*see also* agouti 2) a dasyproctid hystricomorph rodent *Dasyprocta agouti*; it is distinguished from the paca by its long legs and more delicate build

agouti 2 (*see also* agouti 1) a pelt color in mammals produced by the fact that each hair is broadly banded in brown and yellow

[**Agouti**] a mutant (gene symbol *A*) mapped at 0 on linkage group II of the rat. The name refers to the color of the fur

-agra- *comb. form* meaning "field"

agrimony any of several species of the rosaceous herbs of the genus Agrimonia

  bastard agrimony either of the compositaceous herbs *Eupatorium cannabinum* (= **hemp agrimony**) *or Ageratum conyzoides*

Agrionidae a family of zygopteran odonatan insects usually with black or red markings on the wings

Agromyzidae a family of minute myodarian cycloraphous dipteran insects commonly called leaf miner flies because the larvae cut channels through the substance of leaves

agronomy the science of agriculture

Agrostideae a tribe of the plant family Gramineae

-agrosto- *comb. form* meaning "grass"

agujon the belonid fish *Strongyeura acus*

ah *see* [Hoffman's anthocyaninless]

aheneous brassy green, a color particularly common in carabid Coleoptera

aholeholes popular name of kuhliid fish

ai the xenarthran mammal *Bradypus tridactylus* (= three-toed sloth)

-aigial- *comb. form* meaning "seashore" (cf. -psammath-)

aigialium a community of beach-dwelling organisms

aileron sometimes used as synonymous with alula (*q.v.*), but more properly a similar structure, actually a large scale, in front of the base of the forewing of some insects

Ailurinae a subfamily of procynoid carnivorous mammals containing the pandas. Most contemporary workers include the giant panda with the Ursidae

-aio- *comb. form* meaning "eternity" or "everlasting"

AIR = 5-aminoimidazole ribotide

-aitio- *comb. form* meaning "cause" or "causation". Frequently, but confusingly, spelled -etio-

Aizoaceae that family of centrospermous dicotyledons which contains, *inter alia*, the carpet weeds. This family is clearly characterized by the pseudopolypetalous flowers

akalat 1 (*see also* akalat 2) any of very many species of timaliid birds of the genus Malacocincla (*cf.* babbler)

  brown akalat *M. fulvescens*

  black cap akalat *M. cleaveri*

akalat 2 (*see also* akalat 1) a few turdid birds resembling the true akalats

  orange breasted akalat *Sheppardia cyornithopsis*

  thrush akalat *Ptyrticus turdinus*

Aiariaceae a small family of geraniale dicotyledons

Akariota = Monera

akeake *N.Z.* the sapindaceous tree *Dodonaea viscosa*

akepa the drepaniid bird *Loxops coccinea*

akepiro *N.Z.* the compositaceous tree *Olearia fururacea*

Akeridae a family of estuarine gastropod mollusks with a small fragile internal shell from which their popular name of glassy bubbles derives

akiapolaau the drepanid bird *Hemignathus wilsoni*

akiraho *N.Z.* the compositaceous tree *Shawia paniculata* (= golden akeake)

-akm- *comb. form* meaning "point" almost invariably transliterated -acm- (*q.v.*)

akro- *comb. form* meaning "apex" *etc.* almost invariably transliterated -acro- (*q.v.*)

-akt- *comb. form* meaning a "rocky coast", usually transliterated -act- (*q.v.*)

-aktim- *comb. form* meaning "ray" usually transliterated -actin- (*q.v.*)

al *see* [anthocyanin loser] *or* [albescent]

Al *see* [Alopecia]

al *see* [albinism] *or* [aristaless]

al-2 *see* [Albino-2]

ala symbol for a bacterial mutant indicating a requirement for alanine

-ala- *comb. form* meaning "wing"

  ala temporalis a cartilage in the embryonic chondrocranium that appears immediately adjacent, and slightly anterior to the processus alaris

  metala the mesothoracic wing of insects

Alangiaceae a unigeneric family of myrtiflorous angiosperms distinguished by the articulated pedicels and valvate petals

L-alanine L-α-aminopropionic acid $CH_3CH(NH_2)COOH$. An amino acid not essential in rat nutrition

  L-phenyalanine α-amino-β-phenylpropionic acid. $C_6H_5CH_2CH(NH_2)COOH$. An amino acid essential to the nutrition of rats

  L-3, 4-dihydroxyphenylalanine an amino acid, α-amino-β-(3, 4-dihydroxyphenyl) propionic acid, usually derived from leguminous seeds and widely known as dopa

alar pertaining to the wing

alaria the projection from the notum against which the insect wing articulates

alarima the cleft between the paraglossae

alarm a signal indicative of danger or the reaction produced by such a signal

alate winged

  dealate having had the wings removed

  exalate lacking wings

Alaudidae a family of passeriform birds erected to contain the larks. Many have crests and ear tufts.

-alb- *comb. form* meaning "white"

de albate appearing to

albacore the scombrid fish *Thunnus alalunga* (*cf.* tuna)

albatross 1 (*see also* albatross 2) popular name of diomedeid birds. Some species are among the largest of all extant flying birds, one of them, the wandering albatross, having a wing spread of more than 11 feet

black-browed albatross *Diomedia nigripes*

dusky albatross *Phoedetria fuliginosa* (= sooty albatross)

black-footed albatross *D. nigripes*

Laysan albatross *D. immutabalis*

yellow-nosed albatross *D. chlororhynchos*

royal albatross *D. epomophora*

snowy albatross (= wandering albatross)

sooty albatross (= dusky albatross)

short-tailed albatross *D. albatrus*

wandering albatross *D. exulans*

albatross 2 (*see also* albatross 1) any of several large australasian and asiatic pierid lepidopteran insects of the genus Appias. *Austral.* without qualification, usually *A. paulina.* (For *I.P.* species *see* puffin)

albedo the white, spongy mesocarp of a hesperidium

albefaction = blanching

albinism the condition of being an albino

[albinism] a mutant (gene symbol *al*) in the sex chromosome (linkage group I) of the domestic fowl. The albinism of the phenotype is rarely complete

[albinism] a mutant (gene symbol *c*) found at locus 0 in linkage group I of the rat. The name is expressive of the phenotype

albino properly an organism lacking all pigment but often applied to partial albinos

[albino] a mutant (gene symbol *c*) in linkage group I of the mouse. The phenotypic expression is simple albinism

[Albino-2] a mutant (gene symbol *al-2*) in linkage group I of *Neurospora crassa.* The name is descriptive of the phenotype

Albulidae a family of isospondylous fish containing the two species called bonefishes and ladyfishes. They are distinguished by the small mouth and the rear of the tongue and mouth are covered with coarse pavemental teeth

albumin any of numerous water- or saline- simple proteins. This term, particularly in *bot.*, is also used in the sense of "food reserve" or endosperm

albuminous pertaining to, or consisting of, albumin (*see also* seed)

exalbuminous said of a seed lacking a food reserve or endosperm

alburnous pertaining to sapwood

Alcedinidae that family of coraciiform birds which contains the kingfishers. They typically have very large heads and short necks and a heavy pointed bill. Most are very brightly colored

Alcelaphinae a subfamily of antelopid artiodactyle mammals having the horse-shaped body of the hippotragine antelopes but with longer and more slender legs, necks, and heads. They are sometimes called the deer-antelopes. The best known members are the hartebeests and the gnus

alchemist *see* alchymist

alchymist *Brit.* the noctuid lepidopteran insect *Catephia alchymista*

Alcidae a family of charadriiform birds containing the forms commonly called auks, guillemots, and puffins. They are predominantly short, squat birds, black above and white below, with very large heads and short necks. Unlike most webfooted birds they have strong sharp claws

Alcinae a subfamily of cervid artiodactyl mammals erected to contain the moose and *Eur.* elk

Alcippidae a family of abdominale cirripede crustacea with a weak stalk and a large chitinous disc of attachment

Alcyonacea an order of alcyonarian Anthozoa frequently called the soft corals. They occur in the form of fleshy masses from which the oral ends of the polyps protrude

Alcyonaria a class of anthozoan coelenterates containing, *inter alia*, the "soft corals", horny corals, sea fans, and sea pens. The Alcyonaria are clearly distinguished from the Zoantharia by having eight tentacles and eight septa

alder 1 (*see also* alder 2, aphis, beetle, borer, fly catcher, sawfly) originally many of numerous shrubs or trees of the betulaceous genus Alnus

green alder *U.S. A. michelliana—Brit. A. viridis*

smooth alder *A. serrulata*

speckled alder *A. rugosa*

alder 2 (*see also* alder 1) any of many varied shrubs and trees having a similar appearance, or habitat, to the true alders

black alder the aquifoliaceous shrub *Ilex verticillata* or the rhamnaceous shrub *Rhamnus fragula*

dwarf alder either the rhamnaceous shrub *Rhamnus alnifolia* or the hamamelidaceous shrub *Fothergilla gardenii*

white alder any of several shrubs of the coethraceous genus Clethra

witch alder *U.S.* the hamamelidaceous shrub *Fothergilla gardenii*

aldolase the term is usually applied to a widely specific enzyme catalyzing the production of aldehydes from ketose phosphates

oxohydroxybutyrate aldolase catalyzes the production of pyruvate and formaldehyde from 2-oxo-4-hydroxybutyrate

pentosealdolase catalyzes the production of formaldehyde and erythrulose 1-phosphate from ribose 5-phosphate

indoleglyceroephosphate aldolase catalyzes the production of indole and D-glyceraldehyde 3-phosphate from indole 3-glycerolphosphate

deoxyriboaldelase catalyzes the production of D-glyceraldehyde 3-phosphate and acetaldehyde from 2-deoxy-D-ribose 5-phosphate

ketotetrosealdolase catalyzes the production of dihydroxyacetone phosphate and formaldehyde from erythrulose 1-phosphate

threonine aldolase catalyzes the production of glycine and acetaldehyde from threonine.

allothreonine aldolase catalyzes the production of glycine and acetaldehyde from allothreonine

transaldolase = dihydroxyacetonetransferase

aldosterone a hormone secreted by the adrenal cortex. Active in the control of electrolytes, particularly potassium in blood

Alepisauridae a marine family of myctophoid fishes commonly called the lancet fishes. They are distinguished by the extremely elongate body (six feet) and long dorsal fins

Alepocephalidae a marine family of abyssal isospondylous fishes

alethe any of several turdid birds of the genus Alethe

brown-chested alethe *A. poliocephalia*

white chested alethe *A. fuellaborni*

firecrest alethe *A. diademata*

-aleto- *comb. form* meaning "vagrant"

Aleurodidae = Aleyrodidae

aleurone plant proteins, particularly those in the periphery of seeds

[Aleurone color] a mutant (gene symbol *C*) mapped at 26 on linkage group IX of *Zea mays*. The gene controls the color of the aleurone layer

alevin *Brit.* fish fry on which the yolk sac is still apparent. Particularly applied to salmonids

alexander *Brit.* either of two umbelliferous herbs: without qualification usually *Smyrnium elusatrum*

   golden alexander *Zizia aurea*

alexine a term at one time used for a postulated plant antitoxin

Aleyrodidae a family of homopteran insects commonly called whiteflies. They are well known greenhouse pests in the north, and citrus pests in the south, of the *U.S.*. The name derives from the waxy powder covering the wings which also protects them against most insecticides

alfalfa (*see also* aphis, beetle, bug, caterpillar, hopper, looper, midge, weevil, worm) the leguminous forage plant *Medicago sativa* (= lucerne)

   tree alfalfa *Medicago arborea*

alfonsino popular name of berycid fishes

alga any chlorophyll bearing thallophyte

   brown alga popular name of Phaeophyta

   green alga popular name of Chlorophyta

   blue-green alga popular name of Cyanophyta

   yellow-green alga popular name of Chrysophyta

   hymenial alga one in a sporocarp in a lichen

   red alga popular name of Rhodophyta

   soil alga any alga found in soil

alginase an enzyme catalyzing the breakdown of alginate through the hydrolysis of the β-1, 4-mannuronide link

alg^r a yeast genetic marker indicating inhibition of growth by allyl glycine

Alismaceae that helobean family of monocotyledons which contains the water plantains and arrowheads. Distinguishing characteristics are the exalbuminous seeds, numerous carpels, and whorled flowers with a differentiated perianth

alk = auk

alkanet any of numerous species of herbs of the boraginaceous genus Anchusa

   bastard alkanet *Lithospermum arvense* (= corn gromwell

-all- *see* -allo-

heal-all the labiateous herb *Prunella vulgaris*

-allag- *comb. form.* meaning "exchange", frequently written -allax- in compounds

allantoicase an enxyme catalyzing the hydrolysis of allantoate to glyoxylate and urea

allantoid sausage-shaped

allantoinase an enzyme catalyzing the hydrolysis of allantoin to allantoic acid

allantois an extraembryonic membrane growing out from the hind part of amniote embryos and which gives rise to the urinary bladder

   chorio-allantois a extraembryonic membrane formed by the union of the chorion with the allantois

-allax- *see* -allag-

archallaxis 1 (*see also* archallaxis 2) literally, an early or ancestral, change, but particularly one which occurs early in the phylogenetic history of a taxon or an embryo

   archallaxis 2 (*see also* archallaxis 1) the type of development which shows no trace of the recapitulation of ancestral characters

   morphallaxis the reorganization of existing tissues in the course of regeneration (*cf.* epimorphosis *under* morphosis)

trophallaxis the mutual exchange of food

Alleculidae a family of small coleopteran insects closely allied to the Tenebrionidae but distinguished from them by the pectinate tarsal claws, from which derives the name comb-clawed beetle

Allee's *epon. adj.* from W. C. Allee (1885–1955)

allele alternate forms of genetic characters that occur at the same locus on the chromosome are said to be alleles, or allelic to each other (*cf.* allelomorph *under* -morph)

   dominant allele one that determines the phenotypic expression in a heterozygous form

   isoallele one allele that so closely resembles another that it can only be distinguished by special techniques

   pseudoallele a group of closely linked loci once thought to be a single locus. Pseudoalleles do not complement, and recombine only rarely

   recessive allele one that produces the phenotypic expression only when in the homozygous state

allelism the relationship between two characters that are alleles

-allelo- *comb. form* meaning "mutual"

allelositism = syntrophy

allergy an immunological reaction induced by prior sensitization

alliance an obsolete botanical taxonomic rank equivalent to cohort

alligator (*see also* fish, lizard, snapper, turtle) properly any member of the crocodilian genus Alligator but also frequently applied to any crocodilian. In *Amer.* usage the term is also locally applied to any large saurian (e.g. Heloderma) or amphibian (e.g. Cryptobranchus). Even the larva of the Dobson fly (*see* hellgramite) is so named

   American alligator *A. mississipiensis*

alliinase = alliin lyase

Allioniaceae = Nyctaginaceae

-allo- *comb. form* meaning "other" in the sense of "different"

Alloeocoela an order of turbellarian platyhelminths distinguished by a digestive tract that is furnished with lobes or short, blunt diverticula

allometric *see* coefficient

-allotri- *comb. form* meaning "unusual" or "unsuitable"

Allotriognathi an order of deep water osteichthyes, allied to the Acanthopterygii, most possessing a protrusible jaw

allux the penultimate joint of the tarsus in rhynchopheran Coleoptera

almond 1 (*see also* almond 2, moth, willow) any of several species of trees of the rosaceous genus Prunus. Almond, without qualification, usually refers to *P. communis* grown for its edible nuts

   flowering almond *P. triloba*

   oriental almond *P. orientalis*

   Russian almond *P. nana*

almond 2 (*see also* almond 1) other trees resembling the almond, principally:

   Indian almond = tropical almond

   malabar almond = tropical almond

   tropical almond the combretaceous tree *Terminalia catappa* (= myrobalan)

[almondex] a mutant (gene symbol *amx*) mapped at 27.7- on the X chromosome of *Drosophila melanogaster*. The phenotypic expression involves rough narrow eyes and sterile females

alo *see* [alopecia]

aloe a term properly applied to the lileacous genus Aloe also loosely used for several species of agave and a few other plants

   bastard aloe *Agave vivipara*

   hedgehog aloe *Agave humilis*

   water aloe the hydrocharitaceous aquatic herb *Stratiotes aloides*

Aloididae = Corbulidae

[Alopecia] a mutant (gene symbol *Al*) in linkage group VII of the mouse. The phenotypic expression involves the production of bald patches from the second or third month on

[alopecia] a mutant (gene symbol *alo*) mapped at 38.3± on the X chromosome of *Drosophila melanogaster*. The phenotype is "bald" in the sense it lacks microchaetae

Alopiidae a family of sharks, called thresher sharks by reason of the distinctive, long, whip-like tail.

alp a small pasture in high mountain country. Since the European Alps were named for the frequency of such pastures, the term is often misused to mean a high mountain

alpaca any of several species of camelid artiodactyles of the genus Lama (*cf.* llama, huanaco)

alpestrine properly applied to plants growing above tree level but, frequently misused as synonymous with alpine 1 (*q.v.*)

Alpheidae a family of natant decapod crustacea usually confused with the Crangonidae. The Alpheidae have both of the first pairs of legs chelate and the first pair is usually much larger than the other. Commonly called snapping shrimps

alpine 1 (*see also* alpine 2) properly applied to organisms occurring in alps (*i.e.* high mountain meadows). It is frequently, however, used as synonymous with alpestrine and applied to all organisms or habitats of mountainous regions

alpine 2 (*see also* alpine 1) *U.S.* any of several species of satyrid lepidopteran insect of the genus Erebia

Alpineaceae = Zingiberaceae

-als- *comb. form* meaning a "grove

alsad a plant of woody groves

alsike occasional abbreviation for alsike clover (*q.v.*)

Alsinaceae = Caryophyllaceae

alsinaceous having a short claw

Alstroemeriaceae a tribe of amaryllidaceous monocotyledons

-alt- *comb. form* meaning "high"

exaltatous tall

-alter- *comb. form* meaning "other"

   sesquialter said of a flower in which there are one and one half times as many stamens as there are petals or sepals

Alternifoliae = Monocotyledoneae

Althaea properly the malvaceous genus to which the hollyhock (*q.v.*) belongs. Used also by gardeners to describe the rose of Sharon (*q.v.*).

   shrubby althea the malvaceous shrub *Hibiscus syriacus* (rose of Sharon)

-althe- *comb. form* meaning "to increase"

Altingiaceae = Hamamelidaceae

-alto- *comb. form* meaning "high"

altricial said of birds hatched in a condition which requires that they receive parental care

Alu *see* [Alula]

Alucitidae a family of pyraloid lepidopteran insects closely allied to the Pterophoridae but having the wings spread into even more plumes

Aluconidae = Tytonidae

alula 1 (*see also* alula 2) the thumb-like feathered digit on the wing of a bird

alula 2 (*see also* alula 1) the expanded basal area of the wings of some Coleoptera and Diptera, and the posterior basal lobe adjacent to the halteres of some Diptera

[Alula] a mutant (gene symbol *Alu*) mapped at 54.9 on chromosome II of *Drosophila melanogaster*. In the phenotype, the alula is fused to the wing.

alutaceous with a tough, cracked surface, like dried out leather. A buff color is also usually implied

alveolar 1 (*see also* alveolar 2 layer) pertaining to that portion of the ectoplasm of a ciliate protozoan which lies immediately beneath the pellicle

alveolar 2 (*see also* alveolar 1, pertaining to the tooth socket of the mammalian jaw

alveolus 1 (*see also* alveolus 2) a terminal cavity of a hollow lobular structure. When used, as is nowadays rarely the case, in distinction from acinus, the alveolus is conical and the acinus more or less spherical or ovate

alveolus 2 (*see also* alveolus 1) one of the division of the gigantic compound sucker of aspidogastrid trematodes

alyssum properly the name of a genus of cruciferous herbs (*see* goldentuft *under* tuft 2). Also applied by gardeners to species of the allied genus Lobularia

am *see* [Amination-deficient]

-ama- *comb. form* meaning "together"

amadou the dried outer portion of the fungus *Polyporus fomentarus*, used at one time for tinder and now for drying trout flies

amakihi the drepaniid bird *Loxops virens*

Amanitaceae a family of agaricale basidomycete fungi best known for the notoriously poisonous forms found in it. The family is distinguished by the presence of a volva and annulus and the fact that the gills are not attached to the stalk

Amarantaceae that family of centrospermous dicotyledons which contains the amaranths. The single perianth and one-seeded fruits are distinctive

amaranth anglicized form of the name of the amaranthaceous genus of dicotyledonous herbs Amaranthus

   globe amaranth the amaranthaceous herb *Gomphrena globosa*

Amaranthoideae a subfamily of Amaranthaceae distinguished from the Gomphrenoidea by having a four-celled anther

Amaryllidaceae that family of liliflorous monocotyledons which contains not only amaryllis, the tuberose and narcissus, but many forms commonly called "lilies" such as the Guernsey lily, the spider lily, etc. Distinguishing characteristics are the inferior 3-celled ovary, the 6-partite perianth and the 6 stamens with introrse anthers

Amaryllidoideae a subfamily of amaryllidaceous monocotyledons

-amath- *comb. form* meaning "sandy soil"

amathad a plant of sandy places

Amathusiidae a family of large nymphaloid lepidopteran insects confined to the Southern Hemisphere. They are distinguished by broad wings and the strongly arched cluster of the forewing. Many of the Latin American species (*cf.* morpho) are brilliantly colored (*cf.* Morphidae)

Amatidae a family of noctuoid lepidopteran insects many of which have a wasp-like appearance with long transparent wings and a banded body

amatungulu the apocynaceous shrub *Carissa bispinosa*

amazon any tough and aggressive female, but particularly those of the slave-making ant. The term is also frequently applied to the Amazon parrot

shining amazon the ant *Polyerges lucidus*

-amb- *comb. form* meaning "around" or "on both sides"

amb *see* [amber]

ambary the malvaceous herb *Hibiscus cannabinus*

Ambassidae = Centropomidae

ambatjang the anacardiaceous tree *Mangifera foetida* and its fruit

amber fossilized, or partially fossilized, natural resins. Used also to designate a golden-brown color

sweet amber the hypericaceous shrub *Hypericum androsaemum*

[**amber**] a mutant (gene symbol *amb*) mapped at 6.8± on the X chromosome of *Drosophila melanogaster*. The phenotypic expression is a pale yellow body and an abundance of sterile males in the population

ambitus literally, circuit or orbit, but applied particularly in *biol.* to the outline of objects, such as a sea urchin, when viewed vertically from above or beneath, so that a heart shaped urchin is said to have a "cordiform ambitus"

Amblycephalidae a small family of New World colubrid snakes, distinguished by the blunt head and large anterior maxillary teeth

Amblyopsidae a family of cave dwelling cyprinodontoid fishes distinguished by the great development of sensory papillae on the body and sensory tactile organs beneath the epidermis

Amblypygi a small order of arachnid arthropods at one time united with the Uropygi as the Pedipalpi. They are distinguished by the fact that the tarsus of the first walking legs is modified into an extremely long tactile organ

Amblystomatinae a common misspelling of Ambystominae

ambon the fibrous ring around the socket of a ball-and-socket joint

Ambulacralia a zoological taxon erected by Hatschek to contain the Echinodermata and the Enteropneusta

ambulacrum 1 (*see also* ambulacrum 2, 3) a walking leg, particularly of insects and insect larvae

ambulacrum 2 (*see also* ambulacrum 1, 3) one of the, usually five, radiating grooves in which the tube feet of echinoderms are placed

ambulacrum 3 (*see also* ambulacrum 1, 2) an adhesive disc, formed of an aggregate of hooks, that terminates the tarsus of ixodid ticks

ambulation the act of moving about

amby *U.S.* the larva of the odbson fly *Corydalus cornutus* (= hellgrammite)

Ambystomatinae a subfamily of salamandrid urodele amphibia distinguished by the fact that the palatal teeth are restricted to the posterior portion of the vomers

ameboid pertaining to, or resembling, any rhizopod protozoan, particularly Amoeba

amebula diminutive of ameba, usually applied to ameboid stages in protozoan life histories

ament an inflorescence consisting of a closely bracted spike bearing many flowers: commonly called a catkin (*cf.* julaceous)

Amentiferae a now-obsolete plant taxon erected to contain all those forms (*e.g.* Salicales, Fagales, etc.) that bear catkins

Amera a zoological taxon erected by Lamarck to contain the sponges, hydrozoan coelenterates, and some large protozoa, all then considered to be Vermes. (*cf.* Polymera and Oligomera)

[**Ames' waltzer**] a mutant (gene symbol *av*) found in linkage group IV of the mouse. The phenotypic expression is a typical waltzing syndrome

amidase a general term for an enzyme catalyzing the hydrolysis of numerous carboxylic acid amides to the corresponding carboxylic acids and ammonia

formamidase an enzyme catalyzing the hydrolysis of formylamines to formates

muramidase an enzyme catalyzing the hydrolysis of the wall substances of certain bacteria (= lysozyme)

penicillin amidase an enzyme catalyzing the hydrolysis of benzylpenicillin to phenylacetate and penicillin

phosphoamidase an enzyme that catalyzes the hydrolysis of phosphoamides to amides and orthophosphates

amidavad the ploceid bird *Estrioda amandada*

Amiidae the only extant family of a group once known as the Protospondyli (now included in the Holostei) and of which *Amia calva* (= bowfin) is the only extant member.

aminase a general term for enzymes catalyzing reactions that produce urea or ammonia

glutaminase an enzyme catalyzing the hydrolysis of glutamine to glutamate and ammonia

glycocyaminase an enzyme catalyzing the hydrolysis of guanidinoacetate to glycine and urea

transaminase = aminotransferase

deaminase a group of enzymes, more properly called amino hydrolases, that catalyze the liberation of ammonia by hydrolysis

adenine deaminase catalyzes the hydrolysis of adenine to hypoxanthine and ammonia

adenosine deaminase catalyzes the hydrolysis of adenosine to inosine and ammonia

ADP deaminase catalyzes the hydrolysis of ADP to IDP and ammonia

AMP deaminase catalyzes the hydrolysis of AMP to IMP and ammonia

arginine deminase an enzyme catalyzing the hydrolysis of arginine to citrulline and ammonia

formiminotetrahydrofolate cyclodeaminase catalyzes the production of 5, 10-methenyl-tetrahydrofolate and ammonia from 5-formiminotetrahydrofolate

cytidine deaminase catalyzes the hydrolysis of cytidine to uridine and ammonia

cytosine deaminase catalyzes the hydrolysis of cytosine to uracil and ammonia

guanine deaminase catalyzes the hydrolysis of guanine to xanthine and ammonia

serine deaminase = serine dehydratase

[**amination deficient**] a mutant (gene symbol *am*) in linkage group V of *Neurospora crassa*. The phenolypic expression is the necessity of securing nitrogen from ∝-amino

aminidase a general term for an enzyme catalyzing the hydrolysis of terminal links in various mucopolysaccharides

[**p-Aminobenzoic-1**] a mutant (gene symbol *pap-1*) in linkage group III of *Chlamydomonas reinhardi*. The phenotypic expression is a requirement for *p*-aminobenzoic acid

[**p-Aminobenzoic-2**] a mutant (gene symbol *pap-2*) in linkage group I of *Chlamydomonas reinhardi*. The phenotypic expression is a requirement for *p*-aminobenzoic acid

Ammanniaceae = Lythraceae

ammer the *Ger.* word for finch converted to hammer (*q.v.*) by upper class Victorian English

Ammiaceae = Umbelliferae

-ammoch- *comb. form* meaning "sand"

ammochthad a plant of sandbanks

ammocoete *see* larva

Ammon's almost universal misspelling of Amon's (*q.v.*)

ammonifier an organism, particularly bacteria, that liberates ammonia

amnion 1 (*see also* amnion 2, 3) that embryonic membrane (*q.v.*) which encloses the embryos of reptiles, birds, and mammals. In telolecithal, and similar, forms it arises as folds over the head and tail from the extraembryonic blastoderm

    false amnion = chorion

    proamnion the beginnings of an amniotic fold consisting entirely of ectoderm

amnion 2 (*see also* amnion 1, 3, 4) the thick outer wall of the invaginated ectodermal disk in Nemertinea

amnion 3 (*see also* amnion 1, 2, 4) a membrane resembling an amnion that encloses the embryo of many insects and some other invertebrates

amnion 4 *see also* (amnion 1, 2, 3) clear sap surrounding plant ovules

amnios the cast skin of the first larval stage of an insect

amniote šaid of an organism, or group of organisms, that possesses an amnion. Frequently used in the sense of "higher" vertebrate, that is the reptiles, birds, and mammals

    anamniote antithesis of amniote used principally to distinguish the "lower" from the "higher" vertebrates

Amoebida that class of sarcodinous Protozoa which contains the amebas. The method of locomotion with pseudopodia is characteristically amoeboid

amole the liliaceous herb *Chlorogalum pomeridianum*

Amon's *epon. adj.* from Amon (frequently spelled Ammon by anatomists) an Egyptian god who gave his name to ammonia and whose symbol was a horn— hence Amon's horn (*cornu Amonis*) in the hippocampus. The spelling Ammon's may derive from a mistaken association with Freidrick August von Ammon, an early 19th century anatomist

AMP adenosine monophosphate (*see also* nucleosidase)

-ampel- *comb. form.* meaning "vine"

    pterampelid a climbing fern

Ampelidaceae = Vitaceae

Ampelidae = Bombycillidae

Ampeliscidae a family of tube-dwelling gammaridean amphipod crustaceans

Ampharetidae a family of sedentary polychaete annelids closely resembling the Terebellidae but distinguished from them by having a bundle of setae on each side of the head

-amphi- *comb. form* meaning "around" or "double"

Amphibia a class of gnathostomatous craniate chordates commonly called amphibians. They are distinguished by the absence of keratin scales, hairs, or feathers in the skin. Among the frogs, salamanders, caecilians and the like, enough are amphibious to justify the name

amphibious said of an organism that spends, or is capable of spending, part of its life in water and part on land

Amphibrya an obsolete botanical taxon, equivalent to monocotyledon

Amphicoela a small suborder of anuran amphibia possessing amphicoelous vertebra

Amphictenidae a family of sedentary polychaete annelids forming sand tubes which can be moved about with them. There are short tentacular filaments on the head

amphid variously shaped pits, or tubes, found at the anterior end of nematode worms and presumed to be sense organs

Amphidiscophora an order of hexactinellid sponges, with amphidisc spicules but lacking hexasters

Amphilinidea an order of cestodarian cestodes distinguished by a protrusible proboscis, and by the posterior position of the male and female pores

Amphineura a class of Mollusca commonly called chitons in English though this term is properly applied only to the subclass Polyplacophora

amphiont = zygote

Amphioxi the only class of the subphylum Cephalochordata. It contains the lancelets that are distinguished from other acraniates by retaining throughout life the notochord, nerve chord, slit-like mouth and metamerically segmented muscles

Amphiperatidae a family of gastropod mollusks with a long and slender shell resembling that of a cowry, but with a slender thin aperture lacking teeth

Amphipoda an order of pericarid malacostracan crustacea in which there is no distinct carapace and the first thoracic somite is coalesced with the head. They are distinguished from the closely allied Isopoda by the laterally compressed body, a distinct suture between the telson and the last somite, and the biramous first antennae, The group contains those forms commonly known as sand fleas, sand hoppers, and scuds or side swimmers

Amphisbaenidae a family of mainly limbless, burrowing, vermiform saurian reptiles in which the scales are fused into segment-like rings

Amphistomate a division of prosostomate trematodes with the oral sucker at the front end and the acetabulum at the posterior end

Amphiumidae a small family of urodele amphibia distinguished by the presence of four limbs, maxillary bones, a persistent spiracle, and the absence of moveable eyelids

Amphizoidae a small family of aquatic coleopteran insects with the unusual characteristic of being unable to swim

amplexus an embrace, particularly a sexual embrace, and specifically the sexual embrace of anuran Amphibia

ampliate enlarged

amplification the act of increasing the magnitude of a signal, usually applied to electronic devices

Ampulicidae a family of sphecoid hymenopterans, black in color, and with the head on a fairly elongated neck. They are predatory on roaches

-ampull- *comb. form.* meaning a "narrow necked vessel"

ampulla Used in *biol.* for almost any subspherical hollow body including the muscular sac at the base of an echinoderm tube foot, the cavity in the gonophore of a millipore hydrozoan, the flotation bladders of brown algae, etc., etc. Specifically designated ampullae are:

    Lorenzini's ampulla one of the blindly ending slime filled pits, presumably of sensory function, found on the cheek and snout of chondrichthyean fishes

    Savi's ampulla an analog of Lorenzini's ampulla found in Torpedo and some other elasmobranchs

    Vater's ampulla a swelling in the pancreatic duct at the point of junction with the intestine

Ampullariidae a family of ctenobranchiate gastropod mollusks having both gills and lungs. They are the very large fresh water forms popularly called apple snails

amra the anacardiaceous tree *Spondias pinnata* and its edible fruit

amx *see* [almondex]

amyc *see*[Amycelial]

[Amycelial] a mutant (gene symbol *amyc*) in linkage group I of *Neurospora crassa*. The phenotypic expression is the production of dot-like colonies

Amydriidae a family of lepidopteran insects now usually included in the Tineidae

Amygdalaceae a family of rosale dicotyledons that are usually treated as the subfamily Prunoideae of the Rosaceae. The almonds are the best known examples

-amyl- *comb. form* meaning "starch"

amylaceous starchy

amylase an enzyme hydrolyzing 1,4-glucan links, in starch, glycogen, and related polysaccharides

  glucoamylase removes successive glucose units from the non-reducing ends of the polysaccharide chains

  [amylase negative] *either* a mutant (gene symbol *be* mapped at 1.1 on linkage group VIII of *Bombyx mori*. The phenotypic expression is a low concentration of amylase in the hemolymph *or* a mutant (gene symbol *ae*) mapped at 0 on linkage group VIII of *Bombyx mori*. The phenotypic expression is a low concentration of the amylase in the digestive fluid

amylome the starch containing portions of a plant

amylum *Lat.* for starch and so used in some compound terms

  paramylum any carbohydrate granule, not identifiable as glycogen, found in an animal cell

amymone *U.S.* the nymphalid lepidopteran insect *Mestra amymone.*

an *see*[erythrocyte agglutination] or [anemia]

an₁ *see*[anther ear]

-an- *see* -a-

ana- *comb. prefix* confused between three roots and therefore meaning "without", "up", and "double"

Anabantidae a family of percomorph fishes with the characteristics of the suborder. The paradise fish and gourami are well known examples of this family

Anabantoidea a monofamilial suborder of acanthopterygian fishes distinguished by the presence of a labyrinthine breathing apparatus located in a cavity above each gill slit. From this characteristic the name labyrinth fish derives

Anablephidae a monogeneric family of cyprinodontoid fishes commonly called four-eyed fishes by reason of the separation of the eye into a surface and subsurface portion

Anacanthini an order of actinopterygian fishes containing the codfish and their allies. The are distinguished by the fact that the pelvic fins are anterior to the base of the pectoral fins

Anacardiaceae that family of sapindale dicotyledons which contains, *inter alia,* the mangos, the pistachio and the cashews. It is distinguished from the closely related Sapindaceae by the presence of resins. Many also contain irritant oils

Anacardiineae a suborder of Sapindales containing the single family Anacardiaceae

anacerores anal wax- or honey-producing glands of coccid hemipterans

anaconda the bovine pythonid snake *Eunectes murinus.* Popularly supposed to the the largest extant snake

anal (*see also* vein) pertaining to the anus in all the meanings of this word

  adanal in the vicinity of the anus

analogous used in *anat.* to describe structures of similar function but different phylogeny (*e.g.* the vertebrate and invertebrate eye). Also used in *ecol.* to describe organisms of similar habitat or distribution (*cf.* homologous)

analog analogous structure

Anaplasmataceae a family of rickettsiale microtatobiotes of extremely small size found in the erythrocytes of vertebrates. They can be regarded as intermediate between the rickettsias and the viruses

Anapsida a subclass of reptilian chordates containing the tortoises and turtles. They are distinguished by the possession of a bony or leathery shell and by the absence temporal openings in the skull

Anarhichadidae a small family of very slender blennioid fishes lacking pelvic fins. The name wolffish derives from the large canine and molar teeth

[anarista] a mutant (gene symbol *aa*) mapped at 0± on chromosome III of *Drosophila melanogaster*. The aristae of the phenotype are not lacking but are very small

Anaspidacea a small order of syncarid malacostracan crustacea distinguished by lacking a carapace and by the fact that the thoracic limbs bear both exopodites and a double series of lamellar epipoda attached to the outer side of the coxopodites

anastomose for the entomological use of this word see inosculating (under -osculating)

anastomosis a union between two things, originally between two seas. Now used principally of blood vessels, nerves or hyphae

  Huxley's anastomosis an anterior transverse connection between the lateral excretory canal of rotifera

Anatidae that family of anseriform birds which contains the ducks, geese, swans, and magpie geese. They are distinguished by the short legs, broad bill and webbed feet

Anatinae a subfamily of anatidae containing all the true ducks. They are divided into the Anatini, Someteriini, Aythyini, Cairinini, Mergini, Oxyurini, and Tadornini

ancad a plant of canyons (*cf.* -anko-)

anceps two-edged

ancestrula a colony of animals produced by asexual reproduction from a metamorphosed sexually produced larva: particularly the original zooid from which a colony of Ectoprocta is derived

  Tata ancestrula a type of ancestrula with spines round the orifice at one time thought to be a separate genus Tata

anchor the term is used for anchor-shaped ossicles of Echinodermata or spicules of sponges: also for the collectocystophore (*q.v.*) of Stauromedusae

anchoveta the engraulid fish *Centengraulis mysticetus*

anchovy any of many species of isospondylous fishes of the family Engraulidae

anci- *see* anko-

ancipital two edged

ancipitous in the form of a flattened cylinder having two lateral arms

Ancistrocladaceae a small family of parietale dicotyledons

Ancistrocladineae a suborder of parietale dicotyledons containing the single family Ancistrocladaceae

ancistrous barbed

anconal pertaining to the dorsal surface of the wing of a bird

Ancylostomidae a family of nematode worms commonly called hook worms distinguished by the large mouth opening into a buccal cavity and by the possession of two ventrolateral cutting plates, with or without teeth at the entrances of the large buccal capsule

ander male

    micrander a small male

    nannander a dwarf male

-andr- *comb. form* meaning "male

perigynandra the involucre of Compositae

Andraeobrya that subclass of muscose bryophytes which contains the forms commonly called granite mosses. They share many characters both with the sphagnobrya and the eubrya but are distinguished by the longitudinal dehiscence of the capsule into four valves

anandrarious said of a flower lacking stamens

Andrenidae a family of apoid apocritan hymenopteran insects containing the forms commonly called mining bees or burrower bees. They may be distinguished by the presence of two subantennal sutures. They make complex underground nests, with lateral spurs running out from the main shaft

-andria plant or animal taxa having this ending are listed alphabetically

synandria a group of males living together (*cf.* syngynia)

andrium the male portion of a flower

    clinandrium that part of the column in the flower of an orchid, in which the anther is concealed

    synandrium a group of fused anthers

Andropogoneae a tribe of the plant family Gramineae

[Andrus' green stem] a mutant (gene symbol *ag*) mapped at 139 on chromosome 10 of the tomato. The phenotypic expression is a green stem but the cotyledons are purple

-androus pertaining to a male, or male part. The substantive form terminates in -andry and the alternative adjectival form - andric is occasionally used. Compounds so formed are not listed separately

    adynamandrous the condition of being self-sterile

    apandrous said of a male organ that has lost its function, particularly in plants

    diandrous having two stamens

    ergatandrous pertaining to those ants in which the ergates appear to be males

    gynandrous used of a plant in which the stamens are adnate to the pistil

    heterandrous having stamens of various sizes

    homoeandrous having only one kind of stamen

    macrandrous having a large male, though the term should properly be megandrous

    meroandrous having fewer testes than is normal

    metandrous said of a plant in which the female flower is mature before the male

    nanandrous producing dwarf males

    polyandrous literally having many male sexual partners, but also said of a flower having numerous stamens

    protandrous the condition of a hermaphrodite in which the male portion develops first or which is first male, and later sex reversed to female. Also said of a flower in which the pollen matures before the stigma is receptive

    proterandrous said of a flower in which the pollen matures before the stigma is receptive.

    ecoproterandrous having the male flowers mature before the female.

    phytoproterandrous the condition of having ripe pollen before the stigma is receptive

    spanandrous the condition of an organism in populations in which males are only very sparsely found

anellus one of the supporting structures of the aedeagus

Anelytropsidae a monotypic family of limbless lacertillian reptiles distinguished from the Scincidae by the absence of both a pectoral girdle and temporal arches in the skull

-anem- *comb. form* meaning "wind"

anemone 1 *see also* anemone 2) 1. any of numerous ranunculaceous herbs, mostly of the genus Anemone

    rue-anemone *U.S.* the ranunculaceous herb *Anemonella thalictroides*.

anemone 2 (*see also* anemone 1) animals thought to resemble the flower

    sea anemone any non-skeletogenous anthozoan coelenterate

Anemoneae a tribe of the family Ranunculaceae

aner a male ant

    dorylaner a large male ant with long mandibles and a cylindrical abdomen

    ergantaner an apterous male ant

    gynaecaner a male ant which closely resembles a female

    macraner a large male, particularly a male ant

    mermithaner a male ant structurally modified in consequence of being parasitized by mermithid nematodes

    micraner a dwarf aner

    phthisaner a pupal male ant that has wasted away through the attack of an orasemal parasite

-ang- *comb. form* meaning a "hollow container"

ang *see* "angle wing"

angel (*see also* eye) a term fancifully applied to some organisms

    archangel *Brit.* the labiateous herb *Galeobdolon luteum*

    yellow archangel the labiateous herb *Lamium galeobdolon*

    destroying angel the amanitaceous basidomycete *Amanita verna.* There is no recorded case of the survival of an individual who has ingested significant quantities of this fungus. Even very small quantities are frequently lethal

    sun angel any of several trochilid birds of the genus Heliangelus

Angelica a genus of umbelliferous herbs. The candied angelica of commerce comes from Archangelica

-angio- *see* -aggeio-

Angiopteridaceae a taxon which is variously regarded as a separate family of Marattiales (*q.v.*) or a subfamily of the family Marattiateae (*q.v.*)

Angiospermae those spermatophytae (*q.v.*) in which the seeds are contained in an overy (*cf.* Gymnospermae)

Angiospermia in the Linnaean system of classification that order of the didynamia which contains a single many-seeded ovary

angium any container, particularly, in *bot.*, those concerned with reproduction

    anterangium a sporocarp containing both macro- and micro- spores

    gametangium that organ of lower plants in which the gametes are developed

androgametangium = antheridium

gynogametangium an organ producing eggs

progametangium a resting body in Protomyces

gennylangium = anther

gonangium a hydrozoan coelenterate gonotheca with its enclosed blastostyle

goniangium a common term for cystocarp and scyphi

gonidangium an organ producing a sexual spore

microdiodangium = pollen sac

monangium either a sorus containing a single sporangium or a sporangium with a cucullate indusium

ooangium embryo sac (*bot.*)

sporangium the spore-bearing organ of ferns and spermatophytes

megasporanguim that part of the ovule lying inside the integument of spermatophyte

microsporangium that part of the stamen which contains the pollen

spermatangium the male sex organ of red algae

synangium *bot.* a compound sporangium found in some lower plants *zool.* a common sac like trunk from which arteries arise

angle (*see also* wing) in *biol.* a form produced by the junction of two more or less straight lines, particularly applied to insect wings

anal angle the angle of the hind wing of an insect that lies nearest to the abdomen

apical angle the angle of the apex of an insect wing

costal angle the angle at the tip of an insect wing

[**angle wing**] a mutant (gene symbol *ang*) mapped at 10.5± on chromosome II of *Drosophila melanogaster*. The name is descriptive of the slightly raised wings of the phenotype

[**angora hair**] a mutant (gene symbol *l*) on locus 14.3 on linkage group II of the rabbit. The phenotype has unusually long hair

Anguidae a family of lacertilian reptiles, some with greatly reduced limbs, with pleurodont dentition, and a tongue in which the anterior portion can be retracted into the posterior

Anguillidae a family of apodous fish containing the freshwater eels. They are distinguished from other freshwater eels by the presence of small embedded cyloid scales, gill slits on the side of the head, and the possession of pectoral fins

Anguillulidae a large family of free-living, ascaroid nematode worms containing, *inter alia,* the well known vinegar eel which is typical of the group

Anguilliformes = Apodes

Anguinidae = Anguidae

angular pertaining to an angle

acutangular said of a plant, the stems of which, in section, show sharp angles

multangular *see* bone

Anhimidae that family of anseriform birds which contains the forms popularly known as screamers. They are distinguished either by a long frontal crest, or a long frontal spike

anhinga either of two species of anhingid birds *U.S.* without qualification *Anhinga anhinga*

Anhingidae that family of pelecaniform birds which contains the anhingas; closely allied to the cormorants but distinguished from them by the absence of the recurved tip at the end of the bill

ani any of several species of cuculid bird of the genus Crotophaga

groove billed ani *C. aulcirostris*

smooth billed ani *C. ani*

greater ani *C. major*

ania a projection arising from the base of the culmen on a feather

anianiau the drepaniid bird *Loxops parva*

Aniliidae a small family of primitive burrowing snakes, closely allied to the Uropeltidae, with solidified skull bone articulations, compressed teeth, and short, sometimes modified, tails

animal any organism that lacks chlorophyll and that is not thought, as are fungi and some other parasitic plants, to be immediately descended from chlorophyll bearing ancestors. Most animals have locomotor responses to external stimuli

animalcule diminutive of animal. The term was at one time applied to all microscopic animals and is still sometimes used

barrel animalcule any of several species of ciliate protozoan of the genus Coleps

slipper animalcule various species of ciliates of the genus Paramoecium

wheel animalcule now rarely heard term for Rotifera

Animalia (*see also* animal) that kingdom of organisms which contains the animals. There is no single characteristic diŝtinguishing animals from plants

animalization excessive development of animal pole characters (*e.g.* cilia) in the egg of an echinoderm

anise the aromatic umbelliferous herb *Pimpinella anisum*

Japanese star anise the magnoliaceous shrub *Hicium religiosum*

-aniso- *comb. form* meaning "unequal"

Anisochytridiales a small order of phycomycete fungi of uncertain affinities. These marine parasites have two flagellae but a cell wall of cellulose

Anisopodidae = Rhyphidae

Anisoptera a suborder of odonatan insects in which the hind wings are wider than the front wings and the wings are held horizontally when at rest. Commonly called dragonflies

Anisotomidae = Leiodidae

anko- *comb. form* meaning a "hollow" or just possibly a "glen." Used by ecologists, in the transliteration -anci-, in the sense of "canyon"

anlage a *Ger.* word of numerous meanings used by some embryologists in the sense of "precurser" or "rudiment"

annectent pertaining to forms intermediate between other forms

Annelida a phylum of bilateral coelomate animals containing, *inter alia,* the earthworms, bristle worms, and leeches. They are distinguished from other worms by their metameric segmentation

Annonaceae that family of ranale dicotyledons which contains, *inter alia,* the custard-apple and the pawpaw. Distinguishing characteristics are the tripartite calyx and corolla with numerous spiral stamens

annotinous pertaining to the branches produced in the previous year

annual (*see also* -ennial) that which recurs each year

plurannual a cultivated plant which, though commonly perennial, is grown as an annual in an adverse climate

annulet *Brit.* the geometrid lepidopteran insect *Gnophos obscurata*

annulus literally a "ring" and used in *biol.* for numerous annular structures particularly a thin membrane extending from the stalk to the rim of the cap in many agaricale basidomycete fungi

anoa the Celeban buffalo *Anoa depressicornis* (*cf.* tamarou)

Anobiidae a large family of coleopteran insects allied to the Tenebrionidae but clearly distinguished by the

11-jointed antennae and the fact that the head is bent down under the prothorax. Two species are destructive pests of stored vegetable products; these are the drugstore beetle and the cigarette beetle. The death-watch beetles also belong in this family

anodic in *biol.,* means "upwards"

anole *U.S.* widely used anglicized form for the iguanid saurian reptilian genus Anolis (*cf.* chameleon)

Anomalodesmacea an order of pelecypod Mollusca having the mantle lobes more or less completely united with two siphonal and one pedal opening between them

Anomalopsida a marine family of berycoid fishes distinguished by the posession of a band of luminous bacteria that may be intermittently covered

Anomaluridae a family of sciuromorph rodents containing a small number of African flying squirrels. They differ from the holarctic flying squirrels (which are in the Sciuridae) by the presence of a series of scales, known as the climbing organ, at the root of the tail

Anomocoela a suborder of anuran Amphibia containing the toad-like creatures known as spade-foots

Anomiidae a family of pelecypod mollusks commonly called jingle shells. They are distinguished by the thin transparent shell usually permanently attached by a calcified byssus

Anomiopsyllidae = Dolichopsyllidae

Anomura a division of reptant decapod crustacea with the abdomen usually soft or bent under itself though less so than in the Brachyura. This division includes, *inter ali*, the hermit crabs and their allies. They lack a chela on the third pair of legs

anon the annonaceous tree *Annona squamosa* and its edible fruit

anonilla the annonaceous tree *Rollinia jimenizii* and its edible fruits

Anophyta = Bryophyta

Anopla a subclass of Nemertea distinguished by having the mouth posterior to the brain

-anoplo- *comb. form* meaning "unarmed"

Anoplopomatidae a small family of scleroparous actinopterygian fishes commonly called sablefish

Anoplura that order of insects, allied to, and once united with the Hemiptera which contains the forms commonly called sucking lice. They are characterized as wingless mammalian parasites with mouth parts adapted to sucking

Anostraca an order of branchiopod crustacea commonly called fairy shrimps. They are distinguished by their elongate body lacking a carapace and their stalked eyes. They also swim upside down

Anseranatinae a subfamily of Anatidae containing the magpie geese

Anseriformes that order of birds which contains the ducks, geese, swans and screamers. They are distinguished by the flattened bill and webbed feet

Anserinae a subfamily of Anatidae including the whistling ducks, the swans and the geese as opposed to the Anatinae containing the true ducks and the Anseratinae or magpie geese

anserinase = aminoacyl-methylhistidine dipeptidase

L-**anserine** *see* carnosine. N-methylcarnosine. A little-known amino acid isolated from some birds including, as the name indicates, geese

ansulate said of a hoop-shaped structure coiled at the apex

ant- *comb. prefix* meaning "against," "instead" or "opposite." Takes the form anti- before consonants (*cf.* ante-)

ant (*see also* beetle, bird, eater, pipit, pitta, shrike, tanager, thrush vireo, and wren) popular name of hymenopterous insects of the superfamily Formicoidea

agricultural ant = harvester ant

amazon ant = slave-making ant

Argentine ant *U.S. Iridomyrmex humilis*

army ant *U.S.* popular name of doryline ants. Elsewhere, any ant living in very large colonies which move together. Many are of the genus Anomma

black ant *Brit. Lasius fuliginosus.*
  little black ant *U.S. Monomorium minimum*

bulldog ant *Austral.* any of numerous species of the genus Myrmecia

callow ant a worker ant newly emerged from the cocoon

carpenter ant any of numerous formicines of the genus *Camponotus* which burrow in wood.
black carpenter ant *U.S. Camponotus pennsylvanicus*
  Florida carpenter ant *U.S. Camponotus abdominalis floridanus*
  red carpenter ant *U.S. Camponotus ferrugineus*

cornfield ant *U.S. Lasius alienus*

cow ant *Brit.* popular name of mutillid wasps (*cf.* velvet ant)

crazy ant *U.S. Paratrechina longicornis*

leaf-cutting ant any of several species of the genus Atta.
leaf-cutting ant *U.S. Atta texana*

driver ant any of numerous species of the genus *Dorylus*

fire ant *U.S. Solenopsis geminata.*
  imported fire ant *U.S. Solenopsis saevissima*
  little fire ant *U.S. Wasmannia auropunctata*
  southern fire ant *U.S. Solenopsis xyloni*

fungus ant any of numerous species of myrmicine formicids of the genus Trachymyrmex which feed on fungi cultivated in the nest

garden ant *Brit. Lasius niger.*
  fungus garden ant = leaf cutting ant

harvester ant any of numerous myrmicine ants mostly of the genera Pogonomyrmex and Pheidole. They have this name because they store seeds.
California harvester ant *Pogonomyrmex californicus*
Florida harvester ant *U.S. Pogonomyrmex badius*
Red harvester ant, *U.S. Pogonomyrmex barbatus*
Texas harvester ant *Pogonomyrmex barbatus*
Western harvester ant *U.S. Pogonomyrmex occidentales*

big-headed ant *U.S. Pheidole megacephala*

hill ant *Brit. Formica rufa*

honey ant any of numerous formicine ants of the genus Myrmecocystus in which one caste (called a replete) serves as a food reservoir for the other workers

odorous house ant *U.S. Tapinoma sessile*

legionary ant = army ant

meadow ant *Brit. Lasius flavus*

Allegheny mound ant *U.S. Formica exsectoides*

pavement ant *U.S. Tetramorium caespitum*

pharoah ant *U.S.* the myrmicine *Monomorium pharaonis*

pyramid ant *Dorymyrmex pyramidus*

red ant *Brit.* = hill ant, *Brit.*

sanguinary ant *Formica sanguinea*

slave-making ant any of numerous formicines of the genus Polyergus that are totally dependent on slaves

silky ant *U.S. Formica fusca*

soldier ant = dinergate (*see* ergate)

western thatching ant *U.S. Formica obscuripes*

thief ant *U.S. Solenopsis molesta*

tree ant *I.P. Oecophlla smaragdina*

velvet ant *U.S.* popular name of the mutillid wasps (*cf.* cow ant)

white ant popular name of termites

wood ant *Brit.* = hill ant *Brit.*

larger yellow ant *U.S. Acanthomyops interjectus*

smaller yellow ant *U.S. Acanthomyops claviger*

antacava the socket into which the antenna of many arthropods articulates

ante—*comb. prefix* meaning "before"

anteater *see* eater

antelope (*see also* beetle, horn, nilghai and squirrel) popular name for many members of the artiodactyl family Antelopidae and particularly the sub-family Reduncinae. Frequently used for other, similar, forms and in *U.S.* for the antilocaprid *Antilocapra americana*

goat-antelope the term should probably by synonymous, if used at all, with goat-gazelle (*q.v.*) but is frequently misleadingly applied to the rock-goats (Rupicaprinae) particularly the serows (*q.v.*). The term has also been applied to the ovibovine takin (*q.v.*) but never, apparently, to the closely related muskox.

golden goat antelope = takin

Japanese goat antelope *Nemorrhaedus crispus* (*cf.* goral)

harnessed antelope any of several species of striped antelopid of the genus Tragelaphus

four-horned antelope = chousingha

rapier-horned antelope a term sometimes applied to the oryx in contrast to the sabre-horned antelopes of the genus Hippotragus

screw-horned antelope = addax

horse-antelope popular name of antelopids of the sub-family Hippotraginae

roan antelope the hippotragine *Hippotragus equinus*

royal antelope any of several species of neotragine antelopid of the genus Neotragus, particularly *Neotragus pygmaeus*

sable antelope any of several species of hippotragine artiodactyles of the genus Hippotragus particularly *H. niger*. They are sometimes called the sabre-antelope from the general shape of the horn

saiga antelope = saiga

Antelopidae a large family of ruminant artiodactyle mammals containing the antelopes They are distinguished by having bony cores to the horn

-antenn- *comb. form.* properly meaning "yard" or "spar' of a sailing vessel

antenna (*cf.* -corn) an anterior sensory appendage. The term is not confined to the jointed appendage of arthropods but is used also for analagous, unsegmented, structures in polychaete worms and rotifers. The term is also used for antenna-like processes arising from the rostellum of some orchids

angustate antenna one that is very long and narrow

auriculate antenna one in which the basal joint is flattened into a concave plate that covers the rest of the antenna

basantenna = antecoria

breviate antenna one which is of approximately the same length as the head of the insect bearing it

breviorate antenna one which is longer than the head, but shorter than the body of the insect bearing it

caudal antenna a sensory organ found between the toes of the foot of some Rotifera

cirrate antenna pectinate, but with very long, curved lateral branches

compound antenna a capitate antenna with several joints

dorsal antenna a single, mid-dorsal, sensory projection at the anterior end of a rotifer

elbowed antenna = geniculate antenna

fabriate antenna a simple, straight antenna with one long seta on each segment

fractate antenna one with a single, very long segment to which the others are attached at an angle

lateral antenna one of a pair of sensory palps found on the trunk of some Rotifera

nodicorn antenna one in which the apex of each segment is swollen

pectiniform antenna one with the angles of the segments extended to give a comb effect

preantenna a theoretical anterior appendage of insects thought to be represented in some chilopods by embryonic lobes

precephalic antenna = antennule

scopiferous antenna one having one or more bundles of hairs protruding from it

setaceous antenna one that is long, thin and tapering

subulicorn antenna one that is awl-shaped

Antennariidae a family of pediculate fishes distinguished by by their free bait-like first dorsal spine forming a lure and their baloon-shaped bodies covered with loose skin roughened by many small denticles. Commonly called frog fishes

antennula the second of two pairs of antennae in crustacea

antennule the smaller pair, when there are two pairs, of antennae in an arthropod

Anterozoa a rarely used taxon of the animal kingdom erected in contrast to Parazoa (*q.v.*) and which therefore contained all animals except Protozoa and Porifera

-anth- *comb. form* meaning "flower." The *adj. terms.* -anthic and -anthous (*q.v.*) are interchangeable but compounds are recorded in this work only under the latter form. A distinction is sometimes, very properly but rarely, made between -anthous, -anthemous and -antherous

perianth those parts (*e.g.* calyx, corolla) which surround the petals of a flower. Also used for the sheath that, in some lower plants, encases the reproductive organs

pseudoperianth the envelope of an archegonium

anthela (*obs.*) a panicle in which the lateral axes are greater than the main axis

-anthemous (*see also*- antherous, -anthous) pertaining to a flower as a whole

calycanthemous said of a flower in which sepals have been converted into petals or in which the corolla is inserted in the calyx

anthemy a cluster of flowers without regard to its shape or origin

anther properly that expanded portion of a stamen in which the pollen is formed, but sometimes incorrectly used as though synonymous with stamen

didymous anther one in which the two lobes are not, or only very slightly, connected

basithecal anther one in which the pollen-bearing portion lies towards the base usually below a leaf-like extension

basitonous anther one in which the pollen-sac extends to the base

[**anther ear**] a mutant (gene symbol an_1) mapped at 107 on linkage group 1 of *Zea mays*. The phenotypic expression is the appearance of male flowers in the female inflorescence

antherid = antheridium

antheridium that structure which bears the male gametes in the lower plants

-antherous (*see also*-anthemous, -anthus) pertaining to the anther, or stamen

    chasmantherous the liberation of pollen within a cleistogamous flower

    cleisthantherous the condition of a flower which remains partially closed and from which the stamens protrude

    eleutherantherous having the anthers free, not fused together

    gynantherous the condition of stamens that have been metamorphosed into pistils

    phanerantherous with obvious, protruding anthers

    synantherous the condition of anthers that fuse

anthesis the opening of a flower or the condition of being open. Also the period during which a flower is open or "in full bloom" or shedding pollen

    proanthesis either early flowering or the first flower to open of an inflorescence

    synanthesis = synacmy

anthesmus (obs.) = inflorescence

-anthic alternate form of -anthous (*q.v.*)

Anthicidae a small family of coleopteran insects that derive their name of antlike flower beetles from their appearance and habitat

anthine pertaining to flowers

    chrysanthine yellow flowered

anthium a flower, or portion of a flower

    amphanthium a dilated receptacle found in some inflorescences

    clinanthium the receptacle in compositaceous flowers

    hypanthium = a tube-like extension of the receptacle of a flower, partially or wholly enveloping the ovary, and sometimes its apical portions composed also of fused basal portions of corolla and calyx

    scleranthium an achene enclosed in a hardened calyx

anthobian flower-feeding

Anthoboscidae a family of scolioidean apocritan hymenopteran insects usually included with the Tiphiidae

Anthocerotae a phylum of bryophyte plants commonly called hornworts. They differ clearly from the mosses and liverworts in the absence of leaf-like structures

Anthocoridae a family of small predaceous hemipteran insects distinguished by their well defined embolium. They are often called flower bugs.

anthocyanin a general term for blue and red glycoside pigments found dissolved in plant cells

[**anthocyanin**] *either* a mutant (gene symbol a_1) mapped at 103 on linkage group III of *Zea mays*. The phenotypic expression is the presence of anthocyanin pigments *or* a mutant (gene symbol a_2) mapped at 1 on linkage group V of *Zea mays*. The phenotype contains anthocyanin pigment

[**anthocyanin absent**] a mutant (gene symbol *a*) mapped

at 29 on linkage group V of the tomato. The term is descriptive of the phenotypic expression

[**Anthocyanin booster**] a mutant (gene symbol *B*) mapped at 49 on linkage group II of *Zea mays*. As the name indicates the gene causes an increase in anthocyanin pigments

[**anthocyanin loser**] a mutant (gene symbol *al*) mapped at 0 on chromosome 8 of the tomate. The phenotypic plant loses its purple color in about three weeks

anthodium (*obs.*) the capitulum of compositaceous flowers

Anthogamae a no longer acceptable botanical taxon, combining the Bryophytes with the Characeae

Anthomedusae an order of medusae more properly referred to Tubulariae (*q.v.*)

Anthomyidae = Anthomyiidae

Anthomyiidae a family of myodarian cycloraphous dipteran insects closely allied to the Muscidae and frequently fused with them

Anthomyzidae an obscure group of seaside-dwelling myodarian cycloraphous diptera

Anthophoridae a family of apoid apocritan hymenopteran insects usually included with the Apidae

Anthophyta = Phanerogama

-anthous pertaining to flowers. Frequently misused for -antherous (*q.v.*) or -anthemous (*q.v.*)

    aianthous continuously in flower

    ananthous lacking flowers

    cenanthous the condition of a flower lacking either stamens or pistils

    chloranthous 1 (*see also* chloranthous 2) having green flowers

    chloranthous 2 (*see also* chloranthous 1) having many floral parts transformed into leaf-like organs

    anticlinanthous said of the scales around the base of a compositaceous flower

    cryptanthous = cleistanthous

    dianthous the condition of a flower that is pollinated from another flower of the same plant

    dysanthous the condition of being fertilized by the pollen of another plant

    epanthous said of a fungus that parasitizes flowers

    erianthous woolly flowered

    gymnanthous having naked flowers

    hemeranthous flowering only by day

    hysteranthous said of leaves that are produced after flowers

    lasianthous woolly flowered

    leucanthous white-flowered

    nyctanthous night-flowering

    phaneranthous = phanerogamic

    pleuranthous the condition of a symposium of which the inflorescence are borne only on lateral axes

    proteranthous the condition of a plant in which the flowers open before the leaves appear

    pollacanthous said of a plant that flowers many times

    rhizanthous said of a plant that apparently flowers from the root crown

    salpiganthous the condition resulting from the transformation of disk florets in a Compositaceous flower into tubular florets

    speiranthous the condition of having twisted flowers

    symmentranthous having radially symmetrical flowers

    synanthous the condition that results from the fusion of two flower buds

    teleianthous hermaphrodite

Anthozoa that class of coelenterates which contains, *inter alia*, the sea anemones, sea pens, sea fans, and corals. The Anthozoa are clearly distinguished from the other classes of Coelenterata by the lack of a medusoid generation and the strongly developed stomodaeum

anthracinous coal black

Anthribidae a family of curculionoid coleopteran insects commonly called fungus weevils. They do not in general resemble weevils since the beak is short and broad and the antennae are not elbowed. Some are easily confused with cerambycids

Anthriboidea s superfamily of polyphagous coleopteran insects replacing the Anthribidae if this group is considered to have superfamilial rank

-anthrop- *comb. form* meaning "man"

anthropoid man-like

Anthropoidea a suborder of primate mammals containing the apes and monkeys. They are distinguished by an expanded brain case and an abbreviated facial skeleton in which the eyes are set close together and face forward. This term was by some used for Hominoidea

anthurus a cluster of flowers produced at the end of an unusually long stalk

anthus (*obs.*) = corolla

-anthy *subs. suffix* derived from -anth (*q.v.*). Most words listed under -anthous (*q.v.*) can be turned into nouns by substituting -anthy for -anthous

anti *comb. suffix* replacing ant- (*q.v.*) in compounds with words starting with a consonant

antia a short feather at the base of the bill in birds

anticous that part of an organism furthest from the main axis. Used mainly of higher plants

Antigonidae a family of abyssal zeomorphous fishes distinguished by the great depth of the laterally compressed body which makes the general form of a rhomboid. Commonly called boar fish

Antilocapridae a monospecific family of artiodactyl mammals containing the pronghorn (*q.v.*). They resemble the Bovidae in possessing hollow horns but are like the Cervidae in that part of the horns are deciduous

Antilopinae monospecific subfamily of antelopid artiodactyls, containing the blackbuck, widely distributed in India

Antipatharia an order of Zoantharia distinguished by a black horn-like central axis. Usually mistaken for Alcyonarians and popularly called black corals

Antipathidea = Antipatharia

Antirrhinoideae a subfamily of Scrophulariaceae

antler a solid, frequently branched, and usually seasonal, horn found in the Cervidae and in no other family

antlia the spiral proboscis of certain lepidopteran insects

antrum that portion of the gonoduct which, when not otherwise distinguished, lies immediately adjacent to the gonopore

anubis the baboon *Papio anubis* apparently named from a misconception of the animal affinities of the Egyptian god of the same name; the sacred baboon of the ancient Egyptian was hamadryas (*q.v.*)

Anura = Salienta (the *adj. form.* anuran is, however, still widely used)

anus the orifice that terminates the alimentary canal

aorta any large arterial vessel communicating directly with the heart

  dorsal aorta the blood vessel formed by the union of the two aortic roots and which carries blood back along the middorsal line of vertebrates

  ventral aorta that aorta which runs directly forward from the heart along the floor of the pharynx and from which the aortic arches arise

aoudad the Saharan ovine caprid *Ammotragus tragelaphus*. It is sometimes known as the maned sheep in view of the large mane of the male

ap *see* [apodal] and [apterous]

ap- *see* a-, apo-

ap symbol used to indicate taxa that are apomicts

-apag- *comb. form* meaning "once"

apapane the drepaniid bird *Himatione sanguinea*

apar the dasypodid xenarthran mammal *Polypeutes tricinctus* (*cf.* teba, three banded armadillo)

Apatidae = Bostrichidae

ape 1 (*see also* ape 2 and 3) any of several tailless, old world, anthropoid primates

  anthropoid ape a term applied by some to all anthropoids but by others confined to the great apes in contrast to the lesser apes

  great ape the pongid anthropoids in general, *i.e.* chimpanzee, orangutan, gorilla, man, and allied fossil forms

  lesser ape a term reserved for the hylobatid anthropoids by those who divide the great apes from the lesser apes

ape 2 (*see also* ape 1 and 3) any large old world primate with an inconspicuous tail

  barbary ape the cynopithecid *Macaca sylvana*

  black ape any of several cynopithecids of the genus Cynopithecus

  celebesian ape the cynopithecid *Cynopithecus niger*

  gibraltar ape = barbary ape

  holy ape (of India) any of several species of cercopithecid primate of the genus Semnopithecus (=langur)

ape 3 (*see also* ape 1 and 2) any animal thought to have a vaguely primate appearance or habit

  sea ape variously applied to the thresher shark and to the sea otter

apex = tip

Aphasmidea a class of Nematoda distinguished from the Phasmidea by the lack of phasmids and by the fact that the excretory system consists of a single ventral gland

aphelenchoid *see* pharynx

aphid (*see also* wasp, wolf) anglicized form of homopteran family Aphidae. The generic term aphis is frequently misused as a substitute. A few psyllids, noted in the list that follows, are popularly called "aphids"

  woolly alder aphid U.S. *Prociphilus tessellatus*

  spotted alfalfa aphid U.S. *Therioaphis maculata*

  apple aphid U.S. *Aphis pomi*

    rosy apple aphid U.S. *Dysaphis plantaginea*

    woolly apple aphid U.S. *Eriosoma lanigorum*

  pine bark aphid U.S. the psyllid *Pineus strobi*

  bean aphid U.S. *Aphis fabae*

  beech blight aphid U.S. *Procipkilus imbricator*

  tulip bulb aphid U.S. *Dysaphis tulipae*

  cabbage aphid *Brevicoryne brassicae*

  caragana aphid U.S. *Acertosiphon caraganae*

  black cherry aphid U.S. *Myzus cerasi*

  chrysanthemum aphid U.S. *Macrosiphoniella sanborni*

  black citrus aphid U.S. *Toxoptera aurantii*

  brown citrus aphid U.S. *Toxoptera citricidus*

  clover aphid U.S. *Nearctaphis bakeri*

    sweetclover aphid U.S. *Therioaphis riehmi*

    yellow clover aphid U.S. *Therioaphis trifolii*

  cotton aphid U.S. *Aphis gossypii*

  currant aphid U.S. *Cryptomyzus ribis*

  boxelder aphid U.S. *Periphyllus negundinis*

  woolly elm aphid U.S. *Eriosoma americanum*

foxglove aphid *U.S. Acertosiphon solani*

gall aphid any aphid or psyllid which produces a gall

poplar pitiole gall aphid *U.S. Pemphigus populitransversus*

Cooley spruce gall aphid *U.S.* the psyllid *Chermes cooleyi*

eastern spruce gall aphid *U.S.* psyllid *Chermes abietis*

poplar twig gall aphid *U.S. Pemphigus populiramulorum*

golden glow aphid *U.S. Dactynotus rudbeckiae*

apple grain aphid *U.S. Rhopalosiphum fitchii*

English grain aphid *U.S. Macrosiphum avena*

hop aphid *U.S. Phorodon humuli*

ivy aphid *U.S. Aphis hederae*

larch aphid *U.S. Cinara laricis*

corn leaf aphid *U.S. Rhopalosiphum maidis*

elm leaf aphid *U.S. Myzocallis ulmifolii*

crescent-marked lily aphid *U.S. Neomyzus circumflexus*

purple-spotted lily aphid *U.S. Macrosiphum lilii*

waterlily aphid *U.S. Rhopalosiphum nymphaeae*

Norway-maple aphid *U.S. Periphyllus lyropictus*

painted maple aphid *U.S. Drepanaphis acerifoliae*

black-margined aphid *U.S. Monellia costalis*

melon aphid *U.S. Aphis gossypii*

crapemyrtle aphid *U.S. Myzocallis kakawaluokalani*

pea aphid *U.S. Acyrthosiphon pisum*

cowpea aphid *U.S. Aphis craccivora*

black peach aphid *U.S. Brachycaudus persicaecola*

green peach aphid *U.S. Myzus persicae*

woolly pear aphid *U.S. Eriosoma pyricola*

black pecan aphid *U.S. Myzocallis caryaefoliae*

white-pine aphid *U.S. Cinara strobi*

mealy plum aphid *U.S. Hyalopterus pruni*

rusty plum aphid *U.S. Hysteroneura setariae*

potato aphid *U.S. Macrosiphum euphorbiae*

privet aphid *U.S. Myzus ligustri*

root aphid any subterranean aphid infesting roots

corn root aphid *U.S. Anuraphis maidiradicis*

strawberry root aphid *Aphis forbesi*

rose aphid *U.S. Macrosiphum rosae*

yellow rose aphid *U.S. Acyrtosiphon porosum*

snowball aphid *U.S. Neoceruraphis viburnicola*

spirea aphid *U.S. Aphis spiraecola*

spruce aphid *U.S. Elatobium abietinum*

green spruce aphid *U.S. Cinara fornacula*

strawberry aphid *U.S. Chaetosiphon fragaefolii*

sugar-beet root aphid *U.S. Pemphigus populivenae*

yellow sugarcane aphid *U.S. Sipha flava*

thistle aphid *U.S. Brachycaudus cardui*

buckthorn aphid *U.S. Aphis nasturtii*

tuliptree aphid *U.S. Macrosiphum liriodendri*

turnip aphid *U.S. Hydophis pseudobrassicae*

balsam twig aphid *U.S. Mindarus abietinus*

poplar vagabond aphid *U.S. Mordiwilkoja vagabunda*

violet aphid *U.S. Neotoxoptera violae*

viburnum aphid *U.S. Anuraphis viburniphila*

grapevine aphid *U.S. Aphis illinoisensis*

walnut aphid *U.S. Chromaphis juglandicola*

gooseberry witchbroom aphid *U.S. Kakimia houghtonensis*

woolly aphid a term generally applied to a group of aphids having waxed glands which coat them with a wool-like substance. They are sometimes separated from the other aphids into the subfamily Eriosomatinae

Aphididae that family of homopteran insects which contains the forms commonly called aphids. The pear-like shape, and the pair of cornicles at the posterior end of the abdomen, are distinctive

aphodus the narrow canal between the chamber and the excurrent canal in some sponges

Aphredoderidae a monospecific family of fishes containing the freshwater pirateperch (*Aphredoderus sayanus*), distinguished by 3 or 4 dorsal spines, 2 anal spines, and by the fact that the anus is located immediately under the throat in adults

Aphrizidae a family of charadriiform birds usually included with the Scolopacidae

aphrodite *U.S.* the nymphalid lepidopteran *Speyeria aphrodite*. (*c.f.* diana, greater fritillary)

Aphroditidae a family of errant polychaete annelids distinguished by the fact that the back is totally covered with broad overlapping chaetose scales

aphrostase an obsolete term for cellular tissue used at a time when it was thought that many other kinds of tissues existed in organisms

Aphyllae = Thallophyta

apical (*see also* ridge, tuft) pertaining to the apex

Apiaceae = Umbelliferae

apices *plural of* apex

apiculate ending in a short, sharp point

Apidae a very large family of apoid insects divided into numerous sub-families the more important of which are noted in this dictionary

Apinae a subfamily of apid apocritan hymenopterous insects commonly called the social bees since the group includes both the honey bees and the bumble bees

Apioceratidae = Apioceridae

Apioceridae a family of large elongate flower-loving dipteran flies

Aplacophora a small order of aberrant amphineuran Mollusca with a worm-like body, lacking a shell, but with numerous calcareous spicules in the cuticle

Aplantatae = Aflagellatae

Aplondontidae a monospecific family of sciuromorph rodents, distinguished from other sciuromorphs by the absence of post-orbital processes (*cf.* sewellel)

Apneumonomorpha a suborder of araneid arthropods distinguished from all other suborders by lacking lungs

-apo- *comb. form* meaning "away from" or "separate"

Apocrita a suborder of hymenopteran insects in which the basal segment of the abdomen is fused with the thorax and separated from the rest of the thorax by a constricted area often referred to as a wasp-waist. This group contains all of the Hymenoptera except for sawflies and horntails (*cf.* Symphyta)

Apocynaceae that family of contortous dicotyledons which contains, *inter alia*, the dogbanes, oleanders, and periwinkles. The family may be distinguished from the closely related Asclepiadaceae by the absence of a corona and from the Gentianaceae by the milky juice

Apocynales an order of dicotyledonous angiosperms erected to contain the families Apocynaceae and Asclepiadaceae by those who wish to remove these from the Contortae

Apoda 1 (*see also* Apoda 2) an order of holothurioid Enchinodermata distinguished from all others by lacking both ambulacral feet and respiratory trees

Apoda 2 (*see also* Apoda 1) a small order of worm-like Amphibia distinguished by the absence of limbs (= Caecilia)

|apodal| a mutant (gene symbol *ap*) mapped at 0 on linkage group III of *Bombyx mori*. The larva of the phenotype has rudimentary legs

Apodes an order of osteichthyes containing the eels. They are distinguished by their eel-like shape, the absence of pelvic or ventral fins and by passing through a leptocephalus larval stage in their life history

Apodidae a family of apodiform birds containing the swifts. These are distinguished by their thin narrow body, very long and pointed wings and small, de-curved, large-gaped bills

Apodiformes that order of birds which contains the swifts and hummingbirds. They are distinguished, amongst other things, by their very small feet

Apogonidae a marine family of percoid fishes distinguished by the separation of the two dorsal fins and the possession of two anal spines. The males incubate the eggs in their mouths. Commonly called cardinal fish

Apoidea a superfamily of apocritan Hymenoptera containing the forms commonly called bees. They agree with the sphecoids, with which they are frequently included, in all characteristics save that the apoid posterior tarsae are elongate and dilated and that the hairs on the thorax are plumose

apollo *I.P.* Popular name of papilionid lepidopteran insects of the genus Parnassius

apolytic said of tapeworms that shed gravid segments

　anapolytic said of those tapeworms that do not shed gravid segments

　pseudoapolytic said of those tapeworms that shed chains of egg-exhausted segments

Aponogetonaceae the helobian family of aquatic monocotyledons that contains, *inter alia*, the water hawthorns. Distinctive characters are the petaloid perianth, numerous ovules, and straight embryo

Aporidea an order of eucestoid cestodes distinguished by the lack of segmentation

Apporrhaidae a family of gastropod Mollusca distinguished by their high spire and thickened, flaring, wing-like expansion of the lip

Apostasiaceae a family, usually considered to be a subfamily of Orchidaceae, distinguished by the absence of a column or gynostegium

app *see* [approximated]

apparatus a complex of functionally linked parts

　egg apparatus the three cells at the micropylar end in the embryo sac of an angiosperm

　envelope apparatus the sporocarp of ascomycetes

　glottoid apparatus a complex telostom (*q.v.*) found in some nematodes

　Golgi apparatus a mass of lipoidal reticulating fibers, granules and vesicles, present in most animal and many plant cells. The function is thought to be that of an intracellular pump which regulates the movement of fluids in the cell and the expulsion of excretory products from the cell. It may be necessary for the secretion of very large molecules

　Weberian apparatus a series of small bones connecting the swim bladder with the inner ear found in ostariophysid osteichthyes. These bones are modifications of the anterior vertebra

appendage a subordinate part that is appended to a major part. Most frequently used in *zool.* for the articulated structures (*e.g.* limbs) of arthropods and vertebrates

　buccal appendage any mouth part of an arthropod that is articulated and moveable

　intercalary appendage any appendage, usually rudimentary, interposed between those of the regular series

　Simroth's appendages narrow tubules running from the main ring of the water-vascular system through the viscera in some ophiuroid echinoderm

appendent the condition of a sessile ovule

Appendicularia = Larvacea

appendix anything appended to a regular structure as the anal cerci of insects. Without qualification usually refers to the vermiform appendix

　hydatid appendix the rudimentary remains of the oviduct in male mammals

　vermiform appendix a blind tube that terminates the caecum of a few mammals

applanate flattened

apple 1 (*see also* apple 2,3,4,5,6, aphid, borer, bug, chalcid, curculio, hopper, mite, seed, skeletonizer, and sucker) the rosaceous tree *Pyrus malus* and the edible fruit borne by it and its very numerous horticultural varieties. Also the trees and fruits of other species of Pyrus

　crab apple *Brit.* the small fruits of any of several, rarely cultivated species of Pyrus (without qualification) *P. sylvestris*

　　prairie crab apple = western crab apple

　　Siberian crab apple *P. baccata*

　　western crab apple *P. ioensis*

　Chinese flowering apple *P. spectabilis*

apple 2 (*see also* apple 1, 3, 4, 5, 6) a variety of trees and shrubs the fruit of which more or less resembles an apple

　Adam's apple (*see also* apple 5,6) either a horticultural variety of lime *Citrus limetta* or the abocynaceous shrub *Tabernae montana*

　alligator apple the tree, and fruit from *Annona glabra*

　baked apple the shrub *Rubus chamaemorus* or its fruit

　bastard apple *Austral.* any of several trees of the myrtaceous genus Eucalyptus, particularly *E. cambagei*

　blade apple the cactaceous scrambling shrub *Pereskia aculeata*

　Cain's apple the fruit of the tree *Arbutus unedo*. Also applied to the tree itself

　carib apple the annonaceous tree *Fusaea longifolia* and its fruit

　custard apple the annonaceous tree *Annona reticulata* and its fruit

　mammee apple the guttiferous tree *Mammea americana* and its fruit

　pond apple the annonaceous tree *Annona glabra* and its fruit (= alligator apple and monkey apple)

　rose apple the myrtaceous tree *Eugenia jambos*

　Sodom apple the solanaceous herb *Solanum incanum*, the cucurbitaceous vine *Citrullus colocynthis* (= bitter apple) or the asclepadaceous herb *Calotropis procera*

　star apple the sapodaceous tree *Chrysophyllum cainito* and its fruit

　sugar apple the annonaceous tree *Annona squamosa* and its fruit

　tahita apple the anacardiaceous tree *Spondias cytherea* and its fruit

　wood apple the rutaceous tree *Feronia limonia*

apple 3 the edible fruit of several cucurbitaceous vines

　bitter apple the *Citrullus colocynthis* and its fruit (= Sodom apple)

balsam apple the fruit of any of several species of gourd of the genus Mormordica. (*cf.* balsam pear)
wild balsam apple any of several species of vine of the genus Echinocystis
apple 4 a variety of fruits, mostly of the family Solanaceae, of more or less apple-like shape. Many are poisonous
  devil's apple either of the solaneceous herbs *Mandragora officianarum* (= mandrake) or *Datura stramonium* (= jimson weed) or *U.S.* = may apple
  egg apple the fruit of the eggplant (*Solanum melogena*)
  love apple the fruit of the solanaceous vine *Lycopersicum esculentum* (= tomato)
  may apple the berberidaceous herb *Podophyllum peltatum* and its fruit
  thorn apple any of several species of the solanaceous genus Datura
apple 5 (*see also* apple 1,2,3,4,6) a variety of fruits having no relation in shape to the apple
  Adam's apple (*see also* apple 2,6) the fruit of the musaceous tree *Musa paradisiaea (cf.* banana)
  oak apple (*see also* apple 6) *Austral.* the cone-like fruit of several species of Casuarina (*cf.* beefwood)
  Peru apple = Cape gooseberry
  pineapple (*see also* bug, flower, guava, weed, weevil) the bromeliaceous herb *Ananas sativus*, and the fleshy fruit of this plant
  sea apple *U.S.* the fruit of the palm *Manicaria saccifera*
apple 6 (*see also* apple 1,2,3,4,5) a variety of organisms or structures having no obvious affinity to an apple
  Adam's apple (*see also* apple 2,5) the swelling in the human trachea represented by the cricoid, arytenoid, and thyroid cartilages
  oak apple (*see also* apple 5) *U.S.* and *Brit.* any of several insect-induced leaf galls of the oak tree
  shell apple the fringillid bird *Loxia curvirostra* (= crossbill *Brit.*)
apple-of-Peru either the solanaceous herb *Nicandra physalodes* or *U.S. dial. Datura stramonium* of the same family (= Jimson weed)
apposite side by side
apposition the growth of a cell wall, or other structure, by the successive deposition of layers on its outside (*cf.* intussusception)
appressoria hyphae that attach parasitic fungi to surfaces (*cf.* haustoria)
[approximated] a mutant (gene symbol *app*) mapped at 37.5 on chromosome III of *Drosophila melanogaster*. The name refers to the closely approximated crossveins on the wings of the phenotype
apricot (*see also* plum) without qualification, the rosaceous tree *Prunus armeniacā* and its edible fruit. Also other trees with comparable fruit
  black apricot *Prunus dasycarpa*
  Japanese apricot *Prunus mume*
  purple apricot = black apricot
  Santo Domingo apricot the guttiferous tree *Mammea americana* and its fruit (= tropical apricot, mammee apple)
  tropical apricot = Santo Domingo apricot
apron the equivalent of a transverse dewlap in some amphibia
  devil's apron the large broad laminariale alga *Laminaria saccharina*
-apsid- *comb. form* meaning "shield"
  anapsid pertaining to a reptilian skull with no temporal openings

diapsid pertaining to a reptilian skull in which there are two temporal openings
euryapsid a reptile skull in which there is only a single temporal opening
parapsid pertaining to a reptilian skull in which there is one temporal opening above the postorbital-squamosal junction
synapsid pertaining to a reptilian skull in which there is one temporal opening below the post-orbital squamosal junction
parapsis a lateral division of the insect scutellum
Aptenodytidae = Speniscidae
Aptera an order once proposed for all wingless insects (in addition to some other arthropods) then restricted to the Thysanura and Collembola and now abandoned
[apterous] a mutant (gene symbol *ap*) mapped at 55.4 on chromosome II of *Drosophila melanogaster*. Not only the wings but also the balancers are missing from the phenotype
Apterygidae that family of apterygiform New Zealand birds containing the kiwis. The coarse, fur-like plumage, long bill with nostrils at the tip, and vestigial wings and absence of tail are distinguishing
Apterygogenea a generalized term for insects wingless in all instars
Apterygota a subclass erected to contain the assemblage of primitively wingless insects
apyrase a plant enzyme that, when activated by Ca, catalyzes the hydrolysis of ATP to ADP (*cf.* ATPase)
aquatic pertaining to water. In *biol.* usually descriptive of water dwelling organisms
cerebral aqueduct the narrow canal through the mesencephalon of an embryo
Sylvius's aqueduct = iter
Aquifoliaceae that family of sapindalous dicotyledons which contains the hollies. The seemingly exstipulate leaves, polygamodioecious flowers, and the absence of any interstaminal disc are typical
Aquilidae = Accipitridae
aquiprata plant communities of wet meadows
ar *see* [abdomen rotatum]
ar a yeast genetic marker indicating a requirement for arginine
ara a bacterial mutant indicating utilization of arabinose
araA a bacterial genetic marker having a character affecting the activity of L-arabinose isomerase, mapped at 0.25 mins. for *Escherichia coli*
araB a bacterial genetic marker having a character affecting the activity of L-ribulokinase, mapped at 0.25 mins. for *Escherichia coli*
arab *I.P.* any of several species of pierid butterfly of the genus Colotis
  topaz arab *S.Afr.* the pierid lepidopteran insect *Colotis celais (cf.* tip)
aracari any of several species of ramphastid birds of the genus Pteroglossus
  collared aracari *P. torquatus*
Araceae that family of spathiflorous flowering plants which contains *inter alia*, the arums. They are mostly distinguished by the inflorescence being a spadix enveloped by a single spathe
Arachnida a large class of the phylum Arthropoda containing the forms commonly known as scorpions, spiders, mites, ticks, harvestmen and some other lesser known groups. They are distinguished in the division of the body into a cephalothorax and abdomen, the former bearing six pairs of appendages

the first of which are chelicerae and the second palpae. The remaining four pairs of appendages are walking legs

arachnoid properly resembling a member of the Arachnida but used in *anat.* in the sense of "spider web-like" and in *bot.* as cobwebby, an appearance caused by the soft, entangled indumentum

araD a bacterial genetic marker having a character affecting the activity of L-ribose 5-phosphate 4-epimerase, mapped at 0.25 mins. for *Escherichia coli*

Aradidae a family of small flattened hemipteran insects usually with a scalloped flattened lateral extension to the abdomen. They are commonly called flat bugs or fungus bugs

Araeolaimidea a class of Nematoda with spiral amphids, a smooth cuticle, and four cephalic bristles

Araliaceae that family of umbelliflorous dicotyledons which contains, *inter alia*, the ivies and the ginseng. The family may be distinguished from the Umbelliferae proper by the berry-like fruit and the numerous carpels

Aramidae a monospecific family of gruiform birds erected to contain the limpkin. They are distinguished from the closely allied Gruidae by the laterally compressed bill and by the functional hallux

Araneida that order of arachnid arthropods which contains the animals commonly called spiders. In general they are distinguished from other arachnids by the possession of non-chelate chelicerae bearing at their tips the opening of poison glands, and by the production of several kinds of silk

araneous properly, having the form of a spider, though most commonly used as if synonymous with arachnoid

Arantio's common misspelling of Aranzio's

Aranzio's *epon adj.* from Julio Caesar Aranzio (1530–1589)

arataim an osteoglossid fish *Arataima gigas* of Southern American rivers, considered to be the largest freshwater fish in the world

araticu any of several trees or shrubs of the annonaceous genus Rollinia and the edible fruit derived from them

Araucariaceae a family of coniferous gymnosperms containing the trees commonly called kauri pines and monkey puzzlers. The very large, usually winged, seeds are typical as are the persistent leaves on the stem

arawana the osteoglossid fish *Osteoglossum bicirrhosum*

Arber's *epon. adj.* from Agnes Arber (1879–1960)

-arbor- *comb. form* meaning "tree"

arbor vitae any of numerous species and horticultural varieties of the cupressaceous genus Thuja

  common arbor vitae *T. occidentalis*

arboreal pertaining to, or dwelling in, trees

arborescent branched like a tree

arbuscula a shrub with a central stem

arbustive with scattered groups of trees

Arbutoideae a subfamily of Ericaceae distinguished by having a loculicidal capsule and a seed that is not winged

arc properly a portion of the circumference of a circle but used in *biol.* for any curved line limiting an area

  migrarc the zone within which migration takes place

[arc] a mutant (gene symbol *a*) mapped at 99.2 on chromosome II of *Drosophila melanogaster*. The wings of the phenotype are broad and bent down

arch 1 (*see also* -arch-2, -arch-3, arches, and duke) a

curved structure, or a pair of curved structures, which maintains an aperture. In *anat.* extended to structures with the shape, but not the function, of an arch

aortic arch one of six, or rarely 7, vascular channels connecting the ventral aorta or truncus with the dorsal aorta in the region of the pharynx

branchial arch there is great confusion in the literature as to whether this term is synonymous with visceral arch (*q.v.*) whether it applies to all the visceral arches except the maxilla, or whether it should be confined to those visceral arches that bear gills in fish and to their homologues in higher forms

gill arches those visceral arches (*q.v.*) which bear gills

haemal arch that arch which depends ventrally from the centrum of a vertebra and through which run the caudal vein and artery

hyoid arch the branchial arch immediately posterior to the mandibular arch

extrahyoid arch one of a pair of thin cartilages lying backwards from the base of the styloid cartilage under the optic capsule and articulating with the first branchial arch in the cyclostome visceral skeleton

mandibular arch that visceral arch which forms at least the rudiments of the upper and lower jaws

neural arch that arch which rises dorsally from the centrum of a vertebra and through which runs the spinal cord

occipital arch a series of paired cartilages forming the side of the foramen magnum in the chondrocranium

subocular arch one of a pair of arch-shaped cartilages, running downward and forwards from the chondrocranium of Cyclostomes

visceral arch one of a series of U-shaped aggregations of cartilages or bones supporting the pharyngeal cavity and forming the jaw of vertebrates. In all bony forms, the maxillary portion of the mandibular arch, or upper jaw, is attached to the cranium. The lower portion of the mandibular arch forms the lower jaw. The hyoid arch, which lies immediately behind or below the mandibular, varies greatly in structure and function in different classes. Behind the hyoid there are from five (normal) to seven (the shark Heptanchus) branchial arches (*q.v.*) that support the pharynx in water-dwelling forms, but are modified or lost in terrestrial forms. The components of a typical branchial arch are, from the top down, **pharyngobranchial, epibranchial, ceratobranchial,** and **basibranchial** cartilages or bones, as the case may be (*cf.* branchial arch)

zygomatic arch a prominance rising from the squamosal bone and suturing in front with the jugal bone

-arch- 2 (*see also* arch 1, -arch-3, arches and duke) *comb. form* meaning "primitive" or "beginning" but by extension "origin", particularly as applied to seres (*q.v.*)

hydrarch a plant succession taking place in water

litharch an adsere on hard rock

psammarch an adsere originating on sandy soil

xerarch a succession originating in a dry area

-arch- 3 (*see also* arch 1, -arch-2, arches and duke) *comb. form* as -arch-2 but particularly applied to xylem formation

centrarch a mass of xylem surrounding protoxylem

cyclarch the member of a whorl

diarch a stele having two protoxylem groups

endarch the pattern of primary xylem production in a

plant organ in which differentiation progresses in a radial direction from the inner edge of the procambium outwards, *i.e.* the oldest ("archaic") xylem is farthest from the surface.

exarch that pattern of primary xylem production in a plant organ in which differentiation progresses from the outside inwards

mesarch that pattern of primary xylem production in a plant organ in which differentiation progresses inward and outward from the first formed cells

[**arch**] a mutant (gene symbol *arch*) mapped at 60.5± on chromosome II of *Drosophila melanogaster*. The phenotypic wings are downcurved

Archaeopsyllidae = Pulicidae

archangel *see* angel

archangelica the umbelliferous herb yielding the "angelica" of commerce

Archangiaceae a family of myxobacteriale schizomycetes with irregular twisted, fruiting bodies and the resting cells of which are never enclosed in cysts

-archegon- *comb. form* meaning "first ancestor"

menarche the time of appearance of the menstrual cycle in human females or oestral cycle in other mammals (*cf.* menopause)

Archegoniatae a no-longer used plant taxon combining the Bryophyta and the Pteridophyta

arches *Brit.* any of several lepidopteran insects. The plural is the only form of the name

    buff arches the thyatirid *Habrosyne deras*

    black arches any of several species of nolid of the genus Nola and the liparid *Lymantria monacha*

    green arches any of several noctuids mostly of the genera Eurois and Aplecta

-archi- = -arch-

Archiacanthocephala an order of the Acanthocephala distinguished by concentrically arranged spines, and the possession of protonephridia

Archiannelida a small class of annelid worms having many of the characteristics of polychaete larvae

megarchidium = nucellus

Architectonicidae a family of gastropod Mollusca of extremely flattened shape from which their popular name of sundial shells derives

-archo- *comb. form* meaning "chief" or "principal"

Archostemata a suborder of coleopteran insects distinguished by the fact that the apex of the wings is rolled spirally when in repose

Arcidae a family of pelecypod mollusks with a strongly ridged shell having comb-like teeth on both sides of the hinge line

[**arcoid**] a mutant (gene symbol *ad*) mapped at 60.7 on chromosome II of *Drosophila melanogaster*. The phenotypic wings are arched, broad and short

arctate pertaining to an arc

    coarctate (*see also* pupa) crowded or closely pressed together

arctic 1 (*see also* arctic 2) that area of the planet Earth which lies within 23° 30¹ of latitude of the North pole (*cf.* antarctic)

    antarctic that part of the planet Earth which lies within 23° 30′ of latitude from the "south" or lower pole

    holarctic pertaining to those northern areas of the planet Earth which are not tropical and therefore comprising both the nearctic and palaeartic regions

    nearctic pertaining to those northern areas of the New World which are not tropical, *i.e.* Greenland, Canada, Labrador, together with the mountainous and *N.* parts of the *U.S.* and the Mexican Plateau

    palaearctic pertaining to those areas of the Old World which are not tropical, *i.e.* Europe, Africa *N.* of the Sahara, *N.* Arabia and Asia *N.* of the Himalayas

arctic 2 *U.S.* any of several species of satyrid lepidopteran insects of the genus Oeneis

Arctiidae a family of noctuoid lepidopteran insects containing the forms commonly called tiger moths and footman moths

Arctotideae a tribe of tubuliflorous Compositae

[**arctus oculus**] a mutant (gene symbol *at*) mapped at 60.1± on chromosome II of *Drosophila melanogaster*. The phenotypic expression is a reduced number of facets in the eye

arcualia cartilaginous precursors of the neural or hemal arches of the vertebra that persist as simple rods in Cyclostomes

arcule a crossvein between the radius vein and cubitus vein in the wings of some insects (*cf.* vein)

Ardeidae that family of ciconiiform birds which contains the herons. They are in general distinguished from other families of the order by the long filamentous plumes on the back of the head

ardella the apothecia of certain lichens

Ardisiaceae = Myrsinaceae

ardosiacous slate-gray

area in *biol.* a delimited region

    axial area the smooth surface between the margins of diatoms

    central area the clear central space of a diatom frustule

    germinal area a morphologically undifferentiated area in a gastrula, or other early embryonic stage, that has the potency to form a specific organ

    hyaline area that part of a diatom frustule which lacks markings

    interfascicular area the parenchymatous regions between vascular fascicles in a plant stem

    lateral area any area, other than central or axial, of a diatom frustule that lacks markings

    sieve area that on the wall of sieve cells or on the lateral walls of sieve tube members with clusters of perforations or pores, through which run protoplasmic connections (*cf.* sieve plate)

area opaqua the relatively more or less opaque area of the telolecithal blastoderm

area pellucida that clear embryonic area which overlies the blastocoel in the early development of a telolecithal egg

    area vasculosa the extraembryonic blood, and blood vessel, forming area in the telolecithal blastoderm

Areacaceae = Palmae

Arecales = Principes

aren- *comb. form* meaning "sand" (cf. -psam-)

arenaceous pertaining to sand

arend *S.Afr.* = lammergeyer

Arenicolidae a family of large sedentary sand dwelling polychaete worms having no appendages on the head, few chaetae, and rudimentary parapodia

areola 1 (*see also* areola 2–5) literally, a small area, and variously used in *biol.* to refer to limited areas of surfaces, small cells, or even tessalated patterns

areola 2 (*see also* areola 1, 3–5) naked tracts between scales on the feet of birds

areola 3 (*see also* areola 1,2,4,5) a naked area surrounding the nipple on a mamma

areola 4 (*see also* 1–3, 5) the bare area surrounding the boss of an echinoderm tubercle, and to which are attached the muscles operating the spine

areola 5 (*see also* 1–4) a large pseudopore in Ectoprocta

arg-1 to **arg-11** *see* [Arginine-1] to [Arginine 11]

argalis the ovine caprid *Ovis ammon* distinguished by the immense size of its horns

Argasidae a family of acarine arthropods containing those ticks which, in distinction from the Ixodidae, do not possess a scutum

argA a bacterial genetic marker having a character affecting the activity of acetylornithinase, mapped at 77.25 mins. for *Escherichia coli*

argB a bacterial genetic marker having a character affecting the activity of N-acetyl glutamate synthetase, mapped at 55 mins. for *Escherichia coli*

argC a bacterial genetic marker having a character affecting the activity of N-acetyl-γ-glutamokinase, mapped at 77.25 mins. for *Escherichia coli*

argD a bacterial genetic marker having a character affecting the activity of ornithine transcarbamylase, mapped at 5.0 mins. for *Escherichia coli*

argE a bacterial genetic marker having a character affecting the activity of argininosuccinic acid synthetase, mapped at 60.50 mins. for *Escherichia coli*

-argent- *comb. form* meaning "silver"

argent-and-sable the geometrid lepidopteran insect *Eulype hastata* (= mottled beauty)

argentate silvery

argentine anglicized form of the fish family Argentinidae

Argentinidae a marine family of salmonoid fishes

argF a bacterial genetic marker having a character affecting the activity of argininosuccinase, mapped at 77.25 mins. for *Escherichia coli*

argG a bacterial genetic marker having a character affecting the activity of acetylornithine-δ-transaminase, mapped at between 61.0 and 64.0 mins. for *Escherichia coli*

argH a bacterial genetic marker having a character affecting the N-acetylglutamic -γ-semialdehyde dehydrogenase, mapped at 77.25 mins. for *Escherichia coli*

Argidae a small group of symphytan hymenopteran insects readily distinguished by the three segmented antenna of which the two basal joints are together less than one fifth the length of the large elongate ovate terminal joint

-argill- *comb. form* meaning "clay"

argillaceous pertaining to clay

arginase an enzyme catalyzing the hydrolysis of arginine to ornithine and urea

L-arginine 1-amino-4-guanidovaleric acid. $N_2N_3CN \cdot HCH_2CH_2CH(NH_2)COOH$. An amino acid known to be essential in rat nutrition

[Arginine-1] *either* a mutant (gene symbol *arg-1*) in linkage group I of *Neurospora crassa.* The phenotypic expression is a requirement for arginine and the failure to utilize ornithine or citrulline *or* a mutant (gene symbol *arg-1*) in linkage group I of *Chlamydomonas reinhardi.* The phenotypic expression is the requirement for arginine, citrulline, or ornithine

[Arginine-2] *either* a mutant (gene symbol *arg-2*) in linkage group I of *Chlamydomonas reinhardi.* The phenotypic expression is a requirement for arginine and an inability to utilize citrulline or ornithine *or* a mutant (gene symbol *arg-2*) found in linkage group IV of *Neurospora crassa.* The phenotypic expression is a requirement for arginine though citrulline is also used

[Arginine-3] a mutant (gene symbol *arg-3*) in linkage group I of *Neurospora crassa.* The phenotypic expression is a requirement for arginine or citrulline and a failure to utilize ornithine

[Arginine-4] a mutant (gene symbol *arg-4*) in linkage group V of *Neurospora crassa.* The phenotypic expression is the requirement for arginine though the phenotype can also use ornithine or citrulline

[Arginine-5] a mutant (gene symbol *arg-5*) in linkage group II of *Neurospora crassa.* The phenotypic expression is a requirement for ornithine, citrulline, or arginine

[Arginine-6] a mutant (gene symbol *arg-6*) in linkage group I of *Neurospora crassa.* The phenotypic expression is a requirement for ornithine, citrulline, or arginine

[Arginine-7] a mutant (gene symbol *arg-7*) in linkage group V of *Neurospora crassa.* The phenotypic expression is a requirement for arginine though the phenotype may also use ornithine or citrulline

[Arginine-8] a mutant (gene symbol *arg-8*) in linkage group V of *Neurospora crassa.* The phenotypic expression is a requirement for arginine

[Arginine-10] a mutant (gene symbol *arg-10*) on linkage group VI of *Neurospora crassa.* The phenotypic expression is a requirement for arginine and an inability to use ornithine or citrulline

[Arginine-11] a mutant (gene symbol *arg-11*) on linkage group VII of *Neurospora crassa.* The phenotypic expression is a requirement for arginine, adenine, and uridine

Argiopidae a family of dipneumonomorphic araneid arthropods known as orb-web spiders or garden spiders. They are distinguished by the production of the standard "spider web," a spiral thread crossing radial bars

-argo- *comb. form* meaning "passive"

argodromile pertaining to sluggish streams (*cf.* fluvial)

argonaut frequently anglicized name of the genus Argonauta (*see* paper nautilus)

Argonautidae a family of dibranchate cephalopod Mollusca in which the mantle is not united to the head and in which there is a spiral egg case, produced by the female that superficially resembles, and is frequently called, a shell

Arguloidea a group of parasitic crustacea lacking egg sacs and with a strongly depressed body bearing sucking discs in front and lacking egg sacs. They are variously regarded as a separate order of crustacea or a suborder of Copepoda

argus 1 (*see also* argus 2) the scatophagid teleost fish *Scatophagus argus* (*cf.* scat)

argus 2 (*see also* argus 1) *Brit.* the lycaenid lepidopteran insect *Lycaena astrache* (*cf.* silver studded blue) *I.P.* any of numerous satyrid lepidopteran insects of the genus Erebia. Usually, without qualification, *E. nirmala. Austral.* any of several nymphalid lepidopteran insects of the genus Precis

Scotch argus *Erebia aethiops*

argute sharp

Argyresthiidae a family of lepidopteran insects now usually included in the Yponomeutidae

argyrous silvery

Ariciidae a family of sedentary polychaete worms living in sandy burrows and with a proboscis divided into lobes

arietinous having the form of the horns of a ram

Ariidae a family of naked marine catfishes distinguished by the presence of an adipose fin. Most of the ariids incubate their eggs in the mouth of the male

aril a fleshy outgrowth of the funiculus that envelops the integument of an ovule. The term is also applied to reproductive organs posessing an aril such as those of the yew

Arillatae = Scitaminales

arista a sensory bristle or tuft of bristles on the antenna of dipteran insects

[**aristaless**] a mutant (gene symbol *al*) mapped at 0 chromosome II of *Drosophila melanogaster*. The aristae of the phenotype are not absent but are much reduced

aristate awned

  exaristate lacking awns

-aristero- *comb. form* meaning "left" usually abbreviated to -aristo-

aristo- *see* -aristero-

Aristolochiaceae that family of plants which contains, *inter alia*, the Dutchman's pipes, the snake roots and the wild gingers. This peculiar family is principally distinguished by its very simple flowers of showy color and offensive aroma

Aristolochiales an order of dicotyledonous angiosperms characterized by the uniseriate and petaloid perianth with an inferior ovary

ark anglicized name for arcid shells particularly those of the genus Arca (*cf.* blood worm)

  Noah's ark the arcid shell *Arca noae*

arm the anterior limb of pongid anthropoids and, by extension, many other elongate structures

  oral arm a lengthy perradial lobe, descending from the corners of the manubrium in some Scyphozoa

  translator arm the tissue that joins each of the two pollinia of asclepiadaceous plants to the glands

Armadillidiidae a family of oniscoid Isopoda distinguished by their ability to roll the body into a tight ball and hence called "pill bugs"

armadillo (*see also* lizard) popular name of dasypodid mammals distinguished by their armored, and frequently segmented, covering, *U.S.*, without qualification, *Dasypus novemcinctus*

  peludo armadillo *Dasypus sexinctus*

  pigmy armadillo any of several species of six-banded armadillos of the genus Zaedyus

  six-banded armadillo *see* peludo, weasel-headed armadillo, and pigmy armadillo

  three-banded armadillo (*see also* peba) the armadillo *Polypeutes tricinctus* (=apar)

  giant armadillo the five-banded armadillo, *Priodontes giganteus*

  weasel-headed armadillo any of several species of six-banded armadillos of the genus Euphractus

armature the sum total of the spines or spikes on the body of an organism

  apertural armature that around the mouth of some gastropod mollusc shells

  genital armature those portions of the reproductive system of an arthropod which are directly used in copulation

armeniaceous apricot-colored

Armeriaceae = Plumbaginaceae

army term for a large group of the same species, particularly of insects and frogs

old arnold *W.I.* the cuculid bird *Crotophaga ani*

aroA a bacterial genetic marker having a character affecting the activity of shikimic acid to 3-enolpyruvylshikimate-5-phosphate, mapped at 20.25 mins. for *Escherichia coli*

aroB a bacterial genetic marker having a character affecting the activity of shikimic acid to 3-enolpyruvylshikimate-5-phosphate, mapped at 41.7 mins. for *Escherichia coli*

aroC a bacterial genetic marker having a character affecting the activity of shikimic acid to 3-enolpyruvylshikimate-5-phosphate, mapped at 45.0 mins. for *Escherichia coli*

aroD a bacterial genetic marker having a character affecting biosynthesis of shikimic acid, mapped at 32.70 mins. for *Escherichia coli*

Aroideae a subfamily of the Araceae

arolium the central lobe of the pulvillus, sometimes used as synonymous with pulvillus (*q.v.*)

aroM a bacterial mutant indicating a multiple requirement for several aromatic amino acids or aromatic vitamins

arom-1-arom-4 *see* "Aromatic-1" to "Aromatic-4"

[**Aromatic-1**] a mutant (gene symbol *arom-1*) on linkage group II of *Neurospora crassa*. The phenotypic expression requires an aromatic amino acid but can grow on shikimic acid

[**Aromatic-3**] a mutant (gene symbol *arom-3*) on linkage group II of *Neurospora crassa*. The phenotypic expression is a requirement for aromatic amino acids. The phenotype cannot use shikimic acid

[**Aromatic-4**] a mutant (gene symbol *arom-4*) on linkage group II of *Neurospora crassa*. The phenotypic expression is a requirement for an aromatic amino acid but the phenotype cannot use shikimic acid

arrau the pelomedusid chelonian reptile *Podocnemis expansa*

arrect stiffly upright, as distinct from erect which is merely upright (= porrect)

-arrhe- *comb. form* meaning "male"

sea arrow any of several species of ommastrephid mollusks of the genus Ommastrephes (= flying squid)

-arsen- *comb. form* meaning "male" or "masculine"

monarseny a term used by entomologists when polygamy is meant (*cf.* monothely)

Artamidae that family of passeriform birds which contains the wood-swallows of the Far East and Australasian zones. They are stout birds with very long and pointed wings and short stout legs

arteriole diminutive of artery

arterenol = norepinephrine

artery a vessel that conducts blood away from the heart

  arcuate arteries those branches of the renal arteries that run parallel to the surface of the kidney at the level of the cortico-medullary junction

  branchial artery an artery associated with a gill, and derived from the aortic arch

    afferent branchial artery an artery carrying blood to the gill

    efferent branchial artery artery carrying blood from the gill

  carotid artery one of two arteries arising from the first aortic arch. The **internal carotid** goes to the brain; and the **external carotid** supplies the superficial parts of the head

    common carotid artery that portion of the ventral aorta between the third and fourth arches, which carries blood for the carotids alone, in those tetrapods in which the radix disappears between the third and fourth arches

  brachiocephalic artery the common trunk of the subclavian and carotid arteries

  subclavian artery the most anterior of the major arteries arising from the dorsal aorta, and which supplies blood to the forelimbs

  coeliac artery an artery arising from the dorsal aorta near the radix and dividing into the gastric, splenic and hepatic arteries

coronary artery an artery supplying blood to the walls of the ventricle

epigastric artery an artery which extends from the aorta to the ventral side of the body, supplies the muscles of the body wall and gives rise to the femoral artery

femoral artery an artery running down the anterior side of the leg

gluteal artery = sciatic artery

helicine artery the coiled arteries in the flaccid penis

hyaloid artery an embryonic artery supplying blood to the developing vitreous body of the eye

hypogastric artery an artery arising from the dorsal aorta at or near the origin of the iliac artery and that supplies blood to the posterior viscera

innominate artery = brachiocephalic artery

iliac artery the most posterior of the major arteries leaving the dorsal aorta and which supplies blood to the hindlimb

  external iliac artery = femoral artery

ischiadic artery = sciatic artery

mesenteric artery one or more derivatives of the omphalo-mesenteric artery

  omphalo-mesenteric artery a major artery arising near the central region of the dorsal aorta and supplying blood to the principal viscera

peroneal artery a branch of the popliteal artery supplying blood to the muscles of the calf

popliteal artery that portion of the sciatic artery which lies in the lower leg

sciatic artery an artery running down the posterior side of the leg

vertebral artery one of a pair of derivative arteries of the subclavian that run forward to unite as the basilar artery running forward under the medulla

-arthro- *comb. form* meaning "joint" (*q.v.*)

arthrium the concealed tarsal segment in some coleopteran insects that appear to have three or four, but actually have four or five segmented tarsi

syn**arthries** a union between two plates of an echinoderm arm that are joined by elastic fibers (*cf.* syzygy)

Arthromitaceae a family of caryophanale schizomycetes occurring in the form of trichomes having disc-like nuclei alternating with protoplasmic segments but lacking septa. Spores are formed in the distal end of the trichome, the base of which forms a spherical body attached to the intestinal wall of many anthropods and some amphibia

Arthropleona a suborder of collembolan insects distinguished from the Symphypleona by the elongate, clearly segmented, body

Arthropoda an enormous phylum of the animal kingdom distinguished, in all classes except the Onychophora, by a chitinous jointed exoskeleton and jointed appendages. The body cavity is a haemocoel

arthrosis a joint

amphiarthrosis a joint capable of flexion, but not free movement. In human *anat.* a fibrocartilaginous joint

  diarthrosis a moveable joint. In human *anat.* a synovial joint

  synarthrosis an apparent joint, but one that is incapable of movement

arthrous jointed

  anarthrous without joints

arti- *comb. form* meaning "complete"

artichoke either of several completely unrelated plants. Without qualification, either the globe, or the Jerusalem, artichoke is usually meant

Chinese artichoke = Japanese artichoke

globe artichoke the compositaceous plant *Cynara scolymus* and its edible flower bud

Japanese artichoke the subterranean tubers of the labiateous herb *Stachys sieboldii*

Jerusalem artichoke the compositaceous herb *Helianthus tuberosus* and its edible tubers

article = segment

articular pertaining to a joint

Articulata an obsolete zoological taxon once containing all forms showing metameric segmentation

Articulata (Brachiopoda) = Testicardines

Articulatae = Equisetineae

articulate jointed

  inarticulate = unjointed

articulus a segment of a calcareous alga

-artio- *comb. form* meaning "even"

Artiodactyla an order of placental mammals once fused with the Perissodactyla into the order Ungulata. The Artiodactyla are commonly referred to as the cloven-hoofed animals and therefore contain the cattle, sheep, antelopes, deer, and similar forms. They are distinguished by having an even number of digits with the axis of symmetry passing between these digits

Artocarpaceae = Moraceae

Artocarpoideae a subfamily of Moraceae distinguished by having straight stamens and leaves convolute in the bud

arui the wild sheep *Ovis tragelaphus* (= barbary sheep)

arum any of several plants of the family Araceae

  arrow arum *U.S.* any of several ariaceous herbs of the genus Peltandra

  dragon arum *U.S.* any of several araceous herbs of the genus Arisaema. (*cf.* jack-in-the-pulpit and Indian turnip)

  water arum the araceous plant *Calla palustris*

arytenoid (*see also* cartilage) in the shape of a pitcher

as *see* "asynaptic" or "ascute"

as^r a yeast genetic marker indicating inhibition of growth by arsenate

asa a bacterial mutant indicating a requirement for aspartic acid

-asc-, *comb. form* meaning "sac" or "bladder"

Ascalaphidae a family of neuropteran insects strongly resembling Myrmeleontidae both in appearance and habit. They may be distinguished by the antennae that are as long as the body

Ascaphidae = Liopelmidae

Ascaridae a family of typical ascaroid nematode worms containing numerous intestinal parasites of man and domestic animals

Ascaridea a class of parasitic Nematodes, distinguished by three prominent lips

Ascaroidea a suborder of telogonian nematodes distinguished by the three prominent lips and the spirally curled posterior end of the male

ascension term for a group of larks (*cf.* exaltation)

Aschelminthes a taxon of the animal kingdom that may be regarded either as a superphylum containing the phyla Rotifera, Gastrotricha, Kinorhyncha, Nematoda, and Nematomorpha or a phylum containing these taxa as classes. Until recently the Priapulida were also included

Aschiza an assemblage of those families of cyclorrhaphous brachyceran dipteran insects which lack a frontal suture

-ascia- *comb. form* meaning "hatchet"

-ascid- *see* -asc-

Ascidiacea a class of tunicate Chordata containing the sea squirts. They are mostly sedentary forms with a recurved gut and a pharynx modified to form a ciliary-mucoid filter feeding apparatus

ascidium literally a pitcher. Particularly in *biol.*, that of the pitcher plant Nepenthes

Asclepiadaceae that family of contortous dicotyledons which contains, *inter alia*, the milkweeds. The family can be distinguished from the Spocynaceae by the presence of a corona and from the Gentianaceae by possessing a milky juice

-asco- *comb. form* literally meaning "wine skin" but extended in *biol.* to many bladder-like structures

asco *see* [Ascospores colorless]

Ascolichenes group erected to contain those lichens that produce asci

Ascomycetae a phylum of Eumycophyta distinguished by the presence of an ascus (*q.v.*)

asconoid a type of sponge structure consisting of a simple vase-like shape, the simple cavity being lined with choanocytes (*cf.* leuconoid, synconoid, sylleibid)

[**Ascospore lethal**] a mutant (gene symbol *le-l*) in linkage group IV of *Neurospora crassa*. The phenotypic expression is colonial growth

[**Ascospore colorless**] a mutant (gene symbol *asco*) in linkage group VI of *Neurospora crassa*. The phenotypic expressions are a requirement for lysine and a low germination rate

Ascothoracida an order of parasitic cirripede crustacea thought by many to be a separate sub-class. Some are enclosed in a bivalved shell and are ectoparasitic on various echinoderms. Others lose most traces of cirripede appearance and are endoparasites in echinoderms and marine coelenterates

ascus 1 (*see also* ascus 2) the spore sac of ascomycete fungi, typically containing eight ascospores within a tubular, or oval, case produced either from the ascogonium or from specialized hyphae

ascus 2 (*see also* ascus 1) a thin walled sac which permits volume changes in Ectoprocta with calcified zoecia (= compensation sac)

[**ascute**] a mutant (gene symbol *as*) amapped at 46.0± on chromosome IIO of *Drosophila melanogaster*. The phenotypic expression is downwardly held wings

Aselloidea a suborder of free living isopod crustacea having the uropods terminal and the first pair of pleopods modified to form a thin opercular plate that covers the other pleopods. They are further distinguished from the Oniscoidea by their aquatic habit. The very similar Bopyroidea are parasitic

asepsis the condition of lacking microorganisms particularly pathogenic forms

ash 1 (*see also* ash 2, borer, sawfly, sphinx) any of numerous trees of the oleaceous genus Fraxinus. *Brit.* without qualification, *F. excelsior*

Arizona ash *F. velutina*
bastard ash = red ash
blue ash *F. quadrangulata*
European ash *F. excelsior*
flowering ash *F. ornus*
green ash *F. lanceolata*
Oregon ash *F. latifolia*
pumpkin ash *F. tomentosa*
red ash *F. pennsylvanica*
water ash *F. caroliniana*
white ash *F. americana*

ash 2 (*see also* ash 1) any of a variety of trees mostly having foliage, or habits, like those of Fraxinus

bitter ash the samaroubiaceous tree *Picrasma excelsum* (= bitter wood)
Jerusalem ash = wood
mountain ash any of numerous rosaceous trees of the genus Sorbus, *Brit.*, without qualification, usually *S. aucuparia*. *U.S.*, without qualification, usually *S. americana*
prickly ash any of several species of rutaceous shrubs or trees of the genus Zanthoxylum
sea ash = prickly ash
stinking ash the rutaceous tree *Ptelea trifoliata* (= hop tree)

Asilidae a very large group of predatory dipteran insects mostly with densely hairy legs and long conical bodies. Others resemble bumblebees and are frequently confused with the Bombyliidae. Some tropical forms reach a length of two inches and can inflict the most painful insect bite known. They are commonly called robber flies

Asionidae = Strigidae

asity any of four species of philepittid bird

-asko- *comb. form* meaning "wine-skin", usually transliterated -asco-

asp there is endless argument about the meaning of this word unless it is loosely taken to be synonymous with viper, adder or even snake. At one extreme are those who reserve the word for the viperid *Cerastes vipera;* at the opposite end of the scale are those who consider that Cleopatra clasped the elapid *Naja haje* to her bosom. Some reserve the term for the genus Cerastes, others include any viperid. The only agreement appears to be that an asp is a snake

asp *see* [Asparagine] this symbol is also used for a bacterial mutant indicating a requirement for asparagine

Asparaginase an enzyme catalyzing the hydrolysis of asparagine to aspartate and ammonia

[**Asparagine**] a mutant (gene symbol *asp*) on linkage group V of *Neurospora crassa*. The phenotypic expression is a requirement for asparagine

L-**asparagine** α-Aminosuccinamic acid, a common amino acid not essential in rat nutrition

asparagus (*see also* beetle, fern, miner) a large genus of liliaceous herbs best known for the edible spears of *A. officinalis*

diaspasis said of an amitotic division in which the nucleus appears to be crudely torn apart (*cf.* diatmesis)

aspection the seasonal succession of aspects or appearances in a community

aspen (*see also* beetle) properly any of numerous species of the salicaceous genus Populus. (*cf.* poplar and cottonwood)

American aspen *P. tremuloides*
European aspen *P. tremula*
quaking aspen = American aspen
large-toothed aspen *P. grandidentata*
trembling aspen = American aspen
-aspergill- *comb. form.* meaning "brush"

Aspergillales an order of plectomycete ascomycete fungi distinguished by their mat-like hyphae turning to a pulverulent mass when mature. The well-known genera Aspergillus and Penicillium lie in this order

asperous roughened with hairs, like an unshaven human face

asphodel any of several liliaceous herbs, mostly of the genus Narthecium

    bastard asphodel *U.S. N. americanum Brit. N. ossifragrum* (= bog asphodel)

    bog asphodel *U.S.,* usually *N. americanum; Brit. N. ossifragrum*

    false asphodel any of several species of the genus Tofieldia

    marsh asphodel = bastard asphodel

    Scottish asphodel *Brit.* any of several species of the genus Tofieldia

-aspid- *comb. form* meaning "shield"

endaspidean said of the foot of the bird in which the scutes of the tarsal sheath extend around the inner side of the tarsus, their ends meeting on the outer surface

    exaspidean said of the foot of a bird in which the scutes of the tarsal sheath extend around the outside of the tarsus their ends meeting on the inner surface

    holaspidean said of a foot of a bird in which the anterior and rear surfaces of the tarsus are covered by a large rectangular scutes which meet along lines on the inner and outer aspects of the tarsus

    pycnaspidean said of a bird in which the posterior surface of the tarsus is "granulated" with small scales

    taxaspidean said of birds in which the rear surface of the tarsus is covered by two or three series of scales (*cf.* booted, laminoplantar)

Aspidobothria = Aspidocotylea

Aspidobranchia an order of streptoneuran gastropods containing *inter alia,* the marine slipper limpets and a few fresh water forms. They are distinguished by the more or less bilaterally symmetrical appearance of the shell and by the relatively diffuse central nervous system

Aspidochirota an order of holothurioid Echinodermata having numerous ambulacral feet and with the branched tentacles compacted into a circular distal disc

Aspidocotylea an order of digeneous trematode platyhelminthes distinguished by an enormous ventral sucker subdivided into compartments

aspis the propodosomal plate of oribatid mites, in which group it serves to close the cavity of the hysterosoma into which the gnathosoma can be withdrawn

asp-of-Jerusalem the cruciferous herb *Isatis tinctoria* (= woad)

Asporomycetetes = Fungi imperfecti

Aspredinidae a family of *S. Amer.* catfishes referred to as banjo catfish, a term descriptive of their shape

ass any of several species of equid mammal of the genus Equus distinguished principally from the horse by the smaller size, longer ears, tufted tail and in having a longer period of gestation (*cf.* donkey, onager, and kiang)

    asiatic ass *E. onager*

    jackass a male ass or any organism that makes a similar noise

        laughing jackass *Austral.* the alcedinid bird *Dacelo novaeguineae.* (= kookaburra)

    nubian ass *E. africanus*

assemblage any group of organisms taken together without, unless further defined, any other connotation

assembly the smallest community recognized in ecology

assimilation the basic power of living matter to change other things into its own substance

    genetic assimilation the fixation of a genetic character not evident in the original phenotype, by artificial environmental changes. Has been widely confused with fixation of acquired characters

association *see* -sociation

assortment the separation of genes at meiosis

    independent assortment the condition when, in meiosis, genes segregate independently of one another

assumentum the valve of a siliquum

assurgent curved down and then up, as for example, the horns of many bovids

ast *see* ['asteroid']

Astacura a section of reptant decapod crustacea containing the lobsters, crayfish and the like. They are distinguished by the broad tail fan, well developed abdomen, and large chela

Asteiidae an obscure group of myodarian cycloraphous dipteran insects closely allied to the Drosophilidae

-aster- *comb. form* meaning "star" (*see also* -astr-)

aster 1 (*see also* aster 2, aster 3) a large genus of compositaceous herbs. The term is also applied in horticulture to several other plants with star-like flowers

    bog aster *U.S. Aster nemoralis*

    China aster any of several species and numerous horticultural varieties of the Asiatic compositaceous herb Callistephus

    golden aster any of numerous compositaceous herbs of the *Amer.* genus Chrysopsis

    sea aster *Aster trifolium*

    Stokes aster the compositaceous herb *Stokesia laevis* and its numerous horticultural varieties

    tree aster the compositaceous tree *Olearia haastii*

    white topped aster any of several species of compositaceous herb of the genus Sericocarpus

aster 2 (*see also* aster 1, aster 3) a star-like structure, associated with the centrosome and spindle fibers, that appears during cell division

    dyaster = prophase of mitosis or meiosis

    karyaster mitotic spindle

aster 3 (*see also* aster 1, aster 2-acline) a polyaxonic sponge spicule; though the terminal combination aster is used also of some types of monaxonic sponge spicule

    amphiaster a microscleric, monaxonic sponge spicule with spines at each end

    discotaster a microscleric hexactinellid sponge spicule the branches of which are spread in the form of a brush

    hexaster a hexactine sponge spicule with branched rays

        discohexaster a hexactinellid microscleric spicule in which the ends of the rays terminate in disks

        oxyhexaster a microscleric hexactinellid sponge spicule in which the ends of the rays branch

    oxyaster a small centered polyaxonic sponge spicule with pointed rays

    plesioaster a monaxonic, microscleric sponge spicule with spines projecting from a short axis

    sanidaster a short, rod-shaped streptaster

    spheraster a large centered, polyaxonic, sponge spicule with definite rays

    sterraster a sponge spicule, consisting essentially of a sphere with sharp prominances arising from it irregularly

    streptaster a microscleric, monaxonic sponge spicule

    strongylaster a polyaxonic sponge spicule the rays of which have rounded ends

    tylaster a polyaxonic sponge spicule with knobs on the end of the rays

Asteraceae = Compositae

Astereae a tribe of tubuliflorous Compositae

asteroid star-like

[asteroid] a mutant (gene symbol *ast*) mapped at 1.3± on chromosome II of *Drosophila melanogaster*. The phenotypic expression is small rough eyes

Asteroidea that class of eleuthrozoan Echinodermata containing the sea stars and starfish. They are distinguished by their flattened star-like shape with five (rarely more) radiating arms

Asterozoa = Eleuthrozoa

-asthes- *comb. form* meaning "perception" or "sense". Properly, but very rarely, spelled -aesthes- (*cf.* aesthete)

asthesia the condition of perceiving, though sometimes used in *bot.* in the sense of tropism or taxis

   anesthesia the condition of not perceiving

   geoasthesia response to stimulus by gravity. This is given only as an example, other responses being listed under -tropism and -taxis

   hyperasthesia heightened perception

   paresthesia the condition of apparently perceiving

Astiidae = Asteiidae

-astr- comb. form meaning "star" (*see also* -aster-)

Astraeidae an old family including madreporarian Zoantharia typically distinguished by the compact massive colony crowded with zooids

astragaloid dice-shaped, for the reason that the original dice were knuckle bones

[asynaptic] a mutant (gene symbol *as*) mapped at 53 on linkage group I of *Zea mays*. The phenotypic expression is a failure of the chromosomes to pair at meiosis

at *see* [arctus oculus]

At *see* [production of atropinesterase]

-atacto- *comb. form* indicating irregularity

atavism return to an ancestral type

[ataxia] a mutation (gene symbol *ax*) found in linkage group XV of the mouse. The locomotor ataxia of the phenotypic expression leads to death in three or four weeks

Atelopodidae a family of anuran amphibia distinguished by the procoelous vertebrae the fusion of the epicoracoid cartilages, the absence of a sternum and the degeneration and disappearance of some toes

ater dull black

athera = awn

Atherinidae a family of acanthopterygian fishes frequently called smelts but distinguished by the separate spiny and soft dorsal fin, the absence of a lateral line, the slenderness of dorsal spines III to VIII and the anal fin with one spine. Properly called silversides

Athiorhodaceae a family of pseudomodale schizomycetes resembling the Thiorhodaceae in the production of pigment but requiring neither light nor sulphur for their development

atlas the vertebra next to the skull

   preatlas = proatlas

   proatlas a neural arch, of unknown homologies, found anterior to the atlas in some reptiles

atoll a circular, or sometimes horseshoe-shaped coral reef, encircling a lagoon

ATP = adenosine triphosphate

ATPase an enzyme, otherwise called myosin, which, when activated by Ca, acts on ITP, CTP, GTP, UTP, and ATP to yield orthophosphate and the corresponding diphosphate. Another form activated by Mg acts only on ITP and ADP

atramentarious inky black

Atremata an order of ecardine brachiopods distin-

guished by the fact that both valves of the shell are grooved for the passage of the pedicel

atretic *adj.* imperforate

Atrichiidae = Atrichornithidae

Atrichornithidae a family erected to contain the two *Austral.* scrub-birds distinguished by their large bill, small wings and long tail

atrium properly "chamber". In *anat.* refers to the cavity of the auricle of the heart or to the tympanic cavity of the ear. In *zool.* refers to any of numerous cavities, particularly the cavity between the branchial pouch and the body wall of prochordates

Atropidae that family of corrodent insects to which the so-called book-lice belong

atta the annonaceous tree *Annona squamosa* and its fruit

Attagidae = Thinocoridae

attenuate drawn out

attenuation used specifically of the gradual reduction in virulence of a microorganism

Attidae a family of dipneumonomorph Araneae that do not make webs and whose frequent mode of progression and predatory habits cause them to be called jumping spiders

attila any of several cotingid birds of the genus Attila

attingent being in contact with

attire the stamens and pistils of a flower, taken together

Atypidae a family of mygalomorph spiders lacking the rastellum

aubergine the solanaceous shrub *Solanum melongena* and its edible fruit (= egg plant)

auchene *see* hair

Auchenorrhyncha a suborder of homopteran insects containing those forms in which the antennae are short and bristle-like (*cf.* Sternorrhyncha)

auctous enlarged or swollen

aucuparious attractive to birds

Auerbach's *epon. adj.* from Leopold Auerbach (1828–1897)

aufwuchs a German word prefered by some to the English periphyton

augur any of several species of terebrid gastropod mollusks of the genus Terebra

auk any of numerous sea birds of the family Alcidae

   razor-billed auk *Alca torda*

   great auk the extinct *Pinguinis impennis*

   little auk *Plautus alle* (= dovekie)

auklet any of many species of alcid birds in general smaller than auks

   Cassin's auklet *Ptychoramphus aleutica*

   crested auklet *Aethia cristatella*

   least auklet *A. pusila*

   parakeet auklet *Cyclorrhynchus psittacula*

   rhinoceros auklet *Cerorhinca monocerata*

   whiskered auklet *A. pygmaea*

Aulacidae a family of proctotrupoid apocritan Hymenoptera usually fused with the Gasteruptiidae

Aulacogastridae a family of brachyceran dipteran insects commonly included with the Drosophilidae

aulaeum (obs.) = corolla

-aulo- *comb. form* confused from two Greek roots, and meaning "tube" or "dwelling place"

Aulogamae (obs.) = Bryophyta

Aulopidae a small family of fishes strongly resembling the Synodontidae from which they may be distinguished by the shorter head and larger dorsal fin

Aulorhynchidae a family of osteichthyes closely allied to the Gasterosteidae (sticklebacks) from which they differ in their greater length and the fact that there are never less than twenty-five dorsal spines. Commonly called tubenoses

Aulostomidae a family of fishes related to the Fistulariidae but distinguished from them by the notably compressed head and body, rounded caudal fin without a filament and the presence of numerous spines in front of the dorsal fin. Commonly called trumpet fishes.

auntie-effie *W.I.* the icterid bird *Icterus leucopteryx* (= Jamaican oriole)

auntie-katie *W.I.* the icterid bird *Icterus leucopteryx* (= Jamaican oriole)

-aur- *comb. form.* confused from three roots and therefore meaning "ear", "gold" or "air"

aur *see* [Aurescent]

Aurantiaceae = Rutaceae

aurantiaceous orange-colored

auratous with the sheen of metallic gold (*cf.* aureous)

aureous golden yellow (*cf.* auratous)

[Aurescent] a mutant (gene symbol *aur*) in linkage group I of *Neurospora crassa.* The phenotype is at first white but later forms pigmented conidia

auricle almost any ear-shaped structure but most frequently the derivative of the atrium of the vertebrate heart. Also swollen lateral appendages or parts of appendages in many insects, one of a pair of prominent lateral ciliated projections posterior to the corona on some rotifers, an internal projection from the perignathic girdle of some echinoderms, lateral sensory projections from the head of some Turbellaria, a spirally coiled process arising, in some Ctenophora, from the comb row, and an earlike appendage to a leaf or stipule.

Auricula a species of Primula having so many horticultural varieties that the specific name is frequently used as though it were generic

auricularia *see* larva

Auriculariales an order of heterobasidiomycete Basidiomycetes containing those forms commonly known as Jew's ears or monkey's ears. These terms are sufficiently descriptive of their shape to permit their ready identification in the field as saprophytes on dead tree stumps. They are technically distinguished by the fact that the hypobasidium becomes divided into four cells

auroch properly the wild ox *Bos taurus* (= domestic ox) The term is, however, commonly applied to the European bison or wisent

aurora *Afr.* the ploceid bird *Pytilia phoenicoptera*

auroral pertaining to the dawn (*cf.* crepuscular)

auroreous the rosy gold color of a bright dawn

Austroastacidae a family of *Austral.* freshwater crayfishes that differ from the Parastacidae in that no transverse suture occurs on the telson or uropods

autacoid a hormone produced by an endocrine gland

auto- *comb. form* meaning "self"

Autosauri = Lacertilia

Autosauria = Lacertilia

-auxes- *comb. form* meaning "growth"

auxesis literally, growth, but used specifically for an increase in cell size without cell division or the transformation of a small cell into a large one; it has also been used as an expression for the maximum size to which a cell may grow before cell division is inevitable. Sometimes used as synonymous with "swelling" (on a plant stem) or "dilation" (of a diatom frustule)

bradyauxesis the condition of an organism in which the relative growth rate of one part is slower than the growth of the whole

ectauxesis the condition of a plant organ that grows out through the substance of a parent shoot

endauxesis said of an organ which lies proximal to the main axis

heterauxesis differential growth of two sides or parts. In *bot.* usually pertaining to a different shoot, but applied to such diverse conditions as the production of a pseudobulb from a single swollen internode by an orchid, or the production of an adult form not similar to the parent, but an outgrowth from the latter, as is the case in Chara. It is used in *zool.* to express the difference between a larval, or more rarely juvenile, form and the adult condition as well as the differential growth of parts

geoheterauxesis heterauxesis due to gravity

isauxesis the condition in which a part grows at the same relative rate as the whole

auxin a plant hormone or growth factor

-auxo- *comb. form* meaning "increase"

-av- *comb. form.* meaning "bird"

av *see* [Ames' waltzer]

avadat the ploceid bird *Estrilda amandava*

avahi the indriid lemuroid primate *Lichanotus laniger*

-aven- *comb. form.* meaning "oat"

avens *Brit.* any of several species of the rosaceous genus Geum

common avens *Brit.* G. *urbanum*

mountain avens *Brit.* any of several species of rosaceous herb of the genus Octopetala

water avens *Brit.* G. *rosale*

Aves that class of gnathostomatous craniate chordates which are usually called birds. They are distinguished by the presence of feathers on the skin and the habit of laying eggs

avicularia a modified zooid in the shape of a bird's beak found in ectoprocts

avocado (*see also* fly, mite) the lauraceous tree *Persea gratissima* and its edible fruit

avocet (*see also* bill) any of several species of recurvirostrid birds of the genus Recurvirostra *U.S.* (without qualification) *R. americana* (= American avocet). *Brit. R. avocetta* (= European avocet). *Austral. R. novaehollandiae* (= red-necked avocet)

avoset = avocet

aw *see* [without anthocyanin]

aweoweo the priacanthid fish *Priacanthus cruentatus* (= glasseyed snapper)

awl 1 (*see also* awl 2, awl 3, bill, hair) any sharply pointed structure

awl 2 (*see also* awl 1, awl 3, king) *I.P.* and *Austral.* any of several species of hesperid lepidopteran insects, most of the genera Badamia, Bibasio, Chromis, and Hasora

awl 3 (*see also* awl 1, awl 2) any of several organisms bearing awl-like structures, though many of these are simply called cobbler (*q.v.*).

cobbler's awl any bird with an unusually long and sharp beak. *Brit.* usually the avocet (*q.v.*). *Austral.* the meliphagid *Acanthorhynchus tenuirostris*

awn 1 (*see also* awn 2) the bristle that terminates the bract of some grasses, of which barley is a typical example

awn 2 (*see also* awn 1) barbed processes on the end of an intromittent organ, particularly those of some reptiles

ax *see* [ataxia]

Ax *see* [Abruptex]

-ax- *comb. form* meaning "axle" or "axis"

axenic said of organisms isolated in pure culture from

their normal environment or animals sterile both internally and externally, in which case the word is synonymous with gnotobiotic

monaxenic said of an axenic culture to which one *other* kind of organism has been added. Similarly a **diaxenic** culture contains three types of organism and so on

axial relating to an axis (*q.v.*)

abaxial located away from, or further away from the axis

adaxial located near, or nearest to, the axis

epaxial located above, or dorsal to, an axis

hypaxial located beneath an axis, particularly beneath the vertebral column

axiation the state of the relation that exists between organ forming areas and the main axes in developing embryos

axil the angle formed between the axis and an appendage to the axis

-axilla- *comb. form* meaning "arm pit"

axillar a feather arising from the axilla region of a bird

axis 1 (*see also* allaxis, axis 2, axis 3, axis 4, plane) the point or line around, along or across which symmetry is established or gradients measured.

longitudinal axis the axis running the length of a biradially or bilaterally symmetrical object. It differs from the sagittal axis in that it does not divide the object into symmetrical parts

sagittal axis the axis of a biradially or a bilaterally symmetrical object which divides the object into mirror image halves

transverse axis the axis at right angles to the sagittal axis

axis 2 (*see also* axis 1, axis 3, axis 4) similar to axis 1 but with specialized meaning in regard to diatom frustules

apical axis a curved line joining the apices of the frustule, running throughout at an equal distance from the surface

transapical axis that which passes at right angles through the apical axis

isopolar axis those axes of diatom frustules which are identical in their extremities

heteropolar axis the axis of a diatom, of which the ends are different

pervalvar axis that axis of a diatom frustule which runs along the center of the dividing plane, at equal distances from the enclosing walls

axis 3 (*see also* axis 1, axis 2, axis 4) the main part of an elongate body, particularly a plant

ascending axis = stem

descending axis = root system

false axis = sympodium

inflorescence axis that part of a plant axis on which the flowers are borne

primary axis (*bot.*) = main stem

pseudaxis = sympodium

tropaxis that plane in a growing plant from which the epicotyl grows in one direction and the hypocotyl in another

axis 4 (*see also* axis 1, axis 2, axis 3) the second cervical vertebra on which the atlas pivots

axolotl the unmetamorphosed, but sexually mature, larvae of some salamanders, particularly *Ambystoma mexicanum*

axon those nerve fibers that pass from a nerve cell to the side where the impulse is to be conducted. (*cf.* dendrite)

axonic pertaining to the axis

monaxonic the condition of a cell or organ when the two transverse axes are equal

mesaxonic descriptive of the feet of those ungulates in which the axis of symmetry passes through the third digit

paraxonic descriptive of the feet of those ungulates in which the axis of symmetry passes between the third and fourth digits

aye-aye the chiromyid lemur *Chiromys madagascariensis*

ayer ayer the meliaceous tree *Lansium domesticum* (= langsat)

Aythyini a group of Anatinae containing the pochards. They are small short-legged, short-tailed diving ducks

ayu the salmonid fish *Plecoglossus altivelis*. (Sometimes made the type of a monospecific family the Plecoglossidae)

Azotobacteraceae a small family of eubacteriale schizomycetes. A gram negative, obligate aerobe usually in the form of relatively large rods and capable of fixing atmospheric nitrogen

azi a bacterial genetic marker having a character affecting resistance or sensitivity to azide. One has been mapped at 1.0 mins. for *Escherichia coli*.

azure 1 (*see also* azure 2) sky blue

azure 2 (*see also* azure 1 crown) *Austral.* any of numerous lycaenid lepidopteran insects of the genus Ogyris (*cf.* blue)

B

B *see* [Anthocyanin booster], [Bar], or [Barring]

b *see* [black, brown]

b2 *see* [brown egg-2]

ba *see* [balloon]

ba₁ *see* [barren stalk]

Babbler any of very numerous birds of the family Timaliidae

Bombay babbler *Turdoides striatus*

white-browed babbler *Pomatostomus superciliosus*

brown babbler *Turdoides plebejus*

red-capped babbler *Timalia pileata*

capuchin babbler *Phyllanthus atripennis*

rusty-cheeked babbler *Pomatorhinus erythrogenys*

white-crested babbler *Garrulax leucolophus* (*cf.* laughing-thrush)

black-headed babbler *Turdoides reinwardtii*

red-headed babbler *Stachyris ruficeps*

chestnut-headed babbler *Alcippe castaneceps*

hill-babbler any of several species of the genus Alcippe

jungle-babbler any of several species of the genera Pellorneum and Trichastoma

black-lored babbler *Turdoides melanops*

Madras babbler *Turdoides affinis*

arrow-marked babbler *Turdoides jardineii*

Nepal babbler *Alcippe nipalensis*

oriole-babbler = chestnut-headed babbler

pied babbler any of several species of the genus Turdoides, without qualification, usually *T. hypoleucus*

quaker-babbler any of several species of the genus Alcippe

rail-babbler *Eupetes macrocerus*

scimitar-babbler any of several species of the genera Pomatorhinus and Xiphirhynchus

shrike-babbler any of several species of the genus Pteruthius

tit-babbler any of several species of the genus Macronus

tree-babbler any of several species of the genus Malacopteron and Stachyris

yellow-winged babbler *Garrulax erythrocephalus*

wren-babbler any of several species of the genera Rimator, Napothera and Prioepyga

pygmy wren-babbler *P. pusilla*

babirusa the *E. Indian* suinid *Babirusa alfurus*; the male is unique in possessing four large tushes

baboon any large, short-tailed, primate of the genera Papio or Cynocephalus

Arabian baboon *C. hamadryas*

blue-faced baboon = mandrill

sacred baboon = either hamadryas or Arabian baboon

bush-baby popular name of galagid lorisoid primates

babyroussa anglicized form of the genus Babirusa, a genus of suid artiodactyle mammals

-bac- *comb. form.* meaning "berry". The form -bacc- is common in compounds

bacbakiri = bakbakiri

bacca = berry

-bacill- *comb. form.* meaning a "little staff" or "little rod" but frequently used in *biol.* in the derivative sense of "bacillus" (*cf.* -bacul-)

baccate having berries or berry-like fruit

bachang the anacardiaceous tree *Mangifera foetida* and its fruit

Bacillaceae a family of eubacteriale schizomycetes in the form of gram positive, rod-shaped cells capable of producing endospores. The two principal genera are Bacillus and Clostridium. The former are mostly soil dwelling saprophytes though one (*B. popilliae*) is the cause of milky disease in Japanese beetles. The anerobic clostridiums are almost without exception dangerous pathogens, such as *C. tetani*, the cause of tetanus or several species causing gas gangrene. *C. botulinum* produces an exotoxin which is the most poisonous substance yet discovered by man

Bacillariophyceae a class of chrysophyte algae containing those forms commonly known as diatoms. They are distinguished by the possession of a silicified skeleton composed of two overlapping valves

bacillus literally, a rod or staff, and applied not only to bacteria but also to diatoms and some other organisms of that shape

back the dorsal surface of animals, particularly vertebrates. Applied as part of a compound name, to many forms in which this region is distinctive

painted bronze-back the Asian colubrid snake *Dendrelaphis boiga*

canvasback *U.S.* the aythyine duck *Aythya valisineria*

copperback *Austral.* the timaliid bird *Cinclosoma castanotum* (= Chestnut quail-thrush)

diamondback (*see also* moth) *U.S.* either of two crotaline snakes of the genus Crotalus (*cf.* rattlesnake)

featherback popular name of notopterid fish

fiddleback *Austral.* the cetoniid coleopteran insect *Enpocoelia australasiae*

fireback = fireback pheasant

grayback = body louse

atlantic leatherback the dermochelyid chelonian reptile *Dermochelys coriaceae*

olive-back any of several ploceid birds of the genus Nesocharis

34

puffback any of several species of laniid bird of the genus Dryoscopus, usually, without qualification, *D. gambensis*

  black-backed puffback *D. cubla*

  pink-footed puffback *D. angolensis*

quillback the catostomid fish *Carpiodes cyprinus*

redback *U.S.* = red backed sand piper

roachback *W. U.S.* = grizzly bear

saddleback *Brit.* and *U.S.* = the great black-backed gull *U.S.* the cochlidiid lepidopteran insect *Sibine stimulea*. The stinging hairs of which are notorious. *N.Z.* the callaeid bird *Creadion carnunculatus* (= tieke)

sailback the loricariid fish *Panaque nigrolineatus*

sawback *U.S.* any of several species of testudinid chelonian reptile of the genus Graptemys (*cf.* map turtle), in which dorsal projections rise from the dorsal keel

stickleback popular name of stichid fishes, particularly those of the genus Gasterosteus

thickback *Brit.* the soleid fish *Solea variegata*

thornback the rhinobatid ray *Platyrhinoidis triseriata*

yellowback *U.S.* the unionid pelecypod mollusk *Lampsilis anodontoides*. *W.I.* the fringillid bird *Loxipasser anoxanthus* (= yellow-shouldered grass-quit)

bacbakiri = bakbakiri

Bacteria a group of universally distributed, rigid, essentially unicellular microscopic organisms lacking chlorophyll usually appearing as spheroid, or rod-like, or curved entities but occasionally appearaing as sheets or chains or even, most unusually, as branched filaments. Bacteria are usually placed in the Protista but are also frequently, as a matter of convenience, regarded as plants

  budding bacteria bacteria that reproduce by buds formed at the tip of five, thread-like extensions of the mother cell

  stalked bacteria popular name of Caulobacteraceae

  sulphur bacteria any bacteria deriving its energy from sulphur or sulphur compounds

    colorless sulphur bacteria popular name of the family Thiobacteriaceae

    red sulphur bacteria popular name of Phyorodaceae

  chemolithotrophic bacteria those bacteria which are capable of synthesizing all their protoplasmic constituents from inorganic materials and which derive the energy to do this from the oxidation of inorganic materials

  gliding bacteria bacteria that are flexible and show a gliding or creeping movement over solid surfaces. They are not flagellated

  slime bacteria bacteria that produce a thick gelatinous sheath of slime, usually a polysaccharide

Bacteroidaceae a family of gram negative, rod shaped, or occasionally filamentous eubacteriale schizomycetes, mostly found in the intestinal canal of mammals to which a few are pathogenic

bacteriophage virus that invade bacteria

-bacul- *comb. form* meaning a "staff" or "rod" (*cf.* -bacill-)

badger any of several heavy-set fossorial mustelid carnivores sometimes distinguished as a separate family Melidae. *U.S.*, without qualification, *Taxidea taxus*. *Brit.*, without qualification, *Meles taxus*. The distinction between these has been questioned. *Austral.* a notoriously predatory dasyurid *Sarcophilus ursinus* (= Tasmanian devil). The biblical "badger" (Hebrew *tachash*) has never been identified and guesses include seals and dugongs

  ferret-badger = tree badger

  honey badger = ratel

  pig-like badger *Arctonyx collaris*

  sand badger either of two species of the genus Arctonyx

  skunk badger *Mydaus meliceps*

  tree badger any of several species of carnivores of the genus Helictis variously placed in the Melidae or in a separate sub-family

badious chestnut brown

Baer's *epon. adj.* from Karl Ernst von Baer (1792–1876)

Baetidae a family of ephemerotopteran insects with the venation of the wings greatly reduced

[Bag] a mutant (gene symbol *Bg*) mapped at 51.6 on the X chromosome of *Drosophila melanogaster*. The phenotypic expression are shaky, short, blunt, inflated wings

bagre sapo any of several species of batrachoidid fish of the genera Thalassophryne and Thalossophia. All are extremely venomous

Bagridae a family of naked catfishes with two pairs of mental barbels and nasal openings close together. This family is by some fused with the Pimelogidae

bailer = scaphognathite

bait originally called "food"

  whitebait *Austral.* and *N.Z.* the galaxiid fish *Galaxias attenuatus*. *Brit.* Young herring or young sprats from two to three inches long

bakbakiri *S.Afr.* the laniid bird *Telophorus zeylonus*

bal see [balloon]

Balaenicipitidae that family of ciconiiform birds which contains the whale-billed stork. This family of African birds is distinguished by its enormous bill and large broad wings

Balaenidae that family of mystacocetan whales which contains the right whales. They are distinguished from the Balaenopteridae by the large head and absence of a dorsal fin

Balaenopteridae that family of whalebone whales which contains the rorquals. They are distinguished by the small head and presence of a dorsal fin

balaka the palmaceous tree *Balaka seemannii*

-balan- *comb. form.* meaning "acorn"

genic balance the mechanism of phenotypic determination in which numerous genes with opposing actions control a particular character, such as sex, through an interaction balance.

balancer (*see also* haltere, sometimes called balancer) a pair of external villi protruding from just behind the posterior corners of the mouth in the larva of some urodele amphibia

-balano- see -balan-

Balanomorpha a division of thorace cirripede crustacea distinguished by the absence of a stalk. Commonly called rock barnacles

Balanophorales a monotypic order of parasitic dicotyledonous angiosperms containing the family Balanophoraceae

Balanopsidales a monotypic order of dicotyledonous angiosperms containing the family Balanopsidaceae found only in New Caledonia

balao the hemiramphid fish *Hemiramphis balao*. (*cf.* ballyhoo)

balausta the fruit of a pomegranate

bale 1 (*see also* bale 2) term for a group of turtles

bale 2 (*obs.*) (*see also* bale 1) = glume

Balearicidae = Gruidae

baleen the horny substance from which is made the buccal food sieve of plankton-eating whales (= whalebone)

balicassiao the dicrurid bird *Dicrurus balicassius* (*cf.* drongo)

Balistidae a family of plectognathous fish known as the trigger fishes because the first dorsal spine when erected is locked into place by the second dorsal spine moving forward

ball any spherical object but most often used in *biol.* for a solid sphere of cells

  blowball any of numerous compositaceous herbs of the genus Taraxacum (= dandelion)

  buttonball = button wood

  germ ball reproductive cells in some larvae from which other larvae may be produced. Particularly a clump of cells, actually a rudimentary embryo, found in the rear of a miracidium larva

  ovarian ball the dissociated remnants of the ovaries in Acanthocephala

  puffball popular name of lycoperdid basidomycete fungi

    giant puffball *U.S. Calvatia gigantea*

ballonet an anterior, cuticular inflation of some nematodes

[Balloon] a mutant (gene symbol *bal*) in linkage group II of *Neurospora crassa*. The term is descriptive of the swollen hemispherical phenotypic colony

[balloon] a mutant (gene symbol *ba*) mapped at 107.4 on chromosome II of *Drosophila melanogaster*. The name derives from the inflated wings of the phenotype

ballyhoo the hemiramphid fish *Hemiramphis braziliensis*

balm any of numerous horticultural varieties of the labiateous herb *Melissa officinalis*

  bastard balm *Mellitis melissophyllum*

  bee balm the labiateous herb *Monarda didyma* (= fragrant balm)

  field balm the labiateous herb *Satureia nepeta* (= sheep mint)

  fragrant balm = bee balm

  horse balm any of several species of labiateous herb of the genus Collinsonia

  Molucca balm the labiateous herb *Moluccella laevis*

  wild balm *Brit.* the labiateous herb *Melittis melissophyllum* (= bastard balm)

balm-of-gilead *either* the labiateous herb *Cedronella triphylla or* the salicaceous tree *Populus candicans* (= Ontario poplar)

balmony *U.S.* the scrophulariaceous herb *Chelone glabra*

balsam 1 (*see also* balsam 2) a sticky exudate of numerous pinaceous trees consisting essentially of one or more resins dissolved in one or more essential oils

balsam 2 (*see also* balsam 1, apple, midge, beetle, sawfly) any of numerous plants either producing balsams or having a fragrance thought to resemble that of a balsam

  garden balsam *either* the balsaminiaceous herb *Impatiens balsamina or* Peru-balsam (*see* tree)

  seaside balsam the euphorbiaceous tree *Croton eluteria* (= cascarilla)

  she balsam the pinaceous tree *Abies fraseri*

  tolu-balsam *see* tree

Balsaminaceae that family of sapindalous dicotyledons which contains, *inter alia*, the balsams and jewel weeds. The presence of five hypogynous stamens is typical of the family

Balsamineae a suborder of Sapindales containing the single family Balsaminaceae

baltimore *U.S.* the nymphalid lepidopteran *Euphydryas phaeton*

balucalag the euphorbiaceous tree *Aleurites trisperma*

bamboo (*see also* briar woodpecker, wren) any of numerous, large, usually tropical grasses with hollow, woody stems. The term is also in parts of the *U.S.* applied to the liliaceous climber *Smilax laurifolia*

  feathery bamboo *Bambusa vulgaris*

  male bamboo *Dendrocalamus strictus*

Bambuseae a tribe of the plant family Gramineae

banana (*see also* bird, borer, coot, quit, spider) any of several species, and numerous horticultural varieties, of musaceous tree-like herbs of the genus Musa and particularly their fruits

  Abyssinian banana *M. ensete*

  chotda banana *hort. var. M. paradisiaca*

  common banana *M. sapientum*

  cooking banana *M. paradisiaca*

  dwarf banana *M. cavendishii*

  Japanese banana *M. basjoo*

  lady finger banana *M. champa*

  red banana *M. rubra*

band 1 (*see also* disc) a narrow strip

  Casparian band = Casparian strip

  circumapical band a band of long cilia forming the principal ciliated region of the corona of the Rotifera

  cnidoband parallel rows of elongated nematocysts on a tentillum

  flagellated band the two outer bands, parallel to the cnidoglandular band on the septum of Anthozoa

  germ band the thickened area in an arthropod egg from which the embryo is produced

  girdle band (diatom) = girdle (diatom)

  cnidoglandular band a central ridge, flanked by flagellated bands, running along the edge of the septum in Anthozoa

  lateral band a band of cilia running along the edge of the hypobranchial groove in the endostyle

  marginal band a band of cilia running along the outer edge of the endostyle

  median band a band of cilia running along the center of the endostyle

  mesodermic band a strip of mesoderm cells in the larva of annelids and some mollusks

band 2 (*see also* band 1) term for a group of jays

bandicoot popular name of peramelid marsupials

  North Australian bandicoot *Perameles macrura*

  pig-footed bandicoot any of several species of the genus Choeropus

  rabbit bandicoot any of several species of the genus Peragale

  striped bandicoot *Perameles bougainvillii*

bandy-bandy *Austral.* any of several species of elapid snake of the genus Rhynchoelaps

bane anything inimical to something else. Widely used of plants actually, or supposedly, poisonous to various organisms

  bugbane any of several species of ranunculaceous herbs of the genus Cimicifuga

  false bugbane *U.S.* any of several ranunculaceous herbs of the genus Trautvetteria

  cowbane *Brit.* the umbelliferous herb *Cicuta virona*. *U.S.* the rosaceous herb *Potentilla palustris*

  spotted cowbane the umbelliferous herb *Cicuta maculata*

dogbane (*see also* beetle) any of several species of herb of the apocynaceous genus Apocynum

climbing dogbane any of several species of apocynaceous herb of the genus Trachelospermum

fleabane any of numerous species of the compositaceous genus Erigeron, also the compositaceous herb *Pulicaria dysenterica*

marsh fleabane the compositaceous herb *Senecio congestus*

leopard's bane any of several species of compositaceous herbs of the genus Doronicum

sheepbane any of several herbs of the umbelliferous genus Hydrocotyle (= marsh pennywort)

wolfsbane any of numerous herbs of the genus Aconitum (*see also* monkshood)

yellow wolfsbane *A. lycoctonum*

Bangioidaceae a class of the algal phylum Rhodophyta distinguished by the lack of cytoplasmic connections between neighboring cells

banistel the sapotaceous tree *Lucuma nervosa*

bank that part of land which is contiguous to a body of running water (*cf*. shore), or an elevation in the floor of a body of water which nearly reaches the surface. In tidal waters, banks are often exposed at low tide

banner the standard of a papilionaceous flower

bantam a small breed of domestic fowl

banteng a *S.E. Asiatic* bovine bovid (*Bos banteng*) distinguished by a horny casque joining the base of the horns

-bar- *comb. form* meaning "weight" or "heavy" (*cf*. -bary-)

bar 1 (*see also* bar 2) a rod, or strip of material

[**Sanio's bar** a wall thickening above and below the pit in a conifer tracheid (= crassula)

[**terminal bar** a dense rod-like structure outlining an epithelial cell

[**tongue bar** a down growth from the dorsal wall of the pharyngeal slit in Amphioxus and hemichordates, indicated also in some tunicates which divides the slit into anterior and posterior openings

bar 2 (*see also* bar 1) any of several lepidopteran insects. *S. Afr.* without qualification, those of the lycaenid genus Spindasis

broad-bar *Brit.* any of several species of geometrids

dunbar *Brit.* the noctuid *Calymnia trapezina*

silver bar *Brit.* the noctuid *Bankia argentula*

[**Bar**] a mutant (gene symbol *B*) mapped at 57.0 on the X chromosome of *Drosophila melanogaster*. The eye is restricted to a narrow vertical bar in males and homozygous females

[**bar-3**] a mutant (gene symbol *bar-3*) mapped at 79.1 on chromosome III of *Drosophila melanogaster*, but identical in its phenotypic expression with Bar

baracoa the musaceous tree-like herb *Musa rubra* and its fruit (= red banana)

barashinga the Indian cervine artiodactyle *Cervus duvauceli*

barb 1 (*see also* barb 2-8) any projection, spine or bristle that ends in a hook or other device that renders extraction or detachment more difficult than entry or attachment: or the device that renders a spine, or similar structure, a barb

barb 2 (*see also* barb 1, and barb 3-8) branches of a plume that rise directly from the rachis

barb 3 (*see also* barb 1, 2 and 4-8) an appendage around the mouths of certain fishes, particularly catfishes (= barbel)

barb 5 (see also barb 1-4 and 6-8) a variety of domestic pigeon

barb 6 (see also barb 1-5 and 7-8) a breed of *Austral*. dog alleged to have dingo ancestry

barb 7 (*see also* barb 1-6 -nd 8) aquarists' name for cyprinid teleost fishes of the genus Barbus and some of its near allies

Algerian barb *B. setivimensis*

black barb *B. nigrofasciatus*

cherry barb *B. titteya*

clown barb *B. everetti*

flying barb *Esomus danricus*

dwarf gold barb *B. gelius*

rosy barb *B. chonchonius*

spanner barb *B. lateristriga*

African 3-spot barb *B. trispilos*

spotted barb *B. binotatus*

swamp barb *B. chola*

tiger barb *B. partipentazona*, though the name is frequently, but incorrectly, applied to *B. pentazona* and *B. hexazona*

barb 8 (*see also* barb 1-7) a small beard, as that borne by goats

barba literally, a beard

barbacou any of numerous species of capitonid or bucconid birds (*cf*. barbet and puffbird)

barbastelle *Brit*. the vespertilionid bat *Synotus barbastellus*

barbel 1 (*see also* barbel 2) diminuitive of barb, particularly as applied to barbs of fish

barbel 2 (*see also* barbel 1) *Brit*. the freshwater cyprinoid fish *Barbus fluviatilis*

barbet any of many species of bird of the family Capitonidae

bearded barbet *Lybius dubius*

prong-billed barbet *Semnornis frantzii*

tooth-billed barbet *Lybius bidentatus*

yellow-billed barbet *Trachylaemus purpuratus*

brown-breasted barbet *Lybius melanopterus*

hairy-breasted barbet *Tricholaema hirsutum*

yellow-breasted barbet *Trachyphonus margaritatus*

black-collared barbet *Lybius torquatus*

coppersmith barbet *Megalaima haemacephala*

naked-faced barbet *Gymnobucco calvus*

red-fronted barbet *Tricholaema diadematum*

gaudy barbet *Megalaima mystacophanos*

white-headed barbet *Lybius leucocephalus*

great Himalayan barbet *Megalaima virens*

green barbet *Buccanodon olivacea*

little barbet *Megalaima australis*

mountain barbet *Pogoniulus coryphaeus* (*cf*. tinkerbird)

bristle-nosed barbet *Gymnobucco peli*

red-and-yellow barbet *Trachyphonus erythrocephalus*

black-spotted barbet *Capito niger*

red-spotted barbet *Lybius vieilloti*

yellow-spotted barbet *Buccanodon duchaillui*

grey-throated barbet *Gymnobucco bonapartei*

fire-tufted barbet *Psilopogon pyrolophus*

barbicel a projection on the side of a barbule of a feather

barbier any of several species of serranid fishes of the genus Hemanthias

red barbier *H. vivanus*

barbiturase an enzyme catalyzing the hydrolysis of barbiturates to malonates and urea

barbu the polynemid fish *Polydactylus virginicus*

barbudo the polymyxiid fish *Polymixia nobilis*

barbule diminutive of barb, applied specifically to the

inner row of teeth on the peristome of some mosses
and to branches arising from the barb of a feather

bargander *Brit. obs.* = merganser

bark 1 (*see also* bark 2, bark 3, beetle, miner, weevil) a
loosely defined term, usually taken to mean all tissues external to the vascular cambium of a woody
stem

   outer bark the cork layer, and all tissues external to it,
of the periderm (= rhytidome)

   ring bark the structure produced when phellogen occurs as a cylinder around the whole stem

   scale bark that which scales off the tree as the plane

   spent bark exhausted tanbark once used, while fer-
menting, as a source of heat in plant houses

   [**tanbark** any bark rich in tannin

bark 2 (*see also* bark 1, bark 3) any of numerous plants
distinguished by a peculiarity of the bark

   bitter bark = Georgia bark

   black bark the myrtaceous tres *Eucalyptus bicolor*

   Georgia bark the rubiaceous tree *Pinckneya pubens*

   iron bark any of several species of tree of the myrtaceous genus Eucalyptus

      silver leaved iron bark *E. melanophloia*

   Jesuits' bark any of several species of tree of the
rubiaceous genus Cinchona

      bastard Jesuit's bark *U.S.* the compositaceous
shrub *Iva frutescens*. *Brit.* and *U.S.* either of two
caprifoleaceous shrubs *Viburnum opulus* or *V.
lantana*

   lace bark the thymelaeaceous tree *Lagetta linteria*.
The name is sometimes applied to other members
of the same genus and to *N.Z.* either of two species
of malvaceous tree of the genus Hoheria (*cf.* ribbonwood)

   nine bark the rosaceous shrub *Physocarpus opulifolius*

   seven bark the saxifragaceous shrub *Hydrangea
arborescens*

   shagbark (*see also* hickory) the leguminous tree
Pithecolobium microdenium (*cf.* ebony). In *U.S.*
the term is more usually used as an abbreviation of
shagbark hickory (*q.v.*)

      false shagbark the juglandaceous tree *Carya ovalis*

   stringy bark any of several species of myrtaceous tree
of the genus Eucalyptus

      peppermint stringy bark *E. piperita*

      yellow stringy bark *E. muelleriana*

bark 3 (*see also* bark 1, bark 3) a short sharp cry made
by any animals including the dog and sometimes
extended as a part of a compound name

   shear-bark *W.I* the vireonid bird *Vireo crassirostris*

barker *Brit. obs.* = avocet

barley (*see also* jointworm) numerous agricultural varieties of the grass *Hordeum vulgare* and its edible
grains

   wall barley *Brit. Hordeum murinum*

   wild barley *W.U.S. H. murinum. E.U.S. H. jubatum*

barm brewer's yeast

barnacle (*see also* goose, scale) any crustacean of the
groups Cirripedia or Ascothoracica. The term
should properly be restricted to the first group

   acorn barnacle any sessile barnacle of the family Ba-
lanidae

   goose barnacle any stalked barnacle particularly of the
genus Lepas

   ivory barnacle *U.S. Balanus eburneus*

   parasitic barnacle any crustacean of the group Ascothoracida

   rock barnacle popular name of balamonomorph cirripedes

baron *I.P.* any of numerous species of nymphalid lepidopteran insect of the genus Euthalia usually without qualification, *E. garuda* (*cf.* baronet, count,
duchess, duke, earl, marquis)

baronet I.P. the nymphalid lepidopteran insect *Euthalia
nais* (*cf.* baron, duchess, duke, count, earl, marquis)

barracuda popular name of predatory sphyraenid fish

barracudina popular name of paralepidid fish

barramunda the dipnoan fish *Neoceratodus forsteri*
(= Australian lung-fish)

barren 1 (*see also* barren 2, ground) a tract of land,
either in high altitudes or high latitudes on which
there are typically shrubs, with occasional stunted
trees, but no regular trees

barren 2 (*see also* barren 1) term for a group of mules

[**barren stalk**] a mutant (gene symbol ba_1) mapped at
64 in linkage group III of *Zea mays*. The phenotypic expression is the absence of ears

[**Barring**] a mutant (gene symbol *B*) in the sex chromosome (linkage group I) of the domestic fowl. The
term refers to the appearance of bars on the feathers

Bartholin's *epon. adj.* from Caspar Secundus Bartholin
(1655–1738)

Bartonellaceae a family of rickettsiale microtobiotes
occuring as minute bacteria-like cells in vertebrate
erythrocytes

-bary- *comb. form* meaning "weight" or "heavy" (*cf.*
-bar-)

-bas- *comb. form* meaning "bottom"

basal pertaining to, forming, or found at, the base. Also,
by a confusion of meaning, pertaining to basic dyes

   hypobasal behind the basal wall, said particularly of
the parts of a plant embryo so placed

   monobasal possessing a single, short, unbranched root

basale a term variously applied to *either* the distal
pterygiophore *or* the whole pterygiophore complex

phallobase the support of the aedaegus

Basellaceae a small family of centrospermous dicotyledons which contains, *inter alia*, the madeira vine.
The twining habit of growth and the one-celled
ovary producing single seed are distinctive

Basidiolichenes a class of lichenes erected to contain
those forms in which the fungal component is a
Basidiomycete

Basidiomycetae that phylum of Eumycophyta which contains the forms commonly called toadstools and
mushrooms. They are distinguished by clamp cells
(*q.v.* under cell 2) and by a unique type of spore, the
basidiospore, of which four are formed on each
basidium

Basidiorhizae = Basidiomycetes

basidium the condiophore of a basidiomycete fungus

   autobasidium one that is non-septate and typical of
higher basidiomycetes

   endobasidium an enclosed basidium

   hemibasidium one that is septate and typical of lower
basidiomycetes

basil usually, without qualification, the labiateous
potherb *Satureja vulgaris*, but in *U.S.* also any of
numerous labiateous herbs of the genus Pycanthe-
mum

   mountain basil the labiateous herb *Pycnanthemum
virginianum*

   wild basil *Brit.* the labiateous herb *Calamintha
clinopodium*

basilisk any of numerous species of iguanid saurian reptiles of the genus Basiliscus, usually, without qualification, *B. americanus*

basis literally "bottom"

sarcobasis a carcerule

Basitonae a division of the orchidaceous subfamily Monandrae distinguished by the caudicle and viscidia rising from the apices of pollinia

basket a woven container, but in *zool.* specifically a holothurian ossicle, having the shape of a concave, perforated cup

Venus's flower basket any of several species of hexactinellid sponge of the genus Euplectella

pharyngeal basket a food gathering passage held permanently open by a girdle of trichites in certain protozoa

pollen basket = corbiculum

bass 1 (*see also* bass 2, 3, 4) *U.S.* any of numerous marine, and a few freshwater serranid fish

orangeback bass *Prionodes annularis*

chalk bass *Serranus tortugarum*

crimson bass *Anthias asperilinguis*

blackear bass *Paracentropristes pomospilus*

kelp bass *Paralabrax clatbratus*

lantern bass *Prionodes baldwini*

reef bass *Pseudogramma brederi*

saddle bass *Prionodes notospilus*

sand bass *Paralabrax nebulifer*

spotted sand bass *P. maculatofasciatus*

bank sea bass *Centropristes ocyurus*

   black sea bass *C. striatus*

   giant sea bass *Stereolepis gigas*

   pygmy sea bass *Serraniculus pumilio*

   rock sea bass *Centropristes philadelphicus*

   southern sea bass *Centropristes melanus*

streamer bass *Pronotogrammus aureorubens*

striped bass *Roccus saxatilis*

longtail bass *Hemanthias leptus*

splittail bass *H. peruanus*

stone bass *Polyprion americanus*

yellowtail bass *Pikea mexicana*

roughtongue bass *Ocyanthias martinicensis*

white bass *Roccus chrysops*

yellow bass *R. mississippiensis*

bass 2 (*see also* 1, 3, 4 and bug) *U.S.* any of several species of centrarchid fishes mostly of the genus micropterus. The white bass (previously *Lepibema chrysops*) is now regarded as a serranid (see bass 1): the white sea bass (*Cynoscion nobilis*) is a sciaenid.

large mouth bass *M. salmoides*

small mouth bass *M. dolomieui*

bass 3 (*see also* bass 1, 2, 4) *Brit.* any of several marine labrid fish usually, without qualification, *Labrax lupus*

bass 4 (*see also* bass 1, 2, 3) the bast of the lime tree. Sometimes used as synonymous with bast

bassaricyon either of two species of procyonid carnivores of the genus usually called Bassaricyon but properly Corynorhinus

bassarisc the procyonid carnivore *Bassariscus astutus* (= ring rail cat)

basse = bass

bast (*see also* wedge) the phloem tissue of plant stems and roots. The term is commonly used for dried, fibrous strips of phloem (*cf.* bass 4)

margin bast a fiber running around the edge of some leaves

primary bast the outer coat of sieve tubes and parenchyma

bastard that which is not exactly what it appears to be (*e.g.* bastard oak, bastard gemsbok, etc.)

bat 1 (*see also* bat 2, flea) the name commonly applied to cheiropteran mammals

big brown bat *U.S.* the vespertilionid *Eptesicus fuscus*

little brown bat *U.S. Myotis lucifugus*

leafchin bat *U.S.* the phyllostomatid *Mormoops megalophylla*

common bat *Brit. Vespertilio pipistrellus*

big-eared bat *U.S.* either of two species of vespertilionid of the genus Corynorhinus

long-eared bat *Brit. Plecotus auritus U.S. Myotis evotis*

mouse-eared bat *Brit.* the vesperitilionid *Myotis myotis*

evening bat *U.S.* the vespertilionid *Nyticeius humeralis*

flower-faced bat the hipposiderid *Anthops ornatus*

fish bat the emballonurid *Noctilio leporinus*

small-footed bat *U.S.* the vespertilionid *Myotis subulatus*

[**fox-bat** many very large fruit-eating bats of several genera, of which Pteropus is the best known

fringe bat *U.S.* the vespertilionid *Myotis thysanodes*

fruit bat popular name of pteropodids

great bat *Brit. Nyctalus noctula*

silver-haired bat *U.S.* the vespertilionid *Lasionycteris noctivagans*

hammer-headed bat any of several of the West African pteropine bats of the genus Hypsignathus

hoary bat *U.S.* the verpertilionid *Lasiurus cinereus*

Indiana bat the vespertilionid *Myotis sodalis*

javelin bat the phyllostomatid *Phyllostoma hastatum*

long-legged bat *U.S.* the vespertilionid *Myotis volans*

mastiff bat *U.S.* any of several species of molossid of the genus Eumops

Mississippi bat *U.S. Myotis austroriparius*

hognose bat *U.S.* the phyllostomatid *Choeronycteris mexicana*

leafnose bat *U.S.* without qualification the phyllostomid *Macrotus waterhousei*

longnose bat *U.S.* the phyllostomatid *Leptonycteris nivalis*

plainnose bat *U.S.* any vespertilionid

horseshoe-nosed bat any of very numerous species of the families Rhinolophidae and Hipposideridae. The name is derived from the horseshoe shaped mass of naked skin below the nostrils

tube-nosed bat any of several species of Asiatic vespertilionid of the genus Harpiocephalus

pallid bat *U.S.* the vespertilionid *Antrozous pallidus*

plantain bat any of numerous bats but most particularly vespertilionids of the genus Kerivoula

red bat *U.S.* the vespertilionid *Lasiurus borealis*

horseshoe bat *Brit.* either of two species of rhinolophid of the genus Rhinolophus

spotted bat *U.S.* the vespertilionid *Euderma maculatum*

freetail bat *U.S.* any molossid but particularly those of the genus Tadarida

long-tailed bat *N.Z.* the vespertilionid *Chalinolobus morio*

short-tailed bat *N.Z.* the emballonurid *Mystacops tuberculatus*

vampire bat any of several blood lapping phyllostomatid bats of the genus Desmodus, without qualification usually *D. rufus*. The large *S. Amer.* bat *Vampyrus spectrum* is, in point of interest, a predator on birds

false vampire bat the phyllostomatid *Vampyrus spectrum*

Indian vampire bat the nycterid *Megaderma lyra*

whiskered bat *Brit. Myotis mystacinus*

yellow bat *U.S.* either of two species of vespertilionid of the genus Lasiurus

eastern yellow bat *U.S. L. intermedius*

western yellow bat *L. ega*

bat 2 (*see also* bat 1, fish, hawk, ray) any bat-like animal

goore-bat *Austral.* the cracticid bird *Gymnorhina dorsalis* (= western magpie)

[bat] a mutant (gene symbol *bat*) mapped at 71.0 on chromosome II of *Drosophila melanogaster*. The phenotypic expression is extended backwardly directed wings

bateleur *Afr.* the accipitrid bird *Terathopius ecaudatus*

Batesian *epon. adj.* from Henry Walter Bates (1825–1892)

-bath- *comb. form* = -bathy-

eurybath any organism found in a wide range of depths of water

stenobath an organism restricted to a narrow range of depths of water

bathile pertaining to the bottoms of deep lakes (*cf.* chilile, pythmic)

-bathy- *comb. form* meaning "low", "deep" "depth" and "broad"

Bathyergidae a small family of myomorph rodents containing the cape mole rat. They are permanently fossorial forms distinguished from all other rodents by the fact that they have rudimentary limbs and dig with their enormous front teeth

Bathylagidae a small family of bathypelagic smelts

Bathymasteridae the small family of acanthopterygian fishes commonly called ronquils

Bathypteroidae a small family of myctophoid abyssal fishes distinguished by the fact that the upper fin rays of the pectoral extend almost to the tail. Sometimes called spider fishes

Batidaceae *see* Batidales

Batidales a monotypic order of dicotyledonous angiosperms containing a single dioecious species (*Batis maritima*). The plant is readily distinguished by the succulent subterete opposite leaves and unisexual flowers in axillary aments

Batoidea that subclass of elasmobranch fish which contains the rays, and torpedoes, distinguished by the flattened body

Batrachoididae the only family of haplodocid osteichthyes. Their slow moving habit, the broad depressed head, and large mouth have given them the name toad fish

Batrachoidiformes = Haplodoci

bauno the anacardiaceous tree *Mangifera verticillata* and its fruit

bave (*see also* brin) the fluid that is extruded from the spinning glands of caterpillars

bay 1 (*see also* bay 2) any of numerous trees, usually, but not invariably, with dark green shiny leaves

bull bay the magnoliaceous tree *Magnolia grandiflora* (also sometimes called red bay)

loblolly bay the theaceous shrub *Gordonia lasianthus*

red bay the lauraceous tree *Persea borbonia*

rose bay either *U.S.* the apocynaceous shrub *Nerium oleander* or the ericaceous shrub *Rhododendron maximum Brit.* any of several onagraceous herbs of the genus Epilobium

mountain rose bay *U.S.* the ericaceous shrub *Rhododendron catawbiense*

swamp bay the magnoliaceous tree *Magnolia glauca*

sweet bay = swamp bay

white bay = swamp bay

bay 2 (*see also* bay 1) light chestnut brown

baya the ploceid bird *Ploicea phillipinus*

baza any of several accipitrid birds of the genus Aviceda

black-crested baza *A. leuphotes*

bb *see* [bobbed]

Bb *see* [Bubble] or [Beaded]

bd *see* [dilute black] or [branched silkless]

-bdell- *comb. form* meaning "leech"

Bdelloidea a class of Rotifera with completely retractile anterior ends distinguished by the leech-like creeping motion which is employed as an alternative to swimming

Bdellomorpha = Bdellonemertea

Bdellonemertea an order of enoplan Nemertea distinguished by the fact that the proboscis does not bear stylets

be *see* [amylase negative]

beach that area on the shores of an ocean which lies between high and low water mark. The term is also applied to sandy areas, subject to wave action, on the shores of fresh water lakes

bead any small, roundish object

Apache bead the dried roots of the sauruaceous aquatic herb *Anemopsis californica*

black bead the leguminous shrubby tree *Pithecolobium ungis-cati*

coral beads *U.S.* any of several species of menispermaceous herb of the genus Cocculus

cuticle bead a pearl like surface gland on a leaf

[Beaded] a mutant (gene symbol *Bd*) mapped at 93.8 on chromosome III of *Drosophila melanogaster*. The phenotypic expression is a beaded edge to the wing

[Beadex] a mutant (gene symbol *Bx*) mapped at 59.4 on the X chromosome of *Drosophila melanogaster*. The phenotype has long, narrow wings with snips missing from edges

beak 1 (*see also* beak 2) the horny mouth parts of birds and reptiles and the very similar mouth parts of some cephalopod mollusks. The term is also loosely applied to any hardened prominence associated with the mouth parts of any animal and with many beak-like structures, such as the gnathosome of acarines and the pedicel valve of brachiopod valves. Compounds of the word are applied to many animals, particularly birds

grosbeak (*see also* weaver) *U.S.* any of numerous fringillid birds, usually distinguished by their powerful beaks. *Brit.* without qualification, *Coccothraustes coccothraustes* (= hawfinch)

blue grosbeak *Guiraca caerulea*

rose breasted grosbeak *Pheuctichus ludovicianus*

crimson collared grosbeak *Rhodothraupis celaeno*

evening grosbeak *Hesperiphona vesperina*

black-faced grosbeak *Caryothraustes poliogaster*

black-headed grosbeak *Pheugucticus melanocephalus*

hooded grosbeak *Hesperiphona abeillci*

pine grosbeak *Pinicola enueator*

yellow grosbeak *Pheucticus chrysopeplus*

black-and-yellow grosbeak *Mycerobas icterioides*

half beak popular name of hemiramphid fish

beak 2 *Austral.* and *I.P.* any of several species of riodinid lepidopteran insects of the genus Libythea. *Austral.*, without qualification, usually *L. geoffroyi* (*cf.* metal mark)

[beaked fruits] a mutant (gene symbol *bk*) mapped at

36 on chromosome 2 of the tomato. The term refers to a sharp point on the phenotypic fruit

beam 1 (*see also* beam 2) in the sense of wood in compound tree names

hornbeam (*see also* maple) *U.S.* the betulaceous tree *Carpinus caroliniana* European hornbeam *C. betulus*

hop hornbeam the betulaceous tree *Ostrya virginiana*

beam 2 in the sense of light ray in the compound names of several lycaenid lepidopteran insects

moonbeam *Austral. Philiris kurandae*

sunbeam *I.P.* any of several trochilid birds of the genus A glaectis

bean 1 (*see also* bean 2, 3, aphis, lima, roller, skeletonizer, thrip tree, weevil) any of numerous leguminous plants the seeds of which are ovoid or subovoid and particularly the seeds themselves

adzuki bean *Phaseolus acularis*

asparagus bean *Vigna sesquipedalis*

black bean *Castanospermum australe*

broad bean *Vicia faba*

buffalo bean *U.S. Astragalus caryocarpus*

civet bean *Phaseolus lunatus*

common bean = kidney bean

djenkol bean *Pithecolobium lobatum*

English dwarf bean = broad bean

goa bean *Psophocarpus tetragonolobus*

haricot bean = kidney bean

horse bean = broad bean

hyacinth bean *Dolichos lablab*

jack bean *Canavalia ensiformis*

jumping bean any dried bean, and occasionally other seeds, infested with a larva of the olethreutid lepidopteran *Laspeyresia saltitans*

kidney bean any of several species and very numerous horticultural hybrids and varieties of the genus Phaseolus, usually *P. vulgaris*

dutch case-knife bean = scarlet runner bean

lima bean *hort. var.* of civet bean

lyon bean *Stizolobium niveum*

mescal bean *Sophora secundiflora*

metcalf bean *Phaseolus retusus*

moth bean *Phaseolus aconitifolius*

mung bean *Phaseolus boreus*

potato bean *Apios americana* (= ground nut)

rice bean *Phaseolus calcaratus*

sarawak bean *Dolichos hosei*

scarlet runner bean *Phaseolus multiflorus*

screw bean the tree *Prosopis pubescens*

soy bean *Glycine maxima*

sword bean *Canavale gladiator*

velvet bean any of several leguminous herbs of the genus Stizolobium

wild bean *U.S. Phaseolus polystachios.* Also, more rarely, *Apios tuberosa* and several species of *Strophostyles*

windsor bean = broad bean

yam bean *Pachyrhizus erosus*

yokohama bean *Stizolobium hasjoo*

bean 2 (*see also* bean 1,3) several non-leguminous plants with bean-like or supposedly bean-like seeds

buck bean the gentianaceous herb *Menyanthes trifoliata*

castor bean the seed of the castor oil plant (*q.v.*)

coffee bean the seed of the rubiaceous genus Coffea from which the beverage is prepared

Indian bean any of several species of bignoniaceous trees of the genus Catalpa, particularly *C. bignonioides*

sacred bean popular name of nymphaeaceous aquatics of the genus Nelumbo

vanilla bean = vanilla

bean 3 (*see also* bean 1,2) animals, or parts of animals, supposedly bean shaped

coffee bean any of several species of cypraenid gastropod Mollusca of the genus Trivia or the eratid gastropod mollusk *Pusula californiana*

bear 1 (*see also* bear 2, bug, foot, grass) any carnivorous mammal of the family Ursidae, mostly of the genus Ursus

black bear *U.S. U. americanus* (= cinnamon bear)

Tibetan blue bear *U. pruinosus*

brown bear *Brit. U. arctos U.S. U. middendorffi* (= Kodiak bear)

cat bear = panda

cinnamon bear *U.S.* = black bear *U.S.*

great cave bear *U. spelaeus*

Himalayan bear *U. tibetanus*

Kodiak bear = brown bear *U.S.*

moon bear any of several species, or possibly races, of ursid carnivores in the genus Selenarctos, distinguished by a V-shaped white mark on the chest

Peruvian bear = spectacled bear

polar bear *Thalarctos maritimus*

Sloth bear the Indian ursid *Melursus labiatus.* It is distinguished from all other bears by its elongate snout

spectacled bear *Tremarctos ornatus*

sun bear the Asiatic *Tremarctos malayanus*

bear 2 any animal thought, by reason of its shape, gait, or shaggy fur, to resemble a bear

ant bear = ant eater

honey bear = kinkajou

koala bear = koala

native bear (*Austral.*) = koala

sea bear any of several otariid pinniped carnivores of the general Arctocephalus and Callorhinus (= fur seal)

water bear popular name of animals of the phylum Tardigrada

woolly bear popular name of hairy caterpillars particularly those of arctiid moths to some of the adults of which, in the *U.S.*, this term is extended

banded wooly bear *U.S. Isia isabella*

yellow wooly bear *Diacrisia virginica*

beard 1 (*see also* beard 2-4, fish, grass, mystax) a tuft of hair pendant from the chin

beard 2 (*see also* beard 1, 3, 4) any tuft of filaments on any part of an animal, but in vertebrates usually the head or breast. Widely used of the barbels of catfish and pendant tufts of feathers in birds. Not infrequently misused as though synonymous with byssus (*q.v.*) and sometimes even applied to the gills of pelycopod mollusks. In *bot.* sometimes used for awn (*q.v.*)

beard 3 (*see also* beard 1, 2, 4) in compound names of many plants thought to have beard-like structures

Aaron's beard the hypericaceous herb *Hypericum calcycinum*

blue beard the verbenaceous shrub *Caryopteris incana*

crown beard any of several species of compositaceous herb of the genus Verbesina

goat's beard *Brit.* and *U.S.* the compositaceous herb *Tragopogon pratensis U.S.* the rosaceous herb *Aruncus dioicus*

false goat's beard *U.S.* any of several species of saxifragaceous herb of the genus Astilbe

hawks beard any of numerous species of compositaceous herb of the genus Crepis

Jupiter's beard either the leguminous herb *Anthyllis barba-jovis* or the valerianaceous herb *Centranthus ruber*

old man's beard the saxifragaceous herb *Saxifraga sarmentosa*. The term is also applied to lichens of the genus Usnea

turkey's beard any of several liliaceous herbs of the genus Xerophyllum, particularly *X. asphodeloides*

beard 4 (see also beard 1–3) in compound names of "bearded" animals

forkbeard *Brit.* the gadid fish *Phycis blennoides* (*cf.* hake)

beardie *Austral.* the gadid fish *Lotella calerias*

bearer used in compound names both in the sense of "to carry" and "to bear young"

casebearer popular name of those lepidopteran insects, mostly coleophorid and pycitid tinioids, the larvae of which build portable cases similar to those of caddisworms

birch casebearer *U.S.* the coleophorid *Coleophora salmani*

cherry casebearer *U.S.* the coleophorid *Coleophora pruniella*

cigar casebearer the coleophorid *Coleophora fletcherella*

pecan cigar casebearer *U.S.* the coleophorid *Coleophora caryafoliella*

elm casebearer *U.S.* the coleophorid *Coleophora limosipennella*

larch casebearer *U.S.* the coleophorid *Coleophora laricella*

pecan leaf casebearer *U.S.* the phycitid *Acrobasis juglandis*

pecan nut casebearer *U.S.* the phycitid *Acrobasis caryae*

pistol casebearer *U.S.* the coleophorid *Coleophora malivorella*

walnut casebearer *U.S.* the phycitid *Acropasis juglandis*

filament bearer, *U.S.* the geometrid lepidopteran insect *Nematocampa limbata*

live-bearer aquarists' name for any ovoviviparous fish

sack-bearer *U.S.* popular name of lacosomid lepidopteran insects

shield bearer *U.S.* popular name of cycnodoid heliozelid lepidopterans

resplendent shield bearer *Coptodisca splendoriferella*

train-bearer either of two trochilid birds of the genus Lesbia

visor-bearer either of two trochilid birds of the genus Augastes

beaugregory the pomacentrid fish *Eupomacentrus leucosticus*

beauty 1 (see also beauty 2, 3) in compound names of numerous lepidopteran insects. *Brit.,* without qualification, the geometrid genus Boarmia

camberwell beauty *Brit.* the nymphaline nymphalid butterfly *Nymphalis antiopa* (= *U.S.* mourning cloak)

bush beauty *S. Afr.* the satyrid *Meneris dendrophilus*

lilac beauty *Brit.* the geometrid *Hygrochroa syringaria*

marbled beauty *Brit.* the noctuid *Bryophila perla*

mottled beauty *Brit.* the noctuid *Eulype hastata* (= argent-and-sable)

painted beauty *U.S.* the nymphalid *Vanessa virginiensis*

pine beauty *Brit.* the noctuid *Panolis griseo-variegata*

table mountain beauty *S. Afr.* the satyrid *Aeropetes tulbaghia*

willow beauty *Brit.* the geometrid *Boarmia gemmaria*

beauty 2 (see also beauty 1, 3) in compound names of many plants

pine barren beauty the diapensiaceous creeping herb *Pyxidanthera barbulata*

meadow beauty any of several melastomaceous herbs of the genus Rhexia

rock beauty (see also beauty 3) the cruciferous herb *Draba pyreniaca*

rutland beauty the convolvulaceous vine *Convolvulus sepium*

spring beauty any of several species of portulacaceous herb of the genus Claytonia

beauty 3 (see also beauty 1, 2) in compound names of other organisms

rock beauty (see also beauty 2) the chaetodontid fish *Holacanthus tricolor*

beaver (see also tree) either of two castorid sciuromorph rodents of the genus Castor

American beaver *C. canadensis*

European beaver *C. fiber*

mountain beaver *U.S.* the aplondontiid rodent *Aplodontia rufa*

becard any of several species of cotingid birds of the genera Platypsaris and Pachyramphus

cinnamon becard *Pachyramphus cinnamomeus*

grey-collared becard *Pachyramphus major*

rose-throated becard *Platypsaris aglaiae*

white-winged becard *Pachyramphus polychopterus*

beccafico the Italian equivalent of fig eater (*q.v.*) frequently used as a *U.S.* word

bêche-de-mer dried sea cucumbers (*q.v.*) used as an article of food in some Asiatic and many Australasian regions (= trepang)

bed a term for a group of ducks on land (*cf.* paddling, raft)

bee (see also eater, fly, moth, plant) any apoid apocritan hymenopteran insect. The bee is popularly, and by no means inaccurately, distinguished from a wasp by the apparent hairiness of the body due to the fact that in bees many or all of the hairs covering the body, particularly the thorax, are plumose

apple bee = *obs.* wasp

alkali bee the halictid *Nomia melanderi*

bumble bee properly thick bodied bees of the apid subfamily Apinae. The term is, however, in common parlance very usually applied to such forms as carpenter bees and digger bees

earth bumble bee *Brit.* the bombid *Bombus terrestris*

stone bumble bee *Brit.* the bombid *Bombus lapidarius*

carder bee *Brit.* popular name of apids of the genus Anthidium

carpenter bee *U.S.* popular name of xylocopine apids

cuckoo bee a term commonly applied to those anthophorine apids that are parasitic in habit. The term is also applied to nomadids

leaf cutter bee popular name of megachilids

digger bees those anthophorine apids which burrow. They can be distinguished by the great broadening of the hairy legs that adapts them to this habit

yellow-faced bee those colletid apoid hymenopteran insects in which there are yellow markings on the face. They are usually mistaken for wasps since the hind legs do not have pollen brushes and are very sparsely haired

hairy flower bee *U.S.* the apid *Anthophora occidentalis*

honey bee any of numerous species of apine apid of the genus Apis. The domesticated bee is usually *A. mellifera*

long-horned bee *Brit.* the apid *Eucera longicornis*

humble bee = bumble bee

mason bee *Brit.* any of several species of apid hymenopteran of the genera Chalcicodoma and Osmia

mining bee popular name of andrenids

plasterer bee popular name of those colletid hymenopterans which plaster the inside of their burrows with a thin transparent material

social bee popular name of apine apids

sweat bee any of several species of halictid of the genus Halictus that are attracted by human perspiration

beech (*see also* aphis, drops, scale) any of numerous fagaceous trees of the genus Fagus. *U.S.* without qualification *F. grandifolia*. *Brit.* without qualification *F. sylvatica*. In *N.Z.* the various beeches are of the genus Nothofagus and the *Austral.* "beech" is a Eucalyptus

Australian beech the myrtaceous tree *Eucalyptus polyanthemos*

black beech *N.Z. N. solandri*

blue beech the betulaceous tree *Carpinus caroliniana*

copper beech a bronze-leaved horticultural variety of the European beech

hard beech *N.Z. N. truncata*

mountain beech *N.Z. N. cliffortioides*

red beech *N.Z. N. fusca*

silver beech *N.Z. N. menziesii*

beef (*see also* wood) the edible flesh of domestic cattle and, by extension, of some other organisms

sea beef the edible foot of polyplacophoran mollusca

vegetable beef any of several European red or purple "fleshed" polypore fungi

beet (*see also* aphis, beetle, borer, hopper, maggot) any of many horticultural varieties of the chenopodiaceous herb *Beta maritima* and a few plants of similar form

Chilean beet = leaf beet

leaf beet a variety of beet grown for its large leaves (= Swiss chard)

sea kale beet = leaf beet

Sicilian beet = leaf beet

sugar beet = mangel

wild beet *U.S.* either the amaranthaceous herb *Amaranthus retroflexus* or the saxifragaceous herb *Saxifraga pennsylvanica*

beetle 1 (*see also* borer, curculio, weed, weevil) any coleopteran insect typically distinguished from other insects by the hardened forewings, or elytra, which encase the hind wings when at rest

ambrosia beetle = timber beetle

antelope beetle *U.S. the* lucanid *Dorcus parallelus* (*cf.* stag beetle)

antlike beetle popular name of brathinids

golden apple beetle *Brit.* popular name of chrysomelids

asparagus beetle either of two species of chrysomelid of the genus Crioceris. Usually, without qualification, *C. asparagi*

spotted asparagus beetle *U.S. C. duodecimpunctata*

American aspen beetle *U.S.* the chrysomelid *Gonioctema americana*

atlas beetle *U.S.* the scarabaeid *Chalcosoma atlas*

bacon beetle *Brit.* the dermestid *Dermestes lardarius*

bark beetle one living immediately under or in the bark of trees. Many are scolytids.

alder bark beetle U.S. the scolytid *Alniphagus aspericollis*

western balsam bark beetle the scolytid *Dryocoetes confusus*

birch bark beetle *U.S.* the scolytid *Dryocoetes betulae*

western cedar bark beetle *U.S.* the scolytid *Phloeosinus punctatus*

comb clawed bark beetle *U.S.* popular name of alleculids

cylindrical bark beetle popular name of Colydiidae

elm bark beetle *Brit.* the scolytid *Scolytus destructor*

European elm bark beetle *Scolytus multistriatus*

native elm bark beetle *U.S.* the scolytid *Hylurgopinus refipes*

flat bark beetle *U.S.* popular name of cucujids

hickory bark beetle *U.S.* the scolytid *Scolytus quadrispinosus*

broad nosed bark beetles any of numerous curculionids with short beaks and a long curved spine at the apex of each front tibia

peach bark beetle *U.S.* the scolytid *Phloeotribus limenaris*

wrinkled bark beetle *U.S.* popular name of rhysodids

Mexican bean beetle the coccinellid *Epilachna varivestis*

Bess beetle *U.S.* the passalid *Popilius disjunctus*

Betsy-beetle popular name of passalids

blister beetle the term is properly applied to meloids. However the Spanish fly or European blister beetle was for many years known as *Cantharis vesicatoria* and placed in the Cantharidae though it is now a meloid *Lytta vesicatoria*. The active principal is still, however, called cantharidin

black blister beetle *U.S.* the meloid *Epicauta pennsylvanica*

caragana blister beetle *U.S.* the meloid *Epicauta subglabra* (Fall)

clematis blister beetle *U.S.* the meloid *Epicauta cinerea*

European blister beetle the meloid *Lytta vesicatoria* (= Spanish fly)

gray blister beetle *U.S.* = clematis blister beetle

ash-gray blister beetle *Epicauta fabricii*

green blister beetle *U.S.* the meloid *Lytta cyanipennis*

infernal blister beetle *U.S.* the meloid *Lytta stygica*

marginal blister beetle *U.S.* the meloid *Epicauta pestifera*

striped blister beetle *U.S.* the meloid *Epicauta vittata*

three-striped blister beetle *U.S.* the meloid *Epicauta lemniscata*

turnip blossom beetle *Brit.* the nitidulid *Meligethes aeneus*

bombardier beetles popular name of carabid beetles of the genus Brachinus

bone beetle *U.S.* popular name of corynetids

burying beetle popular name of silphids

burying beetle *Brit.* = sexton beetle *Brit.*

sugarcane beetle   *U.S. Euetheola rugiceps*
capricorn beetle   = longhorn beetle
carpet beetle   *U.S.* any of several dermestids, usually, without qualification, *Anthrenus scrophulariae*
  black carpet beetle   *U.S.* the dermestid *Attagenus piceus*
  furniture carpet beetle   *U.S.* the dermestid *Anthrenus flavipes*
  varied carpet beetle   *U.S.* the dermestid *Anthrenus verbasci*
carrion beetle   = burying beetle
roving carrion beetle   *Brit.* any of several silphids of the genus Silpha
carrot beetle   *U.S.* the scarabaeid *Bothynus gibbosus*
cedar beetle   *U.S.* popular name of rhipicerids
cellar beetle   *Brit.* the tenebrionid *Blaps mortisuga* (*cf.* churchyard beetle)
charcoal beetle   *U.S.* the buprestid *Melanophila consputa*
checkered beetle   popular name of clerids
churchyard beetle   *Brit.* the tenebrionid *Blaps mucronata* (*cf.* cellar beetle)
cigarette beetle   *U.S.* the anobiid *Lasioderma serricorne*
short circuit beetle   = lead cable borer
comb-clawed beetle   popular name of alleculids
click beetle   popular name of elaterids
eyed click beetle   *U.S. Alaus oculatus*
clock beetle   *Brit.* the coprid *Geotrupes stercocarius* (= dumble beetle)
Colorado beetle   *U.S.* the chrysomelid *Leptinotarsa decemlineata* (= potato beetle)
fire colored beetle   *U.S.* popular name of pyrochroids
cone beetle   *U.S.* popular name of scolytids (*cf.* bark beetle)
  lodgepole cone beetle   *U.S. Conophthorus contortae*
  Monterey-pine cone beetle   *U.S. Conopthorus radiatae*
  ponderosa-pine cone beetle   *U.S. Conophthorus pondorosae*
  red-pine cone beetle   *U.S. Conophthorus resinosae*
  sugar-pine cone beetle   *U.S. Conophthorus lambertianae*
  white-pine cone beetle   *U.S. Conophthorus coniperda*
  pinon cone beetle   *U.S. Conophthorus edulis*
seed-corn beetle   *U.S.* the carabid *Agonoderus lecontei*
cucumber beetle   any of several species of chrysomelid of the genera Diabrotica and Acalymma
  banded cucumber beetle   *U.S. D. balteata*
  spotted cucumber beetle   *U.S.* the *D. undecimpunctata howardi*
  western spotted cucumber beetle   *U.S.* the chrysomelid *D. U. undecimpunctata*
  striped cucumber beetle   *U.S. A. vittata*
  striped cucumber beetle   *U.S. D. duodecimpunctata* (= southern corn rootworm)
  western striped cucumber beetle   *U.S. A. trivitta*
darkling beetle   popular name of tenebrionids
  false darkling beetle   popular name of melandryidae
death-watch beetle   any of several wood-boring anobiids that maintain a monotonous ticking as they bore through wood
dermestid beetle   properly applied to those dermestid beetles which infest hides but sometimes extended to the whole family Dermestidae

predaceous diving beetle   popular name of dytiscids
dogbane beetle   the chrysomelid *Chrysochus auratus*
drugstore beetle   properly the anobiid *Stegobium panicea* but the term is frequently extended to cover the whole family
dumble beetle   *Brit.* the coprid *Geotrupes stercorarius*
dung beetle   any of numerous cobrophagous forms most belonging to the families Geotrupidae and Scarabaeidae
elephant beetle   any of several very large scarabaeid beetles which lack a pronotal horn (*cf.* rhinoceros beetle, unicorn beetle)
engraver beetle   scolytids that engrave patterns on the underside of the bark of trees (*cf.* bark beetle)
sap feeding beetle   *U.S.* popular name of nitidulids
fiddler beetle   either *Austral.* the cetoniid *Eupocoelia australasiae* (= fiddle back) or *W.I.* the curculionid *Prepodes vittatus*
Douglas-fir beetle   *U.S.* the scolytid *Dendroctomus pseudotsuga* (*cf.* bark beetle)
flea beetles   any of numerous small, usually black, chrysomelid beetles many of which are pests on crops
  alder flea beetle   *U.S. Altica ambiens*
  apple flea beetle   *U.S. Altica foliaceae*
  western black flea beetle   *U.S. Phyllotreta pusilla*
  cabbage flea beetle   *U.S. Phyllotreta albionica*
  corn flea beetle   *U.S. Chaetocnema pulicaria*
  desert corn flea beetle   *U.S. Chaetocnema ectypa*
  echo flea beetle   = tobacco flea beetle
  eggplant flea beetle   *U.S. Epitrix fuscula*
  elongate flea beetle   *U.S. Systena elongata*
  grape flea beetle   *U.S. Altica chalybea*
  hop flea beetle   *U.S. Psylliodes punctulata*
  horseradish flea beetle   *U.S. Phyllotreta armoraciae*
  red-legged flea beetle   *U.S. Derocrepis erythropus*
  potato flea beetle   *U.S. Epitrix cucumeris*
  sweetpotato flea beetle   *U.S. Chaetocnema confinis*
  western potato flea beetle   *U.S. Epitrix subcrinita*
  prairie flea beetle   *U.S. Altica canadensis*
  spinach flea beetle   *U.S. Disonycha xanthomelas*
  three-spotted flea beetle   *U.S. Disonycha triangularis*
  striped flea beetle   *U.S. Systena blanda*
  western striped flea beetle   *U.S. Phyllotreta ramosa*
  strawberry flea beetle   *U.S. Altica ignitia*
  sumac flea beetle   *U.S. Blepharida rhois*
  tobacco flea beetle   *U.S. Epitrix hirtipennis*
  toothed flea beetle   *U.S. Chaetocnema denticulata*
  tuber flea beetle   *U.S. Epitrix tuberis*
flour beetle   *U.S.* any of numerous small dermestids
  confused flour beetle   *U.S. Tribolium confusum*
  depressed flour beetle   *U.S. Palorus subdepressus*
  broad-horned flour beetle   *U.S. Gnathocerus maxillosus*
  red flour beetle   *U.S. Tribolium castaneum*
flower beetle   any beetle popularly associated with flowers
  antlike flower beetle   popular name of anthicids
  bumble flower beetle   *U.S.* the scarabaeid *Euphoria inda*
  shining flower beetle   *U.S.* popular name of phalacrids
  soft winged flower beetle   *U.S.* popular name of melyrids

sunflower beetle the chrysomelid *Zygospila exclamationis*

tumbling flower beetle popular name of mordellids

white-fringed beetle *U.S.* any of numerous species of curculionids

dried-fruit beetle *U.S.* the nitidulid *Carpophilus hemipterus*

fungus beetle one living on or in fungi

hairy fungus beetle *U.S.* popular name of mycetophagids

handsome fungus beetle *U.S.* popular name of endomychids

pleasing fungus beetle *U.S.* popular name of erotylids

shining fungus beetle *U.S.* popular name of scaphidiids

silken fungus beetle popular name of cryptophagids

tooth-necked fungus beetle popular name of derodontids

tree fungus beetles *U.S.* popular name of cisids

fringe-winged fungus beetles popular name of orthoperids (*see also* fringe-winged beetle)

asiatic garden beetle *U.S.* the scarabaeid *Maladera castanea*

bark-gnawing beetle *U.S.* = grain beetle *U.S.*

goliath beetle any of several large cetoniids of the genus Goliatha

grain beetle *U.S.* any of numerous, small, grain infesting beetles, many of the family Cucujidae

flat grain beetle *U.S.* the cucujid *Cryptolestes pusillus*

mexican grain beetle *U.S.* the cryptophagid *Pharaxonotha kirschi*

square-necked grain beetle *U.S.* the cucujid *Cathartus quadricollis*

red rust grain beetle *U.S.* the cucujid *Laemophloeus ferrugineus*

saw-toothed grain beetle *U.S.* the cucujid *Oryzaephilus surinamensis*

peppergrass beetle *U.S.* the chrysomelid *Galeruca browni*

ground beetle any of numerous soil-inhabiting beetles usually of the family Carabidae

ham beetle *U.S.* any of several species of corynetids of the genus Necrobia, particularly *N. rufipes* and *N. ruficollis*

harlequin beetle the giant (3–4 in.) brilliantly colored *S.Amer.* cerambycid *Acrocinus longimanus*

hercules beetle the scarabaeid *Dynastes hercules* (the largest living coleopteran).

Eastern hercules beetle *U.S. D. tityus* (*cf.* unicorn beetle)

hide beetle *U.S.* the dermestid *Dermestes maculatus*

black hills beetle *U.S.* the scolytid *Dendroctonus ponderosae*

hister beetle anglicized form of the family name Histeridae

khapra beetle *U.S.* the dermestid *Trogoderma granarium*

long horn beetle any of numerous beetles with unusually long antennae, properly applied to members of the family Cerambycidae

elderberry long-horn beetle *U.S. Desmocerus palliatus*

storehouse beetle *U.S.* the ptinid *Gibbium psylloides*

Japanese beetle the scarabaeid *Popillia japonica*

june beetle *U.S.* any of numerous scarabaeid leaf eaters

green june beetle *U.S. Cotinis nitida*

lined june beetle *U.S.* any of several species of scarabaeid of the genus Polyphylla

lady beetle *U.S.* popular name of coccinellids (= *Brit.* lady bird and *U.S.* lady bug)

bean lady beetle *U.S.* = Mexican bean beetle

black lady beetle *U.S. Rhyzobius ventralis*

steel-blue lady beetle *U.S. Orcus chalybeus*

convergent lady beetle *Hippodamia convergens*

red lady beetle *U.S. Cycloneda munda*

spotted lady beetle *U.S. Ceratomegilla maculata*

two-spotted lady beetle *U.S. Adalia bipunctata*

twice-stabbed lady beetle *U.S. Chilocorus stigma*

eastern larch beetle *U.S.* the scolytid *Dendroctonus simplex*

larder beetle *U.S.* the dermestid *Dermestes lardarius*

black larder beetle *U.S. D. ater*

leaf beetle any of numerous forms feeding on leaves. Those listed, unless otherwise identified, are chrysomelids

aspen leaf beetle *U.S. Chrysomela crotchi*

bean leaf beetle *U.S. Cerotoma trifurcata*

beet leaf beetle *U.S. Erynephala puncticollis*

western beet leaf beetle *U.S. Monoxia consputa*

cherry leaf beetle *U.S. Galerucella cavicollis*

clover leaf beetle *U.S.* the curculionid *Hypera punctata* (= clover leaf weevil)

southern corn leaf beetle *U.S. Myochorus denticollis*

cottonwood leaf beetle *U.S. Chrysomela scripta*

elm leaf beetle *U.S. Galerucella xanthomelaena*

larger elm leaf beetle *U.S. Monocesta coryli*

gray leaf beetle *U.S. Glytoscelis liebecki*

long-horned leaf beetle any of several chrysomelid beetles superficially resembling cerambycids

long leaf beetle *U.S. Myochrous longulus*

yellow-margined leaf beetle *U.S. Microtheca ochroloma*

rose leaf beetle *U.S. Nodonata puncticollis*

sweetpotato leaf beetle *U.S. Typophorus nigritus*

watercress leaf beetle *U.S. Phaedon aeruginosus*

waterlily leaf beetle *U.S. Galerucella nymphaeae*

gray willow leaf beetle *U.S. Galerucella decora*

Pacific willow leaf beetle *Galerucella carbo*

imported willow leaf beetle *U.S. Plagiodera versicolora*

leather beetle *Brit.* popular name of dermestids

patent leather beetle popular name of passalids

lion beetle *U.S.* the cerambycid *Ulochaetes leoninus*

lizard beetle popular name of languriids

mud-loving beetle *U.S.* popular name of heterocerids

May beetle *U.S.* any of numerous medium-sized scarabaeids (= June beetle *U.S.*, June bug *U.S.*)

mimic beetle *Brit.* popular name of histerids

leaf mining beetle any of many chrosomelids having leaf-mining larvae

musk beetle *Brit.* the cerambycid *Aromia moschata*

mammal-nest beetles popular name of leptinids

bloody nosed beetle the chrysomelid *Timarcha tenebricosa*

oil beetle any of several meloids of the genus Meloe

oriental beetle *U.S.* the scarabaeid *Anomala orientalis*

ox beetle *U.S.* the scarabaeid *Strategus antaeus*

green peach beetle *U.S.* the scarabaeid *Cotinis texana*

pill beetle *U.S.* popular name of byrrhids (*cf.* pill bug)

pinch beetle = stag beetle

pine beetle any of numerous scolytids (*cf.* cone beetle)

Arizona pine beetle *U.S.* = smaller Mexican pine beetle

Colorado pine beetle *U.S. Dendroctonus parallelocollis*

Jeffrey pine beetle *U.S. D. ponderosae*

roundheaded pine beetle *U.S. D. adjunctus*

lodgepole-pine beetle *U.S. D. murrayanae*

larger Mexican pine beetle *D. parallelocollis*

smaller Mexican pine beetle *U.S. D. frontalis*

mountain pine beetle *U.S.* = *Jeffrey pine beetle*

southern pine beetle *U.S.* = *smaller Mexican pine beetle*

western pine beetle *U.S. D. brevicomis*

southwestern pine beetle = western pine beetle

soft-bodied plant beetle *U.S.* popular name of helodids

powder-post beetle *U.S.* popular name of bostrichids

potato beetle *U.S.* the chrysomelid *Leptinotarsa decimlineata*

false potato beetle *U.S.* the chrysomelid *Leptinotarsa juncta*

old-fashioned potato beetle the meloid *Epicauta vittata*

three-lined potato beetle *U.S. Lema trilineata*

rhinoceros beetle any of several very large scarabaeid beetles from the thorax of which arise two horns. (*cf.* unicorn beetle)

riffle beetle popular name of elmids

road beetle *Brit.* popular name of staphylinids

goldenrod beetle *U.S.* the chrysomelid *Trirhabda canadensis*

rose beetle any of several pests of the cultivated roses. Usually, *U.S.*, without qualification, the scarabaeid *Adoretus sinicus* (*cf.* rose chafer)

Fullers rose beetle *U.S.* the curculionid *Pantomorus godmani*

rove beetle *U.S.* popular name of staphylinids

hairy rove beetle *U.S. Creophilus maxillosus*

sacred beetle the scarabaeid *Scarabaeus sacer*

round sand beetle popular name of omophronids (= savage beetle)

sap beetle popular name of nitidulids

corn sap beetle *U.S. Carpophilus dimidiatus*

dusky sap beetle *U.S. Carpophilus lugubris*

savage beetle *U.S.* popular name of omophronids (= round sand beetle)

sawyer beetle *U.S.* any of several species of cerambycid of the genus Monochamus

brown scavenger beetle *U.S.* popular name of lathridiids

water scavenger beetle popular name of hydrophilids

seed beetle popular name of bruchids

corn seed beetle *U.S.* either of two species of carabids of the genus Agonoderus

sexton beetle *Brit.* any of numerous silphids of the genus Necrophorous

shield beetle *Brit.* popular name of chrysomelids

corn silk beetle *U.S.* the chrysomelid *Luperodes brunneus*

skin beetle those dermestids and trogids which feed on stored hides

smut beetle *U.S.* the phalacrid *Phalacrus politus*

snapping beetle *U.S.* = click beetle

snout beetle a term often applied to any weevil (*q.v.*) but sometimes restricted to the Curculionidae

alfalfa snout beetle *U.S.* curculionid *Brachyrhinus ligustici*

bison snout beetle any of several species of curculionids, many of the genus Thecesternus. They are distinguished by the fact that the head and beak can be completely withdrawn into the prosternum

soldier beetle *U.S.* popular name of cantharids

spider beetle *U.S.* popular name of ptinids

brown spider beetle *U.S. Ptinus clavipes*

golden spider beetle *U.S. Ptinus hololeucus*

hairy spider beetle *U.S. Ptinus villiger*

white-marked spider beetle *U.S. Ptinus fur*

spruce beetle those scolytids which infest spruces (*cf.* cone beetle, pine beetle)

Alaska spruce beetle *U.S.* = *eastern spruce beetle*

Allegheny spruce beetle *U.S. Dendroctonus punctatus*

eastern spruce beetle *U.S. Dendroctonus obesus*

European spruce beetle *U.S.* the scolytid *Dendroctonus micans*

Sika-spruce beetle *U.S.* = eastern spruce beetle

squash beetle *U.S.* the coccinellid *Epilachna borealis*

stag beetle popular name of many lucanids

giant stag beetle *U.S. Lucanus elaphus*

giraffe stag beetle the lucanid *Cladognathus giraffa*

oak stag beetle *U.S. Platyceroides agassii*

rugose stag beetle *U.S. Sinodendron rugosum*

stink beetle *U.S.* the carabid *Nomius pygmaeus*

stone beetle *U.S.* popular name of scydmaenids

ant-like stone beetle *U.S.* = stone beetle

jumping sumac beetle *U.S.* = sumac flea beetle

tiger beetle popular name of cincindelid beetles *Brit.*, without qualification, *Cincindela campestris*

giant tiger beetle *Mantichora herculanea*

wood tiger beetle *Brit. Cincindela sylvatica*

timber beetle scolytids that burrow in hardwood

sugar maple timber beetle *U.S. Corthylus punctatissimus*

ship-timber beetle popular name of lymexylids

spruce timber beetle *U.S. Trypodendron bivattatum*

tortoise beetle any of many brilliantly colored chrysomelids with very wide elytra frequently extended as a flattened projection from the body. Others superficially resemble Cincindelids

argus tortoise beetle *Chelymorpha cassidea*

beet tortoise beetle *U.S. Cassida nebulosa*

eggplant tortoise beetle *U.S. Cassida pallidula*

golden tortoise beetle *U.S. Metriona bicolor*

mottled tortoise beetle *U.S. Deloyala guttata*

tule beetle *U.S. Agonum maculicolle*

red turnip beetle *U.S.* the chrysomelid *Entomoscelis americana*

black turpentine beetle *U.S.* the scolytid *Dendroctonus terebrans*

red turpentine beetle *U.S.* the scolytid *Dendroctonus valens*

twig beetle popular name of psoids and also of a few scolytids

apple twig beetle *U.S.* the scolytid *Stephanoderes obscurus*

unicorn beetle any of several very large scarabaeid beetles from the thorax of which arises a single large horn. Without qualification in *U.S.* usually applied to *Dynastes tityus* (*cf.* rhinoceros beetle)

wasp beetle *Brit.* any of numerous cerambycids of the genus Clytus

water beetle any aquatic form

crawling water beetle popular name of haliplids

margined water beetle *Brit.* the dytiscid *Dytiscus marginalis*

silver water beetle *Brit.* the hydrophilid *Hydrophilus piceus*

red milkweed beetle *U.S.* the cerambycid *Tetraopes tetrophthalmus*

whirligig beetle popular name of gyrinids

feather-winged beetles popular name of ptiliids

fringe-winged beetles popular name of clambids (*see also* fungus beetle)

leather winged beetle *U.S.* popular name of cantharids

net winged beetle *U.S.* popular name of lycids

worm beetle one best known from its larva

fruitworm beetle popular name of byturids

beetle 2 (*see also* beetle 1) any organism, other than a coleopteran, which resembles a beetle

black beetle *Brit.* = Cockroach *Brit.*

marsh beetle the typhaceous reed *Typha latifolia* (= cattail)

beggar *U.S.* the geometrid lepidopteran insect *Eudeule mendica*

Beggiatoaceae a family of beggiatole schizomycetes occurring as trichomes capable of gliding over a substrate with flexuous movements, the origin and nature of which is unknown save that flagella and cilia are not involved. They strongly resemble the cyanophycean Oscillatoria and are by many regarded as algae that have lost their photosynthetic pigments. They live in the same environment as Oscillatoria, some species of which, like many of the Beggiatoaceae, are capable of withstanding hydrogen sulphide and depositing granules of sulphur

Beggiatoales an order of schizomycetes occuring as trichomes that frequently show bending or flexing, or in single cells that may glide over the surface of the substrate. They are by some regarded as Sciomophyceae that have lost their pigment

begonia properly a large and varied genus of begoniaceous herbs and shrubs

wild begonia *U.S.* the polygonaceous herb *Rumex venosus*

Begoniaceae that family of parietalous dicotyledons which contains the begonias. The perianth of staminate flowers with two valvate sepals and two petals is typical

Begoniineae a suborder of dicotyledons containing the single family Begoniaceae

behemoth a biblical beast frequently thought to be an elephant but actually (Job 40:15–24) the hippopotamus

[beige] a mutant (gene symbol *bg*) in linkage group XVII of the mouse. The phenotypic expression is a diluted hair color

beira the neotragine antelopid *Dorcatragus melanotis*

beisa the hippotragine antelope *Oryx beisa*

Belemnitidae a family of fossil decapod Mollusca knowns for their straight shells

Belidae a monospecific family of curculionoid coleopteran insects containing the so-called New York weevil

bell 1(*see also* bell 2) any bell-shaped structure particu-

larly the hydrozoan umbrella (*q.v.*) when the rim is bent down and constricted

uterine bell the first part of the female reproductive canal in Acanthocephala

bell 2 (*see also* bell 1, bird, flower, pepper) plants with campanulate flowers

bluebell (*see also* creeper) *U.S.* any of numerous species of the campanulaceous genus Campanula. *Brit.* the liliaceous herb *Scilla nutans*. The term is also applied to several species of the liliaceous genus Scilla

California bluebell the hydrophyllaceous herb *Phacelia whatlavia*

Virginia bluebells the boraginaceous herb *Mertensia virginica* (= Virginia cowslip, Roanoke bells)

canterbury bells the campanulaceous herb *Campanula media*

coral bells the saxifragaceous herb *Heuchera sanguinea*

cowbell the caryophyllaceous herb *Silene latifolia*

crimson bells = coral bells

easter bell the caryophyllaceous herb *Stellaria holostea*

fairy bell any of numerous species of herb of the liliaceous genus Disporum

feather bells *U.S.* the liliaceous herb *Stenanthium gramineum*

California golden bells the hydrophyllaceous herb *Emmenanthe penduliflora*

harebell any of numerous species of the campanulaceous genus Campanula particularly *C. rotundifolia* and also the liliaceous herb *Scilla nonscripta*

New Zealand harebell the campanulaceous herb *Wahlenbergia gracilis*

merry bells *U.S.* any of several species of liliaceous herb of the genus Uvularia (= bell wort)

Roanoke bells the boraginaceous herb *Mertensia virginica* (= Virginia cowslip, Virginia bluebell)

silver bell any of several species of the styracaceous genus Halesia

velvet bells *U.S.* the scrophulariaceous herb *Bartsia alpina*

belle *Brit.* any of several species of geometrid moth of the genus Aspilates

lead belle *Brit.* the geometrid *Ortholitha plumbaria*

Bellini's *epon. adj.* from Lorenzo Bellini (1643–1704)

belly the underside of the abdomen

white belly *W.I.* the columbid bird *Leptotila jamaicensis.* (= white bellied dove)

yellow belly the poeciliid teleost fish *Glaridichtyhs falcatus*

Belonidae a family of synentognathous fishes commonly called needle fishes from the elongated tooth-filled rostrum resembling that of a gar

Beloniformes = Synentognathi

Belostomatidae a family of very large aquatic hemipteran insects. They are distinguished from the closely allied Nepidae which they resemble in form but not size by the possession of a hinged, rather than rotating, coxa. They are commonly called giant water bugs or (*U.S.*) electric light bugs.

belt one of a pair of raised girdles furnished with microscopic tentacles in the mid region of the body of Pogonophora

belted said of an animal particularly a bird having a band of color across the abdomen

[belted] a mutant (gene symbol *bt*) in linkage group

VI of the mouse. The phenotypic expression is a medium band of white

beluga the delphinid cetacean *Delphinapterus leucas* (= beluga whale)

Belutinidae a small family of gastropod Mollusca distinguished by their very thin shells covered with a velvety periostracum from which the name velvet shell is derived

belvedere the chenopodiaceous herb *Kochia scoparia*

Belytidae = Cinetidae

ben the moringaceous tree *Moringa oleifera*

white ben the caryophyllaceous herb *Silene latifolia*

bengali the ploceid bird *Fringilla bengalus*

herb bennet *Brit.* the rosaceous herb *Geum urbanum*

[bent] a mutant (gene symbol *bt*) mapped at 1.4 on chromosome IV of *Drosophila melanogaster*. The phenotypic expression is not only a bent wing but also a knobby leg

[bent] a mutant (gene symbol *Bn*) found in the sex chromosome (linkage group XX) of the mouse. The description refers to the short crooked tail of the phenotype

benth- *comb. form* meaning "deep"

benthic 1 (*see also* benthic 2) pertaining to depths

benthic 2 (*see also* benthic 1) pertaining to organisms living in, or on, the sediments of freshwater and marine habitats

archibenthic pertaining to the zone of the ocean, between approximately fifty fathoms and five hundred fathoms

eurybenthic pertaining to benthic organisms that occur at widely varying depths

benthon There is much confusion in the literature between the use of benthon and benthos, either term being restricted by some to the meaning given under either benthos 1 or benthos 2

benthos 1 (*see also* benthos 2) the sum total of organisms living in or on the sediments of freshwater and marine habitats

abyssobenthos organisms growing on the ocean floor at great depths

holobenthos organisms that pass their entire life at great depths

phytobenthos plants, or communities of plants, of the deep ocean

potamobenthos the population of a river bottom

benthos 2 (*see also* benthos 1, benthon) depths or deep zones of water

geobenthos that portion of the bottom of a freshwater lake that does not support rooted vegetation

phytobenthos that portion of the bottom of a lake or marine area which is covered with vegetation or supports rooted plants

[bent scutellar] a mutant (gene symbol *bsc*) mapped at 1.1 on the X chromosome of *Drosophila melanogaster*. The name is descriptive

ber *see* [berrytail]

Berberidaceae that family of ranalous dicotyledons which contains the barberries. Distinctive of the family are the solitary carpels and the cyclic flowers with stamens opposed to the petals

bergamot 1 (*see also* bergamot 2) any of several fragrant-leaved labiateous herbs

wild bergamot the labiateous herb *Monarda fistulosa*

bergamot 2 (*see also* bergamot 1) the rutaceous tree *Citrus bergamea* and particularly the aromatic rind of the fruit

bernacle *Brit. obs.* = barnacle

Beroidea the only order in the ctenophoran class

Nuda. Clearly distinguished by the absence of tentacles

Berothidae a small family of predatory neuropteran insects mostly with falciform wings

berrin-berrin *Austral.* the meropid bird *Merops ornatus* (*cf.* bee-eater)

berry 1 (*see also* berry 2,3, bug and eater) a fleshy fruit the tissues of which are wholly carpellary in origin. In compound names the term applies equally to the plant bearing the fruit

apple berry *Austral.* the vine *Billardiara scandens*

baked apple berry the rosaceous creeping herb *Rubus chamaemorus*

baneberry any of several species of the ranunculaceous herb Actaea

barberry any of numerous shrubs of the berberidaceous genus Berberis and its near allies

common barberry *B. vulgaris*

ash-leaved barberry = *Mahonia aquifolia*

bearberry the ericaceous shrub *Arctostaphylos uva-ursi*

beauty berry the verbenaceous shrub *Callicarpa americana*

bilberry (*see also* blueberry, deerberry, cranberry, huckleberry) any of numerous species of edible fruited ericaceous shrubs of the genus Vaccinium. Frequently, without qualification, *V. myrtillus*

alpine bilberry the ericaceous shrub *Vaccinium uliginosum* (= bog bilberry)

dwarf bilberry *V. caespitosum*

blackberry any of several long-stemmed rosaceous shrubs of the genus Rubus. *Brit.*, without qualification, *R. fruticosus*

American blackberry numerous horticultural varieties and hybrids mostly derived from *R. allegheniensis, R. argutus, R. floridus* and *R. frondosus*

thornless blackberry *R. canadensis*

evergreen blackberry *R. laciniatus*

cut-leaved blackberry = evergreen blackberry

sand blackberry *R. cuneifolius*

smooth blackberry *U.S. R. canadensis*

sow-teat blackberry *U.S. R. allegheniensis*

blueberry (*see also* bilberry, deerberry, cranberry and huckleberry, mite, and maggot) any of numerous blue or black fruited ericaceous shrubs of the genus Vaccinium

Arctic blueberry *V. uliginosum*

bare blueberry = hairy blueberry

high bush blueberry *V. corymbosum*

Canada blueberry *V. canadense*

hairy blueberry *V. hirsutum*

low blueberry *V. pennsylvanicum* or *V. vacillans*

male blueberry = male berry

swamp blueberry = high bush blueberry

downy swamp blueberry *V. atrococcum*

boxberry the ericaceous trailing shrub *Gaultheria procumbens*

Boysen berry *Rubus loganobaccus*

buffalo berry *U.S.* the elaeagnaceous herb *Shepherdia argentea*

catberry the aquifoliaceous shrub *Nemopanthus mucronata* (= wild holly)

checkerberry the ericaceous trailing shrub *Gaultheria procumbens*

Chinaberry the meliaceous tree *Melia azedarach*

choke berry any of several rosaceous shrubs of the genus Aronia

black choke berry *A. melanocarpa*

purple choke berry   *A. atropurpurea*
red choke berry   *A. arbutifolia*
cloud berry   the rosaceous creeping herb *Rubus chamaemorus* (= molka)
coffee berry   properly applied to coffee (*q.v.*) but also misused for soybean (*q.v.*)
coral berry   the caprifoliaceous shrub *Symphoricarpus orbiculatus* and its edible berries
cowberry   the ericaceous shrub *Vaccinium vitis-idaea* (= mountain cranberry, partridge berry)
crake berry   = crowberry
cranberry   (*see also* bush, blueberry, bilberry, deerberry, huckleberry, and worm) any of several ericaceous shrubs of the genus Vaccinium, or the fruit of these shrubs. The term is also applied to many shrubs bearing fruit resembling that of the cranberry *Brit.*, without qualification, *V. oxycoccus*
high cranberry   (high bush cranberry) the caprifoliaceous shrub *Viburnum americanum*
large cranberry   *U.S. V. macrocarpon*
mountain cranberry   *V. vitis-idaea*
small cranberry   *V. oxycoccus*
crowberry   *U.S.* any of several species of the empetraceous genus Empetrum
broom crowberry   either of two species of empetraceous shrubs of the genus Corema
curlew berry   *U.S.* the empetraceous shrubby herb *Empetrum nigrum*
dangleberry   the ericaceous shrub *Gaylussacia frondosa* (= huckleberry, tangleberry)
deerberry   (*see also* blueberry, bilberry, cranberry, and huckleberry) any of several species of ericaceous edible fruited shrubs of the genus Vaccinium but particularly *V. stamineum*
dewberry   a term applied to those varieties or species of blackberries which habitually trail rather than climb. Many *U.S.* are derived from *Rubus procumbens* and *R. flagellaris. Brit.*, without qualification, *R. caesius*
California dewberry   *R. vitifolius*
southern dewberry   *R. trivialis*
dogberry   *either* the rosaceous tree *Sorbus americana* (= American mountain ash) or the saxifragaceous shrub *Grossularia cynosbati*
two-eyed berry   the rubiaceous herb *Mitchella repens* (*cf.* partridge berry)
farkleberry   *Vaccinium arboreum* (*cf.* blueberry)
foxberry   *N. U.S.* the ericaceous shrub *Vaccinium vitisidaea* (= mountain cranberry, cowberry, partridge berry)
bitter gallberry   the aquifoliaceous shrub *Ilex glabra*
gooseberry   (*see also* moth, sawfly, worm) properly any of several species of shrub of the saxifragaceous genus Ribes. The true gooseberry is distinguished from the currant in having a fruit that is larger, ovate, and frequently hairy. Many other plants with more or less similar fruit are also called "gooseberry"
Barbados gooseberry   the cactaceous scrambling shrub *Pereskia aculeata*
cape gooseberry   the solanaceous herb *Physalis peruviana*
dwarf cape gooseberry   the solanaceous herb *Physalis pubescens*
European gooseberry   *R. grossularia* and numerous horticultural varieties and hybrids of this species
hill gooseberry   the myrtaceous shrub *Rhodomyrtus tomentosa*

West India gooseberry   the euphorbiaceous shrub *Phyllanthus acidus*
Otaheite gooseberry   = West Indian gooseberry
southern gooseberry   the ericaceous shrub *Vaccinium melanocarpum*
star gooseberry   = West Indian gooseberry
swamp gooseberry   *U.S. Ribes lacustre*
hackberry   (*see also* gall, engraver) any of several trees of the ulmaceous genus Celtis
Himalaya berry   *Rubus procerus*
huckleberry   (*see also* blueberry, bilberry, deerberry and cranberry) any of numerous ericaceous shrubs with edible berries of the genera Vaccinium and Gaylussacia
black huckleberry   *G. baccata*
he-huckleberry   the cyrillaceous shrub *Cyrilla racemiflora*
squaw huckleberry   either of the ericaceous shrubs *Vaccinium caesium* or *V. stamineum*
sugar huckleberry   the ericaceous shrub *Vaccinium vacillans* (= low blueberry)
inkberry   either of the two aquifoliaceous shrubs *Ilex glabra* or *I. verticillata*
juneberry   any of several rosaceous shrubs of the genus Amelanchier
limeberry   the rutaceous shrub *Triphasis trifolia*
ling berry   the ericaceous shrub *Vaccinium vitis-idaea*
liver berry   *U.S.* the liliaceous herb *Streptopus amplexifolius*
loganberry   the rosaceous shrub *Rubus ursinus*
maidenhair berry   the ericaceous shrub *Gaultheria hispidula*
male berry   the ericaceous shrub *Lyonia ligustrina*
marble berry   the myristicaceous tree *Ardisia pickeringia*
marsh berry   = cranberry
mayberry   the myricaceous shrub *Myrica carolinensis* or the rosaceous shrub *Rubus palmatus*
mooseberry   the caprifoliaceous shrub *Viburnum edule*
mulberry   (*see also* fly) any of several species and numerous horticultural varieties of the moraceous genus Morus. The term is also applied to other plants with not dissimilar berries including the flowering raspberry (*q.v.*)
American mulberry   *M. rubra*
black mulberry   *M. nigra*
Indian mulberry   the rubiaceous tree *Morinda citrifolia* and its fruit
French mulberry   the verbenaceous shrub *Callicarpa americana*
paper-mulberry   the moraceous tree *Broussonetia papyrifera*
Russian mulberry   *hort. var.* of *Morus alba*
white mulberry   *M. alba*
nanny berry   the caprifoliaceous herb *Viburnum lentago*
nub berry   = cloud berry
partridge berry   any of several species of the rubiaceous genus Mitchella, particularly *M. repens*. Also the ericaceous shrub *Gaultheria procumbens* and the ericaceous shrub *Vaccinium vitis-idaea* (= mountain cranberry, cowberry)
pigeon berry   *U.S.* the phytolaccaceous herb *Phytolacca americana*
raspberry   (*see also* borer, maggot, sawfly, and worm) *U.S.* any of several rosaceous shrubs of the genus Rubus. *Brit.*, without qualification, *R. idaeus*

purple-cane raspberry   *R. neglectus*

black cap raspberry   *R. occidentalis*

dwarf raspberry   *U.S.* any of several species of the rosaceous genus Rubus particularly those belonging to the subgenus Cylactis

European raspberry   *R. idaeus*

fire raspberry   *R. trifidus*

flowering raspberry   *R. odoratus*

red raspberry   *U.S. R. strigosus*

Rocky Mountain raspberry   *R. deliciosus*

strawberry-raspberry   *R. illecebrosus*

red-berry   (*see also* mite) the rhamnaceous shrub *Rhamnus crocea*

rhine berry   the rhamnaceous shrub *Rhamnus cathartica*

rock berry   *U.S.* the empetraceous herb *Empetrum eamesii*

salmonberry   the rosaceous shrub *Rubus spectabilis* and its edible fruit. The term is sometimes improperly applied to the cloudberry (*q.v.*)

sarkleberry   the ericaceous shrub *Vaccinium arboreum* (= sparkleberry)

sarvice berry = service berry

scoot berry   *U.S.* the liliaceous herb *Streptopus amplexifolius*

service berry   *U.S.* any of several species of shrub or tree of the rosaceous genus Amelanchier

sleep berry   the caprifoliaceous herb *Viburnum lentago*

silk berry   the sapindaceous tree *Sapindus drummondii*

silver berry   the elaeagnaceous shrub *Elaeagnus argentea*

snake berry   *U.S.* the ranunculaceous herb *Actaea rubra*

snow berry   any of several caprifoliaceous herbs of the genus Symphoricarpos but particularly *S. albus.* Also several rubiaceous shrubs of the genus Chiococca

creeping snow berry   the ericaceous trailing shrub *Gaultheria hispidula*

soap berry   *U.S.* the elaeagnaceous shrub *Shepherdia canadensis* and any of several trees of the sapindaceous genus Sapindus

sparkleberry   the ericaceous shrub *Vaccinium aboreum*

squash berry   the caprifoliaceous shrub *Vibrunum edule*

squaw berry   the rubiaceous shrub *Mitchella repens*

strawberry   (*see also* aphis, beetle, blite, borer, bush, fern, fly, geranium, mite, moth, roller, weevil, and worm) any of numerous species and very many horticultural varieties of the rosaceous genus Fragaria. The garden strawberry is probably a horticultural variety of *F. chiloensis,* but possibly also with an admixture of *F. virginiana.* The name is also applied to a wide variety of plants with similar fruit, foliage, or habit

alpine strawberry   *F. vesca*

barren strawberry   *U.S.* the rosaceous herb *Waldsteinia fragarioides*

bog strawberry   the rosaceous herb *Potentilla palustris*

Indian strawberry   *U.S.* the rosaceous herb *Duchesnea indica*

wood strawberry   *Brit.* alpine strawberry

yellow strawberry   the rosaceous herb *Duchesnea indica*

sugar berry   the ulmaceous shrub *Celtis occidentalis*

tangle berry   the ericaceous shrub *Gaylussacia frondosa* (= huckleberry, dangleberry)

tea berry   the ericaceous shrub *Gaultheria procumbens*

thimble berry   the rosaceous shrub *Rubus occidentalis*

twin berry   the rubiaceous shrub *Mitchella repens*

wax berry   the caprifoliaceous shrub *Symphoricarpos albus* and its edible fruit (= snow berry) or the ericaceous shrub *Gaultheria hispida*

whinberry = wineberry

whortleberry   frequently, without qualification, *V. myrtillus*

wine berry   *U.S.* the rosaceous shrub *Rubus phoenicolasius N.Z.* the elaeocarpaceous tree *Aristotelia serrata* (= makomako)

winter berry   any of several aquifoliaceous shrubs of the genus Ilex particularly *I. glabra, I. laevigata,* and *I. verticillata*

mountain winter berry   *U.S.* the aquifoliaceous tree *Ilex montana*

wolf berry   the caprifoliaceous shrub *Symphoricarpos occidentalis*

yellow berry   the rosaceous creeping herb Rubus chamaemorus

berry 2   (*see also* berry 1, 3,) the black knob on the bill of a swan

berry 3   (*see also* berry 1, 2) the attached egg masses of large crustacea

[berrytail]   a mutant (gene symbol *ber*) mapped at 52.2± on the X chromosome of *Drosophila melanogaster.* The name refers to the protruded berry-like mass of external genitalia borne on the end of the narrow abdomen of the phenotype

berseem   the leguminous herb *Trifolium alexandrinum* (= Egyptian clover)

Bertin's   *epon. adj.* from Exupere Joseph Bertin (1712–1781)

Berycidae   a family of berycoidous fish strongly resembling the Holocentridae

Beryciformes = Berycoidei

Berycoidei   an order of acanthopterygians distinguished by the fact that the dorsal ray is preceded by a series of spines and that the ventral fins are thoracic

Bethylidae   a small family of apocritan hymenopteran insects closely allied to the Chrysididae but many species have wingless forms that are frequently mistaken for ants

betony   (*see also* moth) the labiateous herb *Stachys officinalis*

wood betony   any of several species of herb of the scrophulariaceous genus Pedicularis (= lousewort)

betrothed   *U.S.* the noctuid lepidopteran insect *Catocala innubens* (*cf.* underwing)

bettong   *Austral.* popular name of rat-kangaroos of the genus Bettongia

Betulaceae   that family of fagalous dicotyledons which contains, *inter alia,* the birches, alders, and hornbeams. Distinguishing characteristics are the cymose group of flowers for each bract, the pistillate flowers in spikes, and the two carpels

Betz's   *epon. adj.* from Vladimir Aleksandrovich Betz (1834–1894)

bevy   term for a group of roedeer, or other shy animals, including quail and swans, and in literary usage, maidens

bezard = bezoar stone

bezoar   an accretion of materials, frequently calcified,

found in the stomachs of many artiodactyle animals and once used as a specific against poisons

bf   *see* [black feet] and [brief]

Bg   *see* [Bag]

bg   *see* [beige]

Bh   *see* [Blotched aleurone]

bharal   the Himalayan ovine *Pseudois nahoor*. It strongly resembles the tur (*q.v.*) and is a graphic demonstration of the impossibility of separating sheep and goats

bhokra = chousingha

-bi-   *comb. form* meaning "two"

bi   *see* [bifid] and [bifurcata inflorescence] the term is also used for a yeast genetic marker indicating a requirement for biotin

biatorine   having, like the lichen Biatora, a conspicuous margin to the apothecium

  pseudobiatorine   having an apothecium lacking a conspicuous margin

bib   *Brit.* the gadid fish *Gadus luscus* (= pout)

Bibionidae   a family of dark colored red and yellow marked medium sized nematoceran dipteran insects commonly called march flies in virtue of their frequency in early Spring

bichir   any of several species of cladistian fish of the family Polypteridae. Usually, without qualification, *Polypterus weeksi*

Bidder's   *epon. adj.* from Friedrich Heinrich Bidder (1810–1894)

biddy   female domestic chicken, though hen is the more common term

biduous   two days long

bifer   a plant producing two crops a season

bifid   cleft

[**bifid**]   a mutant (gene symbol *bi*) mapped at 6.9 on the X chromosome of *Drosophila melanogaster*. The phenotypic expression is the fusion of the wing veins into a bifid stalk

Biflagellata   a class of phycomycete fungi with cellular walls. The name, however, derives from the presence of two flagella on the zoospores

[**bifurcata inflorescence**]   a mutant (gene symbol *bi*) mapped at 0 in linkage group V of the tomato. The phenotypic expression is a branched inflorescence

Bignoniaceae   a family of tubuliflorous dicotyledons, mostly climbers, including the well-known trumpet creeper. They are distinguished as pantropical woody vines or trees and by the usually non-endospermous winged seeds in capsules

Bilateria   a division of the animal kingdom erected to contain all those forms which show bilateral symmetry. It therefore contains all the phyla of the animal kingdom except Protozoa, Coelenterata, Porifera, and Ctenophora

bile   the secretion of the gall bladder of animals

Bilharziidae = Schistosomatidae

bill 1   (*see also* bill 2, 3) a term used interchangeably with beak and used in compound names of many birds

  mountain avocet-bill   the trochilid bird *Opisthoprora euryptera*

  fiery-tailed awlbill   the trochilid bird *Avocettula recurvirostris*

  bent bill   any of several species of tyrannids of the genus Oncostoma usually without qualification *O. cinereagulari*

  black bill   *W.I.* the fringillid *Loxigilla violacea* (= Greater Antillean bullfinch)

bluebill   *Afr.* any of several ploceids of the genus Spermophaga

  Grant's bluebill   *S. poliogenys*

  marsh bluebill   the anserine bird *Aythya affinis* (= lesser scaup)

  red-headed bluebill   *S. ruficapilla*

boat-bill = boat-billed heron (*q.v.*)

bristle bill   any of several *W.Afr.* species of pycnonotids of the genus Bleda, usually without qualification, *B. syndactyla*

broadbill   any of several species of eurylaimids

  African broadbill   *Smithornis capensis*

  banded broadbill   *Eurylaimus javanicus*

  black-and-red broadbill   *Cymbirhynchus macrorhynchos*

  silver-breasted broadbill   *Serilophus lunatus*

  dusky broadbill   *Corydon sumatranus*

  green broadbill   any of several species of the general Calyptomena and Pseudocalyptomena

  long-tailed broadbill   *Psarisomus dalhousiae*

  wattled broadbill   *Eurylaimus steerii*

channel-bill   the cuculid bird *Scythrops novaehollandiae* (= channel-billed cuckoo)

conebill   any of several parulid birds of the genus Conirostrum

crossbill   any of several fringillids of the genus Loxia, usually without qualification, *L. curvirostra*

  two-barred crossbill   *Brit.* = white-winged crossbill *U.S.*

  parrot crossbill   *L. pytyopsittacus*

  red crossbill   *L. curvirostra*

  white-winged crossbill   *U.S. Loxia leucoptera*

flatbill   any of several species of tyrannid of the genus Rhynochocyclus

  eye-ringed flatbill   *R. brevirostris*

hookbill   *Austral.* the psittacid *Purpureicephalus spurius* (= king parrot)

  chestnut-winged hookbill   the furnariid bird *Ancistrops strigilatus*

hornbill   any of numerous coraciiform birds of the family Bucerotidae. They are distinguished by an enormous horny casque on top of the bill

  black hornbill *Anthracoceros malayanus*

  red-billed hornbill   *Tockus erythrorhynchus*

  yellow-billed hornbill *Tockus flavirostris*

  yellow-casqued hornbill *Ceratogymna elata*

  bushy-crested hornbill *Anorrhinus galeritus*

  long-crested hornbill *Berenicornis comatus*

  crowned hornbill *Tockus alboterminatus*

  great hornbill *Buceros bicornis*

  grey hornbill *Afr. Tockus nasutus I.P. T. birostrio*

  ground hornbill *Bucorvus abyssinicus*

  southern ground hornbill *Bucorvus leadbeateri*

  helmeted hornbill *Rhinoplax vigil*

  laughing hornbill *Bycnaistes sharpii*

  pied hornbill *Tockus fasciatus*

  piping hornbill *Bycanistes fistulator*

  rhinoceros hornbill *Buceros rhinoceros*

  tarictic hornbill *Penelopides panini*

  trumpeter hornbill *Bycanistes bucinator*

  wrinkled hornbill *Aceros corrugatus*

ibis-bill   the recurvirostrid bird *Ibidorhyncha struthersii*

ivory-bill = ivory-billed woodpecker

lancebill   either of two species of trochilid bird of the genus Doryfera

longbill   any of several species of sylviids of the genus Macrosphenus (*cf.* warbler)

grey longbill *M. concolor*
yellow longbill *M. flavicans*
openbill either of two species of ciconids of the genus Anastomus
  Asiatic openbill *A. oscitans*
  African openbill *A. lamelligerus*
parrotbill any of numerous paradoxornithid birds of the genera Conostoma and Paradoxornis
  brown parrotbill *P. unicolor*
  black-fronted parrotbill *P. nipalensis*
  great parrotbill *C. oemodium*
razorbill = razorbilled auk
recurvebill either of two furnariid birds of the genus Simoxenops
redbill *N.Z.* the charadriid *Haematopus unicolor* (=*Austral.* sooty oyster-catcher)
saddlebill the ciconiid *Ephippiorhychus senegalensis*
sawbill applied to numerous birds most commonly to mergine water birds and motmotid land birds
scimitarbill either of two phoeniculid birds of the genus Rhinopomastus; without qualification, *R. cyanomelas*
  Abyssinian scimitar-bill *R. minor*
scythebill any of several dendrocolaptid birds of the genus Campylorhamphus
sharpbill the oxyruncid *Oxyruncus cristatus*
sheathbill either of two species of chionidids of the genus Chionis, usually without qualification, *C. alba*
shoebill the balaenicipitid bird *Balaeniceps rex* (= whale-headed stork)
shortbill the cotingid *Phidalura flavirostris*
sicklebill either of two species of trochilid bird of the genus Eutoxeres
silverbill *Afr.* the ploceid *Euodice cantans*
spadebill any of several species of tyrannids of the genus Platyrhinchus
  golden-crowned spadebill *P. coronatus*
  white-throated spadebill *P. mystaceus*
spinebill *Austral.* either of two meliphagids of the genus Acanthorhynchus particularly *A. superciliosus N.Z.* several species of acanthisittids of the genus Acanthisitta (*cf.* wren)
  eastern spinebill *A. tenuirostris*
  western spinebill *A. superciliosus*
spoonbill *W.I.* the anatine *Spatula clypeata* (= shoveler) properly any of several species of threskiornithid bird mostly of the genus Platalea, *Brit.*, without qualification *P. leucorodia*
  African spoonbill the threskiornithid *Platalea alba*
  yellow-billed spoonbill *Platibis flavipes*
  lesser spoonbill *Platalea minor*
  popeler spoonbill any of several species of the genus Platalea
  roseate spoonbill *P. ajaja* (often separated into a genus Ajaia)
  royal spoonbill *P. regia*
  white spoonbill *P. leucorodia*
thornbill *Austral.* any of many species of malurines of the genus *Acanthiza S. Amer.* either of two species of apodiform trochilids Ramphomicron
  brown thornbill *A. pusilla*
  little thornbill *A. nana*
  striated thornbill *A. lineata*
  buff-tailed thornbill *A. reguloides*
  chestnut-tailed thornbill *A. uropygialis*
  yellow-tailed thornbill *Austral. A. chrysorrhoa* (= *Austral.* tomtit)
waxbill *Afr.* any of many species of ploceid bird mostly

of the genus Estrilda. Usually, without qualification, *E. astrild*
  yellow-billed waxbill *Coccopygia melanotis*
  blue waxbill *Uraeginthus angolensis* (*cf.* cordonbleu)
  fawn-breasted waxbill *E. paludicola*
  black-capped waxbill *E. nonnula*
  black-cheeked waxbill *E. erythronotos*
  orange-cheeked waxbill *E. melpoda*
  black-headed waxbill *E. atricapilla*
  ruddy waxbill *Lagonosticta rhodopareia* (*cf.* fire finch)
  black-rumped waxbill *E. troglodytes*
  zebra waxbill *E. subflava*
wedgebill *Austral.* the muscicapid *Sphenostoma cristatum*
weebill *Austral.* either of two malurines of the genus *Smicrornis*, usually without qualification, *S. brevirostris*
  brown weebill *S. brevirostris*
  yellow weebill *S. flavescens*
wrybill *N.Z.* the charadriid *Amarhynchus frontalis* (= wrybill plover)
yellowbill the cuculid *Ceuthmochares aureus* (*cf.* coucal)
bill 2 (*see also* bill 1, 3) any animal with a bird-like bill
  Atlantic hawksbill the cheloniid chelonian reptile *Eretmochelys imbricata*
bill 3 (*see also* bill 1, 2) in compound names of plants usually with bill-like projections on the corolla
  cranesbill any of numerous species of the geraniaceous genus Geranium. *U.S.* without qualification usually *G. maculatum*
  dove's foot crane's bill *G. molle*
  parrot's bill the leguminous vine *Clianthus pumiceus*
  stork's bill properly any of numerous species of the geraniaceous genus Erodium (previously Geranium) and its near allies, but sometimes applied to species of the geraniaceous genus Pelargonium which is the "geranium" of the florist
Billroth's *epon. adj.* from Christian Albert Theodore Billroth (1829–1894)
billy a male goat or other caprid
bilsted the hamamelidaceous tree *Liquidambar styraciflua*
bimous two years long
binary consisting of two parts
binate consisting of one pair
bind in compounds of names of climbing herbs, most usually as an adjective (*e.g.* bindweed)
  corn bind *U.S.* any of several species of polygonaceous herb of the genus Polygonum particularly those of the subgenus Tiniaria
bine = vine
  woodbine *U.S.* any of several species of vitaceous vines of the genus Parthenocissus (*cf.* Virginia creeper) *Brit.* any of several species of the caprifoliaceous genus of climbing shrubs Lonicera (= honeysuckle), most frequently *L. periclymenum*
  Italian woodbine *L. caprifolium*
-bini- *comb. form* meaning "twin"
binjai the anacardiaceous tree *Mangifera caesis* and its fruit
binous paired
binturong the paradoxurine viverrid carnivore *Arctictis binturong*. It is the only carnivore, and one of the very few non-primate mammals, with a prehensile tail
-bio- *comb. form* indicating life. For most compounds

see main root (*e.g.* cenose, for biocenose; climate for bioclimate, etc.)

bio a bacterial genetic marker indicating a requirement for biotin. One has been mapped at 15.5 mins. in *Eschirichia coli*

biome a large community of living organisms having a peculiar form of dominant vegetation and associated characteristic animals

bion a living unit including not only cells but also viruses. By some the term is considered synonymous with "individual" and by others as a variant spelling of biome. There is a further confusion in the literature with "biont"

geobion a plant association of dry land

halobion an association of marine plants

limnobion an association of organisms in a lake

symbion = symbiont

bionomics = ecology

biont a living thing—but see remarks under bion. Also widely used in the sense of a member of a biome. The specific condition of a -biont is -biosis

aerobiont *either* an organism living in air as distinct from water or soil *or* an organism requiring oxygen

anaerobiont an organism capable of anaerobic existence

anabiont = perennial

antibiont antipathetic organisms

anoxybiont an organism incapable of using oxygen as distinct from one that is aerobic. Frequently used of facultative anaerobes

epibiont *either* one organism that lives on the surface of another without any connotation of mutualism, *or* an organism in an isolated small community surviving from a once widely distributed species

geobiont an organism living in soil

hobiont = holophyte

metabiont a many-celled organism (*cf.* metaphyte, metazoon)

diplobiont those red algae with two morphologically distinct diploid phases in the life cycle

haplobiont those red algae with two morphologicially distinct halploid phases in the life cycle

symbiont an individual living in symbiosis

heterosymbiont used of a symbiotic association of several members, particularly of a lichen containing several species of algae

macrosymbiont the larger of two symbionts

microsymbiont applied to one symbiont that is noticeably smaller than the other

obligate symbiont one which cannot exist apart from its partner

troglobiont a cave-dwelling organism

biosis *either* the condition of being alive *or* the conditions pertaining to a specific type of life. An individual in a condition of - -biosis is a - -biont, or - -bion and it is said to be - -biotic

amphibiosis the condition of an organism that spends part of its life on land and part in water

anabiosis the condition of an organism that has passed into a resting stage, which is cyclic or seasonal, but produced by a change in the environment, such as the loss of moisture

antibiosis the method of existence of a microorganism that secretes a substance destroying or inhibiting other microorganisms

archebiosis the original development or origin of life

asthenobiosis *either* the condition of an inactive larva not yet metamorphosed to a pupa *or* autointoxication, particularly among insects

calobiosis a form of symbiosis in which one species becomes temporarily the guest of another

cleptobiosis a form of symbiosis based on theft, best known in ants in which one species will systematically steal the collected food of another laboring species

cryptobiosis the condition of an organism that must in theory have at one time existed, but that has left no fossil traces

ecobiosis the conditions pertaining to a mode of life within a specific habitat

endobiosis *either* the condition of an organism that lives within another (usually with the connotation of parasitism) *or* the condition of those benthic organisms that live in the surface of bottom mud

hemiendobiosis the condition of a parasite that can exist both inside and outside the host

holendobiosis the condition of being an obligate endobiont and specifically that of a parasitic fungus which produces spores within the host tissue

epibiosis *either* the condition of those organisms that live on the surface of another *or* the condition of those benthic organisms that live on the surface of bottom mud

gnotobiosis the condition of animals that are sterile both internally and externally, or of the procedures used to secure and maintain these conditions

hypobiosis the condition of those benthic organisms which live under surface structures such as domes and the like

lestobiosis = cleptobiosis

metabiosis *either* a condition of symbiosis in which one symbiont sets the stage for the arrival of the other *or* a mutual association of two organisms of which one is thought to benefit without detriment to the other

parabiosis *either* the condition of living together, applied variously to mixtures of species of similar habit as man and the rat or to the union of two individual animals (*cf.* twin) *or* the condition of symbiosis between two species of ant in which colonies of neighboring nests are contiguous but do not mingle

phylacobiosis a friendly association between ants and termites

symbiosis (*see also* symphily) the condition of two or more different organisms living together in close association. At one time, and occasionally today, the term is restricted to an association supposed to be to the mutual advantage of both organisms

antagonistic symbiosis a condition in which one symbiont seeks to establish domination over the other. Ocassionally used as synonymous with parasitism

contingent symbiosis the condition of an organism that lives within another, apparently only for shelter and without either causing damage or receiving nutrients. In this sense synonymous with endobiosis

endosymbiosis a form of symbiosis in which one organism lives within another as green algae in Hydra or Convoluta

conjunctive symbiosis the condition, as in lichens, where two symbionts appear to form a single individual

disjunctive symbiosis a temporary symbiotic relationship from which either partner can release itself at will

mutualistic symbiosis symbiosis resulting in mutual advantage

parasymbiosis when one symbiont damages another. It is difficult to distinguish this from parasitism

social symbiosis the condition of one organism that feeds on food stored by another organism

synclerobiosis an adventitious, and frequently temporary, association between two species of ants

thermobiosis = thermophilic

athermobiosis dormancy induced by low temperatures

zoobiosis the condition of a parasitic fungus that infests an animal

biota the sum total of the living organisms of any designated area

-biotic *see* -biosis

biotin *see* vitamin H

biotinidase an enzyme catalyzing the hydrolysis of p-biotinylaminobenzoate to biotin and p-amino-benzoate (= biotin-amide aminohydrolase)

biotope an ecological niche, or restricted area, the environmental conditions of which are suitable for certain fauna and flora. A tree with its associated organisms is a biotope: a forest is a biochore (*q.v.* under -chore)

bipennaria *see* larva

birch (*see also* borer, sawfly, skeletonizer, weevil) any of numerous trees of the betulacean genus Betula, and a few which resemble these superficially

black birch *B. lenta*

canoe birch = silver birch

cherry birch = black birch

dwarf birch *B. nana*

European birch *B. pendula*

grey birch *B. populifolia*

paper birch = silver birch

red birch *B. nigra*

river birch = red birch

scrub birch *B. glandulosa*

silver birch *B. papyrifera*

sweet birch = black birch

water birch *B. fontinalis*

West Indian birch the burseraceous tree *Bursera simaruba*

white birch *B. alba* though occasionally used of the grey birch

yellow birch *B. alleghaniensis*

bird 1 (*see also* bird 2, cherry, eye, fish, grape, louse, moth, rape, shell, and tick) any feathered chordate (an expanded definition is given under Aves)

adjutant bird = adjutant stork

alarm-bird *Austral.* the charadriid *Lobibyx novaehollandiae* (= spur-winged plover). In other parts of the world various birds that utter raucous cries when disturbed

antbird any of numerous myrmecophagous birds but particularly the formicariid passerines. In *W.I.* the term is applied to the parulids *Mniotilta varia* (= black-and-white warbler) and *Dendroica pharetra* (= arrow-headed warbler)

chestnut-backed antbird *Myrmeciza exsul*

bicolored antbird *Gymnopithys leucaspis*

bare-crowned antbird *Gymnochichla nudiceps*

dusky antbird *Cercomacre tyrannina*

spotted antbird *Hylophylax naevioides*

apostle-bird *Austral.* the timaliid *Pomatostomus superciliosus* (= white-browed babbler)

banana-bird *W.I.* the icterids *Icteria leucopteryx* (= Jamaican oriole) and *I. dominicensis* (= black-cowled oriole). Also the parulid *Coereba flaveola* (= banana quit)

black banana-bird *W.I.* the icterid *Nesopsar nigerrimus* (= Jamaica blackbird)

barking-bird the rhinocryptid *Pteroptochos tarnii*

barley bird *Brit.* any of numerous birds, including the siskin, the yellow wagtail, the wryneck, the nightingale, and several finches

wheelbarrow-bird *Austral.* the muscicapid *Sphenostoma cristatum* (= wedgebill)

bay-bird a term loosely applied to any beach-dwelling charadriiform bird

beach-bird *U.S.* = bay-bird

beam bird *Brit.* the muscicapid *Muscicapa striata* = *Brit.* spotted flycatcher)

bellbird almost any bird whose song is thought to have a bell-like sound has received this designation. Most frequently applied to several South American cotingids of the genus Procnias, the *N.Z.* meliphagid *Anthornis melanura* but in *Austral.* to any of several birds

crested bellbird *Austral.* the muscicapid *Oreoica gutturalis*

three wattled bellbird *Procnias tricarunculata*

bishop-bird any of several species of ploceid of the genus Euplectes. Also *U.S.* = indigo bunting *E. orix*

black bishop-bird *E. gierowii*

yellow-crowned bishop-bird *E. afra*

fire-fronted bishop-bird *E. diademata*

yellow bishop-bird *E. capensis*

blackbird any of numerous medium sized black birds. *Brit.*, without qualification, the turdid *Turdus merula*. *U.S.*, without qualification frequently *Sturmus vulgaris* (= starling) Properly *U.S.* this name is applied to many icterids

brewer's blackbird *Euphagus cyanocephalus*

red-breasted blackbird *Leistes militaris*

tricolored blackbird *Agelaius tricolor*

Cuban blackbird *Dives atroviolaceus*

yellow-headed blackbird *Xanthocephalus xanthocephalus*

Jamaican blackbird *Nesopsar nigerrimus*

marsh blackbird *U.S.* = red-winged blackbird

melodious blackbird *Dives dives*

rusty blackbird *Euphagus carolinus*

savanna blackbird *W.I.* the cuculid *Crotophaga ani* (*cf.* ani)

tawny-shouldered blackbird *Agelaius humeralis*

yellow-shouldered blackbird the icterid *Agelaius xanthomus*

red-winged blackbird *Agelaius phoeniceus*

bloodbird any dark red meliphagid particularly those of the genus Myzomela (*cf.* honeyeater)

bluebird *U.S.* any of several turdid birds of the genus Sialia, usually, without qualification, *S. sialis*

eastern bluebird *S. sialis*

mountain bluebird *S. currucoides*

western bluebird *S. mexicana*

boatswain-bird *Austral.* = tropic bird

bosun-bird = boatswain-bird

bowerbird any of many species of ptilonorhynchid bird

golden bowerbird *Prionodura melanura*

great bowerbird *Chlamydera nuchalis*

regent bowerbird *Sericulus chrysocephalus*

satin bowerbird *Ptilonorhynchus violaceus*

spotted bowerbird *Chlamydera maculata*

bristle-bird any of several timaliids of the genus Stachyris (*cf.* babbler) *Austral.* any of several malurines of the genus Daryornis, usually, without qualification, *D brachypterus*

eastern bristle-bird *D. brachypterus*

rufous bristle-bird *D. broadbenti*

western bristle-bird *D. longirostris*

brown bird   *W.I.* the mimid *Cinclocerthia ruficauda* (= trembler)

bugler-bird   the icterid *Icterus icterus* (= troupial)

butcherbird   any laniid bird which impales its prey on thorns (= shrike) *Austral.* properly any of several species of cracticid of the genus *Cracticus*

  grey butcherbird   *C. torquatus*

  pied butcherbird   *C. nigro gularis*

butter-bird   *W.I.* the icterid *Dolichonyx oryzivorus* (= *U.S.* bobolink)

capuchin-bird   the cotingid *Perissocephalus tricolor*

cashew bird   the cracid *Pauxi pauxi* (= helmeted curassow) In *W.I.* the name is applied to the stripe-headed tanager

catbird   *U.S.* the mimid bird *Dumetella carolinensis.* *W.I.* the sylviid *Polioptila caerulea.* (= blue-grey gnatcatcher *U.S.*). *Austral.* the timaliid *Pomatostomus superciliosus* (= white-browed babbler)

  Abyssinian catbird   the muscicapid *Parophasma galinieri*

  tooth-billed catbird   the ptilonorhynchid *Scenopoeetes dentirostris*

  black catbird   the mimid *Melanoptila glabrirostris*

  green catbird   the ptilonorhynchid *Ailuroedus crassirostris*

chapman-bird   *W.I.* the cuculid *Crotophaga ani* (= smooth billed ani)

chew-bird   *W.I.* = catbird *W.I.*

chitty-bird   *W.I.* = sin-bird *W.I.*

cicada bird   *Austral.* the campiphagid *Coracina tenuirostris*

cocoa-bird   *W.I.* = coffee-bird

coconut-bird   *W.I.* the icterid *Icterus dominicensis* (= black-cowled oriole)

coffee-bird   *W.I.* the fringillid *Loxigilla violacea* (= greater Antillean bullfinch)

coffin-bird   *W.I.* the cuculid *Coccyzus minor* (= mangrove cuckoo)

corn-bird   *W.I.* the icterid *Molothrus bonariensis* (= glossy cowbird)

corporal-bird   *W.I.* the icterid *Nesopsar nigerrimus* (= Jamaican blackbird)

cotton-bird   *W.I.* = catbird *W.I.*

cowbird   *U.S.* any of several icterids, particularly *Molothrus ater. Brit.* the motacillid *Motacilla flava* (= yellow wagtail)

  bronzed cowbird   *Tangavius aeneus*

  giant cowbird   *Scaphidura oryzivora*

  glossy cowbird   *Molothrus bonariensis*

  brown-headed cowbird   *Molothrus ater*

cowry-bird   the fringillid *Munia punctulata*

cream-bird   *W.I.* the jacanid bird *Jacana spinosa* (= Brazilian coot)

crocodile-bird   a glareolid (*Pluvianus aegypticus*) that feeds largely on the ectoparasites of the crocodile

crown-bird   any of several species of Musophagidae (= touracos)

cuckoo-bird   (*see also* cuckoo) *W.I.* the strigid *Otus nudipes* (= Puerto Rican screech owl)

long-day bird   the turdid bird *Turdus jamaicensis* (= white-eyed thrush)

death-bird   *W.I.* the tytonid *Tyto alba* (= barn owl)

desert-bird   *Austral.* any of several desert-inhabiting sylviids

devil-bird   either the Asiatic owl *Strix indranee* or the caprimulgid *Caprimulgus kelaarti*

diamond-bird   *Austral.* any of several species of dicaeid bird of the genus *Pardalotus,* particularly *P. punctatus* (= pardalote)

doctor-bird   *W.I.* = hummingbird

doe-bird   = dough-bird

dollar bird   *Austral.* the coraciid *Eurystomus orientalis*

dough-bird   *U.S. dial.* properly the scolopacid *Numenius borealis* (= Eskimo curlew) but frequently applied to other scolopacids

dunbird   the aythyine bird *Aythya ferina* or, more properly the female of this species (*cf.* pochard)

eggbird   = sooty tern

black-eyed bird   *W.I.* the parulid *Geothlypis rostrata* (= Bahama yellowthroat)

fern bird   *N.Z.* the sylviid *Bowdleria punctata*

brainfever-bird   any bird with a piercing repetitive cry particularly the cuculids *Cacomantis merulinus* (*cf.* sand-tailed cuckoo), and *Cuculus varius*

figbird   any of several oriolids of the genus *Sphecotheres* (= fig-eater)

  southern figbird   *S. vieilloti*

  yellow figbird   *S. flaviventris*

flax-bird   *U.S.* = goldfinch *U.S.*

butterfly-bird   *W.I.* the parulid *Setophaga ruticilla* (= redstart *U.S.*)

friarbird   *Austral.* any of several meliphagid birds of the genus *Philemon,* usually, without qualification *P. corniculatus*

  little friarbird   *C. citreogularis*

  noisy friarbird   *C. corniculatus*

frigate bird   any of several pelicaniform fregatids of the genus *Fregata* distinguished by their very long wingspread

  Ascension frigatebird   *F. aquila*

  Christmas Island frigatebird   *F. andrewsi*

  great frigatebird   *F. minor*

  least frigate bird   *F. ariel*

game-bird   any bird that is hunted for pleasure by man. In *Brit.* practice is used to denote the galliform game birds (pheasant, partridge, grouse) in distinction to wild fowl (*q.v.*)

go-away-bird   any of several musophagids of the genus *Crinifer* (*cf.* turaco) without qualification, *C. concolor*

  white-bellied go-away-bird   *C. leucogaster*

  bare-faced go-away-bird   *C. personata*

god-bird   *W.I.* the trochilid bird *Chlorostilbon ricordii* (= Cuban emerald)

gold-bird   *W.I.* the thraupid bird *Spindalis zena* (= stripe-headed tanager)

golden bird   *Austral.* the ptilorhynchid *Prionodura melanura*

grass-bird   any species usually observed rising from grass. *Brit.* = sandpiper. *W.I.* the fringillid *Tiaris olivacea* (= yellowfaced grassquit) and the grass sparrow (*q.v.*) *Austral.* the sylviid *Megalurus gramineus* (= little marsh bird) *S.Afr.* the sylviid *Sphenoeacus afer* (= idle jack)

  yellow-backed grass-bird   *W.I.* the fringillid *Loxipasser anoxanthus* (yellow-shouldered grassquit)

  cape grass-bird   *S.Afr.* the sylviid *Sphenoeacus afer*

grey bird   any of numerous campephagids of the genus *Coracina* (*cf.* cuckoo-shrike)

ground-bird   *Austral.* = quail-thrush *Austral.*

half-bird   any smaller than usual duck such as teal

hangbird   any of numerous icterid birds which produce hanging nests

haybird   a term applied in England to numerous species of Old World warbler

honey bird   a designation of great antiquity and no specific meaning

hard-head bird   *W.I.* = loggerhead kingbird

hummingbird   (*see also* trumpet) any of numerous apodiform birds of the family Trochilidae
    Allen's hummingbird   *Selasphorus sasin*
    Anna's hummingbird   *Calypte anna*
    beautiful hummingbird   *Calothorax pulchra*
    bee hummingbird   *Mellisuga helenae*
    buff-bellied hummingbird   *Amazilia yucatanensis*
    scaly-breasted hummingbird   *Phaeochroa cuvierii*
    bumblebee hummingbird   *Atthis heloisa*
    calliope hummingbird   *Stellula calliope*
    black-chinned hummingbird   *Archilochus alexandri*
    emerald-chinned hummingbird   *Abeillia abeillei*
    cinnamon hummingbird   *Amazilia rutila*
    Costa's hummingbird   *Calypte costae*
    Antillean crested hummingbird   *Orthorhynchus cristatus*
    violet-crowned hummingbird   *Amazilia verticalis*
    white-eared hummingbird   *Hylocharis leucotis*
    black-fronted hummingbird   *Hylocharis xantusii*
    blue-headed hummingbird   *Cyanophaia bicolor*
    violet-headed hummingbird   *Klais guimeti*
    lucifer hummingbird   *Calothorax lucifer*
    magnificent hummingbird   *Eugenes fulgens*
    Rieffer's hummingbird   = rufous-tailed hummingbird
    Rivoli's hummingbird   = magnificent hummingbird
    rufous hummingbird   *Selasphorus rufus*
    broad-tailed hummingbird   *Selasphorus platycercus*
    rufous-tailed hummingbird   *Amazilia tzacatl*
    sparkling-tailed hummingbird   *Tilmatura duponti*
    stripe-tailed hummingbird   *Eupherusa eximia*
    garnet-throated hummingbird   *Lamprolaima rhami*
    ruby throated hummingbird   *Archilschus colubris*
    vervain hummingbird   *Mellisuga minima*

hurricane-bird = frigatebird or (*W.I.*) the sooty tern

incubator bird = megapode

indigo-bird = *U.S.* indigo bunting *Afr.* any of several ploceids of the genus Hypochera
    black indigo bird   *H. nigerrima*
    dusky indigo bird   *H. funerea*
    purple indigo bird   *H. ultramarina*

jack-bird   *W.I.* = paw-paw bird *W.I.*

Jesus-bird = jacana

Judas-bird   *W.I.* the tyrannid bird *Elaenia martinica*

jumbie-bird   *W.I.* = death-bird *W.I.*

jumby-bird   *S.U.S.* = owl or, more rarely, any bird considered to be of "ill omen"

kingbird   any of several species of tyrannid bird of the genus Tyrannus. The name is also occasionally applied to terns
    thick-billed kingbird   *T. crassirostris*
    cassin's kingbird   *T. vociferans*
    eastern kingbird   *T. tyrannus*
    giant kingbird   *T. cubensis*
    grey kingbird   *T. dominicensis*
    loggerhead kingbird   *T. caudifasciatus*
    tropical kingbird   *T. melancholicus*
    western king bird   *T. verticalis*

ladybird   (*see also* bird 2) *W.I.* the vireonid *Vireo altiloquus* (= black whiskered vireo)

laughing bird   (*see also* jackass) *W.I.* the larid *Larus atricilla* (= laughing gull)

leafbird   any of several species of irenids of the genus Chloropus
    orange-bellied leafbird   *C. hardwickii*
    golden-fronted leafbird   *C. aurifrons*
    blue-winged leafbird   *C. cochinchinensis*

locust-bird   *S.Afr.* any of several birds but usually the sturnid *Creatophora carunculata* (= wattled starling). Also the ciconiid *Ciconi alba* (usually great locust bird) and the glareolid *Glareola nordmanni* (usually the little locust bird)

lotus-bird   *Austral.* the jacanid *Irediparra gallinacea* (*cf.* jacana)

lovebird   properly, any of several psittaciform birds of the genus Agapornis; *Austral.*, without qualification, *Melopsittacus undulatus* (= budgerigar)
    grey-headed lovebird   *A. cana*
    red-headed lovebird   *A. pullaria*
    masked lovebird   *A. personata*
    black-winged lovebird   *A. taranta*

lyrebird   either of two species of menurid birds of the genus Menura
    Prince Albert's lyrebird   *M. alberti*
    superb lyrebird   *M. superba*

mackerel-bird   *Brit.* = wryneck or shearwater *Brit.* or razorbill *Brit.* or *U.S.* tern

maize-bird = red-winged blackbird

mallee-bird   the megapodid *Leipoa ocellata*

old man bird   *W.I.* the cuculid *Hyetornis pluvialis*
    rifleman bird = rifle bird

mango-bird   *W.I.* the trochlid *Anthracothorax mango.* *I.P.* the oriolid *Oriolus kundoo* (*cf.* oriole)

man-o'-war bird   frigatebird

maybird   *Brit.* the scolopacid bird *Numenius phaeopus.* (= whimbrel) *U.S.* either the scolopacid *Calidris canutus* (= knot) or the icterid *Dolichonyx oryzivorus* (= bobolink). *W.I.* any of numerous cuculids

millerbird   the Hawaiian sylviid *Acrocephalus familiaris*

miner bird   *Austral.* the meliphagic *Myzantha flavigula* (= yellow-throated miner)

mistle-bird   = mistle thrush

mistletoe-bird   *W.I.* the thraupid *Tanagra musica* (= blue-hooded euphonia)

mockingbird   any of numerous species of passerine birds of the family Mimidae, mostly of the genus Mimus
    Bahama mockingbird   *M. gundlachii*
    common mockingbird   *M. polyglottos*
    blue mockingbird   *Melanotis coaerulescens*
    blue-and-white mocking bird   *Melanotus hypolencus*
    St. Andrews mockingbird   *M. magnirostris*
    tropical mockingbird   *M. gilvus*

moose-bird   the corvid *Perisoreus canadensis* (= Canada jay)

mound-bird   popular name of megapodes

mousebird   *Afr.* popular name of coliids of the genus Colius (= coly)
    red-backed mousebird   *C. castanotus*
    white-backed mousebird   *C. colius*
    red-faced mousebird   *C. indicus*
    white-headed mousebird   *C. leucocephalus*
    blue-naped mousebird   *C. macrourus*
    speckled mousebird   *C. striatus*

mutton-bird   *N.Z.* the procellariid *Puffinus griseus* (= sooty shearwater)
    Kermadec Island mutton bird   *N.Z.* the procellariid *Pterodroma neglecta*

nun-bird   any of several bucconids of the genera Hapaloptila and Monasa

nutmeg bird = cowry bird

October bird   *Austral.* the icterid *Dolichonyx oryzivorus* (= bobolink)

oilbird   the steatornithid *Steatornis caripensis*

orange bird *W.I.* the thraupid *Spindalis zena* (= stripe headed tanager)

organ-bird *Tasmania* the cracticid *Gymnorhina hyporleuca* (= white-backed magpie). *S.Amer.* the troglodytid *Troglodytes aedon* (= house wren)

ovenbird any of very many species of furnariid bird Also *Brit.* several sylviids of the genus Phylloscopus (*cf.* willow wren). *U.S.* the parulid *Seiurus aurocapillus*

ox-bird *Brit.* the scolopacid *Erolia alpina* (*cf.* red-backed sandpiper, dunlin)

paddy-bird any of several ciconiid and ardeid birds frequenting rice fields. Usually, without qualification, *Ardeola grayii*

palm-bird *Afr.* the ploceid *Ploceus cucullatus* (= village weaver)

paradise-bird see bird-of-paradise

parson-bird *N.Z.* the meliphagid *Prosthemadera novaeseelandiae* (= tui) *W.I.* the fringillid *Tiaris bicolor* (= black-faced grassquit)

pasture-bird *W.I.* any of numerous scolopacids, particularly in the course of their migration

paw-paw bird *W.I.* the mimid *Margarops fuscatus* (pearly-eyed thrasher) and the parulid *Coereba flaveola* (= bananaquit)

poe-bird = parson-bird

pilot bird *Austral.* the muscicapid *Pycnoptilus floccosus*

post-bird the muscicapid *Muscicapa striata* (= spotted flycatcher)

prince-bird *W.I.* the thraupid *Tangara cucullata* (= hooded tanager)

puffbird any member of the tropical family Bucconidae

white-necked puffbird *Notharchus macrorhynchos*
pied puffbird *Notharchus tectus*
white-whiskered puffbird *Malacoptila panamensis*

qua-bird the ardeid *Nycticorax nycticorax* (= black crowned night heron)

rafter-bird *Muscicapa striata* (= spotted flycatcher)

rain-bird *N.U.S.* the picid *Dryocopus pileatus* (= pileated woodpecker) *S. Amer.* several cuculids of the genus Piaya. *Brit.* the picid *Picus viridis* (= green woodpecker) *N.Z.* the procellariid *Pterodroma inexpectata* (= scaled petrel) *W.I.* any of numerous cuculids, a variety of apodid birds, the tyrannid *Tyrannus dominicensis* (= grey king-bird) and the hirundinid *Lamprochelidon euchrysea* (= golden swallow)

rainbow-bird *Austral.* the meropid *Merops ornatus* (*cf.* bee-eater)

redbird *U.S.* = cardinal
summer redbird the thraupid *Piranga rubra* (= summer tanager)

reed-bird variously applied to almost any bird frequenting reedy marshes. In particular, *Brit.*, the sylviid *Acrocephalus schoenobaenus* (= sedge warbler)

reel-bird the sylviid *Locustella naevia* (= grasshopper warbler)

regent-bird *Austral.* the ptilonorhynchid *Sericulus chrysocephalus*

rifle bird any of several paradisaeids of the genus Ptiloris, particulary *P. paradiseaus*

rhinoceros-bird = hornbill

rice-bird *U.S.* the icterid *Dolichonyx oryzivorus.* (= bobolink). The term is widely applied in commerce to many small edible birds. The canned rice-

bird of *U.S.* commerce is commonly the English sparrow. The rice-bird of the Orient is frequently *Munia oryzivora* (= Java sparrow)

rock-bird *W.I. Troglodytes aedon* (= house wren)

rudder-bird the oxyurine bird *Oxyura rubida* (= ruddy duck)

sage-bird *W.I.* = Bahama yellowthroat

sand-bird *W.I.* the charadriid bird *Charadrius wilsonia*

sarah-bird *W.I.* the tyrannid bird *Elaenia fallax*

satin-bird the ptilonorhynchid *Ptilonorhynchus violaceus* (= satin bowerbird)

saursop-bird *W.I.* = paw-paw bird *W.I.*

savanna-bird *W.I.* the fringillid *Ammodramus savannarum* (= grasshopper sparrow)

screech-bird *Brit.* the turdid *Turdus viscivorus* (= mistle thrush)

scrub-bird *Austral.* either of two species of atrichornithid of the genus Atrichornis
noisy scrub-bird *A. clamosus*
rufous scrub-bird *A. rufescens*

secretary-bird the sagittariid *Sagittarius serpentarius*

sedge-bird *Brit.* = sedge warbler *Brit. U.S.* the icterid *Dolichonyx oryzivorus* (= bobolink)

oat-seed bird the motacillid *Motacilla flavea* (= yellow wagtail)

silver bird the muscicapid *Empidornis semipartitis*

sin-bird *W.I.* the fringillid *Tiaris bicolor* (= black-faced grassquit)

snake-bird any bird with a long snake-like neck but properly, and usually, any of the species of anhingids (*cf.* anhinga, darter)

snow-bird usually either the snow bunting or the snow finch

soldier-bird (*see also* soldier) *Austral.* the meliphagid *Myzomela sanguinolenta* (= scarlet honeyeater) *W.I.* the charadriid bird *Charadrius vociferus* (= killdear)

soursop-bird *W.I.* the thraupid *Tangara cucullata* (= hooded tanager)

spinifex bird *Austral.* the sylviid *Eremiornis carteri*

sprat-bird *W.I.* the larid bird *Thalasseus maximus* (= royal tern)

stink-bird the opisthocomid *Opisthocomus hoazin* (= hoatzin)

stitchbird *N.Z.* the meliphagid *Notiomystis cincta*

straw-bird *W.I.* = grass-bird *W.I.*

sugar-bird *W.I.* any of several species of parulids particularly *Coereba flaveola*

summer-bird *Austral.* the campephagid *Coracina novaehollandiae* (= black-faced cuckoo-shrike)

sunbird any of very numerous species of nectarinids
amethyst sunbird *Chalcomitra amethystina*
violet-backed sunbird *Anthreptes longuemarei*
beautiful sunbird *Nectarinia pulchella*
bronze sunbird *Nectarinia kilimensis*
Himalayan yellow-backed sunbird *Aethopyga siparaja*
Nepal yellow-backed sunbird *A. nipalensis*
Carmelite sunbird *Chalcomitra fuliginosis*
scarlet chested sunbird *Chalcomitra senegalensis*
cheerick sunbird *Cyanomitra vertecales*
grey-chined sunbird *Anthreptes tephrolaema*
yellow-chinned sunbird *Anthreptes rectirostrio*
collared sunbird *Anthreptes collaris*
copper sunbird *Cinnyris cupreus*
grey sunbird *Anthreptes gabonica*
malachite sunbird *Nectarina famosa*
olive sunbird *Cyanomitra olivacea*

pigmy sunbird *Hedydipna platura*
splendid sunbird *Cinnyris coccinigaster*
violet-tailed sunbird *Anthreptes aurantia*
scarlet-tufted sunbird *Anthreptes fraseri*
variable sunbird *Cinnyris venustus*
golden-winged sunbird *Drepanorhynchus reichenowi*
sunset-bird *W.I.* the tyrannid *Myiarchus tyrannulus* (= rusty-tailed flycatcher)
surfbird *U.S.* the charadriid bird *Aphriza virgata*
tailor-bird any of several species of sylviid bird of the genus Orthotomus, particularly *O. sutorius*
tannia-bird *W.I.* the icterid *Icterus oberi* (= Montserrat oriole)
Java temple-bird = Java sparrow
thistle-bird the fringillid *Carduelis carduelis* (= goldfinch)
thorn bird any of several funariids of the genus Phacellodromus
tick-bird *W.I.* the cuculid bird *Crotophaga ani* (= smooth billed ani). *Afr.* either of the two sturnids of the genus Buphagus (= oxpeeker)
tinker-bird any of several capitonid birds of the genus Pogoniulus (*cf.* barbet)
  red-fronted tinker-bird *P. pusillus*
  yellow-fronted tinker-bird *P. chrysocomus*
  green tinker-bird any of several green species of Pogoniulus; without qualification, *P. simplex*
  golden-rumped tinker-bird *P. bilineatus*
  least tinker-bird *P. subsulphureus*
  speckled tinker-bird *P. scolopaceus*
tropic-bird any of several species of oceanic phaethontid bird of the genus Phaeton
  red-billed tropicbird *P. aethereus*
  yellow-billed tropicbird = white-tailed tropicbird
  red-tailed tropicbird *P. rubricauda*
  white-tailed tropicbird *P. lepturus*
Uncle Sam bird *W.I.* the trogonid *Priotelus temnurus* (= Cuban trogon)
umbrellabird any of several large contingids of the genus Cephalopterus
unicorn-bird the anhimid *Anhima cornuta* (= horned screamer)
waggon-bird *Austral.* = wedgebill
wall-bird *Brit.* the muscicapid bird *Muscicapa striata* (= spotted flycather) *W.I.* = rock-bird
walrus-bird = the scolopacid *Evolia melanotus* (= pectoral sandpiper)
wattle-bird *N.Z.* any of three species of callaeid birds of the genus Callaeas. *Austral.* any of several species of meliphagid bird of the genus Anthochaera
  little wattle-bird *A. chrysoptera*
  red wattle-bird *A. carnuculata*
  yellow wattle-bird *A. paradoxa*
weaver-bird any of numerous passeriform birds of the family Ploceidae. The term "weaver" (*q.v.*) is more commonly used except in parts of the orient and West Indies
whale-bird *Austral.* and *N.Z.* any of several species of procellariid of the genus Pachyptila
whipbird *Austral.* either of two timaliids of the genus Psophodes
  eastern whipbird *P. olivaceous*
  black-throated whipbird *P. nigrogularis*
whistling bird *W.I.* the parulid *Catharopeza bishopi* (= whistling warbler)
white-bird *W.I.* = tropicbird
widow-bird any of many species of long tailed plo-

ceid birds frequently separated as a separate subfamily the Viduinae (*cf.* whydah)
  paradise widow-bird *Steganura paradisea*
wire-bird the charadriid bird *Charadrius sanctaehelenae* peculiar to the island of St. Helena (*cf.* plover)
old woman bird *W.I.* the cuculid *Saurothera vetula* (*cf.* cuckoo)
yam-bird *W.I.* the turdid *Turdus nudigenis* (= bareeyed thrush)
yellow-bird *W.I.* = canary *W.I.*
bird 2 (*see also* bird 1) applied in *Brit.* to an insect
  ladybird *Brit.* popular name of coccinellid coleopteran insects (= lady beetle *U.S.*)
bird-of-paradise (*see also* flower) any of very many species of paradisaeid bird distinguished by their brilliant colors and aberrant plumage. In *W.U.S.* the term is also applied to the tyrannid *Muscivora forficatus* (= scissor tail)
bird-of-prey popular name of falciform and strigiform birds
bird-on-the-wing the polygalaceous herb *Polygala paucifolia*
biriba the annonaceous tree *Rollinia deliciosa* and its edible fruit
biribiri *W. Afr.* the columbid bird *Vinago waalia* (*cf.* green pigeon)
·**birth** the process, and fact, of the emission of young by a viviparous animal
  afterbirth the placental and foetal membranes of the mammal
bis *see* [bistre]
Bis *see* [Biscuit]
[Biscuit] a mutant (gene symbol *bis*) in linkage group V of *Neurospora crassa*. The phenotypic expression is a conidiating colonial growth
bishop (*see also* bird) the poeciliid teleost fish *Brachyrhaphis episcopi*
bison (*see also* beetle) both the English and Latin name of a genus of bovine bovid artiodactyle mammals represented today by the N. American form commonly miscalled the buffalo (*q.v.*) and the European form known as the wisent (*q.v.*)
bistort any of several species of polygonaceous herb of the genus Polygonum particularly those of the subgenus Bistorta. *Brit.* without qualification, usually *P. bistorta*. *U.S.*, without qualification, usually *P. bistortoides*
[bistre] a mutant (gene symbol *bis*) mapped at 19.8± on the X chromosome of *Drosophila melanogaster*. The phenotypic expression involves dark brown eyes and ocelli
bit a word variously meaning a small piece and a bite. It enters into compound plant names in both senses
  devil's bit *either* the dipsacaceous herb *Succisa pratensis or* the lileaceous herb *Chamaelirium luteum*
  frogbit either of two hydrocharitaceous floating aquatic herbs
    American frogbit *Limnobium spongia*
    European frog bit *Hydrocharis morusranae*
  frogsbit *see* frog bit
  hawk-bit *Brit.* any of several species of compositaceous herb of the genus Apargia
  hen bit either of the labiateous herbs *Lamium amplexicaule* or *L. hybridum*
  sheep's bit the campanulaceous herb *Jasione montana*

bitch a female dog, coyote, ferret and occasionally applied to the fox, though vixen is the proper term

devil's **bite** (*see also* bit) any of several species of ranunculaceous herbs of the genus Helleborus (= helibore)

toe **biter** *U.S.* = electric light bug *U.S.*

willow **biter** the parid bird *Parus caerula* (= blue titmouse)

[**bithorax**] a mutant (gene symbol *bx*) at a pseudoallelic locus mapped at 58.8 on chromosome III of *Drosophila melanogaster*. The phenotypic expression is the conversion of the front portion of the haltere into a wing-like structure

bithoraxoid] a mutant (gene symbol *bxd*) mapped at 58.8± on chromosome III of *Drosophila melanogaster*. The phenotypic expression is enlarged, disc-shaped, halteres

Bittacidae that family of mecopteran insects which is commonly called hanging scorpionflies. They differ from other mecopterans in the extremely narrow wings that frequently cause them to be mistaken for craneflies

bittern any of numerous species of ardeid bird, many of the genus Botaurus, most with the habit of ruffling the neck feathers when calling. *Brit.*, without qualification, *B. stellaris. U.S. B. lentiginosus. N.Z. B. poiciloptilus*

black bittern *Dupetor flavicollis*

chestnut bittern *Ixobrychus cinnamomeus*

white-crested bitten *W.Afr. Tigriornis leucolopha*

dwarf bittern *Afr. Ixobrychus sturmii*

green bittern *Austral.* any of several species of the genera Butorides and Botaurus

least bittern *Ixobrychus exilis*

little bittern *N.Z. Ixobrychus minutus*

sunbittern the eurypygid bird *Eurypyga helias*

tiger bittern *Gorsachius melanolophus* (*cf.* tiger heron)

yellow bittern *Ixobrychus sinensis*

bivium literally, a junction of two roads, but, by *biol.* extension the pair of arms, at the junction of which lies the madreporite in asteroid echinoderms (*cf.* trivium)

Bixaceae a small family of parietalous dicotyledons distinguished by numerous stamens and a compound one-cell ovary

bk *see* [broken], [buckled], or [beaked fruit]

bk₂ *see* [brittle stalk]

Bkd *see* [Blackoid]

bl *see* [black]

Bl *see* [Bristle]

bla *see* [bladder wing]

black *U.S.* any of several species of geometrid lepidopteran insect of the genus Euchoeca

waved black *Brit.* the noctuid lepidopteran insect *Parascotia fuliginaria*

[**black**] either a mutation (gene symbol *bl*) in linkage group I of *Habrobracon juglandis*. The name is descriptive of the phenotypic body color *or* a mutant (gene symbol *b*) mapped at 48.5 on chromosome II of *Drosophila melanogaster*. The name is descriptive

blackamoor the characinid teleost fish *Gymnocorymbus ternetzi*

[**black feet**] a mutant (gene symbol *bf*) in linkage group VIII of *Habrobracon juglandis*. The name is descriptive of the appearance of the phenotype

[**Blackoid**] a mutant (gene symbol *Bkd*) mapped at 65.0± on chromosome II of *Drosophila melano-*

gaster. The phenotypic expression is a dark body color

[**black pupa**] a mutant (gene symbol *bp*) mapped at 17.1 on linkage group XI of *Bombyx mori*. The name is descriptive of the phenotype

bladder (*see also* senna) any thin walled expansible structure. Without qualification, usually refers to the urinary bladder of mammals

gallbladder a sac-like structure found in the liver of many vertebrates that serves to accumulate bile

reserve bladder one of a number of peripheral excretory channels in some trematodes

swim bladder a thin-walled, gas-filled, bladder lying along the dorsal wall of the coelom in bony fishes. It is developed as an outgrowth of the alimentary canal and may retain, in some fishes, an attachment either to the esophagus or the pharynx

urinary bladder an outgrowth of the hindgut, or derivative of the allantois, into which the ureters open in amniotic vertebrates. The term is often improperly applied to the water storing bladder of amphibia

[**bladder wing**] a mutant (gene symbol *bla*) mapped at 33.6± on the X chromosome of *Drosophila melanogaster*. There are bladders on the deformed wings of the phenotype

blade any flattened object, usually pointed at one or both ends. Commonly used of leaves, particularly of monocotyledons, or of the flattened portion of leaves

skipping blade one of one or more pairs of flattened spines, paddles or setose bristles, attached to the trunk of some Rotifera

twayblade *U.S.* any of several orchids of the genera Lapris and Histera. Usually, without qualification *Liparis liliifolia*

blanquilo popular name of malacanthid fish

-blast- (*see also* -cyte, -plast) *comb. form* meaning "bud", "shoot" but also used for any developing or developmental structure, particularly cells, (*see also* -cyto) or for cells or structures that are producing something. There is much confusion, apparently not entirely due to misprints with -plast (*q.v.*). For example, blepharoblast and blepharoplast are both used for the same structure but bioblast and bioplast appear to be distinguishable

angioblast mesenchyme cells precursor to blood vessels and cells

antheroblast = androcyte

apoblast a barren plant shoot

auxoblast any part of a plant that can be used for vegetative reproductive purposes

bioblast an independent, or semi-independent living, unit whether it be a single celled form, or an organ, or an individual within a colony of such forms as protozoa or hydrozoa

blepharoblast *see* blepharoplast

brachyblast a short branch

calcoblast a type of scleroblast producing calcareous spicules

chondroblast a cell that secretes the matrix of cartilage

cnidoblast = nematocyte

coeloblast (*bot.*) a syncytial organism

colloblast a cell that secretes a sticky secretion designed to entangle prey, particularly such a cell on the tentacle of a ctenophore

cytoblast = bioblast. The term was also at one time briefly used as synonymous with nucleus

hemocytoblast a free rounded cell, derived directly from a primitive reticular cell as an ancestor of blood cells of all types

endoblast (*see also* hypoblast 1) the lower germ layer in the early development of the telolecithal egg (*cf.* endoderm)

epiblast 1 (*see also* epiblast 2) a rudimentary second cotyledon found in some grasses

epiblast 2 (*see also* epiblast 1) the upper, compact germ layer of involuted endoderm in the first stages of the development of the telolecithal egg

eremoblast a cell once united to, but now separated from, a mass

erythroblast the precursor of an erythrocyte

basophilic erythroblast the descendent of a proerythroblast and the direct precursor of a polychromatophilic erythroblast. Distinguished from a proerythroblast by the greater basophilic staining capacity of the nucleus, the smaller size, and the absence of clearly visible nucleoli

polychromatophilic erythroblast a descendent of a basophilic erythroblast and precursor of a normoblast. It is distinguished by the fact that it takes up both the acidic and basic components of blood stains

proerythroblast a descendent of a hemocytoblast giving rise to basophilic erythroblast and distinguished in general by the large size of the nucleus in which two nucleoli are clearly visible

fibroblast a cell found in areolar connective tissue arising directly from the primitive reticular cell and thought to give rise to fibers

floatoblast a statoblast lacking hooks and that floats

geoblast a developing plant, the cotyledons of which do not rise above the surface of the soil

gonimoblast hair-like filaments arising from the fertilized carpogonium of red algae

hematoblast the first stage in the development of an erythrocyte

histioblast 1 (*see also* histioblast 2, a stage intermediate between a hemocytoblast and a histiocyte

histioblast 2 (*see also* histioblast 1) a small active cell involved in germination of the gemmule of a sponge

histoblast a group of cells that remain unaltered when histolysis occurs in neighboring cells. Also used in the sense of a characteristic cell of a particular tissue

hypoblast 1 (*see also* hypoblast 2) the lower, loose, endodermal cells in the early development of the telolecithal egg. By some used as synonymous with endoderm

hypoblast 2 (*see also* hypoblast 1) the flattened dorsal cotyledon of a grass embryo

idioblast any cell differing markedly from the other cells in the tissue in which it is found. In plants, for example, crystal idioblasts are common. The term is also used in the sense of a hypothetical structural unit of a cell

katablast a plant arising from below ground

lemnoblast a neuroglial cell, of the oligoglia variety, that accompanies peripheral nervous fibers and is differentiated into Schwann's cells

lipoblast a fat-storing cell of the embryo

lymphoblast = plasmablast a derivative of a hemocytoblast ancestoral to a plasm cell

megakaryoblast the first stage in the development of a thrombocyte from a hemocytoblast

megaloblasts erythropoietic cells most commonly found in birds

mesoblast the middle germ layer in the early development of the telolecithal egg (*cf.* mesoderm). The term was at one time used as synonymous with nucleus

metablast = nucellus

monoblast a descendant of a hemocytoblast and direct ancestor of a monocyte

myeloblast a direct descendant of a hemocytoblast, destined to form a myelocyte

nematoblast = cnidoblast

neoblast undifferentiated blastema cells that are used in the regeneration of invertebrate tissues

neuroblast those cells in the embryonic neural tube which will subsequently become neurons

normoblast the descendant of the polychromatophilic erythroblast and the direct precursor of an erythrocyte. In mammals distinguished by a pyknotic nucleus that is subsequently extruded

pronormoblast the second stage in the development of an erythrocyte

odontoblast those mesodermal cells that secrete the dentin in the formation of a tooth

orthoblast an upstanding prothallium

osteoblast a cell that secretes the mineral portion of bone

periblast a mass of incompletely separated yolk cells that unite the blastodisc with the yolk mass in telolecithal eggs

photoblast a shoot developed above the soil

phylloblast a lichen thallus that resembles a leaf

phytoblast an undifferentiated plant cell

plasmablast a large leucocyte thought by some to be the site of antibody production

pleuroblast an outgrowth from a fungus that has a bud-like appearance

poikeloblast a deformed irregular leucocyte

protoblast (*bot.*) a cell lacking a cell wall

scleroblast a sponge mesenchyme cell that produces spicules, or other skeletal elements

sessoblast statoblasts that sink to the bottom or remain attached to the parent tissue (*cf.* floatoblast)

silicoblast a type of scleroblast producing silicous spicules

spheroblast a cotyledon with an apical bulb

spinoblast a floatoblast provided with attaching hooks

spongioblast *either* a cell in the epithelium of the embryonic neural tube that will subsequently become a neuroglia cell *or* a spongin producing scleroblast

sporoblast spore mother-cell (= merispore)

pansporoblast an endogenous bud formed in some sporozoan Protozoa

statoblast a chitinous-covered gemmule found in sponges and a somewhat similar structure, usually provided with hooks for attachment, found in some fresh water Ectoprocta

teloblast a large mother cell that buds off columns of smaller cells at the growing tip of many invertebrate embryos

trichoblast parenchymatous cells of plants that are elongate and branched

trophoblast the outer coat of the implantation stage of the early embryo of a placental mammal

cytotrophoblast = cytotrophoderm

plasmotrophoblast a cell of the plasmatrophoderm (*q.v.*)

syntrophoblast those cells of the trophoblast which invade the maternal uterine wall

Blastaea a hypothetical organism, in general resembling Volvox, from which all metazoa were at one time presumed to be descended (*c.f.* Gastraea)

blastea (*bot.*) a plant structure having the form of a holoblastula

blastema literally a sprout. In *zool.* an area of segregated cells in an embryo that will subsequently develop into a specific organ In *bot.* the radicle and plumule of an embryo before the appearance of the cotyledon

cytoblastema = protoplasm

blastesis the processes of reproduction, particulary those in lichens that involve the conidia

acroblastesis the condition of having a germ tube protruding from the end of a spore

diablastesis an outgrowth from the hyphal layer of a lichen

ecblastesis the proliferation of a flower by internal budding

periblastesis the over-growing of gonidia by neighboring cells

-blastic pertaining to developing structures or parts of developing structures. The form -blastous is occasionally seen, as is the substantive -blasty, particularly in *bot.*

ablastic pertaining to parts of an organism that have failed to develop

asymblastic pertaining to the variation in germination time of a group of seeds from the same plant

diblastic two-celled, used particularly of double spores

dichoblastic a type of dichotomy in which tree or shrub branches lie closely pressed together and, therefore, apparently form an axis

diploblastic consisting of, or possessing, two embryonic layers

enantioblastic the conditon of a plant embryo that lies opposite the hilum

gymnoblastic having a superior ovary

holoblastic *see* cleavage

homoblastic used of organisms of which the larval and adult forms appear practically identical, of orchid pseudobulbs containing several internodes, of which only the last is foliaceous, of that type of development that is direct, that is, which does not pass through a larval condition and of a spore, the whole of which is concerned in embryogeny

meroblastic (*bot.*) said of a spore, only part of which is used in embryogeny

opsiblastic said of an egg, particularly that of a gastropod, with an unusually heavy shell designed to withstand adverse environmental conditions

pleuroblastic *either* pertaining to a pleuroblast *or* pertaining to an early monocotyledonous embryo

tachyblastic said of a normal, as distinct from opsiblastic, egg

triploblastic consisting of, or possessing, three developmental layers

blasticium (*bot.*) = daughter cell

blastidule any asexual reproductive body that is not a spore

Blastobasidae a family of small gelechioid lepidopteran insects with lanceolate wings. The family contains, *inter alia*, the acorn moths

Blastocladiales an order of uniflagellate phycomycetes distinguished by isogamous sexual reproduction

blastos literally a shoot

-blastous *see* blastic

blastula 1 (*see also* blastula 2) that stage which terminates the cleavage of many animal eggs and which usually consists of a hollow sphere of cells

amphiblastula a blastula in which the cells of one pole differ markedly either in size or shape from those of the other pole

archiblastula = coeloblastula

coeloblastula a blastula having a central cavity. This is what is usually meant by the use of the word blastula without qualification

discoblastula a blastula consisting of a disk of cells resting on yolk (= blastodisc)

periblastula a blastula consisting of a membrane one cell thick, enclosing the central yolk

stereoblastula a blastula that fails to develop a central cavity (= morula in mammals)

blastula 2 (*see also* blastula 1) the mother cell of Volvox

-blasty *See* -blastic

Blattariae an order of insects containing the cockroaches

Blattidae that family of blattarian insects to which the roaches belong. They are easily distinguished from the Mantidae by the lack of prehensile forelegs

Bld *see* [blond]

bleak *Brit.* the cyprinid fish *Alburnus lucidus*

heather **bleat** *obs. Brit.* = snipe

bleater = snipe

bleb (obs.) = pith

-blem- *comb. form* meaning "a bed coverlet" and, by extension, something which may be "thrown off" or "shed"

epiblem that portion of a root tip which bears root hairs

periblem the middle of three meristem layers in the root tip of spermatophytes

Blenniidae a family of scaleless blennioid fish sometimes called the combtooth blennies

Blennioidea a very large suborder of acanthopterygian osteichthyes distinguished by the position of the pelvic fins in front of the base of the pectoral fins

blenny popular name of blennioid fishes particularly those of the families Clinidae and Blenniidae

combtooth blenny popular name of blenniid fishes in contrast to clinids

decorated blenny any of several stichid fish of the genus Chirolophus

eelblenny = snake blenny

rockblenny any of several stichid fish of the genus Siphister

scaled blenny popular name of clinid fishes

snakeblenny any of several species of slender sticheid fishes of the genera Eumesogrammus and Lumpenus

viviparus blenny *Brit.* the zoarcid fish *Zoarces vivaparus* (*cf.* eel pout)

-blephar- *comb. form* meaning "eyelid" or "eyelash" and sometimes used as an adjectival form of flagellum

Blephariceridae = Blepharoceridae

Blepharoceratidae = Blepharoceridae

Blepharoceridae a family of small mosquito-like nematoceran dipteran insects commonly called net-winged midges because they have a network of fine lines between the wing veins

blesmol general name for bathyergid myomorph rodents

blinks the portulacaceous herb *Montia fontana*

blight (*see also* blite) any causative agent of a withered condition in plants, or the condition itself

alder blight the aphid *Prociphilus tesellatus*

[blistered] a mutant (gene sybmol *bs*) mapped at 107.3 on chromosome II of *Drosophila melanogaster*.

The small wings of the phenotype are blistered and there are extra veins

[blistery] a mutant (gene symbol *by*) mapped at 48.7 on chromosome III of *Drosophila melanogaster.* The word refers to the condition of the wings of the phenotype

blite any of several chenopodaceous herbs and plants resembling them. Sometimes spelled "blight"

    coast blite the chenopodiaceous herb *Chenopodium rubrum*

    eye blite any of very numerous scrophulariaceous herbs of the genus *Euphrasia*

    sea blite any of several species of chenopodiaceous herb of the genus Suaeda. *N.Z.*, without qualification, *S. novae-zelandiae*

    strawberry blite the chenopodiaceous herb *Chenopodium capitatum*

blo *see* [bloated]

Blo *see* [Blotchy]

[bloated] a mutant (gene symbol *blo*) mapped at 58.8 on chromosome II of *Drosophila melanogaster.* The phenotypic expression is ballooned wings with extra veins

bloater *U.S.* the salmonid fish *Coregonus hoyi. Brit.* a lightly smoked fish, usually a herring

may-blob *U.S.* the ranunculaceous herb *Caltha palustris* (= marsh marigold)

block any relatively well defined mass, frequently more or less square-edged. In *biol.,* without qualification, usually a mass of paraffin, or plastic, in which an object to be sectioned is embedded

    heterochromatic block a thickened condensed area on a chromosome usually best visible in interphase and early prophase stages

[Blond] a mutant phenotype (gene symbol *Bld*) associated with a deficiency on the X chromosome or with a translocation between the X chromosome and chromosome II in *Drosophila melanogaster.* Homozygous lethal; heterozygote has bristles with gleaming yellow tips

blood 1 (*see also* blood 2, fin, island) a fluid connective tissue used for the transport of dissolved or dispersed materials in animals

blood 2 (*see also* blood 1) in compound names but mostly as an adjective

black blood the lythraceous herb *Lythrum salicaria*

bloom 1 (*see also* bloom 2, 3) a flower

    bitter bloom *U.S.* the gentianaceous herb *Sabatia angularis* (*cf.* quinine flower)

bloom 2 (*see also* bloom 1, 3) a sudden increase in the number of algae in lakes. The term is often modified by seasonal adjectives

bloom 3 (*see also* bloom 1, 2) a thin whitish or bluish pruinose coating of minute granules, or microorganisms on the surface of leaves or fruits

blossom 1 (*see also* blossom 2, crown) synonym for flower

blossom 2 (*see also* blossom 1) any of several lepidopteran insects

    peach blossom *Brit.* the thyatirid *Thyatira batis*

    pease-blossom *Brit.* the noctuid *Chariclea delphinii*

[blot] a mutant (gene symbol *blt*) mapped at 55.2± on chromosome II of *Drosophila melanogaster.* The wings of the phenotype are blackened and inflated

[Blotched aleurone] a mutant (gene symbol *Bh*) mapped at 45 on linkage group VI of *Zea mays.* The phenotypic seed has a blotched aleurone layer

[Blotchy] a mutation (gene symbol *Blo*) occurring in the sex chromosome (linkage group XX) of the mouse.

The term Blotchy applies to the appearance of the pelt in the phenotype

blt *see* [blot]

blubber the oily subdermal tissues of cetaceans

blue (*see also* in many compounds entered under the *subs.* [*e.g.* bird, bottle]) any of very many blue or bluish lycaenid lepidopteran insects (*cf.* copper, hairstreak)

    acacia blue *I.P.* any numerous species of the genus Surendra

    azure blue *Brit. Lycaena argiolus*

    babul blue *I.P.* any of numerous species of the genus Azanus

    Bloxworth blue *Brit.* = short-tailed blue *Brit.*

    bush blue *S.Afr. Cacyreus lingeus*

    cape blue *S.Afr. Eicochrysops messapus*

    ciliate blue *I.P.* any of numerous species of the genus Lycaenesthes

    Clifden blue *Brit. Lycaena bellargus*

    clover blue *S.Afr. Zizina antanossa*

    red clover blue *S.Afr. Actizera stellata*

    common blue *Brit. Lycaena icarus S. Afr. Syntarucus telicanus*

    dotted blue *S.Afr. Tarucus sybaris*

    eyed blue any of several species of the genus Hemiargus

    freestate blue *U.S. Lepidochrysops letsea*

    grass blue any of numerous species of the genus Zizeeria

    guava blue *I.P.* any of several species of the genus Virachola

    hedge blue any of numerous species of the genus Lycaenopsis

    chalk hill blue *Brit. Lycaena corydon*

    holly blue *Brit. Cyaniris argiolus*

    king blue *S.Afr. Lepidochrysops tantalus*

    large blue *Brit. Nomiades arion*

    leaf blue *I.P. Horsfieldia anita*

    lime blue *I.P. Chilades laius*

    line-blue *Austral.* any of several species of the genus Nacaduba

    little blue *Lycaena minima*

    Malayan blue *Megisba malaya*

    mandarin blue *I.P. Charana mandarinus*

    orange margined blue any of several species of the genus Lycaeides

    marsh blue *S.Afr. Harpendyreus noquasa*

    mazarine blue *Brit. Nomiades semiargus*

    meadow blue *S.Afr. Cupidopsis cissus*

    mountain blue *S.Afr. Harpendyreus tsomo*

    oakblue *Austral.* and *I.P.* any of several species of the genus Amblypodia

    pigmy blue any of several species of the genus Brephidium

    small blue *Brit. Zizera minima*

    silverstreak blue *I.P.* any of numerous species of the genus Iraota

    silver studded blue *Brit. Lycaena aegon*

    tailed blue *S.Afr. Stugeta bowkeri*

    long-tailed blue *Brit. Lampides baeticus*

    short-tailed blue *Brit. Cupido argiades*

    topaz blue *S.Afr. Azanus jesus*

    thorn tree blue *S.Afr. Azanus moriqua*

    tropical blue any of several species of the genus Leptotes

    walnut blue *I.P. Chaetoprocta odata*

    straightwing blue *I.P.* any of numerous species of the genus Orthomiella

    zebra blue *I.P. Syntarucus plinius*

    zulu blue *S.Afr. Lepidochrysops ignota*

[Blue egg] a mutant (gene symbol O) in the linkage group IV of the domestic fowl. The name is descriptive of the phenotypic expression

bluet 1 (*see also* bluet 2) U.S. any of several species of coenagrionid odonatan of the genus Enallagma

bluet 2 (*see also* bluet 1) any of several plants, usually with small blue flowers. The best known are the rubiaceous herb *Houstonia caerulea*, and the compositaceous plant *Centaurea cyanus* (= cornflower)

  mountain bluet *Centaurea montana*

maiden's blush *Brit.* the geometrid lepidopteran insect *Ephyra punctaria*

bm$_{1,2}$ *see* [brown midrib]

[b$_8$-mottled] a mutant (gene symbol *obt*) mapped at 21.0 on linkage group VII of *Bombyx mori*. The phenotypic expression is a moderate translucency of the larva

Bn *see* [Bent] or [Brown endosperm]

bn *see* [Button]

bo *see* [bordeaux]

boa popular name of boid snakes. Without qualification, usually the boa constrictor (*q.v.*)

  Cuban boa *Epicrates angulifer*

  emerald boa *Boa canina*

  rainbow boa *Epicrates cenchris*

  rosy boa *Lichanura roseofusca*

  rubber boa *U.S. Charina bottae*

  sand boa any of several species of the genus Eryx

boar (*see also* fish, hog, pig) used without qualification, a male domestic pig, though in the case of wild pigs, it is extended to both sexes. It is also used for a male of a number of other animals such as the bear, guinea pig, mink, and skunk

  wild boar *U.S.*, without qualification, *Sus scrofa*

boatswain any of several birds but commonly the skua (*q.v.*)

bob *I.P.* any of several species of hesperiid lepidopteran insect of the genera Iambrix, Arnetta, and Suastus

  chestnut bob *I.P. Iambrix salsala*

  palm bob *I.P. Suastus gremius*

[bobbed] a mutant (gene symbol *bb*) mapped at 66.0 on the X chromosome of *Drosophila melanogaster*. The phenotypic expression is small bristles and irregular sclerites

bobolink *U.S.* the icterid bird *Dolichonyx oryzivorus*

bob-white *U.S.* the phasianid bird *Colinus virginianus*

bocaccio the scorpaenid fish *Sebastodes paucispinis*

bod *see* [bowed]

body 1 (*see also* body 2, 3, corpus, corpuscle) any organ or organelle not possessing a specific name

  Barr body = heteropycnotic body (*q.v.*)

  basal body a small body found at the base of many cilia (= kinetosome)

    endobasal body = intranuclear body

    parabasal body a body, variously shaped, connected with the blepharoplast

  peribiliary body a microbody, thought by some to be a lysosome found in the liver, kidney and adrenal cortex

  ultimobranchial body a glandular structure of unknown function derived from the posterior wall of the fifth branchial pouch

  broodbody leaf gemmae of Bryophytes

  brown body *either* brownish colored fatty masses found in hibernating mammals, *or* a brownish mass produced by the degeneration of an ectoproct or endoproct polyp which subsequently regenerates into a new polyp (= reduction body) *or* a small clump of cells found in the coelom of Holothuria

  central body the nuclear material in blue green algae

  chlorophyll body = chloroplast

  chromatoid body a highly refractile elongate granule in Protozoa

  coccygeal body a small, heavily vascularized body situated just in front of the apex of the coccyx of great apes. The term is sometimes applied to the coccygeal gland (*q.v.*)

  lateral geniculate body that part of the mammalian brain in which the optic tract terminates

  Golgi body *see* Golgi apparatus

  habenular body the nerve center in the epithalamus

  Hassall's body a roundish group of acidophile cells typical of the medulla of the thymus

  mammillary body a small dome-shaped prominence on the underside of the hypothalamus

  microbody a granule disclosed only by electron microscopy

  Mueller's body gland on the base of the petioles of some plants from which ants derive nourishment

  mushroom body *either* one of a pair of pedunculate nervous structures, running forward from the protocerebral lobes in the insect brain *or* a compound of Jacobson's organ (*q.v.*) with a bony column

  Nissl body clumps of granules that stain deeply in alkaline solutions of methylene blue in the cytoplasm of nerve cells

  intranuclear body any micro-body within the nucleus

  pineal body a body developed from the posterior of the two structures known as epiphysis

  heteropycnotic body chromatin of the sex chromosome when it appears as a dark staining mass attached to the nuclear membrane in a resting nucleus (= Barr body)

  reduction body a degenerate, or dedifferentiated, mass of tissue formed in some invertebrates and urochordates from which a new individual can regenerate. In Ectoprocta and Endoprocta it is also called brown body

  interrenal body cells or a discrete body found between the kidneys and corresponding to the medulla of the adrenal gland of higher forms

  suprarenal body cells corresponding to the cortex of the adrenal body in lower vertebrates

  restiform body the fiber tracts of the dorso-lateral funiculus where it bends abruptly inward into the cerebellum

  slime body slime-producing organelles of greatly variable shape, commonly found in the cytoplasm of sieve tubes

  spongy body one of five lymphoid bodies corresponding in position and origin in the echinoid echinoderms to the Polian vesicles of other echinoderms

  paraterminal body that portion of the gray matter of the brain which is stretched out by the development of the corpus callosum in higher forms

  polar body the nuclear material, with a minute amount of accompanying cytoplasm, separated from the egg in the course of meiosis (= polar cell)

  urn body a ciliated, urn-shaped, multicellular body found in the coelomic fluid of some sipunculoids

  intervertebral body a vertebral body, in diplospondylous vertebral columns, lacking spinous processes, lying between bodies bearing processes

  vitreous body the semi-solid content in the retinal chamber of the eye

  Wolffian body = mesonephoros

body 2 (*see also* body 1, 3) any substance, or postulated substance, lacking a specific name

antibody the specific substance produced in blood serum by the injection of a specific antigen

Eberth's body amorphous masses, developed from fine threads, in the epidermis of larval anuran amphibia

interstitial body a mucilaginous disc on a pollen grain

body 3 (*see also* body 1, 2) a whole organism or the major part of an organism

Blochmann's body any intracellular organism in the egg of an arthropod. Most are bacteria, supposedly symbiotic

forebody an anterior body region, sharply distinguished from a hindbody, but not distinguished as a head or other morphologically acceptable division. The term is usually applied to trematode worms

hindbody *see* forebody

bog (*see also* pine) permanently soggy ground, distinguished from a marsh (*q.v.*) by the quantity and permanence of the water, and the types of vegatation

sphagnum bog = moss moor

bogart *U.S.* the larva of the dobson fly *Corydalus cornutus* (= hellgrammite)

Boidae a relatively large family of usually big snakes including the boas, pythons and their allies. They are distinguished from most other snakes by the presence of a pelvic girdle and hind limb vestiges, and from the blind snakes by the presence of differentiated dorsal and ventral scales. Usually divided into several subfamilies including the Pythoninae and Boinae

Boinae a subfamily of boid snakes distinguished from the Pythoninae by the absence of supraorbital bones and, usually, the habit of ovoviviparity

Bojanus's *epon. adj.* from Ludwig Heinrich Bojanus (1776–1827)

bok (*see also* buck) a *S.Afr.* word meaning "buck" but not interchangeable with this word since names such as "gemsbok", which has no English equivalent, are used as much by English-speaking as by Taal-speaking, individuals

blaubok hippotragine antilopid *Hippotragus leucophaeus* (= blue buck)

blesbok the *S. African* alcelaphine antilopid *Damaliscus albifrons* (*cf.* korrigum, topi, sassaby, bontebok)

bontebok alcelaphine antilopid *Damaliscus pygargus* (*cf.* korrigum, topi, sassaby, and blesbok)

boschbok the harnessed antelope *Tragelaphus sylvaticus*

duikerbok any of several species of antelope of the genus Cephalophus

gemsbok the hippotragine antelope *Oryx gazella*
bastard gemsbok the antelope *Egscerus aquinus*

grysbok the neotragine antilopid *Raphiceros melanotus* (*cf.* steinbok) also any of several species of antelope of the genus Raphiceros (*cf.* steinbok)

springbok the gazelle-like antelope *Antidorcas euchore*

steinbok any of several species of antelope of the genus Raphiceros (*cf.* grysbok)

bokmakierie = backbackiri

bokombouli the lemur *Hapalemur griseus* (*cf.* gentle lemur)

-bol- *comb. form* originally meaning "throw" but extended first to mean any kind of "movement" and later "change". The *biol.* compounds are very confused because the substantive terminations -boly and -bolism and the alternative adjectival terminations -bolous and -bolic, have all developed distinct meanings

bolarous brick red

Boldoeae a tribe of the family Nyctaginaceae

bole = tree trunk

xeriobole a plant the seeds of which are dispersed through the drying up of the carpel

Boletaceae a family of basidiomycete fungi containing the forms usually known in English as boletus or, quite frequently, by their French name *cèpe*. They are also known as fleshy-pore mushrooms in virtue of the fact that the head, unlike that of the Polyporacea, is soft and decomposes readily

bolete anglicized form of boletus. A common genus of boletacean basidiomycete fungi

-bolic (*see also* -bolous, -bolism, -boly) *adj. suffixes* from -bol- (*q.v.*) limited in *biol.* to the meaning "change"

anabolic pertaining to those metabolic activities which are concerned with synthesis

amphibolic said of a toe, or animal possessing a toe, that can be reversed at will to point either forwards or backwards

catabolic pertaining to those metabolic activities that are concerned with the breakdown of large molecules to smaller ones

metabolic pertaining to the sum total of the physiological activity of living matter

ametabolic said of plants having shoots that die after the production of fruiting organs

photometabolic pertaining to the utilization in biosynthesis of photon energy

-bolism (*see also* -bolic, -bolous, -boly) *subst. suffix* pertaining either to the adjectival terminations -bolic, or -bolus even though these differ in meaning. Thus metabolic and metabolous have, by usage, developed quite different meanings but both refer to "metabolism" (*i.e.* a "condition of change"). To avoid repetitive definitions, only the adjectival forms are given

Bolitophilidae a family of nematoceran dipteran insects now usually included with the Mycetophilidae

boll (*see also* weevil, worm) a dry pericarp, particularly that of the cotton plant

-bolous (*see also* -bolic, -boly, -bolism) adjectival form from -bol (*q.v.*) but, unlike bolic (*q.v.*), limited in *biol.* to "change" in the sense of metamorphosis

metabolous pertaining to metamorphosis

ametabolous said of those wingless insects the eggs of which hatch into a nymph, differing from the adult principally in size, and in the life history of which there is thus no metamorphosis

hemimetabolous in its broad sense it is said of any insect that is not holometabolic. In its restricted sense (e.g. contrasted with ametabolous and paurometabolous metamorphoses) it is said of those insects in which the egg hatches into a naiad, which, though differing markedly from the adult, does not pass through a pupal stage

holometabolous said of insects, though applicable to other forms, in which the eggs, larva, pupa and imago are morphologically distinct from each other

paurometabolous is said of those insects in which the egg hatches into a nymph living in the same environment as the adult, but which usually lacks the wings found in the adult animal

bolt 1 (*see also* bolt 2) the elongate cylindrical missile fired by a cross bow and hence any animal or plant which resembles this

adder bolt *obs.* for dragonfly

hackbolt *obs.* any of numerous alcid birds

hagbolt = hackbolt

bolt 2 (*see also* bolt 1) any of several ranunculaceous flowers, particularly of the genus Trollius

bolus a globose mass of chewed, undigested food, particularly that found in the rumen of artiodactyls

-boly (*see also* -bolism, -bolic, -bolous) *subst. suffix* derived from -bol (*q.v.*) with no apparent purpose except to avoid -bolism when this, in combination with the desired prefix (e.g. ana-, meta-) is preempted

anaboly the type of development in which the ancestral pattern is followed up to the last ancestral ontogenetic stage, to which further new stages are then added

emboly the formation of the endoderm of a gastrula by invagination

epiboly the process of gastrula formation by which the cells of the animal pole grow down over the cells of the vegetal pole which become enclosed as the endoderm

metaboly a change in shape of an ameboid, or similar, cell or organism

Bolyerunae a subfamily of aberrant boid snakes known only from Mauritius. Distinguished by the absence of a pelvic girdle and hind limb rudiments

sea-bomb *N.Z.* the phaeophyte alga *Adenocystis utricularis*

Bombacaceae that family of malvalous dicotyledons which contains *inter alia* the baobab. The family may be distinguished from the closely allied Malvaceae by the many-celled anthers

Bombidae a family of apoid apocritan hymenopteran insects usually included with the Apidae

Bombycidae a monospecific family of bombycoid lepidopteran insects erected to contain the silkworm moth *Bombyx mori*

Bombycillidae that family of passeriform birds containing the waxwings. They are distinguished by their long pointed wings, short legs, and prominent crests and by the red or yellow waxy tipped wing feathers

bombycinous silky

Bombycoidea a superfamily of lepidopteran insects noted for their ability to produce silk and containing not only the silkworm moths but also the tent caterpillars

Bombyliidae a family of orthoraphous brachyceran dipteran flies with stout hairy bodies and the general appearance of a bee or bumblebee. Most have spotted or blotched wings that are held out-stretched when at rest. They are often seen hovering over flowers and are usually called bee flies

-bome- *comb. form* meaning a "prominence"

nemathybome a pit containing nematocysts usually occuring in vertical rows on the body wall of some Zoantharia

bone 1 (*see also* bone 2, 3, 4, beetle, set) a calcified connective tissue forming the principal portion of the skeleton of most vertebrates

cancellous bone bony tissue in which the trabeculae are joined as in a scaffolding

cartilage bone a bone formed by the replacement of cartilage

chondral bone a bone that arises in or around cartilaginous precursors

endochondral bone one in which ossification commences in the center of the cartilage

perichondral bone one in which the first ossification takes place between the perichondrium and the cartilage

compact bone a bone in which the bony structure is predominant as distinct from cancellous bone

dermal bones bones that ossify directly in connective tissue masses

lamellar bone a bone consisting of parallel layers

membrane bone a bone formed in a membranous area

spongy bone lamellar bone lacking Haversian canals but with a system of interconnecting, blood-filled, lacunae

trabecular bone lamellar bone in which the bone occurs as strands or pillars interconnected across lacunar spaces

bone 2 (*see also* bone 1, 3, 4) a particular mass of bone

acromial bone = clavicle

angular bone one of a pair of dermal bones forming that surface of the inner angle of the lower jaw between the dentary and the prearticular, in many vertebrates other than mammals in which it is modified into the tympanic bone

multangular bones those carpals which lie under the first and second metacarpals.

The **greater multangular bone** lies under the first metacarpal between the radius and the navicular. The **lesser multangular bone** lies under the first metacarpal

supraangular bone one of a pair of dermal bones lying at the posterior end of the lower jaw, between the coronoid and angular bones in many vertebrates other than mammals in which it forms the main part of the malleus

articular bone one of a pair of membrane bones lying at the inner posterior angle of the lower jaw, in vertebrates other than mammals

prearticular bone one of a pair of dermal bones in the posterior region of the lower jaw, immediately ventral to the articular and posterior to the angular, found in many vertebrates other than mammals in which it forms the anterior process of the malleus

retroarticular bone one of a small pair of chondral bones at the posterior medial corner of the lower jaw of actinopterygian fishes

astragalus bone that tarsal bone which lies at the base of the tibia

basale bone *see* pterygiophore bone

cannon bone the elongate metacarpal metatarsal in the leg of perissodactyle mammals, particularly the horse

capitate bone that carpal which lies at the base of the third metacarpal between this and the lunate bone

carpal bone one of a series of small bones lying between the radius and ulna and the metacarpals. They are usually called wrist bones

metacarpal bone one of those bones in the hand which lie between the phalanges and carpals

chevron bone one of the small V-shaped bones that replace the hemal arch in higher vertebrates

clavicle bone a membrane bone in the pectoral girdle of mammals that articulates with the scapula at one end and usually lies free at the other. In crossopterygian fish it is jointed to the lower end of the cleithrum

interclavicle bone a bone jointed to, and between, the

two clavicles in some fishes, amphibians, reptiles and monotremes

cleithrum bone   a dermal bone of the pectoral girdle lying alongside, and jointed to, the chondral bones of that girdle in some bony fishes, amphibia and reptiles. In osteichthyian fishes this bone lies between the supercleithrum above and the clavicle below

postcleithrum bone   one of a pair of bones in the pectoral girdle of osteichthyian fishes, lying between the supracleithrum and the cleithrum

supracleithrum bone   one of a pair of bony elements in the pectoral girdle of most bony fish, lying between the posttemporal and the postcleithrum

coffin bone   that part of the terminal bony structure of the leg of a perissodactyl or artiodactyl mammal which is enclosed by the hoof

coracoid bone   a chondral bone of the pectoral girdle articulating with the scapula and contributing to the glenoid cavity

procoracoid bone   a bone lying ventro-medial to the coracoid in the skeleton of some reptiles and monotremes, identified as perforated by a foramen, this is the only coracoid bone in the bird. The "coracoid" amphibians may represent the coracoid and precoracoid of later forms

coronoid bone   one of a pair, or several pairs, of dermal bones, found in the lower jaws of many vertebrates other than mammals, this tooth bearing bone, or series of bones, lies medial to the dentary and anterior to the prearticular

cuboid bone   a tarsal bone lying at the base of the fourth and fifth metatarsal

cuneiform bone   one of three tarsal bones of the mammal.

The **medial cuneiform bone**   lies immediately at the base of the first metatarsal,

the **intermediate cuneiform bone**   at the base of the second, and the

lateral cuneiform bone   at the base of the third, adjacent to the cuboid

epimeral bone   one of a series of dorso-lateral frequently "Y"-shaped bones, lying between the somites of a fish

epural bone   expanded, flattened neural spines at the posterior end of the vertebral column in fish

ethmoid bone   a chondral bone of the skull forming the nasal part of the chondrocranium

lateral ethmoid bone = preethmoid bone

parethmoid bone   a small chondral bone in the floor of the nasal capsule and articulating with the medial process of the maxilla

preethmoid bone   one of a pair of chondral bones in the anterior wall of the orbit of fishes

sphenethmoid bone   a bone extending from the nasal region into the interorbital region in primitive amphibians and reptiles

fibula bone   the smaller of the two bones (the other is the tibia bone) that lie between the knee and ankle

fibulare bone = calcaneus

frontal bone   a membrane bone forming the anterior region of the roof of the skull, and therefore, lying between the parietal and the nasal bones

postfrontal bone   a small bone at the posterior dorsal edge of the orbit, lying across the anterior end of the postorbital bone and usually running a short distance parallel to the frontal bone, in many vertebrates other than mammals

prefrontal bone   one of a pair of membrane bones

lying in the anter-dorsal of the orbit between the frontal, nasal, and maxillary bones, in many vertebrates other than mammals

entoglossal bone   a tooth-bearing prolongation of the basihyal bone in some fish and a tongue-supporting bone in birds, distinct from the basihyal

gular bone   one of a series of chondral bones, associated with the hyoid arch and which form the bony support of the floor of the mouth in some fishes. In crossopterygians the central pair are the **lateral gular bones**, with the **marginal gular bones** outside them. They come together in front in the unpaired **anterior medial gular bone**

hamate bone   that carpal which lies at the base of the fourth metacarpal bone alongside the capitate bone

heterotopic bone   a bone that is not a part of either the axial or the appendicular skeletons

humerus bone   the bone of that part of the forelimb that lies between the shoulder and the elbow

hyal bone   any bone derived from the hyoid arch

stylohyal bone   that part of the ossified hyoid arch which, in mammals, is not fused to the tympanic bulla

tympanohyal bone   that part of the tympanic bulla which is formed from the hyoid arch

urohyal bone   a derivative of the hyoid arch, lying anterior to and below the copular bone in some fish

hypomeral bone   one of a ventro-lateral series of bones, usually confined to the tail region, corresponding to the dorso-lateral epimeral bones

hypural bone   a series of flattened plates supporting the caudal fin and terminating the vertebral column in actinopterygian fishes. The most posterior plates correspond to the urostyle in other forms

incus bone   the center of the three ossicles that, in mammals, conduct vibration from the tympanum to the inner ear (*cf.* malleus, stapes) and the homolog of the quadrate of lower forms

innominate bone   that part of the pelvic girdle which is formed by the fusion of the illium, ischium and pubis

intercaloid bone   one of a pair of dermal bones overlying the posterior pair of otic capsules. It serves for the ligament of attachment of the pectoral girdle in actinopterygian fishes

jugal bone   one of a pair of membrane bones of the skull that, as part of the zygomatic arch, encloses, or partly encloses, the orbit. In many forms, the anterior end sutures with the maxilla and lachrymal bones

quadratojugal bone   one of a pair of membrane bones lying in the area between the quadrate and the jugal bones. This bone is lost in lizards and mammals

lacrimal bone   one of a pair of membrane bones in the skull that lies in the antero-ventral part of the orbit, usually penetrated by the lacrimal duct. In crossopterygian fishes, the lacrimal bone forms part of the second orbital skeleton

lunate bone   that carpal between the capitate bone and the junction of the radius and ulna

malleolus bone   a projection from the lower end of the fibula

malleus bone   the innermost of the three ossicles, which conduct vibration from the tympanum to the inner ear in mammals (*cf.* incus, stapes)

hyomandibular bone   one of a pair of membrane derived from the hyoid arch of actinopterygian fishes, that lies immediately behind the orbital complex and that, together with the symplectic and quadrate bones, forms the attachment for the lower jaw

marsupial bone   a bone in the pelvic girdle of marsupials that extends forward from the anterior margin of the pubis

maxilla bone   a dermal bone of the splanchnocranium forming the main portion of the upper jaw, and the anterior end of the cranium. In mammals it bears all teeth, except the incisors

premaxilla bone   one of a pair of dermal bones of the splanchnocranium lying at the anterior end of the maxilla and bearing in mammals the incisor teeth

septomaxilla bone   a membrane, or mixed, bone in the floor or sidewall of the nasal passages in some reptiles and amphibia

mental bone   a small chondral bone occuring in some vertebrates at the anterior articulation or synthesis of the mandibles

nasal bone   one of a pair of membrane bones in the anterior region of the skull, lying above the nasal cavity, immediately in front of the frontal and dorsal to the ethmoid

internasal bone   one of one or more membrane bones in the dermocranium of crossopterygian fish, lying between, and partly anterior to, the nasals

navicular bone   that tarsal bone which lies between the base of the medial and intermedial cuneiform, and the astragalus (= radiale)

nobelian bone   one of a pair of supporting rods in the intromittent organ of some anuran amphibia

occipital bone   any of several bones lying about the foramen magnum

basioccipital bone   one of a pair of chondral bones of the skull lying immediately below the foramen magnum

exoccipital bone   one of a pair of chondral bones at the posterior region of the skull, lying on each side of the foramen magnum

supraoccipital bone   one of a pair of chondral bones at the posterior end of the skull, lying immediately above the foramen magnum, adjacent to the postparietal, petrosal, and parietal bones

opercular bone   one of the dorsal pair of the two large membrane bones in the operculum of fish

interopercular bone   one of a pair of chondral bones derived from the hyoid arch in actinopterygian fish, lying between the preopercular and the subopercular

preopercular bone   one of a pair of bones in the cheek skeleton of fish, lying immediately anterior to the junction of the opercular and subopercular

subopercular bone   one of a pair of two large bones, lying immediately under the opercular bone

opisthotic bone   a chondral bone of the posterior part of the otic capsule of Reptilia, Aves, and Amphibia (some), probably fused into the exoccipital bone of most Amphibia

postorbital bone   one of a pair of membrane bones lying immediately posterior to the postfrontal and connected with this bone in front, the jugal below, and the squamosal behind, in many vertebrates other than mammals. In some fishes there are several pairs of postorbitals

palatine bone   one of a pair of dermal bones of the splanchnocranium, lying immediately behind the maxilla and forming the posterior part of the secondary palate of mammals

palpebral bone   one in the upper eyelid of some reptiles and birds

parietal bone   one of a pair of membrane bones, forming the posterior portion of the roof of the skull, lying immediately above the squamosal and posterior to the frontal

interparietal bone   = post parietal bone

postparietal bone   a membrane bone of the rear end of the skull lying immediately above the supraoccipital and between, or behind, the two parietal bones

patella bone   the cup-shaped bone which protects the knee (= knee cap)

petrosal bone   one of a pair of chondral bones of the posterior end of the skull enclosing the inner ear of mammals

pisiform bone   that bone which lies beneath, and usually laterally to, the triangular bone, between the fifth metacarpal and the ulna bone

prepollex bone   the small bone lying between the outer margin of the radius and the outer margin of the navicular bone in some vertebrates and in some mammals

prootic bone   one of a pair of chondral bones forming the anterior part of the inner ear capsule in most vertebrates

pterygiophore bone   one of a group of bones lying immediately above the spinal cord in fishes, and supporting the fin rays. A pterygiophore may be a single bone, in which case it is most commonly called a basale, or composed of distal, middle and proximal portions, in which case either the distal portion alone, or sometimes, all three portions together are known as the **basale bone**

pterygoid bone   a pair of dermal bones of the splanchnocranium. It lies between the palatine bones and forms the lateral walls of the nasal passage in mammals

basipterygoid bone   a bone in the monotreme skull corresponding in position to, and often identified with, the alisphenoid bone

ectopterygoid bone   one of a pair of dermal bones in the palatoquadrate complex of osteichthyan fishes, and some amphibia, lying immediately outside and below the metapterygoid bone

epipterygoid bone   one of a pair of membrane bones extending upwards from the dorsal side of the pterygoid bone to the wall of the cranium in vertebrates other than mammals. It is the homolog of the alisphenoid bone of mammals

metapterygoid bone   one of a pair of bones forming part of the palatoquadrate complex of actinopterygian fishes, lying slightly posterior and dorsal to the pterygoid bone and thought by many to be homologous with the epipterygoid

quadrate bone   one of a pair of chondral bones found at the posterior outer angle of the cranium, in vertebrates other than mammals. It is the homolog of the incus bone of mammals

radiale bone   that carpal which lies next to the radius (= navicular bone)

radius bone   that one of the two bones of the forearm which moves along the arc of a circle when the limb is moved from the pronate to the supinate position

rostral bone   one of a series of dermal bones in the snout skeleton of crossopterygian fishes, lying at the anterior end, below the septomaxilla bone and carrying the infraorbital sensory line

scapula bone   (*see also* scapula) a pair, or series, of dermal bones carrying the occipital transverse sensory canal in bony fishes. These bones lie behind the post parietal bone

sesamoid bone one or more bones in the foot of some mammals, lying below the metatarsal bones and phalanges or the patella, formed in relationship with muscle tendons. A kind of heterotopic bone

sphenoid bone any of several chondral bones forming the base, and part of the lower sides, of the cranium

alisphenoid bone one of a pair of bones of mixed origin, in the splanchnocranium, lying behind and above the orbitosphenoid, and immediately below the lateral edge of the parietal, bone

basisphenoid bone one of a pair of chondral bones on the base of the skull immediately anterior to the basioccipital bones

ethmosphenoid bone a general term for the anterior region of the endocranium in fish

latero-sphenoid bone one of a pair of bones lying, in some vertebrates (crocodilians and birds) immediately on each side of the basisphenoid bones

orbitosphenoid bone one of a pair of chondral bones of the skull lying in the orbit anterior to the basisphenoid bone, perforated often by the optic nerve

parasphenoid bone a dermal bone forming the floor of the cranium of amphibia and the roof of the mouth cavity in fishes. Reduced and fused with the basisphenoid bone in reptiles and birds, a vestige or absent in mammals

presphenoid bone the midline fused area of the orbitosphenoid bones as observed in mammals

postspiracular bone one of a pair of membrane bones in the dermocranium of crossopterygian fishes, lying immediately dorsal to the junction of the opercle and squamosal and immediately below the junction of the tabular and extrascapular bones

splenial bone one of a pair of chondral bones lying along the medial face of the dentary in some amphibians and reptiles. Two pairs of splenials are seen in sarcopterygian fishes along the ventral margin of the mandible

dentosplenial bone a compound bone replacing the dentary in the lower jaw of actinopterygian fishes

squamosal bone one of a pair of membrane bones at the posterior side of the skull, immediately below the parietal bone and from which arises the zygomatic arch that fuses with the jugal

stapes bone the outermost of the three ossicles that conduct vibration from the tympanum to the inner ear (*cf.* malleus, incus)

Sylvius's bone the lenticular process of the malleolus

symplectic bone one of a pair of chondral bones derived from the hyoid arch (epihyal) of actinopterygian fish, lying immediately ventral to the hyomandibular bone

tabular bone one of a pair of membrane bones in the dermal cranium of crossopterygian fishes and early tetrapods. They lie immediately on each side of the postparietal bones, immediately behind the supratemporal bones

supratemporotabular bone one of a pair of membrane bones in the dermocranium of actinopterygian fishes corresponding to the supratemporal and tabular bones of crossopterygian fishes

tarsal bone one of those bones in the vertebrate leg which lies between the metatarsal bones and the tibia. Commonly called ankle bones

metatarsal bone one of those bones of the verte-brate foot which lie between the tarsals and the phalanges

tarsale bones a group name for those tarsal bones (cuneiform and cuboid) which lie at the base of the metatarsal bones

temporal bone a compound bone, or group of bones, in the cranium of some mammals

intertemporal bone one of a pair of membrane bones in the dermo-cranium of fishes, lying on each side posterior to the orbit and immediately anterior to the supratemporal bones

postemporal bone one of a pair of bones of the pectoral girdle of bony fishes articulating anteriorly with the endocranium. Posteriorly it is attached to the supracleithrum bone

supratemporal bone one of a pair of membrane bones lying at the posterior junction of squamosal and parietal bones in many vertebrates other than mammals

tibia bone the larger of the two bones (the other is the fibula) which lie between the knee and the ankle

triangular bone that carpal which lies, together with the pisiforme bone beneath it, opposite the head of the ulna bone

turbinal bone one of several rolled sheets of bone nearly filling the nasal passage

tympanic bone a dermal bone of the splanchnocranium that forms the outer part of the tympanic bulla of the mammal (*cf.* tympanic bulla). It is the homolog of the angular bone of reptiles and amphibians

ulna bone that one of the bones of the forearm which does not pass through the arc of a circle when the hand is rotated from the pronate to the supinate position

ulnare bone = triangular bone

vomer bone one of a pair of cranial bones lying in the floor of the cranium immediately anterior to the parasphenoid or orbitosphenoid bones. In forms lacking a palate it consitutes part of the roof of the mouth and bears teeth

bone 3 (*see also* bone 1, 2, 4) a skeletal tissue called a bone though not composed of bone

cuttle bone the internal calcified shell of the cuttle fish

whale bone = baleen

bone 4 (*see also* bone 1, 2, 3) structures or organisms the name of which is compounded from bone

devil's bones the tubers of the dioscoreacean herb *Dioscorea paniculata*

green bone *Brit.* the belonid fish *Rhamphistoma belone*

boneset the compositaceous herb *Eupatorium perfoliatum*

bongo 1 (*see also* bongo 2) the bombacaceous tree *Cavanillesia platanifolia*

bongo 2 (*see also* bongo 1) the strepsicerosine bovine bifid artiodactyle *Boocercus eurycerus* of African forests

bonitation the condition of well-being of a species as indicated by the numerical development of a population

bonito any of several species of scombrid fishes of the genus Sarda

Atlantic bonito *S. sarda*

Pacific bonito *S. chiliensis*

striped bonito *S. orientalis*

bonnet any headdress with a brim and hence any plant or animal resembling this. Particularly any of numerous species of cassidid mollusks of the genus Cypraecassis (*cf.* helmet shell)

alligator bonnet *U.S.* water lily, particularly Nymphaea

blue bonnet *Austral.* the psittacid bird *Psephotus hoematogaster* (= oak parrot) *U.S.* any of several species of leguminous herb of the genus Lupinus **Texas blue bonnet** *L. texenensis*

bonxie stercorariid bird *Catharacta skua* (= skua)

boobook *N.Z.* the strigiform bird *Ninox novae-seelandiae*

booby properly any of several species of sulid bird of the genus Sula, but in *W.I.* used for tern and also for the larid *Anous stolidus* (= brown noddy). *Austral.*, without qualification, *S. leucogaster*
brown booby *S. leucogaster*
blue-faced booby *S. dactylatra*
blue-footed booby *S. nebouxii*
red-footed booby *S. sula*
dry land booby *W.I.* the procellariid bird *Pterodroma hasitata* (= black-capped petrel)

boodie *Austral.* popular name of rat-kangaroos of the genus Bettongia

Booidea a superfamily of Serpentes containing the very large constrictor snakes and a few other constrictors

book (*see also* lung) any organ consisting of leaves bound together and thus resembling a book
gill book a series of leaf-like pads functioning as gills found in marine arachnids

boomslang a venomous African snake *Dispholidus typus*

boot *zool* an undivided tarsal sheath in birds. *bot* the large sheathing leaf base of arecoid palms
devil's boot any of several species of sarraceniaceous insectivorous herbs of the genus Sarracenia

Bopyroidea = Epicaridea

borage *Brit.* the boraginaceous herb *Borago officinalis*

Boraginaceae that family of tubiflorous dicotyledons which contains, *inter alia*, the borages, the hounds' tongues, the gromwells and the heliotropes. The circinate inflorescence and two-carpeled, four-celled ovary are distinctive

Boragineae a suborder of tubifloral dicotyledonous angiosperms containing the families Boraginaceae and Hydrophyllaceae

Boraginoideae a subfamily of the Boraginaceae

Borboridae = Sphaeroceridae

[**bordeaux**] a mutant (gene symbol *bo*) mapped at 12.5 on the X chromosome of *Drosophila melanogaster*. The term refers to the wine-red eye color

border 1 (*see also* border 2) a conspicuous edge to a flat surface
brush border the appearance of epithelial cells in which microvilli are of a size visible by electron microscopy
red border striations along the distal tip of the sensory cells in the eyes of some Turbellia

border 2 (*see also* border1) any of several lepidopteran insects
broad-border *Brit.* the noctuid *Triphaena ianthina*
dotted border *S.Afr.* any of several species of pierids of the genus Mylothris

-borea- *comb. form* meaning "northwind"
hyperborean northern

Boreidae a family of mecopteran insects commonly distinguished as the snow scorpion fllies, by reason of the frequency that they are found in this environment

borer 1 (*see also* borer 2-4) any of many coleopteran insects that bore holes
branded alder borer *U.S.* the cerambycid *Rosalia funebris*
round headed apple tree borer *U.S.* the cerambycid *Saperda candida*
red-headed ash borer *U.S.* (*see also* borer 2) the cerambycid *Neoclytis acuminatus*
ponderosa-pine bark borer (*see also* borer 2) *U.S.* the cerambycid *Acanthocinus spectabilis*
bronze birch borer *U.S.* the buprestid *Agrilus anxius*
lead cable borer *U.S.* the bostrichid *Scobicia declivis* (= short circuit beetle)
raspberry cane borer *U.S.* the cerambycid *Oberea bimaculata*
red-necked cane borer *U.S.* the buprestid *Agrilus ruficollis*
western cedar borer *U.S.* the buprestid *Trachykele blondeli*
flatheaded cone borer *U.S.* the buprestid *Chrysophana placida*
round-headed cone borer *U.S.* the cerambycid *Paratimia conicola*
red necked cane borer (*see also* borer 2) *U.S.* the buprestid *Agrilus ruficollis*
strawberry crown borer (*see also* borer 2) *U.S.* the curculionid *Tyloderma fragariae*
elm borer *U.S.* the cerambycid *Saperda tridentate*
flatheaded fir borer *U.S.* the buprestid *Melanophila drummondi*
roundheaded fir borer *U.S.* the cerambycid *Tetropium abietis*
lesser grain borer *U.S.* the bostrichid *Rhyzopertha dominica*
California flatheaded borer *U.S.* the buprestid *Melanophila californica*
Pacific flatheaded borer *U.S.* the buprestid *Chrysobothis mali*
hemlock borer *U.S.* the buprestid *Melanophila fulvoguttata*
banded hickory borer *U.S.* the cerambycid *Cerasphorus cinctus*
painted hickory borer *U.S.* the cerambycid *Megacyllene caryae*
pin-hole borer any of several platypodids
shot-hole borer *U.S.* the scolytid *Scolytus rugulosus*
larger shot-hole borer *U.S.* the scolytid *Scolytus mali*
old-house borer *U.S.* the cerambycid *Hylotrupes bajulus*
spotted limb borer *U.S.* the bostrichid *Psoa maculata*
linden borer *U.S.* the cerambycid *Saperda vestita*
locust borer *U.S.* the cerambycid *Megacyllene robiniae*
gall-making maple borer (*see also* borer 2) *U.S.* the cerambycid *Xylotrechus aceris*
sugar-maple borer *U.S.* the cerambycid *Glycobius speciosus*
round head mesquite borer *U.S.* the cerambycid *Megacyllene antennata*
two-lined chestnut borer *U.S.* the buprestid *Agrilus bilineatus*
Australian pine borer *U.S.* the buprestid *Chrysobothris tranquebariaca*
black-horned pine borer *U.S.* the cerambycid *Callidium antennatum*
sculptured pine borer *U.S.* the buprestid *Chalcophora angulicollis*

poplar borer	*U.S.* the cerambycid *Saperda calcarata*
  bronze poplar borer	*U.S.* the buprestid *Agrilus liragus*
banana root borer	*U.S.* the curculionid *Cosmopolitus sordidus*
  clover root borer	*U.S.* the scolytid *Hylastinus obscurus*
  broad necked root borer	*U.S.* the cerambycid *Prionus laticollis*
rustic borer	*U.S.* the cerambycid *Xylotrechus colonus*
oak sapling borer	*U.S.* the cerambycid *Goes tesselatus*
saskatoon borer	*U.S.* the cerambycid *Saperda bipunctata*
potato stalk borer	(*see also* borer 2) *U.S.* the curculionid *Trichobaris trinotata*
  tobacco stalk borer	*U.S.* the curculionid *Trichobaris mucorea*
clover stem borer	(*see also* borer 2) *U.S.* the languriid *Languria mozardi*
currant tip borer	*U.S.* the cerambycid *Psenocerus supernotatus*
flat headed apple tree borer	(*see also* borer 2) *U.S.* the larva of the buprestid *Chrysobothris femorata*
  round head apple tree borer	*U.S.* the cerambycid *Saperda candida*
  spotted apple tree borer	*U.S.* the cerambycid *Saperda cretata*
  cedar tree borer	*U.S.* the cerambycid *Semanotus ligneus*
  sinuate pear tree borer	*U.S.* the buprestid Agrilus *sinuatus*
grape trunk borer	*U.S.* the cerambycid *Cerasphorus albofasciatus*
apple twig borer	(*see also* borer 2) *U.S.* the bostrichid *Amphicerus bicaudatus*
  dogwood twig borer	*U.S.* the cerambycid *Oberea tripunctata*
wharf borer	*U.S.* the oedemerid *Nacerdes melanura*
poplar-and-willow borer	*U.S.* the curculionid *Sternochetus lapathi*
cottonwood borer	(*see also* borer 2) the cerambycid *Plectrodera scalator*
  metallic woodborer	*U.S.* any of several buprestids
borer 2	(*see also* borer 1, 3, 4) any of numerous lepidopteran insects the larvae of which bore holes
ash borer	*U.S.* (*see also* borer 1) the aegeriid *Podosesia syringae*
apple bark borer	(*see also* borer 1) *U.S.* the aegeriid *Thamnosphecia pyri*
blackberry-borer	*U.S.* the aegeriid *Bembecia marginata* (= raspberry crown borer)
strawberry-borer	*U.S.* the aegeriid *Synanthedon rutilans*
bidens borer	*U.S.* the olethreutid *Epiblema otiosanum*
blue cactus borer	*U.S.* the phycitid *Melitara dentata*
maple callus borer	*U.S.* the aegeriid *Sylvora acerni*
sugarcane borer	*U.S.* the crambid *Diatraea saccharalis*
Columbine borer	*U.S.* the noctuid *Papaipema purpurifascia*
European corn borer	(*see also* borer 1) the pyralid *Ostrinia nubilalis*
southwestern corn borer	*U.S.* the crambid *Zeadiatraea grandiosella*
sugar beet crown borer	(*see also* borer 1) *U.S.* the phycitid *Hulstia undulatella*
  raspberry crown borer	*U.S. Bembecia marginata*

currant borer	*U.S.* the aegeriid *Ramosia tipuliformis*
burdock borer	*U.S.* the noctuid *Papaipema cataphracta*
pear fruti borer	*U.S.* the phycitid *Nephopteryx rubizonella*
iris borer	*U.S.* the noctuid *Macronoctua onusta*
lilac borer	*U.S.* the aegeriid *Podosesia syringae*
maple borer	(*see also* borer 1) *U.S.* the aegeriid *Synanthedon acerni*
pitch mass borer	*U.S.* the aegeriid *Vespamima pini*
peach borer	*U.S.* the aegeriid *Sanninoidea exitiosa*
persimmon borer	*U.S.* the aegeriid *Sannina uroceriformis*
American plum borer	*U.S.* the phycitid *Euzophera semifuneralis*
lima-bean pod borer	*U.S.* the phycitid *Etiella zinckenella*
Caribbean pod borer	*U.S.* the phycitid *Fundella pellucens*
pumpkin borer = squash borer
rhododendron borer	*U.S.* the aegeriid *Ramosia rhododendri*
elder shoot borer	*U.S.* the noctuid *Achatodes zeae*
Asiatic rice borer	*U.S.* the crambid *Chilo suppressalis*
graperoot borer	*U.S.* the aegeriid *Vitacea polistiformis*
cotton square borer	*U.S.* the lycaenid *Strymon melinus*
squash borer	the aegeriid *Melittia cucurbitae*
stalk borer	(*see also* borer 1) *U.S.* the noctuid *Papaipema nebris*
  southern cornstalk borer	*U.S.* the crambid *Diatraea crambidoides*
  lesser cornstalk borer	*U.S.* the phycitid *Elasmopalpus lignosellus*
  lined stalk borer	*U.S.* noctuid *Oligia fractilinea*
  rice stalk borer	*U.S.* the crambid *Chilo plejadelus*
potato stem borer	(*see also* borer 1) *U.S.* the noctuid *Hydroecia micacea*
peach tree borer	(*see also* borer 1) *U.S.* the aegeriid *Sanninoidea existiosa*
  lesser peach tree borer	*U.S. Synanthedon pictipes*
  western peach tree borer	*U.S.* the aegeriid *Sanninoidea exitiosa*
boxelder twig borer	(*see also* borer 1) *U.S.* the olethreutid *Proteoteras willingana*
locust twig borer	*U.S.* the olethreutid *Ecdytoeopha insiticiana*
peach twig borer	*U.S.* the gelechiid *Anarsia lineatella*
lima-bean vine borer	*U.S.* the phycitid *Monoptilota pergratialis*
  sweetpotato vine borer	*U.S.* the pyraustid *Omphisa anastomosalis*
  squash vine borer	*U.S.* the aegeriid *Melittia curcurbitae*
ragweed borer	*U.S.* the olethreutid *Epiblema strenuanum*
  smartweed borer	*U.S.* the pyraustid *Pyrausta ainsliei*
dogwood borer	(*see also* borer 1) *U.S.* the aegeriid *Thamnosphecia scitula*
borer 3	(*see also* borer 1,2,4) a few boring insects other than Coleoptera and Lepidoptera
maple petiole borer	*U.S.* the larva of the tenthredinid hymenopteran *Caulocampus acericaulis*
borer 4	(*see also* borer 1–3), drill 1) a few boring animals other than insects
rock borer	popular name of saxicavid pelecypods

Boselaphinae a subfamily of bovid artiodactyl mammals containing the nilghai and the chousingha. They are sometimes called the "deer-oxen"

boss any rounded protuberance, specifically that portion of the echinoderm tubercle which bears the mamelon (*q.v.*) at its tip

-bostrych- *comb. form* meaning "a ringlet"

Bostrichidae a family of wood-boring coleopteran insects closely allied to the Anobiidae but distinguished from them by the enlarged thorax that completely conceals the deflexed head. They are called powder-post beetles from their habit of reducing dried timber to powder

Bostrichoidea a superfamily of polyphagous coleopteran insects containing a number of families of wood-boring beetles

Bostrychidae = Bostrichidae

bostryx a bostrychoid cyme (*q.v.*) in which one of the branches is totally suppressed

Botallo's (usually misspelled Botalli's) *epon. adj.* from Leonardo Botallo (1530–?)

Botallus's occasional misspelling of Botallo's

Bothidae a family of small oval heterostomate osteichthyes in which both eyes are found on the left side of the body and which are therefore usually called "left-eye" flounders. The "right-eye" flounders are the Pleuronectidae (*q.v.*).

-bothr- *comb. form* meaning "hole"

bothridium leaf-like thin flexible structures on the scolex of a tapeworm, (*cf.* bothrium)

Bothriocephaloidea = Pseudophyllida

bothrium a pair of elongated sucking grooves on the scolex of a tapeworm (*cf.* bothridium)

trichobothrium tactile setae or tufts of setae found on some scorpions and hemipteran insects

-bothro- *comb. form* meaning a "hole" or "groove"

-botry- *comb. form* meaning a "bunch of grapes"

botryoidal in the form of a bunch of grapes

bottle in compound names of organisms thought to resemble a bottle in shape, texture, color or function

blue bottle 1 (*see also* blue bottle 2–4) any of several calliphorid dipterans with a blue metallic body

blue bottle 2 (*see also* blue bottle 1,3,4) any of numerous blue plants including the grape hyacinth and the cornflower

blue bottle 3 (*see also* blue bottle 1, 2, 4) *Austral.* any of several species of siphonophoran hydrozoan coelenterate of the genus Physalia (= Portugese man-of-war)

blue bottle 4 (*see also* blue bottle 1–3) *I.P.* any of several species of papilionid lepidopteran insects, usually without qualification, *Zetides sarpedon*

coke bottle *U.S.* the siphonophoran hydrozoan colenterate *Physalia pelagica* (*cf.* bluebottle 3)

green bottle any of several calliphorid dipterans with a green metallic body

scent bottle *U.S.* the orchid *Habenaria dilata*

bottom used as a colloquial synonym for both abdomen and buttock

sulphur bottom = blue whale

-botul- *comb. form.* meaning "sausage"

boubou any of several laniid birds of the genus Laniarius, usually, without qualification, *L. ferrugineus*

slate-colored boubou *L. funebris*

sooty boubou *L. leucorhynchus*

bouncing bet the caryophyllaceous herb *Saponaria officinalis*

bourrelet a conspicuous swelling on the interambulacral plates where they meet the peristome in some echinoid echinoderms

bouton a terminal plate of the labium in Hymenoptera

boutu the amazonian platanistid cetacean *Inia geoffrensis*

Bovidae that family of artiodactyl mammals which contains the oxen, goats and antelopes. They are distinguished by hollow, unbranched, non-deciduous horns. By some, the family is restricted to oxen, in which case *see also* Capridae and Antilopidae

Bovinae that sub-family of bovid artiodactyl mammals which contains the true oxen and cattle. They are principally distinguished from the other sub-families by the diverging horns

[bowed] a mutant (gene symbol *bod*) plotted at 48.3 on chromosome III of *Drosophila melanogaster*. The wings of the phenotype are arched

Bowenioideae a subfamily of Cycadiacea containing the single genus from which the name is derived

virgin's bower (*see also* bird) any of many species of woody shrub of the ranunculaceous genus, Clematis particularly *C. virginiana* (= devil's needle)

Bowman's *epon. adj.* from William Bowman (1816–1892)

box 1 (*see also* box 2, berry, thorn, tree) properly any of several shrubs or trees of the genus Buxus, and the timber derived from them. The term is also applied to numerous shrubs and trees of similar aspect, or to those having fruits in which the seeds rattle

apple box *Austral.* any of several trees of the genus Eucalyptus

bastard box the myrtaceous tree *Eucalyptus goniocalyx*

Brisbane box the myrtaceous shrub *Tristania conferta*

Australian grey box the myrtaceous tree *Eucalyptus hemiphloia*

poplar box the myrtaceous tree *Eucalyptus populifolia*

rattle box any of several species of the leguminous genus *Crotalaria* (= seed box)

devil's rattle box the seed pods of the caryophyllaceous herb *Silene latifolia* (= bladder campion)

running box the rubiaceous herb *Mitchella repens*

seed box the onagraceous herb *Ludwigia alternifolia*

devil's snuff box (= puff ball)

victorian box the pittosporaceous tree *Pittosporum undulatum*

yellow box the myrtaceous tree *Eucalyptus melliodora*

box 2 (*see also* box 1, turtle) a few animals with box shaped bodies, or coverings, but usually as an adjective

jewel box any of several species of chamid mollusks of the genus Chama

boy juvenile male human. Used as a compound in names of several organisms

orange bellboy *S.Afr.* the hesperid lepidopteran insect *Zezonia zeno*

playboy *S.Afr.* any of numerous species of lycaenid lepidopteran insect of the genus Deudorix

plumboy *U.S.* any of several species of the rosaceous genus Rubus particularly those belonging to the subgenus Cylactis

Boyden's *epon. adj.* from Edward Allen Boyden (1886–)

bp *see* [black pupa] or [brown pericarp]

bp *see* [brachypodism]

br *see* [broad], [brown eyes], [brachydactyly] and [brachytic]

brace 1 (*see also* brace 2) two of anything, particularly game birds, trout, and greyhounds running in concert (*cf.* leash)

brace 2   (*see also* brace 1) term for a group of geldings

-brach-   *comb form* meaning "arm" but frequently confused in compounds with -brachy- (*q.v.*)

primibrach   those plates in the arm of a crinoid echinoderm, which lie distal to the branching of the arm

secundibrach   those plates in the arm of a crinoid echinoderm which lie between the first and second fork

tertibrach   those plates in the arm of a crinoid echinoderm which lie between the second and third fork (= palmar)

Brachelytra = Staphylinoidea

brachia   in insects the term is applied either to the aedeagus or to structures closely associated with the aedeagus

brachiate   with widely spreading branches

brachiation   to progress by swinging from hand to hand

brachidium   the internal skeleton of the lophophore in brachiopods

Brachiopoda   a phylum of lophophorian coelomate animals commonly called lampshells. They are distinguished from other lophophorians by the possession of a bivalved shell

brachium   literally "arm" but applied less to these structures in vertebrates than to the arms of such forms as crinoid echinoderms

-brachy-   *comb. form* meaning "short" but frequently confused in compounds with -brach- (*q.v.*)

Brachycera   a suborder of dipteran insects containing those forms with short antennae usually of three, but rarely of five or more, segments (*cf.* Nematocera)

brachychiton   either of two species of stercoulariaceous tree *Sterculia acerfolia* (= flame tree) or *S. diversifolia*

[brachydactyly]   a mutant (gene symbol *br*) at locus 0 on linkage group V of the rabbit. The name is descriptive of the phenotypic expression

Brachygnatha   a division of brachyuran decapod crustacea distinguished by having a square, or at least four-sided, mouth-frame. Almost all well known crabs belong to this group (= Brachyrhyncha)

[brachymacrochaete]   a mutant (gene symbol *brc*) plotted at 0 on the X chromosome of *Drosophila melanogaster*. As the name indicates the phenotypic expression is a reduction in length of the macrochaetes

Brachypodidae = Pycnonotidae

[brachypodism]   a mutant (gene symbol *bp*) found in linkage group V of the mouse. The name is descriptive of the phenotypic expression

Brachypteraciidae   a family of coraciiform birds usually included with the Coraciidae

Brachyrhyncha   a superfamily of brachyuran crustacea with a quadrate mouth frame. Most of the forms commonly called crabs belong in this group (= Brachygnatha)

[brachytic]   *either* a mutant (gene symbol *br*) mapped at 80 on linkage group I of *Zea mays*. The phenotypic stalk has short internodes *or* a mutant (gene symbol *br*) mapped at 0 on chromosome I of the tomato. The phenotypic expression is a short internode

Brachyura   a section of reptant decapod crustacea containing the true crabs, readily distinguished by the great reduction of the abdomen that is flexed under the thorax and by the absence of uropods

[Brachyury]   a mutant (gene symbol *T*) in linkage group IX of the mouse. The phenotypic expression is a short tail

bracken   (*see also* brake) any of several polypodiaceous ferns of the genus Pteridium but partialuly *P. aquilinum*

Braconidae   a large family of apocritan hymenopteran insects closely resembling the Ichneumonidae but differing from them in certain details of wing venation and abdominal segments

bract 1   (*see also* bract 2, 3) a small leaf on a floral stem often closely pressed to the base of the flower and sometimes, as in Poinsettia and Bougainvillea, brilliantly colored and frequently mistaken for a petal

silver-bract   the crassulaceous herb *Cotyledon pachyphytum*

bract 2   (*see also* bract 1,3) a leaf-like protective zooid of a siphonophoran hydrozoan

bract 3   (*see also* bract 1,2) the leaflet-like organs of which the perichaetium of a moss is constructed

bracteody   the condition in which petals are changed into bracts

bracteole   diminutive of bract

bracteose   with conspicuous bracts

-brady-   *comb. form* meaning "slow"

Bradypodidae   that family of xenarthran mammals which contains the sloths. They are distinguished by their arboreal habitat and curious method of up-side-down progression

brain   the large anterior end of the central nervous system. The term was at one time confined to vertebrates but is nowadays often used for the cerebral ganglion of invertebrates, particularly insects

forebrain = prosencephalon

hindbrain = rhombencephalon

midbrain = mesencephalon

brake   any of several species of polypodiaceous fern of the genus Pteridium (= bracken) but sometimes used as synonymous with "fern"

canker-brake   the polypodiaceous fern *Polystichum acrostichoides* (= Christmas fern)

cliff-brake   any of several polypodiaceous ferns of the genus Pellea. The term is also sometimes applied to *Cryptogramma stelleri*

hog brake = pasture brake

pasture brake   any of several species of polypodiaceous fern of the genus Pteridium (= bracken)

rock-brake   any of several polypodiaceous ferns of the genus Cryptogramma

bramble   (*see also* briar, finch, hopper) any rosaceous shrub of the genus Rubus with elongate stems. *Brit.* without qualification, *R. fruticosus*

common bramble   *Brit. R. saxatilis*

brambling   *Brit.* = bramble finch

bran   that part of the coat of a wheat grain which combines the pericarp, the integuments, and the nucellus

-branch-   *comb. form* meaning a "fin" or "gill"

branch 1   (*see also* branch 2) any offshoot leaving a mainstem or trunk

accessory branch   one growing from an accessory bud

embryonic branch   a branch of Chara which breaks off and develops into a new plant

proembryonic branch   a propagative body, like a proembryo, arising from a node of Chara

soredial branch   a branch like structure arising from a lichen due to the development of a soredium that has not become detached

style branch   the free styles of syncarpous flowers in which only the lower portions of the carpel are united

branch 2 (*see also* branch 1) a gill

hemibranch a branchial arch having the gill along one side only

holobranch a gill bar with a row of filaments on each side

marsipobranch a branchial cleft in the form of a pouch. Found only in cyclostomes, which have therefore, been called marsipobranchia

pseudobranch the external gill of a mollusk as contrasted with the true gill or ctenidium. Also a gill-like structure thought to lack respiratory function

[branched silkless] a mutant (gene symbol *bd*) mapped at 96 on linkage group VII of *Zea mays*. The phenotypic expression is branched ears without silks

branching the act or method of producing branches

cymose branching a type of branching in which alternate branches are given off, which grow as fast or faster than the original terminal branch, resulting in a rounded type of growth (*cf.* cyme)

dichasial branching forming two main axes

dichotomic branching the condition in which each branch splits into two at its end to produce a Y-shaped structure each of which itself subsequently produces two branches and so on.

intercalary branching branching from other than the apex

monopodial branching that method of branching in which the main axis dominates the lateral. There is therefore a main stem clearly defined from the lateral branches

polychasial branching having many main axes

sympodial branching that form of branching in which most branches are derived from the growing tip and a very few from the main stem

Branchiobdellidae a family of parasitic oligochaete annelids sometimes still classed with the Hirudinea. They lack setae and have a sucker at the posterior end of the body but not at the anterior

Branchiopoda a subclass of crustacean arthropods containing the animals commonly called fairy shrimps, tadpole shrimps, clam shrimps, and water fleas. They are distinguished from other crustacea by the leaf-like appendages on the thorax that serve both for locomotion and respiration

Branchiostegidae = Malacanthidae

Branchiura a taxon of crustacean arthropods considered by some to be a separate subclass and by others to be an order of the subclass Copepoda which must, in that case, be divided into Branchiura and Eucopepoda with all other copepods included in this last order. The Branchiura are mostly fresh water ectoparasites of fish and are distinguished by possessing compound eyes and usually having a large pair of suckers

brandling the lumbricid oligochaete annelid *Eisenia foetida*

brant the goose *Branta bernicla*

brasiletto any of several leguminous trees of the genus Caesaltania

burnished brass *Brit.* the noctuid lepidopteran insect *Plusia chrysitis*

Brassicaceae = Cruciferae

firebrat the lepismatid thysanuran insect *Thermobia domestica*

Brassolidae a family of shade-loving neotropical lepidopteran insects closely allied to the Satyridae and sometimes included with them. Many are called owl-butterflies

Brathinidae a small family of coleopteran insects closely allied to the Staphylinidae but usually called ant-like beetles in virtue of their shape. They may be distinguished by the absence of hind wings and the very large coxae

Braulidae a monotypic family of pupiparous wingless insects parasitic on the honeybee

brc *see* [brachymacrochaete]

bread (*see also* fruit) the term is applied to almost any edible substance

beef bread the pancreas

Hottentot's bread any of several species of dioscoriaceous shrub of the genus Testudinaria

St. John's bread the edible seed pods of the leguminous tree *Ceratonia siliqua*

sow bread *Brit.* the primulaceous herb *Cyclamen hederaefolium*

sweetbread the thymus gland

buttock sweetbread the testis

bread-and-cheese the leguminous shrubby tree *Pithecolobium ungis-cati*

break a spontaneous, but not of necessity heritable, change appearing in an F-1 generation of plants

chromatid break a fracture of a chromatid

breaking a term applied to the production of algal blooms or the opening of dormant buds

breaking-of-the-meres = algal bloom

break-o-day-boy *Austral.* a cracticid bird *Cracticus nigrogularis* (*cf.* butcher-bird, bell-magpie)

bream (*see also* king-of) *Brit.* without qualification the fresh water cyprinid *Abramis brama*

sea bream any of several species of sparid fishes. *U.S.* without qualification, *Archosargus rhomboidalis Brit. Pagellus centrodontus* (*cf.* sheepshead, porgy)

white bream *Brit.* the fresh water cyprinid *Blicca bjornka*

breast 1 (*see also* breast 2) the human mammary gland

breast 2 (*see also* breast 1) the upper, anterior, body surface particularly of birds

redbreast 1 (*see also* redbreast 2) *W.I.* either of two species of the Antillean bullfinch *Loxigilla*

robin redbreast *Brit.* commonly nowadays called robin (*q.v.*). A small red-breasted turdid bird (*Erithacus rubecula*) which has for many centuries been the most cherished British bird, *W.I.* the todid, *Todus todus* (= Jamaican tody)

redbreast 2 (*see also* redbreast 1) *I.P.* either of two species of papilionid lepidopteran insects but usually, without qualification *Papilio rhetenor*. (*cf.* mormon, swallowtail, peacock, spangle, raven, helen)

mountain velvet breast the trochilid bird *Lafresnaya lafresnayi*

yellowbreast *W.I.* the parulid bird *Coereba flaveola* (= bananaquit)

breath 1 (*see also* breath 2) inspired or expired air

breath 2 (*see also* breath 1) any airy plant

angels breath any of several species of herbs of the caryophyllaceous genus Gypsophila (*cf.* baby's breath) but also, rarely, the liliaceous genus Androstephium

baby's breath any of several species of the caryophyllaceous genus Gypsophila, particularly *G. paniculata*. Also *U.S.* the rubiaceous herb *Galium sylvaticum*

breath-of-heaven the rutaceous shrub *Adenandra fragrans*

bear's breech any of numerous species of the acanthaceous genus Acanthus

Dutchman's **breeches** the fumariaceous herb *Dicentra cucullaria*

breed a variety or race of constant genetic character. The term is mostly used of domestic animals and is by some considered synonymous with race

crossbreed = racial hybrid

mouth **breeder** any aquarium fish that harbors eggs in the mouth, Most are cichlids

inbreeding the mating of close relatives

Breitschneideraceae a small family of rhoeadale dicotyledonous angiosperms found in southwest China

Breitschneiderineae a suborder of Rhoeadales containing the family Breitschneideraceae

Brenthidae = Brentidae

Brentidae a family of brentoid coleopteran insects commonly called primitive weevils. They are extremely long-snouted, the pronotum being as long as the elytra

Brentoidea a superfamily of polyphagous coleopteran insects containing the primitive long-snouted weevils

brephic = larval

Brevibacteriaceae a family of gram positive eubacteriale rod-shaped schizomycetes, commonly found in water and dairy products. One species occurs in the rumen of cattle where it synthesizes relatively large quantities of vitamin B

[**brevis**] *either* a mutant (gene symbol *bv*) mapped at 104.3 on chromosome III of *Drosophila melanogaster*. The phenotypic expression is short stubby bristles *or* a mutant (gene symbol *bv*) mapped at 13 on linkage group V of *Zea mays*. The phenotype is dwarfed.

-bri- *comb. form* meaning "to acquire strength"

bri *see* [bright]

briar 1 (*see also* briar 2,3,tree) any rosaceous shrub with an elongate prickly stem, particularly of the genus Rosa (*cf.* bramble)

Australian briar *R. foetida*

sweet briar *R. rubiginosa* (= eglantine)

briar 2 (*see also* briar 1,3) any of many liliaceous vines of the genus Smilax

bamboo briar *S. hispida*

bull briar *U.S. S. rotundifolia*

cat briar *S. glauca*

China briar *U.S.* the lilaceous herb *S. bona-nox.*

false China briar *S. laurifolia*

green briar *U.S.* any of several species of Smilax

horse briar *S. rotundifolia*

saw briar *U.S. either S. glauca* or *S. bona-nox*

briar 3 (*see also* briar 1,2) a few vines of other affinities

sensitive briar *U.S.* any of several species of leguminous herbs of the genus Schrankia, particularly *S. uncinata*

brick *Brit.* the noctuid lepidopteran insect *Amathes circellaris*

bride *U.S.* the noctuid lepidopteran insect *Catocala neogama* (under-wing)

mourning bride any of numerous species of the dipsacaceous genus Scabiosa

bride-of-California the leguminous vine *Lathyrus splendens*

bridge used by biologists to describe any connection either physical, taxonomic, geographic, or anatomic

Gaskell's bridge = His's bundle

intercellular bridges supposed connections between

the cytoplasm of neighboring cells particularly in epithelia. Those investigated under the electron microscope have proved to be opposed prominences not actual connections

Varolius' bridge (pons Varolii) a transverse band of fibers running across the lower surface of the brain between the two sides of the cerebellum

sweet **bridget** the vireonid bird *Vireo magister*

bridle 1 (*see also* bridle 2) a pair of ridges, or a horse-shoe shaped ridge, on the mesosome of Pogonophora. It is presumed to maintain the body at the anterior edge or bend of the tube

bridle 2 (*see also* bridle 1) protoplasmic strands passing from the nucleus to the periphery of a plant cell

[**brief**] a mutant (gene symbol *bf*) mapped at 95.0± on chromosome III of *Drosophila melanogaster*. The phenotypic expression is a small body with small bristles

brier = briar

eye-**bright** *Brit.* the scrophulariaceous herb *Euphrasia officinalis*

[**bright**] a mutant (gene symbol *bri*) mapped at 54.3± on chromosome II of *Drosophila melanogaster*. The phenotypic expression is bright red eyes

brill *Brit.* the bothid fish *Psetta laevis*

brille a transparent scale fitting closely over the eye of a snake

brilliant 1 (*see also* brilliant 2) *I.P.* any of several species of lycaenid lepidopteran insects of the genus Simiskina

brilliant 2 (*see also* brilliant 1) any of numerous trochilid birds of the genus Heliodoxa

brin the bave (*q.v.*) of the silk worm

brindle *Brit.* any of several noctuid lepidopteran insects

cloud bordered brindle *Xylophasia rurea* (*cf.* arches)

clouded brindle *Xylophasia hepatica*

small clouded brindle *Apamea unanimis*

feathered brindle *Aporophyla australis*

slender brindle *Xylophasia scolopacina*

-brious *Adj. suffix from* -bry- (*q.v.*)

isobrious referring to two organs of equal strength, such as the cotyledons of most dicotyledons, or to a condition of equal growth

anisobrious said of a bilateral organism in which one side is more powerful, or grows more rapidly, than the other

bristle (*see also* bill, bird, front, mouth) a stout, epidermal protrusion. Also used for a feather with a stiff shaft and no web

stilt bristle a large posterior bristle, used in locomotion, on the posterior region of some marine nematodes

[**Bristle**] a mutant (gene symbol *Bl*) mapped at 54.8 on chromosome II of *Drosophila melanogaster*. The bristles of the phenotype are short and fat

brit 1 (*see also* brit 2) the calanoid copepod crustacean *Calanus finmarchicus*

brit 2 (*see also* brit 1) a juvenile clupeid fish

[**brittle endosperm**] a mutant (gene symbol *bt₁*) mapped at 8 on linkage group V of *Zea mays*. The name is descriptive

[**brittle stalk**] a mutant (gene symbol *bk₂*) mapped at 74 on linkage group IX of *Zea mays*. The term is descriptive of the phenotypic plant

[**broad**] *either* a mutant (gene symbol *br*) in linkage group V of *Habrobracon juglandis*. The name refers to the wide thorax of the phenotype *or* a mutant (gene symbol *br*) mapped at 0.6 on the X chromosome of *Drosophila melanogaster*. The phenotypic expression is a broader and shorter wing

brocade *Brit.* any of several noctuid lepidopteran insects

beautiful brocade *Manestra contigua*

dark brocade *Eumichtis adusta*

dusky brocade *Apamea obscura*

flame brocade *Rhizotype flammea*

great brocade *Erois occulta*

light brocade *Mamestra genistae*

pale-shouldered brocade *Mamestra thalassina*

double spot brocade *Micelia bimaculosa*

broccoli *U.S.* a loose branching cabbage of which the green flower stems and unopened buds are eaten. *Brit.* a large, slow growing cauliflower with a slight purplish tinge

-broch- *comb. form* meaning "noose"

brock *Brit.* = badger *Brit.*

brocket any deer with a spikelike horn, but particularly a two-year old European red deer

[broken] a mutant (gene symbol *bk*) in linkage group III of *Habrobracon juglandis*. The phenotypic expression is a broken outer margin to the wing

brolga *Austral.* the gruid bird *Grus rubicunda* (= Australian crane)

Bromeliaceae that family of farinosean monocotyledons which contains the pineapple and other bromeliads. The epiphytic habit of growth, the herbaceous calyx, and the peculiar thick, stiff leaves are easily recognizable characteristics

-brom- *comb. form* meaning "food"

myrmecobromous providing food for ants

-bronch- *comb. form* meaning "wind pipe"

bronchiole the terminal divisions of second order bronchi

bronchus properly, the two tubes into which the trachea divides

dorsal bronchus a series of dorsal connections with the mesobronchus in birds

mesobronchus a bronchus passing down the center of the lung in birds and reptiles

second order bronchus a tube dividing off from the bronchus

parabronchus a transverse connection between dorsal and ventral bronchi in birds

ventral bronchus one of a series of ventral connections with the mesobronchus in birds

[bronze] a mutant (gene symbol *bz*) mapped at 31 on linkage group IX of *Zea mays*. the phenotypic plant is bronze

brood 1 (*see also* brood 2,3, body, bud, cell, gemma) an assemblage of individuals all hatched at the same time from eggs laid by a single parent or any similar conglomeration of synchronously produced offspring

brood 2 (*see also* brood 1,3) term for a group of grouse, though covey is frequently used, particularly in *U.S.*

brood 3 (*see also* brood 1,2) term for a group of young animals, particularly birds, and these mostly when accompanied by the hen

broom (*see also* corn) any of several yellow-flowered leguminous shrubs particularly of the genera Genista and Cytisus. *U.S.*, without qualification, often refers to the grass *Andropogon scoparius*

butcher's broom the liliaceous herb *Ruscus aculeatus*

climbing butcher's broom the liliaceous shrubby climber *Semele androgyna*

indigo broom the leguminous shrub *Baptisia tinctoria*

New Zealand broom the papilionaceaous shrub *Carmichaelia flagelliformis*

pink broom *N.Z.* the leguminous shrub *Notospartium carmichaeliae*

leaflet rush-broom the leguminous herb *Viminaria denudata*

Scotch broom *Cytisus scoparius*

weeping broom *N.Z.* the leguminous shrub *Chordospartium stevensoni*

-brot- *comb. form* meaning "mortal" or "human"

brother *U.S.* the noctuid lepidopteran insect *Raphia frater*

brotion a plant succession due to human agency, as in the rotation of crops

brotula common name of brotulid fish

Brotulidae a small family of ophidioid acanthopterygian fishes commonly called brotulas, related to the eelpouts and codfishes

brow a crest or ridge

eyebrow the ridge or crest over the eye, or in some mammals, the hair growing in this region

brown 1 (*see also* brown 2) any of very numerous, widely distributed satyrid lepidopteran insects

bushbrown *Austral., I.P.* and *S.Afr.* any of numerous species of the genus Mycalesis. *S.Afr.*, without qualification, usually *M. safitza*

evening brown *Austral.* and *I.P.* any of several species of the genus Melanitis *Austral.*, without qualification, *M. leda*

common brown *Austral.* Heteronympha merope

eyed brown *U.S.* Lethe eurydice (*cf.* pearly eye)

sword-grass brown *Austral.* either of two species of the genus Tisiphone

meadow brown *Brit.* Epinephele ianira. *I.P.* any of numerous species of the genus Maniola

rockbrown *I.P.* any of numerous species of the genus Eumenis

shouldered brown *Austral.* Heteronympha penelope

tigerbrown *I.P.* Orinoma damaris

brown 2 (*see also* brown 1) *Brit.* any of several notodontid and noctuid lepidopteran insects

marbled brown *Brit.* any of several notodontids. Usually, without qualification, *Drymonia trimacula*

dusky marbled brown *Gluphisia crenata*

lunar marbled brown *Drymonia chaonia*

pale shining brown the noctuid *Aplecta advena*

[brown] *either* a mutant (gene symbol *b*) mapped at 52 on the linkage group II of the rat. The phenotypic expression involves the replacement of black pigment by brown *or* a mutant (gene symbol *bw*) mapped at 104.5 on chromosome II of *Drosophila melanogaster*. The name refers to the eye color of the phenotype *or* a mutant (gene symbol *b*) found at locus 42.8 on linkage group I of the rabbit. The name is descriptive of the coat color of the phenotype *or* a mutant (gene symbol *b*) in linkage group VIII of the mouse. The phenotypic expression is the production of brown instead of black pigment

[brown egg-2] a mutant (gene symbol *b₂*) mapped at 8 on linkage group VI of *Bombyx mori*. There is a grayish-brown pigment in the serosa of the phenotypic egg

[brown endosperm] a mutant (gene symbol *Bn*) mapped at 60 in linkage group VII of *Zea mays*. The term is descriptive

[brown eyes] a mutant (gene symbol *br*) found in the sex chromosome (linkage group I) of the domestic fowl. The name is descriptive of the phenotypic expression

brownie *I.P.* any of several species of lycaenid lepidopteran insect of the genus Gerydus

[brown midrib] *either* a mutant (gene symbol bm_1) mapped at 7 in linkage group V of *Zea mays, or* a mutant (gene symbol bm_2) mapped at 161 on linkage group I of *Zea mays.* The name is descriptive of the phenotypic leaf

[brown pericarp] a mutant (gene symbol *bp*) mapped at 44 on linkage group IX of *Zea mays.* The term is descriptive of the phenotypic expression

brubru the African laniid bird *Nilaus afer*

Brucellaceae a family of gram negative, coccoid, eubacteriale schizomycetes the great majority of which are pathogenic to man and domestic animals

Bruchelidae = Anthribidae.

Bruchidae = Mylabridae

brumalous pertaining to the winter solstice

Brunelliaceae a small family of South American rosale dicotyledonous angiosperms

bruneolous brownish

bruneous brown

Bruniaceae a small family of rosale dicotyledons. Its heathlike habit and dense flowerheads distinguish it clearly from other rosale families

Brunner's *epon. adj.* from Johann Conrad Brunner (1653–1727)

brush 1 (*see also* brush 2,3,runner) a fruit or flower with protruding stigmas or stamens

brush 2 (*see also* brush 1,3,) a group of hairs or trichomes

  eye brush hairs on the forelegs of insects used in cleaning the compound eyes

  stylar brush the collecting hairs of flowers

brush 3 (*see also* brush 1,2) small shrubs, or shrubby herbs, either individually or as a mass

  bitter brush the rosaceous shrub *Purshia tridentata*

  bottle brush any of several species of myrtaceous shrubs or tree of the genus Metrosideros, and *Callistemon lanceolatus*

  devil's paintbrush almost any orange or red colored compositaceous herb has received this name in various parts of the *U.S. Hieracium aurantiacum* is one of the best known

    Flora's paint brush the compositaceous herb *Emelia flammea*

  sage brush *either* of the compositaceous shrubs *Artemsia arbuscula or A. tridentata*

  wood brush *Brit.* any of several species of the juncaceous genus Luzula

-bry- *comb. form* meaning to "swell", "grow", or "burst forth." Often transliterated -bri-

Phycobrya = Characeae

Bryogam = Bryophyte

bryony the cucurbitaceous vine *Bryonia dioica*

  bastard bryony the vitaceous vine *Cissus sicyoides*

Bryophyta a subkingdom of plants containing those forms commonly called liverworts, hornworts, and mosses. They are distinguished by the absence of a vascular system, of true roots and of true leaves. The alternate generations are distinct

Bryopsidaceae that family of siphonale chlorophyceae in which the thallus is differentiated into a lower rhizome-like tube and erect branch portion

Bryozoa a no longer acceptable taxon at one time containing the forms now divided between the phyla Ectoprocta and Endoprocta

bs *see* [blistered]

bsc *see* [bent scutellar]

bt *see* [bent] and [belted]

bt₁ *see* [brittle endosperm]

bu *see* [bushy]

Bu *see* [Burnt]

bu *see* [bulged]

Bubalornithidae a family of passeriform birds commonly included with the Ploceidae

bubble any of several gastropod mollusks with spheriod translucent shells

  glassy bubble popular name of akerids

  lined bubble popular name of hydatinids

[Bubble] a mutant (gene symbol *Bb*) mapped at 48.0± on chromosome III of *Drosophila melanogaster.* The wings of the phenotype are bubble like

Bubonidae = Strigidae

-bucc- *comb. form.* meaning mouth or cheek

buccal pertaining to the mouth or mouth cavity

  peribuccal around the mouth region

Buccinidae a family of large gastropod mollusks commonly called whelks. They are distinguished by having few whorls and a large aperture notched anteriorly

Bucconidae that family of piciform birds which contains the puffbird. They are distinguished by the strong rounded bill with its hooked tip and conspicuous rictal bristles

Bucerotidae a large family of coraciiform birds commonly called hornbills. The huge bill usually bears a casque on the culmen

buck 1 (*see also* buck 2, mast, moth, root) the male of numerous artiodactyl mammals, particularly cervids. The term is also sometimes applied to caprids, for which the term "billy" is preferable, and ovids, for which the term "ram" is preferable. The term is also applied to male kangaroos. In strict *Brit.* usage the term used to be confined to male fallow deer more than five years old

buck 2 (*see also* buck 1, bok) in compound names of artiodactyls without regard to sex

  blackbuck the antilopine antilopid *Antilope cervicapra*

  bluebuck a hippotragine antelope *Hippotragus leucophaeus*

  bushbuck any of numerous species of strepsicerosine bovid artiodactyles of the genus Tragelaphus

  marsh buck the Afr. antelope *Limnotragus spekei*

  reedbuck any of several species of reduncine antilopids of the genus Redunca. Also applied to the genus Cervicapra

  springbuck an African gazelline antilopid (*Antidorcas euchore*) The female differs from all other female gazellines in bearing horns

  waterbuck either of two species of reduncine antilopid of the genus Kobus

[buckled] a mutant (gene symbol *bk*) mapped at 59.8± on the X chromosome of *Drosophila melanogaster.* The term refers to the misshapen wings of the phenotype

bud (*see also* cone, corm, moth, scale, variation weevil, and worm) a protuberance on an organism that will subsequently grow into an organ, or another organism. Most widely used in *bot.* for the precursors of flowers or shoots. Also, rarely in compound names

  accessory bud one that is additional to those normally formed

  adventitious bud one in an unusual position, *i.e.* not axillary

axillary bud one that is in the axil of a leaf

brood bud (lichen) = soredium

  brood bud (Bryophyta) = bulbil

end bud the tail bud of a developing embryo

imaginal bud = imaginal disc

lateral bud = adventitious bud or axillary bud

limb bud the primordium of a limb

lung bud one of a pair of diverticula from the laryngo-tracheal groove (*q.v.*)

periosteal bud a mass of invading osteogenic cells, osteoblasts and capilliaries in the course of endo-chrondral formation of bone

red bud any of several trees of the leguminous genus Cercis

taste bud one of numerous types of sensory end organ in the tongue and palate

terminal bud one on the end of a stem

budding *zool.* the proliferation of offspring as vegetative outgrowths from the parent. *Bot.* the production of buds

endogenous budding the liberation of buds into an internal brood chamber

budgerigar the psittacid bird Melopsittacus undulatus (*cf.* parrakeet)

budgerygah preferred *Austral.* spelling of budgerigar

budjerigar = budgerigar

Buettneriaceae = Sterculiaceae

buff *Brit.* any of several noctuid lepidopteran insects *S. Afr.* any of numerous species of lycaenid lepidopteran insect of the genus Teriomima

buffalo 1 (*see also* buffalo 2, berry, bean, bur, grass, hopper, weaver, and weed) this term is properly applied to bovine bovid artiodactyls of the genera Bubalus (water buffalo), Syncerus (African buffaloes) and anoa (Indian buffalo). It should not be applied to the American animal, usually so designated, which is a bison (*q.v.*)

cape buffalo the bovine bovid artiodactyle *Syncerus caffer*

forest buffalo the *Afr.* bovine bovid *Syncerus nanus*

water buffalo the asiatic, domesticated bovine artiodactyle *Bubalus bubalus*. Wild, as distinct from feral, stocks exist on some E. Indian islands

buffalo 2 (*see also* buffalo 1) any of several catosomid fishes of the genus Ictiobus

black buffalo *I. niger*

bigmouth buffalo *I. cyprinellus*

smallmouth buffalo *I. bubalus*

Bufonidae a family of anuran amphibians that lack teeth in the upper and lower jaws and have procoelous vertebrae. This family is often defined, particularly in European books, as containing the "toads" as distinct from the "frogs". Actually the term toad (*q.v.*) is applied to the members of many other families

bug this term, without qualification, can mean any small arthropod. Entomologists would like to confine the word to Hemiptera (bug 1, below) though many Homoptera (bug 2, below) are frequently so called. Numerous Coleoptera (bug 3, below) and a few insects of other groups (bug 4, below) are commonly called bugs as are several crustacea (bug 5, below) and one tick (bug 6, below)

bug 1 (*see also* bug, bug 2–6) any of very many hemipteran insects

Abe Lincoln bug the pentatomid *Murgantia histrionica* (= harlequin cabbage bug)

ambush bug U.S. popular name of phymatids. Without qualification, usually *Phymata pennsylvanica*

assassin bug popular name of reduviids

shield-backed bug popular name of scultellerids

bark bug popular name of aradids

bat bug popular name of polyctenids

bed bug popular name of all the family Cimicidae, but usually applied only to the parasite of man *Cimex lectularius*

giant bed bug the reduviid *Triatoma sanguisuga*

burrower bug popular name of cydnids

cannibal bug *Brit.* = assassin bug *U.S.*

caspid bug popular name of mirids (= plant bug, leaf bug)

chinch bug *U.S.* the lygaeid *Blissus leucopterus*

false chinch bug *U.S.* the lygaeid *Mysius ericae*

hairy chinch bug *U.S.* the lygaeid *Blissus hirtus*

western chinch bug *U.S.* the lygaeid *Blissus occiduus*

damsel bug popular name of nabids, particularly *Nabis ferus*

box elder bug the corizid *Leptocoris trivittatus*

flat bug popular name of aradids

flower bug popular name of anthocorids

leaf-footed bug popular name of coreids particularly *Leptoglossus phyllopus*

fungus bug popular name of aradids

gnat bug popular name of enicocephalids

grass bug popular name of corizids

harlequin bug the brightly colored pentatomid *Murgantia histrionica*

broad-headed bug popular name of corixids

unique-headed bug popular name of enicocephalids

lace bug popular name of tingids

azalea lace bug *U.S. Stephanitis pyrioides*

hackberry lace bug *U.S. Corythucha celtidis*

chrysanthemum lace bug *U.S. Corythucha marmorata*

cotton lace bug *U.S. Corythucha gossypii*

elm lace bug *U.S. Corythucha ulmi*

lantana lace bug *U.S. Teleonemia scrupulosa*

oak lace bug *U.S. Corythucha arcuata*

eggplant lace bug *U.S. Gargaphia solani*

rhododendron lace bug *U.S. Stephanitis rhododendri*

sycamore lace bug *U.S. Corythucha ciliata*

hawthorn lace bug *U.S. Corythucha cydoniae*

basswood lace bug *U.S. Gargaphia tiliae*

leaf bug popular name of mirids

ash-gray leaf bug *U.S.* popular name of piesmids

thread-legged bug popular name of ploiariids

electric light bug *U.S.* popular name of belostomatids (= giant water bug)

negro bug *U.S.* the pentatomid *Corimelaena pulicaria*

palm bug popular name of thaumastocorids

third party bug the pentatomid *Murgantia histrionica* (= harlequin cabbage bug)

pirate bug *Brit.* = assassin bug

plant bug popular name of mirids

alfalfa plant bug *U.S. Adelphocoris lineolatus*

ash plant bug *U.S. Neoborus amoenus*

caragana plant bug *U.S. Lopidea dakota*

hickory plant bug *U.S. Neolygus caryae*

hollyhock plant bug *U.S. Melanotrichus althaeae*

hop plant bug *U.S. Taedia hawleyi*

four-lined plant bug *U.S. Poecilocapsus lineatus*

meadow plant bug *U.S. Leptopterna dolabrata*

onion plant bug *U.S. Labopidea allii*

pear plant bug *U.S. Neolygus communis*

phlox plant bug *U.S. Lopidea davisi*

rapid plant bug *U.S. Adelphocoris rapidus*

tarnished plant bug *U.S. Lygus lineolaris*

ragweed plant bug *U.S. Chlamydatus associatus*

yucca plant bug *U.S. Halticotoma valida*

poultry bug *U.S. Haematosiphon inodorus*

red bug popular name of mirid hemipteran insects

apple red bug *U.S. Lygidea mendax*

soldier bug *U.S.* any of several species of pentatomids of the genus Podisus

spined soldier bug *U.S. P. maculiventris*

squash bug *U.S.* the coreid *Anasa tristis*

horned squash bug *U.S.* the coreid *Anasa armigera*

stilt bug popular name of neids

stink bug *U.S.* a term properly applied to pentatomids but extended occasionally to coriscids and corizids. All those here listed are pentatomids

brown stink bug *U.S. Euschistus servus*

dusky stink bug *U.S. Euschistus tristigmus*

green stink bug *U.S. Acrosternum hilare*

southern green stink bug *U.S. Nezare viridula*

rice stink bug *U.S. Oebalus pugnax*

one-spot stink bug *U.S. Euschistus variolarius*

two-spotted stink bug *U.S. Perillus bioculatus*

swallow bug *U.S.* the cimicid *Oeciacus vicarius*

toad bug *U.S.* the gelastocorid *Gelastocoris oculatus*

jumping tree bug popular name of isometopids

turtle bug popular name of podopids

creeping water bug popular name of naucorids

giant water bug *U.S.* the belostomatid *Lethocerus americanus*

velvet water bug popular name of hebrids

milk weed bug either of two lygaeids *Lygaeus kalmii* or *Oncopeltus fasciatus*

wheel bug *U.S.* the reduviid *Arilus cristatus*

bug 2 (*see also* bug 1,3–6) any of many homopteran insects

datebug *U.S.* the issid *Ascarcopus palmarum*

greenbug *U.S.* the aphid *Schizaphis graminum*

mealy bug popular name of psuedococcids

apple mealy bug *U.S. Phenacoccus aceris*

gray sugarcane mealy bug *U.S. Dysmidococcus boninsis*

pink sugarcane mealybug *U.S. Saccharococcus sacchari*

citrophilus mealybug *U.S. Pseudococcus fragilis*

citrus mealybug *U.S. Pseudococcus citri*

coconut mealybug *U.S. Nipacoccus nipae*

comstock mealybug *U.S. Pseudococcus comstocki*

grape mealybug *U.S. Pseudococcus maritimus*

ground mealybug *U.S. Rhizoecus falcifer*

Mexican mealybug *U.S. Phenacoccus gossypii*

pineapple mealybug *U.S. Dysmicoccus brevipes*

striped mealybug *U.S. Ferrisia virgata*

long-tailed mealybug *U.S. Pseudococcus adonidum*

spittle bug popular name of cercopids

alder spittlebug *U.S. Clastoptera obtusa*

dogwood spittle bug *U.S. Clastoptera proteus*

diamond-backed spittlebug *U.S. Lepyronia quadrangularis*

sunflower spittlebug *U.S. Clastoptera xanthocephala*

heath spittlebug *U.S. Clastoptera saintcyri*

lined spittlebug *U.S. Neophilaenus lineatus*

meadow spittlebug *U.S. Philaenus spumarlus*

pecan spittlebug *U.S. Clastoptera achatina*

pine spittlebug *U.S. Aphrophora parallela*

Saratoga spittle bug *U.S. Aphrophora saragotensis*

bug 3 (*see also* bug, bug 1,2,4–6) any of many coleopteran insects

apple bug *U.S.* = penny bug *U.S.*

bessbug popular name of passalids

billbug any of many species of curculionid of the genus Rhynchophorus and its allies

cockle bur billbug *U.S. Rhodobaenus tredecimpunctatus*

clay-colored billbug *U.S. Sphenophorus aequalis*

corn billbug any of several species of the genus Sphenophorus

bluegrass billbug *U.S. Sphenophorus parvulus*

maize billbug *U.S. Sphenophorus maidis*

low-tide billbug *U.S. Sphenophorus setiger*

Timothy billbug *U.S. Sphenophrus zeae*

tule billbug *U.S. Sphenophorus discolor*

curlewbug the curculionid *Calendra callosa*

gold bug *U.S.* the chrysomelid *Metriona bivittata*

june bug *Brit.* the rutelid *Phyllopertha horticola U.S.* = june beetle *U.S.*

ladybug *U.S.* popular name of coccinellid coleopteran insects, better called lady beetle (*q.v.*) and known in *Brit.* as lady bird

lightningbug *U.S.* popular name of lampyrids (= firefly)

penny bug *U.S.* popular name of gyrinids (= whirligig beetle)

pinch bug *U.S.* popular name of lucanids (= stag beetle

potato bug *U.S.* the emeloid *Epicauta vittata*

tumble bug *U.S.* any of several scarabaeid coleopteran insects of the genera Canthon and Deltochilum

bug 4 (*see also* bug 1–3,5,6) a few insects, or insect larvae, neither Hemiptera, Homoptera nor Coleoptera

bass bug *U.S.* larvae of odonata

conniption bug *U.S.* the larva of the dobson fly *Corydalus cornutus* (= hellgrammite

Croton bug a name given to *Blattella germanica* (German cockroach) in New York in consequence of a heavy infestation at the time of the completion of the Croton Aquaduct

doodlebug *U.S.* the larva of myrmeleontid neuropterans (= antlion)

perch bug *U.S.* = bass bug *U.S.*

bug 5 (*see also* bug, 1–4,6) any of several crustaceans

pill bug *U.S.* popular name of armadillidiid isopods

salve bug *U.S.* the cymothoid parasitic isopod *Aega psora*

sand bug the hippid, anomuran, reptant crustacean *Emerita talpoida* (= sand crab)

sow bug any isopod crustacean, though the term pill bug is usually applied to those of the family Armadillidiidae (= *Brit.* wood louse)

watercress sowbug *U.S.* the asellid *Lirceus brachyurus*

bug 6 (*see also* bug, bug 1–5) an argasid tick

miami bug frequent error for miana bug

miana bug the argasid tick *Argas persicus*. The term is also applied to *A. mianensis*

bugle *Brit.* any of several labiateous herbs of the genus Ajuga (= *U.S.* bugleweed)

bitter bugle either of two species of the labiateous genus Lycopus

bugloss 1 (*see* bugloss 2) any of several boraginaceous herbs of the genera Echium, Lycopsis, Asperugo, and Anchusa and the compositaceous herb *Picris echioides*

viper's bugloss (*see also* bugloss 2) any of several species of the boraginaceous genus Echium

bugloss 2 *Brit.* the noctuid lepidopteran insect *Dian-*

thoecia irregularis, also called viper's bugloss (*see* bugloss 1)

mudnest **builder** any of four species of grallinid bird

building term for a group of rooks

bul *see*[bulge]

bulb 1 (*see also* bulb 2) *zool.* any hollow globose organ, frequently with the connotation that a hollow open shaft is attached to the bulb

duodenal bulb a thickened portion of the duodenum in some mammals, containing Brunner's glands

ejaculatory bulb any structure that by compression forces spermatozoa through a penis or aedeagus

end bulb the swollen termination of an axon

flame bulb the excretory organs of Entoprocta which are large compound flame cells

head bulb a swollen band of ballonets in the anterior region of some nematodes

ocellar bulb = tentacular bulb

olfactory bulb one of a pair of evaginations from the antero-dorsal end of the telencephalon that contributes the sensory portion of the nasal organs

penis bulb a muscular body at the base of the penis papilla (*q.v.*) in some Turbellaria

aquapharyngeal bulb the anterior portion of the water-vascular system of holothurian echinoderms

pseudobulb (*see also* bulb 2) a bulb at the base of the pharynx of a nematode that is incapable of closing the lumen

tentacular bulb the bulb on the base of the tentacle of a hydrozoan medusa

bulb 2 (*see also* bulb 1, aphis, fly, mite, scale) *bot.* a globose mass of fleshy leaves growing from a short stem, that serves as the dormant phase of many flowering plants. The term is often, in error, applied to the corm (*q.v.*)

naked bulb one with no outer coat over the modified leaves

plumule bulb a bulb arising directly from a seed

pseudobulb (*see also* bulb 1) the thickened internode of an orchid

runner bulb a bulb formed on a stolon

solid bulb = corm

tunicated bulb one with an outer coat of scaly leaves like the onion

bulbel a new bulb arising from the base of a mother bulb

bulbil 1 (*see also* bulbil 2) a small bulb, produced above ground

bulbil 2 (*see also* bulbil 1) a resting vegetative gemmule produced by some lower plants

spore bulbil an abortive apothecium in a lichen

bulblet a diminutive bulb, not infrequently aerial, produced by some monocotyledonous plants

bulbodium (obs) = corm

bulbul any of very many birds of the family Pycnonotidae (*cf.* green bul)

ashy bulbul *Hypsipetes flavala*

bearded bulbul any of several species of the genus Spinoxos

black bulbul *Hypsipetes madagascariensis*

black collared bulbul *Neolestes torquatus*

chestnut-eared bulbul *Hypsophetes amaurotis*

Gaboon bulbul *Pycnonotus barbatus*

little green bulbul *Andropadus virens*

striated green bulbul *Alcurus striatus*

grey-headed bulbul *Pycnonotus priocephalus*

honey guide bulbul *Baeopogon indicator*

icterine bulbul *Phyllastrephus icterinus*

spotted bulbul *Ixonotus guttatus*

red-vented bulbul *Pycnotus cafer*

red-whiskered bulbul *Pycnotus locosus*

bulbule = bulbil

bulbus arteriosus a strongly muscularized portion in the anterior region of the conus arteriosus in some fishes

[**bulge**] a mutant (gene symbol *bul*) mapped at 43.6 on chromosome III of *Drosophila melanogaster.* The phenotypic eye bulges

[**bulged**] a mutant (gene symbol *bu*) in linkage group I of *Habrobracon juglandis.* The name refers to the eye condition of the phenotype

cellular **bulk** = mesophloem

-**bull-** *comb. form* meaning "blister"

bull (*see also* briar, finch, frog, head, grape) without qualification, a term for the male of domestic cattle, but applied to the male of many other species of animal, such as the alligator, camel, caribou, eland, elephant, giraffe, hartebeest, moose, muskox, terrapin, walrus, whale, yak, and zebu

bulla any hollow knob or disc. Specifically a swollen patch with many veins on the wing of an insect, or a weak spot where a vein is crossed by a furrow

tympanic bulla the bony spheroid which surrounds the inner ear, formed in part from the tympanic bone

bullace *Brit.* the rosaceous shrub *Prunus communis* (= sloe)

bullate blistered

Bullidae a family of gastropod Mollusca almost lacking a cone and with a long flaring aperture greater in length than the body whorl

bumbum the anacardiaceous tree *Mangifera odorata* and its fruit

bumper *U.S.* the carangid fish *Caranx ruber*

bundle 1 (*see also* bundle 2, -arch) in *bot.* a common abbreviation for vascular bundle, that is a group of plant conducting cells, including phloem and xylem, a component of the vascular cylinder

concentric bundle one in which xylem surrounds the phloem (amphivasal vascular bundle) or the phloem surrounds the xylem (amphicribral vascular bundle)

hadrocentric bundle one in which the hadrome is surrounded by the leptome

leptocentric bundle one which has the leptome totally surrounded by the hadrome

vasicentric bundle one in which there is parenchyma around each vessel

cauline bundle a stem bundle which has no direct communication with leaf bundles

closed bundle one in which all the procambrium cells become permanent tissue

collateral bundle one in which the phloem occurs on one side and the xylem on the other side of the bundle

bicollateral bundle one in which the xylem lies between two bundles of phloem

monarch bundle one containing a single strand

bundle 2 *zool.* a large mass of nerve, or other fibers

His's bundle a mass of nerve impulse conducting fibers connecting the auricles and ventricles of the heart

bundy the myrtaceous tree *Eucalyptus camcagei*

bungulan *hort. var.* of the musaceous tree-like herb *Musa paradisiaca* (= banana)

bunting any numerous fringillid birds. *Brit.* without qualification, *Emberiza calandra*

rose-bellied bunting *Passerina rositae*

blue bunting *Cyanocompsa parellina*

golden-breasted bunting *Emberiza flaviventris*
  orange-breasted bunting *Passerina leclancherii*
chestnut bunting *Emberiza rutila*
cirl bunting *Emberiza cirlus*
corn bunting *Emberiza calandra*
black-headed bunting *Emberiza melanocephala*
house bunting *Fringillaria striolata*
indigo bunting *Passerina cyanea*
Lapland bunting *Brit.* = *U.S.* Lapland longspur
lark bunting *Calamospiza melanocorys*
lazuli bunting *Passerina amoena*
little bunting *Emberiza pusilla*
African little bunting *Emberiza forbesi*
McKay's bunting *Plectrophenax hyperboreus*
ortolan bunting *Emberiza hortulana* (= ortolan)
painted bunting *Passerina ciris*
pine bunting *Emberiza leucocephala*
reed bunging *Brit. Emberiza schoeniclus*
rock bunting *Emberiza cia*
rustic bunting *Emberiza rustica*
snow bunting *Plectrophenax nivalis*
varied bunting *Passerina versicolor*
[buo] *see* [burnt orange]
Buphagidae a family of passeriform birds usually included with the Sturnidae
Buprestidae a very large family of coleopteran insects. They are usually medium sized, brilliantly colored beetles tapering to a rather sharp point behind and having very large compound eyes. Many are serious pests of fruit trees
bur = burr
  butterbur *Brit.* the noctuid lepidopteran insect *Hydroecia petasitis*
bur *see* [burgundy]
burbot the freshwater gadid fish *Lota lota*
[burgundy] a mutant (gene symbol *bur*) mapped at 55.7± on chromosome II of *Drosophila melanogaster*. The phenotypic eye is dull, dark brown, not burgundy
Burhinidae a small family of charadriiform birds containing the forms commonly called thick-knees by virtue of the great thickening of the tibiotarsal joint
Burmanniaceae a small family of microspermous monocotyledonous angiosperms with scale-like leaves and numerous minute seeds
burnet 1 (*see also* burnet 2, saxifrage) any of several rosaceous herbs particularly of the genus Sanguisorba
  salad burnet *Brit.* any of several species of the rosaceous genus Poterium
burnet 2 (*see also* burnet 1) Brit. any of numerous species of zygaenid lepidopteran insects
[Burnt] a mutant (gene symbol Bu) mapped at 5.5 on linkage group XI of *Bombyx mori*. The phenotypic expression is a burn-like scar on the larval skin
[burnt orange] a mutant (gene symbol *buo*) mapped at 57.1 on chromosome II of *Drosophila melanogaster*. The name refers to the eye color of the phenotype
burr 1 (*see also* burr 2, fish) any roughened, usually globose, object specifically the seeds and fruits of many plants. In many compound names of plants producing burrs
  buffalo burr the solanaceous herb *Solanum rostratum*
  butter burr *Brit.* the compositaceous herb *Petasites vulgaris* (*P. fragrans* = *U.S.* winter heliotrope)
  clotburr any of several species of compositaceous herb of the genus Arctium (= burdock)
  cockle burr (*see also* billbug, weevil) any of numerous species of compositaceous herb of the genus Xanthium

New Zealand burr any of several species of the rosaceous herbs of the genus Acaena
sand burr *U.S.* any of several species of grass of the genus Cenchrus
sheep burr the compositaceous herb *Acanthospermum australe*
burr 2 (*see also* burr 1) the swollen base of a deer horn
burro *S.W.U.S.* = donkey
burrow (*see also* mite) a tunnel dug by an animal
bursa literally a purse, but used in *biol.* for any pouch or sac. Particularly an eversible pouch used to grasp the rear end of the female in copulating acanthocephalans, a pouch alongside each side of each arm on the oral surface of ophiuroid echinoderms and a copulatory expansion at the hinder end of strongyloid nematodes supported by expanding rays
copulatory bursa a seminal receptacle that houses sperm for a brief space of time (= genital bursa)
Ent's bursa a pouch lying between the pyloric valve and the entrance to the spiral valve in the alimentary canal of lower fish
genital bursa in insects, the pouch in which the female receives and stores sperm
Burseraceae a family of geranealeous dicotyledons that is distinguished from most other families in the order by the presence of resin chambers in the bark. Many of these resins are of economic importance
Bursidae a family of gastropod mollusks with oblong, laterally compressed shells having two rows of continuous varices, one on each side. They are usually called frog shells
egg burster protuberances on the head of an insect larva that rupture the shell to permit hatching
Buryalae an order of ophiuroid echinoderms distinguished by the branching arms and the lack, or least reduction, of superficial plates
bush (*see also* dog, hen, shrike tanager, tit) a low, much branched shrub, or shrubby herb
beaver bush = spice bush
bellyache bush the euphorbiaceous shrub *Jatropha gossypifolia*
Benjamin bush = spice bush
cranberry bush the caprifoliaceous shrub *Viburnum americanum*
European cranberry bush the caprifoliaceous *Viburnum opulus*
strawberry bush the celastraceous shrub *Euonymus americana*
burning bush any of several species of rutaceous plants of the genus Dictamnus. Also the chenopodiaceous bushy herb *Kochia trichophylla*, the urticaceous herb *Pilea microphylla* (= artillery plant), the celastraceous shrub *Euonymus atropurpureus*, and, from time to time, almost any plant with red foliage
butterfly bush any of several shrubs of the loganiaceous genus Buddleia
button bush any of several species of the rubiaceous genus Cephalanthus, particularly *C. occidentalis*
caper bush the capparidaceous shrub *Capparis spinosa*
coral bush the leguminous shrub *Templetonia retusa*
feather bush any of several species of shrub of the ericaceous genus Leucothoe
fetter bush the ericaceous shrub *Pieris floribunda*
fever bush *U.S.* any of several species of lauraceous herb or shrub of the genus Lindera
hobble bush the caprifoliaceous shrub *Viburnum alnifolium*
indigo bush either the leguminous shrub *Amorpha fruticosa* or the tree *Dalea spinosa* of the same family

ivy bush   U.S. the ericaceous shrub *Kalmia latifolia*
(= mountain laurel)
jew bush any of several species of euphorbiaceous
succulent shrub of the genus Pedilanthus
maybush *Brit.* = maythorn *Brit.*
milkbush any of numerous euphorbiaceous shrubs
yielding a milky sap but particularly those of the
genus Synadenium
African milkbush *S. grandii*
minnie bush the ericaceous shrub *Menziesia pilosa*
niter bush any of several shrubs of the zygophyllaceous genus Nitraria but particularly *N. schoberi*
polecat bush the anacardiaceous shrub *Rhus aromatica* (= fragrant sumac)
quinine bush the garryaceous shrub *Garrya elliptica*
salt bush U.S. the chenopodiaceous shrub *Atriplex
argentea* and the verbenaceous tree *Avicennia
nitida*
shad bush any of several rosaceous shrubs of the genus
Amelanchier
skunk bush the anacardiaceous shrub *Rhus trilobata*
soap bush the rhamnaceous shrub *Noltea africana*
spice bush U.S. the lauraceous shrub *Lindera benzoin*
stag bush the caprifoliaceous shrub *Viburnum pruni-
folium*
stagger bush the ericaceous shrub *Lyonia mariana*
steeple bush the rosaceous shrub *Spiraea tomentosa*
tetter bush the ericaceous shrub *Lyonia lucida*
[bushy] a mutant (gene symbol *bu*) mapped at 49 on
chromosome 8 of the tomato. The aspect of the
phenotypic plant is bushy with short internodes and
long petioles
business term for a group of ferrets
bussu the palm tree *Manicaria saccifera*
bustard any of numerous large European, Asiatic, and
Australian gruiiform birds of the family Otididae.
Brit. without qualification, *Otis tarda.* The term
was at one time applied in Canada to the Canada
goose
Australian bustard *Eupodotis australis*
black-bellied bustard *Lissotis melanogaster*
white-bellied bustard *Eupodotis senegalensis*
black bustard *Afrotis atra*
blue bustard *Eupodotis caerulescens*
crested bustard *Lophotis ruficrista*
Denham's bustard *Neotis cafra*
florican bustard any small bustard
great bustard *Otis tarda*
houbara bustard *Chlamydotis undulata* (= houbara)
great Indian bustard *Choriotis nigriceps*
thick-kneed bustard *Brit.* = thicknee
little bustard *Tetrax tetrax*
Nubian bustard *Neotis nuba*
Stanley bustard *Neotis cafra*
buteo the name of a large genus of buteoine hawks, frequently used as an English word particularly in *U.S.*
Buteoninae a subfamily of Accipitridae containing the
so-called buzzard-hawks distinguished from the
Accipitrinae by their soaring habit and long, broad
wings
Butomaceae a family of helobian monocotyledons. Most
distinctive characteristic is the presence of numerous ovules borne between the margins and midrib
of the carpal
Butomales an order of monocotyledonous angiosperms
erected to contain the Hydrocharitaceae and the
Butomoceae

black **butt** the myrtaceous tree *Eucalyptus pilularis*
wooly butt *E. longifolia*

butter-and-eggs the scrophulariaceous plant *Linaria
vulgaris* (= toad flax)
buttercup *see under* -cup.
butterfly *see under* -fly
buttock the swollen area in the posterior region of the
thigh produced by the enlargement of the *gluteus
maximus* muscle
button 1 (*see also* button 2) any button-shaped organism
or part of an organism, particularly an oval, holo-
thurinan ossicle with two rows of holes in it
alligator button the seeds of the lotus
sea button popular name of eratid gastropod mollusks
but in *U.S.* particularly *Pusula solandri*
button 2 (*see also* button 1) any of several plants having
button-shaped flowers
bachelor's button the compositaceous herb *Centaurea
cyanus* (= cornflower) and the amaranthaceous herb
Gomphrena globosa N.Z. the compositaceous herb
Cotula coronopifolia
yellow bachelor's button the polygalaceous herb
Polygala lutea
barbara's button any of numerous species of herb of
the compositaceous genus Marshallia
bitter buttons compositaceous herb *Tanacetum vulgare*
bog button *U.S.* the eriocaulaceous herb *Lachnocaulon anceps*
white-button *U.S.* the eriocaulaceous herb *Eriocaulon
septangulare*
[button] a mutant (gene symbol *bn*) on linkage group
VII of *Neurospora crassa.* The phenotypic expression is a nonconidiating colonial growth
buttons (obs.) = bud
buttress above ground protuberent portions of a trunk or
root of a tree
leaf buttress a lateral prominence on the side of the
shoot apex that will give rise to a leaf
butyrous of the consistency of butter, usually applied to
bacterial colonies
Buxaceae that family of sapindale dicotyledonous angiosperms that contains, *inter alia*, the boxwoods and
Japanese spurges. The family is readily distinguished by the glossy black seeds
buxeous pertaining to the color, or more rarely to some
other property, of box-wood
Buxineae a suborder of Sapindales containing the single
family Buxaceae
buzzard (*see also* eagle, hawk) any large diurnal bird of
prey. *Brit. obs.* = harrier (*Brit.*) In *U.S.* the term is
applied to the turkey vulture (*q.v.*) *Brit.*, without
qualification, *Buteo buteo*
augur buzzard = jackal buzzard
black-breasted buzzard *Hamirostra melanosternon*
desert buzzard *Buteo burmanicus*
bat-eating buzzard = bat hawk
grey-faced buzzard *Butastur indicus*
honey-buzzard *Pernis apivorus*
grasshopper-buzzard *Butastur rufipennis*
long-legged buzzard *Buteo rufinis*
lizard-buzzard *Kaupifalco monogrammicus*
Mexican buzzard = crested caracara (*q.v.*)
moor buzzard *Brit.* = marsh harrier *Brit.*
mountain buzzard *Buteo oreophilus*
red-necked buzzard *Buteo auguralis*
steppe buzzard *Buteo vulpinus*
turkey buzzard *U.S.* the cathartid *Cathartes aura*
(= turkey vulture)
bv *see* [brevis]
bw *see* [brown]
bx *see* [bithorax]

Bx *see*[Beadex]

bxd *see*[bithoraxoid]

by *see* [blistery]

Byblidaceae a small family of Australian rosale angiosperms

Byrrhidae a family of small oval coleopteran insects commonly called pill beetles from their ability to draw their legs flat against the body and remain motionless when disturbed

byssaceous resembling a byssus

byssus literally, freshly combed flax and therefore applied to any loose, long, tuft. Particularly a series of branched projections at the poles of some nematode eggs, a loose stipe of some basidomycetes, and the attaching fibers of sessile pelecypod mollusks

Bytuiidae a small family of coleopteran insects the larvae of which are frequently a pest on fruit trees (*cf.* fruitworm)

bz *see* [bronze]

bz^r a yeast genetic marker indicating growth inhibition by benzimidazole

c   *see* [albinism], [albino], [canteloupe], [curved] and
     [potato leaf]
c   in yeast genetics is a marker for centromere
C   *see* [Aleurone color] and [Golden egg]
Ca   *see* [Caracul]
ca   *see* [claret]
ca[r] a yeast genetic marker for growth inhibition by
     caffeine
cabassou   the dasypodid xenarthran mammal *Xenurus
     unicinctus.* In spite of its Latin name, it has either
     twelve or thirteen movable bands of scutes
cabbage   (*see also* aphis, butterfly, curculio, beetle,
     looper, maggot, palm, and weevil) without qualifica-
     tion, any of numerous horticultural varieties of the
     cruciferous herb *Brassica oleracea,* and in com-
     pound names of plants of similar appearance or
     flavor
   Chinese cabbage   *Brassica chinensis*
   deer cabbage   the leguminous herb *Lupinus diffusus*
   isle-of-man cabbage   *Brit. B. monensis*
   John's cabbage   *U.S.* the hydrophyllaceous herb
     Hydrophyllum virginianum
   St. Patrick's cabbage   the saxifragaceous herb *Saxi-
     fraga umbrosa*
   sea cabbage   *Brit. B. oleracea*
   skunk cabbage   *E.U.S.* the aracean flowering plant
     Symplocarpus foetidus. W.U.S. the aracean flower-
     ing plant *Lysichiton camtschatcense*
   squaw cabbage   the cruciferous annual herb *Caulan-
     thus inflatus*
   yellow cabbage   the araceous herb *Hysichiton ameri-
     canum*
   wild cabbage   *U.S.* = squaw cabbage
cabezon   the cottid fish *Scorpaenichthys marmoratus*
Cabombaceae   a family erected to contain the cabom-
     boid members of the family Nymphaceae by those
     who wish to give them separate familial rank
Cabomboideae   a subfamily of the Nymphaeaceae
     distinguished, by those who wish to treat it as a sub-
     family, by the biseriate perianth
cabrilla   any of several species of serranid fish of the
     genus Epinephelus
   spotted cabrilla   *E. analogus*
cacainous   chocolate brown
cacao   the sterculiaceous tree *Theobroma cacao* and the
     unmanufactured product obtained from it. The man-
     ufactured product is cocoa and its derivative choco-
     late
   alligator cacao   *Theobroma pentagona*
Cacatuidae   a rarely used ornithological taxon at one
     time regarded as a family of psittaciform birds con-

taining the forms commonly called cockatoos. Now
     usually fused with the Psittacidae
cachalot   the physeterid cetacean *Physeter macroce-
     phalus.* (= sperm whale or great sperm whale)
cachalot   any of several furnariid birds of the genus
     Pseudoseisura
cachiman   any of several species of the annonaceous
     herbs of the genus Rollinia and their edible fruits
cacique   any of several species of icterid bird of the
     genera Amblycercus, Cacicus, and Casiculus
   yellow-billed cacique   *Ambylcercus holosericeus*
   scarlet-rumpted cacique   *Cacicus uropygialis*
   yellow-rumped cacique   *Cacicus cela*
   yellow-winged cacique   *Cassiculus melanicterus*
-caco-   *comb. form* meaning "bad"
cacoa   *see* cacao
cacomixtle   any of several species of procyonine carni-
     vore of the genus Bassariscus. In the *S.U.S.* they
     are commonly called ring-tailed cats
Cactaceae   the only family in the order Opuntiales. The
     usually thick fleshy stems, and spines, of the cactus
     distinguish the family from most groups except
     some of the Euphorbiaceae from which, however,
     they may be clearly distinguished by the berry-like
     fruit and watery sap
Cactuidae   a family of birds, once containing the cock-
     atoos, now fused with the Psittacidae
Cactales = Opuntiales
cactus   (*see also* borer) any of many cactaceous plants.
     The term is also applied to euphorbiaceous and
     some other plants of similar habit or appearance
   agave cactus   *Leuchtenbergia principia*
   bird cactus   any of several species of euphorbiaceous
     succulent shrub of the genus Pedilanthus
   red bird cactus = bird cactus
   button cactus   *Epithelantha micromeris*
   Christmas cactus = crab cactus
   crab cactus   *Zygocactus truncatus*
   deerhorn cactus   *Peniocereus greggii*
   easter cactus   *Schlumbergera gaertneri*
   fishhook cactus   *Ancistrocactus scheeri*
   giant cactus   *Carnegia gigantea*
   hatchet cactus   *Pelecyphora aselliformis*
   hedgehog cactus   any of many species of the genus
     Echinocactus
   mistletoe cactus   *Rhipsalis cassutha*
   old man cactus   *Cephalocereus senilis*
   organ-pipe cactus   *Lemaireocereus marginatus*
   sea cactus   *N.Z.* the phaeophyte alga *Splachnidium
     rugosum* (= gummy weed)
   rat-tail cactus   *Aporocactus flagilliformis*

sea urchin catus any of many species of the cactaceous genus Echinopsis

star cactus a species of Astrophytum

vine cactus the fouquieriaceous shrub *Fouquieria splendens*

caddis the larva of a trichopteran insect

cadelle *U.S.* the larva of the ostomatid coleopteran *Tenebroides mauritanicus*. It is a pest of stored grain

caducous literally, withered, but used to describe premature shedding of leaves or components of the floral envelope; or arthropod bristles

caecilian a term popularly applied to any worm-like burrowing amphibian of the order Apoda

black caecilian *Ichthyophis monochrous*

sticky caecilian any of several species of the genus Ichthyophis

caecum any blindly ending pouch. Without qualification, usually refers to the pouch coming off, in mammals, from the junction of the ileum and the colon, at the end of which in some forms, there is the vermiform appendix

enteric caecum any diverticulum of the alimentary canal posterior to the stomach

hepatic caecum the rudimentary pouch at the anterior end of the embryonic alimentary canal from which the liver is subsequently developed

pyloric caecum a caecum arising from the junction of the pyloric stomach and the intestine in some fish

-caeno- *see* -kaino-

Caenolestidae a family of marsupial mammals known from only a few species taken in South America. In structure they far more closely resemble the dasyurids than the didelphids

caeruleous sky-blue

Caesalpinioideae a subfamily of Leguminosae distinguished from the Mimosoideae by the zygomorphic flowers and from the Lotoideae by having the posterior petal (standard) innermost and the wing petals (keel) outermost

caesious pale gray

caespitose tufted

cafta the celastraceous shrub *Catha edulis*

caguan = cobego

cahow the procellariid bird *Pterodroma cahow*

caiman (*see also* lizard) a genus of Central and Southern American alligatorid reptiles. The word is interchangeable in Latin or English

Cairini the group of Anatinae containing the perching ducks. They closely resemble the Mergini but lack the diving habit of these forms

cajupet *see* tree

cal *see* [coal]

calabazilla the cucurbitaceous vine *Cucurbita foetidissima*

calamarian sedge-like

calamistrum a row of curved spines on the metatarsus of spiders

calamondin *see* orange

Calamophyta a phylum of pteridophyte plants containing the forms commonly called horsetails. They are distinguished from other pteridophytes by the presence of a whorl of leaves at each node. The majority of groups are known only as fossils, and are not therefore listed in this dictionary

calamus literally a "shaft" but particularly applied to the bare quill base of a feather

calander the alauidid bird *Melanocorypha calandra* (= calandra lark)

Calanoidea an order, often regarded as a suborder (Calanoida), of copepod crustacea distinguished by the fact that the fifth thoracic segment is firmly attached to the fourth and there is a moveable articulation between the sixth and seventh. They are exclusively marine planktonic forms

intercalary literally "inserted" but referring usually to growth which takes place neither at the base nor at the tip

-calath- *comb. form.* meaning "vase shaped"

calcaneum that tarsal bone which lies between the head of the fibula and the cuboid

calcar a spur

ecalcarate spurless

Calcarea a class of Porifera distinguished by the presence of a calcareous skeleton

calcareous pertaining to chalk or to its color or texture, or, improperly, to other calcium compounds

-calce- *comb. form.* meaning "slipper"

calceous dull white

calces *plural of* calx

calciferol = vitamin D_2

calcification the deposition of calcium salts in tissue (*cf.*) ossification)

calcipete seeking chalky soil

calcospherite calcareous granules found in the fat bodies of some insects

calcule one of numerous tactile scales surrounding the crestals of Shinisaurus

Calenduleae a tribe of tubiflorous Compositae

calf without qualification, the young of domestic cattle though the term is widely applied to other artiodactyls such as the bison, buffalo, camel, hartebeest, etc., and also sometimes to the sea lion and seal, for which, however pup is preferable

Caligoidea an order, often regarded as a suborder (Caligoida), of copepod crustacea almost all of which are parasites of fishes. The temporary ectoparasites may be distinguished by the fact that the fourth thoracic segment forms a moveable articulation with the fifth segment but is firmly attached to the fifth. In some permanent parasites, however, all articulation is lost

caliper the anal forceps of the earwig

caliph *I.P.* any of several species of amathusiid lepidopteran insects of the genus Enispe

calla properly the araceous herb known as the water arum (*Calla palustris*). Commonly used in horticulture for several species and numerous horticultural hybrids of the araceous genus Zantedeschia

black calla the araceous herb *Arum palestinum*

common calla *Z. aethiopica*

golden calla *Z. elliottiana*

pink calla = rose calla

rose calla *Z. albo-maculata*

black throated calla *Z. melanoleuca*

yellow calla *Z. oculata*

Callaeidae a small family of passeriform birds erected to contain the New Zealand wattlebirds that derive their name from the large orange or blue wattles at the corners of the mouth

Callianassidae a family of anomuran decapod crustacea lacking podobranchs on any of the legs. Commonly called ghost shrimps

Calliceratidae = Ceraphronidae

Callichthyidae a family of armored catfishes in which

the armor plates are confined to two longitudinal rows along the sides

elm **calligrapher** *U.S. Calligrapha scalaris*

Callimomidae = Torymidae

Callionymidae a family of acanthopterygian fishes distinguished by the presence of a sharp preopercular spine and/or hook and a very small gill opening. Commonly called dragonets

Callionymoidea a suborder of acanthopterygian fishes containing the single family Callionymidae. The characteristics of the order are those of the family

Calliphoridae a large family of stout myodarian cycloraphous dipteran insects commonly called blow flies. Many are metallic blue (blue bottles) or metallic green (green bottles)

Callithricidae a family of hapaloid primate mammals containing the forms commonly called marmosets and tamarins

Callitrichaceae an order of aquatic geraniale dicotyledonous angiosperms containing the forms commonly called water starworts. They are distinguished by the presence of solitary flowers lacking a perianth and subtended by two horn-like bracteoles in leaf axils.

Callitrichineae a suborder of geraniale dicotyledonous angiosperms containing the single family Callitrichaceae

Calloideae a subfamily of the Araceae

calloo *Brit. dial.* the aythyine bird *Clangula hyemalis* (= oldsquaw)

Callorhynchidae a monogeneric family of Holocephali distinguished by the long flexible nose

callus (*see also* rod, cushion) any hardened or thickened part, usually an external part. *bot.* a parenchymatous tissue formed under wounds

  definitive callus the accumulation of callose around a series of sieve elements, the function of which has usually been lost

Calobatidae = Trepidariidae

Calobryales an order of hepaticate bryophyta distinguished by the presence of erect leafy gametophytes each bearing three vertical rows of similar leaves

-calori- *comb. form* meaning heat. In compounds, -thermo- is more commonly used

calotte 1 (*see also* calotte 2, 3) a rosette of four cells covering the anterior end of Dicyemids

calotte 2 (*see also* calotte 1, 3) the polar cap of a holoblastically cleaving egg

calotte 3 (*see also* calotte 1, 2) the anterior, usually white, tip of a nematomorph worm

calthrop a tetraxonic sponge spicule in which all the rays are of equal length

caltrop *U.S.* the zygophyllaceous herb *Tribulus terrestris*

  water caltrop the hydrocarytaceous aquatic herb *Trapa natans* (= water chestnut)

calvarium the dome of the anthropoid cranium

calx the heel, or the portion of a limb corresponding to the heel in other forms

-calyb- *comb. form.* meaning "cottage"

-calyc- *comb. form.* meaning a "shallow cup" (*cf.* -cyath-)

calyculus dimutive of calyx

calybio a single celled dry fruit produced from an inferior ovary as are many nuts

Calycanthaceae a small family of ranalous dicotyledons distinguished by their aromatic bark and magnolia-like flowers

calycine pertaining to the calyx

  acalycine lacking a calyx

calycle a whorl of bracts outside the true calyx

calyculate pertaining to a structure of bracts imitating a calyx

calyculous having the form of small cups

calymma the extracapsular, usually frothy, protoplasm of radiolarians

calyphyomy the condition in which sepals adhere to petals

calypter the dipterous alula when it is sufficiently large to cover the haltere

-calyptr- *comb. form.* meaning a "blunt, squat, conical hat" or a "veil"

calyptra applied, particularly by botanists, to almost any cap or crowning structure, and particularly to a flat or conical structure covering a flower or fruiting body and dehiscing from it in one piece

Calyptraeidae a family of gastropod Mollusca with a conical uncoiled shell differing from all other similar Mollusca by the possession of a cup-like process on the inner side of the shell

Calyptratae an assemblage of those families of myodarian schizophoran cyclorrhaphous brachyceran dipteran insects which have calypters on the wings

calyptrogen the independent meristem concerned with the formation of a root cap

calyx 1 (*see also* calyx 2) any cup-like or funnel shaped structure (*cf.* infundibulum) particularly the area immediately surrounding the renal papilla in many mammalian kidneys. Also to one of many pockets in the surface of the intromittent organ of many snakes, to that portion of an Entroproct that contains the viscera in distinction to the stalk, to the aboral cup of a crinoid echinoderm, and to a spicule-containing basal portion of the anthocodium of some Alcyonaria

  egg calyx the upper opening of the oviduct

  nectocalyx = nectophore

calyx 2 (*see also* calyx 1) the outer cycle of the appendages of a flower consisting of sepals which may be distinct or marginally connate

  adenocalyx one containing many glands

  caulocalyx = pseudoperianth

  epicalyx an involucre resembling a calyx

  inferior calyx one that is below the ovary

  [macrocalyx] a mutant (gene symbol *mc*) mapped at 46 on chromosome 7 of the tomato. The sepals of the phenotype are unusually large

camass *U.S.* any of several liliaceous herbs of the genus Camassia

  death camass *U.S.* any of several liliaceous herbs of the genus Zigadenus

cambium 1 (*see also* cambium 2) the inner layer of the periosteum

cambium 2 (*see also* cambium 1, zone) that meristem which adds new vascular tissues thereby increasing the girth of a plant stem or root

  cork cambium = phellogen

  fascicular cambium cambium originating within fascicular bundles

  interfascicular cambium cambium originating from intervascicular parenchyma

  procambium partially differentiated derivative of apical meristem that will later give rise to the vascular cambium and the primary vascular system of the plant

  pseudocambium meristem resembling cambium

storied cambium = stratified cambium

stratified cambium　cambium in which the fusiform initials (q.v.) occur in horizontal tiers

nonstratified cambium　cambium in which the fusiform initials are not arranged in horizontal tiers

vascular cambium　that secondary meristem which produces secondary xylem and secondary phloem growth (= vascular meristem)

camel　either of two large camelid artiodactyls of the genus Camelus

Arabian camel　C. dromedarius. distinguished by the presence of one hump

Bactrian camel　C. bactrianus distinguished by the presence of two humps

dromedary camel = Arabian camel

camellia　properly the name of the ternstroemiaceous genus of cultivated shrubs. Used also incorrectly for the theaceous genus Stewartia

Camelidae　a family of ruminant artiodactyle mammals containing the camels and dromedaries. They are distinguished by their long limbs which lack second and fifth toes and are unique in possessing oval red blood corpuscles.

camelinous　dirty yellow-brown (i.e. the color of a camel)

cameos　popular name of cassidid gastropods

camomile　any of numerous species of the compositaceous genus Anthemis

-camp-　comb. form hopelessly confused between two Gr. and one L. roots and therefore variously meaning "caterpillar" (properly -campode-), "marine" and "plain" in the sense of a flat area of ground

Campanulaceae　that family of campanulalous flowers which contains, inter alia, the bellflowers. The gamopetalous epigynous flower with many ovules and frequently united stamens is distinctive

Campanulales = Campanulatae

Campanulariae　a colonial marine hydrozoan coelenterate distinguished from the Tubulariae by the presence of a hydrotheca on the hydranth. The medusoid forms were at one time placed in a separate order the Leptomedusae

Campanulatae　that order of dicotyledonous angiosperms which contains, inter alia, the bell flowers, the lobelias, and all the compositae

Campephagidae　that family of passeriform birds which contains the cuckoo-shrikes. These Asiatic and Australian birds are usually highly colored and frequently transversed with stripes. The tail broadens toward the end and is forked in some

campestral　pertaining to fields

camphor　(see also scale, thrip, tree, weed) any of several fragrant terpenoid ketones

campion 1　(see also campion 2) any of several species of herb of the caryophyllaceous genera Silene and Lychnis

bladder campion　S. latifolia

evening campion　L. alba

moss campion　S. acaulis

red campion　L. dioica

rose campion　L. coronaria

starry campion　S. stellata

campion 2　(see also campion 1) Brit. the noctuid lepidopteran insect Dianthoecia cucubali

-campode-　comb. form. meaning "caterpillar" (cf. -eruc-)

Campodeidae　a family of thysanuran insects distinguished by the absence of styli on the first abdominal segment

Campophagidae = Campephagidae

campt-　comp. form. meaning bend

campylotropous　see ovule

Campynematoideae　a subfamily of amaryllidaceous monocotyledons

can　see [Canavanine]

canʳ　a yeast genetic marker for growth inhibition by canavanine

Canaceidae　a group of minute myodarian cycloraphous dipteran insects with larvae breeding in brackish water

canaigre　the polygonaceous herb Rumex hymenosebalus

canal　(see also raphe) a term commonly used in the biological sciences to denote a narrow tube

alar canal　a horizontal channel through the alisphenoid of some mammals basically for the internal maxillary artery

Bidder's canal = marginal canal

brachial canal　the canal in the oral arm of Schyphozoa

carenal canal　a water canal opposite a ridge of the stem of Equisetum

semicircular canal　those tubes in the inner ear which are concerned with balancing

gynecophoric canal　that in-curved portion of the ventral surface of male schistosomatid flukes in which the female is carried

conjugation canal　a connection between algal gametes and between conjugating bacteria

copulation canal　a canal used for copulation in some hermaphrodite invertebrates, but which is not a vagina in the sense of being confined to the female part of the reproductive system

Cuvier's canal = sinous venousus

intracytoplasmic canal　any canal that passes through the substance of the cytoplasm but particularly such canals when they constitute part of the excretory system of nematodes

festoon canal　a ring canal following the lappets in those medusae which lack a manubrium

girdle canal　intercellular air spaces round palisade cells

Haversian canal　the longitudinal blood vascular channel in bone (cf. osteon)

insemination canal　a canal leading from a seminal bursa to an oviduct

Kupffer's canal　an outgrowth from the mesonephric duct to metanephric units

Laurer's canal　a copulation canal in some trematode platyhelminths

marginal canal　a canal formed by the fusion of mesonephric units growing towards the testes, where they will subsequently become vasa efferentia

neck canal　the elongated portion of the archigonium

neural canal 1　(see also neural canal 2) the cavity of the developing central nervous system

neural canal 2　(see also neural canal 1) the arch of ventral apodemes, protecting the nerve cord, in the thorax of arthropods

neurenteric canal　the connection, that is the remnants of the blastopore, between the archenteron and the neural canal

pit canal　the connection between the lumen of a pit in a plant wall and the pit chamber when the border is unusually thick

podial canal　a branch of the water-vascular system of holothurian echinoderms, that supplies the podia

pore canal 1　(entomol.) (see also pore canal 2) in

insects, the cuticular channel connecting two or more sensory setae

pore canal 2 (*bot.*) (*see also* pore canal 1) a canal joining two pits

radial canal gastrodermal canals radiating from the coelenteran of medusoid coelenterates

ring canal the circum-oral portion of the water-vascular system of echinoderms

stone canal the tube connecting the water-vascular system of echinoderms to the hydropore

stylar canal the loose tissue in a style through which the pollen tube passes

Volkmann's canal a transverse canal in bones joining Haversian canals to each other and to the periostium

Williamson's canal a canal penetrating the bony base of the dermal denticles or ganoid scales

canaliculate with a longitudinal groove

1-canaline an amino acid $NH_2OCH_2CH_2CH(NH_2)$-COOH most commonly known as a hydrolysis product of canavanine

canary 1 (*see also* canary 2, grass) properly any of numerous fringillid birds of the genus Serinus, usually, without qualification, *S. canarius.* The term is also applied in various parts of the world to almost any yellow bird particularly warblers, weavers, and (*U.S.*) goldfinches. Usually without qualification, *W.I.* the parulid *Dendroica petechia* (= yellow warbler), or the fringillid *Sicalis flaveola* (= saffron finch), *Austral.* the meliphagid *Miliphaga penicillata* (*cf.* white-plumed honeyeater)

bastard canary *W.I.* = canary *W.I.*

white-bellied canary *S. dorsostriatus*

brimstone canary *S. sulphuratus*

bush canary *N.Z.* the malurine *Mohoua ochrocephala*

yellow-crowned canary *S. flavivertex*

black-faced canary *S. capistralus*

yellow-fronted canary *S. mozambicus*

grass canary *W.I.* the fringillid *Sicalis luteola* (= yellow grass-finch)

grey canary *S. leucopygius*

grosbeak canary *S. donaldsoni*

guineahen canary *W.I.* the parulid *Dendroica pharetra* (= arrow-headed warbler)

Japanese canary *W.I.* the parulid *Mniotilta varia* (= black-and-white warbler)

mangrove canary *W.I.* = canary *W.I.*

wild canary *W.I. Sicalis flaveola* (= saffron finch)

canary 2 (*see also* canary 1) *S.Afr.* the pierid lepidopteran insect *Dixeia spilleri*

canestero any of numerous funariid birds of the genus Asthenes

1-canavanine an amino acid $NH_2C(= NH)-NH-O\cdot CH_2-CH_2CHNH_2-COOH$ first isolated from jackbean meal

[Canavanine] a mutant (gene symbol *can*) in linkage group I of *Neurospora crassa.* The phenotypic expression is resistance to canavanine.

Cancellariidae a family of gastropod mollusks having solid cone shaped shells markedly ribbed along their surface

cancellate coarsely reticulate

cancellous spongy

Cancridae a family of brachyuran crustacea containing many of the larger edible crabs of the world. They are distinguished by the broadly oval carapace, the anterior margin of which is arched and serrate, and the pointed tips of the last pair of legs

Cancromidae = Cochleariidae

candelabra a tetraxon sponge spicule with branched rays

candiru a trichomycterid catfish (*Vandellia cirrhosa*) found in South America, that enters the genital apertures of humans bathing in rivers. It has for this reason been described as the only vertebrate parasite of man

bog-candle *U.S.* the orchid *Habenaria dilatata*

swamp candle the primulaceous herb *Lysimachia terrestris*

cane (*see also* borer, bug, roller) properly applied to woody grasses of the genus Arundinaria but often used for other forms particularly sugar cane

large cane the grass *Arundinaria macrosperma*

maiden cane *U.S.* the grass *Panicum hemitomon*

scotch cane = small cane

small cane the grass *Arundinaria tecta*

sugar cane the grass *Saccharum officinarum*

switch cane = small cane

Canellaceae a small family of parietale dicotyledonous angiosperms distinguished by the fact that the stamens are joined by the filaments into a tube. The so-called wild cinnamon (= cassia) of some writers is derived from a tree (*Canella alba*) in this family

canellaceous pertaining to cinnamon

canescent becoming gray with old age

Canidae a family of carnivorous mammals containing the dogs, wolves, jackals and their immediate allies. The family is distinguished by the large smooth, auditory bulla and the long and prominent paroccipital process. All are furnished with short, thick, nonretractile claws

Caninae a sub-family of canid, carnivorous mammals, commonly called the "true dogs" and containing the wolves, jackals, and foxes. They differ from the "false dog" (Cymocyoninae *q.v.*) in possessing three or four lower molars in each jaw and five toes on each foot

canna a tall ornamental monocotyledonous plant, with large brightly colored flowers, of the family Cannaceae

bastard canna the cucurbitaceous herb *Curcuma longa* (= turmeric)

Cannabinaceae a family of moraceous plants erected to contain the groups usually contained within the subfamily Cannaboideae

Cannaboideae a subfamily of Moraceae distinguished by the short, straight stamens and the absence of a milky sap

Cannaceae a small family of scitaminous monocotyledons containing the genus Canna. The family is very closely related to the Marantaceae by having many ovules in each cell of the ovary

cannibal an animal that devours individuals of its own species

Canoidea that superfamily of fissipede Carnivora which includes *inter alia* the dogs, bears, raccoons, skunks, and weasels

canon a law (*q.v.*) or rule

Morgan' canon a biological restatement of the principle better known as Ockham's (Occam's) razor

canopy 1 (*see also* canopy 2) a membrane lying between the testa and nucellus in some seeds

canopy 2 (*see also* canopy 1) any part that shades another, as the upper branches of trees in a forest

canous hoary

cantaloupe *U.S.* = muskmelon

[cantaloupe] a mutant (gene symbol *c*) in linkage group

I of *Habrobracon juglandis*. The phenotypic eyes are light pink darkening later to a deep red

canter a three beat method of locomotion, resembling a gallop (*q.v.*) but much slower

canthal pertaining to the canthal ridge

Cantharellaceae a family of cantharellale basidiomycete fungi distinguished by possessing upright vaselike fruiting bodies. They are commonly called by their French name of chanterelle

Cantharellales a class of basidiomycete fungi characterized by a rather stiff upright growth that may be branched like a coral in the Clavariaceae or cup-shaped as in the Cantharellaceae

Cantharidae a large family of soft-bodied beetles commonly called soldier beetles. They are closely allied to the Lampyridae from which they may be distinguished by the fact that the head protrudes beyond the pronotum

canthariden a colorless crystalline strongly vesicant compound extracted from the body fluids of the meloid coleopteran insect *Lytta visicatoria*. It is an extremely dangerous poison at one time misused by the ignorant as an aphrodisiac. The name derives from the fact that the insect was once called *Cantharis vesicatoria* and placed in the family Cantharidae

Cantharoidea a superfamily of coleopteran insects containing the soldier beetles, lightningbugs, and their allies

Canthigasteridae a family of plectognathous fishes distinguished from the closely allied Tetraodontidae by the long sharp nose. Commonly called sharp nose puffers

canthus the partition that divides the eyes of some insects into upper and lower portions

cap 1 (*see also* cap 5–6) a cap shaped mass of cells

    bundle cap a group of parenchyma cells lying between a leaf vein and epidermis and not concerned with conduction

    rootcap a cap of parenchyma cells lying immediately distal to the root meristem covering it

cap 2 (*see also* cap 1, 3–6) a cap like mass of protoplasm

    knee cap (*see also* shell) = patella bone

    polar cap a solid mass formed at the poles of the nuclei in some Protozoa undergoing cell division

cap 3 (*see also* cap 1–2, 4–6) a cap-like organ

cap 4 (*see also* cap 1–3, 5, 6) a fungus having a distinctive pileus

    brick cap *U.S.* the strophariaceous agaricale basidomycete fungus *Naematoloma sublateritium*

    inky cap the coprinaceous agaricale fungus *Coprinus stramentarius*

cap 5 (*see also* cap 1–4, 6) an herb with a cap-shaped flower or a distinctively capped fruit

    bishop's cap any of several species of saxifragaceous herbs of the genus Mitella

    black cap (*see also* cap 6) = black-cap raspberry

    skullcap any of numerous species of labiateous herb of the genus Scutellaria

        Hyssop skullcap *S. integrifolia*

cap 6 (*see also* cap 1–5) a bird with a distinctive "cap" of feathers

    blackcap (*see also* cap 5, akalat) any of numerous birds having a black crown to the head. *U.S.*, local, usually various chickadees. *Brit.* usually the sylviid *Sylvia atricapilla* though the term is also applied to the stone chat and reed-bunting

    bush blackcap *Afr.* the pycnonotid *Lioptilornis nigricapillus*

    blue-cap *Brit.* the parid *Parus caeruleus* (= blue titmouse)

    firecap the parid *Cephalopyrus flammiceps*

    red cap = goldfinch

    snow cap = the trochilid *Microchera albocoronata*

carrying capacity the ability of a habitat to support a population or the maxium level of such a population

capelin the osmerid fish *Mallotus villosus*

caper (*see also* spurge) the edible flower buds of several shrubs of the genus Capparis of the family Capparidaceae

capercailzie a very large European tetraonid bird *Tetrao urogallus*

capercally = capercailzie

capillaceous pertaining to, or having the shape of, a hair

capillary (*see also* capillary 2) those minute blood vessles, lacking any muscular wall, which connect the arteries and veins

capillary 2 (*see also* capillary 1) in *bot.* used in the sense of hair-like, as applied in some roots and to root hairs

-capit- *comb. form.* meaning "head"

capitate terminating in a head or head-like structure

Capitelliformia an order of phanerocephalous polychaete annelids with a conical prestomium and no prestomium processes

Capitonidae that family of piciform birds which contains the barbets. They are distinguished from other piciform birds by the tufts of feathers over the nostril and well developed chin bristles

capitulum 1 (*see also* capitulum 2, 3, 4) the upper of the two articular processes of the rib

capitulum 2 (*see also* capitulum 1, 3, 4) the upper thin-walled region of the "body" of an Anthozoan (*cf.* scapus)

capitulum 3 (*see also* capitulum 1, 2, 4) the gnathosoma of Acarina

capitulum 4 (*see also* capitulum 1–3) an aggregation of small flower heads into a usually dense terminal cluster

Capniidae a family of plecopteran insects with few or no medial and cubital crossveins

-capnod- *comb. form.* meaning "smoke"

capnodous smoky gray

capon a castrated domesticated cock

caponet a capon-like form induced by the injection of hormones

Capparidaceae that family of rheodale dicotyledons which contains *inter alia*, the capers, and spider-plants. They are with difficulty distinguished from the closely related Cruciferae, though the one-celled ovary is characteristic

Capparidineae a suborder of Rhoedales containing the Capparidacea, the Cruciferae, and Tovariacae

capparinous the brownish sage-green color of capers

Caprellidea a suborder of amphipod Crustacea distinguished by the presence of six free thoracic segments, best known from the skeleton shrimps and whale lice

capreolate having tendrils

Capreolinae a subfamily of cervid artiodactyl mammals containing the forms commonly called roe deer

Capridae a family of artiodactyle mammals which contains the goats. It is almost impossible to draw a clear line between the Capridae and the Antilopidae (*cf.* goat-gazelle, gazelle-goat)

caprifier a generalized term for agaontid hymenopterous insects that fertilize figs (= fig insect, fig wasp)

Caprifoliaceae that family of rubialous dicotyledons which contains the honeysuckles, the elders, and the snow berries. The family may be distinguished from the closely related Rubiaceae by the absence of stipules

Caprimulgidae a large family of caprimulgiform birds containing the nightjars. They are distinguished from other caprimulgiforms by the very long rictal bristles and very large eyes. The beak is small and weak but has a very wide gape

Caprimulgiformes that order of birds which contains the goatsuckers and their allies. They are distinguished by their short, weak, somewhat hooked bill with imperforate nostrils and by their loose fluffy plumage

Caprinae a subfamily of caprid artiodactyl mammals containing the "true goats", or at least those animals which more or less resemble the domestic form

Capromyidae that family of hystricomorph rodents which contains, *inter alia,* the hutias, agoutis and coypus, having little in common save that they cannot be reasonably placed in any other family

-caps- *comb. form.* meaning "box"

capsaicin the phenolic amide that causes the "hot" flavor of capsicum peppers (*see* pepper 2)

capsella = achene

capsule 1 (*see also* capsule 2–5) a dry fruit usually produced from a compound ovary (that from a simple ovary is a follicle)

circumscissile capsule one whose line of dehiscence separates the top from the main body as a cap or calyptra

loculicidal capsule one whose line of dehiscence opens into the carpellary cavity

septicidal capsule one whose line of dehiscence is along the point of union of septa with capsule wall

pore capsule (obs.) = poricidal capsule

poricidal capsule a capsule which dehisces through numerous holes

capsule 2 (*see also* capsule 1, 3–5) spherical, or subspherical, portions of the skull

cranial capsule that part of the skull which encloses the brain

nasal capsule that portion of the chondrocranium which encloses the nasal organs

olfactory capsule = nasal capsule

optic capsule the thin cartilagenous capsule that encircles, or lies above, the orbit in chondrocrania

otic capsule the postero-lateral capsule of a chondrocranium containing the inner ear

capsule 3 (*see also* capsule 1, 2, 4, 5) any thin-walled globose structure

Bowman's capsule = glomerular capsule

brood capsule capsules within a hydatid cyst, producing secondary scolices

cartilage capsule the layer of matrix immediately surrounding a cartilage cell

Glisson's capsule the connective tissue sheathing of vessels, both bilary and hepatic, within the liver

glomerular capsule the swollen termination of a metanephric or mesonephric unit containing the glomerulus

uterine capsule a detached portion, containing one to several embryos, broken off from the uterus in cestodes

capsule 4 (*see also* capsule 1–3,5) the spore containing organ of bryophytes

capsule 5 (*see also* capsule 1–4) miscellaneous capsular structures not covered in capsule 1–4 above

adhesive capsule = isorhiza

bacterial capsule a thickened layer, usually polysaccharide, around some bacteria

polar capsule an oval body attached to the spores of cnidosporidia containing a nematocyst-like thread that attaches the spore to the wall of the digestive tract of the host

candelillo the euphorbiaceous herb *Euphorbia antisyphilitica*

candicant shiny white

captaculum a filamentous tentacle arising from the head of scaphopod Mollusca

capuchin (*see also* bird) any of several S. Amer. monkeys of the genus Cebus

Capulidae a family of gastropod Mollusca of conical uncurved shape but with a slight twist at the extreme end of the shell

capybara the hydrochoerid hystricomorph rodent *Hydrochoerus capybara*. It is the largest known rodent sometimes reaching a weight of considerably more than 200 lbs. (= carpincho)

car *see* [carnation]

Carabaeoidea a superfamily of adephagous coleopteran insects containing both the predatory land beetles and the predatory water beetles

Carabidae a very large family of coleopteran insects containing the predatory forms commonly called ground beetles

caracal the felid carnivorous mammal *Lynx caracal* (*cf.* lynx)

Caracanthidae a remarkable family of scleroparous fishes greatly flattened and so covered in spines as to appear hairy

caracara a name commonly applied to all polyborine birds but most properly to those of the genera Caracara, Daptrius and Milvago

Audubon's caracara = crested caracara

crested caracara *C. plancus*

Guadeloupe caracara *C. lutosus*

yellow-headed caracara *M. chimachima*

red-throated caracara *D. americanus*

caracol the leguminous herb *Phaseolus caracalla*

[Caracul] a mutant (gene symbol *Ca*) in linkage group VI of the mouse. The phenotypic expression is short wavy hair and whiskers

carairi = caracara

caraks *S. Afr.* widely used anglicized form of Charaxes, a genus of nymphalid lepidopteran insects

carambola the oxalidaceous tree *Averrhoa carambola* and its fruit

Carangidae a large family of percid acanthopterygii distinguished by the bony scutes or plates that extend along the caudal peduncle and, in some species, along the whole lateral line

carapace any dorsal skeletal shield over an animal. It is applied equally to the chitinous structures of arthropods, the horny structures of armadillos, or the bony structures of turtles

Carapidae a small family of ophidioid fishes distinguished by their extremely narrow shape. The great majority live symbiotically within other animals, particularly sea cucumbers, sea urchins, and tunicates

caraunda the apocynaceous shrub *Carissa carandas*

caraway the umbelliferous herb *Carum carvi*

carboxylase a group of ligase enzymes catalyzing the attachment of carbon dioxide, using energy derived from the breakdown of ATP and, therefore, more properly called ligases

   acetyl-CoA carboxylase catalyzes the production of malonyl-CoA from acetyl-CoA

   methyl crotonoyl-CoA carboxylase catalyzes the production of 3-methylglutaconyl-CoA from 3-methylcrotonyl-CoA

   propionyl-CoA carboxylase catalyzes the production of methylmalonyl-CoA from propionyl-CoA

   phosphoribosyl-amino-imidazole carboxylase catalyzes the production of 5'-phosphoribosyl-5-amino-imidazole from 5'-phosphoribosyl-5-amino-4-imidazolecarboxylate

   pyruvate carboxylase catalyzes the production of oxaloacetate from pyruvate

   phosphopyruvate carboxylase catalyzes the production of GDP and phosphoenolpyruvate from GTP and oxaloacetate

decarboxylase a large group of enzymes, more properly carboxylases that catalyze the removal of CO_2

   acetoacetate decarboxylase catalyzes the production of acetone from acetoacetate

   oxaloacetate decarboxylase catalyzes the production of pyruvate from oxaloacetate

   aconitate decarboxylase catalyzes the production of itaconate from cis-aconitate

   aminomalonate decarboxylase catalyzes the production of lycine from aminomalonate

   arginine decarboxylase catalyzes the production of agmatine from arginine

   carbamoylaspartate decarboxylase catalyzes the production of carbamoyl-β-alanine from carbamoylaspartate

   aminobenzoate decarboxylase catalyzes the production of aniline from aminobenzoate

   malonyl-CoA decarboxylase catalyzes the production of acetyl-CoA from malonyl-CoA

   oxalyl-CoA-decarboxylase catalyzes the production of formic acid and CoA from oxalyl-CoA

   pantothenoylcysteine decarboxylase catalyzes the production of pantetheine from N-(ʟ-pantothenoyl)-ʟ-cysteine

   DOPA decarboxylase catalyzes the production of dihydroxyphenylethylamine from DOPA

   benzoylformate decarboxylase catalyzes the production of benzaldehyde from benzoylformate

   glutamate decarboxylase catalyzes the production of 4-aminobutyrate from glutamate

   hydroxyglutamate decarboxylase catalyzes the production of 3-hydroxy-4-aminobutyrate from ʟ-33-hydroxyglutamate

   histidine decarboxylase catalyzes the production of histimane from histidine

   acetolactate decarboxylase catalyzes the production of acetoin from acetolactate

   lysine decarboxylase catalyzes the production of cadaverine from lysine

   ornithine decarboxylase catalyzes the production of putrescine from ornithine

   oxalate decarboxylase catalyzes the production of formate from oxalate

   orotidine-5'-phosphate decarboxylase catalyzes the production of UMP from orotidine 5 -phosphate

   diaminopimelate decarboxylase catalyzes the production of ʟ-lysine from *meso*-2, 6-diaminopimelate

   pyruvate decarboxylase in the presence of thiamine pyrophosphate, catalyzes the production of an aldehyde from 2-oxo-acids

   cysteinesulphinate decarboxylase catalyzes the production of hypotaurine from ʟ-cysteinesulphinate

   tryptophan decarboxylase catalyzes the production of tryptamine from tryptophan

   hydroxytryptophan decarboxylase catalyzes the production of 5-hydroxytryptamine from 5-hydroxy-ʟ-tryptophan

   tyrosine decarboxylase catalyzes the production of tyramine from tyrosine

   valine decarboxylase catalyzes the production of isobutylamine from valine

   pyrophosphomevalonate decarboxylase catalyzes the production of ADP and orthophosphate and isopentenyl pyrophosphate from ATP and 5-pyrophosphomevalonate

carcerule a dry, indehiscent multilocular superior fruit

Carcharhinidae a large family of sharks commonly called grey, or requiem, sharks

Carchariidae a family of sharks popularly called sand sharks. They are distinguished by the elongated pre-oral rostrum and the reduction, or absence, of a spiracle

carcithium (*obs.*) = mycelium

-card- *comb. form* confused between a *L.* root meaning "hinge" or "pivot" and a *Gr.* root meaning "heart"

cardamon the zingiberaceous herb *Elettaria cardamomum*

   false cardamon *E. granum-paradise* (= grains of paradise)

   large cardamon *E. cardamomum*

   small cardamon the zingiberaceous herb *Amomum cardamom*

cardate hinged. Used particularly of the rotiferan mastax in which the rami are hinged to the fulcrum

cardelle (Ectoprocta) = condyle (Ectoprocta)

wool carden *Brit.* the apid hymenopteran insect *Anthidium manicatum*

cardia the anterior end of the insect midgut (= gizzard)

cardiac pertaining to the heart

   myocardiac pertaining to cardiac muscle

Cardiidae a family of pelecypod mollusks containing the forms known as cockles. The Latin name derives from the heart-shaped shell, and they are sometimes called heart clams

cardinal 1 (*see also* cardinal 2) any of several fringillid birds of the genera Richmondena, Paroaria and Gubernatrix *U.S.*, without qualification. *R. cardinalis*. The word is also applied in various parts of the world to any red bird including tanagers, and weavers

   Brazilian cardinal *P. gularis*

cardinal 2 (*see also* cardinal 1) in *entomol.* pertaining to the cardo (*q.v.*) and in *comp. anat.* to a sinus (*q.v.*)

[cardinal] a mutant (gene symbol *cd*) mapped at 76.2 on chromosome III of *Drosophila melanogaster*. The phenotypic eyes are dull scarlet

cardinalia the sum total of the skeletal structures in the beak region of the dorsal valves of a brachiopod

cardines *plural of* cardo

cardita popular name of carditid pelecypods

Carditidae a family of pelecypod mollusks with equal strongly ridged valves having an erect heavy tooth under the umbones

cardium heart, but used mostly in compounds

   endocardium the inner epithelial layer of the heart

   epicardium *either* a backwardly directed prolongation

from the pharnyx of some urochordates *or* the outermost layer of cells of the heart, or embryonic heart

mesocardium the vertical mesentery that primitively unites the ventral side of the heart in the embryo to the ventral surface of the body

myocardium that part of a developing heart which gives rise to the muscular wall

cardo this term has been variously applied to almost any hinged or pivoted object, particularly to the shells of pelecypods and brachiopods, and to the basal portions of several insect appendages, particularly to the proximal joint of the protopodite and the hinge of the maxilla. It is also applied to a basal ring of some insect genitalia

alacardo the ventro-lateral portion of the cardo of the insect maxilla

Carduales an order of dicotyledonous angiosperms erected to contain the family Compositae when this is not regarded as belonging with the Campanulatae

cardoon the compositaceous herb *Cynara cardunculus.* The blanched midribs of the leaves are eaten

Cardueline a family of passeriform birds commonly included with the Ploceidae or the Fringillidae

carena *see* canal

Carettochelyidae a family of cryptodiran chelonian reptiles in which the neck cannot be retracted and the skin is soft without horny shields

carey = tortoise shell

Cariamidae a family of gruiform birds containing the two species popularly called the cariama or chunga. They are distinguished by their very long legs, naked tibia, semi-palmate back, and hair-like nuchal crest

carib any of several trochilid birds of the genera Eulampis and Sericotes

    green-throated carib *S. holosericus*

    purple-throated carib *E. jugularis*

caribe = piranha

caribou *U.S.* any of three species of cervid artiodactyle of the genus Rangifer of which one, *R. tarandus,* is more commonly spoken of as the reindeer

Caricacea that family of parietalous dicotyledons which contains the pawpaw (papaya) tree. The large, melon-like fruit is typical

Caricoideae a subfamily of Cyperaceae

carina literally, a keel. Particularly, in *bot.*, keeled petals of flowers and in *zool.* the ridge of keel along the bottom of a bird's sternum

Carinatae a taxon of birds erected to include those in which the sternum is keeled to provide for the attachment of flight muscles. It is used in contrast to Ratitae

carinate keeled

carinula diminuitive of carina. Particularly, applied to the ridge along the snout of weevils

carious corroded, moth-eaten, shabby

[carmine] a mutant (gene symbol *cm*) mapped at 18.9 on the X chromosome of *Drosophila melanogaster.* The name refers to the phenotypic eye color

carnation (*see also* maggot) any of very numerous horticultural varieties of the caryophyllaceous herb *Dianthus caryophyllus*

[carnation] a mutant (gene symbol *car*) mapped at 62.5 on the X chromosome of *Drosophila melanogaster.* The name refers to the dark eyes of the phenotype

carneous flesh-colored

Carnidae a family of brachyceran dipteran insects commonly included with the Phyllomyzidae

Carnivora an order of placental mammals distinguished by their carnivorous habit and the adaptation of the teeth for this purpose. Among the better-known Carnivora are the dogs, cats, weasels, bears, walruses, and seals

Carnosa = Homosclerophora

carnosinase = aminoacyl-histidine dipeptidase

carnosine β-alanylhistidine. $[C_3H_3N_3]CH_2CH \cdot [COOH]NHCOCH_2CH_2NH_2$ amino acid common in muscular tissue but not essential to rat nutrition (*cf.* anserine)

carnous fleshy

carob the leguminous tree *Ceratonia siliqua* and particularly its edible seed pods

carotenoid a large class of oil soluble yellow, orange, or red biochromes

-carp- *comb. form* meaning "wrist" or "wrist joint." A frequent, and very confusing, transliteration of "-karp-" (*q.v.*)

carp 1 (*see also* carp 1, 2, 4) properly "karp," with reference to a fruit or part of a fruit of a vascular plant. Any of the derivative substantives here listed can be rendered as adjectives terminating in -carpic, or -carpous (*see* carpous 1)

achenocarp any dry indehiscent fruit

actinocarp a fruit in which the placentas radiate in a star shape

allocarp a fruit produced from cross fertilized flowers

ambleocarp a fruit containing a few seeds even though many ovules were formed

anthocarp the fruit in which asexual parts of the flower contribute to the whole

apocarp a fruit developed from a flower in which the carpels are separate

autocarp a fruit produced from a self-fertilized flower

bastardocarp a fruit produced by hybridization

coenocarp a collective fruit derived from an entire inflorescence

conocarp an aggregate fruit surrounding a conical receptacle

cremocarp a dry fruit consisting of two monospermous carpels surrounded by an epigynous calyx

dialycarp a fruit composed of distinct carpels

discocarp a collection of fruits within a hollow receptacle

endocarp the inner of the three layers into which the pericarp may differentiate

epicarp = exocarp

exocarp the outer of the three layers into which the pericarp may differentiate

geitonocarp a fruit produced in consequence of fertilization between flowers on the same plant

geocarp a fruit that ripens underground even though developed on an aerial flower

holocarp any fruit, irrespective of technical type, that is produced from a number of carpels (*but see* holocarpous 1 and 2)

helicocarp a fruit derived from spirally arranged carpels

hydrocarp said of a fruit that is produced under water from a flower fertilized above water

hypocarp an enlarged peduncle beneath the fruit

heteromericarp a fruit, the various parts of which differ from each other, rather than forming a symmetrical whole

macrobiocarp a fruit retained on the parent over a long period of time

mesocarp the middle of the three layers into which a pericarp may differentiate

monocarp a plant that bears fruit but once in a lifetime. It may be annual, biennial, or perrenial

paracarp an abortive female portion of a flower

parthenocarp a fruit produced without fertilization

pericarp that part of a fruit which develops from the wall of the ovary

podocarp a stipulate fruit (and also the common name of plants of the gymnospermous family Podocarpaceae)

pseudocarp a fruit such as a strawberry, containing parts not derived from the ovary

pyrenocarp = drupe (*see also* carp 2)

sarcocarp a fleshy fruit

schizocarp a pericarp that splits into one-seeded divisions

spermocarp the fruiting body of Characeae

sphalerocarp an accessory fruit

streptocarp a fruit with spiral markings on it

syncarp a multiple or aggregate fleshy fruit

    pseudosyncarp a collective fruit not properly a syncarp

trachycarp a rough coated fruit

xenocarp a fruit produced in consequence of xenogamy

xylocarp a woody fruit

carp 2 (*see also* carp 1, 3, 4) properly "karp," a fruit (*see* carp 2) and hence a reproductive ("fruiting") body of a lower plant. See remarks under carp 1 and -carpous 2)

amphicarp a fruit envelope produced from an archegonium

archicarp the group of cells fertilized in an ascomycete

ascocarp the fruiting body of the ascomycete fungi. It consists of an outer layer of interwoven hyphae and an inner layer of interwoven hyphae among which the asci develop

cleistocarp an ascocarp which does not rupture and from which spores escape after decay of the wall

cystocarp a walled reproductive body in red algae

desmidocarp a cystocarp of Balbiania

discocarp = apothecium

gloiocarp = tetraspore

gonimocarp a type of carposporophyte which is surrounded by the protective gametophytic tissue

oocarp = oospore

phylactocarp a modified hydrocladium, or branches from an hydrocladium, designed to protect the gonangia in some hydrozoan coelenterates

pyrenocarp = perithecium (*see also* carp 1)

sporocarp literally, a "spore bearing fruit." That is, a spore producing body derived from a zygote (*cf.* sporocyst)

podosyncarp a double moss capsule of which only one half is developed

zoocarp = oospore

carp 3 (*see also* 1, 2, 4) the cyprinid fish *Cyprinus carpio*

    leather carp a naturally occurring scaleless variation

    mirror carp a naturally occurring variety with very large scales along the sides of the body

carp 4 a fruit-like body on an animal

carpedeium = cremocarp

carpal pertaining *either* to the wrist *or* to the angle of the last joint of a bird's wing

carpale a genetic term for those bones (*q.v.*) which lie at the base of the metacarpals

carpel the morphological component of the female element of a flower. The entire pistil—ovary, style and stigma—is composed of carpellary tissue

carpellody the reversion of a petal into a carpel

[Carpet] a mutant (gene symbol *cpt*) in linkage group II of *Neurospora crassa*. The term is descriptive of the flat uniform growth of the phenotypic colony

carpet (*see also* beetle) *Brit.* any of very numerous geometrid lepidopteran insects

    least carpet *Brit. Acidalia rusticata*

    satin carpet *Brit.* the thyatirid *Palimpsestis fluctuosa*

    water carpet the saxifragaceous herb *Chrysosplenium*

-carpho- *comb. form* meaning a "splinter" or "twig"

-carpic *adj. suffix* synonymous with -carpous, the form preferred in this work

carpincho the caviid hystricomorph rodent *Hydrochoerus capybara* (= capybara)

-carpium = -carp

Carposinidae a small family of tortricoid lepidopteran insects containing, *inter alia,* the current fruitworm

carpous 1 (*see also* carpous 2) pertaining to the fruit of a vascular plant (*see* carp 1 above) These entries, to avoid duplication, do not include adjectival forms directly derived from the substantives given at carp 1

acarpous said of a plant lacking fruit

acanthocarpous having spiny fruit

acrocarpous bearing fruit on the termination of a shoot or stem

aerocarpous producing fruit above ground

    amphicarpous having two kinds of fruit, of which one may occur both in aerial and subterranean forms

anthocarpous fruits having accessory parts not derived from the ovary

aulacocarpous with furrowed or ridged fruit

bradycarpous bearing fruit in the season following that in which flowers were borne

cladocarpous bearing fruit on the termination of a lateral branch

gymnocarpous having a naked fruit, or a fruit (*e.g.* the sporocarp in the female cone of gymnosperms) having naked, open, carpels

hebecarpous downy fruited

heterocarpous having two or more kinds of fruit

holocarpous (*see also* carpous 2) having an entire pericarp

homocarpous possessing only one kind of fruit

lasiocarpous having a woolly-skinned fruit

monocarpous said either of a plant that bears fruit only once in its existence *or* of a flower containing a single pistil formed from one carpel

pachycarpous with a thick pericarp

paracarpous ovaries in which the carpels are joined by the margins only

phaenocarpous having obvious fruits

platycarpous broad fruited

polycarpous said *either* of a plant bearing fruit many times, *or* of a flower the gynoecium of which is composed of two or more carpels

pterocarpous with winged fruit

rhabdocarpous with very elongate fruit

rhizocarpous (*see also* carpous 2) producing underground fruit

rytidocarpous with wrinkled fruit

spinicarpous with spiny fruit

stegocarpous the condition of a moss capsule having an operculum

stenocarpous having narrow fruit

stephanocarpous having fruit arranged in the form of a wreath

stichocarpous having fruit arranged spirally around an axis

sychnocarpous capable of bearing many successive crops of fruit

symphicarpous *either* with confluent fruits or said of the gynoecium of a flower when the individual simple carpels are only basically fused into a single fruit

syncarpous said of those flowers in which the carpels are fused

carpous 2 (*see also* carpous 1) pertaining to a reproductive ("fruiting") body of a lower plant (*see* carp 2 above) These entries to avoid duplication, do not include adjectival forms directly derived from the substantives given at carp 2

angiocarpous having the reproductive bodies enclosed in a receptacle

dichocarpous having two types of fruiting body, particularly in fungi

hemiangiocarpous an ascocarp that is first closed, but on ripening opens to disclose the asci

cleistocarpous the condition of a fungus perithecium that lacks an aperture

eucarpous 1 (*see also* eucarpous 2) said of an alga that bears, as distinct from transforming itself into, a sporangium

eucarpous 2 (*see also* eucarpous 1) said of a fungus that reproduces several times from the same thallus

hemigymnocarpous the condition of developing spores in closed gynoecia, which subsequently open

holocarpous said of an alga that transforms itself into a sporangium. Sometimes used as synonymous with eucarpous. Also said of a fungus, that reproduces only once from the thallus

pleurocarpous a moss that fruits on a lateral growth

rhizocarpous (*see also* carpous 1) the condition of bearing "fruits" (sporangia) on rhizoids

carpus that part of the wing of some insects at which a transverse fold takes place

carr a graded association between a peaty bog and a dry scrub

swamp carr a carr grading from a swamp rather than a peaty bog

carrageen a rhodophyte alga (*Chondrus crispus*) collected and dried for food in Northern Europe, particularly Ireland

carrot (*see also* beetle) the umbelliferous herb *Daucus carota* and particularly its edible roots

cartilage 1 (*see also* cartilage 2) a connective tissue, usually skeletal, which consists of a matrix of collagen and chondroitin sulfate containing the scattered, usually paired, remnants of the cells that secreted it.

articular cartilage hyaline cartilage of a joint

calcified cartilage cartilage permeated, or even replaced, by calcium salts, but without any organized structure as in bone

elastic cartilage cartilage with elastic fibers in its intercellular substance

fibrocartilage cartilage with collagen fibers in its intercellular substance

hyaline cartilage cartilage consisting of chondrocytes and intercellular substances only

precartilage an aggregate of mesenchyme cells with indeterminate cell boundaries and lamellar ground substance

cartilage 2 *see also* cartilage 1) a cartilagenous skeletal element

annular cartilage the main supporting structure of the buccal funnel of lampreys

apical cartilage an unpaired cartilage in the base of the "tongue" in the rasping organ of Cyclostomes

arytenoid cartilage the posterior of the three laryngeal cartilages

branchial cartilages *see* arch

acrochordal cartilage an unpaired cartilage lying immediately above the anterior end of the parachordals in the embryonic chondrocranium

scapulocoracoid cartilage a cartilage lying between the cleithrum and the clavicle in the pectoral girdle of some fish

supracoracoid cartilage a cartilaginous extension of the coracoid bone

copula cartilage an unpaired cartilage at the anterior end of the hyoid, joined by the first ceratohyals behind and the hypohyals in front

cornual cartilage one of a pair of cartilages, articulated at the rear with the styloid cartilage, and running forward for about one fifth of the length of the piston cartilage in the chondrocranial complex of lampreys

cricoid cartilage the anterior of the three laryngeal cartilages

dentigerous cartilage a cartilaginous plate rising from the floor of the buccal mouth in myxinoids

ensiform cartilage = xiphoid cartilage

hypochiasmatic cartilage a cartilage in the embryonic chondrocranium just below the optic nerve

hyposphyseal cartilage one of a pair of cartilages in the embryonic chondrocranium lying on each side of Rathke's pouch

libial cartilage one of a series of cartilaginous rods associated with the palatoquadrate and mandible in chondrichthian fish

laryngeal cartilage a general name for the cricoid, arytenoid, and thyroid cartilages that form the skeletal support of the larynx

posterior lateral cartilage one of a pair of cartilaginous plates forming the side of the chondrocranium of lampreys

lingual cartilage one of a group of three cartilages (anterior, medial and posterior) lying along the floor of the buccal cavity in myxinoids

Meckel's cartilage the cartilaginous lower jaw of chondrocrania

paranasal cartilage one of a pair of precursors of the nasal capsules in the embryonic chondrocranium

orbital cartilage a cartilage that, in the embryonic chondrocranium, lies just dorso-lateral to the optic nerve and that forms the dorsal region of the orbit in the cyclostome chondrocranium

interorbital cartilage a cartilage rising vertically from the fused trabecula in the embryonic chondrocranium

pedicular cartilage a cartilage rising from the side of the nasal septum in some chondrichthian fish

piston cartilage a long, unpaired cartilage running backward from the apical cartilage in the skeleton of the rasping organ in lampreys

polar cartilage one of a pair of cartilages in the embryonic mammalian chondrocranium lying immediately external to, and on the other side of the carotid artery from, the hypophyseal cartilages

Reichert's cartilage the precursor of the hyoid arch in embryonic chondrocrania

rostral cartilage one of a number of cartilages supporting the rostrum of holocephalan fish

Santorini's cartilage a loose cartilage lying on the median margin of the arytenoid cartilage of ranid anurans

paraseptal cartilage the precursor of the floor of the nasal capsule in the embryonic chondrocranium

spiracular cartilage one of several small cartilages supporting the spiracular opening in chondrichthian fish. They are usually considered to be a portion of the hyoid arch

spinous cartilage one of a pair of ventro-lateral cartilages running backwards from each side of the annular cartilage in the chondrocranial complex of lampreys

styloid cartilage the remnants of the hyoid arch in the lamprey viscera skeleton

anterior tectal cartilage one of a pair of cartilages forming the anterior portion of the orbit and anterior roof of the lamprey chondrocranium

posterior tectal cartilage one of a pair of cartilages in the roof of the chondrocranium of lamprey

tentacular cartilage one of the cartilages supporting the tentacles in myxinoids

thyroid cartilage the center one of the three laryngeal cartilages

tracheal cartilages the cartilagenous rings that support the trachea of air-breathing land vertebrates

nasal tube cartilage one of a series of angular cartilages supporting the nasal tube of myxinoids

ventral cartilage one of a series of cartilages precursor to the ventral portions of the vertebrae

basiventral cartilage the cartilagenous precursor of the lateral wall of the centrum

xiphoid cartilage that which terminates the sternum

caruncle 1 (*see also* caruncle 2, 3) the thick fleshy wattles and combs of certain birds

caruncle 2 (*see also* caruncle 1, 3) an outgrowth from a seed near the micropyle

caruncle 3 (*see also* caruncle 1, 2) a sucker-like organ found between the claws terminating the walking legs of many acarines

-cary- *comb. form* meaning "nut", properly and frequently transliterated -kary-

Caryocaraceae a small family of parietale dicotyledons

caryolite in insects, a single nucleated muscle fiber

Caryophanaceae a family of caryophanale schizomycetes in the form of large trichomes or motile bacilliary structures. Some aquatic forms have motile trichomes

Caryophanales an order of schizomycetes which always occur in trichomes and are characterized by a clear nucleus readily visible in living cells

Caryophyllaceae a very large family of centrospermous dicotyledons which contains, *inter alia*, the pinks, the carnations, the catchflies, the chickweeds, the campions, and the star worts. Distinctive characteristics are the opposite entire leaves and the ten stamens with five separate petals

caryopsis an achene produced from a superior ovary, often restricted to the fruits of the grass family

cascade *see* determination

cascarilla the euphorbiaceous tree *Croton eluteria*

spore **case** = sporangium

cashew *see* nut and bird

casque an excrescence or shield on the upper surface or anterior region of the head. In birds the casque may form the base of the bill

cassabanana the monoecious cucurbitaceous vine *Sicana odorifera* and its edible fruit

cassandra any of several species of shrub of the ericaceous genus Chamaedaphne

cassava a farinaceous foodstuff derived from the tapioca plant (*q.v.*)

butter cassava = tapioca plant

cassena the aquifoliaceous shrub *Ilex vomitor*

cassia a genus of caesalpinaceous leguminous herbs, shrubs, and trees having no relation to the cassia tree (*q.v.*)

cassideous helmet shaped

Cassididae 1 (*see also* Cassididae 2) a family of coleopteran insects now fused with the Chrysomelidae

Cassididae 2 (*see also* Cassididae 1) a family of gastropod mollusks with thick heavy shells. that are commonly called cameo shells in virtue of their frequent use for cutting cameos

cassie the leguminous shrub *Acacia farnesiana*

cassowary any of three species of casuariiform birds of the genus Casuarius. In distribution, they are limited to Northern Queensland, New Guinea, and a few neighboring islands

cast anything that is shed. Any form resembling the original, as the skin of snakes or the molted exoskeletons of arthropod instars

castaneous chestnut-colored

caste a group of specialized function in a social order. Among insects they are often morphologically distinguished (*cf.* gate)

third form caste tertiary reproductives in a termite colony

nasatus caste a subdivision of the soldier caste of termites adapted for chemical warfare

reproductive caste the king and queen of a termite colony

supplementary reproductive caste secondary kings and queens of smaller size in a termite colony whose primary business is nest building but who also reproduce

soldier caste in a termite colony, sterile adults with greatly enlarged heads and mandibles

worker caste a group of termites consisting of nymphs and sterile adults

castor *I.P.* any of several species of nymphalid lepidopteran insect of the genus Ergolis

Castoridae that family of sciuromorph rodents which contains the beavers

castrate a normally sexed organism from which one or more sexual organs have been removed

amphigenous castrate the condition of a male that shows an intermingling of female characters or *vice versa*

Casuariidae that family of ratite birds which contains the cassowaries. They are distinguished from other ratite birds by the highly colored wattles and caruncles on the head and neck, and a bony casque on the crown

Casuariiformes an order of birds containing the Australasian cassowaries and emus. They are distinguished from other flightless birds by the fact that the feathers have an aftershaft as large as the main shaft

Casuarinaceae a family of mostly *Austral.* verticillalous plants containing the single genus Casuarina. Distinguishing characteristics of the family are its catkins of reduced staminate flowers and peculiar woody, cone-like fruits

cat 1 (*see also* bird, cat 2–6, louse, mite, shark) Originally, a carnivore of the family Felidae or genus Felis but in recent years it has become customary to place the great cats (lion, tiger, leopard, and jaguar) in the genus Panthera, to use the genus Profelis for the puma and clouded leopard, and to confine the term Felis to the smaller cats

Bornean bay cat the Bornean *F. badius* apparently closely allied to the flat headed cat (*q.v.*)

bobcat *U.S. Lynx rufus* (= bay lynx)

desert cat a term generally applied to the wild cats of the Near East and Northern India, though, when used without qualification, specifically refers to *F. ornata* of India and Turkestan

domestic cat a cultigen of doubtful origin possibly a hybrid of *F. caffra* and *F. manul.* though many other suggestions have been made. It is certainly not a domesticated variety of the European wild cat *F. catus*

fishing cat *F. viverrina*

golden cat either of two species the Asiatic *F. temmincki* or the African (*F. aurata*). Both are reddish brown

grass cat the Argentinian *F. pajeros*

flat-headed cat the *S.E. Asian F. planiceps*

hunting cat = cheetah

jungle cat *Lynx chaus* (*cf.* lynx)

leopard cat any of numerous felines too large to be called cats and too small to be called leopards. Without qualification, the term is sometimes applied to *F. bengalensis*

Manx cat a geographic race of domestic cat lacking or having a greatly reduced tail

marbled cat the term is generally applied to *F. marmorata* which resembles, in geographic distribution, structure, and color, the clouded leopard (*q.v.*) but which is the size of a house cat

margay cat *U.S. Felis wiedi*

Pallas' cat the Asiatic *F. manul* from which some suppose the Persian domestic breed to be descended

Persian cat a domestic breed thought by some to be a hybrid from Pallas' cat (*q.v.*)

plain cat a term applied to the *S.E. Asian* forms *F. badius* (Bornean bay cat) and *F. planiceps* (flat headed cat) the coats of which are without pattern

tabby cat in popular parlance a domestic cat of indeterminate blackish, yellowish, and whitish markings. The term is also sometimes applied to *F. lybica* from which the Egyptian cat was undoubtedly, and the contemporary domestic cat possibly, descended (*cf.* waved cat)

ring-tailed cat (*see* cat 2) almost any small carnivore with annular markings on the tail has received this designation at one time or another. The term properly belongs to the cacomixtles

tiger cat a term of much the same meaning, or lack of meaning, as leopard cat (*q.v.*) but sometimes applied to the African *F. servalina*

waved cat this is thought by some either to be an independent Asiatic species from which the Asiatic domestic cat is descended or to be a varied descendent of *F. lybica* (*see* tabby cat) introduced to the Far East from Egypt

wild cat *Brit.* without qualification *Felis catus*

cat 2 (*see also* cat 1, 3–6) other mammals, not felids, but having a cat-like habit or appearance

civet cat the term widely applied to all viverrid carnivores but properly to those of the genus Viverra and particularly *V. civetta*

hydrophobia cat the mustelid *Spilogale potorius* (= spotted skunk)

Madagascar cat the lemur, *Lemur catta*

miner's cat *U.S.* the bassariscid carnivore *Bassariscus astutus* (= ringtail)

native cat *Austral.* popular name of dasyurine dasyurids of the genus Dasyurus

polecat (*see also* bush) any of several species of musteline carnivores of the genus Putorius. *Brit.* without qualification *P. foetidus*, also frequently called the European polecat

cape polecat either of two species of the genus Poecilagle (= muishond)

ring tail cat the bassariscid carnivore *Bassariscus astutus* (= bassarisc)

cat 3 (*see also* cat 1, 2, 4–6) in compound names of birds but mostly as an adjective

blue cat *Brit. dial.* the parid *Parus caeruleus* (= blue titmouse)

cat 4 (*see also* cat 1–3, 5, 6) mostly *U.S.* as an abbreviation of catfish (*q.v.*)

blindcat any of several species of blind cave dwelling catfishes

eel cat any of several species of lariid of the genera Gymnallabes and Channalabes

sea cat *Brit.* the anarrhichadid fish *Anarrhichas lupus* (= cat fish or wolf fish *q.v.*)

skunkcat the callichthyid fish *Corydoras arcuatus*

stonecat the ictalurid *Noturus flavus* (*cf.* madtom)

white cat a species of ictalurid catfish, *Ictalurus catus*

cat 5 (*see also* cat 1–4, 6, briar, ear, foot, grape, grass, mint, nip, tail, valerian) in compound names of plants, but mostly as an adjective

blackcat the rosaceous trailing shrub *Rubus occidentalis* (*cf.* berry)

cat 6 (*see also* cat 1–5) in names of fishing baits

white cat *Brit.* popular name of nereid worms

-cata- *comb. form* meaning "down"

Catalase a haemoprotein enzyme, or group of haemoproteins enzymes, which either catalyzes the production of oxygen and water from hydrogen peroxide or of water and aldehyde from hydrogen peroxide and alcohol

Catamblyrhynichidae a monospecific family of passeriform birds erected to contain the plush-capped finch of the Andes. It is a medium sized bird with golden feathers and a large crown of plushlike feathers on the head

Catarrhina a division of primates embracing the Old World monkeys and apes that are distinguished from the Platyrrhina by the fact that the nostrils point downwards. This taxon therefore contains the Simioidea and Anthropodea

catcher a word very commonly used in compound names of birds

caterpillar-catcher *Austral.* any of numerous campephagid birds of the genus Coracina (= cuckoo-shrike)

flycatcher (*see also* shrike) any of numerous passerine birds of the New World family Tyrannidae and the Old World family Muscicapidae. The Ptilogonatidae are the *S.Amer.* silky-flycatchers

alder flycatcher the tyrannid *Empidonax traillii*

ashy flycatcher the muscicapid *Alseonax cinereus*

puffback flycatcher any of numerous muscicapids of the genus Batis. Without qualification, *B. capensis*

beardless flycatcher the tyrannid *Camptostoma imberbe*

ochre-bellied flycatcher the tyrannid *Pipromorpha oleaginea*

sulphur-bellied flycatcher the tyrannid *Myiodynastes luteiventris*

boat-billed flycatcher *Austral.* the muscicapid

Machaerirhynchus flaviventer. U.S. the tyrannid *Megarynchus pitangua*

black flycatcher any of several muscicapids of the genus Melaenornis

black-and-white flycatcher the muscicapid *Bias musicus*

blue flycatcher any of numerous blue-colored muscicapids, usually, without qualification *Erannornis longicauda*

dusky blue flycatcher *Afr. Pedilorhynchus comita*

Nilgiri blue flycatcher *Erranornis albicaudata*

buff-breasted flycatcher the tyrannid *Empidonax fulvifrons*

white-browed flycatcher the muscicapid *Ficidula superciliaris*

brown flycatcher the muscicapid *Muscicape latirostris*

sepia-capped flycatcher the tyrannid *Leptopogon amaurocephalus*

dusky-capped flycatcher the tyrannid *Myiarchus tuberculifer*

collared flycatcher the muscicapid *Ficedula albicollis*

Coues' flycatcher the tyrannid *Contopus pertinax* (= greater pewee)

crested flycatcher any of several species of the tyrannid genus Myiarchus (*U.S.*) or the muscicapid genus Trochocereus (*Afr.*) Without qualification *U.S. Myiarchus crinatus. W. Afr. T. nitens*

great crested flycatcher *U.S.* = crested flycatcher *U.S.*

Wied's crested flycatcher = rusty-tailed flycatcher

dusky flycatcher *U.S.* the tyrannid *Empidonax oberhoseri Afr.* the muscicapid *Artomyias fuligiosa*

white-eye flycatcher the muscicapid *Alseonax cinereus*

grey flycatcher *U.S.* the tyrannid *Empidonax wrightii Afr.* the muscicapids *Alseonax cassini* and *Bradornis microhynchus*

forest flycatcher the muscicapid *Fraseria ocreata*

jungle flycatcher any of several species of muscicapid bird of the genus Rhinomyias

kiskadee flycatcher either of two tyrannids of the genus Pitingus

least flycatcher the tyrannid *Empidonax minimus*

magpie-flycatcher *Austral.* = magpie-lark

olivaceous flycatcher = dusky-capped flycatcher

pale flycatcher the muscicapid *Bradornis pallidus*

paradise flycatcher any of several species of muscicapid birds of the genera Terpsiphone and Tchitrea

Cape paradise flycatcher *Tchitrea suahelica*

pied flycatcher *Brit.* and *Afr.* the muscicapid *Ficedula hypoleuca Austral.* the muscicapid *Arses kangi*

pigmy flycatcher *Afr.* the muscicapid *Alseonax adustus*

pine flycatcher the tyrannid *Empidonax affinis*

piratic flycatcher the tyrannid *Legatus leucophaius*

rufous flycatcher *Afr.* the muscicapid *Stizorhina fraseri*

royal flycatcher any of several tyrannids of the genus Onychorhynchus. Without qualfication, *O. mexicanus*

sulphur-rumped flycatcher the tyrannid *Myiobius barbatus*

shrike-flycatcher the muscicapid *Megabyas flammulatus*

olive-sided flycatcher the tyrannid *Nuttallornis borealis*

silky flycatcher any of numerous species of several genera of the family Ptilogonatidae

social flycatcher the tyrannid *Myiozetetes similis*

sooty flycatcher the muscicapid *Muscicapa sibirica*

spectacled flycatcher *Afr.* the muscicapid *Platysteira cyanea*

spotted flycatcher the muscicapid *Muscicapa striata*

stolid flycatcher the tyrannid *Myiarchus stolidus*

streaked flycatcher the tyrannid *Myiodynastes maculatus*

swamp flycatcher the muscicapid *Alseonax aquaticus*

fantail flycatcher any of several species of muscicapids of the genus Rhipidura

white-browed fantail flycatcher *R. albifrontata*

fork-tailed flycatcher the tyrannid *Muscivora tyrannus*

rufous-tailed flycatcher the tyrannid *Myiarchus validus*

rusty-tailed flycatcher the tyrannid *Myiarchus tyrannulus*

scissor-tailed flycatcher the tyrannid *Muscivora forficata*

ash-throated flycatcher the tyrannid *Myiarchus cinerascens*

tit-flycatcher any of several muscicapids of the genus Parisoma. Without qualification, *P. plumbeum*

banded tit-flycatcher *P. boehmi*

brown tit-flycatcher *P. lugens*

grey tit-flycatcher *P. plumbeum*

tody-flycatcher any of several species of tyrannid bird of the genus Todirostrun

common tody-flycatcher *T. cinereum*

slate-headed tody-flycatcher *T. sylvia*

black-capped gnatcatcher *P. nigriceps*

tyrant flycatcher popular name of tyrannids in general

verditer flycatcher the muscicapid *Muscicapa thalassina*

vermilion flycatcher the tyrannid *Pyrocephalus rubinus*

western flycatcher the tyrannid *Empidonax difficilus*

tufted flycatcher the tyrannid *Mitrephanes phaeocercus*

yellow flycatcher any of several muscicapids of the genera Chloropeta and Chloropetella. Without qualification, *Chloropeta natalensis*

yellow-olive flycatcher the tyrannid *Tolmomyias sulphurescens*

gnatcatcher any of several species of sylviid birds of the genus Polioptila

white-lored gnatcatcher *P. albiloris*

Cuban gnatcatcher *P. lembeyei*

blue-grey gnatcatcher *P. caerulea*

tropical gnatcatcher *P. plumbea*

plumbeous gnatcatcher = black-tailed gnatcatcher

black-tailed gnatcatcher *P. melanura*

western gnatcatcher = blue-gray gnatcatcher

oyster catcher any of several species of charadriiform birds of the family Haematopodidae. *Brit.,* without qualification, *Haematopus ostralegus N.Z.* without qualification *H. longirostris*

American oyster catcher *H. palliatus*

black oystercatcher *Afr. H. moquini. N.Z. H. unicolor. S. Amer. H. ater. U.S. H. bachmani*
pied oystercatcher *H. ostralegus*
sooty oystercatcher *Austral. H. unicolor*
penny-catcher *W.I.* the tyrannid *Elaenia fallax* (*cf.* flycatcher)
spider-catcher an ill-defined term applied to several nectariniid birds (*cf.* spider hunter)
-caten- *comb. form* meaning "chain"
catenulate in the form of a chain
caterpillar (*see also* catcher) the larva of lepidopteran insects and, rarely, of symphytan Hymenoptera. The names given below, unless modified, apply to the adult lepidopteran insect
alfalfa caterpillar the pierid *Colias eurytheme*
saddleback caterpillar *U.S.* the limacodid *Sibine stimulea*
velvetbean caterpillar *U.S.* the noctuid *Anticarsia gemmatalis*
woolly-bear caterpillar any hairy caterpillar but particularly those of arctiids
Florida fern caterpillar *U.S.* the noctuid *Callopistria floridensis*
striped garden caterpillar *U.S.* the noctuid *Polia legitima*
genista caterpillar *U.S.* the pyrausiid *Tholeria reversalis*
clover-head caterpillar *U.S.* theolethreutid *Grapholitha interstinctana*
hedgehog caterpillar any hairy caterpillar that coils up when disturbed
red-humped caterpillar *U.S.* the notodontid *Schizura concinna*
variable oak leaf caterpillar *U.S.* the notodontid *Heterocampa manteo*
salt marsh caterpillar *U.S.* the arctiid *Estigmene acrea*
yellow necked caterpillar *U.S.* the notodontid *Datana ministra* (*cf.* handmaid moth)
ugly nest caterpillar, *U.S.* the tortricid *Archips cerasivoranus*
walnut caterpillar *U.S.* the notodontid *Datana integerrima*
puss caterpillar *U.S.* the cossid *Megalopyge opercularis*
range caterpillar *U.S.* the saturniid *Hemileuca oliviae*
stinging rose caterpillar *U.S.* the limacodid *Parasa indetermina*
pink scavenger caterpillar *U.S.* the cosmopterygid *Sathrobrota rileyi*
slug caterpillar *see* slug larva
tent caterpillar any of several species of lasiocampid building community silken nests.
eastern tent caterpillar *U.S.* *Malacosoma americanum*
California tent caterpillar *M. californicum*
prairie tent caterpillar *U.S. M. lutescens*
western tent caterpillar *U.S. M. pluviale*
unicorn caterpillar *U.S.* the notodontid *Schizura unicornis*
vegetable caterpillar = aveto
water caterpillar the larva of any of several species of South American eupterotids of the genus Palustra
zebra caterpillar *U.S.* the noctuid *Ceramica picta*
Catharactidae = Stercorariidae
Cathartidae a family of falconiform birds containing the American vultures. They are readily distin-

guished from the old world true vultures—and, indeed, from all other Falconiformes—by many details of their internal structure and such superficial features as the perforate external nares and the presence of a distinct membrane between the basal phalanges of the second and third toes
Catitellidae a family of sedentary polychaete annelids living in tubes in muddy sand. They are distinguished by the absence of tentacles or palps but have a pair of ciliated retractile tentacular organs
catkin popular name of an ament
Catostomidae a family of osteriophysous fishes closely allied to the Cyprinidae but distinguished from them by the single row of comb-like pharyngeal teeth. Commonly called suckers
cattle (*see also* grub, tick) this term is properly applied only to bovine bovids of the genus Bos. This includes not only the domesticated ox and the domesticated yak but the wild gaur, banteng, couprey and gayal, though, in some parts of Burma, the latter is partly domesticated
ant cattle aphids tended by ants
-caud- *comb. form* meaning "tail"
caudad moving or turning in the direction of the tail
caudal pertaining to the tail
Caudata an order of amphibian chordates containing the salamanders, newts, etc. They are distinguished by the retention of the tail in the adult form
caudate possessing a tail
-caudic- *comb. form.* meaning the "trunk" of a tree
-caul- *comb. form* meaning "stem" or "stalk"
caul 1 (*see also* caul 2, 3) a plant stem
sarcocaul an extremely fleshly stem like that of a fungus
caul 2 (*see also* caul 1, 3) the fat bodies of an insect larva
caul 3 (*see also* caul 1, 2) = amnion, or amnion + chorion
Caulerpaceae that family of siphonale Chlorophyceae in which the thallus is differentiated into a rhizome-like portion with root-like appendages from which arise shoot-like branches
concaulescene the condition in which plant stems are fused along their lengths
caulescent a condition intermediate between sessile and stalked
haplocaulescent uniaxial
caulidium a leaf of a gametophyte generation (*cf.* caulome)
cauliflower *see* flower
[Cauliflower] a mutant (gene symbol *cfl*) found in linkage group II of *Neurospora crassa*. The term is expressive of the appearance of the phenotypic colony
cauligerous borne on a stem
cauline pertaining to a stem
acauline without a stem, sessile
Caulobacteraceae a family of pseudomonodale schizomycetes strongly resembling the Pseudomonoadaceae but distinguished from them by being attached to the substrate by a stalk that frequently contains iron salts. Most are marine
caulome 1 (*see also* caulome 2, 3) the sum total of the stem structure of a plant
caulome 2 (*see also* caulome 1, 3) the stalk of a sporophyte generation (*cf.* caulidium)
chylocaulome a succulent stem
pericaulome the outer portion of the stem lying immediately below the leaf attachment, and therefore having leaf traces

protocaulome   the first axis

pterocaulome   a winged stem

sclerocaulome   a hard stem

tetraplocaulome   a stem having quaternary axes

caulome 3   (*see also* caulome 1,2) the "stalk" of a hydroid hydrozoan coelenterate

anthocaulome   the stalk from which the anthocyathus larva is cut off

hydrocaulome   the main stalk of a colony of hydroid coelenterates

rhizocaulome   a bundle of vertical stolons imitating a stem in hydroid coelenterates

-caus-   *comb. form* meaning "burn"

cavalla   any of several species of carangid fish

cave   a large cavity in rock or soil

cavernarious   *either* dwelling in caves *or* pertaining to caves

Caviidae   a large family of hystricomorph rodents containing, *inter alia*, the guinea pigs and the capybara. The largest known rodents are in this family which is distinguished by the reduced number of toes and rudimentary tail

cavity   any hollow space not otherwise designated in an organism

atrial cavity   that cavity which, in prochordates, lies between the pharnyx and the body wall

Baer's cavity = blastocoel

pericardial cavity   that portion of the coelom which immediately surrounds the heart

ossicular cavity   a cavity in the tympanic bullae of some mammals separated by the typanic bone from the bullar recess

pit cavity   that part of a pit which is opened internally to the lumen of a plant cell

gastrovascular cavity = coelenteron

cavy   popular name of many hystricomorph rodents, particularly those of the families Caviidae and Dasyproctidae

forest cavy   the dasyproctid *Coelogenys taczanowskii*

Patagonian cavy   the caviid hystricomorph rodent *Dolichotis patachonica* (= mara)

restless cavy   the wild stock of the caviid rodent *Cavia porcellus* (= guinea pig)

spotted cavy   the dasyproctid *Coelegenys paca* (= paca)

cayman = caiman

CCK   cholecystokinin

ccw   *see* [concave wing]

cd   *see* [cardinal]

cd^r   a yeast genetic marker for growth inhibition by cadmium

CDP   cytidine diphosphate

Cebinae   that subfamily of ceboid primates which contains those having prehensile tails, and the squirrel monkeys

Ceboida   a superfamily of primate mammals containing all of the New World forms

Cebrionidae   a family of coleopteran insects closely allied to the Elateridae but readily distinguished from them by the large mandibles

-ceci-   *comb. form* meaning "a gall"

Cecidomyidae = Cecidomyiidae

Cecidomyiidae   a large family of minute, extremely long legged, nematoceran dipteran insects the larvae of which form plant galls so that these flies are usually called gall midges or gall gnats

cecum   *see* caecum

cedar   (*see also* beetle, borer) popularly any tree having fragrant wood. The term has been applied to nu-merous genera of cupressaceous trees including Cedrus, Juniperus, Thuja, Chamaecyparis, and Libocedrus. Also to numerous meliaceous trees and in some areas to a great variety of trees not taxonomically related to these or other cedars. In the West Indies and parts of Latin America various mahogany trees are called cedars as is also a bignoniaceous tree *Tabebuia pallida*. In *U.S.* the term is even applied to a club moss

alaska cedar   *Chamaecyparis nootkatensis*

bastard cedar   any of numerous trees including the sequoia, the Spanish cedar, ribbonwood and guazuma

cigarbox cedar   the meliaceous tropical tree *Cedrella odorata*

ground cedar   *U.S.* the club moss *Lycopodium complanatum*

incense cedar   any of several species of cupressaceous tree of the genus Libocedrus

mountain cedar   *U.S.* any of numerous species of Juniperus, particularly *J. mexicana* (= rock cedar) *N.Z.* the cupressaceous tree *Libocedrus bidwillii* (= kaikawaka)

red cedar   the cupressaceous tree *Juniperus virginiana*

western red cedar   the cupressaceous tree *Thuja plicata*

rock cedar   *U.S. Juniperus mexicana*

spanish cedar   the meliaceous tree *Cedrela odorata*

stinking cedar   the taxaceous tree *Torreya taxifolia*

West Indian cedar   the meliaceous tree *Cedrela odorata*

white cedar   any of several cupressaceous trees including *Chaemaecyparis thyoides* and *Libocedrus decurrens*. The term is also applied to the common arbor vitae (*q.v.*) and to *Tabebuia pallida*

yellow cedar   the pinaceous tree *Chamaecyparis nootkatensis*

cedar-of-lebanon   the pinaceous tree *Cedrus libani*

celandine   (*see also* poppy) any of several species of the papaveraceous genus Chelidonium

lesser celandine   the ranunculaceous herb *Ranunculus ficaria*

tree celandine   any of several species and numerous horticultural varieties of the papaveraceous genus Macleya

Celastraceae   a family of sapindalous dicotyledons that contains *inter alia*, the horticultural Euonymus. In general, the ovary is sunk into the disk and the stamens alternate with the petals.

Celastrineae   a suborder of Sapindales containing the families Cyrillaceae, Aquifoliaceae, Celastraceae, Hippocrataceae, and Staphyleaceae

celeriac   a swollen rooted horticultural variety of celery (*q.v.*)

celery   (*see also* fly, looper) the umbelliferous herb *Apium graveolens*

water celery   *U.S.* any of several species of hydrocharitaceous aquatic herbs of the genus *Vallisneria* (*cf.* eel grass)

cell   this word originally meant one of the smallest rooms of a large house used for the storage of food or, later, prisoners. The reticulated structure seen by early microscopists in plant tissue seemed to be clearly analagous to the rooms in a house and were therefore called cells. In contemporary biology, the word has come to mean a living unit consisting of a nucleus, cytoplasm, and numerous specialized organelles. In the list that follows the types of cell are divided, as a matter of convenience, into animal cells (cell 1), plant cells (cell 2), cell terms pertaining to both plant and animal (cell 3), and other uses of the word (cell 4). The root -cyte, meaning a "pouch" is regarded by many biologists as synonymous with "cell" and numerous entries will be found under this suffix

cell 1 (*see also* cell, cell 2, 3, 4) animal cells (*see* remarks under cell above)

acidophil cell any cell that stains readily in an "acid" dye

acinar cell any secretory peripheral cell in the acinus of the pancreas

centroacinar cell an internal, non-secretory cell in the acinus of the pancreas

alpha cell a chromophil cell of the anterior lobe of the pituitary heavily granulated and with a diplosome

amacrine cell a unipolar nerve cell

arcade cell one of nine cells that run along the pharynx of nematode worms, and are united by arches at the base of the tip

argentaffine cells any cells, particularly those in the intestinal glands, that stain readily by silver techniques

axial cell the single, elongate inner cell of a mesozoan

band cell a granulocyte (*q.v.*) the nucleus of which is not divided into parts

basket cells stellate nerve cells, the axons of which envelope Purkinje's cells (basket cell (Mammalia) = myo-epithelial cell)

basophil cell any cell that stains readily in "basic" dyes

beta cell a chromophile cell of the anterior lobe of the pituitary showing less granulation and fewer mitochondria than the alpha cells

Betz's cell one of the large pyramidal cells in the motor region of the cerebral hemisphere

bipolar cell a nerve cell in the retina with one axon and one dendrite connecting the photoreceptors with the ganglion cells

bladder cell a large cell in the epidermis of cyclostome fishes

C cell = chief cell

castration cell a signet ring type of gonadotroph cell found in castrated animals

chief cell a chromophobe cell in the anterior lobe of the pituitary

neurochord cell a usually large unipolar ganglion cell lying between the ventral nerve chords of nemertines

chromaphile cell a cell of nervous origin, migrant from sympathetic ganglia to many glands, and that possesses a peculiar affinity for chromic acid

chromophobe cell a cells that does not stain under specific circumstances, particularly a cell of the pituitary that is thus contrasted with the acidophils and the basophils

clavate cell one of two long, hollow lamella cells found in the lip pulp of nematodes

cleavage cell = blastomere

club cells cells in the dermis of actinopterygian fishes, that produce the intercellular matrix

collar cell *see* choanocyte

cone cell a cone-shaped cell in the retina of those vertebrates that are capable of perceiving color

crystal cell a hemocyte containing a crystal found in some holothuria

Deiter's cells columnar cells supporting the hair cells in Corti's organ

reticuloendothelial cell a descendant of a primitive reticular cell specialized for phagocytosis and lining blood passageways. The name derives from their ability to make networks of reticular fibers

ependymal cell one in the coat lining the cavity of the central nervous system

myo-epithelial cell a stellate cell underlying the basement membrane of salivary and some other glands. Its contraction is presumed to facilitate the movement of the secretion

fat cell a cell that stores fat

fiber cell a cell with apparently fibrous cytoplasm found in the lip pulp of the dorsal and lateral lips of nematodes

flame cell a primitive excretory cell. The cell is in the shape of an elongate flask, to the inside of the base of the bulbous end of which are attached elongate flagella that, by lashing, force the contained fluid along the neck of the flask either directly to the exterior or to an excretory canal (*cf.* solenocyte)

follicle cell one of those cells immediately surrounding the developing ovum in the follicle of the mammalian ovary. The term is applied to analogous cells in some invertebrates

formative cell *either* in insects, an epidermal cell that produces a seta or hair *or* cells used by some invertebrates in the course of regeneration

ganogene cell one that assists in the production of wax in some insects

Rhode's giant cell one of a group of segmentally arranged cells in the neural tube of Amphioxus

juxtaglomerular cell a granulated cell found in the wall of afferent arterioles of the kidney

goblet cell an epithelial cell that secretes mucous

gonadotroph cell a basophilic cell in the pars intermedia of the pituitary, the grandular content of which varies with the production of gonadotrophic hormone

hair cell a sensory, cilliated cell in Corti's organ

Hensen's cell one of the columnar cells surrounding the hair cells in Corti's organ

Hesse's cell anterior photo-sensory cells in the nerve tube of Amphioxus

His's cells those extraembryonic mesoderm cells which give rise to blood vessels

imaginal cell = a cell of the imaginal disc

iris cell a pigment cell adjacent to a cone cell in ommatidia

Joseph's cell one of the posterior photo-sensory cells in the neural tube of Amphioxus

Kupffer's cells reticulo-endothelial cells of the liver

lasso cell = colloblast

Leydig's cell *either* a secretory interstitial cell in the testes *or* a large secretory cell in the skin of actinopterygian fishes

mast cells large mesenchymal cells containing numerous polychromatic granules

mother cell = teloblast

epitheliomuscular cell a cell containing one or more myofibrils in the epidermis of coelenterates

glandulomuscular cell a glandular cell with a contractile extension such as is found in the pedal disk of many coelenterates

nerve cell any cell adapted to originate or transmit nerve impulses

nettle cell = nematocyst

olfactory cell a bipolar nerve cell in the nasal epithelium

osteogenic cell the precursor of an osteoblast

ovicell an external brood chamber on the surface of an autozooid of some Ectoprocta

acanthostegous ovicell one formed from spines developed from the frontal membrane

hyperstomial ovicell one which is free from the base of the succeeding zooid

peristomial ovicell one formed from hollowed plates, produced from enlarged orificial collars

endozoecial ovicell one embedded in the base of the succeeding zooid

oxyphil cell technically one that stains well in acid dyes, but specifically a heavily granulated cell in the parathyroid gland

Paneth cell a large cell found typically in the bottom of Leiberkuhn's glands in the small intestine

parietal cell a granulated free cell in a stomach gland

pillar cell one of the supporting cells in Corti's organ

pointer cell = deuter cell

polar cell = polar body

propagatory cell in the early cleavage division of the trematode egg, one blastomere remains as a propagatory cell that contributes to the miracidium larvae. The other is a somatic cell that forms a vitelline membrane

Purkinje cell a large flask shaped nerve cell in the cerebellum having much branched dendrites

radial cells specialized neuroglia cells in the retina

reticular cell a cell with numerous processes so that the reticular cells with their processes joined make a spongelike framework or stroma such as that found in lymphatic tissue

primitive reticular cell an undifferentiated mesenchyme cell

rod cell a rod shaped cell in the retina that is sensitive to light and that is the only light sensitive cell in the retina of those animals which cannot perceive color

satellite cell a cell in the capsule of a peripheral nerve ganglion

Schwann's cell a cell that contributes to the formation of a sheath around a myelinated nerve fiber

neurosecretory cell a nerve cell additionally secreting a hormone

Semper's cell the cell of the crystalline eye cone in insects

primary sensory cell a cell, an extension of which acts as a direct receptor of a stimulus rather than receiving the stimulus from another cell

Sertoli cell a supporting cell of the testicular epithelium

somatic cell = *either* propagatory cell *or* body cell

spider cell = astrocyte

stinging cell = nematocyst

sustentacular cell a cell in a taste bud having the form of a segmental slice of a thick-walled hollow sphere. Used by some as synonymous with Sertoli cell

sympathochromaffin cells the precursors of the sympathetic and the medullary cells in the embryonic adrenal gland

thread cell a large cell containing thread-like structures of unknown function in the epidermis of cyclostomes

tooth cell one of the horny cells comprising the "teeth" of anuran larvae

thyrotroph cell one of the basophillic cells found in the thyroid

vibratile cell a flagellated, or pseudoflagellated, coelomocyte of echinoid echinoderms

vitelline cell a cell storing or producing yolk

cell 2 (*see also* cell, cell 1, 3, 4) plant cells (*see* remarks under cell above)

adjective cell in Chara, the cell that lies between the oogonium and the stalk cell

alar cell one at the basal angle of a leaf of a moss

albuminous cell a parenchymous cell in gymnosperms analogous to a companion cell in angiosperms

aleurone cell the outer layer of endosperm cells of grains such as wheat and maize

apical cell a single cell from which, in some plants, all derivative meristem is clearly produced and which is therefore responsible for continued growth

augment cell a modified diatomaceous auxospore that itself produces daughter cells

auricle cells cells of the auricle at the base of a moss leaf

auxiliary cell that algal cell which unites with the conjugation tube

epibasal cell the top cell of a 2-cell plant embryo

bast cell a long fibrous phloem cell

body cell the initial of the sperm cells in conifer pollen tubes

boundary cell = heterocyst

bracket cell a plant secretory cell with projecting papillae

bract cell a projecting cell of Chara

brood cell the first cells produced in the gonidium

intercalary cell a cell that, in the formation of an ascidiospore mother cell is equivalent to a polar cell

neck canal cell a cell in the neck canal of the archegonium of cryptogams

central cell that cell in the archegonium from which the oosphere and ventral canal cell arise

pericentral cell = auxiliary cell

clamp cell one, found in basidiomycete fungi, having a clamp-like branch that connects with another cell for the purpose of maintaining the dikaryotic condition

clathrate cell = sieve tube

collecting cell a round cell lacking chlorophyll, immediately beneath the palisade cells in the leaf

companion cells specialized parenchyma-type cells in the phloem of angiosperms. They arise by division from the sieve tube mother cell and are thought to be connected by plasmodesmata to the sieve tubes themselves

complementary cell a cell produced by the outward division of lenticel phellogen

conjugation cell (alga) = gamete

connecting cell = heterocyst

cotyloid cell a large cell in the developing oat, branches from which ramify through the nucellus

cover cell the apical cell in the neck of an archegonium

daughter cell (*see also* cell 3) a cell produced within the original boundary of another cell

derivative cell a cell of potential specialized function derived from an initiating cell in the meristem

hypodermal cell the sub-apical cell of the nucellus from which the embryo sac is derived

endodermoid cell a plant cell in the position of a endodermis cell, but lacking Casparian strips

Deuter cell one of a row of large cells in the central strand of moss

dislocator cell the cell that liberates the spermatocyte from its attachment in a gymnosperm

embryo cell = oosphere

envelope cell the common envelope of a colony of algae

fertile cell one that gives rise to ascidiospores in uredineous fungi

foot cell a cell in the mycelium of aspergillale fungi from which are derived those hyphae which later bear conidiophores

bulliform cell a large, highly-vacuolated epidermal cell, usually found in rows between the veins of monocotyledonous leaves

cribiform cell = sieve cell

free cell one that is cut off from a plant syncytium

fundamental cell = parenchyma cell

generative cell the cell in a pollen grain that by mitosis produces two sperms

progamic cell = generative cell

grit cell a scleride in fruit

guard cell one of the two bean-shaped epidermal cells that together form a stoma on the surface of a leaf or other plant structure

hinge cell a compressed cell in the upper face of the leaf of a grass, that renders folding easy

hydropic cell an enlarged or swollen somatic cell in a blue-green alga

hygroscopic cell a cell that, through its hygroscopic properties causes a change of shape in a plant structure under conditions of varying atmospheric humidity

initiating cell a cell in the meristem that remains unchanged while budding off derivative cells

latticed cell = sieve tube

limitar cell = heterocyst

libriform cell a thick-walled woody plant cell

lip cell one of a pair of lignified cells marking the line of dehiscence on the sporangia of some ferns

Malpighian cell one of the line of lignin-containing cells in the outer layer of some seeds

mantle cell = tapetal cell

mother cell (*see also* cell 3) one of a group of cells derived from apical initials that give rise to eumeristem in shoot formation in gymnosperms

opening cell one which causes the dehiscense of a plant organ

ophiure cell = astroscleride

palisade cell an individual chloroplast containing cell of palisade parenchyma (*q.v.*)

  arm-palisade cell a palisade cell having numerous arms or branches

passage cell a thin-walled endodermal cell in a monocotyledon root through which water passes readily

antipodal cell one of three cells at the chalazal end of an angiosperm female gametophyte

primordial cell a plant cell before the production of a cell wall

prong cell a parenchymatous cell in some plants containing siliceous deposits

retort cell a funicular cell in Sphagnum having an elongate, recurved apex like the retort of early chemists

seam cell flat cells along the course of which the stomium of fern sporangia rupture

sieve cells sieve elements in the form of long and slender cells with relatively unspecialized sieve areas

sperm cell a male gamete

archesporial cell one of the initials that give rise both to the parietal cells and the pollen mother cells of a stamen

stalk-cell the cell between the antheridial mother

cell and the vegetative cell. The male gametophyte of gymnosperms

pachystichous cell one having a thick wall on one side

stone cell = brachyscleride

swarm cells a term usually applied to the flagellated stage that results from the germination of the spores of a myxomycete

primary tapetal cell the initial of the tapetum in the periblem

unicell a term sometimes applied to organisms, particularly algae, that exist as individuals rather than as aggregates or colonies

vegetative cell that cell in the pollen grain which produces the pollen tube.

water cells suberized, water-retaining cells in the palisade tissue of succulent plants

cell 3 (*see also* cell, cell 1, 2, 4) terms pertaining to both plant and animal cells (*see* remarks under cell above)

couple cell = zygote

daughter cell (*see also* cell 2) one derived from division of a mother cell

isodiametric cell one that has an equal diameter in all directions

generative cell = gamete (but see also cell 2 for specialized botanical usage)

germ cell a gamete

mother cell (*see also* cell 2) one that divides to form a specific group of cells (*cf.* teloblast)

nurse cell one cell that specifically provides nourishment to an adjacent cell as the follicle cells in an ovary

pairing cell = gamete

prolific cells = reproductive cells

resting cell *either* a dormant cell *or* a cell which is not dividing

sister cells the daughters of a mother cell

cell 4 (*see also* cell, cell 1, 2, 3) cell used in other senses (*see* remarks under cell above)

cell an area enclosed by the veins (*q.v.*) of an insect wing

onychocell an avicularium with a conspicuously winged jaw

pollen cell the cavity of the anther in which pollen is formed

cellulase an enzyme catalyzing the breakdown of cellulose through the hydrolysis of β–1,4 glucan links

Celtidoideae a subfamily of Ulmaceae distinguished by the fact that flowers are borne on twigs of the current season (*cf.* Ulmoideae)

cementum the material cementing a bony tooth into its socket

-ceno- *see also* -coeno-

cenosis a community dominated by two distinct, but not of necessity mutually antagonistic, species. Used by some as synonymous with community and by others as synonymous with association and biocenosis

cenote a bottle-shaped cave more or less vertically excavated in the ground and opening through its narrow neck

centaury anglicized name of the gentianaceous genus Centaurium, without qualification usually *C. umbellatum*

center 1 (*see also* center 2) in the sense of mid-point

  central center the midpoint of the pervalvar axis of diatoms

  chromocenter the area of a chromosome immediately on each side of the centromere in salivary gland chromosomes. The term is also applied to Feulgen positive bodies in interphase nuclei commonly thought to be heterochromatic segments of chromosomes

center 2 (*see also* center 1) in the sense of principal area

astrocenter = centrosome

diaphyseal center a center of osteogenesis in the middle of a shaft bone

epiphyseal center a center of ossification at the end of a shaft bone

organ center the point around which the growth of a plant organ takes place

Centetidae = Tenrecidae

-cento- *comb. form* meaning "patchwork"

centonate blotchy

-centr- *comb. form* meaning "middle" *but see* -kentro-

centralium a usually longitudinal cavity, of diagnostic value when found in the seeds of some palms

Centrarchidae a family of North American fresh water fishes. They are distinguished from the closely allied Percidae by the spiny rayed and soft rayed portions forming a continuous dorsal fin. The sun fish, blue gill, crappies, and fresh water bass belong in this family.

centre *see* center

Centrechinoida an order of echinoid Echinodermata distinguished from the Exocycloida by having an apical periproct and from the Cidaroida by the presence of sphaeridia and peristomal gills

centric (*see also* centrous) pertaining to the center

acentric said of chromosomes that lack a centromere

homocentric = concentric

centriole two minute areas of differentiated cytoplasm that arise from the centrosome at the beginning of mitosis. Each forms one pole of the mitotic figure

Centriscidae a family of solenichthous fishes distinguished by their curiously compressed knife-like body with transparent overlapping protective plates

Centrolenidae a family of South American anuran amphibia once fused with the Hylidae but separated from them by the possession of a single tarsal bone. They are closely related to the Leptodactylidae

Centrolepidaceae a family of farinose monocotyledonous angiosperms

Centropomidae a family of acanthopterygian fishes distinguished by the fact that the lateral line extends to the end of the tail fin and that the preorbital is serrated. The best known members of the family are the snook and Nile perch

Centrospermae a large order of dicotyledonous angiosperms that contains *inter alia*, the goosefoots, spinach, the beets and chards, the amaranths, the four o'clocks, the pokeberries, the carpet weeds, the purslanes, the pinks, the chickweeds and the campions. The order is distinguished by the typically biseriate perianth and the coiled or curved embryo

centrous (*see also* centric) pertaining to the centrum

acentrous lacking a centrum

archicentrous said of those vertebrae in which the base of the neural arch contributes to the substance of the centrum

autocentrous said of those vertebrae in which the neural arches chondrify separately from the centrum

centrum 1 (*see also* centrum 2) the basal portion of a vertebra formed as a replacement for the original notochord

ectochordal centrum one in which the center of the

notochord is replaced with cartilage surrounded by a cylinder of bone

holochordal centrum one in which the segment of notochord from which the centrum is derived becomes calcified

stegochordal centrum one in which only the dorsal arch of the perichordal sheath becomes ossified

hypocentrum the anterior equivalent of the pleurocentrum

pleurocentrum that part of the centrum which is derived from the inner portions of the dorsals and ventrals (*q.v.*) In Amniotes, the whole centrum is derived from the pleurocentrum

centrum 2 (*see also* centrum 1) a central body

coenocentrum a mass of granules in the oosphere of certain fungi, thought by some to be the equivalent of yolk

cytocentrum the centriole and its adnexed structures

cepaceous pertaining to garlic

cepe French common name, widely used in English, for boletaceous basidiomycete fungi

-cephal- *comb. form* meaning "head" The adjectival forms -cephalous (here preferred) and -cephalic appear synonymous but -cephaline (*q.v.*) has developed a distinct meaning as has the substantive cephalon

cephalic *see* cephalous

Cephaliidae a family of brachyceran dipteran insects usually included with the Otitidae

cephaline said of those gregarines in which the body is divided into several merites (*q.v.*)

acephaline said of simple bodied gregarines

cephalion an unpaired scaly plate, anterior to the mouth of gastrotrichs (*cf.* hypostomium, pleuron)

cephalium the woody termination of a cactus stem on which flowers are borne

cephalization 1 (*see also* cephalization 2) the process by which the highest degree of specialization became localized in the anterior end of animals

cephalization 2 (*see also* cephalization 1) the simplification of floral elements

Cephalochordata a subphylum of Chordata that contains the single class Amphioxi (*q.v.*)

Cephalodiscida an order of pterobranchiate hemichordates distinguished by the fact that the so-called colony is really an aggregate of discontinuous individuals

cephaloid = capitate

cephalon brain or head as indicated by the nature of the compound

archicephalon the prostomial ganglion of arthropods

diencephalon the portion of the brain which lies between the telencephalon and the mesencephalon

encephalon in insects the brain or that portion of the head containing it

mesencephalon the midbrain. That is, the part which lies between the diencephalon and the metencephalon

metencephalon that portion of the brain which rises immediately behind the mesencephalon

myelencephalon the most posterior portion of the brain

prosencephalon the anterior region of the developing brain that subsequently divides into the telencephalon and diencephalon

rhinencephalon = olfactory bulb

rhombencephalon that portion of the brain which subsequently divides into the metencephalon and myelencephalon

telencephalon the anterior portion of the brain

thalamencephalon the thickened portion of the roof of the diencephalon

Cephalophinae a subfamily of bovid artiodactyl mammals containing the forms commonly called duikers

Cephalopoda a class of siphonopodous Mollusca containing the squids, cuttlefish, and octopus. This class is characterized by the fact that the foot is drawn out into tentacles surrounding the head

Cephalotaceae a family of rosalous dicotyledons containing only one species and genus (*Cephalotus follicularis*) an insectivorous plant of localized distribution in W. Austral.

Cephalotaxaceae a family of coniferous gymnosperms containing the forms commonly known as the plum-yews. They are distinguished by bearing the staminate flowers in globose heads in the leaf axils or, occasionally, in spicate strobili

cephalous pertaining to the head

eucephalous having a true head. Applied particularly to insect larvae in which all of the head appendages are present

homocephalous condition of a flower of which the pollen fertilizes another flower of the same inflorescence

megacephalous large headed. In *bot.*, referring particularly to the capitula of Compositae

pycnocephalous thick-headed

spherocephalous round headed. Used particularly of flowers, such as those of the genus Allium, with a spherical inflorescence

Cephidae a family of symphytan hymenopteran insects with very slender bodies and in many cases a slight waist but still clearly symphytan rather than apocrytan. Many are pests of crops and are commonly called stem sawflies

Cepolidae perciform acanthopterygian fishes distinguished by their very elongate body and continuous median fins, known in England as "snake fish"

-ceptor *abbr. term* for "receptor" (*q.v.*)

acceptor = receptor

baroceptor an organ perceiving weight

enteroceptor a sensory structure or organ located in the visceral mass of an organism, particularly one that provides an organism with information about the interior of its own body

exteroceptor a sensory organ on or near the exterior of the body that permits an organism to perceive its immediate external environment

interoceptor = interoreceptor

nociceptor one that is responsive to possibly injurious stimuli

proprioceptor a structure that enables a complex organism, such as man, to locate one part of his body in relation to another

teloceptor a sense organ, such as the eye or ear, that provides information about stimuli originating at a distance

cera- *comb. form* meaning "horn"

ceraceous waxy

Cerambycidae a very large family of coleopteran insects commonly called the long horn beetles. The antennae of many forms exceeds the body in length. Many are brightly colored and considered to be among the most beautiful insects

Ceramiales an order of tetrasporophytic floridian Rhodophyta in which the procarp has an auxilliary cell produced after fertilization

ceramidium = cystocarp

Ceraphronidae a small family of proctotrupoid apocritan hymenopteran insects, many of which are wingless

ceras the external gill of a nudibranch mollusc

cerate pertaining to horn or horns in the widest sense

adelocerate with the antenna concealed, usually in a groove in the head

Ceratinidae a family of apoid apocritan hymenopteran insects usually included with the Apidae

Ceratiidae a family of pediculate fishes containing the deep sea angler fishes. They are distinguished by the dwarf males that are parasites on the females

ceratium a slender one-celled, two-valved, fruit produced from a superior ovary

Ceratocampidae = Citheroniidae

Ceratodontidae a monospecific family of dipneustian fish containing the Australian lung fish *Neoceratodus forsteri*. It is distinguished from other dipneusti by the presence of only a single lung and the large paddle like fins

Ceratophyllaceae that family of ranale dicotyledonous angiosperms which contains the single genus Ceratophyllum commonly called hornworts. The genus is easily distinguished from other permanent aquatics by the whorled dichotomously divided sessile leaves

Ceratophyllidae = a large family of fleas all parasitic on rodents but readily distinguished from Hystrichopsyllidae by the presence of eyes

Ceratopogonidae a family of minute predatory and blood sucking insects commonly called biting midges, punkies, or no-see-ums

Ceratopsyllidae = Ischnopsyllidae

Ceratopteris = Parkeriaceae

-cerc- *comb. form* meaning "tail"

cercal properly pertaining to the tail but used in *biol.* principally in relation to the caudal fin of fish (*cf.* cercous)

diphycercal said of a caudal fin that is a protocercal fin, but in which equal dorsal and ventral lobes are developed

gephyrocercal said of a caudal fin in which the axial skeleton is truncated and without hypurals. The fin is usually vestigial and without lobes

heterocercal said of a caudal fin in which the dorsal fold is lacking and in which the ventral fold has a much longer posterio-dorsal than antero-ventral lobe

homocercal said of the tail of a fish that has equal ventral and dorsal lobes. Both lobes are derived from the hypochordal fold

hypocercal said of a caudal fin in which the tail of the fish is directed downward, and there is only a single dorsal lobe. It is, in effect, the reverse of a heterocercal type

isocercal said of a caudal fin of a fish that lacks any lobes, and therefore appears as a smooth prolongation of the tail end of the body

protocercal said of a caudal fin in which the ray supported part of the fin is limited to the margin and in which there are both epichordal and hypochordal folds

cercaria (*see also* larva) the basopodite of an anal cercus in insects

-cercid- *comb. form* meaning a "small comb."

Cercidiphyllaceae a unigeneric monospecific family of ranale dicotyledonous angiosperms distinguished by the dimorphic leaves of both palmate and pinnate venation, unisexual flowers, dioeciously disposed, with the pistilate inflorescence having four to six apetalous flowers each with a multi-ovulate pistil

cercidium a fringed or pectinate mycelium

Cercolabidae a family of hystricomorph rodents commonly called tree porcupines or climbing porcupines

Cercoleptinae a monospecific subfamily of procyonid carnivorous mammals containing the kinkajou (*q.v.*)

cercomer the tail of a cercaria larva (*q.v.*)

Cercopidae a family of homopteran insects commonly called froghoppers or spittlebugs in virtue of the bubbly mass with which the nymphs surround themselves. They are closely related to the Cicadellidae from which they are distinguished by their heavily spined tibia

Cercopithecidae a family of simioid primate mammals distinguished from the Colobidae by the large cheek pouches and from the Cynopithecidae, which also have cheek pouches, by the long tail

Cercopithecoidea that superfamily of the primate suborder which contains the Old World monkeys. They are distinguished by their downwardly directed nostrils (*cf.* catarrhine) and by the absence of a prehensile tail

cercous (*see also* cercaria larva) pertaining to horn, in the widest sense, or to a tail (*cf.* cercal)

  acercous lacking projections from the head whether technically horns or not

  brachycercous having short horns or a short antenna

  cladocercous having branched horns or antennae

  helocercous with clavate antennae

  microcercous tiny tailed

  nematocercous having thread-like antennae

cercus a projection from the posterior end of an arthropod (*cf.* cercopod)

  alacercus the median cercus when three are present

cere a soft patch on the proximal end of the upper mandible in birds

cereal any grass seed, or derivative of a grass seed, used for food

cerebellum one of a pair of outgrowths from the latero-dorsal wall of the metencephalon. In birds and mammals these outgrowths form deeply convoluted lobes

-cerebr- *comb. form* meaning "brain"

cerebrum that portion of the brain which consists of paired lobes derived from the telencephalon. It may be a scarcely noticeable swelling, as in fish, or the major portion of the brain, as in mammals. The term is sometimes used as synonymous with brain

  deutocerebrum the second or middle pair of supraoesophageal ganglia of the insect brain

  protocerebrum the first pair of supraoesophageal ganglia of the insect brain

  tritocerebrum the third and posterior of supraoesophageal ganglia of the insect brain

Cereoideae a subfamily of the Cactaceae distinguished by having the leaves reduced to minute scales

Ceriantharia an order of zoantheridian Anthozoa that is distinguished by the elongate form of the anemone-like body which lacks a pedal disk

Cerithiidae a family of gastropod Mollusca having elongate many whorled shells, usually papillate or rugose. Usually called "horn shells"

cernuous with a hanging or nodding head

cero the scombrid fish *Scombromorus regalis*

Ceroplatidae a family of nematoceran dipteran insects now usually included with the Mycetophilidae

cerous pertaining to wax. Very frequently misused for cercous

certation competition between pollen grains as to the speed with which they can grow down the style

Certhiidae that family of passeriform birds which contains the creepers a term derived from their habit of creeping about like mice over rocks, cliffs, walls, and trees. They are distinguished by the slender long laterally compressed decurved beak

cerulean any of numerous species of lycaenid lepidopteran insect of the genus Jamides

Ceruridae = Notodontidae

cerussatous dull white

cerval any of several species of felid carnivore of the genus Leptailurus. They are distinguished by their long legs and short tails

cervical pertaining to the neck

cerviculate having a long neck

Cervidae a very large family of ruminant artiodactyl mammals distinguished from all other ruminants by the possession of antlers

Cervinae a subfamily of cervid mammals containing the European and Oriental deer, frequently called the "true deer." They differ from the New World deer (Odocoileinae) in their generally larger size and either backwardly or vertically projecting antlers

cervine (*bot.*) the color of a stag

cervix the junction of the vagina and uterus in mammals

cespitose pertaining to any dense, short, matted growth, in tufts or coalesced tufts. Sometimes used for turf

-cest- *comb. form* meaning "ribbon" or "girdle"

Cestida = Cestoidea

Cestoda that class of parasitic Platyhelminthes commonly called tapeworms. Distinguished by the complete absence of a digestive system or mouth, and the division of the body into proglottids

Cestodaria a subclass of cestode Platyhelminthes with an undivided body

Cestoidea 1 (*see also* Cestoidea 2) = Cestoda

Cestoidea 2 (*see also* Cestoidea 1) an order of tentaculate ctenophores elongate and laterally compressed. Venus' girdle is typical

Cestreae a tribe of the family Solanaceae

Cetacea that order of placental mammals commonly called whales and porpoises. They are clearly distinguished by the modification of the limbs for swimming

cete the term for a group of badgers

Cetoniidae a family of coleopteran insects closely allied to, and frequently fused with, the Melolonthidae or the Scarabaeidae

cf *see* [crayfish]

Cf₁ - Cf₃ *see* [*Cladosporium resistance*]

cf₁ *see* [Cauliflower]

cg *see* [comb-gap]

cH a term sometimes used for the pH of soil

ch *see* [chocolate], [chubby] and [congenital hydrocephalus]. This symbol is also used a yeast genetic marker indicating a requirement for choline

Ch *see* [Chocolate]

ch *see* [chartreuse petals]

chachalaca any of several cracid birds of the genus *Ortalis*

Chaeluridae a family of gammaridean woodboring amphipod crustacea distinguished from other amphipoda by the flattened abdomen and enlarged uropods

-chaen- *comb. form* meaning "gape"

-chaet- *comb. form* meaning "bristle"

chaeta 1 (*see also* chaeta 2) the sporophore of a moss

chaeta 2 (*see also* chaeta 1) the chitinous structures projecting from some annelid worms and most arthropods. It usually designates a stouter structure than a seta, and in insects refers specifically to a jointed outgrowth from the epidermis (*cf* seta)

ammochaeta one of a group of stiff bristles on the head of some Hymenoptera, particularly one of the bristles that form a beard-like tuft on the underside of the mentum of eremophilous ants. It is used to dust off the strigil before the latter is used to clean the antenna

glandula chaeta a hollow chaeta serving for the outlet of an epidermal gland in some insects (*cf.* glandular seta)

[chaete] numerous mutants with names ending in -chaete are to be found in their alphabetical order

[chaetelle] a mutant (gene symbol *chl*) mapped at 60.8 on chromosome II of *Drosophila melanogaster*. The phenotypic expression is very small bristles

perichaetium the involucre of a moss

Chaetodontidae a large marine family of oval, laterally compressed acanthopterygian fishes with small teeth and no preopercular spine. They are commonly called butterfly fishes in virtue of their brilliant coloration

Chaetognatha a group of coelomate bilateral animals comprising the planktonic organisms commonly called arrowworms. They are mostly torpedo-shaped with large grasping spines at the head and a horizontal tail fin at the posterior end

Chaetonotoidea an order of the phylum Gastrotricha distinguished by having adhesive tubes limited to the posterior end

Chaetophoraceae that family of ulotrichale Chlorophyceae which contains those forms having a branching filamentous thallus

Chaetopteridae a family of large sedentary polychaete annelids living in U-shaped parchment-like tubes buried in sand. The body is divided into three distinct regions, the anterior of which usually carries wing-like parapodia

chaetosema sensory bristles, particularly those on a lepidopteran head

chafer any of numerous plant eating scarabaeid coleopteran insects. The term chafer is sometimes confined in the *U.S.* to the genus Macrodactylus

European chafer *U.S. Amphimallon majalis*

cockchafer *Brit.* any melolonthid coleopteran insect but usually, without qualification, *Melolontha vulgaris*

northern masked chafer *U.S. Cyclocephala borealis*
southern masked chafer *U.S. Cyclocephala immaculata*

pine chafer *U.S. Anomala oblivia*

rose chafer *Brit.* thecetoniid *Cetonia aurata*. The term is occasionally applied to other cetoniids. *U.S. Macrodactylus subspinosus*

green rose chafer *U.S. Dichelonyx backi*
western rose chafer *U.S. Macrodactylus uniformis*

chaff (*see also* finch, seed) the mature, dry bracts on the base of the capitum of the compositaceous flowers. Also used for the glumes of grass infloresences, particularly those detached when grains are thrashed

chain in *biol.* an elongate mass of repetitive structures or a linked function

food chain the transfer of nutrients, and hence energy, from one group of organisms to another

Jamin's chain alternate bubbles of air and drops of water running the length of the stem of a water plant

love's chain the polygonaceous climbing vine *Antigonon leptopus*

chair *see* sella

Chalastrogastra = Symphyta

-chalaz- *comb. form.* meaning "hail stone" or "lump" but mostly often used in *biol.* in the sense of "knot"

chalaza 1 (*see also* chalaza 2–4) the twisted albumen lying between the individual eggs of those amphibia which lay their eggs in strings

chalaza 2 (*see also* chalaza 1,3,4) one of the two apical twists of thick albumen which holds the yolk of a bird's egg in place

chalaza 3 (*see also* chalaza 1,2,4) a pimple-like swelling on an insect body wall bearing a seta

chalaza 4 (*see also* chalaza 1–3) that end of the angiosperm female gametophyte (embryo sac) where the nucellus, integument, and funiculus fuse and which is opposite the micropylar end

chalcid any of several torymid and eurytomid hymenopteran insects

apple seed chalcid *U.S.* the torymid *Torymus druparum*

clover seed chalcid *U.S.* the eurytomid *Bruchophagus gibbus*

grape seed chalcid *U.S.* the eurytomid *Evoxysoma vitis*

Chalcidae = Chalcididae

Chalcididae a family of chalcicoid apocritan hymenopteran insects readily recognized by the enormously swollen hind femur that is usually toothed

Chalcidoidea a superfamily of apocritan hymenopteran insects commonly called chalcid flies or chalcid wasps. They are readily recognized by the presence of a prepectus in front of the mesopleurum and the presence of only a single vein on the front wing. Almost all are parasitic on other insects and are of the greatest economic value. The elbowed antennae are another conspicuous feature

Chalcopariidae a family of passeriform birds usually included with the Nectariniidae

-chali- *comb. form* meaning "gravel" but used by ecologists in the sense of a "gravel slide"

chalicad a gravel slide plant

chalone a hormone exercising an inhibitory action on a metabolic process

chalybeous steel gray

-chamae- *comb. form* meaning "on the ground"

Chamaeidae a family of passeriform birds usually included in the Timaliidae

Chamaeleontes an assemblage of lacertilian reptiles erected to contain the single family Chamaeleontidae (*cf.* Geckones and Lacertae)

Chamaeleontidae a family of bizarre African and Asian lacertilian reptiles, distinguished by the combination of bony skull casque, prehensile tail and zygodactylous grasping feet

Chamaemyidae = Ochthiphilidae

Chamaemyiidae = Ochthiphilidae

chamber 1 (*see also* chamber 2,3) a cavity in an animal or animal organ

anterior chamber that portion of the cavity between the cornea and the lens of the eye which is exterior to the iris

end chamber the gamete-producing area of a closed reproductive system

peribranchial chamber = atrial cavity

posterior chamber that portion of the cavity between cornea and lens which is internal to the iris of the eye

chamber 2 (*see also* chamber 1,3) a cavity in a plant or plant organ

pit chamber that part of a bordered pit which is covered by the border

pollen chamber a cavity in the apex of the style of angiosperms in which pollen accumulates or, in gymnosperms, the cavity made by the integuments of the ovule where the pollen grains come to rest and later start to germinate

stigmatic chamber that part of the orchid rostellum which bears the retinaculum

chamber 3 (*see also* chamber 1,2) in the sense of room

copulation chamber a burrow excavated by certain animals, including many insects, for purposes of copulation

chameleon (*see also* fly) popular name of chamaeleontid lacertilian reptiles. The term is frequently misused

false chameleon *U.S.* any of several species of iguanid of the genus Anolis, particularly *A. carolinensis* *Bri.* the chameleon like agamid lacertilian *Lyrocephalus scutatus*

Chamidae a family of pelecypod mollusks commonly called rock oysters. They are distinguished by their thick irregular shells

chamois the rupicaprine caprid *Rupicapra capra*, best known as the mountain goat of the Alps, but also found in the Carpathians and Caucasus ranges. The biblical chamois (Deuteronomy 14:5) was some other artiodactyle

chamomile any of numerous species of compositaceous herb of the genus Anthemis

false chamomile any of several species of compositaceous herbs of the genera Boltonia and Matricaria

sweet false chamomile *M. chamomilla*

scentless false chamomile *M. inodora*

field chamomile *Anthemis arvensis*

wild chamomile = false chamomile

Chandidae = Centropomidae

changa *U.S.* the gryllotalpid *Scapteriscus vicinus*

Chanidae a monospecific family of isospondylous fish containing the milk fish *Chanos chanos*. It is distinguished by the toothless mouth and the large deeply forked tail

Channidae a family of percomorph fishes with the characteristics of the suborder; commonly called snakeheads

Channoidea a monofamilial suborder of acanthopterygian fish closely related to the anabantids but distinguished from them by having a simple vascular chamber instead of a labyrinth for air breathing.

chanterelle French name of edible cantharellaceous fungi but now adopted as an English word applied to the whole family Cantharellaceae. The term, without qualification, is usually taken in English to mean *Cantharellus cibarius*

chanticleer male domestic chicken, though cock or rooster are more frequent terms

chaparral (*see also* cock) an arid region of dense evergreen, thorny shrubs particularly in *S.W.U.S.*

char *Brit.* any of numerous species or, far more probably, varieties, of the salmonid genus Salvelinus (*cf.* trout)

arctic char *S. alpinus*

Characiaceae that family of chlorococcale Chlorophyceae which consists of sessile elongate cells either solitary or joined in radiate colonies

Characidae a family of ostariophysid fishes closely allied to the Cyprinidae but distinguished from them by the presence of teeth in the jaws and the possession of a small adipose fin

characin anglicized form, sometimes used by aquarists, of characid fishes

character 1 (*see also* character 2) a specific, determinate attribute

acquired character a character acquired by an individual in the course of its life. (*cf.* Lamarckism)

alaesthetic character one perceived only by the aesthetic appreciation of an onlooker

biocharacter a specific individual character of a living organism

characteristic character the ultimate phenotypic expression in the adult of a gene or group of genes

dominant character the expression of a dominant allele

epigamic character = tertiary sexual character

sex-influenced character one that appears recessive in one sex and dominant in the other

sex limited character one in which the heterozygote phenotype appears only in one sex

sex linked character one which is controlled by an allele on the sex chromosome and is therefore heterogametic in one sex and homogametic in the other

recessive character the opposite of dominant character. Therefore, it appears only in individuals homozygous for this character

sexual character one pertaining to, or distinguishing between, male and female

primary sexual character those organs directly concerned with the establishment of sex *i.e.*, the gonads

secondary sexual characters those characters directly concerned with the function of sex, *i.e.* intromittent organ, mammary gland, etc.

tertiary sexual characters those characters that visually, but not functionally, distinguish the sexes *i.e.*, plumage, etc.

somatogenic character = acquired character

taxignomic character anatomical characters visible from the outside

character 2 (*see also* character 1) a unit of writing

Chinese character *Brit.* the drepaniid lepidopteran insect *Cilix glaucata*

Hebrew character *Brit.* the noctuid lepidopteran insect *Noctua cenigrum*

Charadriidae that family of charadriid birds containing the plovers. They are distinguished by their long wings and vestigial or lacking hallux. A few are crested and wattled

Charadriiformes that order of birds which contains, *inter alia*, the gulls, auks, and shore birds such as the oyster catchers and plovers. They are distinguished by the pointed wings with eleven primaries and eleven secondaries. The hallux is, save in the jacanas, small or absent

Swiss chard a subspecies of beet (q.v.) cultivated for its leaves; treated by some as a species (*Beta chilensis*)

charity the polemoniaceous herb *Polemonium caeruleum*

black charles *W.I.* the fringillid bird *Loxigilla violacea* (= Greater Antillean bullfinch)

creeping charlie the primulaceous herb *Lysimachia nummularia* (= money wort). Ther term is also applied to many other low growing, or creeping, herbs such as the stone crop

darkie charlie *Brit.* the dalatiid selachian *Scymnus lichia*

charlock the cruciferous herb *Brassica arvensis*

jointed charlock *U.S.* the cruciferous herb *Raphanus raphanistrum*

charm term for a group of goldfinch

Charophyceae that class of chlorophytan algae which contains the forms commonly known as stoneworts. The thallus, which is often heavily calcified, is divided into segments by a regular succession of nodes

charr = char

chartaceous papery

[chartreuse petals] mutant (gene symbol *ch*) mapped at 65 on chromosome 8 of the tomato. The phenotypic expression is a yellow-green petal

Charybdeidea = Cubomedusae

-chas- *comb. form* meaning "branching" or "separation"

chasial pertaining to the method of branching (*cf.* chastic

    dichasial said of a method of branching that results in a fork shape, or pertaining to a type of dichotomy in which the paired branches arise below a terminal flower bud

    monochasial said of the type of plant branching in which branches are produced only on one side of the main stem

chasium the result of a method of branching

    pleiochasium a cyme in which each main axis produces more than two branches

-chasko- *comb. form* meaning "open", usually transliterated -chasco-

-chasm- *comb. form* meaning "a chasm"

chastic pertaining to a method of separation (*cf.* chasial)

    hygrochastic said of fruits that burst open through the absorption of water

    xerochastic the condition of plants the seeds of which are distributed by the bursting of the desiccated fruit

chat (*see also* shrike, tanager) any of numerous birds, turdids unless otherwise noted below, whose song is considered to contain a "chat" noise. *U.S.* the parulid *Icteria virens Brit.*, without qualification, the sylviid *Acrocephalus schoenobaenus* (= sedge warbler) *I.P.* any of numerous charadriiform birds of the genus Pratincola

    Australian chat popular name of turdids of the genus Ephianthura

    black chat *Myrmecocichla nigra*

    red breasted chat *Oenanthe bottae* (*cf.* red-breasted wheatear)

    yellow-breasted chat the parulid *Icteria virens*

    crimson chat *Epthianura tricolor*

    desert chat *Ashbyia lovensis*

    white-faced chat *Epthianura albifrons*

    grey bush-chat *I.P. Saxicola ferrea*

    Indian bush-chat the charadriiform *Pratincola maura*

    anteater-chat *Myrmecocichla aethiops*

    cliff-chat *Afr.* any of several species of turdids of the genus Thamnolaea (*cf.* mocking-chat). Without qualification, usually *T. cinnamomeiventris*

    fallow-chat = wheatear

    furze chat = stonechat

    hill chat *Cercomela sordida*

    mocking chat *I.P.* = cliff-chat

    orange chat *Epianthura aurifrons*

    palmchat the dulid *Dulus dominicus*

    pied chat *I.P. Saxicola caprata*

    robin-chat any of numerous turdids of the genus Cossypha. Without qualification *C. caffra*

        white-browed robin-chat *C. Heuglini*

        red-capped robin-chat *C. natalensis*

        mountain robin-chat the turdid *Cossypha isabellae*

        snowy-headed robin-chat *C. niveicapilla*

        Ruppell's robin-chat *C. semirufa*

    rock-chat any of several turdids of the genus Cercomela

    sooty chat *Myrmecocichla nigra*

    stonechat *Brit.* any of several turdids of the genus Saxicola. Usually, without qualification, *S. torquata*

    black-tailed chat *I.P. Cercomela melanura*

    red-tailed chat *C. familiaris*

    grey-throated chat the parulid *Granatellus sallaei*

    whinchat *Saxicola rubetra*

    sickle-winged chat *Cercomela sinuata*

    woodchat *Brit.* the laniid bird *Lanius senator* (= woodchat shrike)

chatterer any of numerous birds of chattering habit including the waxwings (*q.v.*). Without qualification, usually refers to usually brilliantly colored, blue or green, tropical American cotingids

chattering term for a group of choughs or starlings (*cf.* murmuration)

Chauliodontidae a small family of abyssal isospondylous fishes sometimes called viper fish by reason of their elongate shape and needle-like teeth

chayote the cucurbitaceous vine *Sechium edule* and its edible squash-like fruit

ch-b *see* [chilblained-b]

cheat *U.S.* the grass *Bromus secalinus*

check *U.S.* the grass *Bromus secalinus*

checkerboard the venerid mollusk *Macrocallista maculata*

cheechak the lacertilian reptile *Hemidactylus frenatus* (*cf.* gecko)

tufted **cheek** any of several funariid birds of the genus Pseudocolaptes

moss-**cheeper** *N.Brit.* the motacillid bird *Anthus pratensis* (= meadow pipit)

cheer the phasianid bird *Catreus wallichi*

cheery-cheery *W.I.* the tyrannid bird *Elaenia martinica*

pink**cheese** *Brit. dial.* the parid bird *Parus caeruleus* (= blue titmouse)

cheetah the felid carnivorous mammal *Acinonyx jubata* (= hunting leopard)

cheetal = chital

-cheil- *comb. form* meaning "lip" usually transliterated -chilo- (*q.v.*)

Cheliostomata an order of gymnolaemate Ectoprocta distinguished from the closely allied Ctenostomata (*q.v.*) by the presence of avicularia

-cheimo- *comb. form* meaning "winter" frequently transliterated -chimo-

-cheir- *comb. form* meaning "hand" frequently transliterated -chiro-

Cheridiidae a family of pseudoscorpionid arthropods distinguished by having the femur of all the legs entire and the legs consisting of only five segments

chela 1 (*see also* chela 2) an arthropod limb in which the ultimate joint is articulated with the base of the penultimate so as to form a claw, like that of the lobster

chela 2 (*see also* chela 1) a microscleric monaxonic sponge spicule with recurved hooks, plates, or flukes at each end

    isochela a chela with both ends alike

    anisochela a chela with unlike ends

chelate possessing chelae, or being in the shape of chelae

    heterochelate said of an arthropod in which the right and left chelae are different

chelicera the most anterior appendage of arachnids, usually chelate and always used in feeding

Chelicerata the term coined to describe those classes of the phylum Arthropoda which lack a true mandible. As so defined the taxon contains the Xiphosura, Pycnogonida, and Arachnida

Chelisodochidae a family of dermapteran insects lacking a dilated lobe on the side of the second tarsal segments

Chelonethida = Pseudoscorpionida

Chelonia the only living order of anapsid reptiles. This order contains those tortoises and turtles which are clearly distinguished from other reptiles by the possession of a bony or leathery shell

Cheloniidae a family of water-adapted marine cryptodiran chelonian reptiles. In *Brit.* the word turtle is usually confined to these forms which are distinguished by their paddle shaped limbs and horny shields

Chelydidae a family of bizarre pleurodiran chelonian reptiles in which the head can be partly retracted

Chelydridae a family of cryptodiran chelonian reptiles containing the forms commonly called snapping turtles, musk turtles and mud turtles. They are distinguished by the long costiform process of the proneural bone, the strong tail, and the powerful hooked beak on the snout

Chenopodiaceae that family of centrospermous dicotyledons which contains, *inter alia*, goosefoots, glassworts and saltbushes. The one-celled one-seeded ovary and the inconspicuous flowers are distinctive. Many are halophytic

-cherad- *comb. form* meaning "silt" but used by ecologists in the sense of sand bars

cheradad a plant of sand bars

cherimoya the annonaceous tree *Annona cherimolia* and its fruit

  wild cherimoya *A. longifolia*

cherlock *Brit.* the cruciferous herb *Brassica arvensis* (= wild mustard)

Chermidae a family of homopteran insects strongly resembling aphids but differing from them in lacking cornicles. There is no vernacular name though the genus Pylloxera is universally known for its depredations on grape vines

Chernetidea = Pseudoscorpiones

cherry 1 (*see also* cherry 2, aphis, beetle, bird, laurel, maggot, orange, picker, sawfly, sucker and weevil) any of many species and horticultural varieties of rosaceous trees and shrubs of the genus Prunus

  bird cherry *U.S. P. pensylvanica Brit. P. padus*

  bitter cherry *P. emarginata*

  wild black cherry *P. serotina*

  choke cherry *U.S.P. virginiana*

  dwarf cherry = ground cherry or sand cherry

  fire cherry *U.S.* = bird cherry *U.S.*

  Japanese flowering cherry *P. serrulata* and *P. sieboldii* and numerous horticultural hybrids and varieties of these

  ground cherry (see *also cherry* 2) *P. fruticosa*

  Islands cherry *P. lyonii*

  mahaleb cherry = St. Lucie cherry

  mazzard cherry = sweet cherry

  morello cherry *P. cerasus*

  mountain cherry *P. angustifolia* (= chickasaw plum)

  pie cherry = morello cherry

  pin cherry *U.S.* = bird cherry *U.S.*

  quinine cherry = bitter cherry

  red cherry *Brit. P. cerasus* (= *U.S.* sour cherry)

    wild red cherry = bird cherry *U.S.*

  rosebud cherry *P. subhirtella*

  St. Lucie cherry *P. mahaleb*

  sand cherry any of several species of the rosaceous genus Prunus, particularly *P. pumula, P. susquehennae, P. besseyi* and *P. depressa*

  western sand cherry *P. besseyi*

  sour cherry = morello cherry

  sweet cherry *P. avium*

  wild cherry *U.S. P. ilicifolia Brit. P. avium* (= *U.S.* sweet cherry)

cherry 2 (*see also* cherry 1) any of numerous trees and shrubs not of the genus Prunus, bearing red, purple, or black fruit with small stones

  Barbados cherry the malpighiacean tree *Malpighia glabra* and its fruit

  bastard cherry the boraginaceous shrub *Ehretia tinifolia*

  bladder cherry = winter cherry

  Australian bush cherry the myrtaceous tree *Eugenia myrtifolia*

  cornelian cherry the cornaceous tree *Cornus mas*

  ground cherry *U.S.* any of several species of the solanaceous herb Physalis. *Brit.* and *U.S.* the solanaceous herb *Physalis pubescens*

  Indian cherry the rhamnaceous tree *Rhamnus caroliniana*

  Jerusalem cherry the solanaceous shrub *Solanum pseudo-capsicum*

  madden cherry the rosaceous tree *Maddenea hypoleuca*

  Spanish cherry = Surinam cherry

  Surinam cherry the myrtaceous tree *Eugenia uniflora*

  winter cherry the solanaceous herb *Physalis alkekengi* (= strawberry tomato)

-chers- *comb. form* meaning "dry land"

chersad a plant of dry wasteland

chervil *Brit.* any of several species of umbelliferous herb of the genera Anthriscus and Chaerophyllum

  hemlock chervil the umbelliferous herb *Torilis anthriscus* (= upright hedge-parsley)

  leaf chervil = salad chervil

  rock chervil *Brit. C. temulentum*

  turnip-rooted chervil *Brit. C. bulbosum*

  salad chervil *Brit. a. cerefolium*

  tuberous chervil *C. bulbosum*

  wild chervil any of several species of umbelliferous herb of the genus Cryptotaenia

crescent chest any of several rhinocryotid birds of the genus Melanopareia (*cf.* tapaculo)

chevron *Brit.* the geometrid lepidopteran insect *Lygris testata*

chevrotain properly any of several species of tragulid artiodactyl mammals, though the term is often applied to other small deer-like forms

chewink *U.S.* the fringillid bird *Pipilo erythrophthalmus* (= towhee)

grey chi *Brit.* the noctuid lepidopteran insect *Polia chi*

-chiasm- *comb. form* meaning "a cross" or, more properly, the shape of the *Gr.* letter "chi"

chiasma 1 (*see also* chiasma 2, terminalization) a point of contact, appearing as a cross, between segments of homologous half chromosomes in meiosis, commonly thought to indicate genetic exchange or crossing over.

  distal chiasma one distal to any marker and the end of the chromosome

  proximal chiasma one between any marker and the centromere

chiasma 2 (*see also* chiasma 1) pertaining to other structures

  optic chiasma the external cross-over of the optic nerves under the lower surface of the brain

chick (*see also* weed) a young bird. Without qualification, usually that of the domestic chicken. Also in numerous compound bird names

dabchick *Brit.* the podicipedid bird *Podiceps ruficollis* (= little grebe)

twopenny chick *either* (*W.I.*) the turdid *Turdus aurantius* (= white chinned thrush) *or* the rallid *Porzana flaviventer* = yellow-breasted crake)

chickadee *U.S.* any of numerous species of parid bird of the genus *Parus* (= *Brit.* titmouse)

chestnut-backed chickadee *P. rufescens*

boreal chickadee *P. hudsonicus*

black-capped chickadee *P. atricapillus*

brown-capped chickadee = boreal chickadee

Carolina chickadee *P. carolinensis*

Mexican chickadee *P. sclateri*

mountain chickadee *P. gambeli*

Siberian chickadee *P. cinctus*

chicaree the sciurid rodent *Tamiasciurus hudsonicus* (= *U.S.* red squirrel)

chicken 1 (*see also* chicken 2, 3, flea, grape, louse, mite and Mother Carey) without qualification, the domestic fowl. Also in compound names of many birds of similar appearance

meadow chicken any of several species of rallid bird

Mother Carey's chicken any black and white storm-petrel of the family Hydrobatidae (*cf.* goose)

Pharaoh's chicken the accipitrid *Neophron perchopterus* (= Egyptian vulture)

prairie chicken *U.S.* any of several species of tetraonid bird of the genus Tympanuchus, particularly *T. cupido* (= **greater prairie chicken**)

lesser prairie chicken *T. pallidicinctus*

chicken 2 (*see also* chicken 1, 3) the young of many animals including turtles and lobsters (*cf.* chick)

chicken 3 (*see also* chicken 1, 2) some other edible animals

mountain chicken *W.I.* the giant leptodactylid anuran amphibian *Leptodactylus fallax*

chick-me-chick *W.I.* the turdid bird *Turdus aurantius* (= white chinned thrush)

chick-of-the-village *W.I.* the vireonid bird *Vireo crassirostris* (= thick-billed vireo)

chicory the compositaceous herb *Cichorium intybus*

chief *S.Afr.* the danaid lepidopteran insect *Amauris echeria*

false chief *S.Afr.* the nymphalid lepidopteran insect *Pseudacraea lucretia*

Chief-John-stirrup *W.I.* the vireonid bird *Vireo altiloquus* (= black whiskered vireo)

chiffchaff *Brit.* the small sylviid bird *Phylloscopus collybita* (*cf.* willow-wren)

chigaree *Austral.* the malurid bird *Acanthiza chrysorrhoa* (= yellow-tailed thornbill)

chigger any of numerous partially, or wholly, parasitic trombidiid mites of the subfamily Trombiculinae. *U.S.*, without qualification, *Trombicula irritans*

tree-toad chigger *Hannemania hylae*

chigoe *U.S.* the chigger *Tunga penetrans*

-chil- *comb. form* meaning "lip" or "margin" more correctly, but rarely, transliterated -cheil-

chil the lip, particularly of a flower

achil lacking a lip

epichil a distinct terminal portion of the labellum of an orchid

hypochil the base of the labellum of orchids

mesochil the central lobe of a tripartite orchid lip

chilarium the boundary of a pit, that by hygroscopic means causes the rupture of the silique of a legume

gnathochilarium a plate, ventral to the mandible formed from the first and second maxilla of diplopod Arthropoda. Homologous to the maxillae and labium of insects

[chilblained-b] a mutant (gene symbol *ch-b*) mapped at 23.8 on the X chromosome of *Drosophila melanogaster*. In the phenotype the tarsi are swollen and adherent to each other

chilidium a plate closing the notothyrium of a brachiopod shell

chilile the bottoms of shallow portions of lakes down to a depth of 6 meters (*cf.* bathile, pythmic)

Chilopoda a class of opisthogoneate Arthropoda, containing those animals commonly referred to as centipedes. The flattened shape and widely separated legs distinguishes them clearly from the millipedes with which they were once associated. In many centipedes the first pair of walking legs are modified as poison claws

-chim- *comb. form* meaning "winter" or, by extension "cold"

chimaera 1 (*see also* chimaera 2) an organism containing tissues derived from at least two genetically distinct parents; most chimaeras are known only from the plant kingdom

diplochlamydeous chimaera one in which the outer component is two cells thick

haplochlamydeous chimaera one in which the epidermis forms one component and all the other tissues form the other component

mericlinal chimaera one in which one of the constituents partially surrounds the other

periclinal chimaera one in which one of the constituents completely surrounds the other

chromosomal chimaera one in which the different tissues do not have the same chromosome number

hyperchimaera one in which the components are arranged in an irregular mosaic

mixochimaera one which contains protoplasmic constituents from two different individuals or species (= mixote)

sectorial chimaera one in which one component is segregated in a small area, usually a sector of a cylindrical stem, and, therefore, the branches derived from that sector

chimaera 2 (*see also* chimaera 1) popular name of Holocephali

elephant chimaera popular name of callorhyncid Holocephali

long nosed chimaera any of several species of rhinochimaerid Holocephali

plow-nosed chimaera = elephant chimaera

Chimaeridae a family of holocephalan fishes distinguished by their cartilaginous skeleton and only one external gill opening

-chiminal *see* chimous

chimney a protrusion of epidermal cells around a stoma

chimous pertaining to cold or winter

brachychimous pertaining to short winters

hydrochimous adapted to a rainy winter

isochimous pertaining to areas of equally low temperatures

brachyxerochimous pertaining to short dry winters

brachytheroxerochimous pertaining to short summers, coupled with dry winters

chimpanzee the simiid primate of the genus Anthropopithecus considered to have only the one species *A. troglodytes*. It is distinguished from the gorilla by its smaller size, friendly disposition, and unpigmented skin; and from the orang-utan by its short hair

blue chin the loricarrid teleost fish *Xenocara dolichoptera*

chinchary *W.I.* the tyrannid bird *Tyrannus domicensis* (= grey king bird)

chinchilla popular name of chinchillid hystricomorph rodents; usually, without qualification, applied to *Chinchilla lanager*; this animal is principally distinguished by the mythical values placed on its fur

Cuvier's chinchilla any of several species of chinchillid rodent of the genus Lagidium

rat-chinchilla popular name of abrocomid hystricomorph rodents

Chinchillidae a family of Andean hystricomorph rodents with large ears and silky fur. They are distinguished from most other hystricomorphs by the presence of a nail instead of a claw on the big toe. The chinchilla and vizcacha are well known examples

[chinese] a mutant (gene symbol *oc*) mapped at 40.8 on linkage group V of *Bombyx mori*. The phenotype has a highly translucent larva

river chink *W.I.* the jacanid bird *Jacana spinosa* (= Brazilian coot)

chinkapin the nymphaeaceous aquatic *Nelumbo lutea*

chino the leguminous tree *Pithecolobium mexicanum*

chinquapin (*see also* oak) the fagaceous trees *Castanea pumila* and *Castanopsis chrysophylla* (= **golden chinquapin**)

chion- *comb. form* meaning "snow"

chionad a plant dwelling in snow

Chionidae = Chionididae

Chionididae a family of charadriiform birds erected to contain the two sheathbills. These oceanic island birds have a black bill with a large horny sheath on the base and a carunculated plate. In other respects, except for the absence of webbed feet, they resemble gulls

chip-chip *W.I.* any of several parulid birds

chipchop = chiffchaff

chipmunk any of numerous *U.S.* terrestrial sciurid rodents of the genera Tamias and Eutamias

  cliff chipmunk *E. dorsalis*

  Colorado chipmunk *E. quadrivittus*

  Eastern chipmunk *T. striatus*

  yellow pine chipmunk *E. amoenus*

  lodge pole chipmunk *E. speciosus*

chir- *see* cheir-

chir = cheer

chirimoya = cherimoya

Chirocentridae a monospecific family of isospondylous fishes containing the form commonly known as a wolf herring. It is distinguished from all other clupeid fishes by the fang-like teeth in the jaws

Chiromyidae a family of lemuroid primates containing the single species *Chiromys madagascariensis* (= aye-aye)

Chiromyzidae a family of brachyceran dipteran insects now usually included with the Stratiomyidae

Chironomidae a family of slender, long-winged, long-legged, small, and very delicate nematoceran Diptera commonly called midges or true midges. They are scavengers not blood suckers

Chiroptera that order of mammals which contains the bats. The adaptation of the limbs for flight is characteristic of the group

chiru the Tibetan saigine caprid *Panthalops hodgsoni*. It is distinguished by a laterally expanded muzzle (*cf.* saiga)

chital the Indian cervine artiodactyle mammal *Axis axis* (*cf.* hog deer)

chitin a substance present in the exoskeleton of all arthropods, in the skeletal elements of many other invertebrate animals, and in the cell wall of many fungi. It is a high molecular weight polymer composed of N-acetylglucosamine residues joined together by β-glycosidic linkages

  actinochitin an optically active chitin found in the setae of certain acarines

  pseudochitin = tectin

chitinase an enzyme catalyzing the breakdown of chitin through the hydrolysis of α-1,4 acetylamino-2-deoxy-D-glucoside links

chitobiase an enzyme catalyzing the hydrolysis of chitobiose into 2,2-acetylamino-2-deoxy-D-glucose

chiton the name popularly applied to amphineuran mollusks

chive 1 (*see also* chive 2, chives) an offset of a bulbous plant

chive 2 (*see also* chive 1, chives) an anther

chives the lileaceous herb Allium schoenoprasum

chl *see* [chaetelle]

Chlaeneae a suborder of malvale dicotyledons containing the single family Chlaenaceae

-chlamy- *comb. form* meaning "cloak" but extended to the mantle of mollusks and the perianth of flowers

achlamydate said of mollusks lacking a mantle

chlamydous pertaining to the perianth but also, without qualification, used to designate a fruit with complete integuments

  achlamydous said of a flower lacking a perianth

  monodichlamydous having either one or both sets of floral envelopes

  dichlamydous said of a flower that has both components of the perianth, that is calyx and corolla

  haplochlamydous having a single perianth

  heterochlamydous having a clearly differentiated calyx and corolla

  homochlamydous having all the perianth leaves identical

  homoichlamydous having a uniform perianth

  monochlamydous having a perianth of a single series

Chlamydiaceae a family of rickettsial microtatobiotes parasitic in invertebrates and vertebrates, including man, but distinguished from the Rickettsiaceae in lacking arthropod vectors. They are the causative agents of, *inter alia*, psittacosis and several types of conjunctivitis

chlamydium a bud-scale

Chlamydobacteriaceae a family of chlamydobacterial schizomycetes in which the trichomes are frequently branched and in which motile swarm cells with a subpolar tuft of flagella are frequently produced

Chlamydobacteriales an order of gram-negative schizomycetes found in both fresh and salt water. They occur in trichomes, sometimes with false branching, and the sheaths frequently contain iron or manganese deposits

Chlamydomonadaceae that family of volvocale Chlorophyceae distinguished as unicellular forms with a definite cell wall and with all, or a portion, of the protoplast adjoining the wall

Chlamydoselachidae a family of sharks distinguished by the possession of six gills, of which the gill cover of the first hyoid is continuous across the body ventrally, and a single dorsal fin

-chled- *comb. form* meaning "rubbish heap' (*cf.* -ruderal-)

chledad a plant of rubbish heaps

-chlor- *comb. form* meaning "grass green" and, by extension, "posessing chlorophyll"

phyto**chlore** (*obs.*) = chlorophyll

Chlorhaemidae a family of sedentary polychaete annelids with many short green tentacular filamentous gills and green blood

chlorinous yellowish-green

DDT-dehydro**chlorinase** an insect enzyme that detoxifies DDT

Chloroanthaceae a small family of piperale dicotyledonous angiosperms distinguished from other Piperales by the opposite stipulate leaves and the inferior ovary

Chlorobacteriaceae a family of pseudomonodale schizomycetes commonly called the green sulphur bacteria. They require hydrogen sulphide and light for their development but produce only green pigments and deposit sulphur in extracellular form

chlorochrous green skinned

Chlorococcaceae that family of chlorococcale Chlorophyceae distinguished by the possesion of more or less globous cells showing unicellular zoospores

Chlorococcales that order of chlorophycean Chlorophyta distinguished by the fact that neither the uninucleate nor the multinucleate cells divide vegetatively

chlorogogue tissue that envelops the alimentary canal and fills the typhlosole of oligochaete annelids

Chloromonadida an order of phytomastigophorous Protozoa containing pale chloroplasts, possessing two flagella, and exhibiting amoeboid movement

Chloromonadina an order of mastigophoran Protozoa with two flagella, numerous chloroplasts and no stigma

Chloroperlidae a family of plectopteran insects with usually an anal lobe on the hind wing, this lobe containing at most three veins

Chlorophyceae that class of chlorophyte algae which contains all green algae except the stoneworts from which they are distinguished in lacking ensheathing structures around the sex organs

chlorophyll *see* -phyll-

Chlorophyta a subphylum of algae containing the forms commonly called green algae. They are distinguished from all other algae by the bright green chloroplasts

Chloropidae a family of small myodarian cycloraphous Diptera living in grassy habitats and frequently rather brilliantly colored in black, orange, and yellow. They are often called frit flies, but their attraction to the human eye, where they act as vectors of pink eye, sometimes causes them to be called eye gnats

-chlorous pertaining to green, or green things

chimonochlorous retaining green leaves throughout the winter

melanochlorous blackish green

choana literally, a funnel and applied to many structures of this shape, but most frequently to the internal nares, particularly of reptiles

Choanichthyes = Sarcopterygii

chocho the cucurbitaceous vine *Sechium edule* and its edible squash-like fruit

chockalott *Austral.* the psittacid bird *Kakatoe leadbeateri* (pink cockatoo, chock-a-lock)

[**Chocolate**] a mutant (gene symbol *Ch*) mapped at 128 on linkage group II of *Zea mays*. The name refers to the color of the pericarp of the phenotype

[**chocolate**] a mutant (gene symbol *ch*) mapped at 0 on linkage group XIII of *Bombyx mori*. The name refers to the color of the newly hatched larva of the phenotype

chok *S.Afr.* the eagle *Aquila rapax* (= tawny eagle)

hog **choker** *U.S.* the soleid fish *Trinectes maculatus*

choko the cucurbitaceous vine *Sechium edule* and its edible squash-like fruit

chol-1 *see* [Choline-1]

Cholecystokinin a hormone secreted in the upper intestinal mucosa. It is active in the contraction and emptying of the gall bladder

choline (β-hydroxyethyl)-tri-methylammonium hydroxide. A water soluble nutritional factor frequently classed as a vitamin

acetylcholine a hormone secreted in the nervous system that affects the conduction of electrical impulses along nerve fibers, and opposes the activity of epinephrin

[**choline-1**] a mutant (gene symbol *chol-1*) in linkage group IV of *Neurospora crassa*. The phenotypic expression is a requirement for choline

-chom- *comb. form* meaning "an aggregation", or "a heap"

icochomous pertaining to branches springing synchronously from the same stem

Chondrichthyes that class of gnathostomatous craniate chordates which contains those fishes which lack a bony, as distinct from an occasionally calcified, skeleton

chondriome the sum total of the granules in a cell

-chondr- *comb. form* meaning "granule," "gristle," and, by extension, "cartilage"

chondrion an intracellular inclusion

desmochondrion = mitochondrion

mitochondrion an organelle, usually sausage-shaped but varying from spherical to filamentous, found in the cytoplasm of all cells except bacteria and Cyanophyceae. The interior is imperfectly partitioned with usually transverse, but sometimes longitudinal or concentric, cristae on which, as on the outer surface of the organelle, are granules that, are source of ATP and hence control the oxidative metabolism of the cell

chondrium cartilage

perichondrium the connective-tissue envelope of cartilage

chondroid the condition of the medulla of the lichen in which the fungal hyphae form a solid axis

chondrome the sum total of the granules in a cell

Chondrophora a division of decapod cephalopod Mollusca distinguished by the presence of a chitinous shell. This group is commonly called squid

chondrosis a cartilagenous joint

synchondrosis immobile "joints" in which the bones are connected with cartilage

Chondrostei a group, variously regarded as an order or a superorder of Osteichthyes with a heterocercal tail, much cartilage in the skeleton, and distinguished by having the maxillary firmly united with the ectopterygoid

chone a sponge canal which penetrates the cortex and is provided with a sphincter of myocytes

Chonotrichida a small order of peculiar euciliate ciliate Protozoa with a spirally coiled body and a few cilia on the anterior end

Choragidae = Anthribidae

-chord- *comb. form* meaning "cat gut", or by extension any straight cylindrical form. Now thoroughly confused with -cord- which is the *Eng.* form of the same word

hypochord a thin rod immediately below the notochord in some embryos

longitudinal chord one of four longitudinal ridges projecting from the epidermis into the pseudocoel of Nematodes

notochord a turgid rod of cells lying immediately beneath and parallel to, the nerve cord in Cephalochordata, and in vertebrate embryos

stomochord a diverticulum of the buccal tube that projects into the protocoel of hemichordates

chordal usually in *biol.* pertaining to the notochord

parachordal one of a pair of cartilages in an embryonic chondrocranium which lies immediately lateral to the anterior end of the notochord

Chordariales an order of heterogenerate Phaeophyta in which the branched filamentous sporophyte does not form a parenchymatous thallus

Chordata a phylum of bilateral coelomate animals containing the vertebrates and their allies. The phylum is characterized by the possession of a hollow dorsal nerve cord, a notochord, and pharyngeal slits in the anterior region of the alimentary canal, at some stage of development. In some groups all of these characters occur only in embryonic or larval stages.

Chordonia a zoological taxon erected by Kowalewski to include the urochordates and vertebrates

-chore- *comb. form* indicating the "agent of dispersion" of plants, or, from another root, a "place" or "region"

chore 1 (*see also* chore 2) in the sense of an agent of dispersal

androchore a plant dependent upon man for its distribution

anemochore a plant distributed by wind. The term is also specifically applied by some to a plant that retains its seeds through the winter for wind distribution in spring

anthropochore a plant that is introduced through the agency of man

autochore *either* a plant with motile spores *or*, a plant that distributes its own seeds as, for example, by projection

blastochore a plant propagated by offshoots

bolochore a plant distributed by mechanic propulsion

brotochore a plant that is dispersed by man

clitochore a plant that is distributed by land slides

gynochore an organism that is dispersed by wandering females

hydrochore a plant distributed through the action of water

indiochore an organism that distributes itself

myrmecochore said of a plant dispersed through the agency of ants

saurochore a plant distributed through the agency of reptiles

zoochore a plant distributed by animals (*cf.* zoidogamae)

epizoochore the dispersal of fruits by casual transport on the outside of animals

endozoochore the dispersion of plants by seeds that pass through the gut of an animal

synzoochore the condition of an organism that is dispersed through the agency of several animals

chore 2 (*see also* chore 1) in the sense of a place or part

allochore an organism occuring in two different habitats in the same geographic region

biochore *either* a group of similar biotopes so large as to form a recognizable habitat, thus forests, deserts and prairies are biochores; a tree with its associated organisms is a biotope (*q.v.*) *or* a climatic boundary indicated by the vegetation

eurychore an organism having a wide distribution

kinetochore = centromere

stenochore a plant of restricted distribution

Choreutidae = Glyphipterygidae

-chori- *comb. form* meaning "separate"

chorion 1 (*see also* chorion 2–5) the outer layer of the embryonic sac of a mammal consisting of both trophoderm and mesoderm elements and the outermost extraembryonic membrane in birds and reptiles

chorion 2 (*see also* chorion 1, 3–5) a membrane around the egg outside the vitelline membrane and produced by the follicle cells in the ovary

chorion 3 (*see also* chorion 1, 2, 4, 5) the vitelline membrane of tunicates and fishes

chorion 4 (*see also* chorion 1–3,5) the pulpy contents of a young ovule

synchorion = carcerule

chorion 5 (*see also* chorion 1–4) the shell of an insect egg

endochorion the membrane lying just within the chorion in insect eggs

exochorion the outer, ectodermal layer of the chorion in insect eggs

chorism division or splitting, particularly, the splitting of a leaf into several parts

anachorism = degeneration

cytochorism the process of division of cells

gonochorism sex determination

choristate unlined or unseparated or undivided

Choristida an order of tetractinellid sponges distinguished by the presence of triaenes

-choro- *comb. form* meaning "place" (*cf.* -chore-)

chorology the study of the distribution of organisms

chorus an aggregate of repetitive noises particularly those of insects and amphibia

Belcher's chorus the muricid gastropod mollusk *Forreria belcheri*

chouchoute the cucurbitaceous vine *Sechium edule* and its edible squash-like fruit

chough any of several, mostly corvid, birds. *Brit.*, without qualification *Pyrrhocorax pyrrhocorax* (= **redbilled chough**)

alpine chough *Pyrrhocorax graculis*

white-winged chough the graculid *Corcorax melanorhampus*

chousingha the boselaphine bovid artiodactyle *Tetraceros quadricornis* a close relative of the nilghai (*q.v.*) but distinguished from all other bovids by the presence of four horns on the male

-chres- *comb. form* meaning "use"

chresard that portion of the water content of soil which is available for plant growth

christophine the cucurbitaceous vine *Sechium edule* and its edible squash-like fruit

-chro- *comb. form* meaning "skin"

pleochroicism variously colored

-chrom- *comb. form* meaning "color". In *biol.* the sustantive "chrome" is usually confined to pigments and "chromatin" for those portions of a fixed nucleus which stain readily. "Chromatism" and "chromatic" usually refer to pigments while the termination "-chromous", though often used as synonymous with "chromatic" pertains to color as distinct from the causative agent of color. "Chromasy" has been used in reference to the nucleus, apparently through some imagined connection with chromatin

Chromadorida an order of aphasmid nematodes distinguished from the Enoplida by the fact that the esophagus is short and divided into three regions

Chromadoridea a class of aquatic nematodes with a heavily ornamented cuticle and spiral amphids

chromasy the condition of a nucleus. (*see* remarks under -chrom-)

achromasy the loss of part of a nucleus

hyperchromasy an increase in the relative size of the nucleus in relation to the cytoplasm

chromatid one of the pair of identical duplicated chromosome strands prior to a nuclear division

sister chromatids those derived from the same chromosome

non-sister chromatids those derived from homologous chromosomes

chromatin those parts of the nucleus which stain densely in nuclear stains

achromatin = linin

euchromatin that portion of the chromatin in a chromosome which contains gene loci

heterochromatin a term once applied to that chromosomal material that remains discreet in the interphase nucleus but nowadays taken to mean that portion of the chromatin of a chromosome which is devoid of genes

idiochromatin the "generative" chromatin postulated in the binuclearity theory (*q.v.*)

prochromatin an early term for the protoplasm of nucleoli

trophochromatin the vegetative chromatin postulated in the binuclearity theory (*q.v.*)

chromatism the condition of being pigmented (*see* remarks under -chrom-)

seasonal amphichromatism the production at different seasons of differently pigmented flowers on the same plant

aptosochromatism the condition in which pigment is augmented without growth

chrome in *biol.* a pigment (*see* remarks under -chrom-)

adenochrome a red pigment found in some cephalopod Mollusca

biochrome any pigment found in a living organism

indigoid biochrome a group of blues and purples derived as the end product of the metabolism of tryptophane. They are best known in the plant kingdom but also account for the mollusk-derived Tyrean purple of the ancients

quinone biochrome a common group of reds and yellows best known as the pigment of the cochineal insect

cytochrome a hemoprotein utilized in biokinetic systems for electron transport by virtue of the reversible valency of heme-Fe

cytochrome A a cytochrome in which the Fe is in a formylporphyrin linkage

cytochrome B a cytochrome in which the Fe is in a protoporphyrin linkage

cytochrome C a cytochrome in which the Fe is in a mesoporphyrin linkage

cytochrome D a cytochrome in which the Fe is in a dehydroporphyrin linkage

endochrome a pigment other than chlorophyll within a cell

entomurochrome the pigment of the arthropod malphigian tubule

ominochrome a brown visual pigment found in insects

phytochrome a chromoprotein that regulates germination, growth and flowering in higher plants

schemochrome a color produced by structure, as the iridescent colors of butterfly wings

chromide popular name of cichlid teleost fish of the genus Etroplus

chromidium 1 (*see also* chromidium 2) a term at one time applied to a granule of "chromatin" but now used for cytoplasmic granules staining with nuclear stains

chromidium 2 (*see also* chromidium 1) the gonidium of a lichen

chromosome *see* -some

-chromous *adj. term* pertaining to color (*see* remarks under -chrom-)

amphichromous said of a plant bearing flowers of two colors on the same stock

coccochromous blotchy with the colored areas granular

homeochromous pertaining to the condition in which there appears to be a tendency for color variations among different species to tend toward the same color in a given geographic area

homochromous when one sex of an insect has two color phases, this term applies to the one that resembles the other sex

heterochromous 1 (*see also* heterochromous 2) when one sex of an insect has two color phases this term applied to the phase that does not resemble the opposite sex

heterochromous 2 (*see also* heterochromous 1) pertaining to the production of different colored flowers in the same inflorescence

leochromous lion colored

metachromous changing color or pertaining to changed color

placochromous with the color in plates or discs, used particularly of diatoms

-chron- *comb. form* meaning "time"

plastochron the interval between the appearance of successive leaf primordia on a shoot

chronic pertaining to time. -chronous is synomous

allochronic a morphological discontinuity which arises simply through the passage of time in contrast to allopatric (*q.v.*)

biochronic the period of time during which mutations have been possible (De Vries)

heterochronic *either* pertaining to the type of development that appears to recapitulate ancestral types, but in the wrong order *or* to an unequal rate of development of parts, usually of an embryo

polychronic occuring several times or originating at several epochs

chronous pertaining to time

synchronous at the same time

dyschronous not at the same time

-chrys- *comb. form* meaning "gold"

chrysalis a popular name for the pupa of lepidoptera

imagochrysalis a dormant nymphal stage in the life history of trombiculid mites

nymphochrysalis a quiescent, as distinct from dormant, nymphal state in the life history of trombiculid mites

pseudochrysalis = semi-pupa

chrysaloideous wrapped and folded as a butterfly pupa

chryseous golden yellow (= aureous, not auratous)

Chrysididae a family of apocritan hymenopteran insects of medium size with brilliant metallic blue and green bodies. They strongly resemble some of the metallic colored bees but may be distinguished by the fact that there is no closed cell in the venation of the hind wing. They are commonly called cuckoo wasps

Chrysidoidea that superfamily of hymenopteran insects which includes the cuckoo wasps

Chrysobalanoideae a subfamily of Rosaceae distinguished by having a superior ovary, indehiscent fruit, and zygomorphic flowers with gynoceium of a single pistil

Chrysocapsales an order of chrysophycean chrysophytes in which the cells are found in gelatinous colonies

Chrysochloridae a family of insectivores containing the South African golden moles. They are more closely allied through their teeth to the Solenodontidae than to the Talpidae from which they also differ in the structure of the pectoral girdle and in possessing only four digits, of which the center two are greatly enlarged, on the forelimb

Chrysomelidae a very large family of coleopteran insects commonly called leaf beetles. They are closely allied to the cerambycids but have much shorter antennae and thicker, shorter, bodies. Many are brilliantly colored

Chrysomeloidea a superfamily of polyphagous coleopteran insects containing most of the foliage-eating beetles

Chrysomonodales an order of chrysophycean chrysophytes containing those form in which the vegetative phases are motile

Chrysomonadida an order of mastigophoran Protozoa distinguished by the presence of yellow or brown chromoplasts. Most biologists regard these forms as algae, in which case, the Chrysomonida of the zoologists corresponds to the Chrysomonadales (*q.v.*) of the botanists. Some unicellular Rhizochrysidales (*q.v.*) are occasionally included in the Chrysomonadida

Chrysophyceae a class of chrysophyte algae containing those forms in which the chromatophores are golden-brown in color

Chrysophyta a sub-phylum of algae distinguished by the fact that the chromatophores are golden since carotene and xanthophylls are sufficiently predominant to mask the color of the chlorophyll

Chrysopidae a large family of green neuropteran insects commonly called lacewings. They are readily distinguishable as the only insect with tent-shaped green transparent wings

Chrysosphaerales an order of chrysophycean chrysophytes containing those forms which are nonmotile, unicellular or, at least, not filamentous colonies

Chrysotrichales an order of chrysophycean chrysophytes distinguished by having branching filamentous thalli

chub *U.S.* any of very many species of fresh water cyprinid fish and also several marine kyphosids. *Brit.* the cyprinid *Leuciscus cephalus*

  creek chub *U.S. Semotilus atromaculatus*
  streamline chub *U.S. Hybopsis dissimilis*
  river chub *U.S. Hybopsis micropogon*
  bigeye chub *U.S. Hybopsis amblops*
  yellow chub *U.S. Kyphosus incisor*

[chubby] a mutant (gene symbol *ch*) mapped at 72.5 on chromosome II of *Drosophila melanogaster*. The phenotypic larva, pupa, and adult are all of short stature

chuck *U.S.* any of several sciurid rodents

  he-chuck a male woodchuck
  rock chuck the sciurid rodent *Arctomys flaviventris* (yellow belly marmot)
  woodchuck the large sciurid rodent *Arctomys monax* (= Quebec marmot ground hog)

chuckar = chukar

chuckwalla *U.S.* the iguanid lacertilian reptile *Sauromalus obesus*

chuck-will's-widow the caprimulgid bird *Caprimulgus carolinensis*

chufa the edible tuber of the sedge *Cyperus esculentus*

chufe the sedge *Cyperus esculentus*

chukar the phasianid bird *Alectoris graeca*

chukkar = chukar

chumpa the musaceous tree-like herb *Musa chumpa* (= ladyfinger banana)

[Chunky] a mutant (gene symbol *chy*) mapped at 18± on chromosome II of *Drosophila melanogaster.* The phenotypic expression is a short, heavy-set body and short wings

chunga the *S. Amer.* cariamid bird *Chunga burmeisteri*

chy *see* [chunky]

chyle *either* lymph rendered milky by emulsified fat *or* partially digested nutrients in the alimentary canal

  perichyle a plant with water-storing tissue between the epidermis and the chlorenchyma

chylema an obsolete term for protoplasm

  cytochylema the sum total of the contents of a plant cell, involving both the protoplasm and the vacuoles

  enchylema the more fluid part of the protoplasm

  nucleochylema = nuclear sap

-chym- *comb. form* meaning "juice" and once used for "protoplasm"

plasomochym protoplasm in the gel condition

-enchyma *comb. suffix* used to indicate plant tissues. The termination -enchyme is preferred for animals

  actinenchyma stellate basic tissue of a plant, corresponding to stellate reticulum in animals

  angienchyma vascular tissue

  atractenchyma = prosenchyma

  bothrenchyma tissue composed of pitted cells or ducts

  ceratenchyma a horny plant tissue developed from sieve tubes

  chlorenchyma chlorophyll-containing cells

  cienchyma intercellular spaces

  cladenchyma branched parenchyma

  collenchyma elongated prismatic cells in seed plants that, though they function as the supporting structures of young leaves and stems, are still capable of growth

    protocollenchyma partially differentiated collenchyma

  colpenchyma plant cells with sinuous walls

  coenchyma conical trichogynes

  critenchyma the nonvascular tissue of the vascular bundles

  cylindrenchyma a tissue made up of cylindrical cells

  cytenchyma a tissue made up of plant cells with large

  daedalenchyma tissue made up of woven hyphae

  diachyma = mesophyll

  hexagonienchyma plant tissue, the cell walls of which appear hexagonal in section

  hyphenchyma a tissue composed of intertwined hyphae

  inenchyma tissue, such as the spiral cells in sphagnum, that have the appearance of being conducting vessels, but are not

koleochyma = kritenchyma

kritenchyma the cell layer immediately ensheathing the vascular bundle

merenchyma tissue composed of spherical cells

mesenchyma that tissue which lies between xylem and phloem in vascular bundles in roots

orthenchyma a parenchyma of upright cells

orthosenchyma = orthenchyma

ovenchyma a parenchyma of oval cells

parenchyma without qualification, refers to the fundamental or supporting tissue of a plant. The term is occasionally used of loose animal connective tissue but all the following entries are botanical

  abaxial parenchyma paratracheal parenchyma in contact with the outer surface of vessels

  banded parenchyma apotracheal parenchyma appearing in bands in the growth ring

  vasicentric parenchyma paratracheal parenchyma surrounding the vessels

  diffuse parenchyma apotracheal parenchyma evenly dispersed throughout the growth ring

  palisade parenchyma parenchyma consisting of elongate cells lying vertical to the surface of the blade of a leaf

  phloem parenchyma parenchyma cells of phloem tissue

  pseudoparenchyma tissue resembling parenchyma but actually not consisting of cells related to each other in function and structure

  sclerotic parenchyma parenchyma cells with lignified secondary walls

  spongy parenchyma that portion of the mesophyll that is not palisade parenchyma

  terminal parenchyma apotracheal parenchyma restricted to the outside of the growth ring

  apotracheal parenchyma secondary xylem parenchyma that is independent of the vessels

  paratracheal parenchyma secondary xylem parenchyma definitely associated in a specific pattern with the vessels

perienchyma irregular cellular tissue in a plant

phycenchyma the ground tissue of large algae

pinakenchyma the tissue of medullary rays

pinenchyma = pinakenchyma

plectenchyma a tissue composed of hyphae

  paraplectenchyma layered or folded mycelium

  prosoplectenchyma hyphal tissue that has the appearance of being plaited

pleurenchyma xylem

porenchyma a tissue of pitted cells

prismenchyma cellular tissue appearing prismatic

prosenchyma a tissue of lengthy cells with overlapping ends

protenchyma ground tissue in general

schlerenchyma coherent masses of skeletal, as distinct from conducting, tissue of a plant

  girder sclerenchyma a sclerenchyma which in section has the shape of a "T" or "H"

  protosclerenchyma colenchyma resembling sclerenchyma

spherenchyma ground tissue consisting of spherical cells

statenchyma tissue formed of statocysts

taphrenchyma = bothrenchyma

perenchymatia any of many types of microscleric hexatine sponge spicules scattered in the trabecular net

chyme the partially digested contents of the stomach

-enchyme termination indicating animal tissue. The botanical equivalent is -enchyma

coenenchyme living tissue occupying the space between the thecae of coral polyps or the "fleshy" part of alcyonarians

collenchyme an embryonic tissue of cells loosely embedded in a gelatinous matrix

  cartilage collenchyme collenchyme with greatly thickened cell walls usually bordering a cavity (*cf.* rift collenchyme)

  rift collenchyme collenchyme on the edge of a cavity with a great thickening of those parts of the cell wall, that abut on the cavity (*cf.* cartilage collenchyme)

mesenchyme those portions of the embryonic mesoderm which are not segregated into layers or blocks

Chyromyidae = Chyromiidae

Chyromyiidae a family of little known cycloraphous myodarian dipteran insects

Chytridiales an order of uniflagellate phycomycetes in which the thallus is never a true myclium

ci *see* [cubitus-interruptus]

cibarium a food pocket or pouch in some insects

cibivia the sucking tube of sucking insects

cicada (*see also* bird, hawk, killer) popular name of large cicadid homopterous insects notorious for their strident evening song. In many parts of the United States the term locust is erroneously applied to these forms

  dog-day cicada *U.S.* any of several species of the genus Tibicen

  periodical cicada *U.S. Magicicada septendecim*

Cicadellidae a family of homopteran insects closely allied to the Cercopidae but having numerous spines extending the whole length of the hind tibia. They are commonly called leafhoppers

Cicadidae an order of homopteran insects containing the forms commonly called cicadas. These are among the largest homopteran insects and may readily be recognized by the general appearance

cicatrice the scar left after the abscission of a leaf or fruit

sweet **cicely** any of several umbelliferous herbs of the genus Osmorhiza. The name is also applied to the umbelliferous herb *Myrrhis odorata*

Cichlidae a large family of freshwater acanthopterygian fishes clearly distinguished by the possession of a single nostril on each side

Cichorieae a tribe of the Compositae distinguished by the ligulate flowers and presence of lactiferous vessels. This tribe is sometimes erected to separate familial rank

Cicindelidae a family of coleopteran insects commonly called tiger beetles. They are usually brightly colored and both run and fly rapidly

Ciconiidae that family of ciconiiform birds which contains the storks. They are distinguished by their long, massive ungrooved bill and long broad wings

Ciconiiformes that order of birds which contains, *inter alia,* the herons, flamingoes, and storks. They are distinguished by having very long bills and long legs

Cidaroida an order of echinoid Echinodermata having an apical periproct but distinguihsed from the Centrechinoida by the absence of sphaeridia and peristomal gills

Ciidae = Cisidae

-cil- *comb. form* meaning "hair," particularly those of the eyelid

cnidocil a projection on a nematocyte that triggers the discharge of the nematocysts

ciliary pertaining to the eyelid, eyebrow or cilia

superciliary either set above the eyes or pertaining to the eyebrow

Ciliata a class of ciliophoran Protozoa distinguished from the Suctoria (*q.v.*) by the presence of cilia throughout the whole life cycle

Ciliophora a subphylum of Protozoa distinguished from the Plasmodroma (*q.v.*) by the possession of cilia at some stage of the life cycle

cilium 1 (*see also* cilium 2) a row of hairs or bristles on arthropods

supercilium literally, the eyebrow but used also in insects of a line of hairs found above some compound eyes

cilium 2 (*see also* cilium 1) a vibratile organelle process found on many cells throughout the animal kingdom. Each cilium consists of nine pairs of peripheral filaments wrapped round a central pair, the whole embedded in a matrix. The vibratory motion may serve to propel the cell or, if the cell is fixed, to move substances over its surface

pseudocilium a structure resembling a cilium in form but not in function

stereocilium long, non-motile "cilium" on the epithelium of the epididymial duct

Cimbicidae a group of large symphytan hymenopteran insects distinguished from other sawflies by the clubbed antenna

Cimicidae that family of parasitic hemipteran insects commonly called bed bugs. They are flattened wingless forms

cimier the head crest on the pupa of pierid lepidopteran insects

cincinnal- *comb. form* meaning "curly or curled"

Cinclidae that family of passeriform birds which contains the dippers, some of the few passerines that dive and swim well under water. Many walk on the bottom of streams searching for insect larvae. The wings are very short and markedly concave

cinclide a pore in the body wall of some Zoantharia

-cinct- *comb. form* meaning "girdle"

circumcinctous the condition of being girded around or bound

-cine- *comb. form* meaning "move" or "movement" frequently and properly, transliterated -kine-

cineraceous ashy

cinereous ash gray

Cinetidae a family of proctotrupoid apocritan hymenopteran insects usually fused with the Diapriidae

cingulum literally, a girdle and used for almost any structure having this appearance including the ciliary zone on the disc of a rotifer, the clitellum of oligochaete annelids, the connecting edges of the frustules of diatoms, a circular ridge on the crown of some teeth, the junction between the stem and the root of a plant, and a colored band on gastroped shells

cinnabar *Brit.* the arctiid lepidopteran insect *Hipocrita jacobaeae*

[cinnabar] a mutant (gene symbol *cn*) mapped at 57.5 on chromosome II of *Drosophila melanogaster*. The phenotypic eye is bright scarlet

cinnamon (*see also* teal, tree, vine) the highly aromatic bark of several lauraceous trees of the genus Cinnammonium, or the trees themselves. *C. zeylanicun* is considered by most to be the "true cinnamon"

bastard cinnamon *Cinammonium cassia*

chinese cinnamon = bastard cinnamon

wild cinnamon the canellaceous tree *Canella winterana* and its aromatic bark

-cino- *see* -cine-

Cinosternidae = Kinosternidae

cinquefoil *see* -foil

Cinura = Thysanura

Cioidea = Cisidae

cion = scion

circadian repeated more or less daily—*i.e.* on a 23 hr to 25 hr cycle (*cf.* diurnal)

circe *I.P.* the nymphalid lepidopteran insect *Hestina nana*

Circinae that subfamily of accipitrid bires which contains the harriers. They are distinguished by the presence of an incomplete feathered ruff around the eye and very long bristle-like feathers on the lores, their tips extending about the outer outline of the cere

circinnate coiled into a ring

apicircinnate with the apex coiled in a ring

Willis' circle the arterial ring formed where the basilar artery splits to pass around on each side of the hypophysis and reunites on the other side

circulation in *biol.* primarily applied to the movement of fluids

pulmonary circulation the circulation of blood from the heart through the lung

respiratory circulation the circulation of the blood from the heart through a respiratory organ either lung or gill

systemic circulation the circulation of blood through those parts of the body not concerned with respiration

-circum- *comb. form* meaning "around" in the sense of surrounding

-cirr- *comb. form* meaning a "curl" of hair, very commonly misspelled -cirrh-

Cirratulidae a family of sedentary polychaete worms lacking head appendages and with very long filamentous dorsal cirri. Many are attached to the underside of stones without burrows

cirrhate having or pertaining to tendrils

Cirrhitidae a family of reef-dwelling acanthopterygian fishes commonly called hawk fishes. They are distinguished by the thickened rays of the pectoral fins

-cirrh- a common misspelling of -cirr-

cirrhous tufted

Cirripedia a subclass of crustacean arthropods containing the forms usually called barnacles as well as the parastic Ascothoracia and Rhizocephala. These last two groups are regarded by many as separate subclasses leaving the Cirripedia to consist only of the order Thoracica

cirrus 1 (*see also* cirrus 2) any flexible, often tactile, projection. In general, a cirrus is shorter than a flagellum (in the sense of flagellum 2), and longer than a papilla. In many invertebrates the term is synonymous with penis and is used also for projections from the stalks of crinoid echinoderms

radicular cirrus one of those crinoid cirri that act as roots

cirrus 2 (*see also* cirrus 1) a plant tendril

cisco any of several species of salmonid fishes, usually of the genus Coregonus

arctic cisco *C. autumnalis*

Bonneville cisco *Prosophium gemmiferum*

blackfin cisco *C. nigripinnis*

longjaw cisco *C. alpenae*

shortjaw cisco *C. zenithicus*

least cisco *C. sardenella*

Nipigon cisco *C. nipigon*

shortnose cisco *C. reighardi*
deepwater cisco *C. johannae*
Cisidae a family of minute coleopteran insects commonly called tree-fungus beetles but that are also sometimes destructive pests of buildings and furniture
Cistaceae that family of parietale dicotyledons containing the rock roses. The family is distinguished by its convolute petals and many hypogynous stamens
Cistelidae = Alleculidae
cisticola any of numerous sylviid birds of the genus Cisticola
croaking cisticola *C. natalensis*
desert cisticola *C. aridula*
red-faced cisticola *C. erythrops*
rattling cisticola *C. chiniana*
singing cisticola *C. cantans*
zitting cisticola *C. juncidis* (= *Brit.* fan-tailed warbler)
Cistineae a suborder of parietale dicotyledons containing the families Cistaceae and Bixaceae
cistron a genetic functional unit thought to control the synthesis of a single product (*cf.* recon, muton, configuration)
Citheroniidae a family of very large saturnioid lepidopteran insects closely allied to the Saturniidae
citrange the fruit of a hybrid between the rutaceous trees *Poncirus trifoliata* and *Citrus aurantium*
citron properly the ruaceous tree *Citrus medica*. But also applied to some melons
citronella (*see also* grass) the labiateous herb *Collinsonia canadensis*
L-**citrulline** α-amino-δ-ureidovaleric acid. H_2-NCO-$NH(CH_2)_3CHNH_2COOH$. An amino acid, not essential for rat nutrition, which, though first indicated from the watermelon, appears most commonly to be synthesized from ornithine by microorganisms
London **city** *W.I.* the cotingid bird *Platypsaris niger* (*cf.* becard)
civet usually civet cat (*q.v.*) except in compounds
anteater-civet the hemigaline viverrid carnivore *Eupleres goudotii* the very peculiar dentition of which, particularly the minute canines, at one time caused it to be placed in the Insectivora
otter-civet the hemigaline viverrid carnivore *Cynogale cynogale*
palm-civet popular name of paradoxurine viverrids
masked palm-civet any of several species of paradoxurine viverrid carnivores of the genus Paguma
ck *see* [crinkled]
cl *see* [clot]
-clad- *comb. form* meaning "branch"
Cladistia an order of Osteichthyes containing the single extant family Polypteridae. They are distinguished by the possession of ganoid scales and numerous dorsal finlets
cladium a branch, frequently anglicized as "clad." The adjectival termination is -cladous
hydrocladium a lateral branch from a hydrocaulus
pericladium the base of a leaf which ensheathes the stem
phyllocladium a flattened branch having the form and function of a leaf
sporocladium a branch bearing reproductive bodies
Cladocera an order of branchiopod crustacea containing the animals normally called water fleas. They are distinguished from the closely allied Conchostraca in the head not being enclosed in the bivalved shell, in posessing only five or six trunk appendages, and in lacking a telson and caudal rami
Cladocopa an order of ostracod Crustacea distingushed by only three pairs of post oral appendages and by their circular valves lacking an antennal notch, with no aperture when closed. They are exclusively marine
cladode a short flattened portion of a stem with limited growth, resembling a leaf and replacing it in photosynthesis (*cf.* phyllode)
Cladophorales that order of chlorophycean Chlorophyta distinguished by their multinucleate cylindrical cells united end to end in simple or branched filaments
[**cladosporium resistance**] any of 3 mutants with gene symbols Cf_1, Cf_2 and Cf_3 mapped respectively at 65 on chromosome 1, 61 on chromosome 6 and 86 on linkage group V of the tomato. The phenotypic expression of Cf_1 is resistance to races 1–3 of *Cladosporium fulvum*. Cf_2 and Cf_3 are resistant to races 1–4.
-cladous pertaining to a branch
acanthocladous having spiny branches
anocladous with out-curved branches
brachycladous having short branches
catocladous having branches bent out and down like a weeping willow
drepanocladous having sickle-shaped branches
eocladous said of a leaf, the initial of which in the meristem is itself branched
homalocladous straight branched
leptocladous with slender branches
macrocladous with long branches
mesocladous possessing branches of medium length
orthocladous straight branched or upright branched
syncladous the condition of having branches produced at the same time and thus forming tufts
clakis *Brit. dial.* = goose, particularly *Branta leucopsis* (= barnacle goose)
clam *U.S.* any of very numerous pelecypod mollusks
awning-clam any of several species of the genus Solemya
basket-clam popular name of corbulids
bean-clam the donacid *Donax gouldi*
black clam popular name of pleurophorids
blood-clam popular name of arcid pelecypods, particularly of the genus Arca
bloody clam the arcid *Arca plexata*
pocketbook-clam the unionid *Lampsillis ventricosa*
boring clam popular name of gastrochaenids
heart-clam popular name of cardids (*cf.* cockle)
hen-clam the mactrid *Spisula solidissima* (*cf.* surf clam)
long clam the myacid *Mya arenaria*
long-necked clam = soft shelled clam
bent-nosed clam the tellinid *Macoma nasuta*
perforated clam popular name of thraciids
pismo clam the venerid *Tivela stultorum*
razor-clam popular name of solenids
soft-shelled clam popular name of myacids
long-siphon clam popular name of sanguinolariids
slender clam popular name of pandorids
steamer-clam = soft-shelled clam
surf-clam popular name of mactrids
veiled clam popular name of solemyids
Washington clam the venerid *Saxidomus nuttalli*
freshwater clam popular name of unioids
wedge-clam popular name of mesodesmatids

Clambidae a curious group of minute beetles distinguished by the fringe of long hairs on the hind wings. They are commonly called fringe-winged beetles

clan in *ecol.*, a local group of organisms of a restricted number of types intermediate between a colony and a society

clapper (Hepaticae) = lobule (Hepaticae)

[**claret**] a mutant (gene symbol *ca*) mapped at 100.7 on chromosome III of *Drosophila melanogaster*. The phenotypic eye is ruby red

Clariidae a family of naked catfishes distinguished by the possession of accessory respiratory dendritic organs in a pocket above the gill arches

clary the labiateous herb *Salvia sclarea*. The term is also sometimes applied to *S. hormium*

-clas- *comb. form* meaning "fracture"

clasp the forked prominence on the underside of collembolan insects that holds the furcula in place (*cf.* trabeculum)

clasper 1 (*see also* clasper 2–4) a modified portion of the pelvic girdle of male elasmobranch fish, serving as an intromittent organ

clasper 2 (*see also* clasper 1, 3, 4) the term is also applied to an organ of doubtful function, rising from the top of the head, just in front of the eyes, and from the pelvic fins of some Holocephalans

clasper 3 (*see also* clasper 1, 4, 2) a modified segmented appendage, part of the genitalia, used by a male to clasp a female, particularly in insects (*cf.* harpé, valve)

clasper 4 (*see also* clasper 1–3) a plant tendril

class a taxon of either the plant or animal kingdom ranking immediately below phylum and above order

subclass a taxon intermediate between class and order

superclass a taxon that is with great difficulty distinguished from subphylum

-clast- *comb. form* meaning "to shatter"

osteoclasts multinucleate cells, often forming a syncytium responsible for the destruction of bone

clathrate latticed

-claus- *comb. form* meaning "closed"

-clav- *comb. form* meaning "club"

Clavariaceae a family of cantharellale basidiomycete fungi commonly called coral mushrooms in virtue of their branched upright growth and the fact that many are brightly colored

clavate club-shaped

obclavate a clavate form attached by its thickened end

Claviceptales an order of pyrenomycete Ascomycetes containing the fungi commonly called ergots. They are easily distinguished as dark-colored compact masses parasitic on the ovaries of various grasses

clavicle *see* bone

clavicularium the antero-lateral dorsal plates of a turtle shell

claviculate *either* club-shaped *or* key-shaped

Clavigeridae a family of coleopteran insects strongly resembling the Pselaphidae (*q.v.*) with which they are frequently fused

clavule 1 (*see also* clavule 2) flattened club-shaped spines found on some echinoid echinoderms

clavule 2 (*see also* clavule 1) a monactine derivative of a hexactine sponge spicule with bulbs at each end

claw 1 (*see also* claw 2–4) any structure terminating an animal limb that is adapted for scratching, clawing, or clutching

dewclaw a vestigial digit or a claw representing such a digit on the foot of a mammal

claw 2 (*see also* claw 1, 3, 4) an animal distinguished by its claws

longclaw any of several species of motacillid bird of the genus Macronyx

pink-throated longclaw *M. ameliae*

yellow-throated longclaw *M. croceus*

claw 3 (*see also* claw 1, 2, 4) a spur on a tree (*see* spur 2)

claw 4 (*see also* claw 1–3) in compound names of plants either claw-shaped or with notable thorns

cat's claw *U.S.* the leguminous herb *Schrankia nuttallii*, the leguminous shrubby tree *Pithecolobium unguis-cati* and the leguminous shrub *Acacia greggii* (= devil's claw)

crab's claw the hydrocharitaceous aquatic herb *Stratiotes aloides*

devil's claw the leguminous shrub *Acacia greggii* (= cat's claw) and the martyniaceous herb *Martynia louisiana* (= unicorn plant)

caly *Brit.* either of two noctuid lepidopteran insects of the genus Noctua

cleat the silicious projection from the secondary hoops of diatoms

cleavage the process by which a fertilized egg subdivides into blastomeres

bilateral cleavage that in which the blastomeres exhibit marked bilateral symmetry

biradial cleavage that in which the tiers of blastomeres are symmetrical with regard to the first cleavage plane

holoblastic cleavage the type of cleavage shown by isolecithical eggs in which the entire egg segments into separate blastomeres

meroblastic cleavage that form of cleavage shown by large yolked eggs in which only the animal pole divides into separate blastomeres

discoidal cleavage that type of cleavage of a telolecithal egg in which the cleavage is restricted to a germinal disc

determinate cleavage holoblastic cleavage in which each cell is destined to form a specific part of the embryo

indeterminate cleavage holoblastic cleavage in which the cleavage pattern bears no definite relation to the embryo

radial cleavage holoblastic cleavage in which the tiers of cells lie on top of each other parallel to the polar axis

spiral cleavage that type of cleavage in which, after the first few divisions, the daughter blastomeres lie at an angle to the longitudinal plane of the cell mass and thus tend to form a spiral pattern

superficial cleavage that type of cleavage of a telolecithal egg in which cleavage is gradually spread over the whole surface

cleavers any of numerous species of herb of the rubiaceous genus Galium but particularly *G. aparine*

anal cleft a cleft produced by scales that extend on each side of the insect anus

cleg = clegg

clegg *Brit.* popular name of tabanid dipterans. *U.S.* any biting dipteran of outstanding unpleasantness

cleithrum *see* bone

-clem- *comb. form* meaning "branch" (*cf.* -clad-) but by many writers used only in the sense of "twin"

clemous pertaining to twigs

pachyclemous with stout twigs

heteroclemous = heterophyllous

leptoclemous with slender twigs

macroclemous with long twigs

Cleonymidae a family of chalcidoid apocritan Hymenoptera closely allied to the Pteromalidae and the Podagrionidae

Cleptidae a small family of apocritan hymenopteran insects

-clept- *comb. form* meaning "theft"

myrmecoclepty the condition of stealing from ants, particularly that of a guest which steals food

Cleridae a family of moderate sized brightly colored densely hairy beetles commonly called checkered beetles. Though most are predatory some (*cf.* ham beetle) are destructive pests

Cleroidea a superfamily of polyphagous coleopteran insects containing the dermestid beetles, the soft-winged flower beetles, and their immediate allies

-cles- *comb. form* meaning "closed"

diclesium an achene with a free perianth

clestine a raphid-containing parenchymatous cell

Clethraceae that family of ericalous dicotyledons which contains the pepperbush. The polypetalous corolla and the three-celled ovary are typical

-clima- *comb. form* meaning "climate"

-climac- *comb. form* meaning "ladder"

Climacorhizae a botanical taxon erected to contain all gymnosperms and dicotyledons except the Nymphaeceae

Climacteridae a family of passeriform birds usually included with the Certhiidae

climate the sum total of the meteorological phenomena of any given area of the planet

bioclimate = microclimate

microclimate a small restricted area, such as that under a decomposing log, which differs markedly from the general surrounding climate (*cf.* microhabitat)

ecoclimatic pertaining to the interaction of organisms with the climatic features of their environment

climax (*see also* community, unit, zone) a plant association which has reached its full development and is likely to remain stable unless disturbed by climatic or other environmental changes

disclimax *either* a climax that has been disturbed through the introduction by man or other agency of new organisms that destroy stability *or* a climax caused and maintained by human interference

edaphic climax *either* as association specifically modified to its environment *or* one that has reached its condition in consequence or the nature of the soil

monoclimax the concept that a given region has only a single potential climax

panclimax the sum total of two closely related climaxes

polyclimax the concept that a region has many potential climax terminations

postclimax either the passing, or alternation, of an existing climax owing to a climatic change or a situation that exists in a region contiguous to a true climax but in which the climate is more favorable and the situation is therefore more advanced

preclimax *either* the vegetation that immediately precedes a climax *or* a situation which exists in an area contiguous to a true climax but in which the climate is less favorable, and the situation is therefore less advanced

prevailing climax a term used to describe a growth form occupying the majority of sites in a given area even though no true climax has been formed

quasiclimax a community, such as that of fresh-water plants, that is a climax in the sense that it is relatively stable but not a climax in that it is more or less independent of climate

serclimax an impermanent sub-climax

subclimax *either* a climax which has not reached a stable level by reason of factors other than climatic *or* a stage which precedes a true climax but which persists for an unusually long time

temporary climax an apparent, but not actual, climax

climber mostly in plant terms

root climber a scandent plant having root-like holdfasts

petticoat-climber *U.S.* the grass *Eragrostis spectabilis* (= tumble grass)

-clin- *comb. form* hopelessly confused between two roots meaning "bed" or "slope" and "bend," the last often extended to "tend towards." The *adj. term* of the "bend" sense is -clinal and that of the "tend" sense -clinic. The *ecol.* "cline" is presumably from "slope" but the connection is not clear

-clinal pertaining to a position or bend

anticlinal said of a division of a meristem cell, that occurs at right angles to the surface of the meristem

periclinal *either* curved around *or*, in *bot.*, said when the line of division of a meristem cell is parallel to the surface

declinate bent downwards

cline 1 (*see also* cline 2) in the sense of a slope or gradient and particularly, in *ecol.*, a group within a population of one species that shows differing characters reflecting a geographic or ecological situation. Particularly used when such characters grade gradually from one to the other

ecocline one reflecting ecological conditions in general

genocline a gradual change of character across a geographical region due to gene flow

geocline one reflecting geographical rather than ecological conditions

hibrid cline one composed of interspecific hybrids

northocline = hybrid cline

thermocline that layer of water in which the temperature changes 1°C. with each meter increase in depth

topocline a geocline extending over an unusually long distance

cline 2 (*see also* cline 1) in the sense of bed. The preferred *adj. term* for this meaning is -ous (*see* clinous)

anthocline the receptacle of a compositaceous flower head

epicline a nectary on a receptacle

pericline the involucre of the compositaceous capitelum

-clinic *adj. term* meaning "tending towards." The termination -ous (*see* -clinous) is used by some in this sense but by others reserved for -clin- in the sense of house (*see* cline 2)

goneoclinic said of a hybrid that shows the phenotypic characters of only one parent

homoclinic the condition of a flower that is self-fertilized

matroclinic a hybrid the phenotype of which resembles that of the female parent

patroclinic a hybrid the phenotype of which resembles that of the male parent

Clinidae a family of blennioid fishes distinguished from the closely allied Blenniidae by the presence of scales. Often called the klipfish or scaled blennies

clinidium the stalk of a stylospore

-clinium preferred by some to cline 2 (*q.v.*)

-clinous the more usual, but not invariable, *adj. suffix* from cline 2 (*cf.* -clinic)

diclinous the condition of a plant in which the male organs are in one flower, and the female in another

heteroclinous the condition of a flower that has the stamens and pistils on separate receptacles

monoclinous hermaphrodite

[**clipped**] a mutant (gene symbol *cp*) mapped at 45.3 on chromosome III of *Drosophila melanogaster*. The phenotypic expression is a clipping of the wing margin

clipper 1 (*see also* clipper 2) *U.S.* the larva of the dobson fly *Corydalus cornutus* (= hellgrammite)

clipper 2 (*see also* clipper 1) *I.P.* the nymphalid lepidopteran insect *Parthenos sylvia*

-clito- *comb. form* meaning a "hillside" or "slope"

Clistogastra = Apocrita

clitellum a glandular anular swelling anterior to the genital apertures of land dwelling oligochaete worms. It is used for the secretion of the egg case

clitoris the vestigial penis of a female mammal

clm *see* [clumpy marginals]

cloaca a posterior involution of the body into which open both the anal and urinary apertures and, in many forms, the sex ducts

mourning cloak *U.S.* the nymphaline nymphalid butterfly *Nymphalis antiopa* (= *Brit.* Camberwell beauty)

biological **clock** a postulated cause of otherwise unexplained endogenous cycles and rhythms in organisms

clonal pertaining to a bud or to a clone

clone a group of organisms descended asexually from a single ancestor known as an ortet

synclopia = cleptobiosis

-closter- *comb. form* meaning "spindle"

closter an elongate, pointed xylem cell

[**clot**] a mutant (gene type *cl*) mapped at 16.5 on chromosome II of *Drosophila melanogaster*. The phenotypic expression is a maroon eye color

cloth a woven fabric

bark cloth *S.Afr.* the bark of any of several species of tree of the genus Brachystegia or the trees themselves

cloud 1 (*see also* cloud 2) term for a group of grasshoppers or locusts (in the old-world meaning of that term) *or* term for a group of foxes, particularly when pursued by the same hunter or hunters (*cf.* skulk, troop)

cloud 2 (*see also* cloud 1) any of several noctuid Lepidoptera

purple cloud *Brit. Cloantha polyodon*

silver cloud *Brit. Xylomania conspicillaris*

clove 1 (*see also* clove 2) a bulblet produced by a bulb

clove 2 (*see also* clove 1, tree) the dried bud of the myrtaceous tree *Eugenia aromatica*

clover 1 (*see also* clover 2 and aphis, borer, caterpillar, chalcid, curculio, hopper, looper, midge and weevil) any of numerous leguminous herbs particularly those of the genus Trifolium

alsike clover *T. hybridum*

bastard clover = alsike clover

bitter clover *Melitotus indica*

buffalo clover *T. reflexum*

running buffalo clover *U.S. T. stoloniferum*

bur clover *U.S.* either of two species of leguminous herb of the genus Medicago, *M. hispida* or *M. minima*

bush clover any of numerous small shrubs or herbs of the leguminous genus Lespedeza

crimson clover *T. incarnatum*

dutch clover = white clover

Egyptian clover *T. alexandrinum*

old field clover = rabbit-foot clover

rabbit-foot clover *U.S. T. arvense*

hop clover *U.S. T. agrarium*

hungarian clover *T. pannonicum*

Japan clover the leguminous herb *Lespedeza striata*

mammoth clover *T. medium*

prairie-clover any of several leguminous herbs of the genus Petalostemum

silky prairie-clover *P. tenuifolium*

violet prairie-clover *P. purpureum*

white prairie-clover *P. candidum*

purple clover *Brit.* = red clover *U.S.*

red clover *U.S. T. pratense*

sand clover the leguminous herb *Anthyllis vulleraria*

scarlet clover = crimson clover

stone clover *U.S.* = rabbit foot clover

strawberry clover *Trifolium fragiferum*

yellow suckling clover *T. filiforme*

Swedish clover = alsike clover

sweet clover any of several species of the leguminous genus Melilotus

tick-clover *U.S.* any of several species of leguminous herb of the genus Desmodium

pea-vine clover = red clover

white clover *T. repens*

yellow clover = hop clover

zigzag clover = mammoth clover

clover 2 (*see also* clover 1) any of several herbs, not of the family Leguminosae, but of similar growth or appearance

stinking clover *U.S.* the capparidaceous herb *Cleome serrula*

pin clover *U.S.* the gernaiaceous herb *Erodium cicutarium*

marsh clover the gentianaceous plant herb *Menyanthes trifoliata* (= buck bean)

clowder term for a group of domestic cats (*cf.* cluster)

club 1 (*see also* club 2) any structure with a bulb or knob on one end, particularly the broad terminal portion of the retractile arm of a decapod mollusk or the swollen terminal portion of an insect antenna

vibratile club a minute club with a ciliated end projecting into the coelom along the longitudinal muscle bands of some Holothuria

club 2 in compound names of plants

devil's club *U.S.* the araliaceous shrub *Oplopanax horridus*

golden club *U.S.* the araceous plant *Orontium aquaticum*

Hercules' club *either* the rutaceous tree *Xanthoxylum clava-herculis or* the araliaceous tree *Aralia spinosa*

[**clumpy**] a mutant (gene symbol *rd*) mapped at 52.1 on linkage group XII of *Bombyx mori*. the phenotypic expression is an irregular egg shape

[**clumpy marginals**] a mutant (gene symbol *clm*) mapped at 32.5± on the X chromosome of *Drosophila melanogaster*. The name refers to the clumping of the marginal wing hairs in the phenotype

Clupeidae a very large family of isospondylous fishes including the herrings and sardines. They are noted for their oily flesh

Clupeiformes = Isospondyli

Clusiaceae = Guttiferae

Clusiidae a family of myodarian cycloraphous dipteran insects closely allied to the Diopsidae (*q.v.*)

cluster term for a group of domestic cats (*cf.* clowder)

clutch the totality of eggs produced by a bird for hatching at one time

-clyp- *comb. form* meaning a "round shield," larger than is indicated by -pelt- (*cf.* -pelt- and -scut-)

clypeolus = anteclypeus

clypeus the shield shaped median anterior plate on an insect head lying below the frons and to which the labrum is attached

anticlypeus the lower or anterior portion of the clypeus in insects when it is divided

paraclypeus a sclerite alongside the clypeus

postclypeus the upper or posterior part of a divided clypeus

-clys- *comb. form* meaning "wave action" or "tide"

epiclysile pertaining to tide pools sufficiently far off the beach to show a significant temperature rise between tides

-clyst- *comb. form* meaning "pipe"

physoclystous the condition of the fish in which the swim bladder is not connected to any portion of the alimentary canal

Clythiidae = Platypezidae

Cm *see* [Crimp]

cm *see* [carmine]

cmp *see* [crumpled]

CMP cytidine 5'-phosphate

cn *see* [cinnabar]

-cnem- *comb. form* meaning "knee." Frequently confused with -nema- (*q.v.*)

metacneme an internal radial portion of anthozoan coelenterates that does not attach to the stomodaeum

protocneme an anthozoan mesentery that is fused to the stomodaeum

cnemial pertaining to the knee or to the tibia

cnemidium that portion of a bird's leg which bears scales and not feathers

epicnemis a synarthrosis at the base of the tibia in many arachnids

-cnid- *comb. form* meaning "nettle" and by extension "sting"

botrucnid a cluster of cnidorhagi

cnida = cnidocyst

Cnidaria a term used to denote a subphylum of Coelenterata (*q.v.*) when the ctenophores are included, the latter then being placed in the Acnidaria

cnidocil *see* -cil-

Cnidosporidia a subclass of sporozoan Protozoa distinguished by their amoeboid movement and by the fact that the spores, with one or two polar capsules, hatch into an amoeboid stage

CNS a widely used abbreviation for central nervous system, meaning, in general, the brain and spinal cord

co- *see* -con-

co *see* [col-4] and [coalescent]

Co *see* [Confluens]

Co a yeast genetic marker indicating grow inhibition by cobalt

Co II *obs.* = Coenzyme II *obs.* = NADP

CoA = Coenzyme A

coadunate fused at the base

coaita any of several species of long-limbed cebid monkeys of the genus Ateles (= spider monkey)

[**coal**] a mutant (gene symbol *cal*) mapped at 59.5± on chromosome II of *Drosophila melanogaster*. The phenotypic body is black

[**coalescent**] a mutant (gene symbol *co*) found in linkage group I of *Habrobracon juglandis*. The phenotypic expression is a coalescence of antennal segments

seedcoat = testa

coatimundi either of two species of long-snouted, procyonine carnivores of the genus Nasua, also called coati

cob 1 (*see also* cob 2,3) a male swan

cob 2 (*see also* cob 1,3) *Brit.* any large gull, particularly *Larus marinus* (= black-back gull)

cob 3 (*see also* cob 1, 2) *U.S.* the receptacle bearing the fruits (kernels) of corn (*Zea mays*)

cobalamin *see* vitamin B_{12}

cyanocobalamin *see* vitamin B_{12}

bessi **coban** *W.E.* the parulid *Coereba flaveola*

cobbler 1 (*see also* cobbler 2) any of numerous fishes furnished with spines thought to resemble a cobbler's awl including: *U.S.* the carangids *Trachinotus carolinus* (= pompano) and *Alectis ciliaris* (= threadfish); *Brit.* the cottid, *Cottus bubalis*) *Austral.* the scorpaenid *Gymnapistus marmoratus* and the photosid *Cnidoglanis microceps*

cobbler 2 (*see also* cobbler 1) *W.I.* the fregatid bird *Fregata magnificens*

cobego a native name applied to various Dermoptera

cobia the rachycentrid fish *Rachycentron canadum*

Cobitidae a family of Old World ostariophysous fish commonly called loaches, distinguished by the large number of barbels around the mouth

cobra any of many hooded elapid snakes. Usually, without qualification *Naja naja* (spectacled cobra)

hooded cobra the elapine *Naja naje*

king cobra *Naja hannah* (= hamadryad)

splitting cobra the elaphid snake *Hemachatus haemachatus* (= ringhal)

coca the erythroxylaceous shrub *Erythroxylum coca*. The term is also applied to the leaves of this shrub which yield the drug cocaine

-cocc- *comb. form* meaning "grain" or "kernel" and, by extension, "nucleus"

Coccidae that family of homopteran insects which contains the scale insects and mealybugs. These very aberrant insects are well described by their English name

Coccidia a group of telosporidian Sporozoa variously regarded as a separate order or as a suborder of Coccidiomorpha (*q.v.*) They are distinguished from the Haemosporidia by the immobile zygote-producing spores containing sporozoites

Coccidiomorpha an order of telosporidian Sporozoa containing, *inter alia*, the malarian parasite. All are intracellular parasites

chlorococcine pertaining to those algae that show only sexual reproduction

Coccinae a subfamily of coccid homopteran insects containing the cochineal insect

Coccinellidae a large family of oval convex brightly colored coleopteran insects commonly called ladybirds or ladybugs; with the exception of the Mexican bean beetle they are of the utmost value in pest control

coccineous the scarlet color of carmine used with a tin lake

pericoccium that portion of the protoplasm which immediately surrounds the nucleus

-cocco- *comb. form* from same root as -cocc- (*q.v.*) but in *biol.* usually meaning "berry" or "chamber"

-coccous pertaining to coccus 2 (*q.v.*)

tetracoccous an aggregation of four closed carpels

dicoccous having two chambers, particularly of fruit

coccus 1 (*see also* coccus 2) a rounded bacterium

coccus 2 (*see also* coccus 1) the individual lobe of a schizocarp

coccyx those vertebrae that lie posterior to the sacrum in forms (*e.g.* great apes and birds) that do not have a bony tail

cochineal (*see also* insect) the commercially available dried bodies of cochineal insects (*q.v.*) The red color is a quinone biochrome

-cochl- *comb. form* meaning either "spoon" or "spiral shell"

cochlea the spiral (shell-like) half of the labyrinth of the inner ear

Cochleariidae that family of ciconiiform birds which contains the boat-billed herons, so called for their typical heavy, thick, clumsy beak and relatively short legs

cochleate pertaining to the form of a snail shell (but, *see* cochleariform)

Cochlidiidae = Limacodidae

Cochlospermaceae a small family of parietale dicotyledonous angiosperms distinguished principally by the oily endosperm of the seed

Cochlospermineae a suborder of parietale dicotyledons containing the single family Cochlospermaceae

cock (*see also* cock 2, 3, 4, chafer, comb, roach) without qualification, the male of the domestic fowl, though also used for most other birds. It is in addition used for the male lobster and the males of many fish

cock 2 (*see also* cock 1,3,4) in compound names of many birds without regard to sex

bastard cock *W.I.* the thraupid *Spindalis zena* (= stripe headed tanager)

bilecock the rallid *Rallus aquaticus* (= water rail)

blackcock the male black grouse (*q.v.*)

chaparral cock = roadrunner

gorcock *Brit. dial.* red grouse

heathcock *Brit.* = blackcock *Brit.*

holmcock *Brit.* = mistlethrush

"Burmese" jungle cock = grey jungle fowl

logcock *U.S. dial.* = pileated woodpecker

mackerel cock = mackerel-bird

maycock *Brit.* the charadriid *Squatarola squatarola* (= grey plover *Brit.*)

moorcock *Brit.* the tetraonid *Lagopus scoticus* (= red grouse)

mountain cock *Gough Island*, the rallid *Porphyriornis comeri* (*cf.* gallinule)

peacock (*see also* cock 3, plant, squid) the notoriously beautiful male galliiform bird *Pavo cristatus*.

snow cock *I.P.* the tetraonid *Tetraogallus himalayensis*

stormcock the turdid *Turdus viscivorus* (mistle thrush)

watercock *I.P.* the rallid *Gallicrex cinerea*

woodcock *U.S.* the charadriiform scolopacid bird *Philohela minor Brit.* the allied *Scolopax rusticola*

cock 3 (*see also* moth, plant) any of severally brilliantly colored lepidopteran insects

peacock (*see also* cock 2) *Brit.* the nymphalid lepidopteran insect *Vanessa io. I.P.* any of several species of papilionid insect of the genus Papilio but usually, without qualification, *P. polyctor.* (*cf.* redbreast, mormon, swallowtail, spangle, raven, helen). *U.S. Junonia coenia*

white peacock *U.S.* the nymphalid lepidopteran insect *Anartia jatrophae*

cock 4 (*see also* cock 1-3, squid) in other compound names

rock-cock *Brit.* the labrid fish *Centrolabrus exoletus*

cockatiel *Austral.* the psittacid bird *Leptolophus hollandicus*

cockatoo commonly applied in popular parlance to any psittacid bird having a marked crest, at one time forming a separate family the Cacatuidae (*cf.* corella, galah)

black cockatoo any of several species of the genera Probosciger and Calyptorhynchus

glossy black cockatoo *C. lathami*

great black cockatoo *P. alterrimus*

red-tailed black cockatoo *C. magnificus*

rose-breasted cockatoo *Kakatoe roseicapilla* (= galah)

lemon-crested cockatoo *Kokatoe sulphurea*

sulphur-crested cockatoo *Kakatoe galerita*

gang-gang cockatoo *Callocephalon fimbriatum*

pink cockatoo *Kakatoe leadbeateri*

cockatrice a biblical beast of doubtful identity but probably a poisonous snake

cockle 1 (*see also* cockle 2) popular name of cardiid pelecypods, particularly those of the genus Cardium, and of a few allied forms

rayed cockle the sanguinolarid *Asaphis beflorata*

cockle 2 (*see also* cockle 1, bur) any of numerous caryophyllaceous herbs

sticky cockle *U.S. Silene noctiflora*

corn-cockle *Agrostemma githago*

cow-cockle *Brit. Silene caccaria*

cock-of-the-plains *U.S.* the large tetraonid bird *Centrocercus urophasianus* (= sage grouse)

cock-of-the-rock any of two species of cotingid birds of the genus Rupicola distinguished by their crimson to scarlet color and extraordinarily laterally compressed crest. They are often placed in a separate family the Rupicolidae

cock-of-the-wood = capercailze

coco *W.I.* any of several species of threskiornithid bird, particularly the ibis

cocoa (*see also* plum) the manufactured and processed product obtained from the seeds of the cacao (*q.v.*)

coco de mer the enormous seed, said to be the largest known, of the palm tree *Lodoicea maldavica*

cocoon 1 (*see also* cocoon 2) the woven silk case that surrounds the pupa of many Lepidoptera, or the eggs of many spiders

cocoon 2 (*see also* cocoon 1) any invertebrate structure in which numerous eggs, capsulated or uncapsulated, are enclosed in a common capsule or shell such as that secreted from the clitellum of some oligochaete annelids

-cod- *comb. form* meaning "head" in the sense of a lump on the end of a stalk but hopelessly confused in compounds with -codi- and -codo-

cod 1 (*obs.*) (*see also* cod 2,3) = pod

cod 2 (*see also* cod 1,3) any of numerous species of anacanthinous fishes *Brit.* without qualification, *Gadus morhua* (= *U.S.* Atlantic cod)

arctic cod *Boreogadus saida*

Atlantic cod *U.S. Gadus morhua*

Greenland cod *Gadus ogac*

Pacific cod *Gadus macrocephalus*

polar cod *Arctogadus glacialis*

poor cod *Brit. G. minutus*

saffron cod *Eleginus gracilis*

tomcod any of several species of gadid fishes, of the genus Microgadus

Atlantic tomcod *M. tomcod*

Pacific tomcod *M. proximus*

cod 3 (*see also* cod 1,2) a few fish not of the family Gadidae

ling cod *U.S.* the hexagrammid fish *Ophiodon elongatus*

coddy-moddy *Brit. obs.* the larid bird *Larus ridibundus* (= black-headed gull)

-codi- *comb. form* meaning "sheepskin" but hopelessly confused in compounds with -cod- and -codo-

Codiaceae that family of siphonale Chlorophyceae in which the "body" is formed of interlacing branches of the thallus

antho**codium** that portion of an alcyonarian polyp which projects above the coenenchyme

-codo- *comb. form* meaning "bell" but hopelessly confused in compounds with -cod- and -codi-

codon 1 (*see also* codon 2) a unit in the cistron (*q.v.*) controlling the synthesis of a single polypeptide unit of the genetic code

codon 2 (*see also* codon 1) the umbrella of a hydrozoan medusa

    entocodon the primordium of the sub-umbrella of a developing medusa produced from the epidermis of the pith of the gonophore

codonic pertaining to codon 2

    adelocodonic said of medusae that remain attached to the hydroid and therefore lack bells

    phanerocodonic said of medusae that are free swimming

Coeciliae = Apoda (Amphibia)

coecum *see* caecum

allometric **coefficient** the slope of a curve obtained by plotting the logarithm of some measurement of an organ or part against the logarithm of measurements of the whole remainder or of another part. This is also termed the heterogonic or heteroausecic coefficient

-coel- *comb. form* meaning "heaven" hence, following the *Gr.* view of this space, "a cavity". Properly, but very rarely, transliterated -koelo-

coel (*see also* -enteron) a cavity

    axocoel a narrow, dorsal evagination from the enterohydrocoel of crinoids

    blastocoel the segmentation cavity or cavity of the blatula

    endocoel that part of the coelenteron of anthozoan coelenterates which is enclosed by septa

    endocoel the space between paired septa in Anthozoa

    enterocoel a coelom derived as the cavity of mesodermal pouches evaginated from the archenteron

    epicoel = metacoel

    exocoel that portion of the coelenteron of an anthozoan which lies between widely separated pairs of septa that themselves include an endocoel between them

    gonocoel the cavity of a gonad

    haemocoel a blood-filled body cavity, particularly that of mollusks and the arthropods. In such a system, the organs are bathed directly in blood and the true coelom occurs only as scattered remnants (*cf.* phlebedesis)

    hydrocoel a ventral sac developed from the enterohydrocoel of crinoids

    enterohydrocoel the anterior of the two primary divisions of the archenteron of a crinoid larva (*cf.* somatocoel)

    mesencoel a coelom produced through the rearrangement of mesodermal cells to enclose a space

    mesocoel the cavity of the mesencephalon and thus the connection between the third and fourth ventricles in lower vertebrates (*cf.* iter)

    metacoel the cavity of the mesencephalon and also

an extension of the fourth ventricle into the cerebellum in lower vertebrates

    myocoel that portion of the coelum which lies within the myotome

    nephrocoel the cavity of a nephrotome

    neurocoel the cavity of the central nervous system

    ovocoel the gonocoel of the ovary

    protocoel that part of the coelom of protochordates which lies in the protosome

    pseudocoel a body cavity, such as that of a nematode, which is not a true coelom but is derived from the blastocoel

    rhynchocoel the cavity containing the rhynchodaeum

    schizocoel a coelom derived from the splitting of a mesodermal band or plate

    sclerocoel that portion of the coelom that lies within the sclerotome

    somatocoel the posterior of the two primary chambers of the crinoid archenteron (*cf.* enterohydrocoel)

    spongocoel the main cavity of a simple sponge

    telocoel the cavity of the telencephalon

Coelenterata a phylum of the animal kingdom distinguished by the presence of nematocysts, a diploblastic body wall in Hydrozoa, and a single internal cavity opening to the exterior only at the oral end. So defined the phylum comprises the hydrozoa, the jellyfish, corals, and sea anemones but excludes the Ctenophora (comb-jellies) that lack nematocysts (*cf.* Cnidaria)

coelenterate (*bot.*) said of insectivorous plants that trap their prey in cavities

coelom (*see also* coel) that body cavity which is limited on all surfaces by the mesoderm. This is also called the "true coelom" to distinguish it from other types of body cavity

    cardiocoelom = pericardium

    septal coelom that portion of the coelom that runs along the outer margin of the pharyngeal bar of prochordates

Coelomata a term coined to define those phyla of the animal kingdom which possess a true coelom. This taxon therefore embraces the Chaetognatha, Hemichordata, Pogonophora, Phoronidia, Ectoprocta, Brachiopoda, Siphunculida, Echinodermata, Prochordata, and Chordata

Coelomocoela the term used in distinction to Enterocoela to group together all those phyla possessing a true coelom or, by some authors, either a true coelom or a pseudocoelom

Coelopidae a family of medium sized myodarian cycloraphous dipteran insects. One of the few groups of insects associated with a marine habitat, its members are commonly called seaweed flies, because the larvae breed in decomposing seaweed. The body and legs of the adult fly are extremely bristly

-coelous (*see also* -gyrous, vertebra) *either* possessing, or in the form of, a cavity *or* pertaining to the coelom

    amphicoelous concave on both sides

    anomocoelous having different kinds of centra in various parts of the vertebral column

    cyclocoelous said of an animal in which the loops of the intestine are spirally aranged

    dicoelous having two cavities

    orthocoelous said of animals in which the loops of the intestine are parallel to each other and the long axis of the body

pericoelous said of an animal in which the second loop of the intestine encloses the third loop

antipericoelous said of those animals in which the second intestinal loop encloses the third loop. The former is left handed and the latter right handed

plagiocoelous said of animals in which one or more coils of the intestine are secondarily coiled

schizocoelous said of an organism in which the coelom is produced as a split in a previously solid band of mesoderm

excoemum the tuft of trichomes at the base of the glume

-coen- *comb. form* properly from a *L.* root meaning "filthy" but widely used as the transliteration of a *Gr.* root, properly -koin-, meaning "sharing" or "togetherness"

coen the sum total of the components of an environment. Occasionally used as an abbreviation for coenosis (*q.v.*)

biocoen the sum total of the living components of an environment

abiocoen the sum total of the non-living components of an environment

Coenagrionidae a family of zygopteran odonatan insects distinguished by the fact that the clear wings are pressed together when at rest

coendou any of several species of erethizontid hystricomorph rodents from *Cent.* and *S. Amer.* They are completely arboreal and with a well-developed prehensile tail

coenecium an aggregate of protective structures such as tubes or capsules in which live an aggregation of such forms as Pterobranchia or Ectoprocta

coenobi- *comb. form* meaning "cloister", properly, but very rarely, transliterated -koenobi-

coenobium literally, a "cloister," but variously used to describe plant or animal organisms (*e.g.* Volvox) occurring in colonies of discrete individuals. The term is also specifically applied to a type of fruit consisting of four distinct fruitlets round a central style

Coenomyiidae a family of brachyceran dipteran insects now usually included with the Xylophagidae

coenose pertaining to distribution

eurycoenose of wide distribution

stenocoenose restricted to narrow limits

coenosis an unstable assemblage of organisms having a common ecological preference

biocoenosis mutualism between plants and animals. The term is also used as a synonym for community

coenosium a community of plants

biocoenosium a coenosium confined to a narrow habitat

isocoenosium a coenosium of isocies (*q.v.*)

permanent coenosium climax

temporary coenosium temporary climax

Coenothecalia an order of alcyonarian Anthozoa containing the blue corals. These differ from the zoantherian corals in containing polyps with eight pinnate tentacles and lacking scleroseptae

coenurus a compound tapeworm cystercus consisting of a large bladder with numerous scolices attached to its inner surface

Coerebidae a family of passeriform birds usually included in the Thraupidae and Parulidae

coetaneous = coaeval

-coeto- *comb. form* meaning a "bedroom" more properly transliterated -koeto-

coffeatous coffee-colored

coffee 1 (*see also* coffee 2, bean 2, bean 3, berry, pea,

senna, tree, and weevil) any of several species of rubiaceous tree or shrub of the genus Coffea and its berries, from which the beverage is extracted

coffee 2 any shrub or tree bearing coffee-like berries

Kentucky coffee the leguminous tree *Gymnocladus dioica* and particularly its seeds

wild coffee *U.S.* any of several species of the caprifoliaceous genus Triosteum particularly *T. perfoliatum* and *T. aurantiacum*

cognate related through the female line (*cf.* agnate)

cohort (*obs.*) a botanical taxonomical rank, consisting of a group of orders

cohosh *U.S.* any of several species of the ranunculaceous genus Cimifuga (= bugbane)

blue cohosh the berberidaceous herb *Caulophyllum thalictroides*

-col- *comb. form* meaning "to dwell" The *adj. terms* -coline and -colous are both used but the former appears to be obsolescent

col a bacterial mutant showing colicin sensitivity or resistance to colicin

col-1–col-4 *see* [Colonial-1] to [Colonial-4]

cola the sterculiaceous tree *Cola acuminata* and the numerous extracts derived from it, or imitating it

Colaciales an order of euglenophyte algae in which the cells are permanently united into sessile colonies

colchicine an alkaloid, derived from the liliaceous herb *Colchicum autumnale,* that inhibits the formation of the mitotic spindle

-cole- *comb. form* meaning "sheath". Better transliterated -kole-

cole a general term for all edible "cabbages" used in the broadest sense to include cauliflower, kale, brussel sprouts, etc.

Coleochaetaceae that family of ulotrichale Chlorophytaceae in which the vegetative cells bear long cytoplasmic setae which are partly or wholly surrounded by a gelatinous envelope

Coleophoridae a family of tinioid lepidopteran insects commonly called casebearers for the reason that the larvae build portable shelters very similar to those of caddisworms

Coleoptera that order of insects containing the forms commonly called beetles. They are distinguished by the metamorphosis of the anterior wings into strongly chitinized elytra or wing covers

colesule a membrane around the outside of the sporangium

colicin any of several surface-produced substances from bacteria that serve to inhibit or kill other bacteria

colets the sturnid bird *Sarcops calvus*

Colignoneae a tribe of Nyctaginaceae with glabrous ovaries, curved embryos, and opposite leaves

Coliidae a small family of coliiform birds commonly called colies. They are distinguished by the red or blue skin surrounding the eyes and the extremely long tail in marked contrast to the short rounded wings

Coliiformes a small order of birds containing, *inter alia,* the colies. They are distinguished by their short legs and long toes

colin any of many small phasianid birds, generally synonymous with quail

-coline *adj. term.* from -col- (*q.v.*) referring to habitat. For compounds see under -colous, the form preferred in this work

-coll- *comb. form* confused from one *L.* and one *Gr.* root and therefore meaning either "neck" or "glue." The prevalence of collagenous connective tissues in animals has led to a further meaning "connective tissue," even though, in plants, these may be cellulosic

mesocolla the middle of three layers postulated in a plant cuticle before the development of electron microscopy

collagen literally, "that which yields glue." It is a protein, usually recurring in fibrous form, and is the chief constituent of white connective tissue and a major constituent of cartilage

collangenase = clostridiopeptidase A

collar 1 (*see also* collar 2–5) in *bot.* *either* the annulus round the stem of some Basidomycetes *or* the neck or cingulum of a plant

collar 2 (*see also* collar 1, 3–5) that area of the hemichordate body which separates the protosome from the metasome

collar 3 (*see also* collar 1, 2, 4, 5) a thin collar shaped lamella of mucus-agglutinated sand produced to hold together the eggs of many marine mollusca, particularly naticid gastropods. Also called sand collar

collar 4 (*see also* collar 1–3, 5) the periphery of the aperture in an ectoproct zoecium

setigerous collar = pleated collar

orificial collar an expanded rim

pleated collar a projection from the diaphragm into the vestibule of Ectoprocta

collar 5 (*see also* collar 1–4) a lepidopteran insect

black collar *Brit.* the noctuid lepidopteran insect *Noctua flammatra*

collard a large leafed herb, particularly a variety of cabbage eaten in *S.U.S.*

marsh collard *U.S.* popular name of nymphaeaceous aquatics of the genus Nuphar (= yellow pond lily)

water collard = marsh collard

collare = pleated collar (*see* collar 4)

collarette the thickened "neck" region of Chaetognatha

Collembola an order of wingless insects distinguished from all other insects by the presence of nine postcephalic segments, the last six of which bear ventral appendages. These characteristics are considered by some to distinguish the Collembola so clearly from all other insects that they should be placed in a separate class of the Arthropoda

rhizocollesy the joining of individuals by their roots

colleter a glandular trichome secreting a sticky substance

Colletidae a family of apoid apocritan hymenopteran insects containing the forms usually called plasterer bees and yellow-faced bees. They may be distinguished by the short tongue, truncate or bilobed at the apex

colliculose covered with hillocks

Collothecacae an order of monogonate rotifers in which the corona is replaced by a lobed funnel

Colobidae a family of large simioid primates mostly arboreal, and with long tails

coloboma an anomaly resulting from failure in closing of the choroid fissure

colocollo the felid carnivorous mammal *Herpailurus colocollo* (*cf.* jaguarondi)

colocynth the cucurbitaceous vine *Citrullus colocynthis* (= bitter apple)

colon 1 (*see also* colon 2) the terminal section of the alimentary canal of vertebrates. The term is often extended to the corresponding region of the invertebrate gut

ascending colon that portion of the colon which curls forward on the right side of the body in certain mammals

descending colon that part of the colon in certain mammals, which runs from the transverse colon directly posterior

ileocolon the undifferentiated anterior portion of the hindgut of arthropods

mesocolon that part of the mesentery which supports the colon

sigmoid colon that part of the colon, in man, which connects the descending colon to the rectum

transverse colon that portion of the colon which, in certain mammals, passes from side to side of the abdominal cavity

colon 2 (*see also* colon 1) a punctuation mark (:) used as descriptive of some animals

white colon *Brit.* the noctuid lepidopteran insect *Mamestra albicolon*

[colonial-1] a mutant (gene symbol *col-1*) in linkage group IV of *Neurospora crassa*. The name derives from the colonial growth of the phenotypic colony

[Colonial-2] a mutant (gene symbol *col-2*) in linkage group IV of *Neurospora crassa*. The phenotypic expression is a nonconidiating colonial growth

[Colonial-3] a mutant (gene symbol *col-3*) in linkage group VII of *Neurospora crassa*. The phenotypic expression is a nonconidiating colonial growth

[Colonial-4] a mutant (gene symbol *col-4*) in linkage group IV of *Neurospora crassa*. The phenotype is colonial with numerous macroconidia

[Colonial-temperature sensitive] a mutant (gene symbol *cot*) in linkage group IV of *Neurospora crassa*. The phenotypic expression is a substitution of colonial for normal growth in 34° C

colony 1 (*see also* colony 2, 3) a natural community of two or more species. The term does not designate any degree of mutual interdependence

autocolony a group of conjoined algae derived from a single mother cell

depauperate colony in ants, one that is dying for lack of nourishment

reciprocal colony = corm

irreciprocal colony one of which some members are independent agents

colony 2 (*see also* colony 1, 3) a visible growth of microorganisms usually on a solid medium. Many colonies are true clones. Most descriptive terms of colonies are self-explanatory (*e.g.* dull colony, lobate colony). Specific colony designations are:

curled colony: a colony in the form of parallel, wavy chains;

erose colony: one having an irregularly shaped border;

punctiform colony: one that is only just apparent to the naked eye;

raised colony: one with a raised edge;

daughter colony: a colony derived, either by transfer or outgrowth from another colony

colony 3 (*see also* colony 1, 2) the term for a group of beavers or gulls

color a sensation induced in the occipital region of the cerebrum by the differential effect of various wavelengths of photon energy on the retina. These sensations are vocalized in conventional terms, (red, blue, etc.) established by custom, and correlated with wavelength. The term is used in *biol.* in the sense of pigment (*see* pigment, chrome) as descriptive of a particular color (color 1, below) as descriptive of the method by which a color is produced (color 2, below,) and the supposed function of color (color 3, below)

color 1 (*see also* color 2, 3) as descriptive of a particular color

discolor having two colors, particularly on the upper and lower surface of a leaf

lignicolor wood colored

hepaticolor a dark brown, slightly tinged with red, bearing little relation to the appearance of a fresh mammalian liver

persicicolor peach colored

tylicolor dark slate gray

vinicolor wine colored

color 2 (*see also* color 1, 3, pigment, chrome) the method by which a color is produced

structural color interference colors produced by reflection from surface structures, also called interference colors or schemochromes

Tyndall color one that is produced by a schemochrome blue over a pigment. Many green feathers, for example, are produced by a blue interference color over a yellow pigment

color 3 (*see also* color 1, 2) color as a functional phenomenon

apatetic color those which make an organism able to mimic either its environment or another organism, or by extension, any color thought to have a protective function

confusing color sematic color that gives a different appearance when the animal is at rest than when it is moving

cryptic color sematic color designed to blend an animal with its background

procryptic color a color designed to assist in concealing an organism

anticryptic color a color or color pattern used for concealment by a predator

dymantic color that which is easily remarkable but the purpose of which is not immediately apparent

epigamic color colors used in or developed for mating displays

sematic color a color that is thought to serve a useful purpose to an organism

antiaposematic color *either* coloration that disguises a predator *or* one which is used as a threat

pseudepisematic color colors that attract

self-colored a florist's term meaning uniform in color

colorific color producing

[colorless fruit skin] a mutant (gene symbol *y*) mapped at 30 on chromosome 1 of the tomato. The term is descriptive of the phenotypic expression

concolorous of a uniform color throughout

colostrum the milk produced by the mammary glands glands immediately following birth

-colous pertaining to the location, or habit, of growth

aigicolous beach-dwelling or dwelling among pebbles

alsocolous dwelling in woody groves

amathocolous dwelling in sandy places

amnicolous growing on river banks

ancocolous living in canyons

arenicolous growing in sand

argillicolous growing in clay

calcicolous growing on chalky soil

caulicolous stem-dwelling, applied particularly to parasitic fungi

cavernicolous cave-dwelling

crenicolous dwelling in spring fed brooks

dendrocolous dwelling on trees

fimicolous growing on dung heaps

foliicolous growing on leaves

gallicolous dwelling in galls

graminicolous growing on grass

pontohalicolous dwelling in salt marshes

hylocolous living in forests

lapidicolous dwelling under or among stones

leimicolous inhabiting wet meadows

lignicolous plants living on wood

limicolous *either* shore inhabiting *or* dwelling in mud (*cf.* luticolous)

limniicolous dwelling in lakes

luticolous dwelling in mud (*cf.* limicolous)

muscicolous growing on mosses

nidicolous nest-dwelling. Used particularly of birds restricted by being hatched in an immature condition

omnicolous living in all places

paludicolous marsh-inhabiting

potamicolous river-dwelling

psicolous dwelling on prairies

radicicolous dwelling in or on a root. The word is also sometimes misused as a synonym for radiciflorous

rubicolous dwelling on brambles

rupicolous dwelling among rocks

sabulicolous sand-dwelling

saxicolous dwelling in stony places

sepicolous dwelling in hedges

silicicolous living on flints

telmicolous living in fresh-water marshes

thinicolous dwelling on shifting sand

tiphicolous pond-dwelling

umbraticolous dwelling in shady places

viticolous living on vines

-colp- *comb. form* meaning "bosom", better transliterated -kolp-

colt (*see also* bur, foot) properly a male foal, though frequently used as synonymous with foal

Colubridae a very large family of snakes containing about 90 per cent of all living species. They are distinguished from all other snakes by the fact that both jaws are toothed, the mandibles lack a coronoid process, and the prefrontals are not in contact with the nasals

Colubrinae a subfamily of aglyphan colubrid snakes that contains the overwhelming majority of living species

colubrine = serpentine

Colubroidea a superfamily of snakes including the family Colubridae and several other small families

colugo = cobego

colulus a papilla that projects between the two anterior spinnerets of aranaeid arthropods

Columbellidae a small family of gastropod mollusks with fusiform shells greatly thickened at the lip

Columbidae a huge family of columbiform birds containing the pigeons and doves. They are distinguished from the Pteroclididae, the other family in the Columbiformes, by the absence of feathered toes

Columbiformes that order of birds which contains the pigeons, doves and their allies. They are distinguished by their small bills, imperforate nostrils, short legs, and dense easily detached plumage

columbine 1 (*see also* columbine 2) any of numerous ranunculaceous herbs mostly of the genus Aquilegia

common columbine U.S. *A. canadensis* Brit. *A. vulgaris*

feathered columbine *Thalictrum aquilegifolium*

columbine 2 (*see also* columbine 1) *I.P.* the riodinid lepidopteran insect *Stiboges nymphidia*

columbo horticultural term for gentianaceous plants of the genus Swertia

columella 1 (*see also* columella 2, 3) the central column of the megasporangium of a cycad

columella 2 (*see also* columella 1, 3) the central, columellar mass of a coral

columella 3 *see also* columella 1, 2) the central sterile parts of the sporangium of some bryophytes

Columelliaceae a small family of tubulifloral dicotyledonous angiosperms

column any tissue or organ or organism in the shape of a column (*cf.* columella) specifically that portion of the orchid flower more properly called gynostegium

    Bertin's columns the connective tissue lying between the lobes of the medulla in polylobular kidneys

    commissural column the central strand of tissue in the fern stem

    Trajan's column the cactus *Pachycereus columna-tragani*

Columniferae = Malvales

coly any of several species of bird of the family Coliidae

    cameroon coly *Colius striatus*

    red-faced coly *Urocolius indicus*

    blue naped coly *Colius macrourus*

Colydiidae a family of coleopteran insects sometimes called the cylindrical bark beetle. Most are predators on other insects

Colymbidae = Podicipedidae

Colymbidae = Gaviidae

Colymbiformes = Podicipediformes

com *see* [Compact]

coma literally, a tuft, particularly those projecting from plant structures, such as seeds and fruits. It is also sometimes used for the entire head of a tree (*cf.* tuft)

comb 1 (*see also* comb 2–4) an erect pectinate caruncle (*q.v.*) on the head of a bird

    pea comb three parallel, and partly fused, combs of which the central is usually the largest

    rose comb a pea comb from the rear end of which projects a large, fleshy spike

    strawberry comb a condition in which the comb is reduced to a fleshy lump

    V comb a greatly reduced comb consisting of two vertical "fingers" in the shape of a V

comb 2 (*see also* comb 1, 3, 4) any structure that has closely apposed teeth along one side

    natal comb the conspicuous comb on the posterior margin of the prothorax of siphonapteran insects

    pollen comb = scopa

    tactile comb patches bearing long stiff hairs on the margins of some medusae

    terminal comb one of many thin triangular projections at the tip of the arm of some crinoids

comb 3 (*see also* comb 1, 2, 4, flower, grass) any of several organisms which, or parts of which, are pectinate

    cockscomb 1 (*see also* cockscomb 2, 3) *Brit.* the scrophulariaceous herb *Rhinanthus cristagalli*

    cock's comb 2 (*see also* cockscomb 2, 3) *Brit.* the amaranthaceous herb *Celosia cristata*

    cockscomb 3 (*see also* cockscomb 1, 2) the stichalid fish *Anoplarchus purpurescens*

    Venus comb *Brit.* the umbelliferous herb *Scandix pectens*

comb 4 (*see also* comb 1–3) the structure of fused hexagonal, waxy cells, in which bees store honey

[comb-gap] a mutant (gene symbol *cg*) mapped at 71.1 on chromosome II of *Drosophila melanogaster*. The phenotypic expression are large sex combs and a gap in wing vein L4

comber *Brit.* the serranid fish *Serranus cabrilla* (*cf.* bass 1)

recombination the rejoining of a broken portion of a chromosome in such a manner as to cause chromosome aberration (*cf.* restitution)

Combretaceae a family of myrtiflorous dicotyledonous trees or shrubs distinguished from other myrtiflorous families by having a drupaceous, often winged fruit

come a sponge spicule

    floricome a plumicome (*q.v.*) with expanded tips

    plumicome a microscleric hexactine siliceous sponge spicule in which the ends of the rays terminate in numerous fine, outwardly directed plumes

comet any of several trochilid birds of the genera Sappho and Polyonymus

comfrey any of several herbs or low growing shrubs of the boraginaceous genera Symphytum and Cynoglossum

    common comfrey *S. officinale*

    prickly comfrey *S. asperum*

    wild comfrey *Cynoglossum virginianum*

-comma- *comb. form* confused between a *L.* and *Gr.* root and therefore variously meaning "ornamentation" (better -komma-) and "septum"

comma 1 (*see also* comma 2) *U.S.* any of several species of nymphalid lepidopteran insects of the genus *Polygonia* (*cf.* angle wing) *Brit.* without qualification, *P. C-album*

comma 2 (*see also* comma 1) in the sense of septum

    myocomma the septum separating two myotomes

commander *I.P.* the nymphalid lepidopteran insect *Limenitis procris* (*cf.* admiral, commodore)

Commelinaceae that family of farinosean monocotyledons which contains the spiderworts. Distinguishing characteristics are the transformed anthers and stamen hairs together with the complete differentiation of the perianth into calyx and corolla

commensalism the condition of two organisms living together for the purpose of sharing each other's food

    ectocommensalism the type of commensalism involving commensals living on the outside of other forms

commiscuum an assemblage, either real or postulated, of all members of a taxonomic group that are capable of interbreeding (*cf.* convivum)

commissure 1 (*see also* commissure 2, 3) literally, a joint or seam. Particularly, in *bot.* where plant organs or segments of organs unite

    palatine commissure the junction of the two subocular arches in Myxine

commissure 2 (*see also* commissure 1, 3) a bundle of nerve fibers connecting two ganglia

    hippocampal commissure the ventral of the two connections between the two cerebral hemispheres in the higher vertebrates

    Jacobson's commissure the connection, in some urodele amphibia, of the seventh cranial nerve with the sympathetic trunk

    pallial commissure the connection between the hippocampal areas on the two sides of the brain in lower vertebrates, which in higher forms becomes differentiated into the corpus callosum and the hippocampal commissure

commissure 3 (*see also* commissure 1,2) the line that is apparent along the side of the closed beak of a bird

commodore *I.P.* any of numerous species of nymphalid lepidopteran insects, mostly of the genus Limenitis (*cf.* admiral, commander)

communication the process of transferring information

biocommunication the process of transferring information between non-human organisms

sexual communication the transfer of information between non-human animals in the course of mating, or in premating courtship

community a group of organisms which may be a clan, a coenosium in the ecological sense, or a climax; the living portion of an ecosystem

climax community a stable climax

closed community one in which no further species can find lodgement either because of unfavorable environmental conditions or because all the ecological niches are already filled

open community one in which the competition for space is not so severe as to render difficult the invasion by a new species

-como- *comb. form* meaning "hair" better transliterated -komo-

comose tufted

[Compact] a mutant (gene symbol *com*) in linkage group III of *Neurospora crassa*. The phenotypic colonies are small and compact

burnet companion *Brit.* the noctuid lepidopteran insect *Euclidia glyphica*

native companion *Austral.* the gruid bird *Crus rubicunda*

company term for a group of widgeons (*cf.* knob)

comparium a biosystematic unit composed of one or more coenospecies that are able to intercross.

compass one of five horizontal aboral components of Aristotle's lantern (*q.v.*)

competence the reactivity of a tissue, or of cells, to an inductor in development

complement a component of blood that reacts with sensitized cells to cause lysis

complex an assemblage of interconnected or interacting parts (*cf.* apparatus)

modifier complex an assemblage of modifier genes

Compositae an enormous family of campanulalous flowering plants. The involucrate head with the gamopetalous flowers and one-seeded dry fruits are typical. It is the largest family of flowering plants containing more than eight hundred genera and about fifteen thousand species. It would be misleading to pick out from this mass "typical" flowers though it must be pointed out that the daisies, thistles and sunflowers are included as well as the goldenrods, asters, and marigolds

[compound inflorescence] a mutant (gene symbol *s*) mapped at 53 on chromosome 2 of the tomato. The phenotypic inflorescence is much branched and carries many flowers

compressed pressed together

obcompressed anteroposteriorly compressed

comptie = coontie

-con- *comb. form* meaning "cone" sometimes transliterated -kon- or a prefix meaning "with". In the latter meaning, the *n* is frequently dropped and sometimes changed to *r*.

concatenate linked together

[concave wing] a mutant (gene symbol *ccw*) mapped at 23.4± on the X chromosome of *Drosophila melanogaster*. The name is descriptive of the phenotypic expression

conceptacle a word that has been applied to almost any type of receptacle in a plant or animal, but that is used in general terms most commonly to refer to a superficial cavity opening outwards, and frequently the one concerned in reproduction

-conch- *comb. form* meaning "shell"

conch 1 (*see also* conch 2, 3) the external ear of mammals

conch 2 (*see also* conch 1, 3) in the meaning of shell in general

dissoconch the shell of a veliger larva

protoconch the shell of a larval gastropod

conch 3 (*see also* conch 1, 2) any of many large gastropod mollusks, usually, without qualification those of the family Strombidae

crown conch any of several species of neptuneid gastropod mollusks of the genus Melongene

horse conch the fasciolariid gastropod mollusk *Fasciolaria gigantea*

pear conch any of several species of the neptuneid genus Busycon but particularly *B. caricum* and *B. canaliculatum* (*cf.* whelk)

pepper conch = horse conch

Conchostraca an order of branchiopod crustacea commonly called clam shrimps. They are distinguished by the body being entirely enclosed within a bivalved shell and by posessing 18–28 pairs of trunk appendages

conchuela *U.S.* the pentatomid hemipteran insect *Chlorochroa ligata*

conchula a protruding, siphon-like extension of the siphonoglyph of some Zoantharia

concrement a term applied to accumulation of crystals, or crystal-like bodies, in animal structures, particularly those found in certain acoele Turbellaria

concrementation the process of getting rid of nitrogenous wastes by converting them to insoluble substances

condor either of two New World cathartid birds. *U.S.*, without qualification, usually *Gymnogyps californianus* (= California condor)

Andean condor *Vultur gryphos*

-condyl- *comb. form* meaning "a knuckle" but extended to any knobby joint. Better transliterated -kondyl-

condyle 1 (*see also* condyle 2, 3) the surface at the posterior end of the skull that articulates with the anterior vertebra

condyle 2 (*see also* condyle 1–3) in insects, any process that articulates consecutive body segments or an appendage to the body and particularly that which articulates the base of the mandible to the head

occipital condyle in insects, the process that articulates the head to the thorax

condyle 3 (*see also* condyle 1, 2) one of a pair of lateral teeth hinging the operculum in the zoecium of some Ectoprocta

acondylose lacking joints or nodes

cone 1 (*see also* cone 2–6) in *bot.* groups of scales, or leaves, particularly those which constitute the gamete bearing organs of coniferous trees

berry cone (*obs.*) a cone with fleshy and fused scales

bud cone an abortive inflorescence

vegetative cone a multibractiate vegetative bud

cone 2 (*see also* cone 1, 3–6) a photoreceptor in the retina functioning principally under conditions of bright illumination

acone lacking a cone. Usually said of a type of compound eye

cone 3 (*see also* cone 1, 2, 4–6) conical prominences

blastocone a segment of a cleaving blastodisc that is not yet separated from the yolk and cannot, therefore, be referred to as a blastomere

fertilization cone a prominence extending from the surface of some eggs at the moment of, or in some cases allegedly shortly before, contact with the sperm

epiproctal cone the central conical prominence, having nothing to do with the digestive tract, found on the aboral surface of some asteroid echinoderms

cone 4 (see also cone 1–3, 5, 6, conid, style) one of the projections from the surface of a mammalian tooth

hypocone the postero-internal cusp on a talonid

metacone the postero-external cusp on a trigon

paracone the antero-external cusp on a trigon

protocone the antero-internal cusp on a trigon

cone 5 (see also cone 1–4, 6) the two ends of a dinoflagellate

epicone that part of a dinoflagellate protozoan which is anterior to the equatorial groove

hypocone that part of a dinoflagellate protozoan which is posterior to the equatorial groove

cone 6 (see also cone 1–5, bill, fish) conical molluscan shells, particularly those of the gastropod family Conidae

thragmocone the chambered terminal portion of a decapod molluscan shell

coney (see also cony) U.S. the serranid fish Cephalopholis fulva

conferted crowded

configuration in biol. used particularly of chromosomes

cis configuration the condition in which two mutants at different sites within a cistron are located on the same chromosome

trans configuration the condition in which two mutants at different sites within a cistron are located on different chromosomes

[Confluens] a mutant (gene symbol Co) mapped at 3.0± on the X chromosome of Drosophila melanogaster. The phenotypic expression is thick wing veins

confoederata = mos

conformist Brit. the noctuid lepidopteran insect Graptolitha furcifera

nonconformist Brit. the noctuid lepidopteran insect Graptolitha landa

[Congenital hydrocephalus] a mutation (gene symbol ch) found in linkage group XIV of the mouse. The phenotypic expression involves a diminution of the cartilaginous skeleton

congestin a toxin obtained by the glycerine extract of nematocyst bearing tentacles (cf. hypnotoxin and thalassin)

conglutinate to cohere, particularly said of joints in insect appendages

congregate clustered together but without crowding

congregation term for a group of plovers

Congridae a family of large marine eels commonly called conger eels. They are distinguished by their large pectoral fins

-coni-, -conid- comb. form (better -koni-, -konid-) meaning "dust"

conic in the shape of a cone

obconic in the shape of a cone standing on its thin end

-conid- (see -coni-)

conid the cusps on a lower molar (cf. cone)

entoconid the postero-internal cusp on a talonid

hypoconid the postero-external cusp on a talonid

metaconid the postero-internal cusp on a trigonid

paraconid the antero-internal cusp on a trigonid

protoconid the antero-external cusps on a trigonid

Conidae a very large family of gastropod mollusks deriving their name from the cone shaped shells that have a large conical body whorl under a short blunt turret. Many tropical conids have a venomous bite

conidium the asexual reproductive body constricted from the tips of special hyphal branches in ascomycete fungi. The term is also used as synonymous with gonidium (q.v.)

acroconidium one that matures at the apex of the conidiopore

cladogynoconidium a branch bearing a female gonidium

conifer a common English term for any gymnosperm spermatophyte particularly of the order Coniferales

Coniferales that order of gymnospermae (q.v.) which contains the yews (Taxaceae) and pines (Pinaceae). The name which means cone bearers, is descriptive

conium the L. form of cone, preferred by some

androconium a patch of modified scales, thought to function as an odor-producing or sexual attractant organ, on the wings of male butterflies

otoconium one of the minute calcareous particles, analogous, or possible homologous, to otoliths found in the semicircular canals of higher vertebrates

statoconium one of numerous small granules acting on the hairs of neurosensory cells in statocysts (cf. statolith)

syconium a closed, fleshy receptacle enveloping many achenes, as in the fruit of the fig

conjugant one of a pair of conjugating protozoans, or a protozoan about to conjugate

exconjugant a freshly separated conjugant protozoan

macroconjugant the larger of the two products of progamous fission

microconjugant the smaller of the two products of progamous fission

conjugation the joining together of organisms, as in the conjugation of Paramoecium, the joining together of living entities such as gametes, and the joining together of parts of entities such as chromosomes

deconjugation precocious separation of chromosome pairs in meiosis

scalariform conjugation the conjugation of two alga filaments as they lie side by side

non-conjunction the failure of chromosomes to pair at the metaphase of mitosis

conjunctiva 1 (see also conjunctiva 2) the epithelium of the inner surface of the eyelid and outer surface of the eye

conjunctiva 2 (see also conjunctiva 1) the membranous connection at the joint between the segments of an insect (= coria)

Connaraceae a small family of rosale dicotyledonous angiosperms, mostly tropical twining shrubs

connascent produced at the same birth

connate said of homologous plant parts when fused together, especially for units of any one whorl of the floral envelope or of the enclosed sex organs (cf. adnate)

connation the congenital growing together of like parts

connective = commissure 2

interstrial connectives conducting fibrils running parallel to and between the interciliary fibrils in Protozoa

connexivum a flattened protruding shelf-like structure

surrounding the abdomen of some hemipteran insects

connivent 1 (*see also* connivent 2) converging

connivent 2 (*see also* connivent 1) pertaining to an eyelid capable of being moved up and down

-conop- *comb. form* meaning "gnat" better transliterated -konop-

conopeous gnat-like

Conopidae a family of cycloraphous myodarian dipteran insects, many so closely resembling small wasps that an examination of the wing is necessary to distinguish them. In addition to the "wasp-waisted" appearance, some have elongate ovipositors like those of saw-flies and ichneumons

Conopophagidae a small family of passeriform birds commonly referred to as antpipits. They are distinguished from the closely allied ant birds by the broad flattened and slightly hooked bill

conservation a term used by biologists to describe their endeavors to preserve, unravaged by man, some areas of the planet Earth

consolidated the result of cohesion between unlike organs or parts

consors one of a group of associated organisms that cannot be characterized by such customary terms as commensal, symbiont, etc.

consort *U.S.* the noctuid lepidopteran insect *Catocala consors* (*cf.* underwing)

consortes plural of consors (*q.v.*)

consortism = symbiosis

consortium a somewhat ill-defined term more or less meaning symbiosis sometimes reserved for the type of mutualism found in lichens

constable *I.P.* the nymphalid lepidopteran insect *Dichorragia nesimachus*

constipate crowded into very compact masses

attachment **constriction** = centromere

constrictor a term applied to many boid snakes (*see* boa) but properly reserved for the boa constrictor (*Constrictor constrictor*)

consutous united by threads, as though sewn together

-cont- *comb. form* meaning a rod or pole

contabsecence the condition of a plant in which male parts have failed to mature

contematosous roughened like an unshaven chin

context the "fleshy" substance of large fungi

continuum the area in which two contiguous populations overlap

Contortae that order of dicotyledonous angiosperms which contains, *inter alia*, the olives, the privits, the jasmines, the lilacs, the buddleias, the gentians, the oleanders, and the milkweeds, The order is distinguished by the actinomorphic gamopetalous flowers with a bicarpellate pistil in each

contractibility in *biol.* the ability of some organisms or cells to alter their shape

contraction in the biological sense either a lessening in length, as in the contraction of muscles, or a withdrawal from a point of contact

biological **control** a reduction in the population of an unwanted species by the intentional introduction of a predator, parasite, or disease

conule literally, a little cone, but specifically used of conical projections on the surface of keratose sponges and of cusps on molars (*cf.* conid, cone)

metaconule any cusp on an upper molar

conulid diminutive of conid (*q.v.*)

entoconulid any cusp on a lower molar

hypoconulid the postero-medial cusp on a talonid

conure a term used for many medium sized psittaciform birds (*cf.* parrot)

conus arteriosus that part of the truncus arteriosus which contains valves

convivum those members of a commiscuum (*q.v.*) that are prevented from interbreeding by geographical barriers

convocation term for a group of eagles

Convolvulaceae that family of tubiflorous dicotyledons which contains, *inter alia*, the morning glories, the bindweeds, and the sweet potato. The distinguishing characteristics of this family include the plaited corolla, and the few erect sessile ovules with axile placentation

Convolvulales an order erected to contain the family Convolvulaceae by those who do not wish to retain it in the Tubuliflorae

Convolvulineae a suborder of tubulifloral dicotyledonous angiosperms containing the families Convolvulaceae, Polemoniaceae, and Fouquieriaceae

cony (*see also* coney) *Brit.* without qualification = rabbit *W.I.* = hutia *U.S.* = pika

cookoo *see* cuckoo

coon 1 (*see also* coon 2) *I.P.* the hesperiid lepidopteran insect *Psolos puligo*

coon 2 (*see also* coon 1) *U.S.* common abbreviation for racoon

coontie the cycadaceous-plant *Zamia floridana*

cooperation an ill defined term rarely used in *biol.*

protocooperation that type of symbiotic association in which either partner can survive without the other

coot (*see also*, bandicoot) any of several water-dwelling birds, mostly rallids, mostly of the genus Fulica *U.S.* without qualification, *F. americana*, *Brit.* *F. atra*

blue bald coot *Austral.* the rallid *Porphyrio porphyrio* (= swamphen)

banana coot *W.I.* = Brazilian coot *W.I.*

Brazilian coot *W.I.* the jacanid *Jacana spinosa* (= American jacana)

red-knobbed coot *F. cristata*

Caribbean coot *F. caribbae*

blue painted coot *W.I.* = plantain coot *W.I.*

plantain coot *W.I.* the rallid *Porphyrula martinica* (= purple gallinule)

red seal coot *W.I.* the rallid *Gallinula chloropus* (= moorhen)

white seal coot *W.I. F. caribaea* (= caribbean coot)

Spanish coot *W.I.* = Brazilian coot *W.I.*

cooter *U.S.* any of several species of testudinid chelonian reptile of the genus Pseudemys (*cf.* slider)

cootie the pediculid anopluran insect *Pediculus humanus* (= body louse or head louse)

cop see [copper]

Copeognatha = Psocoidea

Copepoda a subclass of crustacean arthropods showing striking polymorphism. The females of all save a very few aberrant parasites may be distinguished by bearing the eggs in egg masses, paired or unpaired, attached to the outside of the body through the genital aperture. The free-living planktonic forms are easily recognized by the long first antennae which is used either for swimming or as a balancing organ

copper any of many lycaenid lepidopteran insects, usually orange, brown, or red with black markings. (*cf.* blue, hair-streak)

American copper *Lycaena hypophleas*

colly copper *S.Afr. Phasis chrysaor*

kaffir copper *S.Afr. Phasis taikosama*
king copper *S.Afr. Phasis sardony*
large copper *Brit. Chrysophanus dispar*
samba copper *S.Afr. Phasis clavum*
small copper *Brit. Chrysophanus phlaeas*
large spotted copper *S.Afr. Phasis argyraspis*
water copper *S.Afr. Phasis palmus*
[**copper**] a mutant (gene symbol *cop*) mapped at 43.3±
on the X chromosome of *Drosophila melanogaster*.
The name refers to the eye color of the phenotype
coppice *Brit.* originally a small group of trees planted
near a farmhouse to provide firewood by rotational
cropping. *U.S.* applied to small areas of second
growth timber (*cf.* copse)
-copr- *comb. form* meaning "dung" better transliterated -kopr- (*cf.* -fim-)
Copridae a family of coleopteran insects erected to contain those scarabaeids which are coprophagous, as
distinct from the Melolonthidae which are phytophagous. Nowadays these forms are usually placed
in the subfamily Scarabaeinae
Coprinaceae a family of agaricale basidomycete fungi
readily distinguished by the fact that at maturity
the gills, and in some species the whole cap, becomes liquid. The popular name inky cap derives
from this
coprodaeum the anterior portion of the cloaca (*q.v.*) into
which the anus opens
copse a cluster of trees planted for a specific purpose,
either decorative or utilitarian (*cf.* coppice)
copula literally a connection, specifically an unpaired
cartilage at the anterior end of the hyoid, joined by
the first ceratobranchials behind and the hypohyals
in front
    hypocopula the intermediate cell wall in the smaller
frustules of some diatoms
copulate (*see also* canal, path) to unite in sexual intercourse
    gametangial copulation the fusion of nuclei within a
gametogenic syncytium which subsequently divides
into distinct gametes
phenocopy a phenotype, resembling that of a gene, but
produced by the action of the environment
coquette any of several species of trochilid bird usually
of the genera Lophornis and Eaphosia
    black-crested coquette *L. helenae*
    rufous crested coquette *L. delattrei*
coqui the phasianid bird *Francolinus coqui*
coquilla the palmaceous tree *Attalea funifera* (= Pias-
sava palm)
coquina any of several species of donacid shells of the
genus Donax but usually *D. variabilis*
-cor- *see* -con-
-corac- *comb. form* meaning "raven" better transliterated -korak-
Coraciidae that family of coraciiform birds commonly
called rollers. They are distinguished by their very
short feet on which the second and third toes are
united basally
Coraciiformes that order of birds which contains the
kingfishes and their allies. They are distinguished
by the strong, long bill with imperforate nostrils.
Many are crested
coracinous raven-hued
coracoid (*see also* bone) in the shape of a crow's beak
coral (*see also* bell, bush, drop, flower, plant, root and
tree) any marine coelenterate that produces a calcareous skeleton

astraeid corals a group of families of imperforate
corals which are the chief reef builders
black coral popular name of zoantharians of the suborder Antipatharia
brain coral a coral consisting of a rounded mass with
many longitudinal "folds" containing parallel rows
of polyps, particularly the astraeid *Neandrina sinuosa*
bush coral the alcyonarian *Acanella normani*
fungian corals those which resemble corals of the
genus Fungia in having long laminate septa
hermatypic corals those corals characterized by the
presence of symbiotic unicellular algae. The term is
of ecological, rather than taxonomic, significance
    ahermatypic corals those corals which lack algal
symbionts
blue coral *Hellopora coerulea*
ivory coral popular name of slender branching corals
of the family Oculinidae
mushroom coral popular name of corals of the family
Fungiidae
meandrine corals = brain coral
pepper coral popular name of milliporine coelenterates
perforate coral one in which the polyps are joined by
canals penetrating the thecae
    imperforate corals those in which the polyps are not
joined by canals but only by coenenchyme
red coral the alcyonarian *Corallium nobile*. Also
called precious coral
soft corals a term frequently applied to alcyonarians
stone coral popular name of madreporarian zoantharia
thorny coral = black coral
corallite that portion of the skeleton of a coral which
surrounds one individual polyp (*cf.* corallum)
corallum the total skeleton of an entire coral (*cf.* corallite)
-corb- *comb. form* meaning "basket"
corbel an area of fringed bristles at the distal end of the
tibia in many coleopteran insects
corbina (*cf.* corvina) any of several fish
    California corbina the sciaenid *Menticirrus undulatus*
corbiculum a basket of high curved setae from the dilated hind tibia of bees. Often called pollen basket
corbula a phylactocarp (*q.v.*) with leaflike outgrowths
Corbulidae a family of pelecypod mollusks commonly
called basket clams by virtue of their very unequal
shells
corchorus the rosaceous shrub *Kerria japonica*
corcle a plant embryo
Corcoraciidae a family of passeriform birds commonly
included with the Grallinidae
corculum literally, a little heart, but applied specifically
to the chambers of the dorsal vessel through which
the blood of insects flows
-cord- *comb. form* meaning "heart" but also the *Eng.*
form of -chord- (*q.v.*)
cord 1 (*see also* cord 2,3) in the sense of string or rope
    central cord cells resembling vessels in moss stems
    flagellar cord a flagellum which is attached to the out-
side of the body wall of some mastigophoran Protozoa and which keeps the body in a condition of
constant undulation
    funicular cords cords running through the pore plates
in some Ectoprocta
    pulp cord the red pulp situated between adjacent si-
nusoids in the spleen. The term is also applied to
strings of sex cells in the developing ovary

spermatic cord the vas deferens and its associated ligaments in the mammal

umbilical cord 1 (*see also* umbilical cord 2) the connection between a mammalian embryo and the placenta

umbilical cord 2 (*see also* umbilical cord 1) = funicle

vocal cord 1 (*see also* vocal cord 2) tendons stretched across the glottis, or in some vertebrates within the laryngeal cavity, used in the production of speech and song

vocal cord 2 (*see also* vocal cord 1) in insects, modified thoracic spiracles that produce the typical hum of a mosquito and similar forms

cord 2 (*see also* cord 1, 3) in the sense of heart

obcordate in the shape of a symbolic heart attached by its lobes

cord 3 (*see also* cord 1, 2) in compound names of organisms

nigger's cord the euphorbiaceous tree *Antidesma bunius*

Cordiluridae = Scopeumatidae

Cordioideae a subfamily of the Boraginaceae

cordon rope, cord or strand, but used particularly for strings of eggs of invertebrates and some amphibia. Also synonymous with the nematode epaulet

cordon-bleu *Afr.* any of several ploceid birds of the genus Uraeginthus

  blue-capped cordon-bleu *U. cynocephalus*
  red-cheeked cordon-bleu *U. bengalus*

Cordulegastridae a family of anisopteran odonatan insects distinguished by lacking an oblique cross vein behind the proximal end of the stigma

Corduliidae a family of anisopteran odonatan insects with the hind margin of the compound eyes notched

Cordylidae a small family of *Afr.* saurian reptiles distinguished by the whorls of large keeled scales encircling the tail from which derives the popular name girdle-tail lizard

Cordyluridae = Scopeumatidae

Coregonidae a family of isospondyl fish containing the white fish and cisco when these are not retained in the Salmonidae

Coreidae a family of hemipteran insects many of which are distinguished by their flattened and leaf-like legs from which the family derives the name of leaf-footed bugs. Many members have a repulsive odor

corella *Austral.* any of several species of psittaciform birds of the genus *Kakatoe*, usually, without qualification, *K. tenuirostrio*

-corem- *comb. form* meaning a "broom", better transliterated -korem-

corema scent-tufts on the abdomen of male Lepidoptera

-cori- comb. form meaning "skin"

coria the membrane spanning the joint between segments in arthropods (= conjunctiva)

  antacoria the segmental membrane between each antennal segment of an anthropod

  coxacoria a flexible membrane of the coxal joint
    matacoxacoria that which attaches appendages to the metathorax

  labacoria the membrane between the labium and the margin of the head in insects

  latacoria that which lies between the sternum and tergum

  mandacoria the membrane articulating the mandible to the head

  maxacoria that portion which lies between the maxilla and the head

  metacoria that which joins the metathorax to the mesothorax

  segmacoria that which lies between segments

coriaceous literally leathery, but often used in the sense of tough

Coriariaceae a unigeneric family of sapindale dicotyledonous angiosperms

Coriariaceae a small family of sapindalous shrubs

Coriariineae a suborder of Sapindales containing the single family Coriariaceae

excoriation the shedding of a skin or outer layer

Corimelaenidae small broadly oval black bugs frequently mistaken for beetles from their general appearance. The popular name Negro bug is often applied

Coriscidae a family of hemipteran insects with very broad heads and long narrow bodies. They are commonly called broad-headed bugs. Many have a most offensive odor, which causes them frequently to be called stink bugs, though this term applies properly to the Pentatomidae

corium literally leather, but applied to any leathery structure such as the thickened basal portion of an hemipteran wing. It is also sometimes used for the bark of a tree or the dermis of the skin

Corizidae a family of hemipteran insects strongly resembling the Coreidae but smaller in size, usually lighter in color, and lacking scent glands. They are commonly called grass bugs

cork the spongy outer layers of the bark of woody stems

  pore cork cork cells in lenticels

  storied cork cork formed in monocotyledons by the suberization of parenchyma cells and divided off in layers without a specific phellogen

-corm- *comb. form* meaning "tree trunk" or "stump," better transliterated -korm-. The meaning is extended to "stem" or even "plant"

corm 1 (*see also* corm 2, 3) a swollen underground stem designed for food storage and which serves as a resting phase of the plants involved. Corms, such as those of the crocus and gladiolus, are often erroneously called bulbs (*q.v.*)

  protocorm a primitive corm-like structure produced by seedling orchids

  tubercorm a fleshy root like that of the turnip

  rhizocorm a fleshy rhizome

corm 2 (*see also* corm 1,3) in the sense of an entire plant, or plant organ system

  bud corm the sum total of the root system of an herbaceous plant

  protocorm a plant embryo that is not yet morphologically differentiated

  metacorm a completely differentiated plant

corm 3 (*see also* corm 1, 2, theory) a group of individuals, such as a colony of hydrozoan colenterates, all members of which are both morphologically and physiologically united

cormel a new corm produced from the base of a "mother" corm

cormic pertaining to a tree trunk

  epicormic branches or buds that develop on the trunk of a tree

  monocormic possessing a single trunk

  polycormic having many trunks

cormidium a group of zooids attached to the stem of a siphonophoran hydrozoan

cormorant any of many species of pelecaniform phalacrocoracid birds of the genus Phalacrocorax. They are distinguished by their long neck and skill in diving. The biblical cormorant was probably a pelican

  Brandt's cormorant *P. penicillatus*

white-breasted cormorant *Austral. P. fuscescens Afr.* = great cormorant

double-crested cormorant *P. auritus*

red-faced cormorant *P. urile*

Guanay cormorant *P. bougainvillii*

great cormorant *P. carbo*

olivaceous cormorant *P. olivaceus*

pelagie cormorant *P. pelagicus*

long-tailed cormorant *P. africanus*

-corn- *comb. form* meaning "horn"

corn 1 (*see also* corn 2, beetle, borer, cockle, flag, flower, marigold, hopper, maggot, salad, and worm) in the sense of grain or cereal. *U.S.*, without qualification, *Zea mays* (maize, indian corn); *Brit.*, without qualification, *Triticum spp* (wheat)

barley corn a single grain of barley

broom corn (*see also* millet) *Sorghum halapensis* (= sorghum)

Indian corn = maize

Jerusalem corn *hort. var.* of *Sorghum halapensis* (= sorghum)

squirrel corn the fumariaceous herb *Dicentra canadensis*

sweet corn a horticultural variety of maize having an unusually high sugar content in the immature seed

turkey corn *U.S.* the papaveraceous herb *Dicentra eximia* (= wild bleeding heart)

corn 2 (*see also* corn 1) in the sense of horn

lamellicorn a type of antenna of which the terminal joints are modified as parallel plates. (*cf.* Lamellicornia)

longicorn in insects, possessing antennae as long as, or longer than, the body. At one time applied exclusively to cerambycid Coleoptera

palpicorn a palp so elongate as to resemble an antenna

unicorn (*see also* caterpillar, shell) a mythical forest-dwelling ungulate variously supposed to be the product of bad eyesight, imagination, or a fraudulent trade in narwhal's tusks

Cornaceae that family of umbelliflorous dicotyledons which contains, *inter alia*, the dogwoods. The family is distinguished by the fruit, a 2-4-loculate berry

cornea the outer coat of the front of the eye

cornel *Brit.* any of several species of tree or shrub of the cornaceous genus Cornus (= *U.S.* dogwood)

dwarf cornel *U.S. Cornus canadensis*

cornelian *I.P.* any of several species of lycaenid lepidopteran insect of the genus Deudoryx but usually, without qualification, *D. epijarbas*

corneous horny

cornicle an elevated tube through which insects, particularly aphids, secrete a waxy material

cornification the development of a horny appearance or texture by a cell or cell layer

cornified modified into the form of a horn, or horny tissue

corolla diminutive of corona (*q.v.*) and specifically the whorl of petals in a flower

caryophylleous corolla one in which the petals have a long claw

catacorolla a secondary corolla surrounding the true corolla

gamopetalous corolla with the petals marginally connate, at least basally

polypetalous corolla with the petals distinct

sympetalous corolla = gamopetalous corolla

pericorolla a gamopetalous perigynous corolla

-coron- *comb. form* meaning "crown"

corona 1 (*see also* corona 2–5) in the sense of a crown of cilia, specifically the equatorial girdle of a very long cilia round the early embryo of ectoprocts or the ciliary loop of Chaetognaths or the anterior ciliated area of rotifers

corona 2 (*see also* corona 1, 3–5) in the sense of a crown of calcareous structures. Specifically the body, as distinct from the stalk, of crinoid echinoderms and all that remains of the test of an echinoid echinoderm after the removal of the apical system

corona 3 (*see also* corona 1, 2, 4, 5) in the sense of a crown of spines, specifically those on the cucullus of male Lepidoptera

corona 4 (*see also* corona 1–3, 5) any part of a flower lying between the corolla and the stamens

corona 5 (*see also* corona 1–4) a ring of primary wood in the medullary sheath

corona radiata a layer of follicle cells adhering on the surface of the zona pellucida of a mammalian egg

Coronatae an order of scyphozoan coelenterates with a scalloped margin separated from the bell by a circular furrow

coronet 1 (*see also* coronet 2, 3) the swollen base of a deer horn

coronet 2 (*see also* coronet 1, 3) *Brit.* the noctuid lepidopteran insects *Craniophora ligustri*, and several species of the genus Dianthoecia

coronet 3 (*see also* coronet 1, 2) any of several trochilid birds of the genus Boissonneaua

coronula literally, a small crown. But particularly a crown of spines at the distal end of the insect tibia

Corophiidae a family of tube-dwelling gammaridean amphipod crustacea distinguished from all other amphipods by the fact that the body is flattened and bears small uropods

corosol the fruit of the annonaceous tree *Annona muricata* (= soursop)

corpora quadrigemina two pairs of lobes developed in the mammalian brain from the roof of the mesencephalon corresponding to the single pair of optic lobes of the lower forms

corpus (*see also* corpora, corpuscle, body) literally "body." The use, in English, of corpus, body and corpuscle is dictated by custom

corpus allatum one of a pair of wing-like structures projecting laterally from the hypocerebral ganglion of insects. They are primarily hormonal rather than nervous

corpus callosum the dorsal of the two connections between the cerebral hemispheres in higher vertebrates

corpus luteum the mass of yellow secretory cells formed in the cavity of a ruptured Graafian follicle

corpus pedunculatum = mushroom body

corpus striatum a large ganglionic mass in the lower lateral wall of each of the cerebral hemispheres, usually considered part of the brain stem

corpuscle the diminutive of corpus and thus applied to any small body. The *L.* form corpusculum, unlike corpus, appears to have disappeared from English biological literature

Belt's corpuscle a food body on the leaves of some species of acacia

bridge corpuscle = desmosome

chlorophyll corpuscle = chloroplast

chordal corpuscle free cells lying in the sheath of the notochord in prechordates

cylindrical corpuscle an end organ consisting of a club-shaped nerve termination in a connective tissue sheath

genital corpuscle a specialized type of Vater's corpuscle found in the erotogenic areas

germinal corpuscle = oosphere

Gluge's corpuscle a phagocytic cell produced in pathological states from microglia

Grandry's corpuscle a specialized type of Merkel's corpuscle (*q.v.*) found in the skin of anseriform birds

Hassall's corpuscle concentrically layered globose corpuscles found in the developing thymus

Herbst's corpuscle a type of vagus corpuscle found in birds

Krause's corpuscle an end organ, sensing cold, consisting of a spherical mass of connective tissue with a much branched central nerve ending

Malpighian corpuscle the capsule and glomerulus, taken together, of a mesonephric or metanephric unit

meconium corpuscle a granule in the interstinal epithelium of embryos of higher vertebrates

Meissner's corpuscle a touch sensing end organ in the form of ovoid structures with the central mass of irregular cells penetrated by irregularly curved nerve endings

Merkel's corpuscle an end organ consisting of two or more tactile cells in a connective tissue sheath

Pacini's corpuscle = Vater's corpuscle

red corpuscle = erythrocyte

Ruffini's corpuscle a heat sensing end organ, containing a loose arborization of nerve fibers, ending in flattened expansions in a mass of granular tissue

Stannius' corpuscle a body of possibly endocrine function in the kidneys of actinopterygian fishes

tactile corpuscle an end organ consisting of a single tactile cell in a terminal nerve cup

Vater's corpuscle an end organ sensing pressure, consisting of the central elongated granular mass containing the nerve ending, surrounded by many layers of thin connective tissue

white corpuscle = leucocyte

Corrodentia = Psocoidia

corselet 1 (*see also* corselet 2, 3) an area of scales under the pectoral fins of otherwise scaleless scombrid fish

corselet 2 (*see also* corselet 1, 3) a restraining skeletonic investure such as the nonarticulated bony keels of some chelonian reptiles

corselet 3 (*see also* corselet 2, 3) the coleopteran prothorax

Corsiniaceae a family of marchantiale Hepaticae that are distinguished by having sex organs in a linear series of receptacles lying on either side of the midline of the dorsal surface

-cort- *comb. form* meaning "bark" or, by extension, "periphery"

cortex 1 (*see also* cortex 2, 3) literally, bark, but used for the outer portion of an egg or of an organ, such as the cerebrum, kidney, or adrenal gland in contrast to the inner medulla

exocortex the outermost layer of anything

hippocampal cortex that portion of the brain cortex which runs back from the olfactory bulb to meet at the posterior pole of the hemisphere

neocortex the cerebral cortex of mammals

olfactory cortex the combination of the hippocampal cortex and the pyriform cortex

pyriform cortex that portion of the cortex of the brain which lies superficial and dorsal to the corpus striatum

cortex 2 (*see also* cortex 1, 3) that part of the fundamental tissue of a plant which lies between the epidermis and the vascular region

primary cortex = periblem

cortex 3 (*see also* cortex 1, 2) the outer, frequently spicule bearing, layer of sponges

cortexone = desoxycorticosterone

Corti's *epon. adj.* from Alphonso, Marquis Corti (1822–1888)

-cortic- *see* -cort-

corticate pertaining to the possession of a cortex or of bark

ecorticate lacking bark

decorticate having had the bark removed

cortisol = hydrocortisone

cortisone one of several glucocorticoids (*cf.* corticosterone). Active in glycogen deposition in liver, muscle work performance, potassium-sodium ratio, growth, and the production of secondary sex characters

fluorocortisone similar in origin and action to cortisone

hydrocortisone similar in its effects to cortisone

corticosterone one of several glucocorticoids (cf. cortisone) active in glycogen deposition in liver, muscle work performance, potassium-sodium ratio, growth, and the production of secondary sex characters

deoxycorticosterone a hormone secreted in trace quantities by the adrenal cortex and in general similar in its effects to aldosterone

dehydrocorticosterone similar in its effect to corticosterone

Corvidae a very large family of passeriform birds containing, *inter alia*, the crows, magpie, and jays. They are the largest of the passerine birds; many have crests and extremely long tails. They are principally distinguished by the large heavily scaled tarsi

corvina (*cf.* corbina) *U.S.* any of several species of sciaenid fish of the genus Cynoscion (*cf.* sea trout)

shortfin corvina *C. parvipinnis*

orangemouth corvina *C. xanthulus*

corvinous shiny blue-black

Corydalidae a family of neuropteran insects commonly called dobson flies. They differ from all other Neuroptera except the Sialidae by the fact that the wings are folded flat and overlapping rather than tentwise when at rest and from the Sialidae by the presence of ocelli

Corydirhynchidae = Rhinomaceridae

Corylaceae = Betulaceae

Corylophidae = Orthoperidae

corymb a flat-topped indeterminate cluster of flowers

-coryn- *comb. form* meaning "club"

Corynebacteriaceae a family of eubacteriale schizomycetes in the form of gram positive rods usually showing plentiful metachromatic granules. Many are pathogenic and one is the cause of diphtheria in man

-coryph- *comb. form* meaning "summit"

coryphad a plant of alpine meadows

Coryphaenidae a family of acanthopterygian fishes commonly called dolphins. The family is distinguished by the blunt nose, deeply forked tail, and very long dorsal fins

Coryphaenoididae a family of anacanthinous fishes mostly living in deep water but distinguished by the long rat-like tail fin. Commonly called rat tails or grenadiers

corysterium an accessory female sex gland of insects that secretes a supplementary covering for the eggs

cosmine a dentine-like material underlying the ganoin in some primitive ganoid scales

cosmopolitan *Brit.* the noctuid lepidopteran insect *Leueania loreyi*

cosmopolite *U.S.* the nymphalid lepidopteran insect *Vanessa cardui* (= painted lady)

Cosmopterygidae a family of gelechioid lepidopteran insects with long narrow sharply pointed wings. The larvae of most are leaf miners

Cossidae a family of large and medium sized tortricoid lepidopteran insects including such forms as the leopard moths and the goat moth

costa literally a rib, as much that of a plant leaf as of a chordate thorax. The term is also used for the comb-rows of ctenophorans, the first, or anterior longitudinal veins on an insect wing, and an extension of the scleroseptum of a coral connecting the thecae to the pseudothecae

  antecosta an internal anterior ridge for the attachment of muscles to the plate of an arthropod exoskeleton

[costakink] a mutant (gene symbol *csk*) mapped at 33.4± on the X chromosome of *Drosophila melanogaster*. The costal vein is kinked in an undersized wing in the phenotype

costate ribbed

  curvicostate having curved ribs

costmary the compositaceous herb *Chrysanthemum balsamita*

coster *I.P.* popular name of acraeid lepidopteran insects

cot *see* [colonial-temperature sensitive]

coterie a closed society (*q.v.*) consisting of several species

cotinga anglicized form of the name of the bird family Cotingidae

Cotingidae a family of passeriform birds commonly referred to as cotingas or umbrellabirds, because of the huge umbrella-like crest borne by many species

Cottidae a very large family of scleroparous fishes distinguished by the position of the eyes on top of the large head and the very sharp spines on the preoperculum. They have large fan shaped pectoral fins and the spiny and soft portions of the dorsal fin are separate

cotton 1 (*see also* cotton 2, aphis, hopper, mite, perforator, rose, wick strainer, tree and wood) any of numerous low growing malvaceous shrubs of the genus Gossypium and, particularly, the fiber that is borne on the seeds

  Chinese cotton *G. indicum*

  Egyptian cotton *G. barbadense*

  Indian cotton *G. neglectum*

  sea island cotton = Egyptian cotton

  Japanese cotton = Chinese cotton

  Levant cotton *G. herbaceum*

  tree cotton *G. brasiliense* and *G. peruvianum*

    oriental tree cotton *G. arboreum*

  American upland cotton *G. hirsutum*

cotton 2 (*see also* cotton 1) numerous plants with cottony seeds

  bog cotton *U.S.* any of several species of sedge of the genus Eriophorum (= cotton grass)

  devil's cotton the sterculiaceous shrub *Abroma augusta*

  lavender cotton the compositaceous herb *Santolina chamaecyparissus*

  wild cotton *U.S.* either of the malvaceous herbs *Hibiscus moscheutos* and *Thurberia thespesiodes*

-cotyl- *comb. form* meaning the "cavity of a cup"

cotyl pertaining to a developing plant

epicotyl that part of the axis of a plant embryo that lies above the cotyledons

hypocotyl that part of the axis of a plant embryo which lies below the cotyledons, but above the radicle

mesocotyl the node between the sheath and the cotyledon of seedling grasses

cotylar = cotyledonary

cotyle the socket of a ball-and-socket joint in arthropods

cotyledon the seed leaf: that is the leaf or leaves which first appear when the seed germinates

  accumbent cotyledon one that has its edges pressed against the radicle

  conduplicate cotyledon one of two that are folded together lengthwise

  incumbent cotyledon one of two that are face to face, the back of one against the hypocotyl

  pseudomonocotyledon a dicotyledonous plant that appears to have only a single cotyledon, either through the suppression of one, or the fusion of two

cotyledonous pertaining to the cotyledon

  acotyledonous lacking cotyledons, or at least appearing to lack them

  amphicotyledonous a cotyledon in the form of a cup

  amphisyncotyledonous having cotyledons fused into the form of a funnel

  dicotyledonous having two cotyledons

  heterocotyledonous having unequal cotyledons

  isocotyledonous having cotyledons of equal size

  anisocotyledonous having unequal cotyledons

  monocotyledonous having a single cotyledon

  polycotyledonous having more then two cotyledons, or at least appearing to

Cotyloideae a term used to describe those families of plants, both monocot and dicot, in which the flower is derived by transverse adnations

cotylous = cotyledonous

coucal any of numerous species of cuculid birds mostly of the genus Centropus without qualification usually *C. sinensis*

  black coucal *C. toulou*

  white-browed coucal *C. superciliosus*

  common coucal *C. sinensis*

  Gaboon coucal *C. anselli*

  blue-headed coucal *C. monachus*

  lesser coucal *C. bengalensis*

  pheasant-coucal *C. phasianinus*

  rufous coucal *C. epomidis*

  black-throated coucal *C. leucogaster*

cougar *U.S.* the felid carnivore *Felis concolor* (= mountain lion *U.S.*, panther *U.S.*)

coulter-neb = puffin

count *I.P.* any of numerous species of nymphalid lepidopteran insect of the genus Euthalia. (*cf.* baron, earl, duke, duchess, baronet, marquis)

coupling the condition of two loci when each has a recessive or downward allele (*cf.* repulsion)

coupray a very large and highly bovine bovid (*Bos sauveli*) of Cambodia. It has an exaggeratedly large dewlap, and the sweeping recurved horns of the male terminate in a tuft of fibers

courser any of several species of glareolid birds of the genera Cursorius and Rhinoptilus sometimes placed in a separate family as the Cursoriidae

  two-banded courser *R. africanus*

  cream-colored courser *C. cursor*

  Henglin's courser *R. cinctus*

  Indian courser *C. coramandelicus*

  bronze-winged courser *R. chalcopterus*

courian the aramid bird *Aramus guanura*

courtesan *I.P.* any of numerous species of nymphalid lepidopteran insect of the gneus Euripus

courtier *I.P.* any of numerous species of nymphalid lepidopteran insect of the genus Sephisa

courtship the behavior accompanying or preceding the selection of a mate

cousin offspring of a sib

blackbird's cousin *W.I.* the icterid *Molothrus bonariensis*

coverts feathers that cover or lie between the flight feathers or tail feathers of a bird or are closely applied to other areas

ear coverts feathers protecting the auricle in birds

covey term for a group of partridges, or, in *U.S.*, quail

cow (*see also* bane, bell, bird, fish, grass, herb, itch, pea, ray, shark, and wheat) without qualification, the female of domestic cattle, though applied also to a large number of artiodactyl mammals, other than the Suiformes for which the term sow is preferred. Cow is also used for a female elephant, manatee, seal, terrapin, whale, and for some other animals of cow-like habit or form

ant cow aphids tended by ants

bush cow a geographic race of the forest buffalo (*q.v.*)

sea cow the term popularly applied to sirenians in general. If used specifically, usually refers to *Trichechus manatus* (= manatee)

Steller's sea cow *Hydrodamalis stelleri*

cowl literally an enveloping hood and applied to numerous such structures in animals

cowrie = cowry

cowry (*see also* bird) popular name of cypraeid gastropod Mollusca

money cowry *Cypraea moneta*

cowslip *Brit.* the primulaceous herb *Primula vera U.S.* the primulaceous herb *Dodecatheon meadia*. The name is also sometimes applied to the ranunculaceous herb *Caltha palustris* (= marsh marigold)

American cowslip the primulaceous herb *Dodecatheon meadia* and some similar species

cape cowslip any of numerous liliaceous herbs of the genus Lachenalia

Virgina cowslip the boraginaceous herb *Mertensia virginica*

coxa that segment of the insect leg which articulates with the body

precoxa an additional segment derived by the division of the coxa lying between the coxopodite and the body, in some arthropod appendages

coxella a sclerite in the coxacoria

basicoxite a prominent rib around the proximal end of the coxa in some insects

coxola = coxella

coyote the canid carnivore *Canis latrans* (= prairie wolf)

coypu the large, aquatic capromyid hystricomorph rodent *Myopotamus coypus* with webbed hind feet. The pelt is known in the fur trade as "nutria"

phospho-cozymase *obs.* = NADP

cp *see* [clipped]

Cp *see* [Creeper]

Cpt *see* [Carpet]

cr *see* [crisp], [crescent] or [crinkled]

cr₁ *see* [crinkly leaves]

cr-3 *see* [cream in 3]

Cr *see* [Crest]

crab (*see also* apple, clam, grass, louse, plover and spider) properly any brachyuran decapod crustacean. The term is also applied to many other arthropods resembling a crab in shape or habit

blue crab *U.S.* the portunid *Callinectes sapidus*

California crab the cancrid *Cancer magister*

dwarf crab *U.S. Pelia tumida*

edible crab *U.S.* = blue crab *U.S.*

fiddler crab *U.S.* any of several species of the ocypodid genus Uca

coral gall crab any of several species of brachyrhynchans of the genera Hapolocarcinus and its immediate allies

ghost crab any of several species of brachyrhynchan of the genus Ocypode

green crab *U.S.* the portunid *Carcinides moenas*

woolly-handed crab any of several species of geograpsid of the genus Eriocheir

hermit crab popular name of pagurid thalassinids

horseshoe crab popular name of xiphosurans

jonah crab the cancrid *Cancer borealis*

king crab any of several species of pagurids of the genus Lopholithodes, commercially known in *U.S.* as the **Alaska king crab**. The term is also sometimes applied to Xiphosurans (= horseshoe crab)

land crab any large tropical crab that spends more time out of the water than in it

masked crab *Brit. Corystes cassivellaunus*

mole crab popular name of hippids

mud crab popular name of xanthid brachyuran crustacea particularly (*U.S.*) those of the genus Panopeus. The term is also applied to the grapsid genus Hemigrapsus

mussel crab the pinnotherid *Pinnotheres maculatus*

oyster crab the pinnotherid *Pinnotheres ostreum*

palm crab the pagurid *Birgus latro*

porcelain crab any of several species of porcellanid crustacean of the genus Munida and its immediate allies

purple crab the oxystomate *Randallia ornata*

rock crab *U.S.* any of several species of cancrids of the genus Cancer, particularly *C. irroratus*, *C. productus*, and *C. antennarius*

striped rock crab *U.S.* the grapsid *Pachygrapsus crassipes*

sand crab the hippid *Emerita talpoida* (= sand bug) and the coypodid *Ocypode albicans*. The term is also applied to the mole crab

sheep crab the oxyrhynchan *Loxorhynchus grandis*

shore crab *U.S.* any of several species of grapsid of the genus Hemigrapsus (*cf.* mud crab)

spider crab popular name of oxyrhynchan crabs

stone crab erroneously applied to several species of lithodid of the genus Pylocheles and its immediate allies; but better reserved for the xanthid *Menippe mercenaria*

toad crab *U.S. Hyas coarctatus*

Crabroninae a subfamily of sphecid hymenopteran insects that shares with the Philanthinae the designation digger wasp. They are, however, more commonly called square-headed wasps from their large quadrate head

Cracidae that family of galliform birds containing the forms commonly called curassows and guans. They are distinguished from the Megapodiidae by possessing two carotid arteries and the fact that the uropygial gland is feathered

cracker *Brit. obs.* = pintail

mussel cracker any of several sparid fishes

nutcracker either of two species of corvid bird of the genus Nucifraga. *Brit.*, without qualification, *N. caryocatactes. U.S.* without qualification, *N. columbiana.* (Clark's nutcracker)

Clark's nutcracker *N. columbiana*
European nutcracker *N. caryocatoctes*
Himalayan nutcracker = European nutcracker
seed cracker any of several species of ploceid bird of
the genus Pirenestes
 black-bellied seed-cracker *P. ostrinus*
 large-billed seed-cracker *P. maximus*
 whelk-cracker *W.I.* the haematopodid *Haematopus
 ostralegus* (= oyster-catcher)
Cracticidae that family of passeriform birds which contains the bellmagpies. These large Australasian
birds have big stout bills like the Corvidae but with
a hook in front. The wings are usually long and
pointed
crag neck, used particularly of arthropods
Cragonidae = Crangonidae
crake any of numerous rallid birds
 African crake *Crecopsis egregia*
 banded crake any of several species of the genera
 Rallina and Porzana
 black crake *Afr. Limnocrax flavirostra* (= *U.S.* black
 rail)
 yellow-breasted crake *Porzana flaviventer*
 red-chested crake *Sarothura rufa*
 Chinese crake *Porzana paykullii*
 corn crake *Brit. Crex crex* (= land rail)
 dwarf crake *Porzana pusilla*
 little crake *Porzana parva*
 Malay crake *Rallina fasciata*
 Philippean crake *Rallina eurizonoides*
 ruddy crake *I.P. Porzana fusca U.S. Laterallus ruber*
 chestnut-tailed crake *Sarothrura affinis*
 striped crake *Aenigmatolimnas marginalis*
 buff-spotted crake *Sarothura elegans*
 uniform crake *Amaurolimnas concolor*
Crambidae a small family of lepidopteran insects frequently regarded as a subfamily of the Pyralidae.
The tight wrapping of the wings about the body has
led to the name close wing
cran a *Brit.* unit for the measurement of freshly caught
herring. The size of the cran varies from port to
port, but is usually about fifty gallons, or rather less
Cranchiidae a family of dibranchiate cephalopod Mollusca differing from all other families in the group
by lacking a valve on the siphon
crane (*see also* bill, fly) any of many species of gruiform
birds of the family Gruidae, mostly of the genus
Grus
 Australian crane *Grus rubicunda*
 blue crane = Stanley crane
 little brown crane = sandhill crane
 common crane *Grus grus*
 crowned crane *Balearica pavonina*
 demoiselle crane *Anthropoides virgo*
 hooded crane *Grus monacha*
 sandhill crane *Grus canadensis*
 sarus crane *Grus antigone*
 Stanley crane *Anthropoides paradisea*
 wattled crane *Bugeranus carunculatus*
 whooping crane *Grus americana*
Crangonidae a family of natant decapod Crustacea commonly called shrimps. They differ from the Pa-
laemonidae in lacking a chela on the first pair of
legs
cranial pertaining to the skull or, by improper extension, pertaining to the head or brain
Craniata a division of chordate animals erected to contain those classes that have a skull
cranium that part of a bony skull which surrounds the
brain

branchiocranium those skeletal units which support
the mouth and pharynx of vertebrates
chondrocranium a skull, or skull rudiment, consisting
entirely of cartilage. In some elasmobranchs, it
may be calcified, but is never ossified
endocranium that part of the bony skull which is derived by ossification of parts of the embryonic
chondrocranium
exocranium those bones of the skull which are of
dermal origin
neurocranium that part of the skull which encloses the
brain in contrast to the splanchnocranium
splanchnocranium the jaws and visceral arches of a
skull in contrast to the neurocranium
cranterian pertaining to the spacing of teeth
 diacranterian having enlarged posterior teeth separated from the others by a diastema
 syncranterian having no diastema in the arrangement
 of the teeth
crappie *U.S.* any of several centrarchid fishes of the
genus Pomoxis
crash term for a group of rhinoceros
-craspedo- *comb. form* meaning "border" or "edge"
craspedon = velum
craspedote pertaining to a coelenterate medusa that
possesses a velum
 acraspedote lacking a velum, particularly a medusa in
 this condition
-crass- *comb. form* meaning "thick"
incrassate thickened, particularly applied to the skin
crassula an intercellular thickening of the upper and
lower margins of pit-pairs on tracheids of gymnosperms (= Sanio's bar)
Crassulaceae a small family of rosalous dicotyledons
which include, *inter alia*, the sedums, distinguished
by the gynoecium composed of as many mono-
carpellate pistils as there are petals in the flower
-crater- *comb. form* meaning "cup"
craticular the condition of a diatom in which new
frustules are being formed
crawdad *U.S. dial.* = crayfish or crawfish
crawl a method of locomotion using four limbs with the
body pressed to, or close to, the ground (*cf.* creep)
crawler *U.S.* the larva of the dobson fly *Corydalis
cornuta* (= hellgrammite)
 night crawler *U.S.* any large earthworm
[crayfish] a mutant (gene symbol *cf*) mapped at 11.3 on
linkage group XIII of *Bombyx mori*. The crayfish-
like appearance of the phenotypic pupa is caused by
the swollen and laterally expanded wings
[cream in 3] a mutant mapped at 36.5± on chromosome
III of *Drosophila melanogaster* that is a specific
dilutor of W^e one of the alleles of [white]
creatininase an enzyme catalyzing the hydrolysis of
creatinine to sarcosine and urea
creep very much the same thing as crawl (*q.v.*) but
more usually applied to many legged animals
creeper 1 (*see also* creeper 2) any of numerous birds distinguished by their habit of running up and down
vertical surfaces; in *U.S., Brit.* and *Austral.*, without qualification, those of the family Certhiidae. In
Brit., the term is also applied to juveniie phasianids
 brown creeper *U.S.* the certhiid *Certhia familiaris.
 N.Z.* the malurine bird *Finschia novaeseelandiae*
 earth creeper any of several furnariids of the genus
 Upocerthia
 ground creeper *N.Z.* = the malurine bird *Finschia
 novaseelandiae*
 Hawaiian creeper the drepaniid *Loxops maculata*

honeycreeper popular name of many birds once united into the family Coerebidae now distributed between the Parulidae and Thraupidae

Hawaiian honeycreeper any of many species of the family Drepaniidae

palm creeper the furnariid *Berlepschia rickeri*

spotted creeper the certhiid *Salpornis spilonota*

treecreeper *Brit.* any of several certhiids of the genus *Certhia*. Usually without qualification, *C. familiaris*

red-browed treecreeper *C. erythrops*

white-browed treecreeper *C. affinis*

brown treecreeper *C. picumnus*

white-throated treecreeper *C. leucophaea*

woodcreeper any of numerous dendrocolaptid birds

barred woodcreeper *Dendrocolaptes certhia*

ivory-billed woodcreeper *Xiphorhynchus flavigaster*

strong-billed woodcreeper *Xiphocolaptes promeropirhynchus*

wedge-billed woodcreeper *Glyphorhynchus spirurus*

spot-crowned woodcreeper *Lepidocolaptes affinis*

streak-headed woodcreeper *Lepidocolaptes souleyetii*

olivaceous woodcreeper *Sittasomus griseicapillus*

ruddy woodcreeper *Dendrocincla homochroa*

spotted woodcreeper *Xiphorhynchus erythropygius*

white-striped woodcreeper *Lepidocolaptes leucogaster*

tawny-winged woodcreeper *Dendrocincla anabatina*

wall-creeper the certhiid *Tichodroma muraria*

creeper 2 (*see also* creeper 1) plants of creeping, or climbing habit

Australian blue bell creeper the pittosporaceous shrub *Sollya heterophylla*

rangoon creeper the combretaceous shrub *Quisqualis indica*

trumpet creeper *U.S.* the bignoniaceous herb *Campsis radicans*

Virginia creeper (*see also* sphinx) the vitaceous woody vine *Parthenocissus quinquefolia*

[Creeper] a mutant (gene symbol *Cp*) in linkage group II of the domestic fowl. The name is descriptive of the activities of the phenotype in which the shaft bones fail to ossify

-crem- *comb. form* meaning "to hang" or "to overhang". Frequently confused with -cremno-

cremaster suspensor. The term is used both for the muscles suspending the testes in the scrotum and for the hook on the end of the case by which some insect pupae can be suspended

cremnad a plant dwelling on cliffs

-cremno- *comb. form* meaning "cliff"

-cren- *comb. form* hopelessly confused between *Gr.* -kren- meaning "notch" and *L.* -cren- meaning "spring" in the sense of a source of water

crenad a plant dwelling near springs

crenate notched, the margin scalloped

rheocrene a spring that feeds a stream

crenic pertaining to springs and their immediately derivative waters

Crenotrichaceae a family of chlamydobacterial schizomycetes occuring as trichomes frequently swollen at the free end, and also often encrusted with iron or manganese compounds at the base

crenulate with small notches

-creo- *comb. form* meaning "meat" or "flesh"

-crepi- *comb. form* meaning "shoe" better transliterated -krepi-

Crepidulidae a family of slipper-shaped uncoiled gastropod mollusks commonly called slipper limpets or boat shells. They are distinguished from other uncoiled gastropods by the fact that the inner cavity of the shell is divided by a horizontal platform

crepis a minute irregular sponge spicule

crepuscular pertaining to the dim light of dusk or dawn (*cf.* auroral, matutinal, vespertine)

-cresc- *comb. form* meaning "grow"

crescent 1 (*see also* crescent 2, 3, chest) an area in the shape of a new ("growing") moon

gray crescent an area on the fertilized egg of anuran amphibians exactly opposite the point of entry of the sperm

malaria crescent the gamont of a malarial parasite

crescent 2 (*see also* crescent 1, 3) pertaining to growth

supercrescent growing on something else

crescent 3 (*see also* crescent 1,2) *U.S.* any of numerous species of nymphalid butterfly of the genus Phyciodes

[crescent] a mutant (gene symbol *cr*) in linkage group I of *Habrobracon juglandis*. The phenotype has small eyes with crescent shaped ocelli.

cress any of numerous small cruciferous plants, either edible or supposed to be edible

bastard cress *Lepidium campestris*

bitter cress any of several species of the genus Cardamine

garden cress *Lepidium sativum*

hoary cress *U.S. Cardaria drabe*

Belle-Isle cress *U.S. Barbarea verna*

lake cress *Armoracia aquatica* (*cf.* horse radish)

marsh cress *U.S. Rorippa islandica* (*cf.* yellow cress)

penny cress *Brit. Thlaspi arvense*

rock cress any of several species of the genus Arabis

purple rock cress *Aubrieta deltoidea* and numerous horticultural varieties of this species

stone cress any of several species of the genus Aethionema

swine cress *U.S.* any of several species of the genus Coronopus

thale cress *Brit. Sisymbrium thalianum*

upland cress = winter cress

wall cress = thale cress

wart cress *U.S.* any of several species of the genus Coronopus

watercress (*see also* beetle, bug) *Brit.* and *U.S. Nasturtium officinale. U.S. Cardamine rotundifolia* (= **mountain watercress**)

winter cress any of several species of the genus Barbarea

yellow cress *Brit.* either *Armoracia amphibia* or *Nasturtium sylvestre U.S.* any of several species of the genus Rorippa (*cf.* marsh cress)

crest 1 (*see also* crest 3, 2) a conspicuous tuft of feathers or hair arising from the crown of the head

cnemial crest the flap of skin along the posterior edge of the limbs of crocodilian reptiles

cuticle crest a projection on the lower surface of some leaves

neural crest a band of cells just lateral to the line of closure of the neural folds in an embryo

nuchal crest the transverse crescentic crest drawn out from the supraoccipital and exoccipital bones above the foramen magnum

orbital crest the horizontally flat ridge of cartilage projecting above the orbit

sagittal crest the median dorsal crest of the skull

crest 2 (*see also* crest 1, 3) a ridge, or projection, rising from a surface

crest 3 (see also crest 1, 2) organisms distinguished by the possession of a crest

goldcrest Brit. the sylviid bird Regulus regulus (= golden crested "wren")

golden crest U.S. the amaryllidaceous herb Lophiola americana

bearded helmet-crest the trochilid bird Oxypogon guerinii

[Crest] a mutant (gene symbol Cr) in linkage group III of the domestic fowl. Phenotypic expression are crest and cerebral hernia

crestal one of the keeled and ridged bony scales running along the dorsum of the lizard Shinisaurus

cribellum 1 (see also cribellum 2) the network connecting the cells of Volvox

cribellum 2 (see also cribellum 1) a sieve-like plate such as that found on some insect mandibles or the abdomen of some spiders

-cribr- comb. form meaning "sieve"

cribrose pertaining to a sieve

    endocribrose lying within a sieve tube

Cricetidae A family of myomorph rodents containing all the indigenous New World rats and mice as well as the Old World hamsters and voles. They are distinguished from the Muridae in that the molars have only two rows of cusps

cricket properly any orthopteran insect with reduced forewings, or lacking wings, belonging to the families Gryllacrididae, Gryllidae, Gryllotalpidae, and Tridactylidae. However, the Mormon cricket is a tettigoniid and the water cricket is a hemipteran

camel cricket a name popularly applied to a variety of gryllacrids

cave cricket the gryllacrid Ceuthophilus gracilipes

coulee cricket U.S. the tettigonid Peranabrus scabricollis

field cricket U.S. any of numerous gryllid insects of the genus Acheta, particularly A. assimilis Brit. the gryllid Gryllus campestris

house cricket the gryllid Acheta domesticus

Jerusalem cricket any of several members of the gryllacricoid genus Stenopelmatus

mole cricket popular name of gryllotalpids

    African mole cricket U.S. Gryllotalpa africana

    northern mole cricket U.S. Gryllotalpa hexadactyla

    southern mole cricket U.S. Scapteriscus acletus

Mormon cricket any of several species of the tettigonioid genus Anabrus

pygmy sand cricket popular name of tridactylids U.S. without qualification usually Tridactylus apicalis

greenhouse stone cricket U.S. the gryllacrid Tachycines asynamorus

tree cricket popular name of gryllids, many of the genus Oecanthus

water cricket Brit. the hydrometrid hemipteran Velia currens

[Crimp] a mutant (gene symbol Cm) mapped at 43.5± on chromosome III of Drosophila melanogaster. The name refers to the crimping of the posterior edge of the phenotypic wing

-crin- comb. form meaning "hair". The Gr. derivatives "lily" and "separate" are better transliterated -krin-

crine pertaining to a gland

    apocrine said of a gland that loses part of its cytoplasm in producing its secretion

    ectocrine an external hormone (cf. pheromone)

    endocrine said of a gland that lacks ducts

    exocrine that which secretes externally, specifically a gland which secretes through a duct

enterocrinin a hormone secreted in the intestinal mucosa. It is active in controlling the rate of production and concentration of digestive enzymes in the alimentary canal

[crinkled] either a mutant (gene symbol ck) mapped at 53.0± n chromosome II of Drosophila melanogaster. The wings of the phenotype are crinkled and flimsy. Or a mutant (gene symbol cr) found in linkage group XIV of the mouse. The phenotypic expression involves the absence of guard hairs and zigzags from the pelt

[crinkly leaves] a mutant (gene symbol cf_1) mapped at 0 on linkage group III of Zea mays. The name is descriptive of the phenotype

Crinoidea the only extant pelmatozoan Echinodermata. This group, often called sea lilies, is either attached by an adoral stalk or moves freely by adoral "legs"

[Crisp] a mutant (gene symbol cr) in linkage group I of Neurospora crassa. The phenotypic expression is early uniform conidiation

crissum the area of soft feathers lying under the tail and coverts of a bird, and particularly those immediately surrounding the cloacal aperture

-crist- comb. form meaning "crest"

crista literally, a crest. Without qualification, usually applies to the spike-like, or sac-like, structures that project into the interior of a mitochondrion. The term is also applied to ridge-like membranes on the surface of some bacteria

-crit- comb. form meaning "select", "chosen", "notable" and better transliterated -krit-

crk see [crooked setae]

croaker any of several sciaenid fishes

    Atlantic croaker Micropozon undulatus

    black croaker Cheilotrema saturnum

    blue croaker Vacuoqua sialis

    Catalina croaker Ophioscion thompsoni

    spotfin croaker Roncador sternsi

    yellowfin croaker Umbrina roncador

    reef croaker Odontoscion dentex

    white croaker Genyonemus lineatus

crocatous = croceous

croceous saffron colored

crochet literally, a small hook and applied to numerous parts or organs with this general appearance, particularly some oligochaete setae, and to the sharply curved spines on insects

crock U.S. the larva of the dobson fly Croydalus cornutus (= hellgrammite)

crocker Brit. obs. the bird Larus ridibundus (= black-headed gull) U.S. (obs.) Branta bernicla (= brant)

crocodile (see also bird) properly any crocodilian reptile of the genus Crocodilus. In Brit. usage is also commonly applied to the alligator (q.v.) or, more rarely, caiman (q.v.)

    American crocodile C. acutus

    dwarf crocodile small crocodylid crocodilians of the W.Afr. genus Osteolaemus

    marsh crocodile C. palustris (= mugger)

    Nile crocodile Crocodilus niloticus

Crocodylia an order of diapsid reptiles containing the crocodiles and alligators. They differ from the Squamata by the fact that the cloacal opening is a longitudinal slit

Crocodylidae that family of crocodilian reptiles which contains the crocodiles, dwarf crocodiles, and false gavial

Crocoideae a subfamily of the Iridaceae

crocus properly, a large genus of iridaceous herbs but extended to many other monocotyledons

autumn crocus the liliaceous herb *Colchicum autumnale*

Chilean crocus either of two species of the amaryllidacous herbs of the genus Tecophilaea

clump of gold crocus *Crocus suisianus*

Dutch crocus *C. moesiacus*

saffron crocus *C. sativus*

Scotch crocus *C. biflorus*

crombec any of several sylviid birds of the genus Sylvietta (*cf.* nuthatch-warbler)

long-billed crombec *S. rufescens*

red-faced crombec *S. whytii*

green crombec *S. virens*

[crooked setae] a mutant (gene symbol *crk*) mapped at 60.1± on the X chromosome of *Drosophila melanogaster.* The name is descriptive of the phenotypic expression

crop 1 (*see also* crop 2, 3) a food reservoir, lying between the mouth and the stomach, usually produced as an enlargement of the esophagus, particularly in birds and insects

crop 2 (*see also* crop 1, 3) the yield of organisms produced by a given area

standing crop a measure of the community biomass of an ecosystem

crop 3 in the geologist's sense of "outcrop", applied to saxicolous plants

stonecrop *Brit.* any of numerous species of herb of the crassulaceous genus Sedum, particularly *S. acre*: *U.S.*, particularly *S. ternatum*

ditch-stonecrop the saxifragaceous herb *Penthorum sedioides*

crosnes the labiateous herb *Stachys sieboldii* and its edible root (= Japanese artichoke)

cross 1 (*see also* cross 2, 3) in the sense of chiasma (*q.v.*)

muscle cross connections between the outer and inner circular muscle layers in nemertines and some other forms

cross 2 (*see also* breed, cross 1, 3) used, usually colloquially, in the sense of breeding or its results

cross 3 (*see also* cross 1, 2) in the sense of cross-shaped, in compound names of organisms

Jerusalem cross = maltese cross

maltese cross the caryophyllaceous herb *Lychnis chalcedonica*

St. Andrew's cross the guttiferous herb *Ascyrum hypericoides*

widow's cross the crassulaceous herb *Sedum pulchellum*

crossing cross breeding

diallele crossing all possible crosses between different strains, *i.e.* two reciprocal between and two inbred

crossing-over = cross-over

Crossosomataceae a small family of rosale dicotyledonous trees distinguished by the bisexual flowers with three to five pistils and the multi-staminate androecium

Crossopterygii an order of sarcopterygiian bony fishes containing many fossil forms and the living Latimeria

cross-over a genetic term used to describe that which occurs when homologous regions are interchanged between two members of a pair of chromosomes

somatic cross-over an exchange of genetic material between homologous chromosomes in mitosis

[crossveinless] a mutant (gene symbol *cv*) mapped at 13.7 on the X chromosome of *Drosophila melanogaster.* The name refers to the appearance of the wings

[crossveinless-c] a mutant (gene symbol *cv-c*) mapped at 57.9 on chromosome III of *Drosophila melanogaster.* As the name indicates the phenotype lacks a posterior crossvein

[crossveinless-d] a mutant (gene symbol *cv-d*) mapped at 65.0± on chromosome III of *Drosophila melanogaster.* The phenotypic expression is an absence of the posterior cross vein

crotch the inside of the base of a "v" shaped structure

croton (*see also* bug) properly any of several species of the euphorbiaceous genus Croton. The "Crotons" of horticulture belong to the euphorbiaceous genus Codiaeum (= variegated laurel)

purging croton the euphorbiaceous tree *Croton tiglium*

crotonase = enoyl-CoA hydratase

croup the postero-dorsal region of a horse immediately anterior to the tail

crow 1 (*see also* crow 2, 3, berry, foot, pheasant, poison) properly, any˙of numerous species of passeriform bird of the family Corvidae particularly those of the genus Corvus. Without qualification *U.S. C. brachyrhynchos, Brit. C. frugilegus* (= rook), *Austral. C. cecilae W.I. C. nasicus* (= Cuban crow)

large billed crow *C. macrorhynchos*

slender billed crow *C. enca*

black crow *S.Afr. C. capensis*

carrion crow *C. corone*

common crow *C. brachyrhynchos*

Cuban crow *C. nasicus*

fish crow *C. ossifragus*

hooded crow *C. cornix*

house crow *C. splendens*

Mexican crow *C. imparatus*

little crow *Austral. Corvus bennetti*

jabbering crow = Jamaican crow

Jamaican crow *C. jamaicensis*

white-necked crow *C. leucognaphalus*

northwestern crow *C. caurinus*

palm crow *C. palmarum*

pied crow *Afr. Corvus albus*

crow 2 (*see also* crow 1, 3) numerous birds other than corvids having the appearance or habits of crows. *W.I.*, without qualification, the cathartid *Cathartes aura* (= U.S. turkey vulture)

carr crow *Brit. obs.* for tern

carrion crow *W.I.* = crow *W.I.*

fruit crow any of several large cotingids

bare-headed crow common name of timaliids of the genus Picathartes, at one time placed in the family Picathartidae (*q.v.*)

North Island crow *N.Z.* the callaeid *Callaeas wilsoni*

South Island crow *N.Z. C. cinerarea*

john crow *W.I.* the cathartid *Cathartes aura* (= U.S. turkey vulture)

king-crow the dicrurid *Dicrurus adsimilis*

piping crow any of several species of laniid bird of the genus Gymnorhina

rain-crow *W.I.* any of numerous cuculids

long-tailed crow *W.I.* the cuculid bird *Crotophaga ani*

blue-wattled crow the callaeid *Callaeas wilsoni* (= kakako)

orange-wattled crow the callaeid *Callaeas cimerae* (= kokako)

crow 3 (*see also* crow 1, 2) *Austral.,* and *I.P.* any of numerous species of danaid lepidopteran insects of the genus Euploea

common Australian crow *E. corinna*

brown crow *Austral. E. confusa*

crown 1 (*see also* crown 2) in the sense of a thickened annulus

leaf crown the elongate lobular lips of many strongyloid nematodes

medullary crown = medulla sheath

pharynx crown a thick cuticular ring, projecting from the anterior end of the pharynx of a Kinorhynch into the buccal cavity

crown 2 (*see also* beard, crown 1) an organism in the form of, or bearing a, crown

azure crown any of several trochilid birds of the genus Amazilia

blossom-crown the trochilid bird *Anthocephala flaviceps*

greenbacked firecrown the trochilid bird *Sephanoides sephanoides*

gold crown any of numerous flowers of the carduacean genus Gorteria

Nero's crown the apocynaceous shrub *Erratamia coronaria*

crown-of-thorns the euphorbiaceous shrub *Euphorbia splendens*

crozier literally, a shepherd's crook, but in *bot.,* commonly applied to the unfolding fern's frond

cr-u *see* [cream-underscored]

-cruc- *comb. form.* meaning "cross"

cruciate crossed

Cruciferae that family of rheodalous dicotyledons which contains, *inter alia,* the cabbages, the mustards, the radishes, the wallflowers and woad. The four sepals and four petals arranged in the form of a cross are typical of the family

recrudescence in *bot.,* the production of a new flower-bearing shoot from an inflorescence in fruit

cruentatous blotched with red

cruentous blood red

cruiser *Austral.* the nymphalid lepidopteran insect *Cynthia arsinoe I.P. C. erota*

crumen a pouch

[crumpled] a mutant (gene symbol *cmp*) mapped at 93.0± on chromosome II of *Drosophila melanogaster.* The phenotypic wing is small and crumpled

leaf crumpler *U.S.* the larva of the phycitid lepidopteran insect *Acrobasis indigenella*

-cruor- *comb. form* meaning "blood"

cruorin a general term for "blood pigment" at one time synonymous with hemoglobin

chlorocruorin a green respiratory pigment found in some polychaete annelids. It resembles hemoglobin in having iron linked to a porphyrin prosthetic group

erythrocruorin large hemoglobin-like molecules in the blood of many oligochaete and polychaete annelids and various mollusks

crura properly the plural of crus but specially the brachidium of brachiopods when it consists of a simple fork

quadricrural supported by the four corners, or on four legs

cruralia the brachidium of brachiopods when the arms of the crus are united

bicrurous two legged

-crus- *comb. form* meaning "shell" in the sense of toughened integument

crus a leg, in humans technically confined to portions distal to the knee

Crustacea a large class of the phylum Arthropoda containing, *inter alia,* the crabs, lobsters, shrimps, beach hoppers, sow bugs, barnacles, water fleas, and allied forms. They may be distinguished from other arthropod classes by the possession of two pairs of preoral antennae and most possess at least three pairs of postoral appendages functioning as jaws. Almost all are aquatic

crustose having the appearance of a crust. Said particularly of lichens closely adpressed to a rocky surface in distinction from foliose

ecrustose used of a lichen that lacks a thallus

crustulinous toast-colored

-cry- *comb. form* meaning "cold"

crymad a plant of the "polar barrens"

crymnion = cryoplankton

-crypt- *comb. form* meaning "hidden" and thus, by extension, a "cavity" or "vault". Better, but rarely, transliterated -krypt-

crypt any deep cavity or depression. Specifically in *bot.* the front cavity of a stoma

crypt of Lieberkuhn a tubular epithelial gland in the jejunum and ileum

Crypteroniaceae a small family of myrtiflorous dicotyledonous angiosperms found in Malaysia

cryptic concealed or concealing

allocryptic said of organisms that use a covering of other organisms or inanimate material to conceal themselves

Cryptocephala a subclass of polychaete annelids in which the prestomium is hidden within the peristomium. The tentacles are reduced but the palps are greatly expanded to form a crown of gills. All are tubiculous

Cryptochaetidae a family of brachyceran dipteran insects commonly included in the Ochthiphilidae

Cryptodira an assemblage of chelonian reptiles containing those families in which the retractile neck bends in a vertical S-shaped curve

cryptodirous the motion of a turtle which withdraws its head by bending the neck in a vertical plane

Cryptogam a broad, ill-defined plant taxon embracing all plants other than spermatophytes

Cryptogamia in the Linnaean system of plant classification the group that contained all those plants in which the existence of stamens and pistils could not be detected

Cryptomonadida an order of mastigophorous Protozoa distinguished by the presence of a cuticle and variously colored chromoplasts. Since they produce cellulose many consider them to be algae, in which case they correspond to the order Cryptomonadales in the class Cryptophyceae, a group of uncertain botanical affinities

Cryptonemiales an order of tetrasporophytic floridian Rhodophyta with an auxiliary cell borne in a special filament of the gametophyte

Cryptophagidae a family of small coleopteran insects commonly called silken fungus beetles, in view of their dense pubescence, but many of which occur under bark or on flowers

Cryptoproctinae a monotypic sub-family of viverrid carnivorous mammals erected to contain the fossa (*q.v.*)

Crypturidae = Tinamidae

CSF *see* cerebrospinal fluid

CSH = cysteine

chs *see* [Cushion]

csk *see* [costakink]

ct *see* [cut]

-cten- *comb. form* meaning "comb"

ctene the plate of fused cilia in the form of a comb from which the Ctenophora derive their name

ctenidium any comb-like structure and specifically a comb-shaped spine found in many insects, best demonstrated in the Siphonaptera

Ctenizidae a family of mygalomorph spiders distinguished by the rastellum on the basal joint of the Chelicera

Ctenodactylidae a small family of hystricomorph rodents distinguished by possessing a comb-like mass of strong bristles on the hind feet. They are popularly called gundis

ctenoid comb-like

Ctenomyidae a family of small, fossorial hystricomorph rodents containing the forms popularly called tucotucos

Ctenophora a phylum of mesozoan animals containing, *inter alia*, the sea gooseberries. They were once united with the Cnidaria as the Coelenterata but differ from the Cnidaria in the absence of nematocysts and the presence of mesenchymal muscles. The eight rows of ciliary plates from which their name derives is typical

Ctenostomata an order of gymnolaemate Ectoprocta distinguished by the membranous zoecia the terminal orifice of which has a closing apparatus

Ctingidae a family of pssseriform birds commonly referred to as cotingas or umbrella birds, because of the huge umbrella-like crest borne by many species

CTP = cytidine triphosphate

cu *see* [curled]

Cu *see* [Curl] and [Curly]

cu^r a yeast genetic marker indicating inhibition of growth by copper

cuataquil either of two species of procyonine carnivore of the genus Bassaricyon

cub term applied to the young of many carnivores, particularly those of the seal, bear, wolf, lion, and leopard. The term is also applied to the young of many other animals, such as the shark or even the woodchuck

cubbyu the sciaenid fish *Equetus acuminatus*

cubital (*see also* vein) pertaining to the 4th and 5th secondaries of a bird's wing

 aquincubital said of a bird wing with a gap between the 4th and 5th secondaries

 quincubital said of a bird's wing that lacks a gap between the 4th and 5th secondaries (= eutaxic)

cubitus the fifth longitudinal vein of the insect wing

[cubitus-interruptus] a mutant (gene symbol *ci*) mapped at 0 on chromosome IV of *Drosophila melanogaster*. The phenotypic expression is an interruption of the vein L4

Cubomedusae an order of scyphozoan coelenterates distinguished by a generally cubical form

cuchi the Indo-Burmese osteichthyid fish *Amphitnous cuchia*. Usually placed in the Synbranchidae even though it possesses lung-like air sacs which connect with the gill cavity

cuckold any of numerous species of compositaceous herb of the genus Bidens (= beggar's tick)

cuckoo (*see also* bee, bird, dove, falcon, fish, flower, shrike and wasp) any of numerous species of the family Cuculidae. *Brit.*, without qualification, *Cuculus canorus*. In *Austral.* the strigid *Ninox novaeseelandiae* (= *N.Z.* morepork) is called cuckoo

 African cuckoo *Cuculus canorus*

 bay cuckoo *Penthoceryx sonneratii*

 chestnut-bellied cuckoo *W.I. Hyetornis pluvialis*

 channel-billed cuckoo *Scythrops novaehollandiae*

 black-billed cuckoo *Coccyzus erythropthalamus*

 thick-billed cuckoo *Pachycoccyx audeberti*

 yellow-billed cuckoo *Coccyzus americanus*

 black cuckoo *Cuculus clamosus*

 bronze cuckoo *Chalcites basalis*

 fire-chested cuckoo = red-chested cuckoo *Cuculus solitarius*

 crested cuckoo any of several species of the genus Clamator

 pied crested cuckoo *C. jacobinus*

 brush-winged crested cuckoo *C. coromandus*

 diederik cuckoo *S.Afr. Chrysococcyx cupreus* (= emerald cuckoo)

 drongo cuckoo *Surniculus lugubris* (*cf.* drongo)

 black-eared cuckoo *Misocalius osculaus*

 emerald cuckoo *I.P. Chalcites maculatus Afr. Chrysococcyx cupreus*

 hawk cuckoo any of several species, considered to be more hawk-like than most, of the genus Cuculus. Usually without qualification, *C. sparverioides*.

 Asiatic hawk-cuckoo *C. fugax*

 common hawk-cuckoo *C. varius*

 large-heeled cuckoo any of numerous species of the genus Centropus (= coucal)

 Himalayan cuckoo = oriental cuckoo

 Indian cuckoo *Cuculus micropterus*

 Japanese cuckoo *Cuculus canorus* = *Brit.* cuckoo

 lesser cuckoo *Cuculus poliocephalus*

 lizard-cuckoo either of two species of the genus Saurothera

 mangrove cuckoo *Coccyzus minor*

 oriental cuckoo *Cuculus saturatus*

 pallid cuckoo *Cuculus pallidus*

 pheasant-cuckoo *Dromococcyx phasianellus*

 pied cuckoo *Clamator jacobinus*

 plaintive cuckoo *Cacomantis merulinus*

 shining cuckoo *N.Z. Chalcites lucidus Austral. C. malayanus*

 great spotted cuckoo *Clamator glandarius*

 squirrel cuckoo any of several species of the genus Piaya. Usually, without qualification, *P. cayana*

 striped cuckoo *Tapera naevia*

 fan-tailed cuckoo *Cacomantis pyrrophanus*

 long-tailed cuckoo *N.Z. Urodynamis taitensis W. Afr.* any of several species of the genus Cercococcyx. Usually *C. mechowi*

 sand-tailed cuckoo *Cacomantis variolosus*

 violet cuckoo *Chalcites xanthorhynchus*

Cucujidae a family of extremely flattened beetles that utilize their shape to function as predators on other small arthropods under the bark of trees. Commonly called flat bark beetles

Cucujoidea a huge superfamily of polyphagous coleopteran insects usually regarded as a dumping ground for those families that cannot be placed elsewhere

Cuculidae a large family of cuculiform birds containing such improbable allies as the cuckoo and the road runner. They are distinguished by their strong heavy beaks and bare orbital skin. Almost all have conspicuous eyelashes and very long tails

Cuculiformes that order of birds which contains the cuckoos and their allies. They are distinguished by their more or less decurved bill lacking a cere and with imperforate nares

cucullate hooded

cucullus the terminal part of the valve of the genitalia of some male Lepidoptera

cucumber 1 (*see also* cucumber 2, beetle, root, tree) the cucurbitaceous vine *Cucumis sativus* and its edible fruit

　bitter cucumber the cucurbitaceous vine *Citrullus colocynthus* (= colocynth)

　bur cucumber any of several species of cucurbitaceous vine of the genus Sicyos, particularly *S. angulatus*

　mandera cucumber *Cucumis sacleuxii*

　squirting cucumber the cucurbitaceous vine *Ecballium elaterium*

　wild cucumber any of several species of cucurbitaceous vine of the genus Echinocystis

cucumber 2 (*see also* cucumber 1) organisms resembling cucumbers

　sea cucumber any holothurian echinoderm (*cf.* bêche de mer)

Cucurbitaceae that family of cucurbitale dicotyledons which contains the gourds, squashes, melons and cucumbers. Apart from the well known fruits (*see* pepo) the family is distinguished by the scandant habit, unisexual gametopetalous flowers, tricarpellale parietal placentation and, usually, the presence of tendrils

Cucurbitales an order of dicotyledenous angiosperms distinguished from the related Campanulales by the unisexual flowers, tricarpellate ovary and the exendospermous seeds

Cucurbiteae a subfamily of the Cucurbitaceae

cud (*see also* weed) the bolus of a ruminant animal particularly the domestic cow

cui-ui the catastomid fish *Chasmistes cujus* (*cf.* sucker 2)

Culicidae a large family of nematoceran dipteran insects called mosquitoes when they are blood sucking and phantom midges when they are not

culm a stem, such as that of a grass, which consists of hollow sections interrupted by solid nodes

culmen a ridge, specifically that which runs along the back of a caterpillar and that which runs along the upper border of a bird's maxilla

cultellus literally a dagger, but used specifically of the blades in the mouthparts of biting Diptera

cultigen an organism known only in cultivation or domestic association not yet recognized anywhere as native or indigenous

cultivar a plant variant found only under cultivation and maintained by asexual propogation or controlled breeding

cultrate knife like

culture a laboratory-produced association of organisms under controlled conditions

　stab culture a culture produced by stabbing the innoculating instrument into the solid medium. Descriptive terms for stab cultures are mostly self-explanatory (*e.g.* arborescent, filiform) though **villose** and **plumose** differ only in the thickness of the lateral branches and **echinulate** is used for a rather thick filiform stab culture having areas of prominences coming from it

　streak culture a culture produced by streaking the surface of a solid culture medium without breaking

the surface. Most of the descriptive terms are self-explanatory (*e.g.* filiform, rhizoid) but **effuse** is used for a streak that is diffuse around the edges

-cuma- *comb. form* meaning "a wave" better transliterated -kuma-

Cumacea an order of eumalacostracan crustacea distinguished from the amphipods and isopods by the presence of a carapace and from the Mysidacea by the fact that the eyes are sessile and all the pereiopods are not biramose

-cumben- *comb. form* meaning "to lie down"

decumbent (*see also* cotyledon) lying on the ground with the tip erect

　dorsicumbent = supine

　incumbent said of a digit that is so placed that the whole length rests on the ground when the animal is standing

　procumbent lying along the ground

　ventricumbent prone

cumulus oophorus a mound of follicle cells surrounding the oogonium

-cun- *comb. form* confused from two roots meaning respectively "wedge" and "cradle"

cuneate wedge shaped

cuniculate *zool.* resembling a rabbit. *bot.* resembling the burrow of the European rabbit, that is a tube with an opening only at one end

cuniculine dwelling in burrows

cunner *U.S.* the labrid fish *Tautogolabrus adspersus*

Cunoniaceae a family of planale dicotyledons. They are trees or shrubs of South America or Australasia and, apart from their geographic distribution, are distinguished by the internal carpel margins recurved or distinct, and by the usually arborescent habit

cup 1 (*see also* cup 2) any hemispheroid depression, body or organ. In *bot.* used for hypanthium

　bud cup = cyathus

　dorsal cup the calyx of crinoid echinoderms

　honey cup = nectary

　pigment cup the protruberant pigmented area surrounding the eye of many Mollusca

cup 2 (*see also* cup 1, plant) an organism distinguished by cup-shaped parts

　buttercup (*see also* crowfoot, shell) any of numerous flowering plants of the family Ranunculaceae, and specifically of the genus Ranunculus. A few similar plants of other families are called by this name

　　Bermuda buttercup the oxalidaceous herb *Oxalis cernua*

　　common buttercup *U.S. R. acris*

　　prairie buttercup *U.S. R. rhomboides*

　forefather's cup any of several species of sarraceniaceous insectivorous herbs of the genus Sarracenia

　huntman's cup = forefather's cup

　Indian cup the compositaceous herb *Silphium perfoliatum*

　king cup *U.S.* the ranunculaceous herb *Caltha palustris* (= marsh marigold)

　leafcup any of several species of compositaceous herb of the genus Polymnia

　　Canada leafcup *P. canadensis*

　painted cup *U.S.* any of several species of scrophulariaceous herb of the genus Castilleja

　　marsh painted cup the scrophulariaceous herb *Bartsia viscosa*

　scarlet cup *U.S.* the pezzizale ascomycete fungus *Sarcoscypha coccinea*

　toothcup *U.S.* the lythraceous herb *Rotala ramosior*

cupang the leguminous tree *Parkia timoriana*

Cupedidae = Cupesidae

Cupesidae a small family of coleopteran insects distinguished by the dense scales that cover them

cupid *Austral.* the lycaenid lepidopteran insect *Euchrysops cnejus. I.P.* any of numerous species of lycaenid lepidopteran insect of the genus Everes

Cupidae = Cupesidae

Cupidinidae = Lycaenidae

cup-of-flame the papaveraceous flower *Eschscholtzia californica* (= california poppy)

cup-of-gold the solanaceous shrub *Solandra guttata*

Cupressaceae a family of coniferous spermatophytes containing the forms commonly called cypresses and junipers. They are distinguished by having the cone scales opposite or whorled with small, and frequently scale-like, leaves

cupula 1 (*see also* cupula 2) small, cup shaped, dense cytoplasmic inclusions disclosed by electron microscopy in many Protozoa

cupula 2 (*see also* cupula 1) a domed mass of material but applied particularly to the dome-shaped mucous protuberances from the lateral line sense organs of fish

cupule literally, a little cup. Particularly, that formed of adherent bracts as the cup of an acorn

cupreous with the color and lustre of copper

cupresoid resembling a cypress

-cupul- *comb. form* meaning "cask"

Cur *see* [Curl]

curassow any of several cracid birds of the genera Mitu, Pauxi, and Crax

 razor-billed curassow any of several species of the genus Mitu; without qualification, *M. mitu*

 great curassow *C. rubra*

 helmeted curassow *P. pauxi*

curculio *U.S.* term frequently used to distinguish members of the family Curculionidae from other weevils. However, the distinction is not absolute since many curculionids have the popular name of weevil

 apple curculio *U.S. Tachypterellus quadrigibbus*

 cabbage curculio *U.S. Ceutorhynchus rapae*

 grape curculio *U.S. Craponius inaequalis*

 butternut curculio *U.S. Conotrachelus juglandis*

 cow pea curculio *U.S. Chalcodermus aeneus*

 plum curculio *U.S. Conotrachelus nenuphar*

 quince curculio *U.S. Conotrachelus crataegi*

 rhubarb curculio *U.S. Lixus concavis*

 clover root curculio *U.S. Sitona hispidulus*

 rose curculio *U.S. Rhynchites bicolor*

 cabbage seedstalk curculio *U.S. Ceutorhynchus quadrideus*

Curculionidae a very large family of curculionoid coleopteran insects containing the greater number of the true weevils. Sometimes called snout-beetles

[Curl] *either* a mutant (gene symbol *Cu*) mapped at 38 on linkage group I of the tomato. The phenotypic leaves are curled. *or,* a mutant (gene symbol *Cu*) mapped at 55.2± on chromosome II of *Drosophila melanogaster.* The phenotypic wing is curled and folded *or* a mutant (gene symbol *Cur*) mapped at 66.0± on chromosome III of *Drosophila melanogaster.* The phenotypic expression is curly wings

[curled] a mutant (gene symbol *cu*) mapped at 50.0 on chromosome III of *Drosophila melanogaster.* The phenotypic expression is a dark body with upcurved wings

curlew (*see also* berry, bug) any of several charadriiform birds mostly scolopacids of the genus Numenius (*cf.* sandpiper)

 Australian curlew *N. madagascariensis*

 long-billed curlew *N. americanus*

 slender-billed curlew *N. tenuirostris*

 common curlew *N. arquata*

 Eskimo curlew *N. borealis*

 Hudsonian curlew *N. phaeopus* (= *Brit.* whimbrel, jack curlew)

 jack curlew *Brit.* = *U.S.* Hudsonian curlew

 stone curlew any of several burhinids (= thickknee)

[curlex] a mutant (gene symbol *cx*) mapped at 13.6 on the X chromosome of *Drosophila melanogaster.* The phenotype has upwardly bent wings

blue curls any of several species of labiateous herb of the genus Trichostema

[Curly] *either* a mutant (gene symbol *Cy*) mapped at 7.0 on chromosome II of *Drosophila melanogaster.* The name refers to the upcurled wings of the phenotype *or* a mutant (gene symbol *Cu*) mapped at 4 on linkage group II of the rat. The phenotype is essentially the same as [shaggy] (*q.v.*)

currant 1 (*see also* currant 2, 3, aphis, borer, girdler, mite, moth, sawfly, weevil, worm) properly any of several species of saxifragaceous shrub of the genus Ribes

 alpine currant *R. alpinum* (= *U.S.* mountain currant)

 American black currant *R. americanum*

  California black currant *R. bracteosum*

  European black currant *R. nigrum*

  swamp black currant *U.S. R. lacustre*

 buffalo currant *U.S. R. odoratum*

 fetid currant = skunk currant

 golden currant = *R. aureum.* The term is also sometimes applied to the buffalo currant

 Missouri currant = buffalo currant

 mountain currant *U.S.* = alpine currant

 red currant *R. sativum,* and numerous hybrids and horticultural varieties of this species

  northern red currant *U.S. R. rubrum*

  swamp red currant *U.S. R. triste*

 skunk currant *U.S. R. glandulosum*

currant 2 (*see also* currant 1, 3) any of several small shrubs and trees with berry-like fruits

 Indian currant the caprifoliaceous shrub *Symphoricarpos orbiculatus* and its edible berries

currant 3 (*see also* currant 1, 2) small, dried, usually black grapes

currawong any of several cracticid birds of the genus Strepera

 black currawong *S. fuliginosa*

 grey currawong *S. versicolor*

 pied currawong *S. graculina*

decurrent running down. in *bot.* said of a leaf petiole when basally adnate to the stem

 [excurrent eunning out from or in *bot.,* running up and through the apex

percurrent running from the base to the apex

Cursoria an order of insects combining the Dictyoptera with the Phasmitodea

Cursoriidae a family of charadriiform birds usually included with the Glareolidae

cururo the octodont hystricomorph rodent *Spalacopus poeppigi*

Darwinian curvature bending induced at the apex of the root by mechanical stimulation

 Sach's curvature curvature resulting from a differential growth rate in the two sides of a root

decurved recurved in a downward direction

 incurved bent inwards

 recurved curved back on any axis

[curved] a mutant (gene symbol *c*) mapped at 75.5 on chromosome II of *Drosophila melanogaster*. The term refers to the spread lifted wings of the phenotype

cuscus popular name of nocturnal phalangerine phalangerids of the genus Phalanger

Cuscutaceae a family erected to contain the genus Cuscuta (the dodders) by those who do not wish to retain them in the Convolvulaceae

cushat *Brit. obs.* = pigeon

callus **cushion** a pad covering a pit on the side of a sieve-tube

blue pincushion the brunonaceous herb *Brunonia australis*

Robin's pincushion *Brit.* the gall formed by the cynipid hymenopteran insect *Rhodites rosae*

[Cushion] a mutant (gene symbol *csh*) in linkage group I of *Neurospora crassa*. The phenotypic expression is a diminished colonial growth

cusimanse any of several species of herpestine viverrid carnivores (usually called mongooses) of the genus Crossarchus

cusk the gadid fish *Brosme brosme*

bastard cusk the marine teleost *Cryptacanthodes maculata* (= wrymouth)

cusp a point or pointed structure. In *zool.* usually applied to substructures on teeth and in *bot.* to fronds

cuspid pointed

bicuspid having two sharp points (*cf.* tooth)

Cuspidariidae a family of pelecypod mollusks commonly called dipper shells by reason of the long elongation of the posterior end of the valves

cuspidate used in *bot.* to describe a structure with the apex abruptly rounded and ending in a elongated, sharply pointed, tip

cut in the sense of a slit

gill cut one of the slits in the peristomial test of echinoids that bears a gill

Cut see [Cut]

[cut] *either* a mutant (gene symbol *ct*) mapped at 20.0 on the X chromosome of *Drosophila melanogaster*. The phenotype has sharply pointed scalloped wings *or* a mutant (gene symbol *ct*) in linkage group I of *Habrobracon juglandis*. The outer margin of the phenotypic wing appears to be cut

[Cut] a mutant (gene symbol *Cut*) on linkage group I of *Neurospora crassa*. The term refers to the truncated appearance of a tube culture of the phenotype

cut-and-come-again any of several species and numerous horticultural varieties and hybrids of the compositaceous genus Zinnia

cute *Brit. obs.* = coot

Cuterebridae a family of large mostly tropical myodarian cycloraphous flies closely allied to the Hypodermatidae. The majority are parasitic on rodents

cuticle the dead outer layer of an integument

endocuticle the inner layer of the cuticle

epicuticle the outer layer of the cuticle

Cutleriales an order of isogenerate Phaeophytae distinguished by the posession of a flattened, or blade-like, thallus

cutter in the sense of an organism that cuts

cane cutter *U.S.* the leporid lagomorph *Sylvilagus aquaticus* (= swamp rabbit)

waterlily leaf cutter *U.S.* the larva of the pyraustid lepidopteran *Synchita obliteralis* (Walker)

maple leaf cutter *U.S. Paraclemensia acerifoliella* (Fitch)

plantcutter any of three species of phytotomid birds of the genus Phytoma

Cuverian *epon. adj.* from Georges Leopold Chrétien Frederic, Baron Cuvier (1769–1832)

cv see [crossveinless]

cv-c see [crossveinless-c]

cv-d see [crossveinless-d]

cx see [curlex]

Cy see [Curly]

cyaceous very pale gray

-cyan- *comb. form* meaning "dark blue" better transliterated -kyan-

cyanaecous corn flower blue

hemocyanin *see under* hemo-

Cyanastraceae a small family of farinose monocotyledonous angiosperms

Cyanophyceae = Myxophyceae

Cyanophyta a subphylum of algae containing those forms commonly called the blue-green algae. They are distinguished by the lack of definite chromatophores and by the undifferentiated "nucleus"

-cyath- *comb. form* meaning a "deep cup" or "goblet" (*cf.* -calyc-)

Cyatheaceae a family of filicale Pterophyta containing the forms commonly called tree ferns; the habit of growth indicated by this name is typical. The family is distinguished from the Dicksoniaceae by the superficial sori dorsal to the veins

cybele an ecologically annotated flora

Cycadaceae a family of gymnospermous plants with palm like foliage, short thick usually unbranched woody trunks, and very numerous microsporangia scattered over the microsporophylls

Cycadales an order of Gymnospermae that contains the family Cycadaceae. The characteristics of the only family distinguish the order

Cycadioideae a subfamily of Cycadacea containing the single genus from which the name is derived

-cycl- *comb. form* meaning "a circle" and every possible derivative of this word. Better translated -kvkl-

Cyclanthaceae a family of synanthaceous monocotyledons intermediate between the palms and the arums. The distinguishing characteristics are the palm-like foliage, the pistillate flowers borne abundantly on a fleshy spadix, each with numerous parietally disposed ovules and the infloresence enclosed by many deciduous spathe-like bracts

Cyclanthales = Synanthae

Cyclanthereae a subfamily of the Cucurbitaceae

Cyclarhidae a small family of passeriform birds erected to contain the two pepper-shrikes of Central and South America. They are distinguished by their olive-green loose-webbed plumage and short-hooked, heavy, laterally compressed bill

cycle 1 (*see also* cycle 2–4) a circular arrangement of parts and particularly one complete spiral of a phyllotactic arrangement

contact cycle a phyllotactic cycle in which the members overlap each other

leaf cycle one complete spiral in phyllotaxis

hemicycle partly cyclic, as of plants one part of the flower of which is arranged in whorls and the other in spirals

pericycle the sheath of cells between the endoderm and the stele

cycle 2 (*see also* cycle 1,3,4) any rhythmic phenomenon of more or less uniform periodicity

life cycle the sum total of the changes through which an organism passes between the fertilization of an egg and the production of another mature egg. The term is sometimes used as representing the sum total of the change between any two periods in the life history, such as that which intervenes between one type of larva and the subsequent production of a similar type

oestrous cycle the hormone-induced changes that prepare the uterine epithelium for the implantation of a developing mammal

anoestrous cycle the period between eostrous cycles

dioestrous cycle the period between oestrous cycles

anovulatory cycle the occurrence of menstruation without ovulation

seasonal cycle rhythmic cycles caused by, or associated with, the procession of the equinoxes

sex cycle periodic appearance and regression of sex specific characters, usually associated with the breeding season

cycle 3 (*see also* cycle 1,2,4) the biosphere is divided into three environmental **biocycles** known as the marine, the fresh-water, and the terrestrial

cycle 4 (*see also* cycle 1–3) a repetitive metabolic process

arginine cycle *see* urea cycle

biochemical cycle the circulation through organisms of chemical elements in the biosphere

carbon cycle the cycle through which carbon, changed to a carbohydrate by synthesis, ultimately reappears in the atmosphere as carbon dioxide

citric cycle *see* tricarboxylic cycle

Cori cycle the cycle of blood glucose to muscle glycogen to blood lactic acid to liver glycogen

energy cycle a general term for the successive processes of anabolism and catabolism

glyoxalate cycle a fat to carbohydrate mechanism involving the entry of glyoxalate (from iso citrate) into the tricarboxylic acid cycle. Fatty acids are then utilized to yield malate as the precursor of oxalocetic acid

Krebs cycle *see* tricarboxylic acid cycle

Krebs-Henseleit cycle *see* urea cycle

nitrogen cycle a loose term covering the fixation of atmospheric nitrogen (either by biological or physical means), the synthesis of organic compounds from this nitrogen and the final return of the gas to the atmosphere as the end product of decomposition

ornithine cycle *see* urea cycle

tricarboxylic cycle the complex metabolic cycle starting with the production of citric acid from pyruvic acid through acetyl-CoA ultimately ending in the production of citric acid through oxaloacetic acid

urea cycle the cyclic formation of urea by the hydrolysis of arginine yielding ornithine, following the synthesis of arginine from citrulline

cyclic pertaining to, or consisting of, cycles. In *zool.* specifically that type of foraminiferan shell in which the chambers are arranged in concentric circles. In *bot.* refers to parts disposed in whorls about an axis as opposed to a spiral arrangement

acyclic lacking a cyclic arrangement, used particularly of flowers in which the parts of the corolla are arranged in a spiral

dicyclic having a double cycle, either in time, as biennial plants or in space as when organs occur in two whorls. The term is also applied to those crinoids in which there are *three* cycles of plates in the calyx

cryptodicyclic said of crinoids in which the innermost ring of the dicyclic type (*q.v.*) have been lost or reduced to minute proportions

ectocyclic literally, outside the circle or ring, commonly used in *bot.* to designate structures that are external to, for example, the stele or the corolla

endocyclic said of those echinoid echinoderms in which the periproct is enclosed within the apical system and of crinoids in which the mouth is more or less central on the disc

eucyclic said of flowers with successive cycles that contain equal numbers of parts

exocyclic said of those echinoid echinoderms in which the periproct is not enclosed by the apical system and of those crinoids in which the mouth is displaced to the periphery of the disc

heterocyclic said of a flower when successive whorls of its parts do not consist of equal numbers

holocyclic the condition of a stem that is completely sheathed by the base of the leaf petioles (*cf.* amplexicaul)

isocyclic = eucyclic

mericyclic occupying only a part of the diameter

monocyclic having only a single whorl of each of the members of a floral series. The term is sometimes applied to annual plants. It is also used for those crinoids (*cf.* dicyclic) in which there are *two* cycles of plates in the calyx

pseudomonocyclic = cryptodicyclic

polycyclic said of a flower in which there are many series of cycles or of a stem which has several steles or vascular strands

Cyclobeae a subfamily of Chenopodiaceae distinguished from the Spirolobeae by having an annular embryo

cyclome a ring shaped mass of anthers

Cyclophyllidea = Taenioidea

Cyclopoidea an order, or suborder (Cyclopoida) of copepod crustacea distinguished by the fact that the fifth thoracic segment forms a moveable articulation with the sixth. Many fresh-water and marine planktonic forms have the typical cyclopoid shape but a number are modified as parasites

cyclops *I.P.* the satyrid lepidopteran insect *Erites falcipennis*. The word is also the name of a widely distributed genus of fresh water copepods

Cyclopsittacidae = Psittacidae

Cyclopteridae a family of scleroparous Osteichthyes distinguished from the closely allied liparids by the very much larger size of the tubercules on the back and the globose rather than elongated body. Commonly called lump suckers or lump fishes

Cyclorhidae = Cyclarhidae

Cyclorrhapha an assemblage of those brachyceran dipteran insects the adults of which emerge from the pupa through a circular opening at one end (*cf.* Orthorrhapha)

cyclosis the regular movement along apparently predetermined paths of food vacuoles in some ciliate Protozoa, and the streaming of protoplasm in plant cells

Cyclosporeae an order of phaeophyte algae in which there is no alteration of generations

Cyclostomata 1 (*see also* Cyclostomata 2) an order of gymnolaematous Ectoprocta distinguished from other extant orders by their calcareous skeleton

Cyclostomata 2 (*see also* Cyclostomata 1) = Marsipobranchii

Cycnodoidea a superfamily of lepidopteran insects containing some minute moths most of which are known as leaf miners

Cydippidea an order of tentaculate ctenophores of simple oval form with two branched retractile tentacles

Cydnidae a family of hemipteran insects commonly called burrower bugs. They are distinguished from the Pentatomidae which they strongly resemble by the spiny tibia

cyesis the period, or the length of the period, between fertilization and birth in a viviparous animal

cygnet a young swan

cygneous resembling a swan, particularly when curved like the neck of a swan

Cyladidae a family of brentoid coleopteran insects with an extremely long thin body giving them in general an ant-like appearance

cylinder in the sense of stele (*q.v.*) or vascular bundle (*q.v.*)

    hadrome cylinder that part of the vascular bundle or stele which carries water and supposedly similar tissue in a moss stem

    vascular cylinder a term preferred by some botanists to stele and somewhat broader in its concept

Cylindrocapsaceae that family of urotrichale Chlorophyceae in which the cells of the unbranched filaments have concentric stratified walls

Cylindrotomidae a family of nematoceran Diptera now usually included with the Tipulidae

-cym- *comb. form* meaning "wave" better, and not infrequently, transliterated -kym-

Cymatiidae a family of gastropod mollusks containing those forms commonly called tritons. They are rugged, frequently spiny shelled, distinguished from the closely allied Muricidae by the fact that there are never more than two varices to a volution

cymatium = apothecium

Cymatophoridae a family of lepidopteran insects closely allied to the Geometridae and frequently included with them

-cymbo- *comb. form* meaning "cup" better transliterated -kymbo-

cyme a broad, flattened, determinate floral inflorescence in which each branch ends in a single bloom that matures before the lateral ones

    bostrychoid cyme one in which either the right or left branch is always the more vigorous, and which thus develops a helicoid

    idachasial cyme one of which the secondary members are dichasia (*cf.* dichasial branching)

    monochasial cyme one with a single main axis

    cincinnal cyme one in which successive flowers are on alternate sides of the pseudoaxis

    dicyme one in which the axes branch to form secondary cymes

    dichotomous cyme = dichasium

    helicoid cyme a sympodial inflorescence the lateral branches of which all lie on one side (= bostryx)

    biparacyme = dicotomous inflorescence

    multiparous cyme one that has many axes

    scorpioid cyme one in which the lateral branches occur on opposite sides of a pseudoaxis

Cymocyoninae a subfamily of canid carnivorous mammals commonly called the "false dogs." The best known representatives are the dhoies and hunting dogs. They differ from the "true dogs" (Caninae) either in lacking a molar in the lower jaw or in having only four toes

Cymodoceaceae a tribe of the family Najadaceae when this family includes the Potamogetonaceae

cymose pertaining to a cyme but also used to describe hyphae which produce a branched sporangium

    dicymose doubly cymose

Cymothoidea a suborder of isopod Crustacea distinguished from all other orders except the Ido-

theoidea by having lateral uropods and from the latter by the fact that the uropods form with the telson with a caudal fan

Cymothoidea a suborder of isopod Crustacea in which the first pair of legs are not chelate and the uropods are more or less lamellar in form

-cyn- *comb. form* meaning "dog", shortened from -cyon- and, in any case, better transliterated -kyon-

-cynar- *comb. form* meaning "artichoke"

Cynareae a tribe of tubiflorous Compositae

-cynip- *comb. form* indicating an "insect living under bark"

Cynipidae a large family of cynipoid hymenopteran insects commonly called gall wasps. They may be distinguished from other apocritan hymenopterans by the fact that the abdomen is strongly laterally compressed

Cynipoidea a superfamily of apocritan hymenopteran insects. The majority are gall-producing so that the group in general is referred to as the gall wasps. A few are parasitic. They may readily be told from the Chalcidoidea by the absence of a prepectus and the fact that the antennae are not elbowed

Cynoglossidae a family of pleuronectiform fishes resembling the Bothidae in having both eyes on the left side but differing from them in the pointed tail and absence of ribs. Commonly called tongue-soles or tongue-fishes

Cynomoriineae a suborder of myrtiflorous dicotyledonous angiosperms containing the family Cynormoriaceae

-cyon- *comb. form* meaning "dog" better transliterated -kyon- (*cf.* cyn)

Cynopithecidae a family of simioid primate mammals commonly called the dog-faced monkeys and containing the macaques and the baboons. They differ from the Cercopithecidae in having a stubby tail and from the Colobidae in possessing cheek pouches

Cyperaceae the sedge family of glumifloral monocotyledons. The sedges may be distinguished from the grasses by the solid stem which is often triangular in section

Cyperales a plant order erected to contain the family Cyperaceae if this is not considered to be part of the Glumiflorae

cyphella a "dimple" under a lichen thallus

Cyphioideae a subfamily of the Campanulaceae

Cyphonidae = Helodidae

Cypraeidae that family of gastropod mollusks which contains the cowries. The spire is usually covered by the body whorl giving an uncoiled appearance. The long aperture with teeth on both lips is typical of this group

cypress 1 (*see also* cypress 2) any of several species of cupressaceous trees of the genus Cupressus

    deciduous cypress = bald cypress

    Italian cypress *C. sempervirens*

    mock cypress = summer cypress

    Monterrey cypress *C. macrocarpa*

cypress 2 (*see also* cypress 1, pine, spurge) many evergreen trees, and some herbs that resemble the cypress

    bald cypress the taxodiaceous tree *Taxodium distichum*

    Montezuma cypress *Taxodium mucronatum*

    summer cypress the chenopodiaceous herb *Kochia trichophylla*

    standing cypress the polemoniaceous herb *Gilia rubra*

Cyprinidae a very large family of ostariophysid fishes containing the minnows carps, chubs and their allies. They are distinguished by the toothless jaws and soft-rayed fins

Cypriniformes = Ostariophysi

Cyprinodontidae a family of microcyprinous fishes commonly called top-minnows and killifishes

Cyprinodontiformes = Microcyprini

Cypripediloideae the only tribe of orchids in the subfamily Diandrae

cypsela an achene with an adnate calyx

Cypselidae (Aves) = Apodidae

Cypselidae (Insecta) = Sphaeroceridae

Cyrillaceae a small family of sapindalous dicotyledons distinguished by its racemose inflorescence, the small usually basally gamopetalous corollas, and the 2–4 carpelled ovary with few collateral ovules

Cyrtidae = Acroceridae

cys *either* a yeast genetic marker indicating a requirement for cysteine *or* a bacterial mutant indicating a requirement for cysteine

cysB a bacterial genetic marker having a character affecting the activity of 3′-phosphoadenosine 5′-phosphosulfate to sulfide, mapped at 23.8 mins. for *Escherichia coli*

cysC a bacterial genetic marker having a character affecting the activity of sulfate to sulfide; four known enzymes mapped at 46.5 mins. for *Escherichia coli*

cys-1,2 *see* [Cysteine-1], [-2]

cySH = CSH

-cyst- *comb. form* meaning "bladder" better transliterated -kyst-

cyst in *biol.* any cavity of more or less spherical structure, with a distinct wall, particularly one which encloses the resting stage of an organism

acrocyst a modification of the perisarc in some hydrozoan coelenterates which serves as a brood chamber for the young medusae

adenocyst a group of cells surrounding a plant gland

aerocyst the flotation bladders of plants

ascocyst an empty plant cell with a thick wall

blastocyst the early mammalian embryo following cleavage and during implantation

chlorocyst = chloroplast

chondrocyst a rod-like structure too large to be called a rhabdoid in the ectoderm of turbellarian platyhelminths

cnidocyst the rigid oval capsule containing the eversible thread in the cnidoblast (*q.v.*). For types of cnidocyst see entries under -neme and -phore

coniocyst a tubercular sporangium

cryptocyst a calcareous shelf projecting from a gymnocyst

daughter cyst a secondary cyst within a coenurus

ectocyst the outer wall of a cyst

endocyst the thin, transparent inner wall of a cyst

gametocyst any closed cavity containing more than one gamete

goniocyst = sporangium

gymnocyst the calcified part of the frontal wall in the zoecium of some Ectoprocta

heterocyst large cells in algal filaments separating hormogonia

hormocyst a hormogonium in a thick sheath

hydatid cyst the coenurus of the tapeworm *Echinococcus granularis*

hypnocyst a temporary dormant stage, usually occasioned by lack of water

macrocyst a resting stage of some lower plants in which a mass of cells is enclosed in an outer cell wall

mesocyst the central nucleus of the embryo sac in plants

microcyst a small cyst, though used as a specific technical description of the cysts of Myxomycetes

nematocyst = cnidocyst

olocyst the inner layer of the calcified wall of some ectoproct zoecia (*cf.* pleurocyst, tremocyst)

oocyst an encysted zygote, particularly of a sporozoan protozoan

otocyst a sensory organ of doubtful function found in many invertebrates consisting of a fluid-filled cavity containing an otolith and sensory hairs

paracyst the antheridium of Pyronema. Also used for isogamete

pedal cyst = podocyst

phaeocyst = nucleus

phytocyst a plant cell

pleurocyst the middle layer, commonly granular, in the calcified wall of the zoecium of some Ectoprocta (*cf.* olocyst, tremocyst)

podocyst chitinous cysts in the pedal disk of some schiphistoma larvae. These cysts produce ciliated larvae

resinocyst a cyst on the side of a trichome of Begonia

saggitocyst a rhabdite in certain acoele Tubellaria that contains a central protrusible rod

sarcocyst a many-chambered cyst of a sarcosporidian Protozoa

somatocyst a cavity in the anterior nectophore of many simple siphonophorous coelenterates. It is presumed to assist in floating the colony

spermatocyst the mother cell of a moss antheridium

spirocyst a type of nematocyst, or cell resembling nematocysts, found in Zoantharia. It differs from nematocyst proper in that the contained tube is unarmed and of uniform diameter

sporocyst (*see also* larva) *either* any cyst containing spores, particularly in sporozooan Protozoa, *or* a stage in the life history of trematode worms consisting of a cyst-like body from which the redia are budded off internally *or* a cell from which spores are produced asexually (*cf.* sporocarp)

merisporocyst a branched or divided up sporocyst

statocyst = statoblast

tentaculocyst = lithostyle

tremocyst the outer portion of the calcareous wall of the zoecium of some Ectoprocta (*cf.* olocyst, pleurocyst)

trichocyst a rod, immediately under and at right angles to the pellicle of some ciliates

cnidotrichocyst a type of trichocyst, found in a few ciliates having essentially the same structure as the nematocysts of a coelenterate

protrichocysts rod-like structures which are never discharged like trichocysts in the pellicle of certain ciliate Protozoa

L-cysteine 2-amino-3-mercaptopropionic acid. $HSCH_2 \cdot CH(NH_2)COOH$ an amino acid not essential in the nutrition of rats. Associated principally with hair and keratin

L-homocysteine 2-amino-4-mercaptobutyric acid essentially cysteine with the addition of one HCH group. apparently formed in tissues by the demethylation of methionine

[Cysteine-1] a mutant (gene symbol *cys-1*) on linkage group VI of *Neurospora crassa*. The phenotypic expression is a requirement for either cysteine or methionine

[Cysteine-2] a mutant (gene symbol *cys-2*) on linkage group VI of *Neurospora crassa*. The phenotypic expression is a requirement for either cysteine or methionine

adenosylhomocysteinase an enzyme catalyzing the hydrolysis of S-adenosyl-L-homocysteine into adenosine and homocysteine

polycystic = multicellular

cystid the skeletal portion of an ectoproct zooid (*cf.* polypide)

Cystignathidae a very large family of anuran amphibia serving as a dumping ground for those Anura which do not obviously belong in other families

L-**cystine** 3,3′-dithiobis (2-aminopropionic acid) that is, dicysteine. An amino acid not essential to rat growth formed predominantly from cysteine and forming as much as 10% by weight of some hair structures

L-**homocystine** 4,4′-dithiobis (2-amine butyric acid) essentially cystine with the addition of 2 HCH groups apparently directly derived from homocysteine

encystment the process of forming a cyst or becoming enclosed in (*cf.* excystment)

excystment emergence from a cyst

Cystophorinae a subfamily of phocid pinnipede carnivores containing the crested seals and the sea elephant. They are distinguished by having two incisors in the upper jaw and one in the lower

cyt a bacterial mutant, and a postulated yeast genetic marker, indicating a requirement for cytosine

cyt-1-2 *see* [Cytochrome-1], [-2]

-cyte- *comb. form* meaning "a hollow vessel" or "gourd" but taken by most biologists to mean "cell"

amoebocyte any coelomocyte or hemocyte that exhibits amoeboid movement

amphicytes flattened cells surrounding ganglion cells

androcyte the initial of an antherozoid

archaeocytes large wandering amoebocytes in the wall of sponges. They are thought by some to give rise to sex cells or to control regeneration

astrocyte either a star-shaped neuroglia cell or any star-shaped cell particularly those in stroma tissues

athrocyte a large cell with a conspicuous nucleus and vacuolated cytoplasm arranged in clusters round the nephridial tubules of some platyhelminths

dicaryocyte a cell with two nuclei

syncaryocyte = egg

cementocyte a cell secreting the cementum of teeth

cerodecyte an oenocyte (*q.v.*) which is supposed, in some insects, to be concerned in the production of wax (*cf.* oenocyte)

choanocyte a cell typical of sponges that possesses a thin protoplasmic collar round the flagellum. A similar type of cell is found among certain flagellate Protozoa

chondrocyte a cell that secretes the matrix of cartilage

clasmatocyte a resting histiocyte

coelomocyte cells analogous to hemocytes, but found free in coelomic fluid

pseudocoelomocyte wandering cells in the pseudocoel of nematodes

coenocyte an organism consisting of a multinucleate protoplast within a single cell wall. The botanical equivalent of the zoologist's syncytium

zoocoenocyte (*bot.*) a free swimming coenocyte

colencyte *either* a cell of collenchyme *or* the type of amoebocyte with slender pseudopodia, found in the walls of sponges. At times they unite as a network

cystocyte one of the cells forming a protective covering around germ cells

desmacyte a fibrous cell found in layers of the cortex of some sponges

diplocyte a diploid cell

eleocyte oligochaete coelomocytes derived from chlorogogue

erythrocyte a mature red blood cell

polychromatophilic erythrocyte an erythrocyte that does not stain typically with acid dyes

fibrocyte a cell contributing to the fibrous structure of tissues particularly bone and cartilage

gametocyte a cell ancestral to a gamete

granulocyte a granular leucocyte

haplocyte a cell with a haploid nucleus

hemocyte a cell found in blood fluid. The term is usually applied to invertebrate forms

histiocyte a loose large wandering cell in areolar connective tissue directly descended from the primitive reticular cell, and distinguished from a fibroblast by its large size and ovoid shape (= macrophage)

megakaryocyte a giant blood cell with a large polylobular nucleus directly descended from the hemocytoblast and thought to give rise to platelets

polykaryocyte = osteoclast

leucocytes a white blood corpuscle

adipoleucocyte a haemocyte, particularly in insects containing many fat droplets

lithocyte a single cell functioning as a statoblast

lymphocyte a totipotent type of leucocyte, the dominant component of lymphoid organs

large lymphocyte a lymphocyte with a slightly indented nucleus, much cytoplasm and occasionally small azurophilic granules

small lymphocyte a lymphocyte with a spherical nucleus and little or no visible cytoplasm

macrocyte the larger of two dimorphic unicellular forms

meiocyte a cell in meiosis

merocyte a nucleus in the non-segmenting portion of an egg showing meroblastic cleavage including nuclei formed from supplementary spermatozoa. The term is also sometimes used as synonymous with schizont

monocyte a large leucocyte with a slightly indented nucleus and occasionally very fine neutrophil granules

myelocyte the antipenultimate stage in the development of a polymorph leucocyte

metamyelocyte the penultimate stage in the development of a polymorph leucocyte

basophilic metamyelocyte a descendent of a basophilic myelocyte in which the nucleus is "u" shaped

eosinophilic metamyelocyte a descendent of a eosinophilic myelocyte in which the nucleus is "u" shaped

neutrophilic metamyelocyte a descendent of a neutrophilic myelocyte in which the nucleus is "u" shaped

basophilic myelocyte a descendent of a promyelocyte in which the granules are apparent, but the nucleus is still ovoid

eosinophilic myelocyte a descendent of a promyelocyte in which the eosinophilic granules are apparent but the nucleus is still ovoid

neutrophilic myelocyte a descendent of a promyelocyte in which the granules are apparent but the nucleus is not yet lobed

promyelocyte a direct descendent of a myloblast

containing a few not readily differentiated granules. Directly ancestral to myelocytes

myocyte literally, a muscle cell; but, usually applied to the fusiform contractile cells in certain sponges

nematocyte = cnidocyst

paranephrocyte = athrocyte

oenocyte one of several pairs of clusters of yellow, fat-containing cells in insect bodies

oocyte a cell that is a precursor of an egg which is produced by the first meiotic division

secondary oocyte that which is produced by the first meiotic division

osteocyte a cell in bone characterized by a plum-stone shaped body and numerous processes that are continuous with the processes of neighboring osteocytes

phagocyte a cell that engulfs solid particles

rhorocyte a nurse cell for those eggs which develop in pockets in the manubrium in some medusae

pinacocyte the basic structural cell in the body wall of sponges

pinocyte a cell that engulfs fluids as a phagocyte engulfs solids

pituicytes modified neuroglia cells in the neurohypophysis of man. They are distinguished by their argentaphil granules

plasmocyte = leucocyte

podocyte a cell lying on the outer surface of the glomerulus of vertebrate kidneys

poikelocyte a deformed irregular erythrocyte

polocyte = polar body

polymorphocyte = neutrophil

porocyte the tubular cell of a sponge, the outer end of which terminates in an ostium

reticulocyte the penultimate stage in the development of an erythrocyte

solenocyte an elongated flame cell (*q.v.*) in which one or more long flagella beat through the entire length of the neck

spongiocyte extensively vacuolated cells, often forming part of spongy tissue as in the adrenal gland

sporocyte = spore mother-cell

megasporocyte a cell in the ovule. It normally develops into four megaspores

microsporocyte pollen mother-cell

steatocyte an amoebocytic cell that destroys fat cells particularly in insects

stichocyte very large cells forming the wall of the posterior portion of the pharynx of trichiuroid nematodes

thesocyte an amoebocyte filled with food reserves found in the mesogloea of sponges

thigmocyte an insect blood cell that aids in rapid clotting over a fracture in the exoskeleton

thrombocyte a blood platelet

thymocyte a cell in the cortex of the thymus

trephocyte an invertebrate coelomocyte that transports substances

trophocyte = zygote

cythel *Brit.* any of several species of caryophyllaceous herbs of the genus Cherleria

Cytinaceae = Rafflesiaceae

cytium a term without classical justification used for "cell"

apocytium = syncytium

goniocytium = gonidangium

syncytium multi-nucleate animal tissues in which cell boundaries are not apparent (*cf.* coenocyte)

[Cytochrome-1] a mutant (gene symbol *cyt-1*) on linkage group I of *Neurospora crassa*. The phenotypic expression is a very slow colonial growth

[Cytochrome-2] a mutant (gene symbol *cyt-2*) on linkage group VI of *Neurospora crassa*. The phenotypic expression is very slow growth

cytode the sum of the nuclear material in a blue green alga. The term was also briefly used as synonymous with cell

oenocytoid an insect blood cell resembling an oenocyte (*q.v.*)

Cytophagaceae a group of myxobacteriale schizomycetes differing from all other members of the order in that neither fruiting bodies nor resting cells are produced

cytosis the activities of a cell

emeiocytosis the reverse of pinocytosis. That is, the expulsion of minute particles by a cell

phagocytosis the fact that, or method by which, solid particles may be invaginated by the cell wall and carried into the interior of the cell

autophagocytosis the destruction of tissues by cells originating in, rather than entering, a given tissue

pinocytosis the engulfing of fluids by a cell

-d symbolic suffix appended to bacterial mutant symbols to indicate dependance on the compound indicated (cf. -r, -s)

d see [dachs], [dilute] or [dwarf plant]

d₁ see dwarf

D see [Dichaete] or [Duplex comb]. In yeast genetics indicates homothallism

da see [dark], [Dapple] or [daughterless]

dab a term applied to almost any small flat, fish *Brit.* without qualification the pleuronectid fish *Pleuronectes limanda* (= lemon sole)

  longhead dab *U.S.* the pleuronectid *Limanda proboscidea*

  rough dab *Brit.* the pleuronectid *Hippoglossoides limandoides*

  sand dab *U.S.* any of several species of bothid fish of the genus Citharichthys

    longfin sand dab *U.S.* the bothid fish *C. nanthostigma*

    Pacific sand dab *U.S.* the bothid fish *C. sordidus*

    speckled sand dab *U.S.* the bothid fish *C. stigmaeus*

daboia the viperine snake *Vipera russelli* (= Russell's viper)

dace *U.S.* any of several species of cyprinid fish. Usually applied to the genera Clinostomus, Agosia Rhinichthys and Chrosomus *Brit.* without qualification, *Leuciscus leuciscus*

  spike dace *U.S. Meda fulgida*

  spine dace *U.S. Lepidomeda vittata*

Dacelonidae a family of coraciiform birds now usually included with the Alcedinidae

[dachs] a mutant (gene symbol *d*) mapped at 31.0 on chromosome II of *Drosophila melanogaster*. The phenotype has 4 -jointed tarsae

[dachsous] a mutant (gene symbol *ds*) mapped at 0.3 on chromosome II of *Drosophila melanogaster*. The phenotype has a broad short wing with closely spaced crossveins

Dacromycetales an order of heterobasidiomycete basidiomycetes distinguished by the presence of unseptate basidia. They may be readily distinguished in the field by their gelatinous or waxy texture

-dacry- *comb. form* meaning "tear" or "lachrymal" better transliterated -dakry-

dacryoideous pear-shaped

-dactyl- *comb. form* meaning "finger" or "toe"

dactyl in the sense of finger, or finger shaped object, or pertaining to such structures

  adenodactyl a secondary penis in turbellarian platyhelminthes

  heterodactyl said of the foot of a bird in which the first and second toes are directed backwards

  anisodactyl said of feet on which some toes project forward and others backward as in birds

  monodactyl one fingered. Said also of an insect limb terminating in a chela

  pamprodactyl said of a bird having all four toes turned forwards

  syndactyl said of animals, particularly birds, in which two or more toes are fused for the greater part of their length

  zygodactyl said of birds which have two toes projecting anteriorly and two posteriorly

dactylinous finger-shaped

Dactylopteridae a family of acanthopterygian fishes related to the gurnards. Commonly called flying gurnards by reason of their tremendously enlarged pectoral fins

Dactylopteroidea a monofamilial order of Osteichthyes sometimes regarded as a suborder of Acanthopterygii

Dactylopiidae a family erected to contain the cochineal insects and their immediate allies by those who do not wish to retain them in the Coccidae

Dactyloscopidae a family of acanthopterygian fishes distinguished from the uranscopids by the absence of electric organs

dactylous pertaining to fingers or similar structures

  eleutherodactylous said of a bird the feet of which are not webbed

dactylum a term used by ornithologists for the pedal digits of birds

  hypodactylum the undersurface of the toe

  paradactylum the side surface of the toes, particularly of birds

dactylus = dactylopodite

hyperdactyly = polydactyly

  polydactyly having many, or more than the usual number of, fingers

daddy long legs *U.S.* popular name of phalangid arthropods. (*Brit.* = harvestman) *Brit.* popular name of tipulid diptera (*U.S.* = crane fly)

daedaleous labyrinthiform

-daeum see -deum

daffodil any of several species, and very numerous horticultural varieties, of the amaryllidaceous genus Narcissus

  pheasant's eye daffodil *N. poeticus*

  hoop-petticoat daffodil *N. bulbocodium*

  polyanthus daffodil *N. tazetta*

  sea daffodil any of several species of herb of the amaryllidaceous genus Hymenocallis

trumpet daffodil *N. pseudo-narcissus*
dagger any of several species of lepidopteran insects
marsh dagger *Brit.* the noctuid lepidopteran insect *Acronycta strigosa* (= grisette)
dahoon the aquifoliaceous tree *Ilex cassine*
daisy 1 (*see also* daisy 2, brittle star) almost any compositaceous flower not distinguished by another name
Barbeton daisy *Gerbera jamesonii*
bastard daisy *Bellium minutum*
blue daisy *Felicia anelloides*
Dahlberg daisy *Thymophylla tenuiloba*
English daisy *Bellis perennis*
ox-eye daisy *Chrysanthemum leucanthemum*
giant daisy any of many large compositaceous herbs but particularly *Chrysanthemum uliginosum*
michaelmas daisy = aster
Paris daisy *Chrysanthemum frutescens*
Swan River daisy *Brachycome iberidifolia*
true daisy = English daisy
turfing daisy *Chrysanthemum tchihatchewii*
western daisy *U.S. Astranthium integrifolium*
white daisy the compositaceous herb *Layia glandulosa*. This name is also given the ox-eye daisy (*q.v.*)
yellow daisy *Rudbeckia serotina*
daisy 2 (*see also* daisy 1) a few flowers of other families
mountain daisy *U.S.* the caryophyllaceous herb *Arenaria groenlandica*
marsh daisy the plumbaginaceous herb *Armeria maritima*
Dalatiidae a family of squaloid sharks closely allied to the Squalidae but lacking the spine in the second dorsal fin
Dalceridae a family of zygaenoid lepidopteran insects
dalisier the musaceous tree-like foliage plant *Heliconia bihai*
Dalliidae a monospecific family of haplomous fishes containing the Alaska blackfish commonly placed in the Umbridae but differing from them in its much larger number of pectoral fin rays
dam female parent (*cf.* sire)
chotda **dama** *hort. var.* of the musaceous tree-like herb *Musa paradisiaca* (= banana)
damalisk a general name for alcelaphine antelopes of the general Beatragus (Hunter's hartebeest) and Damaliscus (sassaby, blesbok, etc.)
dame the term for a female goose
Danaidae a family of large nymphaloid lepidopteran insects containing *inter alia* the well known monarch. They are sometimes called milk-weed butterflies because of the food preference of the larvae
dance stylized repetitive motion, usually associated with courtship
combat dance an exhibition between two males either of which may desire to establish dominance over the other, without occasioning him bodily harm
mating dance the pre-copulatory activities of an animal, usually a male
karroo **dancer** *S.Afr.* the hesperid lepidopteran insect *Alenia sandaster*
dandelion any of many low growing, yellow-flowered compositaceous herbs but properly those of the genus Taraxacum
common dandelion *T. officinale*
dwarf dandelion any of several species of the genus Krigia
fall dandelion *Leontopodon autumnalis*
false dandelion *Pyrrhopappus carolinianus*
potato dandelion *Krigia dandelion*

dandy *U.S.* the noctuid lepidopteran insect *Charadra decora*
[Danforth's shorttail] a mutant (gene symbol *Sd*) in linkage group V of the mouse. The phenotypic expression involves not only a shortened tail but also urogenital abnormalities
danio aquarists' name for cyprinid teleost fishes of the genus Brachydanio
dap + hom a bacterial genetic marker having a character affecting the activity of aspartic semialdehyde dehydrogenase, mapped at between 61.0 and 64.0 mins. for *Escherichia coli*
dapA a bacterial genetic marker having a character affecting the activity of dihydrodipicolinic acid synthetase, mapped at between 49.6 and 54.1 mins. for *Escherichia coli*
dapB a bacterial genetic marker having a character affecting the activity of N-succinyl-diaminopimelic acid deacylase, mapped at between 49.6 mins. and 54.1 mins. for *Escherichia coli*
daphne the name of a large genus of thymelaceous plants, often extended to other members of the same family
sand daphne *N.Z.* the thymelaeaceous herb *Pimelea arenaria* (*cf.* rice flower)
New Zealand daphne The thymelaeaceous, low growing shrub *P. prostrata*
Daphniphyllaceae a small family of geraniale dicotyledons
didapper = divedapper
divedapper *Brit.* the podicipedid bird *Podiceps ruficollis W.I.* either of two podiceped birds *Podiceps dominicus* (= least grebe) or *Podilymbus podiceps* (= pied-billed grebe)
[Dapple] a mutant (gene symbol *da*) in linkage group II of *Neurospora crassa.* The term is descriptive of the sparse irregular growth of the phenotypic colony
[dark] a mutant (gene symbol *da*) in linkage group I of the mouse. The phenotypic expression is a band of darkly pigmented hair along the dorsal surface of agoutis and yellows
[dark eye] a mutant (gene symbol *dke*) mapped at $73.0\pm$ on chromosome II of *Drosophila melanogaster.* The phenotypic eye color is dull and dark
darkie *I.P.* any of several species of lycaenid lepidopteran insect of the genus Allotinus
darnel any of several species of grasses of the genus Lolium, particularly *L. temulatum and L. perenne*
darner *U.S.* any of several odonatan insects
common darner *U.S.* common name of aeshnids
gray darner popular name of petalurids
green darner *U.S. Anax junius*
darr *Brit. obs.* = tern
dart 1 (*see also* dart 2-5, moth) *U.S.* any of very large numbers of species of noctuid lepidopteran insects
deep-brown dart *Brit. Aporophyla lutulenta*
dart 2 *Austral. I.P.* and *Afr.* any of numerous hesperid lepidopteran insects. *S.Afr.*, without qualification, *Andronymus neander*
grass dart *I.P.* and *Austral.* any of several hesperiid lepidopteran insects of the genus Taractrocera. *Austral.*, without qualification, *T. papyria*
palmdart *Austral.* the hesperid insect *Cephrenes augiades*
dart 3 the sting of an insect
dart 4 a calcareous, hormone-impregnated, spike transferred between some land snails as a preliminary to copulation

dart 5 a short, sharp, restricted motion of an entire organism

darter 1 (*see also* darter 2,3) any of very numerous small fresh water percid fishes. The largest genus Etheostoma contains upwards of seventy North American species

banded darter *E. zonale*

bluebreast darter *E. camurum*

greenbreast darter *E. jordani*

fantail darter *E. flabellare*

johnny darter *E. nigrum*

rainbow darter *E. caeruleum*

sand darter any of several species of the genus Ammocrypta

blackside darter *Percina maculata*

greenside darter *E. blennioides*

darter 2 (*see also* darter 1,3) popular name of birds of the family Anhingidae

African darter *Anhinga rufa*

Indian darter *A. melanogaster*

darter 3 (*see also* darter 2,3) *Austral.* any of numerous hesperid lepidopteran insects

dartlet (*see also* dart) *I.P.* any of several species of hesperiid insects of the genus Oriens

Darwinian *epon. adj.* from Charles Robert Darwin (1809–1882)

darwinism properly the views expressed on evolution by Charles R. Darwin who stressed the struggle for survival and the "survival of the fittest" (*i.e.* those best fitted to their environment). Among laymen this term is sometimes thought to be synonymous with "evolution"

neodarwinism that view which explains the mechanics of evolution on the basis of mutation and selection

Dascilloidea a superfamily of polyphagous coleopteran insects containing the soft-bodied plant beetles

dash *U.S.* any of several species of hesperid lepidopteran insects

black dash *Atrytone conspicua*

broken dash *Wallengrenia otho*

long dash *Polites mystic*

dassy *S.Afr.* popular name for the Hyracoidea

-dasy- *comb. form* meaning "thick"

Dasyatidae a family of rays commonly called sting rays from the presence of a tapered venomous spine at the base of the tail

Dasychadales an order of chlorophycean Chlorophyta containing those forms in which the thallus consists of a central axis bearing whorls of branches

Dasypodidae that family of xenarthran mammals which contains the armadillos, distinguished from all other mammals by the possession of an anterior and posterior shield of fused scutes, and a number of movable bands of scutes separating the shield

Dasypogoneae a subfamily of liliaceous monocotyledons containing some Australian xerophytic genera

Dasyproctidae that family of hystricomorph rodents which contains the pacas and agoutis. They are distinguished from other hystricomorphs by the very large jugal arch and by the great development of the three middle toes of the hind foot

dasyure anglicized form of Dasyurus (*q.v.*)

Dasyuridae a family of marsupial mammals distinguished by the presence of eight incisors in each jaw and five-toed hind feet. There are many analogues of structure and habit with the eutherian carnivores

Dasyurinae a subfamily of dasyurid marsupials containing the Australian "native cat" and the Tasmanian devil

Dasyurus a genus of dasyurid marsupials closely paralleling the viverine cats in their general form and commonly called dasyures

date 1 (*see also* date 2, bug, scale) the palmaceous tree *Phoenix dactylifera* and its edible fruit

date 2 (*see also* date 1) any of several species of mytilid mollusk of the genus Lithophaga

Datiscaceae a small family of parietale dicotyledonous angiosperms characterized by the unisexual flowers and dioeceous habit

Datiscineae a suborder of parietale dicotyledons containing the single family Datiscaceae

Datureae tribe of the family Solanaceae

mud-dauber *U.S.* popular name of tryposcylonine spheooid hymenopteran insects

daughter properly a female offspring, but frequently used to designate one part developed from another, *e.g.* daughter chromosome, daughter cyst

[daughterless] a mutant (gene symbol *da*) mapped at 39.3± on chromosome II of *Drosophila melanogaster*. The term is descriptive

davie the larid bird *Sterna dougallii* (= roseate tern)

daw *Brit. obs.* = jackdaw

black daw *B.W.I.* the cuculid bird *Crotophaga ani* (= smooth billed ani)

jackdaw either of two corvid birds of the genus Corvus. Usually, without qualification *C. monedula*

dayal = dhyal

dd *see* [displaced[2]]

de *comb. prefix* used for "from" in the privative sense

de *see* [droopy ear]

de₁ *see* [defective endosperm]

de₁₆ *see* [defective endosperm]

deaminase see aminase

death the removal of the phenomenon of life from an organic complex of otherwise known structure

Decandria that class of Linnaean classification of plants distinguished by the possession of ten stamens

Decapoda (Crustacea) a large order of eucarid malacostracan crustacea containing the shrimps, prawns, lobsters, crayfish, and crabs. They are distinguished from the Euphausiacea by the fact that the first three pairs of thoracic limbs are specialized as maxillipedes

Decapoda (Mollusca) a suborder of cephalopod mollusks distinguished by the presence of ten arms, four pairs of which are shorter than the other pair

decomposers that group of organisms (bacteria and fungi) which degrade dead organic material in ecosystems

decussate arranged in pairs at alternate right angles

[deep orange] a mutant (gene symbol *dor*) mapped at 0+ on the X-chromosome of *Drosophila melanogaster*. The name refers to the color of the eyes of the phenotype in which the females are also sterile

deer (*see also* cabbage, grass, kill, oxen) any of numerous cervid artiodactyle mammals distinguished by the solid deciduous antlers rising from boney bases on the skull. They are divided among the subfamilies Moschinae, Muntiacinae, Cervinae, Odocoileinae, Alcinae, Rangiferinae, Hydropotinae and Capreolinae

axis deer either of two species of cervines of the genus Axis known respectively as the chital (*q.v.*) and hog-deer (*q.v.*)

Chilean deer the odocoileine *Odocoileus chilensis* (*cf.* mule deer)

fallow deer the cervine *Dama dama,* best known for its decorative sporting

hog deer the cervine *Axis porcinus*

marsh deer any of several species of *S. Amer.* odocoileins of the genus Blastocerus

mule deer *U.S.* the odocoilein *Odocoileus hemionus* (*cf.* blacktail deer)

musk deer a small moschine *Moschus moschiferus.* It is distinguished not only by the musk sac but also by the very large canines

pampas deer any of several species of odocoileins of the genus Ozotoceros inhabiting the savannahs and pampas

Père David's deer the cervine mammal *Elaphurus davidiana.* They are distinguished by the huge straight, but angularly branched, antlers

red deer any of several species and numerous races of the cervine genus Cervus. *Brit.* without qualification, *Cervus elaphus.* (*cf.* wapiti) Phillipines, without qualification, *C. alfredi*

reindeer any of several species of the genus Rangifer, usually, without qualification, *R. tarandus*

roe deer any of several capreolines of the genus Caprellus

silka deer The cervine *Cervus nippon*

swamp deer *I.P.* without qualification, the cervine *Cervus duvauceli* (= barashinga)

blacktail deer *U.S.* a geographic race of the mule deer *U.S.*

white-tailed deer any of numerous species of odocoileins of the genus Odocoileus, *U.S.,* without qualification, *O. virginianus* (= Virginia deer)

Thorold's deer the Far Eastern cervine *Cervus albirostris*

Virginia deer *U.S.* = whitetail deer *U.S.*

water deer the cervid *Hydropotes inermis*

deerlet = chevrotain

[defective endosperm] *either* a mutant (gene symbol *de₁*) mapped at 0 in linkage group IV of *Zea mays, or a* mutant (gene symbol *de₁₆*) mapped at 74 in linkage group IV of *Zea mays.* The name is descriptive in both instances

deficiency the loss or inactivation of a section of a chromosome. In *Drosophila melanogaster* a general term (symbol *Df*) for all mutants associated with such a chromosomal anomaly

defixed immersed

defoliate to remove leaves from

sagebrush **defoliator** *U.S.* the gelechiid lepidopteran *Aroga websteri*

degeneration see generation

[Deformed] a mutant (gene symbol *Dfd*) mapped at 47.5 on chromosome III of *Drosophila melanogaster.* The name is descriptive

-dehisc- *comb. form* meaning "to yawn" or "gape"

dehiscence spontaneous opening of a ripe plant structure, such as a fruit, anther or sporangium

circumscissile dehiscence dehiscence at right angles to the long axis of a fruit or anther

extrorse dehiscence dehiscence from the inside outward

loculicidal dehiscence dehiscence along the center of a carpel

introrse dehiscence dehiscence from the outside inward

lateral dehiscence splitting along the side

marginicidal dehiscence = septicidal dehiscence

poricidal dehiscence dehiscence through pores

septicidal dehiscence dehiscence along the line between fused carpels

septifragal dehiscence dehiscence along the line of juncture of the wall and a septum

valvate dehiscence dehiscence through an aperture formed by the partial detachment of a flap

deirid one of a pair of papillae in the region of the nerve ring in some nematodes

Deiter's *epon. adj.* from Otto Friedrich Karl Deiter (1834–1863)

del *see* [delicate]. In yeasts indicates a deletion marker

Del *see* [deletion]

[deletion] the remainder of a chromosome after the occurence of a non-terminal deficiency. In *Drosophila melanogaster* a general term (symbol *Del*) for all mutants associated with such a chromosomal anomaly

[delicate] a mutant (gene symbol *del*) on linkage group VI of *Neurospora crassa.* The phenotypic expression is a slow rate of rather sparse growth

queen's **delight** *U.S.* the euphorbiaceous herb *Stillingia sylvatica*

Delphacidae a small family of homopteran insects closely allied to the Fulgoridae

-delph- *comb. form* meaning "womb" but extended in *biol.* to the whole female reproductive system in both plants and animals. Frequently confused with -adelph- (*q.v.*)

delphic pertaining to female parts

diadelphic possessing two sets of stamens

didelphic possessing a pair of ovaries and oviducts

heteradelphic possessing two fused carpels, one of which atrophies

monodelphic possessing a single ovary, and oviduct

polydelphic possessing many ovaries and oviducts

Delphinidae that family of odontocetan cetacea which contain the dolphins, porpoises, killer whales, etc. They have numerous teeth both in the upper and lower jaw

delphous *see* delphic

[Delta] a mutant (gene symbol *Dl*) mapped at 66.2 on chromosome III of *Drosophila melanogaster.* The veins of the phenotypic wing are thickened at the margin

deliquescent 1 (*see also* deliquescent 2) rapid degeneration of tissues or organs into an amorphous, often semiliquid, mass as some flower perianths or fungi

deliquescent 2 (*see also* deliquescent 1) said of some tree habits lacking a continuous central axis (as the elm); antonym = excurrent (*q.v.*)

[deltex] a mutant (gene symbol *dx*) mapped at 17.0 on the X chromosome of *Drosophila melanogaster.* The phenotypic expression involves thickened wings with delta shaped vein margins

deltidium a plate closing the delthyrium

deltoid said of foliar organs of plants when triangular in form with the base more or less truncate

-deme- *comb. form* hopelessly confused between "demos" ("people") and demas ("body" or "structure")

deme 1 (*see also* deme 2) in the sense of "people." A population, usually meaning, without qualification, an aggregate of similar cells or similar species. Also used for an isolated population within a species

gamodeme an isolated community of interbreeding organisms

topodeme a fraction of a population restricted to a specific area

deme 2 in the sense of "structure"

apodeme those hollow, inwardly projecting, portions

of an arthropod exoskeleton which form an internal framework for the support of muscles (*cf.* apophysis)

articulatory epideme the thickened membrane that articulates the insect wing to the thorax

stipodeme the apodeme of the labrum in certain dipteran insects

demersal inhabiting the lowest layer of a lake

demersed submersed

-demic *adj. term* from deme in the sense of "people"

endemic confined to, but normally inhabiting, a limited area

epidemic pertaining to a rapid, great, temporary increase in the population of a given species

pandemic epidemic over a wide area

polydemic occurring in several, separated areas

demissous hanging downwards

demography the field of human population analysis, also applied to the study of other animal populations

demoiselle 1 (*see also* demoiselle 2) any of several gruid birds (cranes) but particularly *Anthropoides virgo*

demoiselle 2 (*see also* demoiselle 1) = damsel fly

demon *I.P.* and *Austral.* any of several species of herperiid insect of the genera Notocrypta and Udaspes

banded demon *Austral.* the hesperid lepidopteran insect *N. waigensis*

Demospongiae a class of Porifera having skeletons either of horny fibers or of silicious spicules which are never triaxonic

Jack **Dempsey** the cichlid teleost fish *Cichlasoma biocellatum*

-dendr- *comb. form* meaning "tree"

dendrad a plant of orchards

dendrite those nerve fibers which extend from the nerve cell body to the source of stimulus (*cf.* axon, telodendron)

dendritic branched or tree like

Dendrobatidae a subfamily of ranid anuran Amphibia commonly called arrow poison frogs in view of the use to which the virulent exudate of their skins is put by South American Indians. They are distinguished from other anurans by the absence of teeth or (Phyllobates) a reduction of teeth to a microscopic size

Dendrochelidonidae = Hemiprocnidae

Dendrochirota an order of holothurioid Echinodermata with numerous ambulacral feet distinguished from the Aspidochirota by the fact that the tentacles are irregularly branched

Dendrocolaptidae that family of passeriform birds which contains the wood-creepers. They are distinguished by the strong rigid sharp pointed lower tail feathers which assist them to maintain a vertical position on tree bark

dendrogram the type of diagram commonly referred to as "a family tree" designed to show postulated relationships between taxa

Dendromyinae a subfamily of murid rodents distinguished by their long stilt-like legs

gonodendron a branched stalk bearing gonophores

telodendron one of the fine branches that unites to form an efferent axon (*cf.* dendrite)

denigrate blackened

-dent- *comb. form* meaning "tooth"

dentary that bone in the lower jaw, or mandible, which carries teeth

dentate (*see also* -dont) toothed. In *bot.* specifically applied to that form of marginal dentition, usually of foliar parts, with the teeth at right angles to the primary axis (*cf.* crenate, serrate)

double dentate a dentate margin the teeth of which are themselves dentate

maxadentes teeth on the insect lacinia

denticle diminutive of tooth

dermal denticles the tooth-like "scales" found in the skin of elasmobranch fish

edentate smooth edged, that is without teeth

dentine the layer between the enamel and the pulp cavity in teeth and placoid scales. A hard tissue, usually devoid of cells and with parallel, perpendicular to the surface, tubules

circumpulpar dentine the inner dentine surrounding the pulp cavity

fibrodentine a term applied to the calcified fibrous material which corresponds to dentine in the dermal denticles of elasmobranch fish

hyodentine a modified vitreous dentine representing or replacing the enamel layer in some fish denticles

mesodentine a form of semidentine in which the odontoblasts lie in interconnected lacunae

orthodentine the regular dentine structure consisting of an outer layer of pallial dentine and an inner layer of circumpulpar dentine

osteodentine dentine in the form of a network of tubes of dentine, analogous to the Haversian system of bone. Bound together by a matrix of bony substance

pallial dentine the outer dentine of the tooth where it is not covered by enamel

plicidentine dentine which is folded in a series of longitudinal ridges penetrating the pulp cavity

semidentine contains fine, parallel tubules, with thin branches coming off at right angles, and which contains odontoblasts embedded in that part which lies next to the pulp cavity

tubulodentine consists of parallel tubes of dentine with an interstititial matrix of bone

vasodentine dentine in which there are radiating, anastomosing, capillary channels, but no tubules

vitrodentine the vitreous (non tubular) material covering the teeth and placoid scales of elasmobranch fish

dentition (*see also* dental formula) the sum of, or the arrangement of, the teeth in a given organ, or the presence or type of marginal teeth on foliar or other plant organs

edentulate lacking teeth

deodar (*see also* weevil) the pinaceous tree *Cedrus deodara* (*cf.* cedar)

depauperation said of a plant growing on impoverished soil and therefore having a starved appearance

depend to hang down

deposit the term is sometimes used for the ossicles of holothurian echinoderms

Depressariidae = Oecophoridae

depressed sunk down

der *see* [deranged]

[deranged] a mutant (gene symbol *der*) mapped at 57.2 on the X chromosome of *Drosophila melanogaster*. The derangement is in the thoracic bristles

derbio *Brit.*, the carangid fish *Lichid glauca*

-derm- *comb. form* meaning "skin"

derm 1 (*see also* derm 2, dermis) in the sense of the surface layers of animals and their embryos. The term -dermis (*q.v.*) is identical in meaning but is rarely used except for words epidermis and hypodermis

anoderm literally lacking a skin, but usually lacking a cuticle

ectoderm the outer layer of an animal, particularly an embryo

mesectoderm mesenchyme cells supposedly derived from the ectoderm as those budded off from the primitive streak of an avian egg

endoderm the innermost layers of an animal

　mesendoderm the layer of cells over the archenteron, produced by the involution of cells through the blastopore of the gastrula

enteroderm that part of the endoderm which gives rise to the gut itself

entoderm = endoderm

gastroderm the lining of the coelenteron of coelenterates

mesoderm (*see also* derm 2) *zool.* the middle of the three layers of triploblastic animals. In coelomate forms it is divided into an outer (somatic) and an inner (splanchnic) layer (*cf.* -pleure)

　ectomesoderm mesodermal, usually invertebrate, mesenchyme cells budded off from the ectoderm

　endomesoderm mesoderm derived from endoderm

paraderm the membrane that encloses a pronymphal insect

periderm (*see also* derm 2) the outermost layer of the epidermis in the mammalian embryo, from which the cornified layer is derived, and also the chitinous tube protecting the hydranths in certain hydrozoan coelenterates

plasmoderm = ectoplasm

somatoderm the outer, or peripheral, cells of a mesozoan

trophoderm the outermost layer of cells in the blastula stage of a placental mammal

　cytotrophoderm the thickened portion of the trophoblast from which the plasmotrophoderm is produced

　plasmotrophoderm the outermost layer of a mammalian embryo before the formation of the chorion

derm 2 (*see also* derm 1) in the sense of the surface layers of plants

carpoderm = pericarp

cytioderm the cell wall of diatoms (*cf.* cytoderm)

cytoderm that part of a plant cell protoplast immediately adjacent to the wall (*cf.* cytioderm)

endoderm (*see also* derm 1) the sheath of cells surrounding the vascular cylinder of a root, or a conifer leaf or the stem of a vascular cryptogram

epiderm the outer cell-layer of a plant

exoderm thickened, frequently suberized, cells immediately under the epidermis of a root

mesoderm (*see also* derm 1) The center of the three layers of tissue in the theca of a moss

periderm (*see also* derm 1) the outer layer of a plant replacing the epiderm and comprising phellem, phellogen and phelloderm

phelloderm that tissue in a plant stem which is derived from phellogen and lies internal to the latter

protoderm partially differentiated meristem which will later give rise to the epidermal system of a plant

rhizoderm the epiderm of a root (= epiblem)

sarcoderm a fleshy layer between the exopleura and endopleura of a seed coat

spermoderm the seed coat

sporoderm the coat of a spore

dermalia spicules on the outer surface of a sponge (*cf.* gastralia)

Dermanyssidae a family of parasitic acarines infesting birds and mammals

Dermaptera that order of insects containing the forms commonly called earwigs. They are distinguished by having wings which are folded in a complex manner under anterior elytra and in almost all species by the presence of a pair of forceps at the posterior end of the abdomen

Dermatemydidae a primitive family of Central American chelonian reptiles distinguished by their very short tail and the separation of the plastral scutes from the marginals by a complete row of inframarginal scutes. The only extant species is *Dermatemys marvii*

Dermatoptera = Dermaptera

Dermestidae a very large family of destructive and economically important coleopteran insects. The family contains not only the forms commonly called dermestids which feed on stored hides and museum specimens but also the carpet beetles, buffalo moths and numerous other destructive pests

dermis skin. The term is usually replaced by -derm (*q.v.*) except for the two words that follow

　epidermis the cells covering the surface of an organism

　hypodermis the innermost layer of the skin of an animal

Dermochelyidae a monotypic family of chelonian reptiles containing the leathery turtle, so called from the fact that the leathery skin lacks the epidermal shields found in other Chelonians

Dermoptera that group of placental mammals which contains the forms commonly called "flying" lemurs. They are distinguished by a double furred skin parachute that extends between the neck and front paws and between the front and hind paws as well as between the hind feet and tail. The unique feature is that the lower incisors are very wide and pectonate

dermotism the ability to perceive light in those areas of skin not furnished with known photoreceptors

pachydermous with a thick skin. The term pachyderm is journalese for elephant

Derodontidae a small group of fungus-eating coleopteran insects readily distinguished by the notches along each edge of the elongate pronotum from which the name tooth-necked fungus beetles is derived

Deroptera = Orthoptera

dertrum the hook of the bill of a bird

Descemet's *epon. adj.* from Jean Descemet (1732–1810)

descent term for a group of woodpeckers

desert 1 (*see also* desert 2) from the ecological point of view any area in which less than 20% of the ground surface is covered with permanent vegetation. Deserts are commonly classified according to the reasons that have caused them to exist as low rainfall, cold, low nutrients in the soil, etc.

desert 2 (*see also* desert 1) term for a group of lapwings

Desfontainaceae a small family of contortate dicotyledonous angiosperms

-desm- *comb. form* meaning "bundle" or "chain" and, by extension, "tie" or "ligament"

desma 1 (*see also* desma 2) a sponge spicule with irregular, tree-like branches, developed through the deposition of silica on a crepis

desma 2 (*see also* desma 1) a bond, chain, link or group

　kinetodesma one of the numerous fine strands forming

a network connecting the kinetosomes of ciliate protozoa

peridesma the layer of cells between the endoderm and a vascular bundle when the vessels of the latter are not combined into a stele

plasmodesma a connection between plant cells probably formed in part by endoplasmic reticulum

sporidesma a multicellular body, each of the cells of which is, in effect, a spore

desman popular name of talpid insectivorous mammals of the genera Desmana and Galemys. They resemble a mole in general shape but are aquatic and have long flexible snouts

Desmarestiales an order of heterogenerate Phaeophyta distinguished by the presence of a single filament at each growing apex

desmic pertaining to a stele (*q.v.*)

monodesmic possessing a monarch stele

dialydesmic possessing a stele composed of separate bundles

desmid popular name of algae of the family Desmidiaceae

Desmidiaceae that family of zygonematale Chlorophyceae containing the forms commonly referred to as desmids. The variously shaped cells which may be solitary, united end to end, cr associated in amorphous colonies are distinguished by the possession of a median constriction dividing them into two distinct halves

Desmodontidae a family of chiropteran mammals erected to contain the two bloodsucking, or vampire, bats *Desmodus rufus* and *Diphylla ecaudata*. The teeth adapted to "shaving" the skin of the victim, and the stomach adapted to solely liquid food are typical

Desmognathinae a subfamily of plethodontid urodele Amphibia distinguished by having the tongue attached at the front but free behind in adults, and by hypertrophy of the temporal muscles

desmoid said of jointed plant organs, as the fruit of the legume Desmodium

Desmokontae = Desmophyceae

desmone a hormone at one time postulated as an inductor of mitosis

Desmophyceae a class of pyrrophyte algae differing from the Dinophyceae in that the flagella are borne at one end

Desmoscolecidea a class of marine nematodes with a heavily ringed and frequently bristled cuticle, and with crescentic amphids

desmose a single, thickened fiber, connecting the divided centrioles in mitotic division of some Protozoa

centrodesmose the axial spindle of the achromatic spindle

desmosis a joint

syndesmosis a form of amphiarthrosis in which the bones are joined by a ligament, but do not articulate freely

polydesmous pertaining to a scyphistoma larva or other strobilating form that produces more than one disc at one time

destruction term for a group of wild cats (*cf.* rout)

det *see* [detached]

[detached] a mutant (gene symbol *det*) mapped at 72.5 on chromosome III of *Drosophila melanogaster*. The phenotypic expression is broken crossveins

[Detached] a mutant (gene symbol *Dt*) mapped at 10.0± on chromsome II of *Drosophila melanogaster*. The

phenotypic expression is in vein L2 which does not reach the margin of the wing

detritus dead organic tissues and organisms in an ecosystem, usually including the live micro organisms engaged in the decomposition of the material

detectous naked

determinant that which determines anything, as a chromosome determines specific heritable characters

determinate said of an inflorescence, or part of an inflorescence, the axis of which is terminated by a flower

cascade determination the concept that embryos or regenerating tissues depend on determinants that become serially independent

-deum suffix indicative of a "way" or "route" or "cavity through which things pass"

proctodeum a posterior invagination, which subsequently becomes either the cloacal or anal aperture

rhynchodeum the cavity into which a protrusible proboscis is retracted, particularly in the Nemertina

stomodeum an anterior opening into the archenteron or, in anthozoan coelenterates a short inturned tube projecting into the coelenteron

urodeum the posterior portion of the cloaca into which open the ureter and sex ducts

deustous appearing scorched

-deutero- *comb. form* meaning "second" or "secondary" but often misused for "double"

Deuteromycetae a botanical taxon erected to contain those fungi in which the formation of zygotes or spores has not been observed. The group is also referred to as fungi imperfectae

Deuterostomia a term coined to describe those bilaterally symmetrical animals in which the embryonic blastopore becomes the anus. This includes the Chaetognatha, Echinodermata, Hemichordata and Chordata (*cf.* Protostomia)

development the sum total of the changes occuring in the life history of an organism

regulative development a pattern of development in which capacities of parts of the organism are greater (*e.g.* more diverse) than their normal fate

schizogenic development development resulting from cell division as distinct from increase in cell size

devil (*see also* appie, bite, bones, claw, club, cotton, fig, flax, fish, grass, gut, hair, hand, horn, horse, ivy, milk, needle, paintbrush, rattlebox, root, toenail, tongue, and wood) the antithesis of a beneficient god and, by extension, organisms with grossly undesirable qualities

hell devil *U.S.* the larva of the dobson fly *Corydalus cornutus* (= hellgrammite)

hickory horned devil *U.S.* the caterpillar of the citheroniid moth *Citheronia regalis* (= regal moth)

king devil the compositaceous herb *Hieracium floribundum* and *H. pratense*. The term is also sometimes applied to *H. florentinum*

pine-devil *U.S.* the ceratocampid lepidopteran insect *Citheronia sepulchralis*

Tasmanian devil (*Austral.*) a notoriously predatory dasyurine dasyruid *Sarcophilus ursinus* (= badger *Austral.*)

devil-in-a-bush any of several species of ranunculaceous plant of the genus Nigella (= love-in-a-mist)

devil-in-a-mist = devil-in-a-bush

deviling *Brit. obs.* = swift

devil's-head-in-a-bush the malvaceous herb *Hibiscus trionum* (= flower-of-an-hour)

dew moisture condensed at dusk and, therefore, organisms actually or apparently producing droplets

    daily-dew *U.S.* any of several species of the droseraceous insectiverous genus Drosera (= sun dew)

    honey dew a sweet secretion of aphid insects, utilized as a food source by ants

    sundew any of numerous insectivorous herbs of the genus Drosera

        great sundew *Brit. D. anglicum*

dewlap loose skin on the ventral side of the necks of mammals

Dexiidae a family of brachyceran dipteran insects commonly included with the Tachinidae

-dextr- *comb. form* meaning "right hand"

dextrinase an enzyme catalyzing the hydrolysis of α-1,6-glucan links

limit **dextrinase**–oligo-1,6-glucosidase

Df *see* [deficiency]

Dfd *see* [Deformed]

DFP di-isopropyl phosphorofluoridate

DFPase an enzyme that catalyzes the hydrolysis of di-isopropylphospho-fluoridate to di-isopropylphosphate and hydrofluoric acid

Dh *see* [Dominant hemimelia]

dhal the leguminous herb *Cajanus indicus* (= pigeon pea)

dhoie any of several species of Asiatic cymocyonine canid carnivores of the genus Cuon. They differ from "true dogs" in having only two mandibles

dhyal = dyal

-di- *comb. form* meaning "two" (*cf.* dis-)

Di *see* [Dirty]

di *see* [divergens]

-dia- *comb. form* meaning "through"

adiabatic that which cannot be transferred

diad one of a pair of chromosomes resulting from the separation of two homologous members of a tetrad in meiosis

Diadelphia that class in the Linnaean system of plant classification distinguished by the possession of stamens combined by their filaments into two sets

dusky **diadem** *I.P.* the satyrid lepidopteran insect *Anadebis himachala*

Diadocidiidae a family of nematoceran dipteran insects now usually included with the Mycetophilidae

floral **diagram** a diagrammatic section across a flower, on which the position of the various parts is indicated as though they lay in one plane

-diaire- *comb. form* meaning "division" frequently transliterated -diere-

sundial 1 (*see also* sundial 2) the architectonicid gastropod *Architectonica granulata* (*cf.* sundial shell)

sundial 2 (*see also* sundial 1) the leguminous herb *Lupinus perennis*

-dialy- *comb. form* meaning "to disband"

dialysis *see* -lysis

[**dialytic stamens**] a mutant (gene symbol *dl*) mapped at 40 on chromosome 8 of the tomato. The name is descriptive of the phenotypic expression

diana *U.S.* the nymphalid lepidopteran insect *Speyeria diana* (*cf.* greater fritillary)

Diandrae a subfamily of Orchidaceae distinguished by the suppression of the inner whorl of stamens and the fact that the two remaining functional stamens are situated on either side of the column

Diandria that class of the Linnaean classification of plants distinguished by the possession of two stamens

Diapensiaceae small family of ericalous dicotyledons distinguished from other families in the order by having three carpels and five stamens

Diapensiales a monofamilial order of dicotyledonous angiosperms. The characteristics of the family are those of the order

diaphanous translucent

diaphery the condition of having two flowers within one calyx

diaphragm *see* -phragm

Diapriidae a large family of minute black proctotrupoid apocritan hymenopteran insects with an unusual head best described by saying that the antennae arise from a shelf in the middle of the face

Diapsida an order of reptilian chordates containing all extant reptiles except the anapsid chelonians from which they are distinguished by the absence of a shell

diarinous = diandrous

Diaspididae a family of homopteran insects regarded by many as a subfamily of the Coccidae. They are distinguished by the possession of a heavy scale formed from a combination of wax and molted skins from which they derive their popular name of armored scales

Diastatidae a family of brachyceran dipteran insects commonly included with the Drosophilidae

-diastem- *comb. form* meaning "gap or "interval"

diastema 1 (*see also* diastema 2) the gap between teeth

diastema 2 (*see also* diastema 1) the hyaline area of protoplasm that indicates the plane of cell division in telophase

diastole the dilation of a contractile organ or vesicle

diatmesis *see under* -tmes

diatom common name of Bacillariophyceae

Dibamidae a family of vermiform limbless lacertilian reptiles covered with scales but lacking osteoderms

dibatag an Abyssinian gazelline antilopid strongly resembling a small gerenuk (*q.v.*)

Dibranchia an order of siphonopodous Mollusca containing the squids, octopus and cuttlefish. They are distinguished by the presence of a rudimentary shell which is covered by the integument

Dicaeidae that family of passeriform birds which contains the flowerpeckers of Asia and Australasia. They are small birds with thin curved serrate beaks and short necks with long wings and short tails

-dich- *comb. form* meaning "to disunite"

[**Dichaete**] a mutant (gene symbol *D*) mapped at 40.4 on chromosome III of *Drosophila melanogaster*. The phenotypic expression is widely spread wings

Dichapetalaceae a family of geranale dicotyledonous angiosperms

Dicholophidae = Cariamidae

Dichondraceae a family erected to contain the single genus Dichondra by those who do not hold it to be a member of the Convolvulaceae

Dichotomosiphonaceae that family of siphonale Chlorophyceae in which the thallus consists of dichotomously branched tubes interwoven into a cloth-like layer

dick a frequent component of popular names of organisms

    jumping-dick *W.I.* = popping-dick *W.I.*

    mountain-dick *W.I.* the cotingid bird *Platypsaris niger* (*cf.* becard)

    popping-dick *W.I.* the turdid *Turdus aurantius* (= white-chinned thrush)

    long-day-popping-dick *W.I.* the turdid bird *Turdus jamaicensis* (= white-eyed thrush)

sleepy dick *Austral.* the malurine bird *Gerygone fusca* (= western warbler *Austral.*)

slippery dick the labrid fish *Halichoeres bivittatus*

dickcissel *U.S.* the fringillid bird *Spiza americana*

Dicksoniaceae a family of filicine Pteridophyta containing those tree ferns which bear marginal sori at the vein tips. By some treated as a subfamily of the Cyatheaceae (*q.v.*)

Diclidantheraceae a small family of ebenale dicotyledons

Dicotylae = Dicotyledoneae

Dicotyledoneae one of two taxa of angiosperms distinguished by seedlings typically with two cotyledons, the leaves with netted pinnate or palmate venation. (*cf.* Monocotyledoneae)

Dicotylidae = Tayassuidae

Dicruridae that family of passeriform birds which contain the drongos. In many species the tail is deeply forked and racket-tipped and all possess elongate hair-like feathers in the crest

dictum sometimes used in the sense of "rule" or "law"

dictyonine said of a silicious sponge skeleton in which the spicules are cemented together

Dictosiphanoales an order of heterogenerate Phaeophyta in which growth is initiated by a single apical cell at the tips of a profusely branched cylindrical thallus

isodictyal = renierine

-dictyo- *comb. form* meaning "net" better transliterated -diktyo-

archedictyon the irregular network-like veins in the wings of primitive insects

Dictyoptera an order of insects containing the forms commonly called cockroaches, and the praying mantis. They were at one time, together with the Grylloblattodea and Phasmida, included with the Orthoptera, a group which, in this system of classification, is restricted to the grasshoppers, locusts and crickets. The Dictyoptera possess ten evident abdominal segments but are distinguished from the Grylloblatodea in not possessing a prominent ovipositor

Dictyotales an order of isogenerate Phaeophyta distinguished by the fact that growth of the erect thalli is initiated by an apical cell or a row of apical cells at the tip of each branch

Didelphyidae that family of polyprotodont marsupials which contains the forms commonly called opossums

Dididae = Rhapidae

Didiereaceae a small family of cactus-like sapindale dicotyledons

Didiereineae a suborder of Sapindales containing the single family Didiereaceae

Didunculidae a monospecific family erected to contain the tooth-billed pigeon *Didunculus strigirostris* now usually included in the Columbidae

didymous twinned

Didynamia that class of the Linnaean classification of plants distinguished by the possession of four stamens, two longer than the other two

diel a chronological day (24 hours) as distinct from the daylight portion of varying duration (*cf.* circadian, diurnal)

intradiel occuring with one twenty-four hour period

-diere- *comb. form* meaning "division" sometimes transliterated -diaire-

cytodieresis = mitosis or meiosis

plasmodieresis cell division

Dietesiae low growing perennial plants

differentiation the process by which cells, usually those

of an embryo, develop specific characters or patterns

dedifferentiation the reversion of cells, particularly in the life history of many invertebrates, from a specific to a generalized form. This process differs from degeneration in conveying the implication that redifferentiation will take place

redifferentiation the process by which dedifferentiated cells return to specific forms or patterns

diffract said of a coarse, net-like surface, some or all of the walls between the holes being themselves double and open

diffusion the intermingling of atoms, ions, or molecules in solutions or gasses

Digenea an order of trematode Platyhelminthes distinguished by the presence of an anterior sucker circling the mouth, a ventral sucker and an absence of hooks

digestion the enzymatic breakdown of large molecules into smaller ones more readily absorbed for food

gold-digger *Austral.* the meropid bird *Merops ornatus* (*cf.* bee-eater)

-digit- *comb. form* meaning finger

digit the jointed terminations or termination of a limb, *i.e.* fingers, thumbs and toes

digitate fingered, or in the form of fingers. In *bot.* = palmate

digitellus one of the tentacle-like projections from the gastric septa of Schyphozoa

Digynia those orders of the Linnaean system of plant classification distinguished by the possession of two pistils

dik-dik any of numerous species of neotragine antelopid of the genera Modoqua and Rhynchotragus

dikkop the *S.Afr.* burhinid bird *Burhinus capensis* (= spotted thickknee)

water dikkop *S.Afr.* the burhinid bird *B. vermiculatus*

-diktyo- *comb. form* meaning "a net" frequently transliterated -dictyo-

dil *see* [specific dilutor] and [dilute]

Dilaridae a small family of neuropteran insects readily identified by possessing a very long ovipositor

dilation *bot.* the cleavage of the xylem in wood through an increase in the parenchyma

Dilleniaceae a small family of parietalous dicotyledons distinguished by the polypetalous flowers with very numerous unicarpellate pistils and distinct or fasiculate stamens

wild **dilly** the sapotaceous tree *Mimusops parvifolia*

[dilute] *either* a mutant (gene symbol *dil*) mapped at 17 on chromosome 2 of the tomato. The phenotypic expression is a light green leaf *or* a mutant (gene symbol *d*) found in linkage group II of the mouse. Phenotypic expression involves the clumping of pigment granules in the hair

[dilute black] a mutant (gene symbol *bd*) mapped at 6.7 on linkage group IX of *Bombyx mori*. The phenotypic expression is a dilute black larva

[Dilution] a mutant (gene symbol *Sd*) found in the sex chromosome (linkage group I) of the domestic fowl. The phenotypic expression is a dilution of the pigment to blue

dimidiate divided into two unequal halves

diminution the loss of chromosomal elements in somatic, as distinct from germ, cells

[diminutive] *either* a mutant (gene symbol *dm*) mapped at 4.6 on the X chromosome of *Drosophila melanogaster*. The name is descriptive of the phenotype *or* a mutant (gene symbol *dm*) found in linkage group V of the mouse. The phenotypic expression involves not only small size, but also abnormal vertebrae and ribs

dimotous remote from

dingle a small wooded valley

dingo a wild dog (*Canis dingo*) found in *Austral.* but very doubtfully indigenous

[Dingy] a mutant (gene symbol *dn*) in linkage group IV of *Neurospora crassa*. The name derives from the dingy lumpy appearance of the phenotypic colony

Dinocapsales that order of dinophysian pyrrophytes which are pyrroid in shape and not motile even for short periods

Dinococcales that order of dinophysian pyrrophytes which lack transverse grooves and flagella and are commonly attached to some base by a short stalk

Dinoflagellata = Dinophyceae

Dinoflagellida an order of phytomastigophorous Protozoa the non parasitic forms of which are distinguished by the presence of two flagella, one trailing in the body axis, the other transverse to the axis. They may also be considered as algae, in which case they are included in the class Dinophyceae (*q.v.*) of the phylum Pyrrophyta (*q.v.*)

Dinomyidae a monospecific family of hystricomorph rodents containing the Pacarana

Dinophyceae that class of pyrrophyte algae which is distinguished by the possession of a transverse groove in which lie the paired flagellae. They are regarded by some zoologists as animals and, in this case, form the order Dinoflagellida of phytomastigophoran Protozoa

Dinotrichales that order of dinophysian pyrrophyta in which the cells are joined together in branching filaments

dioch any of several species of ploceid bird of the genus Quelea

Dioctophymoidea a class of parasitic nematodes. Usually, somewhat elongate, the males having a bursa without rays, and the mouth lacking lips but with numerous papillae

-diod- *comb. form* meaning "a passage"
micro**diode** pollen grain

Diodontidae a family of plectognathous fishes closely allied to the Tetraodontidae, and also capable of puffing themselves up, but distinguished by the presence of spines all over the outer surface of the body. Popularly called porcupine fish or burr fish

Diodyrhynchidae = Rhinomaceridae

Dioecia that class in the Linnaean system of plant classification distinguished by having the stamens and pistils in separate flowers and on different plants

Diomedeidae that family of procellariiform birds which contains the albatrosses. They are readily distinguished from other families in the order by the large size, exceptional powers of sailing flight and extremely long wings

Dionaeaceae a family erected to contain the single genus Dionaea more usually placed in the Droseraceae

Dioonioideae a subfamily of Cycadaceae containing the single genus Dioon from which the name is derived

Diopsidae a family of myodarian cycloraphous dipteran insects easily distinguished by the fact that the eyes are situated on the ends of long stalks. Hence the

popular name of stalk-eyed fly applied to these largely tropical insects

Dioptidae a family of noctuoid lepidopteran insects

Dioscoreae a subfamily of the Dioscoreaceae distinguished from the Stenomerideae by possessing dioeceous plants with unisexual flowers

Dioscoriaceae that family of liliflorous monocotyledons which contains the yams. Distinctive characteristics are the inferior 3-celled ovary, the presence of two albuminous seeds and usually, the climbing habit of growth and the net-veined leaves

Diospyrineae a suborder of Ebenales containing the following families Ebenaceae, Diclidantheraceae, Symplocaceae, Styracaceae, and Lissocarpaceae

Diphyllidae an order of eucestode cestodes distinguished by having two bothria on the scolex and a spiny head stalk

Diplasiocoela a suborder of anuran Amphibia distinguished by having ten vertebrae assorted between procoelous, amphicoelous and acoelous. The common frog Rana is included in this group

Dipleurula a postulated common ancestor of echinoderms (*cf.* Pentacula)

diploe = mesophyll

Diplogangliata = Arthropoda

Diploglossata an order of insects, usually combined with the Dermaptera, containing a single genus of ectoparasites of the banana

Diplopoda a class of progoneate arthropods containing the forms commonly referred to as millipedes. The cylindrical, or hemicylindrical body, bearing from forty to a hundred pairs of legs is typical of the class

diplotegia a dry fruit with an adnate calyx, produced from an inferior ovary

Diplozoa an obsolete taxon once combining the sponges with the coelenterates (*see also* -zoon)

Diplura = Aptera

Dipneuomonomorphae a suborder of araneid arthropods distinguished from the other three suborders by possessing a single pair of lungs

Dipneusti an order of osteichthyes popularly referred to as lung fishes by virtue of the presence of one or more sac like lungs

Dipnoi = Dipneusti

Dipodidae a family of myomorph rodents containing the forms commonly known as jerboas. They are distinguished by the enormous development of the hind legs in relation to the front legs so that they progress by means of long jumps

dipper 1 (*see also* dipper 2) any of several species of cinclid bird of the genus Cinclus *U.S.*, without qualification, usually *C. mexicanus* (*U.S.* water ouzel). *Brit. C. cinclus*

   American dipper *C. mexicanus*

   brown dipper *C. pallasii*

dipper 2 (*see also* dipper 1) any of many cuspidariid mollusks

Diprionidae a small family of relatively large symphytan hymenopteran insects differing from all other sawflies in the serrate antennae of the female and the deeply pectinate antennae of the males

Diprotodontia a main division of marsupial mammals containing the kangaroos and wombats. Distinguished by the presence of one strong pair of incisors in the lower jaw

Dipsacaceae that family of rubialous dicotyledons which contains the teasels. It is distinguished by the bibracteate epicalyse that in some genera becomes scariously spiney in fruit

Dipsadinae = Amblycephalidae

Dipsadomorphinae a subfamily of colubrid snakes erected to contain many of the opisthoglyphan forms, now considered an unnatural assemblage

Diptera that order of insects containing the forms commonly called houseflies and mosquitoes. They are distinguished by possessing a single pair of wings, the place of the second pair being taken by halteres which function as gyroscopic stabilizers

Dipteridaceae = a Polypodiaceae

dirolent said of those forms which smell by moving their heads backwards and forwards (*cf.* latolent)

[Dirty] a mutant (gene symbol *Di*) mapped at 0 on linkage group XIV of *Bombyx mori*. The phenotypic larva is irregularly spattered with black

-dis- *comb. form* meaning "apart" and also a frequent form of -di- meaning "two"

disc a thin circular object. Specifically the central area of an insect wing, the receptacle of the flower head in Compositae and the elevated nectaries in some flowers

A-disc the darker of the alternate light and dark discs of which striated muscle appears to be composed

amphidisc a type of sponge spicule with a disc or cup at each end. Found mostly in the walls of gemmules

adhesive disc the modified termination of the tendril of many climbing plants

cement disc (Orchidaceae) = retinaculum (Orchidaceae)

ectodermal disc a series of invaginated paired discs, cut off from the larval ectoderm in the development of nemertines

epiphyseal disc a transverse disc of cartilage which separates epiphyseal bone from diaphyseal bone

fixation disc that by which the free-swimming larvae of sessile invertebrates form their first attachment or by which the larvae of some motile invertebrates, particularly echinoderms, attach themselves before undergoing metamorphosis (*cf.* fixation papillae)

germinal disc that area of yolk free protoplasm on the upper suface of a telolethical egg, to which early development is confined

I-disc the lighter of the alternate light and dark discs of which striated muscle appears to be composed

imaginal disc those histoblasts, or new formations, in the larvae and pupa of holometabolous insects from which the parts of the imago are formed. Also called imaginal cells and imaginal buds

intercalated disc any disc interposed between two other objects, particularly between vertebrae, and between the anastomoses of myocardial cells

intervetebral disc a disc between two vertebrae

oral disc the flattened upper survace of anthozoan coelenterate

pedal disc the adhesive base of a coelenterate polyp

squamodisc a disc near the trematode opisthohaptor bearing concentric circles of spines

trochal disc one of a pair of semicircular ciliated areas into which the corona of bdelloid rotifers is divided

Discocephali an order of osteichthyes distinguished by the conversion of the "spiny" dorsal fin into a sucking disc. Commonly called remoras or sucker fish

Discoglossidae a family of phaneroglossan anuran amphibia distinguished by having a non-protrusible tongue in the form of a round disc

Discomedusae a term once used to describe an order of scyphozoan coelenterates. The order is now divided into the Semaesstomeae and the Rhizostomeae

Discomycetes the class of ascomycete fungi in which the asci lie in disclike or potato-like ascocarps. The former, having the general appearance of toadstools, are often mistaken for basidomycetes. The tuber-like forms are the well-known truffles

pleurodiscous lateral to a disc

discrete separate

washdish = the motacillid bird *Motacilla lugubris* (= pied wagtail)

disk = disc

dispersal the manner in which organisms are dispersed (*cf.* dissemination)

[displaced²] a mutant (gene symbol *dd²*) mapped at 27.2 on the X chromosome of *Drosophila melanogaster*. The phenotypic expression includes sunken antennae in a deformed head

character **displacement** the situation which results when the differences between two species are exaggerated by natural selection in the areas where they overlap but not in areas where each exists alone

display an activity for the purpose of attracting attention

epideictic display the communal, or group displays of large populations

epigamic display mating display

dissected deeply cut

dissemination the methods by which individual offspring or seeds are liberated from the parent (*cf.* dispersal)

disseminule a characteristic or device leading to widespread dissemination of a species

dissepiment partitions dividing a space, as outgrowths from the wall of an ovary or the calcareous skeleton in the septa of a coral

spurious dissepiment a similar partition produced by any other means

dissilient flying apart

distad = distal

distaff of or pertaining to descent from the female side (*cf.* spear)

distal that which lies further from. In general biological usage the anterior end and the main axis are used as the points of departure so that the stomach is distal to the esophagus and the arm is distal to the shoulder (*cf.* proximal)

distinct separate. In *bot.* said of parts of the same series when neither coherent nor connate

-disto- *comb. form* meaning "to stand apart" or by extension "to be remote from"

Distomata a division of prosostomatous trematodes with the oral sucker at the front end and with the acetabulum on the ventral surface

distractile spread widely apart

Ditomyiidae a family of nematoceran dipteran insects now usually incuded with the Mycetophilidae

dittany any of several species of rutaceous plants of the genus Dictamnus (particularly *D. albus*) and the labiate herb *Cunila origanoides*

bastard dittany the labiateous herb *Ballota pseudo-dictamnus*

cretan dittany the labiateous herb *Origanum dictamnus*

Maryland dittany the labiateous herb *Cunila mariana*

diurnal pertaining to daylight hours either as a rhythmic cycle or as an activity opposed to nocturnal or crepuscular (*cf.* circadian, diel)

dive to proceed downwards in a controlled manner either through air, from air to water, or through water

diver 1 (*see also* diver 2) a term applied to almost any aquatic bird capable of diving but in Britain frequently restricted to the gaviids

   hell-diver (*see also* diver 2) *W.I.* either of two podiciped birds *Podiceps dominicus* (= least grebe) or *Podilymbus podiceps* (= pied billed grebe)

   mackerel-diver = mackerel-bird

   magpie-diver *Afr.* = tufted duck

   great northern diver *Brit. Gavia immer*

   red diver *W.I.* = ruddy duck *W.I.*

   red-throated diver *Brit. Gavia stellata*

diver 2 (*see also* diver 1) any of several burrowing organisms

   helldiver (*see also* diver 1) *U.S.* the larva of the dobson fly *Corydalus cornutus* (= hellgrammite)

   mud diver the pelobatid anuran amphibian *Pelodytes punctatus*

   sand diver popular name of krameriid fish and also applied to the syndontid *Sinodus intermedius*

ecological **divergence** the production of races and subspecies adapted to varying ecological conditions

[divergens] a mutant (gene symbol *di*) mapped at 20 on chromosome 4 of the tomato. The phenotype has a gray-green cast

[divergent] a mutant (gene symbol *dv*) mapped at 20.0 on chromosome III of *Drosophila melanogaster*. The name refers to the spread wings of the phenotype

[divers] a mutant (gene symbol *dvr*) mapped at 28.1 on the X chromosome of *Drosophila melanogaster*. The phenotype has short dark wings

diverticulum any blindly ending pouch and specifically in *bot.* the connection between the procarp cells and the placenta in some algae

   hepatic diverticulum either a diverticulum running anteriad from the junction of the esophagus and gut in Amphioxus or an outgrowth from the primitive gut, precursor to the liver

division the process of breaking into two or more parts

   cell division frequently taken as synonymous with mitosis even though this term applies properly to the nucleus

   equational division = maturation division

   maturation divisions that meiotic division in which the chromsome number is not reduced

   reduction division that meiotic division in which the number of centromeres, and usually the number of chromosomes, is reduced to half

Dixidae a small family of minute mosquito-like nematoceran dipteran insects. They do not bite and are one of the many groups to which the word midge is loosely applied

dke *see* [dark eye]

Dl *see* [Delta]

dl *see* [dialytic stamens]

dm *see* [diminutive]

dn *see* [Dingy] and [doughnut]

DNA deoxyribonucleic acid

DNP = dinitrophenol

proto**docha** = prisere

dock 1 (*see also* dock 2, sawfly) any of several polygonaceous herbs of the genus Rumex

   alpine dock *R. alpinus*

   bitter dock *R. obtusifolius*

   curly dock *R. crispus*

   golden dock *R. maritimus*

   patience dock *R. patientia*

   sour dock *R. crispus*

   spinach dock *R. patientia*

   water dock *R. altissumus*

     great water dock *R. hydrolathum*

   yellow dock = curly dock

dock 2 any of several other plants having a general similarity to Rumex

   burdock any of several species of the compositaceous genus Arctium

   prairie dock the compositaceous herb *Silphium terebinthinaceum* (= rosin weed)

   splatter dock any of several species of nymphaeaceous aquatic herbs of the genus Nuphar but particularly *N. advena*

   wendock the nymphaceous aquatic herb *Drasenia schreberi*

dockmackie the caprifoliaceous shrub *Viburnum acerifolium*

doctor any of several organisms superstitiously thought to have curative powers for themselves or others

   fish doctor (*see also* doctor fish) *U.S.* the zoarcid fish *Gymnelis viridis*. *Brit.* the cyprinid fish *Tinca vulgaris* (= tench)

   snake doctor *U.S.* either the larva of the dobson fly *Corydalus cornutus* (= hellgrammite) or any of several odonatan insects

doda = chousingha

dodder (*see also* weevil) any of many species of convolvulaceous parasitic herbs of the genus *Cuscuta*

-dodeca- *comb. form* meaning "twelve"

Dodecandria that class in the Linnaean classification of plants distinguished by the possession of twelve to eighteen stamens

blue **dodger** *U.S.* the cicadellid homopteran insect *Oncometopia undata*

   grass dodger *W.I.* the fringillid bird *Ammodramus savannarum* (= grasshopper sparrow)

dodlet the columbid bird *Didunculus strigirostris* (= tooth-billed pigeon) at one time made the type of a special family the Didunculidae

dodo the extinct bird *Didus ineptus*. Sometimes placed with the pigeons in the Columbidae but more usually placed in a separate family the Raphidae

doe the female of many cervid mammals, and also sometimes applied to the hare, (though puss is preferable), kangeroo and ferret

dog 1 (*see also* dog 2–4) specifically any male canine, either domestic or wild, in contrast to bitch. The term is also used for the male otter

dog 2 (*see also* dog 1, 3, 4, face, fish, grass, kennel, louse, mint, moth, rose, tick, whelk and wood) any of numerous, rather diverse, animals forming the carnivorous family Canidae, containing also the foxes and wolves. The origin of domestic dogs, sometimes called *Canis familiaris*, is completely unknown but they are probably polyphyletic

   Azara's dog *Dusicyon azarae* (*cf.* jackal)

   bushdog the cymocyonine canid carnivore *Speothus venaticus*. The term is sometimes also applied to the kinkajou

   coydog a relatively frequent natural hybrid between a domestic dog, usually a German shepherd, and a coyote

   crab-eating dog *Dusicyon cancrivora* (*cf.* jackal)

   hunting dog the cymocyonine canid carnivore *Lycaon pictus*. It differs from a canine canid in having only four toes on the front foot. Also called Cape hunting dog

raccoon dog the Japanese canine *Nyctereutes procyonides*

dog 3 (*see also* dog 1, 2, 4) any of several other vertebrates

blue dog *U.S.* = porbeagle

hedgidog obs. = hedgehog

prairie dog a sciurid rodent of the genus Cynomys

black tail prairie dog *C. ludovicianus*

white tail prarie dog *C. gunnisoni*

waterdog *U.S.* any of several species of proteid urodele amphibian of the genus Necturus. The term is also variously applied in parts of the *U.S.* to almost any large urodele, particularly Cryptobranchus

dog 4 (*see also* dog 1–3) any of several lepidopteran insects

hop dog *Brit.* the lymantriid lepidopteran insects *Dasychiar pudibunda*

orange dog *U.S.* the papilionid *Papilio cresphontes* (= giant swallowtail)

dohle the canid carnivore *Cyon dukkunensis*

-dolabr- *comb. form* meaning a "hatchet" or "mattock"

-doli- *comb. form* meaning a "cask"

-dolicho- *comb. form* meaning "along"

Dolichoderinae a subfamily of formicid hymenopteran insects with a single segmented pedicel and no constriction between the first and second segments of the gaster. They are particularly distinguished by possessing glands secreting a foul-smelling fluid

Dolichopidae = Dolichopodidae

Dolichopodidae a large family of small orthoraphous brachyceran dipteran insects mostly distinguished by their brilliant metallic colors. They are frequently called long-legged flies because of the long legs, often curiously ornamented, of the males

Dolichopsyllidae = Ceratophyllidae

dolichosis the condition of being stunted

Doliolida an order of thaliacian tunicates that resembles the Salpida in being either solitary or aggregated in chains, but which are furnished with numerous transverse gills

sand **dollar** the clypeastrine echinoid echinoderm *Echinarachnius parma.* The term is sometimes applied to other flattened echinoids

dolly varden the salmonid fish *Salvelinus malma*

dolphin 1 (*see also* dolphin 2) popular name, particularly in *U.S.*, of coryphaenid fishes. *U.S.*, without qualification, usually *Coryphaena hippurus*

dolphin 2 (*see also* dolphin 1) any of many small cetacean mammals particularly of the delphinid genus Delphinus. *Brit.*, without qualification, *D. delphis*

Amazon dolphin the platanistid *Inia geoffrensis*

Ganges dolphin the platanistid *Platanista gangeticae* (= susu)

La Plata dolphin the platanistid *Pontoporia blainvillii*

Risso's dolphin the delphinid *Grampus griseus*

-dome- *comb. form* meaning "house" frequently confused with the totally unrelated root -deme- (*q.v.*)

apodome the internal portions of an exoskeleton and therefore comprising both apodeme (*q.v.*) and apophysis (*q.v.*). Used by many as synonymous with apodeme

cladome the three short rays of a sponge spicule attached to a rhabdome

myodome a cavity between the base of the cranium and the parasphenoid in actinopterygian fishes

plastidome the sum total of all the plastids in an organism

rhabdome the longest of the four axes of a tetraxon sponge spicule

rhytidome the cork layer, and all tissues external to it, of the periderm (= outer bark)

domin an organism showing weak dominance in an association

vedomin a minute domin

dominance 1 (*see also* dominant 1, dominance 2) in the genetic sense, the condition of a character that appears both in the heterozygote and the homozygote. The character is said to be "dominant" and the effect to result from a dominant allele

alternate dominance an obsolete theory of sex determination which presumed that sex depended on male or female determinates in sexually heterozygous individuals

conditional dominance said at one time of a gene which was known from the heterozygote only

conditioned dominance the situation in which a dominant gene can be modified by another gene

incomplete dominance the incomplete masking of a recessive character by a dominant character in the F_1 generation

overdominance a condition in which the phenotype of a heterozygote is more extreme than that of either homozygote

partial dominance a condition in which the phenotype of the heterozygote is intermediate between the two parents

pseudodominance the appearance of a recessive character in a heterozygous individual owing to the absence of the dominant allele from that individual as the result of a deficiency

semi-dominance = partial dominance

dominance 2 (*see also* dominant 2, dominance 1) in the ecological sense, the condition of an organism which dominates a community either in virtue of its habits or of the sheer weight of its numbers

dominant 1 (*see also* dominant 2, dominance 1) an allele (*q.v.*) which dominates another allele. For variations of this last condition *see* dominance 1

delayed dominant an allele the phenotypic expression of which does not appear until quite late in development

double dominant the condition in which a phenotypic expression is dependent on the presence of both of two dominant alleles

dominant 2 (*see also* dominance 2, dominant 1) in the ecological sence, an organism or several organisms in a community which so behave, either passively or actively, as to dominate or control the whole habitat

subdominant an organism which becomes dominant in areas not controlled by the regular dominant

[**Dominant chocolate**] a mutant (gene symbol *l-a*) mapped at 5.9 in linkage group IX of *Bombyx mori.* The phenotype expression is the same as chocolate

[**Dominant hemimelia**] a mutant (gene symbol *Dh*) found in linkage group XIII of the mouse. The phenotypic expression involves not only hemimelia but also the absence of a spleen

[**Dominant spotting**] a mutant (gene symbol *W*) in linkage group III of the mouse. In addition to white spotting of the coat the phenotypic expression includes macrocytic anemia and sterility

[**Dominant white**] a mutant (gene symbol *I*) in linkage

group III of the domestic fowl. The name is descriptive

dominule the dominant of a microhabitat

-domit- *comb. form* meaning "tamed"

domous "dwelling in the house of" or "providing a house for"

   monodomous dwelling in a single house

   myrmecodomous providing housing for ants

   polydomous having many houses, as birds that occupy two or more nests in the same season, or ants that have several interconnecting colonies

Donacidae a family of pelecypod mollusks commonly called wedge shells, butterfly shells, or coquina. They are small wedge-shaped clam-like shells with the posterior ends prolonged and rounded

donkey a long eared, domesticated ass (*q.v.*) possibly, but not certainly, derived from *Equus africanus*

bella**donna** (*see also* lily) either of two species of herbs of the solanaceous genus Atropa

donor 1 (*see also* donor 2) the individual from which a tissue transplant is removed to a receptor

donor 2 (*see also* donor 1) a molecule from which atoms are removed in the course of an enzyme catalyzed reaction

-dont- *comb. form* meaning "tooth" (*cf.* -dent-)

   acrodont said of a dentition in which the teeth are attached by their sides to a bony ridge in the jaw

   amphidont = mesodont

   angulodont having bluntly angular teeth

   aulodont said of those echinoderms in which the teeth lack a keel and the epiphyses of Aristotle's lantern are widely separated

   brachydont said of a form in which the teeth grow only for a limited period. Also said of a mammal the teeth of which have short crowns but well developed roots

   bunodont having tubercles on the crown of the molar teeth

   camarodont said of those echinoid echinoderms in which the teeth are keeled and the epiphyses of Aristotle's lantern are joined together

   catadont having teeth only in the mandible

   coelodont hollow-toothed

   coryphodont possessing a dentition in which there is a gradual diminution in tooth length from front to back

   curvidont having curved teeth

   diphyodont said of a form with two generations of teeth

   haplodont said of a form with simple conical teeth

   heterodont a term used of a form possessing several kinds of teeth

   homodont said of a form in which all of the teeth are of a similar size and shape

   hypsodont said of a mammal in which the teeth grow over a long period

   kumatodont = mesomegadont

   lophodont having ridges on the crowns of the teeth

   megadont possessing some teeth markedly larger than others

     mesomegadont having the largest teeth in the central area of the maxilla with smaller teeth anterior and posterior to them

     opisthomegadont with the largest teeth at the rear

     diaopisthomegadont having the large teeth separated from the small by a diastema

     synopisthomegadont having the megadont teeth grading into each other without a diastema

   promegadont with the largest teeth in the anterior region

   mesodont a form intermediate between a priodont and a teleodont

   notodont having teeth or tooth-like structures along the back, as in the larvae of many moths

   oligodont having few and widely separated teeth

   oinododont a type of dentition in which there is a gap somewhere in the maxillary series of teeth

   oligophyodont having successive series of teeth replacing each other

   polyphodont having successive generations of teeth replacing each other

   pleurodont said of teeth that rise from the top of a bony ridge in the jaw or that are attached to the outer wall of the alveolar groove

     eupleurodont that form of dentition in which the teeth are replaced from beneath

     subpleurodont that form of dentition in which the teeth are replaced by intercalation

   polyodont having many functional teeth at one time

   priodont said of a male lucanid coleopteran insect with very small mandibles (*cf.* mesodont, teleodont)

   scaphiodont = promegadont

   statodont a term used of a form in which lost teeth are not replaced

   teleodont said of a male lucanid coleopteran insect with unusually large mandibles (*c.f.* mesodont, priodont)

   thecodont said of a form in which the teeth are set in sockets

     subthecodont said of forms in which the teeth are fused to a bony sheath

dooja the rutaceous tree *Microcitrus australis*

DOPA *see* 1-3,4-dihydroxyphenylalanine and carboxylase

dor *Brit.* the coprid coleopteran *Geotrupes stercorarius* (= dumble beetle)

   dumble dor = cockchafer

dor *see* [deep orange]

dorab the chirocentrous fish *Chirocentrus dorab*

doré *see* dory

Doradidae a family of *S.Amer.* armored catfish. In some the overlapping armour plates are festooned with hooks

dorca *Austral.* any of several species of macropodid marsupial of the genus Dorcopsis (= dorca kangaroo)

Dorilaidae = Pipunculidae

-dorm- *comb. form* meaning "sleep" but used, by extension, for words descriptive of any structure or organism which remains inactive

dormancy a condition in a plant or animal in which the life processes are slowed down, usually for the purpose of surviving a temporarily inclement environment

Dorosomidae a family of isospondylous fishes closely allied to the Clupeidae and usually included with them. If separated it contains the genus Dorosoma known as the gizzard shads

dorr (*see also* hawk) *Brit.* the scarabaeid beetle *Melontha solstitialis* (*cf.* dor, chafer)

dorsal pertaining to the back. In *bot.* the dorsal surface of a leaf is the lower surface which was originally the outer surface in the bud (*cf.* sulcal)

dorsalia = arcualia

dorse *obs. Brit.* codling

dory any of numerous fishes of a golden color, the name

being a corruption of *Fr.* doré. *U.S.* occasionally, particularly in the form doré, applied to the wall eyed pike

bastard dory *N.Z.* the zeid teleost *Cythus australis*

john dory the laterally compressed, golden yellow ("jaune doré") zeid teleost *Zeus faber*

Doryceridae a family of brachyceran dipteran insects usually included with the Otitidae

Dorylaidae = Pipunculidae

Dorylaimidea a class of nematodes, with a smooth cuticle, cyathiform amphids, and two circles of labial papillae

dorylamoid *see* pharynx

Dorylinae a subfamily of Formicidae containing the forms commonly called legionary ants or army ants from their nomadic habit

dot 1 (*see also* dot 2) a small object

   Casparian dot = Casparian strip

   germinal dot the centrosome of diatoms

dot 2 (*see also* dot 1) any of several lepidopteran insects *Brit.*, without qualification, the noctuid *Mamestra persicariae*

   green brindled dot *Brit.* the noctuid *Valeria oleagina*

   polka dot *S.Afr.* the acraeid *Pardopsis punctatissima*

   straw dot *Brit.* the noctuid *Rivula sericealis*

[dot] a mutant (gene symbol *dot*) in linkage group I of *Neurospora crassa* The phenotypic expression is a restricted colonial growth

dot *see* [dot]

Dothidiales an order of pyrenomycete ascomycetes distinguished by the production of a massive dark-colored stroma on the surface of the plant which is parasitized

Dotryllidae a family of ptychobranchiate ascidians that form rock-encrusting colonies

[Dotted] a mutant (gene gymbol *Dt*) mapped at 0 on linkage group IX of *Zea mays.* The gene controls the mutability of the gene d_2 in linkage group V

dotterel *Brit.* the charadriid bird *Eudromias morinellus.* *N.Z.* the charadriid *Pluviorhynchus obscurus.* *Austral* the glareolid *Peltohyas australis*

   Asiatic dotterel *Eupoda asiaticus*

   red-keed dotterel *Erythrogonis cinctus*

double used specifically of flowers having an increase in the number of units in one or more series of the floral envelope

douc a brightly colored colobid primate *Pygathrix nemaea* usually classed with the langurs (*q.v.*)

[doughnut] a mutant (gene symbol *dn*) mapped at 47.0± on chromosome III of *Drosophila melanogaster.* The phenotypic expression of a sepia-doughnut cross is an eye with a light central spot

Douglasiidae a small family of leaf mining cycnodoid lepidopterans

douroucoulis popular name of pithecid primates of the genus Aotes. They are distinguished by their very large eyes from which their other name of owl-monkey is derived

dout term for a group of wild cats (*cf.* destruction)

dove (*see also* shell, tree) any of numerous columbiform birds of the family Columbidae. There is no real distinction between doves and pigeons though the former term is usually applied to smaller birds. The *U.S.* "prairie dove" is a gull

barbary dove *W.I.* = bridled quail dove

barred dove *Geopelia striata*

white-bellied dove *W.I. Leptotila jamaicensis*

blue dove *W.I.* = crested quail-dove

Caribbean dove *Leptotila jamaicensis*

cinnamon dove *Afr.* = lemon dove

cuckoo-dove any of several species of the genus Macropygia

diamond dove *Geopelia cuneata*

violet-eared dove *Zenaidura auriculata*

emerald dove *Chalcophaps indica*

red-eyed dove *W.Afr. Streptopelia semitorquata*

fruit dove any of several species of the genera Ptilinopus and Leucotreron

ground dove any of many species predominantly ground dwelling doves particularly those on the genera Colombigallina and Claravis. *W.I.*, and *U.S.* without qualification, *Columbigallina passerina.*

   blue ground-dove *Columbigallina minuta*

   eastern ground-dove *U.S. Columbigallina passerina*

   ruddy ground-dove *Columbigallina talkacoli*

   Trinidad ground-dove = violet-eared dove

pink-headed dove *W.Afr. Streptopelia roseogrisea*

Inca dove *Scardafella inca*

laughing dove *W.Afr.* = *Streptopelia senegalensis*

lemon dove any of several species of the genus Aplopelia, particularly *A. larvata*

marmy dove *W.I.* = bridled quail-dove

morning dove frequent misspelling of mourning dove

mountain dove *W.I.* = zenaida dove

mourning dove *U.S. Zenaidura macroura Afr. Streptopelia decipiens*

namaqua dove *Oena capensis*

ring-necked dove *Streptopelia capicola*

prairie dove *U.S.* the larid bird *Larus pipixcan* (= Franklin's gull)

quail-dove any of numerous species of columbid mostly bird of the genus Geotrygon

   bridled quail-dove *G. mystacea*

   crested quail-dove *G. versicolor*

   blue-headed quail-dove *Starnoenas cyanocephalus*

   grey-headed quail-dove *G. caniceps*

   ruddy quail-dove *G. montana*

   Key West quail-dove *G. chrysia*

ring dove *Brit. Columba palumbus* (= wood pigeon) *I.P. Streptopelia decaocto*

rock dove *Columba livia* (= domestic pigeon)

seaside dove *W.I.* = zenaida dove

Senegal dove *W.Afr.* = laughing dove

blue-spotted dove *W.Afr. Turtur afer* (= blue-spotted wood dove)

Malay spotted dove *Streptopela chinensis*

stock dove *Columba oenas*

stone dove = ground dove

longtailed dove *Oena capensis*

tambourine dove *W.Afr. Tympanistria tympanistria*

tobacco dove *W.I.* = ground dove *W.I.*

torpedo dove *W.I.* = violet eared dove

town dove *W.Afr.* = Senegal dove *W.Afr.*

turtle dove *Brit. Streptopelia decaocto. U.S.S.R. risoria.* *W.I. Zenaidur macroura*

   adamawa turtle dove *Streptopelia hypopyrrha*

   Saharan turtle dove *Streptopelia turtur*

bronze winged dove *Chalcophaps indica*

   white winged dove *Zenaida asiatica*

wood dove *W.I.* = zenaida dove

zenaida dove *Zenaida aurita*

dovkie the alcid bird *Plautus alle*

dow *see* [downy]

dowitcher any of several scolapacid birds of the genus *Limnodromus griseus*

down 1 (*see also* down 2, 3) any soft hairy material such as the underfeathers of ducks or the pappus on seeds like those of the dandelion

down 2 (*see also* down 1, 3) term for a group of hares (*cf.* husk, drove, trip)

down 3 (*see also* down 1, 2) in compound names of organisms

john down = fulmar

lookdown the carangid fish *Selene vomer*

[**downy**] a mutant (gene symbol *dow*) mapped at 8.0 on the X chromosome of *Drosophila melanogaster*. The phenotype has fuzz rather than bristles

Doyère's *epon. adj.* from Louis Doyère (1811–1863)

Dp *see* [Duplication]

dp *see* [dumpy]

DPN *obs.* = diphosphopyridine nucleotide *obs.* = NAD

dr *see* [droopy], [dreher]

drab *Brit.* any of numerous species of noctuid lepidopteran insect of the genus Taeniocampa

Dracaeneae a subfamily of amaryllidaceous monocotyledons. Considered by many properly to belong in the Liliaceae

Dracenioideae a subfamily of liliaceous monocotyledons erected to contain, *inter alia*, the agaves

Dracunculidae a family of very elongate narrow filarioid nematode worms containing the well-known guinea worm (*q.v.*)

Dracunculidea a class of parasitic nematodes, lacking lips or even a cuticularized buccal capsule. There is no bursa

dragon 1 (*see also* dragon 2) properly a large mythical reptile but also applied to many large lizards and some other animals. The biblical dragon was at times a jackal (Micah 1:8) and at others (Ezekiel 29:3) a crocodile. The term dragon is also used in U.S. for the larva of the dobson fly (= hellgrammite)

flying dragon the agamid lacertilian reptiles *Draco volans*

Komodo dragon the varanid lacertilian *Varanus komodoensis*

white dragon the hylobiid urodele amphibian *Batrachuperus karlschmidti*

dragon 2 (*see also* dragon 1, head) any of several plants

flying dragon a horticultural variety of the rutaceous tree *Poncirus trifoliata*

green dragon *U.S.* the araceous plant *Arisaema dracontium*

snapdragon any of several species and numerous horticultural varieties of the scrophulariaceous genus Antirrhinum

dwarf snapdragon any of several species of herb of the scrophulariaceous genus Chaenorrhinum

dragonet popular name of callionymid fish, particularly *Callionymus drago*

drake 1 (*see also* drake 2) the male of any duck or duck-like bird, and as a compound in names of ducks without regard to sex

sheld drake the anatine *Tadorna tadorna* (= sheld duck)

drake 2 (*see also* drake 1) in compound names having nothing to do with ducks

mandrake the solanaceous herb *Mandragora officinale*. The term is sometimes erroneously applied to the may apple (*q.v.*)

green drake the ephemerid ephemeropteran insect *Ephemera vulgata*

dray term for a group of squirrels or for a young squirrel

dreen the shrubby, or sometimes woody, swamp which occupies the "channel" between a partially detached island and the mainland

[**dreher**] a mutant (gene symbol *dr*) found in linkage group XIII of the mouse. The phenotypic expression is a typical waltzing syndrome

Dreisseniidae a family of pelecypod mollusks commonly called platform mussels. They are distinguished from the Mytilidae by the presence of a plate under the beak

-drepan- *comb. form* meaning "a sickle"

Drepaniidae a family of drepanoid lepidopteran insects containing the forms commonly called hook-tip moths

Drepaniidae a family of passeriform birds commonly called honeycreepers and confined to the Hawaiian islands. They are distinguished by the narrow pointed wings with only nine functional primaries, and their strong flight

drift a term for a group of wild pigs or swine when feeding peacefully (*cf.* sounder) and for a group of domestic cattle, though drove and herd are preferable for the latter

genetic drift the irregular change in gene frequencies in a population from generation to generation as a result of random processes

drill 1 (*see also* drill 2, borer 4) any of numerous hole-drilling mollusks. Most operate on the shells of other mollusks. Without qualification, usually the muricid gastropod *Eupleura caudata*

oyster drill the muricid pelecypod mollusk *Urosalcinx cimereus*

drill 2 (*see also* drill 1) popular name of such cynopithecid primates of the genus Mandrillus as are not otherwise specifically designated. Without qualification, usually refers to *M. leucophaeus* (*cf.* mandrill)

-drim- *comb. form* meaning "pungent" but used by ecologists in the sense of "alkaline"

drimad a plant of alkaline formations

drimium an alkali plain formation

drinker *Brit.* the lasiocampid lepidopteran insect *Cosmotriche potatoria*

driodad a plant of dry thickets

-drom- *comb. form* meaning "a course" or "direction" usually in the form -dromo-

peridroma the rhachis of a fern

Dromadidae a monospecific family of charadriiform birds erected to contain the crabplover. This is distinguished by its strong coarse laterally compressed bill

Dromaeidae = Dromiceiidae

Dromaiidae = Dromiceiidae

dromedary a large one-humped artiodactyl better called the Arabian camel (*q.v.*)

Dromiceiidae that family of casuariiform birds which contains the emu. Readily distinguished by the absence of rectrices

dromous pertaining to a direction

anadromous said of marine fish that enter fresh waters to spawn

antidromous said of the branch of a plant the stems of which twist in diverse directions

catadromous descriptive of fish which pass from fresh water to salt water for reproductive purposes

didromous twice twisted

heterodromous said of a spiral that changes the direction of its twist

homodromous having all the spirals running in the same direction

opisthodromous a genetic spiral that turns once between the bract and the first floral segment

drone a male bee

drongo (*see also* cuckoo) any of many species of bird of the family Dicruridae

  ashy drongo *Dicrurus longicaudatus*

  glossy-backed drongo *Dicrurus adsimilis*

  cuckoo drongo = drongo cuckoo

  velvet mantled drongo *Dicrurus modestus*

  shining drongo *Dicrurus atripennis*

  square tailed drongo *Dicrurus sharpei*

[droopy] a mutant (gene symbol *dr*) mapped at 71.2± on chromosome II of *Drosophila melanogaster*. The term refers to the appearance of the wide spread wings of the phenotype

[droopy ear] a mutant (gene symbol *de*) found in linkage group XVI of the mouse. The laterally projecting ears of the phenotype are set low on the head

[droopy wing] a mutant (gene symbol *drw*) mapped at 52.3± on the X chromosome of *Drosophila melanogaster*. The phenotypic expression involves small flies with droopy wings. The males of the population are sterile

drop widely used frequently in the plural, in compound names of plants having drop-shaped flowers

  beech drops the orobranchaceous herb *Epifagus virginiana*

  false beech drops the pyrolaceous parasitic herb *Monotropa hypopithys* (= pinesap)

  coral drops the liliaceous herb *Bessera elegans*, often called Mexican coral drops

  golden dewdrop the verbenaceous herb *Duranta plumieri*

  golden drop any of several species of boraginaceous herb of the genus Onosma

  golden eardrops the fumariaceous herb *Dicentra chrysantha*

  lady's eardrop any of several species of onagraceous shrubs of the genus Fuchsia and also the polygonaceous herb *Brunnichia cirrhosa*

  pine drops the pyrolaceous root-parasitic herb *Pterospora andromedea*

  snowdrop any of several species of the amaryllidaceous genus Galanthus particularly *G. nivalis*

  giant snowdrop *G. elwesii*

  sundrops the onagraceous herb *Oenothera linearis*

dropping term for a group of sheld drakes

Droseraceae that family of sarraceniales dicotyledons which contains the sundews. The insectivorous sensitive leaves are characteristic of the family

-droso- *comb. form* meaning "dew"

Drosophilidae a family of small yellowish myodarian cycloraphous dipteran insects which derive their importance from the contribution made by *Drosophila melanogaster* to the study of genetics. The name fruit fly applied to these forms is a misnomer and should properly be used with the Tephritidae the larva of which actually eat fruit. The Drosophilidae eat yeasts and fungi associated with decomposing fruit and are sometimes called pomace flies

drove a term for a group of asses (*cf.* pace), domestic cattle (though herd is more common), or hares (*cf.* husk, down, trip)

drug any reagent used in the treatment of disease but often specifically applied to habit forming alkaloids

drum 1 (see also drum 2) a tightly stretched membrane

  ear drum = tympanic membrane

drum 2 (*see also* drum 1) any of several sciaenid fishes

  banded drum *Larimus fasciatus*

  black drum *Pogonias cromis*

  red drum *Sciaenops ocellata*

  sand drum *Umbrina coroides*

  spotted drum *Equetus punctatus*

  star drum *Stellifer lanceolatus*

  striped drum *Equetus pulcher*

  freshwater drum *Aplodinotus grunniens*

-drup- comb. form meaning "an olive" in the sense of the fruit (*cf.* -elaeo-)

drupe a one-seeded fruit, such as a plum, cherry or peach, in which there is a thick stony endocarp and a fleshy mesocarp

  false drupe a nutlike fruit with a fleshy lower part

  spurious drupe any fleshy body, not a true drupe, but which encloses a stone, and therefore appears to be a drupe

druse a stellate crystal in a plant cell

drusy covered with a bloom like a ripe grape

drw *see* [droopy wing]

-drya- *comb. form* meaning "woodnymph"

  yellow dryad *I.P.* the amathusiid lepidopteran insect *Aemona amathusia*

drymium a formation of woody plants

  halodrymium a mangrove formation

  therodrymium the leafy forest formation

  xerodrymium a dry thicket

Dryomyzidae a small family of myodarian cycloraphous dipterans closely allied to the Sciomyzidae

Dryopidae a small family of aquatic coleopteran insects. They are, together with the closely allied Psephenidae and Elmidae, commonly called long-toed water beetles a name descriptive of their general appearance. The Dryopidae are distinguished from the others by being densely hairy

Dryopoidea a superfamily of polyphagous coleopteran insects containing most of the non-predatory aquatic beetles

ds *see* [dachsous]

Dsd a bacterial genetic marker having a character affecting the activity of D-serine deaminase mapped at 43.9 mins. for *Escherichia coli*

Dt *see* [Detached] *or* [Dotted]

du *see* [ducky] or [dutch pattern]

grand duchess *I.P.* the nymphalid lepidopteran insect *Euthalia patala* (*cf.* duke, baron, earl, count, baronet, marquis)

duck 1 (*see also* duck 2, 3, foot, head, louse, meat, weed and wheat) any of numerous anseriform birds. The distinction between a duck and a goose is largely one of size and local custom

  acorn duck *U.S.* = wood duck *U.S.*

  white-backed duck the oxyurine *Thalassornis leuconotus*

  bay duck *U.S.* = diving duck

  spoonbill duck *Austral.* the anatine *Spatula rhynchotis* (= Australian shoveler)

  blue-billed duck *Austral.* the oxyurine *Oxyura australis*

    knob-billed duck *W.Afr.* the cairine *Sarkidiornis melanotos* (= comb duck Afr.)

    yellow-billed duck the anatine *Anas undulata*

    red-billed duck the anatine *Anas erythrorhyncha*

    spot-billed duck the anatine *Anas erythrorhyncha*

  black duck *U.S.* the anatine duck *Anas rubripes*. *W.I.* any of several species of aythyine mostly of the genus Aythya. *Austral.* the anatine *Anas superciliosa* (= grey duck) *Afr.* the anatine *Anas sparsa*

  blue duck *N.Z.* the aythyine *Hymenolaimus malacorhynchos*

brown duck *N.Z.*the anatine *Anas chlorotis* (= brown teal)

Carolina duck the anatine *Aix sponsa* (= *U.S.* wood duck)

comb duck either of two species of cairinine duck of the genus Sarkidiornis

dabbling duck any of numerous anatine ducks including such forms as the domestic duck, the mallards and the teals

diving duck popular name for ducks of the subfamily Aythyinae

eider duck any of several species of aythyine duck of the genera Somateria and Polysticta, noted for their soft down

pink-eared duck *Austral.* the anatine *Malacorhynchus membranaceus*

white-eyed duck any of several athyine ducks of the genus Aythya

white-faced duck *W.Afr.* = white-faced tree-duck

flightless duck the anatine *Nesonetta aucklandica*

freckled duck the anatine *Stictonetta naevosa*

grey duck the anatine *Anas superciliosa*

harlequin duck *U.S.* the aythyine duck *Histrionicus histrionicus*

Hawaiian duck the anatine *Anas wyvilliana*

black-headed duck the anatine *Heteronetta atricapilla*

loggerhead duck = steamer-duck

pink-headed duck the anatine *Rhodonessa caryophyllacea*

white-headed duck the oxyurine *Oxyura leucocephala*

Labrador duck the aythyine Camptorhynchus labradorius

Mallard duck = mallard

mandarin duck a brilliantly colored, crested, anatine duck *Aix galericulata*

maned duck *Austral.* the cairinine *Chenonetta jubata* (= maned goose)

mangrove duck = tree duck

masked duck the oxyurine *Oxyura dominica* = ruddy duck

Mexican duck the anatine *Anas diazi*

mottled duck the anatine *Anas fulvigula*

mountain duck *Austral.* the tadornine *Tadorna tadornoides* (= Australian sheldrake)

muscovy duck a large blue-black cairinine duck *Cairina moschata*

muck duck the oxyurine *Biziura lobata*

ring-necked duck *U.S.* the aythyine *Aythya collaris*

night duck *W.I.* = tree duck

paradise duck *U.S.* the tadornine *Tadorna tadorna* (= shelddrake) *N.Z.* the tadorine *Casarca variegata* (= paradise sheldrake)

perching duck the common name of the subfamily cairini containing such forms as the muscovy duck

quail-duck *W.I.* = masked duck

rubber duck *W.I.* the oxyurine *Oxyura jamaicensis* (= ruddy duck)

rudder duck = ruddy duck

ruddy duck any of several species of oxyurines of the genus Oxyura but, without qualification, usually *Oxyura jamaicensis*

scale duck = shelddrake

sea duck *U.S.* = diving duck *U.S.*

sheld-duck *U.S.* any of several ducks but particularly *Tadorna tadorna* (= shelddrake)

smee duck any of several ducks but particularly the wigeon and pintail

spirit duck = goldeneye

squat duck *W.I.* = masked duck

oldsquaw duck an arctic migrant aythyine duck *Clangula hyemalis*

steamer-duck any of several species of aythyine duck of the genus Tachyeres, without qualification usually *T. brachyptera*

summer duck *U.S.* = wood duck *U.S.*, *W.I.* the anatine *Anas bahemensis*

spine-tailed duck = old squaw duck

long-tailed duck the aythyine *Clangula hyemalis* (= old squaw)

torrent duck the mergine *Merganetta armata*

tree-duck any of several species of dendrocygnine duck of the genus Dendrocygna *U.S.*, without qualification, *D. bicolor*

tufted duck *W.Afr.* the aythyine *Aythya fuligula*

whistling duck the common name of the subfamily Dendrocygninae. They are large long-legged grazing ducks. In many areas the term is applied to tree ducks

white-winged duck *N.Z.* the aythyine *Aythya australis* (= Australian pochard)

wood duck *U.S.* the anatine *Aix sponsa.* In *Austral.* the term is applied to the maned goose (*q.v.*)

duck 2 (*see also* duck 1, 3) various aquatic birds not of the order Anseriformes

blue mountain duck *W.I.* the procellariid bird *Pterodroma hasitata* (= black capped petrel)

duck 3 (*see also* duck 1, 2) various animals, other than birds, particularly mullusks

Bombay duck the harpodontid fish *Harpodon mehereus*, and particularly the dried fillets of this fish found in commerce

channeled duck the mactrid pelecypod mollusk *Anatina canaliculata*

gooey duck *U.S.* the very large saxicavid pelecypod *Panope generosa*

[ducky] a mutant (gene symbol *du*) in linkage group II of the mouse. The phenotypic expression is a duck-like waddle

duct any tubular vessel carrying either fluids or gasses in an organism

aeriduct almost any duct concerned with respiration in insects, including the internal trachea of most forms, and the breathing tubes of aquatic larvae

alveolar duct the duct connecting alveoli in the lungs to bronchioles

Aranzio's duct = ductus venosus

Bellini's duct collecting tubules of the metanephros

bile duct the duct leading from the gall bladder to the small intestine

Botallo's duct = ductus arteriosus

anterior cephalic duct a lymph duct, parallel to the jugular vein

coelomoduct any duct which connects the coelom to the exterior, usually applied to the terminal tubule of nephridia

pharyngocutaneous duct a duct running from the last branchial pouch on the left side of myxinid cyclostomes to an opening some distance posterior

dot duct = pitted vessel

ejaculatory duct that portion of the male gonoduct which by contraction forces out the sperm

Gartner's duct the degenerate remnant in the female of portions of the mesonephric duct

gonoduct any duct connected to the reproductive system, but frequently used as synonymous with coelomoduct

gum duct an intercellular canal bearing gum in dicotyledonous wood

hermaphrodite duct a duct which carries, usually in alternate seasons, either sperm or eggs

lacteal ducts lymph ducts carrying chyle

endolymph duct the primitive connection between the auditory vesicle and the exterior

Müller's duct the embryonic duct which in the adult female becomes the oviduct

oviduct the duct carrying eggs to the exterior. It is frequently divided into distinct portions (*cf.* uterus, vagina)

common oviduct the terminal duct resulting from the union of two or more oviducts or oviductules

ovovitelline duct a duct coming from a germovitellarium

pneumatic duct connection of the swim bladder to the alimentary canal

resin duct duct for transporting resin in gymnosperms

respiratory duct the passage through which air for respiration is drawn through the nose. In some urodele amphibia, it is separated from the olfactory pocket, which is lined with sensory epithelium

Santorini's duct an accessory pancreatic duct opening posterior to the main duct

scalariform duct a vessel with ladder-like markings on the side

Steno's duct the duct of the parotid gland

Wirsung's duct = pancreatic duct

Wharton's duct the duct of the submaxillary gland

ductule diminutive of duct

oviductule small collecting ducts, coming from a follicular (in the sense of scattered) ovary

ductus *L.* for duct

vasiductus (*bot.*) = raphe

ductus arteriosus the rudimentary remnants of the sixth aortic arch when the latter is primarily, but not entirely, converted to a pulmonary artery

duff the top layer of soil, consisting of partially decomposed vegetable matter and lacking any admixture of mineral soil from the lower layers

duffer *I.P.* any of several species of amathusiid lepidopteran insects of the genus Discophora

dug an abdominal mammary gland

dugong both the *Eng.* and *L.* name of the sirenian which inhabits the Western Pacific, the Indian Ocean and the Red Sea

duiker (*see also* cormorant) any of numerous species of several genera of cephalophine bovid artiodactyles

duke *I.P.* any of numerous nymphalid lepidopteran insects of the genus Euthalia (*cf.* baron, earl, count, duchess, baronet, marquis)

archduke *I.P.* any of several species of nymphalid lepidopteran insect of the genus Adolias, usually, without qualification, *A. dirtea*

dul a bacterial gene marker indicating utilization of dulcitol

dule term for a group of doves

Dulidae a monospecific family of passeriform birds erected to contain the palmchat of the West Indies. This is a thrush-like bird with a long bill laterally compressed and with the upper mandible decurved

dulosis the condition of slavery among ants

dulse any of several edible rhodophyte algae particularly of the genus Rhodymenia

dulwilly *Brit. dial.* the charadriid bird *Charadrius hiaticula* (*cf.* plover)

dumetose bushy

dumose shrubby

[dumpy] a mutant (pseudoallelic locus, gene symbol *dp*) mapped at 13.0 on chromosome II of *Drosophila melanogaster*. The phenotypic expressions are truncated wings, vortices and whorls of bristles and hairs on the thorax, and recessive lethality in some alleles

dun the sub-imago of an ephemerpteran insect

dunker = pochard

dunlin the charadiiform scolopacid bird *Erolia alpina*. There are several races with various names, one of which is also known as the red-backed sand-piper

dunnock the prunellid bird *Prunella modularis* (= Brit. hedge sparrow)

dunnock *Brit. dial.* = sparrow

dunter = eider duck

duodenum that part of the small intestine which is immediately adjacent to the stomach

mesoduodenum that part of the mesentery which supports the duodenum

[Duplex comb] a mutant (gene symbol *D*) in linkage group VI of the domestic fowl. The name is descriptive of the phenotypic expression

duplicate *see* -plicate

dura mater the outer of the two meninges

dura spinalis the outer of the two meninges covering the spinal cord

duramen heartwood, additionally hardened by an inorganic deposit

durgon any of several species of balistid fish of the genus Melichthys

durian the bombacaceous tree *Durio zibethinus* and its fruit

durra *hort. var.* of *Holcus halepensis* (= sorghum)

[dusky] a mutant (gene symbol *dy*) mapped at 36.2– on the X chromosome of *Drosophila melanogaster*. The dark phenotype has small wings and cannot be distinguished from [miniature]

Dussumieriidae a family of isospondylous fishes closely allied to the Clupeidae and known as the round herrings

[dutch pattern] a mutant (gene symbol *du*) at locus 0 on linkage group II of the rabbit. The phenotypic expression is a white transverse central bar.

feather duster the popular name of marine tubiculous polychaete worms of the genus Spirobis

dv *see* [divergent] and [dwarf virescent]

dvr *see* [divers]

dw *see* [dwindling] and [dwarf]

dw-24F *see* [dwarf in salivary chromosome section 24F]

dwale *Brit.* the solanaceous herb *Atropa belladonna* (= deadly night shade)

dwarf any form which is significantly smaller than normal

[dwarf] *either* a mutant (gene symbol *dw*) on the sex chromosome (linkage group I) of the domestic fowl. The name is descriptive of the phenotypic expression; *or* a mutant (gene symbol *dw*) at locus 14.7 in linkage group IV of the rabbit. The name is descriptive of the phenotypic expression which also leads to early death; *or* a mutant (gene sumbol *d₁*) mapped at 18 on linkage group III of *Zea mays*. The name is descriptive of the phenotype

[dwarfex] a mutant (gene symbol *dwx*) mapped at 33.2 on the X chromosome of *Drosophila melanogaster*. In addition to small size the phenotypic expression involves coarse wings

[dwarf in salivary chromosome section 24F] a mutant (gene symbol *dw-24F*) mapped at 13.0± on chromosome II of *Drosophila melanogaster*. The phenotype has a dwarf body with dull eyes

dwarfism the condition of being a dwarf particularly when it results from a single gene mutation

[**dwarf plant**] a mutant (gene symbol *d*) mapped at 5 on chromosome II of the tomato. The name is descriptive

[**dwarf virescent**] a mutant (gene symbol *dv*) mapped at 0 on chromosome II of the tomato. The phenotypic expression is a stunted plant

rock **dweller** popular name of petricolid pelecypods

[**dwindling**] a mutant (gene symbol *dw*) occuring in linkage group II of *Habrobracon juglandis*. The phenotypic expression involves irregularity of antennal segments

dwx *see* [dwarfex]

dx *see* [deltex]. Also used as a yeast genetic marker indicating a requirement for dextrin

dy *see* [dusky]

dyad a subdivision of a tetrad into single elements

dyal the madian turdid bird *Copsychus saularis*

Dyctyobranchia an order of ascidiacean urochordates distinguished by their translucent tunic and simple tentacles

schin**dylesis** the joint in which one plate of bone lies between two plates of another bone or plates of two other bones

spon**dylium** a "u" shaped ridge on the dental plate of a brachiopod valve

-dynam- *comb. form* meaning "power"

dynamic Pertaining to energy, power, or importance in *bot.*, said of a tissue which swells on one side but not on the other

androdynamic possessing unusually prominent stamens

dichodynamic said of hybrids which display the phenotypic appearance of both parents to an equal extent

gynodynamic applied to an hermaphrodite organism in which the female portion is the more important

homodynamic said of hybrids displaying equally the phenotypic characters of both parents

isodynamic equally well developed

anisodynamous = anisobrious

monodynamic having one stamen particularly prominent

oligodynamic said of waters containing sufficient impurities to kill delicate, but not sufficient to kill tough, organisms

poikilodynamic said of a hybrid showing a phenotypic expression almost entirely derived from one parent

trophodynamic pertaining to energy transfer between food levels in an ecosystem

-dynania plant or animal taxa having this ending are listed alphabetically

-dys- *comb. form* meaning "bad"

-dysis- *comb. form* meaning "to clothe"

ecdysis the shedding of the outer layer or molting

metecydysis the period immediately following ecdysis in arthropods

endysis the growth of a new cuticle

dystrophy *see* -trophy

dyticon a community dwelling in ooze

Dytiscidae a considerable family of aquatic predatory beetles usually of large size. The males of several species are distinguished by the presence of a large adhesive pad on the front tarsus. They are commonly called predaceous diving beetles

e *see* [ebony], [elongate] and [entire leaves]

E *see* [Plain supernumerary legs] and [Extension]

E1 a bacterial mutant indicating that the mutant is bacteriocinogenic for calicin E1

E2 a bacterial mutant indicating that the mutant is bacteriocinogenic for calicin E2

e- privative meaning "without" or "destitute of", but not in the sense of having had something removed which is better expressed by de-

E–S *see* [Enhancer of Star]

eagle (*see also* hawk, owl, ray) any of numerous large diurnal birds of prey of the family Acciptridae. The distinction between eagle and hawk is a matter of size and varies from place to place. The "eagle" of the *W.I.* is an osprey

  bald eagle *U.S. Haliaeetus leucocephalus*

  black eagle *Ictinaetus malayensis*

  Bonelli's eagle *Hieraeetus fasciatus*

  booted eagle *Hieraeetus pennatus*

  crowned eagle *Afr. Stephanotus coronatus*

  fish-eagle *Cucuma vocifer*

  fishing-eagle any large piscivorus eagle but properly those of the genus Ichthyothaga

  golden eagle *Aquila chrysaetos*

  harrier-eagle any of several species of the genus Circaetus

  hawk eagle any of several large hawks

    African hawk-eagle *Hieraeetus spilogaster*

    mountain hawk-eagle *Spizaetus nipalensis*

    variable hawk eagle *Spizaetus cirrhatus*

  imperial eagle *Aquila heliaca*

  little eagle *Austral. Hieraeetus morphnoides*

  martial eagle *Polemaetus bellicosus*

  monkey-eating eagle *Pilkecophaga jefferyi*

  sea eagle any marine eagle but properly those of the genus Haliaeetus

    African sea eagle *Cucuma vocifer*

    white-bellied sea-eagle *H. leucogaster*

    Cape sea eagle = *African sea eagle*

    Kamchatkan sea-eagle *H. pelagicus*

    Pallas's sea-eagle *H. lecoryphus*

    Steller's sea eagle *Thallasoaeetus pelagicus*

  serpent eagle any of several species, mostly of the genus Spilornis

  snake eagle any of several species, mostly of the genus Circaetus

    forest snake eagle *Dryotriorchis spectabilis*

  solitary eagle *Urobitornis solitaria*

  tawny eagle *Aquila rapax*

  steppe eagle *Aquila nipalensis*

  short-toed eagle *Circaetus gallicus*

  whistling eagle *Austral. Haliastur sphenurus*

[eagle] a mutant (gene symbol *eg*) mapped at 47.3 on chromosome III of *Drosophila melanogaster.* The phenotypic expression is that the wings are held like those of an eagle raised and spread

ear 1 (*see also* ear 2–6, tick, wig) a complex vertebrate phonoreceptor. The term properly applies to the whole apparatus but in mammals is often loosely applied to the pinna

ear 2 (*see also* ear 1, 3–6) any object shaped like an ear, *e.g.* paired tufts of feathers on a bird's head

ear 3 (*see also* ear 1, 2, 4–6) any of several ear shaped mollusks

  baby's ear any of several species of maticid gastropods of the genus Sinum (*cf.* ear shell)

  Venus' ear any haliotid gastropod (*cf.* abalone, ormer)

ear 4 (*see also* ear 1–3, 5, 6, fern, flower, weed) in many compound names of plants, some of which have ear shaped organs

  cat's ear any of several species of the compositaceous genera Antennaria and Hypochaeris

  elephant ear any of several large-leaved herbs of the begoniaceous genus Begonia

  hare's ear *Brit.* either of two species of umbelliferous herb of the genus Bupleurem

  jew's ear = monkey's ear

  lamb's ears the labiateous herb *Stachys olympica*

  lion's ear the labiateous shrub *Leonotis leonurus*

  monkey's ear any of several edible gelatinous fungi of the basidiomycete order Auricularialaes, particularly of the genus Auricularia. The name is descriptive both of the texture and shape

  mouse ear the compositaceous herb *Hieracium pilosella.* The term is also sometimes applied to the forget-me-not

    bastard mouse ear *Hieracium tenoreanum*

  pig's ears the cantharellaceous basidomycete fungus *Cantharellus clavatus*

ear 5 (*see also* ear 1–4, 6) a spike of grain

ear 6 (*see also* ear 1–5) in compound names of organisms, frequently for no apparent reason

  feathered ear *Brit.* the noctuid lepidopteran insect *Pachetra leucophaea*

  pink ear *Austral.* the atherinid teleost fish *Melanotaenia nigrans* (*cf.* rainbow fish)

  violet ear any of several trochilid birds

  wheatear any of several species of turdid birds of the genus Oenanthe. Usually, without qualification, *O. oenanthe*

    black wheatear *O. leucura*

    capped wheatear *O. pileata*

    common wheatear *O. oenanthe*

desert wheatear *O. deserti*
black-eared wheatear *O. hispanica*
Greenland wheatear = common wheatear
red-headed wheatear *O. bottae*
hooded wheatear *O. monacha*
isabelline wheatear *O. isabellina*
pied wheatear *O. pleschanka*
white-rumped wheatear *O. leucopyga*
white-underwinged wheatear *O. lugens*

earl *I.P.* any of several species of nymphalid lepidopteran insect of the genus Euthalia (*cf.* count, baron, duke, duchess, baronet, marquis)

earwig *see* -wig

heart's ease *U.S.* the polygonaceous herb *Polygonum persicaria* and the violaceous herb *Viola tricolor* (= wild pansy)

eater he who or that which eats. The *adj. form* is either -phagous (*q.v.*) or -vorous (*q.v.*)

anteater (*see also* chat) any myrmecophagous animal, most particularly xenarthran mammals. Used without qualification, usually refers to the great, or giant, anteater *Myrmecophaga jubata*
Australian anteater the dasyurid marsupial *Myrmecobius fasciatus* sometimes separated from the Dasyuridae as a separate family. The name is sometimes applied to the monotreme *Echidna aculeata*, better termed **spiny anteater**
cape enteater the tubulidentate mammal *Orycteropus capensis* (= aardvark)
giant anteater the xenarthran *Myrmecophaga jubata*
great anteater = giant anteater
lesser anteater the xenarthran *Tamandua tetradactyla* (= tamandua)
marsupial anteater the perameloid *Myrmecobius fasciatus* (= Australian anteater). The term is sometimes, without any reasonable justification, applied to the wombat (*q.v.*)
scaly anteater = pangolin

bee-eater any of numerous coraciiform birds of the family Meropidae The term used without qualification refers to *M. apiaster* in Europe and *M. ornatus* in *Austral.*
blue-beared bee-eater *Nyctyornis athertoni*
black bee-eater *Melittophagus gularis*
blue-breasted bee-eater *Melittophagus variegatus*
carmine bee-eater *Merops nubicus*
southern carmine bee-eater *Merops nubicoides*
cinnamon-chested bee-eater *Melittophagus oreobates*
black-crowned bee-eater *Merops albicollis*
European bee-eater *Merops apiaster*
white-fronted bee-eater *Melittophagus bullockoides*
green bee-eater *Merops orientalis*
blue-headed bee-eater *Melittophagus muelleri*
chestnut-headed bee-eater *Merops leschenaulti*
little bee-eater *Melittophagus pusillus*
rosy bee-eater *Merops malimbicus*
Somali bee-eater *Melittophagus revoilii*
blue-tailed bee-eater *Melittophagus bulocki*
swallow-tailed bee-eater *Dicrocercus hirundineus*
red-throated bee-eater *Melittophagus bulocki*
white-throated bee-eater *Aerops albicollis*

beef-eater the African sturnid bird *Buphagus africanus* (= oxpecker)

berry-eater either of two cotingid birds of the genus Carpornis

caterpillar-eater *Austral.* the campephagid bird *Lalage sueurii* (= white-winged triller)

chicken-eater *W.I.* the buteonine bird *Buteo platypterus* (= broad-winged hawk)

fig-eater any of numerous small birds, particularly those common to southern European orchards. There is a convention among lexicographers, but not ornithologists, that the sylviid bird *Sylvia borin* (= *Brit.* garden warbler) is the true "fig eater." In *U.S.* the term is used for the scarabaeid coleopteran insect *Cotinis nitida*

fly-eater *Austral.* any of several malurine birds of the genus Gerygone (*cf.* warbler) Malay the sylviid bird *Gerygone sulphurea* (*cf.* warbler)

fruiteater any of several cotingid birds of the genera Pipreola and Ampelioides

gnateater any of several conopophagid birds of the genera Conopophaga and Corythopis

grape-eater *Austral.* popular name of zosteropid birds (= white-eyes) Also used for the meliphagid *Meliphaga virescens* (= black-faced honey-eater)

honeyeater any of very numerous species of meliphagid birds
brown honeyeater *Gliciphila indistincta*
spiny-cheeked honeyeater *Acanthagenys rufogularis*
white-cheeked honeyeater *Meliornis niger*
black-chinned honeyeater *Melithreptus gularis*
crescent honeyeater *Phylidonyris pyrrhoptera*
tawny-crowned honeyeater *Gliciphila melanops*
dusky honeyeater *Myzomela obscura*
white-eared honeyeater *Meliphaga leucotis*
black-faced honeyeater *Meliphaga virescens*
blue-faced honeyeater *Entomyzon cyanotus*
yellow-faced honeyeater *Meliphaga chrysops*
yellow-fronted honeyeater *Meliphaga plumula*
fuscous honeyeater *Meliphaga fusca*
brown-headed honeyeater *Melithreptus brevirostris*
mallee honeyeater *Meliphaga ornata*
white-naped honeyeater *Melithreptus lunatus*
painted honeyeater *Grantiella picta*
pied honeyeater *Certhionyx variegata*
white-plumed honeyeater *Meliphaga penicillata*
regent honeyeater *Zanthomiza phrygia*
scarlet honeyeater *Myzomela sanguinolenta*
yellow-tufted honeyeater *Meliphaga melanops*
yellow-winged honeyeater *Meliornis novaehollandiae*
yellow honeyeater *Meliphaga flava*

plaintain-eater any of several musophagid birds (= touracos and go-away birds)
blue plantain-eater *Corythaeola cristata* (= great blue touraco)
grey plantain-eater *Crinifer africanus*

seedeater any of several species of fringillid bird in *U.S.*, particularly those of the genus Sporophila or in *Afr.*, ploceid birds of the genus Serinus (*cf.* canary)
yellow-bellied seedeater *S. nigricollis*
thick-billed seedeater *Serinus burtoni*
blue-black seedeater *Volatinia jacarina* (= blue-black grassquit)
blue seedeater the fringillid *Amaurospiza concolor*
slate-blue seedeater the fringillid *Amaurospiza relicta*
ruddy-breasted seedeater *Sporophila minuta*
white-collared seedeater *Sporophila torqueola*
black-eared seedeater *Serinus mennelli*
streak-headed seedeater *Serinus gularis*
white-rumped seedeater *Serinus leucopygius*
yellow-rumped seedeater *Serinus atrogularis*
streaky seedeater *Serinus striolatus*

variable seedeater *Sporophila aurita*

snail-eater the ciconiid bird *Anastomus oscitans* (= openbill) or any of several colubrid snakes of the genus Sibon

ebe-ebe = iiwi

Ebenaceae that family of ebenale dicotyledonous angiosperms which contains *inter alia*, the true ebonies. The family is distinguished from the closely related Sapototaceae by the absence of a milky sap

Ebenales that order of dicotyledonous angiosperms which contains, *inter alia*, the sapodilla, ebony, the persimmon, and styrax. A tree-like habit of growth and unisexual flowers are distinctive to the family

ebeneous ebony black

Eberth's *epon. adj.* from Karl Joseph Eberth (1835–1926)

Ebner's *epon. adj.* from Victor Ebner, Ritter von Rosenstein 1842–1888

ebony any of several trees having dark wood particularly those of the ebenaceous genus Diospyros, and the leguminous *Pithecolobium flexicaule*

Macassar ebony the ebenaceous tree *Diospyros ebenum* and the lumber obtained from it

mountain ebony any of numerous leguminous trees of the genus Bauhania

[ebony] a mutant (gene symbol *e*) mapped at 70.7 on chromosome III of *Drosophila melanogaster*. The term refers to the body color of the phenotype

eborinous ivory-white

eburneous off white, in the direction of yellow

ec *see* [echinus]

ecad an organism frequently but not of necessity, sessile which is specifically, and sometimes uniquely, adapted to the environment in which it is found

megecad a group of closely related ecads

Ecardines a class of Brachiopoda distinguished by the fact that the shells are not hinged together

eccrine secretory. Used specifically to distinguish a gland that produces what its name indicates (*e.g.* an eccrine sweat gland produces sweat) as distinct from a gland which does not (*e.g.* an apocrine sweat gland that produces a milky fluid)

ecdyson a hormone secreted in the prothoracic gland of insects which controls molting

ecesis the establishment of a plant in a new location which it reaches as a seed

Echeneidae the only family of discocephalous fishes. The order is recognized by a sucking disc on top of the head

Echeniformes = Discocephali

Echidnidae that family of monotreme mammals which contain the spiny anteaters

Echidnophagidae = Tungidae

Echimyidae a very large family of hystricomorph rodents commonly referred to as "spiny-rats."

-echin- *comb. form* meaning "spine"

echinate prickly

echinating said of sponge spicules that project towards, but do not penetrate, the surface

Echinodera a small phylum of microscopic marine pseudocoelomatous bilateral animals distinguished by the superficial spines and a body superficially segmented into thirteen segments. This phylum is taken by some as a class of the Aschelminthes

Echinodermata that phylum of the animal kingdom which contains the sea urchins, the sea stars, the sea cucumbers, and the sea lilies. They are distinguished from other enterocoelous coelomate animals by a radial symmetry of five and the presence of a calcareous skeleton often consisting of external plates bearing spines

[echinoid] a mutant (gene symbol *ed*) mapped at 11.0 on chromosome II of *Drosophila melanogaster*. The phenotypic expression is rough large eyes

Echinoidea that class of eleuthrozoan echinoderms which contains the sea urchins. They may be spheroidal, disk-shaped, or heart-shaped, but are distinguished from other echinoderms by the large number of moveable spines on the surface which is covered with a test of contiguous calcareous plates

Echinophthiriidae a family of anopluran insects parasitic on marine animals and distinguished by the very spiny, and sometimes scaly, body

Echinorhinidae a family of sharks containing the bramble shark *Echinorhinus brucus*. It is distinguished by the very large spiny dermal denticles

Echinosoricinae a subfamily of erinacid insectivorous mammals confined to Southeast Asia and the adjoining islands

Echinozoa at one time a subphylum of the phylum Echinodermata embracing the classes Echinoidea and Holothuroidea. Now included in the Eleuthrozoa

echinulate with small spines, but specifically applied to a streak or stab bacterial culture in which the line of growth is spiny

[echinus] a mutant (gene symbol *ec*) mapped at 5.5 on the X chromosome of *Drosophila melanogaster*. The phenotypic expression is large-faceted, large, rough eyes

Echiurida a small phylum of coelomate bilaterally symmetrical animals closely allied to the Annelida and by some placed as a class in that phylum. Most possess a pair of setae but they lack metameric segmentation. An outstanding characteristic is the very large spatulate prostomium

echymenine = extine

ecize = colonize

eclosion coming forth from an egg or a pupal case

-eco- *comb. form* meaning a "dwelling" (*cf.* -oec)

ecology a study of organisms in relation to each other and to their environment

autecology the relation of an individual organism, as distinct from an association of organisms, to its habitat

synecology the relation between an association and its environment or the ecology of communities

dynamic synecology the study of the interaction of organisms within a community

geographic synecology the relation of environmental factors to the distribution of communities

morphological synecology the relation of the morphology of organisms to their position in the community in which they live

ecronic *ecol. jarg.* for estuarine

-ecto- *comb. form* meaning "outside"

Ectocarpales an order of isogenerate Phaeophyta containing those forms which have a branched filamentous thallus

Ectoprocta a phylum of lophophorian coelomate animals at one time combined with the phylum Entoprocta into the taxon Bryozoa. They are distinguished from other lophophorians by their exoskeleton of small horny or calcareous cases or gelatinous masses. They have often been referred to as moss animalcules from their habit of growth

ed *see* [echinoid]

edaphic (*see also* ecotype) pertaining to the influence of soil upon organisms growing in or on it

-edapho- *comb. form* meaning "soil" (*cf.* -ge-)

edaphon the fauna and flora of soils

phytoedaphon plants (*i.e.* fungi and bacteria) living entirely underground (*cf.* Phytodyte)

Edentata an assemblage, once regarded as an order, of mammals lacking true teeth. Nowadays usually divided into the orders of Pholidota, Xenarthra, and Loricata

hoary edge *U.S.* the hesperioid lepidopteran insect *Achalarus lyciades*

Edoliidae = Dicruridae

eel (*see also* blenny, cat, goby, grass, trout and worm) any of numerous fishes of the order Apodes particularly those of the genus Anguilla. The name is also used for any of numerous elongate sinuous animals

conger eels popular name of marine eels of the family Congridae, *Brit.* without qualification, *Conger conger. U.S.* frequently used in error for congo eel (*q.v.*)

congo eel *U.S.* the amphiumid urodele amphibian *Amphiuma means.* The term is also used in various parts of *U.S.* for siren (*q.v.*)

cusk eel popular name of ophidiid fishes

ditch eel *U.S.* = congo eel *U.S.*

electric eel the gymnotid fish *Electrophorus electricus*

gulper eel popular name of lyomerous fishes (= gulper fish)

horn eel *Brit.* the syngnathid *Syngnathus acus*

lamper eel *U.S.* = congo eel *U.S.*

leopard eel the cobitid fish *Acanthophthalmus kuhlii*

moray eel = moray

mud-eel *U.S.* the sirenid urodele amphibian *Siren lacertina*

snub-nosed eel the simenchelid *Simenchelys parasiticus*

rice eel the synbranchid *Monopterus albus*

sand eel *Brit.* popular name of ammodytid fish (= *U.S.* sand lances)

slime eel = hagfish

snake eel popular name of eels of the family Ophichthidae

snipe eel popular name of nemichthid eels

spiny eel popular name of mastacembelid fishes

swamp eel popular name of synbranchid fishes

vinegar eel popular name of the anguillulid ascaroid nematode worm *Anguillula aceti*

wolf eel the anarhichadid fish *Anarrhichthys ocelatus*

worm eel = snake eel

eelpout any of several marine fish resembling blennys but of the family Zoracidae

effect (*see also* law) the result of an action, or interaction, so constant and so frequently observed that it has been recorded in the literature under a specific name. The terms "effect" and "law" (*q.v.*) are, in *biol.* literature used almost interchangeably

cis-trans effect the condition in which two alleles at different sites within a cistron, produce a mutant phenotype in the trans configuration (one on each chromosome) or a wild type phenotype in the cis configuration (both on the same chromosome)

edge effect the tendency for variety and density to increase where communities join

Pasteur effect the inhibition of fermentation by oxygen

position effect the circumstance that a gene may exercise a different phenotypic effect in relation to its position in the chromosome

Sewall Wright effect the result of random genetic drift

effete of an age at which reproduction is no longer possible

ecological **efficiency** an expression of the energy required by a given organism to produce a given unit of protoplasm

effuse properly over-flowing, but used by botanists in the sense of diffuse (*see also* streak culture)

eft *Brit. obs.* newt

eg *see* [eagle]

egg 1 (*see also* egg 2) properly the female gamete but generally applied to the entire female reproductive unit including extracellular food reserves and the case of shell. By some the term ovum is used for the gamete and egg for the unit

holoblastic egg one which is completely divided into blastomeres during cleavage

opsiblastic egg the resting or winter egg of a gastrotrich

tachyblastic egg one that hatches rapidly and is not adapted to withstand adverse conditions

cleidoic egg one, like that of a bird, that is enclosed in a more or less moisture-proof box

dormant egg an egg, in invertebrates usually thickshelled, that is destined to remain for some considerable time before developing (*cf.* subitaneous egg)

ephippial egg the winter egg of Cladocera

isotropic egg one that lacks a predetermined axis

mictic egg an egg that may, or may not, be fertilized

amictic egg an egg that cannot be fertilized, and therefore can only develop parthenogenetically into a female

mosaic egg one in which developmental patterns are established before cleavage so that induced modification of cleavage patterns result in modification of development (*cf.* regulative egg)

pseudo-egg (Mesozoa) = Infusorigen

regulative egg one in which the developmental pattern is not established before cleavage so that induced modifications of cleavage do not modify subsequent development (*cf.* mosaic egg)

subitaneous egg a thin-shelled egg, destined to hatch rapidly in contrast with a thick-shelled egg, intended to remain dormant for some time (*cf.* dormant egg)

winter egg one produced by some invertebrates, particularly Cladocera, which are adapted to resist freezing temperatures and which usually will not hatch until after they have been frozen

egg 2 (*see also* egg 1, fruit, plant, shell) an organism, or part of an organism, having an egg like appearance

ant egg a cocoon containing an ant pupa

golden eggs the onagraceous herb *Oenothera ovata*

goose-egg *Brit.* the drepanid lepidopteran insect *Cilix glaucata* (= Chinese character)

eggar *Brit.*, popular name of lasiocampid lepidopteran insects (*cf.* lackey)

grass eggar *Lasiocampa trifolii*

oak eggar *Lasiocampa guerca*

pale oak eggar *Trichura crataegi*

small eggar *Eriogaster lanestris*

eglantine the rosaceous shrub *Rosa rubiginosa* (= sweet briar)

egret any of very numerous species of heron like bird of of the family Ardeidae

cattle egret *Bubulcus ibis*
Chinese egret *Egretta eulophotes*
common egret *Casmeroduis albus*
large egret = common egret
lesser egret *Egretta intermedia*
little egret *Egretta gazzetta*
reddish egret *Dichromonassa rufescens*
snowy egret *Egretta thula*

Ehretioideae a subfamily of the Boraginaceae

eider the popular name of ducks of the group Aythyini
common eider *Somateria mollissima*
king eider *Somateria spectabilis*
spectacled eider *Lampronetta fischeri*
Steller's eider *Polysticta stelleri*

-eidos- *comb. form* meaning "resemblance". Frequently abbreviated in compounds to the termination -id

figure-of-eighty *Brit.* the thyatirid lepidopteran insect *Palimpsestis octogessima*

ovejector that part of the invertebrate oviduct which, by its contraction, expels the egg

-ekto- *comb. form* meaning "outside" usually transliterated -ecto-

ekwalat *Afr.* any of several species of turdid bird of the genus Neocossyphus

el *see* [elbow] and eyeless]

Elacatidae = Othniidae

Elachistidae a small family of leaf mining cycnodoid lepidopterans

Elachistodontinae a subfamily of opisthoglyphan colubrid snakes with teeth on the palatines

Elachistoidae = Cycnodidea

Elaeagnaceae a relatively small family of myrtiflorous plants containing *inter alia* the oleasters and buffalo berries. The family is easily distinguished by the peculiar silvery scales which cloak the plants

-elaeo- *comb. form* meaning "olive" either in the sense of its color or of its oily nature (*cf.* -drup-)

Elaeocarpaceae that family of malvalous dicotyledons which contains *Elaocarpus grandis* (Brisbane quandong). It differs from the closely allied Tiliacae by the hairy petals

elaeodochon the oil gland of birds

elaeodous olive green

eland either of two species of strepsicerosine bovid artiodactyles of the genus Taurotragus (Arias). Without qualification *T. canna* is usually meant

Elaninae that subfamily of accipitrid birds commonly called the white tailed kites a term descriptive of their appearance

elaphinous tawny

Elapidae a subfamily of proteroglyphan snakes mostly extremely venomous and including the cobras and mambas

Elasipoda an order of holothurian echinoderms with numerous podia and peltate oral tentacles

Elasmidae a family of small chalcidoid apocritan hymenopteran insects readily distinguished by their very expanded, flattened, often almost discoidal hind coxae

Elasmobranchii a group, variously regarded as a class or subclass, of gnathostomatous craniate chordates containing those forms commonly known as sharks and rays. They are distinguished by the possession of a cartilaginous skeleton which in larger forms sometimes becomes calcified but never ossified

-elat- *comb. form* meaning "drive" in the sense of causing movement

elater that which causes an organism to jump as the furcula of Collembola, or the prosternal process of

elaterid beetles. In *bot.* any of numerous structures, which by their sudden expansion, serve to disseminate seeds or spores

Elateridae a large family of coleopteran insects commonly called click beetles in view of their ability to right themselves when turned over by a jump-like motion produced through a prosternal spine that fits into a groove on the mesosternum

Elateroidea a superfamily of polyphagous coleopteran insects containing the click beetles and their immediate allies

Elatinaceae that family of parietale dicotyledonous angiosperms containing the forms commonly called waterworts. The family may be distinguished from other permanent aquatics by the opposite or whorled leaves and paired spicules

elbow in primates, particularly anthropoids, the joint between the humerus and the radio-ulna (*cf.* knee)

[elbow] a mutant (gene symbol *el*) mapped at 50.0 on chromosome II of *Drosophila melanogaster*. The wings of the phenotype are bent and the alulae and balancers are small

elder 1 (*see also* elder 2, aphis, borer, bug) any of numerous caprifoliaceous shrubs and trees of the genus Sambucus
American elder *S. canadensis*
red-berried elder *S. racemosa*
European elder *S. nigra*
sweet elder = American elder

elder 2 (*see also* elder 1) a variety of shrubs having the habit or appearance of elder 1
box elder *U.S.* the maple *Acer negundo*
marsh elder *U.S.* the compositaceous shrub *Iva fructescens*. *Brit.* and *U.S.* either of two caprifoleaceous shrubs *Viburnum opulus* or *V. lantana*
poison elder *U.S.* the anacardiaceous shrubby tree *Rhus verix* (= poison sumac)
water elder *Brit.* the caprifoliaceous shrub *Viburnum opulus*
wild elder the araliaceous shrub *Aralia hispida*
yellow elder the bignoniaceous shrub *Tecoma stans*

elecampane the compositaceous herb *Inula helenium*

electrinous amber colored

element in *biol.* used for a component, or group of components, of a larger structure
kinetic element that part of a cell which is active in mitotic division and thus usually referring to the centromere
sieve elements individual units which, joined end to end, make a sieve tube
tracheary element a conducting element of the xylem
sieve tube element a relatively large cell with sieve plates on its end walls
vessel element (*bot.*) = vessel member (*bot.*)

Eleotridae a family of gobioid acanthopterygian fishes lacking a functional sucking disc and with the inner rays of the pelvic fins longest. Commonly called sleepers

elepaio the Hawaiian muscicapid bird *Chasiempis sandwichensis*

elephant 1 (*see also* elephant 2, 3, chimaera, ear, foot, moth, nose, seal and shell) either of two gigantic proboscidean mammals
African elephant *Loxodonta africana,* distinguished from *E. maximus* by the possession of larger tusks, particularly in the female, and by the larger pinnae
Indian elephant *Elephas maximus*

elephant 2 various large mammals, though in compounds the word is usually used as an adjective

sea elephant either of two species of cystophorine phocid pinniped of the genus Mirounga

elephant 3 *Brit.* any of several sphingid lepidopteran insects, usually without qualification, *Chaerocampa elpenor*

small elephant *Brit. Metopsilus porcellus*

spotted elephant *Brit. Deilephila galii* (= bedstraw hawk)

-eleuthero- *comb. form* meaning "free"

Eleutherata = Coleoptera

Eleuthrozoa a subphylum of Echinodermata distinguished from the Pelmatozoa by the absence of an adoral stem for attachment

eleutriation the process of the separation of different sized, or different shaped, particles by differential flotation

elf *S.Afr.* either of two species of hesperid lepidopteran insect of the genus Eretis

elfin *U.S.* any of numerous lycaenid lepidopteran insects of the genus Incisalia. *S.Afr.* any of several species of hesperid lepidopteran insect of the genera Sarangesa and Eagris

elk 1 (*see also* elk 2) in the Old World applied to the alcine artiodactyl *Alces americanus* (= *U.S.* moose) of circumpolar distribution but in *U.S.* usually applied to the cervine *Cervus canadensis*

elk 2 (*see also* elk 1) the cygnine bird *Cygnus musicus* (*cf.* swan)

Ellobiidae a family of gastropod mollusks with a spiral shell having an elongate aperture and covered with a heavy horny periostracum. They are commonly called salt-marsh snails

elm (*see also* aphis, beetle, bug, minor, sawfly, scale, sphinx, weevil) any of numerous ulmaceous trees, almost all of the genus Ulmus

American elm = white elm

bastard elm any of several trees of the ulmaceous genus Celtis

cedar elm *U. crassifolia*

Chinese elm *U. parvifolia*

cork elm *U. racemosa*

Dutch elm *U. hollandica* but probably a horticultural hybrid

English elm *U. campestris*

smooth-leaved elm *U. foliacea*

red elm = slippery elm

rock elm = cork elm

Scotch elm = wych elm

slippery elm *U. rubra*

wahoo elm = winged elm

water elm *U.S.* the ulmaceous tree *Planera aquatica*

white elm *U. americana*

winged elm *U. alata*

wych elm *U. glabra*

Elmidae a small family of aquatic coleopteran insects closely allied to the Dryopidae and Psephenidae and like them commonly called long-toed water beetles. The Elmidae are also called riffle beetles in virtue of their customary habitat

[elongate] a mutant (gene symbol *e*) mapped at 36.4 on linkage group I of *Bombyx mori*. The phenotype expression involves an elongation of the first and second abdominal segments of the larva

Elopidae a family of isospondylous fishes distinguished by a long, deeply forked, tail and a single dorsal fin placed about the center of the body. The best known member of the family is the tarpon

elver a young eel, newly metamorphosed from a leptocephalus larva, and which is ascending a river

-elytra- *comb. form* meaning "sheath"

elytron a sheath or shield particularly the modified horny anterior wings of beetles and some other insects, or the modified chaetae in the form of scales or plates found in some polychaete worms

hemielytron the anterior wing of hemipterous insects. One half of this wing is thickened

homoelytron one which is of the same texture throughout

emarcid withered

embal the euphorbiaceous shrubby tree *Phyllanthus emblica* (myrobolan)

Emballonuridae a family of microchiropteran mammals lacking a nose leaf but distinguished from the Vespertilionidae by the fact that the tail is either free or partly free. Several are piscivorous

emerald any of numerous trochilid buds, mostly of the genera Chlorostilbon and Amazilia

Embiodea an order of minute elongate flattened insects distinguished by the enlarged front tarsae used for spinning the silken webs in which the insect lives

Embioptera = Embiodea

Embiotocidae a large family of acanthopterygian fishes commonly called surf perches or viviparous perches

emblic = myrobolan

Embolemidae a family of sphecoid apocritan hymenopteran insects usually included with the Dryinidae

embolium the costal part of the hemielytron

embryo in animal development, a stage incapable of supporting a separate existence in contrast to a larva or nymph. In plant development, a stage in which specific organs or organ systems are not visibly differentiated, particularly the partially developed sporophyte in a seed

endoscopic embryo a plant embryo, usually of a spermatophyte, in which the apical pole of the embryo is directed inwards, that is towards the base of the archegonium or embryo sac

exoscopic embryo a plant, usually bryophyte, embryo in which the apical pole of the embryo is directed outwards and the basal region is in contact with the gametophyte tissues from which nutrients will be drawn

fixed embryo (*bot.*) a leaf bud

holoblastic embryo (*bot.*) one which is in the product of the entire ovum

proembryo 1 (*see also* postembryo 2) the massive cells developed from the oospore of Chara and from which the plant develops as a lateral bud

proembryo 2 (*see also* proembryo 1) a spermatophyte embryo before histodifferentiation and organ formation

embryonic pertaining to the embryo

curvembryonic a plant embryo which is curved into any form except the atropous

extraembryonic pertaining to structures connected to, but outside the main body of, a developing organism

monoembryonic producing one embryo from each egg

polyembryonic having several embryos in one ovule or producing several embryos from one egg

embryony the condition of producing or having an embryo

adventitious embryony a form of apomixis in plants in which an embryo is formed directly by the outgrowth of a cell of the parent sporophyte

epoembryony the condition of a plant where the embryo is apparently suppressed

parembryum the coat round the plumule and radicle of a monocotyledonous embryo

Embyoptera = Embioidea

emerald 1 (*see also* emerald 2) any of numerous trochilid birds of the genus Chlorostilbon

emerald 2 (*see also* embryo 1) *Brit.* any of rather numerous green species of geometrid lepidopteran insects

emeu = emu

emigrant *I.P.* any of numerous yellow pierid lepidopteran insects of the genus Catopsilia

Doyères **eminence** a bump formed at the point where a motor nerve ending penetrates the sarcolemma

emmet = ant

empaled surrounded by

empalement = calyx

emperor *I.P.* any of numerous nymphalid lepidopteran insects of the genera Helcrya, Eulaceura, Dilipa, Eriboea and Apatura

New Guinea emperor *Apaturina erminea*

purple emperor *Brit. Apatura iris*

tawny emperor *U.S. Asterocampa clyton*

Empetraceae that family of sapindalous dicotyledons which contains the crowberries. The heathlike growth, the one-seeded ovary with an external micropyle and the absence, or great reduction of, the corolla are typical

Empetrineae a suborder of Sapindales containing the single family Empetraceae

-emphys- *comb. form* meaning "breathe on"

Empidae = Empididae

Empididae a family of small orthoraphous brachyceran dipteran insects commonly called dance flies because they are so frequently seen in huge swarms in which they fly up and down. They are predatory on smaller insects

empress *I.P.* the nymphalid lepidopteran insect *Sasakia funebris*

emu either of two species of dromiceiid Australian and Tasmanian flightless birds

en *see* [engrailed]

En *see* [English]

-en- *comb. form* meaning "in"

enamel the outer layer of a tooth consisting of overlapping prismatic scale of apatite cemented in an organic matrix

-enant- *comb. form* meaning "opposite"

Enantioblastae a suborder of monocotyledonous angiosperms containing the families Restionaceae, Centrolepideae, Xyridaceae and Eriocaulaceae

enate = cognate

enation the growth of one organ from another

-enaulo- *comb. form* meaning "water course" but used by ecologists in the sense of "sand draw"

-enchym- *comb. form* meaning "poured" or "molded"

-enchyma- *see* -chyma-

-enchyme- *see* -chyme-

Enchytraeidae a vary large family of slender oligochaete annelids mostly from one to three centimeters long and usually less than a millimeter thick. Many inhabit fresh waters, though some are found on ocean shores and in decaying vegetation

Encyrtidae a family of small chalcidoid hymenopteran apocritan insects readily distinguished by their broad convex mesopleuron and one of the classic examples of polyembryony in the animal kingdom with recorded instances of a thousand embryos developing from a single egg

encystment the process of becoming enclosed in a cyst

-end-, -endo- *comb. form* meaning "inner"

-endeca- *comb. form* meaning "eleven"

endiviaceous cornflower blue

endive properly the compositaceous herb *Cichorium endivia*

French endive the etiolated leaves of chicory (*q.v.*)

-endo- *comb. form* meaning "within"

endome the inner layer of pachyte, roughly equivalent to the pheloderm

Endomycetales an order of hemiascomycete Ascomycetae distinguished by the fact that plasmogamy is immediately followed by karyogamy

Endomychidae a family of small brightly colored fungicolous coleopteran insects commonly called handsome fungus beetles. They are closely allied to the Coccinellidae but separated from them by the absence of teeth on the tarsal claws

Endorhizae = Monocotyledonae

Endosphaeraceae that family of chlorococcale Chlorophyceae distinguished as unicellular forms with large irregularly-shaped cells

Endosporeae a subclass of the Myxomycophyta that produce their spores internally within a fruiting body of definite shape

Endromididae a small family of bombycoid lepidopteran insects

ependyma the sheet of neuroglial tissue lining the cavities of the central nervous system

-eneilm- *comb. form* meaning "a wrapper"

energesis = anabolism

energid a unit consisting of a nucleus and the cytoplasm that it directly controls. Also used for "cell" as a unit of living matter

monoenergid a cell containing a single nucleus

polyenergid an individual containing several energid units, as a multinucleate protozoan

[English] a mutant (gene symbol *En*) at locus 1.2 on linkage group II of the rabbit. The pelt of the phenotype is spotted on a white ground

[engrailed] a mutant (gene symbol *en*) mapped at 62.0 on chromosome II of *Drosophila melanogaster*. The phenotypic expression are broken veins, a scutellar notch and an extra sex comb

Engraulidae a family of isospondylous fishes, commonly called anchovies, distinguished by the small size of the lower jaw

engraver *U.S.* popular name of scolytid coleopteran insects of the genus Scolytus, which engrave the bark of trees (*cf.* bark beetle)

hackberry engraver *S. muticus*

fir engraver *S. ventralis*

Douglas-fir engraver *S. unispinosus*

Engystomatidae a family of anuran Amphibia distinguished by having a solid sternum and a dilated sacral diapophyses. They are commonly called narrow-mouthed toads

enhancer a gene that intensifies the effect of a mutant, making it more extreme in its departure from wild type

[Enhancer of am] a mutant (gene symbol *i*) in linkage group V of *Neurospora crassa*. The phenotypic expression is inhibition of the utilization of glutamic acid by amination-deficient genotypes

[Enhancer of Star] a mutant (gene symbol *E–S*) mapped at 6.0± on chromosome II of *Drosophila melanogaster*. The name is descriptive

[enhancer of white-eosin] a mutant (gene symbol en-w^e) mapped at 32.0 on the X chromosome of *Drosophila melanogaster*. In combination with w^e

alleles the phenotype reverts to a white eye. This mutant also suppresses the pseudoalleles [forked]

Enicocephalidae a small family of slender predaceous hemipteran insects distinguished from all other hemipterans by possessing entirely membraneous wings

Enicuridae a family of passeriform birds usually included in the Turdidae

-ennea- *comb. form* meaning "nine"

Enneandria that class of the Linnaean classification of plants distinguished by the possesssion of nine stamens

-ennial pertaining to a period of a year

biennial an event which occurs once very two years or a plant which blooms two years after the seeds are sown and then dies

perennial that which occurs in more than two successive years, particularly an herb which blooms in this manner

biperennial living two years but reproducing indefinitely

triennial lasting for three years

Enopla a subclass of the phylum Nemertea distinguished by having the mouth anterior to the brain

Enoplida an order of aphasmid nematodes distinguished from the Chromadorida by the fact that the esophagus is elongate and divided into two regions

Enoplidae a large family of minute, free-living nematode worms distinguished by their hairy or bristly mouth

Enoplidea a class of nematode worms with a bristly cuticle, cyathiform amphids and six labial papillae

-ens- *comb form* meaning a "sword"

-entell- *comb form* meaning to "order" in the sense of command

entellechy the postulated condition that vital functions can be suspended by a hypothetical agent which cannot be perceived or measured (*cf.* vitalism)

Enterobacteriaceae bacteria occuring as short rods frequently motile with peritrichous flagellae. A few are plant parasites or even saprophytic but the great majority are dangerous pathogens including such genera as Escherichia, Klebsiella, Proteus, Salmonella, and Shigella

Enterocoelia = Deuterostomia

-enteron- *comb. form* meaning "gut"

archenteron 1 (*see also* archenteron 2) the cavity, precursor to the gut, formed by the invagination of the blastula

archenteron 2 (*see also* archenteron 1) the cavity of Volvox

coelenteron the single body cavity of the Coelenterata

Enterocoela a term used in distinction to Coelomocoela to group together those phyla having only a single body cavity

Enteropneusta a class of the phylum Hemichordata containing the acorn worms. They are distinguished by the acorn-shaped proboscis

entire in *bot.,* having a smooth edge

[entire leaves] a mutant (gene symbol *e*) mapped at 0 on chromosome 4 of the tomato. The phenotypic leaves are either entire or at least very broad

Entomobryidae a family of collembolan insects with a reduced prothorax and without bristles, distinguished from the Sminthuridae by possessing cylindrical bodies

Entomophthorales an order of aflagellate phycomycete fungi distinguished from the Mucorales by the presence of transverse walls in the mycelia

Entomostraca a no-longer acceptable zoological taxon at one time embracing all of the Crustacea except the Malacostraca. As so defined it contained the Branchiopoda, Ostracoda, Copepoda, and Cirripedia, each now regarded as a separate subclass

Entoprocta a phylum of pseudocoelomate bilateral animals distinguished by a distal circlet of ciliated tentacles and with the anus opening inside the circlet. These sessile forms were once combined with the Ectoprocta into the no-longer valid phylum Bryozoa

Entotrophi an order of insects erected to contain the japygids and their allies by those who do not hold them to belong in the Thysanura

envelope in *Biol.* any enveloping structure

cellular envelope = mesophloem

floral envelope the sum total of the perianth, calyx and corolla

paraphysial envelope the peridium of Uredineae

environment the physical matrix in which organisms exist

microenvironment = microclimate

en-we *see* [enhancer of white-eosin]

enzyme a protein or conjugated protein which, because of its configuration, both lowers the energy of activation of, and directs the stepwise pathway taken by, chemical reactions in a living organism. It has been estimated that a cell may require a thousand or more enzymes to maintain the vast number of cyclic reactions which characterize "living" systems (*see also* -zyme)

antienzyme = enzyme inhibitor

apoenzyme an enzyme that cannot function without a coenzyme

co-enzyme a non-protein component of many enzymatic reactions to which it is essential, but from the other components of which it may be separated (*cf.* activator)

coenzyme I *obs.* = NAD

coenzyme II *obs.* = NADP

coenzyme A a most important coenzyme in biokinetic systems. The systematic name is 3'-phosphoadenosine diphosphate-pantoyl -β-alanine

coenzyme Q = ubiquinone

coenzyme R *obs.* = biotin

holoenzyme a complete function enzyme, that is the apoenzyme and the coenzyme taken together

isoenzyme one of two or more molecular forms of the same enzyme

Old Yellow enzyme (*obs.*) = NADPH diaphorase

PR-enzyme = phosphorylase phosphotase

-eo- *comb. form* meaning "dawn" and used by biologists in the sense of "early" or "earliest" in contrast to "-nes-" and "-neo-"

Eoacanthocephala an order of Acanthocephala distinguished from the Archiacanthocephala by lacking protonephridia and from the Palaeacanthocephala by the radiai arrangement of the proboscis hooks

eosin *see* [white-eosin]

Epacridaceae a family of ericalous dicotyledons which may be distinguished from the closely related Ericaceae by having only a single whorl of stamens

Epacridineae a suborder of Ericales containing the single family Epacridaceae

epaulet 1 (*see also* epaulet 2) a heavy, cuticular, cordlike thickening at the anterior region of some nematodes

epaulet 2 (*see also* epaulet 1) a hairy scale at the base of the anterior wing of some Diptera

ependyma cells lining the lumen of the neural tube

-ephattom- *comb. form* meaning "grasp"

-epheb- *comb. form* meaning "adult"

ephebic pertaining to the imago of a winged insect

phyloephebic = paracme

Ephedraceae a family of gnetale spermatophytes having the characteristcs of the order, containing the joint-firs

-ephem- *comb. form* meaning "short lived"

ephemer 1 (*see also* ephemer 2) an organism which is reproductively mature only for one day

pseudoephemer an ephemer that lasts more than one day

ephemer 2 (*see also* ephemer 1) organisms that are introduced to a new environment but soon disappear from it

Ephemeridae a family of ephemeroteran insects having a vein M_2 of the forewing strongly bent toward Cu basally and with the hind tarsae 4-segmented

Ephemeroptera that order of insects containing the forms commonly called mayflies. They are distinguished by four pairs of membranous wings of which the anterior pair are usually triangular in shape

ephemerous living, or appearing, for a short time

euephemerous said of flowers lasting only one day

Ephippidae a family of deep bodied acanthopterygian fishes distinguished by the complete separation of the soft rayed and spiny rayed portions of the dorsal fins. Commonly called spade fishes

ephippium the outer, extra, shell on the winter egg of a cladoceran

Ephydridae a very large group of myodarian cycloraphous dipteran insects commonly called shore flies, from their habit of occurring in vast numbers both on the shores of lakes and of seas. It is in this group that the famous petroleum fly *Psilopa petrolei,* the larvae of which live in pools of crude petroleum, belongs. Other larvae occur in brackish water or, in *W. U.S.,* in pools so alkaline as to be poisonous to other forms, and are therefore known as brine flys

-epi- *comb. form* properly meaning "upon" but often used in the sense of "uppermost" or "outermost"

Epibiotica a general term for "living fossils." That is, scattered survivors of a group most of which are extinct

-epiblem- *comb. form* meaning "a cloak"

epiblem the epidermis of a root (= rhizodermis)

epiboly the growth of one layer of cells over another, as in the overgrowth of endoderm by mesoderm in the development of the frog

Epicaridea a suborder of isopod Crustacea. They are aberrant forms modified for parasitic existence

epidiymis that portion of the sperm duct of the male vertebrate which is derived from the remants of the mesonephros

Epimachidae a family of passeriform birds commonly included with the Paradisaeidae

epimer a stereoisomer which differs from the other isomer in possessing more than one asymmetrical carbon atom only one of which, with its attached groups, contributes to the stereoisomerism

epimerase a group name for enzymes that catalyze the production of one epimer from another

aldose 1-epimerase catalyzes the production of β-D-glucose from α-D-glucose

3-hydroxybutyryl-CoA epimerase catalyzes the production of D-3-hydroxybutyryl-CoA- from L-3-hydroxybutyryl-CoA

UDPglucose epimerase catalyzes the production of UDPgalactose from UDPglucose

ribulose phosphate epimerase catalyzes the production of D-xylulose 5-phosphate from D-ribulose 5-phosphate

diaminopimelate epimerase catalyzes the production of meso-diaminopimelate from 2,6-L-diaminopimelate

threonine epimerase catalyzes the production of D-threonine from L-threonine

epinasty an outward and downward bending movement of a plant

pseudoepinasty = geotropism

epinephrine a hormone secreted by the adrenal medulla and active in glucose metabolism and vaso-constriction

norepinephrine a hormone similar in origin and function to epinephrine, but with greater effect on arteriolar constriction

Epipaschiidae a small family of lepidopteran insects usually regarded as a subfamily of the Pyralidae

Epiplemidae a family of uranioid lepidopteran insects

Epipyropidae a family of zygaenoid lepidopteran insects principally distinguished by the fact that the larvae are parasitic on fulgorid homoptera

epithelium see -thelium

epithem thin-walled paranchyma at the ends of conducting vessels in leaves. In most plants, they under lie hydathodes

Epitoniidae a family of gastropod mollusks commonly called wentletraps or staircase shells in virtue of the marked ribbing transverse to the high spiral whorls

epizous = epizoochory

eponyms see *under* organ, or system, not under name of individual

Epthianuridae a family of Australian birds usually included with the Muscicapidae

-equi- *comb. form* meaning "equal"

Equidae the family of perissodactyle mammals containing the horses and their immediate allies, All except Grevy's zebra (*q.v.*) are of the genus Equus

equilenin a hormone having properties similar to estradiol

Equisetaceae that family of Pterydophyta which contains the horsetails. The single genus Equisetum is distinguished by the erect axis with joints from which rise scale-like leaves

Equisetales the only order of the Equisetinae containing the single family Equisetaceae the characteristics of which are those of the order

Equisetinae that class of the Pterydophyta which contains the horsetails. The characteristics of the family Equisetaceae are those of the class

equitant set astride

ecological equivalent an animal occupying an ecological niche, and having in the broadest sense the same general appearance as, another animal to which it is not closely related

Eratidae a small family of gastropod Mollusca resembling the Cypraeidae but easily distinguished from them by the fact that the shell is not polished. They are sometimes called sea buttons

erectile capable of being erected either by distension or movement

-erem- *comb. form* confused from three roots and therefore variously meaning "desert," "solitary" and "gentle"

eremad a desert plant

Eremialectoridae = Pteroclidae

-eremu- *comb. form* meaning "hermit"

Erethrizontidae that family of hystricomorph rodents which contains, *inter alia,* the New World porcupines. They differ from the Old World porcupines (Hystricidae) in numerous adaptations to arboreal life

-ergas- *comb. form* meaning "labor"

ergastic said of non-living material within a cell

ergate (*see also* -andromorph, -androus, -aner, -gyne, -gynous) a worker ant

desinergate a "sub-soldier" caste

dinergate that caste of ant with enlarged heads and jaws commonly called the soldier caste

gynergate a fertile female worker ant

lateral gynergate a lateral gynandromorphic ant

macrergate a giant worker ant

mermithergate a worker ant, structurally altered in consequence of being parasitized by mermithid nematodes

micrergate a dwarf ergate

pleurergate = replete

perergate = replete

phthisergate a worker which has wasted away through the attack of a parasite

pteregate a worker ant with vestigial wings

ergatoid a fertile ergate

ergid an abbreviation of energid (*q.v.*)

synergid one of the two nuclei of the upper end of the plant embryo sac

ergot a dark, spongy, parasitic mass found on the ovaries of various grasses. It is the sclerotium of Claviceps, a genus of claviseptate ascomycetes

Ericaceae a very large family of ericalous dicotyledons containing *inter alia,* the barberries, the heaths, the heathers, the trailing arbutus, the mountain laurel, the cranberry, the huckleberry, the azaleas and the rhododendrons

Ericales that order of dicotyledonous angiosperms which contains *inter alia,* the pepperbushes, the azaleas and rhododendrons, the mountain laurel, the cranberries, the heaths. The order is characterized by the pentamerous flowers with distinct and free petals

Ericinae a suborder of Ericales containing all the families except the Epacridaceae

Ericoideae a subfamily of Ericaceae distinguished by having a superior ovary but lacking the winged seed of the Rhododendroideae

erigous said of a prostrate plant stem with a vertically directed tip

Erinacidae a family of insectivores containing the European hedgehog and the gymnures

Erinacinae a sub-family of erinacid insectivores containing the hedgehogs. Their most typical characteristic is the fusion of hairs into thin narrow spines

erineum a dense mass of fluffy hairs produced on a plant by some gall mites

Erinnidae = Xylophagidae

erinous resembling a hedgehog

-erio- *comb. form* meaning "wool"

Eriocaulaceae a family of farinose monocotyledonous angiosperms containing the forms commonly called pipeworts. They are grass-like, often with pelucid, and sometimes membranous leaves

Eriocaulales a rarely used order of monocotyledonous angiosperms erected to contain the single family Eriocaulaceae

Eriocephalidae = Micropterygidae

Eriococcinae a subfamily of coccid homopteran insects commonly called mealy bugs. The name derives from the floury appearance

Eriocraniidae a family of jugate lepidopteran insects

Eriophyidae a family of acarine arthropods with only four legs and feeding on plants. Members of this family cause some types of gall

Eriosomatinae a subfamily of aphid Homoptera containing the woolly aphids

Erisiphales an order of euascomycete Ascomycetae containing those forms commonly called "powdery mildews." They are distinguished by the globose ascocarp and their superficial parasitic habit

-erko- *comb. form* meaning "fence" usually transliterated -herco-

ermine 1 (*see also* ermine 2) the white winter color phase of *Mustela erminea* (= stoat)

ermine 2 (*see also* ermine 1, moth) *Brit.* any of several species of arctiid lepidopteran insects usually, without qualification, those of the genus Spilosoma

seven spot ermine *Brit.* the arctaeid lepidopteran insect *Diaphora mendica* (= muslin)

ermineous off white with yellow streaks or blotches

erose having the appearance of having been gnawed

Erotylidae a family of medium-sized brightly colored fungicolous coleopteran insects commonly called pleasing fungus beetles

-erp- *comb. form* meaning "creep" commonly transliterated -herp-

Erpopdellidae a family of gnathobdellid hirudinean annelids, distinguished from the Hirudinidae by having the jaws reduced to muscular ridges

errantia an assemblage, sometimes regarded as an order, of those polychaete worms which are, with the exception of the Myzostomidae, free living and predatory (*cf.* Sedentaria)

-eruc- *comb. form* meaning "caterpillar" (*cf.* -camp-)

erumpent being about to break through, or having such an appearance

Erycinae a small, possibly heterogenous, assemblage of small, short-tailed, ovoviviparous boid snakes

Erycinidae = Riodinidae

Eryonidea a superfamily of macruran Crustacea containing a few deep water forms

hemeerythrin an iron-protein respiratory pigment found in brachiopods, sipunculids and some annelids

-erythro- *comb. form* meaning red

[erythrocyte agglutination] a mutant (gene symbol *an*) found at locus 36.8 on linkage group V of the rabbit. The name expresses the phenotypic reaction

Erythroxylaceae that family of geranealous dicotyledons which contains the coca shrub. The peculiar non-capsular fruit, of considerable economic importance, is typical of the family

Escallonoidea a subfamily of Saxifracaceae considered by some to be of familial rank

escape cultivated plant that has reverted to wild plant, corresponding to the zoological term feral

-escens *comb. term* corresponding to the *Engl.* -ish

escolar properly the gempylid fish *Lepidocybium flavobrunneum* but also applied the Mediterranean scombrid *Ruvettus pretiosus*

esculent edible

escutcheon literally, a shield and therefore used sometimes in place of scutum or scutellum and of many other shield-shaped structures or areas

Esocidae the pike or pickerel family of haplomous fishes readily distinguished by their duck-like shovel mouths and posteriorly placed dorsal and anal fins. The English names of members of this family are much confused

esophagus *see* oesophagus

esoteric developing, or being produced, within an organism

-ess *comb. term* indicative of the female sex, as leopardess, lioness, etc.

esterase a hydrolase that acts on ester bonds

acetylesterase catalyzes the hydrolysis of acetic esters into acetic acid and alcohols

arylesterase catalyzes the hydrolysis of phenyl esters into phenol and acids

carboxylesterase catalyzes the hydrolysis of carboxylic esters, into alcohols and carboxylic acids

cholinesterase catalyzes the hydrolysis of choline esters into choline and the appropriate acid

acetylcholinesterase catalyzes the hydrolysis of acetylcholine to choline and acetic acid

benzoylcholinesterase catalyzes the hydrolysis of benzoylcholine to choline and benzoic acid

glycerophosphorylcholine diesterase catalyzes the hydrolysis of L-3-glyceryl-phosphorylcholine to yield choline and glycerol 1-phosphate

cholesterol esterase catalyzes the hydrolysis of esters of cholesterol to cholesterol and the appropriate acid

phosphodiesterase catalyzes the hydrolysis of phosphoric diesters to yield phosphoric monoesters and alcohols

pectinesterase catalyzes the hydrolysis of pectin to methanol and pectic acid

glutathione thiolesterase catalyzes the hydrolysis of S-acylglutathione to reduced glutathione and an acid

tropinesterase catalyzes the hydrolysis of tropine esters to tropine and the appropriate acid

vitamin A esterase catalyzes the hydrolysis of vitamin A acetate to vitamin A and acetic acid

esthete *see* aesthete

estradiol a hormone, secreted by the ovary, necessary for the maintenance of pregnancy and for the development of female secondary sex-characters

di-OH-estrin = estradiol

keto-OH-estrin = estrone

tri-OH-estrin = estriol

estriol a hormone, secreted by the ovary, having properties similar to estradiol

estrone a hormone secreted by the ovary having properties similar to estradiol

estrous *see* oestrous

et *see* [etched endosperm]

-etaer- *comb. form* meaning "companionship"

etaerio an aggregate drupe

[etched endosperm] a mutant (gene symbol *et*) mapped at 115 on linkage group III of *Zea mays*. The name is descriptive of the phenotypic endosperm

-etes- *comb. form* meaning "annual"

etesial pertaining to herbaceous perennials

Ethmiidae a small family of gelechioid lepidopteran insects almost all of which are strikingly marked in black and white

ethology the reaction of an organism to its environment

etiolated said of a seedling which is drawn out and bleached through the absence of light

etrog a horticultural variety of citron (*q.v.*)

-eu- *comb. form* meaning "true" or "real" or "genuine"

Euascomycetae a class of ascomycete fungi in which the asci are formed on ascogenous hyphae

Eubacteriales an order of schizomycetes. If motile, they have peritrichous flagellae, and most are either spherical or rod shaped. Both gram-positive and gram-negative forms occur. This is the largest order of schizomycetes and contains most of those forms commonly thought of as "bacteria."

Eubrya that class of muscose Bryophyta which contains the forms usually referred to as mosses. They are distinguished by the fact that the gametophores have leaves with a midrib more than one cell in thickness

Eucarida a superorder (or division) of malascostracous crustacea distinguished from all others by the fact that the carapace colaesces dorsally with all the thoracic segments

Euceridae a family of apoid apocritan hymenopteran insects usually included with the Apidae

Eucestoda a subclass of cestode Platyhelminthes distinguished by the ribbon like body divided into many segments

Eucharidae = Eucharitidae

Eucharitidae a group of medium sized chalcidoid apocritan hymenopteran insects with a petiolate abdomen and a prominent, sometimes spined, scutellum. The eggs hatch into a unique type of larva called a planidium

Euchromiidae = Amatidae

Euciliata a subclass of ciliate Protozoa distinguished from the Protociliata (*q.v.*) by the presence of a "mouth"

Eucleidae = Limacodidae

Eucnemidae a family of coleopteran insects closely allied to the Elateridae and frequently fused with them

Eucopepoda a taxon of crustacean arthropods erected to contain all copepods other than the Branchiura if the latter are to be regarded as copepods

Eucryphiaceae a unigeneric family of parietale dicotyledonous angiosperms best characterized as evergreen resinous shrubs

eudoxid a cormidium which has broken loose from the parent colony and lives a separate existence

Eufilicales an order of filicine Pteridophyta containing the majority of ferns

Euglenales an order of euglenophyte algae containing all those forms which are motile cells capable of euglenoid movement. This order corresponds to the zoologist's protozoan order Euglenida

Euglenida an order of phytomastigophorous Protozoa distinguished by their amoeboid like movements and their ability to exist either holophytically or saprozoically. As holophytes they are placed in the algal phylum Euglenophyta (*q.v.*)

Euglenidae (Insecta) a family of coleopteran insects closely resembling the Anthicidae in appearance and habit

Euglenocapsales = Colaciales

Euglenophyta a phylum of algae containing both motile unicellular forms exhibiting euglenoid movement and a few sessile forms. The fomer are regarded as animals by zoologists who class them in the protozoan order Euglenida (*q.v.*)

Eulabetidae a family of passeriform birds usually included with the Sturnidae

eulachon the osmerid fish *Thaleichthys pacificus* (= candle fish)

eulalia properly the name of a genus of grasses but also frequently misapplied to the grass *Miscanthus sinensis*

Eulamellibranchia an order of pelecypod Mollusca containing, *inter alia* the oysters, clams, cockles and their allies. They are distinguished by the fact that the branchial filaments of the gills are united at regular intervals by vascular junctions

Eulebetidae a family of passeriform birds usually included with the Sturnidae

Eulophidae a family of small chalcidoid apocritan hymenopteran insects closely allied to the Trichogrammatidae but distinguished from them by the four segmented tarsae. Sometimes used in the broad sense to include four or five subfamilies which would otherwise have familial rank

Eumelacostraca a large taxon, frequently referred to as a series, of malacostracan crustacea. They are distinguished from the Leptostraca by the absence of an adductor muscle of the carapace and by possessing typically six (sometimes less) abdominal somites

Eumenidae a family of vespoid apocritan hymenopteran insects usually included with the Vespidae

Eumeninae a large subfamily of vespid Hymenoptera commonly called mason or potter wasps because those of the genus Emenes construct vase-like nests of mud which are attached to twigs. They should not, however, be confused with the mud daubers which are sphecoids

Eumetazoa a taxon erected to contain all metazoa except Porifera

Eumycophyta a subkingdom or phylum of plants containing those forms popularly called fungi. That is, the molds, mildews, toadstools, mushrooms, and the like. All save a few bacteria—which many do not regard as belonging in the Eumycophyta—lack chlorophyll. They differ from the Myxomycophyta, which some would include in the fungi, in possessing a definite cell wall throughout the stages of vegetative development

Eunicidae a family of errant polychaete annelids distinguished by the fact that the prostomium is not annulated and that there are many jaws

Eupelmidae a small family of chalcidoid hymenopteran apocritan insects many of which are wingless and are capable of leaping. The females posess a broad convex mesopleuron similar to that of the allied Encyrtidae

Eupetidae a family of passeriform birds usually included in the Timaliidae

Euphausiacea an order of eucarid malacostracan Crustacea distinguished by the fact that none of the thoracic limbs are specialized as maxillipedes

Euphorbiaceae that family of geranealous dicotyledons which contains, *inter alia,* the castor oil plant, the croton oil plant, the rubber tree and a number of poisonous plants from the sap of which arrow poisons are made in South America. The family is remarkably polymorphic but differs from most of the other geranealous families in having albuminous seeds

euplantula = pulvillus

Eupteleaceae a family of ranale dicotyledonous angiosperms frequently included with the Trochodendraceae from which it is principally distinguished by having the stamens in a single whorl

Eupterotidae = Zanolidae

-euri- *comb. form* meaning "broad," better written -eury-

Euribiidae = Trypetidae

-euro- *comb. form* meaning "mold" but used by ecologists in the specific sense of "leaf mold"

euroky =euryoky

-eury- *comb. form* meaning "broad" or "wide"

Euryalae an order of ophiuroid Echinodermata distinguished from the Ophiurae by the branched arms and irregular or absent superficial plates

Eurycerotidae a family of passeriform birds usually included with the Vangidae

Eurylaimidae that family of passeriform birds commonly called broadbills, a name derived from the broad, flattened, wide-gaped, hooked bill

euryoky the condition of being able to sustain a wide range of variation in environmental condition

Eurypygidae a monospecific family of gruiform birds containing the sun-bittern distinguished by its big head and long slender neck

Eurystethidae a family of coleopteran insects most unusual in their marine habitat in clefts in rocks below high tide mark. Apart from this habitat, they are distinguished by their very long legs

eurytele a type of heteroneme nematocyst in which the summit of the butt is dilated (*cf.* rhabdoid, stenotele)

Eurytomidae a family of chalcicoid apocritan hymenopteran insects with head and thorax often coarsely punctate. The pronotum is large, the posterior femora are not enlarged

Eusporangiate a class of pterophyte plants containing all living ferns as well as a number of extinct types. They are distinguished by the fact that the spores are either borne in sori on leaf blades or on a special fertile spike

Eustachian *epon. adj.* from Bartolomeo Eustachi (1513?–1574)

Eutardigrada a division of the Tardigrada distinguished by the absence of buccal appendages

eutele the condition of an animal that retains the same number of nuclei in the adult throughout its life, except in the gonad, and which therefore "grows" only through increase in cytoplasmic mass

eutely = eutele

Eutheria that group of mammalian chordates which contains all the orders except Marsupialia and Monotremata. This group is sometimes referred to as the placental, or true, mammals

-euthy- *comb. form* meaning "immediately"

Euthyneura a subclass of gastropod Mollusca containing, *inter alia,* the snails, slugs and their allies. They are distinguished by the detortion of the visceral mass, particularly the nerve commissures, though tortion is frequently present in the shell

Eutrichosomatidae a small family of chalcidoid apocritan Hymenopteran having the body thickly covered with scale-like hairs

Evaniidae a family of proctotrupoid apocritan hymenopterans with an extremely small abdomen attached by a long stalk to a very large thorax. The abdomen appears to be carried as a flag which gives to this group their popular name of ensign wasps

Evanioidea a superfamily of apocritan hymenopteran insects erected to contain the family Evaniidae by those who hold that these should not be retained in the Proctotrupoidea

everlasting *see* flower

eversible said of a portion of an organism, particularly the pharynx of an invertebrate, that is protruded and turned inside out at the same time (*cf.* protrusible)

evocator a chemical substance involved in induction (*q.v.*)

evolute unfolded

evolution a process of gradual change by which one form of something slowly changes into a similar, but significantly different, form

accidental evolution that which occurs in consequence of a mutation which does not appear to improve survival value

clandestine evolution that which takes place in the larva and is transmitted by neoteny but which subsequently forms part of the genetic composition of the adult

macroevolution the evolution of broad groups as distinct from individuals

allomorphotic evolution that leading rapidly to specialization

aromorphotic evolution that which leads to different, but not highly specialized, forms

quantum evolution a transition from one form to a very different one under such adaptive pressure that intermediate forms do not survive for significant periods

regressive evolution the appearance in a taxon of organisms having characters usually considered to be associated with more primitive forms

saltatory evolution that which is supposed to proceed by leaps and bounds

ew³ *see* [extended wings]

ewe a female sheep or other ovid, through the term is sometimes applied to other artiodactyls, such as the impala

ex *see* [expanded]

-ex- *comb. form* variously meaning "out," "without" or "out of"

exaltation term for a group of larks (*cf.* ascension)

exarate grooved. It is also said of those insect pupae in which the legs and wings lie free of the body and therefore groove the outer surface of the pupal case (*cf.* obtect)

exasperate roughened with prickly points, as the fruit of the chestnut

exciple literally, a basin, but particularly the cuffed ring round the base of an apothecium

excretion the process of removing unwanted materials from a cell, organ, or organism (*cf.* secretion)

exerted protruded

exi *see* [exiguous]

[exiguous] a mutant (gene sybmol *exi*) mapped at 51.5± on the X chromosome of *Drosophila melanogaster*. The phenotypic expression is a small dark body

eximious outstandingly beautiful

exine the outer wall of an angiosperm microspore (*cf.* extine)

ektexine the outer coat of the exine frequently decorated with structures or patterns of taxonomic value

endexine the smooth inner coat of the exine

intexine the inner of the two layers of a two-layered exine

exite an outer lobe on the appendages of eubranchipod crustacea

-exo- *comb. form* meaning "out of" (*cf.* -ex-)

Exobasidiales a small group of blister and gall producing hymenocete basidiomycete fungi distinguished by the presence of a hymenium which covers the infected portion of the host

Exocoetidae that family of synentognathous fish which contains the flying fishes. The development of the fins for this purpose is typical

Exocycloida an order of echinoid Echinodermata distinguished from the Cidaroida and the Centrechinoida by the fact that the periproct is posterior to the apex

Exosporeae a subclass of Myxomycophyta producing external spores on an erect fruiting body

exostosis the root nodule of a legume, and less frequently other types of plant nodules or even burrs

exoteric arising from outside the organism

exotic foreign

expanded spread (*cf.* explanate)

[expanded] a mutant (gene symbol *ex*) mapped a 0.1 on chromosome II of *Drosophila melanogaster*. The name refers to the broad wings of the phenotype

expressivity the extent to which a gene exercises a phenotypic effect (*cf.* penetrance)

reduced expressivity the condition of an individual showing a lesser phenotypic expression of the character than is customary in the species

Ext *see* [Extras]

[extended wings] a mutant (gene symbol *ea³*) in linkage group VII of *Habrobracon juglandis*. The name is descriptive of the appearance of the phenotype

extensacuta a sclerite in the mandacoria assisting to articulate the mandible to the head segment

extension that movement of a jointed appendage which results in opening the angle subtended by the two segments

sheath extension (*bot.*) = vein extension (*bot.*)

vein extension an extension of the bundle sheath towards the epidermis

[Extension] a mutant (gene symbol *E*) at locus 0 on linkage group VI of the rabbit. The extension refers to the dark pigment in the pelt of the phenotype

pennexterna = epicarpium

extine the outer coat of a pollen grain (*cf.* exine)

-extra- *comb. form* meaning "outside"

[Extras] a mutation (gene symbol *Ext*) mapped at 15.2± on the X chromosome of *Drosophila melanogaster*. The name refers to extra veins in the wings

exutive applied to naked seeds

exuvia that which is cast by molting, particularly the sloughed cuticle of insects

ey *see* [eyeless]

eyas a fledgling hawk

eye 1 (*see also* eye 2–7) a photoreceptor organ particularly one in which there is a lens and retina, or equivalents

cerebral eye an eye embedded in the substance of the brain of rotifers

compound eye an eye typically found in arthropods consisting of numerous discrete, small, eyes joined together

converse eye *see* ocellus

eucone eye a compound eye in which the cone is crystalline

exocone eye a compound eye in which the crystalline cone is replaced by an ingrowth from the corneal facet

pseudocone eye a compound eye in which the cone is semi-liquid within the cone cells above their nuclei

day eye an eye specifically adapted to diurnal habit, particularly among insects

inverse eye *see* ocellus

parietal eye an eye developed from the anterior part of of the epiphysis and commonly confused with the pineal eye

pineal eye an eye-like structure, apparently functional in lampreys and some Amphibia, derived from the epiphysis

eye 2 (*see also* eye 1, 3-7) in compound names of many birds, usually those with conspicuous eyes

bare-eye any of several formicariids of the genus Phlegopsis

bull's eye *U.S.* any of several aythyines of the genus Bucephala (= goldeneye). The term is also locally applied to many charadriids

fire-eye any of several formicariids of the genus Pyriglena

fish eye *W.I.* the turdid *Turdus jamaicensis* (= white-eyed thrush)

glass eye any of several species of sylviids of the genus Camaroptera. In parts of W.I. the term is synonymous with fish-eye (*q.v.*)

goldeneye (*see also* eye 4 and 5) any of several species of aythyine ducks of the genus Bucephala

ox eye (*see also* eye 7) the parid *Parus major* (= great titmouse)

shine eye *W.I.* the turdid *Turdus jamaicensis*

silver eye *Austral.* and *N.Z.* any of several zosteropids of the genus Zosterops

wattle eye popular name of muscicapid birds of the genera Dyaphorophyia and Platysteira. Usually, without qualification, P. cyanea

　yellow-billed wattle eye *D. ansorgei*

　blackthroat wattle-eye *P. peltata*

white-eye any of numerous zosteropids of the genus Zosterops *I.P.,* without qualification, *Z. palpebrosa*

eye 3 (*see also* eye 1, 2, 4-7) in compound names of some reptiles

cat-eye (*see also* eye 6) *U.S.* the colubrid snake *Leptodeira septentrionalis*

eye 4 (*see also* eye 1-3, 5-7) in compound names of many fish, either with conspicuous eyes or with an eye-like marking on the body

barrel eye common name of opisthoproctids but particularly *Macropinna microstona*

big eye popular name of prinacanthids

glow-eye = lamp-eye

goldeneye (*see also* eye 2, 5) the hiodontid *Hiodon alosoides*

lamp-eye the cyprinodont *Aplocheilichtys macrophthalmus*

lantern-eye = lamp-eye

mooneye the hiodontid *Hiodon tergisus*

opal eye the girellid *Girella nigricans*

pearl eyes (*see also* eye 5) common name of scopelarchids

popeye any of several species of priacanthids but particularly *Pristigenys serrula*

prickle eye the cobitid *Acanthophthalmus kuhlii*

tube eye popular name of stylephorids

walleye *U.S.* the percoid fish *Stizostedion vitreum*

eye 5 (*see also* eye 1-4, 6-7) in compound names of insects, mostly lepidoptera with eye-like markings on the wings

black eye *S.Afr.* either of two species of lycaenid lepidopterans of the genus Leptomyrina

bright-eye (*see also* eye bright) *Brit.* the noctuid lepidopteran *Leucania conigera*

brown-eye *Brit.* the noctuid lepidopteran *Memestra oleracea. Austral.* the hesperid *Chaetocneme porphyropsis*

buckeye (*see also* eye 7) *U.S.* the nymphalid lepidopterans *Precis lavinia*

catseye (*see also* eye 6) *I.P.* any of numerous satyrid lepidopterans, many of the genus Zipoetis

　scarce catseye *I.P. Coelites nothis*

Enfield eye *Brit.* nymphalid lepidopteran *Pararge-egeria*

golden eye (*see also* eye 2, 4, bird, fish) *Brit.* poular name of chrysopid Neuroptera (*cf.* lacewing)

pearly eye (*see also* eye 4) *U.S.* any of several species of satyrid lepidopteran of the genus Lethe

red-eye *Austral.* any of several species of hesperid lepidopterans (*cf.* skipper). *I.P.* any of several species of hesperiid lepidopterans of the genera Matapa and Gangara. Without qualification, usually *Matapa aria*

eye 6 (*see also* eye 1-5, 7) in compound names of Mollusca

cat's eye (*see also* eye 5) the naticid gastropod *Natica maroccana* and the operculum of the turbinid gastropod *Turbo petholatus*

sharp eye popular name of naticid gastropod (*cf.* moon shell)

eye 7 (*see also* eye 1-6, bright) in compound names of plants, mostly compositaceous flowers with a marked annular pattern

angel's eyes the scrophulariaceous herb *Veronica chamaedrys*

bird's eyes the scrophulariaceous herb *Veronica chamaedrys* and (*U.S.*) *V. persica*

buckeye (*see also* eye 5) *U.S.* any of several species of tree of the hippocastanaceous genus Aesculus (*cf.* horse chestnut)

　Mexican buckeye the sapindaceous herb *Ungnadia speciosa*

　Ohio buckeye *A. glabra*

　red buckeye *A. pavia*

　sweet buckeye *A. octandra* = yellow buckeye

　yellow buckeye = sweet buckeye

doll's eyes *U.S.* the ranunculaceous herb *Actaea pachypoda*

oxeye (*see also* eye 2) any of several compositaceous flowers including *Chrysanthemum leucanthemum* = oxeye daisy, *Anthemis arvens* = field chamomile and several species of the genera Buphthalmum and Heliopsis

sea ox-eye the compositaceous herb *Borrichia frutescens*

pheasant's eye the amaryllidaceous herb *Narcissus poeticus* or, more rarely, the ranunculaceous herb *Adonis aestivalis*

[eye-gone] a mutant (gene symbol *eyg*) mapped at 35.5 on chromosome III of *Drosophila melanogaster.* Both the eyes and head of the phenotype are greatly reduced

[eyeless] *either* a mutant (gene symbol *ey*) mapped at 2.0 on chromosome IV of *Drosophila melanogaster.* The name is descriptive of the phenotype *or* a mutant (gene symbol *el*) found in linkage group I of *Habrobracon juglandis.* The name is descriptive

eyg *see* [eye-gone]

eyra the felid carnivorous mammal *Herpailurus eyra.* (*cf.* jaguarondi with which some consider this species to be identical)

eyrie the nest of a hawk or eagle

eyry = eyrie

ezi the anacardiaceous tree *Spondias cytherea* and its edible fruit

F abbreviation for filial, itself meaning "pertaining to children." The characters F_1, F_2, etc. are used to denote the first, second, etc. generations of offspring from a given mating. Also used in yeast genetics as an indicator for flocculence. *See also* [Flesh] and [Frizzling]

f *see* [fawn], [fasciated fruit], [flexed tail], [forked] and [furless]

f_1 *see* [fine striped]

fa *see* [facet]

-fab- *comb. form* meaning "bean," particularly the "broad bean" (*Vicia faba*)

face the anterior surface of the head of a mammal together with the sense organs there located and in names of some organisms

dogface *U.S.* the pierid lepidopteran insect *Colias cesonia* (*cf.* sulphur)

pigface *N.Z.* the aizoaceous herb *Disphysma australe* (= ice plant)

whiteface *Austral.* any of several malurine birds of the genus Aphelocephla

[facet] one of four pseudo alleles loci (gene symbol *fa-* the others are notchoid split and notch) mapped at 3.0± on the X chromosome of *Drosophila melanogaster*. The phenotypic expression is mixed wings and rough eyes

faciation an association within which one or more dominants has been replaced, thus differing from a consociation in which a dominant has dropped out but not been replaced

facies used in *bot.*, in the sense of "over-all shape"

facilitation the alleged property of some nerves to have the admission of subsequent stimuli made easier by earlier stimuli

artifact an appearance, or structure, produced by treatment of material in the course of preparation and which is not present in the original material before the manipulation

factitious artificial

factor in biology usually applies to an unidentified substance involved in, or the causative agent of, a specific reaction or process (*cf.* principle)

anti-acrodynia factor = pyridoxine

additive factors = cumulative factors

anti beriberi factor = vitamin B_1

complementary factor = complementary gene

cumulative factor one of several factors that are not allelic but each of which enhances the effect of the other

chick antidermatitis factor = pantothenic acid

edaphic factor the contribution of the inhabitants of the soil to an ecosystem. Also any of numerous physical-chemical soil conditions that influence the structure and function of an ecosystem

growth factor plant or animal hormones involved in the initiation, or maintenance, of growth of higher forms or any substance required for the growth of a microorganism

antihemorrhagic factor = vitamin K_1, K_2

anti-egg-white injury factor = vitamin H

lactation factor = vitamin L

Lactobacillus casei factor = folic acid

Lactobacillus lactis Dorner (LLD) factor = vitamin B_{12}

latency factor = latency

antineuritic factor = vitamin B_1

pellagra preventing (pp) factor = nicotinic acid

anti-rachitic factor = vitamin D_2

antiscorbutic factor = vitamin C

anti-sterility factor = vitamin E

antixerophthalmia factor = vitamin A

facultative optional, or, more accurately, capable of exercising an option to

FAD = flavin-adenine dinucleotide

faeces *see* feces

Fagaceae that family of fagalous dicotyledons which contains, *inter alia* the chestnuts, beeches and oaks. Distinctive characteristics are that only the staminate flowers are borne in catkins and that there are three carpels and a one-seeded fruit

Fagales that order of dicotyledonous plants which contains, *inter alia* the beeches and chestnuts. The order is distinguished by the inferior unilocular ovary containing two or more ovules

fairy any of several apodiform trochiid birds of the genus Heliothryx

grassland fairy the hesperid lepidopteran insect *Isoteinon inornatus*

-falc- *comb. form* meaning "sickle"

falces *plural* of falx

falcon any of numerous hawks of the falconiform family Falconidae. They are distinguished from the tree hawks in having a weak foot. The term falcon, used without qualification, usually refers to the peregrine falcon and is often used only for the female, tercel being used for the male

aplomado falcon *Falco femoralis*

bat falcon *F. rufigularis*

carrion falcon popular name of polyborine birds (*cf.* caracara)

red footed falcon *F. vespertinus*

forest falcon popular name of micrasturine birds

gerfalcon = gyrfalcon

gyrfalcon a large Arctic falcon (*F. rusticolus*) so

difficult to secure and so powerful in flight, that its use in medieval falconry was restricted to kings

hobby falcon *F. subbuteo*

lagger falcon *F. jugger*

lanner falcon *F. biarmicus*

laughing falcon popular name of herpetotherine birds

New Zealand falcon *F. novaeseelandiae*

peregrine falcon a large falcon *F. peregrinus*, regarded by medieval falconers as ranking third in the aristocracy of birds, being inferior only to the eagle and the gyrfalcon

pigmy falcon *Afr. Polihierax semitor quatus*

prairie falcon *F. mexicanus*

saker falcon *F. cherrug*

shahin falcon = peregrine falcon

slight falcon = peregrine falcon

sore falcon a bird of the first year that has not molted

falconet any of several small falconid birds, particularly those of the genera Microhierax and Neohierax

Falconidae a family of falconiform birds containing the falcons. They are most readily distinguished from the Accipitridae by the sharply pointed wings and the fact that the foot is not as powerful

Falconiformes that order of birds which contains the diurnal birds of prey. They are distinguished by the hooked bill, with a cere in the center of which the nostrils are located

Falcininae that subfamily of falconid birds which contains the true falcons. They are distinguished by their compact bills, long wing, and conspicuously toothed and notched bill

Falculidae a family of passeriform birds usually included with the Vangidae

Falcunculidae a family of passeriform birds usually included with the Muscicapidae

fall 1 (*see also* fall 2) term for a group of mixed wildfowl, in the *Brit.* sense of this word

fall 2 (*see also* fall 1) *U.S.* = autumn *Brit.*

Fallopian *epon. adj* from Gabriele Fallopio (1523–1563)

falx literally, a sickle, (*cf.* falciform) and applied to almost any biological object of that shape, from a piece of the genitalia of Lepidoptera to the septum extending vertically between the cerebral hemispheres of vertebrates

family 1 (*see also* family 2, 3) a biological taxon ranking immediately above genus and below order. In *zool.* all families terminate in -idae and in *bot.* most terminate in -aceae

subfamily a division of a family containing several distinct but allied genera. In *bot.* the names of subfamilies terminate in -oideae and in *zool.* invariably in -inae

superfamily a zoological taxon erected to comprise a group of related families. In *zool.* the name of a superfamily properly ends in -oidea

family 2 (*see also* family 1, 3) a group of common parentage living as a unit

cell family a group of cells having a common origin

family 3 (*see also* family 1, 2) a term for a group of beavers whether known to be related or not (*cf.* colony)

famulus a microsensory seta found on the limbs of some mites

fan (*see also* tail) a structure, or organism, consisting of, or supported by, radiating parts

anal fan a fan shaped structure protruding from the anal angle (*q.v.*) in insects

caudal fan the fan of breathing tubes at the posterior end of a mosquito larva

sea fan popular name of gorgoniid alcyonarian coelenterates, in *U.S.* particularly *Gorgonia flabellum*

tail fan the telson and uropods of crustacea taken together

fanaloka the Madagascan hemigaline viverrid *Fossa daubentoni* also known as the fossane. These names have caused endless confusion with the fossa (q.v.)

fang a popular term for a large tooth, particularly of carnivorous mammals, or venomous snakes; the word is also applied to venom bearing Chelicerae of spiders. For types of the former *see* -glyph

Faniidae = Anthomyiidae

blue fanny *Austral.* the papilionid lepidopteran insect *Papilio sarpedon* (= blue triangle)

farctate stuffed full

infarctate turgid

field fare *Brit.* the turdid bird *Turdus pilaris*

farewell-to-spring the onagraceous herb *Godetia amoena*

farina literally "meal" in the sense of coarsely ground cereal. The term has been applied to almost anything of this consistency, including some larger pollen grains

farinaceous starchy

Farinosae that order of monocotyledonous plants that contains, *inter alia*, the bromeliads, spiderworts and pickerel weeds

farinose mealy

-farious *comb. form* meaning "placed in rows"

bifarious arranged in two rows

multifarious ranged in many ranks

trifarious facing in three directions

farrow the bearing of young by a pig

farumfer *S.Afr.* either of two species of bathyergid rodent of the genus Heterocephalus

fasciated mutiloculate

[fasciated fruit] a mutant (gene symbol *f*) mapped at 2 in linkage group V of the tomato. The phenotypic expression is multiloculate fruit

fascicle a bundle, particularly of vessels in a plant stem (*see* vascular bundle.) The term is also used for bundles of fused zoecia of some Ectoprocta

Fasciolariidae a family of large snail-like marine gastropod mollusks with a sharply pointed spire and no umbilicus. Commonly called tulip shells though this term is sometimes confined to the genus Fasciolaria

fasciole a tract of clavules (*q.v.*) in an echinoid Echinoderm

acid-fast said of bacteria from which an initial stain is not removed by treatment with acid solutions

hold fast any large adhesive organ, or structure, particularly those of some trematode platyhelminthes, the terminal discs on the rhizoid of an Ectoproct or the thickenings of the wall of those sipunculids which inhabit abandoned gastropod shells. In *bot.* used specifically for the rhizoids of large marine algae

fastigiate literally gabled, but used to describe shrubs having erect clusters of branches

fat a popular term for the esters of glycerol and fatty acids stored as a food reserve by animals in adipose tissues

winter fat the chenopodiaceous *W. U.S.* desert shrub *Eurotia lanata*

[fat] a mutant (gene symbol *ft*) mapped at 12.0 on chromosome II of *Drosophila melanogaster*. The name refers to the short squat body of the phenotype

fate in *embryol.* the anticipated end of a developing part (e.g. the production of one lateral half of the whole from one blastomere after the first division of an echinoderm egg *cf.* potency)

fatima *U.S.* the nymphalid lepidopteran insect *Anartia fatima*

fatiscent cracked or gaping

-fauc- *comb. term* meaning "throat"

faucial situated in, or pertaining to, the throat, particularly that of flowers

[faulty chaete] a mutant (gene symbol *fc*) mapped at 0.9± on the X chromosome of *Drosophila melanogaster*. The phenotypic expression is short thin bristles

faun *O.P.* any of several species of amathusiid lepidopteran insect of the genus Faunis

fauna the animal population of a given region

    epifauna that which lives on the surface of the bottom of the ocean or, less specifically, any encrusting fauna

    infauna those bottom-dwelling forms that burrow

    interstitial fauna animals living within crevices, particularly those between sand grains

    meiofauna the smaller, invertebrate fauna of sea bottoms

defaunated deprived of animals, used specifically of the removal of intestinal symbiotic protozoa from insects

faunula the population of a micro-environmental niche

faveolate honey-combed

favose = faveolate

fawn a young cervid, in *Brit.* usage particularly a young fallow deer

[fawn] a mutant (gene symbol *f*) mapped at 44.6 on linkage group V of the rat. The name refers to the color of the coat which, however, has a distinct shade of blue

fb *see* [fine bristle]

fc *see* [faulty chaete]

feather 1 (*see also* feather 2, 3, louse) a unit of the external covering of birds

    contour feather a complete feather which lies on the outside of the feather mass of a bird and serves to establish the apparent shape of the animal

    down feather a soft underfeather

        powder-down feathers down feathers which disintegrate at the tip

    flight feather a stiff feather in a bird's wing. (*cf.* remex and rectrix)

    pin feather a juvenile contour feather

feather 2 (*see also* feather 1, 3, bells, grass, geranium) any of several feather like plants

    parrot's feather the haloragidaceous aquatic herb *Myriophyllum proserpinacoides*

    prince's feather (sometimes princess feather) any of several species of the amaranthaceous genus Amaranthus, particularly *A. hypochondriacus*

    princess feather (sometimes prince's feather) the polygonaceous herb *Polygonum orientale*

    water feather the haloragaceous aquatic *Myriophyllum brasiliense*

feather 3 (*see also* feather 1, 2 back) any of several feather-like or feathery, animals

sea-feather popular name of pennatulid alcyonarian Coelenterata (*cf.* sea-pen)

feces the solid or semi-solid excreta of higher animals

fecula literally "dregs" or the "lees of wine." The term is sometimes applied to the excrement of insects

fecundation = fertilization

fecundity the quality of being able to produce many offspring

federion a mos of several species

snake feeder *Brit.* = dragonfly

filter-feeding a mechanism common in invertebrates, but found also in the baleen whales, by which planktonic food or food particles are filtered from large volumes of water

feeler any tactile organ from the antenna of a lobster to the barbel of a catfish

felical pertaining to ferns

Felidae that family of fissipede carnivorous mammals which contains the true cats. They are distinguished by having a greatly inflated auditory bulla and in possessing a carnassial tooth of the upper jaw with three lobes to the blade. The commonly given characteristic of retractile claws applies to all save the cheetah

felleous bitter as gall

Feloidea that suborder of fissipede carnivores which contains, *inter alia* the cats, the civet cats and the hyenas

female that form of a dioecious organism which produces eggs

    metafemale one having additional female sex determinants, such as a XXX chromosome

    neofemale the female produced by sex reversal from the male

    super female = metafemale

[female-sterile] a mutant (gene symbol *fes*) mapped at 5.0± on chromosome II of *Drosophila melanogaster*. The name is descriptive

femur in vertebrates the proximal bone of the leg skeleton, or thigh bone, and in arthropods that joint of the leg which is third from the articulation of the body, and which therefore, in insects, articulates with the trochanter at its proximal end and with the tibia at its distal end

    basifemur that segment of the arachnid pedipalp which lies between the trochanter and the telofemur

    telofemur that segment of the arachnid pedipalp that lies between the basifemur and the genu

fen a peaty, moist tract, usually derived from the ageing or draining of a swamp

flower fence the leguminous tree *Caesalpinia pulcherrima*

fenestrate windowed

fennec a huge-eared, fluffy, fox-like canine *Fennucus zerda*

fennel (*see also* tick) the umbelliferous herb *Foeniculum officinale*

    dog fennel either of two compositaceous herbs *Anthemis cotula* or *Eupatorium capillifolium*

    Florence fennel *Foeniculum dulce*

    giant fennel any of several species and horticultural varieties of umbelliferous herbs of the genus Ferula

    hog's fennel *Brit.* any of several species of umbelliferous herb of the genus Peucedanum

fenugreek the leguminous herb *Trigonella foenumgraecum*. The dried and ground herb is in commerce as a culinary seasoning

-fer- *comb. form* meaning "to bear" (*cf.* -phor-)

    alifer projection of the pleurum against which the insect wing moves

    antennifer the socket like portion of a ball-and-socket-like joint by which some insect antennae are inserted

    lacticifer a cell that produces milk (*cf.* laticifer)

    laticifer a cell that produces latex (*cf.* lacticifer)

    palpifer a palp-bearing segment particularly a sclerite bearing the insect maxillary palp (*cf.* palpiger)

pilifer the sclerite-resembling rudimentary mandible subtending the clypeus in lepidopteran insects

feral a domestic animal, such as a cat, which has adopted a "wild existence" (*cf.* escape)

biferate having two perforations

fer-de-lance the crotalid snake *Bothrops atrox*

afferent to convey towards

deferent to convey downward

efferent to convey away from

fermentation anaerobic metabolism in which both the electron donor and acceptor are organic compounds

fern (*see also* bird, scale, tongue, owl) any member of the plant phylum Pterophyta and popularly applied to a few other fern-like forms (*e.g.* asparagus "fern")

adder's fern the polypodaceous fern *Polypodium vulgare*

asparagus fern any of numerous species of the liliaceous genus of climbing herbs Smilax or *Asparagus plumosus*

beach fern the polypodiaceous fern *Dryopteris hexagonoptera*

bladder fern any of several species of polypodiaceous ferns of the genus Cystopteris

bracken fern = bracken

snuff box fern the polypodiaceous fern *Dryopteris thelypteris*

chain fern any of several polypodiaceous ferns of the genus Woodwardia

Christmas fern the polypodiaceous fern *Polystichum acrostichoides*

cinnamon fern the osmundaceous fern *Osmunda cinnamonea*

cliff fern the polypodiaceous fern *Dryopteris fragans*

climbing ferns the term is usually applied to members of the schizaeaceous genus Lygodium

cloak fern any of several polypodiaceous ferns of the genus Notholaena

cotton fern the polypodiaceous fern *Notholaena newberry*

dagger fern = Christmas fern

elephant-ear fern the polypodiaceous fern *Elaphoglossum crinitum*

fancy fern the polypodiaceous fern *Dryopteris spinulosa*

filmy fern popular name of ferns of the family Hymenophyllaceae

floating fern the ceratopteridaceous fern *Ceratopteris thalictroides*

fragile fern the polypodiaceous fern *Cystopteris fragilis*

grape fern any of several species of osmundaceous fern of the genus Todea or of ferns of the family Ophioglossaceae

grass fern the name popularly applied to many members of the fern family Schizaeaceae

maiden hair fern any of numerous species of the polypodiaceous genus Adiantum

Hartford fern the schizaeaceous fern *Lygodium palmatum*

holly fern either of two polypodiaceous ferns *Polystichum lonchitis* or *Cyrtonium falcatum*

stag horn fern any of several species of fern of the polypodiaceous genus Platycerium

interrupted fern the osmundaceous fern *Osmunda clatoniana*

lace fern the polypodiaceous fern *Notholaena parryi*

lady fern the polypodiaceous fern *Athyrium felixfemina*

lip fern any of several polypodiaceous ferns of the genus Cheilanthes

male fern the polypodiaceous fern *Dryopteris filixmas*

marsh fern the polypodiaceous fern *Dryopteris thelypteris*

meadow fern *U.S.* the myricaceous shrub *Myrica gale*

ameristic fern one in which antheridia, but no meristem are produced from a frond

New York fern the polypodiaceous fern *Dryopteris novoboracensis*

oak fern = beech fern

ostrich fern any of several species of polypodiaceous fern of the genus Pteretis

royal fern the osmundaceous fern *Osmunda regalis*

hay-scented fern the polypodiaceous fern *Dennstaedtia punctilobula*

shield fern any of several species of polypodiaceous fern of the genus Dryopteris

strawberry fern the polypodiaceous fern *Hemionitis palmata*

sun fern = beech fern

sword fern any of numerous species of the polypodiaceous genus Nephrolepis but particularly *N. exaltata*

tree fern a fern belonging to the pterophyte families Cyatheacea or Dicksoniacea distinguished by their large tree like forms

adder's tongue fern the ophioglossaceous fern *Ophioglossum vulgatum*

hart's tongue fern the polypodaceous fern *Phyllitis scolopendrium*

walking fern any of several species of the polypodiaceous genus Camptosorus

wall fern the polypodiaceous fern *Polypodium vulgar*

wood fern any of several species of the polypodiaceous genus Dryopteris

-ferous *comb. term* meaning "bearing" (*cf.* -fer, -phor, -phorous)

aliferous wing-bearing

calamiferous bearing a hollow stem

ceriferous wax producing

trichiferous bearing hairs

coniferous bearing cones

criniferous hairy or bearded

foliferous leaf-bearing

frondiferous frond-bearing

fructiferous fruit-bearing

gemmiferous reproducing by means of of buds

graniferous = monocotyledonous

hiliferous having a hilum on the surface

lactiferous milk carrying (*cf.* laticiferous)

laticiferous said of a stem bearing laticifers (*cf* lactiferous)

ligniferous wood bearing. *i.e.,* a branch that produces only wood but no flowers

nectariferous bearing nectar

nuciferous nut-bearing

ovuliferous bearing ovules

papilliferous producing nipples

pappiferous having a pappus

pediferous having a stalk

piliferous hairy

pomiferous bearing pomes

proliferous (*bot.*) producing offspring as offshoots

propaguliferous said of a plant bearing offsets

pupiferous pertaining to a generation, particularly in aphids, which produces sexual individuals

racemiferous having racemes

radiciferous root-bearing, particularly applied to organs such as stems, that do not commonly do so

ramiferous bearing branches

sacchariferous bearing sugar

sarmentiferous bearing sarments

sebiferous waxy

seminiferous seed bearing or spermatozoa bearing

soriferous bearing sori

spiniferous thorny

strobiliferous cone-bearing

succiferous bearing sap

surculiferous bearing suckers

tergiferous the condition of bearing sporangia on the dorsal surface of a leaf

vaginiferous sheath-bearing

veneniferous carrying venom, though sometimes used in the sense of carrying poison

ferret (see also badger) a musteline carnivorous mammal variously regarded as a domesticated breed of the European polecat or as a separate species *Putorious furo*. It is used in Europe to flush rabbits from their burrows. The biblical ferret (Hebrew anakah) was probably a lizard of some kind

black-footed ferret *U.S. Mustela nigripes*

ferrugineous rust-colored

ferruminate joined or fused

conferruminate adherent by adjacent faces

fertilization the union of a male and female gamete

closed fertilization the condition of a plant which is fertilized by its own pollen

double fertilization when nuclei in the pollen tube fuse with both the egg nucleus and with the polar nucleus

external fertilization the production of a zygote outside the body

legitimate fertilization fertilization by pollen from another flower of the same species

postfertilization in *bot.*, the period between fertilization and the ripening of the seed

prefertilization in *bot.*, the development of the ovule before fertilization

fertility the possibility, or a measure of the possibility of a female becoming fertilized

differential fertility in population genetics a term used to describe the result of genotypic selection

infertility inability to reproduce

fertilizin a substance, produced by an egg, causing agglutination of spermatozoa

anti-fertilizin a supposed opposite of fertilizin on the surface of spermatozoa

ferulaceous having the form of a hollow reed

arcifery the condition of having the epicoracoids overlap, as in some amphibia, rather than fuse as in others (*cf.* firmisterny)

fes see [female-sterile]

festucine straw-colored

festoon *Brit.* the limacodid lepidopteran insect *Cochlidion limacodes*

Festuceae a tribe of the monocotyledonous family Gramineae

fetticus any of several species of the valerianaceous herb *Valerianeola* but particularly *V. olitoria* (= corn salad, lamb's lettuce)

fetus = foetus

Fevilleae a subfamily of the Cucurbitaceae

feverfew the compositaceous herb *Chrysanthemum parthenium*

bastard feverfew *Parthenium histerophorus*

corn feverfew *Brit.* the compositaceous herb *Matricaria inodora*

[Few fruit locules] a mutant (gene symbol *Lc*) mapped at 67 on chromosome 2 of the tomato. The name is descriptive of the phenotypic fruit

fi see [fidget]

fiber 1 (*see also* fiber 2) in plants, without qualification, an elongate sclerenchyma cell (*cf.* sclereid) and usually distinguished as an easily separable thread of high tensile strength

bast fiber the term is sometimes used as synonymous with extraxylary fiber, but see [bast]

chambered fiber one which is septate

labriform fiber extremely long, woody fibers, occuring in phloem

gelatinous fiber (*bot.*) = mucilaginous fiber (*bot.*)

mucilaginous fiber (*bot.*) a fiber containing hydroscopic walls

phloic fiber an extraxylary fiber originating in phloem

perivascular fiber an extraxylary fiber located on the periphery of a vascular cylinder

extraxylary fiber one of the fibers outside the xylem in a plant stem

xylem fiber one of the fibers in the xylem

fiber 2 (*see also* fiber 1) in animals any fine, threadlike structures

cholinergic fiber one which liberates acetylcholine when activated

collagenic fibers intercellular fibers of collagen

elastic fibers fibers possessing considerable elasticity and mechanical strength and which are bound with other connective tissues as branched filaments without organized pattern

gamma fiber a motor nerve fiber supplying a neuromuscular spindle

nuclear fiber *obs.* = chromosome

Purkinje fiber atypical, impulse-conducting, cardiac muscle fibers

radicular fiber (Ectoprocta) = rhizoid (Ectoprocta)

reticular fibers small branched fibers closely allied to collagen fibers but smaller in diameter and differing also from collagen in that reticular fibers can be demonstrated by silver staining techniques

septate fiber a fiber with transverse partitions

Sharpey's fibers fibers in connective tissue, thought by some to be nervous, running from the periostium through bone, or from the dentine of a tooth into the pulp, or from a bony scale or skin plate into the connective tissue below

cytostomal fiber an extension of the blepharoplasts wrapped round the cytostome

spindle fibers 1 (*see also* spindle fiber 2) fibers constituting the mitotic spindle

spindle fibers (*see also* spindle fiber 1) fibers found in smooth muscle

fibrils diminutive of fiber and, in *zool.*, used amost interchangeably with fiber

argyrophilic fibrils = reticular fibers

interciliary fibril the conducting fibrils connecting the basal bodies of cilia in protozoa

microfibril diminutive of fibril

myofibrils the unit fibril of muscle

neurofibrils protoplasm adapted to conduct impulses

protofibril diminutive of microfibril

Tomes' fibrils fibrils in dentine

tonofibril one of a series of fibrils, at one time thought to interconnect cells, but now usually considered to be an artifact

spirofibril one of the spirally coiled fibrils of which protoplasm was, at one time, suppose to consist

fibrous composed of fibre

fibula (*see also* bone) (*bot.*) a cylindrical podetium ending in an apothecium

ficin an enzyme of similar function to papain (*q.v.*)

Ficoideae = Aizoaceae

-fid- *comb. form* meaning "cleft"

fidelity the extent to which a species is confined to a given set of conditions

[**fidget**] a mutant (gene symbol *fi*) in linkage group V of the mouse. The phenotypic expression is the waltzing syndrome frequently accompanied by polydactyly

field 1 (*see also* field 2, 3, cricket, fare, wren) an area of ground in which crops are grown, or, by extension, any restricted area of land

fellfield a stony area with dwarf, scattered plants

field 2 (*see also* field 1, 3) a localized area.

apical field the flat, central region at the anterior end of a rotifer, which is surrounded by the corona

basal field the anterior, non-annulated portion of a teleost scale

buccal field the large oval area, covered in short cilia, in the ventral region of the corona of rotifera

embryonic field a theoretical area in, or around, an embryo which may exercise an effect on the behavior of cells within its influence

polar fields the sensory floor of a statocyst

regeneration field the restricted area, from which new organelles arise after the division of some Protozoa

field 3 (*see also* field 1, 2) term for a group of horses racing

fieldfare the large European turdid bird *Turdus pilaris*

fierasfer a name sometimes applied to caparid and fierasferid fishes

Fierasferidae a family of elongate fishes, often combined with the Caparidae, dwelling in holothurians and therefore called cucumber fishes

fig 1 (*see also* fig 2, bird, eater, mite, scale) any of several species and numerous horticultural varieties of the moraceous genus Ficus but particularly those of *F. carica*

Moreton Bay fig *F. macrophylla*

caprifig the male fig producing the pollen distributed by the caprifier (*q.v.*)

cluster fig *F. glomerata*

creeping fig *F. pumila*

sycamore fig *F. sycomorus*

little-toe fig *F. diversifolia*

fig 2 (*see also* fig 1) any of numerous plants bearing more or less fig-like fruit

Adam's fig the musaceous tree-like herb *Musa paradisiaca* and its fruit (= plantain, common cooking banana)

bastard fig fruit of the cactus *Opuntia*

devil's fig any of several species of papaveraceous herbs of the genus Argemone, particularly *A. mexicana* (= prickly poppy) and also several cacti of the genus Opuntia

Indian fig *U.S.* any of several species of cactus of the genus Opuntia

fighter *W.I.* either of two tyrannid birds *Tyrannus dominicensis* (= grey kingbird) and *T. caudifasciatus*

night fighter *S.Afr.* any of several species of hesperid lepidopteran insects of the genus Artitropa

tom fighter *W.I.* the tyrannid bird *T. caudifasciatus* (*cf.* kingbird)

effigurate having a definite outline

figure used mostly in *biol.* in descriptions of mitosis

achromatic figure all those visible structures, except chromosomes, appearing in the course of mitotic division

amphiastral figure all those structures, except chromosomes, appearing in a mitotically dividing cell at the metaphase

anastral figure the appearance presented by a cell in mitosis when only the chromatic figure is visible

chromatic figure the arrangement of the chromosomes in a mitotically dividing cell (*cf.* amphiastral figure)

mitotic figure the general appearance of a cell in mitosis at some particular phase (*cf.* -phase)

filament a thread-like structure

acrosomal filament a filament produced by the acrosome of an activated spermatozoan, usually in response to an egg

gonimoblast filament the structure bearing the carpospores in floridiophycidean Rhodophyceae

gastric filament a nematocyst containing thread on the septa of Scyphozoa

lateral filament filamentous appendages arising from the side of the abdominal segments in some aquatic insect larvae

paranematal filaments = paranemata

septal filament a sinuous, thick filament containing gland cells and nematocysts, running along the free edge of the septa in Anthozoa

filander the original name applied by the early Dutch explorers to kangaroos and wallabies

Filaridea a class of parasitic nematodes, lacking either a pharyngeal bulb or a bursa, and with the buccal capsule small or rudimentary

Filariformes a dividison of nematode worms lacking lips on the mouth and a bulb on the esophagus. The adults live in vertebrate tissues

Filariidae a family of very long thin filarioid nematodes including many which infest man, particularly *Filaria bancrofti* the causative agent of elephantiasis and the notorious loa-loa worm (*q.v.*)

Filarioidea a sub-order of telogonian nematode worms with thread-like bodies The notorious guinea worm, loa-loa worm, and Filaria, the causative agent of elephantiasis are all in this group

filarious on the form of a thread

filator the spinnerets of caterpillars

filbert *Brit.* the betulaceous shrub *Corylus maxima*. *U.S.* any of several species of the genus Corylus

-fili- *comb. form* meaning "thread" frequently confounded with -filia- and -filic-

-filia- *comb. form* meaning "daughter" frequently confounded with -fili- and -fific-

Filibranchia an order of prorhipidoglossomorphic pelecypod Mollusca containing *inter alia*, the mussels and scallops. They are distinguished by the possession of gills formed of parallel, ventrally directed and reflected filaments

-filic- *comb. form* meaning "fern" frequently confounded with -fili- and -filia-

Filicales an order of leptosporangiate Pteridophyta containing the great majority of living ferns. They are distinguished from the Marsiliales and the Salviniales by being homosporous

Filicinae that class of Pteriodophyta which contains the true ferns. They are distinguished from all other pteriodophytes by their typically large leaves

fillip any of several species of geometrid lepidopteran insects of the genus Heterophleps

filly a female foal

fillymingo *W.I.* = flamingo

Filopteridales an order of isogenerate Phaeophyta distinguished by the freely-branched thalli

filose terminating in a thread

circumfilus one of the numerous intersegmental whorls of long, thin hairs on the antennae of cecidomyiid dipterans

-fim- *comb. form* meaning "dung" (*cf.* -copr-, -kopr-)

fimbria = pilus

fimetarious pertaining to dung heaps

fimbriate fringed

-fin- *comb. form* meaning "limit"

fin 1 (see also fin 2, foot) an appendage, used for swimming, balancing, or steering found in aquatic animals, particularly fishes

adipose fin a small posterior, dorsal fin, containing much fatty matter, typical of salmonid fishes

anal fin a ventral, unpaired, usually posterior fin of fishes

caudal fin the terminal posterior unpaired fin of fishes, commonly called the tail or tail fin; or the expanded margin of the tail in larval salamanders

dorsal fin one or more median-dorsal unpaired fins of fishes

plesodic fin a fin in which the radials extend nearly, or quite, to the margin of the fin

aplesodic fin a fin in which the radial does not extend to the fin margin

tail fin = caudal fin

fin 2 (*see also* fin 1) in many compound names of fishes

bloodfin the characinid *Aphyocharax rubropinnis*

glass bloodfin the characinid *Prionobrama filigera*

bowfin the popular name of the protospondylid *Amia calva*

dragon fin any of several species of characinid of the genus Pseudocorynopoma

featherfin the characinid *Hemigrammus unilineatus*

hairfin the anabantid *Trichogaster trichopterus* (*cf.* gourami)

mudfin = bowfin

sailfin the poeciliid *Molliensia velifera* (*cf.* molly)

threadfin popular name of polynemids

woundfin the cyprinid *Plagopterus argentissimus*

finch 1 (*see also* finch 2, lark) any of numerous passeriform birds properly of the family Fringillidae, though the name is applied, in various parts of the world, to members of other families

banded finch the ploceid *Steganopleura bichenovii*

bramble finch *Brit. Fringilla montifringilla* (= brambling)

red-browed finch *Austral.* the ploceid *Aegintha temporalis*

brush finch any of several fringillid birds of the genus Atlapetes

bullfinch *Brit.* the fringillid *Pyrrhula europaea*

Antillean bullfinch any of many fringillids of the genus Loxigilla

Greater Antillean bullfinch *L. violacea*

Lesser Antillean bullfinch *L. noctis*

brown bullfinch the fringillid *Pyrrhula nitalensis*

Cuban bullfinch the fringillid *Melopyrrha nigra*

Puerto Rican bullfinch the fringillid *Loxigilla portoricensis*

chaffinch *Brit. Fringilla coelebs*

plush-capped finch the catamblyrhynchid *Catamblyrhynchus diadema*

chestnut-breasted finch the ploceid *Donacola castaneothorax*

citril finch the fringillid *Serinus citrinella*

crimson finch the ploceid *Neochmia phaeton*

Darwin finch any of several Galapagos Island fringillids, placed by some in a separate family the Geospizidae. Their remarkable specialized feeding adaptations were a major argument used by Darwin in support of his theories

fire-finch *Afr.* any of several species of ploceids of the the the genus Lagonosticta

firetail finch any of several species of ploceid birds of the genus Zonaeginthus. *Austral.*, without qualification, *Z. bellus*

goldfinch *U.S.*, without qualification, the fringillid *Spinus tristis* (*cf.* siskin). *Brit.* the fringillid *Carduelis carduelis*. *W.I.* the thraupid *Spindalis zena* (= striped headed tanager). *Afr.* the *Muscicapa hypoleuca* (= pied flycatcher)

Himalayan goldfinch the fringillid *Carduelis caniceps*

lesser goldfinch the fringillid *Spinus psaltria*

grass finch any of several species of fringillids of the genus Sicalis

yellow grass finch *Sicalis luteola*

greenfinch any of several of fringillids of the genus Chloris, usually, without qualification, *C. chloris*

Himalayan greenfinch *C. spinoides*

hawfinch any of several fringillids of the genera Eophona and Coccothraustes. *Brit. Coccothraustes vulgares*

house finch the fringillid *Carpodacus mexicanus*

indico finch *Afr.* any of several ploceids of the genus Hypochera

Laysan finch the drepaniid *Psittirostra cantans*

masked finch the ploceid *Poephila personata*

melba finch the ploceid *Pytilia melba*

oriole finch the fringillid *Linurgus olivaceus*

painted finch the ploceid *Emblema picta*

purple finch the fringillid *Carpodacus purpureus*

quail finch the ploceid *Ortygospiza atricollis*

rose finch any of several Old World fringillids of the genus Carpodacus; without qualification, usually *C. erythrinus*

red-mantled rose finch the fringillid *Carpodacus rhodochlamys*

rosy finch any of several fringillids of the genus Leucosticte; without qualification, *L. tephrocotis*

saffron finch the fringillid *Haematospiza sipahi*

scarlet finch = saffron finch

seed finch any of several species of fringillids of the genus Oryzovorus

snow finch any of several species of ploceids of the genus Montifringilla, particularly *M. nivalis*

star finch the ploceid *Taeniopygia castanotis*

thistle finch *U.S.* = goldfinch *U.S.*

weaver-finch a general term for ploceids but particularly those of the genus Ploceus

finch 2 a fish

water goldfinch the characinid teleost fish *Pristella riddlei* (= X-ray fish)

[fine bristle] a mutant (gene symbol *fb*) mapped at 1.0± on the X chromosome of *Drosophila melanogaster*. The name is descriptive

[fine striped] a mutant (gene symbol f_1) mapped at 85 on linkage group I of *Zea mays*. The phenotypic expression is fine longitudinal green and white stripes on the leaves.

finger 1 (*see also* finger 2, 3) one of the four terminal digits—the fifth is the thumb—of the primate limb

finger 2 (*see also* finger 1, 3, fish) in compound names of finger-shaped animals

deadman's fingers (*see also* finger 3) either a large, fleshy, alcyonarian coelenterate (*Alcyonium digitatum*) or any of several monaxonid sponges of the genus Haliclona, particularly *H. arbuscula*

finger 3 (*see also* finger 1, 2, grass) in compound names of plants having finger-like parts

deadman's fingers (*see also* finger 2) the orchid *Habenaria conopsea* and several other orchids having finger-shaped roots. Also, rarely, the basidiomycete fungus *Xylaria polymorpha*

five finger any of numerous rosaceous herbs and shrubs of the genus Potentilla

ladies fingers *Brit.* and *U.S.* the leguminous herb *Anthyllis vulneraria* (= kidney vetch *Brit.*)

fingerling a post-larval teleost less than one year old, differing from a fry in that it can readily be identified through its resemblance to the adult form

fir 1 (*see also* fir 2, borer, engraver, sawfly and sawyer) any of numerous coniferous trees. If this term is used to distinguish among coniferous trees, it should be applied to the genus Abies

sub-alpine fir *A. lasiocarpa*
balsam fir *A. balsamea*
Caucasian fir *A. nordmanniana*
grand fir *A. grandis*
Grecian fir *A. cephalonica*
Nikko fir *A. homolepio*
red fir *A. magnifica*
sacred fir *A. religiosa*
silver fir *A. picea*
  Pacific silver fir *A. amabilis*
spanish fir *A. pinsapo*
white fir *A. concolor*

fir 2 (*see also* fir 1) a fir shaped organism

sea fir the sertulariid hydroid coelenterate *Abietinaria abietina*

floral **firecracker** the liliaceous plant *Brevoortia idamaia*

fire-on-the-mountain the euphorbiaceous shrub *Euphorbia heterophylla*

fiscal *Afr.* any of several laniid birds particularly *Lanius collaris*

fish 1 (see also fish 2, 3, geranium, grass, hawk, owl) a popular name for all permanently gill-breathing craniate chordates and therefore applied indiscriminately to cyclostomes, elasmobranchs and osteichthyans; the plural, in ichthyological usage is "fish" when applied to a single species and "fishes" when applied to several. "Fish" is used throughout the following list in order to avoid alphabetic confusion

alligator fish *U.S.*, without qualification, usually the agonid *Aspidophoroides monopterygius*

angelfish any of several ray-like sharks particularly of the genus Squatina and almost any laterally compressed brightly colored fish found in tropical waters.

  fresh water angelfish any of several chichlids of the genus Pterophyllum particularly *P. scalare*, common among aquarists

anemone fish any of several species of pomacentrid fish of the genus Amphiprion. They derive their name from their custom of living among the tentacles of large anemones

archerfish common name of toxotid fishes, particularly those of the genus Toxotes

balloonfish locally applied to any tetraodontid (= puffer toad) or diodontid but usually, without qualification, *Diodon holacanthus*

bandfish *Brit.* popular name of cepolids

barbfish (*see also* barb) the scorpaenid *Scorpaena brasiliensis*

barrelfish *Brit.* either of two species of stromateids of the genus Lirus. *U.S. Palinurichthys perciformis*

batfish common name of ogocephalid fishes

beaconfish = head-and-tail-light fish

breadfish popular name of polymyxiids

bumble-bee fish the gobiid *Brachygobius xanthozona*

bellows fish *U.S.* popular name of macrorhamphosids (= snipe fish) *Brit.* = trumpet fish *Brit.*

billfish popular name of istiophorids

birdfish any of several species of labrids of the genus Gomphosus

blackfish (*see also* fish 2) *either Brit.* popular name of stromateids (= *U.S.* butter fish), without qualification, usually *Centrolophus niger* or *Brit.* a female salmon which has spawned and is descending the rivers on her way back to the sea (= kelt)

  Alaska blackfish the daliid *Dallia pectoralis*
  Sacramento blackfish the cyprinid fish *Orthodon microlepidotus*

bluefish the pomatomid *Pomatomus salatrix*

boar fish *Brit.*, without qualification, *Capros aper.* *U.S.* any of several species of antigonids of the genus Antigonia

bonefish the albulid *Albula vulpes*

  threadfin bonefish the albulid *Dixonia nemaoptera*

bony fishes popular term for teleosts

box fishes popular name for ostraciontids

breadfish any of several species of carangid fish of the genus Alectis and also the scorpaenid *Sebaste marinus*

buffalo fishes any of several species of catostomid fish of the genus Ictiobus, the cottid fish *Enophrys bison* and the cichlid fish *Symphysodon discus* (= pompadour fish)

burrfishes popular name of diodontids

butterfish *U.S.* popular name of stromateids. *Brit.* = gunnel (*Brit.*)

butterfly fish popular name of chaetodontids and also the pantodontid *Pantodon buchholzi*

candlefish the osmerid fish *Thaleichthys pacificus* (= eulachon)

cardinal fishes popular name of apogonids

catfish any of many ostariophysous fishes identified by the presence of numerous barbels, and of spines in front of the dorsal and pectoral fins. *Brit.*, without qualification, the anarchichadid *Anarrhicas lupus*

  armored catfish any catfish of the families Doradidae, Callichthyidae and Loricariidae

  banjo catfish popular name for catfish of the family Aspredinidae

  blue catfish *either* the ictalurid *Ictalurus furcatus* *or* the callichthyid *Corydoras nattereri*

  bronze catfish the callichthyid *Corydoras aeneus*

  upside-down catfish any of numerous mochocids of the genus Synodontis

  electric catfish the malapterurid *Malapterurus electricus* (*cf.* eel)

  glass catfish the silurid *Kryptopterus bicirrhis* (*cf.* glass fish)

  harlequin catfish the pimelodid *Microglanis parahybae*

  leopard catfish the callichthyids *Corydoras julii* and *C. leopardus*

naked catfish any catfish not possessing armor
peppered catfish the callichythid *Corydoras paleatus*
spiny catfish the doradid *Acanthodoras spinosissimus*
sucking catfish any of several species of loricariid of the genus Otocinclus
talking catfish the doradid *Acanthodoras spinosissimus*
cavefish any cavernicolous fish particularly those of the family Amblyopsidae and the characinid *Anoptichthys jordani*
chameleon fish the cichlid *Cichlasoma facetum*
cherubfish the chaetodontid *Centropyge argi*
cigarfish any of several species of stromateid of the genus Cubiceps
clingfish popular name of gobiesocids
white cloud fish the cyprinid *Tanichthys albonubes*
coalfish the gadid *Gadus virens*. There is considerable confusion between the coalfish and the pollak in various dialects
codfish *see* cod
conchfish the apogonid fish *Apogon stellatus*
pinecone fish *Austral.* the monocentrid *Monocentris gloriae-maris*
cornetfish popular name of aulostomids
cowfish (*see also* fish 2) properly the ostraciontid *Lactophrys quadricornis*, though the term is occasionally applied to other ostraciontids
cramp fish the torpedinid ray *Hypnarce monopterygium*
creole fish *U.S.* the serranid *Paranthias furcifer*
crestfish popular name of lophotids
cuckoo fish *Brit.* = boar fish *Brit.*
cucumber fish popular name of fierasferids that dwell in holothurian echinoderms
cutlassfish popular name of trichurids
damselfish popular name of pomacentrids
dealfish the trachipterid *Trachipterus arcticus*
devilfish (*see also* fish 3) any of numerous large batoid elasmobranchs particularly those of the genera Manta and Mobula. The name is also sometimes applied by aquarists to the cichlid *Geophagus jurupari*
doctorfish (*see also* fish doctor) the acanthurid *Acanthurus chirurgus*
dogfish *Brit.* popular name of scyliorhinid selachians (= *U.S.* cat shark) In parts of *U.S.* the term is applied to the bowfin. *S.Afr.* the gadid fish *Merluccius capensis*
    alligator dogfish = bramble shark
    smooth dogfish *U.S.* the scyliorhinid shark *Mustelus canis*
    spineless dogfish popular name of sharks of the family Dalatiidae
    spiny dogfish any of numerous species of shark of the genus Squalus
dollar fish = harvest fish
driftfish any of several species of stromateids of the genus Psenes
grideye fish popular name of ipnopids
    lanterneye fish popular name of anomalopids
foureyed fish popular name of anablepids
fallfish *U.S.* the cyprinid fish *Semotilus corporalis*
fighting fish the anabantid fish *Betta splendens* (= Siamese fighting fish)
filefish any of several species of balistid fish of the genus Alutera. *Brit.* without qualification *Balistes capriscus*. (= *U.S.* gray trigger fish)

fingerfish popular name of monodactylids
flagfish the cyprinodont *Jordanella floridae*
flamefish *either* the characinid *Hyphessobrycon flammeus or* the apogonid *Apogon maculatus*
flatfish popular name of any fish of the order Heterostomata
flying fish popular name of exocoetids
    freshwater flyingfish = butterflyfish
forcepsfish the chaetodontid *Forcipiger longirostris*
frogfish *Brit.* the lophiid *Lophius piscatorius* (*cf. U.S.* goosefish) *U.S.* properly any antennariid but sometimes applied, copying British custom, to lophiids
frostfish = scabbard fish
garfish *Brit.* = gar pike *Brit.*
ghostfish the silurid *Kryptopterus bicirrhis* (= glass catfish)
glassfish (*see also* catfish and tetra) any more or less transparent fish. Most usually applied to centropomids of the genus Chanda
    Indian glassfish the centropomid *Chanda rariga*
globe fish any plectognath that can puff up its body to a globular form. *U.S.* cyclopterid *Cyclopterichthys glaber*. *Brit.* the tetraodontid *Lagocephalus lagocephalus* (= *U.S.* oceanic puffer)
goatfish popular name of mullids
goldfish the cyprinid *Carassius auratus*, occurring in many domesticated varieties
goosefish *U.S.* popular name of lophiids but particularly, without qualification, *Lophius americanus*
guitarfish popular name of rhinobatid rays
gulper fish popular name of Lyomeri
hagfish the popular name of myxinoid cyclostomes
harlequinfish (*see also* catfish) the cyprinid *Rasbora heteromorpha*
harvestfish any of several species of stromateids of the genus Peprilus
hatchetfish popular name of sternoptychids and of several characinids, particularly *Gasteropelecus levis*
hawkfish popular name of cirrhitids
serpent-head fish any of several species of channids of the genus Chanda
head-and-tail-light-fish the characinid *Hemigrammus ocellifer*
hogfish popular name of labrids. *U.S.*, without qualification, usually *Lachnolaimus maximus*
    red hogfish *Decodon puellaris*
    Spanish hogfish *Bodianus rufus*
    spotfin hogfish *Bodianus pulchellus*
hedgehog fish = porcupine fish
go-home fish *S.Afr.* any of several species of sparids
keyholefish the cichlid *Aequidens maronii*
    portholefish the poeciliid *Poecilistes pleurospilus*
houndfish the belonid *Strongylura raphidoma*
jawfish popular name of opisthognathids
jewfish *U.S.* the serranid *Epinephelus itajara*
jugfish the tetraodontid *Lagocephalus pachycephalus*
kelpfish any of several species of clinids of the genus Gibbonsia
killifish any of numerous species of cyprinodonts particularly those of the genus Fundulus
    least killifish the poeciliid *Heterandria formosa* (= dwarf topminnow)
kingfish *U.S.* any of several sciaenids of the genus Menticirrhus. *Brit.* = sunfish *Brit.*
    gulf kingfish *M. littoralis*
    northern kingfish *M. saxatilis*
    southern kingfish *M. americanus*
klipfish = kelpfish

knifefish any of numerous freshwater gymnotids, particularly *Gymnotus carapo* and also the characinid *Ctenobrycon spilurus* (= silver tetra)

African knife fish popular name of notopterids (= feather-back)

jackknife fish the sciaenid *Equetus lanceolatus*

ladyfish the elopid *Elops saurus*

lancetfish popular name of alepisaurids

lantern fish popular name of myctophids

leaffish any of several nandids but particularly *Monodirrhus polycanthus* and *Polycentrus scomburgki* named from its startling resemblance to a leaf

lionfish the scorpaenid *Scoraena grandicornis*. In some parts of the world this name is applied to the turkeyfish

lizardfish popular name of synodontids

lumpfish popular name of cylopterids. *U.S.*, without qualification, usually *Cyclopterus lumpus*

lungfish popular name of dipneustian fishes. The *Australian lungfish* (*cf.* barramunda) is in the family Ceratodontidae, the *S. Amer.* in the family Lepidosirenidae, and the African in the family Protopteridae

man-of-war fish the stromateid *Nomeus gronowi*

medusafish any of several species of stromateids that live among the tentacles of scyphozoan coelenterates

milk fish the chanid *Chanos chanos*

millionsfish any of several species of poeciliids of the genus Phalloceros

minkfish *U.S.* the sciaenid *Menticirrhus focaliger*

monkfish *U.S.* the lophiid *Lophius piscatorius* (= anglerfish or frogfish.) *Brit.* the rhinobatid batoid *Rhina squatina* (*cf.* guitarfish)

moonfish any of numerous laterally compressed fishes of both fresh and marine waters. Used without qualification, it usually applies to the carangid genus Selene or to the poecilid minnow *Platypoecilus maculatus* (*cf.* platy). The name is also applied to the lamprid *Lampris regius*

mosquito fish any of several species of poeciliids but particularly (*U.S.*) *Gambusia affinis*

mudfish any of several poeciliids of the genus Limia. *N.Z.* the galaxiid *Neochanna apoda*

needlefish *U.S.* popular name of belonids. *Brit.* the syngnathid *Syngnathus acus*

spotted needlefish the loricariid *Farlowella acus*

numbfish torpedinid rays of the genera Narcine and Torpedo

oarfish *U.S.* the regalecid *Regalecus glesne*. *Brit.* popular name of trachypterids

oilfish the gempylid *Ruvettus pretiosus*

opalescentfish the cyprinid *Brachydanio albolineatus*

paddlefish either of two species of polyodontid Chondrostei

paradisefish any of several species of anabantid of the genus Macropodus. Without qualification, usually *M. opercularis*

parrotfish any marine teleost of the family Scaridae. The term is also locally applied to some globe fishes that possess a parrot-like beak

pearl fish popular name of fierasferids particularly those seeking shelter in holothurians. *U.S.*, without qualification, usually *Fierasfer bermudensis*.

Argentine pearlfish the cyprinodontid *Cynolebias bellotti*

pencilfish any of several elongate characids many of the genus Poecilobrycon

petticoatfish the characinid *Gymnocorymbus ternetzi*

pigfish the pomadasyid *Orthopristis chrysopterus*

pilotfish the carangid *Naucrates ductor*

pinfish the sparid *Lagodon rhomboides*

pipefish *Brit.* popular name of sygnathids

ghost pipefish popular name of solenostomids

pompadourfish the cichlid *Symphysodon discus*

porcupinefish popular name of diodontids

porkfish the pomadasyid *Anisotremus virginicus*

puffer-fish = balloon fish

queen-fish the sciaenid *Seriphus politus*

rabbitfish *Brit.* the holocephalan *Chimaera monstrosa*. The term is also sometimes applied to balloon fishes, and, in the indo-pacific region, to several species of siganids

ragfish popular name of icosteids

rainbowfish *Austral.* any of several species of atherinid of the genus Melanotaenia

ratfish any of several species of chimaerid holocephali, particularly *Hydrolagus colliei*

razorfish *U.S.* any of several species of labrids of the genus Xyrichthys

redfish *Brit.* the male equivalent of a blackfish

reedfish popular name of polypterids

ribbonfish *U.S.* popular name of trachipterids. *Brit.* without qualification, *Regalecus glesne* (= *U.S.* oarfish)

Japanese ricefish the cyprinodont *Oryzias latipes*

rockfish popular name of scorpaenids and sometimes applied to the Scleroparei

roosterfish the carangid *Nematistius pectoralis*

rosefish any of several species of scorpaenids particularly *Helicolenus dactylopterus*

rudderfish kyphosids in general and specifically the carangid *Seriola zonata* (*cf.* amberjack)

sailfish any of several istiophorids of the genus Istiophorus but particularly

Atlantic sailfish *I. albicans*

sailorfish the cottid *Nautichthys oculofasciatus*

sandfish (*see also* fish 3) popular name of trichodontids and of the gonorhynchid *Gonorhynchus gonorhynchus*

Atlantic sandfish *U.S.* the malacanthid *Malacanthus plumeri*

belted sandfish the serranid *Serranellus subligarius*

sable fish the anaplomatid *Anaploma fimbria*

sargassumfish the antennariid *Histrio histrio*

sawfish any of several marine elasmobranchs of the batoid genus Pristis. Often confused with swordfish

scabbardfish the trichurid *Lepidopus xantusi*

scaldfish *Brit.* the bothid *Arnoglossus laterna*

scorpionfish any of numerous teleosts of the family Scorpaenidae. Many have venom-bearing spines

shee fish the salmonid *Stenodius mackenzii* (= inconnu)

shrimpfish popular name of centriscids. The term snipefish is also applied to this family

Siamese fish a brilliantly colored anabantoid *Betta splendens* (= fighting fish)

silverfish (*see also* fish 3) the monodactylid *Monodactylus argenteus*

skilfish the anoplopomatid *Eriletsis zonifer*

skillet fish *U.S.* the gobiesocid *Gobiesox strumosus*

snailfish common name of liparids

snakefish *U.S.* the synodontid *Trachinocephalus myops*. *Brit.* popular name of cepolids

snipe fish *U.S.* popular name of centriscids of the genus Macrorhamphosus. *U.S.*

long-spine snipe fish *M. sclopax* (= *Brit.* trumpet fish)

soapfish *U.S.* any of several serranids of the genus Rypticus, usually, without qualification, *R. saponaceus*

freckled soapfish *R. arenatus*

spotted soapfish *R. subbifrenatus*

soldierfish popular name of holocentrids

spear fish any of several species of istiophorids of the genus Tetrapturus

longbill spearfish *T. belone*

shortbill spearfish *T. angustirostris*

spiderfish popular name of bathypteroids

spikefish popular name of triacanthodids

springfish *U.S.* the amblyopsid *Chologaster agassizi* (= spring cavefish)

squawfish any of several species of cyprinids of the genus Ptychocheilus

squirrelfish popular name of holocentrids

head-standing fish the characinids *Chilodus punctatus* and *Abramites microcephalus*

stockfish *Brit.* dry, salted cod

stonefish the scorpaenid *Synanceja verrucosa*. The most venomous of all fishes, having caused death to a human in two hours

studfish any of several species of cyprinodontids of the genus Fundulus but particularly *F. catenatus* and *F. Stellifer* (*cf.* killifishes)

suckerfish popular name of discocephalous fishes

sunfish the term includes many freshwater centrarchids of the genus Lepomis. In *Brit.*, the name is used both for the molid *Mola mola* and occasionally for the lamprid *Lampris luma*

ocean sunfish popular name of molids. *U.S.*, without qualification, usually *Mola mola*

surgeonfish any of numerous acanthuroids of the family Acanthuridae distinguished by the possession of a sharp "knife" just in front of the tail

sword fish a large xiphiid with a sword-like beak (*Xiphias gladius*)

talkingfish the doradid *Acanthodoras spinosissimus* (= talking catfish)

tasselfish popular name of polynemids

threadfish the carangid *Alectis ciliaris*

tigerfish the characid *Hydrocyon golias.*, an African fish reaching a weight of 125 pounds. The term is also applied generally to scorpaenids of the genus Pterois

tilefish any of several malacanthids but usually, without qualification, *Lopholatilus chamaeleonticeps*

toadfish any of numerous haplodocids of the family Batrachoididae distinguished by their large mouths, broad heads, and the body tapering to a long slender tail

tonguefish any of numerous cynoglossids

topfish any of several species of cyprinodonts of the genus Fundulus (= topminnow)

treefish the scorpaenid *Sebastodes serriceps*

triggerfish popular name of balistids

trumpet fish *U.S.* the aulostomid *Aulostomus maculatus*. *Brit.* the centriscid *Macrorhamphosus scolopax* (= *U.S.* long-spine snipe-fish)

trunkfish popular name of ostraciontids *U.S.*, without qualification, usually *Lactophrys trigonus*

elephant trunkfish the mastocembelid *Macrognathus aculeatus*

tubfish the triglid *Trigla hirundo*. *Brit. T. lucerna* (*cf.* gurnard)

turkeyfish any of numerous species of scorpaenids of the genus Pterois

unicornfish any of several species of anthurids

viperfish *U.S.* popular name of chauliodontids

walkingfish the anabantid *Anabas testudineus*. It is capable of walking by the use of its paired fins from one body of water to another

Chinese walkingfish the channid *Channa asiatica*

waspfish = bumble-bee fish

weakfish *U.S.* the sciaenid *Cynoscion regalis* (*cf.* sea trout)

bastard weakfish *C. nothus* (= bastard trout)

weatherfish the cobitid *Cobitis taenia*. The term is sometimes also applied to other cobitids

weaverfish any of several trachinids of the genus Trachinus, noted for their poisonous opercular spines

whalefish luminous abyssal fish of the order Xenoberyces distinguished by the absence of a swim bladder

whitefish any of numerous species of salmonids of the genera Coregonus and Prosopium

humpback whitefish *C. pidschian*

bonneville whitefish *P. spilonotus*

broad whitefish *C. nasus*

lake whitefish *C. clupeaformis*

Bear Lake whitefish *P. abyssicola*

mountain whitefish *P. williamsoni*

pygmy whitefish *P. coulteri*

round whitefish *P. cylindraceum*

wolffish popular name of anarhichadids

woodcockfish *Brit.* = trumpetfish *Brit.*

wreckfish the serranid *Polyprion americanus*

X-rayfish the characinid *Pristella riddlei* (= water goldfinch)

zebrafish the cyprinid *Brachydanio rerio*

fish 2 (*see also* fish 1, 3) a term sometimes applied in error to cetaceans

blackfish (*see also* fish 1) *U.S.* any of numerous cetacean mammals of the genus Globiocephala

cowfish (*see also* fish 1) any of numerous species of ziphiid cetacean mammal of the genus Mesoplodon. *N.Z.* the delphinid cetacean *Tursiops tursio*

fish 3 (*see also* fish 1, 2) a term applied to a number of invertebrates for reasons not always apparent

basketfish the euryale ophiuroid echinoderm *Gorgonocephalus arcticus*

crawfish *Brit.* popular name of palinurid decapods. (= *U.S.* sea crawfish) *U.S.* crawfish = crayfish. *W.U.S.* the callianassid *Callianassa californiensis* (= ghost shrimp)

crayfish common name of astacid and parastacid decapod crustacea. They are distinguished from the Homaridae (lobsters) by the sternum between the last pair of legs not being fused with those anterior to it and by posessing a flattened rostrum. *Brit.* without qualification, *Austropotamobius pallipes*, usually miscalled *Astacus fluviatilis*, the correct name of which is *A. astacus* and which does not occur in Britain

cuttlefish any of many members of the cephalopod molluscan genus Sepia distinguished by the presence of a calcified internal shell called the cuttle bone

devilfish (*see also* fish 1) *U.S.* any of many species of octopodid cephalopod Mollusca (= octopus)

jelly fish any pelagic medusoid coelenterate, but particularly the larger Scyphozoa. Locally the term is sometimes also applied to Ctenophora

sandfish (*see also* fish 1) any of several species of

scincid saurian reptile of the genus Scincus (see skink)

silverfish (see also fish 1) any thysanuran insect but usually, without qualification, *Lepisma saccharina*

fisher the *N. Amer.* musteline carnivore, *Martes pennanti* (= pekan)

kingfisher 1 (see also kingfisher 2) any of very many species of coraciiform bird of the family Alcedinidae, frequently brilliantly colored. *U.S.*, without qualification, *Megaceryle alcyon. Brit.*, *Alcedo athis N.Z. Halcyon vagans*

banded kinfisher *Lacedo pulchella*
belted kingfisher *Megaceryle alcyon*
stork billed kingfisher any of several species of alcedinid bird of the genus Pelargopsis
shining blue kingfisher *W.Afr. Alcedo gradribrachys*
crab-eating kingfisher *Clytocerys rex*
dwarf kingfisher *Afr. Myiocey lecontei*
green kingfisher *U.S. Chloroceryle americana*
malachite kingfisher *W. Afr. Alcedo cristata*
pied kingfisher *Ceryle rudis* and *C. lugubris*
pigmy kingfisher *Afr. Ispidina picta*
sacred kingfisher *Halcyon sancta*
striped kingfisher *Afr. Halcyon chelicuti*

kingfisher 2 (see also kingfisher 1) *Brit.* = demoiselle *U.S.*

fission used in *biol.* in the sense of division or splitting
binary fission division into two
progamous fission that fission which in some ciliate protozoa divides the organism unequally into macro and micro conjugants
multiple fission a division of a single cell into many cells, as in spore formation in protozoa

Fissipedia that suborder of carnivora which contains the land-dwelling forms

fissure a space or split
choroid fissure 1 (see also choroid fissure 2) the temporary groove on the ventral side of the embryonic eye cup
choroid fissure 2 (see also choroid fissure 1) a fissure associated with the choroid plexus
orbital fissure the gap between the alisphenoid and the orbitosphenoid bones
rhinal fissure the fissure which separates the olfactory lobes from the rest of the cerebrum in higher forms
sclerotomic fissure a space free of cells that lies between successive sclerotomes
intersegmental fissure = sclerotomic fissure
Sylvius' fissure the deep, vertical, lateral fissure marking the posterior margin of the temporal lobe of the cerebral hemisphere

Fissurellidae a family of gastropod mollusks often called volcano shells. They have uncoiled conical shells like a limpet but with perforation at the tip. Many are called limpets

Fistulariidae a monogeneric family of solenichthous fishes distinguished by the long tubular snouts and a filament extending from the center of the tail fin. Popularly called cornet fishes

fistulous hollow

fitch = european polecat

basifixed attached by the base
dorsifixed attached at the dorsal side, as an anther to its filament

fj see [four-jointed]

fl see [fluffy], [fluted] or [wingless]

fl₁ see [floury endosperm]

Fl see [Flightless]

fla see [flat eye]. Used also for a bacterial mutant affecting flagellation. One has been mapped at 32.5–38.6 mins. for *Escherichia coli*

flabellate fan-shaped, or bearing a fan-shaped structure

flabellum literally, a small fan and applied to numerous organisms and structures of this general shape, particularly fan-shaped extensions to appendages of arthropods

flabs = labella

flaccid limp

Flacourtiaceae that family of parietalous dicotyledons which contains *inter alia*, Gynocardia from which chaulmugra oil is obtained. The regular flower with numerous stamens and an enlarged receptacle is typical

flag popular name for many monocotyledonous plants of the family Iridaceae, also applied to a fish *Brit.*, without qualification, *Iris pseudacorus* or *I. germanica*
blue flag *Iris virginica* or *I. versicolor*
corn flag *Brit.* = common flag *Brit.*
crimson flag the iridaceous herb *Schizostylis coccinea*
Spanish flag the serranid fish *Gonioplectrus hispanus*
spiral flag any of several species of the zingiberaceous genus Costus
sweet flag the araceous *Acorus calamus*

-flagell- comb. form meaning "whip"

Flagellariaceae a small family of farinose monocotyledonous angiosperms

Flagellata = Mastigophora

flagellate literally furnished with whips or anything in the form of a whip

flagellum 1 (see also flagellum 2, 3) a motor organelle found on many free-living cells in both the plant and animal kingdom. A flagellum differs from a cilium (*q.v.*) only in being larger and occurring singly or in small groups (*cf.* tractellum)
flagellum 2 (see also flagellum 1, 3) any long tactile arthropod appendage or part of an appendage
flagellum 3 (see also flagellum 1, 2) a whip-like prolongation of the tail of the larvae of some anuran Amphibia

flake a nectary. Used also in compound names of many plants
snowflake many white flowers particularly those of several species of amaryllidaceous herb of the genus Leucojum and (*U.S.*) of the labiateous herb *Lamium album*
spring snowflake *Leucojum vernum*
water snowflake the gentianaceous aquatic herb *Nymphoides indicum*

flame 1 (see also flame 2) the noctuid lepidopteran insect *Axylia putris* (*cf.* flame shoulder) and *Brit.* the geometrid lepidopteran insect *Anticlea rubidata*
flame 2 (see also flame 1, cell) a flame-shaped structure
lateral flame a tuft of cilia along the course of a protonephridial tubule

flamingo (see also tongue) any of several birds of the ciconiiform family Phoenicopteridae, most of the genus Phoenicopterus distinguished by the long legs and the thick bill curved at a right angle
American flamingo *P. ruber*
greater flamingo *P antiquorum*
lesser flamingo *Phoeniconaias minor*

flank the side of a bilateral body

rhachilla-flap an outgrowth from the internode of the spikelet axis of grasses
flipflap *U.S.* the larva of the dobson fly *Corydalus cornutus* (= hellgrammite)

flapper a young duck

[**flap wings**] a mutant (gene symbol *flp*) mapped at 31.0± on the X chromosome of *Drosophila melanogaster.* The phenotypic expression is concave wings and bulging rough eyes

flash *I.P.* any of numerous species of lycaenid lepidopteran insects of the genus Rapala and its allies

green flash *I.P. Artipe eryx*

indigo flash *Austral. Rapala varuna* (*cf.* blue)

flat (*see also* bill) *I.P.* and *Austral.* any of numerous hesperid lepidopteran insects (*cf.* skipper)

clouded flat *S.Afr. Tagiades flesus*

pied flat *I.P. Coladenia dan*

small flat *I.P.* any of several species of the genus Sarangesa

snow flat *I.P.* any of several species of the genus Tagiades

[**flat eye**] a mutant (gene symbol *fla*) mapped at 0.3± on the X chromosome of *Drosophila melanogaster.* The eyes of the phenotype are both flat and small

Flatidae a small family of homopteran insects closely allied to the Fulgoridae

flavedo the oil-gland bearing exocarp of a hesperidium

lactoflavin = vitamin B$_2$

ovoflavin = vitamin B$_2$

riboflavin = vitamin B$_2$

riboflavinase an enzyme catalyzing the hydrolysis of riboflavin to ribitol and lumichrome

flavous bright yellow

flax (*see also* bird, seed, thrip, worm) properly any of numerous species of the linaceous genus Linum. The flax from which the commercial fiber is obtained is *L. usitatissimum.* Many other blue-flowered, or fiber producing, plants are also called flax

big flax *N.Z.* the liliaceous herb *Phormium tenax*

devil's flax = toadflax

false flax *U.S.* any of several species of cruciferous herb of the genus Camelina

flowering flax *Linum grandiflorum*

mountain flax *U.S.* the polygalaceous herb *Polygala senega. N.Z.* the liliaceous herb *Phormium colensoi*

toadflax the scrophulariaceous plant *Linaria vulgaris*

bastard toadflax any of several species of santalaceous herbs of the genus Comandra (*U.S.*) or Thesium (*Brit.*)

fairy flax *Linum catharticum*

New Zealand flax any of several species of liliiaceous herb of the genus Phormium

fld *see* [fluffyoid]

flea 1 (*see also* flea 2, bane, beetle, hopper) any of many ectoparasitic insects of the order Siphonaptera

cat flea *Ctenocephalides felis*

European chicken flea the ceratophyllid *Ceratophyllus gallinae*

western chicken flea *U.S.* the ceratophyllid *Ceratophyllus niger*

chigoe flea the tungid *Tunga penetrans* (= jigger)

dog flea the pulicid *Ctenocephalides canis*

human flea the pulicid *Pulex irritans*

mouse flea the leptopsyllid *Leptopsylla segnis*

plague flea the pulicid *Xenopsylla cheopis* (= Oriental rat flea)

common rat flea the ceratophyllid *Nosopsyllus fasciatus*

northern rat flea *U.S.* the ceratophyllid *Nosopsyllus fasciatus*

oriental rat flea = plague flea

squirrel flea any of several species of pulicids of the genus Hoplopsyllus

sticktight flea the tungid *Echidnophaga gallinacea*

flea 2 (*see also* flea 1) various hopping and jumping arthropods other than siphonapteran insects

beach flea popular *U.S.* name of talitrid amphipod crustacea more commonly called sand hoppers in *Brit.*

lucerne flea the sminthuran collembolan insect *Sminthurus viridis*

sand flea *U.S.* sometimes synonymous with sand hopper but also applied to hippid anomuran crustacea

turnip flea *Brit.* popular name of sminthuran collembolan insects

-flect *comb. term* meaning bent

deflected bent upwards

exflected bent away from the main axis

genuflected bent at a sharp angled bend, as that of a bent knee

inflected bent towards the main axis

reflected in *biol,* has the mening of bent backwards

fledge to develop adult feathers

fledgling a bird lacking adult plumage and not yet able to fly

fleece the pelt of an ovid or caprid mammal or any organism thought to resemble one

mountain fleece the polygonaceous herb *Polygonum amplexicaule*

fleet term for a group of coots (*cf.* pod)

flesh (*see also* fly) soft, usually edible, tissues of both plants and animals

[**Flesh**] a mutant (gene symbol *F*) mapped at 13.6 in linkage group VI of *Bombyx mori.* The term refers to the color of the phenotypic cocoon

fleur-de-lys the iridaceous herb *Iris germanica*

[**flexed tail**] a mutation (gene symbol *f*) found in linkage group XIV of the mouse. The phenotypic expression involves not only the condition described by the name but also anemia

flexion that movement of a jointed appendage which results in closing the angle about the joint

flexuose bent in a zigzag, more abruptly than is connoted by sinuous and less so than is connoted by serpentine

flexure a bend, particularly one in the central nervous system

nuchal flexure the bend of the embryonic brain in the region of the medulla oblonga

pontal flexure the flexure of the developing brain in the region of the cerebellum which is in the reverse direction to the primary and nuchal flexures

primary flexure the initial bend in the embryonic brain by which the procencephalon and its derivatives are bent at right angles to the remainder

flicker any of several species of picid bird of the genus Colaptes. *U.S.* without qualification, *C. auratus* (= yellow-shafted flicker)

gilder flicker *C. chrysoides*

red-shafted flicker *C. cafer*

yellow-shafted flicker *C. auratus*

flier the centrarchid fish *Centrarchus macropterus*

flight 1 (*see also* flight 2) continued supported motion in air

mating flight = nuptial flight

nuptial flight that flight of hymenopteran insects in the course of which the queen is fertilized

flight 2 (*see also* flight 1) term for a group of a number of wild birds, including the cormorant, dove, many accipters, many wildfowl (in the *Brit.* sense of this term) and swallows

[**Flightless**] a mutant (gene symbol *Fl*) in linkage group V of the domestic fowl. The name derives from the fact that the flight feathers of the phenotype break off with their first use

flipper 1 (*see also* flipper 2) a mammaliam limb lacking digits and therefore adapted to swimming

flipper 2 (*see also* flipper 1) sometimes used for seal
  square flipper = bearded seal

tree **flitter** *I.P.* the hesperiid lepidopteran insect *Hyarotis adrastus*

flotsam *see,* for information as to its ecological usage, jetsam

flocculus a small lateral outgrowth from the base of the vermis
  paraflocculus a latero-ventral subdivision of the flocculus in some mammals

flock a general term for a group of domesticated birds or for those artiodactyle mammals for which specific group designations do not exist

-flor- *comb. form* meaning "flower"
  nudiflor a plant on which the flowers appear before the leaves

flora the sum total of the plants in any given area or environment
  microflora either the sum of those plants in any locality that are too small to be distinguished by the naked eye or the dwarf plants of high mountains

floral pertaining to a flower
  thalamifloral when the hypogynous parts of a flower are separately inserted on the torus

de**florate** the condition of a plant after the flowers have fallen

ef**florescence** = anthesis

in**florescence** the grouping of flowers set apart from the foliage into racemes and like
  acroblastic inflorescence one which arises from a terminal bud
  acropetalous inflorescence a type of inflorescence in which the flowers open from the base upwards
  botryoidal inflorescence one of racemose botryoidal clusters
  centrifugal inflorescence an inflorescence in which the flowers develop from the tip down or the inside outwards
  centripetal inflorescence an inflorescence of which the flowers open from the base up or the outside in
  compound inflorescence one which is itself composed of secondary inflorescences
  cymose inflorescence a type of inflorescence in which the flowers open from the apex downwards
  determinant inflorescence when the inflorescence ends in a bud
  helicoid inflorescence an inflorescence of flowers in a single row
  intercalary inflorescence an inflorescence through the center of which the main axis continues to grow
  polyaxial inflorescence one in which the flowers are borne on secondary and tertiary branches
  simple inflorescence any inflorescence with a single axis
  strobile inflorescence one made up of overlapping scales as a pine cone

florescent a period of flowering

floret diminutive of flower, usually used for the components of composite flowers
  tubiform floret the disk floret of a compositaceous flower
  ray floret an outer floret of a compositaceous disk

florican any of several charadriiform otitid birds of the genera Houbaropsis and Sypheotides
  lesser florican *Sypheotides indica*

Floridaceae a class of the algal subphylum Rhodophyta distinguished by the presence of cytoplasmic connections between the cells

floriken = florican

-florous *comb. form* meaning "flowered" in the sense of uniflorous, etc. The substantive termination, compounds of which are not recorded in this work, is -flory
  apertiflorous = chasmogamic
  cauliflorous the continued lateral production of flowers or fruit on old wood
  diversiflorous having polymorphic flowers
  geminiflorous with either twinned, or double flowers. The former meaning is more common
  labiatiflorous having lipped flowers
  neutriflorous having neutral flowers or florets, particularly in a compositaceous capitulum
  nodiflorous bearing flowers arising from nodes
  radiciflorous said of a flower which apparently arises directly from the crown of a root
  ramiflorous flowering on branches
  scapiflorous having flowers on scapes
  secundiflorous having the flowers all turned in one direction
  similiflorous having identical flowers
  squamiflorous having a perianth of scale-like bracts
  tubuliflorous having flowers in the shape of little pipes
  umbelliflorous bearing flowers in umbels

-flos- *comb. form* meaning "flower"

floscelle a flower-like appearance in the oral region of some echinoid echinoderms which is formed by the combination of bourrelets (*q.v.*) and phyllodes (*q.v.*)

floscular pertaining to florets

Flosculariacae an order of monogonate Rotifera with a greatly enlarged circular, and frequently lobed, corona

floscule = floret

flounder any of very numerous species of fishes of the families Bothidae and Pleuronectidae. *Brit.*, without qualification, the pleuronectid *Pleuronectes flexuosa*
  arctic flounder *U.S.* the pleuronectid *Liopsetta glacialis*
  arrowtooth flounder *U.S.* the pleuronectid *Atheresthes stomias* (= arrowtooth halibut)
  Bering flounder *U.S.* the pleuronectid *Hippoglossoides robustus*
  broad flounder *U.S.* the bothid *Paralichthys squamilentus*
  channel flounder *U.S.* the bothid *Syacium micrurum*
  dusky flounder *U.S.* the bothid *Syacium papillosum*
  three-eye flounder *U.S.* the bothid *Ancylopsetta dilecta*
  eyed flounder *U.S.* the bothid *Bothus ocellatus*
  spotfin flounder *U.S.* the bothid *Cyclopsetta fimbriata*
  fringed flounder *U.S.* the bothid *Etropus crossotus*
  gray flounder *U.S.* the bothid *Etropus rimosus*
  gulf flounder *U.S.* the bothid *Paralichthys albigutta*
  Gulf Stream flounder *U.S.* the bothid *Citharichthys arctifrons*
  Mexical flounder *U.S.* the bothid *Cyclopsetta chittendeni*
  smallmouth flounder *U.S.* the bothid *Etropus microstomus*

ocellated **flounder** *U.S.* the bothid *Ancylopsetta quadrocellata*

pelican **flounder** *U.S.* the bothid *Chascanopsetta lugubris*

peacock **flounder** *U.S.* the bothid *Bothus lunatus*

sash **flounder** *U.S.* the bothid *Trichopsetta ventralis*

shoal **flounder** *U.S.* the bothid *Syacium gunteri*

shrimp **flounder** *U.S.* the bothid *Gastropsetta frontalis*

slim **flounder** *U.S.* the bothid *Monolene antillarum*

smooth **flounder** *U.S.* the pleuronectid *Liopsetta putnami*

southern **flounder** *U.S.* the bothid *Paralichthys lethostigma*

spring **flounder** *U.S.* the bothid *Engyophrys senties*

fourspot **flounder** *U.S.* the bothid *Paraliehthys oblongus*

starry **flounder** *U.S.* the pleuronectid *Platichthys stellatus*

summer **flounder** *U.S.* the bothid *Paralichthys dentatus*

yellowtail **flounder** *U.S.* the pleuronectid *Limanda ferruginea*

deepwater **flounder** *U.S.* the bothid *Monolene sesnilicauda*

winter **fllounder** *U.S.* the pleuronectid *Pseudopleuronectes americanus*

witch **flounder** *U.S.* the pleuronectid *Glyptocephalus cynoglossus*

[**floury endosperm**] a mutant (gene symbol *fl₁*) mapped at 58 on linkage group II of *Zea mays*. The name is descriptive of the phenotypic endosperm

gene-**flow** the wide distribution of genes within a population by interbreeding

flower 1 (*see also* flower 2, beetle, bug, pecker, piercer, thrip) a short, stem bearing, appendage specialized for sexual reproduction in an angiospermous plant

aggregate flower one with spherical or sub-spherical components (*cf.* capitulate)

euacranthic flower one arising from the apex of a shoot which has already produced leaves

butterfly flower (*see also* flower 2) one with a perianth so shaped that the nectaries can only be reached by an insect having a long proboscis

capitulate flower a flattened, compound flower, as in the Compositae

dichotomal flower one which arises in the form of a dichotomous branch (*cf.* dichasium)

complete flower one which has a calix, a corolla, stamen and pistils

incomplete flower a flower which lacks one or more of the cyclic series of organs found in a complete flower

fertile flower = female flower

self-fertile flower one which does not require assistance from an outside agency in pollination

polygamous flower one characterized by the presence of both bisexual and unisexual flowers

neutral flower properly a flower, or rather the termination of a floral axis, in which both sex organs have been surpressed. The term is sometimes improperly applied to staminate flowers

perfect flower one which has both male and female sex organs

imperfect flower one which does not contain both male and female sex organs

entomophilous flower one which attracts insects

pistillate flower a unisexual flower possessing only a gynoecium

pollen flower one which lacks nectar and which is therefore visited by bees for the sake of its pollen

prison flower one which temporarily detains an insect

separate flowers those having distinct sexes

macrosporangiate flower female flower

staminate flower a unisexual flower possessing only an androecium

flower 2 (*see also* flower 1) a plant named for its flowers

Adder's flower any of several *spp.* of the caryophyllalous genus Lychnis

balloon flower the campanulaceous herb *Platycodon grandiflorum* (= Chinese bellflower)

basket flower the amaryllidaceous herb *Hymenocallis calathina* and the compositaceous herb *Centaurea americana*

beard flower *U.S.* popular name of the genus of orchids Pogonia

bellflower any of numerous plants with bell shaped flowers, many of the campanulaceous genus Campanula

Chilean bellflower any of several species of herb of the liliaceous genus Lapageria

Chinese bellflower the campanulaceous herb *Platycodon grandiflorum*

giant bellflower the campanulaceous herb *Ostrowkia magnifica*

Japanese bellflower = Chinese bellflower

marsh bellflower *Campanula aparinoides*

bird-of-paradise flower the musaceous herb *Strelitzia reginae*

blanket flower any of several species of compositaceous herbs of the genus Gaillardia

breeches flower the papaveraceous herb *Dicentra cucullaria* (= Dutchman's breeches *cf.* bleeding heart)

bunch flower the liliaceous herb *Melanthium virginicum*

butterfly flower (*see also* flower 1) any of several species of solanaceous herb of the genus Schizanthus

calico flower the aristolocheous herb *Aristolochia elegans*

cardinal flower the campanulaceous herb *Lobelia cardinalis*

blue cardinal flower *L. siphalitica*

carrion flower the liliaceous herb *Smilax herbacea*

cauliflower (*see also* organ) a large flowered horticultural variety of cabbage (*cf.* cauliflorous)

chigger flower the asclepiadaceous herb *Asclepias tuberosa* (= butterfly weed)

Christmas flower the euphorbiaceous shrub *Euphorbia pulcherrima* (= Poinsettia)

combflower the compositaceous herb *Echinacea purpurea*. Also variously called the **hedgehog combflower** and the **purple combflower**

coneflower any of several species of the compositaceous genus Rudbeckia

prarie coneflower any of several species of the compositaceous genus Ratibida

purple coneflower any of many species of the compositaceous genus Echinaceae, probably in error for combflower

corkscrew flower = snail flower

cornflower the compositaceous herb *Centaurea cyanus*

cuckoo flower the scrophulariaceous herb *Lychnis floscuculi* (= ragged robin) and the cruciferous herb *Cardamine pratensis*

cup flower the loasaceous flower *Scyphanthus elegans* and any of several species of the solanaceous genus Nierembergia

blue-dawn flower *Ipomoea learii*

day flower *U.S.* any of several species of commelineaceous herb of the genus Commelina

dove flower the orchid *Peristeria elata*

duck flower = pelican flower

sacred-ear flower the annonaceous shrub *Cymbopetalum penduliflorum*

Easter flower the euphorbiaceous shrub *Euphorbia pulcherrima* (= "Poinsettia")

everlasting flower any flower which retains its shape and color when dried. The term is most commonly applied to the genera Antennaria, Gomphrena, Anaphalis, Helychrysum, Ammobium and Gnaphalium

fame flower *U.S.* any of several species of portulacaceous herb of the genus Talinum

fennel flower any of several species of the ranumculaceous genus Nigella

flamboyant flower = peacock flower

flame flower the liliaceous herb *Kniphofia uvaria*

foam flower any of several species of the sacifragaceous genus Tiarella, particularly *T. cordifolia*

frost flower any of several species of the compositaceous genus Aster and the liliaceous herb *Milla diflora*

garland flower any of several herbs of the zingiberaceous genus Hedychium

gillie flower = gilly flower

gilly flower *Brit. obs.* = carnation. *Brit.*, usually means wallflower but sometimes stock. *U.S.* usually means stock but sometimes wallflower

marsh gillyflower the caryophyllaceous herb *Lychnis floscuculi* (= ragged robin, cuckoo flower)

globe flower any of several species of ranunculaceous herbs of the genus Trollius and the rosaceous shrub *Kerria japonica*

gold flower the hypericaceous shrub *Hypericum moseranum*

holy-ghost-flower the orchid *Peristeria elate*

goose flower = pelican flower

inside-out flower the berberidaceous herb *Vancouveria parviflora*

leather flower *U.S.* any of several species of ranunculaceous herb of the genus Clematis

leopard flower the iridaceous herb *Belamcanda chinensis*

lobster flower the euphorbiaceous shrub *Euphorbia pulcherrima.* (= "Poinsettia", *cf.* Easter flower)

marshflower *U.S.* the gentianaceous waterplant *Nymphoides lacunosum*

mayflower any of so many species of spring flowering herbs, vines and shrubs that the name has no specific value

mist flower the compositaceous herb *Eupatorium coelestinum*

moccasin flower any of numerous orchids of the genus Cypripedium

monkey flower any of several scrophulariaceous herbs of the genus Mimulus

moon flower any of several climbing vines of the convolvulaceous genus Ipomoea

bush moon flower *I. leptophylla*

musk flower the scrophulariaceous herb *Mimulus moschatus*

pasque flower *U.S.* the ranunculaceous plant *Anemone patens* Brit. *A. pulsatilla*

passion flower any of numerous species of climbing vines of the passifloraceous genus Passiflora

peacock flower the leguminous tree *Poinciana regia*

pelican flower the aristolocheous herb *Aristolochia grandiflora* (= swan flower, goose flower, and duck flower)

pencil flower *U.S.* any of several species of leguminous herb of the genus Stylosanthes

pineapple flower the liliaceous herb *Eucomis punctata*

pinxter flower the ericaceous shrub *Rhododendron nudiflorum*

proboscis flower the martyniaceous herb *Martynia louisiana*

prophet flower the boraginaceous herb *Arnebia echioides*

quinine flower the gentianaceous herb *Sabatia angularis* (= bitter bloom)

red-white-blue flower the lythraceous herb *Cuphea llavea*

rice flower any of numerous species of herb of the phymelaeaceous genus Pimelea

side-saddle flower = side-saddle plant

safflower the compositaceous herb *Carthamus tinctorius*

St. Agnes' flower the amaryllidaceous herb *Leucojum vernum* (= spring snowflake)

satin flower *U.S.* any of several iridaceous herbs of the genus Sisyrinchium and the cruciferous herb *Lunaria annua. Brit.* the caryophyllaceous herb *Stellaria holostea*

shell flower the labiateous herb *Moluccella laevis* and the iridaceous flower *Tigridia pavonia* (= tiger flower)

shrimp flower the acanthaceous shrub *Belloperone guttata*

snail flower the leguminous herb *Phaseolus caracalla*

star flower either of two species of primulaceous herb of the genus Trientalis, particularly *T. borealis*

spring-star flower the liliaceous herb *Brodiaea uniflora*

sunflower (*see also* bug, maggot, midge, moth) any of very numerous species of the compositaceous genus Helianthus. Without qualification, and particularly in horticulture and agriculture, the common sunflower is usually meant

common sunflower *H. annuus*

purple disc sunflower *H. atrorubens*

cucumber-leaved sunflower *H. devilis*

Russian sunflower horticultural variety of common sunflower

showy sunflower *H. laetiflorous*

wild sunflower *U.S. H. decapetalus*

swan flower = pelican flower

tassel flower the compositaceous herbs *Emelia flammea* and *Brickelia grandiflora*

tiger flower the iridaceous flower *Tigridia pavonia* (= shellflower)

trumpet flower *U.S.* the bignoniaceous herb *Campsis radicans*

tube flower the verbenaceous herb *Clerodendron siphonanthus*

twin flower any of several species of herb of the caprifoliaceous genus Linnaea, particularly *L. borealis*

Mexican twin flower the amaryllidaceous herb *Bravoa geminiflora*

urn flower any of several species of amaryllidaceous herb of the genus Urceolina

wallflower the cruciferous herb *Cheiranthus cheiri*

western wallflower either of two cruciferous herbs

of the genus Erysimum, *E. asperum* and *E. arkansanum*

wand flower the diapensiaceous herb *Galax aphylla* and any of several species of iridaceous herbs of the genus Sparaxis

windflower any of numerous species of herb of the ranunculaceous genus Anemone

bastard windflower the gentian *Dasystephana linearis*

zephyr flower any of many species of amaryllidaceous herb of the genus Zephyranthes

flower-de-luce *Brit.* any of many species of monocotyledonous herbs of the genus Iris (*cf.* fleur-de-lys)

flower-of-an-hour the malvaceous herb *Hibiscus trionum*

flower-of-Jove the caryophyllaceous herb *Lychnis flos-jovis*

flower-of-the-gods the orchid *Disa grandiflora*

flowers-of-tan the myxomycete *Aethalium septicum*

flowers-of-wine the floating yeast *Saccharomyces mycoderma* which produces the secondary fermentation of wines like sherry (*cf.* mother-of-vinegar)

flp *see* [flap wings]

fluellen *Brit.* the scrophulariaceous herb *Linara elatine* (*cf.* toad-flax) or the scrophulariaceous herb *Veronica officinalis* (= common speedwell)

fluellin = fluellen

circum**fluence** a flowing together of protozoan pseudopodia around a food particle (*cf.* cirvumvallation)

dif**fluent** readily soluble

in**fluent** a motile organism which moves into a stable community which it disturbs by the destruction of organisms or parts of organisms

in**fluents** those organisms in a community which are not dominant but are nonetheless very numerous

sub**influent** one which is seasonal or transitory and therefore exercises very little effect

ve**fluent** an influent exercising even less effect than a subinfluent

[**fluffy**] a mutant (gene symbol *fl*) on linkage group II of *Neurospora crassa*. The phenotype lacks macroconidia and has relatively few microconidia

[**fluffyoid**] a mutant (gene symbol *fld*) on linkage group IV of *Neurospora crassa*. The phenotypic colonies lack conidia

fluid in *biol.* any liquid

amniotic **fluid** the contents of the amniotic pouch

cerebrospinal **fluid** the fluid in which the central nervous system is bathed

stigmatic **fluid** the sticky secretion on a stigma to which pollen grains adhere

synovial **fluid** the fluid of the synovial joint, consisting of tissue fluid enriched with mucin

tissue **fluid** a fluid that diffuses through the endothelial walls of capilliaries

fluke 1 (*see also* fluke 2, 3) originally any small paddle-shaped object, particularly the limbs of aquatic mammals

fluke 2 (*see also* fluke 1, 3) popular name of trematode platyhelminths

liver**fluke** the distomate trematode *Fasciola hepatica*

Chinese liver **fluke** *Clonorchis sinensis*

giant liver **fluke** *Fasciolopsis buski*

fluke 3 (*see also* fluke 1, 2) a flatfish

white **fluke** *N. Brit.* the pleuronectid fish *Pleuronectes flexuosa* (= flounder)

-flumen- *comb. form* meaning "a river"

flumineous pertaining to running water

flush term for a group of mallards, though flock and flight are frequently used

[**fluted**] a mutant (gene symbol *fl*) mapped at 59.9 on chromosome III of *Drosophila melanogaster*. The term refers to the appearance of the wing of the phenotype

flutterer the apodid trochilid bird *Klais guimeti*

fluvial pertaining to streams (*cf.* argodromile)

Fluviales = Helobiae

fly 1 (*see also* fly 2, catcher, eater, poison, trap) any dipterous insect. Also, loosely, any winged insect not obviously a bee or wasp

adder fly *Brit.* = dragon fly

alder fly any of numerous megalopterous insects of the family Sialidae

ant fly the winged stage of any ant

apefly *I.P.* the lycaenid lepidopteran insect *Spalgis epius*

armed fly *Brit.* popular name of stratiomyid diptera of the genus Stratiomys

humpbacked fly popular name of phorid dipterans

bat fly pupiparous dipteran parasites of bats belonging to the families Nycteribiidae and Streblidae

bee fly popular name of bombyliid dipterans

grasshopper bee fly *U.S. Systoechus vulgaris*

black fly popular name of simuliid nematoceran dipterans. Also, any of several aleyrodid homopterans which cannot be called whiteflies in virtue of their color

citrus blackfly *U.S.* the aleyrodid homopteran *Aleurocanthus woglumi*

blow fly popular name of calliphorid dipterans

black blow fly *U.S. Phormia regina*

bot-fly a term properly applied to oestrid diptera but often extended to any dipterous insects the larva of which are parasitic in a vertebrate

horse bot-fly *U.S.* the gasterophilid dipteran *Gasterophilus intestinalis*

human bot-fly *U.S.* the cuterebrid dipteran *Dermatobia hominis*

nose bot-fly *U.S.* the gasterophilid dipteran *Gasterophilus haemorrhoidalis*

sheep bot-fly oestrid *Oestrus ovis* the larvae of which live in the cranial sinuses of sheep

throat bot-fly *U.S.* the gasterophilid dipteran *Gasterophilus nasalis*

breeze fly *Brit.* popular name of tabanid dipterans

brine fly a popular name of many Ephydridae

bulb fly popular name of certain syrphid dipterans

lesser bulb fly *U.S. Eumerus tuberculatus*

narcissus bulb fly *U.S. Merodon equestris*

onion bulk fly *U.S. Eumerus strigatus*

butterfly (*see also* bird, bush, fish, flower, lily, orchid, pea, ray, shell, tulip) a term commonly applied to the papilionoid and hesperioid lepidopteran insects but properly belonging only to the former. The vast majority of butterflies are diurnal slender-bodied forms (the hesperioids are thick bodies) with thread-like antennae terminating in a club (*cf.* skipper, moth)

alfalfa butterfly *U.S.* the pierid *Colias eurytheme*

ambrax butterfly *Austral.* the papilionid *Papilio ambrax*

brown argus butterfly *Brit.* the lycaenid *Lycaena astrache*

cabbage butterfly the pierid *Pieris rapae*

Christmas butterfly *S.Afr.* the papilionid *Papilio demodocus*

comma butterfly *Brit. Vanessa c-album*

brush-footed butterfly popular name of nymphalids

four-footed butterfly popular name of nymphalids

hunter's butterfly *U.S.* the nymphaline nymphalid butterfly *Vanessa virginiensis* (= painted beauty)

kite butterfly (*see also* kite) *Austral.* any of numerous papilionids lacking tails on the wing

leaf butterfly any of several species of vanessid of the genus Kallima

lime butterfly *I.P.* the papilionid *Papilis demoleus*

map butterfly any of several nymphalids of the genus Cyrestis

monarch butterfly the danaid *Danaus plexippus*

moth butterfly *I.P.* and *Austral.*, the lycaenid *Liphyra brassolis*, one of the very few butterflies possessing a carnivorous larva

Blue Mountain butterfly *Austral.* the papilionid *Papilio ulysses*

noctula butterfly *Austral.* the satyrid *Harsiesis hygea*

owl butterfly general name for amathusiids

patched butterfly *U.S.* any of several species of nymphalid of the genus Chlosyne

pine butterfly *U.S.* the pierid *Neophasia menapia*

snout butterfly popular name of Libytheids

thistle butterfly term used in *U.S.* to distinguish nymphalids still remaining in the genus Vanessa (red admiral, painted lady) from those transferred to Nymphalis (tortoiseshells)

wall butterfly *Brit.* the satyrid *Parage megaera* (*cf.* speckled wood)

milkweed butterfly = monarch

zebra butterfly *U.S.* the heliconiine nymphalid *Heliconius charithonius* (*cf.* zebra swallowtail)

caddisfly popular name of trichopterans

snail-case caddis fly popular name of helicopsychids

finger-net caddis fly popular name of philopotamids

trumpet-net caddis fly popular name of psychomyiids

net-spinning caddis fly popular name of hydropsychids

chameleon fly *Brit.* the stratiomyid dipteran *Stratiomys chameleon*

cheese fly = cheese skipper

cluster fly *U.S.* the calliphorid dipteran *Pollenia rudis*

crane fly (*see also* orchid) *U.S.* popular name of tipulid insects (= *Brit.*, but not *U.S.* daddy long legs)

phantom crane fly popular name of ptychopterid dipterans

primitive crane fly popular name of tanyderid dipterans

range crane fly *U.S.* the tipulid *Tipula simplex*

winter crane fly popular name of trichocerid dipterans

cuckoo fly *Brit.* popular name of chrysidid hymenopteran insects (= ruby wasp *Brit.*, cuckoo wasp *U.S.*)

damselfly popular name of zygopteran odonatan insects

black-winged damselfly *U.S.* the agrionid *Agrion maculatum*

broad-winged damselflies popular name of damselflies of the family Agrionidae

dance fly popular name of empidid dipterans

deer fly *U.S.* popular name of tabanid diptera

dobson fly *U.S.* a term properly applied to the huge neuropteran *Corydalus cornutus*. Sometimes extended to incluse all corydalids

dragonfly popular name of anisopteran odonatan insects

club-tailed dragonflies popular name of gomphid anisopteran odonatan insects

drone fly the syrphid *Eristalis tenax*

dung fly *Brit.* popular name of scopneumatid Diptera

eggfly *Austral.* and *I.P.* any of numerous species of large brilliantly colored lepidopterans of the genus Hypolimnas. Usually, without qualification, *H. bolina*

stalk-eyed fly popular name of diopsid dipterans

fairy fly mymarid hymenopterans

firefly *U.S.* popular name of lampyrid coleopterans (= *U.S.* lightningbug)

fishfly *U.S.* variously used for all corydalid Neuroptera or more frequently for all corydalid Neuroptera except for the dobson fly

flesh fly popular name of sarcophagid dipterans

flower fly *U.S.* popular name of syrphid diptera

flat-footed fly popular name of platypezid diptera

forest fly *Brit.* the hippoboscid dipteran *Hippobosca equina*

frit fly popular name of chloropid diptera, particularly *U.S. Oscinella frit*

fruit fly a term properly applied to the trypetid dipterans the larvae of which are well known pests of fruit. The term is often incorrectly applied to the Drosophilidae which, however, feed primarily on the yeasts and fungi in decomposing fruit

bumelia fruit fly *U.S. Pseudodacus pallens*

cherry fruit fly *Rhagoletis cingulata* (Loew)

black cherry fruit fly *U.S. Rhagoletis fausta*

currant fruit fly *U.S. Epochra canadensis*

West Indian fruit fly *U.S. Anastrepha mombin-praeoptans*

Mediterranean fruit fly *Ceratitis capitata*

Mexican fruit fly *U.S. Anastrepha ludens*

olive fruit fly *U.S. Dacus oleae*

oriental fruit fly *U.S. Dacus dorsalis*

papaya fruit fly *U.S. Toxotrypana curvicauda*

gad fly *Brit.* popular name of tabanid dipterans

gall fly any of several gall making trypetid dipterans

eupatorium gall fly *U.S. Procecidochares utilis*

lantana gall fly *U.S.* Eutreta xanthochaeta

greenfly *Brit.* = aphid *U.S.*

harlequin fly *Brit.* any of numerous species of the chironomid genus Chironomus

big-headed fly popular name of pipunculid diptera

small-headed fly popular name of acrocerid dipterans

thick-headed fly *Brit.* popular name of conopid Diptera

hessian fly the cecidomyiid dipteran *Mayetiola destructor*

horn fly the muscid dipteran *Haematobia irritans*

horse fly (*see also* weed) popular name of tabanid dipterans

black horse fly *U.S. Tabanus atratus*

striped horse fly *U.S. Tabanus lineola*

house fly any of several small dark colored nonbiting dipterous insects particularly of the genus Musca and specifically *Musca domestica*

little house fly *U.S. Fannia canicularis*

hover fly a name commonly given to any medium sized dipteran with the habit of hovering over flowers, particularly (*U.S.*) syrphids

walnut husk fly *U.S.* the trypetid dipteran *Rhagoletis completa*

lantern fly any of numerous homopterous insects

having elongate prominences on the head and with the antennae attached to the cheek below the eyes

latrine fly *U.S.* the muscid dipteran *Fannia scalaris* (*cf.* house fly)

long-legged fly *U.S.* popular name of dolichopodid dipterans

stilt-legged fly popular name of micropezid dipterans

louse fly popular name of hippoboscid dipterans though occasionally extended to all pupiparous dipterans

march fly popular name of bibionid dipterans

marsh fly popular name of schizomyid dipterans

mayfly popular name of ephemeropteran insects
    burrowing mayfly popular name of ephemerids
    stream mayfly popular name of heptageniids

melon fly *U.S.* the trypetid dipteran *Dacus cucurbitae*

leaf miner fly popular name of agromyzid dipterans (*cf.* miner)

moth fly popular name of psychodid dipterans

mydas fly popular name of mydaid dipterans

Narcissus fly = Narcissus bulb fly

onion fly the anthomyiid *Hylemya cepetorum*

orchidfly *U.S.* the eurytomid hymenopteran *Eurytoma orchidearum*

owlfly popular name of ascalaphid neuropterans

palmfly *I.P.* any of several species of satyrid lepidopteran insect of the subfamily Elymniinae

petroleum fly the ephydrid dipteran *Psilopa petrolei* the larva of which lives in pools of crude petroleum

pigeon fly *U.S.* the hippoboscid *Pseudolynchia canariensis*

pomace fly popular name of drosophilid dipterans (*cf.* fruit fly)

robber fly popular name of asilid dipterans

cabbage root fly the anthomyiid dipteran *Anthomyia brassicae* (= cabbage maggot)

Russian fly = Spanish fly

rust fly popular name of psilid Diptera
    carrot rust fly *U.S. Psila rosae*

sand fly popular name of blood sucking psychodid dipteran insects of the genus Plebotomus (= Phlebotomus). They are the vectors of many serious diseases

sawfly popular name of those hymenopteran insects (the suborder Symphyta) in which the body is directly fused to the thorax without the intervention of a wasp-waist. By some the term is confined to a superfamily, the Tenthredinoidea
    striped alder sawfly *U.S.* the tenthredinid *Hemichroa crocea*
    European apple sawfly *U.S.* the tenthredinid *Hoplocampa testudinea*
    black-headed ash sawfly *U.S.* the tenthredinid *Tethida cordigera*
    brown-headed ash sawfly *U.S.* the tenthredinid *Tomostethus multicinctus*
    mountain-ash sawfly *U.S.* the tenthredinid *Pristiphora geniculata*
    gooseberry sawfly *Brit.* the tenthredinid *Nematus ventricosus*
    raspberry sawfly *U.S.* the tenthredinid *Monophadnoides geniculatus*
    birch sawfly *U.S.* the argid *Arge pectoralis*
    dusky birch sawfly *U.S.* the tenthredinid *Croesus latitarsus*
    cherry sawfly *U.S.* = hawthorn sawfly *U.S.*, imported currantworm

    currant sawfly *Brit.* the tenthredinid *Nematus ribesii*
    dock sawfly *U.S.* the tenthredinid *Ametastegia glabrata*
    elm sawfly *U.S.* the cimbicid *Cimbex americana*
    balsam-fir sawfly *U.S.* the diprionid *Neodiprion abietis*
    poplar leaf-folding sawfly *U.S.* the tenthredinid *Phillocolpa bozemani*
    cherry fruit sawfly *U.S.* the tenthredinid *Hoplocampa cookei*
    willow red-gall sawfly *U.S.* the tenthredinid *Pontania proxima*
    grape sawfly *U.S.* the tenthredinid *Erythraspides vitis*
    grass sawfly *U.S.* the tenthredinid *Pachynematus extensicornis*
    hemlock sawfly *U.S.* the diprionid *Neodiprions tsugae*
    larch sawfly the tenthredinid *Pristiphora erichsonii*
    raspberry leaf sawfly the tenthredinid *Priophorus morio*
    European pine sawfly *U.S.* the diprionid *Neodiprion certifer*
    introduced pine sawfly *U.S.* the diprionid *Diprion similis*
    red-headed pine sawfly *U.S.* the diprionid *Neodiprion lecontei*
    red-pine sawfly *U.S.* the diprionid *Neodiprion nanulus*
    white-pine sawfly *U.S.* the diprionid *Neodiprion pinetum*
    lodgepole sawfly *U.S.* the diprionid *Neodiprion burkei*
    curled rose sawfly *U.S.* the tenthredinid *Allantus cinctus*
    willow shoot sawfly *U.S.* the cephid *Janus abbreviatus*
    plum web-spinning sawfly *U.S.* the pamphiliid *Neurotoma inconspicua*
    European spruce sawfly *U.S.* the diprionid *Diprion hercyniae*
    green-headed spruce sawfly *U.S.* the tenthredinid *Pikonema dimmockii*
    yellow-headed spruce sawfly *U.S.* the tenthredinid *Pikonema alaskensis*
    stem sawfly popular name of cephids
    currant stem sawfly the cephid *Janus integer* (= currant stem girdler)
    black grain stem sawfly *U.S. Cephus tabidus*
    wheat stem sawfly the cephid *Cephus cinctus*
    European wheat stem sawfly the cephid *Cephus pygmaeus*
    honeysuckle sawfly *U.S.* the cimbicid *Zaraea inflata* Norton
    hawthorn sawfly *U.S.* the tenthredinid *Profenusa canadensis*
    violet sawfly *U.S.* the tenthredinid *Ametastegia pallipes*
    willow sawfly *U.S.* the tenthredinid *Nematus ventralis*

scavenger fly popular name of sepsid dipterans
    black scavenger fly popular name of scatopsid dipterans

scorpion fly popular name of mecopterans, particularly panorpids
    earwig scorpion fly the meropeid *Merope tuber*
    hanging scorpion fly popular name of bitacid mecopteran

snow scorpion fly popular name of boreid mecopterans

lantana seed fly *U.S.* the agromyzid dipteran *Ophiomyia lantanae*

shadefly *S.Afr.* any of several species of lepidopteran of the genus Coenyra

shore fly popular name of ephydrid dipterans

skipper fly popular name of piophilid dipterans (*but see* skipper)

snakefly popular name of raphidiid neuropterans

snipe fly popular name of rhagionid dipterans

soldier fly popular name of stratiomyid dipterans

Spanish fly the meloid coleopteran *Lytta vesicatoria* (*cf.* Cantharidae)

spongilla fly popular name of sisyrid neuropterans

stable fly *U.S.* the muscid dipteran *Stomoxys calcitrans*

false stable fly *U.S.* the muscid dipteran *Muscina stabulans*

stiletto fly popular name of therevid Diptera

stink fly *Brit.* popular name of chrysopid Neuroptera (*cf.* lacewing, goldeneye)

stonefly the popular name for members of the insect order Plecoptera *U.S.* without qualification, usually applied to Perlidae

giant stonefly *U.S.* popular name of pteronarcid plecopterans

green stonefly *U.S.* popular name of Chloroperlidae

roach-like-stone fly *U.S.* popular name of Peltoperlidae

spring stonefly *U.S.* popular name of Nemouridae

green-winged stonefly *U.S.* popular name of Isoperlidae

rolled-winged stonefly *U.S.* popular name of Leuctridae

winter stonefly *U.S.* popular name of Taeniopterygidae

small winter stonefly *U.S.* popular name of Capniidae

suckfly *U.S.* the mirid hemipteran insect *Cyrotopeltis notatus*

tsetse fly any of several species of muscid flies of the genus Glossina particularly *G. morsitans* well known as vectors of trypanosames

typhoid fly = house fly

tangle-veined flies popular name of nemestrinid Diptera

vinegar fly a term sometimes applied in *U.S.* to the family Drosophilidae (*cf.* pomace fly, fruit fly)

warble fly popular name of hypodermatid dipterans

seaweed fly popular name of coelopid dipterans

white fly popular name of aleyrodid homopterans (*cf.* blackfly)

avocado whitefly *U.S. Trialeurodes floridensis*

azalea whitefly *U.S. Pealius azaleae*

mulberry whitefly *U.S. Tetraleurodes mori*

strawberry whitefly *U.S. Trialeurodes packardi*

citrus whitefly *U.S. Dialeurodes citri*

grape whitefly *U.S. Trialeurodes vittata*

greenhouse whitefly *Trialeurodes vaporariorum*

sweetpotato whitefly *U.S. Bemisia tabaci*

rhododendron whitefly *U.S. Dialeurodes chittendeni*

cloudy-winged whitefly *U.S. Dialeurodes citrifolii*

woolly whitefly *U.S. Aleurothrixus floccosus*

window fly Popular name of scenopinid diptera

picture-winged fly popular name of otitid dipterans

spear-winged fly popular name of lonchopterid dipterans

screw-worm fly the calliphorid dipteran *Cochliomya hominovorax*

yamfly *I.P.* any of several species of lycaenid lepidopterans of the genus Loxura

fly 2 (*see also* fly 1) in compound names of plants

catchfly (*see also* grass) any of numerous species of the caryophyllaceous genus Silene (*cf.* campion)

bladder catchfly *S. latifolia*

German catchfly the scrophulariaceous herb *Lychnis viscaria*

royal catchfly *S. regia*

sweet william catchfly *S. armeria*

highflyer *Brit.* any of several species of geometrid lepidopteran insect of the genus Hydriomena. *S.Afr.* the lycaenid lepidopteran *Aphnaeus hutchinsoni*

fly-up-the-creek *U.S.* heron

flyway the established route of migratory birds

FMN = flavin mononucleotide

fo *see* [folded]. This symbol is also used as a yeast genetic marker indicating a requirement for folic acid

foal a young equid animal

meadow foam the limnanthaceous herb *Limnanthes douglasii*

fody any of several ploceid birds of the Madagascan genus Foudia (= fodi)

foetalization the persistence of certain foetal or immature characters of an ancestor in the adult stages of a descendant

superfoetation the fertilization of an ovary by two kinds of pollen

foetid stinking

foetus an early mammalian embryo and specially applied to the 4-month human embryo

foil (*see also* foyle) a leaf, usually in compound names of plants

cinquefoil any of numerous rosaceous herbs and shrubs of the genus Potentilla

marsh cinquefoil *Brit.* the rosaceous plant Potentilla *palustris*

milfoil the compositaceous herb *Achillea millifolium*

quadrifoil = quadrifoliate leaf

rockfoil any of numerous saxifragaceous herbs of the genus Saxifraga

rush foil any of several species of euphorbiaceous herb of the genus Crotonopsis

Saint-foin *Brit.* any of several species of leguminous herb of the genus Onobrychis

fol *see* [folded wings]

-fol- *comb. form* meaning "leaf"

epichordal fold the dorsal of the two fin-folds, from which the caudal fin of the fish may be derived

hypochordal fold the ventral of the two fin-folds from which the caudal fin of the fish may be derived

costal fold a slit or pocket in the wing of some lepidopteran insects that can be opened to expose the androconia

head fold that area at the anterior end of the embryo which rises from the blastoderm in the early development of a telolecithal egg

metapleural fold *see* metapleure

tail fold that part of a posterior end of an embryo which rises above the blastoderm in the early development of a telolecithal egg

[folded] a mutant (gene symbol *fo*) mapped at 63.± on the X chromosome of *Drosophila melanogaster*. The phenotypic expression is a failure of the wings to unfold

[folded wings] a mutant (gene symbol *fol*) mapped at

39.0± on chromosome II of *Drosophila melano-gaster*. The wings of the phenotype overlap when folded

grape-leaf folder *U.S.* the larva of the pyraustid lepidopteran insect *Desmia funeralis*

-folia- *comb. form* meaning "leaf"

foliaceous pertaining to, or resembling, a leaf

foliage (*see also* gleaner) the sum-total of the leaves

-foliate having or pertaining to leaves. The form folious, not used in this work, is synonymous

  asperifoliate having rough leaves

  angustifoliate having narrow leaves

  centifoliate literally having a hundred leaves, but used in the sense of having very many leaves

  crispifoliate with curled leaves

  defoliate the condition of a plant after the leaves have fallen

  exfoliate to shed the outer covering in leaf-like scales

  graminifoliate having grass-like leaves

  integrifoliate with simple leaves

  latifoliate broad leaved (*but see* laterifoliate)

  laterifoliate growing on the side of a leaf

  serraefoliate having serrate leaves

  serrifoliate = serraefoliate

  spinifoliate with spiny leaves like those of the cactus or holly

  varifoliate either having leaves of variable form or of blotchy color

foliolate covered with leaflets

  efoliolate lacking leaf scales or squamae

foliole a leaflet

foliose having the appearance of a leaf, said particularly of lichens in contradistinction to crustose

-folious an alternative termination for -foliate, the form preferred in this work

follicle 1 (*see also* follicle 2) literally, a small container, but used in histology for a more or less spherical aggregate of cells surrounding a single cell in fluid. Particularly the follicle containing the egg in the ovary

  atretic follicle a primordial follicle of an ovary that degenerates *in situ*

  Graffian follicle the functional follicle in which the mammalian ovum develops in the ovary

  hair follicle glands secreting the hair in mammals

  lymph follicle a patch of reticular tissue in the intestinal wall crowded with lymphocytes (= Peyer's patches)

  primordial follicle a group of cells cut off from, and sinking below the surface of, a mammalian ovary

follicle 2 (*see also* follicle 1) *bot.*, a monocarpic capsular fruit, or any other bag shaped structures

follicular possessing follicles. Also used in platyhelminthes of diffuse gonads, consisting of numerous little follicles

folliculin = estrone

  dihydrofolliculin = estradiol

fontanel literally a little fountain, but in biology applied to shallow depressions such as that which surrounds the frontal and the opening of the frontal gland on the head of termites, or the gaps between the dorsal paired dermal bones in the posterior region of the vertebrate skull

fontanous pertaining to fresh-water springs

tom fool *W.I.* any of several species of tyrannid bird of the genus Myiarchus. Usually, without qualification, *M. stolidus*

  big tom fool *M. validus*

  little tom fool *M. tuberculifer*

foot 1 (*see also* foot 2–6, man) the terminal structure of all four limbs of terrestrial vertebrates except birds, in which the only hind limb is furnished with a foot and primates in which the front foot is called a hand

  heterodactyl foot a foot, characteristic of trogoniform birds, in which the inner two toes are turned backwards

  mesaxonic foot one in which the axis of symmetry passes through the middle of the third digit. This situation pertains in the even-toed ungulates or Perissodactyla

  paraxonic foot one in which the axis of symmetry passes between the third and fourth digit. This situation pertains in the even-toed ungulates or Artiodactyla

  palmate foot a foot, characteristic of many swimming birds, in which the front three toes are entirely webbed

  pamprodactyl foot a foot, characteristic of coliiform and apodid birds, in which all four toes are capable of being turned forwards

  raptorial foot a foot, characteristic of birds of prey, which is modified for grasping prey

  semipalmate foot a foot in which the front three toes are partially webbed

  syndactyle foot a foot, characteristic of coraciiform birds, in which two toes are fused for part of their length

  totipalmate foot a foot, characteristic of pelecaniform birds, in which all four toes are joined by webbing

  zygodactyl foot a foot, such as that found on parrots and woodpeckers, in which the middle two toes point forwards and the other two backwards

foot 2 (*see also* foot 1, 3–5, podium and podite) the locomotory appendage of some invertebrates

  anal foot the holdfast at the posterior end of larval chironomid dipterans

  tube foot the extrusible portion of the water vascular system of echinoderms

foot 3 (*see also* 1,2,4–6) any of several birds with distinctive feet

  finfoot any of three species of heliornithid birds, and also either of the two species of birds of the family Anhingidae (= snakebird)

   African finfoot the heliornithid *Podica senegalensis*

   American finfoot the heliornithid *Heliornis fulica* (= sungrebe)

   masked finfoot the heliornithid *Heliopais personata*

foot 4 (*see also* foot 1–3, 5, 6) any of several Mollusca the shells of which resemble feet

  blue-foot *U.S.* the unionid pelecypod *Quadrula costata*

  duck foot the aporrhaid gastropod *Aporrhais occidentalis*

  pelican's foot the aporrhaid gastropod *Aporrhais pespelicani*

foot 5 (*see also* foot 1–4, 6) any of numerous plants having foot-like leaves, flowers or roots

  bear's foot the compositaceous herb *Polymnia uvedalia*

  bird's foot (*see also* trefoil) *Brit.* common name of the leguminous genus Ornithotus

  cat foot the compositaceous herb *Gnaphalium obtusifolium*

  coltsfoot any of several species of the compositaceous genus Tussilago, particularly *T. farfara*

   sweet coltsfoot any of several species of herb of the

compositaceous genus Petasites, particularly *P. fragrans*

crowfoot (*see also* buttercup, grass) *Brit.* and *U.S.* any of several species of ranunculaceous herb of the genus Ranunculus

bristly crowfoot *R. pennsylvanicus*

marsh crowfoot *R. sceleratus* (= cursed crowfoot)

elephant's foot any of several species of herb of the compositaceous genus Elephantopus and any of several species of the dioscoriaceous shrub Testudinaria

goosefoot any of several species of the chenopodiaceous genus Chenopodium

bastard goosefoot *C. hybridum* (*cf.* pigweed)

sheepfoot the leguminous herb *Lotus corniculatus* (= bird's foot trefoil)

stone foot any of several species of labiateous plants of the genus Collinsonia

thread foot *U.S.* the podostemaceous herb *Podostemum ceratophyllum*

whiteman's foot *U.S.* the plantaginaceous herb *Plantago major*

foot 6 (*see also* foot 1–5) any of several organisms not birds (foot 3) mollusca (foot 4) or plants (foot 5), distinguished by their feet, or foot-like appearance

fan-foot *Brit.* any of several species of noctuid lepidopteran insect of the genera Zanclognatha, Herminia and Pechipogon. Usually, without qualification *Z. tarsipennalis*

flap-foot *see* lizard

scaly-foot *Austral.* any of several species of pygopodid saurians of the genus Pygopus (= salt-bush snake)

spadefoot *U.S.* any of several species of pelobatid anuran Amphibia of the genus Scaphiopus

footman *see* moth

for *see* [Formate]

foramen a hole, particularly one in a bone through which a nerve or blood vessel passes

foramen magnum 1 (*see also* foramen magnum 2) the posterior opening of the skull, through which the spinal cord passes

foramen magnum 2 (*see also* foramen magnum 1) the opening connecting the cavity of the insect head with the cavity of the thorax

Monro's foramen one of the two foramina which connect the third ventricle of the brain to the first ventricles

Panizza's foramen an opening between the two sides of the aortic trunk in Crocodilia

posterior foramen (insect) = foramen magnum 2

foramen ovale the posterior of the two foramina which may pierce the alisphenoid bone for the mandibular branch of the trigeminal nerve

foramen rotundum the anterior of the two foramina which may pierce the alisphenoid for the maxillary branch of the trigeminal nerve

Foraminifera a large order of sarcodinous Protozoa characterized by a calcareous shell through pores in which anastomizing pseudopodia protrude

-forb- *comb. form* meaning "fodder"

forb any plant in a meadow or prairie which is not a grass. Used by ecologists in the sense of herb

anal **forceps** forciculate projections of the male genitalia of some Coleoptera. The term is also used for the modified anal cerci of Dermaptera

forcipate forked like a carpenter's pincers, or the caudal appendages of an earwig. The use as "forcep-like" (i.e. with straight tips) is incorrect

Forcipulata an order of asteroid echinoderms lacking the prominent marginal plates of the Phanerozonia and distinguished from the Spinulosa by the numerous pedunculate pedicellariae

-fore- *comb. term* meaning "anterior"

forest (*see also* fly) any considerable area of land covered with a heavy growth of trees

monsoon forest a tropical deciduous forest fluctuating seasonally between wet and dry

pure forest one which consists of only a single species of tree

rain forest one composed of plants which require an unusually high level of precipitation

high rain forest one having more than seventy-two inches of rainfall annually

tropical rain forest an equatorial rain forest

forester 1 (*see also* forester 2) any of many lepidopteran insects. *U.S.* any of numerous agaristid moths of the genus Alypia and Alypiodes. *Brit.* any of several species of zygaenid lepidopteran insect of the genus Ino. *S.Afr.* the nymphalid lepidopteran insect *Euphaedra neophron*

Christmas forester *S.Afr.* the hesperid *Celaenorrhinus mokeezi*

eight-spotted forester *U.S.* the noctuid *Alypia octomaculata*

forester 2 (*see also* forester 1) *Austral.* the macropodid marsupial mammal *Macropus giganteus* (= great grey kangaroo)

forfex a scissor-like structure at the posterior end of some insects

forficatous scissor-shaped

Forficulidae a large family of dermapterous insects containing most of the common earwigs

forget-me-not 1 (*see also* forget-me-not 2) any of several species of boraginaceous herb of the genus Myosotis but particularly *M. scorpioides*

Chatham Island forget-me-not the boraginaceous herb *Myosotidium hortensia*

creeping forget-me-not the boraginaceous herb *Omphalodes verna*

forget-me-not 2 (*see also* forget-me-not 1) *I.P.* the lycaenid lepidopteran insect *Catochrysops strabo*

biforine a raphid-containing cell open at each end

[forked] a pseudo allelic locus (gene symbol *f*) mapped at 56.7 on the X chromosome of *Drosophila melanogaster*. The phenotype strongly resembles [kinky] but has shorter bristles. (*see also* [enhancer of white eosin])

-form 1 (*see also* -form 2) *adj. suffix* meaning "in the shape of"

abietiform in the shape of a pine tree (particularly used of trichomes)

acetabuliform saucer-shaped

aciform needle-shaped (= acicular)

acinaciform scimitar-shaped

aculeiform in the shape of a thorn

adeniform gland-shaped

aliform wing-shaped

ampulliform flask-shaped

anguilluliform eel-shaped

antenniform in the shape of an antenna

ascidiform pitcher-shaped

asciform hatchet-shaped

auriform ear-shaped

bacciform in the shape of a berry

bacilliform rod-shaped

baculiform in the shape of a cane or reed

beloniform needle-shaped

biform existing in two different shapes

biscoctiform in the shape of an oblong constricted in the middle. This is a common pattern of German cracker (= *Brit.* biscuit)

botuliform sausage-shaped

bufoniform toadlike

bulliform in the shape of a blister

bursiform purse-shaped

cactiform having the shape of a cactus

calathiform having the shape of a deep cup

calcariform spur-shaped

calceiform slipper-shaped

calciform in the shape of a calyx (*cf.* calyciform)

calyciform in the shape of a shallow cup

calyptriform in the shape of a blunt nosed cane

campaniform bell-shaped

campodeiform caterpillar shaped (*cf.* erucaeform)

capitiform head-shaped

cateniform in the shape of a chain with very small links

caudiciform in the shape of the stem of a plant

cauliform stalk-shaped

cerebriform brain-shaped, in the sense that a brain coral or the kernel of a walnut are brain-shaped

cestiform girdle-shaped

cirriform in the shape of a tendril

claviform club-shaped or, rarely, key-shaped

clintheriform platter-shaped

cocciform in the shape of a coccineal insect

cochleariform spoon-shaped (*but see* cochleate)

columelliform in the shape of a pillar

conchiform in the shape of a gastropod shell. The shape of a snail shell is either heliciform or cochleate

cordiform heart-shaped

coroniform in the shape of a coronet

corticiform having the shape or appearance of bark

cotyliform in the shape of a dished wheel

crateriform in the form of a volcanic crater and its tube. Used specifically in bacteriology to describe the liquified portion of a stab culture

cribriform (*see also* organ) having the appearance of a sieve

cristiform in the shape of a crest or ridge but usually used in the latter sense (*cf.* lophiform)

cruciform in the shape of a cross

cucumiform in the shape of a cucumber

cultiform (*see also* -form 2) in the shape of a coulter or colter, that is, of a heavy, blunt-backed, round-ended knife

cuneiform wedge-shaped

cupuliform in the form of a cupola

curviform = curved

cyathiform in the shape of a deep cup

cydariform orange-shaped

cymbiform boat-shaped

dentiform in the shape of a tooth

digitaliform finger-shaped

disciform in the shape of a disc

dolabriform hatchet-shaped

doleiform barrel-shaped

drepaniform sickle-shaped

elytriform in the shape of a beetle wingcase

ensiform sword-shaped

eoliiform = foliaceous

erucaeform caterpillar-shaped (*cf.* campodeiform)

fabiform in the shape of a broad bean

falciform sickle-shaped

fibriform having the shape of a fiber

filiform thread like. Often used in distinction from capitate of a tentacle, or the like, that does not terminate in an expanded portion

flabelliform in the shape of a fan

flagelliform in the shape of a whip

floriform in the shape of a flower

fragariform in the shape of a strawberry

frondiform in the shape of a fern frond

fungiform having a mushroom shape

fusiform cigar-shaped

gemmiform in the shape of a bud

graniform having the shape of a cereal grain

gregariniform *zool.* in the form of a gregarine protozoan usually in the trophoblastic condition. *bot.* pertaining to spores that move as though they were gregarines

guttiform drop-shaped

gyalectiform = urceolate

hastiform halberd-shaped

heliciform snail-like

hippocrepiform horseshoe-shaped

hypocrateriform in the shape of a very flat bowl, or disc with a raised rim

indusiform in the shape of an indusium

infundibuliform funnel-shaped, but used specifically by microbiologists to describe the liquified area of a stab culture when this extends across the whole of the top of the medium and reaches down to the bottom (*cf.* saccate)

juliform in the shape of a catkin

labyrinthiform having internal sinuous chambers or external sinuous markings

laciniform fringed

lacrimiform tear-shaped

lageniform bulbous with a long neck

lentiform = lenticular

leptiform properly long and slender but usually used in the sense of campodeiform, or erucaeform

liguliform strap-shaped

limaciform having the shape of a slug

linguiform tongue-shaped

lophiform in the shape of a crest or bridge, but usually used in the former sense (*cf.* cristiform)

luniform in the shape of the moon, usually taken to mean a crescentic new moon

madreporiform = cribiform

mammiform in the shape of a breast or rounded hillock, but not of necessity having a teat-shaped apex (*cf.* mammilliform)

mammilliform in the form of a breast having a teat

maniform hand-shaped

mitriform mitre-shaped

modioliform in the shape of the nave of a spoked wheel

moniliform having the appearance of a string of beads

muriform arranged in the manner of bricks in the wall

musciform moss-like

napiform in the shape of a turnip. Used specifically in bacteriology to describe the liquified portion of a stab culture

nautiform ship-shaped

nodiform knotted

nuciform nut-like

nymphaeform *either* water-lily shaped *or* chrysalis-shaped

onisciform in the shape of a wood-louse
operculiform lid-shaped
orculaeform cask-shaped
orculiform = orculaeform
oviform in the shape of a hen's egg
paleaform in the shape of small scales
palmatiform in the shape of a hand with the fingers
  extended
palmiform = palmatiform
pampiniform in the shape of a tendril
panduriform fiddle-shaped
paniculiform panicle-shaped
panniform having the texture or appearance of felted
  cloth
papilioform in the shape of a butterfly, or, more usually, a butterfly wing
papilliform shaped like a nipple
pappiform in the shape of pappus
patelliform in the shape of a small dish
pateriform = patelliform
paxilliform in the shape of a tent peg
peltiform in the form of a small, round shield
pelviform basin-shaped
pencilliform shaped like an artist's pencil (cf. peni-
  cill-)
penniform feather-like
piciform resembling a woodpecker
pileiform having the form of a cap
pinniform having the shape of a wing but frequently
  used for penniform
pisiform pea-shaped
placentiform in the shape of a flat cake
pliciform plaited
plumiliform plume-shaped
poculiform goblet-shaped
podetiform shaped like a foot or pedicel
pomiform apple-shaped
poriform pore-shaped
proteiform = polymorphic
pugioniform dagger-shaped
pulviniform cushion-shaped
punctiform literally point-shaped, but used by bacteri-
  ologists to describe colonies on agar plates below
  the limits of resolution of the human eye
pyriform = pear-shaped
racemiform having the appearance of a bunch of
  grapes
radiatiform said of the capitula of compositaceous
  flowers in which the ligulate florets of the outer
  circle are conspicuously the largest
radiciform root-shaped
ramiform (see also -form 2) branch-like
reniform kidney-shaped
rhachidiform having the shape, appearance, or function of a backbone
rhizomatiform having the shape of a rhizome
rimiform cleft-shaped
rostriform beak-shaped
rotiform wheel-shaped
sacciform bag-shaped
sagittiform in the form of an arrow head
scalariform ladder-like
scalpelliform in the shape of a scalpel
scalpriform chisel-shaped
scapiform either boat shaped or in the form of a
  stem lacking leaves
scobiform having the appearance of sawdust
scrotiform pouch-shaped
scutiform = scutelliform

scutelliform in the shape of an oval dish with a round
  edge, but frequently confused with, and often used
  as synonymous with, scutellate (q.v.)
scyphiform cup-like
securiform in the shape of an ax with a broad edge
  and a narrow base
sellaeform saddle-shaped
selliform = sellaeform
septiform having the appearance of a septum, usually
  in the sense that the appearance is deceptive
setiform in the shape of a bristle
setuliform thread-like
siliquiform in the shape of a silique
soleaeform slipper-shaped
soliform = soleaeform
spiciform spike-shaped
squamelliform scale-shaped
stelliform star-shaped
stigmatiform properly in the shape of a point, but
  often used to mean in the shape of the stigma of an
  anther
stipuliform in the shape of a stipule, or of a long rod
stratiform crateriform topped by a cylinder. Used
  specifically in bacteriology to describe the liquified
  portion of a stab culture
strobiliform cone-shaped
strombuliform spirally twisted
styliform in the shape of an elongate column
subuliform awl-shaped
sulciform in the form of a groove
thyrsiform in the shape of a thyrse
trichiform bristle-shaped
trochleariform pulley-shaped
trulliform in the shape of a bricklayer's trowel, or a
  small basin
tubaeform trumpet-shaped
tubiform in the shape of a pipe
turbiniform in the form of a peg-top
tympaniform in the shape of a drum head
umbraculiform in the shape of a parasol
unciform hook-shaped
unguiform in the shape of a claw
utriform = utriculiform
utriculiform having the appearance of a bladder
uviform either in the shape of a grape, or in the shape
  of a bunch of grapes
vagiform of indeterminate shape
variciform in the shape of a ridge
vasiform vessel-or duct-shaped
vergiform rodlike
vermiform worm-shaped
verruciform wart-shaped
versiform (see also form 2) of varying shape
villiform in the form of a villus either in the botanical
  or zoological sense
vulviform in the shape of a vulva i.e., of a cleft with
  lips
-form 2 (see also -form 1) subst. suffix meaning "being
  a type of"
anteform a postulated ancestral type
avoform a form thought to be the original type of a
  ramiform or to resemble a praeform
cultiform-1 (see also cultiform 2) a cultivated variety
  of a wild plant (cf. cultigen)
cultiform-1 (see also cultiform 1) a form of plant
  which has arisen through cultivation
domitoform = cultigen
finiform a taxon, usually plant, all of the known relatives of which are fossils

gregiform a polymorphic finiform

locogregiform a secondary ramiform

growth form the habit of growth of an organism as distinct from a genetic character

hemiform said of fungi in which the teleutospores germinate only after a resting period

hybridiform a hybrid between finiforms

locoform an organism which differs from other organisms through adaptation to its particular environment

lusiform a cultivated form which can be reproduced only by vegetative means

microform 1 (*see also* microform 2) that one of a heteroecious fungus which produces teleutospores

microform 2 (*see also* microform 1) = Jordanon

mistoform a hybrid derived from hybrids markedly differing from the original parent

noviform a cultigen of recent origin

praeform the original ancestor of a taxon either known from the fossil record or postulated

proliform one which is capable of reproducing

hybridoproliform a fertile hybrid of hybridoforms

mistoproliform fertile hybrid of mistoforms

ramiform a markedly variable or polymorphic finiform, or a monophyletic gregiform

satiform a noviform that reproduces by seed

seminiform a cryptogamic reproductive body that is not part of the fructification

sobriniform a versiform belonging to a subgregiform

versiform a form that diverges from the stem form

forma the smallest category commonly used in botanical taxonomy and applying to trivial variations occurring among a population of any species

[formate] a mutant (gene symbol *for*) on linkage group VII of *Neurospora crassa*. The phenotypic expression is a requirement for either formate plus adenine or formate plus methionine

formation a word which is defined in ecology as, or even more, variously than association (*q.v.*). It cannot, at present, be usefully defined and the sense must be determined from the context. Used in *bot.* of an assemblage in the sense that a meadow is an assemblage of grasses

closed formation an assemblage of plants so closely placed in the soil, that others cannot obtain a foothold among them

complex formation a group of plants which is approaching a climax

edaphic formation one dependant on soil

mixed formation an assemblage of several kinds of vegetation, such as a veldt

open formation one which leaves room for additions

panformation a plant community of several combined formations, each, or at least one of which, is dominated by a genus or family

secondary formation one which results from the activity of humans

formicarian pertaining to ants

Formicariidae a very large family of passeriform birds containing those commonly called antbirds. A very variable group most having in common short rounded wings and a strong, slightly recurved, bill

Formicidae a very large family of apocritan hymenopteran insects commonly called ants. They are social insects with three or more castes in the colony with short-lived males, sterile females that lose their wings after a mating flight. Many ants can sting as well as bite. The family is divided into numerous subfamilies the more important of which are noticed in this dictionary (*cf.* aner, ergate)

Formicinae a large subfamily of formicid Hymenoptera containing many well known ants including the carpenter ants and the slave-makers

Formicoidea a monofamilial superfamily of apocritan Hymenoptera containing the Formicidae or ants. The characteristics of the superfamily are those of its only family

formion a group of federions or mores with one, or at the most a few, predominant species to one or more of which some or all of the other species appear to have a social connection

panformion a synusium of several formions each dominated by a distinct taxonomic formation

dental **formula** a method of expressing mammalian dentition as a series of fractions, with maxillary teeth as the numerator and mandible teeth as the denominator, representing successively the incisors, canines, premolar and molars

formylaspartate de**formylase** an enzyme catalyzing the hydrolysis of N-formyl-L-aspartate to formate and aspartate

formyltetrahydrofolate deformylase an enzyme catalyzing the hydrolysis of 10-formyl-tetrahydrofolate to formate and tetrahydrofolate

fornicate literally "arched," but used in *bot.*, for scale-like appendages on the corolla tube

-foss- *comb. form* meaning "ditch"

fossa 1 (*see also* fossa 2) literally a ditch but also used in *biol.* for a groove or trough

antafossa an oval or elongate antacava

cerebral fossa the large anterior portion of the brain case, separated from the cerebellar fossa by the tentorium

Hatschek's fossa an asymmetrical hollow ciliated tube in one of the languets of the wheel organ in Amphioxus just to the left of the notochord

fossa 2 (*see also* fossa 1) the Madagascan cryptoproctine carnivore *Cryptoprocta ferox* which is everlastingly becoming confused with the fanaloka (*q.v.*)

difossate said of those tapeworms which have saucer-like or trench-like bothria on the scolex

tetrafossate said of those tapeworms which have four suction cups on the scolex

fosse literally a ditch (*cf.* fossa). Specifically a groove separating the scapus and capitulum in some Anthozoa

infossous sunk in but leaving a visible channel, as the central nervous system in the neurula stage of an embryo

fossulate having a ditch or ditch-shaped depression on the surface

Fouquieriaceae that family of parietalous dicotyledons which contains the candle woods. The family may be distinguished from the closely allied Tamaricaceae by the hairy stamens and sturdier leaves

[four-jointed] a mutant (gene symbol *fj*) mapped at 81.0± on chromosome II of *Drosophila melanogaster*. The tarsae of the phenotype are 4-jointed

four-o-clock 1 (*see also* four-o-clock 2) popular name of flowering plants of the family Nyctaginaceae, particularly those of the genus Mirabilis, especially *M. jalapa*

four-o-clock 2 (*see also* four-o-clock 1) *Austral.* the meliphagid bird *Philemon corniculatus* (= friar bird)

fovea literally a small pit, and applied in *biol.* to numerous structures of this description

fovilla the contents of a pollen grain

fowl 1 (*see also* fowl 2, grass) *U.S.* synonymous with bird but particularly applying to the domestic. *Brit.* as *U.S.* but more commonly to distinguish wild birds from tame birds

   wildfowl *Brit.*, without qualification, refers to shore and marsh birds

   domestic fowl *Gallus domesticus* (chicken)

   gare-fowl the extinct alcid bird *Pinguinis impennis* (= great auk)

   guinea fowl (*see also* fowl 2) any of several species of the family Numididae

      black guineafowl *Phasidus niger*

      crested guineafowl any of several species of the genus. Usually, without qualification, *Guttera Edouardi*

      helmet guineafowl *Numida meleagris*

      plumed guineafowl *G. plumifera*

      white-ruffed guineafowl *Agelastes meleagrides*

      vulturine guineafowl *Acrylium vulturinum*

   jungle-fowl any of numerous phasianine galliform birds of the genus Gallus having the general appearance of a domestic fowl which is possibly descended from them

      Ceylon jungle fowl *G. lafayetii*

      green jungle-fowl *G. varius*

      grey jungle-fowl *G. sonneratii* distinguished by its "wax" spotted hackles and known to fishermen as the "Burmese" jungle-fowl

      Malayan jungle-fowl *G. bankiva*

      red jungle-fowl *G. gallus*

   mallee-fowl *Austral.* the megapodiid *Leipa ocellata*

   mayfowl = maybird

   moor fowl = moor cock

   oat fowl the fringillid *Plectrophenax nivalis* (= snow bunting)

   peafowl large phasianine galliform birds of the genus Pavo noted for the large eyed displaying tail of the male

      common peafowl *P. cristatus*

      green peafowl *P. muticus*

      scrub fowl popular name of megapodiids

   spur fowl any of several phasianid birds of the genera Galloperdix and Pternistis

   waterfowl 1 (*see also* waterfowl 2) *W.I.* the rallid *Gallinula chloropus* (= moorhen)

   waterfowl 2 (*see also* waterfowl 1) any bird living on or closely associated with water. In *U.S.* the term is usually confined to anseriform birds but in *Brit.* includes the charadriiform and gruiform birds

   wildfowl literally any wild bird. In *Brit.* practice often used to designate birds that are commonly hunted but are not thought of as game birds

fowl 2 (*see also* fowl 1) a lepidopteran insect

   guinea fowl (*see also* fowl 1) *S.Afr.* the nymphalid *Hamanumida daedalus*

fox 1 (*see also* fox 2, 3, grape, glove, moth, squirrel, tail) any of numerous species of small, canine carnivores mostly placed in the genus Vulpes but formerly included with Canis. *Brit.*, without qualification, *Vulpes vulpes* (= red fox)

   Arctic fox *Alopex eagopus*

   silver-backed fox *Vulpes chama*

   fennec fox = fennec (*q.v.*)

   gray fox *U.S. Urocyon cinereoargenteus*

   king fox *U.S. Vulpes velox*

   kit fox *U.S. Vulpes macrotis*

   red fox (*see also* fox 3) *Vulpes vulpes*

   swift fox *U.S. Vulpes velox*

   Virginian fox = red fox

fox 2 (*see also* fox 1, 3) some other fox-like mammals

   bat-eared fox the otocyonine canid carnivore *Otocyon megalotis*

   flying fox the pteropid chiropteran mammal *Pteropus poliocephalus* (*cf.* fox bat)

fox 3 (*see also* fox 1, 2) in compound names of various organisms

   night fox *Brit.* the lasiocampid lepidopteran insect *Bombyx rubi*

   red fox (*see also* fox 1) the myrtaceous tree *Eucalyptus polyanthemos*

interfoyles things that are not quite leaves (*i.e.* bracts, scales, and stipules)

fr see [fray], [fringed], [frizzy] and [frost]

fracidous with the texture of a ripe pear

infracted incurved

phyllotactic fraction one in which the numerator represents the number of times the stem must be encircled before reaching a leaf vertically under the one from which the count starts; while the denominator indicates the number of leaves which must be passed in so doing

-fract- *comb. form* meaning "to break"

anfractuose sinuous, but not of necessity spiral

-frag- *comb. form* meaning "to break"

nucifragal nutcracking

-fragar- *comb. form* meaning "strawberry"

francolin any of numerous phasianine birds of the genus Francolinus and, less frequently, of the *Afr.* genus Pternistis

   natal francolin *F. natalensis*

   ring-necked francolin *F. streptophorus*

   scaly francolin *F. squamatus*

frangipani any of several herbs of the apocynaceous genus Plumeria but particularly *P. rubra* and *P. acutifolia*

Frankeniaceae a small family of parietale dicotyledonous angiosperms distinguished as much for their preference for saline or arid habitat as by the heath like foliage

[fray] a mutant (gene symbol *fr*) in linkage group III of the domestic fowl. The name refers to the frayed appearance of the defective wing and tail feathers

freak *I.P.* the nymphalid lepidopteran insect *Calinaga buddha*

freemartin an intersex in cattle induced in a female twin by the sex hormones of a male twin

Fregatidae that family of pelecaniform birds commonly called frigatebirds. They are distinguished from other members of the order by the long, deeply forked, tail and strong, hooked beak

frenal = loreal

Frenatae a suborder of lepidopteran insects containing the great majority of Lepidoptera in which the hindwing is smaller than the frontwing and the front and hindwings on each side are united by a frenulum when present or by the expanded humeral angle of the hind wing itself

frenulum 1 (*see also* frenulum 2) diminutive of frenum (*q.v.*) Specifically bristles arising from the base of the hindwing in certain insects, especially Lepidoptera, used to lock the wings together in flight

frenulum 2 (*see also* frenulum 1) one of the gelatinous folds supporting a velarium

frenum literally, reins, or a strap, particularly any fold of tissue supporting an organ, such as that under the tongue, or a ridge on insects extending from the scutellum to a base of the anterior wing

friar (*see also* bird) *S.Afr.* the danaid lepidopteran insect *Amauris dominicanus*

widow's **frill** the caryophyllaceous herb *Silene stellata*

extrabrillar **fringe** one round the eye of some saurian reptiles

> **mountain fringe** the fumariaceous vine *Adlumia fungosa*

[**fringed**] a mutant (gene symbol *fr*) mapped at 80.0± on chromosome II of *Drosophila melanogaster*. The wing margins of the phenotype are ragged

Fringillidae a large family of passeriform birds containing those commonly called finches, grosbeaks, buntings and sparrows. They are mostly small birds with short and pointed bills usually thick and very rarely hooked. Many are brightly colored

fritillary 1 (*see also* fritillary 2) any of numerous liliaceous herbs of the genus Fritillaria

fritillary 2 (*see also* fritillary 1) any of numerous species of lepidopteran insects, mostly nymphalids of the genera Argynnis and Melitaea

> **Australian fritillary** *Argynnis hyperbius*
>
> **pearl-bordered fritillary** *Brit. Argynnis euphrosyne*
>
> **high brown fritillary** *Brit. Argynnis adippe*
>
> **Duke of Burgundy fritillary** *Brit.* the riodinid *Nemeobius lucina*
>
> **Glanville fritillary** *Brit. Melitaea cinxia*
>
> **greasy fritillary** *Brit. Melitaea aurinia*
>
> **greater fritillary** *U.S.* any of several species of the genus Speyeria (= silverspot)
>
> **dark green fritillary** *Brit. Argynnis aglaia*
>
> **gulf fritillary** *U.S.* the nymphalid *Agraulis vanillae*
>
> **heath fritillary** *Brit. Melitaea athalia*
>
> **lesser fritillary** *U.S.* any of numerous species of nymphalids of the genus Boloria
>
> **silver washed fritillary** *Brit. Argynnis paphia*

[**frizzled**] a mutant (gene symbol *fz*) mapped at 41.7± on chromosome III of *Drosophila melanogaster*. The bristles and thoracic hairs of the phenotype point towards the median line

[**Frizzling**] a mutant (gene symbol *F*) in linkage group III of the domestic fowl. The name refers to the frizzled appearance of the feathers

[**frizzy**] a mutant (gene symbol *fr*) in linkage group I of the mouse. The phenotypic expression is thin hair and curled whiskers

frog 1 (*see also* frog 2, fish, hopper, mouth, shell) the distinction between frog and toad can only be maintained in countries such as Britain where ranids (frogs) and bufonids (toads) are the only anuran amphibians. In other parts of the world there is a general tendency to regard predominantly aquatic and marshland Anura as frogs and primarily terrestrial forms as toads but there is no general agreement as to such terms as "tree frog" and "tree toad." *Brit.* frog, without qualification, is the ranid *Rana temporaria*

> **arum frog** *S.Afr.* the rhacophorid *Hyperolius horstocki*
>
> **banana frog** *Afr.* any of several species of rhacophorid of the genus Megalixalus
>
> **barking frog** *U.S.* the leptodactyid *Eleutherodactylus augusti*
>
> **mouth-breeding frog** the rhinodermatid *Rhinoderma darwini* (= vaquero)
>
> **bronze frog** *U.S.* = green frog *U.S.*
>
> **brown frog** *Brit.* = grass frog *Brit.*
>
> **bullfrog** *U.S.* the ranid *Rana catesbeiana*. The term is also applied in South America to the giant leptodactylid *Leptodactylus pentadactylus*
>
> **carpenter frog** *U.S.* the ranid *Rana virgatipes*
>
> **Catholic frog** *Austral.* the leptodactylid *Notaden bennetti*

chirping frog *S.Afr.* the ranid *Arthroleptella lightfooti*

chorus frog *U.S.* any of several species of hylids of the genus Pseudacris

clawed frog the pipid *Xenopus laevis*

cliff frog *U.S.* the leptodactylid *Syrrhophus marnocki*

crawfish frog *U.S.* = gopher frog *U.S.*

cricket frog *U.S.* general name for hylids of the genus Acris

flying frog a misnomer frequently applied to the ranid *Rhacophorus pardalis*

ghost frog *S.Afr.* the heleophrynie leptodactylid *Heleophryne rosei*

Goliath frog the African ranid *Rana goliath*

gopher frog *U.S.* the ranid *Rana areolata*

grassfrog *Brit.* the ranid *Rana temporaria*

> **little grassfrog** *U.S.* the hylid *Hyla ocularis*

green frog *U.S.* the ranid *Rana clamitans*

hairy frog the ranid *Astylosternus robustus*

bony-headed frog hylids of the genera Aparaspheno-don or Trachycephalus

> **flat-headed frog** *Austral.* the leptodactylid *Chiroleptes platycephalus*

horned frog any of several species of leptodactylid of the genus Ceratophrys

greenhouse frog *U.S.* the leptodactylid *Eleutherodactylus ricordi*

leopard frog *U.S.* the ranid *Rana pipiens*

white-lipped frog *U.S.* the leptodactylid *Leptodactylus labialis*

marsh frog the ranid *Rana palustris*

marsupial frog any of several hylids of the genus Gastrotheca

mink frog *U.S.* the ranid *Rana septentrionalis*

New Zealand frog the liopelmid *Liopelma hochstetteri*

rice-paddy frog any of several species of ranid of the genus Ooeidozyga

pickeral frog *U.S.* the ranid *Rana palustris*

pig frog *U.S.* the ranid *Rana grylio*

arrow-poison frog popular name of dendrobatine ranids

pouched frog = marsupial frog

rain frog *S.Afr.* the microhylid *Breviceps gibbosus*

river frog *U.S.* the ranid *Rana heckscheri*

mush frog = sedge frog

sedge frog any of several species of rhacophorid of the genus Hyperolius

sheep frog any of several microhylids of the genus Hypopachus, particularly *H. cuneus*

smith frog the hylid *Hyla faber*

speckled frog *S.Afr.* the ranid *Rana adspersa*

tailed frog *U.S.* the liopelmid *Ascaphus truei*

treefrog *U.S.* general name for hylids of the genus Hyla

> **barking treefrog** *H. gratiosa*
>
> **Cuban treefrog** *H. septentrionalis*
>
> **golden treefrog** *Austral. H. aurea*
>
> **gray treefrog** *S.Afr.* the rhacophorid *Chiromantis xerampolina*
>
> **green treefrog** *U.S. H. cinerea. Brit. H. arborea.* The term is also applied to *H. annectens*
>
> **Mediterranean treefrog** *H. arborea meridionalis*
>
> **White's treefrog** *Austral. H. caerulea*

water-frog *Brit.* the ranid *Rana esculenta*

woodfrog *U.S.* the ranid *Rana sylvatica*

frog 2 (*see also* frog 1) an annular structure set on a surface and thus sometimes applied to the ring on the body of an arthropod with which a limb articulates

-**frond**- *comb. form* meaning "leaf" but usually used in the sense of fern frond

frond (*see also* acrorrhagus) the aerial branch of a pteridophyte plant

frondescence any condition applying to fronds as distinct from leaves

frondome the abstract entity of a frond

frons literally forehead (*see* frontal and bone) but also used for the anterior region of an insect head

   bombifrons having an extruded blister pouch in the region of the frons particularly in insects

front in compound names of organisms

   bristlefront any of several rhinocryptid birds of the genus Merulaxis

   jewelfront the apodiform trochilid bird *Polyplancta aurescens*

frontal 1 (*see also* frontal 2–4, bone gland) pertaining to the forehead

frontal 2 (*see also* frontal 1, 3, 4) pertaining to the anterior end

frontal 3 (*see also* frontal 1, 2, 4) descriptive of a sectioned plane parallel to the long axis of the body but at right angles to the sagittal plane

frontal 4 (*see also* frontal 1–3) a subterminal orifice on an ectoproct zoecium

star **frontlet** any of several apodiform birds of the genus Coeligena

[**frost**] a mutant (gene symbol *fr*) found in linkage group I of *Neurospora crassa*. The term describes the appearance of the phenotypic colony

-**fruct**- *comb. form* meaning "fruit" (*cf.* -frug-)

fructescence the condition or period of mature fruit

   infructescence an inflorescence in the fruiting stage

fructification the production of fruit

   double fructification the production of dimorphic reproducing bodies as in certain algae

-**frug**- *comb. form* meaning "fruit" (*cf.* -fruct-)

frugiferous = fructiferous

fruit 1 (*see also* fruit 2, crow, eater, fly, mite, wall, worm) the reproductive body of a seed plant in general

   aggregate fruit a fruit derived from many free carpels produced by one flower

   blastocarpelous fruit one which germinates within the pericarp

   choricarpelous fruit one derived from the flower having free carpels (= aggregate fruit)

   syncarpelous fruit one derived from the flower having the carpels united (= unit fruit)

   epichlamytous fruit one derived from a hypogynous flower (= free fruit)

   hypochlamytous fruit one derived from the perigynous or epigynous flower (= cupped fruit)

   false fruit = pseudocarp

   multiple fruit one formed from several flowers but appearing to grow as a single fruit. The pineapple is an example

   simple fruit one which is derived from a single pistil

   split fruit = cremocarp

fruit 2 (*see also* fruit 1) in compound names of specific, usually edible, fruits

   bread fruit the moraceous tree *Artocarpus incisa* and its fruit

   egg fruit the sapotaceous tree *Lucuma nervosa*

   fog fruit the verbeniaceous herb *Lippia lanceolata*

   grape fruit the rutaceous tree *Citrus paradisi* and its fruit

   jack fruit the moraceous tree *Artocarpus integrifolia*

   Jove's fruit the lauraceous shrub *Lindera melissaefolium*

   marmelade fruit the sapotaceous tree *Lucuma mammosa* (= sapote)

   monkey fruit the annonaceous shrub *Rolliniopsis discreta* and its edible fruit

   passion fruit the edible fruit of several species of passifloraceous vine of the genus Passiflora but particularly that of *P. edulis* and *P. quadrangularis* (= granadilla)

   star fruit *Brit.* the alismaceous aquatic *Actinocarpus damasonium*

fruitlet a unit in an aggregate fruit formed from a single carpel

frumentaceous pertaining to grain

frustule the two silicious plates that enclose a diatom

   intercalary frustule a diatom frustule the bands on which have longitudinal divisions

   discigerous frustule the circular frustule of a diatom

suffrutescent more or less shrubby

frutex = shrub

fruticose 1 (*see also* fruticose 2) shrubby

fruticose 2 (*see also* fructicose 1) bearing fruiting bodies, said particularly of lichens having colored prominences

fry a young fish, particularly those in a stage too immature to permit ready identification

FSH = follicle-stimulating hormone

ft *see* [fat]

fu *see* [fused]

Fu *see* [Fused]

fucaceous literally, resembling brown algae of the genus Fucus, but frequently used in the sense of resembling algae in general

fuchsia any of several species of onagraceous shrubby plants, particularly those of the genus Fuchsia, and some plants of similar aspect

   California fuchsia the onagraceous herb *Zauschneria californica*

   cape fuchsia the scrophulariaceous shrub *Phygelius capensis*

   tree fuchsia *N.Z. Fuchsia excorticata* (= kotukutuku)

fugacious transitory

-**fugal** pertaining to a movement away from. The adjectival termination -fugous is also frequent

   acrofugal = basipetal

   basifugal developing from the base up

   calcifugal avoiding chalky soil, though the term is extended to mean avoiding alkaline soils in general

   frigofugal said of organisms that avoid cold conditions

   lucifugal fleeing from light

   nidifugal nest-fleeing; commonly used as the antithesis of nidicolous (*q.v.*)

-**fugous** *see* -fugal

fulcra spines heavier then fin rays, supporting the anterior edges of the fins of primitive actinopterygian fishes

fulcrate having the form of, possessing, or acting as, a fulcrum. Used particularly of a kind of rotiferan mastax, in which the fulcrum is unusually large

   efulcrate lacking a bract

fulcrum literally, a prop, and used for numerous plant and animal structures of this function

Fulgoridae a family of homopteran insects commonly called planthoppers. They are distinguished from the Cicadellidae (leafhoppers) and Cercopidae (froghoppers) by the placement of the antennae

fuligineous sooty

fulmar any of several procellariid birds of the genera

Macronectes, Fulmarus and Priocella. Usually, without qualification, *F. glacialis*

Fulmaridae = Procellariidae

fulmineous light yellow-brown (*cf.* fulvous)

fulvous dark yellow-brown (*cf.* fulmineous)

Fumariaceae that family of rheodalous dicotyledons which contains the, *inter alia,* bleeding hearts, the dutchman's breeches, and the fumitories. They are easily distinguished from their near relatives, the Papaveraceae by the six anthers on the two filaments and the absence of milky juice

fumeous smoky or smoke-colored

fumitory any of several species of fumariaceous herb of the genus Fumaria particularly *F. officinalis*

   climbing fumitory the fumariaceous vine *Adlumia fungosa*

-fund- *comb. form* meaning "depth"

   profundal the deep basin of a lake in which physical conditions, except at overflow, are at a constant level

fundatrix the founder of a line of descent, and specifically the aphis which emerges in spring from a resting egg as an apterous, viviparous, parthenogenetic female

fungaceous = fungiform

fungal pertaining to fungi

Fungi *see* Eumycophyta

Fungi imperfectae = Deuteromycetae

Fungivoridae a family of nematoceran dipteran insects now usually distributed between the Mycetophilidae and the Sciaridae

fungose spongy

fungus (*see also* beetle, gnat, weevil) any thallophyte plant that lacks chlorophyll. In popular usage the term is generally applied to large, fleshy Basidomycetes and a few Ascomycetes, also variously called mushrooms and toadstools. The lower fungi are commonly called molds, and mildews.

   artist's fungus *U.S.* the polyporaceous basidomycete *Canoderma applanatum*

   beefsteak fungus *U.S.* the polyporaceous basidomycete *Fistulina hepatica*

   jelly fungus a term commonly applied to the Tremellalea

   jack-o-lantern fungus the tricholomataceous basidomycete *Clitocybe illudens*

   bird's nest fungus the lycoperdaceous basidomycete *Crucibulum levis* (*cf.* puffball)

   saddle fungus any of several species of the ascomycete genus Helvella

   shelf fungus popular name of polyporacean Basidomycetes

   tinder fungus the polyporaceous basidomycete *Polyporus fomentarius* (*cf.* amadou)

-funic- *comb. form* meaning "rope"

funicle (*see also* funiculus) literally, a rope. In *zool.* specifically applied to that part of a clavate antenna which lies between the club and the articulation with the head, and to the strand that anchors an ectoproct in the zooecium. In *bot.,* a strand of fungal hyphal tissue, or a chord connecting the ovule to the placenta

funiculus (*see also* funicle) any cord, rope, or stalk-like body particularly those composed of a bundle of fibers or vessels. Specifically, the stalk by which a developing seed is attached to the ovarian wall; a band of tissue attaching the stomach of an ectoproct to the body wall; a ligament which, in certain vespoid Hymenoptera, connects the petiole to the

propodium; the slender unswollen part of a clavate insect antenna

funnel a structure in the form of an open cone attached to a tube. Specifically, the lower expanded chamber of a pneumatophore

   buccal funnel in insects that part of the foregut which lies in the head

   coelomic funnel a ciliated funnel, draining fluid from the coelom (= nephrostome)

   oral funnel the cavity anterior to the velum in Amphioxus

   peritoneal funnel = coelomic funnel.

   subumbrellar funnel one of the interradial pits of scyphozoan coelenterates (= peristomial pits)

fur the hair of nonhuman mammals or, by analogy, any fuzzy coating of an organism

fructofuranosidase an enzyme catalyzing the hydrolysis of fructofuranoside to alcohol and fructose

-furca- *comb. form* meaning "fork"

furca a fork-shaped apodeme arising from the sternum in many insects

   antefurca an internal fork coming forwards from the prosternal process in many insects

   mesofurca the furca of the insect mesothorax

furcate forked

furcula a fork or forked structure, particularly such structures in chordates

furfuraceous scurfy

[furless] a mutant (gene symbol *f*) at locus 28.3 on linkage group V of the rabbit. The name is descriptive of the phenotypic expression though some fur is found on the ends of the limbs

Furnariidae a large family of passeriform birds commonly called ovenbirds which derive their name from the habit of many species of building a domed nest of mud

furrow a shallow groove or ditch

   Mayrian furrow a "Y" shaped groove on the top of the thorax of an ant

[furrowed] a mutant (gene symbol *fw*) mapped at 38.3 on the X chromosome of *Drosophila melanogaster.* The term refers to the condition of the phenotypic eye

furvous swarthy

furze (*see also* chat) the leguminous shrub *Ulex europaeus*

-fus- *comb. form* meaning "spindle" in the sense of the cigar-shaped spindle used in spinning and weaving

[fused] *either* a mutant (gene symbol *fu*) mapped at 59.5 on the X chromosome of *Drosophila melanogaster.* The fusion refers to the wing veins of the phenotype which also lack ocelli *or* a mutant (gene symbol *fu*) in linkage group I of *Habrobracon juglandis.* The phenotypic expressions are fused segments in the tarsae and antennae

[Fused] a mutant (gene symbol *Fu*) in linkage group IX of the mouse. The name refers to the condition of some caudal and other vertebrae

fursula the knob on the end of a spider's spinneret

[fuzzy] *either* a mutant (gene symbol *fy*) mapped at 33.0± on chromosome II of *Drosophila melanogaster.* The term applies to the fuzzy thoracic hair of the phenotype *or* a mutant (gene symbol *fz*) in linkage group XIII of the mouse. The phenotypic expression is sparse crinkled hair and crinkled whiskers

fw *see* [furrowed]

fy *see* [fuzzy]

fz see [frizzled] or [fuzzy]

G *see* [Gull]

g *see* [garnet]

ga a yeast genetic marker indicating a requirement for galactose

Ga$_1$, ga$_6$, ga$_7$ *see* [gametophyte factor]

Gadidae a family of anacanthinus fish, typical of the order, containing the codfish and their allies

Gadiformes = Anacanthini

gadwall the anatine bird *Anas strepera*

gady *W.I.* the larid bird *Thalasseus maximus* (= royal tern)

-gae- *comb. form* meaning "earth" and therefore identical with -ge- and -geo-, but preferred in compounds denoting geographical zones. The *adj.* suffix -gaen is simularly used in contrast to -gean

  amphigaean pertaining to both the Old World and New World

  arctogaea one of the primary zoogeographic zones comprising most of the northern hemisphere and those parts of Asia that lie to the north and west of Wallace's Line (*q.v. under* line 2)

  dendrogaea the neotropical region

  gerontogaea the "old world" as applied to the distribution of plants (*cf.* neogaea)

  neogaea the area comprising Central and Southern America, though sometimes used to include the whole "new world", particularly in regard to the distribution of plants

  notogaea the area that combines the neotropical and Australasian regions

  ornithogaea Polynesian region including New Zealand

  palaeogaea the region of relict faunas including Arctic Europe, parts of North Africa, India and the Australian region

gafftopsail the ariid catfish *Bagre marinus*

gag *U.S.* the serranid fish *Mycteroperca microlepis* (*cf.* grouper)

gaggle term for a group of geese on the water (*cf.* skein)

gait the type of motion demonstrated by an animal

gal a bacterial mutant showing a utilization of galactose

galA a bacterial genetic marker having a character affecting the activity of galactokinase, mapped at 16.0 mins. for *Escherichia coli*

galB a bacterial genetic marker having a character affecting the activity of galactose-1-phosphate uridyl transferase, mapped at 16.0 mins. for *Escherichia coli*

galC a bacterial genetic marker having a character affecting the activity of operator mutants, mapped at 16.0 mins. for *Escherichia coli*

galD a bacterial genetic marker having a character affecting the activity of uridinediphosphogalactose 4-epimerase, mapped at 16.0 mins. for *Escherichia coli*

galactin = lactogenic hormone

α -and β-galactosidases enzymes catalyzing respectively the hydrolysis of α and β -D-galactosides into alcohols and D-galactose

polygalacturonase an enzyme catalyzing the breakdown of pectate through the hydrolysis of α-1,4-D-galacturonide links

Galagidae a family of lorisoid primates readily, but superficially, distinguished from the Lorisidae by the large ears and long bushy tails. They are commonly called bush-babies

galago a term rather generally applied to African lemurs though properly and specifically to members of the genus Galago

galah *Austral.* the psittacid bird *Kakatoe roseicapilla* (*cf.* corella)

Galatheidea a superfamily of anomuran Crustacea distinguished by the depressed body, the flexed abdomen, the chelate first legs and reduced fifth legs

galax the diapensiaceous herb *Schizocodon soldanelloides*

Galaxiidae a family of isospondylous fishes of the southern hemisphere. They are distinguished by the single dorsal and single anal fins set far back on the body

Galbulidae a small family of piciform birds commonly called jacamars. Distinguished from all other piciform birds by the long, attenuated bill

-gale for birds terminating in this suffix of doubtful etymology *see under* initial letter

  galingale 1 (*see also* galingale 2) any of several zingiberaceous plants yielding a pungent rhizome, particularly *Kaempferia galanga*

  Chinese galingale *Alpinia officinalis*

  galingale 2 (*see also* galingale 1) *U.S.* any sedge of the genus Cyperus particularly *C. longus*

  sweet gale popular name for plants of the family Myrtaceae

galea literally a helmet, but most usually applied to the outer of the two lobes that terminate an insect maxilla (*cf.* lacinia)

galericulate bearing a cap

gall 1 = bile

gall 2 (*see also* gall 1, 3, berry) an abnormal growth induced in a plant. Those induced on oaks by cynipid hymenoptera were at one time the principal commercial source of tannin

  Aleppo gall an oak gall, peculiar to *Quercus infec-*

toria, produced by the cynipid hymenopteran *Cynips tincotoria*

blue gall = green gall

crown gall one induced by the bacterium *Agrobacterium tumifaciens*

green gall an immature oak gall

oak gall (*see also* gall 3) any cynipid hymenopteran-induced gall on an oak

gall 3 (*see also* gall 1, 2) an insect causing a plant gall

    elm cockscomb gall *U.S.* the aphid homopteran *Colopha ulmicola*

    hackberry nipple gall *U.S.* the psyllid homopteran *Pachypsylla celtidismamma*

    oak gall (*see also* gall 1) any of numerous cynipid hymenopterans

        bullet oak gall *Disholcaspis globulus*

        giant oak gall *Andricus californicus*

    vagabond poplar gall the aphid homopteran *Mordwilkoja vagabunda*

    spiny witch hazel gall the aphid homopteran *Hamamelistes spinosus*

    witch hazel cone gall the aphid homopteran *Hormaphis hamamelidis*

Galleriidae a small family of lepidopteran insects, frequently regarded as a subfamily of the Pyralidae, erected principally to contain the wax moth (*q.v.*)

Galliformes that order of birds which contains the fowls both wild and domestic. They are distinguished by the large feet always bearing a hallux and the short rounded concave wings

gallinule any of several rallid birds of the general Gallinula, Porphyrula and Porphyrio

    purple gallinule *U.S. Porphyrula martinica—Old World* any of several species of the genus Porphyrio (= swamp hen)

gallito the rhinocryptid bird *Rhinocrypta lanceolata*

gall-of-the-earth *U.S. either* the gentianaceous herb *Gentiana quinque-folia or* the compositaceous herb *Prenanthes trifoliata*

gallop a method of rapid quadruped locomotion. In most quadrupeds (*e.g.* rodents, carnivores) the motion consists of rapidly repeated springs or jumps in which the hind feet are pushed far forwards between the front feet and all four feet are thus off the ground for considerable periods. In horses, first one hind foot then the diagonally opposite forefoot and then the other hindfoot almost synchronously touch the ground in propelling the animal, all four legs of which are off the ground only for the briefest period

galo- *comb. form* meaning "milk"

Galton's *epon. adj.* from Francis Galton (1822–1911)

gam term for a group of whales or porpoises, though school is more frequently used and pod is better usage

-gam- *comb. form* meaning "marriage" and thus, by extension, anything connected with sexual reproduction

-gam termination used in classifying plants according to their method of reproduction or to distinguish parts of plant reproductive systems

    aerogam (obs.) = phanerogam

    asiphonogam any plant fertilized by antherozoids

    carpogam the female organ of a procarp

    cryptogam a plant which reproduces by means other than flowers (*cf.* phanerogam)

    monogam a plant with the anthers united in a simple flower

    phanerogam a flowering plant that reproduces from seeds (*cf.* cryptogam)

    phenogam = phanerogam

    thallogam an obsolete term for vascular cryptogam

    zoidogam a plant in which pollination is effected through the agency of animals

    zoogam a plant having motile reproductive cells and hence sometimes used as synonymous with alga

-gamae *comb. suffix* synonymous with -gam (*q.v.*)

gambet *U.S.* any of several scolopacid birds *Brit.* = ruff

gameon (*cf.* -gam) an organism classified by its reproductive habits

    agameon a species reproducing asexually (*cf.* agamospecies)

    apogameon an organism that is capable both of sexual reproduction and parthenogenesis

hypergamesis the nourishment of the female, or female gamete, by excess spermatozoa

-gamet- *comb. form* meaning "a spouse"

gamete one of two cells that fuse and develop into one or more organisms. Where the sexes can be distinguished, the male gamete is a sperm and the female an egg

    agamete any product of reproductive multiple fission that is not a gamete

    apogamete one which is formed by apomixis (*q.v.*)

    coenogamete a syncytium that will subsequently fragment into gametes

    diplogamete a double gamete produced in one cell

    gynogamete = egg

    heterogamete 1 (*see also* heterogamete 2) one of two gametes that can be distinguished from each other

    heterogamete 2 (*see also* heterogamete 1) that gamete produced by the sex with the potential of producing either sex

    homogamete that produced by the sex with only one sex producing potential

    isogamete one of a group of gametes that cannot be distinguished from each other

    anisogamete = heterogamete 1

    karyogamete the nucleus of a gamete

    prokaryogamete the nucleus of a progamete

    macrogamete = megagamete

    megagamete the larger of a pair of heterogametes and, by convention, usually regarded as female

    merogamete a gamete, which is smaller than the organism, usually a protozoan, from which it is derived by fission

    microgamete the smaller, and by convention male, of a pair of heterogametes

    obligate gamete a gamete that cannot develop parthenogenetically

    parthenogamete a gamete capable of parthenogenesis

    planogamete a motile gamete

    aplanogamete a non-motile gamete

    isoplanogamete a motile gamete the sex of which cannot be distinguished particularly in phycomycete fungi

    anisoplanogamete a motile gamete which differs in size from another planogamete produced by the same organism usually in a phycomycete fungus

    progamete that cell which gives rise directly to gametes either by a single division or by metamorphosis of itself

    spermatogamete male gamete, particularly of a plant

    syngamete = zygote

    zoogamete = planogamete

-gametism suffix indicating the condition of having special kinds of gametes (*q.v.*). For example, heterogametism is the condition of an organism that produces heterogametes

-gamety suffix indicating the condition of producing special gametes (*q.v.*) or producing gametes by special means. For example, apogamety (*see* apogamete) is the formation of gametes by apomixis

　prokaryogamety = meiosis

[gametophyte factor] *either* a mutant (gene symbol *ga₆*) mapped at 15 on linkage group I of *Zea mays*. The gene controls the viability of the gametophyte *or* a mutant (gene symbol *ga₇*) mapped at 121 on linkage group III of *Zea mays*. The gene controls the viability of the gametophyte

[Gametophyte factor] a mutant (gene symbol *Ga₁*) mapped at 35 on linkage group IV of *Zea mays*. The gene controls gametophyte viability

-gamia a termination synonymous with -gamy (*q.v.*) the form preferred in this work. Plant or animal taxa with this termination are listed alphabetically

-gamic (*see also* color) an *adj. suffix* derived from the *subs. suffix* -gamy, under which definitions are given in this work. The alternative form -gamous (*q.v.*) is usually interchangeable though in a few cases (*e.g.* cryptogamic botany) the -ic form appears fixed

Gamicae = Algae

polygamist U.S. the noctuid lepidopteran insect *Catocala polygama* (*cf.* underwing)

-gamous an *adj. term* derived from the substantive form -gamy under which definitions are given in this work. The alternative form -gamic (*q.v.*) is usually interchangeable though in a few cases (*e.g.* polygamous humans) the -ous form appears fixed

perigamium the branchlet of a moss which contains the archegonium

gamma *I.P.* the colubrid snake *Boiga trigonata*

Gammaridae a family of gammaridean amphipod Crustacea common in marine habitats and in all the freshwaters of the world where they form a major item in the diet of fish. It is a very large and varied family and more than half of the species are confined to Lake Baikal. The larger forms are popularly called scud

Gammaridea a suborder of amphipod crustacea distinguished from the Caprellidea by the presence of a fully developed abdomen and usually the presence of seven free thoracic segments and from the Hyperiidea by the usual presence of coxal plates and the legs, by the small head and eyes and the absence of palps on the maxillipedes. The beach hoppers and the fresh water gammarids are the best known

-gamo- *comb. form* meaning "marriage"

gamobium the sexual of two alternate generations and hence sometimes used as synonymous with gametophyte

　gamont that form of an organism which produces gametes

　　agamont that form of an organism that does not produce gametes. Occasionally used as synonymous with schizont

Gamosporae a botanical taxon, at one time including all those algae which produced zygospores

-gamy *comb. suffix* indicating the type or method of reproduction

　adelphogamy mating of siblings, though also used in *bot.* for fertilization between neighboring plants

　adichogamy the condition of a plant in which the male and female organs mature synchronously

homodichogamy = autoallogamy

allogamy the process of cross fertilization

　autoallogamy the condition in which some individuals of a species are adapted to cross-fertilization and others to self-fertilization (*cf.* allautogamy)

　karyallogamy the process or reproduction in which two morphologically identical gametes fuse

amphigamy the condition of lacking sexual organs

anemogamy the condition of a plant fertilized through the agency of wind

apolegamy selective breeding

androgamy the impregnation of a male gamete by a female gamete

　cytoplasmic androgamy the condition where the male gamete is fertilized by the cytoplasm of the female gamete

apogamy the direct production of a plant by budding from the gametophyte without the customary sexual generation or the condition of being asexual in general

　diploid apogamy = euapogamy

　euapogamy apogamy in plants in which the gametophyte is diploid

　generative apogamy = meiotic apogamy

　meiotic euapogamy the condition in which the mother cells of a sporophyte plant have the haploid chromosome number

　generative apogamy the production by a haploid gametophyte of a sporophyte plant by asexual means (*cf.* generative parthenogenesis)

　haploid apogamy = meiotic apogamy

　meiotic apogamy the condition in which a sporophyte plant is developed from the oosphere

　obligate apogamy = parthenapogamy

　ooapogamy parthenogenetic apogamy

　parthenapogamy the fusion of the nuclei of diploid cells

　pseudapogamy the conjugation of an ovum with a somatic nucleus or, more rarely, the conjugation of two somatic nuclei

　somatic apogamy the reproduction of a haploid sporophyte plant

autogamy 1 (*see also* autogamy 2, 3) the fusion of two reproductive nuclei within a single cell derived from a single parent and, particularly in ciliate Protozoa, the division of the micronucleus to eight or more, of which two fuse. From these fused nuclei a new macronucleus arises

autogamy 2 (*see also* autogamy 1, 3) a type of reproduction, apparently confined to Heliozoa, in which a single cell encysts, undergoes a reduction division, and the two autogametes then fuse again

autogamy 3 (*see also* autogamy 1, 2) the condition of a flower fertilized by its own pollen

　allautogamy the state of having an alternative to the normal method of pollination (*cf.* autallogamy)

　paedogamous autogamy the condition in which the nucleus, but not the cytoplasm is concerned in the formation of the zygote

　double autogamy = cytogamy

　xenautogamy the condition of an organism in which cross-fertilization is normal but self-fertilization possible

axogamy the condition of a plant having sex organs on a leafy stem

basigamy the condition of a plant in which the oosphere and synergids are at the lower end of the mother-cell of the endosperm

chasmogamy the condition of a flower in which the perianth expands

dichogamy the condition of an hermaphrodite which cannot fertilize itself, because the two sexes mautre at different times

chasmodichogamy the condition of having cleisto-gamic and chasmogamic flowers in the same inflorescence (cf. cleisto-chasmogamy)

ecodichogamy the condition of dioecious organisms with different maturation times

diecodichogamy the condition of groups of plants in some of which the male flowers mature first and in others of which the females mature first

homodichogamy the condition of a species in which homogamous and dichogamous individuals occur

chromidiogamy the fusion of chromidia (q.v.)

synchronogamy the condition of having the male and female sex organs mature at the same time

cleistogamy the condition of having fertilization take place within closed flowers

archicleistogamy the condition of having diminished sex organs in permanently closed flowers (cf. archocleistogamy)

archocleistogamy the condition of having closed flowers containing ripe sexual organs (cf. archicleistogamy)

chasmocleistogamy the condition of a sporophyte in which some flowers open and others do not

hemicleistogamy the condition of having flowers that are only partly open

hydrocleistogamy the condition of a closed flower which is pollinated through immersion or that of a flower which does not open by reason of the fact that it remains under water

ombrocleistogamy the condition of a closed flower that fertilizes itself through the agency of rain

photocleistogamy the condition of a sporophyte in which flowers remain closed because of insufficient illumination or that of flowers in which cleistogamy is induced in consequence of photo-hyponastic growth of one side

psychrocleistogamy cleistogamy produced by unusually cold weather

thermocleistogamy the fertilization of flowers which have failed to expand through lack of warmth

xerocleistogamy the condition of flowers remaining closed by virtue of drought

cytogamy the condition which results when both male nuclei remain in one of two conjugating animals and therefore both fuse with the female nucleus of the same animal; or the fusion of two cells into a haploid zygote

digamy with both sexes closely together

ditopogamy = heterogamy

endogamy the condition of an individual that fertilizes itself

entomogamy the condition of flowers that are fertilized by insects

epigamy the type of behavior, such as the mating dance, which precedes or hastens copulation

exogamy reproduction between groups or organisms not usually interbreeding or the state of a cross produced from different plants

geitonogamy allogamy from the same plant but a separate flower

gnesiogamy fertilization between two individuals of the same species

gymnogamy fertilization

cytoplasmic gymnogamy when the female gamete is allegedly fertilized by the cytoplasm of the male gamete

nuclear gymnogamy normal fertilization by nuclear fission

hercogamy the condition of a perfect flower which is unable to fertilize itself without the assistance of an insect visitor

heterogamy the condition in which there is a reversal of function between male and female flowers, or that of a plant having two kinds of flowers, or simply a synonym for anisogamy

hologamy reproduction through the fusion of the nuclei of gametes or of two mature Protozoa

anisohologamy the union of gametes which are not quite identical in form

isohologamy = isogamy (only, apparently, more so, though how this is possible is not clear)

homogamy possessing only one kind of gamete on one kind of flower or the condition of a perfect flower in which pollen and stigmax mature at the same time

monoecious homogamy fertilization from another flower on the same plant

homoeogamy the condition in which an antipodal cell and not the oospore is fertilized

homiogamy the fusion of two gametes of the same sex

hylogamy the fusion of a haploid with a diploid nucleus

hypohydrogamy the condition of being fertilized under water

isogamy the fusion of morphologically identical gametes

anisogamy the condition of reproducing by the fusion of dissimilar gametes

hyperanisogamy the condition of having a large, active female gamete

exoisogamy the condition in which an isogamete will only fuse with another isogamete from a different brood

karyogamy the fusion of the nuclei of gametes

endokaryogamy = endogamy

macrogamy = hologamy

malacogamy the condition of a plant which is fertilized by a mollusk

merogamy fertilization through the union of mero-gametes

diamesogamy fertilization by an outside agent

heteromesogamy the condition in which various forms of fertilization are available to different members of the same species

misogamy = reproductive isolation

monogamy the condition of having only one sexual partner

mychogamy self fertilization

neogamy precocious syngamy

nothogamy heteromorphic xenogamy

nyctigamy the condition of a flower which opens by night and closes by day

oogamy a term, usually confined to the discussion of algae, which involves reproduction by anisogametes sufficiently different to be distinguished as "sperm" and "eggs"

ornithogamy the condition of a plant that is fertilized by birds

paedogamy the production of the zygote from gametes of the same gametangium

paragamy the production of zygote nuclei by the fusion of nuclei in a syncytium

apocytial paragamy a similar fusion resulting in an oospore on Saprolegnieae

parthenogamy parthenogenetic development of a diploid cell

phaenogamy = phanerogamy

phanerogamy reproducing by obvious means, specifically the condition of a plant with manifest flowers

phytogamy cross-fertilization of plants

plasmogamy the condition of two cells of which the cytoplasm but not the nuclei have fused, particularly in rhizopod Protozoa

plastogamy the fusion of cells into a plasmodium

pleogamy the condition of a sporophyte having flowers of various degrees of maturity

    gynopleogamy said of a plant which exists in three conditions, some having pistillate flowers, others perfect flowers, and the third staminate flowers

polygamy 1 (*see also* polygamy 2, 3) the condition of having many sexual partners

polygamy 2 (*see also* polygamy 1, 3) the term is sometimes applied in *bot.* to intersexes

polygamy 3 (*see also* polygamy 1, 2) having perfect and imperfect flowers on the same individual

    dioeciopolygamy the condition when some individuals of a species are unisexual and others hermaphrodite

    monoecious polygamy the condition of having single-sexed, and perfect flowers on the same specimen

porogamy the condition of a plant in which the pollen tube passes through the micropyle

    aporogamy the condition in which the pollen-tube does not pass through the micropyle

progamy the condition of an organism before fertilization

protogamy the fusion of gametes to produce a binucleate zygote

pseudogamy 1 (*see also* pseudogamy 2, 3) the situation which results when a spermatozoan enters and activates an egg but degenerates without its nucleus fusing with that of the egg

pseudogamy 2 (*see also* pseudogamy 1, 3) the fusion of two hyphal cells from two different thalli

pseudogamy 3 (*see also* pseudogamy 1, 2) parthenogenesis

siphonogamy plants fertilized by means of pollen tubes

    protosiphonogamy the condition that involves the fertilization of pollen on the ligule of a cone scale

staurogamy cross fertilization

sporogamy the production of spores by an organism derived from a zygote

syngamy (*see also* synchronogamy) the fusion of two gametes

    asyngamy the condition of organisms that are prevented from crossing by reason of having different breeding or flowering seasons

    xenogamy allogamy from the flower of another plant

zoidiogamy the condition of a plant that produces motile antherozoids

zoogamy the condition of plants having motile sex cells

gander the term for a male goose, though stag is sometimes used

gang term for a group of elks or buffaloes though herd is equally common, and also the term for a group of dogs, hunting or roaming without specific purpose being apparent (*cf.* kennel, pack)

ganglion 1 (*see also* ganglion 2) literally a swelling but most commonly in *biol.* applied to a group or cluster of associated nerve cells

    autonomic ganglion one that sends the axons of its cells to cardiac and smooth muscles and to glandular epithelium. They lie in a paired segmental chain along the lumbar spinal chord

    cardiac ganglion a prevertebral ganglion near the origin of the pulmonary artery

    cerebral ganglion the dorsal, *i.e.* supra-esophageal, ganglion of invertebrates divided, in insects into an anterior **protocerebral ganglion,** a middle **deuto cerebral ganglion** and a posterior **trito cerebral ganglion**

    hypocerebral ganglion in arthropods lies on top of the oesophagus immediately posterior to the cerebral ganglion

    chain ganglion one of a series of metameric ganglia lying near the vertebrate aorta, communicating both with each other and with the visceral branches of the spinal nerve

    cheliceral ganglion the arachnid equivalent of the tritocerebral ganglion of insects

    coeliac ganglion a prevertebral or subvertebral ganglion connected with the chain ganglia of the thoracic region

    collar ganglion the principal dorsal ganglion in the central nervous system of pterobranchs

    frontal ganglion in insects a dorsal, unpaired ganglion connected by paved commissures to the tritocerebral ganglia

    Gasserian ganglion = semilunar ganglion

    geniculate ganglion 1 (*see also* geniculate ganglion 2) a ganglion at the base of the facial nerve of mammals

    geniculate ganglion 2 (*see also* geniculate ganglion 1) one of a pair of large ganglia, from which the principal nerves of the body derive in the Rotifera

    semilunar ganglion a large ganglion at the base of the trigeminal nerve

    Meckels ganglion = sphenopalatine ganglion

    nodose ganglion a ganglion on the intestinal branch of the vagus nerve

    petrosal ganglion a ganglion on the ninth cranial nerve which in lower forms receives Jacobson's commissure

    rostral ganglion the arachnoid and crustacean equivalent of the insect frontal ganglion

    sensory ganglion = cerebrospinal ganglion

    cerebrospinal ganglion a ganglion either in the brain or spinal chord that contains the cell bodies of peripheral nerve fibers

    sympathetic ganglion a ganglion which receives its primary impulses from the lateral horns of the spinal chord and forms part of the thoraco-lumbar autonomic nervous system

    parasympathetic ganglion a ganglion which receives its impulses from autonomic components of cranial nerves. They are usually imbedded in the walls of organs and form the cranio-sacral autonomic nervous system

    ventricular ganglion in arthropods lies on the lateral walls of the foregut

    prevertebral ganglion one of a series of ganglia, some single and some paired, lying ventral to the aorta and connected to the chain ganglia

    Wrisberg's ganglion = cardiac ganglion

    X and Y ganglion transitory ganglia in embryonic arachnids equivalent to the protocerebral ganglion of insects

ganglion 2 (*see also* ganglion 1) any knob or expansion in a hypha

ganglioneous (*bot.*) pertaining to trichomes with plumose articulation

gannet any of several large sulid birds of the genus

Morus or, less frequently, Sula. Used without qualification, *U.S.* and *Brit.* refers to *M. bassanus, N.Z.* and *Austral.* to *M. serrator, Afr.* to *M. capensis.* In *S.E.U.S.* the wood ibis (*q.v.*) is referred to as a gannet

ganoin an enamel like substance found on the scales of many actinopterygain fishes

[gap] a mutant (gene symbol *gp*) mapped at 74.0± on chromosome II of *Drosophila melanogaster.* The gap is a break in vein L4 of the phenotype

leaf **gap** = leaf lacunae

gape the extent of the opening in an opened jaw

gaper *W. U.S.* the mactiid pelecypod mollusk *Schizothaerus nuttallii*

gar *U.S.* any of numerous lepidosteiform fish of the genus Lepidosteus. The spelling here given is almost universal but, by strict nomenclatural rules should be Lepisosteus

　　alligator gar *L. spatula*

　　florida gar *L. platyrhincus*

　　longnose gar *L. osseus*

　　shortnose gar *L. platostomus*

　　spotted gar *L. oculatus*

garbanzo the leguminous herb *Cicer arietinum* (= chickpea)

gardener any of several ptilorhynchid birds of the genus Amblyornis

garganey the anatine bird *Anas querquedula* (= *Brit.* summer teal)

garget the phytolacaceous herb *Phytolacca americana* (= poke weed)

garibaldi the pomacentrid fish *Hypsypops rubicunda*

Garidae a small family of pelecypod Mollusca distinguished by their extremely long siphons and thin, nearly transparent, shells

garlic (*see also* mustard) the liliaceous plant *Allium sativum*

　　false garlic *U.S.* the liliaceous herb *Nothoscordum bivalve* (= streak leaved or yellow garlic)

　　wild garlic any of many species of liliaceous herbs of the genus Allium

[garnet] a pseudoallelic locus (gene symbol *g*) mapped at 44.4 on the X chromosome of *Drosophila melanogaster.* The name applies to the phenotypic eye color

Garryaceae the only family of the Garryales with the characteristics of the order

Garryales an order of dicotyledonous angiosperms distinguished from other catkin-bearing plants by the unilocular biovulate ovary

Gartner's *epon. adj.* from Hermann Treschow Gartner (1785–1827)

Gaskell's *epon. adj.* from Walter Holbrook Gaskell (1847–1914)

gaskin *see under* hough

Gasserian *epon. adj.* from Johann Laurentius Gasser (1757–1765)

-gaster- *comb. form* meaning "stomach" or "abdomen"

gaster an abdomen, and particularly one which is swollen, and specifically that portion of the hymenopteran insect abdomen which lies behind the petiole

　　epigaster in various arthropods, either the ventral surface of the posterior segments of the thorax or that of the anterior segments of the abdomen

　　mesogaster the mesentery which supports the stomach of vertebrates

Gasterophilidae a family of large myodarian cycloraphus Diptera the larvae of which are parasitic

within the alimentary tract of horses. They are commonly called the horse bot fly

Gasterostomata a suborder, or order if Digenea is held to be a subclass, of trematode platyhelminths distinguished by the fact that the mouth is in the middle of the ventral surface

Gasterosteidae a family of osteichthyes commonly called stickle backs, distinguished by a series of bony plates along the sides and free spines along the back. Their taxonomic position is greatly debated since they are variously placed in the Thoracostei which may itself be either a separate order or an order of Acanthopterygii while others place the stickle backs with the solenichthyes in a separate order, usually placed between the Anacanthini and Allotriognathi

Gasterosteiformes = Thoracostei

Gasteruptiidae a family of proctotrupoid apocritan hymenopteran insects with a long ovipositor that usually causes them to be mistaken for ichneumons

Gasteruptionidae = Gasteruptiidae

Gastraea a hypothetical, two-layered organism postulated as the ancestor of all Metazoa (*cf.* Blastea)

　　metagastraea a hypothetical ciliated, bottom-feeding gastraea

gastralia 1 (*see also* gastralia 2, 3) sternal rib-like bones found in the ventral abdominal wall between the last true rib and the pelvis in Crocodilia, Sphenodon and some fossil reptiles (= Parasternalia)

gastralia 2 (*see also* gastralia 1, 3) dermal ossifications that contribute to the plastron of Chelonia

gastralia 3 (*see also* gastralia 1, 2) sponge spicules in the lining of the cavity (*cf.* dermalia)

amphigastria stipule-like, stem-enveloping organs in Hepaticae

gastric pertaining to the stomach or abdomen

　　clitogastric a term applied to those Hymenoptera having a petiolate abdomen

　　gamogastric pertaining to a pistil formed by the complete union of ovaries which therefore form a "belly" at its base

　　physogastric having a swollen abdomen

　　stenogastric having a slender abdomen

gastrin an enzyme secreted by the gastric mucosa. It is active in the control of hydrochloric acid production in the stomach

Gastrochaenidae a family of rock-boring or burrowing myocean Mollusca differing from the Saxicavidae in having equal sized valves

Gastromycetes a subclass of homobasidiomycete Basidiomycetes containing the forms commonly called puffballs and stinkhorns. They are distinguished by the fact that the basidiocarp remains permanently closed

enterogastrone a hormone secreted by the duodenal mucosa. It is active in controlling motor activity and acid secretion of the stomach

　　urogastrone a hormone secreted in the gastro intestinal tract. It is active in controlling the function of the stomach

Gastrophilidae = Gasterophilidae

Gastropoda that class of prorhipidoglossomorphan Mollusca which contains, *inter alia,* the snails, slugs and whelks. They are distinguished by their well-developed head and normally by an elongate conical shell which is almost invariably coiled in a spiral

Gastrotricha a phylum of pseudocoelomatous bilateral animals. All are microscopic and have an unsegmented cuticle furnished with spines or scales. The group is held by some to be a class of the phylum Aschelminthes

gastrous alternate, but rarely used, form of gastric

gastrula that stage in the early development of an embryo in which the rudimentary enteron is established but the nervous system is not yet apparent. This stage therefore lies between the blastula and the neurula

archigastrula that type of gastrula in which the endoderm is produced by invagination (emboly). This is usually what is meant by the term gastrula used without qualification

coelogastrula a gastrula derived from a coeloblastula (*q.v.*)

discogastrula one which is produced, usually after meroblastic cleavage, by cells that turn under and grow beneath the ectoderm to form the endoderm

stereogastrula = planula larva (*q.v.*)

gatekeeper *Brit.* the satyrid lepidopteran *Epinephele tithonus*

firewood gatherer the funariid bird *Anumbius annumbi*

gaulin *W.I.* = heron

gaur a large asiatic, bovine bovid (*Bos gaurus*) with a narrow hump extending over the front part of the body and white legs

gavial the gavialid crocodilian *Gavialis gangeticus*

false gavial the crocodylid crocodilian *Tomistoma schlegelii*

Gavialidae that family of crocodilian reptiles which contains the gavial and its immediate allies. They are distinguished from the Crocodilidae by the narrow elongate snout

Gaviidae that family of gaviiform birds which contains the loons. They are distinguished by their small pointed wings, long tapering beaks and necks, well developed tails and webbed feet

Gaviiformes an order of birds containing the forms commonly called loons. The long straight sharp pointed bill is typical of these swimming and diving birds

gayal a semi-domesticated Asiatic bovine bovid (*Bos frontalis*). It greatly resembles a small size gaur with straight horns

gazelle (*see also* goat) the term, though widely applied to all members of the sub-family Gazellinae (*q.v.*) should properly be reserved for the very numerous species of the genus Gazella

goat-gazelle either of two or possibly three species of gazelline antelope of the genus Procapra. It is almost impossible to make a distinction between some of the gazelline antelopids—the zeren (*q.v.*) and the goa (*q.v.*) are good examples—and the subfamily Saiginae (*q.v.*) of the Capridae

Gazellinae a sub-family of antelopid artiodactyle mammals containing not only the forms called gazelles as such, but also *inter alia,* the impalla, gerenuk and springbuck

star **gazer** popular name of uranoscopid fish

sungazer the cordylid saurian reptile *Cordylus giganteus*

Gb *see* [Green b]

Gc *see* [Green c]

GDP = guanosine diphosphate

Ge *see* [Giant egg]

-ge- *comb. form* meaning "earth," both in the sense of the soil (for which -edapho- is a better combining form) and of the planet. *See also* -gae- and -geo-

-gean *comb. term* usually employed to indicate "earth" (soil) as distinct from "Earth" (planet) for which -gaean (*q.v.*) is preferred in *biol.*

amphigean said of a plant which has both aerial and underground flowers

diagean said both of a plant producing a subterranean stolon, and of one the shoots of which rise directly from the soil

epigean dwelling on the surface of the land and particularly the condition of a plant the shoots of which lie on the surface of the soil

hypogean living under the surface of the earth

hypocarpogean producing underground fruit

gecko properly any gekkonid lacertilian

day gecko any of several species of the genus Phelsuma

"flying" gecko any of several species of the genus Ptychozoon

African house gecko *Hemidactylus mabouia*

mourning gecko = scaly-toed gecko

oceanic gecko *N.Z. Gehyra oceanica*

Pacific gecko *Peropus mutilatus*

reef gecko *U.S. Sphaerodactylus notatus*

stone gecko *Austral. Diplodactylus vittatus* (= stone adder)

fat-tailed gecko any of several species of the genus Oedura

leaf-tailed gecko any of several species of the genus Uroplatus

bent-toed gecko any of several species of the genus Cyrtodactylus

half-toed gecko *Hemidactylus frenatus*

scaly-toed gecko *Lepidodactylus lugubris*

waif gecko any of several species of the genus Hemiphyllodactylus

Gekkones an assemblage of lacertilian reptiles erected to contain the family Geckonidae the members of which differ from other lacertilians in having amphicoilous rather than procoelous vertebrae

Gekkonidae a family of lacertilian reptiles having amphicoelous vertebrae but better known for the adhesive lamellae on their digits. This, in spite of popular opinion, is not a universal character. Commonly called geckos

geelbec *S.Afr.* the anatine bird *Anas flavirostris*

-geic *comb. term* used either for -gaean (*q.v.*) or -gean (*q.v.*)

Geissolomataceae a small family of myrtiflorous dicotyledonous angiosperms

-geito- *comb. form* meaning "neighbor"

gekko = gecko

gelada the cynopithecid primate *Theriopithecus gelada* of Abyssinia. It posseses a huge head of hair but a naked pink chest and buttocks

Gelastocoridae a family of hemipteran insects commonly called toad bugs, a name deriving from the large eyes and hopping movements

gelation the transformation of a colloid from the sol to the gel condition

Gelechiidae a very large family of lepidopteran insects, distinguished by the long upcurved labial palps. Most of the larvae are leaf miners though some are leaf-rollers and gall-formers. A few such as the Angoumois grain moth and the pink bollworm are serious pests

Gelechioidea a superfamily of lepidopteran insects containing the Gelechidae and a number of little-known small moths

Gelidiales an order of tetrasporophytic floridian Rhodophyta in which the carposporophyte develops directly from the carpogonium

gelinotte = hazel hen

gem 1 (*see also* gem 2) any of several lepidopteran insects. *Brit.* without qualification, the geometrid *Percnoptilota fluviata. I.P.* any of several species of lycaenids of the genus Poritia

brilliant gem *S.Afr.* the lycaenid *Chiorozelas pseudozeritis*

purple gem *S.Afr.* the lycaenid *Desmolycaena mazoensis*

gem 2 (*see also* gem 1) any of several brilliantly colored organisms

coral-gem the leguminous herb *Lotus bertholetii*

mountain-gem any of several trochilid birds of the genus Lampornis

sun gem the trochilid bird *Heliactin cornuta*

geminate doubled

-gemm- *comb. form* meaning "bud"

gemma literally a little bud, particularly a globose bud becoming detached from the parent and thus being an initial stage of asexual development, particularly in the lower plants

brood gemma a propagative body found in some pterydophytes in which there is a brood cell on one side of the frond and a bulbil on the other

spermogemma = archegonium

sporogemma the oogonium of Chara

gemmaceous pertaining to gemma

gemmidium = tetraspore

gemmule 1 (*see also* gemmule 2) a bud-like body of any plant except an angiosperm. The term has also variously been applied to the plumule and the ovule of angiosperms

gemmule 2 (*see also* gemmule 1) an asexual propagative body found in some aquatic animals. It consists normally of a mass of undifferentiated tissue surrounded by a protective case

Gempylidae a small family of acanthopterygian fishes commonly called snake-mackerals by virtue of the elongate body and projecting lower jaws with large teeth

-gen- *comb. form* combined or abbreviated from numerous Greek roots, meaning "bring forth," "parent," "beginning," "origin," "birth," "ancestor" "nation," or "pertaining to birth." The *Engl.* derivatives are confused and frequently have arbitrary meanings. Entries in this dictionary are under all the following forms: -gen (1–4), gene, gener, generation (1, 2) generic (1, 2), -genesis (1, 2), -geneous, -genetic (1, 2, 3), genetics, -genia, -genic (1, 2), -genous (1, 2), genus, -geny (1, 2). Many adjectival forms (*e.g.* generative) are not separately defined in this place but will be found with the substantives (*e.g.* nucleus) with which they combine in *biol.* usage

-gen 1 (*see also* -gen 2, 3, 4) *comb. suffix* in the sense of "that which produces"

aerogen a gas producer, particularly a bacteria

androgen a general term for male hormones

calyptrogen the initials of a root cap

dermatogen the meristem of the epidermis

florigen the hormone that initiates the production of flowers

primary desmogen = procambium

mucigen the precursor of mucous

mutagen an agent that causes mutations

myogen an obsolete term for sarcoplasm

pangen one of the particles concerned in pangenesis (*q.v.*)

pathogen a parasitic organism, particularly a microorganism, that causes damage to its host

phellogen that type of cambium which produces cork

lenticel phellogen an area developed beneath a stoma which gives rise to a lenticel

sclerogen totally lignified cells such as those which form the shell of some nuts

sporogen any spore-bearing plant

thallogen = thallophyte

trichogen a cell from which an insect seta is derived (*cf.* trichophore, trichopore)

urobilinogen a precursor of urobilin (= stercobilinogen)

zymogen an inactive enzyme, or enzyme precursor, commonly observed in the form of granules

-gen 2 (*see also* -gen 1, 3, 4) *comb. suffix* in the sense of "sequential growth"

acrogen an obsolete word for fern, deriving from the terminal growth of these forms

anagen that stage in the hair cycle of mammals in which the hair is actively growing (*cf.* telogen, catagen)

biogen = biophore

catagen that stage in the growth cycle of hair which is intermediate between the telogen and the anagen

exogen a plant which grows by the addition of wood to the outside

telogen that stage in the hair cycle of mammals in which the hair is not growing (*cf.* anagen, catagen)

-gen 3 (*see also* gen 1, 2, 4) *comb. suffix* in the sense of "ancester" or "ancestry"

cultigen a plant known only under cultivation, the wild ancestor being extinct, or at least unknown

fundatrigen the viviparous parthenogenetic female aphid that is produced by the fundatrix (*q.v.*)

infusorigen a ball of cells inside a rhombogen which developed into the infusoriform larva of Mesozoa

nematogen that stage of a mesozoan in which the acial cell produces agametes by endogenous division (*cf.* rhombogen and infusorigen)

rhombogen a stage in the development of Mesozoa intermediate between the nemetogen and the infusorigen

syngen a genetically isolated variety of ciliate protozoan which cannot interbreed fruitfully with another syngen of the same species

-gen 4 (*see also* -gen 1, 2, 3) *comb. suffix* in the sense of "offspring"

heterogen the offspring of heterozygous parents

homogen the offspring of homozygous parents

gena literally cheek and applied to the lateral facial area of many animals, particularly arthropods. It is also used of the basal, feathered portion of a bird's bill or jaw

genal pertaining to the cheek or an equivalent area

transgenation an early term for a mutation at a single locus

gender the specification of sex (*see also* name)

gene a functional unit of heritable information occupying a specific locus on a chromosome (*cf.* allele)

allogene an early term for recessive allele

buffering gene = modifier gene

chromogene a term once used for a gene locus when this was thought to be a point to which cytogenes were attached

complementary gene a gene which is unable to produce a phenotypic expression in the absence of one or more other genes

cumulative genes genes which, acting together, accentuate a character such as skin color

dominigene a term at one time applied to a gene that appears to modify a dominant

duplicate gene one of several genes at different loci all affecting the same phenotype in the same manner

independent gene non-allelic genes on different chromosomes

inhibitor gene one which masks or inhibits the effect of a gene at another locus

major gene one of which the effects are readily identifiable

mimic gene one which produces a similar effect to another gene to which it is nonallelic

modifier gene one which affects the expression of another loci

modifying gene one which has no function other than to modify the expression of another gene

oligogene one having a major effect on qualitative phenotypic characters, as distinct from a polygene

plasmagene a genetic unit found in the cytoplasm (*cf.* episome)

plastogene a nonnuclear genetic locus associated with the plastids of plants and in part responsible for the traits of these plastids

polygene one of several genes that individually exercise a relatively minor effect, but in combination effect quantitative characters

protogene an early term for dominant allele

triplicate gene one of three non-allelic, non-cumulative genes showing the same effect

wild type gene the allele normally found at any given locus on a chromosome

heterogeneous said of a mixture containing one or more unlike components. (*cf.* heterogenous with which the word is often confused)

gener an organism, particularly one identified by reproductive criteria

bigener a sterile plant resulting from an intergenetic cross

heterogener a plant, particularly a phaeophyte alga, in which the zygote grows into a diploid plant that is different in size and shape from the parent

isogener said of a marine alga, particularly a phaeophyte, in which the zygote grows into a diploid plant that is similar to the parent

microgener = variety

generation 1 (*see also* generation 2) the act of generating or producing

degeneration the breakdown of specific cells or organs to an indeterminate form. It differs from dedifferentiation in carrying the connotation that the process will not reverse

regeneration the ability of an organism to reproduce a part that has been lost; or, to reorganize dedifferentiated material

homoetic regeneration an abnormal form of regeneration in which a serial homologue replaces the lost structure

spontaneous generation the assumption, at one time prevalent, that living things sprang directly from dead things by an instantaneous mutation of inorganic matter into complex living forms

generation 2 (*see also* generation 1) a group of offspring descended from common parents, or a group of parents

anithetic generation either of two morphologically distinct generations one of which haploid and the other diploid

filial generation offspring of a cross, the first being indicated by F_1 the second, or grandchild, by F_2 etc.

generic 1 (*see also* generic 2) pertaining to a genus

congeneric belonging within the same genus

monogeneric said of a taxonomic unit, such as a family, containing only one genus

generic 2 (*see also* generic 1) pertaining to a gener

.genesia plant or animal taxa having this ending are listed alphabetically

-genesis 1 (*see also* genesis 2) *comb. suffix* used in the sense of "a type of reproduction"

alliogenesis a type of reproduction in which there is an alternation of generations

amphigenesis development inaugurated by the fusion of two unlike gametes

androgenesis the development of an egg having only a paternal nucleus

anthogenesis the production of both males and females by parthenogenesis

blastogenesis any method of asexual reproduction that does not involve spores (*but see* blastogenous)

cenogenesis developmental processes which are not supposed to repeat

cytogenesis the production of cells or cellular structure

digenesis alternation of generations

diplogenesis an abnormal doubling of parts

ectogenesis the development of an embryo *in vitro*

embryogenesis reproduction through the medium of embryos

direct embryogenesis in plants, when the product of the spore resembles the adult form

phylembryogenesis the relation between phylogeny and ontogeny

etheogenesis the production of an organism without fertilization from a male gamete (*cf.* parthenogenesis)

gametogenesis the cytoplasmic and nuclear processes involved in the production of gametes

gamogenesis sexual reproduction

agamogenesis = asexual reproduction

geneagenesis = parthenogenesis

gynogenesis = pseudogamy

heterogenesis alternation of generations. The term was at one time also applied to the appearance of mutants

histogenesis the production and differentiation of tissues, particularly reorganization after histolysis

homogenesis the condition in which successive generations resemble each other. That is, the opposite of alternation of generations

hypogenesis = agamogenesis

metagenesis a term preferred by some to alternation of generations when applied to polymorphic coelenterates

monogenesis asexual reproduction

ologenesis the concept that morphogenesis in phylogeny, as well as in ontogeny, is due to an intrinsic mechanism of which the process of development follows the law of differentiation with a division of physiological labor between the parts

organogenesis the production of organs and organ systems

orthogenesis development along clearly defined lines. At one time used to distinguish development controlled by heredity from that enforced by the environment

paedogenesis sexual reproduction by larval or apparently immature organisms. The term is used not only of animals but also of the flowering of plants, particularly trees, before they reach their full development

palaeogenesis = neoteny

paragenesis reproduction induced by crossing a hybrid, otherwise sterile, with a parent

parthenogenesis the reproduction of an organism from one gamete (*cf.* etheogenesis)

    artificial parthenogenesis the laboratory-induced development of an egg using a stimulus other than a spermatozoan

    autoparthenogenesis = artificial parthenogenesis

    automatic parthenogenesis the condition when a haploid gamete divides without chromosome reduction thus giving a diploid offspring

    diploid parthenogenesis = parthenogamy

    female parthenogenesis that form of parthenogenesis in which the female gamete produces a new individual (*cf.* etheogenesis)

    generative parthenogenesis parthenogenesis by a haploid cell and particularly the production by asexual means of a sporophyte from a haploid germ cell of a gametophyte (*cf.* generative apogamy)

    heteroparthenogenesis the condition of producing parthenogenetically either offspring that themselves reproduce parthenogenetically or, alternatively, offspring that reproduce sexually

    male parthenogenesis = etheogenesis

    meiotic parthenogenesis parthenogenesis from a gamete which has become haploid by a normal process of meiotic reduction divisions

    ameiotic parthenogenesis parthenogenesis from a gamete which has become haploid through a single mitotic division

    somatic parthenogenesis parthenapogamy

    zygoid parthenogenesis asexual reproduction from an egg that remains diploid or becomes diploid again in the course of its development

    hemizygoid parthenogenesis asexual reproduction from haploid eggs

patrogenesis the development of an enucleated egg induced by fusion with a normal sperm

protogenesis reproduction by budding

rhexigenesis the separation of structures by mechanical rupture

schizogenesis reproduction by splitting (*cf.* schizogenous, under -genous 1)

sporogenesis the total process of the production of spores or seeds

syngenesis = embryogenesis

tachygenesis abbreviated development, as when one or more larval stages are coalesced

xenogenesis literally means strange reproduction and has therefore been applied to alternation of generations, spontaneous generation and a number of similar things

zygogenesis that method of reproduction in which a male and female gamete nuclei fuse

-genesis 2 (*see also* genesis 1) *comb. form* used in the sense of "origin" or "origination"

angiogenesis the origin of blood vessels

archegenesis the concept that life can evolve stepwise over a long period of time from inanimate precursors

biogenesis the production, or origin, of life

    abiogenesis the concept that life can arise from non-living material in a relatively short space of time (*cf.* archegenesis and spontaneous generation)

    neobiogenesis the theory that life may have been synthesized several times and that there is a continuous recurring possibility of such synthesis taking place

    symbiogenesis the development or origin of symbiotic relationships

caenogenesis the production of characters through the influence of the environment

catagenesis the evolution of a complex to a simpler form

coenogenesis genesis deriving from the action of the environment

ectogenesis the origin of variations from causes outside the organism

epigenesis the theory that the egg is structureless and that the adult develops from it by a process of structural elaboration. The antithesis is preformation

phytogeogenesis the origin of plants on earth

haplogenesis the origin of new forms by evolution

hologenesis the development of species by multiple mutation from a single lost ancestor

lactogenesis the initiation of the secretion of milk in the mammalian mammary gland

morphogenesis the development of shape and form

nomogenesis *either* the concept that life is produced by "natural" rather than by "miraculous" causes *or* the Lamarckian concept that the exterior environment produces heritable adaptive modifications in all individuals of the same species that are subjected to it

palingenesis *either* the doctrine of simple descent *or* that part of the development of an individual that is supposed to repeat the phylogeny

pangenesis that theory of evolution which held that heritable characters are transmitted through minute gemmules that develop in each cell, and lie dormant until the moment of reproduction

phylogenesis the origin of a race or taxon

physiogenesis the development and evolution of physiological processes

phytogenesis the origin and development of a plant

dipleurogenesis = bilateral symmetry

polygenesis the origin of something at several places either in time or space

strophogenesis the process by which an organism once having single generations may have evolved into the condition of having alternation of generations

genet any of several species of viverrine carnivorous mammal of the genus Genettae. Their domestication by Egyptians seems to have preceded the domestication of the cat

-genetic 1 (*see also* genetic 1, 2) *comb. suffix* used in the sense of "reproduction" and thus the *adj. form* of genesis 1 where most definitions of this form are given

basidiogenetic anything borne on a basidium

digenetic *see* digenous

-genetic 2 (*see alo* -genetic 1, 3) *comb. suffix* used in the sense of "origin" and thus the *adj. form* of -genesis 2 (q.v.) where most definitions are given

allogenetic said of organisms transported to the area where found

autogenetic said of organisms originating where found

coenogenetic said of adaptive as distinct from ancestral characters

palingenetic said of ancestral as distinct from adaptive characteristics

-genetic 3 (*see also* -genetic 2, 3, mosaic) *comb. suffix* used in the sense of "pertaining to heredity." Definitions are under the modified substantive (*e.g.* drift, marker)

genetics the study of the causes and effects of heritable characteristics

cryptogenetics the study of genotypic transfer as distinct from phenotypic transfer

cytogenetics the study of the behavior of chromosomes in mitosis and meiosis and the correlation of this behavior with the transmission of heritable characters

syngenetics the study of changes in the structure of associations

genia an assemblage of organisms of common origin

syngenia an assemblage of organisms formed by asexual reproduction

sysgenia an assemblage formed by the fusion of syngenia

genial pertaining to the chin

-genic 1 (*see also* genic 2, 3) pertaining to genes

isogenic belonging to the same genotype

-genic 2 (*see also* -genic 1, 3) *comb. suffix* used in the sense of "production" (*cf.* -genetic 1, -genesis 1)

cyanogenic gas-making, particularly of insects which will repel their enemies by this means

cystigenic *see* valve

erotogenic producing sexual desire. Usually said of areas causing this effect under tactile stimulation

gynecogenic = parthenogenetic

-genic 3 (*see also* -genic 1, 2) *comb. suffix* used in the sense of "causation" and therefore synonymous with -genous 1 (*q.v.*) the form mostly preferred in this work

cacogenic pertaining to the deterioration of a race through constant interbreeding of unsuitable partners

dysgenic pertaining to the degradation of a race

geniculate being in the possession of or having the form of a knee or knee-joint

genip the sapindaceous tree *Melicocca bijuga*

Genista properly a genus of leguminous plants but commonly applied to the broom (*q.v.*) (*Cytisus canariensis*)

genitalia reproductive organs

external genitalia those reproductive organs, either male (*e.g.* penis) or female (*e.g.* ovipositor) which are, or can be, protruded from the body

gennyleion = antheridium

genome one complete haploid set of chromosomes and its effect (*cf.* phenome)

-genous 1 (*see also* -genous 2, 3) *comb. suffix* used in the sense of "causation." The synonymous form -genic (*see* -genic 3) is often used

aetigenous caused by external forces (*cf.* exogenous)

arthrogenous structures which develop from a portion ("joint") of a cell rather than by the division of the cell into two or more or less equal parts

ascogenous that which gives rise to an ascus (*q.v.*)

autogenous said of anything which is self-generated or derived from, or influenced by, internal causes (*cf.* endogenous)

endogenous that which is generated from within as, for example, supposed endogenous rhythms

exogenous arising from causes outside the organism, or, sometimes in *bot.*, from superficial tissue

heterogenous said of something which originates from causes outside the body in contrast to autogenous. (*cf.* heterogeneous with which the word is often confused)

lysigenous derived through lysis. The term is commonly applied to intercellular spaces in plants (*cf.* schizogenous)

hysterolysigenous said of a cavity formed by the lysis of cells

monogenous = endogenous

mutagenous that which produces mutations

photogenous light-producing

protogenous the condition of a structure in which differentiation has must begun

schizogenous derived through splitting. The term is commonly applied to intercellular spaces in plants (*cf.* schizogenous [under -genous 3], schizogenesis [under genesis 1] and lysigenous)

sporogenous producing spores

-genous 2 (*see also* -genous 1, 3) *comb. form* used in the sense of "carrying" "bearing" or "associating"

acrogenous borne on the tip of a hypha

amphigenous growing over or around

biogenous a term combining the meanings of epiphytic and epizoic

carpogenous bearing or producing fruit

amphicarpogenous the production of aerial fruit which is subsequently buried

hypocarpogenous fruit which is produced in the soil from underground flowers

cauligenous arising from the stalk

cyclogenous said of a stem showing annual rings

dysgaeogenous used of plants growing on rocks

entomogenous said of fungi that parasitize insects

hypogenous produced beneath anything, but not of necessity beneath the ground (*cf.* hypogeam)

monogenous = monocotyledonous

pantogenous said of a parasitic fungus which can attack numerous hosts

phyllogenous growing on leaves

-genous 3 (*see also* -genous 1, 2) *comb. form* used in the sense of "sex" or "sex organs"

agenous = neuter

ambigenous said of a flower in which the sepals cannot be readily distinguished from petals

antherogenous said of a double flower in which the additional petals replace anthers

blastogenous an early term applied to heritable changes originating in gametes, and thus equivalent to the modern concept of mutation

cenogenous said of an organism which is sometimes oviparous and at other times viviparous

digenous bisexual, though also used in the sense of a sexually produced offspring to which the term digenetic is also applied

schizogenous (*see also* -genous 1) failing to breed true

somatogenous *see* character

spermatogenous (*bot.*) the development of any male gamete form

gentian any of numerous flowering herbs of the family Gentianaceae and particularly the genus Gentiana

barrel gentian = soapwort gentian

bastard gentian *G. acuta*

blind gentian = bottle gentian

bottle gentian *G. andrewsii* or *G. clausa*

closed gentian = bottle gentian

fringed gentian *G. crinita*

horse gentian any of several species of the caprifoliaceous genus Triosteum, particularly *T. perfoliatum* (= feverroot)

marsh gentian *Dasystephana pneumonanthe*

Plymouth gentian *Sabatia kennedyana*

spurred gentian any of several herbs of the genus Halenia

soapwort gentian *G. saponaria*

Gentianaceae that family of contortous dicotyledons which contains, in addition to the gentians, the buck beans and other plants. The presence of a bitter principle in the herbs and a one-celled ovary are together typical of the family

Gentianales = Contortae

Gentianineae a suborder of Contortae containing all the families of this order except the Oleaceae

Gentianoidea a subfamily of the Gentianaceae distinguished from the Menyanthoideae by having the leaves opposite

genu that segment of the acarine pedipalp which lies between the telofemur and the tibia

genus an assemblage of species considered to be more closely related to each other than they are to members of another genus

autogenus = monotypic genus

daughter genus *see* mother genus

monotypic genus one which contains only one species

mother genus one which splitters (*q.v.*) have furnished with many daughters

subgenus a taxon immediately below genus, the members of which are thought to be associated by genetic rather than geographical factors

supergenus a taxon immediately above genus

-geny *comb. form* used in the sense of a "condition or type of reproduction" and of a "condition or type of inheritance." Many more meanings are given under -genic and -genous

apogeny the loss of function of male and female reproductive organs

arrhenogeny the condition of producing exclusively male offspring

imperfect arrhenogeny the condition of producing predominantly, but not quite exclusively, male offspring

cyclogeny the life cycle of a microorganism

dissogeny the condition of becoming sexually mature both as a larva and as an adult

embryogeny the production of embryos

continued embryogeny the concept that the apical meristem of a plant is in point of fact an embryonic structure

homogeny the inheritance of a part similar in origin even though the function be lost or changed

lysogeny the condition of a microorganism that has a prophage in its genome

ontogeny properly the production of eggs but used in the classic phrase "ontogeny repeats phylogeny" in the sense of the successive developmental stages through which an egg develops into an adult

periphyllogeny the condition of a plant that produces many leaflets around the edge of a leaf

phylogeny properly the development of races but used in contrast to ontogeny (*q.v.*) in the sense of the sucessive evolutionary forms through which an organism has evolved from a remote ancestor

pleiogeny the increase of a plant by branching or budding

pleogeny mutability of function

poecilogeny the condition, found in some Diptera, of having polymorphic larvae some of which develop normally and others of which reproduce paedogenetically

-geo- *see* -ge- and -gae-

geometer *U.S.* larva of geometrid moths. The term is also extended to some adult moths (= measuring worm)

chickweed geometer *Haematopsis grataria*

crocus geometer *Xanthotype sospeta*

chain-spotted geometer *U.S. Cingilia catenaria* = chain-streak moth

notch-wing geometer *Ennomos magnarius*

Geometridae a family of geometroid lepidopteran insects commonly called measuring-worms or geometers from the habit of their larvae which progress in a looping manner (*cf.* looper)

Geometroidea a superfamily of lepidopteran insects containing the large-winged, smallbodied moths commonly called geometers

Geomyidae a family of New World pounched sciuromorph rodents commonly called pocket gophers

Geomyzidae = Opomyzidae

Geospizidae a family of passeriform birds erected to contain the Darwin finches by those who do not hold them to belong in the Fringillidae

Geotrupidae a family of coleopteran insects closely allied to the Scarabaeidae and frequently included in that family. They are among the numerous forms called dung-beetles

-geous *adj. term* synonymous with -gean, the form preferred in this work

Gephyrea at one time a phylum containing the Echiurida, Priapulida and Sipunculoidea. This taxon is no longer acceptable to most zoologists

Geraniaceae that family of geranalous dicotyledons which contains, *inter alia,* the geraniums, the stork's bills, and the florists "geraniums." The arrangement of the sepals and petals in five pairs with ten stamens and five carpels is distinctive of the family

Geraniales that order of dicotyledonous angiosperms which contains, *inter alia,* the geraniums, nasturtiums, the flaxes, the coco plant, the rues, the mahogonies, the milkworts, and the euphorbias

Geraniineae a suborder of geraniale dicotyledonous angiosperms containing families Oxalidaceae, Geraniaceae, Tropaeolaceae, Linaceae, Erythroxylaceae, Zygophyllaceae, Cneorace, Rutaceae, Simaroubaceae, Burseraceae, Meliaceae and Achariaceae

geranium properly the genus of geraniaceous herbs known collectively as cranes' bills or wild geraniums (*q.v.*). The florist's geranium is any of numerous horticultural hybrids of the genus Pelargonium of the same family. The term is loosely applied to many red or pink flowers

apple geranium = nutmeg geranium

beefsteak geranium any of numerous species and horticultural varieties of the begoniaceous genus Begonia

California geranium the compositaceous herb *Senecio petasitis*

feather-geranium the chenopodiaceous herb *Chenopodium botrys*

fish-geranium any of several species of dwarf Pelargonium used for bedding but particularly *P. inquinans*

mint-geranium the compositaceous herb *Chrysanthemum balsamita*

nutmeg-geranium *P. odoratissimum*

rock-geranium *U.S.* the saxifragaceous herb *Heuchera americana*

strawberry-geranium the saxifragaceous herb *Saxifraga sarmentosa*

gerbil properly the cricetid rodent *Gerbillus aegyptius* though the term is applied locally to many of its relatives

gerbille = gerbil

gerenuk an E. African gazelline antelopid (*Litocranius walleri*) with very long legs and an almost giraffe-like neck

gerfalcon = gyrfalcon

germ literally, a bud. The term is colloquially applied to bacteria, and is also frequently used as an adjective, where germinal would be proper, as in germ cell

cousin German *U.S.* the noctuid lepidopteran insect *Mamestra congermana*. *Brit*. the noctuid lepidopteran insect *Noctua sobrina*

germander any of numerous species of labiateous herbs or shrubs of the genus Teucrium

germen the same word as germ which is an abbreviation of it. In the present form it is commonly used for the capsule of mosses or the ovary of plants

germination growth resulting from the breaking of dormancy in a seed

geron-, geront- *comb. form* meaning "old man" but in biology extending to mean "old world"

Gerrhosauridae a family of lacertilian reptiles having pleurodont dentition and a long papillate, but not bifid, tongue. Now usually included in the Cordylidae

Gerridae a family of acanthopterygian Osteichthyes closely allied to the Leiognathidae. Popularly called mojarras

Gerridae a family of hemipteran insects living on the surface of fresh waters and commonly called water striders. Their long hind femora, visible antennae and antiapical tarsal claws are typical

Gesneriaceae a family of tubuliflorous dicotyledons which may be distinguished from the closely allied Scrophulariaceae, Orobanchacae and Begoniaceae by the one-celled ovary

gestation the period between fertilization and birth in oviparous animals

gf *see* [green flesh]

gg² *see* [goggle²]

gh *see* [ghost plants]

GH = growth hormone

gharial = gavial

gherkin the cucurbitaceous vine *Cucumis anguria* and its edible fruit

[ghost plants] a mutant (gene symbol *gh*) mapped at 54 on linkage group V of the tomato. The phenotypic plants lack chlorophyll

[giallo ascoli] a mutant (gene symbol *og*) mapped at 7.4 on linkage group IX of *Bombyx mori*. The phenotypic expression is a highly translucent larva

[giant] a mutant (gene symbol *gt*) mapped at 0.9 on the X chromosome of *Drosophila melanogaster*. The term refers to the size of the larvae

[giant-4] a mutant (gene symbol *gt-4*) mapped at 24.0 on chromosome II of *Drosophila melanogaster*. The name is descriptive

[Giant egg] a mutant (gene symbol *Ge*) mapped at 14.0 in linkage group I of *Bombyx mori*. The term is descriptive of the phenotypic expression

[giant larva] *see* [lethal (2) giant larva]

gib a male cat, though the term tom is more common

gibbon general term for those hylobatid primate mammals of the genus Hylobates which are not otherwise specifically designated

agile gibbon = unka-puti

silvery gibbon = wow-wow

gibbous hump backed, though applied to any more or less cylindrical body with a lump on one side

gidgee the leguminous tree *Acacia homalophylla*

bastard gidgee any of numerous *Austr*. Acacias (= myall)

gie-me-me-bit *W.I.* the caprimulgid bird *Chordeiles minor* (= night hawk)

Giganturoidea an order of small abyssal osteichtyes

Gigartinales an order of tetrasporophytic floridan Rhodophyta in whch the auxiliary cell is a vegetative cell of the gametophyte

-ginglym- *comb. form* meaning "a hinge" usually translitered -gingly-

gill 1 (*see also* gill 2, 3) a thin-walled process, or series of processes, designed to promote osmotic exchange between the blood and the environment in water-dwelling animals

external gill a protuberance, finger-like or filamentous process, or series of processes arising from the body wall, and functioning as a gill

gill 2 (*see also* gill 1, 3) the spore-bearing plate of a basidiomycete fungus

gill 3 (*see also* gill 1, 2) in compound names of organisms

bluegill *U.S.* the centrarchid fish *Lepomis macrochirus*

gill-over-the-ground *U.S.* the labiateous creeping herb *Glechoma hederacea*

gilt a castrated young pig

gilvous properly, yellow. The word derives directly from the Latin, *gilvus* which is the root of the German *gelb*. It has, for some unknown reason, been applied by botanists to almost any color between silver gray and bluish pink

ginger (*see also* lily) the zingiberaceous herb *Zingiber officinale* and its edible root

wild ginger any of several species of the aristolochiaceous genus Asarum particularly *A. canadense* and *A. caudatum*

-gingiv- *comb. form* meaning "gum"

gingko popular misspelling of ginkgo

-gingly- *see* ginglym-

Ginglymodi an order of osteichthyes containing the single family Lepisosteidae, commonly called gars

ginglymoid having the appearance of a hinge

ginglymoidy the condition of a vertebra in which the articular surfaces are doubled or asymmetrical

ginkgo a monotypic genus of coniferous trees and the sole genus in the family Ginkgoaceae

Ginkgoaceae the family of plants containing the ginkgo tree, the only extant species. The fan-shaped deciduous leaves and fertilization of the macrosporangia by motile sperm are distinctive

Ginkgoales an order of gymnosperm spermatophytes containing the single family Ginkgoaceae and the single genus Ginkgo

stupid Ginny *W.I.* the tyrannid bird *Myiarchus stolidus*

ginseng the araliaceous herb *Panax quinquefolia* and particularly its root used widely in Chinese pharmacy

gipsy *see* gypsy

giraffe (*see also* beetle) the African, long-necked, giraffine artiodactyle mammal *Giraffa camelopardalis*

Giraffidae that family of artiodactyl mammals which contains the giraffes and the okapi. They are distinguished by the elongate skull which results in a very large diastema between the incisors and the premolars

Giraffinae the subfamily of the Giraffidae in which is classified the giraffe. It differs from the Palaeotraginae in possessing short bony horns covered with unmodified skin, and of course, an almost incredibly elongated neck in the only extant example

girdle any object structure or association of objects and structures which surrounds or partially surrounds another structure. Without qualifications, the term is used for the connecting zone or hoop of the diatom frustule, and as synonymous with the bridle (*q.v.*) of Pogonophora

ciliary girdle (Ectoprocta) = corona (Ectoprocta)

perignathic girdle a calcareous ridge on the inner surface of the peristomial edge of the test of an echinoid, which carries the insertion of muscles from Aristotle's lantern

pectoral girdle the sum total of the skeletal elements to which the anterior limbs of vertebrates are attached

pelvic girdle the sum total of skeletal units to which the hind limbs of vertebrates are attached

Venus's girdle the ctenophoran coelenterate *Cestum veneris*

girdler any of several insects or insect larvae which bore circular galleries girdling (*i.e.* destroying the cambium of) woody plants

cranberry girdler, *U.S.* the crambid lepidopteran *Crambus topiarius*

currant stem girdler *U.S.* the cephid hymenopteran *Janus integer*

rose stem girdler *U.S.* the buprestid coleopteran *Agrilus rubicola*

twig girdler *U.S.* the cerambycid coleopteran *Oncideres cingulata* (*cf.* twig pruner)

oak twig girdler *U.S.* the buprestid coleopteran *Agrilus angelicus*

Girellidae a family of acanthopterygian fishes. The girellid *Girella nigricans* has the habit of pecking at everything, from which the name "nibbler" has been applied to the whole family

githagineous dark green, blotched or streaked with purple or red

-gito- *see* geito-

gizzard (*see also* shad) in birds a muscular sac with sclerotized internal teeth immediately behind the proventriculus (*q.v.*). In insects the proventriculus is sometimes called the gizzard

gl *see* [glass] or [golden]

gl₁–gl₁₇ *see* [glossy seedling]

Gl *see* [glued]

Gla *see* [glazed]

glabrous shiny and smooth

glade an open space in, or open passage through, fairly dense woodland. Also (*S.U.S.*) a sphagnum bog or a swamp interspersed with hummocks

gladiate sword-shaped

gladiator any of numerous species of laniid bird of the genus Malaconotus (= bush-shrike)

water gladiole the campanulaceous herb *Lobelia dortmanna*

gladwin the iridaceous herb *Iris foetidissima*

gland 1 (*see also* gland 2) a group of cells, frequently globular or flask-shaped, which secrete a specific substance or group of substances. In invertebrates the term is also extended to similar structures of excretory function

abdominal gland a tubular accessory sex gland in male urodele amphibia

acid gland any of numerous invertebrate glands secreting acid, such as the sulfuric acid-secreting salivary gland of shell, and limestone, boring gastropod Mollusca, the formic acid-secreting glands of stinging Hymenoptera, or the HCN-secreting glands of certain diplopods

acinous gland = alveolar gland

adhesive gland any of numerous invertebrate glands which secrete a sticky substance

marginal adhesive glands the eosinophilous glands arranged in a ring encircling the ventral surface of some triclad Turbellaria, and which sometimes give rise to adhesive papillae

adrenal gland a gland, adjacent to the kidney in mammals, formed from the fusion of the suprarenal body and the interrenal body of lower forms

alveolar gland a much branched type of gland, with each branch terminating in a secretory capsule

anal gland a term loosely applied to any gland adjacent to or associated with the anus, in many fish and birds. In insects a gland opening in an area adjacent to the anus. The term is often used where rectal gland would be better

angular gland the buccal scent gland of some reptiles

apocrine gland *see below* under -crine gland

areola gland 1 (*see also* areola gland 2, 3) the sebaceous glands in the area (areola) immediately surrounding the nipple of a mammary gland (*cf.* Montgomery's gland)

areola gland 2 (*see also* areola gland 1, 3) the area of the iris which immediately borders on the pupil of the eye

areola gland 3 (*see also* areola gland 1, 2) the interstices between the veins of a leaf

axial gland a mass of tubules, probably endocrine in function, lying inside the axial sinus

Bartholin's glands mucous secreting glands in the vestibule of the vagina of some mammals, probably homologous with Cowper's glands

Batelli's gland = froth gland

belly gland adhesive secreting epidermal glands aiding in the amplexus of some anuran amphibia (= abdominal gland)

Bowman's glands small, scattered glands in the nasal cavity of mammals

Brunner's glands enzyme secreting glands in the submucosa of the anterior region of the duodenum

calciferous gland one of several oesophageal outfoldings in many oligochaete annelids. They control the *Ca* content of the blood (*cf.* chalk glands)

cardiac gland a gland of unknown function occurring at the rear end of the pharynx in some nematodes

carotid gland a spongy swelling at the base of the internal and external carotid artery in some amphibia

caudal glands literally glands of the tail region, but usually applied, without qualification, to the adhesive tip of Turbellaria and some other invertebrates

cement gland any gland secreting a substance which hardens for the attachment of an organism. Used without qualification, generally refers to the frontal gland of the larvae of cirripede crustacea (*cf.* colleterial gland)

cephalic gland (Platyhelminthes) = penetration gland

post cerebral glands the salivary glands of insects

cervical gland (Nematoda) = ventral gland (Nematoda)

chalk-gland a gland depositing and voiding excess calcium in some plants. (*cf.* calciferous gland, Mettenian gland)

ciliary glands modified sweat glands in the eyelids

cloacal gland = anal gland

coccygeal gland properly, the uropygeal gland (*qv.*) but sometimes applied to a coccygeal body (*q.v.*)

colleterial gland glands in the wall of the oviduct of some arthropods which secrete a sticky material holding the eggs in masses

compound gland any gland of which the duct is

branched and terminates in separate glandular elements

Cowper's gland an accessory sex gland of unknown function in male animals

coxal gland a protrusible gland in the coxa of many insects

apocrine gland one in which part of the cytoplasm of the cell is lost together with the secretion

endocrine gland a gland without ducts, the secretory products of which are distributed by the blood and known as hormones

exocrine gland a gland the secretion of which is collected into a duct for transportation to the site of action

holocrine gland a gland in which the secretion is produced by the complete breakdown of some of the glandular cells

merocrine gland a gland the product of which is secreted without the loss of any part of a cell, except the secretory granules

crural gland glands lodged in the base of the legs in Onycophora. That in the seventeenth pair of legs is greatly enlarged and is thought to be an accessory sex gland.

digitiform gland a synonym for the rectal gland of elasmobranch fish

ductless gland = endocrine gland

Dufour's gland a pheromone-producing gland in the abdomen of ants: in other Hymenoptera it contributes to the venom

von Ebner's gland albuminous glands associated with the circumvallate papillae in the mammalian tongue

femoral gland one of the glands constituting the femoral organ (*q.v.*) of lizards

Filippi's gland one of several accessory glands of the silk-producing apparatus of lepidopteran larvae

frontal gland any gland imbedded in the antero-dorsal region of an invertebrate though usually only specifically so designated in Nemertea and Platyhelminthes and termites. In this last case the gland produces a milky secretion through the frontal pore. The term is also applied to the hatching gland of amphibian larvae

froth gland glands which produce the "spittle" of cercopid homopteran insects (= Batelli's glands)

gas gland 1 (*see also* gas gland 2) the gland which secretes gas into the pneumatophore of a siphonophoran

gas gland 2 (*see also* gas gland 1) a mass of capillaries lying at the anterior end of the swim bladder of teleost fishes

Gene's gland an accessory gland of the acarine female reproductive system functioning only during the period of egg laying

cystogenous gland a gland for secreting a cyst. Best known in the cercariae larva of trematode platyhelminthes

Gilson's gland thoracic glandular excretory organ in trichopteran insects

Harderian gland a tear-producing gland found in many reptiles and birds differing from the lacrymal gland of mammals in being situated on the inner side of the orbit

hatching gland a gland or group of glandular cells in the head region of larval amphibians that is supposed to secrete an enzyme to dissolve the gelatinous envelope

holocrine *see above* under -crine

hedonic gland one that is concerned in the stimulation of sexual desire

hibernating gland a gland found in some hibernating mammals possessing some of the characteristics both of fat glands and hemolymph glands

labial gland the salivary gland of insects

lachrymal gland a lubricating (tear-producing) gland on the outer side of the vertebrate orbit

lepal gland a nectar secreting gland developed from a modified stamen (*cf.* nectary)

Leiblein's gland a gland, presumably digestive in function, opening into the base of the esophagus in some gastropod Mollusca

Lieberkuhn's glands simple tubular epithelial glands of the vertebrate intestine

lingual gland in insects, very loosely any structure, even those of doubtful glandular function, found associated with the tongue

lymph gland aggregates of connective tissue, leucocytes and lymph vessels

hemolymph gland a lymph gland distinguished by the presence of erythrocytes

mammary gland one of the prominences, known as breasts when pectoral, dugs when abdominal and udders when lying between the hind legs. They house the mammalian milk glands (*cf.* milk gland 1)

mandibular gland the term applied to any gland which opens at the base of the mandible in insects. In some cases these glands are salivary

internomandibular gland a salivary gland situated at the internal angle of the mandible in some insects

maxillary gland any arthropod gland in the same segment as or opening on the maxilla

Mehlis's gland the shell gland of some trematode platyhelminthes

Meibomian gland a modified sebaceous gland in the mammalian eyelid secreting a fatty lubricant

merocrine *see above* under -crine

Mettenian gland cells which secrete a chalky material in some plumbaginaceous plants (*cf.* chalk gland)

milk gland 1 (*see also* milk gland 2, 3) the milk-producing portions of a mammary gland

milk gland 2 (*see also* milk gland 1, 3) the enlarged sweat glands providing nourishment to the newly hatched protherian mammal

milk gland 3 (*see also* milk gland 1, 2) any gland which produces a nutrient for the nourishment of young, specifically applied by entomologists to such a gland in hippoboscid dipterans

mixed gland one which secretes both mucous and serous material

Moll's gland = ciliary gland

Montgomery's gland one of the areolar glands around the nipple in the mammalian breast

Morren's gland = oesophageal gland (Oligochaetae)

mucous gland one which secretes mucous

mushroom gland a mushroom-shaped assemblage of accessory sex glands in male insects

Nakamura's gland the nuchal venom gland of snakes

oesophageal gland (Oligochaetae) = calciferous gland

oil gland (birds) = uropygial gland, preen gland

ovoid gland (Crinoid) = axial gland (Crinoid)

Pavlovsky's gland the salivary gland of pediculid anopluran insects

penetration gland a gland opening into the terebratorium of a miracidium larva

cyanophilous gland one of those sub-epidermal glands of Turbellaria which stain blue with haematoxylin

eosinophilous gland one of those sub-epidermal glands of Turbellaria which stain deeply in eosin

pituitary gland a compound endocrine gland, derived

in part from the stomodeum (adenohypophysis) and in part from the diencephalon (neurohypophysis)

poison gland any gland secreting a toxin, sometimes intended for injection as those of the venomous insects, fish and some reptiles

preen gland = uropygial gland

preputial gland a gland of unknown function, lying at the anterior end of the penis in mammals

prostate gland a male accessory sex gland of mammals, the secretion of which dilutes spermatic fluid, and possibly activates spermatozoa

prostatoid glands a term applied to some accessory sex glands in the Turbellaria, sometimes also called prostate glands

pygidial glands in certain coleopteran insects, anal glands producing repellant fluids

rectal gland a digitiform gland specifically arising from the rectum of elasmobranch fish. It functions as a salt gland. The term is also used for anal glands

repugnatorial gland = stink gland

rosette gland one of numerous glands in the head of isopod crustacea recently shown to secrete mucopolysaccharides

salivary gland any of several glands, adjacent to the buccal cavity of mammals, that produce saliva. The term is extended to glands of similar location and function in other organisms

salt gland a compound tubular gland found about external nares in marine reptiles and birds. The gland excretes large quantities of sodium chloride thus permitting the animal to utilize sea water

sebaceous gland a small oil gland opening into the hair follicle

secretory gland a gland that secretes, as distinct from excretes

serous gland one that secretes a clear, watery fluid

serific gland one that secretes brin

shell gland any gland that secretes the outer coat of eggs

accessory shell gland a mucus-secreting gland coating the eggs of many gastropods as they are laid

short gland (*bot.*) one of those that, on some saxifrages, exude salt which dries to a chalky appearance (*cf.* chalk gland)

simple gland in *zool.* any unbranched gland or, in *bot.* a unicellular gland

sinus gland a gland in Crustacea located between the two basal optic ganglia and that secretes a hormone that increases the molting rate

Stensen's gland a compact gland in the nasal cavity, lying in the lateroventral wall

stink gland any gland producing a mephitic secretion, but more usually applied to such structures in insects than in vertebrates

sweat gland any gland in the skin of mammmals which is not a sebaceous gland

apocrine sweat gland any sweat gland producing a secretion other than sweat. These glands are produced from, or associated with, hair follicles

eccrine sweat glands sweat glands producing simple sweat and having no apparent relation to the hair follicle

tarsal gland a modified sebaceous gland on the edge of the eyelid more commonly called a Meibomian gland (*q.v.*)

tear gland = lachrimal gland

thymus gland an endocrine gland arising from evaginations of the dorsal wall of pharyngeal pouches

thyroid gland one of a pair of endocrine glands lying

on each side of the thyroid cartilage in mammals, that secretes thyroxin

accessory thyroid gland = ultimobranchial body

parathyroid gland an endocrine gland arising from pockets developed on the anterior surface of pharyngeal pouches

toral gland = lepal gland

tubular gland a simple gland of which the secreting cells form a cylindrical tubule

bulbourethral gland = prostate gland

uropygial gland that integumentary gland of birds found immediately above the tail which produces a secretion used by the bird for dressing or preening its feathers. The gland is very prominent in anseriform and some other aquatic birds (= preen gland, oil gland)

ventral gland the unicellular, excretory gland, found in the vicinity of the pharynx in free-living marine nematodes

Versonian gland one of many epidermal glands in lepidopteran larvae, at one time supposed to control or facilitate ecdysis

gland 2 (*see also* gland 1) in *bot.* used as synonymous with glans (from which the word is indeed derived) as well as for the usual secreting structures. The term is also used for the tissue to which the two translator arms are joined in asclepiadaceous plants

glandulaceous acorn-colored

glandular the gland in the orchid flower, to which the pollen adheres

glans an acorn

glareal pertaining to gravel

Glareolidae a family of charadriiform birds containing the pratincoles distinguished from all other non-web footed charadriiform birds by the conspicuously forked tail

glasoogje *S.Afr.* popular name of zosteropid bird *Zosterops virens* (*cf.* white-eye)

[glass] *either* a mutant (gene symbol *gl*) in linkage group I of *Habrobracon juglandis.* The phenotypic expression is small eyes *or* a mutant (gene symbol *gl*) mapped at 63.1 on chromosome III of *Drosophila melanogaster.* The phenotypic expression is small eyes with fused facets

[glass like] a mutant (gene symbol *gl-l*) mapped at 64± on chromosome III of *Drosophila melanogaster.* The phenotype has roughened small eyes with the central portions pigmented orange

glaucinous bluish green

Glauconiidae = Leptotyphlopidae

glaucus *Brit.* the carangid fish *Lichia glauca*

[Glazed] a phenotypic glass-like eye associated with the inversion in (*2LR*) *Gla* in *Drosophila melanogaster*

-glea *see* -gloea

glead *Brit. obs.* = kite

glean a term for a school of herring

foliage-gleaner any of several species of furnariid bird of the genera Anabacerthia, Automolus, Philydor and Syndactyla

gleba the mass of interwoven mycelia which form the interior of the immature fruiting head of a puffball

glebe a specific plot of land, particularly one assigned for its value, and hence, by extension, any particularly fertile area

glebula chambered, spore-producing tissue in some fungi and algae, or a rounded bump on the thallus of some lichens

gled = glede

glede *Brit. obs.* for buzzard or kite

Gleicheniaceae a family of filicale pterophytes containing the single genus Gleichenia distinguished by the subsessile sporangia with small sori and pseudodichotomous multiforked leaves

-gli- *comb. form* meaning "glue" and thus applied to some forms of connective tissue

glia a general term applied to connective tissue of the vertebrate central nervous system

astroglia macroglial cells with large nuclei and many ramifying fibers (= astrocyte)

macroglia a general term for glial cells of ectodermal origin

microglia a general term for glial cells of mesodermal origin

oligoglia macroglial cells having only a few fibers (= oligodendrocyte)

Rio-Hortega's glia = microglia

glider *S.Afr.* either of two species of nymphalid lepidopteran insects of the genus Cymothoe. *C. coranus* (blonde glider) or *C. alcimeda* (battling glider)

Gliridae that family of myomorph rodents containing the dormouse. The absence of a caecum is distinctive

Glisson's *epon. adj.* from Francis Glisson (1597–1677)

GLG a yeast genetic marker indicating a requirement for glycogen

gl–l *see* [glass-like]

glm a bacterial mutant showing a requirement for glutamine

conglobate balled

Globidia an order of sarcosporidian sporozoan Protozoa distinguished by fusiform spores

-globin (*see also* hemo-) a general term for the protein portion of respiratory pigments

Globulariaceae a small family of tubuliflorous dicotyledons that is distinguished from the Scrophulariaceae by the solitary ovule and one-celled ovary

globule apart from the usual meaning, is applied to the antheridium of Characeae

polar globule = polar body

globulet a secretory trichome

globulin general term for a group of water insoluble simple proteins

euglobulin a simple protein insoluble either in ammonium sulfate solution or water

pseudoglobulin a simple protein soluble in ammonium sulphate solution but insoluble in water

thyroglobulin a secretion of the thyroid gland, normally stored in thyroid follicles

glochidium (*see also* larva) a barbed bristle used as an organ of attachment

-gloe- *comb. form* meaning "glue" and applied generally to apparently structureless colloid layers in organisms

mesogloea a transparent gelatinous matrix found in the walls of sponges, and between the ectoderm and endoderm of coelenterates

pelogloea the inorganic material adhering to aquatic organisms and submerged objects

perigloea the gelatinous investment of a diatom

Gloeophyta = Thallophyta

-gloia- = gloea-

-glom- *comb. form* meaning "ball" or "sphere"

Glomandreae a subfamily of Liliaceae

glome a spherical flower head

conglomerate clustered

glomerule a cluster of flower heads in a common involucre

glomerulus literally a "little ball" but applied in *anat.* to a ball-shaped network of blood vessels projecting into a cavity. Without qualification, in vertebrate anatomy, usually refers to the glomerulus of a kidney unit

glory used in compound names of spectacular organisms

Bhutan glory *I.P.* the papilionid lepidopteran insect *Armandia lidderdalei*

jungle glory *I.P.* the amathusiid lepidopteran insect *Thaumantis diores*

Kentish glory *Brit.* the endromidid lepidopteran insect *Endromis versicolor*

morning glory (*see also* miner) any of several climbing vines of the convolvulaceous genus Ipomaea

Brazilian morning glory *I. setosa*

tall morning glory *I. purpurea*

glory-of-the-snow any of several species of the liliaceous genus Chionodoxa

-gloss- *comb. form* meaning "tongue"

glossa literally a tongue, but specifically applied to the inner of the two terminal lobes of an insect labium

paraglossa one of a pair of structures on the insect labium corresponding to the galea of the insect maxilla

bolitoglossal having a mushroom-shaped tongue

detoglossal with the tongue attached to the lower jaw along part of its length in contrast, to the boletoid tongue

Glossata = Lepidoptera

Glossinidae a family of brachyceran dipteran insects now usually included in the Muscidae

Glossiphoniidae a family of rhynchobdelloid hirudinean annelids in which the posterior sucker is pedunculate but the anterior fused with the body

glossum a word without classic justification but apparently an attempt to latinize the *Gr.* glossa (*q.v.*)

entoglossum the skeleton of the tongue in birds and some turtles

hypoglossum a cartilage found in the floor of the mouth of some chelonian reptiles which is not a part of the hyoid apparatus

paraglossum the anterior, usually cartilaginous, skeletal element of the tongue of birds

[glossy seedling] *either* a mutant (gene symbol gl_1) mapped at 26 in linkage group VII of *Zea mays*. The term is descriptive *or* a mutant (gene symbol gl_3) mapped at 111 on linkage group IV of *Zea mays*. *or* a mutant (gene symbol gl_{17}) mapped at 0 on linkage group V of *Zea mays*. *or* a mutant (gene symbol gl_2) mapped at 38 on linkage group II of *Zea mays*. The name is descriptive

glottid a variant form of glottis, presumably derived from the plural glottides but usually used as though singular

proglottid one of the divisions, improperly called segments, into which the body of a tapeworm is divided (*cf.* proglottis)

craspedote proglottid a proglottid having a velum

acraspedote proglottid a proglottid lacking a velum

glottis poroperly the back of the tongue but now used only for the opening of the pharynx to the trachea

epiglottis a raised flap which folds back to assist in closing the glottis in mammals

hypoglottis in insects, a sclerite between the mentum and labium

proglottis the tip of the tongue (*cf.* proglottid)

foxglove (*see also* aphid) any of several species of herbs of the scrophulariaceous genus Digitalis, usually without qualification *D. purpurea*

false foxglove any of several species of herb of the scrophulariaceous genus Gerardia, particularly those of the subgenus Panctenis

mullein foxglove *U.S.* the scrophulariaceous herb *Seymeria macrophylla*

glt a bacterial mutant showing a requirement for glutamic acid

glucagon a hormone secreted in the α-cells of the islets of Langerhans in the pancreas. Its action is antagonistic to that of insulin (*q.v.*)

glucanase general term for enzymes catalyzing the hydrolysis of α-1,4-glucan links

cycloheptaglucanase an enzyme catalyzing the production of heptaglucan from cycloheptaglucan through the hydrolysis of an α-1,4-glucan link

cyclohexaglucanase an enzyme catalyzing the production of hexaglucan from the cyclohexaglucan from the hydrolysis of an α-1, 4-glucan link

glucosidase a general term for enzymes catalyzing the hydrolysis of glucosides

amylo-1,6-glucosidase an enzyme catalyzing the breakdown of amylopectin and glycogens through the hydrolysis of the α-1, 6-glucan links

oligo-1,6-glucosidase an enzyme catalyzing the breakdown of dextrins, isomaltose and panose through the hydrolysis of α-1,6-glucan links

β-acetylaminodeoxyglucosidase an enzyme catalyzing the hydrolysis of β-phenyl-2-acetylamino-2-deoxy-D-glucoside into phenol and 2-acetylamino-2-deoxy-D-glucose

thioglucosidase an enzyme catalyzing the hydrolysis of a thioglycoside into a thiol and a sugar

α-1,3-glucosidase an enzyme catalyzing the hydrolysis of 1,3 glucosides into glucose

β-glucuronidase an enzyme catalyzing the hydrolysis of β-D-glucuronide into an alcohol and D-glucuronic acid

[Glued] a mutant (gene symbol *Gl*) mapped at 41.4 on chromosome III of *Drosophila melanogaster*. The phenotypic expression is rounded facets in small eyes

-glum- *comb. form* meaning "husk"

Glumales = Glumiflorae

glume a bract, particularly one of a grass

empty glume the bract of a grass not containing a flower

floral glume that glume which contains the flower

outer glume a flower-enclosing glume at the base of a grass spikelet

Glumiflorae that order of monocotyledons which contains the grasses and sedges. These groups are closely related to but clearly distinguished from the rushes by the nature of the infloresence

L-glutamine an amino acid analog, glutamic acid monoamide

glutinant = isorhiza nematocyst

glutton a large palaearctic musteline carnivore *Gulo gulo* (*cf.* wolverine)

Glyceridae a family of errant polychaete annelids distinguished by the fact that the prostomium is annulated

glycine aminoacetic acid. NH_2CH_2COOH. A widely dispersed amino acid not necessary for the growth of rats

glycolysis properly the breakdown of glycogen and the derived sugar

glyoxalase II = hydroxy acyl glutathione hydrolase

-glyph- *comb. form* properly meaning "carved" or

"notched." The common usage as "fang" (of a snake) is difficult to justify

aglyph said of a snake having solid conical teeth

ectoglyph = opisthoglyph

kinetroproglyph = solenoglyph

opisthoglyph said of snakes in which the fangs are one to four rear teeth of the maxilla enlarged and grooved medially

proteroglyph said of a snake in which the fangs are anterior and immobile—that is they cannot be erected

siphonoglyph a groove down one side of the pharynx of anthozoan coelenterates

solenoglyph said of snakes having hollow fangs that may be folded back

Glyphipterygidae a family of yponomeutoid lepidopteran insects

glypholecine having a surface marked with wavy grooves

GMP guanosine 5′ phosphate

gnat (*see also* catcher, eater, wren) a rather meaningless term applied to a variety of dipteran insects many capable of inflicting a "bite." In *Brit.* usage extended to the mosquito

buffalo gnat popular name of simuliids

southern buffalo gnat *U.S. Cnephia pecuarum*

eye gnat popular name of chloropids

fungus gnat popular name of mycetophilids

dark winged fungus gnat popular name of sciarids

gall gnat popular name of cecidomyiids

Clear Lake gnat *U.S.* the cecidomyiid *Chaoborus astictopus*

potato scab gnat *U.S.* the mycetophilid *Pnyxia scabiei*

turkey gnat *U.S.* the simuliid *Simulium meridional*

water gnat *Brit.* the hydrometrid heteropteran insect *Hydrometra stagnorum* (= water measurer)

wood gnat popular name of rhyphids

-gnath- *comb. form* meaning "jaw"

endognath the modified endopodite of a crustacean mouth part

paragnath 1 (*see also* paragnath 2, 3) properly anything that lies alongside a jaw and applied to various palps and the like on the jaws of various anthropods and annelids

paragnath 2 (*see also* paragnath 1, 3) one of the two lobes of the metastoma (*q.v.*)

paragnath 3 (*see also* paragnath 1, 2) rudimentary or embryonic appendages sometimes appearing between the mandible and the maxilla of Thysanura and Collembola, and of Crustacea

-gnathal a synonym of -gnathous, the form preferred in this work

gnathite diminutive of gnath

scaphognathite the second maxilla of decapod Crustacea that fans water over the gills

Gnathobdellida an order of hirudinian annelids which lack a proboscis but usually have jaws

gnathochilarium *see* -chilarium

Gnathostomata a division of craniate chordate animals erected to contain those classes possessing a hinged lower jaw. Thus defined, this group contains all craniates except the Marsipobranchii

Gnathopoda = Arthropoda

gnathos a pair of rods in the genitalia of male Lepidoptera which, when fused together, form the scaphium

-gnathous pertaining to a jaw (*cf.*-gnathal)

aegithognathous said of the skull of those birds, such as the passeriformes, which resemble the schizognathous type except that the vomer is broad extending into the adjacent region of the nasal capsule

desmognathous said of the skulls of birds, such as the hawks, parrots and ducks in which the vomer is small or wanting and the palatine portions of the maxilla meet in the middle line

dromaeognathous said of the skull of a bird such as the Ratitae in which the pterygoid usually articulates with the vomer medial to the palatine

entognathous = entotrophous

epignathous said of birds in which the upper bill is longer than the lower and particularly when the tip is bent down over the lower

exognathous = exotrophous

hypognathous *either* having the mouth pointed directly backwards or downwards *or*, in birds, those in which the lower jaw is longer than the upper

pseudohypognathous having the head so tilted that a prognathous jaw appears hypognathous

metagnathous 1 (*see also* metagnathous 2) said of insects having different types of mouth parts in the larva and in the adult

metagnathous 2 (*see also* metagnathous 1) said of birds more commonly called cross-billed

neognathous used in contrast to palaeognathous and therefore containing aegithognathous, desmognathous, schizognathous and saurognathous types of skull

palaeognathous synonymous with dromaeognathous in definition but used in contrast to neognathous when the latter is used in its broad sense

paragnathous said of birds in which the upper and lower bill are equal in length

prognathous 1 (*see also* prognathous 2) with jaws directed forward, used particularly of insects with a horizontal head and projecting jaws

prognathous 2 (*see also* prognathous 1) said of flowers in which the anthers project forward at the base

saurognathous said of the skulls of birds, such as those of the woodpeckers, in which the delicate rod like vomers and maxillo palatines scarcely extend inwards from the maxillae

schizognathous said of the skulls of those birds such as the gulls, gamebirds, etc., in which the maxillo-palatines do not meet the vomer or each other

gnaur a burr on a tree trunk

Guetaceae the only family of Gnetales. The characteristics of the order apply. It closely approaches the angiosperms in the development of the endosperm and differs from the Coniferae by the presence of a perianth and the absence of resin tubes

Gnetales an order of gymnospermous spermatophytes distinguished by possessing compound staminate strobili. They possess many characteristics of angiosperms

gnomonical abruptly bent or angled

phytognosis = botany

-gnot- *comb. form* meaning "well known"

gnow *Austral.* the megapodiid bird *Leipoa ocellata* (= mallee-fowl)

gnu any of several species of alcelaphine antilopids of the genera Connochaetes and Gorgon

goa a Tibetan "goat-gazelle" *Procapra picticaudata* (*cf.* zeren)

goal-of-pleasure the cruciferous-herb *Camelina sativi*

goat (*see also* beard, fish, grass, louse, mite, moth, rue, and sucker) any of numerous artiodactyle mammals of the caprid subfamily Caprinae. The goats grade directly into the antelopes through the

Saiginae (*q.v.*) and are almost impossible, in fact as in myth, to separate from the sheep

gazelle-goat popular name of saigine caprid artiodactyles (*cf.* goat-gazelle)

Rocky Mountain goat the rupicaprine caprid *Oreamnos americanus* which is the American representative of the rupicaprines corresponding to the chamois of Europe and the serows and gorals of Asia

rock-goat popular name of rupicaprine caprids. This group is distributed over the whole N. hemisphere, the goral and the serow being the Asiatic representatives, the chamois the European, and the Rocky Mountain goat, the American

Persian wild goat *Capra aegagrus*. This is one of the principal ancestors of the domestic goat

gobbler a male turkey, though tom is the more usual term

Gobiesocidae a family of xenopterygious fishes, commonly called clingfish, distinguished by the large ventral sucker by which they attach to rocks

Gobiesociformes = Xenopterygii

Gobiidae a large family of gobioid acanthopterygian fishes possessing a sucker formed by the fusion of the inner edges of the pelvic fins but distinguished from the Taenioidedae by the presence of two dorsal fins

Gobioidea a suborder of acanthopterygian fishes distinguished by the possession of one spine and three to five soft rays in the pelvic fins, the inner rays being the longest. The suborder usually includes the families Apocryptcidae, Eleotridae, Gobiidae, Periophthalmidae and Taenioididae

goblet (*see also* cell) a cavity or organism in the shape of a wineglass without the stem

Neptune's goblet the monaxonid sponge *Poterion neptuni*

goby popular name of gobiid fishes

eel goby popular name of taenioid fishes

godwit any of several species of scolopacid bird of the genus Limosa, and occasionally of other genera. *N.Z.*, without qualification, *L. lapponica*

bar-tailed godwit *L. lapponica*

black-tailed godwit *L. limosa*

Hudsonian godwit *L. haemastica*

marbled godwit *L. fedoa*

snipe-billed godwit = dowitcher

go-go *W.I.* the cuculid *Coccyzus minor* (= mangrove cuckoo)

[goggle²] a mutant (gene symbol *gg²*) mapped at 23.1± on the X chromosome of *Drosophila melanogaster*. The phenotypic expression is bulging eyes

goggleboy *U.S. dial.* = hellgrammite

-goggyl- *comb. form* meaning "round" usually transliterated -gongylo-

goffered having reticulated ridges on the surface

gold *see also* marigold

purple-bordered gold *Brit.* the geometrid lepidopteran insect *Hyria muricata*

[golden] a mutant (gene symbol *g₁*) mapped at 43 on linkage group X of *Zea mays*. The phenotypic expression is an overall golden shade on the adult plant

[Golden egg] a mutant (gene symbol *C*) mapped at 14.0 on linkage group XII of *Bombyx mori*. The phenotypic expression is a golden cocoon

Golgi *epon. adj.* from Camillo Golgi (1834–1926)

Goliath *see* beetle, frog, heron

Gomortegaceae a small family of ranale dicotyledonous angiosperms

gom-paauw *S.Afr.* the otid bird *Otis cristata.* (*cf.* bustard)

-gomph- *comb. form* meaning "club" or "club shaped" and, since bolts are club-headed, extended to mean "bolt" or "fasten"

Gomphidae a family of anisopteran odonatan insects with well separated eyes, an unnotched labium, and usually with yellowish markings on the wings

Gomphrenoideae a subfamily of Amaranthaceae distinguished from the Amaranthoideae by having a two-celled anther

gomphosis the joint between the peg-like roots of the teeth and the walls of the sockets into which they fit

urogomphus a backwardly directed projection from the terminal segment of many insect larvae

-gon- *comb. form* used in a great number of compound words concerned in reproduction but also, from another root, in the meaning "angle." Entries in this work will be found under gonad, -gonalis, -gonangium, -gone, -gonel, -goneutic, -gonia, -gonic, -gonidium, gonimium, -gonium, -gonous and -gony

gonad an organ producing gametes. The male gonad is the testis, the female is the ovary

follicular gonad one which consists of numerous small parts scattered through the tissues of the body, particularly in platyhelminths

gonalis the central portion of the genital ridge

epigonalis the posterior portion of the genital ridge

progonalis the anterior portion of the genital ridge

gonangium *see under* angium

gone the asexual equivalent of gamete

-gone *comb. suffix* usually used in the sense of "a reproductive structure" (*cf.* -gonidium, -gonium)

androgone those portions of an antheridium which do not give rise to reproductive cells

carpogene that part of a procarp which produces the sporocarp

epigone a cell layer covering a sporophore. The term is also sometimes applied to the nucleus of Chara

heterogone a flower possessing more than two lengths of stamen or style

hormogone a filamentous alga which reproduces by breaking off terminal portions

merogone a developmental stage produced entirely by cleavage (*cf.* blastula, morula)

parthenogenetic merogone properly a merogone produced from an unfertilized egg but frequently used specifically for one produced from an enucleated egg

nematogone an asexual gemma produced by the protonema

paraphysagone the base of a paraphysis, in the sense of paraphysis 3 (*q.v.*)

parthenogone an organism produced by parthenogenesis

perigone anything surrounding a reproductive structure, and therefore variously applied to the perianth, the perichaetium and the involucre of the inflorescences in mosses

trophogone a branch of an ascomycete fungus specialized for nutritive functions

gonel the floral reproductive apparatus

amphigonel the reproductive structure of a plant which has polycentric axes

anthogonel the condition of having a flower with a well developed corolla

-goneutic *comb. suffix* used in the sense of the "frequency of reproduction"

digoneutic breeding twice a year or having two broods in one year, used particularly of insects; for other forms *see* divoltine (under -voltine)

polygoneutic laying several clutches of eggs in one season

gongylous knob-like

-goni- *see* -gon-

-gonia *comb. suffix* frequently used as synonymous with -gone (*q.v.*) but better confined to the meaning "angle"

metagonia the anal angle of an insect wing

protogonia the apical angle of the insect forewing

-gonic *comb. suffix* used both in the sense of "pertaining to reproduction" and pertaining to "reproductive structures". Most definitions are given under the *subst.* forms -gony (*q.v.*) and -gonium (*q.v.*). The alternative form -gonous (*q.v.*) appears to be used mostly in a rather different sense

heterogonic (*see also* coefficient) said of the life cycle of a parasite, in which a free-living form is hatched from the egg, and subsequently develops into a parasitic form. The term is most frequently found in accounts of Nematodes

hologonic said of a sac-shaped ovary in which germ cells are proliferated only from the blind proximal end (*cf.* telogonic)

homogonic said of the life cycle of a parasite, in which the parasitic form develops directly from the egg

telogonic said of a sac-shaped ovary in which germ cells are proliferated over the whole of the interior surface of the sac (*cf.* hologonic)

gonidema the gonidial layer of lichens

gonidium the diminutive of gonium (*q.v.*) generally used for a reproductive cell, or structure of a plant, but also applied to the algal component of a lichen

acrogonidium one which forms at the apex of a gonidiophore

androgonidium one concerned with the production of male gametes

cladoandrogonidium an androgonidium borne on a special branch

androgyno-cladogonidium a branch carrying male and female reproductive structures

chlamydogonidium a thick-walled resting gemma of some fungi

deuterogonidium a gonidium in the second generation of a transitorial series

gynogonidium one concerned in the reproduction of female gametes but also used as synonymous with oospore (*q.v.*)

hologonidium = soredium

leptogonidium = microgonidium

metrogonidium = heterocyst

parthenogonidium a protozoan cell that gives rise to a colony by continued fission

phytogonidium a gonidium capable of independent germination

stylogonidium a gonidium derived from spore producing hyphae

teleutogonidium = teleutospore

zoogonidium = zoospore

macrozoogonidium the larger of the two kinds of zoospores of Ulothrix

gonimium the gonidium of a lichen

hormogonimium one of a group of gonimia arranged like beads on a string

gonimon = gonidial layer

asincronogonism = dichogamy

-gonium *comb. suffix* used in the widest sense for "reproductive structure" The *adj. suffix* -gonic (*q.v.*) has a few specialized meanings

amphigonium = archegonium

archegonium that structure which bears the female gemete in lower plants

ascogonium that structure which gives rise to the sexual reproductive organs in ascomycete fungi (= archicarp)

carpogonium that organ in the lower plants, particularly red algae, which gives rise to a female gamete

gametogonium a common term embracing both oogonium and spermatogonium

oogonium = carpogonium

ovogonium the animal oocyte or, rarely, the female gametangium of plants (= carpogonium, oogonium)

spermatogonium the male gametangium of plants or the spermatocyte of animals

sporogonium the sporocarp of a moss

-gonous *comb. suffix* logically synonymous with -gonic (*q.v.*) but rarely used except in the sense of "breeding" as distinct from reproduction. Occasionally found in the sense of "angled (*e.g.* "digonous")

digonous two angled

homogonous said of a race or variety that breeds true

isogonous used of hybrids which combine the parents' characters equally

anisogonous the condition of an hybrid which does not derive characters equally from both parents

gonochrism said of an organism in which the two sexes are distinct, in contrast with hermaphroditism

gonolek *Afr.* the laniid bird *Laniarius barbarus* and, occasionally, other species of Laniarius

Gonorhynchidae a monospecific family of isospondylous fishes containing the sandfish or beaked salmon (*Gonorhynchus gonorhynchus*), readily distinguishable by the single conspicuous barbel and beady ctenoid scales

Gonostomatidae a small family of deep sea isospondylous fishes mostly with luminescent organs. Sometimes called bristlemouths

-gony *comb. suffix* used in the sense of "a method of reproduction." The *adj. suffix* -gonic (*q.v.*) and -gonous (*q.v.*) can be derived from any of these words but both have other specialized uses

agamogony = schizogony

amphigony any type of reproduction involving two individuals

archigony spontaneous generation

dissogony the condition of an organism that reproduces sexually both as a larva and as an adult

gamogony *either* reproduction by gametes, and specifically the process of multiple fission, when the resulting cells are sex cells *or* sporogony

heterogony the alternation of parthenogenetic and zygogenetic generations

merogony development of an enucleated egg initiated by a sperm (*cf.* -gone)

opsigony the production of proventitious buds

schizogony the process of multiple fission when the end products develop directly into adults, not into spore or gametes (*cf.* gamogony, sporogony)

sporogony the process of multiple fission, following sexual fusion of gametes, and when the product is spores

syngony used of some invertebrates, particularly nematodes, when hermaphroditism is meant

telegony the supposed transfer by a female of characters derived from one mate to the offspring of a subsequent mate

gonys the ridge on the anterior tip of the lower bill of a bird

goober *U.S.* = peanut

Goodeniaceae a small family of campanulate dicotyledonous angiosperms closely allied to the Campanulaceae from which they differ in possessing an indusiate pollen-collecting cup under the stigma

goosander = merganser

goose (*see also* barnacle, foot, grass, louse, plum mother-carey, teal) usually any of very numerous large waterfowl of the anserid subfamily Anserinae, though the word is also used for some allied forms

Alaska goose *Chen hyperborea* (= lesser snow goose)

barnacle goose *Branta leucopsis* (*cf.* goose barnacle)

Cape Barren goose *Cereopsis novaehollandiae*

Canada goose *Branta canadensis*

carr goose *Brit. obs.* grebe

comb goose = knob billed duck

African dwarf goose the cairine *Nettapus auritus*

Egyptian goose the tadornine *Alopochen aegyptiaca*

ember goose *Brit.* = great northern diver

grey-lag goose *Anser anser*

magpie goose an Australian bird, *Anseranas semipalmata*, the only member of the subfamily Anseranatinae of the Anatidae. It differs from all other geese in having semipalmate feet

Mother Carey's goose the procellariid bird *Macronectes gigantens* (= giant petrel) (*cf.* goose)

marsh goose *Brit. Anser anser. U.S. Branta canadensis*

pigmy goose = *Afr.* dwarf goose. *I.P.* and *Austral* = cotton teal

quinck goose = brant

rain goose = rain bird

rode goose = brant

snow goose *Anser caerulescens*

lesser snow goose *Chen hyperborea* and, less frequently, *C. caerulescens*

solan goose = gannet

swan goose the Far Eastern and Siberian goose *Anser cygnoides*, supposed to be the ancestor of domestic geese

wild goose *W.I.* = snow goose *U.S.*

spur-winged goose *Afr. Plectropterus gambensis*

gopher (*see also* frog, mouse, plant) any of several species of *Amer.* ground squirrel; without qualification, frequenty *Citellus tridecemlineatus*

grey gopher *Citellus franklini* (= Franklin's ground squirrel)

pocket gopher *U.S.* any of numerous species of geomyid sciuromorph rodents of the genera Thomomys, Geomys and Cratogeomys

goral the Far Eastern rupicaprin caprid *Naemorhedus goral* (*cf.* serow)

Gordiaceae = Nematomorpha

Gordioidea a class of the phylum Nematomorpha distinguished from the Nectonematoidea by the lack of swimming bristles. All are fresh water or terrestrial

gorge = throat

gorgeret the barbed sting of hymenopteran insects

gorget a throat patch, particularly used of a distinctively colored patch on the throat of a bird

gorgon *I.P.* either of two species of papilionid lepidopteran insect of the genus Meandrusa

Gorgoracea that order of alcyonarian Anthozoa which contains the horny corals, the gorgonians, the sea fans, and the sea feathers. The axial skeleton, made either of horny materials or calcareous spicules, is typical

gorilla the simiid primate distinguished from the chimpanzee and oran-utan by its naked face and black skin

gorse the leguminous shrub *Ulex europaeus*

gosling 1 (*see also* gosling 2) a young goose

gosling 2 (*see also* gosling 1) = catkin

gossander = merganser

gothic *Brit.* any of several species of noctuid lepidopteran insects usually, without qualification, *Naenia typica*

   beautiful gothic *Brit.* the *Heliophobus hispidus*

   bordered gothic *Neuria reticulata*

Gotte's *epon. adj.* from Alexander Wilhelm Gotte (1840–1922)

plum gouger *U.S.* the curculionid coleopteran insect *Anthonomus scutellaris*

goujon properly the French word for gudgeon (*q.v.*) but applied by early settlers in the Mississippi Valley to the ictalurid catfish *Pylodictis olivaris*

Gouldiidae a family of pelecypod mollusks commonly called triangular shells in vitue of their shape

goura any of several large Australasian columbid birds of the genus Goura, once separated as the family Gouridae

gourami properly the anabantid fish *Osphronemus gorany*. The term is, however commonly applied by aquarists to other anabantid fishes

   dwarf gourami *Colisa lolia*

   kissing gourami *Helostoma temmincki*

   mosaic gourami *Trichogaster leeri* (= pearl gourami)

   talking gourami *Trichopsis vittatus* (= purring gourami)

   snakeskin gourami *Trichogaster pectoralis*

   skipping gourami *Helostoma temmincki*

gourd any of numerous flask-shaped fruits of several genera of cucurbitaceous vines

   Arizona gourd = calabazilla

   bitter gourd = colocynth

   cranberry gourd *Abobra tenuifolia*

   gooseberry gourd *Cucumis anguira* (= gherkin)

   dish cloth gourd = rag gourd

   hedgehog gourd *Cucumis dipsaceus*

   rag gourd *Luffa cylindrica*

   snake gourd *Trichosanthes anguina*

   tallow gourd *Benincasa hispida*

   wax gourd = tallow gourd

Gouridae a family of columbiform birds commonly called gouras and now usually included in the Columbidae

gowrie *W.I.* the apodid bird *Streptoprocne gonaris.* (= collared swift)

goggle goy *U.S.* the larva of the dobson fly *Corydalus cornutus* (= hell-grammite)

gp *see* [gap]

Gr *see* [Gray egg]

gra a bacterial mutant showing gramicidin resistance

Graafian *see* Graffian

Gracilariidae a very large family of minute tineid lepidopteran insects the larvae of which are all leaf miners

grackle *U.S.* any of several starling-sized icterid birds of the genera Cassidix, Quiscalus and Scaphidura. The term is also applied, in various parts of the world, to sturnid birds. *I.P.*, without qualification, the sturnid *Gracula religiosa* (= myna)

   Antillean grackle any of several West Indian icterids of the genus Quiscalus

   boat-tailed grackle the icterid *Cassidix major*

Graculidae = either Phalacrocoracidae or Sturnidae

-grad- *comb. form* meaning "to walk"

   digitigrade to walk on the toes with the heels clear of the ground

   orthograde to walk with the body vertical

   plantigrade to walk on the flat of the foot with the heel touching the ground

   pronograde to walk with the body horizontal

gradient a gradual change of slope or intensity of effect

   embryonic gradient any embryonic phenomen that differs in intensity along any axis

Graffian *epon. adj*: from Regnier de Graaf (1641–1673). The omission of the second "a" and introduction of the second "f" are almost universal in contemporary literature

graft the action of inserting one part of an organism into another; or the part so inserted

   autologous graft one of which donor and recipient are the same individual

   heterograft one made between two organisms of different species

   homograft one made between two organisms of the same species

grain 1 (*see also* grain 2, aphis, beetle, mite, moth and thrip) a fruit in which the testa and pericarp are fused (= caryopsis)

   Kefir grain = Kefir seed

grains of paradise *see* cardamon

grains of selim the seeds of the annonaceous tree *Xylopia aethiothipa*

grain 2 (*see also* grain 1) the diminutive granule (*q.v.*) is more frequent in *biol.*

   chlorophyll grain = chloroplast

grakle = grackle

grallatorial adapted for wading

Grallinidae a small family of passeriform birds erected to contain the four mudnest-builders. They are distinguished by preparing a large communal nest of mud lined with grass and feathers. Known in Australia as magpie-larks

gramicidin an antibiotic derived from *Bacillus brevis*

graminaceous pertaining to grass or grain

Graminales = Glumiflorae

Gramineae the grass family of glumifloral monocotyledons. The grasses are easily distinguished from the sedges by the mostly hollow, usually cylindrical stem

gramineous = graminaceous

grammicous marked as though with letters or writing

gram-negative said of bacteria that do not retain certain dyes after washing with an iodine solution

   gram-positive said of bacteria that do retain certain dyes after washing with an iodine solution

grampus properly a genus of delphinid whales (*see* Risso's dolphin) but often applied to other cetaceans particularly the killer whale

   water grampus *U.S.* the larva of the dobson fly *Corydalis cornutus* (= hellgrammite)

-gran- *comb. form* meaning "grain" in the sense of a cereal grain

granadilla (*see also* tree) the edible fruit of the passifloraceous vine *Passiflora quadrangularis*

granatinous scarlet-pink

granule any small particle (*cf.* grain)

   Altmann's granule = mitochondrion

   azurophilic granule one that selectively stains with

azure dyes in azure-eosin combinations (*cf.* eosinophilic granule)
basal granule a tubular swelling on the base of a
cilium. It is usually considered that new cilia arise
by the division of basal granules (= basal body)
chlorophyll granule = chloroplast
metachromatic granules the variously staining granules of some bacteria particularly those of *Co-
rynebaclerium diphtheriae*
eosinophilic granule one that selectively stains with
eosin, particularly in eosin-azure, or eosin-methy-
lene blue combination (*cf.* azurophilic granule)
Kuhne's granule one located between the meshes of
the nerve net which terminates a motor nerve
ending
miliary granule a small, solid ossicle of indegerminate
shape in the body wall of holothurian echinoderms
neck granule one visible under the optical microscope
in the neck region of spermatozoa
nissl granule a granule demonstrable in the cytoplasm
of nerve cells stained with an alkaline solution of
methylene blue
Palade's granule = ribosome
polar granule one at the pole of a bacterial cell
"sand" granule a phosphate granule appearing in the
pineal gland
secretory granule one which appears in cells of
exocrine glands that are actively secreting
trichohyalin granule granular material in the cells of
the inner root sheath of a hair
zymogen granule *see* zymogen
grape (*see also* grape 2, beetle, borer, bug, chalcid,
curculio, fly, fruit, midge, mite, moth, sawfly, scale,
skeletonizer, worm) any of numerous species and
very many horticultural varieties and hybrids of the
vitaceous genus Vitis and a few near allies
gray-bark grape *U.S. V. cinerea*
river-bank grape *V. viparia*
bird grape = mustang grape
catbird grape = cat grape
blue grape *V. bicolor.* Also, *N.U.S.* called summer
grape. The term blue grape is also applied horti-
culturally to numerous dark fruited grapes
bull grape *V. rotundifolia*
bullace grape = bull grape
bullet grape = bull grape
bunch grape = blue grape
bush grape *U.S. V. acerifolia* (occasionally called
mountain grape)
Southern California grape = valley grape
canon grape *V. arizonica*
cat grape = red grape
chicken grape *U.S. V. vulpina* (occasionally called
winter grape)
currant grape = mountain grape
everbearing grape = mustang grape
evergreen grape *U.S.* the vitaceous vine *Cissus ca-
tensis*
everlasting grape = mustang grape
fall grape *V. berlandieri.* (Also sometimes called
mountain or winter grape)
fox grape = skunk grape
southern fox grape = bull grape
frost grape *U.S. V. vulpina S.W.U.S. V. riparia* and
V. vulpina
mountain grape *V. rupestris* (occasionally called bush
grape sand grape, or sugar grape)
sweet mountain grape *V. monticola*
muscadine grape *U.S. V. rotundifolia*

mustang grape *V. munsoniana.* The term is also applied to *V. candicans*
opossum grape *V. baileyana*
pigeon grape *U.S. V. cinerea* (also occasionally
called summer grape)
pinewood grape = turkey grape
possum grape *U.S.* the vitaceous vine *Cissus incisa*
raccoon grape = winter grape
red grape *V. rubra* (the term is also generally applied to all red fruited species of Vitis)
rock grape = mountain grape
sand grape = mountain grape
scuppernong grape *U.S. V. rotundifolia*
skunk grape *V. labrusca*
Spanish grape = fall grape
sugar grape = mountain grape
summer grape *V. aestivalis*
turkey grape *V. linsecomii*
valley grape *V. girdinana*
wine grape *V. vinifera*
winter grape *V. cordifolia* (occasionally called
chicken grape)
sweet winter grape *V. cinerea*
grape 2 (*see also* grape 1) in compound names of other
organisms
sea-grape the polygonaceous tree *Coccoloba uvifera*
bastard sea-grape the chenopodeaceous herb *Salsola kali* (= saltwort)
shore grape = sea grape
tail grape the annonaceous woody climbing shrub
Artobotrys odoratissima
Grapsidae a family of free living brachyuran arthropods
with a square flattened carapace the margins of
which are usually parallel and with the last legs as
well developed as those anterior to them
grass 1 (*see also* grass 2, 3, bird bug, land, mite, quit
moth, sawfly, scale, sparrow, sponge, spider, thrip,
veldt, and worm) any plant of the family Gramineae
alkali grass (*see also* grass 2) *U.S.* any of several
species of Puccinellia
creeping alkali grass *P. phyganodes*
triple awned grass *U.S.* any of several species of the
genus Aristida (= needle grass)
barnyard grass *U.S. Echinochloa crusgalli*
beach grass *U.S.* any of several species of the genus
Ammophila, particularly *A. arenaria*
Palm Beach grass *Zoysia japonica*
beard grass *U.S.* any of several species of the genera
Andropogon, Gymnopogon, and Polypogon
silver beard grass *Andropogon argenteus*
woolly beard grass any of several species of the
genus Erianthus
Bengal grass *Setaria italica*
bent grass *U.S.* any of several species of genus
Agrostis
marsh bent grass *A. stolonifera*
purple bent grass any of several species of the genus
Calamovilfa, particularly *C. brevipilis*
reed-bent grass *U.S.* any of several species of genus
Calamagrostis
Rhode Island bent grass *Agrostis canina*
purple bent grass *Calamovilfa longifolia*
rough bent grass *Agrostis hyemalis*
white bent grass *Agrostis alba*
bermuda grass *Cynodon dactylon*
blue grass (*see also* bilbug) any of several species of
the genus Poa
Canada blue grass *P. compressa*
English blue grass = Canada blue grass

Kentucky blue grass *P. pratensis*
Texas blue grass *P. arachnifera*
bottle grass *U.S. Setaria viridis*
brome grass any of several species of the genus Bromus
awnless brome-grass *B. inermis*
bottle brush grass *U.S.* any of several species of genus Hystrix, particularly *H. patula*
buffalo grass *U.S. Buchloe dactyloides*
false buffalo grass *U.S. Munroa squarrosa*
bunch grass *U.S. Andropogon scoparius*
bur grass *U.S.* any of several of the genus Cenchrus
canary grass *U.S.* any of several species of genus Phalaris, particularly *P. canariensis*
reed canary grass *P. arundinacea*
citronella grass any of several species of the genus Cymbapogon
cloud grass *Agrostis nebulosa*
Colorado grass *Panicum texanum*
finger-comb grass any of numerous species of the genus Dactyloctenium
cord grass *U.S.* any of several species of genus Spartina
fresh-water cord grass *S. michauxiana*
couch grass *Agropyron repens* (= witch grass)
blue couch grass *Cynodon incompletus*
crab grass *U.S.* any of several species of genera Digitaria and Eleusine. Usually, without qualification, *D. sanguinalis*
cup grass any of several species of the genus *Eriochloa*. *U.S.*, without qualification, usually *E. contracta*
cut grass *U.S.* any of several species of Leersia
dallis grass *Paspalum dilatatum* (sometimes called Dallas grass)
devil grass = Bermuda grass
billion dollar grass *U.S. Echinochloa frumentacea*
wooly-eared grass *U.S.* any of several species of the genus Erianthus
esparto grass *Stipa tenacissima*
Himalaya fairy grass *Miscanthus nepalensis*
feather grass *U.S.* any of several species of genera Stipa and Leptochloa, particularly *S. pennata*
fescue grass any of numerous species of the genus Festuca. The name is also applied to *Bromus unioloides*
sheep's fescue grass *Festuca ovina*
finger grass any of numerous species of grass of the genera Chloris and Digitaria
floating grass any of numerous grasses of marshy habitat. *U.S.* particularly *Glyceria fluitans*
catchfly grass *U.S. Leersia lenticularis*
flyaway grass *Agrostis hyemalis*
flyback grass *Danthonia compressa*
cocksfoot grass *Dactylis glomerata*
crow's foot grass *Dactyloctenium aegyptium*
gama grass *U.S.* any of several species of the genus Tripsacum (= sesame grass), particularly *T. dactyloides*
goat grass *U.S. Aegilops cylindrica*
goosegrass (*see also* grass 2) *U.S.* usually *Eleusine indica* but also applied to several species of the genus Puccinellia
grama grass *U.S.* any of several species of the genus Bouteloua
guinea grass *Panicum maximum*. Also, erroneously, used for *Sorghum halepensis* (Johnson grass)
hair grass almost any thin hairy grass. In *U.S.* the term is applied to members of the genera Agrostis

(particularly *A. hyemalis* and *A. scabra*) Aira, Deschampsia and Muhlenbergia (particularly *M. capillaris* and *M. expansa*)
tufted hairgrass *Deschampsia caespitosa*
water hair grass *U.S. Catabrosa aquatica*
wood hair grass *Deschampsia flexuosa*
hard grass *U.S. Pholiurus incurvus*
hassock grass *Deschampsia caespitosa*
heath grass *U.S. Sieglingia decumbens*
hedgehog grass (*see also* grass 2) any of numerous species of the genus Cenchrus
hemp-grass any of numerous species of the genus Agrostis
herb-grass *Agrostis alba*
herd's grass *U.S.* and *Brit. Phleum pratense* (*cf.* timothy), though in some *U.S.* localities *Agrostis alba* is so called
holy grass *U.S.* any of several species of Hierochloa, particularly *H. odorata* and *H. borealis*
Hungarian grass *Setaria italica*
Indian grass *U.S. Sorghastrum nutans* and *Hierochloe odorata.*
Johnson-grass *Sorghum halepensis*
blue-joint grass *Calamagrostis canadensis*
june grass *U.S. Danthonia spicata* (= poverty grass) and *Poa pratensis* (= Kentucky blue grass)
knot grass (*see also* grass 2, grass 3) *U.S. Paspalum distichum*
lace grass *U.S. Eragrostis capillaris*
lawn grass loosely, any grass grown on lawns but sometimes restricted to members of the recently introduced genus Zoysia
Japanese lawn grass *Z. pungens*
Korean lawngrass *Z. japonica* (= Palm-Beach grass)
palm-leaved grass *Panicum sulcatum*
lemon-grass *U.S. Ctenium aromaticum* (*cf.* toothache grass)
love grass *U.S.* any of many species of the genus Eragrostis, particularly *E. elegans*
lyme grass (*see also* grass 3) any of several species of the genus Elymus. The "lyme grass" of the upholstery trade is *Deschampsia caespitosa*
Manila grass *Zoysia matrella*
manna grass *U.S.* any of several species of Glyceria
marram grass *Ammophila arenaria*
marsh grass *U.S.* any of several species of genus Spartina
mascarene grass *Zoysia tenuifolia*
mat grass *U.S. Nardus stricta*
may grass *U.S. Phalaris caroliniana*
meadow grass Any of numerous grasses of the genus Poa. *Brit. P. pratensis* (= *U.S.* Kentucky bluegrass) *U.S.* grasses of the genus Glyceria
fowl meadow grass *G. nervata* or *P. triflora*
reed meadow grass *G. grandis*
roughish meadow grass *P. trivialis*
salt meadow grass = salt grass
rough stalked meadow grass *P. trivialis*
wood meadow grass *P. nemoralis*
melic grass any of several species of the genus Melica
mesquite-grass *U.S.* any of several species of genus Bouteloua
millet-grass any of several species of the genus Milium
molasses grass *Melinis minutiflora*
moor grass *U.S. Molinia caerulea*
myrtle grass the araceous herb *Acorus calamus*

needle grass *U.S.* any of several species of genus Aristida

oat grass (*see also* oat) any of several species of the genus Arrhenatherum. Without qualification, usually *A. elatius*

tall meadow oat grass *Arrhenatherum elatius*

wild oat grass *U.S.* any of several species of genus Danthonia

orange grass *U.S. Ctenium aromaticum* (*cf.* toothache grass)

orchard grass *U.S. Dactylis glomerata*

palm grass *Panicum palmifolium*

pampas grass *Cortaderia argentea*

hardy pampas grass any of several species of the genus Erianthus but particularly *E. ravennae*

panic grass *U.S.* any of several species of the genus Panicum

para grass *Panicum barbinode*

pigeon grass *U.S.* the grass *Setaria glauca*

plume grass any of several species of the genus Erianthus but particularly *E. ravennae* (= hardy pampas grass)

pony grass *Calamagrostis stricta*

porcupine grass *Stipa spartea*

poverty grass (*see also* grass 2) *U.S. Sporobolus vaginiflorus, Aristida dichotoma* and *Danthonia spicata* are all so called

quack grass *Agropyron repens*

quaking grass any of the numerous species of the genus Briza

large quaking grass *B. maxima*

little quaking grass *B. minor*

quick grass *U.S.* = witch grass

quitch grass *U.S.* = witch grass

ray grass *Lolium perenne*

redtop grass *Agrostis alba*

reed grass any of several species of the genera Arundo and Bambusa

wood-reed grass *U.S.* any of several species of genus Cinna

rescue grass *Bromus unioloides*

Rhode's grass *Chloris gayana*

rice grass *N.Z. Microlaena stipoides*

rye grass any of several species of the genus Lolium, particularly *L. perenne*

giant rye grass *Elymus condensatus*

Italian rye grass *L. multiflorum*

perenniel rye grass *L. perenne*

salt grass any of several species of the genus Distichlis

sand grass *Calamovilfa longifolia* (= purple bent grass) and *Triplasis purpurea*

silvery sand grass *N.Z. Spinifex hirsutus*

sweet scented grass *Hierochloe odorata*

scratch grass (*see also* grass 2) *U.S. Muhlenbergia asperfolia*

scutch grass *U.S. Cynodon dactylon*

feather sedge grass *Andropogon saccharoides*

bird seed grass *U.S.* any of several species of the genus Phalaris

Seneca grass *Hierochloe odorata*

sesame grass *U.S.* any of several species of the genus Tripsacum, particularly *T. dactyloides*

sheep grass = Bermuda grass

sickle grass *Parapholis incurva*

silk grass (*see also* grass 2) *U.S. Oryzopsis hymenoides* and *Agrostis hyemalis* (= hair grass)

skunk grass *U.S. Eragrostis megastachya*

slough grass *U.S. Beckmannia syzigachne* and *U.S. Spartina cectinata*

smut grass *U.S. Sporobolus poiretii*

snake grass *Eragrostis megastachya* (= skunk grass)

rattlesnake grass *Briza maxima,* or *Glyceria canadensis*

spangle grass *U.S.* = spike grass *U.S.*

spear grass (*see also* grass 2) *U.S.* any of several species of the genus Stipa and a few species of Poa

spike grass *U.S.* any of several species of the genera Distichlis, Demazeria and Uniola, particularly *Demazeria sicula* and *U. latifolia*

marsh spiked grass any of several species of the genus Distichlis

stake grass *U.S.* = skunk grass *U.S.*

blue-eyed stem grass *Calamagrostis canadensis, Andropogon furcatus,* and rarely *Agropyron smithii* (= western wheat grass)

stink grass *Eragrostis megastachya*

sweet grass *U.S. Hierochloe odorata*

cat-tail grass *Phleum pratense*

dog's tail grass any of several species of the genus Cynosurus

crested dog's tail grass *C. cristatus*

fox-tail grass *W.U.S. Hordeum murinum* or *H. jubatum* and any of several species of the genera Alopecurus and Setaria

bristly fox-tail grass *Setaria magna*

hare's tail grass *Lagurus ovatus*

squirrel-tail grass *Hordeum jubatum*

tear grass (*see also* grass 2) *Coix lachrymajob*

terrell grass *U.S. Elymus virginicus* (*cf.* lyme grass)

tickle grass *Agrostis hiemalis* and *A. scabra* (= hair grass). In *W.U.S. Hordeum jubatum* (= squirrel tail grass)

timothy grass any of several species of the genus Phleum but particularly *P. pratense*

toothache grass *U.S.* any of several species of genus Ctenium

golden-top grass *Lamarckia aurea*

tumble grass *U.S. Eragrostis spectablis*

umbrella grass (*see also* grass 2) *Austral. Aristida ramosa*

usar grass *Sporobolus orientalis*

uva grass *Gynerium saccharoides*

vanilla grass *Hierochloe odorata* and *H. borealis*

vasey grass *U.S. Paspalum urvillei*

velvet grass *U.S. Holcus lanatus*

sweet vernal grass *U.S. Anthoxanthum odoratum*

wheat-grass any of numerous species of the genus Agropyron (*cf.* blue-eyed stem grass)

Western wheat grass *A. smithii*

white grass *U.S.* any of several species of Leersia

windmill grass *U.S.* any of several species of the genus Chloris, particularly *C. verticillata*

wire grass many wiry grasses including *Poa compressa, Sporobolus junceus, Cynodon dactylon, Andropogon scoparius* and *Eleusine indica*

witch grass *U.S. Agropyron repens* (= couch grass, quick grass)

old witch grass *Panicum capillare*

wood grass *U.S. Sorghastrum nutans*

yard grass any of several grasses of the genus Eleusine, particularly *E. indica*

zebra grass *Miscanthus sinensis*

grass 2 (*see also* grass 1, 3) any of numerous plants more or less resembling a true grass

adder's grass the fern *Polypodium vulgare*

alkali grass (*see also* grass 1) *U.S.* the liliaceous herb *Zigadenus elegans*

arrow grass popular name of helobe monocotyledonous herbs of the family Scheuchzeriaceae

marsh arrow grass *Triglochin palustris*

artifical **grass** sometimes used for certain forage plants, as sorghum, and also for leguminous plants, as clover, lucerne, sainfoin

bayonet grass *U.S.* the sedge *Scirpus paludosus*

bear grass *U.S.* any of several species of the liliaceous genus Yucca

bog grass = sedge

China grass the urticaceous herb *Boehmeria nivea*

coco grass *U.S.* the sedge *Cyperus rotundus*

cotton grass *U.S.* any of several species of sedge of the genus Eriophorum

cow grass the leguminous herb *Trifolium pratense* (= red clover)

deer grass any of several species of melastomataceous herb of the genus Rhexia

ditch grass any of several species of potamagetonaceous herbs of the genus Ruppia

duck grass *U.S.* the eriocaulaceous herb *Eriocaulon septangulare*

eel grass the hydrocharitaceous aquatic herb *Vallisneria spiralis*

blue-eyed grass *U.S.* any of several species of the iridaceous genus Sisyrinchium

yellow-eyed grass *U.S.* popular name of monocotyledonous plants of the family Xyridaceae

fish grass the nymphaeaceous aquatic plant *Cabomba caroliniana*

goosegrass (*see also* grass 1) *U.S.* the rubiaceous herb *Galium aparine*. *Brit.* the rosaceous herb *Potentilla anserina* (= silver weed *U.S.*)

hedgehog grass (*see also* grass 1) *Brit.* the sedge *Carex flava*

knot grass (*see also* grass 1, 3) *U.S.* any of several species of polygonaceous of the genus Polygonum. *Brit.* any of several species of the caryophyllaceous genus *Illecebrum*

lead grass *U.S.* the chenopodiacous herb *Salicornia virginica*

four-leaved grass the liliaceous herb *Paris quadrifolia*

nut grass *U.S.* the sedge *Cyperus rotundus*

parnassus grass *Brit.* the droseraceous herb *Parnassia palustris*

pepper grass any of several species of cruciferous herbs of the genus Lepidium and the marciliaceous herb *Pilularia globulifera*

poverty grass (*see also* grass 1) *U.S.* the cistaceous herb *H. tomentosa*

pudding grass *U.S.* the labiateous herb *Hedeoma pulegioides*

ribgrass the plantaginaceous herb *Plantago lanceolata*

ripple grass = rib grass

scorpion grass any of several species of boraginaceous herbs of the genus Myosotis (= forget-me-not)

scratch grass (*see also* grass 1) *U.S.* any of several species of polygonaceous herbs of the genus Polygonum particularly those of the subgenus Echinocaulon

scurvy grass *Brit.* any of several species of herb of the cruciferous genus Cochlearia, particularly *C. officinals*. *N.Z.* any of several species of herb of the commelinaceous herb Commelina

silk grass (*see also* grass 1) any of several species of compositaceous herb of the genus Chrysopsis but particularly *C. graminifolia* and *C. nervosa*

spear grass *N.Z.* (*see also* grass 1) the umbelliferous herb *Aciphylla squarrosa* (= spaniard *N.Z.*)

star grass *U.S.* any of several species of herbs of the liliaceous genus Aletris, the amaryllidaceous genus Hypoxis and the callitrichaceous genus Callitriche

water star grass the pontederiaceous herb *Heteranthberea dubia*

sword grass (*see also* grass 3) *U.S.* the sedge *Scirpus americanus*

tape grass = eel grass

tear grass (*see also* grass 1) *U.S.* any of several species of polygonaceous herbs of the genus Polygonum particularly those of the subgenus Echinocaulon.

tongue grass any of several species of cruciferous herbs of the genus Lepidium

umbrella grass *see also* grass 1) *U.S.* any of several species of sedge of the genus Fuirena

viper's grass the compositaceous herb *Scorzonera hispanica* (= black salsify)

whitlow grass any of several species of cruciferous herb of the genus Draba, particularly *D. verna* and the saxifragaceous herb *Saxifraga tridactylites*

worm grass any of several species of loganiaceous herb of the genus Spigelia. Also the crassulaceous herb *Sedum album*

grass 3 (*see also* grass 1, 2) in compound names of various organisms other than plants

cutting-grass the tyronomyid hystricomorph rodent *Thyronomys swindernianus*

knot grass (*see also* grass 1, grass 2) *Brit.* the noctuid lepidopteran insect *Acronycta rumicis*

lyme grass (*see also* grass 1) *Brit.* the noctuid lepidopteran insect *Tapinostola elymi*

sword grass (*see also* grass 2) either of two species of noctuid lepidopteran insect of the genus Calocampa but usually, without qualification, *C. exoleta*

grass-of-parnassus *Brit.* the droseraceous herb *Parnassia palustris*

gravel rock particles between 2 mm and 8 mm in size. The term, though loosely used, indicates a size larger than sand and smaller than pebble

graveolence foetid in virtue of the strength, not the quality, of the aroma

gravid the condition of a pregnant mammal

[Gray egg] a mutant (gene symbol *Gr*) mapped at 6.9 on linkage goup II of *Bombyx mori*. The egg of the phenotype is milky white with a serosal pigment

grayling 1 (*see also* grayling 2) popular name of salmonid fishes of the genus Thymallus

Artic grayling *Thymallus arcticus*

grayling 2 (*see also* grayling 1) any of several satyrid lepidopteran insects *U.S. Cercyonis pegala* (= wood nymph *U.S.*) *Brit. Satyrus semele*

graysby *U.S.* the serranid fish *Petrometopon cruentatus*

king **greasy** *Austral.* the papilionid lepidopteran insect Eurycus cressida

greaved = ochreate

grebe 1 (*see also* grebe 2) any of several species of aquatic bird of the family Podicipediadae. *Brit.*, without qualification, *Podiceps ruficollis*

pied-billed grebe *Podilymbus podiceps*

great crested grebe *Afr.* = *Podiceps cristatus*

eared grebe *U.S. Podiceps caspicus*

horned grebe *U.S. Podiceps auritus*

little grebe *Podiceps ruficollis N.Z.P. rufopectus*

spotted grebe *W.I.* the mimid *Margarops fuscus*

tippet grebe great crested grebe

western grebe *U.S. Aechmophorus occidentalis*

grebe 2 (*see also* grebe 1) occasionally used for a few birds not of the family Podicipedidae

sungrebe the heliornithid *Heliornis fulica* (= American finfoot)

green 1 (*see also* green 2, 3, finch, shank) any of several lepidopteran insects brindled green *Brit.* the noctuid *Eumichtis protea*

frosted green *Brit.* the thyatirid *Polyploca ridens*

marbled green *Brit.* the noctuid *Bryophila muralis*

mottled green *S.Afr.* the nymphalid *Euryphura achlys*

green 2 (*see also* green 1, 3) in compound names of plants

wintergreen the ericaceous trailing shrub *Gaultheria procumbens.* This term is also applied to several herbs of the pyrolaceous genus Pyrola

chickweed-wintergreen either of two species of primulaceous herb of the genus Trientalis, *T. europaea* and *T. americana*

flowering wintergreen the polygalaceous herb *Polygala paucifolia*

cliff green *U.S.* the celastraceous shrub *Pachistima canbyi*

sour green *U.S.* the polygonaceous herb *Rumex venosus*

green 3 (*see also* green 1, 2) in compound names of other organisms

billy green *W.I.* the tyrannid bird *Myiarchus stolidus*

[Green b] a mutant (gene symbol *Gb*) mapped at 0.7 on linkage group VII of *Bombyx mori.* The color refers to the phenotypic cocoon

[Green c] a mutant (gene symbol *Gc*) mapped at 7.8 on linkage group XV of *Bombyx mori.* The phenotypic cocoon is green

[green flesh] a mutant (gene symbol *gf*) mapped at 23 on chromosome 1 of the tomato. The phenotypic fruit remains green

[green stiped] a mutant (gene symbol *gs₁*) mapped at 134 on linkage group I of *Zea mays.* The name refers to stripes intermediate between the vascular bundles on the phenotypic leaf

greenie *Austral.* the zosteropid bird *Zosterops gouldi* (= silvereye)

greenlet any of several species of vireonid birds of the genus Hylophilus

greenling popular name of hexagrammid fishes

-gregar- *comb. form* meaning flock or herd

Gregarinida an order of telosporidian Sporozoa distinguished by a small case containing eight sporozotes

gregarious pertaining to small groups of individuals gathered together for apparently other than reproductive reasons

vixgregarious sparsely vegetated

grenadier common name of macrurid fish

purple grenadier the ploceid bird *Granatina ianthinogaster*

gressorial having legs adapted for walking

grey (*see also* bird, tail) *Brit.* any of several species of lepidopteran insect. Without qualification, the noctuid *Dianthoecia caesia*

early grey *Brit.* the noctuid *Xylcompa areola*

poplar grey *Brit.* the noctuid *Acronycta megacephala*

silver-studded grey *S.Afr.* the lycaenid *Crudaria leroma*

Greyioidea a subfamily of Saxifragaceae considered by some to be of familial rank

gribble any of several wood boring isopod crustaceans of the genus Limnoria, particularly *L. lignorum*

griffon any of several Old World, vulture-like accipitrid birds of the genus Gyps. Usually, without qualification, *G. fulvus* (= griffon vulture)

Ruppell's griffon *Afr. G. ruppellii*

grilse that form of the salmon which develops as from the smolt (*q.v.*) on its first return from ocean to rivers

grinder *Austral.* The muscicapid bird *Seisura inquieta* (= restless flycatcher)

scissors-grinder any of several birds with a call resembling the noise of steel on a grindstone, including the *Austral.* grinder, several nightjars and several flycatchers

grindle the popular name of the protospondylid fish *Amia calva*

grisette *Brit.* the noctuid lepidopteran insect *Acronycta strigosa* (= marsh dagger)

grison any of several species of *S. Amer.* terrestrial musteline carnivores of the genus Grison. The tayra (*q.v.*) is sometimes called grison

grive the French word for thrush but used as an English word in parts of the West Indies

bastard grive *W.I.* the vireonid bird *Vireo altiloquus*

yellow-eyed grive the turdid *Turdus nudigenis* (= bare-eyed thrush)

mountain grive *Turdus fumigatus* (= cocoa thrush)

spotted grive the mimid *Margarops fuscatus* (= scaly-breasted thrasher)

grivet = vervet

gromwell *U.S.* any of several boraginaceous herbs of the genus Lithospermum. In *Brit.* the term is applied to the boraginaceous herb *Mertensia maritima* (*cf.* Virginia bluebell)

bastard gromwell *L. arvense*

gentian-blue gromwell *L. fruticosum*

corn gromwell *L. arvense*

false gromwell any of several species of boraginaceous herbs of the genus Onosmodium

groove any fold or channel

atrial groove an ill-defined term often used for the epibranchial groove of Amphioxus in contrast to the endostyle

epibranchial groove a groove running along the dorsal edge of the pharynx in prochordates

hypobranchial groove the groove along the ventral surface of the atrium, particularly in Amphioxus where its ciliated floor and glandular walls constitute the endostyle

peribranchial groove a groove connecting the hypobranchial to the epibranchial grooves

coronal groove a groove separating the exumbrellar surface of some Scyphozoa into upper and lower portions

branchiogenital groove a groove along the side of some Hemichordata that contains the openings both of the gill and genital pores

laryngotracheal groove a ventral embryonic diverticulum from the posterior floor of the pharynx which gives rise to the lung

nasolabial groove a prominent groove leading from the nostril to the upper lip in plethodontid urodeles

oral groove a depressed area leading to the cytopharynx in some Protozoa

oronasal groove a groove leading from the corners of the angle of the mouth to the internal nares in elasmobranch fishes, precursor to nasal passages of higher forms

vestibular groove a ciliated groove on the internal side of the base of the tentacle of endoprocts. It conveys material to the mouth

[grooved] a mutant (gene symbol *gv*) mapped at 36.2 on chromosome III of *Drosophila melanogaster.* The groove is a longitudinal medial one in the thorax

[grooveless] a mutant (gene symbol *gvl*) mapped at 0.2 on chromosome IV of *Drosophila melanogaster.* The phenotypic expression is a diminution of the scutellar groove

Grossularioidea a subfamily of Saxifragaceae considered by some to be of familial rank

ground-hele the scrophulariaceous herb *Veronica officinalis* (= common speedwell)

groundling *U.S.* any of several species of noctuid lepidopteran insect of the genus Perigea

groundsel (*see also* tree) any of several species of compositaceous herb of the genus Senecio. *Brit.* without qualification, *S. vulgaris*

velvet groundsel *S. petasitis*

grouper *U.S.* any of very numerous serranid fishes. (*cf.* bass, hamlet)

    black grouper *Mycteroperca bonaci*
    yellowfin grouper *Mycteroperca venenosa*
    gulf grouper *Mycteroperca jordani*
    marbled grouper *Dermatolepsis inermis*
    Nassau grouper *Epinephelus striatus*
    red grouper *Epinephelus morio*
    snowy grouper *Epinephelus niveatus*
    broomtail grouper *Mycteroperca xenarcha*
    tiger grouper *Mycteroperca tigris*
    warsaw grouper *Epinephelus nigritus*
    yellowedge grouper *Epinephelus flavolimbatus*

grouse (*see also* sandgrouse) any of numerous galliform birds of the subfamily Tetraoninae, *Brit.* without qualification, *Lagopus scoticus* (= red grouse)

    black grouse *Lyrurus tetrix*
    blue grouse *Dendragapus obscurus*
    dusky grouse = blue grouse
    hazel grouse *Tetrastes bonazia*
    red grouse *Lagopus scotticus*
    ruffed grouse *Bonasa umbellus*
    sage grouse *Centrocercus urophasianus*
    sand grouse *see next entry*
    sooty grouse = blue grouse
    spruce grouse *Canachites canadensis*
    sharp-tailed grouse *Pedioecetes phasianellus*
    willow grouse *Brit. Lagopus lagopus* (= willow ptarmigan)

sandgrouse (*see also* grouse) any of numerous species of columbiform pteroclid birds

    black-faced sandgrouse *Pterocles decoratus*
    pin-tailed sandgrouse *P. alchata*
    yellow-throated sandgrouse *P. gutturalis*

growth an increase in size

    appositional growth growth by deposition on the surface of the tissue

    determinant growth that type of plant growth, in which the season's growth ends in a bud

    interstitial growth growth within the substance of the tissue

    intrusive growth that type of growth of plant cells in which the walls of a growing cell intrude between the walls of others

    symplastic growth the growth of plant cells in such a manner that the cell walls remain in their original relationship without cavities developing or intrusion occurring

    interpositional growth = intrusive growth

    primary growth that growth of a plant which derives directly from apical meristem and results in elongation

    secondary growth that plant growth which derives from lateral meristems and results in thickening

    amphitrophic growth that in which the lateral shoots and buds of a plant grow more rapidly than the terminal

grub any worm-like larva not obviously a caterpillar though these were once included in the term

    common cattle grub *Brit.* and *U.S.* the larva of the the hypodermatid dipteran *Hypoderma lineatum*

    northern cattle grub, *U.S.* the larva of the hypodermatid dipteran *Hypoderma bovis*

    grugru grub the edible larva of several species of curculionid coleopteran insects of the genus Rhyncophorus

    white grub *U.S.* any earth-dwelling larva of a scarabaeid beetle

    yellow grub the metacercariae of trematodes encysted in fish and frogs

Grubbiaceae a small family of santanale dicotyledons

grubby the cottid fish *Myoxocephalus aeneus*

Gruidae that family of gruiform birds which contains the cranes. They are distinguished from other families of the order by their long legs with very short elevated hind toes

Gruiformes that order of birds which contains, *inter alia* the cranes and rails. They are distinguished by the long bill, perforate nostril, and long legs with the lower part of the tibia bare

gruinalous shaped like a crane's bill

grumixameira the myrtaceous tree *Eugenia brasiliensis*

grumous having the form of clusters of granules

grundy *Austral.* the motacillid bird *Anthus novaeseelandiae* (= Australian pipit)

grunion the atherinid fish *Leuresthes tenuis*

grunt any of numerous pomadasyid teleost fishes (*cf.* margate)

    barred grunt *Conodon nobilis*
    black grunt *Haemulon bonariense*
    burro grunt *Pomadasys crocro*
    Caesar grunt *Haemulon carbonarium*
    French grunt *Haemulon flavolineatum*
    smallmouth grunt *Brachgenys chrysargyrea*
    blue-striped grunt *Haemulon sciurus*
    white grunt *Haemulon plumieri*

Gryllacrididae that family of orthopteran insects containing the forms commonly called cave-crickets and camel-crickets. Their very long antennae are typical

Gryllidae that family of orthopteran insects which contains the true crickets or tree crickets

Grylloblattida a small order of wingless insects once included in the Orthoptera which they resemble in the structure of the female ovipositor

Grylloblattodea an order of insects at one time included, together with the Dictyoptera and Phasmida in the Orthoptera, a group restricted by this definition to grasshoppers, locusts and crickets. The Grylloblattodea have ten evident abdominal segments but are distinguished from the Dictyoptera by lacking ocelli

Gryllotalpidae that family of orthopteran insects which contains the mole-crickets. The shovel-shaped tibia of the first legs are typical

gs₁ *see* [green striped]

GSH = reduced glutathione

GSSG = oxidized glutathione

gt *see* [giant]

gt-4 *see* [giant-4]

GTH = adrenoglomerulotropin

GTP = guanosine triphosphate

    deoxyGTPase any enzyme catalyzing the hydrolysis of deoxy-GTP to deoxyguanosine and triphosphate

gu a yeast genetic marker indicating a requirement for guanine

gua a bacterial genetic marker indicating a requirement for guanine, mapped at between 41.6 mins. and 43.2 mins. for *Escherichia coli*

guacharo the steatornithid bird *Steatornis caripensis* (= oil bird)

guaguanche the sphyraenid fish *Sphyraena guachancho* (*cf.* barracuda)

guaibero the psittacid bird *Bolbopsittacus lunulatus*

gualilla = paca

guan any of several cracid birds of the genera Penelope, Chaemaepetes, Pipile, Aburria and Oreophasis, with their near allies the general Ortalis and Penelopina often placed in a separate family the Penelopidae (*cf.* chachalaca)

creasted guan *Penelope purpurascens*

horned guan *Oriophasis derbianus*

piping guan any of several cracids of the genus Pipile

purplish guan = crested guan

wattled guan *Aburria aburri*

sickle-winged guan *Chamaepetes goudotii*

guanine 2-aminohypoxanthine   a breakdown product of nucleic acids present in fish scales and the excrement of fish-eating birds

guano the accumulated droppings of sea birds, particularly those of the guanay cormorant on islands off the coast of Peru. The term is also applied loosely to the accumulated droppings of cavernicolous bats

guard *see also* hair

horse guard *U.S.* a large nyssonine sphecid hymenopteran (*Bombix carolina*) so called from its habit of hunting flies near horses

guava any of several species of myrtaceous tree of the genus Psidium. Without qualification, usually *P. guajava*

Brazilian guava *P. araco*

Chilean guava the myrtaceous shrub *Myrtus ugni*

Costa Rican guava *P. friedrichsthalianum*

pineapple guava either of two species of the myrtaceous genus Feijoa and their fruits

strawberry guava *P. cattleianum*

guayaya the myrtaceous tree *Psidium molle*

guaymochil the leguminous tree *Pithecolobium dulce* and its edible fruit

gubernaculum literally, a rudder; particularly a rudder-shaped thickening of the body wall immediately posterior to the copulatory spicule in nematodes and the ligament anchoring the testes to the scrotum

gudgeon *Brit.* the cyprinid fish *Gobio gobio* (*cf.* gujon)

guemal any of several species of odocoilein cervid artiodactyles of the genus Hippocamelus (= huemal)

guenon the general term for those cercopithecid monkeys of the genus Cercopithecus which are not otherwise specifically named. Often, without qualification, refers to *C. callitrichus*

guereza colobid simioid primates of the genus Colobus. They are distinguished by the absence of a thumb. Many have a remarkable fringe of long white hair, giving the impression of a white cloak

honey guide 1 (*see also* honeyguide 2) a stripe of color on a petal pointing to the nectary

honeyguide 2 (*see also* honey guide 1) any of several species of bird of the family Indicatoridae. Usually, without qualification, *Indicator indicator* (= black-throated honeyguide)

greater honeyguide = black-throated honeyguide

lesser honeyguide *Indicator minor*

spotted honeyguide *I. maculatus*

lyre-tailed honeyguide *Melichneutes robustus*

black-throated honeyguide *I. indicator*

scaly-throated honeyguide *I. variegatus*

guillimot any of several charadriiform alcid birds particularly of the genus Cepphus, less frequently of the genus Uria (= murre)

black guillemot *C. grylle*

pigeon guillemot *C. columba*

sooty guillemot *C. carbo*

guira the cuculid bird *Guira guira*

guisaro the myrtaceous tree *Psidium molle*

guit-guit any of several species of South American coerebid birds particularly those of the genera Coereba and Dacnis

-gula- *Comb. form* meaning "throat"

gula literally, the throat, but in insects applied to that part of the ventral portion of the head which extends from the submentum to the posterior margin

gular pertaining to the throat or gula

gularis aquarists' name for the cyprinodont teleost fishes of the genus Aphyosemion

gull 1 (*see also* gull 2, 3) properly any long winged, swimming, bird of the charadriiform family Laridae

black-backed gull *Brit.* and *U.S. Larus marinus. N.Z. L. dominicanus*

great black-backed gull *L. marinus*

lesser black-backed gull *L. fuscus*

southern black-backed gull *L. dominicanus*

slaty backed gull *L. schistisagus*

black-billed gull *N.S. L. bulleri*

red billed gull *N.Z.* and *Austral. L. novae-hollandiae. N.Z. L. scopulinus*

ring-billed gull *L. delawarensis*

short-billed gull = new gull

Bonaparte's gull *L. philadelphia*

Caifornia gull *L. californicus*

common gull = new gull

franklin gull *L. pipixcan* (= prairie dove)

glaucous gull *L. hyperboreus*

black-headed gull *Austral. Thalasseus bergii*

herring gull *L. argentatus*

Iceland gull *L. glaucoides*

ivory gull *Pagophila eburnea*

laughing gull *L. atricilla*

little gull *L. minutus*

mackerel gull = mackerel bird

mew gull *L. canus*

parson gull = black-backed gull

Ross's gull *Rhodostethia rosea*

Sabine's gull *Xema sabini*

sea gull a meaningless term applied to any gull. In *W.I.* confined to *L. atricilla* (= laughing gull)

black-tailed gull *L. crassirostris*

swallow-tailed gull *Creagrus furcatus*

western gull *L. occidentalis*

glaucous-winged gull *L. glaucescens*

gull 2 (*see also* gull 1, 3) birds, not of the family Laridae, mostly but not all gull-like

wocken gull *Austral.* the psittacid bird *Polytelis anthopeplus* (= regent parrot)

gull 3 (*see also* gull 1, 2) *I.P.* any of several white lepidopteran insects of the genus Cepora

Australian gull *Austral.* the pierid lepidopteran insect *Cepora perimale*

[Gull] a mutant (gene symbol *G*) mapped at 12.0 on chromosome II of *Drosophila melanogaster*. The wings of the phenotype are large and widely spread

gullie *W.I.* almost any gull or tern (*cf.* sea gull)

gum 1 (*see also* gum 2, 3) those portions of the mucous membrane of the mouth that cover the base of the teeth

gum 2 (*see also* gum 1, 3) any water soluble, or water miscible, exudate of trees. Water insoluble exudates are properly resins (*q.v.*)

acacia gum from numerous species of leguminous trees of the genus Acacia. Often called gum arabic

gum arabic = gum acacia

gum brea from any of numerous species of the caesalpineaceous genus Caesalpina

gum ghatti a term applied to numerous gums, and very frequently mixtures of gums, of Asiatic origin. The so called "genuine" ghatti gum is derived from the combretaceous tree *Anogeissus latifolia*

gum karaya from the sterculiaceous tree *Sterculia urens*

gum tragacanth from numerous leguminous trees of the genus Astragalus

gum 3 (*see also* gum 1, 2, succory, weed) any of numerous trees which may, or may not, yield gums in the sense of gum 2

apple gum *Austral.* the tree *Eucalyptus stuartiana* and the lumber derived from it

black gum the nyssaceous tree *Nassa sylvatica*

blue gum the myrtaceous tree *Eucalyptus globulus*

bastard blue gum *Austral.* the myrtaceous tree *Eucalyptus leucoseylon*

cider gum the myrtaceous tree *Eucalyptus gunnii*

cotton gum the nyssaceous tree *Nyssa aquatica*

desert gum the myrtaceous tree *Eucalyptus rudis*

flooded gum the myrtaceous tree *Eucalyptus tereticornis*

grey gum the myrtaceous tree *Eucalyptus tereticornis*

hickory gum the myrtaceous tree *Eucalyptus punctata*

juniper gum the pinaceous tree *Tetraceinis articulatum*

white mahogony gum the myrtaceous tree *Eucalyptus acmenioides*

manna gum the myrtaceous tree *Eucalyptus viminalis*

mountain gum the myrtaceous tree *Eucalyptus goniocalyx*

peppermint gum the myrtaceous tree *Eucalyptus amygdalina*

red gum the myrtaceous tree *Eucalyptus rostrata*

apple-scented gum the myrtaceous tree *Eucalyptus stuartiana*

sour gum the nyssaceous tree *Nyssa sylvatica*

spotted gum the myrtaceous tree *Eucalyptus maculata*

sugar gum the myrtaceous tree *Eucalyptus corynocalyx*

sweet gum the hamamelidaceous tree *Liquidambar styraciflua*

tupelo gum the nyassaceous tree *Nyssa aquatica*

tallow wood gum the myrtaceous tree *Eucalyptus mycrocorys*

gumbo the malvaceous herb *Hobiscus esculentus* and its edible parts (= okra)

gumbo-limbo the burseraceous tree *Bursera simaruba*

gumi the elaeagnaceous shrub *Elaeagnus multiflora*

gummosis the pathological production of gums by plants that are not normally gum producers

gundi popular name of ctenodactyl rodents. Without qualification, usually *Ctenodactylus gundi*

gunnel popular name of pholid fish. *Brit.* without qualification, *Pholis gunnellus* (= *U.S.* rock gunnel)

Gunneraceae a family of myrtiflorous dicotyledonous angiosperms erected to contain the single genus Gunnera more usually placed in the Haloragidaceae

guppy the poeciliid fish *Lebistes reticulatus*. The name is derived from the Rev. J. L. Guppy, who introduced this fish to aquarists

gurnard *Brit.* popular name of triglid fish (= *U.S.* sea robin)

armed gurnard *Brit.* the triglid *Peristethus cataphractun*

flying gurnard popular name of dactylopterid fishes, particularly *U.S. Dactylopteris volitans*

mailed gurnard = armed gurnard

yellow gurnard *Brit.* the triglid *Trigla hirundo*

-gust- *comb. form* meaning "taste" and applied by extension to anterior regions of the alimentary canal

epigusta the cental portion of the propharynx in arthropods

subgusta the coria of the parapharynx

gut 1 (*see also* gut 2, 3) any tubular portion of the alimentary canal

foregut an ill-defined term. In vertebrate embryos, applied principally and usually to that portion of the alimentary canal, which runs from the mouth to the pyloric stomach. In arthropods, it is delimited to that anterior region which is lined with chitin

hindgut that posterior region of the arthropod alimentary canal which is lined with chitin

midgut that region of the arthropod alimentary canal which is not lined with chitin

gut 2 (*see also* gut 1, 3) translucent threads, flexible when wet, obtained by stretching and drying the silk glands of silk worms. Also called catgut (*i.e.* caterpillar gut)

gut 3 (*see also* gut 1, 2) in compound names of several plants

cat gut (*see also* gut 2) the leguminous herb *Tephrosia virginiana*

devil's gut 1 (*see also* devil's gut 2) the convolvulaceous parasitic herb *Cascuta epithymum* (= dodder)

devil's gut 2 (*see also* devil's gut 1) the equisetaceous plant *Equisetum arvense* (*cf.* horsetail)

-gutt- *comb. term* meaning "a drop"

Guttales an order of dicotyledonous angiosperms containing *inter alia*, those families more commonly placed in the suborders Theineae and Tamaricineae of the Parietales

guttate *either* spotted *or* drop-shaped

guttation the voiding of drops of water onto the surface of a leaf

Guttiferae a family of parietale dicotyledonous angiosperms, containing *inter alia*, the mangosteen, so closely related to the Hypericaceae that many include them in this family. The only significant difference is that the Guttiferae lack the pellucid or black dots on the leaves that are typical of the Hypericaceae

Guttiferales an order of dicotyledonous angiosperms erected to contain those families more usually placed in the suborder Theineae of the Parietales

guttule oil droplets in a plant structure

gv *see* [grooved]

gvl *see* [grooveless]

gwyniad *Brit.* the salmonid fish *Coregonus pennantii*

GY *see* [Gyro]

gy *see* [gynoid]

-gyalec- *comb. form* meaning an "urn"

Gymnocerata a group term for those insects having conspicuous, as distinct from concealed, antennae

Gymnodiniales that order of dinophycean Pyrrophyta which is distinguished by the lack of cellulose walls

Gymnolaemata a class of the phylum Ectoprocta distinguished by the circular shape of the lophophore

Gymnophiona = Apoda (Amphibia)

Gymnoptera a group term for those insects having membranous wings

Gymnospermae that division of the Spermatophyta

(*q.v.*) in which the seeds are not contained in an ovary (*cf.* Angiospermae). The group contains in addition to the Coniferae, the Cycadales, Gink-goales and Gnetales

Gymnospermia in the Linnaean system of classification of plants that order of the Didynamia which contains plants with four single seeded ovaries

Gymnosomatidae = Phasiidae

Gymnotidae a family of ostariophysid fishes closely related to the Characidae in their anatomy but mostly eel-like and lacking both dorsal and ventral fins

gymnure popular name of echinosoricine erinaceid insectivoran mammals

-gyn- *comb. form* meaning "female"

Gynandreae = Orchidales

gynandria in the Linnaean system of plant classification that class distinguished by possessing stamens growing in the pistil and united with it

gyne any female organism but, without qualification, usually refers to a fertile female or queen in a colony of social insects (*cf.* aner)

deutogyne the distinct female which characterizes one of the alternate generations of those trombidiform acarines which exhibit this phenomenon

dichthadiigyne a blind ant with a large gaster and ovaries

ergatogyne a fertile female ant lacking wings and therefore resembling an ergate

macrogyne a giant female ant or termite

mermithogyne a female ant, structurally modified in consequence of being parasitized by mermithid nematodes

microgyne a dwarf queen ant

phthisogyne a female pupal ant which has wasted away through the attack of a parasite

protogyne the male-like female that distinguishes one of the alternate generations of those trombidiform acarines which exhibit this phenomenon

pseudogyne a parthenogenetic female, particularly in insects

trichogyne *either* a filament arising from the ascogonium of ascomycete fungi *or* the receptive filament in the cross fertilization of some algae

gynecoid a worker ant that lays eggs

-gynia plant or animal taxa having this ending are listed alphabetically

syngyngia a group of females living together (*cf.* syn-andria)

-gynic *see* -gynous

-gynicous a rare variant of -gynous (*q.v.*) the form preferred in this work

-gynism *comb. term* meaning "a female characteristic"

heterogynism the condition in which the female of the species is more variable than the male in its phenotypic expression

podogynium = gynophore

gynoecy the condition of a species of which only the females are known

[gynoid] a mutant (geney symbol *gy*) found in linkage group I of *Habrobracon juglandis*. The phenotypic expression involves antennae in the males as short as those in the female

-gynous *comb. term* pertaining to "females," "female functions" and "female structures." -gynic and -gynicous are rare variants. The *subst.* form is "-gyny" (*cf.* -androus)

acrogynous bearing female organs at the apex of the stem

adynamogynous pertaining to the loss of function of female sex organs

agynous said of a plant ovary that lacks a pistil, or of a stamen that does not contact the ovary

androgynous = hermaphrodite (but, usually used only of plants having male and female flowers in the same inflorescence)

apagynous = monocarpic

apogynous pertaining to the loss of function in female reproduction organs

apodogynous the condition of a disk which is not adherent to the plant ovary

epigynous the condition in which antheridia lie above the oogonia in lower plants or with a superior perianth and androecium in flowering plants

ergatogynous pertaining to those ants in which the females resemble the ergates

exogynous the condition of a flower in which the style is extruded

gymnogynous having a naked ovary

hypogynous the condition of a flower in which the perianth and androecium attachments are below and free from the ovaries

isogynous having similar females, or female structures (*e.g.* pistils)

anisogynous *either* having dissimilar females or female parts *or* possessing flowers having fewer carpels than sepals

metagynous condition of an organism in which the male matures before the female

monogynous the condition of having a single fertile female as in the colony of many social insects or, in higher forms, having only one female sexual partner

paragynous said of a fungal hypha that grows alongside the oogonium

perigynous anything surrounding a female reproductive organ

symphytogynous the condition of a flower in which the fused calyx and pistil lie above an inverior ovary

progynous the condition of a hermaphrodite in which the female portions mature first (*cf.* protogynous)

proterogynous having the female parts mature before the male parts

ecoproterogynous having the female flowers mature before the male flowers

phytoproterogynous having the styles of a flower mature before the pollen ripens

protogynous frequently used as synonymous with progynous but is also used to describe the condition of being a hermaphrodite that is first female and later becomes a male

syngynous = epigynous

gynum a female structure

epigynum a flap over the genital pore in Araneae

-gyny *comb. form* pertaining to "a female condition." Definitions are given under the *adj. form* -gynous

Gypogeranidae = Sagittariidae

gypseous chalky

-gypsoph- *comb. form* meaning "chalk"

gyrate = circinate

-gyrate *adj. term* synonymous with -gyrous (*q.v.*) the form preferred in this work

gyre one turn of a spiral form. Thus it can be said of a spiral that the gyres are numerous

laevogyrinid said of an anuran amphibian larva with a spiracle on the left side only

mediogyrinid said of anuran amphibian larvae which have a single median spiracle

Gyrinidae a family of small oval black beetles commonly called whirligig beetles from their custom of swimming on the surface. Readily distinguished as having the compound eyes divided into an aerial and an aquatic component

-**gyro-** *comb. form* meaning "round"

[**Gyro**] a mutant (gene symbol *Gy*) found in the sex chromosome (linkage group XX) of the mouse. The phenotypic expression is the waltz syndrome accompanied by abnormally long bones in the male

Gyrocotylidea an order of cestodarian cestodes distinguished by an eversible proboscis and the anterior position of the male and female pores

Gyropidae a family of mallophagan insects parasitic on rodents. They are distinguished from the Trichodectidae by the possession of clavate antennae

gyrose curved backwards and forwards

Gyrostemonaceae a small family of centrospermous dicotyledons

-**gyrous** pertaining to a turn

  isogyrous forming a complete spire

  meiogyrous with the edges curled in

  mesogyrous said of an animal in which the second loop of the intestine is spirally arranged with the caecum at its apex (*cf.* -coelous)

  oxygyrous sharply twisted

  pachygyrous having thick coils

  pleurogyrous said of a fern sporangium with a horizontal annulus

  telogyrous said of an animal in which the distal intestinal loops are coiled into a spiral (*cf.* -coelous)

gyrus a fold or convolution in the cerebral hemisphere

gyttja organic detritus, so finely divided as to seem slimy, on fresh water lake bottoms (*cf.* herpon)

H

h *see* [hairy] and [silkie]

H *see* [Hairs absent] and [Hairless]. Used also to indicate a bacterial genetic marker having a character affecting the activity of flagellar antigen mapped at 32.6 and 38.6 mins. for *Escherichia coli*

H- symbolic prefix indicating that the serotypic response of a bacterial is not somatic

H-1-4 *see* [Histocompatibility-1-4]

ha *see* [hair bristles]

habitat the sum total of the environment of a particular species

   microhabitat the condition produced by the existence of a microclimate (*q.v.*)

habu the crotalid snake *Trimeresurus flavoviridis*

hard hack the rosaceous shrub *Spiraea tomentosa*

   golden hard hack the rosaceous herb *Potentilla fruticosa*

hackle the slender feathers on the neck of a male gallinaceous bird used to simulate the legs of an insect in an artificial fly used for angling

hackmatack the pinaceous tree *Larix laricina* (= tacamahac)

hadada *Afr.* the threskiornithid bird *Hagedashia hagedash*

hadal the sum total of free swimming and planktonic organisms living in depths greater than 2000 fathoms. If this term is employed bathyplankton (*q.v.*) is applied to those forms living between 500 fathoms and 2000 fathoms

haddie *Brit. dial.* for haddock

haddock any of several fishes. Without qualification, always the gadid *Melanogrammus aeglefinus*

   Jerusalem haddock *Brit.* the lamprid *Lampris luna* (= opah)

   Norway haddock *Brit.* the scorpaenid *Sebastes norvegicus*

hadrome = hadromestome

   protohadrome = protoxylem

Hadromerina an order of monaxonic demospongious Porifera in which the spicules are tylostylic

Haeckel's *epon. adj.* from Ernst Heinrich Haeckel (1834–1919)

haekaro *N.Z.* the pittosporaceous tree *Pittosporum umbellatum*

Haematopinidae a family of anopluran insects parasitic on mammals other than man and lacking eyes

Haematopodidae a family of charadriiform birds containing the oystercatchers. Many are distinguished by their black and white plumage, bright red bills and pink legs. The neck is shorter than in most other families of the order

Haemodoraceae that family of liliiflorous monocotyledonous angiosperms which contains the forms commonly called bloodworts. The name derives from the orange or red sap typical of the family

Haemosporidia a group of telosporidian Sporozoa variously regarded as a separate order or as a suborder of Coccidiomorpha (*q.v.*). They are distinguished from the Coccidia by the motile zygotes producing naked sporozoites

hair 1 (*see also* hair 2, 3, bell, grass, mite, slug) a thin strand of keratin secreted by follicles in the mammalian skin and typical of that order

   anchene hair a hair with a single kink found in the pelts of many mammals particularly rodents

   awl hair a hair found in the pelts of many mammals which is intermediate in length between the underfur and the guard hairs

   camel's hair a term commonly applied to the hair of the Malabar squirrel from which paint brushes are made

   compound hair one which is branched

   guard hair a hair found in the pelts of most mammals. Guard hairs are longer and stiffer than the underfur

   zigzag hair a hair in the coat of some rodents, particularly mice, which is normally of a zig-zag shape. If straightened it would be about the length of a guard hair

hair 2 (*see also* hair 1, 3) hair-like structures other than those forming the pelt of mammals (*cf.* seta)

   bracket hair a sickle shaped trichome

   collecting hair a hair on a style, destined to entrap pollen

   crop hair a knobbed trichome

   gathering hair forked, hooked, setae on the tongue of hymenopteran insects

   glandular hair a secretory trichome

   latex hair a trichome contiguous with a laticifer

   peltate hair a terminally flattened trichome

   plant hair a hair-like projection which may be either unicellular or multicellular from the surface of a plant. Better called a trichome

   root hair a hair-like extension of a root cell

   tenent hair *either* an empodium, modified for grasping, found in some acarines *or* a sticky seta found on the tarsi, or post-tarsi of some insects and used in clinging or grasping

   eel-trap hair an insect entangling trichome, in some plants

   tuft hair a trichome which branches from the base upwards

hair 3 (*see also* hair 1, 2) in compound names of organisms devil's hair the ranunculaceous shrub *Clematis virginiana* (= virgin's bower)

vegetable horse hair the bromeliaceous plant *Tillandsia usneoides* (= Spanish moss)

maiden hair *see* fern

[hair bristles] a mutant (gene symbol *ha*) mapped at 22.7± on the X chromosome of *Drosophila melanogaster*. The phenotypic expression is fine short bristles

[hairless] *either* a mutant (gene symbol *hl*) mapped at 49 on linkage group V of the tomato. The name is descriptive of the phenotypic expression *or* a mutant (gene symbol *hr*) mapped at 34.7 on linkage group III of the rat. The phenotypic expression is loss of hair after about a month *or* a mutant (gene symbol *hr*) in linkage group III of the mouse. The name is descriptive of the phenotypic expression

[Hairless] a mutant (gene symbol *H*) mapped at 69.5 on chromosome III of *Drosophila melanogaster*. The phenotype is not "hairless" but the bristles are greatly reduced

[hair-loss] a mutant (gene symbol *hl*) in linkage group VI of the mouse. The phenoytpic expression differs from [hairless] (*q.v.*) in that complete nakedness is rarely reached before the age of three months

[Hairs absent] a mutant (gene symbol *H*) mapped at 61 on chromosome 10 of the tomato. The phenotype lacks hairs on the stem

[hairy] a mutant (gene symbol *h*) mapped 26.5 on chromosome III of *Drosophila melanogaster*. The name is descriptive of the phenotype

[Hairy-wing] a mutant (gene symbol *Hw*) mapped at 0 + on the X-chromosome of *Drosophila melanogaster*. The name is descriptive of the phenotypic expression

hake any of numerous species of gadid fish particularly those of the genera Merluccius and Urophycis. *Brit.* without qualification, *M. vulgaris*

Caroline hake *U. earlli*
Gulf hake *U. cirratus*
longfin hake *Phycis chesteri*
Pacific hake *M. productus*
silver hake *M. bilinearis*
southern hake *U. floridanus*
spotted hake *U. regius*
squirrel hake *U. chuss*
white hake *U. tenuis*

hakoakoa *N.Z. either* the procellariid bird *Puffinus gavia* (= shearwater *N.Z.*) *or* the stercorariid bird *Megalestris antarctica* (= sea hawk *N.Z.*)

hakura *N.Z.* the ziphiid cetacean *Ziphius cavirostris* (= goose-beak whale)

-hal- *comb. form* meaning "salt" or "marine"

Halacaridae a family of acarine arthropods with usually flattened, elliptical bodies, a large rostrum and rudimentary tracheae. Most occur in salt water but many live in freshwater and a few are parasitic

Halcyonidae = Alcedinidae

Haldane's *epon. adj.* from John Burdon Sanderson Haldane (1892–1965)

Halecomorphi = Protospondyli

halibut any very large pleuronectid or bothid fish and particularly *Hippoglossus hippoglossus*

arrow-tooth halibut the pleuronectid *Atheresthes stomias* (= arrow-tooth flounder)

Atlantic halibut *U.S.* the pleuronectid fish *Hippoglossus hippoglossus*

bastard halibut *U.S.* the bothid *Paralichthys californicus*

California halibut = bastard halibut

Greenland halibut *U.S.* the pleuronectid fish *Reinhardtius hippoglossoides*

Pacific halibut *U.S.* the pleuronectid *Hippoglossus stenolepsis*

Halichondrina an order of monaxonic demospongious Porifera in which the megascleres are mostly of two or more kinds and which lack, or have raphid, microscleres

Halictidae a family of apoid apocritan hymenopteran insects that share with the Andrenidae the designation mining bee through burrowing bee would be better. They are mostly small forms, some brilliantly metallic in color

Halicystidaceae that family of siphonale Chlorophyceae in which the vegetative portion of the thallus and the gametangium are not divided by a cellulose wall

alkylhalidase an enzyme that catalyzes the hydrolysis of alkylhalides to formaldehyde and the appropriate halide acids

haline pertaining either to the sea or to salt

euhaline said of waters containing between 30 and 40 parts per thousand of dissolved salts. That is, in most cases, normal sea water

mixoeuhaline said of esturine waters that contain more than 30 parts per thousand of dissolved salts but less than the concentration of the adjacent seas

euryhaline said of organisms capable of withstanding widely varying concentrations of salt in the environment

hyperhaline said of water containing more than 40 parts per thousand of dissolved salts

mixomesohaline said of brackish water containing from 5 to 18 parts per thousand dissolved salts

mixohaline said of all waters that contain anywhere from 0.5 to 30 parts per thousand of dissolved salts

mixooligohaline said of brackish waters containing from 0.5 to 5 parts per thousand of dissolved salts

mixopolyhaline said of brackish water containing from 18 to 30 parts per million dissolved salts

stenohaline properly used of organisms capable of tolerating only slight variations in salt concentrations but also used of those capable of tolerating only low concentrations

Haliotidae a family of gastropod mollusks distinguished by the flattened depressed shell with a row of round or oval holes along the left margin. Commonly called abalones or ear shells

Haliplidae a family of small, usually yellowish or brownish coleopteran insects commonly called crawling water beetles. They may be readily distinguished by the massive hind coxae

Haller's *epon. adj.* from Albrecht von Haller (1708–1777)

hallux the first digit on the hind limb which is the hind toe of birds or the great toe of mammals

haltere a peg-like organ modified from the hindwings of dipterous insects. It vibrates at the same speed as the wings and thus is generally thought to serve as a gyroscopic stabilizer

prehaltere a scale immediately in front of the haltere

halm = haum

Halmidae = Elmidae

-halo- *see* -hal-

halonate having a halo

Haloragaceae that family of myrtiflorous dicotyledons which contains, *inter alia*, the water milfoils. The family may be distinguished from the closely related Onagraceae by having a single ovule in each cell of the ovary

Haloragidaceae = Haloragaceae

hamadryad 1 (*see also* hamadryad 2) the elapid snake *Ophiophagus hanna* (= king cobra)

hamadryad 2 (*see also* hamadryad 1) *Austral.* the danaid lepidopteran insect *Tellervo zoilus*

hamadryas the cynopithecid primate *Papio hamadryas* (= sacred baboon)

Hamamelidaceae that family of rosalous dicotyledons which contain the witchhazels and sweetgums. The fruit, which is a woody two-valved capsule, is distinctive of the family

hamate hooked

hamlet any of several fishes of the families Serranidae and Muraenidae

butter hamlet the serranid *Hypoplectrus unicolor*

mutton hamlet *U.S.* the serranid *Alphestes afer*

hammer (*see also* head, pop) a Victorian English genteelism for ammer (*q.v.*)

yellowhammer the fringillid bird *Emberiza citrinella*

hamster any of numerous pouched murid rodents

common hamster *Cricetus frumentarius*

crested hamster the large, *E.Afr.* myomorph rodent *Lophiomys imhausi*. It is distinguished by the possession of a large, erectile mane

-ham- *comb. form* meaning "hook"

hamulate bearing a small hook

hamulose furnished with numerous small hooks

hamulus a hooked bristle and particularly one that rises from the barbule of a feather

hand 1 (*see also* hand 2) the terminal structure of the front limb of primates

hand 2 (*see also* hand 1) in compound names of organisms

devil's hand the sterculiaceous tree *Chiranthodendron pentadactylon*

monkey's hand the sterculiaceous tree *Chiranthodendron platanoides*

hangul a W. Himalayan race of the cervine artiodactyl *Cervus elaphus* (= red deer)

hanna = hoatzin

hanuman the colobid primate *Simnopithecus entellus*

Hapalidae a family of platyrrhine primates containing the marmosets and their allies. They are distinguished from most other families of monkeys by possessing claws rather than nails

Hapaloidia a suborder of primate mammals containing the *S. Amer.* marmosets and their immediate allies. They differ from the monkeys and apes in possessing claws instead of nails

-hapax- *comb. form* meaning "once"

Haplodoci a monofamilial order of Osteichthyes commonly called toad fishes

Haplondontidae a family of sciuromorph Rodentia all extinct except for the mountain beaver

haploid = *see* -ploid

Haplomi an order of osteichthyes closely allied to the Isospondyli but differing from them in lack a mesocoracoid; by some included with the Isospondyli. The order contains the pike and mud minnow groups

Haplomitriaceae the only family of the hepaticate order Calobryales. The characteristics of the family are those of the order

Haplosclerina an order of monaxonic demospongious Porifera with diactinal megascleres and a considerable skeleton of spongin

haplosis (*see also* -ploid) the reduction of the diploid

number (2N) of chromosomes to the haploid number (N) at meiosis

Haplosporidia a subclass of sporozoan Protozoa distinguished by the possession of oval spores lacking polar capsules but opening by a lid

Haplotaxidae a small family of very long, slender, filamentous oligochaete annelids

-hapto- *comb form* meaning "contract" or, by extension, "togetherness' or even "binding"

-haptor- *comb. term* meaning "sucker"

opisthohaptor the posterior sucker of a trematode platyhelminth

prohaptor the anterior sucker of a trematode platyhelminth

harbinger-of-spring the umbelliferous herb *Erigenia bulbosa*

Harder's *epon. adj.* from Johann Jacob Harder (1656–1711)

Harderian = Harder's

hardun the agamid saurian reptile *Agama stellio*

hare 1 (*see also* hare 2, bell, ear) the confusion between rabbit and hare is similar to that between rat and mouse since the distinction is, in popular parlance, one of size. All are, of course leporid lagomorphs. The European rabbit is *Oryctolagus cuniculus*, usually miscalled *Lepus cuniculus*. European hares belong to the genus Lepus as do most American hares and many large American rabbits. The majority of small American rabbits belong to the genus Sylvilagus. *Brit.*, without qualification, *L. europaeus*

alpine hare *Brit. L. timidus* (= variable hare)

artic hare the white phase of the European hare (*L. timidus*)

blue hare *Brit.* = alpine hare

brown hare *Brit. L. europaeus* (= European hare *U.S.*)

European hare *U.S.* = brown hare *Brit.*

jumping hare *Afr.* the pedetid rodent *Pedetes caffer*

marsh hare *U.S. Sylvilagus palustris*

mountain hare *Brit.* = alpine hare *Brit.*

mouse hare any of several species of ochotonid lagomorphs of the genus Ochotona

piping hare any of several species of ochtonid lagomorphs of the genus Ochotona (= pika)

snowshoe hare *L. americanus* (= *U.S.* varying hare)

tundra hare *L. othus*

variable hare *Brit. L. timidus*

varying hare *U.S.* = horseshoe hare *U.S.*

hare 2 (*see also* hare 1) any organism fancifully resembling a hare

sea hare any of several large nudibranch Mollusca of the family Tethyidae

harle = merganser

harlequin 1 (*see also* harlequin 2, beetle, bug, duck) *Austral.* either of two riodinid lepidopteran insects *Praetaxila segecia* or *Eicallaneura ribbei*. Without qualification the latter is usually intended. *I.P.* the riodinid lepidopteran insect *Taxila haquinus*

harlequin 2 (*see also* harlequin 1) either of two papaveraceous herbs of the genus Corydalis

rock harlequin *C. sempervirens*

yellow harlequin *U.S. C. flavula*

epharmonic said of a shape which is developed in response to environmental, not genetic factors

harmosis the response of an organism to its environment

photoharmosis response to light

epharmosis *ecol. jarg.* for adaptation

Harpacticoidea an order, or suborder Harpacticoida,

of free-living copepod Crustacea distinguished by the fact that the fifth thoracic segment forms a moveable joint with the sixth segment and that the anterior and posterior regions of the body are usually of about the same width. Most have a single egg mass instead of the paired egg masses usually found in copepods

harpago a clasping organ of male insects derived from the styli of the ninth abdominal segment

harpé literally, a scimitar, but applied to each member of the claspers of a lepidopteran (*cf.* valve 3) in the sense that harpago is used for other insects

Harpodontidae a small family of iniomous fish readily distinguished from the closely allied Synodontidae by the presence of extemely long pectoral and pelvic fins. The best known harpodontid is the "Bombay duck"

harpy the acciptrid bird *Phrasaepus harpyia*

harrier any of numerous circine birds. *U.S.*, without qualification, = marsh hawk. *Brit.* without qualification, the circine bird *Circus cyaneus*. *N.Z.* the circine bird *Circus approximans*

Australian harrier *Circus approximans*

hen harrier *Brit. Circus cyaneus* = (*U.S.* marsh hawk)

marsh harrier *Brit.* the *Circus aeruginosus*

spotted harrier *Austral. Circus assimilis*

swamp harrier *Austral.* = Australian harrier

rest harrow 1 (*see also* rest, harrow 2) any of several species of leguminous herb or shrub of the genus Ononis but particularly *O. natrix* (= goat root)

rest harrow 2 (*see also* rest harrow 1) *Brit.* the geometrid lepidopteran insect *Aplasta ononaria*

hart a term commonly used to mean deer but properly reserved for a male red deer in its sixth year

royal hart a hart with more than three anterior tines, at one time reserved for the hunting of kings

hartebeest a large *S.Afr.* antelope *Alcelaphus caana*. The term, usually modified, is applied to many other large antelopes the horns of which are curved backwards

bastard hartebeest *S.Afr.* the antelope *Damaliscus lunatus* (= sassaby)

hunter's hartebeest a rare *E.Afr.* antelope (*Beatragus hunteri*) more properly classified with the damalisks than with the hartebeests

harvester any of numerous species of lycaenid butterfly (*cf.* blue, copper, hair streak). *U.S.*, without qualification, *Feniseca tarquinius*

harvestman popular name of phalangid arthropods, in *U.S.* less frequent than daddylonglegs

Hassal's *epon. adj.* from Arthur Hill Hassal (1817–1894)

hassar the callichthyid teleost fish *Callichthys callichtys*

-hasta- *comb. form* meaning "halberd"

hastate halberd shaped, that is, in the form of an arrow head with the basal lobes recurved

Chinese hat the calyptraeid gastropod *Calyptraea fastigiata*

hatch a term for a group of flying insects that have hatched or emerged at the same time

reed-haunter the funariid bird *Limnornis curvirostris*

wood-haunter any of several species of funariid bird, particularly of the genus Hyloctistes

nuthatch (*see also* warbler) any of numerous passeriform birds of the families Sittidae, Neosittidae and Hypositttidae. *Brit.* without qualification, *Brit. Sitta europaeus*. *N.Z. Xenicus longipes* (= bush wren)

Australian nuthatch any of several speices of neosit-tid birds. Usually, without qualification, *Neositta pileata*

coral-billed nuthatch the hyposittid *Hypositta coral-lirostris*

blue nuthatch *Sitta azurea*

red-breasted nuthatch *Sitta canadensis*

white-breasted nuthatch *Sitta carolinensis*

brown-headed nuthatch *Sitta pusilla*

pigmy-nuthatch *Sitta pygmaeus*

white-tailed nuthatch *Sitta himalayensis*

stone hatch the charadriid bird *Charadrius hiaticula.* (= ringed-plover)

haulm a plant stem usually, but not necessarily, dried particularly one from which an edible seed or fruit has been removed

haumakaroa the araliaceous tree *Nothopanax simplex*

haunch the musculature and bony structures lying between the loin and the hip in artiodactyl and perissodactyl mammals

haustellate adapted to sucking

haustoria hyphae of parasitic fungi that penetrate tissues and absorb nourishment

Haustoriidae a family of beach dwelling, burrowing, gammaridean amphipod Crustacea

Haversian *epon. adj.* from Clopton Havers (1650–1702)

haw 1 (*see also* haw 2, 3, finch) the fruit, actually a pome, of various species of rosaceous shrub of the genus Crataegus (*cf.* hawthorn)

haw 2 (*see also* haw 1, 3) any of numerous shrubs bearing haws or haw-like fruit

apple haw the rosaceous shrub *Cretaegus aestivalis*

black haw the caprifoliaceous shrub *Viburnum pruni-folium*

mayhaw the rosaceous shrub *Crataegus aestivalis*

possum-haw *U.S. either* the aquifoliaceous shrub *Ilex decidua or* the caprifoliaceous shrub *Viburnum nudum*

red haw *U.S.* any of several species of the rosaceous genus Crataegus

swamp-haw = possum-haw

sweet haw = black haw

haw 3 (*see also* haw 1, 2) nictatating membrane

hawk 1 (*see also* hawk 2, 3, beard, cuckoo, eagle, moth, owl, weed, wing) almost any diurnal falconiform bird too small to be called an eagle or a vulture bears this designation

bastard hawk the falconine *Falco sparverius* = sparrow hawk *U.S.* kestrel *Brit.*

bat hawk the accipitrine *Machaerhamphus alcinus*

black hawk the buteonine *Buterogallus anthracinus*

brown hawk the falconine *Falco berigora*

bush hawk *N.Z.* the falconine *Falco novoseelandiae*

buzzard hawk popular name of buteonine birds in general

chicken hawk a colloquial term applied to many large hawks, particularly of the genus Accipiter *W.I.* the buteonine *Buteo jamaicensis*

Cooper's hawk *U.S.* the accipitrine *Accipiter cooperii*

crab hawk *W.I.* = black hawk

duck hawk *U.S.* the falconine *Falco peregrinus* (= peregrine falcon) *Brit.* = marsh harrier

fish hawk = osprey

gabar hawk the falconine *Melierax gabar*

goose hawk = goshawk

goshawk any of numerous accipitrine birds distinguished by their heavy bills. Without qualification, particularly in terms of falconry, properly applied to the falconine *Accipiter gentilis*

Australian goshawk *A. fasciatus*

Chinese goshawk *A. soloensis*
crested goshawk *A. trivirgatus*
grey hawk the buteonine *Buleo nitidus*
killyhawk *W.I.* = sparrow hawk *U.S.*
rough-legged hawk *U.S. Buteo lagopus*
ferruginous rough-legged hawk *B. regalis*
marsh hawk *U.S.* the circine *Circus cyaneus*
molly hawk *see under* molly
mountain hawk *W.I. Chondrohierax uncinatus* (= hook-billed kite)
mullet hawk = osprey
musket hawk = male sparrow hawk *Brit.*
White Nile hawk the accipitrine *Melierax metabates*
pigeon hawk *U.S.* = merlin
quail hawk *N.Z.* the falconine *Falco novaeseealandiae*
red hawk = sore hawk
sharp-shinned hawk the accipitrine *Accipiter striatus* and its many varieties and subspecies
red-shouldered hawk the buteonine *Buteo lineatus*
sore hawk a bird of the first year that has not molted
sparrow hawk *U.S.* the falconine *Falco sparverius* (*cf.* kestrel). *Brit.*, without qualification, the accipitrine *Accipiter nisus. N.Z.* the falconine *Falco novaeseealandiae*
black sparrow hawk the accipitrine *A. melanoleucus*
Japanese sparrow hawk *A. virgatus*
long-tailed hawk the accipitrine *Urotriorchis macrourus*
red-tailed hawk the buteonine *Buteo jamaicensis*
short-tailed hawk *B. brachyurus*
white-tailed hawk *B. albicaudatus*
zone-tailed hawk *B. albinotatus*
white hawk the buteonine *Leucopternis albicollis*
broad-winged hawk *Buteo platypterus*
hawk 2 (*see also* hawk 1, 3) any of several birds other than falconiforms having hawk-like characters
cuckoo hawk any of numerous species of cuculids of the genus Hierococcyx (= hawk cuckoo)
dorr hawk *Brit.* = nightjar
mosquito hawk the caprimulgid *Chordeiles minor* (= nighthawk)
moth hawk = nighthawk
nighthawk *U.S.* any of several species of caprimulgids of several genera, notably Chordeiles, usually without qualification *C. minor.* (*cf.* nightjar)
sea hawk *N.Z.* the stercorariid *Catharacta skua*
hawk 3 (*see also* moth) any of several insects of predatory habit
cicada hawk *U.S.* the sphecid nyssonine hymenopteran insect, *Sphecius speciosus* (= cicada killer)
tarantula hawk *U.S.* the pompilid hymenopteran insect *Pepsis formosa*
hay the dried mixture of grasses and herbaceous plants obtained by cutting pasture land
hazel any of numerous betulaceous shrubs of the genus Corylus. *Brit.* without qualification, *C. maxima. Austral.* any of several rhamnaceous shrubs of the genus Pomaderris, particularly *P. atetals.*
beaked hazel the betulaceous shrub *Corylus cornuta*
Chile hazel the proteaceous shrub *Gevuina aveolana*
witch-hazel any of several species of hamamelidaceous trees of the genus Hamamelis, particularly *H. virginiana. Brit.* both this tree and the ulmaceous tree *Ulmus glabra* (=wych elm)
Hb *see* [Hemoglobin pattern]
HCG = chorionic gonadotropin

hdp *see* [heldup]
head 1 (*see also* head 2, 3) the anterior end, or the complex of organs forming the anterior end, of an organism
false head the gnathosoma of Acarina
head 2 (*see also* head 1, 3, kop) in compound names of numerous animals
bonyhead any of several species of hylid anuran Amphibia of the genera Trachycephalus or Aparasphenodon
bowhead the whale *Balaena mysticetus*
blackhead *W.I.* the aythyine bird *Aythya affinis* (= scaup)
bluehead the labrid fish *Thalassoma bifasciatum*
bonnethead the sphyrnid shark *Sphyrna tiburo*
buffle head the aythyine bird *Bucephala albeola* (*cf.* golden-eye)
bullhead 1 (*see also* bullhead 2, 3, 4) *Brit.* popular name of cottid fishes (= *U.S.* sculpin) and of many heterodontid sharks (Heterodontidae)
alligator fish particularly *Agonus cataphractus*
bullhead 2 (*see also* bullhead 1, 3, 4) *U.S.* any of several species of ictalurid catfishes of the genus Ictalurus
brown bullhead. *I. nebulosus*
yellow bullhead *I. natalis*
bullhead 3 (*see also* bullhead 1, 2, 4) *U.S.* either of several aythyine birds of the genus Bucephala (= goldeneye)
bullhead 4 (*see also* bullhead 1–3) *U.S.* (local) various charadriid birds
copperhead 1 (*see also* copperhead 2) *U.S.* the crotaline snake *Agkistrodon contortrix*
copperhead 2 (*see also* copperhead 1) the sciurid rodent *Citellus lateralis* (=golden mantled squirrel)
fathead *U.S.* the cyprinid fish *Pimephales promelas.* The name is also sometimes used for the labrid *Pimelometopon pulchrum* (= California sheephead)
flathead popular name of percophidid fishes and of the ictalurid *Pylodictus olivaris*
fringehead any of several species of clinid fishes of the genus Neoclinus
gilt head *Brit.* the labrid fish *Crenilabrus melops* and the sparid fish *Pagrus auritus*
hammerhead (*see also* shark) the scopid bird *Scopus umbretta*
hardhead (*see also* head 3) the *U.S.* cyprinid fish *Mylopharodon conocephalus*
lancehead the crotalid snake *Trimeresurus macrolepis*
leatherhead *Austral.* any of several species of Australian meropid birds, usually of the genus Merops
loggerhead (*see also* duck, kingbird, turtle) *B.W.I.* the tyrannid bird *T. caudifasciatus* (*cf.* kingbird)
mountain loggerhead *B.W.I.* the tyrannid *Myiarchus tyrannulus* (= rusty-tailed flycatcher)
markhead *W.I.* the thrupid bird *Spindalis zena* (= stripe-headed tanager)
pikehead popular name of luciocephalid anabantid
redhead *Brit.* any of several species of duck particularly the pochard and widgeon. *U.S.* = red-headed woodpecker and the aythyine bird *Aythya americana*
rockhead the agonid fish *Bothragonus swani*
sheepshead the sparid fish *Archosargus probatocephalus*
California sheepshead the labrid fish *Pimelometopon pulcher*

freshwater sheepshead the cichlid teleost fish *Chichlasoma cyanoguttatum* (= Rio Grande perch)

silverhead *W.I.* the thrupid bird *Spindalis zena* (= stripe-headed tanager)

snakehead (*see also* head 3) popular name of channid fishes

steelhead the migratory sea-run stage of adult rainbow trout

thickhead any of several species of muscicapid bird of the genus Pachycephala and its close allies, at one time forming a separate family the Pachycephalidae (*cf* whistler) *S.Afr.* the burhinid *Burhinus capensis*

whitehead *W.I.* the columbid bird *Columba leucocephala* and the anatine bird *Anas bahamensis*. *N.Z.* the sylviid bird *Mohoua ochrocephala*

head 3 (*see also* head 1, 2) in compound names of numerous plants

arrowhead any of several alismaceous herbs of the genus Sagittaria

giant arrowhead *S. montevidensis*

burhead any of several species of the alismataceous genus of Echinodorus

curly head *U.S.* the ranunculaceous plant *Clematis ochroleuca*

dragonhead any of several species of labiateous herb of the genus Dracocephalum

false dragonhead any of several species of labiateous herb of the genus Physostegia

fiddle heads the osmundaceous fern *Osmunda cinnamonea* (= cinnamon fern)

hard head (*see also* head 2) the compositaceous herb *Centaurea nigra* (= matweed)

medusa's head the euphorbiaceous shrub *Euphorbia caputmedusae*

negro head the annonaceous tree *Annona purpurea* and its fruit

niggerhead the compsoitaceous herb *Rudbeckia serotina*

snake's head (*see also* head 2, iris) the liliaceous herb *Fritillaria meleagris*

sticky head the compositaceous herb *Grindelia squarrosa*

turtle head any of several herbs of the scrophulareaceous genus Chelone

[head-streak-in-down] a mutant (gene symbol *ko*) in the sex chromosome (linkage group I) of the domestic fowl. The name is descriptive of the phenotype

self-heal the labiateous herb *Prunella vulgaris*

heart 1 (*see also* heart 2, 3) any organ that circulates a fluid within an animal

accessory heart in invertebrates, one or more pumping organs accessory to the main heart

cephalic heart a pulsating organ which breaks the shell around young odonate insect larvae

lymph heart a pulsating portion of a lymph vessel

heart 2 (*see also* heart 1, 3, pea, seed) any of numerous plants, mostly with heart-shaped flowers or leaves

bleeding heart any of several members of the funariacean plant genus Dicentra

blue heart *U.S.* any of several species of scrophulariaceous herb of the genus Buchnera

bullock's heart the annonaceous tree *Annona reticulata* and its fruit

bursting heart the celestraceous shrub *Eunonymus americanus*

floating heart any of several species of gentianaceous aquatic herbs of the genus Nymphoides

heart 3 (*see also* heart 1, 2) in compound names of other organisms

black heart *S.Afr.* the lycaenid lepidopteran insect *Uranothauma nubifer*

golden heart any of several columbid birds of the genus Gallicolumba

heart-and-club *Brit.* the noctuid lepidopteran insect *Agrotis corticea*

heart-and-dart *Brit.* the noctuid lepidopteran insect *Agrotis exclamationis*

heartsease the violaceous herb *Viola tricolor* (= wild pansy)

heat = estrus

heath 1 (see also heath 2, 3, and) a tract of land, worthless for cultivation and typically covered in heather

heath 2 (*see also* heath 1, 3) any of many ericaceous shrubs, mostly of the genus Erica, and a few plants of other families. The name is frequently but erroneously used as synonymous with heather (*q.v.*) except in *Brit.* where *E. tetralix*, (= *U.S.* cross-leaved heath) is universally called bell heather. The term heather is also common for members of the ericaceous genera Cassiope (*U.S.*) and Calluna (*Brit.*)

Cornish heath *E. vagans*

Corsican heath *E. stricta*

cross-leaved heath *E. tetralix*

mountain heath any of several ericaceous shrubs of the genus Phyelodoce

sea heath *Brit.* any of several species of the frankeniaceous genus Frankenia

Spanish heath *E. lusitanica*

heath 3 (*see also* heath 1, 2) any of several lepidopteran insects. *Brit.*, without qualification applies to species of both the geometrid genus Ematurga and the satyrid genus Caenonympha

Baluchi heath *I.P.* the satyrid *Coenonympha myops* (*cf.* ringlet)

heather (*see also* bleat) any of numerous cistaceous shrubs of the genus Hudsonia and, by confusion with heath (*q.v.*) a few ericaceous shrubs

beach heather any of several cistaceous shrubs of the genus Hudsonia, particularly (*U.S.*) *H. tomentosa*

bell heather either of several ericaceous shrubs. *Brit.* either *Erica tetralix* or *E. cinerea*——*U.S.* *Cassiope mertensiana* (*cf.* heath)

Arctic bell-heather *C. tetragona*

golden heather *U.S.* the cistaceous shrub *Hudsonia ericoides*

heaven see breath-of, tree-of, etc.

[heavy vein] a mutant (gene symbol *hv*) mapped at 104 on chromosome II of *Drosophila melanogaster*. The name is derived from the thick oblique posterior cross veins on the wing of the phenotype

-hebe- *comb. form* meaning "puberty"

hebetate blunt

hebrew (*see also* character) *U.S.* the noctuid lepidopteran insect *Polygrammata hebraicum*

Hebridae a small family of aquatic hemipteran insects distinguished from all other Hemiptera by the pilose body. Commonly called velvet water bugs

Hectopsyllidae = Tungidae

Hederaceae = Arabaceae

heel 1 (*see also* heel 2) the posterior region of the foot, particularly in plantigrade mammals

heel 2 (*see also* heel 1) (Hymenoptera) = spinula

heifer a virgin cow

heketara *H.Z.* the compositaceous shrub *Olearia rani*

[heldout] a mutant (gene symbol *ho*) mapped at 4.0 on

chromosome II of *Drosophila melanogaster*. The name refers to the extended wings of the phenotype

[heldup] a mutant (gene symbol *hdp*) mapped at 59.6± on the X chromosome of *Drosophila melanogaster*. The wings of the phenotype are held vertically

Heleidae = Ceratopogonidae

helen *I.P.* either of two species of papilionid lepidopteran insects of the genus Papilio. (*cf.* raven, peacock, redbreast, mormon, swallowtail, spangle)

Helenieae a tribe of tubiflorous Compositae

Heleomyzidae a little known family of myodarian cycloraphous dipteran insects

Heleophryninae a monogeneric subfamily of anuran leptodactylid amphibians

Heliantheae a tribe of tubiflorous Compositae

-helic- *comb. form* meaning "twisted" but often used in *biol.* in the sense of snail

helical having a spiral form

Heliconidae a family of nymphalid lepidopteran tropical insects and distinguished from the Nymphalidae by the elongate and narrow front wings

Helicopsychidae a monogeneric family of trichopteran insects the larvae of which build snail-like cases. They are therefore commonly called the snail-case caddisflies

Heliodinidae a family of small yponemeutoid lepidopteran insects distinguished by very long and narrow hindwings with a broad fringe

Heliornithidae a small family of gruiform birds containing those forms commonly called finfoots by reason of the broadly lobed toes

heliotrope (*see also* tree) anglicized form applied to flowering plants of the boraginaceous genus Heliotropium

garden heliotrope the valerianaceous herb *Valeriana officinalis*

winter heliotrope the compositaceous herb *Petasites fragrans*

Heliotropioideae a subfamily of the Boraginaceae

Heliozelidae a family of cycnodoid lepidopteran insects commonly called shield bearers. The larvae are almost invariably leaf miners

Heliozoa an order of actinopodous Protozoa distinguished by their spherical shape and stiff radiating pseudopodia

helium a marsh formation

helix a spiral

phyllotactic helix a spiral which is formed by the origin of successive leaves on nodes as viewed from the apical end of the shoot (= genetic spiral)

hellbender *U.S.* the cryptobranchid urodele amphibian *Cryptobranchus alleganiensis*

hellebore any of numerous ranunculaceous herbs of the genus Helleborus

bastard hellebore any of several orchids, also *H. viridis*

false hellebore a term sometimes applied to liliaceous herbs of the genus Veratrum though these are often called hellebore without qualification

white hellebore *Brit. Veratrum alba. U.S. V. viride*

Helleboreae a tribe of the family Ranunculaceae

helleborine any of several orchids of the genus Epipactis, particularly *E. helleborine*

hellgrammite *U.S.* the larva of the neuropteran insect *Corydalus cornutus* (= Dobson fly)

helminthoid in the shape of a worm (= vermiform)

[Helminthosporium susceptibility] a mutant (gene symbol *hm*) mapped at 66 on linkage group I of *Zea mays*. The name is descriptive of the phenotype

-helo- *comb. form* meaning "a marsh"

Helobiae an order of monocotyledonous angiosperms containing many aquatic and marshy plants including, *inter alia*, the pond weeds, arrow grasses, flowering quillworts, water plantains, flowering rushes and frogbits. They are distinguished from other monocotyledonous orders by the absence of endosperm in the seed and by the aquatic habit

helobious pertaining to or living in marshes

helodad a marsh plant

Helodermatidae a monogeneric family of lacertilian reptiles having a pleurodont dentition with the lower teeth grooved, and carrying poison glands

Helomyzidae = Heleomyzidae

Heloridae a small family of proctotrupoid apocritan hymenopteran insects

helostadion a community of plants of which only the base is submerged (*cf.* emersiherbosa)

helotism a type of relationship, frequently regarded as a form of symbiosis, in which one of the two organisms is definitely subordinate to the other

Hellvellaceae a family of ascomycete fungi containing the forms commonly called morels or sponge mushrooms. They are distinguished by the spongy, frequently wrinkled, appearance of the head

helvolous pale yellow

helvous very light reddish brown

hem a bacterial genetic marker indicating a requirement for heme

heme a general term for the iron-porphyrin group in hemoglobin (*q.v.*) and some other iron-protein respiratory pigments (*see* heme erythrin, under -erythrin)

-hemero- *comb. form* meaning "cultivated"

Hemerobiidae a family of predatory neuropteran insects strongly resembling the Chrysopidae but differing from them in possessing two or more radial sectors in the front wing. In *Brit.* they are called aphis lions

-hemi- *comb. form* properly meaning "half" but widely misused in biology to mean "partly"

Hemiascomycetae a class of ascomycete fungi that lacks ascocarps

Hemichordata a phylum of enterocoelous coelomate animals. At one time they were thought to be closely allied to the chordates but are now not so regarded by some workers. The phylum contains the classes Enteropneusta (acorn worms), Pterobranchia and Planctosphaeroidea

hemifuse spread over the ground (*cf.* humistratous)

Hemigalinae a sub-family of the viverrid carnivorous mammals containing *inter alia*, the otter-civet, the fossa, and the anteater-civet

hemimelia the condition of having stunted extremities

Hemimyaria = Salpida

hemipode any of several species of megapodiid bird of the genus Megapodius

collared hemipode the pedionomid bird *Pedionomus torquatus*

Hemiprocnidae a small family of apodiform birds commonly called crested-swifts. They are distinguished by the crest which gives them their name and a patch of silky feathers on each flank

Hemiptera an order of insects distinguished by the fact that the basal portion of the front wing is thickened while the apical portion is membranous. The term "bug" used without qualification, usually refers to a hemipteran. The Hemiptera are closely allied to, and are sometimes fused with, the order Homoptera, then taking the subordinal name of Heteroptera

Hemiramphidae a peculiar family of synentognathous fishes with an extremely long lower jaw and a short upper jaw

cerebellar **hemisphere** a large hemisphere on the cerebellum developed in mammals lateral to the vermis

    cerebral hemisphere one of a pair of thickened evaginations from the dorsal wall of the dienciphalon. The cerebral hemispheres form the major part of the adult brain of birds and mammals

hemlock 1 (*see also* hemlock 2, borer, looper, parsley, scale) *U.S.* any of numerous trees of the pinaceous genus Tsuga

    Carolina hemlock *T. caroliniana*

    Chinese hemlock *T. chinensis*

    eastern hemlock *T. canadensis*

    Douglas hemlock = Douglas fir

    Japanese hemlock *T. sieboldii*

    western hemlock *T. heterophylla*

hemlock 2 (*see also* hemlock 1) *Brit.* the poisonous umbelliferous herb *Conium maculatum*

    water hemlock any of several species of umbelliferous herb of the genus Cicuta, particularly *C. virosa* and *C. maculata*

hemocyanin a blue, copper containing, respiratory pigment found in crustacean blood. The copper is directly attached to an amino-acid and not, as the iron in hemoglobin, to a porphyrin

hemoglobin a general name for a group of respiratory pigments in which ferrous iron, linked to a protoporphyrin, is conjugated to a protein. All hemoglobins are about 94% protein but the molecular weight varies from about 17,000 to a high of 2,750,000 in some invertebrates

    carbohemoglobin (= carbaminohemoglobin) the compound formed between hemoglobin and carbon dioxide

hemovanadin a green, vanadium containing, blood pigment found in urochordates

[**hemoglobin pattern**] a mutant (gene symbol *Hb*) in linkage group I of the mouse. The phenotype is demonstrable only through electrophoresis of the hemoglobin

hemp (*see also* nettle, weed) any of several species of the moraceous genus Cannabis, particularly *C. sativa*, and *C. indica*

    ambary hemp the malvaceous herb *Hibiscus cannabis*

    bastard hemp any of several fiber producing plants, especially the hemp nettle. Also any of several species of Datisca

    Indian hemp *U.S.* any of several species of apocynaceous herb of the genus Apocynum

    manilla hemp the musaceous tree *Musa textilis* and the fibers derived from it

    bow-string hemp any of several liliaceous herbs of the genus Sanseviera

    water hemp *U.S.* any of several species of amaranthaceous herbs of the genus Acnida

hen 1 (*see also* hen 2) without qualification, the female domestic chicken, but the term is also used for female birds in genera, the female lobster and many female fishes

hen 2 (*see also* hen 1) any of many birds without regard to sex

    bush hen the rallid *Amaurornis olivacea*

    clucking hen *W.I.* the aramid *Aramus guarauna* (= limpkin)

    daker hen *Brit. obs.* = land rail

    false egmont hen the stercorariid *Catharacta skua* (great skua)

    grey hen female blackcock

hazel hen the tetraonid bird *Tetrastes bonasia* (*cf.* hazel grouse)

island hen (of Tristan da Cunha) the rallid *Porphyriornis nesiotis*

mallee hen = mallee fowl

mangrove hen *W.I.* any of several species of rallid bird

marsh hen *U.S.* any of numerous gruiform birds of the family Rallidae. Also (rarely) the bittern. *Brit.* (rare) the rallid *Gallinula chloropus* (= moorhen) *W.I.* the rallid *Rallus longirostris*

meadow hen = meadow chicken

moorhen *Brit.* the rallid bird *Gallinula chloropus*.

    African moorhen *Gallinula chloropus*

native hen *Austral.* the rallid *Tribonyx ventralis* (= waterhen)

reed hen = gallinule

sage hen = sage grouse

scrub hen = megapode

swamp hen *N.Z.* the rallid *Porphyrio poliocephalus*

water hen = gallinule

wood hen *N.Z.* any of several species of rallid bird of the genus Gallirallus (= weka)

woodhen *W.I.* the columbid bird *Geotrygon mystacea* (= bridled quail dove)

hen-and-chickens any of several species of crassulaceous herbs of the genus Sempervivum but particularly *S. tectorum* and *S. soboliferum*

Henle's *epon. adj.* from Friedrich Gustav Jakob Henle (1809–1885)

henna the lythraceous shrub *Lawsonia inermis* and the yellow dyestuff extracted from it

[**Henna**] a mutant (gene symbol *Hn*) mapped at 23.0 on chromosome III of *Drosophila melanogaster*. The phenotypic eye color is dull brown

hen-of-the-woods *U.S.* the polyporaceous basidomycete fungus *Polypilus frondosus*

Henopidae = Acroceridae

good king **Henry** the chenopodiaceous herb *Chenopodium bonus-henricus*

Hensen's *epon. adj.* from Viktor Hensen (1835–1924)

Henslovian *epon. adj.* from John Stevens Henslow (1796–1861)

heparinase an enzyme catalyzing the breakdown of heparin through the hydrolysis of α-1, 4-links between 2-amino-2-deoxy-D-glucose and D-glucuronic acid

Hepaticae that phylum of bryophyte plants which contains the forms commonly called liverworts. The group is of very diverse form and habitat and differs from other bryophytes principally in lacking the leaf-like structures of the mosses and the large pyrenoid containing chloroplasts of the hornworts

[**hepatic fusion**] a mutant (*hf*) in linkage group I of the mouse. The phenotypic expression is the fusion of the median and lateral lobes of the liver

-hepato- *comb. form* meaning liver

Hepialidae a family of jugate lepidopteran insects commonly called ghost moths or, in *U.S.*, swifts. These insects are frequently mistaken for sphingids

Heptageniidae a family of ephemeropteran insects with the hind tarsae 5-segmented

-hepta- *comb. form* meaning "seven"

Heptandria that class of the Linnaean classification of plants distinguished by the possession of seven stamens

herald U.S. the noctuid lepidopteran insect *Scoliopteryx libatrix*

herb 1 (*see also* herb 2) any green flowering plant not a shrub or tree and specifically, such a plant used in the seasoning of food. Rarely, an annual, as distinct from a perennial plant

pot herb any fragrant or aromatic herb used as a seasoning in cookery; or any green herb boiled for food

sweet herb (*Brit.*) = pot herb *U.S.* in first meaning

herb 2 (*see also* herb 1) in compound names of plants

bitter herb the compositaceous herb *Centaurium umbellatum*

cow-herb the caryophyllaceous herb *Saponaria vaccaria*

St. George's herb the valerianaceous herb *Valeriana officinalis*

willow herb any of several onagraceous herbs of the genus Epilobium

great willow herb the onagraceous herb *Epilobium angustifolium* (= fireweed *U.S.*)

Herbaceae a class of dicotyledonous plants erected to contain those families the members of which are primarily herbaceous and some clearly related woody forms

herbaceous said of a plant that lacks a woody stem and is therefore neither a shrub nor a tree

herbarium (*see also* moth) a collection of dried plants arranged taxonomically for reference purposes

herb-christopher the ranunculaceous shrub *Actaea spicata*

herb-of-grace the rutaceous herb *Ruta graveolens*

herbosa *ecol. jarg.* for vegetation

altherbosa a community of tall, usually perennial, herbs

emersiherbosa a community of herbs with their roots, and usually the lower part of their stems, submersed (*cf.* helostadion)

herb-patience the polygonaceous herb *Rumex patienta*

herb-Robert the geranaceous herb *Geranium robertianum*

Herbst's *epon. adj.* from Ernst Friedrich Gustav Herbst (1803–1893)

-herco- *see* -erko-

herd a group of almost any perissodactyle or artiodactyl mammal of species for which no specific group name exists. The term is also used as a group name for cranes (*cf.* siege), curlews, seals (*cf.* pod) and swans

heredity the transmission of characters from parent to offspring

hermaphrodite an organism which has both male and female primary sex characters

pseudohermaphrodite an animal which produces functional gametes of one sex but exhibits some of the secondary sex characters of the other sex

primary somatic hermaphrodite an organism having the gonads of one sex while either all or some part of the secondary sexual characters of both sexes

secondary somatic hermaphrodite an organism having the gonads and secondary sexual characters of one sex together with parts of the secondary sexual characters of the other sex

agamohermaphroditism the condition of having both neuter and perfect flowers on the same plant

Hermelliformia an order of cryptocephalous polychaete annelids in which the peristomium is enormously developed forming a trilobed hood

hermit (*see also* thrush) any of numerous apodiform trochilid birds of the genera Glaucis, Threnetes and Phaethornis

Hernandiaceae a small family of ranale dicotyledonous shrubs distinguished from the Lauraceae by its winged fruits

herniary any of several species of the caryophyllaceous genus Herniaria

heron any of very many species of long-legged, long-necked wading birds of the family Ardeidae, and a few others birds of similar aspect. *Brit.*, without qualification, *Ardea cinerea*

green-backed heron *Butorides striatus*

boat-billed heron the cochleariid bird *Cochlearius cochlearius*

black heron *Melanophoyx ardesiaca*

blind heron = pond heron

blue heron *N.Z.*, without qualification *Demiegretta sacra* (= reef heron)

great blue heron *Ardea herodias*

cattle heron *Bubulcus ibis*

white-fronted heron *N.Z. Notophoyx novaehollandiae*

goliath heron *Ardea goliath*

green heron *Butorides virescens*

grey heron *Ardea cinerea*

Louisiana heron *Hydranassa tricolor*

mangrove heron *Austral. Butorides striatus*

white-necked heron *Ardea pacifica*

night heron any of several species of the genera Nycticorax and Nyctanassa

black-crowned night heron *Nycticorax nycticorax*

yellow crowned night heron *Nyctonassa violacea*

pond heron any of several species of the genus Ardeola

purple heron *Ardea purpurea*

reef heron *Demigretta sacra*

W. Afr. reef heron *Demigretta gularis*

squacco heron *Ardeola ralloides*

tiger heron any of several species of the genera Tigrisoma and Heterocnus

white heron *N.Z.* = common egret *U.S.*

great white heron *Brit.* = common egret *U.S. U.S. Ardea occidentalis*

-herp- *see* -erp-

-herpet- *comb. form* meaning "creep" or "creeping thing"

Herpestinae a subfamily of viverrid carnivorous mammals containing, *inter alia*, the mongooses, cusimanses and the meerkat

Herpetotherinae that subfamily of Falconidae commonly called laughing falcons. They are distinguished by their large head with a relatively large, circular, bony-rimmed nostril lacking a central tubercule

herpism the condition of creeping or being a creeper

Herpobdellidae = Erpopdellidae

herpon the sum total of those microorganisms that live on the surface of muddy or slimy bottoms of fresh water lakes (*cf.* gyttja)

herptile a word used by some as synonymous with herpetologist or herpetology (*see* -logy)

herring (*see also* gull, king-of) *U.S.* any of numerous fishes of the isospondylous family Clupeidae and a few other fish of like aspect. *Brit. Clupea harengus*

Atlantic herring *Clupea harengus*

blueback herring *Alosa aestivalis*

lake herring = cisco

dwarf herring *Jenkinsia lamprotaenia*

flatiron herring *Harengula thrissina*

Pacific herring *Clupea harengus pallasi*

round herring any of several species of dussumierid fish

Atlantic round herring *Etrumeus sadina*

California round herring *E. acuminatus*

skipjack herring *Alosa chrysochloris*

thread herring any species of the genus Opisthonema

Atlantic thread herring *O. oglinum*
Pacific thread herring *O. libertate*
wolf herring the chirocentrous fish *Chirocentrus dorab*

-heaia *comb. suffix* presumably derived by extension from *adhaesere,* used by ecologists to indicate a "swarm"

synhesia a group of animals gathered together during the breeding season (*cf.* synhesma)

Hesionidae a family of errant polychaete annelids distinguished by the fact that the prostomium is not annulated and that there are no jaws on the proboscis

-hesma *comb. form* presumably related to -hesia, *q.v.,* used by ecologists for "reproductive swarm"

synhesma a group of organisms gathered together in consequence of a reproductive drive

amphoterosynhesma such a group of males and females mixed

androsynhesma such a group of males

gynsynhesma such a group of females

hesperidium a berry-like fruit with a leathery rind of oily tissue derived from the ovary and many membranous juice-filled segments derived from the endocarp. Citrus fruits are an example

Hesperiidae the largest family of hesperioid lepidopteran insects. They are commonly called skippers, and are also known in various parts of the world as flats, angles, awls, bobs, redeyes, darts, scrub hoppers, aces, hedge hoppers, swifts and demons

Hesperioidea a superfamily of Lepidoptera containing the Hesperiidae and their near relatives the Megathymidae. They differ from the Papilionoidea in having 5 branches to the radial vein, in possessing thicker bodies, and in some cases in having recurved tips to the antennae

het-2 *see* [Heterocaryon formation]

Heterantherae a tribe of the Pontederiaceae

Heteraxonia the zoological taxon erected by Hatschek to contain all metazoa other than those in his taxon Protaxonia

-hetero- *comb. form* meaning "different"

Heterobasidiomycetes a class of basidiomycete fungi containing the parasites commonly called molds and rusts. They are distinguished by the presence of phragmobasidia that are either vertically or transversely septate

[Heterocaryon formation] a mutant (gene symbol *het-2*) on linkage group II of *Neurospora crassa.* This gene determines heterocaryon compatibility

Heterocera a term at one time used to divide the "moths" from the "butterflies" (*see* Rhopalocera)

Heteroceridae a small family of mud-dwelling beetles readily recognized by the dilated and spiny tibia of the front legs

Heterochloridales an order of xanthophycean Chrysophyta including all the forms having flagellated vegetative cells

Heterococcales an order of xanthophycean Chrysophyta in which the immobile vegetative cells have a thickened wall

Heterocoela an order of calcareous Porifera in which the choanocytes are found in chambers in the body wall, so that the prosopyles open from incurrent canals into these chambers rather than directly from the exterior as in the Homocoela

Heterocotylea = Monogenea

Heterodontidae a family of sharks distinguished by the possession of a heavy spine at the forward margin of each dorsal fin. Popularly called horn sharks

Heterogeneratae that class of phaeophyte algae which show a heteromorphic alternation of generations in which the sporophyte is always larger than the gametophyte

Heterogyna = Formicidae

Heterokontae = Xanthophysaceae

heteromalous spreading about in all directions

Heteromera a zoological taxon at one time containing those beetles in which the number of tarsi differ on the front and back legs. They are now divided among the superfamilies Meloidea, Mordelloidea and Tenebrionoidea

Heteromi an order of Osteichthyes erected to contain the spiny eels of the family Notacanthidae

Heteromyidae a family of sciuromorph rodents, mouse- or rat-like in shape, with fur lined cheeks pouches and containing, *inter ali,* the pocket mice, kangaroo mice and kangaroo rats

Heteronemertea an order of anoplate Nemertea distinguished by the possession of a fibrous dermis

Heteroneura = Frenatae

Heteroneuridae = Clusiidae

Heteroptera the subordinal name taken by the Hemiptera when they are included in the Homoptera

Heteropyxidaceae a small family of myrtiflorous dicotyledonous angiosperms

Heterosiphonales an order of xanthophycean Chrysophyta containing multinucleate unicellular forms with rhizoid-like protuberances

heterosis a group of economically desirable characters associated with inter-variety crosses in domestic plants and animals, leading to increased vigor of hybrid populations

Heterostomata an order of Osteichthyes containing the "flat fish": that is, those fish which, in the adult, have both eyes on one side of the body that thus becomes the upper side. The soles, flounders and halibuts are well-known representatives

Heterostylaceae = Lilaeaceae

Heterotardigrada a group of Tardigrada distinguished by the possession of several pairs of short cirri beside the mouth

Heterothripidae a family of terebrantian thysanopteran insects closely allied to the Thripidae from which they are distinguished by the antennal sense cones being short and triangular, or absent

Heterotricha = Holotrichida

Heterotrichales an order of xanthophycean Chrysophyta containing those forms in which the cells are united to form filaments

Heterocapsales an order of xanthophycean Chrysophyta distinguished by the possession of immobile vegetative cells in amorphous gelatinous colonies

hewel = hewhole

wood hewer = wood creeper

hewhole *Brit. obs.* = woodpecker

-hexa- *comb. form* meaning "six"

Hexagrammidae a family of small-scaled scleroparous fishes usually having two or more lateral lines. Commonly called greenlings because the flesh of some species is light greenish

Hexagynia those orders of the Linnaean system of plant classification distinguished by the possession of six pistils

Hexanchidae a family of sharks distinguished by the possession of six or seven gill slits, commonly called the cowsharks

Hexandria that class of the Linnaean classification of plants distinguished by the possession of six stamens of equal length

Hexapoda = Insecta

Hexasterophora an order of hexactinellid Porifera with hexasters but without amphidisks

hexatine a six-rayed spicule such as that found in silicious sponges

hf *see* [hepatic fusion]

hi a yeast genetic marker indicating a requirement for histidine

hiant gaping wide

-hibern- *comb. form* usually thought to mean "winter" (*cf.* -hiemal-) but actually meaning "winter quarters"

hibernacle a winter bud

hibernaculum literally, winter quarters and applied to any of numerous structures that assist organisms to withstand cold weather

hibernal pertaining to winter or winter quarters

hibernation the condition of physical activity and decreased metabolic activity in which some organisms pass the winter months (*cf.* aestivation)

  artificial hibernation a condition showing physical and psychological resemblance to hibernation induced in an organism that does not normally hibernate

hickory (*see also* beetle, borer, bug, gum, worm) any of many juglandaceous trees of the genus Carya

  shagbark hickory *C. ovata*

  shellbark hickory either of two species: *C. laciniosa* is often called the **big shellbark hickory** and *C. ovata* the **little shellbark hickory**

  big-bud hickory *C. tomentosa*

  bitternut hickory *C. cordiformis*

  pignut hickory *C. glabra*

  nutmeg hickory *C. myristicaeformis*

  red hickory *C. ovalis*

  swamp hickory *C. cordiformis*

  water hickory *C. aquatica*

-hidro- *comb. form* meaning "sweat" but by extension excretion (*cf.* hydro-)

-hiemal- *comb. form* meaning "winter," frequently spelled -hyemal- in compounds

hierarchy the ranks established within a group of organisms by means of the peck order (*q.v.*) or some such method. The term caste is more usually confined to those forms, such as the social insects, in which there are morphological differences between the social ranks

[High tail] a mutant (gene symbol *Ht*) in linkage group VI of the mouse. The phenotypic expression is a short thick erect tail

hihi = ihi

  hile = hilum

rose hill the Australian psittacid *Platycercus eximius*

axon hillock a prominence on a nerve cell from which the axon arises

  sense hillock = neuromast

hilum 1 (*see also* hilum 2) literally, "a trifle" but used by anatomists both plant and animal, for a reentrant portion of an organ (such as a kidney, or the seed of a bean) through which blood vessels or other structures penetrate to the interior

hilum 2 (*see also* hilum 1) a central point, round which rings, not of necessity either circular or symmetrical, radiate. A hilum is to be observed on a starch grain, a teleost scale, the valves of pelecypods, and on the scar left on a seed by the detachment of the funicle

Himantandraceae a small family of ranale dicotyleconous angiosperms

hinau *N.Z.* the elaeocarpacean tree *Elaeocarpus dentatus*

hind 1 (*see also* hind 2) a female red deer

hind 2 (*see also* hind 1) *U.S.* several species of the genus serranid fish Epinephelus

  red hind *E. guttatus*

  rock hind *E. adscensionis*

  speckled hind *E. drummondhayi*

Hiodontidae a small family of freshwater herring-like fish of the north central and eastern United States

hip 1 the area immediately surrounding the articulation of the head of the femur with the pelvic girdle. The term is also applied to the coxa of insects

hip 2 (*see also* hip 1) = cynarrhodium

Hippidea a superfamily of anomuran Crustacea distinguished from the Paguridea by the symmetrical abdomen and from most other groups by the cylindrical thorax

hippo popular form of hippopotamus

  wild hippo *U.S.* the euphorbiaceous herb *Euphorbia corollata*

Hippoboscidae a large family of parasitic pupiparan dipteran insects commonly called louse flies. Their flat shape and leathery appearance is unmistakable. The wingless so-called sheep tick is a member of this group

Hippocastanaceae that family of sapindalous dicotyledons which contains the horse-chestnuts. The large seeds are typical of the family

Hippocrateaceae a family of sapidale dicotyledonous angiosperms distinguished by their climbing habit, mostly opposite leaves and often winged seeds

-hippocrep- *comb. form* made up from -crep- and -hippos- to indicate "horseshoe"

Hipponicidae a family of gastropod Mollusca closely allied to the Capulidae in possessing a hooked but uncoiled shell but distinguished from them by the presence of a shelly plate between the true shell and the point of attachment

Hippopotamidae family of suine artiodactyles containing the hippopotamus and pigmy hippopotamus

hippopotamus the *Afr.* mammal *Hippopotamus amphibius*. Apart from its general appearance it is distinguished by possessing four toes and lacking a caecum

  pigmy hippopotamus the *W.Afr.* hippopotamid *Choeropsis liberiensis*

-hippos- *comb. form* meaning "horse"

Hipposideridae a family of chiropteran mammals containing those horseshoe-nosed bats which have only two phalanges on all the toes (*cf.* Rhinolophidae)

Hippotraginae a subfamily of antelopid artiodactyles sometimes called the horse-antelopes because the general shape of the body strongly resembles that of a horse

Hippuridaceae a monogeneric, monospecific family of myrtiflorous dicotyledonous angiosperms. It is easily distinguished by the linear entire leaves in whorls of six or more

Hippuridineae a suborder of myrtiflorous dicotyledonous angiosperms containing the families Hippuridaceae and Thelygonaceae

hircinous pertaining to goats, particularly with regard to the odor

hirsel term for a group of sheep, through flock is more frequently used

hirsute covered in long, but not shaggy, hairs

hirtellous diminuitive form of hirsute

Hirudinea a class of annelid worms commonly called leeches. They are distinguished by the absence of bristles and the presence of suckers

Hirudinidae a family of gnathobdelloid hirudinean annelids, containing, *inter alia,* the well known medicinal leech. They are distinguished from Erpobdellidae by possessing three-toothed jaws

Hirundinidae that family of passeriform birds which contains the swallows. They have short, broad-gaped, and flattened bills. They are, however, principally distinguished by the short neck, slender body, and very long and pointed wings

his a bacterial genetic marker indicating a requirement for histidine. The mutant affects the activity of ten known enzymes, and an operator, in *Salmonella typhimurium* and has been mapped at 38.6 mins. for *Escherichia coli*

His's *epon. adj.* from Wilhelm His (1831–1904) [cells] or Wilhelm His (1863–1934) [bundle]

hispid bristly, in the sense of an unshaven chin

hist-1-hist-7 *see* [Histidine-1]—[Histidine-7]

histaminase *see* diamine oxidase

Histeridae a family of coleopteran insects lying between the Scarabaeidae and the Hydrophilidae. Commonly called hister beetles

[histidine-1] a mutant (gene symbol *hist-1*) on linkage group V of *Neurospora crassa.* The phenotypic expression is a requirement for histidine

[histidine-1] a mutant (gene symbol *hist-2*) in linkage group I of *Neurospora crasa.* The phenotypic expression is a requirement for histidine

[histidine-3] a mutant (gene symbol *hist-3*) in linkage group I in *Neurospora crassa.* The phenotypic expression is a histidine requirement

[histidine-4] a mutant (gene symbol *hist-4*) on linkage group IV of *Neurospora crassa.* The phenotypic expression is a requirement for histidine

[histidine-7] a mutant (gene symbol *hist-7*) found in linkage group III of *Neurospora crassa.* The phenotypic expression is a requirement for histidine

1-histidine α-amino -4-imidazolepropionic acid. A common amino acid known to be essential in the nutrition of rats

-histo- *comb. form* meaning "web" and like -istem of the same meaning, has come to be extended to cellular structures particularly when seen in section

[Histocompatability-1] a mutant (gene symbol *H-1*) in linkage group I of the mouse. The phenotypic expression is an ability to receive tissue grafts

[Histocompatibility-2] a mutant (gene symbol *H-2*) in linkage group IX of the mouse. The phenotypic expression is the same as Histocompatibility-1

[Histocompatibility-3] a mutant (gene symbol *H-3*) found in linkage group V of the mouse. The phenotypic expression is the same as Histocompatibility-1 and -2

[Histocompatibility-4] a mutant (gene symbol *H-4*) in linkage group I of the mouse. Phenotypic expression is the same as Histocompatibility-1, -2 and -3

histology the study of tissues

histone any of several proteins complexed, at one time or another, with DNA

hitch *U.S.* the cyprinid fish *Lavinia exilicauda*

hiulcous split

hk *see* [hook]

Hk *see* [Hook]

hl *see* [hair-loss] and [hairless]

hm *see* [Helminthosporium susceptibility]

Hn *see* [Henna]

ho *see* [honey] and [holdout]

hoactzin = hoatzin

hoary rendered gray by a sparse pubescence

hoatzin a *S. Amer.* galliform bird (*Opisthocomus hoazin*) which is the only representative of the family Opisthocomidae. The outstanding peculiarity is the presence, in the young, of well-developed claws of the first and second toes of the wings. The young therefore climb about in trees in a markedly reptilian manner

witch hobble the caprifoliaceous shrub *Viburnum alnifolium*

hobby any of several falconine birds of graceful flight and extremely long wings. *Brit.,* without qualification, *Falco subbuteo*

African hobby *F. curierii*

Oriental hobby *F. severus*

hock the tarsal joint of the hind limb of perissodactyl and artiodactyl mammals. In horses, the proximal region is differentiated as the gaskin

[Hoffman's anthocyaninless] a mutant (gene symbol *ah*) mapped at 1.5 on chromosome 9 of the tomato. The phenotypic plants lack anthocyanin

hog 1 (*see also* hog 2, fennel, mite) properly any suid artiodactyl but, in technical agricultural parlance, a domestic pig weighing more than 120 pounds

forest hog a giant Bornean *Hylochoerus meinertzhageni*

pygmy hog *Sus salvania*

Red River hog any of several species of the genus Potamochoerus

wart hog any of several species, or possibly races, of the genus Phacochoerus. The name derives from the grotesquely warted head

hog 2 any of several heavily built animals not suids

ground hog *U.S.* the large sciurid rodent *Marmota monax* (= woodchuck)

hedgehog (*see also* mushroom) any of several species of insectivorous mammal of the erinaceid genus Erinaceus. *Brit.,* without qualification, *E. europaeus*

-homal- *comb. form* meaning "equal" (*cf.* -homo- and -homoe-)

hoiho *N.Z.* the spheniscid bird *Megadyptes antipodes.* (= yellow-eyed penguin)

hojack *U.S.* the larva of the dobson fly *Corydalus cornutus* (= hellgrammite)

holdfast that portion of a thallophyte plant body which attaches the organism to a base. Used, less specifically, for many attaching structures

holly 1 (*see also* holly 2, fern, miner, oak, scale) any of very many aquifoliaceous shrubs and trees and numerous horticultural varieties of the genus Ilex and its close allies. *Brit.,* without qualification, *I. aquifolium*

American holly *I. opaca*

English holly *I. aquifolium*

mountain holly (*see also* holly 2) *U.S.* the aquifoliaceous shrub *I. montana* or *Nemopanthus mucronata* (= catberry, wild holly)

wild holly *Nemopanthus mucronata*

holly 2 (*see also* holly 1) any of several prickly leaved herbs and shrubs

mountain holly *U.S.* the rosaceous tree *Prunus ilicifolia N.Z.* the oleariaceous tree *Olearia ilicifolia*

seal-holly any of several species of herb of the umbelliferous genus Eryngium

hollyhock (*see also* bug, weevil) any of several malvaceous herbs, mostly of the genus Althaea. Without qualification, *A. rosea*

antwerp hollyhock *A. ficifolia*

sea hollyhock the malvaceous herb *Hibiscus palustris*

trailing hollyhock the malvaceous herb *Hibiscus trionum*

holm *Brit. obs.* = holly

-holo- *comb. form* meaning "whole." The correct transliteration, sometimes still used, is -olo-

Holocentridae a family of berycoidous fishes distinguished by their pointed noses and very large brown eyes from which the name squirrel fish is derived

Holocephali a group of gnathostomatous craniate chordates containing the forms usually called rabbit-fishes and ratfishes. They are distinguished by the presence of a cartilaginous skeleton but differ from the elasmobranchs in that the upper jaw is fused to the cranium

Hologonia an order of Nematoda, distinguished from the Telogonia by the fact that genital tract in both sexes is a single unbranched tube

Holostei a superorder of actinopterigyiian bony fishes best characterized as being intermediate between the Chondrosteii and the Teleostei

Holostomata a division of prosostomate trematodes possessing both an oral sucker and an acetabulum and, in addition, a ventral adhesive disc posterior to the acetabulum

Holothurioidea a class of eleuthrozoan Echinodermata commonly called sea cucumbers. They are distinguished by the elongate shape and the reduction of the skeleton to spicules or plates embedded in the leathery body wall

Holotrichida an order of euciliate ciliate Protozoa containing *inter alia*, the well-known Paramecium and its allies. Distinguished by the relatively uniform distribution of even-sized cilia over most of the body

hom a bacterial mutant indicating a requirement for homoserine

Homalopsinae a subfamily of opisthoglyphan colubrid snakes. They are without exception aquatic and mostly estuarine. They are adapted to this habit by having valvular nostrils

Homalopteridae a small family of Asiatic ostariophysous fishes adapted for dwelling in mountain torrents by the presence of ventral suckers

homeosis the alteration, by mutation, of one organ or appendage into another or the transfer, also by mutation, of an appendage to a segment on which it does not usually appear

Hominoidea a suborder of primate mammals erected to contain man and the great apes, now usually regarded as the family Pongidae

homing the process by which certain organisms, such as pigeons, return to a point of departure without of necessity reproducing the method or direction by which they left

Homininae a subfamily of primates erected to contain *Homo sapiens* by those who refuse to place him in the Ponginae

-homo- *comb. form* meaning "identical" (*cf.* -homal-, -homoe-). The h is sometimes omitted from all these forms

Homobasidomycetae a class of basidiomycete fungi that possess a holobasidium

Homocoela an order of calcareous Porifera possessing a simple central cavity lined with choanocytes and with prosopyles that open directly from the exterior to this cavity

-homoe- *comb. form* meaning "similar to" (*cf.* -homal-, -homo-)

homoeosis = metamorphosis

homolog a homologous structure

homologous *adj.* used in *biol.* to describe structures of similar phylogeny but different appearance (*e.g.* the flipper of a whale and the wing of a bird) (*cf.* analogous). In cytology the term is used for a matching pair of chromosomes

Homoneura = Jugatae

homonunculus the name given to the minute preformed individual at one time thought to exist in the head of spermatozoans

Homoptera that order of insects which contains, *inter alia*, the cicadas, spittle bugs, tree hoppers, scale insects, and white flies. They are distinguished except for the scale insects, by the presence of four similar membranous wings typically roofed over the body when at rest.

Homosclerophora an order of tetractinellid sponges mostly lacking triaenes and with the macroscleres and microscleres not differentiated

[Homoserine] a mutant (gene symbol *hs*) in linkage group I of *Neurospora crassa*. The phenotypic expression is a homoserine requirement

homotene having a more primitive, or ancestral-like form, than is customary in the group concerned

honesty the cruciferous herb *Lunaria annua*

honey (*see also* bee, bird, creeper, dew, eater, guide, locust, mushroom, pot, stomach, stopper, streak) the viscous sugary fluid produced by bees as a food reserve

[honey] a mutant (gene symbol *ho*) in linkage group I of *Habrobracon juglandis*. The phenotypic expression is a lack of pigment in the body

honeysuckle (*see also* sawfly, roller) properly any of very numerous species of caprifoliaceous shrubs of the genus Lonicera. The term is also applied to the ranunculaceous herb *Aquilegia canadensis* (*cf.* columbine), to deciduous ericaceous shrubs of the genus Rhododendron (*cf.* Azalea) and, in compound names, to some other tubuliflorous plants. *N.Z.*, without qualification, the large proteaceous tree *Knightia excelsa*

bush honeysuckle any of several species of the caprifoliaceous genus Dierdilla

cape honeysuckle the bignoniaceous climbing shrub *Tecomalia capensis*

French honeysuckle the leguminous herb *Hedysarum coronarium*

Jamacia honeysuckle the passifloraceous vine *Passiflora laurifolia*

purple honeysuckle *U.S.* the ericaceous shrub *Rhododendron nudiflorum*

swamp honeysuckle *U.S.* the ericaceous shrub *Rhododendron viscosum*

trumpet honeysuckle either *L. sempervirens* or the bignoniaceous woody vine *Campsis radicans*

hood 1 (*see also* hood 2) any structure that tends to cover the head, or even a part of the head, so that galea (*q.v.*) is sometimes referred to as the hood of the maxilla. It is specifically used for a fold of the body wall that can be drawn over the head in Chaetognatha, for a patch of distinctively colored feathers on the head of a bird, and for the bell-shaped opening at the anterior end of Amphioxus (= oral funnel)

hood 2 (*see also* hood 1) in compound names of organisms

monkshood any of numerous ranunculaceous herbs of the genus Aconitum

true monkshood *A. napellus*
wild monkshood *A. uncinatum*
hoodie = crow
hoof (*see also* shell) a horny covering protecting the digits of perissodactyl and artiodactyl mammals (*cf.* coffin bone)
[**hook**] a mutant (gene symbol *hk*) mapped at 53.9 on chromosome II of *Drosophila melanogaster*. The phenotypic expression is hook-bristles
[**Hook**] a mutation (gene symbol *Hk*) found in linkage group XVIII of the mouse. The phenotypic expression is a short tail
silver hook *Brit.* the noctuid lepidopteran insect *Hydrelia uncula*
hoolock the gibbon *Hylobates hoolocki*
hoopkoop the leguminous herb *Lespedeza striata*
hoopoe (*see also* lark) any of several coraciiform birds of the families Upupidae and Phoeniculidae, many of the latter being called woodhoopoes. Used without qualification, *Upupa epops* is usually meant
African Hoopoe *U. africana*
cape red-billed hoopoe *Afr. Phoeniculus purpureus* (= green woodhoopoe)
Indian hoopoe *U. orientalis*
black woodhoopoe *P. aterrimus*
forest woodhoopoe *P. castaneiceps*
green woodhoopoe *P. purpureus*
white-headed woodhoopoe *P. bollei*
hop (*see also* beetle, bug, moth, tree, trefoil) any of several moraceous vines of the genus Humulus but most particularly *H. lupulus*
Hoplestigmataceae a small family of ebenale dicotyledons
Hoplocarida a superorder of malacostracous Crustacea resembling the Peracarida in that the carapace leaves four of the thoracic somites distinct but differing from the Peracarida in that two moveable segments are separated from the anterior part of the head
Hoplopleuridae a family of anopluran insects commonly called small mammal sucking lice
Hoplonemertea an order of enoplate Nemertea distinguished by the fact that the proboscis is armed with stylets
hopper 1 (*see also* hopper 2, 3) any of numerous hopping insects
three-cornered alfalfa hopper *U.S.* the membracid homopteran *Spissistilus festinus*
fleahopper *U.S.* any of several mirid hemipterans, many of the genus Halticus
cotton fleahopper *U.S. Psallus seriatus*
garden fleahopper *U.S.* the mirid hemipteran *Halticus bracteatus*
froghopper popular name of cercopid homopterans
grasshopper (*see also* fly, maggot) any of numerous orthopteran insects of the families Acrididae, Tetrigidae, and Tettigoniidae (*cf.* locust)
American grasshopper *U.S.* the acrid *Schistocerca americana*
Carolina grasshopper *U.S.* the acridid *Dissosteira carolina*
devastating grasshopper *U.S.* the acridid *Melanoplus devastator*
differential grasshopper *U.S.* the acridid *Melanoplus differentialis*
long-horned grasshopper popular name of tettigonid orthopterans (= meadow grasshopper)

short-horned grasshopper popular name of acridid orthopterans
red-legged grasshopper *U.S.* the acridid *Melanoplus femurrubrum*
lubber grasshopper *U.S.* the acridid *Brachystola magna*
eastern lubber grasshopper *U.S.* the acridid *Romalea microptera*
meadow grasshopper *U.S.* = long-horned grasshopper
migratory grasshopper *U.S.* the acridid *Melanoplus biliteratus*
Rocky Mountain grasshopper *U.S.* the acridid *Melanoplus spretus*
Packard grasshopper *U.S.* the acridid *Melanoplus packardii*
high plains grasshopper *U.S.* the acridid *Dissosteira longipennis*
Nevada sage grasshopper *U.S.* the acridid Melanoplus rugglesi
green-striped grasshopper *U.S.* the acridid *Chortophaga viridifasciata*
two-striped grasshopper *U.S.* the acridid *Melanoplus bivittatus*
clear-winged grasshopper *U.S.* the acridid *Camnula pellucida*
leafhopper popular name of cicadellid homopteran insects
apple leaf-hopper *U.S. Empoasca maligna*
white apple leafhopper *U.S. Typhlocyba pomaria*
red-banded leafhopper *Graphocephala coccinea*
three-banded leafhopper *U.S. Erythroneura tricincta*
beet leafhopper *U.S. Circulifer tenellus*
bramble leafhopper *U.S. Ribautiana tenerrima*
sugarcane leafhopper *U.S.* the delphacid homopteran *Perkinsiella saccharicida*
clover leafhopper *U.S. Aceratagallia sanguinolenta*
blunt-nosed cranberry leafhopper *U.S. Scleroracus vaccinii*
Virginia-creeper leafhopper *U.S. Erythroneura ziczac*
southern garden leaf-hopper *U.S.* the cicadellid homopteran *Empoasca solana*
yellow-headed leafhopper *U.S. Carneocephala flaviceps*
mountain leafhopper *U.S. Colladonus montanus*
painted leafhopper *U.S. Endria inimica*
plum leafhopper *U.S. Macropsis trimaculata*
potato leafhopper *Empoasca fabae*
rose leafhopper *U.S. Edwardsiana rosae*
saddled leafhopper *U.S. Colladonus clitellarius*
six-spotted leafhopper *U.S. Macrosteles fascifrons*
corn planthopper *U.S.* the delphacid homopteran *Peregrinus maidis*
treehopper popular name of membracid homopteran insects
Buffalo treehopper *U.S. Stictocephala bubalus*
two-marked treehopper *U.S. Enchenopa binotata*
quince treehopper *U.S. Glossonotus crataegi*
hopper 2 (*see also* hopper 1, 3) any of several hopping Crustacea
beachhopper any of numerous arenicolous gammarid amphipod crustacea
sandhopper popular name of talitrid amphipod crustacea more commonly called beach fleas in the *U.S.*
hopper 3 (*see also* hopper 1, 2) any of numerous hesperid

lepidopteran insects. *S.Afr.*, without qualification, those of the genus Platylesches. (*cf.* skipper)

bush hopper *I.P.* any of several species of the genus Ampittia

hedge hopper *I.P.* any of several species of the genus Baracus

scrub hopper *I.P.* any of several species of the genus Aeromachus

-hor- *comb. form* confused from two roots and therefore meaning "boundary" or "limit" as well as "hour" or "time of day." In compounds this form is frequently confused with -horamo-

-horamo- *comb. form.* meaning "what is perceived" or "noticed"

horarious extending over a period measured in hours

hordeaceous in the shape of a barley grain

Hordeae a tribe of the plant family Gramineae, containing, *inter alia*, the barleys

horizon a horizontal level of layer (*cf.* zone)

A horizon the uppermost level of soil, consisting mostly of decomposing organic detritus

B horizon the layer of soil immediately under the A horizon containing considerable organic matter carried to it by rainfall from above or ground water from below

C horizon the soil layer under the B horizon consisting principally of unconsolidated inorganic matter

D horizon the bottom layer of soil lying below the C horizon and consisting of consolidated, unweathered, inorganic material

-hormo- *comb. form* confused from three *Gr.* roots and variously meaning "necklace," "base" and "excitation"

hormon marine plants held in position by holdfasts

hormone (*see also* pheromone, desinone) a substance which exercises an effect on organs that do not secrete it

ectohormone = pheromone

growth hormone one produced by the adenohypophysis that stimulates growth, particularly of bones. In adults, its action is antagonistic to insulin. The term is sometimes also applied to auxins (*q.v.*)

lactogenic hormone one produced in the adenohypophysis that stimulates the production of milk

luteinizing hormone one produced by the adenohypophysis that stimulates ovulation and the growth of the corpus luteum

melanophore hormone = melanocyte-stimulating hormone

molting hormone an arthropod neurohormone inducing ecdysis

neurohormone one produced in a nerve cell

oxytocic hormone = oxytocin

parahormone a hormone not produced by a gland

pupation hormone an insect neurohormone inducing pupation

interstitial-cell-stimulating hormone = luteinizing hormone

follicle-stimulating hormone one produced by the adenohypophysis which stimulates growth of ovarian follicles

melanocyte-stimulating hormones secreted in the pars intermedia of the pituitary of the lower vertebrates. The name is descriptive of the action

parathormone the secretion of the parathyroid gland

parathyroid hormone one secreted by the parathyroid gland. Active in the control of the calcium-phosphorus ratio in the blood

sociohormone = pheromone

adrenocorticotropic hormone = adrenocorticotropin

thyrotropic hormone one produced in the adenohypophysis which stimulates secretion by the thyroid gland

chromatophorotropic hormone = melanocyte-stimulating hormones

horn 1 (*see also* horn 2–4) a keratinous protruberance from the top of the head of most artiodactyls and many perissodactyl mammals. In contrast to the antler of the Cervidae, the horn of the Bovidae is hollow and rarely seasonal. The "horn" of the narwhal is a modified tooth: that of the rhinocerous is a fibrous mass of fused hairs

horn 2 (*see also* horn 1, 3, 4) any horn shaped structure projecting from an organism or forming part of an organ. The arthropod antenna, particularly when unusually long, are frequently miscalled horns. The term is also applied to the posterior extension of the proboscis skeleton of Enteropneusta

Ammon's horn enlarged and inwardly rolled portion of the hippocampus in mammals

anterior horn one of the ventral (anterior in terms of human anatomy) arms of the H of gray matter seen in a transverse section of the spinal cord

posterior horn one of the dorsal (posterior in terms of human anatomy) branches of the H of gray matter seen in a transverse section of the spinal cord

stigmal horn = peritreme

horn 3 (*see also* horn 1, 2, bill fly, mouth, shark, shell) in compound names of animals

bighorn *U.S.* = bighorn sheep

long horn *see* long-horn beetle

pronghorn the North American antilocaprid *Antilocapra americana*

yellow horns *Brit.* the thyatirid lepidopteran insect *Polyploca flavicornis*

horn 4 (*see also* horn 1–3, bean, fern) in compound names of plants

antelope horn *U.S.* the asclepiadaceous herb *Asclepiodora viridis*

buckhorn the osmundaceous fern *Osmunda cinnamonea* (= cinnamon fern) the name is also applied to rib grass

devil's horn = stink horn

huntsman's horn = huntsman's cap

ram's horn the martyniaceous herb *Proboscidea louisianica*

stink horn any of numerous species of ascomycete fungus of the genus Phallus and its near allies

hornero any of several species of funariid bird of the genus Funarius

hornet (*see also* moth) any very large stinging vespine hymenopteran. *Brit.*, without qualification, *Vespa crabro* (*U.S.* giant hornet)

bald-faced hornet *U.S. Vespula maculata*

giant hornet *U.S. Vespa crabro*

horn-of-plenty the cantharellaceous basidomycete fungus *Craterellus cornucopioides* (= trumpet-of-death)

hornotinous pertaining to plant growth of the current year

horoeka *N.Z.* the araliaceous tree *Pseudopanax crassifolium* (= lancewood *N.Z.*)

horopito *N.Z.* the winteraceous tree *Wintera axillaris*

horse 1 (*see also* horse 2) either of two species of large, equid perissodactyle mammals of the genus Equus. The domestic horse is any of numerous varieties of races of *E. caballus;* the wild horse of Siberia and Mongolia is *E. przewalskii*

horse 2 (*see also* horse 1, balm, briar, conch, fly, guard, louse, mint, radish, saffron, sponge, stinger, sugar, tail, tick, weed) in compound names of a great variety of organisms

devil's coachhorse *Brit.* the staphylinid coleopteran insect *Ocypus olens,* or more rarely, other large staphylinids

devil's horse *U.S. dial.* preying mantis

race horse the tadornine bird *Tachyeres brachypterus* (= steamer duck)

redhorse any of several catastomid fishes of the genus Moxostoma

copper redhorse *M. hubbsi*

Carolina redhorse *M. coregonus*

black redhorse *M. duquesnei*

smallfin redhorse *M. robustum*

golden redhorse *M. erythrurum*

gray redhorse *M. congestum*

greater redhorse *M. valenciennesi*

shorthead redhorse *M. breviceps*

V-lip redhorse *M. collapsum*

suckermouth redhorse *M. pappillosum*

Neuse redhorse *M. lachrymale*

northern redhorse *M. macrolepidotum*

river redhorse *M. carinatum*

silver redhorse *M. anisurum*

blacktail redhorse *M. poecilurum*

river horse = hippopotamus

seahorse any of several species of sygnathid fish of the genus Hippocampus. The term is sometimes also applied to the walrus

hospitating "acting as host" Said particularly of a plant that harbors ants

host 1 (*see also* host 2) any organism that harbors parasites

paratenic host an intermediate or transfer host in the life history of a parasite

host 2 (*see also* host 1) term for a group of sparrows

hottentot *S.Afr.* either of several species of hesperiid lepidopteran insect of the genus Gegenes usually, without classification, *G. hottentota*

houbara the otidid bird *Chlamydotis undulata*

hough *N.Brit.* = hock

houhere *N.Z.* the malvaceous tree *Hoheria populnea*

hound 1 (*see also* hound 2, 3) any of many varieties of domestic dog, particularly those used in hunting

hound 2 (*see also* fish, hound, 1, 3) in compound names of animals

nursehound *Brit.* the scyliorhinid selachian *Scyliorhinus catulus*

smoothhound *Brit.* the carcharhinid selachian *Mustelus vulgaris. U.S.* any of several carcharhinid sharks of the genus Mustelus

hound 3 (*see also* hound 1, 2) in compound names of plants

hoarhound = horehound

horehound any of several labiateous herbs many of the genus Marrubium. Usually without qualification *M. vulgare*

bastard horehound = black horehound

black horehound *Brit. Ballota nigra*

fetid horehound any of several species of herb of the labiateous genus Ballota

marsh horehound *Lycopus virginicus*

water horehound any of several species of the genus Lycopus

houpara *N.Z.* the araliaceous tree *Pseudopanax lessonii*

meeting **house** *U.S.* the ranunculaceous herb *Aquilegia canadensis*

houting *Brit.* the coregonid fish *Coregonus oxyrhynchus*

hover 1 (*see also* hover 2) said of a flying organism that remains more or less in one place in the air

wind hover *Brit.* = kestrel *Brit.*

hover 2 (*see also* hover 1) term for a group of trout

howl, howlet = owl, owlet

Howship's *epon. adj.* from John Howship (1781–1841)

hr *see* [hairless]

hs *see* [homoserine]

hst a bacterial marker indicating utilization of histidine

Ht *see* [High tail]

huahillo the leguminous tree *Pithecolobium brevifolium*

huamuchil = guaymochil

huanaco the camelid ruminant artiodactyle *Lama huanacos* (*cf.* llama)

Huguier's *epon. adj.* from Pierre Charles Huguier (1804–1874)

huia *N.Z.* the callaeid bird *Heteralocha acutirostris*

huisache the leguminous shrub *Acacia farnesiana*

hull the outer shell of grain

humantin the squalid selachian *Centrina salvini*

humeral pertaining to the humerus bone but specifically used for a flight feather attached to the humerus

humicular = saprophytic

humification the process by which plant detritus is turned into humus

hummer = hummingbird

humor an old term for fluids or semi-fluids

aqueous humor the fluid filling the anterior and posterior chambers of the eye

vitreous humor the transparent colloid filling the posterior chamber of the eye

[humpy] a mutant (gene symbol *hy*) mapped at 93.3 on chromosome II of *Drosophila melanogaster.* The phenotypic expressions is a truncated wing and a ridged thorax

humus (*see also* necron) decomposing organic matter in the soil

hunter a term used in compound names of predatory organisms and of some others having predator-like shapes or habits. *W.I.* a cuculid bird *Hyetornis pluvialis*

black hunter *U.S.* the phloeothripid thysanopteran insect *Leptothrips mali.* It is, as the name indicates, a predatory species

caterpillar hunter any of numerous carabid coleopteran insects

emmet-hunter *Brit. dial.* = wryneck

fiery hunter *U.S.* the carabid coleopteran insect *Calosoma calidum*

king hunter *Austral.* the alcedinid bird *Dacelo novaeguineae* (= kookaburra)

masked hunter *U.S.* the reduviid hemipteran insect *Reduvius personatus*

spider hunter any of several species of nectarinid birds of the genus Arachnothera

tree hunter any of several species of funariid bird of the genus Thripadectes

hurtle term used for a group of sheep, though flock is more frequent

sweet **hurts** the ericaceous shrub *Vaccinium angustifolium* (= low sweet blueberry)

husk 1 (*see also* husk 2) the dried, detached coat of many seeds

husk 2 (*see also* husk 1) term for a group of hares, or jackrabbits (*cf.* down, drove, trip)

hutia *W.I.* any of several species of tailless, rabbit-like capromyid rodents of the genus Geocapromys.

S.Afr. the capromuid hystricomorph rodent *Capromys melanurus*

hutu *N.Z.* the chloranthaceous tree *Ascarina lucida*

Huxley's *epon. adj.* from Thomas Henry Huxley (1825–1895)

hv *see* [heavy vein]

Hw *see* [Hairy-wing]

hwa-mei the timaliid bud *Garrulax canoris*

hy *see* [humpy]

hyacinth (*see also* squill) anglicized form of the liliaceous genus Hyacinthus. Many species and hybrids are in cultivation but *H. orientalis* is the common hyacinth of the florist. The term is applied to some other members of the same family

  feathered hyacinth *hort. var.* of grape hyacinth

  grape hyacinth any of many liliaceous herbs of the genus Muscari, but particularly *M. botryoides*

  musk hyacinth = nutmeg hyacinth

  nutmeg hyacinth the liliaceous herb *M. moschatum*

  Peruvian hyacinth the liliaceous herb *Scilla peruviana*

  star hyacinth the liliaceous herb *Scilla amoena*

  starry hyacinth the liliaceous plant *Scilla autumnalis*

  summer hyacinth any of several species of herb of the liliaceous genus Galtonia

  water hyacinth the pontederiaceous aquatic *Eichornia crassipes*

  wild hyacinth any of numerous liliaceous herbs of the genus Scilla. *U.S.* the liliaceous herb *Camassia scilloides*

hyaena (*see also* poison) any of three species of two genera of the fissipede carnivore family Hyaenidae. They are distinguished by the presence of four toes with non-retractile claws and a furtive appearance due to the lowered carriage of the rear legs and hind body

  brown hyaena *Hyaena brunnea*

  spotted hyaena *Crocuta crocuta*

  striped hyaena *H. striata*

Hyaenidae a family of carnivorous mammals distinguished by having the hind legs significantly shorter than the front legs, and by having four toes with non-retractile claws

Hyaeninae a subfamily of hyaenid carnivorous mammals containing the true hyaenas which differ from the Protelinae in having four toes on each of the four feet

hyaline translucent

-hyalo- *comb. form* meaning "crystal" or "crystalline"

hyalom = hyaloplasms

Hyalospongiae = Hexactinellida

hyaluronidase = hyaluronate lyase

hybrid (*see also* swarm) an offspring resulting from the union of two different forms which may be varietal, racial, or specific

  amphiclinous hybrid the condition that some hybrids resemble one parent and the other the other parent

  combined hybrid a hybrid between an existing hybrid, and another hybrid or species or race

  cryptohybrid a hybrid which exhibits unexpected characters

  derivative hybrid hybrids derived from hybrids

  dihybrid an individual heterozygous for two characters

  heterodynamic hybrid one of a group showing various proportions of phenotypic characters from both parents

  homodynamic hybrid one which shows the phenotypic characters of the parents in combination with each other

  false hybrid a hybrid which exhibits the phenotypic characters of only one parent

  graft hybrid one resulting from the interaction of scion and stock

  monohybrid an individual heterozygous for one pair of alleles

  monogenodifferent hybrid one which differs from the parent in a single genotype

  mosaic hybrid one which reproduces the colors of the parents in patches

  triphyletic hybrid one which is derived from three species

  pseudohybrid a hybrid which shows the phenotypes of only one parent

  reciprocal hybrid one obtained from the same parents, but with the sexes transposed

    sesquireciprocal hybrid one between an F_1 individual and one of its parents

  secondary hybrid one produced by crossing other hybrids

  bixexual hybrid a hybrid that shows the characters of both parents distinctly and not mingled

    unisexual hybrid one in which a specific phenotype of one parent is not visible

  ternary hybrid a hybrid derived by crossing a hybrid with a species not related to either of its parents

  twin hybrid identical reciprocal hybrid

  variety hybrid one which is produced by crossing two varieties

hybridization the process of producing hybrids

  introgressive hybridization that type of hybridization in which there is a partial transfer of genetic characters from one type to the other during the initial overlap period before a complete barrier to crossing builds up

carboxymethylhydantoinase an enzyme catalyzing the hydrolysis of L-5-carboxymethylhydantoin to L-3-ureidosuccinate

hydathode the structure through which water is passed in the process of guttation

Hydatinidae a family of gastropod mollusks with a thin shell practically lacking an umbilicus, and with an involute spire. Commonly called painted bubble-shells

Hydnaceae that family of basidiomycete fungi which contains the forms commonly called hedgehog mushrooms. They are characterized by the needle-like projections that hang from the underside of the cap

hydr-, -hydro- *comb. form* meaning "water" or "hydrogen"

Hydracarina a loose assemblage of trombidiform acarines distinguished by their aquatic habitat

Hydrachnellae = Hydracarina

hydrad a water loving plant

  helohydrad a marsh forest plant

Hydrangea properly a genus of saxifragaceous shrubs but used also for near relatives

  climbing hydrangea *Decumaria barbara*

  wild hydrangea *U.S. Hydrangea arborescens*

Hydrangeoidea a subfamily of Saxifragaceae held by some to be of familial rank

Hydrariae an order of hydrozoan coelenterates containing the solitary elongate freshwater forms typified by the well known Hydra

carbonic anhydrase an enzyme catalyzing the breakdown of carbonic acid into carbon dioxide and water

hydratase a group of enzymes which catalyze the addition of water to a double bond, hence the opposite of hydrolase

aconitate hydratase catalyzes the production of aconitate from citrate or isocitrate

enoyl-CoA hydratase catalyzes the production of 2-3-*trans*-enoyl-CoA from 1-3-hydroxyacyl-CoA and vice versa

methylglutaconyl-CoA hydratase catalyzes the production of *trans*-3-methyl-glutaconyl-CoA from 3-hydroxy-3-methylglutaryl-CoA

phosphopyruvate hydratase catalyzes the production of phosphoenoepyruvate from phosphoglycerate

dehydratase essentially the same term as hydratase

dihydroxyacid dehydratase catalyzes the production of oxoisovalerate from 2,3-dihydroxyisovalerate

altronate dehydratase catalyzes the production of 2-oxo-3-deoxy-D-gluconate from D-altronate

arabonate dehydratase catalyzes the production of 2-oxo-3-deoxy-D-arabonate from D-arabonate

carbonate dehydratase = carbonic anhydrase

citrate dehydratase catalyzes the production of aconitate from citrate but does not act on isocitrate

galactonate dehydratase catalyzes the production of 2-oxo-3-deoxy-D-galactonate from D-galactonate

phosphogluconate dehydratase catalyzes the production of 2-oxo-3-deoxy-6-phospho-D-gluconate from D-6-phosphogluconate

mannonate dehydratase catalyzes the production of 2-oxo-3-deoxy-D-gluconate from D-mannonate

imidazoleglycerolphosphate dehydratase catalyzes the production of imidazoleacetol phosphate from D-*erythro*-imidazoleglycerol phosphate

5-dehydroquinate dehydratase catalyzes the production of 5-dehydro-shikimate from 5-dehydroquinate

serine dehydratase catalyzes the production of pyruvate and ammonia from L-serine

homoserine dehydratase catalyzes the production of 2-oxobutyrate from L-homoserine

threonine dehydratase catalyzes the production of oxobutyrate from L-threonine

aminolaevulinate dehydratase = porphobilinogen synthase

fumarate hydratase an enzyme catalyzing the production of fumarate from malate

Hydrellidae = Ephydridae

hydric *ecol. jarg.* for "wet." The word hydric actually means "pertaining to" or "containing" hydrogen

-hydro- *comb. form* meaning "water" (*cf.* -hidro-)

Hydrobatidae that family of procellariiform birds which contains the storm-petrels. They are distinguished from other families in the order by the erratic and fluttering flight and generally black color

Hydrocaryaceae an aquatic family of myrtiflorous dicotyledons containing *inter alia* the water chestnuts. The aquatic habit, and the fruit are distinctive

Hydrocharitaceae that family of helobian monocotyledons which contains (*inter alia*) the frog's bits and eel grasses. Characteristic of these aquatic plants are the differentiation of the flower into calyx and corolla, the numerous stamens and the exalbuminous seeds

Hydrochoeridae a monogeneric family of hystrichomorph rodents containing the capybaras. They are distinguished from all other rodents, apart from its great size, by the relatively huge posterior upper molar

Hydrocorallina an order of hydrozoan coelenterates

distinguished from all other hydrozoans by forming a coral-like stock. Now usually divided into the Milliporina and Stylasterina

Hydrodictyaceae that family of chlorococcale Chlorophyceae that contains the forms in which the cells are united into free-floating colonies

Hydrogam = Cryptogam

hydrogenase a little known bacterial enzyme, or group of enzymes, which utilizes molecular hydrogen in the reduction of a variety of materials

dehydrogenase a general term for a group of enzymes that catalyze the removal of hydrogen from molecules (*cf.* transhydrogenase)

acetoin dehydrogenase catalyzes the production of diacetyl from acetoin

lipoamide dehydrogenase catalyzes the production of dihydrolipoamide from oxidized lipoamide using NADH as the hydrogen donor

L-**aminoacid dehydrogenase** catalyzes the production of 2-oxo-acid and ammonia from aliphatic L-amino acids

D-**2-hydroxyacid dehydrogenase** catalyzes the production of pyruvate from lactate using cytochrome c as the receptor

alanine dehydrogenase catalyzes the production of pyruvate and ammonia from L-alanine

alcohol dehydrogenase catalyzes the production of aldehyde or ketone from primary or secondary alcohols

aldehyde dehydrogenase catalyzes the production of the corresponding acids from various aldehydes using either NAD or NADP as the hydrogen receptor

benzaldehyde dehydrogenase catalyzes the production of benzoate from benzaldehyde using either NAD or NADP as the receptor

betaine aldehyde dehydrogenase catalyzes the production of betaine from betaine aldehyde

formaldehyde dehydrogenase catalyzes the production of formate from formaldehyde

aspartate semialdehyde dehydrogenase catalyzes production of L-β-aspartylphosphate from L-aspartate β-semialdehyde + phosphate using NADP as the hydrogen receptor

L-**arabinose dehydrogenase** catalyzes the production of L-arabone-γ-lactone from L-arabinose

L-**arabitol dehydrogenase** catalyzes production of L-ribulose from L-arabitol D- and L-arabitol dehydrogenases catalyze the production of D-and L-xylulose respectively from D- and L-arabitol

α-**hydroxycholanate dehydrogenase** catalyzes the production of 3 oxocholanate from 3-α-hydroxycholanate

choline dehydrogenase catalyzes the production of betaine aldehyde from choline using indophenol as the hydrogen acceptor

acyl-CoA dehydrogenase catalyzes the production of 2, 3-dehydroacyl -CoA from acyl-CoA using cytochrome c as a hydrogen receptor

butyryl-CoA dehydrogenase catalyzes the production of crotonyl-CoA from butyryl-CoA using cytochrome c as a hydrogen acceptor

3-hydroxyacyl-CoA dehydrogenase catalyzes the production of 3-oxo-acyl-CoA from 3-hydroxyacyl-CoA

3-hydroxybutyrate dehydrogenase catalyzes the production either of acetoacetate from 3-hydroxyisobutyrate

cysteamine dehydrogenase catalyzes the production of cystamine disulphoxide from cysteamine

hydroxymethyltetrahydrofolate dehydrogenase catalyzes the production of anhydroleucovorin from 10-hydroxymethyltetrahydrofolate using NADP as the receptor

dihydrofolate dehydrogenase catalyzes the production of folate from 7_x8-dihydrofolate using NADP as the receptor

tetrahydrofolate dehydrogenase catalyzes the production of 7, 8-dihydrofolate from 5, 6, 7-8-tetrahydrofolate using either NAD or NADP as a receptor

formate dehydrogenase catalyzes the production of CO_2 from formate

galactitol dehydrogenase catalyzes the production of D-tagatose from galactitol

galactose dehydrogenase catalyzes the production of D-galactono-γ-from D-galactose

gluconate dehydrogenase catalyzes the production of 2-oxo-D-gluconate from D-gluconate using pyocyanine as the hydrogen acceptor

oxo-gluconate dehydrogenase catalyzes the production of 2, 5-dioxo-D-gluconate from 2-oxo-D-gluconate using phenazines and indophenol as the hydrogen acceptors

phosphogluconate dehydrogenase catalyzes the production of 6-phospho-2-oxo-D-gluconate from 6-phospho-D-gluconate, using NAD as an acceptor of the production of D-ribulose 5-phosphate using NADP as the receptor

glucose dehydrogenase catalyzes the production of D-glucono-δ-lactone from β-D-glucose using either NAD or NADP as the hydrogen receptor

glutamate dehydrogenase any of three enzymes catalyzing production of 2-oxoglutarate + ammonia from L-glutamate. One enzyme uses NAD only, one only NADP and the third either NAD or NADP as hydrogen receptors

2-hydroxyglutarate dehydrogenase catalyzes the production of 2-oxo-glutarate from 2-hydroxyglutarate using pyocyanine and phenazines as hydrogen acceptors

oxoglutarate dehydrogenase catalyzes the production of 6-S-succinylhydrolipoate from 2-oxoglutarate using oxidized lipoate as the hydrogen receptor

glutathione dehydrogenase catalyzes the production of oxidized glutathione and ascorbate from 2-reduced glutathione and dehydroascorbate

glycerol dehydrogenase catalyzes the production of dihydroxyacetone from glycerol

glycerate dehydrogenase catalyzes the production of hydroxypyruvate from D-glycerate

butyleneglycol dehydrogenase catalyzes the production of acetoin from 2,3,-butyleneglycol

L-gulonate dehydrogenase catalyzes the production of 3-keto-L-gulonate from L-gulonate

histidinol dehydrogenase catalyzes the production of L-histidine from L-histidinol

homoserine dehydrogenase catalyzes the production of L-aspartate β-semialdehyde from L-homoserine

D- and L-iditol dehydrogenase catalyze the production of D- and L-sorbose respectively from D- and L-iditol

inositol dehydrogenase catalyzes the production of 2-oxo-*myo*-inositol from *myo*-inositol

isocitrate dehydrogenase catalyzes the production of 2-oxoglutarate from threo-D-isocitrate using either NAD or NADP as the hydrogen acceptor

lactate dehydrogenase catalyzes the production of pyruvate from lactate

galactonolactone dehydrogenase catalyzes the production of ascorbate from galactono-γ-lactone using cytochrome C as the hydrogen receptor

malate dehydrogenase catalyzes the productinon of either oxaloacetate or pyruvate from L-malate

mannitol dehydrogenase catalyzes the production of D-fructose from D-mannitol using cytochrome as a receptor

dihydro-orotate dehydrogenase catalyzes production of orotate from 4,5-L-dihydro-orotate using either oxygen or NAD as the hydrogen receptor

glucose-6-phosphate dehydrogenase catalyzes the production of D-glucono-δ-lactone 6-phosphate from D-glucose 6-phosphate using NADP as the hydrogen receptor

glycerolphosphate dehydrogenase catalyzes the production of dihydroxyacetone phosphate from L-glycerol 3-phosphate

glyceraldehyde-phosphate dehydrogenase catalyzes the production of 1,3-diphosphoglyceric acid from D-glyceraldehyde 3-phosphate and ortho-phosphate

mannitol-L-phosphate dehydrogenase catalyzes the production of D-fructose 6-phosphate from D-mannitol 1-phosphate

inosine monophosphate dehydrogenase catalyzes the production of xanthosine 5'-phosphate from D-monophosphate

propanediolphosphate dehydrogenase catalyzes the production of hydroxyacetone phosphate from 1, 2-propanediol 1-phosphate

pyruvate dehydrogenase catalyzes the production of acetate from pyruvate using cytochrome as a hydrogen acceptor

quinate dehydrogenase catalyzes the production of 5-dehydroquinate from quinate

shikimate dehydrogenase catalyzes the production of 5-dehydroshikimate from shikimate using NADP as the hydrogen receptor

3-α-hydroxysteroid dehydrogenase catalyzes production of androstane-3,17-dione from androsterone using either NAD or NADP as the hydrogen receptor

transhydrogenase an enzyme that catalyzes the transfer of hydrogen from one molecule to another

gluathione-homocystine transhydrogenase catalyzes the production of 2 homocysteine and oxidized glutathione from 2 reduced glutathione and homocystine

NAD (P) transhydrogenase catalyzes the production of NADP + NADH from NADPH + NAD

hydroger a thread in a spiral vessel. The name derives from an obsolete thought that the threads themselves conveyed water

Hydroida an order of hydrozoan coelenterates with well developed, frequently branched, hydroid colonies or solitary individuals and in which the medusae are free swimming

hydrolase a group term for those enzymes which catalyze a hydrolytic reaction. Among common trivial names for hydrolases are lipase, esterase, lactonase phosphatase, deaminase and many others. Most of the enzymes with other than standard names (*e.g.* pepsin, trypsin, papain etc.) are hydrolases

amino hydrolase = deaminase

acetyl-CoA hydrolase catalyzes the hydrolysis of acetyl-CoA to CoA and acetic acid

3-hydroxyisobutyryl-CoA hydrolase catalyzes the hydrolysis of 3-hydroxyisobutyryl-CoA to Co and 3-hydroxyisobutyric acid

hydroxymethyglutaryl-CoA hydrolase catalyzes the hydrolysis of 3-hydroxy-3-methylglutaryl-Co to Co and 3-hydroxy-3-methylglutaric acid

palmitoyl-CoA hydrolase catalyzes the hydrolysis of palmitoyl-CoA to CoA and palmitic acid

succinyl-CoA hydrolase catalyzes the hydrolysis of succinyl-CoA to CoA and succinic acid

methenyltetrahydrofolate cyclohydrolase an enzyme catalyzing the hydrolysis of 5, 4-methenyltetrahydrofolate to 5, 10-formyltetrahydrofolate

IMP cyclohydrolase catalyzes the hydrolysis of IMP to 5-phosphoribosyl-5-formamido-4-imidazole-carboxamide

hydroxyacylglutathione hydrolase catalyzes the hydrolysis of S-2-hydroxyacylglutathione to reduced glutathione and a 2-hydroxyacid

acetoacetylglutathione hydrolase catalyzes the hydrolysis of S-aceto-acetyl-glutathione to reduced glutathione and acetoacetic acid

acetoacetylhydrolipoate hydrolase catalyzes the hydrolysis of S-aceto-acetylhydrolipoate to a reduced lipoate and acetoacetic acid

hydrome the sum total of the water carrying system of a vascular plant

Hydrometridae a family of long slender usually wingless hemipteran insects found on the surface of water or aquatic vegetation. They are commonly called water measurers, marsh treaders, and water crickets

Hydromynea a sub-family of murid rodents containing the Australian water-rats. They differ from other murids except the Phloeomynae in possessing three molars in each jaw and from this group in their adaptation to aquatic life in the form of webbed hind feet

Hydrophidae a family of squamatan reptiles containing the true sea snakes distinguished by the modification of the tail region as an expanded fin or paddle. All are venomous

Hydrophilidae a family of water beetles, some very large, which may be readily distinguished from the Dytiscidae, which they superficially resemble, by the long maxillary palpi, and short, clubbed antennae. Their popular name of water scavenger beetles is descriptive of their habit

Hydrophinae a subfamily of hydrophid snakes distinguished by the reduced ventral plates and viviparous mode of reproduction

Hydrophyllaceae a small family of tubiflorous dicotyledons distinguished from its close relatives the Boraginaceae, by the one-celled ovary

Hydropotinae a sub-family of cervid artiodactyl mammals containing the Asiatic forms commonly known as water-deer. They are distinguished by the large canines and the absence of antlers

Hydropsychidae a very large family of trichopteran insects commonly called net-spinning caddis flies because the larvae construct a pebble case with a cup-shaped net leading to it from upstream

Hydropteridales an order of ferns containing two families of water ferns. They are distinguished from the Eufilicales by the presence of sporocarps enclosing the usually heterosporous sporangia

Hydropteridineae = Hydropteridales

Hydroptilidae a family of very small trichopteran insects the larvae of which make crest-shaped cases

Hydrosporae an assemblage of plants the seeds of which are distributed by water

Hydrostachyales an order erected to contain the family

Hydrostachyaceae containing a single genus of African aquatic herbs

Hydrozoa a class of the phylum Coelenterata containing *inter alia* the hydroids, millepores, and Portuguese man-of-war. The group is distinguished in that both the hydroid and the medusoid forms are known for almost all

-hyema- *comb. form* meaning "winter"

hyena *see* hyaena

isohyet *ecol. jarg.* for "rainfall" or "areas of equal rainfall"

hygric pertaining to moisture (*cf.* hydric)

euryhygric pertaining to an organism capable of sustaining a wide range of relative humidity

stenohygric pertaining to an organism tolerating only a narrow humidity range

-hygro- *comb. form* meaning "moisture" as distinct from -hydro- meaning "water"

hygroscopic capable of alternately absorbing and loosing atmospheric moisture

-hyl- *comb. form* meaning "forest." Precisely the same combining form is used for "material"

Hylactidae = Rhinocryptidae

hylad a forest plant

xerohylad a plant of dry forests

-hyli- *see* -hyl-

Hylidae a family of procoelous anuran amphibia distinguished from the other members of the suborder by the presence of maxillary teeth, intercalary cartilages, an arciferal pectoral girdle, and by the absence of Bidder's organ. The members of this family are commonly called tree frogs

hylile pertaining to forests

carpohylile pertaining to a dry forest community

orohylile pertaining to subalpine forests

hylion a forest climax

helohylium a swamp forest formation

Hylobatidae that family of anthropoid primates which contains the gibbons and the siamang. The family differs from the Pongidae (great apes) on the very doubtful qualification of having enlarged canines

hylodad a plant of open woodlands

Hylophilidae = Euglenidae

Hylotomidae = Argidae

-hymen- *comb. form* meaning "membrane"

hymen a membrane usually, without qualification, the first of the membranes partially closing the vagina of some virgin mammals

hymenium a continuous layer of spore mother-cells on a sporophore of a fungus

pseudohymenium a hymenium like structure covering a sporidium

hymenodous having a membranous texture

Hymenomycetes a subclass of homobasidiomycete Basidiomycetae having a hymenium that is fully exposed before the basidiospores are mature

Hymenophyllaceae that family of eufilicale filicine pteridophyta containing the forms usually called the filmy ferns. They are distinguished by possessing leaves only one cell in thickness

Hymenoptera that order of insects which contains the forms commonly called ants, bees, and wasps. They are distinguished by possessing at some stage of their life history two pairs of membranous wings

Hynobiidae a small family of primitive urodele amphibia closely allied to the Cryptobranchidae but which undergo more complete metamorphosis

hyoid (*see also* arch) in the form of a "Y" with recurved ends to the fork as in the capital Greek letter upsilon. Also, specifically, pertaining to the second visceral arch the residual derivative of which in mammals is of the shape described

hypantra vertical articulations below the level of the zygapophyses

Hypecoideae a subfamily of Papaveraceae distinguished by the actinomorphic flowers with four stamens

-hyper- (sometimes, in error, -hyp-) *comb. form* meaning "upper," "above," "more than" (*cf.* hypo)

Hypericaceae that family of periatalous dicotyledons which contains the St. John's worts. The leaves dotted with semitransparent areas are quite typical

Hyperiidea a suborder of pelagic amphipod crustacea distinguished from the Caprellidea by the presence of seven free thoracic segments and from the Gammaridea by possessing palps on the maxillipedes and usually by the very large head and huge eyes

Hypermastigida an order of zoomastigophorous Protozoa possessing numerous flagella and many parabasal bodies

Hypermastigina = Hypermastigida

Hyperoartii = Petromyzontia

Hyperotreti = Myxiniformes

-hyph- *comb. form* meaning a web

hypha a fungus thread. Numerous, matted hyphae form a mycelium

 ascogenous hypha diploid threads in some fungi in which a second nuclear fusion may take place immediately prior to spore formation

 Woronin's hypha a coiled hypha in the area from which a sporocarp will develop in some Ascomycetes

hyphema the fungus layer in algae

 gonohyphema the fungal layer of lichens (*cf.* gonidial layer, *under* layer)

hyphidium = spermatium

Hyphomicrobiaceae a family of hyphomicrobiale schizomycetes distinguished by the fact that the buds are borne on filaments

Hyphomicrobiales an order of schizomycetes found in freshwater mud. They reproduce by budding, frequently into branched filaments of several hundred cells, that are often attached to the substrate

-hypno- *comb. form* meaning "sleep"

hypnote an organism in the dormant stage

hypnotic applied to seeds which are dormant but not dead

-hypo- (sometimes in error, -hyp-) *comb. form* meaning "lower," "under," "less than" (*cf.* -hyper-)

Hypochilomorphae a suborder of araneid arthropods possessing two pairs of lungs, six spinnerets, and cheliceral fangs that move transversely

Hypochitriales = Anisochitridiales

Hypocoliidae a monotypic family of passeriform birds related to the waxwings

Hypodermatidae a family of myodarian cycloraphous Diptera closely allied to the Oestridae and distinguished from them larges by the fact that the larvae are subdermal rather than internal pests. They are commonly called warble flies

Hyponeuria = Coelomata

Hyponomeutidae = Yponomeutidae

hypophysis *see* -physis

Hypoptidae a family of lepidopteran insects now usually included in the Cossidae

hypopus the second nymphal stage in the life history of some acarines. Most hypopi have suckers or claspers for grasping other animals for dispersion

Hyposittidae a monotypic family erected to contain the coral-billed nuthatch of Madagascar. This small, nuthatch-like, bird is of greenish-blue color with a bright red slightly hooked bill

Hypostomides a monofamilial order of osteichthyes regarded by some as a suborder of Actinopterygii. The characteristics of the family Pegasidae (sea moths) are those of the order

hypoteran = axillar

hypothesis *see* thesis

Hypoxideae a tribe of amaryllidaceous monocotyledons

-hyps- *comb. form* meaning "high" or "lofty"

Hypsidae a family of lepidopteran insects now usually included in the Pericopidae

Hypsiprymnodontinae a subfamily of phalangerid marsupials commonly called the muskrat-kangaroos. They are distinguished from other phalangerids by the naked, scaly tail and musky odor

Hyracoidea a small order of placental mammals containing the small African animals known as dassies or coneys. They are distinguished by the possession of one upper and two lower pairs of enlarged incisors, which, unlike those of the rodents, are prismatic in section and have no enamel on the lingual surface

hysginous dusky pink

hyssop the labiateous herb *Hyssopus officinalis*. Also (rarely *U.S.*) any of several species of the compositous genus Artemisia

 bastard hyssop the labiateous herb *Teucrium polium*

 hedge hyssop any of several species of herb of the scrophulariaceous genus Gratiola

 water hyssop any of several species of aquatic or marshy herb of the scrophulariaceous genus Bacopa

-hystero- *comb. form* meaning "following"

Hystrichopsyllidae a family of siphonapteran insects found parasitic only on rats, mice and shrews

Hystricidae that family of hystricomorph rodents which contains the Old World porcupines. They differ from the New World porcupines (Erethizontidae) by their terrestrial, as distinct from arboreal, habit and by the absence of all those characters such as a prehensile tail and an enlarged sternum, which are associated with climbing

Hystricomorphae an assemblage of rodents so named because it contains both the Old World (hystricid) and New World (efethizonid) porcupines. It also includes, *inter alia,* most of the large S. American rodents

I

i *see* [Enhancer of am]

I *see* [Yellow inhibitor] and [Dominant white]

I-a *see* [Dominant chocolate]

ianthinous violet-colored

Iapygidae = Japygidae

iarovization = vernalization

ibex any of several very widely distributed caprine caprid artiodactyl mammals with very large, swept back horns and long tufted chin-beards. Without qualification, the Spanish ibex is usually intended

    Sinaitic ibex *Capra sinaitica*

    Spanish ibex *Capra pyrenaica*

Ibidae = Threskiornithidae

Ibididae = Threskiornithidae

ibis (*see also* bill) any of numerous long-legged birds with decurved bills of the family Threskiornithidae and, occasionally, the family Ciconiidae

    black ibis *Pseudibis papillosa*

    bald ibis *Afr. Geronticus calvus Eur. G. eremita*

    white faced ibis *Plegadis chihi*

    glossy ibis any of several species of threskiornithid bird of the genus Plegadis, usually, without qualification, *P. falcinellus*

    hadada ibis *S.Afr. Hagedashia hagedash*

    sacred ibis *Threskiornis aethiopiea*

    scarlet ibis *Eudocimis ruber*

    white ibis *U.S. Eudocimis albus*

    wood ibis *U.S.* the ciconiid *Mycteria americana Afr.* the ciconiid *Ibis ibis*

Icacinineae a suborder of Sapindales containing the families Icacinaceae, Aetoxicaceae, Aceraceae, Hippocastanaceae, and Sapindaceae

icaco the rosaceous shrub *Chryobalanus icaco* and its fruit

-ich- *see* -ichth-

ichneumon 1 (*see also* ichneumon 2) a term variously applied to many herpestine viverrid carnivores and therefore roughly synonymous with mongoose. Members of the herpestine genus Ichneumia are commonly called white-tailed mongooses. Ichneumon, without qualification, usually refers to *Herpestes ichneumon*

ichneumon 2 (*see also* ichneumon 1) popular name of ichneumonid hymenopteran insects

Ichneumonidae an extremely large class of apocritan hymenopteran insects commonly called ichneumons. They are typically wasp-like in form but may be distinguished from the true (stinging) wasps in lacking a costal cell and by possessing very long antennae and two segmented trochanters. Many are furnished with a long, and some with a fantastically long, ovipositor

Ichneumonoidea a superfamily of apocritan hymenopteran insects containing the majority of the larger parasitic Hymenoptera. The Gasteroptiidae are sometimes included but this family possesses a costal cell and has only fourteen antennal segments instead of the twenty to sixty or more of the other families

-ichth- *comb. form* meaning "fish"

Ichthyobdellidae = Piscicolidae

-icosa- *comb. form* meaning "twenty"

Icosandria that class of the Linnean classification of plants distinguished by the possession of twenty or more stamens placed on the calyx

Icosteidae a small family of Osteichthyes distinguished by the almost complete lack of hard bones in the skeleton. The only two known species live in the open ocean. Commonly called ragfishes

ICSH = luteinizing hormone

Ictaluridae a family of naked catfish possessing four pairs of short barbels around the mouth, an adipose fin, and spines in front of the pectoral fins

ictericous dirty yellow

Icteridae that family of *Amer.* passeriform birds which contains the troupials, blackbirds, orioles and their relatives. They are small to medium sized black birds often with a metallic gloss

-id- *see* -eido-

id a term used for the heredity unit before gene or allele

Id *see* [Inhibitor]

-idae suffix indicating familial rank in animal taxonomy. The *bot.* equivalent is -aceae

-idio- *comb. form* meaning "peculiar"

Moorish idol the xanclid fish *Zanclus canescens*

Idotheoidea a suborder of isopod crustacea distinguished from all other orders except the Cymothoidea by having lateral uropods and from the latter by the fact that the uropods are ventral in position covering the pleopods

IDP = inosinediphosphate

I-f *see* [Intensifier of forked]

if *see* [inflated]

igneous flame-colored

igniarious having the consistency of a puff ball, at one time used for tinder

iguana a term properly applied to iguanid lacertilian reptiles of the genus Iguana but sometimes by extension to other large lizards

    black iguana any of several species of Ctenosaura

    common iguana *Iguana iguana*

    desert iguana *Dipsosaurus dorsalis*

    ground iguana any of several species of Cyclura

    marine iguana *Amblyrhynchus cristatus*

Iguanidae a family of lacertilian reptiles with pleurodont dentition, a short thick tongue, and no osteoderms

ihi *N.Z.* the meliphagid bird *Notiomystis cincta* (= stitchbird)

iiwi *N.Z.* the drepaniid bird *Vestiaria coccinea*

ij *see* [iojap]

ilama the annonaceous tree *Annona diversifolia* and its fruit

ile a bacterial genetic marker having a character affecting the activity of threonine deaminase, mapped at 72.2 mins. for *Escherichia coli*. This symbol is also used for a bacterial mutant indicating a requirement for isoleucine

ileum that portion of the small intestine which lies between the jejunum and the large intestine, also the anterior part of the hindgut of insects between the ventriculus and colon

ilium the more dorsal bone of the pelvic girdle of tetrapods. It participates in the attachment of that girdle to the vertebral column

Illadopsidae a family of passeriform birds usually included in the Timaliidae

illadopsis any of several timaliid birds of the genus *Trichastoma*

Illecebraceae a family of monospermous dicotyledonous angiosperms usually included within the Caryophyllaceae

Illiciaceae a small family of ranale dicotyledonous angiosperms. It is distinguished from the closely allied Schisandraceae by its shrubby or tree-like form and bisexual flowers

arginine deiminase an enzyme catalyzing the hydrolysis of arginine to citrulline and ammonia

Ilysiidae = Aniliidae

imaginal (*see also* disc) pertaining to the imago

imago the terminal, or sexually mature, stage in the life history of an arthropod

   pseudoimago = sub-imago

   sub-imago a stage in some insects, particularly Ephemoptera, intercalated between the pupa and the perfect imago (*cf.* dun)

imbricate overlapping, as the scales on a butterfly wing

imbu = umbu

-imgioc- *comb. form* meaning "claw" or "nail"

aminoimidazolase an enzyme that catalyzes the hydrolysis of 4-aminoimidazole with the liberation of ammonia

ilva a bacterial mutant indicating a requirement for isoleucine plus valine multiple

ilvA a bacterial genetic marker having a character affecting the activity of α-hydroxy β-keto acid reductoisomerase of *Salmonella typhimurium*, mapped at 72.3 mins. for *Escherichia coli*

ilvB a bacterial genetic marker having a character affecting the activity of α, β-dihydroxyisovaleric dehydrase of *Salmonella typhimurium*, mapped at 72.3 mins. for *Escherichia coli*

ilvC a bacterial genetic marker having a character affecting the activity of transaminase B, mapped at 72.3 mins. for *Escherichia coli*

immunity the condition of being resistant to an infection

   acquired immunity that which is derived by an individual from a specific immunization through exposure to the infectious agent or a derivative of it

   active immunity that which is the result of the possession of the appropriate antibodies

   congenital immunity that which results from the transfer of antibodies from the mother to the offspring

   humoral immunity = active immunity

   innate immunity that which results from the presence of antibodies not acquired by exposure to the infectious agent

   natural immunity = innate immunity

   passive immunity that which results from the injection of serum or the like

   racial immunity that which appears to be associated with a geographically limited community

   species immunity complete immunity of one species to a pathogen affecting other species in the same genus

IMP = inosine monophosphate

impala = impalla

impalla (*see also* lily) a large African gazelline antelopid *Aepyceros melampus*. The name impalla is also used in the jargon of fly-tyers to indicate calf tail

Impennes a superorder of birds, rarely used, erected to contain the Spenisciformes. The characteristics of the order are those of this group

imperial *I.P.* any of several lycaenid lepidopteran insects of the genera Neocheritra, Suasa, Cheritrella, Cheritra and Ticherra. Without qualification, usually *Cheritra freja*

   crown imperial the liliaceous herb *Fritillaria imperialis*

impregnation = fertilization

   hypodermic impregnation a term used for the fertilization process in those invertebrates in which the sperm are injected through the epidermis

   pseudoimpregnation the fusion of nuclei of two teleutospores

in *see* [inturned], [incisorless], and [intensifier of aleurone color]. This symbol is also used as a yeast genetic marker indicating a requirement for inositol

inaka *N.Z.* the epacridaceous shrub *Dracophyllum longifolium*

inane = empty

-inae suffix indicating subfamiliar rank in animal taxonomy or sub-tribal rank in plant taxonomy. The name of plant subfamilies terminate in-oidea

Inarticulata (Brachiopoda) = Ecardines

inca any of several trochilid birds of the genus Coeligena

incanous hoary

inchoate incomplete

incised sharply cut

[incisorless] a mutant (gene symbol *in*) mapped at 28 on linkage group II of the rat. The name is descriptive of the phenotypic expression

included *In bot.,* used as antithetic to protruded

inclusion a general term for non-living substances found in a cell (*cf.* metaplasm)

   cytoplasmic inclusion any structure in the cytoplasm which has not so far been identified as an organelle

inconnu the salmonid fish *Stenodus mackenzii* (= shee fish)

incubate to maintain embryos, eggs, or organisms at a constant temperature, usually above that of the environment

incudate having the form of an anvil. Used particularly of a rotiferan mastax with an unusually large incus

incunabulum = cocoon

Incurvariidae a small family of incurvarioid leaf mining lepidopteran insects. The females are unusual in possessing a leaf-piercing ovipositor

Incurvarioidea a superfamily of lepidopteran insects containing some small leaf miners as well as the yucca moths

incus literally, an anvil. Particularly the combination of ramus and fulcrum in the mastax of a rotiferan and the middle of the three bones of the mammalian inner ear

ind a bacterial genetic marker having a character affecting the activity of tryptophanase. One has been mapped at 71.8 mins. for *Escherichia coli*

Indicatoridae a family of piciform birds commonly called honeyguides in virtue of their custom of leading man and other mammals to stores of wild honey. The nostrils have raised rims and the tail is tapered

indigo any of several species of leguminous shrubs and of herbs of the genus Indigofera and the dyestuff prepared from them. Without qualification usually means *I. tinctoria* or *I. anil*

   bastard indigo *U.S.* any of several species of leguminous herb of the genus Lotus, and shrubs of the genera Amorpha and Tephrosia

   false indigo any of several species of leguminous herbs of the genus Baptisia. The term is also used as synonymous with bastard indigo

      fragrant false indigo *U.S.* the leguminous herb *Amorpha nana*

   wild indigo *Baptisia tinctoria*

individual a single specific organism

individualism a type of symbiosis, in which the aggregate differs from any of its components. A lichen is a case in point

indri any of several indriid lemuroid primates usually, without qualification, *Indris brevicaudata*

Indriidae a family of lemuroid primates distinguished by the broad muzzles, widely spaced eyes and naked face

induction the production of a part under the influence of another as the induction of the embryonic lens by the embryonic eyecup

inductor a substance thought to cause embryonic induction

-indus- *comb. form* meaning "clothing" or "covering"

indusium literally, a petticoat, but used for any structure of much that shape, such as the epidermal outgrowth round the sorus of a fern, or the ring of collecting hairs below a stigma. It is also used for the annulus of some basidiomycetes, the case of a caddis fly larva and the inner of the membranes that surround an arthropod larva

induviate with the cast skin, or with withered leaves adhering

-ineae suffix indicating sub-ordinal rank in plant taxonomy

inermous without spines

infant *U.S.* the geometrid lepidopteran insect *Brephos infans*

inferior = below

[inflated] a mutant (gene symbol *if*) mapped at 55.0± on the X chromosome of *Drosophila melanogaster*. The inflation refers to the wings with thickened veins of the phenotype

inflorescence *see* -florescence

influent *see* -fluent

infundibulum literally, a funnel and used in *biol.* for any organ of this shape including the calyx of a kidney, the ostium of an oviduct, the pouch that grows down from the diencephalon and, joining with Rathke's pouch, forms the pituitary gland, and the outer chamber of the coronal funnel in Rotifera

infuscate dusky

Infusoria the zoological taxon erected by Lamarck to contain those microscopic organisms such as Protozoa and Rotifera found in hay infusions. The

term has gone through many changes, and most recently (middle 19th. century) was confined to ciliate Protozoa

infusorigen the cluster of cells, or psuedo-egg, produced by cleavage from the axial cell of the rhombogen stage of a mesozoan (*cf.* rhombogen and nematogen)

ingluvial pertaining to the crop

ingression in *biol.*, the movement of cells from an outer layer towards the interior

   multipolar ingression the production of an endoderm by the budding off of cells from all parts of the blastula wall

   unipolar ingression the production of endoderm from vegetable pole cells which detach themselves and rearrange themselves under the ectoderm

inguinal pertaining to the groin

inheritance (*see also* character) the sum total of genetically transmitted characters

   allosomal inheritance the inheritance of characters influenced by genes in an allosome (*e.g.* sex-linked characters)

   blending inheritance a term at one time used to describe the condition in which phenotypic characters were expressed as a blend of parental characters such as a pink color derived from red and white parents

[Inhibitor] a mutant (gene symbol *Id*) on the sex chromosome (linkage group I of the domestic fowl). The inhibition is of melanin

Iniomi a small order of actinopterygian fishes distinguished by soft rays and the presence of a small adipose fin on the upper surface of the caudal peduncle. Many lack an air bladder and are abyssal or bathypelagic

initial a word used in *bot.* to designate a cell or group of cells destined to produce a specific type of cell, tissue or organ

   apical initial one of a group of cells no one of which can be designated as a specific apical cell

      subapical initial the initial from which the internal tissue of some leaf axes is produced

   fusiform initial those cells of a vascular cambium which will give rise to all those cells of the xylem and phloem which have their axes parallel to the direction of growth

   marginal initial one of the line of cells from which marginal meristem is developed, giving rise to the outer surface of the blade of a leaf

      submarginal initial one of a line of initials lying immediately under the marginal initials, giving rise to the internal tissues of the lamina of a leaf

   ray initial a cell of the vascular cambium which gives rise to xylem and phloem lying at right angles to the direction of growth

inl a bacterial gene marker indicating utilization of inositol

innate borne on the apex of the support

innocence *U.S.* the rubiaceous herb *Houstonia caerulea* (= bluets)

innovation (*bot.*) a shoot which, passing out from the parent stem, develops an independent existence after severing the connection

inos *see* [Inositol]

inosinase an enzyme catalyzing the hydrolysis of inosine into hypoxanthine and ribose

inositol (hexahydroxycyclohexane) a water soluble nutrient frequently classed as a vitamin and acting as a growth factor in some animals and microorganisms. No human need has been demonstrated

[inositol] a mutant (gene symbol *inos*) in linkage group V of *Neurospora crassa*. The phenotypic expression is a requirement for inositol

inquiline one who dwells in the house of another. Specifically, one organism which shares the quarters of another or lives on another without damage or help to either, and particularly an insect developing in a gall produced by another species

insect the popular name of any member of the arthropod class Insecta

cochineal insect any of several species of either of two genera of dactylopiid homopteran insects which produce a crimson dye from which carmine is manufactured. *U.S.*, without qualification, *Dactylopius coccus*

fig insect any of several species of agaontid chalcicoid hymenopteran (= fig wasp)

kermes insect Mediterranean kermisiine homopterans of the genus Kermes widely used as a source of red dyestuffs before the introduction of cochineal

lac insect any of several species of coccid homopteran which secrete lac

Indian lac insect *Laccifer lacca* which produces the shellac of commerce

scale insect any of numerous coccid homopteran insects which derive their name from the waxy scale, in part composed of molted skins, with which they protect themselves. The common specific names of most scale insects are abbreviated to scale (*q.v.*)

armored scale insect any of numerous coccid insects with the scale separated from the body. They are often placed in a separate family the Diaspidinae

pit scale insect any of several genera of scale insect in which the scale is in the form of a waxy film or mass

tortoise scale insect any of several genera of scale insects in which the female body is tortoise shaped

wax scale insect any of numerous genera of coccid homopteran insects which secrete a waxy cover. The term is sometimes restricted to a group of genera in a subfamily (Coccinae)

social insect one that lives in a large group or colony in which there is a division of function and distribution of labor. Frequently, supported by polymorphic types known as castes

stick insect *U.S.* any of numerous phasmid insects. Usually, *U.S.*, without qualification, *Diapheromera femorata* (= walking stick)

water stick insect *Brit.* the nepid heteropteran *Ranatra linearis*

Insecta a class of opisthogoneate Arthropoda very variously defined. If classed as possessing antennae and six legs it must include the Collenbola and Protura which many omit. To exclude these groups it is necessary to add to the definition of insects that they possess fourteen post-cephalic segments which develop without adding body segments in post-embryonic stages

Insectivora a heterogenous order of mammals containing, *inter alia*, the solenodons, the tenrecs, the moles, the hedgehogs, and the shrews. They may be defined as plantigrade placental mammals with primitive sharp-cusped teeth, a long snout, and a primitive skull

garden inspector *S.Afr.* the nymphalid lepidopteran insect *Precid archesia*

instar any of the several stages through which a metamorphosing arthropod passes before reaching the adult or imago

instinct a behavior pattern which is not based on learning

insulin a hormone secreted by the β-cells of the islets of Langerhans in the pancreas. It is active in the control of glucose metabolism (*cf.* glucagon)

integument the outer skin of an organism or organ. In *biol.* specifically that portion of the ovule which surrounds the nucellus

intelligence an attribute by which man distinguishes himself from other animals

[intensifier of aleurone color] a mutant (gene symbol *in*) mapped at 4 in linkage group VII of *Zea mays*. The term is descriptive of the phenotypic expression

[intensifier of forked] a mutant (gene symbol *I–f*) mapped at 86.5 on chromosome II of *Drosophila melanogaster*. The name is descriptive

gene interaction the mutual effect of non-allelic genes

interchange translocation between non-homologous chromosomes

segmental interchange an obsolete term for reciprocal translocation

interferon a postulated substance, apparently a protein of molecular weight 20,000–100,000 elaborated by cells under the influence of a virus, and which is antagonistic to another virus

intermedin = melanocyte-stimulating hormones

interstitial that which lies in small places between cells

Intestina a subdivision of the Linnaean Vermes. It contained a variety of worms, both free-living and parasitic

intestine a term looosely applied to any portion of the alimentary canal which lies posterior to the stomach

large intestine that portion of the alimentary canal which runs from the small intestine to the anus (= colon)

small intestine that portion of the intestine which runs from the stomach to the large intestine

terminal intestine a term used in place of large intestine for those fishes in which the diameter of the posterior portion of the alimentary canal is less than that of the anterior portion

-intra- *comb. form* meaning "within"

intricate literally entangled, and so used in *bot.*

introvert the tentacle-bearing, anterior region of a holothurian, which can be withdrawn completely into the body

[inturned] a mutant (gene symbol *in*) mapped at 46.9 on chromosome III of *Drosophila melanogaster*. The inturning refers to the thoracic bristles

inulase an enzyme catalyzing the breakdown of inulin through the hydrolysis of β-1,2-fructan links

Inuleae a tribe of tubiflorous Compositae

invagination the pushing of a layer of cells into a cavity as in the formation of a gastrula from a simple blastula (*cf.* involution)

inversion the condition of a portion of a chromosome in which the gene loci run in the reverse direction to their normal position

acentric inversion an inversion of any part of the chromosome that does not involve the centromere

paracentric inversion one that does not involve the centromere

pericentric inversion chromosome inversion around the centrosome

overlapping inversion one which is superimposed on a previous inversion

invertebrates a loose term, not an acceptable taxon, embracing all animals except the chordates

involucre a ring of bracts surrounding a cluster of flowers

io *see* io moth

iodinase an enzyme that catalyzes the production of iodine and water frome iodide and hydrogen peroxide. This is part of a complex system once thought to be a single enzyme ("tyrosine iodinase") which produces monoiodotyrosine from tyrosine

[iojap] a mutant (gene symbol *ij*) mapped at 42 on linkage group VII of *Zea mays*. The leaves are longitudinally striped in green and white. (*cf*. [japonica striping])

iora any of several species of irenid birds of the genus Aegithina

ioterium a gland that produces the venom for the sting of Hymenoptera

ipecac a rubiaceous trailing herb *Cephaelis ipecacuanha* and the dried roots of the same

American ipecac the rosaceous herb *Gillenia stipulata*

bastard ipecac the caprifoliaceous herb *Triosteum perfoliatum* (= fever root, horse gentian)

Ipidae a family of coloepteran insects now usually included in the Scolytidae

Ipnopidae a small family of abyssal fishes distinguished by the peculiar eye with a flattened cornea and an enlarged retina. Commonly called grideye fish

ipomoea properly a genus of convolvulaceous vines but also applied as a popular name to some other vines

star ipomoea the convolvulaceous climber *Quamoclit coccinea*

Irenidae that family of passeriform birds which contains the leafbirds and fairy bluebirds. These are small, brilliantly colored, Asiatic birds with short wings, small feet, and a very swift flight

Iridaceae that family of liliflorous monocotyledons which contains not only Iris, but also, *inter alia*, the gladiolus, the crocus and the shell-flowers. Distinguishing characteristics are the three extrose stamens, the inferior 3-celled ovary and the typical showy flower

Iridales an order of liliiflorous monocotyledonous angiosperms erected to contain the single family Iridaceae by those who do not hold that this family belongs with the Liliiflores

Iridineae a suborder of Liliiflorae containing the single family Iridaceae

Iridoideae a subfamily of the Iridaceae

iris 1 (*see also* iris 2) an anterior projection of the ciliary process, lying immediately in front of the lens of the eye and which may be opened or closed

iris 2 (*see also* iris 1) a very large genus of iridaceous plants

English iris *I. xiphioides*

German iris any of numerous species, especially *I. germanica*, and very numerous horticultural hybrids and varieties derived from them, of Iris which have the outer segments of the perianth bearded with multicellular hairs

beachhead iris *U.S. I. hookeri*

snake's head iris the iridaceous herb *Hermodactylus tuberosis*

Japanese iris numerous horticultural varieties of *I. laevigata* more usually referred to by horticulturalists as *I. kaempferi*

lamance iris *U.S. I. brevicaulis*

mourning iris *I. susiana*

prairie iris *U.S.* the iridaceous herb *Nemastylis geminiflora*

red iris *I. fulva*

sand iris *N.Z.* the iridaceous herb *Libertia peregrinans*

Spanish iris *I. xiphium*

vesper iris *I. dichotoma*

yellow iris *Brit. I. pseudacorus*

Irrisoridae = Phoeniculidae

irritability in *biol.* the ability of organisms or parts of organisms to respond to stimuli

irrorate freckled

is a yeast genetic marker indicating a requirement for isoleucine

isabelline dirty gray

Ischnopsyllidae a family of siphonapteran insects found only parasitic on bats

isidiosous having powdery branched excrescences like some lichens

isidium the branched globose erection from the surface of some lichens

blood **island** an extraembryonic aggregate of blood-forming cells

Reil's island the portion of the brain immediately under Sylvius's fissure

islay the rosaceous tree *Prunus ilicifolia*

pancreatic **islets** = islets of Langerhans

islets of Langerhans groups of cells in the pancreas which produce insulin

-iso- *comb. form* meaning "equal"

Isoetaceae that family of isoetine Pteridophyta containing the quillworts. The characteristics of the class are those of the family

Isoetales an order of lepidophyte plants distinguished from the Lycopodiales by the presence of ligules on the leaves and from the Selaginales by possessing a massive rhizophore at the base of the stem

Isoetinae that class of Pteridophyta which contains the single order Isoetales and family Isoetaceae. They are distinguished from other classes of Pteridophyta by the possession of large chambered eusporangia

Isogeneratae a class of phaeophyte algae distinguished by having an isomorphic alternation of generations

isolation the segregation of organisms

ecobiotic isolation segregation by virtue of habitat, as the head louse and the body louse

ecoclimatic isolation segregation by reason of the climate

genetic isolation isolation through mutual sterility

ecogeographic isolation isolation by virtue of geographic location

ecological isolation isolation of populations by virtue of living conditions

reproductive isolation the isolation of species inhabiting the same environment but which are unable to interbreed

[isoleucine-valine-1] a mutant (gene symbol *iv-1*) in linkage group V of *Neurospora crassa*. The phenotypic expression is a requirement for both isoleucine and valine

[isoleucine-valine-2] a mutant (gene symbol *iv-2*) in linkage group I of *Neurospora crassa*. The phenotypic expression is a requirement for both isoleucine and valine

isomer any two compounds of identical composition but different structure are isomeric (*cf*. epimer)

stereoisomer either of a pair of isomers (*e.g.* D-, L-glucose) which are mirror images of each other

isomerase a group of enzymes which catalyse reactions leading to the internal reorganization of molecules. Among trivial names in this group are isomerase, mutase and racemase

maleylacetoacetate isomerase catalyzes the production of 4-fumarylacetoacetate from 4-maleylacetoacetate

arabinose isomerase catalyzes the production of D-ribulose from D-arabinose

vinyacetyl-CoA isomerase catalyzes the production of crotonyl-CoA from vinylacetyl-CoA

muconate cycloisomerase catalyzes the change of 4-carboxymethyl-4-hydroxyisocrotonalactone to *cis-cis*-muconate

erythrose isomerase catalyzes the production of D-erythrulose from D-erythrose

glucuronate isomerase catalyzes the production of D-fructuronate from D-glucuronate

maleate isomerase catalyzes the production of fumarate from maleate

mannose isomerase catalyzes the production of D-fructose from D-mannose

glucosephosphate isomerase catalyzes the production of D-fructose 6-phosphate from D-glucose 6-phosphate

acetylaminodeoxyglucosephosphate isomerase catalyzes the production of D-fructose 6-phosphate, ammonia and acetate from 2-acetylamino-2-deoxy-D-glucose 6-phosphate

aminodeoxyglucosephosphate isomerase catalyzes the production of D-fructose 6-phosphate and ammonia from 2-amino-2-deoxy-D-glucose 6-phosphate

mannosephosphate isomerase catalyzes the production of D-fructose 6-phosphate from D-mannose 6-phosphate

isopentenylpyrophosphate isomerase catalyzes the production of isopentenyl pyrosphosphate from dimethylallyl pyrophosphate

ribosephosphate isomerase catalyzes the production of D-ribulose 5-phosphate from D-ribose 5-phosphate

triosephosphate isomerase catalyzes the production of dihydroxyacetone phosphate from D-glyceraldehyde 3-phosphate

retinene isomerase catalyzes the production of 11-*cis* retinene from 11-*trans*-retinene

xylose isomerase catalyzes the production of D-xylulose from D-xylose

Isometopidae a family of hemipteran insects commonly called jumping tree bugs

isopedine lamellar bone forming the base of a ganoin scale

Isoperlidae a family of plecopteran insects readily distinguished by their greenish wings

Isopoda that order of pericaridan malacostracous crustacea containing, *inter alia*, the forms commonly referred to as wood lice or sow bugs. The order is distinguished from all except the amphipods by the fact that there is no distinct carapace but the first thoracic somite is coalesced with the head. Most are flattened dorsoventrally which usually distinguishes them clearly from the laterally compressed Amphipoda. The order includes many more or less aberrant parasitic forms, mostly marine.

Isoptera that order of insects which contains the termites. The wingless forms are easily distinguished by their white larva-like bodies. Winged forms have equal-sized wings with indistinct veins and a prothorax smaller than the head

Isospondyli an order of osteichthyes distinguished by the fact that the anterior and posterior vertebrae are essentially similar and that the fins are soft-rayed. The order includes, *inter alia*, the tarpons, herrings, anchovies, trout and pike

isostic said of a root which has more than two xylem bundles

isozyme = isoenzyme

Issidae a small family of homopteran insects closely allied to the Fulgoridae

-istem- *comb. form* meaning a "web." The root derives from -isto-, spelt by zoologists histo-. Both usages refer to the web-like appearance of cells in crude sections studied with primitive microscopes

meristem those plant tissues, from which growth originates

adaxial meristem the meristem in a developing leaf lying under the adaxial protoderm

apical meristem the meristem at a growing tip either shoot or root

endomeristem the initial of the central strand in mosses

eumeristem meristem consisting of cells which are unquestionably undifferentiated, unvacuolated, thin-walled, and more or less cylindrical

exomeristem that part of the meristem of a moss, which produces all tissue except the central strand

file meristem = rib meristem

ground meristem partially differentiated meristem which will later give rise to the fundamental system of a plant

indicatory meristem meristem occurring on the axis, but not at the apex of a root or stem

lateral meristem meristematic tissue running parallel to the stem or root as distinct from the apical meristem

marginal meristem meristem, lying in a band along each side of a developing leaf axis from which the blade is produced

mass meristem a meristem that grows by division within all planes to produce a spheroid

plate meristem a meristematic mass that produces a flattened plate like a leaf

promeristem initiating cells and undifferentiated derivative cells

protomeristem = apical meristem

residual meristem meristem from which the vascular system of a shoot develops

rib meristem a meristem that gives rise to a complex of longitudinal cell rows

primary thickening meristem a zone of meristem immediately below and beneath the apical meristem in some monocotyledons

vascular meristem that meristem which produces secondary growth (= vascular cambium)

peristem young meristem

isthmus a sharp constriction in the brain separating the mesencephalon from the rhombencephalon

Istiophoridae a family of scombroid acanthopterygian fishes all of which possess a round bill projecting in front. The roundness of the bill distinguishes them from the sword fish with flattened bills. Commonly called billfishes or marlins

-isto- *see* -histo-

cow **itch** *either* the bignoniaceous herb *Campsis radicans or* the leguminous herb *Stizolobium pruriens*

-ite *comb. term* abbreviated from -podite (*q.v.*)

Iteacoidea a subfamily of Saxifragaceae considered by some to be of familial rank

iter literally a road but used by anatomists in the sense of a small canal. Used without qualification it refers to the connection between the third and fourth ventricles in higher vertebrates (*cf.* mesocoele)

iter chordae anterior the canal by which the tympanic nerve leaves the tympanic cavity (= Huguier's canal)

iter chordae posterior the canal by which the typanic nerve enters the typanic cavity

Ithyceridae = Belidae

Itonididae = Cecidomyiidae

ITP = inosinetriphosphate

-iul- *see* -jul-

iv-1 *see* [isoleucine-valine-1]

iv-2 *see* [isoleucine-valine-2]

ivory (*see also* nut, gull, shell) the modified, fibrous dentine of the tusk of elephants. The dentine of several other large teeth, particularly those of the walrus and sperm whale are also called ivory

vegetable ivory = ivory nut

ivy (*see also* aphis, lettuce) any of numerous species and very numerous horticultural varieties of the araliaceous genus Hedera and many other plants that resemble ivy in their scandent habit. The term is also applied in parts of the *U.S.* to the ericaceous herb *Kalmia latifolia* (= mountain laurel)

Boston ivy the vitaceous woody vine *Parthenocissus tricuspidata*

cape ivy the compositaceous herb *Senecio macroglossus*

devil ivy any of several species of araceous plants of the genera Scindapsus and Rhaphidophora (= ivyarum)

English ivy *H. helix*

German ivy the compositaceous herb *Senecio mikanioides*

ground ivy *U.S* the labiateous creeping herb *Glechoma hederacea. Brit.* the labiateous herb *Nepeta cataria.* (= catnip)

Japanese ivy = Boston ivy

Kenilworth ivy the scrophulariaceous trailing vine *Linaria cymbalaria*

poison ivy properly the anacardiaceous climbing vine *Rhus toxicodendron.* In some parts of the *U.S.* this name is also applied to the poison oak (*q.v.*) and poison sumac (*q.v.*)

ivy-arm any of several species of araceous plant of the genera Scindapsus and Rhaphidophora (= devil ivy)

Ixioideae a subfamily of the Iridaceae

Ixodidae a family of acarine arthropods containing the forms commonly called ticks. They are distinguished from the ticks of the family Argasidae by the presence of a scutum

Ixodides a division of acarine arachnida containing the ticks. They are distinguished by having the hypostome modified as a piercing organ and usually provided with recurved teeth

ixous sticky

J

j *see* [jaunty] or [jittery]

J *see* [Jammed]

j₁ *see* [japonica striping] or [jointless pedicel]

j₂ *see* [japonica striping]

jabiru *U.S.* and *S.Amer.* the ciconiid bird *Jabiru mycteria* *Asia* the cicionid bird *Xenorhynchus asiaticus* *Afr.* the cicionid bird *Ephippiorhynchus senegalensis* (= saddlebill)

jaboticada the myrtaceous tree *Myrciaria cauliflora*

jacama the climbing leguminous herb *Pachyrhizus tuberosus*

jacamar any of numerous metallic green or bronze tropical birds of the piciform family Galbulidae

jacana any of several species of jacanid birds distinguished by their habit of running about over the surface of aquatic vegetation

African jacana *Actophilornis africana* (= lily-trotter)

American jacana *Jacana spinosa*

comb-creasted jacana *Irediparra gallinacea* (= lotusbird)

pheasant-tailed jacana *Hydrophasianus chirurgus*

Jacanidae a small family of charadriiform birds containing the jacanas distinguished by their extremely long toes and claws

subjacent lying just below

jacinth any of several species of liliaceous herb of the genus Scilla

Peruvian jacinth *S. peruviana*

Spanish jacinth *S. hispanica*

jack 1 (*see also* jack 2, 3) a male ass or donkey. Jackass, frequently used, is more properly applied to the mule. The term is occasionally applied to the male of other animals, particularly hawks and ferrets, and to the female hare (*cf.* jenny 1)

jack 2 (*see also* jack 1, 3) in compound name of many animals. *U.S.*, without qualification, applies particularly to carangid fishes of the genus Caranx (*cf.* jenny 2)

amberjack any of several species of carangid fishes of the genus Seriola

idle jack *S.Afr.* the sylviid bird *Sphenoeacus africanus* (= grass birds)

natterjack *Brit.* the bufonid anuran amphibian *Bufo calamita*

skipjack *U.S.* popular name of elaterid coleopteran insects (= click beetle) and any of several species of scombrid fishes of the genus Euthynnus (*cf.* tuna)

union jack *Austral.* any of several species of pierid lepidopteran insects of the genus Delias (*cf.* jezebel)

whisky jack the corvid bird *Perisoreus canadensis* (= Canada jay)

jack 3 (*see also* jack 1, 2) in compound names of many plants (*cf.* jenny 2)

black jack *U.S.* the oak *Quercus marilandica*

flapjack *N.Z.* the phaeophyte alga *Carpophyllum maschalocarpum*

slippery jack *U.S.* the boletaceous basidiomycete fungus *Suillus luteus*

supple jack the rhamnaceous vine *Berchemia scandens*

jackal any of many species of canine carnivores of the genus Thos, though many do not separate this genus from Canis

Indian jackal *T. aureus*

South American jackal any of several species of S. American canine carnivore of the genus Dusicyon (*cf.* dog)

jackass a male mule, though the term is improperly equated with jack

derwent jackass *Austral.* the cracticid bird *Cracticus torquatus* (= butcher bird)

laughing jackass the alcedinid bird *Dacelo novaeguinae* (= laughing kookaburra)

Jack-by-the-hedge the cruciferous herb *Alliaria officinalis* (= garlic-mustard).

jackdaw the corvid bird *Corvus monedula*

leatherjacket *Brit.* the larva of several species of tipulid diptera *U.S.* the carangid fish *Oligoplites saurus*

yellowjacket *U.S.* any of numerous yellow or yellow and brown wasps of the family Vespinae

jack-in-the-pulpit the araceous herb *Arisaema triphyllum.* Also, (*W.U.S.*) the scrophularaceous herb *Scrophularia californica.* The term has also been applied to numerous other flowers including various members of the genera Nigella and Trillium and to a large sepalled variety of the common primrose

jack-sail-by-the-wind the siphonophoran hydrozoan Velella

white-necked jacobin the trochilid bird *Florisuga mellivora*

Jacobson's *epon. adj.* from Ludwig Levin Jacobson (1783–1843)

jacquot *W.I.* the psittacid bird *Amazona arausiaca* (= red-necked parrot)

jaculator a springy hook at the base of some fruits which aids in expelling the seeds

jaeger popular name of predatory gulls of the stercorariid genus Stercorarius

parasitic jaeger *S. parasiticus*

pomerine jaeger *S. pomarinus*

long-tailed jaeger *S. longicaudus*

Jag *see* [Jagged]

[Jagged] a mutant (gene symbol *Jag*) mapped at 54.9 on chromosome II of *Drosophila melanogaster*. The wings of the phenotype are notched and the eyes rough

jaguar *U.S.* the felid carnivore *Felis onca*

jaguarondi the *Amer.* felid carnivore *Felis (Herpailurus) yagouroundi* (*cf.* colocollo)

jalap the convolulaceous herb *Exogonium purga*

wild jalap *U.S.* the berberidaceous herb *Podophyllum peltatum* (= mayapple)

jambeau the triacanthodid fish *Parahollardia lineata*

[Jammed] a mutant (gene symbol *J*) mapped at 41.0 on chromosome II of *Drosophila melanogaster*. The phenotypic expression is that the wings are reduced to a narrow strip

blue **jane** the turdid bird *Mimocichla plumbea* (= red-legged thrush)

Janthinidae a small family of pelagic gastropod Mollusca distinguished by their purple shells and habit of ejecting a purple flluid

[japonica striping] *either* a mutant (gene symbol *j₁*) mapped at 28 on linkage group VIII of *Zea mays*. The phenotypic leaves are longitudinally striped in green and white *or* a mutant (gene symbol *j₂*) mapped at 105 on linkage group IV of *Zea mays*. The phenotypic leaves are longitudinally striped in green and light yellow

Japygidae a small family of thysanurid insects distinguished by the possession of forcep-like cerci and sometimes placed in a separate order, the Entotrophi

jararacussu the erotalid snake *Bothrops jararacussu*

jarovization = vernalization

jarrah *either* of two myrtaceous trees *Eucalyptus marginata or E. rostrata*

bastard jarrah the myrtaceous gree *Eucalyptus botryoides*

sweet **jarvil** *U.S.* the umbelliferous herb *Osmorhiza claytoni*

jasmine anglicized name of the oleaceous genus of woody plants Jasminum, and properly applied to that genus though used for many other fragrant herbs and shrubs

Arabian jasmine *J. sambac*

bastard jasmine any of several solanaceous shrubs of the genus Cestrum

blue **jasmine** *U.S.* the ranunculaceous climber *Clematis crispa*

cape jasmine the rubiaceous shrub *Gardenia jasminoides*

Catalonian jasmine = Italian jasmine

common jasmine *J. officinale*

crape jasmine the apocynaceous shrub *Tabernaemontana coronaria*

West Indian jasmine the apocynaceous shrub *Plumeria rubra* (= frangipani)

Italian jasmine *J. grandiflorum*

rock jasmine any of many species of the primulaceous genus Androsace

royal jasmine = Italian jasmine

Spanish jasmine = Italian jasmine

star **jasmine** the apocynaceous shrub *Trachelospermum jasminoides*

Jasminoideae a subfamily of the Oleaceae containing the genera Jasminum, Mendora, and Nyctanthes

jaspideous having a pattern of multi-colored spots

Jassidae = Cicadellidae

[jaunty] a mutant (gene symbol *j*) mapped at 48.7 on chromosome II of *Drosophila melanogaster*. The phenotypic expression is upturned wings

[javelin] a mutant (gene symbol *jv*) mapped at 19.2 on chromosome III of *Drosophila melanogaster*. The phenotypic expression is the cylindrical nature of the bristle and hairs

[javelin-like] a mutant (gene symbol *jvl*) mapped at 56.7 on chromosome III of *Drosophila melanogaster*. The phenotypic bristles are cylindrical but crooked

javelina *U.S.*, without qualification, the dicotylid artiodactyle *Pecari angulatus* (= peccary)

jaw 1 (*see also* jaw 2) the articulated skeletal elements margining the mouth of gnathostome vertebrates

jaw 2 (*see also* jaw 1, fish) in compound names of organisms

hammerjaw the omosudid fish *Omosudis lowii*

whitejaw *W.I.*, the anatine bird *Anas bahamensis* (= summer duck)

jay 1 (*see also* jay 2, 3) any of numerous small corvid birds, frequently with blue coloration. *Brit.*, without qualification, *Garrulus glandarius*. *U.S.*., without qualification, *Cyanocitta cristata*

Arizona jay = Mexican jay

black and blue jay *Cissilopha sanblasiana*

blue jay (*see also* jay 2) *U.S. Cyanocitta cristata*. *Brit. Garrulus glandarius*. *I.P. Coracias indica*. *Burma Coracias affiniris*

brown jay *Psilorhinus morio*

Canada jay = grey jay

crested jay *Malay Platylophus galericulatus*

Florida jay = scrub jay

green jay *Cyanocorax yncas*

grey jay *Perisoreus canadensis*

Himalayan jay *Garrulus bispecularis*

azure-hooded jay *Cyanolyca cucculata*

magpie jay *Calocitta formosa*

Mexican jay *Aphelocoma ultramarina*

pinyon jay *Gymnorhinus cyanocephalus*

scrub jay *Aphelocoma coerulescens*

Steller's jay *Cyanocitta stelleri*

black-throated jay *Garrulus lanceolatus* or *Cyanolyca pumilo*

tufted jay *Cyanocorax dickeyi*

jay 2 (*see also* jay 1, 3) any of several birds, not corvids, but having a resemblance to true jays in color, shape or voice. *Austral.*, without qualification, the cracticid *Strepera versicolor* (=grey bell-magpie)

blue jay (*see also* jay 1) *Austral.* the campephagid *Coracina novaehollandiae* (*cf.* cuckoo-strike) *I.P.* the coraciid *Coracias benghalensis* (= Indian roller)

popinjay (*see also* jay 3) originally parrot, but nowadays *Brit. dial.* for green woodpecker

jay 3 (*see also* jay 1, 2) any of several brightly colored papilionid lepidopteran insects. *I.P.* any of numerous species of the genus Zetides but usually, without qualification, *Z. doson* (*cf.* bluebottle)

popinjay (*see also* jay 2) *I.P.* the nymphalid lepidopteran insect *Stibochiona nicea*

je *see* [jelly] or [jerker]

jejunum that part of the small intestine which lies between the duodenum and the ileum

[jelly] a mutant (gene symbol *je*) mapped at 46.0± on chromosome III of *Drosophila melanogaster*. The phenotypic eye color is dark pink

royal **jelly** a secretion of the pharyngeal glands of the honey-bee that nurtures all larvae at their earliest stage and the larva of the queen throughout its existence

jennet = jenny

jenny 1 (*see also* jenny 2) a female, donkey, or wren (*cf.* jack 1)

jenny 2 (*see also* jenny 1) in compound names of organisms (*cf.* jack 2, 3)

  creeping jenny the primulaceous herb *Lysimachia nummularia*

  silver jenny the gerrid fish *Eucinostomus gula*

jerboa popular name of dipodid rodents. Without qualification, usually applied to *Dipus hirtipes*

  pounched jerboa any of several species of phascogaline dasyurid marsupials of the genus Antechinomys

[jerker] a mutant (gene symbol *je*) found in linkage group XII of the mouse. The phenotypic expression is a waltzing syndrome accompanied by deafness

jessamine frequently used as synonymous with jasmine

  Caroline jessamine the loganaceous woody vine *Gelsemium sempervirens*

  orange jessamine the rutaceous shrub *Chalcas exotica*

  yellow jessamine = Carolina jessamine

jester *I.P.* any of several species of nymphalid lepidopteran insect of the genus Symbrenthia

jetsam considered by some ecologists to be that area of the beach on which floating material is washed up and moreover to be synomous with flotsam. The correct meaning of these words is given in standard English dictionaries.

wandering jew the commelinaceous herb *Zebrina pendula*

jewel *Austral.* any of many species of lycaenid lepidopteran insect of the genus Hypochrysops

  blue jewel *H. delicia*

  fiery jewel *H. ignita*

  rex jewel *H. polycletus*

  western jewel *H. halyaetus*

jezebel *Austral* and *I.P.* any of numerous species of pierid lepidopteran insects of the genus Delias

  northern jezebel *D. argenthona*

  orange jezebel *D. aruna*

ji *see* [jittery]

jigger the tungid siphonapteran insect *Tunga penetrans*

jill a female ferret (*cf.* jack 1)

stinking-jim *U.S.* popular name of kinosternine chelydrid turtles

jimbling the euphorbiaceous shrub *Phyllanthus acidus* (= West India goose-berry)

[jittery] *either* a mutant (gene symbol *j*) on the sex chromosome (linkage group I) of the domestic fowl. The name is descriptive of the phenotypic expression which is lethal *or* a mutant (gene symbol *ji*) in linkage group X of the mouse. The phenotypic expression is a wobbly gait and early death

joey a young kangaroo

john dory *Brit.* the zeid fish *Zeus faber.* The name is a corruption of the French *jaune dorée* (golden yellow) *U.S.* the zeid fish *Zenopsis ocellata*

johnny *Brit. obs.* penguin

John Philip *W.I.* the vireonid bird *Vireo altiloquus* (= black-whiskered vireo)

Johnny-jump-up *W.I.* the fringillid *Volatinia jacarina* (= blue-black seedeater) and occasionally applied in *U.S.* to certain violets

john-to-whit *W.I.* the vireonid bird *Vireo altiloquus*

joint an inseparable junction, usually but not necessarily moveable, between two parts of anything (*see also* arthrosis, which is synonymous). The term has also become confused with segment

coffin joint the joint articulating the coffin bone (*q.v.*) to the rest of the leg in perissodactyl and artiodactyl mammals

diarthrodial joint = synovial joint

fibrocartilagenous joint those joints in which the skeletal elements are united by fibrocartilage during some stage of their existence

fibrous joints a joint between two bones united by fibrous tissues

synovial joint a joint between two bones which possesses a cavity and is specialized to permit more or less free movement

joint 2 (*see also* joint 1) in compound names of organisms

  blue joint *U.S.* the grass *Calamagrostis canadensis*

[jointless pedicel] a mutant (gene symbol j_1) mapped at 69 on linkage group V of the tomato. The name is typical of the phenotypic expression

jojoba the buxaceous shrub *Simmondsia californica* and its edible fruit

joker *U.S.* the noctuid lepidopteran insect *Feralia jocosa*. *S.Afr.* either of two species of nymphalid lepidopteran insect of the genus Byblia. *B. ilithyia* (**common joker**) or *B. acheloia* (**scarce joker**) *I.P. Byblia ilithyia*

jonquil the amaryllidaceous herb *Narcissus jonquilla*

  campernelle jonquil the amaryllidaceous herb *N. odorus*

jonquilleous bright yellow

Jordanism an early botanic term for splitter (*q.v.*) derived from Alexis Jordan (1814–1897)

jowl any gular prominance such as a dewlap or wattle and applied by some to swollen areas lateral to the mouth of certain insects

jp *see* [jumpy]

juba a loose panicle

jubatous maned

judy 1 (*see also* judy 2) *I.P.* any of several species of riodinid lepidopteran insect of the genus Abisara

judy 2 (*see also* judy 1) *W.I.* the cotingid bird *Platypsaris niger* (*cf.* becard)

Jugatae a suborder, by no means universally recognized, of lepidopteran insects that possess a small basal lobe on the posterior margin of the front wing which unites the two wings. The ghost moths (Hepaliidae) are the only well known forms in this suborder

jugate yoked

  conjugate coupled together as Protozoa in conjugation, or paired as two leaflets

  multijugate many paired

conjugation properly the temporary fusion of two individuals, but frequently used for permanent union

Juglandaceae the family of dicotyledonous plants containing the walnuts and their allies. The indehiscent fruit and the pinnate leaves are distinctive of this family

Juglandales a monotypic order of dicotyledonous angiosperms. The single family Juglandaceae has the characteristics of the order

jugulum that area of a bird's plumage which lies between the neck and the breast

jugum literally, a joke and used in *biol.* for many structures having this function. Specifically in Hemiptera, two lateral lobes of the head bordering the tylus (*q.v.*) and in Lepidoptera and Trichoptera a lobe of the forewing that overlaps the anterior margin of the hindwing and serves to lock the wings together

jujube any of several rhamnaceous shrubs of the genus Zizyphus. Usually, without qualification, *Z. sativa*

-jul- *comb. form* meaning "catkin." Properly, but rarely, spelled -iul-

julaceous bearing catkins (= amentaceous)

julia *U.S.* the nymphalid lepidopteran insect *Dryas julia*

Julianales a monotypic order of dicotyledonous angiosperms found only in Mexico and Peru

jump the word is variously used in the sense of spring (*q.v.*) or for any method of progression in which all feet are off the ground for a considerable time

grey **jumper** *Austral.* the grallinid bird *Struthidea cinerea* (= apostle bird)

rock-**jumper** *E.Afr.* popular name for macroscelid insectivores of the genus Petrodromus (*cf.* elephant-shrew)

orange breasted rock jumper *S.Afr.* the timaliid bird *Chaetops aurantius*

[**jumpy**] a mutant (gene symbol *jp*) in the sex chromosome (linkage group XX) of the mouse. The phenotypic expression is a twitching, staggering, gait leading to death in the first month

Juncaceae that family of liliflorous monocotyledons which contains the rushes. The rushes are clearly distinguished from their nearest allies the lilies by their grass-like habit of growth

juncaceous rush-like

Juncaginaceae = Scheuchzeriaceae

Juncineae a suborder of liliiflorae containing the single family Juncaceae

junco *U.S.* any of several small fringillid birds of the genus Junco

slate-colored junco *J. hyemalis*

grey-headed junco *J. caniceps*

Mexican junco *J. phaeonotus*

Oregon junco *J. oreganus*

white-winged junco *J. aikeni*

tight **junction** = nexus

abjunction the delimiting of part of an organ, such as the cutting off of spores by septi

Jungermanniales an order of hepaticate Bryophyta sometimes called the "leafy liverworts." Most are therefore foliose but the distinctive character is the presence of a globose antheridium borne on a long stalk

Jungermannineae a suborder of jungermanniale Hepaticae with a generally foliose appearance and with the archigonia borne on the apices of the stem

jungle (*see also* fowl) an association of trees interspersed with brush, particularly in India

disjunctor connection between the gonidia in some fungi

juniper (*see also* scale) anglicized form of the cupressaceous genus Juniperus containing numerous species and many horticultural varieties

alligator juniper *J. pachyphaloaea*

common juniper *J. communis*

one-seed juniper *J. monosperma*

Mexican juniper *J. mexicana*

Rocky Mountain juniper *J. scopulorum*

Utah juniper *J. utahensis*

Juniperoideae a subfamily of Cupressaceae

jute several tiliaceous plants of the genus Corchorus and the fiber produced from them

bastard jute the malvaceous herb *Hibiscus cannabinus* (= ambary)

juvenal the plumage of a young bird

jv *see* [javelin]

jvl *see* [javelin-like]

Jyngidae a family of piciform birds erected to contain the two wrynecks. They are distinguished from other piciform birds by the short pointed rather feeble beak

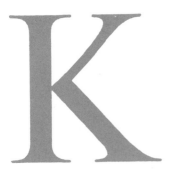

k *see* [kidney] or [kinky]
K *see* [Knobbed] or [slow feathering]
kae *N. Brit.* = jackdaw
kagu the rhynochetid bird *Rhynochetos jubatus*
kahikatea *N.Z.* the podocarpaceous tree *Podocarpus dacrydioides* (= white pine *N.Z.*)
kahu *N.Z.* the circine bird *Circus approximans* (= harrier *N.Z.*)
kaikawaka *N.Z.* the cupressaceous tree *Libocedrus bidwillii* (= mountain ceder *N.Z.*)
kaikomako *N.Z.* the icacinaceous tree *Pennantia corymbosa*
kaiku *N.Z.* the apocynaceaeous herb *Parsonia heterophylla*
-kaino- *comb. form* meaning "recent" usually transliterated -caeno-
kaiser *I.P.* the nymphalid lepidopteran insect *Penthema lisarda*
kaiser-i-hind *I.P.* the papilionid lepidopteran insect *Teinopalpus imperialis*
kaka *N.Z.* the psittacid bird *Nestor meridionalis*
kakapo *N.Z.* the psittacid bird *Strigops habroptilus*
kakariki 1 (*see also* kakariki 2) *N.Z.* either of two species of geckonid lacertilian of the genus Naultinus
kakariki 2 (*see also* kakariki 1) *N.Z.* either of two species of psittacid bird of the genus Cyanoramphus. *C. novaezelandiae* (= red-fronted parakeet) and *C. auriceps* (= yellow-fronted parakeet)
kakelaar 2 *S.Afr.* popular name of phoeniculid birds (= wood hoopoe)
  white headed kakelaar *Phoeniculus bollei*
  Nigerian kakelaar *P. purpureus* (= green wood-hoopoe)
kaki *N.Z.* the recurvirostrid bird *Himantopus himantopus* (*cf.* stilt)
-kako- *comb. form* meaning "bad" usually transliterated -caco-
kale a tough-leaved horticultural variety of cabbage (*q.v.*)
  Roman kale a variety of Swiss chard (*q.v.*)
  sea kale the cruciferous herb *Crambe maritima*
kalidion = cystocarp
kalij = kaleege
kaleege any of several species of phasianid bird of the genus Gennaeus
kalkoentje *S.Afr.* any of several species of motacillid bird of the genus Macronyx (*cf.* longclaw)
kallege = kaleege
Kalotermitidae a family of isopteran insects distin-

guished by the absence of the fontanelle and the presence of an empodium
-kalyb- *comb. form* meaning "cottage" usually transliterated -calyb-
-kalyptr- *comb. form* meaning either "a veil" or a "squat conical hat" usually transliterated -calyptr-
kamahi *N.Z.* the large cunoniaceous tree *Weinmannia racemosa*
kamala the euphorbiaceous tree *Mallotus philippinensis*
kamila the dyestuff obtained from the monkey-face tree (*q.v.*)
-kampt- *comb. form* meaning "bend" frequently transliterated -campt-
kangaroo any of several jumping marsupial mammals of the family Macropodidae. The term wallaby is applied to smaller species of the same family
  brush-kangaroo *Tasmania Macropus ruficollis* (= red-necked wallaby *Austral.*)
  dorca kangaroo any of several species of the genus Dorcopsis (= gazelle-faced wallaby)
  great grey kangaroo *M. giganteus*
  hare kangaroo any of several species of the genus Lagorchestes (= hare wallaby)
  musk-kangaroo *Hypsiprymnodon moschatus*
  rat-kangaroo any of numerous species of several genera of potoroine phalangerids
  muskrat-kangaroo = musk-kangeroo
  rufous rat-kangaroos any of several species of the genus Aepyprymnus
  prehensile-tailed rat-kangaroo any of several species of the genus Bettongia
  great red kangaroo *M. rufus*
  tree kangaroo popular name of macropodine marsupials of the genus Dendrolagus. They differ from other macropodines in having heavy-clawed front limbs
  woolly kangaroo = great red kangaroo
-kapnod- *comb. form* meaning "smoke" usually transliterated -capnod-
kapok the bombacaceous tree *Ceiba pentandrum* and the fibers derived from its seed pods
-kaps- *comb. form* meaning "box" frequently transliterated -caps-
kar *see* [karmoisin]
karaka *N.Z.* the large corynocarpaceous tree *Corynocarpus laevigata*
karakahia *N.Z.* the aythyine bird *Aythya australis*
karamus *N.Z.* the rubiaceous shrub *Coprosma robusta*
karengo *N.Z.* the rhodophyte alga *Porphyra columbina*

kare-wa-rewa *N.Z.* the falconnine bird *Falco novaeseelandiae*. (= quail-hawk N.Z.)

karic pertaining to a nucleus

monokaric the condition of possessing a single nucleus

polykaric = multinucleate

synkarion = syncaryon

[**karmoisin**] a mutant (gene symbol *kar*) mapped at 52.0 on chromosome III of *Drosophila melanogaster*. The eye color of the phenotype is a dull scarlet

karo the pittosporaceous shrubby tree *Pittosporum crassifolium*

karoro *N.Z.* the larid bird *Larus dominicanus* (= black-backed gull *N.Z.*)

-karp- *comb. form* meaning "fruit." Frequently transliterated -carp- (*q.v.*) and thus confused with "wrist"

-karpho- *comb. form* meaning a "splinter" or "twig"

karri the myrtaceous tree *Eucalyptus diversicolor*

karvo- *comb. form* meaning "nut" or, by extension, "nucleus" sometimes transliterated -caryo-

karyon a term for nucleus now rarely used except in compound words (*cf.* caryon)

amphikaryon a nucleus containing two haploid genomes

heterokaryon a cell, or an individual containing cells, formed by the coalescence of the cytoplasm, but not the nuclei, of hyphal cells (*cf.* heterocaryon formation)

monokaryon variously a nucleus with one centriole or a centrosome

perikaryon cytoplasm but particularly the cytoplasm of a neuron

synkaryon a nucleus formed by the fusion of two others, particularly the early fusion nucleus in a zygote

karyosis the condition of a nucleus

heterokaryosis the association in a reproductive structure of nuclei of varied genetic content, a circumstance common in some fungi (*cf.* heterokaryotic)

Karyota = Protista

karyotic pertaining to nuclei

akaryotic the condition of lacking a nucleus

eukaryotic the condition of having a well defined, membrane limited, nucleus (*cf.* prokaryotic)

heterokaryotic the condition of a spore that contains both make and female nuclei (*cf.* heterokaryosis)

homokaryotic said of spores that contain both male and female nuclei

polykaryotic said of an organism that multiplies only by binary division

prokaryotic the condition, such as that in bacteria, and Cyanophyceae, of having an ill-defined nucleus (*cf.* eucaryotic)

karyotin the term at one time applied to the supposed nuclear net

-kata- *comb. form* meaning "down" usually transliterated -cat-

kathodic (*bot.*) applied to the side of a leaf facing in the opposite direction to the turn of the genetic spiral

katydid *U.S.* any of numerous species of tettigonid grasshopper mostly of the genus Scudderia

fork-tailed bush katydid *U.S. S. furcata*

angular-winged katydid *U.S. Microcentrum retinerve*

broad-winged katydid *U.S. M. rhombifolium*

Japanese broadwinged katydid *U.S. Holochlora japonica*

kauri *N.Z.* the araucariaceous tree *Agathis australis*

-kaus- *comb. form* meaning "burn" usually transliterated -caus-

kawaka *N.Z.* the cupressaceous tree *Libocedrus plumosa*

kawakawa *N.Z.* the piperaceous tree or shrub *Macropiper excelsum*

kawau *N.Z.* the phalacrocoracid bird *Phalacrocorax carbo* (= black shag *N.Z.*)

kawri = kauri

kea *N.Z.* the psittacid bird *Nestor notabilis* (*cf.* kaka)

ked the hippoboscid dipteran insect *Melophagus ovinus* (= sheeptick)

neutral **keel** a medullary plate, which is deeper than it is wide

kekeno *N.Z.* the otariid mammal *Arctocephalus forsteri* (= fur seal *N.Z.*)

-keki- *comb. form* meaning "a gall" frequently transliterated -ceci-

kelp any large phaeophyte alga of the order Laminariales

bladder kelp *N.Z.* = butterfish kelp *N.Z.*

bull kelp *N.Z.* any unusually large kelp

butterfish kelp *N.Z.* the *Macrocystis pyrifera*

giant kelp *W.U.S. Macrocystis pyrifera*

kelt *Brit.* a female salmon that has spawned and is descending the rivers on her way back to the sea. (= blackfish)

kennel term for a group of dogs when housed (*cf.* pack, gang)

kentragon that group of larval cells derived from the cypris larva of a parasitic cirripede, which penetrates into the tissues of the host

-kentro- *comb. form* meaning "a sharp point" frequently transliterated -centro- and thus confused with -centr- meaning "middle"

-kera- *comb. form* meaning "horn" frequently transliterated -cera-

Keratosa at one time regarded as an order of non-calcareous sponges but now usually classified as a subclass of the Demospongiae. These are the horny sponges with a skeleton composed of spongin fibers. The "bath" sponges of commerce are typical of the skeletons of members of this subclass

Kerckring's *epon. adj.* from Theodor Kerckring (1640–1693)

-kerkid- *comb. form* meaning "a small comb" frequently transliterated -cercid-

kermes the Southern European equivalent of cochineal prepared from the coccid insect *Kermes ilicis*

Kermesiinae a subfamily of coccid homopterans containing the kermes insects

kernel that part of an ovule that lies within the seed coat

Kerosphaereae a tribe of orchids of the division Acrotonae of the subfamily Monandrae. They are distinguished from the Polychondreae by the waxy pollen

kestrel in general, any hawk particularly those with a hovering flight. *Brit.*, without qualification, *Falco tinnunculus*. *U.S.* the falconine *Falco sparverius* (= sparrow hawk). *Austral. F. cenchroides*

fox kestrel *F. alopex*

greater kestrel *F. rupicoloides*

grey kestrel *F. ardosiaceus*

lesser kestrel *F. naumanni*

Nankeen kestrel *F. cenchroides* (= *Austral.* kestrel)

bladder **ketmia** the malvaceous herb *Hibiscus trionum*

phospho**ketolase** an enzyme catalyzing the production of alcetylophosphate and glyceraldehyde 3-phosphate from D-xylulose 3-phosphate and orthophosphate

transketolase = glycolaldehydetransferase

key a table listing the characteristics of a group of organisms arranged in such a manner as to facilitate the identification of its members

khat the celasteraceous shrub *Catha edulis*, yielding khat tea

kid a young goat, though the term is sometimes applied to the young of other artiodactyls, such as the antelope

kiddaw = guillemot

kidjeroekan the rutaceous tree *Pleiospermum dubium* and its fruit

kidney 1 (*see also* kidney 2) any organ serving to remove water, or water soluble substances, from the blood (*cf.* nephros)

  polylobular kidney a metanephros which is divided into lobes, each entering the pelvis through a separate branch

  unilobulat kidney a metanephros in which the medulla is undivided

kidney 2 (*see also* kidney 1) in compound names of organisms

  double kidney *Brit.* the ₙoctuid lepidopteran insect *Plastenis retusa*

  [kidney] *either* a mutant (gene symbol *k*) found in linkage group II of *Habrobracon juglandis or* a mutant (gene symbol *k*) mapped at 64± on chromosome III of *Drosophila melanogaster*. In both cases, the name refers to the eye shape of the phenotype

kigerukkan the rutaceous shrub *Merope angulata*

lambkill *U.S.* the ericaceous shrub *Kalmia angustifolia*

  pigkill = lambkill

  sheepkill = lambkill

killdeer *U.S.* the charadriid bird *Charadrius vociferus* (*cf.* plover)

kill-'em-polly *W.I.* the larid bird *Sterna albifrons* (= least tern)

bee-killer any of numerous species of philanthine sphecid wasps that provision their nests with bees (= bee-killer wasp). The term is also locally applied to many asilid Diptera

  cicada-killer *U.S.* a very large (close to 2 inches long) nyssonine sphecid hymenopteran insect (*Sphecius speciosus*) which is predatory on cicadas (= cicada hawk)

killie = killifish, in the broad sense of a cyprinodont

killigrew *Brit.* the corvid bird *Pyrrhocorax pyrrhocorax* (*cf.* chough)

killy-killy the falconine bird *Falco sparverius* (= sparrow hawk *U.S.* kestrel, *Brit.*)

killy-kadick *W.I.* the caprimulgid bird *Chordeiles minor* (= nighthawk)

-kinase a large group of kinases, more properly called ATP: phosphotransferases, catalyse phosphorylation with energy derived from the conversion of ATP to ADP. These are all named with simple compound words descriptive of their function (*e.g.* xylulokinase yields xylulose; 5-phosphate protein kinase yields a phosphoprotein, etc.). These enzymes are not listed separately save for the following few which might cause confusion

  acetol kinase catalyzes the production of hydroxyacetone phosphate from hydroxyacetone

  dephospho-CoA kinase catalyzes the production of CoA from dephospho-CoA

enterokinase = enteropeptidase

phosphofructokinase catalyzes the production of D-fructose 1,6-diphosphate from D-fructose 6-phosphate

phosphoglucokinase catalyzes the production of D-glucose, 1,6-diphosphate from D-glucose 1-phosphate

ribosephosphate pyrophosphokinase catalyzes the production of 5-phospho-α-D-ribosylpyrophosphate and AMP from D-ribose 5-phosphate and ATP

triokinase catalyzes the production of D-glyceraldehyde 3-phosphate from D-glyceraldehyde

-kine- *comb. form* meaning "move" or "movement" frequently transliterated -cine-

kinesis 1 (*see also* kinesis 2) movement between parts of a skull

  amphikinesis showing both meso- and meta-kinesis

  mesokinesis movement of the skull about a transverse line anterior to the orbit

  metakinesis movement of the skull about a tranverse line posterior to the orbit

  monokinesis the condition in which the skull moves only at one place

  prokinesis (*see also* kinesis 2) the condition in which the movement is between the nasals and frontals

kinesis 2 (*see also* kinesis 1) movements of cells, or parts of cells

  akinesis = amitosis

  blastokinesis activity resulting in the reorientation of an embryo within an egg

  chemokinesis the attraction of zoospores by chemotaxis

  dytokinesis cell-division, though confined sometimes to the division of the cytoplasm as distinct from nuclear elements

  diakinesis a terminal stage of the prophase in meiosis immediately following diplonema. It is characterized by the disappearance of the nucleolus and the even distribution of the bivalents throughout the cell

  dictyokinesis the splitting of the Golgi apparatus as dictyosomes at cell division

  homoeokinesis = homotypic meiosis

  interkinesis the pause between two mitotic divisions or the interphase between the first and second divisions of meiosis

  karyokinesis an obsolete term once used to denote mitosis, and later the role played by the chromosomes in mitosis

  metakinesis (*see also* kinesis 1) movement of chromosomes to the metaphase plate in mitosis

  prokinesis (*see also* kinesis 1) = prophase of mitosis

kinete a term occasionally used for "cell"

  akinete a single algae cell separated from the thallus

  ookinete a motile zygote, particularly of a sporozoan protozoan, in contrast to an oocyst (*q.v.*)

kinetic pertaining to movement

  ectokinetic said of a sporangium which dehisces through a rupture of the epidermis

  endokinetic said of a sporangium which dehisces by internal rupture spreading outward

kineticism the ability of bones of the skull to move in relation to each other (*see* kinesis 1)

king 1 (*see also* king 2) the male reproductive individual in a colony of hymenopteran or isopteran insects

king 2 (*see also* king 1, bird, fishes) in compound names of organisms considered to be regal in habit or appearance antking any of several species of formicariid bird of the genus Grallaria (= antpitta)

awlking *I.P.* the hesperiid lepidopteran insect *Choaspes benjaminii*

jungleking *I.P.* the amathusiid lepidopteran insect *Thauria lathyi*

little king the paradisaeid bird *Diphyllodes respublica*

palmking *I.P.* the amathusiid lepidopteran insect *Amathusia phidippus*

King-Harry *Brit. obs.* = goldfinch

kinglet *U.S.* any of numerous species of sylviid bird of the genus Regulus. *Brit.* without qualification, *R. regulus* (= goldcrest)

golden-crowned kinglet *R. satrapa*

ruby-crowned kinglet *R. calendula*

king-of-the-breams the sparid fish *Sparus erythrinus* (= pandora)

king-of-the-herrings the regalecid fish *Regalecus glesne* (= oarfish)

king-of-the-meadow *U.S.* the ranunculaceous herb *Thalictrum polygamum*

king-of-the-salmon the trachipterid fish *Trachipterus trachypterus* (= weever)

villikinin a hormone liberated in the intestine that increases the rate of movement of the villi

kinkajou the cercoleptine procyonid mammal *Potos flavus*. It is distinguished from all other procyonids by the prehensile tail

[kinky] *either* a mutant (gene symbol *k*) mapped at 0 on linkage group IV of the rat. The name refers to the condition of the hair and whiskers *or* a mutant (gene symbol *kk*) mapped at 42.0± on the X chromosome of *Drosophila melanogaster*. The name refers to the bent, and sometimes bifurcate, bristles of the phenotype

-kino- *comb. form* meaning "movement" frequently transliterated -cino- (*cf.* -kine-)

Kinorhyncha = Echinodera

Kinosternidae a family of cryptodyran chelonian reptiles containing the forms commonly called mudturtles, now usually regarded as a subfamily of the Chelydridae

[kinshirlyu] a mutant (gene symbol *ok*) mapped on linkage group V of *Bombyx mori*. The phenotypic expression is a highly translucent larva

kioriki *N.Z.* the ardeid bird *Ixobrychus minutus* (= little bittern *N.Z.*)

kipper properly a male salmon. The kipper of commerce is a smoked herring

kiskadee any of several tyrannid birds of the genus Pitangus; without qualification, *P. sulphuratus*

kiss-me-over-the-garden-gate the polygonaceous herb *Polygonum orientale*

kit a young beaver

kite 1 (*see also* kite 2) a term loosely applied to any accipitrid bird with a forked tail usually those of the subfamilies Elaninae, Milvinae and Perminae

yellow billed kite

African black kite = pariah kite

brahminy kite the milvine *Haliatur indus*

hooked-billed kite any of numerous species of pernines of the genus Chondrohierax. *W.I.*, without qualification, *C. uncinatus*

Cayenne kite the pernine *Leptodon cayanensis*

Cuban kite *W.I.* the pernine *Chondrohierax wilsonii*

everglade kite = snail kite

Mississippi kite the milvine *Ictinia misisippiensis*

pariah kite the milvine *Milvus migrans*

pear kite the elanine *Gampsonyx swainsonii*

plumbeous kite the milvine *Ictinea plumbea*

red kite the milvine *Milvus milvus*

black-shouldered kite *Austral.* the elanine *Elanus notatus*

snail kite the milvine *Rostrhamus sociabilis* (= everglade kite)

square-tailed kite the milvine *Lophoictinia isura*

swallow-tailed kite any of several species of pernine birds. They are, as the name indicates, distinguished by the very long, deeply forked, tail. *U.S.* Without qualification, *Elanoides forficatus. Afr. Chelictinia riocourii*

white-tailed kite popular name of elanine birds in general *U.S. Elanus leucurus*

double-toothed kite the milvine *Harpagus bidentatus*

black-winged kite *Elanus careuleous*

kite 2 (*see also* kite 1, butterfly) any of several lepidopteran insects

bush kite *S.Afr.* the papilionid *Papilio euphranor*

kitten 1 (*see also* kitten 2) any young feline. The term is also applied to the young of several other carnivores, particularly the skunk

kitten 2 (*see also* kitten 1) *Brit.* any of several species of notodontid lepidopteran insects of the genus Cerua

kittiwake any of several species of larid bird of the genus Rissa. Without qualification, usually *R. tridactyla*

black-legged kittiwake *R. tridactyla*

red-legged kittiwake *R. brevirostris*

kiwi any of three species of wingless birds of the family Apterygidae found only in New Zealand

brown kiwi *A. australis*

great spotted kiwi *A. haastii*

little spotted kiwi *A. owenii*

kiyi *U.S.* the salmonid fish *Coregonus kiyi*

kk see [kinky]

-klado- *comb. form* meaning "branch" almost invariably transliterated -clado-

-klas- *comb. form* meaning "fracture" usually transliterated -clas-

-klept- *comb. form* meaning "thief"

-klesi- *comb. form* meaning "closing" frequently transliterated -clesi-

-klima- *comb. form* meaning "climate"

-klimac- *comb. form* meaning "a ladder." Frequently transliterated -clima-

-klin- *comb. form* hopelessly confused between two roots meaning "bed" and "blend." Frequently transliterated -clin-

-klino- *comb. form* meaning a bed, almost invariably transliterated -clin- or -klin-

klipspringer any of several species of neotragine antelopid of the genus Oreotragus

klip-sweet *S.Afr.* the dried urine of dasseys

-klito- *comb. form* meaning a "hillside" or "slope" frequently transliterated -clito-

klon = clone

-kloster- *comb. form* meaning "spindle" almost invariably transliterated -closter-

kn see [knot]

Kn see [knotted leaves]

knawel *Brit.* any of several species of caryophyllaceous herb of the genus Scleranthus

knee in most mammals the joint between the femur and the tibiofibula of the hindlimb or between the humurus and the radio-ulna of the forelimb. In perissodactyles, however, the knee is the joint between the tibia and the metatarsal or the tibia and the metacarpal. In primates, particularly anthropoids the knee of the forelimb is called the elbow. In *bot.* an abruptly bent root or branch

thick knee popular name of burhinid birds (*cf.* stone-curlew). Usually, without qualification, *Burhinus oedicnemus*
  reef thickknee *Orthorhamphus magnirostris*
  spotted thickknee *Burhinus capensis*
  double-striped thickknee *B. bistriatus*
  water thickknee *B vermiculatus*
knight *I.P.* the nymphalid lepidopteran insect *Lebedea martha*
kno *see* [Knobbyhead]
knob term for a group of widgeon (*cf.* company)
[Knobbed] a mutant (gene symbol *K*) mapped at 0 in linkage group XI of *Bombyx mori*. Knobs appear on the dorsal surface of larva, pupa and moths in the phenotype
knobber a two-year old red deer
[knobby head] a mutant (geney symbol *kno*) mapped at 63.9± on the X chromosome of *Drosophila melanogaster*. The name is descriptive of the phenotype
knot 1 (*see also* knot 2, 3) a swelling on an elongate structure. In bot. synonymous with node
  net-knot = karyosome
knot 2 (*see also* knot 1, 3) any of several scolopacid birds of the genus Calidris, particularly *C. canutus*
  great knot *C. tenuirostris*
  red knot *C. canutus*
knot 3 (*see also* knot 1, 2) in compound names of other organisms
  shoulder knot *Brit.* the noctuid lepidopteran insect *Graptolitha ornithopus*
  topknot *Brit.* any of several species of bothid fish of the genus Zeugopterus
  true-lover's knot *Brit.* the noctuid lepidopteran insect *Agrotis strigula*
[knot] a mutant (gene symbol *kn*) mapped at 72.3 on chromosome II of *Drosophila melanogaster*. The phenotypic expression is an unusually short space between wing veins L3 and L4
[Knotted leaves] a mutant (gene symbol *Kn*) mapped at 128 linkage group I of *Zea mays*. The name refers to irregular lumpy growths on the leaves of the phenotype
ko *see* [head streak in down]
koa the large leguminous tree *Acacia koa*
koala an arboreal phascolarctid marsupial *Phascolarctos cinereus*. Clearly distinguished from the phalangerids by the absence of a tail and from the wombat by the presence of a canine in the upper jaw (= koal bear)
kob the reduncinine antelopid *Adenota kob* (= puka)
-kodo- *comb. form* meaning "fleece" frequently transliterated -codio-
koekoea the *N.Z.* cuculid bird *Urodynamis taitensis* (= long-tailed cuckoo)
koel *I.P.* the cuculid birds *Eudynamays scolopacea* and *E. honorata*
-koelo- *comb. form* meaning "hollow" usually transliterated -coel-
-koeno- *see* -koino-
kohekohe *N.Z.* the meliaceous shrub *Dysoxylum spectabile*
koh-i-noor *I.P.* the amathusiid lepidopteran insect *Amathuxidia amythaon*
kohlrabi a tuberous-stemmed variety of cabbage (*q.v.*)
kohuhu *N.Z.* the pittosporaceous tree *Pittosporum tenuifolium*
-koino- *comb. form* meaning "sharing" or "togetherness" usually transliterated -coen- and therefore confused with a *L.* derivative properly so spelled

-koinobio- *comb. form* meaning "cloister" usually transliterated -coenobio-
koitareki *N.Z.* the rallid bird *Porzana affinis* (= marsh rail)
-koito- *comb. form* meaning a "bed chamber" usually transliterated -coeto-
kokako *N.Z.* the callaeid bird *Callaeas cinerae*
-kokko- *comb. form* meaning "berry" usually transliterated -cocco-
kokla the columbid bird *Sphenurus sphenurus* (= green pigeon)
-kole- *comb. form* meaning "sheath" frequently transliterated -cole-
-koll- *comb. form* meaning "glue" frequently transliterated -collo-
Kollicker's *epon. adj.* from Rudolph Kollicker (1817–1905)
-kolp- *comb. form* meaning "bosom" frequently transliterated -colp-
-komo- *comb. form* meaning "hair" frequently transliterated -como-
-kon- *comb. form* meaning "cone" frequently transliterated -con-
-kondyl- *comb. form* meaning "a knuckle" but extended to any knobby joint. Frequently transliterated -condyl-
-koni- *comb. form* meaning "dust" frequently used in the sense of "spore" and also transliterated -conid-
-konop- *comb. form* meaning "gnat," frequently transliterated -conop-
-kont- *comb. form* meaning "a rod" or "pole," sometimes, but rarely, transliterated -cont-
  chondriokont = mitochondria
  gonotokont a common term for gamete mother-cells before meiosis
  isokont having two flagella of equal length
    anisokont having one large and one small flagellum
  monokont having a single flagellum
  polykont having many flagella
  tetrakont having four equal flagella
koodoo any antelope of the genus Strepsiceros (= kudu)
kookaburra *Austral.* any of several species of alcedinid bird of the genus Dacelo
  laughing kookaburra *Austral. D. novaeguinae* (= laughing jackass
kookacheea the caprimulgid bird *Caprimulgus ridgwayi*
kookavurra = kookaburra
kop a taal word meaning "head" used in compound names of several organisms
  hammer kop the scopid bird *Scopus umbretta*
  dikkop any of several burhinid birds of the genus Burhinus (*cf.* thickknee)
-kopr- *comb. form* meaning "dung" frequently transliterated -copr- (*cf.* -fim-)
-korac- *comb. form* meaning "raven" frequently transliterated -corac-
korbugo = cobego
koreke *N.Z.* the phasianid bird *Coturnix novaezealandiae* (= New Zealand quail)
-korem- *comb. form* meaning "a broom" frequently transliterated -corem-
korhaan *S.Afr.* any of several species of otidid birds (= bustard)
-korm- *comb. form* meaning "a tree trunk" frequently transliterated -corm-
korokio *N.Z.* the cornaceous shrub *Corokia cotoneaster*
korora *N.Z.* the spheniscid bird *Eudyptula minor* (= blue penguin)

korrigum *W.Afr.* the alcelaphine antelope *Damaliscus korrigum* (*cf.* topi, sassaby, blesbok and bontebok)

kotare *N.Z.* the alcedinid bird *Halcyon sancta* (= sacred kingfisher)

kotuku *N.Z.* the ardeid bird *Casmerodus albus* (= white heron *N.Z.*)

kotukutuku the onagraceous tree *Fuchsia excorticata* (= tree fuchsia *N.Z.*)

-kotyl- *comb. form* meaning "tube" almost invariably transliterated -cotyl-

kowhai *N.Z.* the leguminous tree *Edwardsia microphylla*

  red kowhai the leguminous vine *Clianthus puniceus* (= parrot's bill)

kr *see* [Kreisler]

kraai *S.Afr.* = crow

Kraemeriidae a small family of Indo-pacific and Australian glass-like acanthopterygian fishes commonly called sand divers

krait any of many elapid snakes of the genus Bungarus. Particularly *B. coeruleus*

Krameriaceae a family erected to contain those forms usually considred to be a tribe (Kramerieae) of the leguminous subfamily Caesalpinioideae

-kraspedo- *see* -craspedo-

krassang the rutaceous tree *Feroniella oblatta* and its fruit

-krater- *comb. form* meaning "cup" usually transliterated -crater-

Krause's *epon. adj.* from Wilhelm Krause (1833–1909)

[Kreisler] a mutant (gene symbol *Kr*) found in linkage group V of the mouse. The phenotypic expression is a typical waltzing syndrome accompanied by deafness

-kremno- *comb. form* meaning "cliff" almost invariably transliterated -cremno-

-kren- *comb. form* meaning "notch" but see -cren-

-kreo- *comb. form* meaning "meat" or "flesh" usually transliterated -creo-

-krepi- *comb. form* meaning "shoe" often transliterated -crept-

Krikobranchia an order of ascidiacean urochordates with an elongate body divided into two or three regions

-krin- *comb. form* meaning "lily" or "separate" frequently transliterated -crin- and thus becoming confused with hair

krishna bor the moraceous shrub *Ficus krishnae*

-krit- *comb. form* meaning "chosen" frequently transliterated -crit-

Krogh's *epon. adj.* from Schack August Steenberg Krogh [commonly called August Krogh] (1916–1945)

krypt- *comb. form* meaning "hidden" and thus, by extension, "cavity" or "vault." Usually transliterated -crypt-

kuaka *N.Z.* the scolopacid bird *Limosa lapponica* (= bar-tailed godwit *N.Z.*)

kudu any of several antelopes of the genus Strepsiceros (= koodoo)

Kuhliidae a monogeneric family of Indo-Pacific percoid fishes commonly called aholeholes in Hawaii

Kuhne's *epon. adj.* from Willy Kuhne (1837–1900)

kukri any of several species of colubrid snakes of the genus Oligodon

kuku *N.Z.* the columbiid bird *Hemiphaga novaeseelandiae* (= wood pigeon *N.Z.*)

kukuruatu *N.Z.* the charadriid bird *Thinornis novaeseelandiae* (= shore plover *N.Z.*)

-kuma *comb. form* meaning "a wave" frequently transliterated -cuma-

kumquat any of several species and numerous horticultural varieties of tree of the rutaceous genus Fortunella

  Australian kumquat the rutaceous tree *Eremocitrus glauca*

  Hong Kong kumquat *F. hindsii*

  oval kumquat *F. margarita*

  round kumquat *F. japonica*

Kupfer's *epon. adj* from Karl Wilhelm von Kupfer (1829–1902)

kuruwhengi *N.Z.* the anatine bird *Spatula rhynchoris* (= shoveller *N.Z.*)

[kurz] a mutant (gene symbol *kz*) mapped at 0.7 on the X chromosome of *Drosophila melanogaster*. The phenotypic expression are short, fine bristles

kuwini the anacardiaceous tree *Mangifera odorata* and its fruit

kyah the phasianid bird *Francolinus gularis*

-kyan- *comb. form* meaning "cornflower" and giving directly -cyan- meaning "blue"

-kykl- *comb. form* meaning "circle" almost invariably transliterated -cycle-

-kym- *comb. form* meaning "a wave" frequently translated -cym-

-kymbo- *comb. form* meaning "cup" frequently transliterated -cymbo-

-kyo- *comb. form* meaning "contain" frequently transliterated -cyo-

-kyon- *comb. form.* meaning "dog" frequently transliterated -cyn- or -cyon-

-kypy- *comb. form* meaning "humpbacked"

Kyphosidae a small family of acanthopterygian fishes oval in shape and with small mouths and fine teeth. They are known as rudder fishes from their habit of following closely after ships but in the *U.S.* are better known as sea chubs

-kyst- *comb. form* meaning "a bladder" almost invariably transliterated -cyst-

kz see [kurz]

l *see* [angora hair], [lethal] or [long]. This symbol is also used in yeast genetics to indicate a recessive lethal mutant

l₁ *see* [lutescent foliage]

l₂ *see* [lutescent-2]

l₈ *see* [luteus seedling]

L *see* [Lobe] or [Mulilunar]

La *see* [Lanceolate]

label the terminal segment of a fern-frond

labellum 1 (*see also* labellum 2) the term is variously applied to extensions of the labrum in insects, and also, in honeybees, to the spoon-like tip of the glossa

labellum 2 (*see also* labellum 1) the enlarged third petal of an orchid which is twisted normally from the posterior to the anterior position

labia plural of labium but also used for that part of the exoskeleton of an insect which immediately surrounds the spiracle

 labia majora the outer lips of the vulva of primate mammals

 labia minora the inner lips of the vulva of primate mammals

labial pertaining to the labium

Labiatea that family of tubifloral dicotyledonous angiosperms which contains a very large number of pharmaceutical and culinary aromatics. The family is distinguished by the two-lipped shape of the flower, the quadrangulate stem and opposite leaves

labiate lipped

labides *pl.* of labis

Labiduridae a family of dermapteran insects distinguished by the unusually large number of segments in the antenna

labiella the insect hypopharynx or "tongue"

Labiidae a family of dermapteran insects distinguished by an eleven to fifteen segmented antenna in which the fourth, fifth, and sixth segments together are longer than the first

labiose heavily lipped, particularly a polypetalous flower of which two petals are unusually large

labis one of a pair of hairy appendages in the male genitalia of lycaenoid lepidopteran insects. It is probably homologous with the uncus

labium 1 (*see also* labium 2) in *bot.*, lip of a labiate flower

labium 2 (*see also* labium 1) a compound structure forming the "lower lip" or floor of the mouth in insects but most commonly applied to the most posterior paired appendages of the head, homologous with the second maxillae

 eulabium the terminal division of the labium

 postlabium the basal region of the labium

labor 1 (*see also* labor 2) term for a group of moles

labor 2 (*see also* labor 1) common name for the act of parturition

Laboulbeniales an order of pyrenomycete ascomycetes principally distinguished by their habitat as minute ectoparasites in the chitin of living insects

labraria the lip of the epigusta. The term is also sometimes used as a synonym for epipharynx

Labridae a large family of marine acanthopterygian fishes commonly called wrasses or hogfish. They are distinguished by their extremely well developed projecting incisor teeth

labrum literally upper lip, and specifically that portion of the insect head which covers the base of the mandible and forms the roof of the mouth in mosquitos and other Diptera; it is essentially a cephalic sclerite immediately above the mandibles. The term is also applied to the thickened edge round the mouth of a gastropod shell and to the lip-like projection of the posterior interambulacrum over the peristome in some echinoderms

 epilabrum a lateral lobe on each side of the labrum in chilopds

 paralabrum a toothed moveable plate derived from the labrum in some acarines or a medium flap found immediately in front of the mandibles of many arthropods and thus corresponding roughly to an upper lip

lac 1 (*see also* lac 2) any alcohol-soluble resin-like substance produced by certain coccid insects

 shellac the lac produced by *Laccifer lacca*

lac 2 (*see also* lac 1) *see* [lacquered]. The symbol is also used for a bacterial mutant showing a utilization of lactose

lac O a bacterial genetic marker having a character affecting the activity of operator mutants, mapped at 10.0 mins. for *Escherichia coli*

lac Y a bacterial genetic marker having a character affecting the activity of galactoside permease, mapped at 10.0 mins. for *Escherichia coli*

lac Z a bacterial genetic marker having a character affecting the activity of β-galactosidase. Mapped at 10.0 mins. for *Escherichia coli*

lacapan *hort. var.* of the musaceous tree-like herb *Musa paradisiaca* (= banana)

laccer = lappet 2

lacerous pertaining to a lappet in the sense of lappet 1

Lacertae an assemblage of lacertilian reptiles erected to contain those families having procoelus, rather than amphicoelous, vertebrae (*cf.* Geckones) and which differ from the Chamaeleontes in lacking a prehensile tail

Lacertidae a very large family of Old World lacertilian reptiles, distinguished by the combination of long and strong tails, no reduction of limbs, complete lack of body osteoderms and a long, deeply forked tongue

Lacertilia an assemblage of squamatan reptiles containing those forms commonly called lizards in which the right and left half of the mandibles are fused in front in distinction from the Ophidia (or more properly Serpentes) in which a ligament, not a suture, connects the two halves

lachrymal of, or pertaining to, tears

-lacin- *comb. form* meaning "torn"

lacina a term used by many acarologists in error for lacinia (*q.v.*)

lacinella the lateral lobe of a two-lobed arthropod lacinia (*cf.* lacinoidea)

licinia literally a lappet in the sense of lappet 1 or a slender blade, but typically applied to the inner of the two lobes that terminate an insect maxilla (*cf.* galea) though used also for the narrow, incised segment of a leaf

laciniate cut into narrow lobes

lacinoidea the medial lobe of a two-lobed arthropod lacinia (*cf.* lacinella)

lackey *Brit.* popular name of lasiocampid lepidopteran insects (*cf.* eggar)

Lacosomidae a family of uranioid lepidopteran insects distinguished by the fact that many of the caterpillars make cases resembling those of caddisfly larvae. They are accordingly called sack-bearers

[lacquered] a mutant (gene symbol *lac*) mapped at 7.3± on the X chromosome of *Drosophila melanogaster*. The phenotype is pale and shiny as though lacquered

-lacrim- *comb. form* meaning "tear." Also spelled -lachrim-

lactation the production of milk

lacteal pertaining to milk

lacteous milky, either as to texture or color

lactic pertaining to milk (*cf.* laticic)

prolactin = lactogenic hormone

Lactobacillaceae a family of gram positive eubacterial Schizomycetes having the form of rods that may divide into chains and even become filamentous. Usually divided into two tribes, the Streptococceae almost all of which are dangerous pathogens and the Lactobacilleae a few of which are pathogenic and many of which occur in the production of cheese and other milk products

lactonase general name for a group of enzymes that catalyze the hydrolysis of lactones

aldonolactonase catalyzes the hydrolysis of gulono-γ-lactone to gulonic acid

arabonolactonase catalyzes the hydrolysis of L-arabono-γ-lactone to L-arabonic acid

gluconolactonase catalyzes the hydrolysis of D-glucono-δ-lactone to D-gluconic acid

4-carboxymethyl-4-hydroxy-isocrotonolactonase catalyzes the hydrolysis of 4-carboxymethyl-4-hydroxyisocrotonolactone to 3-oxoadipic acid

uronolactonase catalyzes the hydrolysis of glucurono-δ-lactone to glucuronic acid

lacuna a hole, particularly, in *bot.*, an air-space in tissue. In *zool.* lacuna is frequently used for an invertebrate structure which, if vertebrate, would be called a sinus

Howship's lacuna a pit on the surface of bone in which osteoclasts are clustered

leaf lacuna the gap left in a vascular cylinder where the leaf trace has parted from it

Lacunidae a small family of gastropod Mollusca commonly called chinkshells in virtue of the lengthened groove which runs along the side of the columella

lacunose pitted or, more rarely, perforate

lacuster the middle of a lake

lacustrine pertaining to lakes

Jacob's **ladder** *U.S.* the liliaceous herb *Smilax herbecea Brit.* the polemoniaceous herb *Polemonium caeruleum. U.S. P. vanbruntiae*

lady 1 (*see also* lady 2) any of several lepidopteran insects

old lady *Brit.* the noctuid *Mania maura*

painted lady the nymphalid *Vanessa cardui*

white lady *S.Afr.* either of two species of papilionid lepidopteran insect *Papilio mórania* or *P. pylades*

lady 2 (*see also* lady 1) in compound names of plants

quaker lady the rubiaceous herb *Houstonia caerulea*

Laemobothriidae a small family of mallophagan insects parasitic on water birds and birds of prey. There is a conspicuous swelling front of the eye

-laev- *comb. form* confused from three roots and therefore various meaning "to the left," "smooth" or "slippery," and "nimble"

laevigate polished

-lagen- *comb. form* meaning "flask"

lagena a side pouch from the sacculus which gives rise to the cochlea

Lagenidiales an order of biflagellate phycomycete fungi possessing a simple mycelial thallus

lagger the falconid bird *Falco jugger*

Lagoidae = Megalopygidae

Lagomorpha that order of placental mammals which contains the rabbits. They are distinguished from the rodents, with which they were once fused, by the possession of two pairs of upper incisors and a single relatively short pair of lower incisors

Logomyidae = Ochotonidae

lagopous densely hirsute

lair the semipermanent resident of a predator (*cf.* lochetic)

lak *Brit.* the courting dance of the capercailzie

lake any large body of fresh water

bog lake a stage intermediate between a dystrophic lake and a bog

oxbow lake a crescentic lake formed from an oxbow (*q.v.*) which has been cut off from its parent waters

reservoir lake a lake resulting from the artificial impoundment of water

dystrophic lake an old, heavily sedimented, vegetation-filled lake, often the first stage in the production of a peat bog and sometimes used as synonymous with bog

eutrophic lake a highly productive lake characterized by abundant plankton and fairly high turbidity

mesotrophic lake intermediate in productivity between oligotrophic and mesotrophic conditions

oligotrophic lake a weakly productive lake with little plankton

laker = lake trout

lalang the grass *Imperata cylindrica*

lama = llama

Lamarckian *epon. adj.* from Jean Baptiste Pierre Antoine de Monet, Chevalier de Lamarck (1744–1829)

Lamarckism a view of evolution derived from the unproven postulate that characters acquired through exposure to the environment can be inherited by the offspring of the individual acquiring these characters

lamb 1 (*see also* lamb 2) without qualification, the young of the domestic sheep, though applied to many other ovids

lamb 2 (*see also* lamb 1, lettuce) in compound names of plants

Scythian lamb the cyatheaceous herb *Cibitium barometz*

tartarian lamb = Scythian lamb

lamel = lamella

lamella a little plate, or structure resembling a little plate. Among *biol.* structures so designated are the gill of an agaricale fungus, the gill of a pelecypod mollusc, the thin projections found on the pygidium of coccid insects and the calcareous platelets surrounding the Haversian canals in bone

mesolamella a term suggested to replace mesogloea when it occurs in the form of a thin sheet as in Hydra

primary lamella the outermost coat of a spore

Lamellibranchia = Pelecypoda

Lamellicornia = Scarabaeoidea

lamia a half-grown shark

Lamiaceae = Labiatae

Lamiales an order erected to contain the families Verbenaceae and Labiatae by those who do not hold that these families belong in the Tubiflorae

lamina a sheet. Used particularly in *bot.* to describe the flat portions of leaves and thalli

sensitive lamina one of the numerous sheets of vascularized innervated tissue in the dermal portion of the hoof of an ungulate

terminal lamina that portion of the floor of the embryonic brain that lies anterior to the preoptic recess

vertebral lamina one, of the two, sides of the neural arch

Laminariales an order of heterogenerate Phaeophyta commonly called the "kelps." They are clearly distinguished by the blade-like thallus of the sporophyte provided with a holdfast at its lower end

laminarinase an enzyme catalyzing the breakdown of laminarin through a hydrolysis of β-1,3-glucan links

delamination the process of separating a solid mass into sheets and specifically the production of flattened cells from a layer

lammergeier the accipitrid falconiform bird *Gypaetus barbatus* (= lammergeyer)

lamnadens the plate bearing the prestomal teeth in Diptera

Lamnidae a family of sharks distinguished by the symmetrical tail

lampern *Brit. obs.* freshwater, as distinct from marine, cyclostomes

lamprey the popular name for any pteromyzontid cyclostome

Lamprididae a family of allotriognathous fishes distinguished by its great lateral compression. Commonly called opahs

Lampridiformes = Allotriognathi

Lampyridae a large family of coleopteran insects commonly called lightning-bugs or fireflies, in virtue of the luminescent segments near the end of the abdomen

lana the timber of the rubiaceous tree *Genia americana*

lanate with a woolly texture

sand lance popular name of ammodytid fishes, though the term is sometimes applied to the sand divers (Kraemeridae)

lancelet popular name for amphioxus and its allies

lanceolate shaped like a blade sharpened to a narrow point at one end

[lanceolate] a mutant (gene symbol *ll*) mapped at 106.7 on chromosome II of *Drosophila melanogaster*. The term refers to the shape of the phenotypic wing

[Lanceolate] a mutant (gene symbol *La*) mapped at 0 on linkage group B of *Zea mays*. The term is descriptive of the phenotypic undivided leaves

land any area of the planet not covered with water

badlands an area in *W.U.S.* of fantastically wind-eroded rocks, sandy-top desert soil, and dried up water courses occasionally subject to flash floods (*cf.* hydrotribophilous, tirad)

grassland a large category of territories in which the vegetation consists of herbacious perennials and grasses, the latter being dominant

heathland an area covered with low growing shrubs interspersed with occasional areas of grass

langaha the snake *Langaha nasuta*

langarai the rhizophoraceous tree *Bruguiera parviflora*

langouste decapod crustaceans of the genus Palinurus (= rock lobster)

langsat the meliaceous tree *Lansium domesticum* and its edible fruit

languets literally, a little tongue. Used of any soft projection, particularly the ciliated processes that project into the oral funnel in Amphioxus or in the pharynx of tunicates

langur any of numerous long-limbed, long-haired colobid primates of the Asiatic region. Without qualification, frequently those of the genus Semnopithecus

Languriidae a small family of medium-sized beetles closely allied to the Cucujidae. They are readily distinguished by their elongate shape, red pronotum and black elytra from which the name lizard-beetle derives

Laniidae that family of passeriform birds which contains the shrikes. These are agressive predatory birds and many form larders by impaling their prey on sharp thorns. They have a large head and a strong, hooked, toothed bill

lanner a general term applied to those falconine birds which are of a bluish grey tint. Usually, without qualification, *L. biarmicus* (= Abyssinian lanner)

lanose woolly

lanseh the meliaceous tree *Lansium domesticum* and its edible fruit (= lanoon)

Aristotle's lantern the teeth and chewing apparatus of echinoid echinoderms

Lanthanotidae a monotypic family of lacertilian reptiles having pleurodont dentition, a short bifid tongue, and ridges of tubercles on the back that give it the appearance of a miniature crocodile

Lanthionine cis(2-amino-2carboxyethyl)sulfide. An amino acid not known to be necessary to the nutrition of rat and not definitely established as a protein precursor

languinose with a woolly texture

lanugo the prenatal fur of some mammalian embryos and by extension, any downy growth of hair. The term is also sometimes used for fine, single "hairs" on insects

lanvon the meliaceous tree *Lansium domesticum* and its edible fruit (= lanseh)

lap to secure fluid by repeated scoops of the tongue

-lapid- *comb. form* meaning "stone"

lapideous stony

lapidose dwelling among stones

lapon the scorpaenoid teleost fish *Scorpaena mystes*

lappaceous hamate

lappet 1 (*see also* lappet 2) a flap of tissue, specifically the lateral wattle of a bird and the protuberant portion of the scalloped edge of a medusa

lappet 2 (*see also* lappet 1) any of several lasiocampid lepidopteran insects usually, without qualification, *Gastropacha quercifolia*

   small lappet *Epicnaptera ilicifolia*

lar the gibbon *Hylobates lar*

larch (*see also* beetle, sawfly) properly any of several species of the pinaceous genus Larix, though used for a few other pinaceous trees

   eastern larch *U.S. L. laricina*

   European larch *L. decidua*

   golden larch the pinaceous tree *Pseudolaryx kaempferi*

   western larch *U.S. L. occidentalis*

Lardizabalaceae a small family of ranalous dicotyledons that contains a few genera of edible fruited shrubs. The several-seeded fruits are characteristic

[Large] a mutant (gene symbol *Lg*) mapped at 27.0± on the X chromosome of *Drosophila melanogaster*. The name refers to the size of the adult phenotype

Laridae a large family of charadriiform birds containing the gulls and terns. Almost all have black or grey backs with white undersides with long pointed wings and square or forked tails

lark (*see also* cuckoo, sparrow, spur) originally (*Brit.*) any of very many species of passeriform birds of the family Alaudidae. In *U.S.* the term is applied to icterids mostly of the genus Sturnella as well as to an alaudid and in *Austral.* and *N.Z.* often to motacillids, mostly of the genus Anthus

   antipodes lark *Austral. Anthus novoseelandiae* (= New Zealand pipit)

   bushlark a general term applied to numerous alaudids of the genus Mirafra. Without qualification *Austral. M. javanica*

      red-tailed bush-lark *Afr. M. nigricans*

   flappet lark *Mirafra rufocinnamonea*

   groundlark *N.Z.* the motacillid *Anthus novaeseelandiae* (= New Zealand pipit)

   hoopoe-lark *Afr.* the alaudid *Alaemon alaudipes*

   horned lark *U.S.* the alaudid *Eremophila alpestris*

   magpie-lark any of several species of grallinid bird of the genus Grallina

   meadowlark *U.S.* any of several icterid birds of the genus Sturnella usually, without qualification *S. magna. Brit.* the alaudid *Alauda arvensis*

      eastern meadow lark *U.S. S. magna*

      western meadow lark *U.S. S. neglecta*

   mudlark = magpie-lark

   sand lark the alaudid *Ammomanes deserti*

   sea-lark the charadriid *Charadrius hiaticula* (= ringed plover)

   shore lark *Brit.* the alaudid *Eromophila alpestris*

   skylark *Brit.* = meadowlark *Brit.*

   brown songlark *Austral.* The sylviid *Cinclorhamphus cruralis*

      roufous songlark *Austral. C. mathewsi*

   sparrow-lark any of several species of alaudid of the genus Eremopteryx

   sunlark the alaudid *Galerida modesta*

   titlark = *Brit.* pipit

   short-toed lark the alaudid *Calandrella cinerea*

   woodlark *Brit. Lullula arborea*

larkspur *see* -spur

Larrinae a subfamily of sphecoid hymenopterans commonly called sand-loving wasps owing to the frequency with which they are found burrowing in dry sand

larva a development stage differing from an embryo in being able to secure its own nourishment. Almost all larvae differ markedly from the adult in appearance and attain the adult shape by metamorphosis. Larvae owe their fanciful names to early misidentification as separate generic entities. In *U.S.* entomologists confine the term to the prepupal stage of holometabolous insects, but European entomologists use the word for the prenymphal stage of any metabolizing insect. In *bot.* larva refers to an early plant form which differs markedly from the adult

acanthella larva an acanthocephalan larva developed from an acanthor after the latter has lost its shell, and crossed through the gut wall into the intermediate host's haemocoel

acanthor larva a shell-enclosed larva, strongly resembling a hexacanth, which is the infectious stage of acanthocephalans

acronurus larva the larva of acanthuroidean fishes distinguished by the vertical ridges on the body

actinotroch larva the tentaculate larva of Phoronida

actinula larva a larva developed from the planula in some hydrozoan coelenterates. It resembles a stalkless polyp and creeps about on its tentacles

adolescaria larva a term embracing both cercaria and metacercaria

alima larva the free-swimming larva of the stomatopod crustacean Squilla

ammocoetes larva the freshwater larva of the lampreys (Petromyzontidae)

amphiblastula larva the free-swimming larva of some calcareous sponges consisting of an anterior hemisphere of flagellated micromeres and a posterior hemisphere of non-flagellated macromeres

antlered larva caterpillars with prominent branched protuberances

arachnactis larva the modified, creeping, pseudomesusoid larva of certain arenicolous Anthozoa particularly of the genus Cerianthus

auricularia larva the free-swimming larva of holothurian Echinodermata

bipinnaria larva the free-swimming larva of asteriod Echinodermata which preceeds the brachiolaria

brachiolaria larva free-swimming larva of asteroid Echinodermata which develops from the bipinnaria

caddis larva the larva usually case bearing, of a trichopteran insect

calyoptis larva the zoaea larva of euphausiadid Crustacea in which the carapace overlaps and conceals the eyes

decacanth larva = lycophore larva

caraboid larva the second instar in the life history of meloid coleopterans immediately following the triungulin

cephalotroch larva a trochopore-like stage in the development of polyclad turbellaria distinguished from a trochophore by the possession of one preoral and one postoral ciliated band

cercaria larva the free-swimming larva of trematode platyhelminthes, that develops from the redia

cotylocercous cercaria larva one with a sucker or adhesive gland on the tail

cystocercous cercaria larva one in which the body can be withdrawn into a cavity in the tail

furcocercous cercaria larva one with a forked tail

macrocercous cercaria larva a cercaria with a long thick, but not flattened tail

microcercous cercaria larva one with a very small tail

pleurolophocercous cercaria larva one with a finned tail

rhopalocercous cercaria larva one with a very thick tail

trichocercous cercaria larva one with a bristle tail

xiphidiocercaria larva a cercaria having a boring spine at the anterior end

mesocercaria larva a stage intermediate between a cercaria and a metacercaria

metacercaria larva a cercaria which is encysted and has metamorphosed into a juvenile fluke

strobilocercus larva a terminal larval stage of a cestode in which the posterior portion of the cystircercus begins strobilation

plerocercoid larva a larval stage in the development of a cestode platyhelminth consisting of a scolex, some strobila and sometimes even genital organs. (*cf.* plerocercus, cysticercus, cysticercoid)

procercoid larva an elongate development of a coracidium or onchosphere in which the hooks pass through the posterior and form an organ of attachment

cerinula larva *either* a floating larva of ceriantharian coelenterates which floats on the surface by spreading its tentacles or swims by flapping them *or* a sessile larva intermediate between the arachnactis (*q.v.*) and the adult Cerianthus

coarctate larva a resting stage in the life history of a meloid coleopteran between the scarabaeoid larva and the scolytoid larva

codonocephalus larva = neascus larva

coenurus larva a type of cysticercus in which the inner wall of the bladder proliferates groups of other cysticerci which do not become detached

conaria larva the first, spherical, free-swimming larva of the siphophore hydrozoan Velella

coracidium larva the membrane enclosed onchosphere of diphyllobothrid cestodes

pentacrinoid larva that stage in the development of a crinoid immediately following the cystidean stage. It is distinguished by the presence of arms on the calyx

anthocyathus larva a disk cut off from the top of a young polyp of the developing coral Fungia in much the manner that strobila are released in the life history of Aurelia

cyphonautes larva a tent-shaped free-swimming larva of some Ectoprocta

cypris larva the free-swimming, ostracod-like, larva of cirripedia

cystocercus larva a larval stage in the development of a cestode platyhelminth consisting of a scolex attached to a bladder. There may be more than one scolex and one or more may be invaginated. (*cf.* plercocercus, plerocercoid, cysticercoid)

cystidean larva that stage in the development of a crinoid in which the stalk first appears, but in which the arms are not developed

Desor's larva an oval ciliated postgastrula stage in the life history of the nemertine worm Lineus

dipleurula larva that stage in the development of an echinoderm that lies between the gastrula and the pluteus or brachiolaria. It is often considered to represent the ancestral echinoderm type

discinid larva a planktonic bivalve larva of brachiopods

doliolaria larva a free-swimming ciliated postgastrular stage in the development of crinoids and a post auricularia holothurian larva. It possesses a large apical tuft and either four or five ciliated bands

metadoliolaria larva a holothurian doliolaria larva in which the tentacular lobes are apparent

edwardsia larva the stage in the development of anthozoan coelenterates possessing eight pairs of septa. It is usually sessile but occasionally free-swimming

ephyra larva a primitive medusoid stage budded off from the scyphistoma of scyphozoans

ephyrula larva = ephyra larva

erichthoidina larva that stage in the development of a stomatopod crustacean which hatches from the egg. There are five pairs of biramous thoracic appendages and one abdominal appendage

erichthus larva a stage in the development of a stomatopod crustacean having abdominal appendages, and raptorial limbs, but lacking appendages on the posterior six thoracic segments

campodeiform larva a caddis larva lacking a case (*cf.* campodeiform)

eucoliform larva that stage which emerges from a hymenopteran egg (*cf.* teleaform larva)

infusoriform larva a larva produced from the infusorigen of a mesozoan

teleaform larva that larval stage which follows a eucoliform larva from which it is distinguished by possessing a cephalic process and a girdle of setae

glaucothoe larva the pre-adult stage of a hermit crab corresponding to the megalopa of other crabs but with a precociously developed symmetrical abdomen

glochidium larva a bivalved larva of freshwater clams which attaches by its valves to fish for distribution

Gotte's larva a type of Müller's larva, produced by Stylochus, which has four, instead of eight, lobes

hallocampoides larva the stage immediately following the edwardsia larva in the development of zoantharian coelenterates in which two further pairs of septa have been added

hexacanth larva a spherical or ovoid six-hooked larva hatching directly from the egg of many cestode platyhelminthes

hydratuba larva = scyphistoma larva

leptocephalus larva the marine larva of the freshwater eel Anguilla

Loven's larva = trochophore

lycophore larva a ciliated cestode larva with large frontal glands and ten hooks. It remains in the shell until eaten by a secondary host

megalopa larva the pre-imago stage in the development of a crab in which the eyes and chelae are both very prominent

miracidium larva the free-swimming, conical ciliated larva that hatches from the egg in digenean trematodes

mitraria larva a post-trochophore larva of the polychaete Owenia distinguished by the presence of three extremely long ventrally directed flotation chaetae

Müller's larva an elongate post-trochophore stage in the development of polyclad Turbellaria distinguished by the presence of eight, backwardly directed post oral lobes

nauplius larva the one-eyed larva with three anterior pairs of appendages hatched from the egg of most crustacea

metanauplius larva the stage immediately following the nauplius in which three further rudimentary pairs of appendages are apparent

neascus larva a type of metacercaria with a cup-shaped forebody and a well developed hindbody

nectochaeta larva a post-trochophore larva of polychaete annelids in which one or more pairs of parapodia are developed

olynthus larva that stage in the development of a sponge, in which the parenchymula becomes hollowed out to form a spongocoel

onchosphere larva the first, ciliated, spherical larva of a cestode platyhelminth

pentactula larva the post-doliolaria larva of a holothurian, distinguished by the presence of the five primary tentacles

phyllosoma larva the flattened, leaf-like, transparent larva of scyllaridean crustacea

pilidium larva a free-swimming stage in the development of many nemerteans. It is distinguished by its conical shape, large apical tuft of cilia and a pair of flap-like projections hanging down on each side

planidium larva the active larva hatching from the egg of some chalcicoid Hymenoptera particularly those of the families Eucharitidae and Perilampidae

planocera larva a flattened oval postgastrular ciliated stage in the development of polyclad turbellaria

planula larva a free-swimming ovoid or pear-shaped post-blastula stage in the development of many coelenterates

pluteus larva a general term for the bilaterally symmetrical free-swimming echinoderm larva in which the ciliated bands are prolonged over anteriorly directed arms

ophiopluteus larva a pluteus larva of ophiuroid echinoderms, distinguished by the fact that one pair of arms is much longer than the others

oligopod larva an insect instar with functional thoracic limbs

polypod larva an insect instar with a completely segmented abdomen, each segment bearing a set of functional legs

protopod larva an arthropod instar not yet differentiated

protaspis larva the minute, megacelaphic, almost unsegmented, larva of trilobites

protonymphon larva the larva which hatches from the egg of Pycnogonida. It has three pairs of appendages, of which the anterior are chelate, and a well-developed proboscis

prototroch larva a pre-trochophore larva possessing one to three equatorial bands of pre-oral cilia

protrochula larva a free-swimming larva of some polyclad turbellaria and which is supposed by some to be a precursor of the trochophore larva of other types

rataria larva an hourglass-shaped, free-swimming larva with an anterior disc-like collar which develops from the conaria larva

redia larva a larval stage of a trematode developed within a sporocyst. The redia possesses a distinct mouth and pharynx and produces cercaria by internal budding

rhagon larva a larval stage of many sponges having the form of a cone with a single exhalant aperture at the tip

scarabaeoid larva the third instar in the life history of a meloid insect immediately following the caraboid larva

schizopod larva the larval stage of astacuran, and late larval stage of many natantian, crustacea

scolytoid larva the active larva in the life history of a meloid beetle which follows the coarctate larva and immediately precedes the pupa

scyphistoma larva the sessile hydroid stage of scyphozoan coelenterates

scyphula larva = scyphistoma larva

Semper's larva a cylindrical larva produced by some zoanthid corals with a hole at each end and a longitudinal band of cilia

slug larva any slimy insect larva

sparganum larva = plerocercoid larva

sporocyst larva a hollow cyst developed from the miracidium in the intermediate host of a trematode. Germinal cells in the sporocysts produce either further sporocysts or redia

stereogastrula larva = planula larva

strobila larva a scyphistoma which has begun to split off ephyra from its upper end

rat-tailed larva the aquatic larva of surface Diptera furnished with an elongate breathing tube

tetrathyridium larva a plerocercoid larva with an introverted scolex

tornaria larva the free-swimming larva of Enteropneusta. Its general appearance, and arrangement of ciliated bands is reminiscent of some echinoderm larva

tentaculate tornaria larva a tornaria larva of the genus Ptychodera in which the ciliary loops are fringed by small tentacles

trilobite larva the larva of the Xiphosura

triungulin larva a six-legged, compound-eyed, active larva hatched from the egg of either strepsipteran insects, or meloid coleopteran insects, and which serve as the distributive phase

atrocha larva a type of trochophore uniformly ciliated and without a preoral band

cephalotrocha larva a young Müller's larva

trochophore larva a top-shaped post-gastrula larva, common to the Annelida and Mollusca. It is distinguished by the presence of an apical tuft of cilia and three preoral ciliated bands

metatrochophore larva a trochophore larva possessing secondary ciliated bands in the post-trochal hemisphere

polytrochula larva = nectochaeta larva

veliger larva a late, free-swimming larva of marine gastropods distinguished by two large ciliated lobes. The shell, if present, is a dissoconch

Wagener's larva a free-swimming larva stage of the mesozoan Mycrocyema

zoaea larva that stage in the development of higher Crustacea with compound eyes, a carapace overlapping the thorax and segmented abdomen

pseudozoea larva the zoaeal stage of stomatopod crustacea

zonanthella larva a larva of a zoanthid coral, in the shape of an inverted pegtop with a girdle of very long cilia near the oral pole

zoanthina larva = zoanthella larva

Larvacea a class of tunicate Chordata distinguished by the retention throughout life of a "tail" with metameric muscle segments

Larvaevoridae = Tachinidae

larvarium the housing of a larva

larvina an insect larva, particularly a dipterous larva, lacking appendages

larvula *obs.* the first instar of a larva, particularly that of a mayfly

lary *Brit.* the charadriiform alcid bird *Cephas grylle* (= black guillemot)

larynx that proximal portion of the trachea which is modified in some animals to special functions

lascar *I.P.* any of several species of nymphalid lepidopteran insect of the genus Neptis (*cf.* sailer)

father lasher *Brit.* the cottid fish *Cottus scorpius*

-lasi- *comb. form* meaning "shaggy"

Lasiocampidae a family of bombycoid lepidopteran insects containing *inter alia* the well known *U.S.* tent caterpillars and lappet moths, and the *Brit.* eggars and lackeys

lasion a collection of sessile organisms crowded on a substrate and mutally interdependent (*cf.* epiphyton, periphyton)

lasset *Brit. obs.* = ermine

lasulinous ultramarine blue

latchet *Brit.* the triglid fish *Trigla lucerna*

latebra a flask shaped mass of white yolk, extending from the center of the yolk mass to Pander's nucleus in telolecithal eggs

latebrosous concealed

latency a term once applied to a form showing genotypic, as distinct from phenotypic, characters

laterad moving, or pointing in, a lateral direction

lateral pertaining to the side

bilateral having two sides

isobilateral = bilaterally symmetrical

latericious brick red

latex a milky fluid, found in many plants. Rubber is present in the latex of many genera though it is highest in Hevea

Laticaudinae a subfamily of hydrophid snakes distinguished by the broad ventral plates and an ovoviviparous mode of reproduction

laticic pertaining to latex (*cf.* lactic)

latolent said of those forms that smell by moving their head from side to side (*cf.* dirolent)

Latreille's *epon. adj.* Pierre Andre Latreille (1762–1833)

lattice an abortive sieve plate

lauan any of many dipterocarpaceous trees

laugher *U.S.* the noctuid lepidopteran insect *Charadra deridens*

lauia the scarid teleost fish *Callyodon lauia*

launce = lance (*cf.* sand-eel, sand-diver)

laund *Brit. obs.* a grass carpeted open space in woodlands

Lauraceae that family of ranalous dicotyledons which contains *inter alia*, the fragrant laurel, the cinnamon, and the sassafras. Distinguishing characteristics are the one-celled, one-seeded ovary and the undifferentiated perianth with numerous stamens

laurel (*see also* oak) any shrubs and trees with shiny dark green elliptic leaves are called laurel. Used without qualification, the term usually, but by no means always, applies to the lauraceous shrub *Laurus nobilis* or, occasionally to other members of the Lauraceae

Alexandrian laurel the liliaceous shrub *Danae racemosa*

California laurel the lauraceous shrub *Umbellularia californica*

cherry-laurel the rosaceous shrubby tree *Prunus laurocerasus*

Chilean laurel the monimiaceous tree *Laurelia sempervirens*

Chinese laurel the euphorbiaceous tree *Antidesma ˈbunius*

English laurel the rosaceous tree *Prunus laurocerasus*

great laurel *U.S.* the ericaceous shrub *Rhododendron maximum*

ground laurel *U.S.* the ericaceous shrub *Epigaea repens* (= trailing arbutus)

Japan laurel the cornaceous shrub *Aucuba japonica*

marsh laurel the theaceous shrub *Gordonia lasianthus*

mountain laurel any of several species of the ericaceous genus Kalmia, particularly *K. latifolia*. In *W.U.S.* *Umbellularia californica*

Portugese laurel the rosaceous shrub *Prunus lusitanica*

savino laurel the magnolaceous tree *Magnolia splendens*

seaside laurel the euphorbiaceous shrub *Phyllanthus speciosus*

sheep laurel the ericaceous shrub *Kalmia angustifolia* (*cf.* mountain laurel)

spurge laurel the thymelaeaceous shrub *Daphne laureola*

varigated laurel the euphorbiaceous shrub *Codiaeum variegatum* and its numerous horticultural varieties

West Indian laurel the rosaceous tree *Prunus occidentalis*

laurestinus = laurustinus

Lauroidea a subfamily of the Lauraceae distinguished from the Perseoideae by the two-celled anthers, dehiscing by two valves

laurustinus horticultural name for the caprifoliaceous shrub *Viburnum tinus*

Lauxaniidae a family of small myodarian cycloraphous dipterans the larvae of which are leaf miners

lavanga the rutaceous woody vine *Luvunga scandens*

lavender (*see also* cotton) properly the small labiateous shrub *Lavendula vera,* but applied to many plants of similar habit or appearance

cut-leaved lavender *Lavendula multifida*

hollow-leaved lavender any of several species of sarraceniaceous insectivorous herbs of the genus Sarracenia

sea lavender *U.S.* any of several species of plumbaginaceous herbs of the genus Limonium

laver any of several rhodophycean algae, particularly of the genus Porphya

green laver any of several ulvate chlorophyte algea of the genus Ulva

Lavernidae = Cosmopterygidae

laverock *N.Brit.* = lark

lavrock = laverock

lavy *Brit. dial.* = guillemot

law a statement of biological principle that appears to be without exception at the time that it is made. The terms "rule" and "effect" are, in contemporary biological literature used, almost interchangeably with "law." Many of the laws here cited should more properly be called hypotheses or theories

Allee's law any given habitat has an optimal population level for any given species

Allen's law in poikilothermic animals, the relative size of appendages diminishes in a colder environment

Arber's law any structure disappearing from the phylogenetic line in the course of evolution is never again shown by descendants of that line

Barfuth's law the axis of the regenerated tail will be perpendicular to oblique cuts made on amphibian tails

Bergman's law the average size of individuals of a given species is larger in those inhabiting colder climates than those inhabiting warmer ones

biogenetic law the concept of Haeckel that "ontogeny recapitulates phylogeny"

Fraser-Darling law the relative number of breeding individuals and young in a population of birds increases as the size of the population increases. There is a corresponding shortening of the breeding season

disharmony law the logarithm of a dimension of any part of an animal is proportional to the logarithm of the dimension of the whole animal

Dollo's law structures lost in the course of evolution can never be regained (*cf.* Arber's Law) but structures gained may none the less be lost

Dyar's law the increase in head width between successive instars of Lepidoptera show a regular geometrical progression

Galton's law the primitive genetic concept that the characters of an individual are half innate, the other half being derived equally from each parent. Hence, a sixteenth would be derived from each grandparent, a sixty-fourth from each great-grandparent and so on

Gause's law two species occupying the same ecological niche cannot survive in the same geographic location

Gloger's law pigmentation is related to climate; cold reduces melanins and warmth and humidity increases melanins, but yellow and brown pigments predominate in dry climates

Haeckel's law = biogenetic law

Haldane's law in those F_1 hybrids in which one sex does not occur, or at least is sterile, that sex is the heterogametic

Hardy-Weinberg law the genetic constitution of a population tends to remain stable, even in the absence of selection, since the frequency of the pairs of allelic genes is an expansion of a binomial equation

Hofacker and Sadler's law an older male parent crossed with a younger female will produce more male offspring than a young male parent crossed to an older female

homonym law a rule of nomenclature which requires that two different taxa cannot have the same name

Hopkin's law spring, and its accompanying biological phenomena, in the eastern U.S. is four days later for each degree of northward latitude, five degrees of eastward longitude and four hundred feet of elevation

Jordan's law 1 the nearest relatives of the species are found immediately adjacent to it but isolated from it by a barrier

Jordan's law 2 fish of a given species develop more vertebrae in a cold environment than in a warm one

Knight-Darwin law the postulate that "nature abhors perpetual self-fertilization"

Krogh's law the rate of a biological process is directly correlated to the temperature

Leibig's law the minimal requirement, in respect to some specific factor, is the ultimate determinate in controlling the distribution or survival of a species

Mendel's first law (law of segregation) there are pairs of factors in a sexual organisms which are segregated in the parent but reunited in offspring

Mendel's second law (laws of independent assortment) pairs of factors segregate from each other independently. Mendel also stated, but did not promulgate as a law, that characters may be dominant or recessive

Murphy's law the bills of birds dwelling on islands are longer than those of the same species inhabiting the mainland

priority law the proper name of an organism is that, subject to certain modifications of the international rules, is the first under which it was described

Rensch's Laws in cold climates the races of birds have larger clutches of eggs and mammals larger litters than races of the corresponding species in warmer climates; in warmer climates birds have shorter wings and mammals have shorter fur; in snails, races of land snails in cold climates have brown shells and those in hot have white shells; the thickness of the shell is positively associated with strong sunlight and arid conditions

Sachs's law a cell wall always tends to set itself at right angles to another cell wall

Sewall Wright's law = genetic drift

Waller's law the axis cylinder of a nerve fiber remains intact only so long as it is in connection with its parent cell. It degenerates along its sheath beyond the point where its connection with the parent cell is severed

Weber-Fechner law sensation intensity is a linear function of the logarithm of the stimulus intensity of a sense organ

law of recapitulation = Baer's law

law of toleration distribution of organisms is limited by their tolerance of the fluctuation of a single factor

lawyer any organism thought to be stiff in its manner or ponderous in its appearance or to have any other character from time to time associated in the popular mind with practitioners of the law; the American usage of this word for the gadid fish *Lota lota* (= burbot) is not clear

lax in *biol.* used in the sense of "loose"

layer one, of several, sheet-like structures

abscission layer the tissue separating the leaf from the stem

argenteal layer a whitish layer of calcium salts along the outer edge of the retina of some fishes

basidial layer that portion of the mycelium in fungi from which basidia arise

prickle cell layer a layer of cells immediately above the Malphighian layer in the skin

closing layer a periodically produced layer of firm cells under a lenticel which is for most of the time surrounded by loose complementary cells

comidial layer the fringe-like outer layer of the cuticle of some cestodes

corneagen layer a layer of transparent cells immediately underlying the cornea of the insect eye

cornified layer the outer layer of the epidermis in adult mammals

dictyogenous layer the root meristem of monocotyledons

discontinuity layer a term sometimes used for

thermocline when the conditions do not meet the rigid definition of this word

fibre layer a surface layer in some fungi

germ layer the three layers (ectoderm, mesoderm, endoderm) into which most animal embryos differentiate

germinative layer (of skin) = Malphighian layer

gonidial layer properly applied to a layer of gonidiophores, but widely used for the algal layer in a lichen thallus (*cf.* gonohyphema)

growth layer a discrete layer, produced in one period by a form or organ showing periodic growth (*cf.* ring)

Henle's layer the outer of the three layers of which the internal hair sheath is composed

Huxley's layer the middle of the three layers of which the internal hair sheath is composed

hymenial layer = hymenium

lentigen layer = corneagen layer

lignified layer that part of the abscission layer which remains attached to the branch

Malpighian layer the basal layer of the epidermis

mantle layer a layer of tapetal cells

marginal layer an envelope of longitudinal nerve fibers around the outside of the spinal cord

palisade layer a layer of one or more tiers of columnar cells containing chloroplasts forming the adaxial portion of the mesophyll of a leaf

papillary layer the outer layer of the dermis

piliferous layer that which produces root hairs

inner plexiform layer that layer of the retina in which the axons of the bipolar cells synapse with the dendrites of the ganglion cells

outer plexiform layer that layer of the retina in which the axons of the rods and the cones synapse with the dendrites of the bipolar cells

prismatic layer the layer that surrounds the xylem cylinder in Isoetes

protective layer the layer immediately on each side of the separation layer in an abscission zone

reticular layer the layer of the dermis immediately underneath the papillary layer

separation layer the region in the abscission zone where separation occurs

spore layer a layer of spore mother cells in certain fungi

Lc *see* [Lurcher] or [Few fruit locules]

ld *see* [loboid]

le *see* [lemon] or [light ears]. This symbol is also used as a yeast genetic marker indicating a requirement for leucine

le-1 *see* [Ascospore lethal]

leaden a mutant (gene symbol *ln*) in linkage group XIII of the mouse. The phenotypic expression is a dull silvery-grey pelt produced by the aggregation of pigment granules

leaf 1 (*see also* leaf 2, 3, bird, cladode, crown, cup, lacunae, love, -phyll-, scraper, trace) a lateral appendage born on the stem of plants

binate leaf one of two leaflets at the end of one petiole

climax leaf the most extensively developed, but not of necessity the oldest, leaf

compound leaf one which possesses several blades

decompound leaf (*bot.*) leaves twice or more compounded as some ferns and legumes

dew leaf one which is sloped and cupped so as to collect dew

digitate leaf compound leaf in which all the leaflets are borne on the apex of the petiole

peltatodigitate leaf a digitate leaf with a swollen end to the petiole on which the leaflets are inserted

bifacial leaf one in which palisade parenchyma is found only on one side, usually regarded as the adaxial or ventral side (*see explanation under entry* ventral)

floral leaf = bract

foliage leaf typical leaf, not metamorphosed for other functions

quadrifoliate leaf one in which four leaflets arise from the end of a petiole, also called **quadrifoliolate leaf**

hastate leaf one having the form of an arrow head

lanceolate hastate leaf a hastate leaf with the main lobe lanceolate

bijugate leaf a pinnate leaf with two pairs of leaflets

sejugate leaf a pinnate leaf having six pairs of leaflets

paucijugate leaf a pinnate leaf with few leaflets

quadrijugate leaf one which has four pairs of leaflets

lanceolate leaf one which is narrow, tapering at both ends

lanceovate leaf one which is too narrow to be called ovate, and too broad to be called lanceolate

isolateral leaf one in which palisade parenchyma is present on both sides (*cf.* bifacial leaf)

palmatilobate leaf one which resembles a hand in shape

pinnatilobate leaf one having lobes arranged in a pinnate manner

pedatilobate leaf palmate, with supplementary lobes at the base

nepionic leaf the first seedleaf

angulinerved leaf with the veins coming off the midrib at an angle

pinninerved leaf with parallel veins springing from a central rib and running directly to the margin (*see also* venation)

nest leaf a leaf with a base adapted for the accumulation of humus

conjugate-palmate leaf a leaf having two separate arms, each palmate

parabolic leaf an oblong-ovate leaf becoming thinner towards the tip

bipinnatiparted leaf = bipinnatifid leaf

pedate leaf a palmately divided leaf with the lateral divisions two-cleft

peltate leaf a leaf attached by its abaxial (*i.e.* lower) surface, rather than by its base, to a petiole

pinnate leaf with leaflets arranged on each side of a common petiole

abruptly pinnate leaf a pinnate leaf which ends in a pair of leaflets

alternipinnate leaf a pinnate leaf in which the leaflets are not opposite

bipinnate leaf one in which both the primary and secondary divisions are pinnate

decreasingly pinnate leaf one in which the terminal leaflets are smallest

decursively pinnate leaf one in which the leaflets are prolonged beyond their insertion

digitate pinnate leaf a digitate leaf with pinnate leaflets

equally pinnate leaf one with no terminal leaflet

even-pinnate leaf = abruptly pinnate leaf

interruptedly pinnate leaf one which lacks a terminal leaflet

oddly pinnate leaf = imparipinnate leaf

paripinnate leaf a pinnate leaf with no terminal leaflets

imparipinnate leaf a pinnate leaf with an odd terminal leaflet

ternate pinnate leaf when the secondary petioles arise in threes from the summit of the main petiole

bipinnatifid leaf when the divisions of a pinnatifid leaf are themselves pinnatifid

bipentaphyllous leaf one having from two to five leaflets

polytomous leaf a leaf so cut, as to resemble a pinnate leaf, but lacking a central petiole

primordial leaf the first leaf formed after the cotyledons

quadrinate leaf a digitate leaf having four leaflets arising from the petiole

rain leaf one which is adapted to shed rain from its surface

root leaf one springing from a subterranean portion of the stem

rough leaf the first leaf to be produced on a stem after the cotyledon

sagittate leaf one in the shape of an arrow head

lanceolate saggitate leaf a sagittate leaf with the middle lobe lanceolate

scale leaf cataphyllary leaf

seedleaf = cotyledon

sessile leaf one lacking a petiole

simple leaf one with a single blade

amphistomatous leaf one with stomata on both surfaces

biternate leaf = compound ternate leaf

water leaf (*see also* leaf 2) the submerged type of leaf in those aquatic plants exhibiting dimorphophyll

leaf 2 in compound names of plants

blood leaf *U.S.* the amaranthaceous herb *Iresine rhizomatosa*

broadleaf *N.Z.* the cornaceous tree *Griselinia littoralis* (= papaumu)

copper leaf any of several species of euphorbiaceous herbs or shrubs of the genus Acalypha

Mexican flame-leaf the euphorbiaceous shrub *Euphorbia pulcherrima* (= poinsettia)

lace leaf the aponogetonaceous aquatic herb *Aponogeton fenestralis* (= lattice leaf)

lattice leaf *U.S.* any of several species of the orchidaceous genus Goodyera

leather leaf any of several species of shrub of the ericaceous genus Chamaedaphne

letter leaf the enormous orchid *Grammatophyllum speciosum*, the largest orchid known

liver leaf any of several species of the ranunculaceous genus Hepatica, particularly *H. americana*

long leaf either the pinaceous tree *Pinus palustris* or the fern *Phyllitis scolopendrium* (= hart's tongue)

oyster leaf the boraginaceous herb *Mertensia maritima* (= lungwort)

painted leaf *U.S.* the euphorbiaceous herb *Euphorbia heterophylla*

quiver leaf the salicaceous herb *Populus tremuloides*

satin leaf the sapotaceous tree *Chrysophyllum oliviforme*

shin leaf any of several species of pyrolaceous herbs of the genus Pyrola

stick leaf *U.S.* the loasaceous herb *Mentzelia oligosperma*

sweet leaf any of several species of plant of the

symplocaceous genus Symplocos, particularly the tree *S. tinctoria*

twin leaf *U.S.* the berberidaceous herb *Jeffersonia diphylla*

umbrella leaf the berberidaceous herb *Diphylleia cymosa*

velvet leaf *U.S.* the malvaceous herb *Abutilon theophrasti*

dorsiventral leaf = bifacial leaf

walking leaf = walking fern

water leaf (*see also* leaf 1) any of several species of the hydrophyllaceous aquatic genus Hydrophyllum

leaf 3 (*see also* leaf 1, 2, beetle, bird, bug, crumpler, fish, folder, hopper, roller, tier) in compound names of animals

autumn leaf *I.P.* the nymphalid lepidopteran insect *Doleschallia bisaltide*

dry leaf *S.Afr.* the nymphalid lepidopteran insect *Precis tugela*

oakleaf *I.P.* any of several species of leaf mimicing lepidopteran insect of the genus Kallima

[potato leaf] a mutant (gene symbol *c*) mapped at 0 on chromosome 6 of the tomato. The name is descriptive of the phenotypic expression on the adult plant

leafage = foliage

leaflet (*see also* leaf 1) each separate blade of a compound leaf

opposite pinnate leaflets those in the same plane at right angles to the common petiole

leap term for a group of leopards (= lepe)

leash a term for three, though sometimes used of a larger group, of hounds and hawks used in hunting, particularly three greyhounds running in concert (*cf.* brace)

lebia aquarists' name for the cyprinodont teleost fishes genus Aphanius

lebrancho the percomorph mugilid fish *Mugil brasiliensis*

lecama = hartebeest

Lecanicephaloidea an order of eucestoid Cestoda distinguished by the presence of a scolex with four acetabula and found only in the intestines of elasmobranch fish

Lecaniinae a subfamily of coccid homopteran insects commonly called tortoise scale-insects or, rarely, wax scale-insects. The name derives from the cleft triangular scales, or plates, covering the adult

lechwe either of two species and several geographic races of reduncine antelopids of the genus Onotragus, particularly *O. leche*

-lecith- *comb. form* meaning "yolk"

lecithal pertaining to yolk

alecithal the term sometimes inaccurately used for isolecithal

mesolecithal = medialecithal

centrolecithal said of a heavily yolked egg, such as that of an insect, in which the yolk is concentrated at the center

medialecithal said of an egg with a considerable amount of yolk but still capable of holoblastic cleavage

megalecithal = telolecithal

oligolecithal = said of an egg with a small amount of yolk

telolecithal said of an egg containing such large quantities of yolk that it is only capable of meroblastic cleavage

lecotropal horseshoe-shaped

Lecythidaceae a large family of myrtiflorous dicotyledons containing *inter alia* the Brazil nut. The woody capsule and the hard-shelled nut, of most members of the family, is typical

leech popular name of hirudinean Annelida
  medicinal leech *Hirudo medicinalis*
  oyster leech the acotyl polyclad turbellarian *Stylochus frontalis*
leek 1 (*see also* leek 2, 3) any of several liliaceous potherbs of the genus Allium, without qualification always *A. porrum*
  lily leek *A. moly*
  wild leek *U.S. A. tricoccum* (= ramp)
leek 2 (*see also* leek 1, 3) any of several plants neither potherbs nor Alliums
  houseleek any of several species of crassulaceous herbs of the genus Sempervivum but particularly *S. tectorum* and its horticultural varieties
    cobweb houseleek *S. arachnoideum*
    spiderweb houseleek = cobweb houseleek
leek 3 (*see also* leek, 1, 2) in compound names of animals
  greenleek *Austral.* the psittacid bird *Colytelis barreanda*
leg 1 (*see also* leg 2, 3) a limb designed for walking or, frequently in insects, modified for other specialized functions
  anal leg that on the tenth abdominal segment of a lepidopteran larva
  false leg = proleg
  proleg an unjointed arthropod appendage used in locomotion, as are the abdominal appendages of many caterpillars
leg 2 (*see also* leg 1, 3) in compound names of animals
  puffleg any of several apodiform trochilid birds of the genera Eriocnemis and Haplophaedia
  red leg the testudinid chelonian reptile *Clemmys insculpta* (wood turtle)
  woolly legs *S. Afr.* either of two species of lycaenid lepidopteran insect of the genus Lachnocnema, usually without qualification, *L. bibulus*
  yellowlegs any of several charadriiform scolopacid birds mostly of the genus Totanus
    bastard yellowlegs *U.S. Micropalama himantopus* (= stilt sandpiper)
    greater yellowlegs *T. melanoleucus*
    lesser yellowlegs *T. flavipes*
leg 3 (*see also* leg 1, 2) in compound names of plants
  tangle legs the caprifoliaceous shrub *Viburnum alnifolium*
leguan = iguana
legume a fruit consisting of a shell-like pod containing a number of seeds. The pod splits in two halves as it ripens. Also the popular name for the family Leguminoseae, having such fruits and also the seed of a leguminous plant
Leguminosae a large family of dicotyledons clearly distinguished from all other families by the leguminous fruits, many of which are of economic importance
lehua the myrtaceous tree *Metrosideros polymorpha*
-leimi- *comb. form* meaning "meadow" used by ecologists in the sense of a "wet meadow"
Leiodidae a small family of beetles occuring under decaying bark. They are frequently fused with the Silphidae
Leiognathidae a family of acanthopterygian fishes distinguished by their laterally compressed bodies and the remarkable protrusible mouths. Many house symbiotic luminescent bacteria

Leiotrichidae a family of passeriform birds usually included in the Timaliidae
leipoa the galliform megapodiid bird *Leipoa ocellata* (= mallee-fowl)
Leitneriaceae that family of leitneriale dicotyledons containing the form commonly called the corkwood, readily distinguished from other catkin-bearing plants by the erect pistillate spikes
Leitneriales a monotypic order of dicotyledonous angiosperms. The characters of the family are those of the order
lem see [lemon]
-lemma- *comb. form* meaning "bark" or "rind" and, by extension, "sheath"
lemma any sheath or coating but, without qualification, specifically the lower bract of a grass floret (*cf.* palea)
  eneilemma the inner skin of a seed
  neurolemma the protoplasmic sheath of nerve fiber
  plasmolemma fertilization membrane
  sarcolemma the sheath of a striated muscle fiber
lemmal pertaining to a sheath
  epilemmal literally, "above the sheath" specifically used of that part of a motor nerve ending which lies outside the sarcolemma
  hypolemmal literally "under the sheath," specifically used of that part of a motor nerve ending which lies immediately under the sarcolemma
  intralemmal literally "that which lies inside the sheath" specifically used of the extreme tip of a motor nerve ending which lies within the sarcolemma
lemming *U.S.* any of numerous species of cricetid rodents
  banded lemming *Cuniculus torquatus*
  brown lemming *Lemmus sibiricus*
  collared lemming any of several species of the genus Dicrostonyx
  Scandinavian lemming *Lemmus lemmus*
Lemnaceae that family of spathiflorous monocotyledons which contains the duckweeds. The duckweeds are clearly distinguished by their small size and habit of growth as well as by the rarely occurring minute flowers born in pits on the edge or upper surface of the leaf
-lemnisc- *comb. form* meaning "ribbon"
lemniscate ribbon-like
lemniscus one of a pair of ribbon-shaped extensions of the hypodermis extending backwards from the neck of Acanthocephala into the body cavity
lemon (*see also* mint, lily) properly, the rutaceous tree *Citrus limonia* and its fruit but applied to some other aromatic plants
  garden lemon a horticultural variety of musk melon (*q.v.*)
  water lemon the passifloraceous vine *Passiflora laurifolia* and its edible fruit
[lemon] *either* a mutant (gene symbol *le*) in linkage group I of *Habrobracon juglandis*. The name refers to the phenotypic body color *or* a mutant (gene symbol *lem*) mapped at 22.8 on linkage group III of *Bombyx mori*. The name refers to the skin color of the phenotypic larva
Lemoniidae = Riodinidae
lemur popular name for members of the mammalian suborder Lemuroidea and order Dermoptera
  flying lemur popular name for members of the mammalian order Dermoptera
  gentle lemur any of several lemurids of the genus Hapalemur

mouse lemur any of several lemurids of the genus Microcebus

ruffed lemur the lemurid *Lemur varius*

sportive lemur any of several small lemurids of the genus Lepilemur

bat-tailed lemur either of two species of lemurids of the genus Opolemur

weasel-lemur any of several lemurids of the genus Lepilemur

Lemuridae that family of primates which contains all the lemuroids except the aye -aye and tarsiers

Lemuroidea a suborder of primate mammals containing the lemurs and their allies. They are distinguished by their fox-like muzzle, large eyes and crepuscular habit

lena the cyrillaceous tree *Cyrilla racemiflora* (= firewood)

-lend- *comb. form* meaning "nit," *i.e.* the egg of the head louse

lendigelous nit-like

lenitic pertaining to rapidly flowing waters (*cf.* lentic, lotic)

Lennoaceae a small family of parasitic, chlorophyll-lacking, tubifloral, dicotyledonous, angiosperms

Lennoineae a suborder of tubulifloral dicotyledonous angiosperms containing the single family Lennoaceae

lens a transparent lenticular mass in the eye, designed to produce images. In *bot.* any object resembling a biconvex lens in shape

-lent- *comb. form* meaning a "lentil" and, by extension, first a lentil-shaped (plano-convex) lens and, finally, lenses in general

Lentibulariaceae that family of tubuliflorous dicotyledons which contains *inter alia,* the bladderworts. Almost all members of this family are insectivorous, aquatic, plants with a distinctive insect-trapping bladder

lentic pertaining to standing water (*cf.* lenitic, lotic)

lenticel a lenticular area of plant periderm in which the cells are not suberized and in which there are many intercellular spaces. They are thought to serve for gas exchange

lenticular in the shape of a biconvex lens (*but see* -lent-)

lentil the leguminous vine *Lens esculenta* and its edible seeds

leopard 1 (*see also* leopard 2, 3, bane, eel, plant) any of three species of fissipede carnivoran mammal of the genus Panthera (to which Felis is preferred by many). Without qualification, invariably *P. pardus.* The cheetah (*q.v.*) is sometimes called the **hunting leopard**

clouded leopard *P. nebulosa*

snow-leopard *P. uncia*

leopard 2 (*see also* leopard 1, 3, seal) other spotted carnivores

sea-leopard properly the leopard-seal (*q.v.*) but sometimes also applied to Weddell's seal (*Leptonychotes weddelli*)

leopard 3 (*see also* leopard 1, 2, moth) any of several nymphalid lepidopteran insects. *I.P.* and *Austral.* any of several species of the genus Atella, particularly *A. phalanta* and (*S. Afr.*) *A. columbina.* In *I.P.* the term is applied to *Neurosigma doubledayi*

Lepadomorpha a division of thorace cirripedes containing those forms which possess a stalk and are for the most part called goose-barnacles

lepal a nectary produced by modification of a stamen

lepe *see* leap

chorisolepideus the condition when the scales of the compositaceous flower are distinct from one another

-lepido- *comb. form* meaning "scale"

lepidomorium a minute ovoid denticle, frequently growing by concrescence, found in the skin of primitive elasmobranch fishes

Lepidophyta a phylum of pteridophyte plants containing the forms commonly called club-mosses. The name derives from the club shaped strobili which are typical

Lepidoptera that order of insects containing the forms commonly called butterflies and moths. They are distinguished from all insects except Trichoptera by their scaly wings and from this group by the absence of biting mouth parts (*cf.* macrolepidoptera, microlepidoptera)

Lepidosauria a subclass of reptilian chordates distinguished by the double, lateral, temporal fenestrae and palatal teeth

Lepidosirenidae a monotypic family of dipneustian fishes containing the South American lung fish *Lepidosiren paradoxa.* It is distinguished from the Protopteridae by the possession of five gill arches and four gill clefts

Lepidosteidae a very frequent misspelling of Lepisosteidae

lepidote scaly

alepidote lacking scales

dilepidous two scaled

polylepidous many scaled

lepisma a membranous scale enclosing the ovary in some Ranunculaceae and that is derived from a stamen

Lepismatidae that family of thysanuran insects which contains the forms commonly called silverfish

Lepisosteidae the only family of the osteichthian order Ginglymodi distinguished by the armor-plate-like ganoid scales and a long snout used in predation

-lepo- *comb. form* meaning "scale"

Leporidae a family of Lagomorpha containing the rabbits and hares

leporine having the form of a hare

leprous applied to insects with loose irregular scales

-lepsi- *comb. form* meaning "receive" derived from the *Gr. leptes* (*cf.* lepto)

amphilepsis the condition of a hybrid when the phenotypic characters derived from both parents are equally apparent

monolepsis the condition of a hybrid showing the characters of only one parent

-lept- *see* -lepto-

Leptaleinae a subfamily of Formicidae containing very slender ants that live in small colonies in cavities in plants. Most are tropical

proleptic precocious

Leptinidae a small family of coleopteran insects which live either in the nests of small mammals or more rarely bumblebees. Commonly called mammal-nest beetles

-lepto- *comb. form* of hopelessly varied meaning resulting from the confusion of two real and one imaginary *Gr.* words. *Leptos* means "small" or any of the attributes of smallness, such as "weakness" or "thinness." *Leptes* means an individual, or institution which "receives" or "accumulates." The meaning "solid" as in the word leptom (*q.v.*) has no etymological justification

Leptocardii a division, sometimes referred to as a subphylum of the acraniate chordates, containing the single class Amphioxi. The characteristics of the division are those of the division

Leptoceratidae a family of brachyceran dipteran insects commonly included in the Sphaeroceridae

Leptoceridae a family of trichopteran insects called the long-horned caddisflies since the antennae are often twice as long as the body. The larvae make conical cases of twigs or sand grains

Leptodactylidae a subfamily of ranid anuran Amphibia distinguished by the absence of parotid glands, the arcuferal pectoral girdle and unexpanded sacral diapophyses

leptom the soft-walled parts of phloem (*cf.* hadrom)
  perileptomatic surrounding the leptome
  protoleptome = protophloem

Leptomedusae an order or suborder of hydrozoan coelenterates, now properly referred to the Campanulariae (*q.v.*)

Leptomitales an order of biflagellate phycomycete fungi distinguished by the possession of a mycelial thallus with a distinct holdfast

Leptonidae a small family of pelecypod Mollusca frequently called kelly shells. They are distinguished by the small smooth equal-sized shells

Leptopsyllidae = Hystrichopsyllidae

Leptosomatidae a monospecific family of coraciiform birds containing the Madagascan cuckoo-roller

Leptosporangiatae a class of pterophyte plants containing those ferns in which the sporangia develop individually from a single cell and possess a jacket one cell in thickness

Leptostraca a subdivision of malacostracan Crustacea uniquely possessing eight abdominal segments (including the telson). All are marine

Leptotyphlopoidea an assemblage of serpentes no longer accepted as a valid taxon

Lernaeopodida an order of parasitic copepod Crustacea in which all trace of segmentation is lost from the adult body. The most readily recognizable copepod affinity are the larval stages and the presence of paired egg masses often of considerable length

lerot any of several murine rodents of the genus Eliomys

lespedeza a genus of leguminous herbs and shrubs commonly called the bush clovers

Lestidae a family of zygopteran odonatan insects usually distinguished by the fact that the wings are held in a V-position when at rest

-let- *comb. term* meaning "small" or "juvenile" *e.g.* owlet

letch *Brit. obs.* a pond

lethal said of anything that produces death

[lethal] *either* a general term for lethal mutants (gene symbols *l(1)* for X chromosome, *l(2)* for chromosome 2, *l(3)* for chromosome III and *l(4)* for IV in *Drosophila melanogaster, or* a mutant (gene symbol *xl*) on the sex chromosome (linkage group I), of the domestic fowl. The phenotypic expression is early death *or* a mutant (gene symbol *l*) mapped at 24.3 on linkage group I of the rat

Lethrinidae a small family of acanthopterygian fishes distinguished by the absence of scales on the head

letondol *hort. var.* of the musaceous tree-like herb *Musa paradisiaca* (= chotda banana)

lettuce properly any of several species and numerous horticultural varieties of compositaceous herbs of the genus Lactuca, particularly *L. sativa*, but also extended to a few other plants
  lamb's lettuce any of several species of herbs of the valerianaceous genus Valerianella but particularly *V. olitoria* (= corn salad)

ivy-leaved lettuce *Brit. L. muralis*

mountain lettuce the saxifragaceous herb *Saxifraga micranthidifolia*

water lettuce the araceous aquatic herb *Pistia stratiotes*

white lettuce the compositaceous herb *Prenanthes alba* (= rattlesnake root)

leu a bacterial genetic marker indicating a requirement for leucine and affecting the activity of three known enzymes and an operator of *Salmonella typhomurium*. One has been mapped at 0.5 mins. for *Escherichia coli*

leu (37501) *see* [Leucine (37501)]

leu-1, 3, 4 *see* [Leucine-1], [-3], [-4]

-leuc- *see* -leuco-

Leucastereae a tribe of the family Nyctaginaceae distinguished by the hairy ovaries

L-leucine α-aminoisocaproic acid. $(CH_3)_2CHCH_2 \cdot CH(NH_2)COOH$. An amino acid isomeric with isoleucine and necessary for the nutrition of rats
  isoleucine α-amino-β-methylvaleric acid $CH_3CH_2 \cdot CH(CH_3)CH(NH_2)\text{-}COOH$. An amino acid, necessary to the nutrition of rats and an isomer of L-leucine
  norleucine α-amino-caproic acid. $CH_3(CH_2)_3CH \cdot (NH_2)CO_2H$. An amino acid not essential to rat nutrition. It is isomeric with leucine

[leucine-1] a mutant (gene symbol *leu-1*) found in linkage group III of *Neurospora crassa*. The phenotypic expression is a requirement for leucine

[leucine-3] a mutant (gene symbol *leu-3*) in linkage group I of *Neurospora crassa*. The phenotype expression is a requirement for leucine

[leucine-4] a mutant (gene symbol *leu-4*) in linkage group I of *Neurospora crassa*. The phenotypic expression is identical with [Leucine-3]

[leucine (37501)] a mutant (gene symbol *leu(37501)*) in linkage group IV of *Neurospora crassa*. The phenotypic expression is a requirement for leucine

leucite a term at one time applied to colorless organelles and inclusions
  clasileucite that portion of the cytoplasm of a cell which is physically differentiated in mitosis
  directing leucite = centrosome
  hydroleucite a fluid-containing vacuole in protoplasm
  tinoleucite = centrosome

leuco- *comb. form* meaning "white," properly, but rarely, transliterated -leuko-

leuconoid a grade of sponge structure in which the choanocytes are in many small chambers that communicate with the central chamber through canals (*cf.* asconoid, syconoid, sylleibid)

Leucopsidae a small family of chalcidoid apocritan hymenopteran insects readily distinguished as the only chalcidoids possessing a longitudinal fold in the forewings. Many are marked with transverse bands of brown and yellow and strongly resemble small vespid wasps

leucosin a carbohydrate occurring as granules in some Protozoa

Leucospididae = Leucospidae

Leucotrichaceae a family of beggiatoale schizomycetes in the form of short cylindrical cells arranged in unbranched, nonmotile trichomes attached by a holdfast to substrates in the environment of decomposing algae. Like most other Beggiatoales, they differ from Cyanophaceae only in the lack of photosynthetic pigments

leucous whitish

   erileucous lustrous white

Leuctridae a family of plecopteran insects lacking an apical crossvein in the wing and having tent-shaped wings when at rest

lev a bacterial mutant showing a utilization of levulose

-levat- *comb. form* meaning "to lift up"

levator applied to muscles which raise a part

leveret a young hare

levers *Brit. obs.* any of several species of the iridaceous genus Iris

leviathan a biblical beast best known as a whale (Psalm 104:25–26) though also used (Psalm 74:14) for the crocodile

levigatous slippery

-levo- *comb. form* properly spelled -laev- (*q.v.*) and usually used in the sense of "to the left"

Leydig's *epon. adj.* from Franz von Leydig (1821–1908)

lg *see* [light green foliage]

lg₁ *see* [liguleless]

Lg *see* [Large]

Lg₃ *see* [Liguleless]

LH = luteinizing hormone

li *see* [lineate]

Li *see* [light down]

liʳ a yeast genetic marker indicating growth inhibitor by lithium

liana any climbing plant that has roots in the ground. Most frequently used of woody tropical climbing vines

Libellulidae a family of anisopteran odonatan insects without a notch on the hind margin of the eyes and with the anal loop in the hind wing foot-shaped

liber *obs.* = bast

Libytheidae a family of nymphaloid lepidopteran insects commonly called snout butterflies in virtue of their long projecting labial palps

lice *pl.* of louse but used in several compound names of plants having small seeds or fruits

   beggar's lice any of several species of boraginaceous herb of the genus Cynoglossum or of the rosaceous genus Agrimonia

   harvest lice any of several species of rosaceous herb of the genus Agrimonia

lichen (*see also* moth) popular name of members of the order Lichenes (*q.v.*)

   homoeolichen one with gonidia evenly distributed throughout the thallus

   leaf-lichen one with a leaf-like thallus

   letter-lichen one with letter-like markings on its surface

   pseudolichen one which is parasitic on another lichen

lichenase an enzyme catalyzing the breakdown of lichenin through the hydrolysis of β-1, 4-glucan links

Lichenes a plant taxon containing the forms commonly called lichens which are an association between a fungus and an alga

lichi the sapindaceous tree *Litchi chinensis* and its fruit

licinula diminutive of lacinia (*q.v.*). Applies specifically to the inflexed point of petals in some cruciferous plants

licorice the leguminous plant *Glycyrrhiza glabra*

   hedgehog-licorice *G. echinata*

   wild licorice the rubiaceous herbs *Galium circaezans* and *G. lanceolatum*

lid a moveable closure

   eyelid a retractile flap of skin covering the eyes of many animals

   germinal lid that portion of a pollen-grain through which the pollen tube comes out

Lieberkuhn's *epon. adj.* from Johann Nathaniel Lieberkuhn (1711–1756)

lienal pertaining to the spleen

-ligul- *comb. form* meaning a little tongue or strap

life an evanescent phenomenon dependent for its continued existence, and perpetuation, on cyclic enzymatic reactions in an environment consisting principally of protein and water

life-of-man the araliaceous herb *Aralia racemosa* (= spikenard)

ligament specifically a mass of white fibrous connective tissues holding the articulations of bones together but used also for any structure that supports an organ or parts of an organ including the springy chitinous tissue which serves to open the valves of pelecypod Mollusca

   falciform ligament the term used to describe the ventral mesentery (ligament) connecting the liver with the ventral body wall

   Treitz's ligament the suspensor of the mammalian duodenum

adligant attached with bonds

ligas the anacardiaceous tree *Semecarpus perrottetii*

ligase ligases, also called synthetases, are enzymes which catalyze the combination of two molecules using energy derived from the breakdown of phosphate bond usually in ATP. Most of the trivial names are synthetases save for the carboxylases

   acetyl ligases a group of enzymes more commonly known by their trivial name of synthetases, which attach CoA to an acid or an ester with energy derived from the breakdown of ATP. The names are all typical (*e.g.* acetyl -CoA synthetase catalyzes the production of acetyl-CoA from acetate and CoA), and they are not listed separately

   aminoacid-RNA ligases a group of enzymes, more commonly known by their trivial names of synthetases, which synthetize the attachment of RNA to aminoacids using energy derived from the change of ATP to AMP. The names are all completely descriptive (*e.g.* leucyl-sRNA synthetase catalyzes the production of L-leucyl-sRNA from L-leucine) and they are not listed separately

[light] a mutant (gene symbol *lt*) mapped at 55.0– on chromosome II of *Drosophila melanogaster*. The phenotypic expression is a yellowish pink eye

[Light down] a mutant (gene symbol *Li*) on the sex chromosome (linkage group I) of the domestic fowl. The name is descriptive of the down color of newly hatched chicks

[light ears] a mutant (gene symbol *le*) found in linkage group III of the mouse. The phenotypic expression is a dilution of the color of the coat

[light green foliage] a mutant (gene symbol *lg*) mapped at 8 on chromosome 10 of the tomato. The phenotypic foliage is light green

scarlet lightning the caryophyllaceous herb *Lychnis chalcedonica*

[lightoid] a mutant (gene symbol *ltd*) mapped at 56.0± on chromosome II of *Drosophila melanogaster*. The phenotypic expression is a translucent pink eye color

lights usually means "lungs" but sometimes extended to other viscera

ligneous woody

Lignosae an assemblage of the Dicotyledoneae erected to contain those families the members of which are predominantly trees and shrubs. (*cf.* Herbaceae) lignum vitae usually the zygophyllaceous tree *Guaiacum officinale* and its lumber but used in *Austral.* for numerous other hardwoods

bastard lignum vitae *G. sanctum*

ligula literally a little tongue, including in the original Latin usage the tongue of shoes; specifically applied to the terminal lobes of the insect labium

aligular on the side opposed to the ligule

ligulate strap-shaped or having a ligule

ligule used of almost any ligulate structure including a fleshy prominence on the notopodium of polychaete Annelida, a small outgrowth round the adaxial surface of the leaf of lycopodiale Lycopsida, a small leaf-like structure at the junction of the blade and petiole of a leaf, and the small tongue-shaped corolla of compositaceous flowers

[liguleless] a mutant (gene symbol lg_1) mapped at 11 in linkage group II of *Zea mays*. The name is descriptive of the leaf of the phenotype

[Liguleless] a mutant (gene symbol Lg_3) mapped at 38 on linkage group III of *Zea mays*. The name is descriptive of the leaf of the phenotype

Liguliflorae a subfamily of Compositae containing the tribe Cichorieae distinguished from all the other Compositae (in the subfamily Tubiflorae) by having all ligulate flowers and anastomizing lactiferous vessels

likh *I.P.* the otitid bird *Sypheotides indica* (= lesser florican)

lilac (*see also* borer) any of numerous species and very many horticultural varieties and hybrids of the oleaceous genus Syringa

wild lilac *U.S.* the rhamnaceous shrub *Ceanothus sanguineus*

Lilaeaceae (not Liliaceae) that family of helobe monocotyledonous angiosperms which contains the forms commonly called flowering quillworts. They are most readily distinguished from the Scheuchzeriaceae, with which they were once joined, by having fibrous roots without tubers

Liliaceae an enormous family of liliiflorous monocotyledonous angiosperms containing the lilies, tulips, trilliums, asparagus, the onions, and many other forms. They are distinguished from all other liliiflorous families, except the Juncaceae, by the possession of a superior ovary and from the Juncaceae by the general habit of growth and large showy flowers

Liliales = Liliiflorae

Lilliflorae a large order of monocotyledons containing not only the true lilies but also, *inter alia,* the rushes, the iris and the yams. All are distinguished by possessing trimerous flowers with a biseriate perianth usually undifferentiated into calyx or corolla

Liliineae a suborder of Liliiflorae containing all the families except the Juncaceae and the Iridaceae

lily 1 (*see also* lily 2, 3, 4, thrip, trotter, weevil) strictly, a liliaceous herb of the genus Lilium

annunciation lily = Madonna lily

golden-banded lily *L. auratum*

orange bell lily *L. grayi*

Bourbon lily = madonna lily

candlestick lily *L. dauricum*

Turk's cap lily properly *L. michauxii,* but also loosely applied to any lily having strongly recurved petals such as *L. superbum* (= swamp lily) and some others

American Turk's cap lily = swamp lily

European Turk's cap lily *L. martagon*

golden Turk's cap lily *L. hansonie*

Japanese Turk's cap lily = golden Turk's cap lily

little Turk's cap lily *L. pontonium*

scarlet Turk's cap lily *L. chalcedonicum*

yellow Turk's cap lily *L. pyrenaicum*

Carolina lily *L. michauxii*

coral lily *L. tenuifolium*

Japanese Easter lily = trumpet lily

fern-leaved lily = coral lily

Lent lily (*see also* lily 2) = Madonna lily

leopard lily *L. pardalinum*

Madonna lily *L. candidum*

meadow lily *U.S. L. canadense*

Nankeen lily *L. testaceum*

orange lily *L. bulbiferum*

panther lily = leopard lily

pine lily *L. catesbaei*

golden-ray lily = golden-banded lily

wild red lily *L. canadense*

wild orange-red lily *L. philadelphicum*

southern red lily *L. catesbaei*

St. Joseph's lily = madonna lily

showy lily *L. speciosum*

spotted lily *U.S. L. medeoloides*

star lily *L. concolor*

swamp lily *L. superbum*

southern swamp lily *L. carolinianum*

tiger lily *L. tigrinum*

tiny lily = coral lily

Tom Thumb lily = coral lily

trumpet lily *L. longiflorum*

turban lily = European Turk's cap lily

wheel lily = spotted lily

long-tube white lily = trumpet lily

wood lily (*see also* lily 2) *L. philadelphicum* (= wild red lily)

speckled wood lily *U.S.* the liliaceous herb *Clintonia umbellulata*

wild yellow lily *L. canadense* (= meadow lily)

lily 2 (*see also* lily 1, 3, 4) any of very many monocotyledonous herbs, mostly of the families Liliaceae and Amaryllidaceae, but not of the genus Lilium

African lily the liliaceous herb *Agapanthus umbellatus*

Amazon lily any of several amaryllidaceous herbs of the genus Eucharis

arum lily the araceous herb *Richardia africana*

Atamasco lily the amaryllidaceous herb *Zephyrantes atamasco* (*cf.* fairy lily)

Barbados lily the amaryllidaceous herb *Hippeastrum equestre*

belladonna lily the amaryllidaceous herb *Amaryllis belladonna*

blackberry lily the iridaceous herb *Belamcanda chinensis*

blood lily any of several species of the amaryllidaceous genus Haemanthus

blueberd lily *U.S.* the liliaceous herb *Clintonia borealis* (= corn lily)

Brisbane lily the amaryllidaceous herb *Eurycles sylvestris*

butterfly lily any of several herbs of the zingiberaceous genus Hedychium

calla lily the araceous herb *Zantedeschia aethiopica*

celestial lily popular name of the iridaceous genus Nemastylis

checkered lily the liliaceous herb *Fritillaria meleagris*

climbing lily the liliaceous climbing vine *Gloriosa superba*

corn lily *U.S.* the liliaceous herb *Clintonia borealis*

Cuban lily the liliaceous herb *Scilla peruviana*

day lily any of several species and very numerous horticultural varieties of the liliaceous genus Hemerocallis, though the term is locally applied to some species of the genus Funkia

desert lily any of several species of the liliaceous genus Eremurus

fairy lily any of many species of amaryllidaceous herb of the genus Zephyranthes

fawn lily any of several species of liliaceous herb of the genus Erythronium

ginger lily any of several herbs of the zingiberaceous genus Hedychium

ground lily any of many species of liliaceous herb of the genus Trillium (= wake-robin)

Guernsey lily the amaryllidaceous herb *Nerine farmiensis*

jacobean lily the amaryllidaceous herb *Sprekelia formosissima*

Kaffir lily any of several species of the amaryllidaceous genus Clivia

lemon lily the liliaceous herb *Hemerocallis flava*

Lent lily (*see also* lily 1) the amaryllidaceous herb *Narcissus pseudo-narcissus* (= trumpet daffodil)

liver lily the iridaceous herb *Iris versicolor*

mariposa lily any of numerous species of the liliaceous genus Calochortus

white mountain lily the liliaceous herb *Leucocrinum montanum*

plantain lily any of numerous species of liliaceous herb of the genus Funkia

plume lily the liliaceous herb *Smilacina racemosa*

giant prairie lily the amaryllidaceous herb *Cooperia pedunculata*

queen lily the zingiberaceous herb *Curcuma petiolata*

rain lily the amaryllidaceous herb *Cooperia drummondii*

rengarenga lily *N.Z.* the liliaceous herb *Arthropodium cirratum*

rush lily any of several iridaceous herbs of the genus Sisyrinchium (= blue-eyed grass)

St. Bernard's lily the liliaceous herb *Anthericum liliago*

St. Bruno's lily the liliaceous herb *Paradisea liliastrum*

sand lily the liliaceous herb *Leucocrinum montanum*

Scarborough lily the amaryllidaceous herb *Vallota purpurea*

sego lily the liliaceous herb *Calochortus nuttallii* (*cf.* mariposa lily)

Solomon's lily the araceous herb *Arum palaestimum*

spider lily any of several species of herb of the amaryllidaceous genus Hymenocallis

golden spider lily the amaryllidaceous herb *Lycoris aurea*

sword lily any of several species and numerous horticultural hybrids of the iridaceous genus Gladiolus or (*Austral.*) the amaryllidaceous genus Anigozanthos

toad lily any of several species of liliaceous herb of the genus Tricyrtis

torch lily the liliaceous herb *Kniphofia uvaria*

trout lily *U.S.* the liliaceous herb *Erythronium americanum*

white wood lily (*see also* lily 1) any of several species

of liliaceous herb of the genus Trillium (= wake-robin)

zephyr lily any of numerous species of the amaryllidaceous genus Zephyranthes

lily 3 (*see also* lily 1, 2, 4) a few dicotyledonous herbs with lily-like flowers

cow lily *U.S.* any of several species of the nymphaeaceous aquatic herbs of the genus Nuphar (=yellow pond lily) but particularly *N. advena* (=spatter-dock)

bull-head lily the nymphaeaceous herb *Nuphar variegatum*

impala lily the African apocynaceous shrub *Adenium multiflorum*

lotus lily the nymphaeaceous water plant *Nymphaea odorata*

yellow pond lily any of several species of nymphaeaceous aquatic herbs of the genus Nuphar. *Brit.* without qualification, *N. luteum*

water lily (*see also* aphis, beetle) any of numerous species and very numerous horticultural hybrid aquatic herbs of the nymphaeaceous genus Nymphaea

Cape blue water lily *N. capensis*

royal water lily the giant nymphaeaceous aquatic *Victoria regia*

lily 4 (*see also* lily 1, 2, 3) in compound names of lily-shaped animals

sea-lily popular name of crinoid echinodermata

lily-of-the-Nile either the araceous plant *Zantedeschia aethiopica* (= calla) or the liliaceous herb *Agapanthus umbellatus*

lily-of-the-palace the amaryllidacean herb *Hippeastrum aulicum*

lily-of-the-valley the liliaceous plant *Convallaria majalis* and its numerous horticultural varieities

false lily-of-the-valley *U.S.* the lilaceous herb *Maianthemum canadense*

lima = lima bean

Chickasaw lima the leguminous creeper *Canavalia ensiformis*

-limac- *comb. form* meaning slug

limacine pertaining to, or resembling, slugs

Limacodidae a family of zygaenoid lepidopteran insects distinguished by their short, fleshy and slug-like larvae which gives to the family the name slug-caterpillar

limb 1 (*see also* limb 2, borer) *bot.* the side branch of a tree as distinct from the trunk, the lamina of a leaf or petal as distinct from the petiole or stalk and the expanded border of a gamopetalous corolla as distinct from the tube

limb 2 (*see also* limb 1) a jointed locomotory appendage

limbus literally a border and specifically the border round the basilar membrane of the ear in lizards and the constriction between the columnar body and base of some zoantharian coelenterates

lime 1 (*see also* lime 2, 3) the rutaceous tree *Citrus aurantifolia* and its fruit

lime 2 (*see also* lime 1, 3, berry, butterfly) any of several species of tiliaceous tree of the genus Tilia. (*cf.* linden, bass wood, white wood)

large-leaved lime *T. platyphyllos*

lime 3 (*see also* lime 1, 2) in compound names of other plants

brooklime *Brit.* the scrophulariaceous herb *Veronica beccabunga* (*cf.* speedwell)

American brooklime *V. americana*

finger lime any of several species of rutaceous tree of the genus Microcitrus

Russell River lime the rutaceous tree *Microcitrus inodora*

Spanish lime the sapindaceous tree *Melicocca bijuga*

limer a hunting dog kept on the leash until the quarry is at bay

Limidae a family of pelecypod mollusks commonly called file-shells. They are distinguished by the obliquely oval winged shells

[limited] a mutant (gene symbol *lm*) mapped at 50.0± on chromosome II of *Drosophila melanogaster*. The phenotypic expression is small sternites

-limn- *comb. form* meaning "lake" frequently confused with -limno-

limnad a lake plant

Limnanthaceae a family of sapindale dicotyledonous angiosperms sometimes united to the Geraniaceae but with ovules in reversed position

Limnephilidae the largest family of trichopteran insects containing the great majority of those occurring in the northern hemisphere and hence called northern caddisflies

limnetic pertaining to lakes

bathylimnetic the condition of plants which may either float or be rooted

halolimnetic pertaining to salt lakes, though the term is sometimes incorrectly applied to the marine environment

proschairlimnetic the condition of a plant which occasionally turns up as a constituent of limnoplankton

tycholimnetic fortuitously limnetic

halolimnetic said of typically marine organisms capable of surviving in fresh waters

limnion a horizontal division of the waters of a lake

epilimnion the upper region of a thermally stratified lake

hypolimnion the lower zone of a thermally stratified lake. This non-circulating layer is usually depleted of oxygen

mesolimnion the middle layer of a thermally stratified lake

metalimnion = discontinuity layer

-limno- *comb. form* meaning "marshy" frequently confused with -limn-

Limnobiidae = Limoniidae

limnocrene a spring feeding a pool with no outflow

Limoniidae a family of nematoceran Diptera now usually included with the Tipulidae

limonile deeply subterranean

limpet any of numerous uncoiled conical gastropod mollusks. The term limpet is properly applied to the families Acmaeidae and Lepetidae though it is often also extended to the Fissurellidae, and when modified, to many other groups

cup-and-saucer limpet popular name of calyptraeids, usually without qualification, *Crucibulum striatum*

false limpet popular name of siphonarids, usually, without qualification, *Siphonaria normalis*

key-hole limpet popular name of fissurellids (*cf.* volcano shells)

slipper limpet any of several of the crepidulids, particularly *Crepidula fornicata* (*cf.* boat shell)

limpkin the aramid gruiform bird *Aramus guarauna*

limu any Hawaiian edible marine alga

Linaceae that family of geranealous dicotyledons which contains, *inter alia*, the flaxes. The family is most readily distinguished from the other geranealous

plants by the symmetrical pattern of five that applies to all the parts of the flower

linacel the internal shell of limacid gastropods

linaloa lumber from the burseraceous tree *Bursera alexylon*

boblincoln = bobolink

Port **Lincoln** the psittacid bird *Barnardius zonarius* (= twenty-eight parrot)

linden (*see also* borer, looper) any of several species of tiliaceous tree of the genus Tilia. (*cf.* lime, basswood, whitewood)

American linden *T. americanum*

common linden *T. vulgaris*

Crimean linden *T. euchlora*

large-leaved linden *T. platyphyllos*

small-leaved linden *T. cordata*

white linden *T. tomentosa*

weeping white linden *T. petiolaris*

line 1 (*see also* line 2–5) a distinct thread-like structure, or linear group of structures

lateral line in fish, a chain of lateral sense organs found in cyclostomes, fishes and aquatic amphibians In insects, a line running along the side of caterpillars and some other larvae

luminous line the Malphighian cells on the outside of seeds

Schreger's line a refractive line caused by the spiraling of enamel rods in the teeth

line 2 (*see also* line 1, 3–5) a narrow area of demarcation or the structure forming the demarcation

abscission line a point of future breakage for organs that are shed, such as leaves and horns

Baillarger's line a thickened layer of tangential fibers along the boundary between the superradial and interradial networks in the cortex of the forebrain

cervical line the line of junction between the enamel and cementum in a mammalian tooth

contact line = parastichie

Gennari's line = Baillarger's line

Retzius line the growth lines of the enamel on the tooth

Vicz d'Azyr's line that part of Baillarger's line which abuts on the cuneus

Wallace's line the imaginary line that separates the Australasian from the Asiatic faunas; the important part of the line runs north-northeast between the island of Bali and Lombok and then between Borneo and Celebes; Bali and Borneo thus have an Asiatic fauna while Celebes and Lombok represent the limits of the Australasian

z-line a line which appears to bisect an I-disc. (*q.v.*)

line 3 (*see also* line 1, 2, 4, 5) in the sense of lineage

pure line = clone

line 4 (*see also* line 1–3, 5) any of several noctuid lepidopteran insects

double line *Brit. Leucania turca*

silver lines *Brit.* either of two species of the genus Hylophila

treble line *Brit. Meristis trigrammica*

line 5 (*see also* line 1–4) an obsolete measure of length equivalent to one twelfth of an inch

lineate lined

[lineate] a mutant (gene symbol *li*) mapped at 28 on linkage group X of *Zea mays*. The phenotypic leaves have fine longitudinal stripes

lineolate delicately lined

ling 1 (*see also* line 2, -ling and cod) any of several fish *Brit.* the marine gadoid *Molva molva*. *U.S.* the fresh water gadoid *Lota lota* (= burbot). *Can.*

marine gadoids of the genus Urophycis (*cf.* Lake). *N.Z.* the perciform ophidiidid *Genypterus blacodes*.

ling 2 (*see also* line 1 and -ling) any of several plants. *Brit.* the ericaceous shrub *Calluna vulgaris* (= heather). *U.S.* and *Brit.* the sedge *Eleocharis tuberosa* (= Chinese waterchestnut)

-ling *comb. term* meaning "juvenile" *e.g.* codling and by extension appended to many names of small forms

bitterling the cyprinid fish *Rhodeus sericeus*

rockling any of several species of gadid fishes particularly those of the genus Enchelyopus

lingen = ling berry

linget *Brit.* the linaceous herb *Linum usitatissimun* (= flax)

lingo the fabaceous tree *Lingoum indicum*

-lingu- *comb. form* meaning "tongue" (*cf.* -ligul-)

lingua tongue. In certain insects the maxilla or hypopharynx especially when modified for lapping or sucking

sublingua a horny pointed or serrated plate found beneath the tongue of, and peculiar to, the Lemuroidea

superlingua (insecta) = paragnatha (insecta)

lingual pertaining to the tongue and surface of oral structures facing the tongue

acutilingual with a sharp-pointed tongue

lingula literally, a little tongue, and so loosely used in biology as to be almost meaningless

lingy pertaining to heather

linin a term at one time applied to chromatin (*q.v.*) reduced to an apparently thread-like form by acid fixatives

unlining (*bot.*) the separation of parts originally joined along the line marking the point of juncture

linkage the type of inheritance exhibited by genes located in the same chromosome and therefore having a tendency to remain together in passing from one generation to the next

autosomal linkage the linkage of alleles on the same autosome

Linnaean *epon. adj.* from Karl von Linné (1707–1778) Karl's father Nils, as a peasant had no heritable last name and assumed Linnaeus, after a large linden tree that stood by his house, when he entered the university to study for the ministry. Karl continued to use this, rather than the conventional Nilsson, and derived von Linné from this when he was enobled in 1761. Thus "Linnaeus" and "von Linné" are both correct but "Linne" often seen, is an error

linneanon = linneon

linneon superspecies

linnet *U.S.* the fringillid bird *Carpodacus mexicanus* (= house finch). *Brit.* the fringillid bird *Cardualis cannabina*

green linnet *Brit.* = greenfinch

mountain linnet *Brit.* the fringillid bird *Cardualis flavirostris* (= twite)

Linognathidae a family of anopluran insects commonly called smooth sucking lice

linsang viverrine carnivorous mammals of the genus Prionodon and their immediate allies. They are the Asiatic equivalents of the African genets

lintie *Brit. obs.* = linnet

lion 1 (*see also* lion 2, 3, beetle, monkey, tail) properly the felid carnivorous fissipede mammal *Felis leo* but locally extended to other felids

American lion = puma

mountain lion *U.S.* = American lion

lion 2 (*see also* lion 1, 3, beetle, monkey, tail) various other carnivores

sea-lion any of several species of otarid pinnepede Carnivora mostly of the genus Otaria

Aukland Island sea-lion *O. hookeri*

California sea-lion *Zalophus californicus*

Cape sea-lion *O. pusilla*

northern sea-lion *U.S.* = steller sea-lion

Patagonian sea-lion *Eumetopias jubata*

Steller's sea-lion *Eumetopias stelleri*

lion 3 (*see also* lion 1, 2, beetle, monkey, tail) various other organisms of predatory habit

ant-lion any of several species of planipennid insects, the larvae of which trap ants in funnel-shaped excavations in loose soil

spotted-winged ant-lion *U.S. Dendroleon obsoletus*

aphis-lion *Brit.* popular name of hemerobiid neuropteran insects (*cf.* brown lacewing). *U.S.* often also used of chrysopid neuropterans (*cf.* aphis-wolf)

lioness a female lion, though also applied in *U.S.* to the female of the bobcat and the cougar

Liopelmidae a family of primitive anuran Amphibia once placed with the Discoglossidae but distinguished from them and all other extant anurans by the presence of free ribs in the adult, and by the undivided intervertebral bodies

Liotheidae = Menoponidae

Liotrichidae a family of passeriform birds usually included in the Timaliidae

-lip- *comb. form* meaning "fat"

lip 1 (*see also* lip 2) the free edge of any opening particularly the fleshy flap of skin lying anterior to the teeth in mammals, the area immediately dorsal to the yolk plug in the late amphibian gastrula and the lobes surrounding the mouth in nematode worms. In both *zool.* and *bot.* the word is used interchangeably with labium (*q.v.*)

lip 2 (*see also* lip 1, fern) in compound names of organisms

blue lip the scrophulariaceous flowering plant *Collinsia grandiflora*

oxlip *Brit.* the primulaceous herb *Primula elatior*

sweet lip any of several Indo-Australian pomadasyid fish

Lipariidae a family of scaleless scleroparous fishes with the pelvic fins modified into sucking discs. Frequently called snailfishes. The family is frequently fused with the Cyclopteridae

Liparidae a family of noctuoid lepidopteran insects commonly called tussock moths

lipase an enzyme which catalyzes the hydrolysis of triglyceride into diglycerides and a fatty acid

phospholipase A catalyzes the hydrolysis of lecithin into lysolecithin and an unsaturated fatty acid

phospholipase B = lysophospholipase

phospholipase C catalyzes the hydrolysis of phosphatidyl-choline to yield a 1,2-diglyceride and choline phosphate

phospholipase D catalyzes the hydrolysis of phosphatidyl-choline to yield choline and a phosphatidic acid

lysophospholipase catalyzes the hydrolysis of lysolecithin to glycerolphosphocholine and a fatty acid

Liphistiomorphae a suborder of aranean arthropods distinguished from all other spiders by possessing a segmented abdomen

-lipo- *comb. form* meaning "to depart from" frequently confused with -lip-

lipoid a fat, or fat-like substance

Lipoptera = Mallophaga

liquefaction used specifically by microbiologists to indicate the hydrolysis of a gel medium by a microorganism

colliquescent to become fluid, through some action other than the direct absorption of water (*cf.* deliquescent 1)

deliquescent 1 (*see also* deliquescent 2) to become liquid through the absorption of water (*cf.* colliquescent)

deliquescent 2 (*see also* deliquescent 1) profusely branching so that the original stem is lost ("dissolved") in the numerous branches

liquorice (*Brit.*) = licorice

lirate said of molluscan shells having a thread-like ridge

lirella an elongated apothecium found in lichens

Liropeidae = Ptychopteridae

-liss- *comb. form* meaning "smooth"

Lissocarpaceae a small family of ebenale dicotyledons

litchi = lichi

literate said of an animal or plant ornamented with markings resembling letters

-lith- *comb. form* meaning "stone"

coccolith one of variously shaped pieces of calcium carbonate imbedded in the surface of some marine flagellates

cystolith an outgrowth from the cellulose wall of a plant cell, impregnated with calcium carbonate

otolith a solid particle within a balancing organ

statolith a mineral mass acting on the hairs of the neurosensory cells in statocysts (*cf.* statoconium)

lithi the anacardiaceous shrub *Lithraea caustica*

lithic pertaining to rocks

epilithic growing on rocks

hypolithic pertaining to organisms living under stones

lithistid a network of crepid (*see* crepis) sponge spicules, forming a reticulated skeletal body

Lithophyta a division of the Linnaean group Vermes. It contained the corals and similar forms

litter a group of animals brought forth at one time, usually, but not invariably, from a multiparous mammal, and particularly a carnivore

littling *N. Brit.* one member of a litter (*cf.* sibling)

littoral pertaining to the shore and, in fresh waters, confined to those zones in which rooted vegetation occurs (*cf.* riparian)

elittoral the extension of the shoreline to the point under water where photosynthesis is no longer possible

psammolittoral a sandy shore

sublittoral pertaining to areas adjacent to the littoral zone. In fresh waters the sublittoral zone commences where rooted vegetation ceases

Littorinidae a very large family of littoral gastropod Mollusca commonly called periwinkles. They are distinguished by their sturdy shell of few whorls and dull color

live-forever the crassulaceous herb *Sedum telephium*

liver (*see also* leaf) an organ developed as an outgrowth of the alimentary canal primarily concerned with the storage and metabolism of foodstuffs

livid plumbaceous

liza the mugilid fish *Mugil liza*

lizard 1 (lizard 2) a term commonly applied to any lacertilian reptile having legs. The family Lacertidae or even the genus Lacerta is sometimes referred to as containing the "true lizards"

alligator lizard *U.S.* any of several species of anguids of the genus Gerrhonotus. It is unusual for an anguid in having well developed legs

armadillo lizard the cordylid *Cordylus cataphractus*

bearded lizard *Austral.* the agamid *Amphibolurus barbatus*

caiman lizard either of the two species of teiid genus Dracaena

collared lizard *U.S.* either of two species of iguanid of the genus Crotaphytus

common lizard *Brit.* the lacertid *Lacerta vivipara*. *N.Z.* the scinid *Lygosoma moco*

earless lizard *U.S.* any of several iguanids of the genera Holbrookia and Cophosaurus

emerald lizard *Brit.* = green lizard *Brit.*

eyed lizard *Brit. Lacerta ocellata*

fence lizard *U.S.* without qualification the iguanid *Sceloporus undulatus* (*cf.* spiny lizard)

flap-footed lizard *Austral.* popular name of pygopodids

frilled lizard *Austral.* the agamid *Chylamydosaurus kingi*

glass lizard *U.S.* any of several species of limbless anguid reptiles of the genus Ophisaurus

bunch-grass lizard the iguanid *Sceloporus scalaris*

green lizard *Brit. Lacerta viridis*. *N.Z.* the geckonid *Naultinus grayi*

lyre-headed lizard the agamid *Lyriocephalus scutatus*

horned lizard *U.S.* any of several species of iguanid of the genus Phrynosoma

jeweled lizard the lacertid *Lacerta lepida*

mesquite lizard *U.S.* the iguanid *Sceloporus grammicus* (*cf.* spiny lizard)

Arizona night lizard the xantusiid *Xantusia arizonae*

hook-nosed lizard any of several species of agamid of the genus Harpesaurus

plated lizard any of several species of lacertid of the genus Psammodromus or gerrhosaurid of the genus Gerrhosaurus

prairie lizard *U.S.* the iguanid *Sceloporus undulatus* (*cf.* spiny lizard)

ringed lizard popular name of amphisbaenids

rock lizard *N.Z.* the scincid *Lygosoma grande*

rough lizard *N.Z.* the geckonid *Naultinus rudis*

sand lizard *Brit.* the lacertid *Lacerta agilis*

scrub lizard *U.S.* the iguanid *Sceloporus woodi*

Sita's lizard the agamid *Sitana ponticeriana*

snake lizard *S.Afr.* any of several species of cordylid of the genus Chamaesaura. The term is also applied to members of the teiid genus Ophiognomon

spiny lizard any of numerous lizards with projecting scales but particularly iguanids of the genus Sceloporus, and the lacertid *Lacerta echinata*

spotted lizard *N.Z.* the geckonid *Naultinus elegans*

whip-tail lizard any of numerous species of the teiid genus Anemidophorus

fringe-tailed lizard the lacertid *Holaspis guentheri*

girdle-tailed lizard popular name of cordylids

gridiron-tailed lizard *U.S.* the iguanid *Callisaurus draconoides*

spiny tailed lizard *I.P.* any of several species of agamid of the genus Uromastix

zebra-tailed lizard the iguanid *Callisaurus draconoides*

tegu lizard the teiid *Tupinambis nigropunctatus*

fringe-toed lizard any of several species of lacertid of the genus Acanthodactylus

long-toed lizard *N.Z.* the geckonid *Dactylocnemis granulatus*

short-toed lizard *N.Z.* the geckonid *Dactylocnemis maculatus*

tree lizard *U.S.*, without qualification, the iguanid *Urosaurus ornatus* *N.Z.* the geckonid *Dactylocnemis pacificus*

wall lizard *Brit. Lacerta muralis*

water lizard (*see also* lizard 2) any of several species of giant agamids of the genus Hydrosaurus

　whip lizard *S.Afr.* any of several species of gerrhosaurid of the genus Tetradactylus

　worm lizard *U.S.* the amphisbaenid *Rhineura floridana*

lizard 2 (*see also* lizard 1) in error, some urodele amphibians

　ghost lizard *U.S.* the plethodontid urodele amphibian *Typhlotriton spelaeus* (= grotto salamander)

　spring lizard *U.S. dial.* urodele amphibians in general

　water lizard (*see also* lizard 1) *U.S.* any of several aquatic urodele Amphibia particularly Triturus and Necturus

ll *see* [lanceolate]

llama any of several species of *S. Amer.* camelid artiodactyle of the genus Lama. Without qualification, usually *L. huanacos* (= huanaco)

llano a large flat area of land

llanos South American prairie

lliana = liana

llyn a lake

lm *see* [limited]

ln *see* [leaden]

loa an abbreviation used for "length over-all" and frequently printed, as here, without separating periods

loa-loa the African parasitic nematode *Filaria loa*

loach popular name of cobitid, and occasionally other, fishes

　clown loach *Botia macracantha*

　coolie loach *Acanthophthalmus kuhli*

　spatula loach the pimelodid *Sorubim lima*

　spined loach *Cobitis taenia*

　stone loach *Cobitis barbatula*

　striped loach *Acanthophthalmus kuhlii*

loaf-long *Brit.* = live-forever *U.S.*

Loasaceae a small family of parietalous dicotyledons distinguished by the often grotesque flowers and the peculiar hairs, frequently venemous, borne on the stems and leaves

Loasales an order of dicotyledonous angiosperms erected to contain the families Datiscaceae, Begoniaceae, and Ancistrocladaceae, more usually regarded as belonging in the Parietales

Loasineae a suborder of parietale dicotyledons containing the single family Loasaceae

Lobata an order of tentaculate Ctenophora distinguished by the presence of two large oral lobes

lobe any blunt prominence arising from a surface, or any of several parts into which an organ is divided by constrictions or indentations

cephalic lobe the expanded anterior region of a nemertine

frontal lobe the most anterior region of the cerebral hemisphere

limbic lobe a cerebral lobe surrounding the corpus callosum and therefore forming an arch under the medial borders of the cerebral hemisphere in mammals

mantle lobe an extension of the wall of the brachiopod body over the anterior regions of the valve

middle lobe the small tongue projecting between the two lobes of a fern prothallus

occipital lobe the most posterior lobe of the cerebral hemisphere

optic lobe one of a pair of thickened outgrowths from the dorsal wall of the mesencephalon. In mammals, there are two pairs of lobes known as the corpora quadrigemina, the unterior of which are the homologues of optic lobes. In insects, the term is applied to a lateral lobe of the protocerebrum

parietal lobe the lobe of the cerebral hemisphere that lies between the occipital and the frontal

pyriform lobe the backwardly directed lobe arising from the anterior region of the cerebral cortex

temporal lobe that portion of the lower outer part of the cerebral hemisphere that lies below Sylvius' fissure

[Lobe] a mutant (gene symbol *L*) mapped at 72.0 on chromosome II of *Drosophila melanogaster*. The phenotypic expression is small eyes nicked at the anterior edge

Lobelioideae a subfamily of the Campanulaceae sometimes raised to separate familial rank

postlobin-O = oxytocin

　postlobin-V = vasopressin

loblolly any of several pinaceous trees, particularly *P. taeda*

-lobo- *comb. form* properly meaning the "lobe of the ear" but extended to any "lobe"

lobo *U.S.* the canine fissipede carnivorous mammal *Canis lupus* (= timber wolf)

Lobodontinae a subfamily of phocid pinniped carnivores of exclusively antarctic distribution. They are distinguished from the Phocinae by having two incisors in each jaw

[loboid] a mutant (gene symbol *ld*) mapped at 100.0± on chromosome III of *Drosophila melanogaster*. The term applies to the appearance of the phenotypic eye

Lobosa = Amoebida

Lobotidae a small family of acanthopterygian osteichthyes commonly called tripletails. They are easily recognized by their lateral compression and extended dorsal and anal fins

lobster (*see also* flower, moth) a term loosely applied to almost any large, edible, decapod crustacean not obviously a crab. *U.S.*, without qualification, *Homarus americanus*. *Brit. H. vulgaris*

　locust lobster *Brit.* the scyllarid *Scyllarus arctus*

　Norway lobster *Nephrops norvegicus* (= Dublin Bay prawn)

　rock lobster = spiny rock lobster

　spiny rock lobster any of several species of scyllarid crustacea of the genus Palinurus. Without qualification, *Brit. P. vulgaris*, *W.I. P. argus*, *S.Afr. P. lalandii*

lobule diminutive of lobe, specifically (Hepaticae) the minor lobe at the leaf base (= auricle)

germinal localization the specific localization of parts of the embryo in the egg or in the cleaving egg

locellate divided into secondary compartments

locellus literally any small place or compartment but used specifically of the cavity of a pollen sac or a secondary ("false") compartment of a single-celled ovary

loch a lake or (*N.Brit.* and *N.Z.*) a deep inlet of the sea in mountainous districts

lochan diminutive of loch

lochetic said of the activity of a predator lying in ambush

-lochm- *comb. form* meaning a "thicket"

lochmad a plant of thickets

-lochmod- *comb. form* meaning "bushy" but used by

ecologists in the sense of "dry thicket," as distinct from -lochm- used for thickets in general

lociation a modification of an association which, in contrast to a faciation, has been produced by the removal or addition of other than dominants

locies the term is to socies as lociation is to association

lock a tuft

fetlock a posterior projection, usually bearing a tuft of hair, immediately behind the hoof of many perissodactyl mammals

forelock the tuft of hair or crest of a horse lying anterior to the ears

goldylocks the compositaceous herb *Linosyris vulgaris*

marshlocks the rosaceous herb *Potentilla palustris* (*cf.* cinquefoil)

-loco- *comb. form* meaning "place"

locomotion directed movement

locular pertaining, or having cavities

polaribilocular pertaining to a two celled spore having a tube through the connecting wall

multilocular consisting of several chambers

unilocular possessing, or consisting of, a single chamber

locule any small cavity particularly one of several associated cavities, as the cells of a compound plant ovary, or the cavities in the shell of a foraminiferan protozoan

anther locule = pollen sac

loculus a cavity. In *bot.*, particularly that of an ovary or anther

proloculus the first chamber to be formed of a multiloculate foraminiferan shell

locus *gen. jarg.* for place or site at which a gene is located on a chromosome

locust 1 (*see also* locust 2, borer, miner, roller) *U.S.* any of several trees or shrubs of the leguminous genus Robinia

bastard locust the *W.I.* sapindalous tree *Ratonia apetala* (*cf.* bastard mahogany)

black locust *R. pseudacacia*

bristly locust *R. hispida*

clammy locust *R. viscosa*

honey locust any of several species of tree of the leguminous genus Gleditsia and particularly *Gleditsia triacanthos*

swamp locust the leguminous tree *Gleditsia aquatica*

sweet locust *Gleditsia triacanthos* (= honey locust)

water locust *U.S.* the leguminous tree *Gleditsia aquatica*

yellow locust = black locust

locust 2 (*see also* locust 1, borer, miner, roller) a term properly applied to grasshoppers of the families Acrididae and Tetrigidae (*cf.* grasshopper). In *E. U.S.* however, the term is widely applied to the cicada

grouse locust *U.S.* = pygmy locust *U.S.*

Rocky Mountain locust *Melanoplus spretus*

pygmy locust popular name of grasshoppers of the family Tetrigidae. *U.S.* without qualification, usually *Tetrix ornatum*

17 year locust *U.S.* the homopteran *Cicada septendecim* and two closely related species

lodge a communal dwelling place

lodicule a small scale functioning as a subordinate bract (the main bract is a glume) in the inflorescence of a grass

protolog an original written description of a genus or species

Loganiaceae a family of contortous dicotyledons which is distinguished from other families of the order by the two-celled superior ovary. The drugs curare and nux vomica are obtained from members of this family

Loganiales an order of dicotyledonous angiosperms erected to contain the families Loganiaceae and Oleaceae by those who do not hold these to belong in the Contortae

login *U.S.* that portion of a river that runs smooth

-logy *comb. suffix* literally meaning "wordy" but universally used in compound words in the sense of "the study of" or "the science of"

acarology the study of mites

bacteriology the study of bacteria, particularly the practical applications of this study. The academic aspects are more usually called microbiology

biology the study of life and living organisms from any aspect

behavioral biology the study of the behavior and psychological processes of non-human animals

cryobiology literally the biology of cold and originally used for the study of living forms in cold environments. At present it is more commonly used for the study of the effect of superfreezing on living material

developmental biology the study of the development of plants and animals from the broadest aspect as distinct from embryology

environmental biology = ecology

exobiology the theoretical consideration of the postulated existence of life in other parts of the universe

gnotobiology the study of organisms raised in pure culture, that is free from all other forms of life. This therefore includes the study of mammals raised under such sterile conditions that they lack intestinal flora

microbiology the study of those forms, particularly from an academic aspect, that are usually classified as Protista (*cf.* bacteriology)

molecular biology the study *in vitro* of those molecules and molecular reactions that *in vivo* produce the phenomenon called life.

photobiology the study of the utilization of photon energy by living organisms

radiation biology the study of the effects of radiation on living forms, but particularly radiation other than that in the visible spectrum

bryology the study of mosses

conchology the study of shells of Mollusca as distinct from the Mollusca themselves (*cf.* malacology)

cytology the study of the structure of cells

ecology the study of the interrelationships between living organisms and their environment

embryology the study of the stages by which specific organisms develop as distinct from developmental biology which is more concerned with processes than structures

endocrinology the study of the endocrine gland and their products

entomology the study of insects

enzymology the study of enzymes

epiontology concerning the origin of individuals

ethology the study of animal behavior, particularly as it relates to the interactions of more than one species

herpetology the study of amphibia and reptiles

histology the study of the cellular structure of organs

and organisms as distinct from cytology which is the study of the structure of the cells themselves.

ichthyology the study of fish

immunology originally the study of immunity but now extended to cover all chemical phenomena associated with host-parasite relationships

limnology literally the study of lakes but commonly used in the sense of the study of fresh waters

malacology the study of Mollusca as distinct from the study of their shells which is conchology

mammalogy the study of mammals

morphology the study of the shape of organisms and parts of organisms as distinct from structure, the study of which is better called anatomy

mycology the study of fungi

helminthology properly the study of flatworms but usually used in the sense of the study of parasitic forms

ornithology the study of birds

osteology the study of bone

paleoeology the use of fossils for the study of community structure and succession

paleontology the study of extinct organisms

palynology the study of fossil pollens

parasitology the study of parasitic relationships from the biological point of view. The chemical point of view is more commonly called immunology

phenology the study of rhythmic or periodical events as they are influenced by climatic changes

phycology the study of algae

physiology the study of the function of organisms from an aspect so broad that it cannot be defined as being restricted to biophysics or biochemistry

teratology the study of monsters and other abnormalities

virology the study of virus

zoology the study of animals from any aspect, but, if separated from biology, is most commonly used in the sense of morphological, anatomical, or taxonomic studies

loin the dorso-lateral areas of the posterior epaxial and hypaxial musculature and bony structure of mammals, particularly those used for food

Loliginidae a family of dibranchiate cephalopod Mollusca with the fins at the end of the body and a valve on the siphon. Distinguished from the Ommastrephidae in possessing a cornea. Commonly called squid

loma a seasonal prairie in Latin America

loment a leguminous pod which contracts between the seeds and breaks up on drying, onto one-seeded joints

lomma a plate-like lobe or flap

lommatine said of the feet of birds in which the toes have lateral flaps projecting from them

lon a bacterial genetic marker having a character affecting the activity of filament formation and radiation sensitivity. One has been mapped at 11.8 mins. for *Escherichia coli*

Lonchaeidae a small family of minute flat shiny blackish myodarian cycloraphous dipteran insects

Lonchopteridae a family of aschizan cycloraphous brachyceran dipteran insects with spear-shaped wings. Sometimes called spear-winged flies

[long] a mutant (gene symbol *l*) in linkage group I of *Habrobracon juglandis*. Both the antennae and leg segments are longer in the phenotype than in the wild type

longe U.S. the fresh water salmonid teleost fish *Cristivomer namaycush* (= namaycush)

-longi- *comb. form* meaning "long" (*cf.* -macro-)

loon any of several gaviiform birds of the genus Gavia
  Arctic loon G. *arctica* (= black-throated loon)
  yellow billed loon G. *adamsii*
  common loon G. *immer*
  red-throated loon G. *stellata*

ciliary loop an oval loop of ciliated cells on the anterodorsal region of Chaetognatha. It is variously regarded as a chemoreceptor or as an excretory organ

  Henle's loop that portion of a metanephron which runs from Bowman's capsule to the convoluted tubule

looper U.S. the caterpillars of those noctuid and geometrid moths which move with a looping motion. The name looper is sometimes used by entomologists, but rarely by others, to distinguish the larvae of noctuids from the larvae of geometrids (measuring worms)

  alfalfa looper U.S. the larva of the noctuid *Autographia californica*

  cabbage looper U.S. the larva of the noctuid *Trichoplusiani*

  celery looper U.S. the larva of the noctuid *Anagrapha falcifera*

  clover looper U.S. the larva of the noctuid *Caenurgina crassiurcula*

  fir-cone looper U.S. the larva of the geometrid *Eupithecia spermaphaga*

  forage looper U.S. the larva of the noctuid *Caenurgina erechtea*

  grapevine looper U.S. the larva of the geometric *Lygris diversilineata*

  hemlock looper U.S. the larva of the geometrid *Lambdina fiscellaria lugubrosa*

    false hemlock looper U.S. the larva of the geometrid *Nepytia canosaria*

    phantom hemlock looper U.S. the larva of *N. phantasmaria*

    western hemlock looper U.S. the larva of the geometrid *Lambdina fiscellaria*

  hop looper U.S. the larva of the noctuid *Hypena humuli*

  linden looper = linden moth

  western oak looper U.S. the larva of the geometrid *Lambdina fiscellaria somniaria*

[Loop tail] a mutant (gene symbol *Lp*) found in linkage group XIII of the mouse. The name is descriptive of the phenotypic expression

loosestrife any of several species of herbs of the closely related primulaceous genera Lysimachia and Steironema and of the lythraceous genus Lythrum

  hyssop loosestrife the lythraceous herb *Lythrum hyssopifolia*

  ice loosestrife = purple loosestrife

  purple loosestrife the lythraceous herb *Lythrum salicaria*

  swamp loosestrife the lythraceous herb *Decodon verticillatus*

  yellow loosestrife the primulaceous herb *Lysimachia vulgaris*

-loph- *comb. form* meaning "crest," "mane" or "ridge"

lophad a hill plant

Lophiidae a family of pediculatus fishes distinguished by the enormous mouth fringed with flaps of skin. Commonly called goosefishes in the U.S. and anglerfish in G. Brit.

Lophiiformes = Pediculati

Lophobranchii a suborder of solenichthous fishes containing the pipefishes and sea horses. Frequently regarded as a separate order. These fishes are distinguished by the lack of an operculum and the bony surface plates

lophophore *see* -phore

Lophophoria a term coined to designate those animals possessing a lophophore. This taxon therefore comprises the phyla Phoronida, Ectoprocta and Brachiopoda

Lophopoda = Phylactolaemata

Lophotidae a small family of deep water allotriognathous fishes commonly called crestfishes

Lophyridae = Diprionidae

loppe *Brit. obs.* = spider

loquat the rosaceous tree *Eriobotrya japonica*

Loranthaceae that family of santalalous dicotyledons which contains the mistletoes. Almost all of the family are aerial parasites and are additionally distinguished by the cup-shaped receptacle and the one-celled, one-seeded fruits

Loranthineae a suborder of Santalales containing the single family Loranthaceae

Loranthoideae a subfamily of Loranthaceae distinguished from the Viscoideae by the presence of a calyculus below the perianth

lorate strap-shaped

lorchel popular name of helvellacean fungi of the genus Helvella (= false morel)

Irish **lord** any of several species of cottid fish of the genus Hemilepidotus

lords-and-ladies *Brit.* the araceous herb *Arum maculatum*

lore one of two anterolateral areas of a bird's head that lie between the eye and the bill

loreal an elongate scale lying between the nares and the eye in snakes

lorica a hardened or thickened cuticle forming a shell or case. There is no clear distinction between cuticle, lorica and exoskeleton in invertebrates. In *bot.* the term is applied to both frustles of a diatom taken together

Loricariidae a family of *S.Amer.* armored catfish completely covered with overlapping plates

Loricata (Mammalia) an order of placental mammals, or a suborder of Xenarthra, which contains the armadillos. This group is distinguished by the transverse plates of fused hairs which cover the body

Loricata (Reptilia) an obsolete order once containing the extant crocodiles and alligators

illoricate lacking a lorica

Loriidae = Psittacidae

lorikeet any of numerous species of psittaciform birds of the genera Loriculus, Trichoglossus and some closely related forms

Malay lorikeet *L. galgulus*

loriot *Brit.* the passerine oriolid bird *Oriolus oriolus* (= golden oriole)

loriquet = lorikeet

loris properly a genus of lemurs but usually without qualification refers to *Loris gracilis*. The name is however applied to other lemurs

slender loris *L. gracilis*

slow loris *Nycticebus tardigradus*

Lorisidae a family of lorisoid primate mammals containing the forms commonly called lorises and pottos. They are superficially, but readily, distin-guished from the Galagidae by the small ears and greatly reduced tails

Lorisoidia a suborder of primates erected to contain the families Lorisidae and Galagidae by those who do not hold these to belong in the Lemuroidia. They are principally distinguished from the Lemuroidia by the relatively large number (14 to 16) of dorsal vertebrae

loro the psittaciform bird *Nyiotsitta monachus* (= monk-parrot)

lorum any small chitinous band or process particularly those in the mouth parts of insects. The term is used for so many assorted strap, bridle or whip-shaped structures as to defy definition

lory *Austral.* almost any psittaciform bird of the Australasian region, particularly those of the genera Domicella and Vini *S.Afr.* any of numerous cuculiform birds of the family Musophogidae (= plantain-eaters)

lot = burbot

lotic pertaining to rapidly flowing streams (*cf.* lenitic, lentic)

Lotoideae a subfamily of Leguminosae containing those forms having the typical papilionaceous flower

lotus 1 (*see also* lotus 2, 3 bird) any of numerous Asiatic and African flowering water plants, of the family Nymphaceae

African lotus *Nymphaea lotus*

American lotus *Nelumbo lutea*

African blue lotus *Nymphaea capensis*

Egyptian blue lotus *Nymphaea caerulea*

Indian blue lotus *Nymphaea stellata*

Egyptian lotus = African lotus

Indian lotus *Nelumbo indica*

lotus 2 (*see also* lotus 1, 3) any of numerous leguminous herbs of the family Fabaceae

lotus 3 (*see also* lotus 1, 2) a variety of trees having edible fruit particularly the rhamnaceous *Zizythus lotus*

lough = loch

loulu any palmaceous tree of the genus Pritchardia

lourie any of several species of musophagid birds (= touraco)

louse (*see also* lice) any small parasitic insect, particularly those of the order Anoplura ("true lice") and Mallophaga ("biting lice") Without qualification, usually refers to the human body louse *Pediculus humanus*

bark louse = any outdoor winged corrodentian

bird louse popular name of theose Mallophaga that parasitize birds (*cf.* biting louse)

biting louse popular name for Mallophaga

cattle biting louse *U.S.* the trichodectid *Bovicola bovis*

dog biting louse *U.S.* the trichodectid *Trichodectes canis*

goat biting louse *U.S.* the trichodectid *Bovicola caprae*

angora-goat biting louse *U.S.* the trichodectid *Bovicola limbata*

horse biting louse *U.S.* the trichodectid *Bovicola equi*

sheep biting louse *U.S.* the trichodectid *Bovicola ovis*

human body louse the anopluran pediculid *Pediculus humanus*. (the head louse is usually considered a geographic race)

chicken body louse *U.S.* the menoponid mallophagan *Menacanthus stramineus* goose body louse *U.S.* the menoponid mallophagan *Trinoton anserinum*

book louse common name for members of the order Corrodentia particularly applied to the family Atropidae

cat louse *U.S.* the trichodectid mallophagan *Felicola subrostrata*

long-nosed cattle louse *U.S.* the linognathid anopluran *Linognathus vituli*

   short-nosed cattle louse *U.S.* the haematopinid anopluran *Haematopinus eurysternus*

brown chicken louse *U.S.* the philopterid mallophagan *Goniodes dissimilis*

   large chicken louse *U.S. G. gigas*

clothes louse = body louse

crab louse the phthiriid anopluran *Phthirus pubis*

large duck louse *U.S.* the menoponid mallophagan *Trinoton querquedulae*

   slender duck louse *U.S.* the philopterid mallophagan *Anaticola crassicornis*

elephant louse the haematomzyid mallophagan *Haematomyzus elephantus* sometimes placed in a separate order, the Rhyncothirinida

guinea feather louse *U.S.* the philopterid mallophagan *Goniodes numidae*

   fish louse ectoparasitic copepod Crustacea and, in fresh waters, the closely allied Argulus

fluff louse *U.S.* the philopterid mallophagan *Goniocotes gallinae*

slender goose louse *U.S.* the philopterid mallophagan *Anaticola anseris*

slender guinea louse *U.S.* the philopterid mallophagan *Lipeurus numidae*

human head louse *see* body louse

   chicken head louse *U.S.* the philopterid mallophagan *Cuclotogaster heterographa*

hog louse *U.S.* the haematopinid anopluran *Haematopinus suis*

horse louse the haematopinid anopluran *Haematopinus asini*

human louse either the body louse or the head louse or, rarely, crab louse

pear louse the psylliid homopteran insect *Psylla pyricola*

oval guinea-pig louse *U.S.* the gyropid mallophagan *Gyropus ovalis*

   slender guinea-pig louse *U.S.* the gyropid mallophagan *Gliricola parcelli*

slender pigeon louse *U.S.* the philopterid mallophagan *Columbicola columbae*

   small pigeon louse *U.S.* the philopterid mallophagan *Campanulotes bidentatus*

plant louse a term properly applied to aphid homopterans but sometimes used for other small homopterans

jumping plant louse popular name of psyllid homopterans

potato louse *U.S.* = tomato louse *U.S.*

public louse = crab louse

rabbit louse *U.S.* the hoplopleurid anopluran *Haemodipsus ventricosus*

spined rat louse *U.S.* the hoplopleurid anopluran *Polyplax spinulosa*

   tropical rat louse *U.S.* the hoplopleurid anopluran *Hoplopleura oenomydis*

sea louse any of several species of echinophthiriid anopluran of the genus Echinophthirius

shaft louse *U.S.* the menoponid mallophagan *Menopon gallinae*

dog sucking louse *U.S.* the linognathid anopluran *Linognathus setosus*

goat sucking louse *U.S. L. stenopsis*

horse sucking louse *U.S.* the haematopinid anopluran *Haematopinus asini*

small mammal sucking louse popular name of hoplopleurid mallophagans

smooth sucking louse popular name of homognathid anoplurans

cattle tail louse *U.S.* the haematopinid anopluran *Haematopinus quadripertusus*

tomato louse the psylliid homopteran *Paratrioza cockerelli*

turkey louse the philopterid mallophagan *Virgula meleagridis*

   large turkey louse *U.S.* the philopterid mallophagan *Chelopites meleagridis*

   slender turkey louse *U.S.* the philopterid mallophagan *Oxylipeurus polytrapezius*

whale louse parasitic amphipod Crustacea forming the family Cyamidae of the suborder Caprellidea

wing louse *U.S.* the philopterid mallophagan *Lipeurus caponis*

wood louse *Brit.* any land dwelling oniscoid isopod crustacean though the term pill bug is usually applied to those of the family Armadillidiidae (= *U.S.* sow bug). Also sometimes used for the cassidid gastropod mollusk *Morum oniscus*

louvar the scrombroid luvarid fish *Luvarus imperialis*

lovage *Brit.* any of several species of umbelliferous herb of the genus Ligusticum

leaf love = aphid. The term is also applied to any of several pycnonotid birds of the genus Phyllastrephus, usually, without qualification P. scandens

true-love the liliaceous herb *Paris quadrifolia*

love-entangle the crassulaceous herb *Sedum acre* (= stone crop)

love-in-a-chain *Sedum reflexum*

love-in-a-mist any of several species of the ranunculaceous genus Nigella. *U.S.* the passifloraceous vine *Passiflora foetida*

love-in-idleness *U.S.* the violaceous herb *Viola tricolor* (= wild pansy)

love-in-winter the ericaceous herb *Chimaphila corymbosa*

love-lies-bleeding the amaranthaceous herb *Amaranthus caudatus* and the fumariacean herb *Dicentra spectabilis* (= bleeding heart)

loveman *Brit.* the rubiaceous herb *Galium aparine*

Loven's *epon. adj.* from Sven Ludwig Loven (1809–1895)

mountain lover *U.S.* the celastraceous shrub *Pachistima canbyi*

low a deep sound made by cattle

lowan *Austral.* the galliform megapodiid bird *Leipoa ocellata*

-lox- *comb. form* meaning "slanting"

[lozenge] a pseudo-allelic seriies of mutants (gene symbol *lz*) mapped at 27.7 on the X chromosome of *Drosophila melanogaster*. The phenotypic expression is narrow eyes with abnormally-shaped facets

Lp *see* [Loop tail]

lt *see* [light]

ltd *see* [lightoid]

LTH = lactogenic hormone

lu *see* [luxoid]

lubricous slippery

Lucanidae a family of large coleopteran insects commonly called stagbeetles or pinchingbugs. In many species the male mandibles are half as long as the body. They are closely allied to the Scarabaeidae but differ in having the lamellicorn antenna elbowed

lucban the rutaceous tree *Citrus grandis* (= grapefruit)

luce an obsolete, principally heraldic, term for the teleost fish *Esox lucius* (= pike)

lucerne the leguminous forage plant *Medicago sativa* (= alfalfa)

Lucernariidea = Stauromedusae

lucid shining

luciferase an enzyme which catalyzes the oxidation of luciferin

luciferin a flavin related compound, found in many light producing organisms, which emits light when oxidized

lucine anglicized name of the lucinid pelecypod genus Lucina

Lucinidae a family of pelecypod Mollusca with round and compressed shells

Luciocephalidae a small family of teleost fishes usually included with the Anabantidae

luggar = lagger

Lujanidae = Lutianidae

lulula *S.Afr.* the lycaenid lepidopteran insect *Pseudiolaus poultoni*

lumbang the euphorbiaceous tree *Aleurites moluccana* (= candlenut)

lumbar literally, pertaining to the loins and, especially of vertebrae, pertaining to those which lie between the thoracic and the sacrum

Lumbricidae a family of terrestrial oligochaete annelids commonly called earthworms, though this term is equally well applied to Megascolicidae from which they are distinguished by having the clitellum posterior to the eighteenth segment

lumbricous shaped like an earthworm

Lumbriculidae a small family of minute fresh-water mud-dwelling oligochaete annelids

luminescence self-produced, as distinct from reflected, light

   bioluminescence luminescence produced by living organisms

lumme *Brit.* the podicipediform bird *Colymbus arcticus*

lumper a derogatory term applied by taxonomists to those of their number who tend to throw many small taxa into one large taxon (*cf.* splitter)

lunar pertaining to the moon

lunate moon-shaped, usually meaning crescentic

lunda the charadriiform alcid bird *Fratercula arctica* (= common puffin)

lung 1 (*see also* lung 2, wort) heavily vascularized, usually loculated, sacs into which air for respiration is drawn by terrestrial vertebrates

lung 2 (*see also* lung 1) any device used for gas exchange by animals

   book-lung a series of leaf-like respiratory plates found on the posterodorsal surface of some arachnids, particularly scorpions

lunge = longe

lungie *Brit.* any of several species of birds of charadriiform birds of the alcid genus Cepphus (= guillemots)

lunula the crescentic white area at the base of the nail

lunule a crescent, or new-moon-shaped body. Specifically such an area at the base of the antennae in certain Diptera, or just in front of the umbo on the shells of many pelycepod mollusks

   frontal lunule one round the base of the antenna of some dipterous insects

lunulet diminutive of lunule

lupine any of numerous species of leguminous herbs of the genus Lupinus

   bastard lupine the leguminous herb *Trifolium lupinaster*

   blue lupine *L. hirsutus*

   wild lupine *L. perennis*

lurcher 1 (*see also* lurcher 2) *Brit.* a dog, commonly partially of greyhound ancestry, used by poachers

lurcher 2 (*see also* lurcher 1) *Austral.* any of several species of nymphalid lepidopteran insect of the genus Yoma

[Lurcher] a mutant (gene symbol *Lc*) in linkage group XI of the mouse. The name is descriptive of the gait of the phenotypic expression

lure any device used by a predator to attract its prey. Many fish and some chelonian reptiles use such devices

lurg any polychaete annelid of the genus Nephthys (*cf.* lug worm and lurg worm)

lurid in *bot.*, dingy brown

lutaceous pertaining to mud

luteal pertaining to the corpus luteum

luteous deep yellow

lutescent yellowish

[lutescent-2] a mutant (gene symbol *l₂*) mapped at 79 on chromosome 10 of the tomato. The phenotypic leaves are yellowish

[lutescent foliage] a mutant (gene symbol *l*) mapped at 65 on chromosome 8 of the tomato. The leaves and unripe fruits of the phenotype are yellowish

[luteus seedling] a mutant (gene symbol *l₈*) mapped at 38 on linkage group X of *Zea mays*. The phenotypic expression is a yellow seedling

luth the chelonian dermochelyid reptile *Sphargis coriacea* (= leathery turtle)

Lutianidae a very large family of acanthopterygian osteichthyes commonly called snappers.

Lutrinae a subfamily of mustelid carnivorous mammals containing the otters and their immediate allies. They are all aquatic and one is marine

Luvaridae a monotypic family of scombroid acanthopterygian fish containing the louvar (*Luvarus imperialis*) distinguished by having the vent covered by the ventral fins at the base of the pectoral fins well anterior to the anal fins

[luxate] a mutant (gene symbol *lx*) found in linkage group III of the mouse. The phenotypic expression is indistinguishable from luxoid

[luxoid] a mutant (gene symbol *lu*) found in linkage group II of the mouse. The phenotypic expression is anterior limb polydactyly

lx see [luxate]

ly a yeast genetic marker indicating a requirement for lysine

Ly see [Lyra]

lyase a group of enzymes that catalyze the removal of groups, by other than hydrolysis, in such a manner as to leave a double bond or to add groups to double bonds. Among the better-known trivial names of lyases are decarboxylase, dehydratase, lyase and synthase (*cf.* synthetase)

   alliin lyase catalyzes the production of 2-aminoacrylate and an alkylsulphenate from an S-alkyl-L-cysteine-sulphoxide

   aspartate ammonia-lyase catalyzes the production of fumurate from L-aspartate

methylaspartate ammonia-lyase catalyzes the production of mesaconate and ammonia from L-*threo*-3-methylaspartate

histidine ammonia-lyase catalyzes the production of urocanate and ammonia from L-histidine

isocitrate lyase catalyzes the production of succinate and glyoxylate from isocitrate

lactoyl-glutathione lyase catalyzes the production of glutathione and methylglyoxal from S-lactoyl-glutathione

hyaluronate lyase catalyzes the production of 3(β-D-gluco-3, 4-en-urono)-2-acetylamine-2-deoxy-D-glucose from hyaluronate

hydroxynitrile lyase catalyzes the production of benzaldehyde and HCN from mandelonitrile

adenylosuccinate lyase catalyzes the production of fumarate and AMP from adenylosuccinate

argininosuccinate lyase catalyzes the production of fumarate and arginine from argininosuccinate

Lycaenidae a large family of small lycaenoid lepidopteran insects containing, *inter alia*, the forms commonly called blues, coppers, hair-streaks and (*U.S.*) harvesters

lychnis *Brit.* the noctuid lepidopteran insect *Dianthoecia capsincola* (*cf.* campion)

striped lychnis *Brit.* the noctuid lepidopteran insect *Cucullia lychnitis*

lychnisc a hexatine sponge spicule having a latticework at the point where it joins to another hexatine in a dictyonine mesh

Lycidae a family of soft-bodied coleopteran insects closely allied to the Lampyridae in shape but easily distinguished from them by the network of raised lines on the elytra from which their name net-winged beetle is derived

-lyco- *comb. form* meaning "wolf"

Lycoperdaceae that family of basidiomycete fungi which contains the forms commonly called puffballs. The spherical or ovoid stalkless fruiting head is typical

Lycophen a term coined to designate the Equisetinae, Lycopodinae, Psilotinae and Isoetinae in contrast to the Filicinae

Lycopodiaceae that family of lycopodine Pteridophyta which contains the true club mosses. They are distinguished from the Selaginaceae by the fact that the leaves lack a ligule

Lycopodiales that order of lycopodine Pteridophyta which contains, as its only family, the Lycopodiaceae distinguished by being homosporous and in having leaves without ligules

Lycopodiinae that class of Pteridophyta which contains the club-mosses and the small club-mosses. They are distinguished by the scale-like leaves and biciliate sperm

Lycoriidae = Sciaridae

Lycosidae a family of dipneumonomorph spiders which do not build webs though many line their lairs with silk. They derive their name from their predatory habit and customs and are sometimes called wolf-spiders

Lyctidae a family of coleopteran insects closely allied to the Bostrichidae and frequently fused with them

Lydidae = Pamphilyidae

Lygaeidae a very large family of phytophagous hemipteran insects. They have no popular name as a whole though they contain such well known forms as the chinch bug

-lym- *comb. form* meaning "ruinous"

Lymantriidae = Liparidae

Lymexylidae a considerable family of wood-boring coleopteran insects commonly called ship-timber beetles. They are distinguished by the presence of a short obvious neck and serrate antennae

Lymexylonidae = Lymexylidae

Lymnaeidae a large family of fresh water pulmonate gastropods with dextral shells, commonly called pond-snails (*cf.* Physidae)

lymph literally meaning water and so used as a *comb. form*. More usually refers, either alone or as a *comb. form* to the fluid found in lymph vessels, which is a diluted blood plasma containing a few lymphocytes. In *bot.* is occasionally used for sap

cytolymph protoplasm in the *sol* condition

endolymph the fluid within the semicircular canal

karyolymph nuclear sap

perilymph the fluid between the semicircular canals and their skeletal supports

lynx a term generally applied to the jungle cat, caracals, lynxes and bobcats which are placed in the genus Lynx if this can, indeed, be distinguished from Felis by reason of the tufted ears and heavier hind quarters. *Brit.*, without qualification, usually *L. lynx*. *U.S.*, without qualification, usually *L. canadensis*. *Felis pardina* is sometimes called the **Spanish lynx**

bay lynx *U.S. Lynx rufus* (= bobcat)

lyo a bacterial mutant showing a requirement for lysine. In yeast indicates a lysine petite

Lyomeri an order of osteichthyes containing a few fish commonly called gulperfishes by reason of their enormous mouths. All are abyssal

Lyonetiidae a family of tineiform lepidopteran insects with very narrow wings the hindwing being often linear. The larvae are leaf-miners

Lyonsiidae a family of pelecypod mollusks commonly called papershells, by reason of the fragile pearly valves

lyra the stridulating apparatus of a spider

[Lyra] a mutation (gene symbol *Ly*) mapped at 40.5 on chromosome III of *Drosophila melanogaster*. The phenotypic wings are narrow and cut in at the ends

lyrate shaped like a lyre. Used particularly of irregularly-lobed leaves the tip of which is broader than the base

lyrule a median tooth in the posterior rib of the zoecium of some Ectoprocta

lys-1 to lys 4 *see* [Lysine-1] to [Lysine 4]

Lysianassidae a family of gammaridean amphipod Crustacea with a three-jointed palp on the mandible. Mostly large freshwater forms found in deep lakes

heterohemolysin an antibody that will lyse a blood cell from another individual

L-δ-hydroxylysine α-diamino β-hydroxy M caproic acid. An amino acid not essential to the nutrition of rats

[lysine-1] a mutant (gene symbol *lys-1*) on linkage group V of *Neurospora crassa*. The phenotypic expression is a requirement for lysine

[lysine-2] a mutant (gene symbol *lys-2*) on linkage group V of *Neurospora crassa*. The phenotypic expression is a requirement for lysine

[lysine-3] a mutant (gene symbol *lys-3*) in linkage group I of *Neurospora crassa*

[lysine-4] a mutant (gene symbol *lys-4*) in linkage group I of *Neurospora crassa*. The phenotypic expression is a requirement for lysine

-lys- *comb. form* meaning to "separate" or "loosen" and thus, by extension, to "liquify." The *subs. suffix* -lysis and the *adj. suffix* -lytic do not, in compounds, always have identical meanings

autolysis the degradation after death of cell contents by the contained enzymes

antholysis degeneration of a flower

biolysis death due to or accompanied by lysis

catalysis a change in rate of a reaction

autocatalysis (*bot.*) = self-fertilization

dialysis 1 (*see also* dialysis 2) the separation of large from small molecules by their passage through a membrane of suitable pore size (*cf.* dialytic)

dialysis 2 (*see also* dialysis 1) a separation of parts of a plant usually associated

histolysis literally, the breakdown of cells or cellular structure, but specifically the initial phase of the reorganization of material in, for example, the pupa of an insect or the brown body of an ectoproct

karyolysis the condition of a necrotic nucleus which becomes lysed

plasmolysis the shrinking of plant protoplasm from the cell wall

photolysis the arrangement of chloroplasts within the cell under the action of light

lyssacine said of the type of framework of siliceous sponge spicules that is formed by the interlacing of the elongated rays of hexactines

Lythraceae a large family of myrtiflorous dicotyledons containing the loosestrifes. The family may be distinguished from the Onagraceae by the superior ovary and from the Melastomaceae by the normal stamens

-lyte- *comb. term* indicating a separated part

sarcolyte a fragment of degenerating muscle circulating in the blood, particularly of insects

biolytic pertaining to the destruction of life

dialytic said of the parts of a plant (*e.g.* stamens) which are usually fused but which, in a particular case, are separated (*cf.* dialysis)

ethymolytic said of those echinoid echinoderms in which the genital plates are not completely encircled by the terminal plates (*cf.* ethmopract)

schizolytic applied to gemmae which are detached by splitting of cells

seirolytic pertaining to the separation of heritable characters

lytta a longitudinal cartilaginous rod found in the tongue of many mammals

Lyttidae = Meloidea

lz *see* [lozenge]

m *see* [miniature] [misty], [modifier] and [mottled leaves]

m-1 *see* [Modifier of *vis (3717)*]

m-2 *see* [Modifier of *vis (3717)*]

M *see* [Multiple spurs] and [Tetra molting]

M(1)Bld *see* [Minute (1) Blond]

M(1)n to M(1)36f *see* Minute (1)n to Minute 1(36f)

M(2)1 and M(2)1² *see* [Minute (2)1] and [Minute(2)1²]

M(2)B *see* [Minute (2) Bridges]

M(2)C *see* [Minute (2) Curry]

M(2)D *see* [Minute(2)D]

M(2)e *see* [Minute(2)e]

M(2)S1 to M(2)S13 *see* [Minute(2) Schultz'] to [Minute (2)Schultz' 13]

M(2)z *see* [Minute(2)z]

M(2)33a to M(2)173 *see* [Minute(2)33a] to [Minute(2) 173]

M(3)B *see* [Minute(3) Burkart]

M(3)B² *see* [Minute(3)Bridges]

M(3)be *see* [Minute(3)beta]

M(3)f *see* [Minute (3) f]

M(3)Fla *see* [Minute(3)Florida]

M(3)h–M(3)l *see* [Minute(3)h] to [Minute(3)l]

M(3)S31 to M(3)S39 *see* [Minute(3)Schultz' 31] to [Minute(3)Schultz' 39]

M(3)w and M(3)y *see* [Minute(3)w] and [Minute(3)y]

M(3)124 *see* [Minute(3) 124]

M-4 *see* [Minute-4]

ma *see* [marron] or [marbling]

MA a yeast genetic marker indicating a requirement for maltose

mabuya any of several species of scincid saurian reptile of the genus Mabuya (*cf.* skink)

mac see [Methionine-adenine-cysteine]

macaque a general term for any cynopithecinid primate of the genus Macaca which is not otherwise specifically named

macaw any of numerous brilliantly colored psittaciform birds of the genus Ara *W.I.* the buteonine bird *Buteo jamaicensis* (= red-tailed hawk)

blue-and-yellow macaw *A. ararauna*

scarlet macaw *A. macao*

macaroni the spheniscid bird *Eudyptes chrysolophus.* This designation derives from the use of "macaroni" to describe a well-dressed foppish Londoner of the 18th century

maccaw = macaw

mace the ruminated aril of the nutmeg (*q.v.*)

reed-mace *Brit.* any of several species of aquatics of the typhaceous genus Typha (= cat tail)

Machaerodontidae that family of extinct fissipede carnivores which contains the sabre-toothed tigers

machete the elopid fish *Elops affinis*

Machilidae a family of thysanuran insects commonly called jumping bristle-tails. They are distinguished from the Lepismatidae by the large size of the compound eyes

mackerel (*see also* bird, shark) any of numerous fishes, the majority of the scombrid genus Scomber. *Brit.* without qualification, *Scomber scombrus*

atka mackerel hexagrammid *Pleurogrammus monopterygius*

Atlantic mackerel the *S. scombrus*

chub-mackerel *S. colias*

frigate mackerel any of several species of the genus Auxis. *U.S.* without qualification, usually *A. thazare*

horse-mackerel *Brit.* popular name of carangid fish, without qualification, usually *Caranx trachurus*

king-mackerel *Scomberomorus cavalla* (*cf.* Spanish mackerel)

Pacific mackerel *S. japonicus*

snake-mackerel popular name of gempylid fish

Spanish mackerel any of several species of the genus Scomberomorus, particularly *S. meanlatus* (*cf.* king-mackerel). *Brit., Scomber colias* (= chub-mackerel)

Monterey Spanish mackerel *S. concolor*

macpalxochiquahuitl the sterculiaceous tree *Chiranthodendron platanoides*

-macro- comb. form meaning "long," but almost universally used as a substitute for -mega- which properly means "big"

Macroceratidae a family of nematoceran dipteran insects now usually included with the Mycetophilidae

Macrodasyoidea an order of the phylum Gastrotricha distinguished by having adhesive tubes along the whole body

[macrofine] a mutant (gene symbol *mf*) mapped at 5.5± on the X chromosome of *Drosophila melanogaster.* The phenotype has a small body and fine macrochaetae

Macrolepidoptera an obsolete, unrealistic, division of the Lepidoptera containing those forms large enough to be set on pins and which were therefore of interest to the average amateur collector (*cf.* Microlepidoptera)

Macropodidae that family of diprotodont marsupial mammals which contains the kangaroos and wallabies

Macropodinae a subfamily of macropodid marsupials containing the kangaroos, wallabies and their near relatives. The are distinguished by the long stiff tail and short front legs and well-developed hind legs

Macropterygidae = Hemiprocnidae

Macrorhamphosidae a family of solenichthous fishes distinguished by the long snout and very long single dorsal fin spines. They are commonly called snipe-fishes unless they are fused with the Centrescidae (q.v.)

Macroscelidae a family of insectivorous mammals commonly known as elephant-shrews. They have long pointed snouts and are distinguished from other insectivores by the presence of a caecum

Macrouridae = Coryphaenoididae

Macrura a section of reptant Crustacea with a long abdomen and large uropods containing the lobsters and their allies, differing from the Anomura in the symmetrical abdomen and from the Brachyura in not having the abdomen reduced

Mactridae that family of pelecypod mollusks commonly called surf-clams. They are distinguished by the large spoon-shaped cavity containing the inner cartilaginous ligament

macula a spot, specifically in *bot.* the pit of Coniferae, and in *zool.* the area of an ectoproct colony between the monticules

maculate spotted

immaculate without markings

madder *Brit.* the rubiaceous herb *Rubia peregrina* and the dyestuff derived from it

field madder *Brit.* the rubiaceous herb *Sherardia arvensis*

wild madder the rubiaceous herb *Galium mollugo* (= hedge bed straw)

madge *Brit. obs.* the barn owl or the magpie

-madre- *comb. form* meaning "mother"

Madreporaria an order of Zoantharia distinguished from the Actiniaria and the Antipatharia by the possession of a calcareous skeleton. They are commonly called stone corals

madrona the ericaceous tree *Arbutus menziesii*

madtom *U.S.* any of several species of ictalurid catfishes of the genus Noturus

maggot an insect larva, typically of a dipteran, characterized by the absence of well-defined mouth parts or legs

apple maggot the larva of the tephritid dipteran *Rhagoletis pomonella*

blueberry maggot *U.S.* the larva of the tephritid dipteran *Rhagoletis mendax*

cabbage maggot *U.S.* the larva of the muscid dipteran *Hylemya brassicae*

raspberry cane maggot *U.S.* the larva of the muscid dipteran *Pegomya rubivora*

carnation maggot *U.S.* the larva of the muscid dipteran *Hylemya brunnescens*

cherry maggot *U.S.* the larva of the tephritid dipteran *Rhagoletis cingulata*

seed-corn maggot *U.S.* the larva of the muscid dipteran *Hylemya platura*

grasshopper maggot *U.S.* the larva of the sarcophagid dipteran *Sarcophaga kellyi*

onion maggot *U.S.* the larva of the muscid dipteran *Hylemya antiqua*

pepper maggot *u.s.* the larva of the tephritid dipteran *Zonosemata electa*

sugar-beet root maggot *U.S.* the larva of the otitid dipteran *Tetanops myopaeformis*

wheat stem maggot *U.S.* the larva of the chloropid dipteran *Meromyza americana*

sunflower maggot *U.S.* the larva of the tephritid dipteran *Strauzia longipennis*

rat-tailed maggot the aquatic larvae of several species of syrphid dipterans of the genus Eristalis. *Brit.*, without qualification, that of *E. tenax*

carnation tip maggot *U.S.* the larva of the muscid dipteran *Hylemya echinata*

turnip maggot *U.S.* the larva of the muscid dipteran *Hylemya floralis*

Magnoliaceae that family of ranalous dicotyledons which contains, *inter alia*, the magnolias and tulip trees. Aside from the showy flowers the spiral structure of the hypogynous flowers and separate carpels are distinctive

Magnoliineae a suborder of ranale dicotyledonous angiosperms containing all those families of this order which are not included within the Nymphaeineae, Trochodendrineae and Ranunculineae

magpie (see -pie)

maguey any of several species of Agave (q.v.) particularly *A. atrovirens*

mah *see* [mahogany]

mahoe *N.Z.* the violaceous tree *Melicytus ramiflorous* (= whiteywood)

mahogany any of several trees of meliaceous trees of the genus Swietinia, particularly *S. mahogoni*. The term is also applied to numerous trees the lumber of which is similar in appearance to true mahogany

bastard mahogany usually the *Austral.* tree *Eucalyptus botryoides* but also occasionally *E. marginata*. Also, rarely, the *W.I.* sapindaceous tree *Ratonia apetala*

mountain mahogany any of several trees of the rosaceous genus Cercocarpus

red mahogany the myrtaceous tree *Eucalyptus resinifera*

swamp mahogany the myrtaceous tree *Eucalyptus robusta*

white mahogany *see* gum

[mahogany] *either* a mutant (gene symbol *mg*) in linkage group V of the mouse. The color of the phenotype is well described by the mutation name *or* a mutant (gene symbol *mah*) mapped at 88.0± on chromosome III of *Drosophila melanogaster*. The name is descriptive of the phenotypic eye color

mahseer *I.P.* any of several species of large cyprinid fish of the genus Puntius

old-maid *U.S.* the noctuid lepidopteran insect *Catocala coelebs* (*cf.* underwing)

red maid the portulacaceous herb *Calandrina menziesii*

maile the apocynaceous shrub *Gynopogon olivaeformis*

maina = mynah

maiosis = meiosis

maiotic = meiotic

maire (*see also* taioake) *N.Z.* the santalaceous tree *Mida myrtifolia* = sandalwood *N.Z.*) and also any of several species of oleaceous tree of the genus Olea

maize (*see also* bilbug, bird) the popular name of the edible seeded grass *Zea mays* (= U.S. corn)

milo maize *Hort.* var. of *Holcus halepensis* (= sorghum)

sergeant-major *I.P.* the nymphalid lepidopteran insect *Albrota ganga*

makamaka *N.Z.* the saxifragaceous tree *Ackama rosaefolia*

makaw = macaw

maker frequent in compound names

grape cane gallmaker *U.S.* the curculionid coleopteran insect *Ampeloglypter sesostris*

shoemaker the procellariid bird *Procellaria aequinoctialis*

popular tentmaker *U.S.* larva of the notodontid lepidopteran insect *Ichthyura inclusa*

birch tubemaker *U.S.* the larva of the phycitid lepidopteran insect *Acrobasis betulella*

mako the lamnid shark *Isurus oxyrinchus*

makomako 1 (*see also* makomako 2) *N.Z.* the elaeocarpaceous tree *Aristotelia serrata* (= winberry *N.Z.*

makomako 2 (*see also* makomako 1) *N.Z.* the melaphagid bird *Anthornis melanura* (= bell-bird)

mal a bacterial mutant showing utilization of maltose

mal B a bacterial genetic marker having a character affecting the activity of amylomaltase. Mapped at between 83.25 mins and 88.0 mins. for *Escherichia coli*

ma-l *see* [maroon-like]

mal the lateral area of bird's head immediately behind the jaw but posterior to the ear coverts

Malacanthidae a small family of percid fishes distinguished by the many rayed dorsal and anal fins. Commonly called blanquilos

Malaceae a family of rosale dicotyledonous angiosperms erected to contain those forms considered by many to belong in the rosaceous subfamily Pomoideae

Malacichthys a monofamilial order of Osteichthyes, regarded by many as a suborder of Acanthopterygii, erected to contain the family Icosteidae

-malaco- *comb. form* meaning "soft," but by extension often used to mean "slug," "snail," or even "molluscan"

Malacocotylea = Digenea

malacoid mucilaginous

Malacopsyllidae a small family of siphonapteran insects distinguished by the absence of a pronotal comb and a row of three bristles in front of the eye

Malacostraca a widely used zoological taxon erected to contain those Crustacea having differentiated abdominal appendages

malagash *S.Afr.* the sulid bird *Morus capensis* (= cape gannet)

malanga the araceous herb *Xanthosoma sagittifolium* and its edible corms

Malapteruridae a monogeneric family of ostariophyan fishes erected to contain the electric catfish *Malapterurus electricus*

Malasidae = Eucnemidae

malaxation literally the reduction of a hard substance to a soft mass by pounding or kneading. In *biol.*, specifically the type of kneading applied to its prey by predatory Hymenoptera to render it fit for larval food

Malayan *I.P.* the lycaenid lepidopteran insect *Megisba malaya*

Malcoha = malkoha

Maldanidae a family of sedentary polychaete annelids living in sandy tubes. The head is formed of a fused protomium and peristomium. There are no gills

male (*see also* fern) that one, of two, heterosexual forms of the same species, which normally yields motile gametes called sperm

complemental male a dwarf, frequently bizarre, male, usually associated with, and sometimes parasitic, on a larger female

super male the male genetic equivalent of a metafemale

[**male sterile**] *either* a mutant (gene symbol ms_{17}) mapped at 25 on linkage group I of *Zea mays*. The name is descriptive of the phenotype *or* a mutant (gene symbol ms_8) mapped at 14 in linkage group VIII of *Zea mays*. The name is descriptive of the phenotype

[**male sterile-10**] a mutant (gene symbol ma_{10}) mapped at 35 on chromosome 2 of the tomato. The term is descriptive of the phenotypic expression

malic pertaining to apples

malimbe any of several ploceid birds of the genus Malimbus

Malimbidae = Ploceidae

malkoha any of several species of cuculid birds

red-billed malkoha *Zanclostomus javanicus*

chestnut-breasted malkoha *Ramphococcyx curvirostris*

red-faced malkoha *Phaenicophaeus pyrrocephalus*

Raffles' malkoha *Rhinortha chlorophaea*

mallard the anatine duck *Anas platyrhynchos*

malleate hammer shaped, used particularly for the chewing type of mastax in a rotifer

mallemuch = fulmar

malleolus literally, a little hammer. Specifically *either* in insects the halter *or* in Solpugida one of the stalked leaf-like organs of unknown function found on the fourth pair of legs

malleus literally a hammer. Particularly that formed by the combination of the uncus and the manubrium in the mastax of rotifers and the outermost of the three bones (the others are the stapes and incus) in the mammalian ear

Mallophaga that order of insects containing the forms commonly known as biting lice. They are distinguished as wingless parasitic insects with biting mouth parts save in some bird lice in which the chewing mandibles are adapted to piercing. Most are parasitic on birds

chaetomallous thick maned

homomallous curved towards a common center

mallow 1 (*see also* mallow 2, rose) any of numerous malvaceous herbs particularly those of the genus Malva

common mallow *M. rotundifolia*

curled mallow *M. crispa*

false mallow any of numerous malvaceous herbs particularly those of the genus Malvastrum and in *U.S.*, Sphaeralcea

glade-mallow any of several species of herb of the malvaceous genus Napaea

globe mallow any of several species of malvaceous herb of the genus Sphaeralcea (= false mallow)

Indian mallow *U.S.* any of several species of herb of the malvaceous genus Abutilon

marshmallow the malvaceous herb *Althaea officinalis*

musk-mallow *M. moschata*

rose mallow any of several species of herbs of the malvaceous genus Hibiscus but particularly *H. moscheutos* and *H. oculiroseus*

seashore mallow *U.S.* any of several species of malvaceous herb of the genus Kosteletzkia

tree-mallow *Brit.* the malvaceous herb *Lavateria arborea*

Virginia mallow *U.S.* the malvaceous herb *Sida hermaphrodita*

mallow 2 *see also* mallow 1) *Brit.* the geometrid lepidopteran insect *Ortholitha cervinata*

Malpighian *epon. adj.* from Marcello Malpighi (1628–1694)

Malpighiaceae a family of geranialous dicotyledons containing a considerable number of trees and shrubs none of which, however, are of economic importance. Distinctive to the family are the glandular calyx, clawed petal, and the outer stamens opposite the petals

Malpighiineae a suborder of geraniale dicotyledonous angiosperms containing the family Malpighiaceae, Trigoniaceae, and Bothysiaceae

amylomaltase = maltose 4-glucosyltransferase

Maluridae a family of passeriform birds usually included with the Sylviidae

Malvaceae that family of malvalous dicotyledons which contains, *inter alia*, the mallows and the hibiscus. The stamens united in a column are typical of the family

Malvales that order of dicotyledonous angiosperms which contains, *inter alia*, vegetable ivory, the bass woods, the mallows, the baobabs, cacao and cola. The order is distinguished by the presence of bisexual pentamerous flowers

Malvinae a suborder of malvale dicotyledons containing the Tiliaceae, Malvaceae, Bombacaceae, and Sterculiaceae

mamangi *N.Z.* the rubiaceous tree *Coprosma arborea*

mamba any of several elapid snakes of the genus Dendroaspis

mamelon 1 (*see also* mamelon 2) that portion of the spine-bearing tubercle of an echinoid test which articulates with the spine

mamelon 2 (*see also* mamelon 1) floral axis

mamey the sapotaceous tree *Lucuma mammosa* (= sapote)

mamie = mammee apple

mamin = mamon

-mamm- *comb. form* meaning "breast," specifically a human breast, and thus, by extension, any mound bearing a prominence

mamma = mammary gland

Mammalia a class of gnathostomatous craniate chordates distinguished by the presence of hair on the skin and by having milk-secreting glands

mammary pertaining to the mammalian breast

mammee = mammee apple

mammie = mammee apple

mammillate breast-shaped; specifically having teat-like prominences

mammose = mammiform

mamo any of several drepaniid buds of the genus Drepanis, usually, without qualification *D. pacifica*

mamon the annonaceous tree *Annona reticulata* and its fruit

man- *comb. form* meaning "hand"

man 1 (*see also* man 2, 3) the hominine pongid primate *Homo sapiens*

man 2 (*see also* man 1, 3, mandrake) in compound names of many organisms

water boatman popular name of corixid hemipteran insects

footman popular name of numerous arctiid lepidopteran insects at one time placed in a separate family the Lithosiidae

layman *S.Afr.* the danaid lepidopteran insect *Amauris albimaculata*

midshipman *U.S.* any of several species of batrachoid fishes of the genus *Porichthys*

old man *Austral.* the macropodid marsupial mammal *Macropus canguru* (= great grey kangaroo). *W.I.* the cuculid bird *Hyeopornis pluvialis*

old-plainsman *U.S.* the compositaceous herb *Hymenopappus corymdosus*

peterman *W.I.* the larid bird *Sterna albifrons* (= least tern)

policeman *S.Afr.* any of several species of hesperid lepidopteran insect of the genus Coliades

rifleman *N.Z.* the acanthisittid bird *Acanthisitta chloris* (*cf.* wren)

sandman *S.Afr.* any of numerous species of hesperid lepidopteran insect of the genus Spialia

sheriff's man the fringillid *Carduelis carduelis* (= goldfinch)

water watchman *S.Afr.* the hesperid lepidopteran insect *Parnara monazi*

wenchman the lutianid fish *Pristipomoides andersoni*

yeoman *I.P.* any of several species of nymphalid lepidopteran insects of the genus Cirrochroa

man 3 (*see also* man 1, 2) a bacterial mutant showing utilization of mammose

manakin any of numerous species of birds of the family Pipridae

red-capped manakin *Pipra mentalis*

white-collared manakin *Manacus candei*

thrush-like manakin *Schiffornis turdinus*

long-tailed manakin *Chiroxiphia linearis*

manatee any of the three species of trichechid sirenian mammals. One of these (*Trichechus manatus*) inhabits coastal waters and river estuaries in the Caribbean, as far north as Florida. Another species (*T. inunguis*) is known from the Amazon and Orinoco rivers of South America, and a third (*T. senegalensis*) from some of the large rivers of tropical West Africa.

manatu *N.Z.* the malvaceous tree *Plagianthus betulinus.* (= lowland ribbonwood *N.Z.*)

manawa *N.Z.* the verbenaceous tree *Avicennia officinalis* (= mangrove *N.Z.*)

manchineel the euphorbiaceous tree *Hippomane mancinella*

mandarin 1 (*see also* mandarin 2, 3) *U.S.* any of several species of the liliaceous genus Streptopus

rose mandarin *S. roseus*

white mandarin *S. amplexifolius*

mandarin 2 (*see also* mandarin 1, 3) the anatine duck *Aix galericulata*

mandarin 3 (*see also* mandarin 1, 2) the rutaceous tree *Citrus reticulata* and its edible fruit, often called tangerine

mandible in vertebrates, the lower jaw as a whole. Also the most anterior of three pairs of mouthparts in many arthropods

Mandibulata a term coined to describe those classes of the phylum Arthropoda which have a true mandible. As so defined this taxon contains the Crustacea, Pauropoda, Symphyla, Diplopoda and Insecta

mandoris the aperture between the anterior and posterior halves of the insect pharynx

mandrake the solanaceous herb *Mandragora officinarum* and particularly its root used in pharmacy. *U.S.* (local) = may apple

mandrill the cynopithecid primate *Mandrillus mormon.* It has blue cheeks, a bright red nose and similarly brilliantly colored buttocks and genitalia. It is quite frequently referred to as the blue-faced baboon

mane a crest of long hair arising from the cervical, and sometimes thoracic, region of many mammals

shaggy mane the coprinaceous agaricale fungus *Coprinus comatus*

mangabey any of several species of cercopithecid monkeys of the genus Cercocebus

mangeao *N.Z.* the lauraceous tree *Litsea calicaris*

mangel = sugar beet

mango 1 (*see also* mango 2, weevil) the anacardiaceous tree *Mangifera indica* and its fruit

mango 2 (*see also* bird) any of numerous species of trochilid bird of the genus Anthracothorax

green-breasted mango *A. prevostii*

black-throated mango *A. nigricollis*

mangosteen the guttiferous tree *Garcinia mangostana*

mangrove (*see also* cuckoo, hen) any of several species of trees of the rhizophoraceous genus Rhizophora, most commonly *R. mangle*. *N.Z.* the verbenaceous tree *Avicennia officinalis* (= manawa)

black mangrove the verbenaceous tree *Avicennia nitida* (= white mangrove and honey mangrove)

manicate refers to a dense, cohesive pubescence on a plant form

Manidiidae a family of noctuoid lepidopteran insects

manioc the euphorbiaceous herb *Manihot utilissima* (*see* tapioca plant)

maniplies = manyplies, omasum

manisoba the euphorbiaceous tree *Manihot glaziovii*

mannakin *Austral.* the dicaeid bird *Pardalotus substriatus* (*cf.* diamond bird, mannikin)

mannikin *Afr.* any of several species of ploceid bird of the genera Spermestes and Amauresthes (*cf.* mannakin)

blue-billed manikin *S. bicolor*

bronze-manikin *S. cucullatus*

magpie manikin *A. fringilloides*

α and β -mannosidase enzymes respectively catalyzing the hydrolysis of α and β -D-mannosides into alcohols and mannose

man-of-the-earth the convolvulaceous climbing vine *Ipomoea pandurata*

Portuguese man-of-war any of several species of siphonophoran coelenterate of the genus Physalia

man-o'-war-bird = frigatebird

yerba mansa the saururaceous squatic herb *Anemopsis californicus*

mantid a very misleading term when applied to the Mantispidae and causing much confusion between these neuropterans and the mantidae

Mantidae that family of dydictyopteran insects which contains the preying mantids. The predatory reflexed first legs are typical. (*cf.* Mantispidae)

mantis any of several mantid dictyopteran insects characterized by the fact that the tibia is reflexed against the femur to form a grasping organ. "Praying" mantis is the original term—the *Eur.* form is *M. religiosa*—but "preying" is more common in *U.S.* The term mantid is preferred by some

Carolina mantis *Stagmomantis carolina*

Chinese mantis *U.S. Tenodera aridifolia*

praying mantis *Eur.* (but now acclimated in *E.U.S.*) *Mantis réligiosa*

preying mantis *U.S. Stagmomantis carolina*

narrow-winged mantis *U.S. Tenodera augustipennis*

Mantispidae a family of neuropteran insects so strongly resembling the Mantidae (*q.v.*) as frequently to be mistaken for them. The forewings of Mantispidae are entirely membranous and those of Mantidae parchment-like

mantle 1 (*see also* mantle 2, lobe) an envelopment of the body usually meaning, without qualification, the outer soft coat of Mollusca and brachiopods

myoepicardial mantle the walls of the embryonic tube which give rise to the muscular and epicardial walls of the heart

mantle 2 (*see also* mantle 1) in compound names of organisms

ladies mantle *Brit.* any of several rosaceous herbs of the genus Alchemilla

royal mantle *Brit.* the geometrid lepidopteran insect *Anticlea cucullata*

Mantodea an order of insects containing the mantids now usually united with the cockroaches in the Dictyoptera

manubrium literally, a handle. Specifically, one of a pair of trophi in the mastax of a rotifer, the handle-like extension on the end of which is the mouth of coelenterate medusae, the projection from the mesosternum of elaterid Coleoptera which fits into a corresponding cavity in the prothorax and enables them to "click," and the base of the furcula in Collembola

pseudomanubrium a very elongated peduncle (*q.v.*) on the end of which is the true manubrium

manucode any of several species of birds-of-paradise (*q.v.*) particularly those of the genus Manucodia

manuka *N.Z.* the myrtaceous tree *Leptospermum scoparium* (= red tea-tree)

manu-mea = dodlet

manyplies = omasum, maniplies

manzanillo the euphorbiaceous tree *Hippomane mancinella*

map *I.P.* any of several species of nymphalid lepidopteran insect of the genus Cyrestis. Usually, without qualification, *C. thyodamas*

maple 1 (*see also* maple 2, aphis, beetle, borer, mite, scale, skeletonizer) any of numerous species of trees of the aceraceous genus Acer

bark-maple *A. glabrum*

common maple *Brit. A. campestre*

hornbeam-maple *A. carpinifolium*

Japan maple *A. palmatum*

ash-leaved maple *A. negundo*

big-leaf maple *A. macrophyllum*

mountain maple *A. spicatum*

Norway maple *A. platanoides*

red maple = swamp maple

river maple = silver maple

rock maple = sugar maple

scarlet maple = swamp maple

silver maple *A. saccharinum* (*cf.* sugar maple)

soft maple swamp maple or silver maple

striped maple *A. pennsylvanicum*

sugar maple *A. saccharum* (*cf.* silver maple)

southern sugar maple *A. barbatum*

swamp maple *A. rubrum*

sycamore maple *A. pseudoplatanus*

white maple = silver maple

maple 2 (*see also* maple 1) a few shrubs not of the genus Acer

flowering maple any of several species of the malvaceous genus of shrubs Abutilon

maplet *I.P.* any of several species of nymphalid lepidopteran insect of the genus Chersonesia. Usually, without qualification, *C. risa*

mapou *N.Z.* the myrsinaceous tree *Suttonia australis*

maquis an association of hard-leaved shrubs, principally on the north shore of the Mediterranean

mara the caviid hystricomorph rodent *Dolichotis pata-chonica.* It is an inhabitant of *S. Amer.* arid regions and resembles a small, thick-set deer in appearance and habit (= Patogonian cavy)

marabou the ciconiid bird *Leptoptilos crumeniferus*

maral a Caspian race of the cervine artiodactyl *Cervus elaphas* (= red deer)

Marantaceae a family of scitaminous monocotyledons containing the arrowroots. The family is distinguished by the presence of rhizomes, the one-seeded cells of the ovary and the presence of a joint at the summit of the petiole

Marattiaceae a family of filicine Pteridophyta distinguished from the Ophioglossales by the fact that the leaves are circinate in venation

Marattiales an order of filicine Pteridophyta containing the single family Marattiaceae

purple marbled *Brit.* the noctuid lepidopteran insect *Thalpochares ostrina*

rosy marbled *Brit.* the noctuid lepidopteran insect *Monodes venustula*

[**marbling**] a mutant (gene symbol *ma*) in linkage group IV of the domestic fowl. The name is descriptive of a pattern appearing in the down of the chick

marcescent said of a plant organ which withers, but does not fall

Marcgraviaceae a small family of parietale dicotyledons

Marchantiaceae a family of marchantiale Hepaticae containing the well-known genus Marchantia which, in having the archigonia borne on stalked vertical branches, is typical of the family

Marchantiales an order of hepaticate bryophytes distinguished by their prostrate habit of growth

mare a female horse

bastard margaret *U.S.* a marine teleost *Haemulon parra* (*cf.* grunt)

Margarodidae a small family of scale insects closely allied to the Ortheziidae

margate any of several acanthopterygian fishes, *U.S.*, without qualification, usually *Haemulon album*

black margate the pomadasyid *Anistremus surinamensis*

margay the feline carnivore *Felis wiedi*

antemarginal used of organs, particularly sori, which lie just within the margin of a leaf

marginate furnished with a brim

emarginate said of any structure not having a smooth margin as the slightly forked tail of a bird

Marginellidae a family of gastropod mollusks almost cylindrical in shape in view of the large body whorl and short spire. A long narrow aperture runs almost the whole length of the body. Commonly called rim shells. Many small procellaneous species are known as rice shells

marguerite almost any daisy has from time to time been given this name, but it is most frequently applied to *Chrysanthemum leucanthemum* and *C. frutescens*

blue marguerite the compositaceous herb *Felicia anelloides*

golden marguerite the compositaceous herb *Anthemis tinctoria*

glaucous marguerite *Chrysanthemum anathifolium*

marigold (*see also* beetle) properly any of numerous species and very numerous horticultural varieties and hybrids of herbs of the compositaceous genus Tagetes, but the term is applied to a wide variety of Compositae and a few yellow plants of other families

African marigold *T. erecta*

bur-marigold any of numerous species of the genus Bidens

Cape marigold any of numerous species of the genus Dimorphotheca

corn-marigold *Chrysanthemum segetum*

fetid marigold *Dyssodia papposa*

fig-marigold any of numerous species of aizoaceous herbs of the genus Mesembryanthemum

French marigold *T. patula*

marsh-marigold the ranunculaceous herb *Caltha palustris*

pot-marigold *Calendula officinalis*

sweet-scented marigold *T. lucida*

water-marigold *Megalodonta beckii*

marijuana the cannabinaceous herb *Cannabis indica*

marine pertaining to the sea (*cf.* maritime)

marita a term once employed for the adult fluke in contrast to parthenitae and adolescariae (*q.v.*)

maritime pertaining to regions bordering on the sea (*cf.* marine)

submaritime used of plants typically occurring on seashores but also found elsewhere

maritzgula the apocynaceous shrub *Carissa bispinosa*

marjoram any of several culinary herbs of the labiateous genus Origanum (*see also* orégano)

annual marjoram = sweet marjoram

bastard marjoram the labiate herb *Origanum vulgare* (= wild marjoram)

greek marjoram *O. onites*

sweet marjoram *O. majorana*

wild marjoram *O. vulgare*

metal-mark popular name of riodinid papilionoid lepidopteran insects

question mark *U.S.* the nymphalid lepidopteran *Polygonia interrogationis* (*cf.* angle-wing)

spear-mark *U.S.* the geometrid lepidopteran insect *Rheumaptera hastata*

markary *obs. Brit.* = mercury (the plant)

genetic marker a symbol used in microbial genetics, corresponding to the gene symbols used in higher forms. Genetic markers cannot, for obvious reasons, be mapped as are genes on chromosomes but those of *Escherichia coli* can be arranged in minutes of elapsed time before transfer is complete

pie marker *U.S.* the malvaceous herb *Abutilon theophrasti*

markhor any of several geographic races of the caprine caprid *Capra falconeri.* They are distinguished by their large, straight, spirally-twisted horns

marlin 1 (*see also* marlin 2, 3, spike) any of several species of istiophorid fishes of the genus Makaira

black marlin *M. marlina*

blue marlin *M. nigricans*

striped marlin *M. audax*

white marlin *M. albida*

marlin 2 (*see also* marlin 1, 3) *U.S. dial.* any of several scolapacid birds particularly those of the genus Numenius (*cf.* curlew)

marlin 3 (*see also* marlin 1, 2) *Brit. obs.* = merlin

marlion = merlin

marmoratous veined with color

marmoress *Brit.* = marmoris *Brit.*

marmoris *Brit. obs.* the satyrid lepidopteran insect *Melanargia galatea* (= marbled white)

marmoset popular name of diurnal South American callithricid hapaloid primates

bald marmoset any of several species of the Amazonian genus Marikina

lion-marmost = maned marmoset

maned marmoset any of several species of the genus Leontocebus

pigmy marmoset any of several species of the genus Cebuella

plumed marmoset any of several species of the genus Callithrix

ruffed marmoset any of several species of the genus Hapale

marmot any of numerous ground dwelling sciurid rodents of the genus Marmota

Alpine marmot M. marmotta

yellow-belly marmot M. flaviventris

hoary marmot M. caligata

prairie marmot = prairie dog

Quebec marmot M. monax (= woodchuck)

Siberian marmot M. bobac

[maroon] either a mutant (gene symbol ma) in linkage group IV of Habrobracon juglandis. The name refers to the phenotypic eye color or a mutant (gene symbol ma) mapped at 49.7 on chromosome III of Drosophila melanogaster. The phenotypic eye is not maroon but dull ruby color

[maroon-like] a mutant (gene symbol ma-l) mapped at 67.2 on the X chromosome of Drosophila melanogaster. The name is descriptive of the phenotype

marquis I.P. any of several speices of nymphalid lepidopteran insects of the genus Euthalia (cf. duke, baron, earl, baronet, duchess, count)

marram U.S. any of several species of grass of the genus Amophila (= sand reed)

marrow 1 (see also marrow 2) the vascularized tissue that fills the cavities of bones

marrow 2 (see also marrow 1) many oval-fruited horticultural varieties of the cucurbid Cucurbita pepo (U.S. = squash)

vegetable marrow Brit. an enormous green fruited variety of marrow

marsh (see also bird, hawk, marigold, wort, wren) an area of normally wet ground differing from a bog in being less soggy and in frequently possessing transitory dry areas and also distinguished from a bog by the type of marginal vegatation

salt marsh a marsh the waters of which are brackish

marshlocks the rosaceous herb Potentilla palustris (cf. cinquefoil)

Marsiliaceae a small family of leptosporangiate Pteridophyta distinguished by being heterosporous and producing their sporangia with sporocarps each of which contains numerous sori

Marsiliales a small order of leptosporangiate Pterophyta containing the Marsiliaceae

Marsipobranchia = Cyclostomata

marsupial literally "pouched." Applied in bot. to pouched fruits as well as in zool. to pouched mammals

Marsupialia an order of metatherian mammals in which the prematurely born young are raised in a brood pouch (the marsupium) containing the mammary glands. The placenta is usually lacking though a placenta-like organ is found in Parameles

marsupium 1 (see also marsupium 2) the broad pouch on the abdomen of marsupials

marsupium 2 (see also marsupium 1) in some insects a pouch near the anus in which eggs are carried

marten any of several species of mustelid carnivore of the genus Martes U.S., without qualification, M. americana. Brit., without qualification, M. martes

beech marten M. foina

pine marten Brit. M. martes. U.S. M. americana

martin any of several passeriform hirundinid birds (cf. swallow). Brit. without qualification, Delichon urbica

grey-breasted martin Progne calybea

crag-martin any of several species of the genus Ptyonoprogne, usually, without qualification, P. rupestris

freemartin see under free

house martin Delichon urbica

purple martin U.S. Progne subis

sand-martin any of several species of the genus Riparia, usually, without qualification, R. riparia

African sand-martin R. paludicola

banded sand-martin R. cincta

tree-martin Austral. Petrochelidon nigricans

Martineta the tinamid bird Eudromia elegans

Martyniaceae that family of tubifloral dicotyledonous angiosperms which contains the plants commonly called unicorn plants by reason of the long horn which projects from the mature fruit. This characteristic is typical of the family

cloaked marvel U.S. the noctuid lepidopteran insect Chytonix palliatricula

green marvel U.S. the noctuid lepidopteran insect Dipthera fallax

marvel-of-Peru the nyctaginaceous herb Mirabilis jalapa

blue-eyed Mary the scrophulariaceous herb Collinsia verna

costmary the compositaceous herb Chrysanthemum balsamita

masir = mahseer

mask the extensible portion of the nymphal mouth in Odonata

masked (bot.) = personate (bot.)

maskinonage = muskellunge

arcual mass those cells which are the presursors of the neural arches of the vertebra

biomass the total weight of all living organisms on the planet or in specific habitats on the planet, or of a specific taxon

central mass the lump of cells remaining at one pole when the morula of a mammal develops an asymmetrical cavity. It produces both blastodics and endoderm

ground mass = woody tissue

pollen mass a coherent mass of pollen

massasauga U.S. the crotalid snake Sistrurus catenatus (= swamp rattler)

massula literally, a little lump. Particularly applying to coherent lumps of pollen grains of orchids or of microspores of some ferns

-mast- comb. form meaning "breast" or "breast shaped"

mast 1 (see also mast 2) in the sense of breast, lump or hillock

gynecomast a male human with female-type mammary glands (see gynandromorph)

neuromast sensory cell. The term is applied both to the cells of invertebrate receptors and of the lateral line system of fish

mast 2 (see also mast 1) Brit. nuts used as forage

buckmast the nut of the beech tree

-mastac- comb. form meaning "jaw"

Mastacembelidae a monotypic family of opisthomous fishes commonly called spiny eels but not related to the true eels. They are distinguished by their elongate shape and particularly by the long sensitive snout supported by cartilage. They are further distinguished by the very numerous depressed spines preceding the dorsal fin

mastax the complex pharynx of Rotifera. The term is sometimes used as synonomous with trophi

bushmaster the crotalid snkae *Lachesis muta*

rattlesnake master the compositaceous herb *Eryngium aquaticum*

schoolmaster the lutjanid fish *Lutjanus apodus*

mastic *see also* tree

-mastig- *comb. form* meaning "whip"

Mastigophora a class of Protozoa distinguished by the presence of one or more flagella at all stages of the life history. At one time universally, and still occasionally, called the Flagellata

mastigote = flagellate

heteromastigote having an anterior functional tractellum and a posterior functionless flagellum

isomastigote possessing several flagella of equal length

monomastigote = monokont

paramastigote an organism having one large and one small flagellum

polymastigote = polykont

mastoid nipple-like

water **mat** the saxifragaceous herb *Chrysosplenium americanum*

mat *see* [mat]

[mat] a mutant (gene symbol *mat*) on linkage group IV of *Neurospora crassa*. The phenotypic expression is colonial growth

matagouri *N.Z.* the rhamnacaeous herb *Discaria toumatou*

mata hari the lethrinid fish *Lethrinus medulosus* (scavenger)

matai *N.Z.* the podocarpaceous tree *Podocarpus spicatus* (= black pine *N.Z.*)

matamata the chelid chelonian reptile *Chelys fimbriata*

matata *N.Z.* the sylviid bird *Bowdleria punctatus* (= fernbird)

mate one, of two, heterosexual partners

phycomater a jelly surrounding some germinating spores

pia mater the inner of the meninges

mating the sum total of those activities of a hererosexual couple which terminate in coitus, and the act of coitus itself

assortive mating the system under which mates are selected on the basis of a character thought to be desirable and which therefore appears with increasing frequency in the population

sib mating mating between brother and sister

[mating type] a mutant (gene symbol *mt*) in linkage group VI of *Chlamydomonas reinhardi*. The gene controls the plus or minus linkage type

Matoniaceae a small family of filicale Pterophyta distinguished by the simultaneous development of sporangia in which the sori are protected by an umbrella-shaped indusium

matricary any of several species of herb of the compositaceous genus Matricaria

grey **matter** that part of the central nervous system which consists principally of cell bodies

white **matter** that part of the central nervous system which consists principally of medullated fibers

mattulla plant fiber, particularly that of palm trees

matuku *N.Z.* the ardeid bird *Demigretta sacra* (= reef heron)

matuki-hureopo *N.Z.* the ardeid bird *Botaurus poiciloptilus* (= bittern *N.Z.*)

matuki-moana *N.Z.* the ardeid bird *Notophoyx novaehollandiae* (= white-fronted heron)

matutinal referring to the morning (*cf.* crepuscular, auroral)

mavis = thrush

maxilla literally the upper jaw. In arthropods applied to the mouth parts immediately posterior to the mandibles

second maxilla in Crustacea the second of the two pairs of appendages so named, but in insects the labium

maxillaries = maxillary bones

Mayacaceae that family of farinose monocotyledonous angiosperms containing the forms commonly called bog-mosses. They are aquatic plants easily recognized by the one-nerved, narrow, bidentate leaves

Maydeae a tribe of the plant family Gramineae

Mayrian *epon. adj.* from Gustav Mayr (1830–1908)

mazew *Brit.* the cruciferous herb *Brassica campestris*

mb *see* [minusbar]

mc *see* [microchaete] and [macrocalyx]

md *see* [mottled-2]

md- symbolic prefix used before a bacterial marker to indicate that the mutant is a modifier of the mutant the symbol which follows

Me *see* [Moiré]

me-1 to me-9 *see* [Methionine-1] to [Methionine-9]

Me *see* [Mouse ears]

meadow (*see also* chicken, hen, king-of, lark, parsnip) an association of a few dicotyledonous herbs with grasses, the former being restrained by grazing or cutting

meagre *Brit.* the sciaenid fish *Sciaena aquila*

water-**meal** *U.S.* any of several species of aquatic of the lemnaceous genus Wolffia

water **measurer** any of many species of hydrometrid hemipteran insects (= marsh treader). *Brit.*, without qualification, *Hydrometra stagnorum*

meat the flesh of mammals, or rarely other vertebrates, used as food. By extension, in compound names of many organisms supposed, often erroneously, to be eaten by specific forms

adder's meat *Brit.* the caryophyllaceous herb *Stellaria holostea*. *Brit. obs.* the aracaceous herb *Arum maculatum*

duck's meat any of several species of minute floating aquatic plants of the lemnaceous genus Lemna

meatus a passage, usually in *biol.* one in the form of a tube

external auditory meatus the tube, at the base of which lies the tympanic membrane, in those organisms in which this last is sunk beneath the surface

mechanism the doctrine that all living processes van be explained in terms of inorganic concepts. The doctrine is the antithesis of vitalism

Meckel's *epon. adj.* from Johann Friedrich Meckel (1781–1833) [cartilage] or Johan Friedrich Meckel (1714–1774) [ganglion]

meconium (*see also* corpuscle) the excreta of a fetus in the uterus or an insect in the pupal case

Mecoptera that order of insects which contains the scorpion-flies and their allies. Their distinguishing characteristic is the prolongation of the head into a trunk-like beak and, in most groups, the possession of four equal-sized wings frequently mottled

medaka the cyprinodont fish *Oryzias latipes* (= Japanese ricefish)

mediastinum a cavity between the two pleural sacs of a mammal

medick any of several species of leguminous herb of the genus Medicago

black medick *M. lupulina*

hop-medick = black medick

mediterran- *comb. form* meaning "inland" but almost invariably used in the sense of the "inland sea," *i.e.* the Mediterranean

mediterranean (*but see* mediterranean, *under* -terr-) occurring in the area of the Mediterranean Sea

medlar the rosaceous tree *Mespilus germanica* and its edible fruit

medulla literally "narrow" or "thick." Used generally to distinguish the central ("medullary") from the outer ("cortical") portions of an organ (*cf.* cortex.) In *bot.* frequently used as synonymous with pith

medulla oblonga the thickened floor of the myelencephalon

medusa (*see also* fish, head) the free-swimming jellyfishlike form of a polymorphic coelenterate in contrast to the polyp

meelalla the palmaceous tree *Borassus flabellifer*

meerkat the herpestine viverrid carnivore *Suricata penicillata*. The term is sometimes applied to other herpestines

meg *see* [megaoculus]

-mega- *comb. form* meaning "large." The term -macro-, which properly means "long," is frequently substituted for -mega- in biological combinations

Megachilidae a family of large stout-bodied apoid apocritan hymenopteran insects commonly called leafcutter bees from the fact that some line their cells with neatly cut pieces of leaves

Megachiroptera a monotypic suborder of chiropteran mammals, commonly large in size and frugivorous in habit. They are distinguished by the fact that the molars are not tubercular (*cf.* Microchiroptera)

Megalaemidae = Capitonida

Megaloptera an order of insects erected to contain the families Sialidae and Corydalidae by those who do not hold them to belong in the Neuroptera

Megalopygidae a family of zygaenoid lepidopteran insects in which elongate scales resembling hairs occurs on the wings so frequently as to give a woolly appearance. The name flannel-moth is sometimes used

Megalornithidae = Gruidae

dimegaly the condition of having two sizes of macrogamete

[megaoculus] a mutant (gene symbol *meg*) mapped at 61.9± on the X chromosome of *Drosophila melanogaster*. The phenotypic expression is large rough eyes

megapode any galliform bird of the family Megapodiidae, usually, without qualification, *Megapodius freycinet*

Megapodiidae that family of galliform birds which contains the forms commonly called mound-builders. The are distinguished by having only one carotid artery and a naked uropygial gland

Megapsyllidae = Malacopsyllidae

Megascolicidae a family of large American terrestrial oligochaetes readily distinguished from the Lumbricidae by the fact that the clitellum of the megascolicids is anterior to segment fifteen

Megathymidae a family of hesperioid lepidopteran insects commonly called giant skippers. The larvae of many species feed on agave and are the edible maguey-worms of commerce

megrim *Brit.* the bothid fish *Arnoglossus megastoma*

Meibomian *epon. adj.* from Heinrich Meibom (1638–1700)

-meio- *comb. form* meaning "less"

meiosis two sequential cell divisions which together bring about segregation and reduction of the chromosome number from the diploid to the haploid. The process ultimately leads to gamete formation

brachymeiosis *either* meiosis resulting in half the usual haploid number of chromosomes *or* the second reduction division

initial meiosis meiosis immediately following fertilization

pseudomeiosis = pseudo-reduction

ameiotic said of a division of a gamete in which the diploid number is not produced

Meissner's *epon. adj*: from Georg Meissner (1829–1903)

mel *see* [melanized]. This symbol is also used for a bacterial mutant showing utilization of melibiose and for a yeast genetic marker indicating a requirement for melibiose

mel-3 *see* [Melon-3]

-melan- *comb. form* meaning "black"

Melandryidae a family of coleopteran insects commonly called false darkling beetles. They have the same five-five-four tarsal formula as the tenebrionids but can be distinguished by the two pits found near the posterior border of the pronotum

melanin a general term for a group of indole biochromes best known for the black melanin but also occuring as yellow, orange and brown compounds

[melanized] a mutant (gene symbol *mel*) mapped at 64.1± on the X chromosome of *Drosophila melanogaster*. The darkened body of the phenotype is accompanied by dull red eyes

[melanoscutellum] a mutant (gene symbol *msc*) mapped at 52.6± on the X chromosome of *Drosophila melanogaster*. The phenoytpic expression is a dark scutellum

Melanotaenidae = Atherinidae

Melastomaceae a very large family of myrtiflorous dicotyledons. The indehiscent fruit, the peculiar leaf venation and the stamen appendages distinguish them from most of the families of the order

melatonin a hormone secreted by the pineal in lower vertebrates which is antagonistic to melanocyte-stimulating hormones

Meleagrididae that family of galliform birds which contains the turkeys. Their large size and prominent wattles are distinctive

Meliaceae that family of the geraniale dicotyledons which contains numerous trees of economic importance including the true mahogany. The family differs from the closely allied Rutaceae and Burseraceae in lacking both aromatic oil glands in the leaves and resin chambers in the bark, and is further distinguished by the peculiar staminal tubes and the viscoid, capitate stigmas

Melianthaceae a small family of sapindalous dicotyledons. They may be distinguished from the closely allied Sapindaceae by the peculiar flowers and, in general, the presence of more endosperm in the seeds

Melianthineae a suborder of Sapindales containing the single family Melianthaceae

melilot *Brit.* anglicized version of the name of the leguminous plant genus Melilotus

Melinae a subfamily of mustelid carnivorous mammals containing the badgers

melinous pertaining to quinces

Meliphagidae a large family of passeriform birds commonly called honeyeaters, almost all of which are confined to Australasia. Many have brightly colored bare skin on the head and carry wattles or casques

Melithreptidae = Meliphagidae

Micromelittophilae flowers fertilized by small insects

melleous pertaining to honey, either in regard to color, texture or taste

Mellivorinae a subfamily of mustelid carnivorous mammals containing the ratels

Meloidae a family of medium to large coleopteran insects with soft and flexible elytra and with a pronotum markedly narrower than the head. They are commonly called blister-beetles because of the strongly vescicant properties of the body fluids of many of them. Great confusion is occasioned by the fact that the common European blister-beetle or Spanish fly was for years called *Cantharis vesicatoria* and placed in the family Cantharidae. The insect in question is now known as *Lytta vesicatoria* and placed in the Meloidae. However, the active principal, once notorious as a dangerous aphrodisiac, is still called cantharidin

Meloidea a superfamily of polyphagous coleopteran insects containing the blister beetles and their immediate allies

Melolonthidae a family erected to contain those scarabaeids which are phytophagous in contrast to the Copridae which are coprophagous. This distinction, when retained, is usually given subfamilial status

melon (*see also* aphis, fly, pear, spurge, worm) any of many vines with sweet-fleshed gourd-like fruit of the cucurbitaceous genus Cucumis. Most horticultural varieties are derived from *C. melo*

  gourd-melon the fruit of *Benincasa hispida*

  muskmelon supposed to be the "original" *C. melo*

  orange-melon a horticultural variety of muskmelon

  watermelon the cucurbitaceous vine *Citrullus vulgaris* and its fruit

[melon-3] a mutant (gene symbol *mel-3*) in linkage group III of *Neurospora crassa*. The phenotypic colonies are hemispherical

Melothrieae a subfamily of the Cucurbitaceae

primary **member** *bot.* the hypocotyl or radicle

  sieve-tube member = sieve tube element

  vessel member a xylem element differing from a tracheid in being hollow

Membracidae a family of homopteran insects readily distinguished by the large and frequently distorted pronotum. Commonly called treehoppers

membrane any thin sheet, living or dead, of organic material

  anal membrane the opaque portion at the posterior end of the wing of anisopterous insects

  arachnoid membrane an inner subdivision of the pia mater

  articular membrane the flexible membrane between the segments of arthropods

  basement membrane the layer that lies between epithelial cells and the underlying connective tissue but in insects applied also to the inner surface of the eye

  Bowman's membrane a clear membrane immediately underlying the cornea of the eye

  cell membrane the external limiting membrane of the cytoplasm of the cell

  cloacal membrane a temporary ectodermal-endodermal plate closing the opening of the cloaca to the proctodaeum of the embryo

  cytomembrane a general term proposed for the endoplasmic reticulum and all other membraneous structures found in the cytoplasm

  dermal membrane the outer surface of some sponges consisting of epidermis with a thin underlying mesenchymal layer

  Descemet's membrane an membrane immediately above the inner endothelium of the cornea

  embryonic membrane a name early applied to the amnion, chorion allantois, and yolk sac taken together

  fenestrated membrane the elastic structural unit in the wall of an artery

  fertilization membrane the vitelline and perivitelline membranes taken together; the plasmolemma in fertilized eggs of some species

  frontal membrane the thin, ventral cuticle and underlying body wall of some ectoprocts

  Henslovian membrane cuticle

  nictitating membrane a transparent sheet which may be drawn horizontally across the front of the cornea from the inner angle of the eye, or beneath the lower lid

  nuclear membrane the limiting membrane of the nucleus. It is thought by many to be continuous with the endoplasmic reticulum

  permeable membrane one which permits the passage of substances from one side to the other

    differentially permeable membrane one which permits the passage of some materials but not others

    semipermeable membrane an unfortunate name for differentially permeable membrane

  pit membrane the membrane which closes a pit in a plant cell wall at the line of junction of two cells

  plaited membranes folded, and interlaced, membranes rising from the floor of the stomach wall in Scyphozoa

  plasma membrane the surface boundary of a cell (= cell membrane)

  Schneiderian membrane the mucous membrane lining the respiratory portion of the nasal cavities

  synovial membrane the inner layer of the joint capsule of a synovial joint

  tentacular membrane a membrane round the base of the tentacles of entoprocts forming, as it were, a lip to the calyx

  tympanic membrane a stretched vibrating membrane and thus the usual functional portion of a scolopophore or phonoreceptor. In higher vertebrates the membrane separating the tympanic cavity from the outer ear and against which impinges the lower arm of the malleus

  undulating membrane a delicate membrane supported by several rows of cilia in the cytopharynx of some Protozoa

  vitelline membrane the external surface of an egg

    perivitelline membrane a membrane formed beneath the vitelline membrane immediately after fertilization

membranelle a triangular plate, composed of two or more rows of fused cilia in Protozoa

memnonious dark brownish black

menarche the time of appearance of the first menstrual cycle in human females or oestral cycle in other mammals (*cf.* menopause)

Mendelian *epon. adj. from* Gregor Mendel (1822–1884)

menhaden any of several species of clupeid fish of the genus Brevoortia

  Atlantic menhaden *B. tyrannus*

  finescale menhaden *B. ginteri*

  largescale menhaden *B. patronus*

meninges (*sing.* menix) the connective tissue envelope of the brain

menix *sing. form* of meninges

Menispermaceae a small family of ranalous dicotyledons

containing the moonseeds. The two whorls of sepals, petals and the curved seeds are distinctive

-meno- *comb. form* usually meaning "remain" (*but see* -men-)

Menognatha a group term for those insects provided with mandible in both the larval and adult instars (*cf.* Menorhyncha)

-men- *comb. form* hopelessly confused from many roots and variously meaning "moon," "crescent," "month," "courage," "vigor," "unchanging," "permanent" and "wrath"

-menon- *see* -men-

    autophaptomenon an autotrophic plant

    ephaptomenon all those life forms which live more or less permanently in contact with each other

    heterephaptomenon an organism which is more or less parasitic

    planomenon the sum total of those forms of life capable of movement

    rhizomenon those living organisms which are rooted

menopause the time of the cessation of the menstrual cycle in human females or the oestral cycle in other mammals (*cf.* menarche)

Menoponidae a large family of mallophagan insects parasitic on domestic poultry and a few marsupials. They are distinguished by a large triangular head

Menorhyncha a group term for those insects in which both the imagal and larval instars feed by sucking (*cf.* Menognatha)

epimenous having a superior perianth

-mens- *see* -men-

mentum literally chin, but in insects usually applied to the distal segment of the labrum

    postmentum the median unpaired basal portion of the insect labium

    prementum a median unpaired plate of the insect labium lying immediately in front of the postmentum

Menuridae a family of passeriform birds erected to contain the lyrebirds. They are so called from the unmistakable long and elaborate lyre-shaped tail

Menyanthoideae a subfamily of the Gentianaceae distinguished from the Gentianoidea by having the leaves alternate

Mephitinae a subfamily of mustelid carnivorous mammals containing the skunks

-mer- *comb. form* meaning "a part of." The substantive forms mer, mere, mera, merism, merite and the adjectival forms meric, meristic have all developed more or less specialized meanings

phytomer any of the units from which a plant is built

epimera that part of the arthropod body from which a leg originates

    external paramera = male external genitalia of insects

hop **merchant** *U.S.* nymphalid lepidopteran insect *Polygonia comma*

mercury the chenopodiaceous herb *Chenopodium bonushenricus* and any of numerous plants once thought to have astrological associations

    three-seeded mercury any of several species of the euphorbiaceous herbs of the genus Acalypha

-mere- *see* -mer-

    actinomere one complete segment of a radially segmented organism

    antimere either of the two halves of a bilaterally symmetrical object or a homologous part repeated in segments arranged round the axis of a radially symmetrical organism

    aphanimere amitosis

arthromere a somite or metamere of an arthropod sometimes limited to one bearing a walking leg

branchiomere that division of mesoderm from which the mesodermal components of a visceral arch are developed

centromere that point at which two chromatids are fused along the course of a chromosome, usually indicated by a transparent constriction from which the spindle fibers usually originate during mitosis (*cf.* kinetochore)

cephalomere those segments of a metamerically segmented animal, particularly an arthropod, which are considered to belong to the head

chondriomere that section (the middle section) of a spermatozoan in which the mitochondria are found

chromatomere = chromosome

chromomere The term is not specific but is applied to any division recognizably constant in size and position, such as the bands on salivary gland chromosomes of diptera, or the lumps from which loops project in amphibian lamp-brush chromosomes

telochromomere = telomere

cryptomere a gene the phenotype of which is not known

cytomere cytoplasm of a spermatozoan. The term is also sometimes used as synymous with chondriomere

epimere *either* a dorsal projection from the phallobase *or* the dorsal mesoderm of a vertebrate embryo before the differentiation of myotome and sclerotome

ethomere the condition of having the normal number of chromosomes

gonomere the distinction between male and female

hypomere the lower segment of mesoderm which divides into somatic and splanchnic layers in a vertebrate embryo

idiomere structures thought to be sexual units which were supposed to be developed in the nucleus during its resting stage

kinomere = centromere

metamere one division of a metamerically segmented organism

myomere the dorsal, metamerically segmented, portion of the mesoderm from which the segmental muscles of chordates are derived

myriomere the process of pseudomitosis (*q.v.*)

neuromere = rhombomere

paramere either one peradius with half of the interadius on each side of it or one of two lateral lobes of the phallobase

plastomere = chondriomere

podomere = podite

prosthomere a preoral segment in an arthropod embryo

rhabdomere that portion of a retinal cell that contains the photosensitive pigment (*cf.* rhabdom)

rhombomere one of a pair of ventral prominences in the embryonic rhombocephalon

sarcomere that portion of a myofibril which lies between two Z-lines (*q.v.*)

teleomere the end portion of a chromosome

uromere an abdominal segment of an arthropod

merganser any of numerous fish-eating ducks of the subfamily Merginae, mostly of the genus Mergus

    American merganser *M. serrator*

    common merganser *M. merganser* (= *Brit.* goosander)

    hooded merganser *Lophodytes cucullatus*

    southern merganser *N.Z. M. australis*

Mergini a group of Anatinae containing the scoters,

goldeneyes and mergansers. They are distinguished as hole-nesting, long-tailed diving ducks

-meric pertaining to parts

dimeric having two divisions or parts

isomeric having similar parts in similar numbers

metameric (*see also* segmentation) having parts in company with one another

polymeric having many members of each series of a flower

heteromeric stratified

-mericous = -meric

merid an assemblage of plastids formed by fission from other plastids

merism the division of something into parts. Specifically a botanical term roughly corresponding to the zoological metamerism, but differing from it in that plant parts may be reproduced by increasing the number of radial axes as distinct from longitudinal divisions

meiomerism the condition of lacking some parts

metamerism the condition of having many parts joined together as the segments of a polychaete annelid

locomotory metamerism the theory that metameric segmentation arose in consequence of a sinuous method of progression

pseudometamerism false segmentation

oligomerism the condition of having few segments as distinct from metamery. The term is not in general use nor well defined

trimerism the condition of those animal phyla (Echinodermata, Stomochorda, and Pogonophora) in which the coelom is divided into three parts

meristem *see* -istem

-meristic pertaining to merism

allomeristic said of organisms which differ in the number of the parts of any organ from that which is customary in the group to which they belong. For example, Pycnogonida commonly have four pairs of legs, but a species with twelve pairs of legs has been recorded

isomeristic having the same number of parts

Meristomata a class of the phylum Arthropoda containing, among living forms, only the order Xiphosura

-merite generally used in *biol.* in the sense of a specific segment

caryomerite = idiomere

deutomerite the terminal section of a cephaline gregarine

epimerite the anterior division of a cephaline gregarine. It is often furnished with hooks and teeth and is separated from the reminder of the "body" by an ectoplasmic partition

protomerite that "segment" of a cephaline gregarine which lies immediately behind the epimerite

Merkel's *epon. adj.* from Friedrich Sigmund Merkel (1845–1919)

merle *Brit. obs.* = blackbird *Brit.*

merlin the falconine bird *Falco columbarius* (= *U.S.* pigeon hawk). *Brit.* the falconid bird *Falco aesalon* (= *F. columbarius aesalon*)

false **mermaid** any of several species of herb of the limnanthaceous genus Floerkea

mermi- prefix applied to indicate the alteration of the caste of an ant through infection with mermithid nematodes (*see* aner, ergate)

Mermithidae a family of mermithidean nematode worms admirably described by their popular name of hairworm

Mermithidea a class of nematodes, parasitic in insects in juvenile stages, with cyathiform amphids, a long and tenuous pharynx and a blindly ending intestine

Mermithiformes a taxon with the rank of superfamily preferred by some to the class Mermithidea

mermithoid *see* pharynx

-meron- *see* -mer-

epimeron that area of the arthropod integument which lies between the tergum and the insertion of the appendages and in some forming a posterolateral extension of the exoskeleton. In insects, specifically the posterior part of the pleuron of a thoracic segment

proepimeron the epimeron of the prothorax

sternepimeron the ventral portion of the epimeron

Meropeidae an extremely small archaic family of mecopteran insects originally erected to contain *Merope tuber*. The posterior end of the male of this species possesses forcep-like cerci resembling those of the earwigs

Meropidae a family of coraciiform birds commonly called bee-eaters. They have long, slender, laterally compressed, decurved bills with both mandibles coming to a sharp point. They are among the more brilliantly colored of all birds

Merostomata a class of aquatic Chelicerata containing the horseshoe crab (*cf.* Xiphisura)

Merothripidae a family of Thysanopteran insects commonly called large-legged thrips. They are distinguished by the greatly thickened front and hind femora

-merous a termination synonymous with -meric but rarely used interchangeably

anisomerous = asymmetrical

bimerous said of a bilocular fruit with the locules widely separated

cryptomerous possessing parts which show only in the offspring

embolomerous having an added part thrust in, *i.e.* equal development of centrum and intercentrum

heteromerous when there is no uniformity in the number of parts between organisms of the same species or organs on the same individual. For example, in many beetles the number of tarsal subsegments differ on the first and third legs

homeomerous said of insects having the same number of tarsal subsegments on all legs

homoeomerous the condition of a lichen in which the algal and fungal components are uniformly mixed and not layered

homoimerous the condition of a lichen in which the algal and fungal components have about the same volume, but are not of necessity uniformly distributed

monomerous formed from one pod as, for example, a single carpel that may give rise to a fruit

cryptotetramerous said of insects which have apparently three tarsal subsegments, but with a fourth minute subsegment concealed between two larger ones

cryptopentamerous said of insects having apparently four tarsal subsegments, but with a minute fifth subsegment concealed between two larger ones

merveille-du-jour *Brit.* the noctuid lepidopteran insect *Agriopis aprilina*

-mery termination synonymous with -merism, the form preferred in this work

-mes- *comb. form* meaning "middle"

Mesembryaceae = Aizoaceae

mesenchyme *see* -chyme

mesentery 1 (*see also* mesentery 2) the fold of mesoderm in which the viscera of coelomate animals is suspended

mesentery 2 (*see also* mesentery 1) the septa of anthozoan coelenterates

primary mesentery one of the mesenteries of an anthozoan that extends into, and fuses with, the stomodeum (= protocneme)

secondary mesentery one of the mesenteries of an anthozoan coelenterate that does not reach or fuse with the stomodeum

mesic having a moderate rainfall

mesidium the thickened portion of the mesochil of orchids

mesite either of two species of mesoenatid birds of the genus Mesoenas

white-breated mesite *M. variegata*

brown mesite *M. unicolor*

Mesitidae = Mesoenatidae

Mesitornithidae = Mesoenatidae

-meso-, *see* -mes-

mesod one of the two cells in the angiosperm embryosac in which the polar nuclei are housed

Mesodesmatidae a family of pelecypod mollusks commonly called wedge-clams in virtue of their shape

Mesoenatidae a small family of gruiform birds containing the mesites. They are distinguished by the long operculate slit nostrils

Mesoenidae = Mesoenatidae

mesome that portion of a vascular plant which lies between forkings or branchings and which was itself once a telome

meson the medial or sagittal plane of an animal

ventrimeson an imaginary line running down the ventral surface of an animal

Mesostigmata a suborder of acarine arachnids distinguished by posessing one pair of stigmata lateral to the legs, in lacking Haller's organ, and in posessing a hypostome that is not modified for piercing. In some the hypostome may be lacking

Mesotaeniaceae that family of zygonematale Chlorophyceae containing the forms of sometimes called saccoderm desmids. They consist of variously shaped nucleate cells either solitary or united into simple filaments

Mesoveliidae a small family of hemipteran insects somewhat resembling the Gerridae but clearly distinguished from them by the absence of predatory adaptions to the front legs. They are commonly called water-treaders

mesquite (*see also* borer) the leguminous shrubby tree *Prosopis juliflora*

-mesto- *comb. form* meaning "filled"

mestome *see* -tome

bimestris two months long

met a bacterial mutant indicating a requirement for methionine. This symbol is also used as a yeast genetic marker indicating a similar requirement.

metA a bacterial genetic marker having a character affecting the synthesis of the succinic ester of homoserine in *Salmonella typhimurium*, mapped at 79.9 mins. for *Escherichia coli*.

metB a bacterial genetic marker having a character affecting the activity of the succinic ester of homoserine + cysteine to cystathionine in *Salmonella typhimurium*, mapped at 76.0 mins. for *Escherichia coli*

metF a bacterial genetic marker having a character probably affecting the activity of 5,10-methylene

tetrahydrofolate reductase, mapped at 76.0 mins. for *Escherichia coli*

-meta- *comb. form* which properly denotes "among." Biologists have extended this meaning to "behind," "between," "adjacent," "later" and numerous other fanciful ideas, none of which, however, are as bad as the chemist's "having least water." The use of meta- as the third of a series commencing with pro-, meso- is common, though without classical justification

Metachlamydeae = Compositae

Metagnatha a group term for those insects in which the larval instars have jaws but the imagos feed by suction

Metatheria a term used variously for a superorder of mammals or a subclass of the same group, in both cases containing the single order Marsupialia (*q.v.*)

Metazoa a rarely used taxon of the animal kingdom into which are grouped all animals other than Protozoa

meteoric pertaining to weather. Particularly to those flowers the opening and closing of which is dependent upon atmospheric conditions

Methanomonadaceae a family of pseudomonodale schizomycetes containing those forms which are rod shaped, gram negative, and derive their energy from the oxidation of simple hydrogen and carbon compounds

methionine α-amino-γ-methylmercaptobutyric acid. $CH_3SCH_2CH_2CH(NH_2)COOH$. An amino acid essential to the nutrition of rats. It is important not only as a source of sulfur in the biosynthesis of other sulfur-containing amino acids, but is also a common methyl donor in transmethylation reactions

[methionine-1] a mutant (gene symbol *me-1*) in linkage group IV of *Neurospora crassa*. The phenotypic expression is a requirement for methionine

[methionine-2] a mutant (gene symbol *me-2*) in linkage group IV of *Neurospora crassa*. The phenotypic expression is a requirement for methionine

[methionine-3] a mutant (gene symbol *me-3*) in linkage group V of *Neurospora crassa*. The phenotypic expression is a requirement for methionine

[methionine-5] a mutant (gene symbol *me-5*) on linkage group IV of *Neurospora crassa*. The phenotypic expression is a requirement for methionine

[methionine-6] a mutant (gene symbol *me-6*) on linkage group I of *Neurospora crassa*. The phenotypic expression is a requirement for methionine

[methionine-7] a mutant (gene symbol *me-7*) on linkage group VII of *Neurospora crassa*. The phenotypic expression is a requirement for methionine

[methionine-8] a mutant (gene symbol *me-8*) in linkage group III of *Neurospora crassa*. The phenotypic expression is a requirement for methionine

[methionine-9] a mutant (gene symbol *me-9*) on linkage group VII of *Neurospora crassa*. The phenotypic expression is a requirement for methionine

[methionine-adenine-cysteine] a mutant (gene symbol *mac*) on linkage group I of *Neurospora crassa*. The phenotypic expression is a requirement for methionine and a preference for all three

Methochidae a family of scolioidean apocritan hymenopteran insects usually included with the Tiphiidae

dimethylpropiothetin dethiomethylase an enzyme that catalyzes the production of acrylate and dimethylsulphide from S-dimethyl-β-propiothetin

metochy the relation of a host ant to an inquiline (*cf.* symphily, synechtry)

Metopiidae a family of brachyceran dipteran insects

now usually divided between the Calliphoridae and the Sarcophagidae, the genus Metopia being placed in the latter

-metr- *comb. form.* meaning "uterus"

-metri- *comb. form* meaning "measurement"

basi**metrical** the distribution of organisms on the seabottom, either vertically or horizontally

endo**metrium** the endothelium of the uterus

allo**metron** a genetic change in the proportion of an existing character, such as an increase of depth of color in a flower

allo**metry** disproportionate size, particularly of an organ

bio**metry** biological statistics

Metzgerineae a suborder of jungermaniale Hepaticae in which the sporophytes lie some distance back from the growing apex of the gametophyte

meu *Brit.* the umbelliferous herb *Meum athamanticum* (= spignel)

mew *Brit. obs.* = gull

kirr-mew *N. Brit.* = tern

mf *see* [macrofine]

mg *see* [mahogany]. This symbol is also used for a yeast genetic marker indicating a requirement for α-methyl glucoside

mgt *see* [midget]

mi *see* [microphthalmia] or [minus]

Mi *see* [Nematode resistance]

miam *W.Afr.* any of several species of bucerotid birds of the genus Bycyanistes (*cf.* hornbill)

micelle a unit of a colloid

Micrasturinae that subfamily of Falconidae which contains the forest falcons. They are distinguished by their slender hawk-like bills, short wings, relatively long tail, and bare orbital region

microbe popular term for organisms, especially bacteria, too small to be seen with the naked eye

[**microchaete**] a mutant (gene symbol *mc*) mapped at 54.0 on the X chromosome of *Drosophila melanogaster*. The phenotypic microchaetes are small and irregular

Microchiroptera a suborder of chiropteran mammals containing all those forms in which the molars are multicuspid with sharp cusps (*cf.* Megachiroptera)

Micrococcaceae a family of gram positive, eubacteriale schizomycetes usually spherical in shape and frequently occuring in tetrads, irregular masses or chains. Frequently found on the skin of vertebrates. Many are responsible for food poisoning and some members of the genus Staphylococcus are pathogenic

Microcyprini an order of Osteichthyes, commonly called topminnows or killifishes, distinguished by the absence of a lateral line or adipose fin, the presence of an air bladder and the fact that the mouth opens in a dorsal direction

Microcrustacea a no longer acceptable zoological taxon at one time meaning much the same thing as Entomostraca

Microhylidae a family of anuran amphibians distinguished by their short limbs, pointed heads, and lack of a precoracoid bone. Curiously called narrow-mouthed toads or narrow-mouthed frogs

Microlepidoptera an obsolete division of the Lepidoptera containing all those not then classed as Microlepidoptera

Micromalthidae a monogeneric family of coleopterous insects distinguished by the fact that they possess paedogenetic larvae

Micromyiophilae flowers fertilized by small flies

micron micromillimeter (0.001 mm)

chylomicron a fat droplet in the bile

[**microoculus**] a mutant (gene symbol *mo*) mapped at 6.7± on the X chromosome of *Drosophila melanogaster*. The phenotypic expression is small eyes and narrow wings

Micropezidae a family of elongate, extremely longlegged, myodarian cycloraphous dipteran insects often called stilt-legged flies. They are mostly tropical

[**microphthalmia**] a mutant (gene symbol *mi*) in linkage group XI of the mouse. The phenotypic expression is small eyes and reduced pigment

Microphysidae a small family of hemipteran insects distinguished by their small size, black color, and the presence of a cuneus on the hemelytron

Micropodidae = Apodidae

Micropodiformes = Apodiformes

[**microptera**] a mutant (gene symbol *mp*) mapped at 0 on chromosome III of *Drosophila melanogaster*. As the name indicates the phenotypic wings are small but they are also ballooned and the tarsi are 4-jointed

[**micropterous**] a mutant (gene symbol *mp*) mapped at 24.0 on linkage group XI of *Bombyx mori*. The phenotypic expression is small wings

Micropterygidae a family of jugate lepidopteran insects differing from all other lepidoptera in possessing well-developed functional mandibles

Microsclerophora = Homosclerophora

Microspermae that order of monocotyledonous plants which contains the single family Orchidaceae. The characteristics of the family apply to the order

Microsporaceae that family of ulotrichale Chlorophyceae containing those forms which have unbranched filaments composed of uninucleate cells containing a lobed chloroplast

Microsporidia an order of cnidosporidian Sporozoa. Distinguished from the Actinomyxidia and the Myxosporidia by possessing a single polar capsule

Microtatobiotes a class of Protophyta containing the rickettsias and viruses. They are distinguished as obligate parasites of minute size

micrurgy surgical manipulations at the cellular level

mict that which is produced by mixing

amphi**mict** an organism which reproduces by amphimixis

apo**mict** a plant produced from seeds that have developed from unfertilized eggs or from somatic cells associated with the egg

amphiapo**mict** an organism which reproduces both sexually and parthenogenetically

haplo**mict** a hybrid of which the genome is derived from chromosomes and portions of chromosomes from various sources

mictium a heterogenous association of organisms, particularly plants, to which no other ecological definition can be readily applied

micturation urination

midge popular designation of any small dipteran insect. Usually, without qualification, refers to the Chironomidae

biting **midge** popular name of Ceratopogonids

grape-blossom **midge** *U.S.* the cecidomyiid *Contarinia johnsoni*

catalpa **midge** *U.S.* the cecidomyiid *Cecidomyia catalpae*

curcurbit **midge** *U.S.* the cecidomyiid *Cecidomyia citrulii*

splay-footed midge *Brit.* any of several species of the chironomid genus Tanypus

gall midge popular name of cecidomyiid Diptera

    alfalfa gall midge *U.S. Asphondylia websteri*

    balsam gall midge *Dasineura balsamicola*

    willow beaked gall midge *U.S. Mayetiola rigidae*

    chrysanthemum gall midge *U.S. Diarthronomyia chrysanthemi*

    dogwood club-gall midge *U.S. Mycodiplosis alternata*

horned midge *Brit.* the chironomid *Ceratopogon bicolor*

cloverleaf midge *U.S.* the cecidomyiid *Dasyneura trifolii*

pear midge *U.S.* the cecidomyiid *Contarinia pyrivora*

phantom-midge a small group of culicid nematoceran dipteran insects of the family Culicidae but distinguished from the mosquitos by the fact that they are not blood sucking

Monterey-pine resin midge *U.S.* the cecidomyiid *Cecidomyia resinicoloides*

rose midge *U.S.* the cecidomyiid *Dasyneura rhodophaga*

clover seed midge *U.S.* the cecidomyiid *Dasyneura leguminicola*

sunflower seed midge *U.S.* the cecidomyiid *Lasioptera murtfeldtiana*

solitary midge popular name of thaumaleid Diptera

sorghum midge *U.S.* the cecidomyiid *Contarinia sorghicola*

wheat midge *U.S.* the muscid *Sitodiplosis mosellana*

net-winged midge popular name of blepharocerids

midget *U.S.* either of two species of noctuid lepidopteran insect of the genus Oligia

[midget] a mutant (gene symbol *mgt*) mapped at 49.6± on the X chromosome of *Drosophila melanogaster*. The name is descriptive of the phenotype

mignonette any of several species, and numerous horticultural varieties of resedaceous herb of the genus Reseda

migrant 1 (*see also* migrant 2) any organism that wanders, but particularly those that move seasonally between fixed places

weather migrant an animal which migrates in response to changes, usually seasonal, in the climate

migrant 2 (*see also* migrant 1) *Austral.* any of several species of pierid lepidopteran insect of the genus Catopsilia

    common migrant *C. pyranthe*

    lemon migrant *C. pomona*

    yellow migrant *Austral. C. scylla*

lysolecithin migratase = lysolecithin acylmutase

migration the activity of a migrant

migrule a unit of migration

mildew any white fungal growth just visible to the eye and on which spores are not apparent

powdery mildew any erysiphale Ascomycete

miliarious covered with small glands

Milichiidae a family of brachyceran dipteran insects commonly included with the Phyllomyzidae

milk 1 (*see also* milk 2) the nutritious emulsion fed to young animals and specifically the emulsion produced in the mammary gland of mammals

pigeon's milk a fluid produced in the crop of a pigeon and fed to its young

witch milk milk induced in the mammary glands of many newly born mammals through a carry-over of hormones from the maternal bloodstream

milk 2 (*see also* milk 1, bush, mushroom, thistle and weed) in compound names of plants with a milky sap

devil's milk the milky sap of numerous euphorbiaceous plants particularly *Euphorbia peplus*

wolf's milk the euphorbiaceous herb *Euphorbia purpurea*

windmill *I.P.* any of several species of papilionid lepidopteran insect of the genus Tros (*cf.* batwing, clubtail, rose)

millepore popular name of milliporine hydrozoan coelenterates

Milleporina an order of hydrozoan coelenterates often called the false corals. The hydroid colony lies on the surface of a massive calcareous deposit in special cavities in which the medusae, subsequently becoming free, are formed

miller 1 (*see also* miller 2) an animal with a white, floury, appearance. *Brit.* almost synonymous with moth

dusty miller (*see also* miller 2) any of several white or light grey lepidopteran insects particularly those of the noctuid genus Acronycta

molly miller the bleniid fish *Blennius cristatus*

miller 2 (*see also* miller 1) a plant with a white, dusty appearance

dusty miller (*see also* miller 1) the compositaceous herb *Artemisia stelleriana* and the caryophyllaceous herb *Lychnis coronaria* (= rose campion)

millet any of several grasses with edible seeds. *U.S.*, without qualification *Setaria italica*, also called **German millet** and **Hungarian millet**. *Brit.*, without qualification, *Panicum miliaceum*

    African millet *Eleusine coracana*

    Japanese barnyard millet *Echinochloa frumentacea*

    broom-corn millet = true millet

    hog-millet = true millet

    Indian millet *U.S. Oryzopsis hymenoides* (*cf.* mountain rice)

    Japanese millet *Echinochloa frumentacea* or *Setaria italica*

    pearl-millet *Pennisetum americanum*

    Texas millet *Panicum texanum*

    true millet *Panicum miliaceum*

milo *see* maize

milt 1 (*see also* milt 2) the spermatozoa, or sometimes the entire testes, of fishes

milt 2 (*see also* milt 1) the spleen

milter a male fish

Milvinae that subfamily of accipitrid birds which contains the true kites

mime *I.P.* any of numerous species of papilionid lepidopteran insect of the genus Chilasa all of of which closely mimic Danaids

mimic *U.S.*, without qualification, the nymphalid butterfly *Limenitis archippus* (= viceroy). *S.Afr.* any of several species of nymphalid insect of the genus Hypolimnas

mimicry the assumption of the appearance of one organism by another

Batesian mimicry the mimicry of an inedible form (the model) by an edible one (the mimic)

Mullerian mimicry *either* that shown by several species having the same aposomatic colors in common *or* that form which appears unnecessary since both the mimic and the model are inedible

Mimidae that family of passeriform birds which contains the mockingbirds. The name derives from the habit of some species of imitating the sounds

and calls of other birds. The base of the middle toe
is adherent to the outer toe and the legs are longer
than those of most passerines

mimosa properly, any of several species of shrubs of the
leguminous genus Mimosa. The term is extended
to many other flowering shrubs including in *U.S.*,
the leguminous *Albizzia julibrissin* (*cf.* silktree)

   prairie mimosa *U.S.* the leguminous shrub *Desmanthus illinoensis*

   Texas mimosa the leguminous shrub *Acacia greggii*

Mimosoideae a subfamily of Leguminoseae distinguished by the actinomorphic flowers

mine the burrow made by leaf miners. The various
descriptive terms for these (*e.g.* serpentine) are
self-explanatory

miner 1 (*see also* miner 2) a term applied to numerous
insects which burrow in plant material

   asparagus miner *U.S.* the larva of the agromyzid
dipteran *Melanagromyza simplex*

   apple-bark miner *U.S.* the gracilariid lepidopteran
Marmara elotella

   strawberry crown-miner *U.S.* the larva of the gelechiid lepidopteran *Aristotelia fragariae*

   applefruit-miner *U.S.* the larva of the gracilariid
lepidopteran *Marmara pomonella*

   leaf-miner a term popularly applied to those insects
the larvae of which burrow through leaves

     European alder leaf-miner *U.S.* the tenthredinid
hymenopteran *Fenusa dohrnii*

     arborvitae leaf-miner *U.S.* the yponomeutid lepi-
dopteran *Argyresthia thuiella*

     azalea leaf-miner *U.S.* the gracilariid lepidopteran
Graciliaria azalealla

     birchleaf-miner the tenthredinid hymenopteran
Fenusa pusilla

     blotch leaf-miner popular name of gracilariid lepi-
dopteran insects

     aspen blotch leaf-miner *U.S. Lithocolletis tremu-
loidiella*

     corn blotch leaf-miner *U.S.* the larva of the
agromyzid dipteran *Agromyza parvicornis*

     chrysanthemum leaf-miner *U.S.* the agromyzid
dipteran *Phytomyza atricornis*

     columbine leaf-miner *U.S.* the agromyzid dipteran
Phytomyza minuscula

     cottonleaf-miner *U.S.* the nepticulid lepidopteran
Nepticula gossypii

     eggplant leaf-miner *U.S.* the gelechiid lepidopteran
Keiferia glochinella

     elm leaf-miner the tenthredinid sawfly *Fenusa ulmi*

     hollyleaf-miner *U.S. Phytomyza ilicis*

     native hollyleaf-miner *Phytomyza ilicicola*

     lantana leaf-miner *U.S.* the gracilariid lepidop-
teran *Cremastobombycia lantanella*

     larkspur leaf-miner *U.S.* the agromyzid dipteran
Phytomyza delphiniae

     lilacleaf-miner *U.S.* the gracilariid lepidopteran
Gracilaria syringella

     locust leaf-miner *U.S.* the chrysomelid coleopteran
Xenochalepus dorsalis

     morning-glory leaf-miner *U.S.* the lyonetiid lepi-
dopteran *Bedellia somnulentella*

     gregarious oakleaf-miner *U.S.* the gracilariid lepi-
dopteran *Cameraria cincinnatiella*

     solitary oakleaf-miner *U.S.* the gracilariid lepidop-
teran *Cameraria hamadryadella*

     pealeaf-miner *U.S.* the agromyzid dipteran
Liriomyza bryoniae

     privetleaf-miner the gracilariid lepidopteran
Gracilaria cuculipennella

     serpentine leaf-miner *U.S.* the agromyzid dipteran
Liriomyza brassicae

     sweetpotato leaf-miner *U.S.* the lyonetiid lepidop-
teran *Bedellia orchilella*

     spinach leaf-miner *U.S.* the larva of the an-
thomyiid dipteran *Pegomya hyoscyami*

     tupelo leaf-miner *U.S.* the heliozelid lepidopteran
Antispila nysaefoliella

     basswood leaf-miner *U.S.* the chrysomelid cole-
opteran *Baliosus ruber*

     boxwood leaf-miner *U.S.* the cecidomyiid dipteran
Monar thropalpus buxi

   white-fir needle-miner *U.S.* the olethreutid lepidopteran insect *Epinotia meritana Heinrick*

     lodgepole needle-miner *U.S.* the gelechiid lepidop-
teran *Coleotechnites milleri*

     pine needle-miner *U.S.* the gelechiid lepidopteran
Exoteleia pinifoliella

     spruce needle-miner *U.S.* the olethreutid lepidop-
teran *Taniva albolineana*

   grass-sheath miner *U.S.* the agromyzid dipteran
Cerodontha dorsalis

   appleleaf trumpet-miner *U.S.* the tischeriid lepidopteran *Tischeria malifoliella*

miner 2 (*see also* miner 1) in compound names of birds.
S.Amer. any of several funariids of the genus
Geositta. *Austral.* any of several meliphagids of the
genera Manorina and Myzantha and, ocassionally,
other birds.

   bell-miner *Manorina melanophrys*

   dusky miner *Myzantha obscura*

   gold-miner *Austral.* the meropid *Merops ornatus*
(*cf.* bee-eater)

   yellow-throated miner *Myzanthus flaxigula*

mineate red lead colored

[miniature] *either* a mutant (gene symbol *m*) in linkage
group II of *Habrobracon juglandis*. The name is
descriptive of the phenotypic body *or* a mutant
(gene symbol *m*) mapped at 36.1+ on the X
chromosome of *Drosophila melanogaster*. The
dark phenotypes have small wings and cannot be
distinguished from [dusky]

minivet *Ind.* any of several species of campephagid bird
of the genus Pericrocotus

   scarlet minivet *P. flammeus*

   small minivet *P. cinnamomeus*

mink (*see also* fish) any of numerous domesticated races
and mutations of the musteline carnivorous mammal *Mustela vison*

minnow any small fish frequently, but not always, freshwater cyprinids. *Brit.*, without qualification the
cyprinid *Phoxinus phoxinus*. *Austral.* some small
galaxiid fish are locally so called

   sheep-head minnow the cyprinodont *Cyprinodon
variegatus*

   mudminnow popular name for umbrid fish

   pikeminnow the poeciliid *Belonesox belizanus*
(= dwarf pike)

   topminnow any of numerous small, fresh water, surface feeding fish. In many parts of *U.S.*, it is specifically applied to the poeciliid *Gambusia affinis*,
but there are many cyprinodonts so called

     dwarf topminnow the cyprinodont *Heterandria
formosa*

     Gila topminnow the poeciliid fish *Poeciliopsis
occidentalis*

minor *Brit.* any of numerous species of noctuid lepidopteran insect of the genus Miana

mint (*see also* geranium) any of numerous species and
horticultural varieties of labiateous herbs, many of

the genus Mentha. *Brit.*, without qualification, *M. officinalis* (= *U.S.* garden mint)

applemint *M. rotundifolia*

bergamot-mint *M. citrata*

black mint a horticultural variety of peppermint

calamint *Brit. Calamintha officinalis*

catmint *Brit* = *U.S.* catnip

dogmint the labiateous herb *Satureia neogaea*

garden mint *U.S. M. officinalis* which is held by many to be a *hort. var.* of peppermint

horny mint *Acanthomintha ilicifolia*

horsemint *U.S.* any of several species of the genus Monarda *Brit., Mentha sylvestris* or *Mentha longifolia*

Japanese mint *M. arvensis*

round-leaved mint *M. rotundifolia*

lemon-mint *Monarda citriodora*

mountain mint any of numerous species of the genus Pycnanthemum particularly *P. virginianum*

peppermint (*see also* bark, gum) *M. piperita*

sheepmint *Satureia nepeta* (= field-balm)

spearmint *M. spicata*

stonemint the labiateous herb *Cunila mariana*

white mint *hort. var. M. piperita*

American wild mint *M. canadensis*

woodmint the labiateous herb *Blephilia hirsuta*

[**minus**] a mutant (gene symbol *mi*) mapped at 104.7 on chromosome II of *Drosophila melanogaster*. The phenotype has a small body with hair like bristles

[**minusbar**] a mutant (gene symbol *mb*) mapped at 43.4± on chromosome III of *Drosophila melanogaster*. It modifies [bar] to an almost normal eye condition

[**minute**] a series of genetic factors in *Drosophila melanogaster* which are dominant in their phenotype effect and homozygous lethal. All [minute] heterozygotes have characteristic minute (small, short, fine) bristles, and a longer developmental period than the wild type. There is usually an accompanying complex of other secondary effects

[**Minute (1) Blond**] a mutant (gene symbol *M(1)Bld*) mapped at 0.1± on the X chromosome of *Drosophila melanogaster*. The name is descriptive of the color of the minute bristles of the phenotype

[**Minute (1) 3E**] a mutant (gene symbol *M(1)3E*) mapped at 5.0+ on the X chromosome of *Drosophila melanogaster*. The phenotype shows a slight [Minute] effect

[**Minute (1)36f**] a mutant (gene symbol *M(1)36f*) mapped at 62.0± on the X chromosome of *Drosophila melanogaster*. The phenotype resembles M(1)o but has a better viability

[**Minute(1)n**] a mutant (gene symbol *M(1)n*) mapped at 62.7 on the X chromosome of *Drosophila melanogaster*. The phenotypic expression is a [Minute] effect with low fertility and viability

[**Minute (1) o**] a mutant (gene symbol *M(1)o*) mapped at 56.6 on the X chromosome in *Drosophila melanogaster*. The phenotypic expression involves low viability

[**Minute (2) Bridges**] a mutant (gene symbol *M(2)B*) mapped at 13.0 on chromosome II of *Drosophila melanogaster*. It has a medium [Minute] phenotype

[**Minute (2) Carry**] a mutant (gene symbol *M(2)C*) mapped at between 11.0 and 12.0 on chromosome II of *Drosophila melanogaster*. The phenotype shows a strong [Minute] effect with poor viability and fertility

[**Minute(2)D**] a mutant (gene symbol *M(2)D*) mapped at 55.0± on chromosome II of *Drosophila melano-*

gaster. The phenotype is light in body color and with pale bristles

[**Minute (2)e**] a mutant (gene symbol *M(2)e*) at 46.0± on chromsome II of *Drosophila melanogaster*. The phenotype shows a medium [Minute] effect

[**Minute(2)40c**] a mutant (gene symbol *M(2)40c*) mapped at 65.0± on chromosome II of *Drosophila melanogaster*.

[**Minute(2) 38b**] a mutant (*gene symbol M(2)38b*) mapped at 57.0± on chromosome II of *Drosophila melanogaster*. The phenotype shows an extreme [Minute] effect

[**Minute (2)z**] a mutant (gene symbol *M(2)z*) mapped at 12.9± on chromosome II of *Drosophila melanogaster*. The phenotype shows a medium to extreme [Minute] phenotype

[**Minute(2)33a**] a mutant (gene symbol *M(2)33a*) mapped at 108.0± on chromosome II of *Drosophila melanogaster*. The phenotype shows an extreme [Minute] effect

[**Minute)2)1²**] a mutant (gene symbol *M(2)1²*) mapped at 101.2 on chromosome II of *Drosophila melanogaster*. The phenotype shows a slight [Minute] effect

[**Minute(2)1**] a mutant mapped between 99 and 102.2 on chromosome II of *Drosophila melanogaster*. The phenotype shows an extreme [Minute] effect

[**Minute (2)173**] a mutant (gene symbol *M(2)173*) mapped at 92.3 on chromosome II of *Drosophila melanogaster*. The phenotype shows a moderate [Minute] effect

[**Minute (2) Schultz' 1**] a mutant (gene symbol *M(2)S1*) mapped at 15.0 on chromosome II of *Drosophila melanogaster*. The phenotype shows an extreme [Minute] effect

[**Minute(2) Schultz' 2**] a mutant (gene symbol *M(2)S2*) mapped at 55.1 on chromosome II of *Drosophila melanogaster*.

[**Minute(2)Schultz' 4**] a mutant (gene symbol *M(2)S4*) mapped at 55.1 on chromosome II of *Drosophila melanogaster*. The phenotype shows a moderate [Minute] effect

[**Minute(2)Schultz' 5**] a mutant (gene symbol *M(2)S5*) mapped at 53.5 on chromosome II of *Drosophila melanogaster*. The phenotype shows a moderate [Minute] effect

[**Minute(2)Schultz' 7**] a mutant (gene symbol *M(2)S7*) mapped at 77.5 on chromosome II of *Drosophila melanogaster*. The phenotype shows a strong [Minute] effect

[**Minute(2)Schultz' 8**] a mutant (gene symbol *M(2)S8*) mapped at 55.1 on chromosome II of *Drosophila melanogaster*. The phenotype shows a slight [Minute] effect

[**Minute(2)Schultz' 10**] a mutant (gene symbol *M(2)S10*) mapped at 55.1 on chromosome II of *Drosophila melanogaster*. The phenotypic shows a slight [Minute] effect

[**Minute (2) Schultz' 11**] a mutant (gene symbol *M(2)S11*) mapped at 43.0± chromosome II of *Drosophila melanogaster*. The phenotype shows a slight [Minute] effect

[**Minute(2)Schultz' 12**] a mutation (gene symbol *M(2)S12*) mapped at 56.0± on chromosome II of *Drosophila melanogaster*. The phenotype shows a slight [Minute] effect

[**Minute (2) Schultz' 13**] a mutant (gene symbol *M(2)S13−*) mapped at 50.0± on chromosome II of *Drosophila melanogaster*. The phenotype shows a strong [Minute] effect

[Minute(3)f] a gene (gene symbol M(3)f) mapped at 62.4 on chromosome III of Drosophila melanogaster.

[Minute(3) 124] an allele of M(3)w

[Minute(3) beta] a mutant mapped at 87.0± on chromosome III of Drosophila melanogaster. The phenotype shows a medium [Minute] effect

[Minute (3) Burkart] an allele of M(3)w

[Minute(3)36e] a mutant (gene symbol M(3)36e) mapped at 84.5 on chromosome III of Drosophila melanogaster. The phenotype shows a medium [Minute] effect

[Minute(3)Florida] a allele of M(3)w

[Minute(3)h] An allele of M(3)33j

[Minute(3)j] a mutant (gene symbol M(3)j) mapped at 90.2 on chromosome III of Drosophila melanogaster. The phenotype shows an extreme [Minute] effect

[Minute(3)33j] a mutant (gene symbol M(3)33j) mapped at 40.2 on chromosome III of Drosophila melanogaster. The phenotype shows a medium [Minute] effect

[Minute(3)l] a mutant (gene symbol M(3)l) mapped at 101.0 on chromosome III of Drosophila melanogaster. The phenotype shows a medium [Minute] effect

[Minute(3)Schultz' 31] a mutant (gene symbol M(3)S31) mapped at 50.0 on chromosome III of Drosophila melanogaster. The phenotype shows a medium [Minute] effect

[Minute(3)Schultz' 34] a mutant (gene symbol M(3)S34) mapped at 49.0± on chromosome III of Drosophila melanogaster. The phenotype shows a slight [Minute] effect

[Minute(3)Schultz' 37] a mutant (gene symbol M(3)S37) mapped at 39.7± on chromosome III of Drosophila melanogaster. The phenotype shows an extreme [Minute] effect

[Minute(3)Schultz' 38] a mutant (gene symbol M(3)S38) mapped at 44.0± on chromosome III of Drosophila melanogaster. The phenotype shows a strong [Minute] effect

[Minute(3)Schultz' 39] a mutant (gene symbol M(3)S39) mapped at 47.0± on chromosome III of Drosophila melanogaster. The phenotype shows a strong [Minute] effect

[Minute(3)y] An allele of M(3)33j

[Minute(3)w] a mutant (gene symbol M(3)w) mapped at 79.7 on chromosome III of Drosophila melanogaster. The phenotype shows a strong [Minute] effect

[Minute-4] a mutant (gene symbol M-4) mapped at between 0 and 0.2 on chromosome IV of Drosophila melanogaster. The phenotype shows a medium [Minute] effect

minute effect see [minute]

Mirabileae a tribe of the family Nyctaginaceae with glabrous ovaries, straight embryos, and large cotyledons

mire originally mud though in some places used for bog or marsh

pickmire Brit. dial. numerous birds frequenting muddy swamps, particularly the black-headed gull and the lapwing

Miridae a very large family of hemipteran insects commonly called leaf bugs, plant bugs or capsid bugs. They are distinguished by the possession of a cuneus

mirliton the cucurbitaceous vine Sechium edule and its edible squash-like fruit

miro N.Z. the podocarpaceous tree Podocarpus ferrugineus

miro-miro N.Z. the Muscipapid bird Petroica macrocephala (= pied tit)

mirror the brightly metallic colored area in the wing of some birds (e.g., a drake mallard)

mis see [misproportioned]

Miscogasteridae = Pteromalidae

[misformed] a mutant (gene symbol msf) mapped at 55.6 on chromosome II of Drosophila melanogaster. The misformation applies to the eyes and wings of the phenotype

[misproportioned] a mutant (gene symbol mis) plotted at 1.3± on the X chromosome of Drosophila melanogaster. The abdomen of the phenotype is abnormal both in size and shape

[misheld wings] a mutant (gene symbol mwi) mapped at 0.4± on the X chromosome of Drosophila melanogaster. The phenotypic expression are upright, widely divergent, wings

missey-moosey U.S. the rosaceous tree Pyrus americana (= mountain ash)

mist comb. form meaning "mingled"

Scotch mist U.S. the rubiaceous herb Galium sylvaticum

mistletoe (see also bird, thrush) any of numerous species of parasitic dicotyledonous plants of the family Loranthaceae. U.S., without qualification, usually Phoradendron flavescens. Brit., without qualification, Viscum album

dwarf mistletoe U.S. the loranthaceous parasite Arceuthobium pusillum

false mistletoe U.S. the loranthaceous parasitic herb Phoradendron flavescens

[misty] a mutant (gene symbol m) mapped in linkage group VIII of the mouse. The phenotypic expression is a pale coat-color, occasionally spotted

Major Mitchell Austral. the psittacid bird Cacatua leadbeateri (= pink cockatoo)

mite 1 (see also mite 2) any acarine arachnoid arthropod which is not a tick (q.v.)

beetle-mite popular name of galumnid acarines. They are distinguished by the thick hard integument and the presence of a pseudostigmatic organ

dryberry mite U.S. the trombidiform Phyllocoptes gracilis

rebberry mite U.S. the trombidiform Aceria essigi

bird mite any parasitic acarine of birds

cotton blister-mite U.S. the trombidiform Acalitus gossypii

pear leaf blister-mite U.S. the trombidiform Eriophyes pyri

walnut blister-mite U.S. the trombidiform Aceria erinea

broadmite U.S. the trombidiform Hemitarsonemus latus

brown mite = clover mite

big-bud mite Brit. the trombidiform Vasates fockeui (= plum rust-mite U.S.)

blueberry bud-mite U.S. the trombidiform Aceria vaccinii

citrus bud-mite the trombidiform Aceria sheldoni

currant bud-mite U.S. the trombidiform Cecidophyes ribis

filbert bud-mite U.S. the trombidiform Phytoptus avellanae

pine bud-mite U.S. the trombidiform Phytoptus pini

bulb-mite U.S. sarcoptiform Rhizoglyphus echinopus

cheese-mite the tyroglyphid *Tyroglyphus siro*

chicken-mite *U.S.* the dermanyssid *Dermanyssus gallinae*

chigger mite *see* chigger

clover-mite the tetranychid *Bryobia praetiosa*

wheat curl-mite *U.S.* the trombidiform *Aceria tulipae*

cyclamen-mite *U.S.* the trombidiform *Steneotarsonemus pallidus*

depluming mite *U.S.* the sarcoptid *Knemidokoptes gallinae*

tip-dwarf mite *U.S.* the trombidiform *Eriophyes thujae*

grape erineum-mite *U.S.* the trombidiform *Eriophyes vitis*

fig-mite *U.S.* the trombidiform *Aceria ficus*

citrus flat-mite *U.S.* the trobidiform *Brevipalpus lewisi*

housefly mite the dermanyssid *Macrochelus muscae*

follicle-mite *U.S.* the tromibidiform *Demodex folliculorum*

cat follicle-mite *U.S.* the trombidiform *Demodex cati*

cattle follicle-mite *U.S.* the trobidiform *Demodex bovis*

dog follicle-mite *U.S.* the trombidiform *Demodex canis*

goat follicle-mite *U.S.* the trombidiform *Demodex caprae*

hog follicle-mite *U.S.* the trombidiform *Demodex phylloides*

horse follicle-mite *U.S.* the trombidiform *Demodex equi*

sheep follicle-mite *U.S.* the trombidiform *Demodex ovis*

northern fowl-mite *U.S.* the dermanyssid *Ornithonyssus sylviarum*

tropical fowl-mite *U.S.* the dermanyssid *Ornithonyssus bursa*

dried-fruit mite *U.S.* any of several sarcoptiforms of the genus Carpoglyphus, particularly *C. lactis*

gall mites popular name of eriophyids

grain-mite *U.S.* the sarcoptiform *Acarus siro*

winter grain-mite *U.S.* the trombidiform *Penthaleus major*

Banks grass-mite *U.S.* the trombidiform *Oligonychus pratensis*

red grasshopper-mite *U.S.* the trombidiform *Eutrombidium trigonum*

harvest-mite a group of large trombidiids distinguished by their large size and usually red color. However, they are not red spiders (*q.v.*).

horny mite = galumnid mite

itch-mite any of numerous sarcoptiform acarine skin parasites, particularly *Sarcoptes scabei*

straw itch-mite *U.S.* the trombidiform *Pyemotes ventricosus*

pecan leafroll mite *U.S.* the trombidiform *Aceria caryae*

scaly-leg mite *U.S.* the sarcoptiform *Knemidokoptes mutans*

mole-mite the parasitid *Haemogamasus litonyssoides*

mushroom-mite *U.S.* the *sarcoptiform Tyrophagus putrescentiae*

privet-mite *U.S.* the trombidiform *Brevipalpus obovatus*

rat-mite any of several species of parasitids of the genera Euchinolaelaps and Liponyssus

tropical rat-mite *U.S.* the dermanyssid *Ornithonyssus bacoti*

avocado red mite *U.S.* the trombidiform *Oligonychus yothersi*

citrus red mite *U.S.* the trombidiform *Panonychus citri*

European red mite *U.S.* the trombidiform *Panonychus ulmi*

southern red mite *U.S.* the trombidiform *Oligonychus ilicis*

tomato russet mite *U.S.* the trombidiform *Vasates lycopersici*

apple rust-mite *U.S.* the eriophyid *Aculus schlechtendali*

buckeye rust-mite the trombidiform *Oxypleurites aesculifoliae*

citrus rust-mite *U.S.* the trombidiform *Phyllocoptruta oleivora*

grain rust-mite *U.S.* the trombidiform *Abacarus hystrix*

grape rust-mite the trombidiform *Calipitrimerus vitis*

pear rust mite *U.S.* the trombidiform *Epitrimerus pyri*

plum rust-mite *U.S.* the trombidiform *Vasates fockeui* (= *Brit.* (big-bud mite)

scab-mite *U.S.* the sarcoptiform *Psoroptes equi*

sheep scab-mite *U.S.* the sarcoptiform *Psoroptes equi ovis*

bulb-scale mite *U.S.* the trombidiform *Steneotarsonernus laticeps*

silver mite the trombidiform *Vasates cornutus*

peach silver mite *U.S.* the trombidiform *Aculus cornutus*

sitting mite = spider-mite

snout-mite any of numerous trombidiforms of the family Bdellidae

spider-mite any of numerous tetranychids mostly of the genus Tetrarhynchus (*cf.* red spider)

desert spider-mite *U.S. T. desertorum*

Pacific spider-mite *U.S. T. pacificus*

Schoene spider-mite *U.S. T. schoenei*

four-spotted spider-mite *U.S. T. canadensis*

two-spotted spider-mite *U.S. T. urticae*

spruce spider-mite *U.S. Oligonychus ununguis*

strawberry spider-mite *U.S. T. atlanticus*

turnip spider-mite *U.S. T. tumidus*

yellow spider-mite *U.S. Eotetranychus carpini borealis*

Yuma spider-mite *U.S. Eotetranychus yumensis*

six-spotted mite *U.S.* the trombidiform *Eotetranchus sexmaculatus*

water-mite those trombidiform mites which live in water. The term is of more ecological than taxonomic value

saltwater mite popular name of halacarids

brown wheat-mite *U.S.* the trombidiform *Petrobia latens*

mite 2 (*see also* mite 1) a small object

karyomite *obs.* = chromosome

chondriomite a group of mitochondria arranged in a straight line (*cf.* chondriosphere)

mithan = gayal

atelomitic pertaining to a chromosome the spindle of which is not attached to the end

mitochondrion *see under* chondrion

mitom a term coined to describe the network of threads in protoplasm when that substance was thought to have a reticular structure

paramitom the supposed interstitial substance between the supposed mitoms (= paraplasm)

mitosis the sum total of the nuclear and chromosomal activities in those cell divisions in which continuity of the chromosome complement is retained

amitosis cellular division in which mitotic activity has not been observed

c-mitosis one which has been arrested by the action of colchicine or some similar reagent

cryptomitosis a type of mitosis found in some Protozoa in which the achromatic figure is evident, but chromosomes are not apparent

endomitosis the division of the chromosomes within the bounds of the nucleus, thus resulting in polyploidy

eumitosis mitosis involving a complete series of mitotic figures

haplomitosis that form of mitosis in which the whole endosome acts as a centriole as in some euglenoid Mastigophora

intermitosis the period between two cell divisions

karyomitosis = mitosis

mesomitosis *either* mitosis taking place within a complete nuclear membrane *or* that form of division in which the endosome furnishes the chromosomes, but not the centriole

metamitosis complete mitosis

paramitosis a type of mitosis found in some Protozoa in which the chromosomal behavior is abnormal

promitosis that form of division in which both centrioles and chromosomes come from the endosome

pseudomitosis an early term for a type of nuclear division in which discrete chromosomes were not visible

antimitotic a mitosis inhibitor

-mitr- *comb. form* originally meaning any "headdress" but later confined to a bishop's mitre

papal mitre *U.S.* the mitrid gastropod shell *Mitra papalis*

Mitridae a family of gastropod Mollusca with spindle-shaped thick shells having a sharply pointed spire. Commonly called mitre-shells

-mix- *comb. form* meaning "mixing" and by extension, "breeding" and "fertilization"

amixia the condition of two individuals which are mutually infertile

panmixia random reproduction in the sense of absence of selection (= panmixis)

-mixis- = -mix-

amixis the absence of fertilization. Occasionally extended to mean specifically absence of fertilization by reason of the absence of gonads

amphimixis the union of two gamets from separate parents

apomixis the replacement of sexual reproduction by any form of propagation which, by avoiding meiosis and by syngamy, permits the reproduction unchanged of a given genotype

automixis *either* self fertilization *or* automictic parthenogenesis (*q.v. under -genesis*)

cytomixis the exchange of nuclear material between pollen mother cells

endomixis the reorganization of nuclear material within a protozoan, through the replacement of an existing macronucleus with the products of division of one or more micronuclei (*cf.* hemimixis)

endomixis the dissolution of the macronucleus and its reformation from the micronucleus in Protozoa without conjugation taking place

hemixis = hemimixis

hemimixis the break-up and refusion of the parts of the macronucleus of a protozoan cell without contributions from the micronuclei (*cf.* endomixis)

karyomixis the fusion of the two nuclei of a teleutospore

panmixis = random mating in the sense of absence of selection (= panmixia)

parthenomixis parthenogamy

pseudomixis = pseudapogamy

mixium a term coined to denote structures of dual function

nephromixium a unit consisting of a coelomostome and a protonephridial tubule connected to the same duct (*cf.* nephridium)

metanephromixium a nephridium which serves also as a genital duct (*cf.* nephridium)

mixote = mixochimaera

mk *see* [murky]

mlg a bacterial mutant showing a utilization of melizitose

mnioid moss-like

mo *see* [microoculus]

Mo *see* [Mottled]

moano the mullid fish *Parpeneus multiasciatus*

mob term for a group of kangaroos, though troop is equally common. In *Austral.* extended to domestic cattle

Mobulidae a family of rays commonly called mantas

moccasin (*see also* flower) *U.S.* any crotalid snake of the genus Agkistrodon

highland moccasin *U.S. A. contortrix* (= copperhead)

water-moccasin *U.S. A. piscivorus* (= cottonmouth)

mocha *Brit.* any of several species of geometrid lepidopteran insects of the genus Ephyra. Without qualification, *E. annulata* is usually meant

Mochocidae a family of naked catfish commonly referred to as "upside down catfish" by reason of their peculiar swimming behavior

dauermodification one which is supposed to be produced in an organism by the action of the environment and transmitted with slowly diminishing effect to several successive generations of offspring

[modifier] a mutant (gene symbol *m*) in linkage group II of the guinea pig. The phenotypic expression is a modification of *R*(*q.v.*)

[modifier of thi-lo] a mutant (gene symbol *thi-lo*) in linkage group III of *Neurospora crassa*. This mutation removes from *thi-l* the ability to use thiamine derivatives but not thiamine itself

[modifier of vis (31717)] a mutant (gene symbol *m-1*) in linkage group I of *Neurospora crassa*. The name is descriptive

-modiol- *comb. form* meaning the "nave" of a spoked wheel

mohopereru *N.Z.* the rallid bird *Rallus philippensis* (= banded rail)

mohoua *N.Z.* the malurine bird *Mohoua ochrocephala* (= bush canary)

[Moiré] a mutant (gene symbol *Mé*) mapped at 20.0± on chromosome III of *Drosophila melanogaster*. The phenotypic eye color is flecked-brown

mojarra any of many species of gerrid and cichlid fishes

mokomoko *N.Z.* the scincid lacertilian *Lygosoma moco*

mola popular name of molid fish

mold any fungus forming a matted or felted mycelium in damp places

black mold any mucorale phycomycete

bread-mold any mucorale phycomycete, generally Rhizopus

fish-mold any saproligniale phycomycete

slime-mold popular name for the myxomycetes

mole (*see also* cricket, mite, plant, rat, salamander, snake) any of numerous burrowing insectivores. *Brit.*, without qualification, *Talpa europaea*

California mole *Scapanus latimanus*

eastern mole *U.S. Scalopus aquaticus*

golden mole popular name of South African chrysochlorid insectivores

marsupial mole popular name of parameloid notoryctid marsupials

Pacific mole *U.S. Scapanus orarius*

shrew-mole *U.S. Neurotrichus gibbsi*

starnose mole *U.S. Condylura cristata*

strand mole = strand rat

hairytail mole *U.S. Parascalops breweri*

biomolecule a "living" molecule (*cf.* biomonad, biomore)

molendinaceous having wings like the sail of a windmill

Molgulidae a family of ptychobranchiate ascidians closely allied to the Pyuridae and distinguished from them by the lobes of the oral siphon

Molidae a family of very large plectognathous fishes commonly called ocean sunfish. They are distinguished by the vertically compressed body, the greatly reduced tail, and the absence of pelvic bones and pelvic fins

molka the rosaceous creeping herb *Rubus chamaemorus* and its edible fruit (= cloud berry)

Moll's *epon. adj.* from Jacob Antonius Moll (1832–1914)

Mollusca a phylum of bilateral coelomate animals containing, *inter alia* the clams, the oysters, the snails, the whelks, the squids and octopus. The group is distinguished by the presence of a mantle enclosing a visceral mass and a calcareous skeleton which normally lies outside the body

mollusk a spelling vehemently defended as the only possible one by those who do not regard mollusc as an English word

molly aquarists' name of poeciliid fish of the genus Mollienisia

mollyhawk = mollymawk

mollymauk = mollymawk

mollymawk *N.Z.* any of several species of diomedeid birds (= albatross)

moloch *Austral.* the agamid saurian reptile *Moloch horridus*

Molossidae a family of chiropteran mammals distinguished by having a tail that protrudes beyond the uropatagium

Molpadonia an order of holothuroid Echinodermata distinguished by the absence of ambulacral feet but having a respiratory tree

molt the shedding of feathers by a bird, the shedding of the epidermis by a reptile and the shedding of the outer chitinous case by an arthropod or, occasionally, other invertebrates (= ecdysis)

molula the articulation of the insect femur with the tibia

molybdeous lead-colored

mombin the anacardiaceous tree *Spondias lutea* and its edible fruit. Sometimes applied to Spanish plum (*q.v.*)

Momotidae that family of coraciiform birds which contains the motmots. Most are readily distinguished by the extremely long tail terminating in two raquet-shaped feathers

-mon- *comb. form* meaning "one" or "single"

Monacanthidae a family of plectognathous fishes erected to contain the file-fish (*q.v.*) by those who do not hold that they belong in the Balistidae

Monachinae a subfamily of phocid pinniped carnivores commonly called the monk-seals. They are closely allied to the Phocinae from which they are distinguished by the fact that the first and fifth toes of the hind feet are significantly the longest and terminate in rudimentary nails

monad a general term applied to minute colorless flagellates, frequently showing amoeboid motion, and to zoospores

biomonad a symbiotic system of biomores (*cf.* biomolecule, biomore)

Monadelphia that class in the Linnaean system of plant classification distinguished by the fact that the stamens are united by their filaments into one set

monal any of several species of brilliantly colored phasianid birds of the genus Lophophorus usually, without qualification, *L. refulgens* (*cf.* impeyan pheasant)

Monandrae a subfamily of Orchidaceae distinguished from the Diandrae by the fact that there is only one functional anther situated terminally on the gynostegium

Monandria a class in the Linnaean classification of plants distinguished by the possession of a single stamen

monarch 1 (*see also* monarch 2) the danaid lepidopteran insect *Danaus plexippus*

African monarch *D. chrysippus*

monarch 2 (*see also* monarch1) any of several muscicapid birds, particularly those of the genus Monarcha

Monarchidae a family of passeriform birds, usually included in the Muscicapidae

Monaxonida a subclass of demospongious Porifera in which the spicules are monaxonic

Monera a taxon erected to contain those acellular organisms (*i.e.* bacteria, Cyanophyta) that have a haploid nucleus lacking a nuclear membrane

baldmoney *Brit.* the umbelliferous herb *Meum athamanticum* (= spignel)

bastard baldmoney *Gentiana acuta* (= bastard gentian)

mongabey any of several cercopithecid primates of the genus Cercocebus

mongol *I.P.* the nymphalid lepidopteran insect *Araschnia prorsoides*

mongoose any of about 100 species of herpestine viverrid carnivores, distributed in about half a dozen genera. Some of those in the genus Herpestes, sometimes called the "true mongooses" do kill snakes, but the diet of the group as a whole is very varied

Monhysteridea a class of Nematoda with a smooth or slightly ringed cuticle, sometimes bristly and with circular amphids

monias the mesoenatid bird *Monies benschi*

-monili- *comb. form* properly meaning "necklace" but usually, by extension, "in the form of a string of beads"

Monimiaceae a family of ranale dicotyledonous angiosperms distinguished from its close allies the Calycanthaceae by the transversely dehiscing stamens and separate carpels scattered over the receptacular cup

monitor any of several species of varanid lacertilian reptile of the genus Varanus

earless monitor popular name of Lanthanotidae

false monitor the teiid *Callopistes flavipunctatus*

monjita any of several tyrannid birds of the genux Xolmis

monk *Brit.* = male bullfinch. *Austral.* = friar-bird

monkey (*see also* ear, hand) any primate mammal possesing a tail

bear-monkey the Chinese cynopithecid *Macaca arctoides*

blue monkey = vervet

capuchin monkey cebid primates of the genus Cebus. They are the commonest of all monkeys in captivity

diadem monkey any of several guenons living in the Congo and in which the hair of the head, frequently vividly colored, is raised to a peak

Diana monkey the cercopithecid *Cercopithecus diana* (*cf.* guenon)

douroucouli monkey any of several species of cebid of the genus Nyctipithecus

green monkey = vervet

half monkey popular name of pithecids

howler monkey cebid primates of the genus Alouatta

lion-monkey properly, the tamarin *Leontocebus leoninus* though the term is also applied to the cynopithecid *Macaca albibarbata* (*cf.* wanderoo)

military monkey any of several cercopithecids of the genus Erythrocebus

putty-nosed monkey the guenon *Cercopithecus mona* and some closely allied species

ouakari monkey any of several cebid monkeys of the genus Brachyurus

owl-monkey = douroucouli monkey

proboscis-monkey the Bornean colobid *Nasalis larvatus* which, as the name indicates, is distinguished by the enormous tubular nose of the male

rhesus monkey the cynopithecid *Macaca rhesus* of India, widely used in contemporary biological research

sacred monkey any primate thought by natives to have supernatural powers or by dealers to bring a higher price if this name be applied to it

spider-monkey cebid primates of the genus Ateles. The name derives from the long limbs and very long prehensile tails

squirrel-monkey cebid primates of the genus Saimiri. They are long-legged monkeys with club-shaped tails and large ears

lion-tailed monkey the cercopithecid *Macacus leoninus*

pig-tailed monkey the cercopithecid *Macacus nemestrinus*

vervet monkey the cercopithecid *Cercopithecus lalandii* (*cf.* guenon)

woolly monkey any of several cebid primates of the genus Lagothrix. The name derives from the short thick fur

lanceolated monklet the piciform bucconid bird *Micromonacha lanceolata*

-mono- *see* -mon-

monoao *N.Z.* the podocarpaceous tree *Dacrydium kirkii* (*cf.* pine)

Monoblepharidales an order of uniflagellate Phycomycetes distinguished by the fact that the naked zygote emerges from the oogonium before forming a cell wall

Monocentridae a monogeneric family of berycoid fishes distinguished by the fact that the spines of the dorsal fin are alternately inclined to one side and the other

Monocleaceae a family of marchantiale Hepaticae distinguished by the presence of a hood-like sheath posterior to the female receptacle

Monocotyledoneae a class of Angiospermae named from the fact that the germinating seed produces a single cotyledon. They are also distinguished by the fact that vascular bundles are scattered through the stem but not arranged in a cylinder. The leaves are very commonly parallel veined and the flower parts have a triradial symmetry

Monodactylidae a small family of percoid teleost fish with a very deep and strongly compressed body and elongated, thickly scaled, dorsal and anal fins

Monoecia in the Linnaean system of plant classification the order of the class Polygamia in which all three kinds of flowers are on the same plant

Monogenea an order of trematode Platyhelminthes distinguished by a posterior adhesive disk commonly provided with hooks. There may be anterior adhesive structures but the oral sucker is usually absent. If the Monogenea are regarded as a subclass, rather than as an order, it will comprise the Monopisthocotylea and the Polyopisthocotylea

Monogononta a class of the phylum Rotifera distinguished by the possession of a single germovitellarium

Monogynia those orders of the Linnaean system of plant classification distinguished by the possession of one pistil

Monommatidae = Monommidae

Monommidae a small family of coleopteran insects closely allied to the Rhizophagidae

Monophlebinae a subfamily of coccid homopteran insects commonly called giant coccids or cottony cushion-scales

Monopisthocotylea an order of monogeneous trematode Platyhelminthes distinguished by the presence of a single posterior adhesive organ

Monostomata a division of prosostomate trematodes in which the oral sucker is the only adhesive organ

Monotomidae a small family of bark-dwelling coleopteran insects

Monotremata the only order of prototherian Mammalia. They are represented today by the Australasian duck-billed platypus and the spiny anteater. Reproduction is by means of eggs and the mammary glands are reduced to enlarged sweat glands

Monotropaceae a small family of ericalous plants which are all saprophytic and lack chlorophyll. This characteristic is distinctive

Monotropoideae a subfamily of the Pyrolaceae containing the chlorophyll-less saprophytic genera. This subfamily is often given familial rank

Monro's *epon. adj.* from Alexander Monro (1733–1818)

Gila monster *U.S.* the helodermatid lacertilian reptile *Heloderma suspectum*

Monsteroideae a subfamily of the Araceae

Monstrilloida a small order of aberrant composed Crustacea distinguished by the fact that the mouth and alimentary canal are missing from the adult. The newly hatched young and adults are marine planktonic forms while the intermediate stages are parasitic in polychaete annelids

montane pertaining to mountains

Montgomery's *epon. adj.* from William Fetherston Montgomery (1797–1859)

monticule tubercle-bearing orifices on the surface of some Ectoprocta

moonaul = monal

moor (*see also* hen) an area of predominantly peaty soil
  high moor an area rich in peat but which no longer has bog water avilable to it
  low moor a peaty swamp
  moss moor a boggy moor, which is sufficiently dry to grow sphagnum

mooruk = cassowary

moose (*see also* bird, wood) the very large alcine cervid artiodactyle *Alces alces*. The European race is called elk, a term applied in *N. Amer.* to the cervine *Cervus canadensis*

Moraceae that family of urticalous dicotyledons which contains, *inter alia*, the mulberries and bread fruits. The family is easily distinguished from the closely allied Urticaceae by the milky juice and from the elms by its elastic stamens

moray popular name of muraenid teleost fish

Mordellidae a family of coleopteran insects with a hump-backed wedge-shaped laterally compressed appearance. Commonly called tumbling flower-beetles

bio**more** an aggregate of similar biomolecules (*cf.* biomolecule, biomonad)

morel (*cf.* morrel) popular name of helvellaceous asco-mycete fungi. Without qualification the term morel is usually applied to the widely eaten *Morchella esculenta*
  beefsteal morel *Helvella esculenta*
  black morel *U.S. Morchella angusticeps*
  early morel *U.S. Verpa bohemica*
  false morel popular name of helvellacean fungi of the genus Helvella (= lorchel)
  thick-footed morel *U.S. Morchella crassipes*
  half-free morel *U.S. Morchella hybrida*

morepork *Austral.* and *N.Z.* the strigiform bird *Ninox novaeseelandiae*

mores plural of mos, but most used in reference to the habits or customs of the mos

Morgagni's *epon. adj.* from Giovanni Battisa Morgagni (1682–1771)

morgan a unit of genetic recombination. Thus a centi-morgan refers to a one per cent recombination rate (*cf.* strane)

Morgan's *epon. adj.* from Thomas Hunt Morgan (1866–1945)

moribund on the point of death

morillon = female goldeneye

Moringaceae a family of rhoeadale dicotyledonous angiosperms containing the single genus Moringa distinguished by the alternate decompound leaves with opposite pinnae, and the pentamerous zygo-morphic flowers

Moringincae a suborder of Rhoeadales containing the family Moringaceae

morinous deep purple-black (= morulose)

bipoly**morious** consisting of not less than two parts

mormon *I.P.* any of several species of papilionid lepi-dopteran insects of the genus Papilio (*cf.* swallow-tail, redbreast, peacock, spangle, raven, helen)

Mormyridae a large family of isospondylous fishes distinguished by an enormously elongated snout

Moroideae a subfamily of Moraceae distinguished by having the stamens incurved and the leaves folded in the bud

-morph- *comb. form* meaning "shape." The *subst. forms* -morph, -morphism, -morphosis and -morphy have specialized meanings but the *adj. forms.* -morphis and -morphous appear to be synonymous

morph a form or type of something
  amorph a mutant allele which produces little apparent change
  allelomorph two antagonistic, but closely parallel, genetic characters (*e.g.* smooth or wrinkled skin in peas) were said to be allelomorphs (*cf.* allele). The term is sometimes, but nowadays improperly, used as a synonym for allele which is, however, directly derived from it
  andromorph a male form
    ergatandromorph an ergate showing some male characters
    gynandromorph an organism, of a type not normally hermaphroditic, which is partly perfect male and partly perfect female. Most commonly, and particularly in birds, the two lateral halves are so modified (*cf.* gynecomast)
  antimorph a mutant allele which inhibits the produc-tion of an ancestral or wild type structure
  ergatomorph an ant with male genitalia but an ergate-type body
  gynaecomorph an organism with primary sexual characters of the male and tertiary sexual charac-ters of the female
  hypermorph a mutant allele which has an overriding effect over the ancestral allele
  hypomorph a mutant allele the effect of which is over-ridden by the normal ancestral allele
  isomorph something which appears identical but is not
  micromorph = jordanon
  neomorph a mutant allele influencing development in a manner markedly different from that of the ancestral allele
  nothomorph in plants one of several different hybrid forms of the same parentage
  pleimorph one stage in a polymorphic cycle
  polymorph that which exhibits several forms. Also a common abbreviation of polymorphocyte (*q.v.*)
  rhizomorph the analog of a root produced from fungus hyphae

-morphic *comb. suffix* meaning "having the shape of" or "like." Many of these terms duplicate in meaning words ending in -form (*q.v.*)
  actinomorphic = radially symmetrical. Commonly used of flowers
  adelomorphic of obscure structure
  allotriomorphic having an abnormal or unexpected shape
  alphitomorphic having the shape of coarse meal
  dimorphic having two shapes, particularly (*bot.*) having two forms of the same organ on one plant
  clinomorphic formed or directed without relation to the plane of the earth
  cymbomorphic cup-shaped
  dioecio-dimorphic = heterogonic
  geomorphic a form shaped by the soil, or by gravity
  heteromorphic having different forms, but pertaining particularly to cases where the difference is a function of the life history, as distinct from poly-morphic where the different forms occur in the same instar
  homomorphic with a uniform shape or having the same shape
  mesomorphic a plant which is not adapted to dry climates

monomorphic having only one form, but particularly applied to organisms, such as many cladoceran Crustacea, in which only one sex is known

monoicodimorphic cleistogamic

orthomorphic upright

plesiomorphic of much the same shape

polymorphic having numerous shapes (*cf.* heteromorphic)

rhachimorphic used of the zigzag axis of some grasses

zygomorphic bilaterally symmetrical

Morphidae a family of nymphaloid lepidopteran insects usually included in the Amathusiidae

-morphism the condition of having a shape

cyclomorphism a cyclic change in body form such as that found in some Cladocera

dimorphism having two forms as when the sexes of an insect are markedly different

antidimorphism dimorphism in usually identical parts of the same organism

gynodimorphism the existence of female flowers on a gynodioecious plant

syndimorphism the condition of having two forms of the same organ in one individual

haptomorphism a shape produced by contact stimulus

heteromorphism the condition of being heteromorphic

hygromorphism a change in form induced by moisture

pleimorphism the occurence of several forms in the course of development

pleiomorphism polymorphism, but by some restricted to polymorphism exhibited as successive stages in a life cycle

polymorphism the condition of having one organism occur in several forms, as the polyp and medusoid forms of some Coelenterata

sympolymorphism = polymorphism

Morpho a genus of brilliant iridescent blue *S. Amer.* amathusiid lepidopteran insects

morphosis the manner of development

aitiomorphosis a change in shape forced on an organism by external causes

allomorphosis the rapid development of specialized organs or of general specialization of the organism

anamorphosis 1 (*see also* anamorphosis 2) *either* a process of slow, steady evolution not involving immediately apparent, gross, mutant variations (*cf.* saltation) *or* any gradual change in anything

anamorphosis 2 (*see also* anamorphosis 1) in insect life cycles refers to a form that hatches from the egg with fewer abdominal segments than characterize the adult. (*cf.* metamorphosis 2)

aromorphosis a generalized evolutionary change as distinct from a specialized change

automorphosis = mutation

barymorphosis change in shape occasioned by gravity

cyclomorphosis a condition found, for example, in some rotifers and Cladocera, in which there is cyclic polymorphism in the adult

cytomorphosis a change in the shape of a cell, particularly a bacterial cell, induced by environmental factors

epimorphosis the replacement or partial replacement of tissue removed from an organism (*cf.* heteromorphosis, morphallaxis, regeneration). In insect life cycles synonymous with metamorphosis as distinct from anamorphosis

gerontomorphosis a supposed evolution of new groups from neotenic larvae of other groups

heteromorphosis *either* aitomorphosis *or* the formation, in regeneration, of an organ different from the one lost

homoemorphosis having an analogous shape due to environment, not to heredity

hypermorphosis if and when ontogeny repeats phylogeny, hypermorphosis refers to those stages later in evolution that the last recognizable ancestral type

mechanomorphosis those changes in structures, under the influence of environment, which result in analogous, but not homologous, structures

metamorphosis 1 (*see also* metamorphosis 2) transformation of one part into another, particularly in plants

descending metamorphosis the metamorphosis of reproductive endosomatic parts in plants (*e.g.* the substitution of petals for stamens, etc.)

progressive metamorphosis the change of vegetative into sexual organs

regressive metamorphosis in *bot.*, when a female structure is metamorphosed into a male or when a reproductive structure is metamorphosed into a vegetative one

metamorphosis 2 (*see also* metamorphosis 1) change of shape, particularly applied to the changes of one larval form to another larval form or of a larval form to a nymph or adult. In insect life cycles refers specifically, in contrast to anamorphosis 2, to those forms in which embryonic segmentation is complete before hatching

biaimetamorphosis an exogenously produced change disadvantageous to the organism

complete metamorphosis in insects, a metamorphosis in which successive instars are completely polymorphic

direct metamorphosis (= incomplete metamorphosis)

hypermetamorphosis a form of life history, such as that found in the blister beetles or mayflies, in which there are numerous active larval instars but with very different forms (*e.g.* the triungulin larva of blister beetles) or in which a sub-imago is interpolated (*e.g.* the dun of mayflies)

incomplete metamorphosis an insect life history in which the immature stages bear a recognizable likeness to the adult

secondary metamorphosis an alteration in the life pattern and form of an adult the first metamorphosis of which was the change from the larval to the adult condition

simple metamorphosis = incomplete metamorphosis

neomorphosis the replacement, usually through regeneration, of one part by another different part

thigmomorphosis a change in structure due to contact stimulus

-morphous *adj.* suffix synonymous with -morphic, the form preferred in this work

morphy the condition of having a shape

actinomorphy = radial symmetry

hydromorphy having a shape occasioned by submersion

stasimorphy a shape caused by arrested development

zygomorphy = bilateral symmetry

petty morrel (*cf.* morel) the araliaceous herb *Aralia racemosa* (= spikenard)

morse = walrus

praemorse with the end bitten off

premorse = praemorse

morula the solid ball of cells which results from cleavage of most mammalian eggs

amphimorula a morula derived from an amphiblastula

[morula] a mutant (gene symbol *mr*) mapped at 106.7±

on chromosome II of *Drosophila melanogaster*. The phenotypic expression is small bristles and rough eyes

Moruloidea = Mesozoa

morulose dark purple black (= morinous)

mos an assemblage of organisms "held together by the will and humor of each other" (Publius Terence, 185 B.C.). Used by contemporary ecologists in much the same sense to describe a community of one or more species living in amity but without mutual interdependence (*cf.* mores)

genetic **mosaic** (*see also* egg) the occurence of two types of genetically homozygous tissue in a heterozygote as the result of a mitotic chromosomal exchange

moschate musky

moschatell *Brit.*, the araliaceous herb *Adoxa moschatellina*

Moschinae a subfamily of cervid artiodactyl mammals containing the musk-deer

Moses-in-the-bullrushes the commelinaceous herb *Rhoeo bicolor*

mosquito (*see also* hawk, plant) any of numerous biting dipterous insects of the family Culicidae. A mosquito differs from a gnat partly as a matter of size and partly as a matter of dialectic difference

northwest-coast mosquito *U.S. Aedes aboriginis*

yellow-feaver mosquito *Aedes aegypti. S. Amer. Stegomyia fasciata*

crab-hole mosquito *U.S. Deinocerites cancer*

western tree-hold mosquito *U.S. Aedes sierrensis*

northern house mosquito *U.S. Culex pipiens*

malaria mosquito any of several species of the genus Anopheles

common malaria mosquito *U.S. A. quadrimaculatus*

salt marsh mosquito *U.S. Aedes sollicitans*

brown salt marsh mosquito *U.S. A. cantator*

California salt-marsh mosquito *U.S. A. squamiger*

New Jersey mosquito *U.S.* = salt marsh mosquito *U.S.*

pitcher-plant mosquito *U.S. Wyeomyia smithii*

tiger mosquito = yellow fever mosquito (*U.S.*)

floodwater mosquito *U.S. Aedes sticticus*

spotted-wing mosquito *Anopheles maculipennis*

moss 1 (*see also* moss 2, cheeper) popular name of muscine bryophyte plants

apple moss any of several species of moss of the genus Bartramia

bog moss any of several species of the genus Sphagnum

peatmoss = bog moss

urn moss any urn-shaped moss, particularly those of the genus Physcomitrium

moss 2 (*see also* moss 1) any of several moss-like plants

black moss *U.S.* the bromeliaceous herb *Tillandsia usneoides*

Corsican moss the dried rhodophycean alga *Gracilaria helminthochorton* (= worm moss)

ditch-moss the hydrocharitaceous aquatic herb *Elodea canadensis*

floating moss any of several species of the marsiliaceous fern Salvinia

Florida moss = Spanish moss

flowering moss the diapensiaceous creeping herb *Pyxidanthera barbulata*

Irish moss = carragreen

long moss *U.S.* the bromeliaceous herb *Tillandsia usneoides*

reindeer-moss any of several artic and subarctic foliaceous lichens, but most particularly *Cladonia rangiferina*

rock-moss the crassulaceous herb *Sedum pulchellum*

rose-moss the portulacaceous herb *Portulaca grandiflora*

Spanish moss the bromeliaceous epiphytic herb *Tillandsia usneoides*

spike-moss *U.S.* general name for selaginellaceous pteridophytes

worm moss = Corsican moss

mot a bacterial genetic marker indicating the presence of flagella but absence of motility

mot-28 *see* [mottled-28]

Motacillidae that family of passeriform birds which contains the pipits and the wagtails. These are mostly small to medium sized birds with a short neck and a slender body with an unusually long tail which has given, to some, the name of wagtail

moth 1 (*see also* moth 2, mullein) a term used to describe all lepidopteran insects except the true butterflies (Papilionoidea and Nymphaloidea) and the skippers (Hesperioidea). In general the moths, which are divided among nearly a score of superfamilies, are thick bodied nocturnal forms and most of them have plumose antennae. At one time they were united as the Heterocera (*q.v.*)

acorn moth the gelechioid blastobasid *Valentinia glandulella*

alder moth *Brit.* the noctuid *Acronycta alni*

almont moth *U.S.* the phycitid *Cadra cautella*

antler moth *Brit.* the noctuid *Cerapteryx graminis*

Atlas moth *Austral.* the saturnid *Coscinocera hercules*

autumnal moth *Brit.* the geometrid lepidopteran insect *Oporabia autumnata*

black-banded moth *Brit.* the noctuid *Polia xanthomista*

jumping-bean moth the olethreutid *Laspeyresia saltitans*

bee-moth = wax-moth

bella moth *U.S.* the arctiid *Utetheisa bella*

gooseberry moth *U.S.* = currant moth *Brit.*

grapeberry moth *U.S.* the olethreutid *Paralobesia viteana*

water betony moth *Brit.* the noctuid *Cucullia scrophulariae*

broom-moth *Brit.* the noctuid *Mamestra pisi*

buck-moth *U.S.* any of several species of saturniid of the genus Hemileuca

lesser bud-moth *U.S.* the gelechiid *Recurvaria nanella*

pecan bud-moth *U.S.* the olethreutid *Gretchena bolliana*

raspberry bud-moth *U.S.* the incurvariid *Lampronia rubiella*

eye-spotted bud-moth *U.S.* the olethreutid *Spilonota ocellana*

spruce bud-moth *U.S.* the olethreutid *Zeiraphera ratzeburgiana*

sugarcane bud-moth *U.S.* the tineid *Ereunetis flavistriata*

verbena bud-moth *U.S.* the olethreutid *Endothenia hebesana*

bugong moth *Austral.* the noctuid *Agrotis spina*

burnet moth *Brit.* popular name of zygaenids

cabbage moth *Brit.* the noctuid *Barathra brassicae*

cactus moth the pyralid *Cactoblastis cactorum*. Well known for its successful introduction into Australia for the control of cactus

carpet moth any of numerous geometrids, so called for their pattern, and tineids, so called from their feeding habits. In this last category the term, without qualification, usually refers to *Trichophaga tapetzella*

cecropia moth *U.S.* the saturniid *Hyalophora cecropia*

cinnabar moth *Brit.* the arctiid *Euchelia jacobaeae*

short-cloaked moth *Brit.* the nolid *Nola cucullatella*

clothes moth any of several species of tineid insect infesting fabrics and furs, without qualification usually refers to the *Tineola bisselliella*

    case-making clothes moth *Tinea pellionella*

    tapestry clothes moth *Trichophaga tapetzella* (*cf.* carpet moth)

clover-moth *Brit.* either of two species of noctuid of the genus Heliothis

    spotted clover-moth the noctuid *Heliothis scutosus*

codling moth *U.S.* the olethreutid *Carpocapsa pomonella*

confused moth *Brit.* the noctuid *Hama furva*

crescent moth *Brit.* the noctuid *Helotropha leucostigma*

strawberry-crown moth *U.S.* the aegeriid *Ramosia bibionipennis*

currant-moth *Brit.* the geometrid *Abraxas grossulariata*

    dried-currant moth *U.S.* the pyralid *Cadra cautella*

dagger-moth *U.S.* any of several species of noctuid of the genus Acronicta

    American dagger-moth *U.S. A. americana*

    cottonwood dagger-moth *U.S. A. lepusculina*

    smeared dagger-moth *U.S. A. oblinita*

dart-moth *U.S.* any of several species of noctuids of the genera Rhynchagrotis and Adelphagrotis (*cf.* dart)

December moth *Brit.* the lasiocampid *Poecilocampa populi*

delicate moth *Brit.* the noctuid *Leucania vitellina*

dew-moth *Brit.* the arctiid *Endrosa irrorella*

diamondback moth *U.S.* the yponomeutid *Plutella maculipennis*

ear-moth *Brit.* the noctuid *Hydroecia nictitans*

western elder-moth *U.S.* the noctuid *Zotheca tranquila*

emperor moth *Brit.* the saturniid *Saturnia pavonia*

engrailed moth *Brit.* the geometrid *Tephrosia bistortata*

ermine-moth popular name of yponomeutids though the term used without qualification usually refers to members of the genus Yponomeuta and particularly *Y. padella*

fairy moth *U.S.* popular name of adelids

fern-moth *Brit.* the geometrid *Phibalapteryx tersata*

figure-of-eight moth *Brit.* the noctuid *Diloba caeruleocephala*

flannel-moth *U.S.* any of several species of megalopygids

    crinkled flannel-moth *U.S. Megalopyse crispata*

flour-moth any of numerous moths parasitic on stored grain products. Particularly *U.S.* the phycitid *Anagasta kuehniella* (= **Mediterranean flour moth**)

sunflower moth *U.S.* the phycitid *Homoeosma electellum*

    banded sunflower moth *U.S.* the phaloniid *Phalonia hospes*

footman moth any of several species of arctiid moths

    striped footman moth *U.S. Hypoprepia miniata*

forester moth popular name of agaristids

apple fruit-moth *U.S.* the yponomeutid *Argyresthia conjugella*

    dried-fruit moth *U.S.* the phycitid *Vitula edmandsae serratilinella*

    oriental fruit-moth *U.S.* the olethreutid *Grapholitha molesta*

fox-moth *Brit.* the lasiocampid *Macrothylacia rubi*

fur-moth the tineid *Tinea pellionella*

gall moth any gall-forming lepidopteran but particularly tortricoids and gelechioids

ghost-moth *U.S.* popular name of hepialids. *Brit.*, without qualification, *Hepialus humuli*

goat-moth *Brit.* the cossid *Cossus cossus*

Angoumois grain-moth the gelechiid *Sitotroga cerealella*

    European grain-moth *U.S.* the tineid *Nemapogon granella*

granite moth *U.S.* the geometrid *Sciagraphia granitata*

grass-moth *Brit.* and *U.S.* any of numerous small pyralid moths commonly whitish or pale yellow, which hold their narrow wings folded along the body. They are also therefore called close-wing moths. In *Brit.* the term is commonly restricted to the genus Crambus

gypsy moth *U.S.* the lymantriid *Porthetria dispar*

hagmoth *U.S.* the limacodid *Phobetron pithecium*

handmaid moth *U.S.* any of several species of notodontid of the genus Datana

hawkmoth (*see also* sphinx) popular name of sphingids

    bedstraw hawkmoth *Brit.* the sphingid *Deilephila galii* (= spotted elephant)

    bee-hawkmoth *Macroglossa fuciformis*

    convolulus hawkmoth *Sphinx convolvuli*

    elephant hawkmoth *Brit.* either of two species of the genus Chaerocampa

    death's head hawkmoth *Brit. Acherontia atropos*

    humming bird hawkmoth *Brit. Macroglossa stellarum*

    popular hawkmoth *Brit. Smerinthus populi*

    privet hawkmoth *Brit. Sphinx ligustri*

    spurge hawkmoth *Brit. Deilephila euphorbiae*

heart-moth *Brit.* the noctuid *Dicycla oo.*

herbarium-moth *U.S.* the geometrid *Eois ptelearia*

yellow-horned moth *Brit.* the cymatophorid *Asphalia flavicornis*

hornet-moth *U.S.* the aegeriid *Aegeria apiformis*

    lunar hornet-moth *Brit.* the aegerid *Trochilium crabroniformis*

brown housemoth *U.S.* the oecophorid *Hofmannophila pseudospretella*

    white-shouldered housemoth *U.S.* the oecophorid *Endrosis sarcitrella*

three-humped moth *Brit.* the notodontid *Notodonta phoebe*

imperial moth *U.S.* the citheroniid *Eacles imperialis*

io moth *U.S.* the saturniid *Automeris io*

lappet moth any of several species of lasiocampids *U.S.*, without qualification, *Epinaptera americana.* *Brit. Gastropacha quercifolia*

leopard-moth the cossid *Zeuzera pyrina*

lichen-moth *U.S.* the amatid *Lycomorpha pholus* and any of numerous species of lithosiids

linden-moth *U.S.* the geometrid *Erannis tiliaria*

double-lobed moth *Brit.* the noctuid *Apamea ophiogramma*

lobster-moth *Brit.* the notodontid *Stauropus fagi*

luna moth any of several saturniid moths with very long tail-like prominences from the hindwing and a delicate yellow-green color. *U.S.*, without qualification *Actias luna*

magnet-moth *U.S.* the geometrid *Leptomeris magnetaria*

magpie-moth *Brit.* = currant moth *Brit.*

maple-moth *U.S.* any of several citheroniids of the genus Anisota

white-marked moth *Brit.*, the noctuid *Pachnobia leucographa*

marsh moth *Brit.* the noctuid *Hydrilla palustris* and *Noctua subrosea*

meal-moth *U.S.* the pyralid *Pyralis farinalis*
    Indian meal-moth *U.S.* the phycitid *Plodia interpunctella*

mountain moth *Brit.* any of several species of geometrid, particularly *Psodos coracina*

mullein-moth *Brit.* the noctuid *Cucullia verbasci*

November moth *Brit.* the geometrid *Oporabia dilutata*

scalloped oakmoth *Brit.* the geometrid *Crocallis elinguaria*

orache moth *Brit.* the noctuid *Trachea atriplicis*

oriental moth *U.S.* the limacodid *Cnidocampa flavescens*

owlet moth popular name of noctuids

pandora moth *U.S.* the saturniid *Coloradia pandora*

pea-moth *U.S.* the olethreutid *Laspeyresia nigricana*

peacock moth *Brit.* the geometrid *Semiothisa notata*

pepper-and-salt moth *U..S.* the geometrid *Biston cognataria*

Zimmerman pine-moth *U.S.* the phycitid *Dioryctria zimmermani*

Nantucket pinetip-moth *U.S.* the olethreutid *Rhyacionia frustrana*

Douglas-fir pitch-moth *U.S.* the aegeriid *Vespamima novaroensis*
    sequoia pitch-moth *U.S.* the aegeriid *Vespamima sequoiae*

plum-moth *U.S.* the geometrid *Eustroma prunata*

plume-moth popular name of pterophorids. The term is also applied to the closely allied Alucitidae
    artichoke plume-moth *U.S.* the pterophorid *Platyptilia carduidactyla*
    grape plume-moth *U.S.* the pterophorid *Pterophorus periscelidactylus*
    lantana plume-moth *U.S.* the pterophorid *Platyptilia pusillodactyla*

polyphemus moth *U.S.* the saturniid *Antheraea polyphemus*

Portland moth *Brit.* the noctuid *Agrotis praecox*

potato-moth *U.S.* the gelechiid *Phthorimaea operculella*

privet-moth *U.S.* the pyralid *Glyphodes quadristigmalis*

processionary moth the eupterotid *Cnethocampa processionea*

promethea moth *U.S.* the saturniid *Callosamia promethea*

purslane-moth *U.S.* the agaristid *Copidryas gloveri*

puss-moth *Brit.* the notodontid *Dicranura vinula*

raisin-moth *U.S.* the phycitid *Cadra figulilella*

regal moth *U.S.* the citheroniid *Citheronia regalis*

sand-dune moth *U.S.* the agaristid *Tuerta sabulosa*

satin-moth *U.S.* the lymantriid *Stilpnotia salicis*
    lesser satin-moth *Brit.* the thyatririd *Palimpsestis duplaris*
    white satin-moth *Brit.* the liparid *Stilpnotia salicis*

scape-moth *U.S.* any of several species of syntomid of the genus Scepsis

fir seed-moth *U.S.* the olethreutid *Laspeyresia bracteatana*
    spruce seed-moth *U.S.* the olethreutid *Laspeyresia youngana*

sheep-moth *U.S.* saturniid *Pseudohazis eglanterina*

scallop-shell moth *U.S.* the geometrid *Hydria undulata*

shoot-moth any olethreutid with a shoot boring larva
    European pine shoot-moth *U.S.* *Rhyacionia buoliana*

silkmoth *U.S.* popular name of saturniids (*cf.* silkworm)
    ceanothus silkmoth *U.S.* *Hyalophora euryalus*

skiff-moth *U.S.* the cochlidiid *Prolimacodes scapha*

smoky moth popular name of zygaenids and pyromorphids

four-spotted moth *Brit.*, the noctuid *Acontia luctuosa*

twin-spotted moth *Brit.* the noctuid *Nonagria geminipuncta*

star-moth *U.S.* the noctuid *Derrima stellata*

cotton-stem moth *U.S.* the gelechiid *Platyedra vilella*

chain-streak moth *U.S.* the geometrid *Cingilia catenaria* (= chain-spotted geometer)

suspected moth *Brit.* the noctuid *Dysochorista suspecta*

swift moth *Brit.* pipular name of hepialids

sycamore-moth *Brit.*, the noctuid *Acronycta aceris*

brown-tail moth *U.S.* the lymantriid *Nygmia phaeorrhoea*
    cat-tail moth *U.S.* the cosmopterygid *Lymnaecia phragmitella*
    swallow-tail moth *Brit.* the geometrid *Ourapteryx sambuccaria*

tiger-moth any of numerous arctiid moths with spotted or striped yellow, red, white and black coloration
    buff tiger-moth *Spilosoma fuliginosa*
    garden-tiger moth *Brit.* *Arctia caia*
    scarlet tiger-moth *Callimorpha dominula*
    cream-spot tiger-moth *Brit.* the arctiid *Arctia villica*
    silver-spotted tiger-moth *U.S.* *Halisidota argentata*
    virgin tiger-moth *U.S.* *Apantesis virgo*

hook-tip moth popular name of drepanids

tobacco moth *U.S.* the phycitid *Ephestia elutella*

pine tube-moth *U.S.* the tortricid *Argyrotaenia pinatubana*

turnip moth *Brit.* the noctuid *Agrotis segetum*

tussock-moth popular name of many liparids, arctiids and lymantrids
    Douglas-fir tussock-moth *U.S.* the lymantrid *Hemerocampa pseudotsugata*
    hickory tussock-moth *U.S.* the artiid *Halisidota caryae*

white-marked tussock-moth *U.S.* the lymantriid *Hemerocampa leucostigma*
    pale tussock-moth *U.S.* the arctiid *Halisidota tessellaris*
    pine tussock-moth *U.S.* the lymantriid *Dasychira plagiata*
    rusty tussock-moth *U.S.* the lymantriid *Orgyia antiqua*
    spotted tussock-moth *U.S.* the arctiid *Halisidota maculata*
    sycamore tussock-moth *U.S.* the arctiid *Halisidota harrisii*
    western tussock moth *U.S.* the lymantriid *Hemerocampa vetusta*

pitch twig-moth *U.S.* the olethreutid *Petrova comstockiana*

V-moth *Brit.* the geometrid *Thamnonoma wauaria*

vaporer moth *Brit.* the lymantrid *Orgyia antiqua*

royal walnut-moth *U.S.* = regal moth *U.S.*

wasp-moth *U.S.* popular name of syntomids

wax-moth the galleriid *Galleria mellonella* which feeds on the wax in beehives (= **greater wax-moth** *U.S.*)

lesser wax-moth *U.S.* the galleriid *Achroia grisella*

chickweed moth *U.S.* the geometrid *Haematopsis grateria*

milkweed moth *U.S.* the arctiid *Euchaetias egle*

clearwing moth popular name of Aegeriidae

closewing moth = grass moth

underwing moth popular name of noctuids of the genus Catocala

clear-winged moth *U.S.* popular name of any transparent-winged moth particularly sphingids and thyridids

window-winged moth *U.S.* popular name of thyridids

winter moth *U.S.* the geometrid *Operophtera brumata*. *Brit.* either of two geometrids of the genus Cheimatobia

woolen moth the tineid *Tinea pellionella*

cottonworm moth *U.S.* the noctuid *Alabama argillacea*

silkworm-moth the bombycid moth *Bombyx mori*

giant silkworm-moth popular name of saturnids in general

yucca-moth any of four species of prodoxid of the genus Tegeticula. The female is specifically adapted to the fertilization of plants of the genus Yucca

bogus yucca moth any of several species of prodoxids of the genus Prodoxus which, though predatory on the yucca plant, are unable to fertilize it

moth 2 (*see also* moth 1) in compound names of a few animals not lepidopteran insects

buffalo-moth *U.S.* the dermestid coleopteran *Anthrenus scrophulariae*

sea-moth popular name of pegasiid fish

mother properly, a female parent, but widely used in *biol.* in the sense of "parental"

devil's grandmother the compositaceous herb *Elephantopus tomentosus*

stem mother the original ancestor of a parthenogenetically produced clone

mother-of-pearl 1 (*see also* mother-of-pearl 2) a thin layer of iridescent aragonite forming the inner surface of the shell of many Mollusca

mother o' pearl 2 (*see also* mother-of-pearl 2) a thin either of two species of nymphalid lepidopteran insect of the genus Salamis usually, without qualification, *S. parnassus*

mother-of-the-evening the cruciferous herb *Hesperis matronalis* (= rocket)

mother-of-thousands the scrophulariaceous trailing vine *Linaria cymbalaria* and the saxifragaceous herb *Saxifraga sarmentosa*

mother-of-thyme the labiateous herb *Satureia acinos*

mother-of-vinegar an association of yeast and bacteria, predominantly Acetobacter, which forms a mat over the surface of wine which is turning to vinegar

motmot any of numerous coraciiform birds of the family Momotidae

broad-billed motmot *Electron platyrhynchum*

keel-billed motmot *Electron carinatum*

turquoise-browed motmot *Eumomota superciliosa*

blue-crowned motmot *Momotus momota*

russet-crowned motmot *Momotus mexicanus*

rufous motmot *Baryphthengus ruficapillus*

blue-throated motmot *Aspatha gularis*

tody motmot *Hytomanes momotula*

moto-oi the annonaceous tree *Canagium odoratum*

-motor- *comb. form* meaning "movement"

pilomotor pertaining to the movement of hair

vasomotor concerned with the movement—actually wall contraction—of blood vessels

motorium a body, in which many neurofibrils are united, close to the cytopharynx of some Protozoa

mottle *I.P.* any of several species of lycaenid insect of the genus Logania

[mottled] a mutant (gene symbol *odk*) mapped at 10.7 on linkage group XIV of *Bombyx mori*. The phenotypic expression is the low translucency of the larva

[Mottled] a mutant (gene symbol *Mo*) found in the sex chromosome (linkage group XX) of the mouse. The general appearance resembles [Blotchy] but males die before birth

[mottled-2] a mutant (gene symbol *md*) mapped at 25 on chromosome 6 of the tomato. The phenotypic leaves have yellowish spots on them

[mottled-28] a mutant (gene symbol *mot-28*) mapped at 46.0 on chromosome III of *Drosophila melanogaster*. The phenotypic expression is a brown mottled eye

[mottled leaves] a mutant (gene symbol *m*) mapped at 3 on chromosome 2 of the tomato. The term is descriptive

[mottled translucent] a mutant (gene symbol *oa*) mapped at 26.2 on linkage group II of *Bombyx mori*. The name is descriptive of the larva of the phenotype

mouflon either of two ovine caprid mammals which are the wild sheep of the Mediterranean region. They are distinguished by backwardly swept horns of the male

Corsican mouflon *Ovis musimon*

Cyprus mouflon *O. ophion*

moula the proximal end of the insect tibia

mould *Brit.* = mold *U.S.*

mourner any of several cotingid birds of the genus Rhytipterna

half-mourner *Brit.* the satyrid lepidopteran insect *Melanargia galatea* (marbled white)

bearded-mountaineer the apodiform trochilid bird *Oreonympha nobilis*

mouse 1 (*see also* mouse 2, 3, bird, flea, tail, weed) the terms mouse and rat are indiscriminately applied in all parts of the world to any long-tailed rodent. The distinction between rat and mouse is one of size only and is not uniform in various parts of the world. Old World rats and mice belong to the family Muridae and three of them (the black rat, the Norway rat and the house mouse) are now to be found in almost every inhabited part of the planet. There are no indigenous murid rodents in the New World. (*cf.* rat, vole). The laboratory mouse, of which many mutants are listed in this work, is *Mus musculus* (= house mouse)

barbary mouse any of several murids of the genus Arviacanthis

brush-mouse *U.S.* the cricetid *Peromycsus boylei*

cactus-mouse the cricetid *Peromyscus eremicus*

canyon-mouse the cricetid *Peromyscus crinitus*

cotton-mouse *U.S.* the cricetid *Peromyscus gossypinus*

deer-mouse *U.S.* the cricetid *Peromyscus maniculatis*

dormouse (*see also* phalanger) any of many glirid rodents. *Brit.*, without qualification, *Muscardinus avellanarius* edible dormouse *Glis glis*

spiny dormouse popular name of platacanthomyids

squirrel-tailed dormouse = edible dormouse

Florida mouse *U.S.* = gopher-mouse *U.S.*

white-footed mouse *U.S.* the cricetid *Peromyscus leucopus*

gliding mouse any of several species of anomalurids of the genus Idiurus, furnished, as the name indicates with gliding membranes

golden mouse *U.S.* the cricetid *Peromyscus nuttalli*

gopher-mouse *U.S.* the cricetid *Peromyscus floridanus*

grasshopper mouse *U.S.* either of two species of cricetids of the genus Onychomys

harvest mouse *U.S.* any of several species of cricetids of the genus Reithrodontomys. *Brit. Micromys minutus*

hazelmouse = dormouse (Brit.)

house mouse the murid *Mus musculus*

jumping mouse general name for the zapodids

kangaroo mouse *U.S.* either of two species of heteromyids of the genus Microdipodops

pocket mouse (*see also* mouse 2) *U.S.* any of numerous species of heteromyids, mostly of the genus Perognathus

pygmy mouse *U.S.* the cricetid *Baiomys taylori*

shrew-mouse *see* mouse 2

spiny mouse any of several species of mice in which the hair is fused into porcupine-like spines. Usually, without qualification, the murid *Acomys cahirinus* or, in *U.S.* heteromyid *Liomys irroratus*

wood mouse *Brit.* the murid *Apodemus sylvaticus*

mouse 2 (*see also* mouse 1, 3) mouse-like mammals not rodents

flittermouse *U.S.* the not uncommon anglicized form of *Fledermaus*, the *Ger.* word for bat

pocket-mouse *Austral.* (*see also* mouse 1) popular name of phascogaline dasyurid marsupials

pouched mouse any of several species of perameloid marsupials of the genus Phascologale

long-legged pouched mouse any of several species of Anthechinomys

shrew-mouse (more properly shrew-like mouse) the insectivore *Rhynchomys soricioides* (*cf.* shrew)

mouse 3 (*see also* mouse 1, 2) any of several organisms other than mammals thought to be mouselike. *Brit.*, without qualification the noctuid lepidopteran insect *Amphipyra tragopogonis*

coal-mouse the parid bird *Parus ater* (= coal tit)

sea-mouse any of several species of aphroditid polychaete annelids of the genus Aphrodite

titmouse this term for small parid birds is interchangeable with, but less usual than, tit (*q.v.*) in *Brit.* In *U.S.* the term titmouse is preferred for birds of the genus Parus.

bridled titmouse *P. wollweberi*

black-creasted titmouse *P. atricristatus*

plain titmouse *P. inornatus*

tufted titmouse *P. biclor*

[Mouse ears] a mutant (gene symbol *Me*) mapped at 38 on chromosome 2 of the tomato. The phenotypic expressions are compoundly pinnate mouse-ear-like leaves

mouth 1 (*see also* mouth 2, 3) any aperture leading to a cavity or tube, particularly the anterior aperture of the alimentary canal. The term is frequently improperly used for the buccal cavity

mouth 2 (*see also* mouth 1, 3) any of numerous animals distinguished by the appearance of their mouths

bristle-mouth popular name of gonostomatid fish

chisel-mouth the cyprinid fish *Acrocheilus alutaceous*

cotton-mouth *U.S.* the crotalid snake *Agkistrodon piscivorus*

firemouth the cichlid teleost fish *Cichlasoma meeki*

flute-mouth popular name of fistulariid fish (= cornet fish)

frogmouth popular name of birds of the family Podargidae and, less frequently of the family Aegothelidae

marbled frogmouth *Podargus ocellatus*

tawny frogmouth *Podargus strigoides*

golden mouth the turbinid mollusk *Turbo chrysostomus*

horn mouth any of several species of muricid gastropod of the genus Pterorytis

peamouth the cyprinid fish *Mylocheilus caurinus*

redmouth the olivid gastropod mullusk *Oliva erythrostoma* (*cf.* olive)

shovel-mouth *W.I.* the anatine bird *Spatula clypeata* (= shoveler)

slipmouth popular name of leiognathid fishes

warmouth the centrarchid fish *Chaenobryttus gulosus*

wrymouth the stichid fish *Cryptacanthoides maculatus*

mouth 3 (*see also* mouth 1, 2) in compound names of plants

adder's mouth *U.S.* any of several orchids of the genus Malaxis

snakemouth the orchid *Pogonia ophioglossoides*

mp *see* [microptera] or [micropterous]

mr *see* [morula]

ms$_8$ *see* [male sterile]

ms$_{10}$ *see* [male sterile-10]

ms$_{17}$ *see* [male sterile]

msc *see* [melanoscutellum]

msf *see* [misformed]

MSH = melanocyte-stimulating hormones

mt *see* [Mating type]

mtl a bacterial genetic marker having a character probably affecting the activity of mannitol dehydrogenase, mapped at 70.5 mins for *Escherichia coli*

mu *see* [mussed]

muc a bacterial genetic marker having a character affecting the regulation of capsular polysaccharide production, mapped at 12 mins for *Escherichia coli*

mucilaginous slimy

mucket *U.S.* the unionid mollusk *Actinonais ligamentina*

chondromucoid a glycoprotein forming a considerable portion of the matrix of hyaline cartilage

Mucorales an order of aflagellate phycomycete fungi containing those forms commonly called breadmolds or black molds. They are distinguished by the possession of a loose thallus the mycelia of which are without septa

mucous thick, slimy, gelatinous, usually used in distinction from serous (*q.v.*)

mucro any short, stiff process whether pointed or not. It has been applied to too many parts of too many organisms to have any valid meaning. It has occasionally been used as synonymous with micron

mucronate sharply and stiffly pointed

mucus any slimy secretion of animal origin

mugger the Indian crocodilid reptile *Crocodylus palustris* (= marsh crocodile)

Mugilidae a family of mugiloid acanthopterygian fishes commonly called mullets. They are distinguished by possessing a muscular gizzard-like stomach

Mugiloidea a suborder of acanthopterygian Osteichthyes distinguished by the five-rayed and one-spined abdominal pelvic fins, and the separation between the spiny and soft portions of the dorsal fins. The families Sphyraenidae, Atherinidae, and Mugilidae are usually included in this suborder

muishond either of two species of S.Afr. mustelid carnivore (= cape polecat)

mule (*bot.*) = hybrid

mulga the leguminous shrub *Acacia aneura*

mullein (*see also* foxglove, moth, thrip) any of very many species of scrophulariaceous herbs of the genus Verbascum, and a few other plants

  common mullein *U.S. V. thapsus*

  dark mullein *Brit. V. nigrum*

  great mullein *Brit.* = common mullein *U.S.*

  moth mullein *V. blattaria*

  purple mullein *V. phoeniceum*

  rosette mullein the gesneriaceous herb *Raymondia pyrenica*

Muller's *epon. adj.* from Heinrich Muller (1820–1864) Johannes Peter Muller (1801–1858), Heinrich Jacob Ferdinand Muller (1825–1896), Fritz Muller (1821–1897) or Otto Friedrich Otto Muller (1730–1784)

mullet *U.S.* popular name of mugilid fishes. *Brit.* either mugilid fishes (**grey mullets**) or mullid fishes (**red mullets** = *U.S.* goatfishes)

Mullidae a family of acanthopterygian Osteichthyes commonly called goatfishes in *U.S.* and red mullets in *Eur.* The name derives from the two long tactile barbels which hang down from the underside of the mouth

mulm amorphous organic detritus in fresh water

multi- *comb. prefix* meaning many

[Multilunar] a mutant (gene symbol *L*) mapped at 0 on linkage group IV of *Bombyx mori*, the phenotype have large round spots on the thorax and abdominal segment

[Multiple spurs] a mutant (gene symbol *M*) found in linkage group VI of the domestic fowl. The name is descriptive of the phenotypic expression

mumia = pupa

mummichog the cyprinodontid fish *Fundulus heteroclitus*

mumruffin the parid bird *Aegithalos caudatus* (= long-tailed tit)

munia any of several species of ploceid birds of the general Munia, Lonchura and Estrilda

Muntiacinae a subfamily of cervid mammals containing the forms commonly called muntjacs

muntjac any of several species of Oriental cervid mammals of the subfamily Muntiacinae

-mur- *comb. form* meaning either "mouse" or "wall"

mur *see* [murrey]

Muraenidae a family of marine shallow-water reef-dwelling eels commonly called morays. They are principally distinguished by the absence of pectoral fins

mural pertaining to walls

murder term for a group of crows

murex a genus of muricid gastropod mollusks the name of which has passed into English. The Tyrian purple dye commonly meant by this word is, however, derived from a mollusk nowadays placed in the family Thaisidae

muricate with a texture like a roughly plastered wall

Muricidae a family of gastropod mollusks with thick shiny shells having three varices to a volution

Muridae a very large family of myomorph rodent mammals containing the rats, mice and their immediate relatives (*cf.* Cricetidae)

murinous mouse-colored (*cf.* murine, myochrous)

[murky] a mutant (gene symbol *mk*) mapped at 0.8± on the X chromosome of *Drosophila melanogaster*. The phenotypic female is sterile and both male and female have dark bodies and eyes

murmuration term for a group of starlings (*cf.* chattering)

murre any of several species of alcid birds of the genus Uria

  thick-billed murre *U. lomvia*

  common murre *U. aalge*

murrelet any of several alcid birds of the genera Brachyramphus, Endomychura, and Synthliboramphus

  ancient murrelet *S. antiquus*

  Kittlitz's murrelet *B. brevirostris*

  marbled murrelet *B. marmoratus*

  Xantus' murrelet *E. hypoleuca*

[murrey] a mutant (gene symbol *mur*) mapped at 14.3 on the X chromosome of *Drosophila melanogaster*. The phenotypic expression is reddish purple eyes with a reduced body size

murry *Brit.* the muraenid fish *Muraena helena* (*cf.* moray)

Musaceae that family of scitaminous monocotyledons which contains, *inter alia* the bananas and the bird-of-paradise flower. The outstanding characteristics of the family are the five fertile stamens and the very slight difference between the calyx and the corolla

musang any of several species of paradoxurine viverrid carnivores of the genus Paradoxurus. When used without qualification, usually refers to *P. hermaphrodytus*

-musc- *comb. form* meaning either "moss" or "fly," the latter usually in the sense of a dipteran

muscadine the vitaceous vine *V. rotundifolia* (= bull grape)

muscarian said of seeded flowers which attract coprophilous Diptera

Musci a phylum of bryophyte plants containing the form commonly called mosses. The distinctive character, apart from the mosslike appearance, is the simple thallose protonema bearing upright gametophores

Muscicapidae a large family of passeriform birds commonly called the Old World flycatchers. They have short legs and short rounded wings

Muscidae a very large family of myodarian cycloraphous dipteran insects containing the houseflies, stable-flies, tsetse flies, etc. The mouth parts vary greatly, some as in the housefly, being adapted to the absorption of fluid while others such as the stable-fly are piercing

muscle 1 (*see also* muscle 2) the contractile tissue of animals

  cardiac muscle the peculiarly branched striated muscle of which the chordate heart is composed

  diffuse muscle one in which the fibers are widely separated as in some Zoantharia

  involuntary muscle = smooth muscle

  circumscript muscle used as the antithesis of diffuse muscle in those forms in which the latter occurs

smooth muscle muscle which lacks striations and which is associated with involuntary (*e.g.* peristaltic) movement

striated muscle muscle, the fibers of which appear striated and which is associated with voluntary (*e.g.* limb) movement

voluntary muscle = striated muscle

muscle 2 (*see also* muscle 1) a mass of muscular tissue of specific function

ambiens muscle the muscle used to contract the toes by perching birds. It originates on the pelvis and is inserted on the end of the toe

appendicular muscle a muscle moving a limb

divaricator muscle that which opens the anterior margin of the valve in brachiopods

ciliary muscle the muscle in the ciliary process of the eye which moves the lens

coronal muscle the circular muscle band in the subumbrella of Schyphozoa

epaxial muscle that part of the muscular system of a vertebrate which lies, or originates, dorsal to the horizontal septum

masseter muscle the large adductor muscle of the jaw

papillary muscles muscular braces for the valves of the heart

skeletal muscle that portion of the muscular system which is attached to bony elements

somatic muscle muscle associated with the body wall and its appendages and which arises from myotomes

spindle muscle a muscle running down the center of the coils of the intestine of some Sipunculids

trunk muscle a muscle associated with the body wall of a higher vertebrate

visceral muscle muscles associated with the walls of the alimentary canal

Muscoidea = Myodaria

muscovy the cairinine duck *Cairina moschata* (= muscovy duck)

mushroom (*see also* body, mite) an edible basidiomycete or, rarely, ascomycete fungus. In *U.S.* the term is more widely used than in *Brit.* where it is commonly confined to the common mushroom (*cf.* fungus)

common mushroom the agaricaceous agaricale *Agaricus campestris*. In the *U.S.* the commercially grown "button mushroom" is usually a descendant of *A. bisporus*

hedgehog-mushroom popular name of hydnaceous basidiomycete fungi, particularly those of the genus Hydnum

honey-mushroom the tricholomataceous basidiomycete fungus *Armillaria mellea*

meadow mushroom = common mushroom

milk-mushroom a term quite frequently applied to the family Russulaceae though it should properly belong only to the genus Lactarius in that family

oyster-mushroom the tricholomataceous basidiomycete *Pleurotus ostreatus*

parasol-mushroom the agaricaceous basidiomycete *Leucoagaricus procerus*

pore-mushroom common name of Polyporaceae

fleshy-pore mushroom common name of Boletaceae

fairy-ring mushroom *U.S.* the trichlomataceous basidiomycete *Marasmitus oreades*

snow-mushroom *Helvella gigas* (*cf.* false morel)

sponge-mushroom common name of helvellaceans (= morel)

winter mushroom *U.S.* the tricholomataceous basidiomycete *Flammulina velutipes*

Musidoridae = Lonchopteridae

musk (*see also* beetle) an odorous secretion produced by many animals of which the musk deer (*q.v.*) is the best-known example

muskeg sphagnum moss bogs of the northern Canadian region

muskellunge *U.S.* the esocid fish *Esox masquinonge* (*cf.* pike)

muslin *Brit.* the arctiid lepidopteran insect *Nudaria mundana*

Musophagidae a family of cuculiform birds commonly called touracos or plantain-eaters. They have strong serrate broad bills, are almost invariably brightly colored and the head carries a conspicuous crest

musquash = musk rat

[mussed] a mutant (gene symbol *mu*) mapped at 50.0± on chromosome III of *Drosophila melanogaster.* The phenotypic expression is crumpled thin wings

mussel (*see also* crab) usually, without qualification, mytilid mollusks, though the term is also applied to members of some other molluscan families

edible mussel *Mytilus edulis*

platform-mussel popular name of dreisseniid pelecypods

sea-mussel *U.S. Mytilus californianis*

mustang (*see also* grape) *U.S.* a feral horse

mustard 1 (*see also* mustard 2) any of several species and numerous horticultural varieties of the cruciferous genus Brassica

black mustard *B. nigra*

Chinese mustard *B. juncea*

tuberous Chinese mustard *B. natiformis*

pot-herb mustard *B. japonica*

wild mustard *B. alba*

mustard 2 (*see also* mustard 1) any of several other cruciferous herbs

ball-mustard any of several species of the genus Neslia

hare's ear mustard *Conringia orientalis*

garlic-mustard *Brit. Alliaria officinalis*

hedge-mustard *Brit. Sisymbrium officinale*

mithridate mustard *Thlaspi arvense* (= penny cress)

wormseed mustard *Erysimum cheiranthoides*

tansy-mustard any of several species of the genus Descurainia

treacle-mustard any of several species of the genus Erysimum, particularly (*Brit.*) *E. cheiranthoides*

tumble-mustard *Sisymbrium altissimum*

Mustelidae a family of low-slung carnivorous mammals containing the weasels, minks, badgers, skunks, otters and their immediate allies. They are distinguished by having only one molar in each upper jaw, and at the most two in the lower jaw, and in lacking an alisphenoid canal

Mustelinae a subfamily of mustelid carnivorous mammals containing the weasels and their allies. They are for the most part small, vicious predators

muster term for a group of peacocks (*cf.* pride)

mustering term for a group of storks

musurana the colubrid snake *Clelia clelia*

mutant the result of a mutation (*q.v.*)

aldose mutarotase = aldose 1-epimerase

mutase a general term for those enzymes which catalyze intramolecular transfers

lysolecithin acylmutase catalyzes the change of 2-lysolecithin into 3-lysolecithin

methylaspartate mutase catalyzes the production of L-glutamate from L-threo-3-methylaspartate

phosphoglucomutase catalyzes the production of D-glucose 6-phosphate and D-glucose 1,6-phosphate from D-glucose 1,6-phosphate and D-glucose 1-phosphate

phosphoglyceromutase catalyzes the production of D-3-phosphoglycerate and D-2,3-diphosphoglycerate from D-2,3-diphosphoglycerate and D-2-phosphoglycerate

diphosphoglyceromutase catalyzes the production of D-3-phosphoglycerate and D-2-3-diphosphoglycerate from D-1,3-diphosphoglyceric acid and D-3-phosphoglycerate

glucose phosphomutase = phosphoglucomutase

acetylaminodeoxyglucose phosphomutase catalyzes the production of 2-acetylamino-2-deoxy-D-glucose 6-phosphate and 2-acetylamine-2-deoxy-D-glucose 1,6-diphosphate from 2-acetylamine-2-deoxy-D-glucose 1,6-diphosphate and 2-acetylamino-2-deoxy-D-glucose 1-phosphate

glycerate phosphomutase = phosphoglyceromutase

phosphoglycerate phosphomutase catalyzes the change of 2-phospho-D-glycerate to 3-phospho-D-glycerate

phosphoglycerate phosphomutase = diphosphoglyceromutase

mutation a heritable change in a genetic character resulting either from a change in the gene at a locus, or alterations in chromosomal structure

back-mutation the mutation of a mutant back to the form from which it was derived

block mutation a mutation of a number of neighboring loci

fractional mutation a mutation in one part of an organism occuring through the production of a dominant mutation in an early division

lethal mutation one which causes death

micromutation one which does not markedly alter the phenotype though the ultimate result of many micromutations may be a new geographic race or subspecies

point mutation one which appears to involve only a single locus

praemutation a term of early genetics used to account for the unexpected appearance of a phenotypic character which was held to have been "premutated" within the organism

retrogressive mutation an early term for a genotypic character which failed to appear as a phenotypic expression in a hybrid

reverse mutation one which transforms a mutant back to its original state

mute term for a group of hounds

muticous with the point or spike cut off

Mutillidae a family of scolioid apocritan hymenopteran insects commonly called velvet-ants because the wingless females, which can sting very severely, are often mistaken for ants from which, however, they can be readily distinguished by the downy pubescence covering the entire body

Mutiseae a tribe of tubiflorous Compositae

muton the smallest unit of mutation (a single nucleotide pair) within a cistron (*q.v.* and *cf.* recon)

mutualism at one time a synonym of symbiosis, later extended to include both symbiosis and commensalism and now frequently used for any association between two organisms including parasiticism

paramutualism = facultative symbiosis

muzzle that portion of the head of an animal which protrudes from the rest. Deprived of the lower jaw, a muzzle is a snout

mwi *see* [misheld wings]

Myacidae a family of pelecypod mollusks commonly called soft-shelled clams because of the relatively small size of the widely gaping shell

myall *Austral.* any of numerous trees and shrubs of the leguminous genus Acacia

bastard myall any of several species of Acacia

weeping myall *A. pendula*

myarian pertaining to a muscle (*cf.* myodian)

coelomyarian said of those nematodes in which the longitudinal muscle strip, with its fibrillar zone, bulges into the pseudocoel

holomyarian said of those nematodes in which the muscle layer does not show typical muscle-cells, but is divided into two zones by the lateral chords

meromyarian said of those nematodes in which there are from two to five longitudinal rows of muscle-cells

platymyarian said of those nematodes in which the muscle-cells are flattened with the fibrillar region limited to a basal zone

polymyarian said of those nematodes in which there are a large number of muscle-cells in each strip

mycele = mycelium

mycelium a matted mass of fungus hypae

promycelium the tube which projects from the fungus spore

pseudomycelium short chains of fungal cells such as those produced by some yeasts and bacteria

racquet mycelium one consisting of biscotiform segments

myceloid said of a bacterial culture having a filamentous appearance as though it were composed of mycelia

Mycetobiidae a family of nematoceran dipteran insects now usually included with the Rhyphidae

Mycetophagidae small family of brilliant red, yellow and black fungus beetles commonly called the hairy fungus-beetles by virtue of their densely pilose surface

Mycetophilidae a family of small slender mosquito-like nematoceran dipteran insects with greatly elongated coxae. They are commonly called fungus-gnats in virtue of their feeding habits

mycetous pertaining to fungus

hymenomycetous said of fungus which has the hymenium exposed at maturity

Mycetozoa a peculiar order of sarcodinous Protozoa possessing at one stage of their life history a form strongly reminiscent of the lower fungi. They have from time to time been regarded as a separate phylum (*cf.* myxomycophyta)

mycina an apothecium in the form of a bulb on the end of a stalk

Mycobacteriaceae a family of actinomycetale schizomycetes most of which occur as spherical cells rarely branching. The majority are pathogenic to fish, but *Mycobacterium tuberculosis* is responsible for the disease of that name in man

Mycophyta *see* Eumycophyta

Mycoplasmataceae the only family of mycoplasmatale schizomycetes. The characteristics of the order are those of the family

Mycoplasmatales an order of schizomycetes containing the single genus Mycoplasma. These are highly pleiomorphic organisms most of which have a filterable stage in the life history. They were originally called pleuropneumonia-like organisms

Myctophidae a family of iniomous fish commonly called lantern-fishes by reason of the numerous large photophores along their sides

Myctophiformes = Iniomi

Mydaidae a family of very large dipteran insects, usually resembling wasps, closely allied to the asilids both in structure and habit. They are commonly called mydas flies

Mydasidae = Mydaidae

-myel-, *comb. form* meaning "narrow" or by a curious extension "spinal cord"

myelin a variable mixture of lipoid and proteinaceous materials investing nerve fibers

-myelo- *see* -myel-

Mygalomorphae a suborder of araneid arthropods with two pairs of lungs and with cheliceral fangs that move up and down instead of transversely

-myi- *comb. form* meaning "fly" often improperly rendered -myo-

Myiadestidae a family of passeriform birds usually included in the Turdidae

Myidae = Myacidae

Mylabridae a family of coleopteran insects commonly called pea- and bean-weevils. They are short-snouted beetles closely allied to the Chrysomelidae. The antennae are so deeply serrate as to almost appear pectinate

Myliobatidae a family of flattened rajiform Chondrichthyes distinguished by a fleshy pad extending round the front end of the head and commonly called eagle-rays or cownose-rays. Many are venomous

Mymaridae a family of tiny chalcidoid apicritan hymenopteran insects commonly called fairy flies

mynah any of numerous sturnid birds, mostly of the genera Sternus and Acridotheres. *U.S.* without qualification, *Acridotheres cristatellus* (= **crested mynah**)

  common mynah *Acridotheres tristis*

  golden-crested mynah *Ampeliceps coronatus*

  jungle mynah *Acridotheres fuscus*

-myo- *comb. form* meaning "to contract" or to "close" but widely used for "muscle" which is properly -mys- and "fly" which is properly -myi-

Myodaria an assemblage of those schizophorous cyclorraphous brachyceran dipteran insects which have well-developed wings and are strong fliers in contrast to the Pupipara (*q.v.*)

myodian pertaining to muscle or muscles particularly those of the syrinx of birds (*cf.* myarian)

  acromyodian said of those birds in which the muscles of the syrinx are inserted at the dorsal and ventral ends of the bronchial semirings

  anacromyodian said of those birds in which the muscles of the syrinx are inserted on the dorsal end of the semirings

  catacromyodian said of those birds in which the muscles of the syrinx are inserted on the ventral end of the semirings

  diacromyodian said of a bird in which some of the muscles of the syrinx are inserted on the dorsal and some on the ventral side of the bronchial semirings

  anisomyodian said of birds in which the muscles of the syrinx are inserted irregularly on the ends or in the middle of the bronchial semirings

  mesomyodian said of those birds in which the muscles of the syrinx are inserted in the middle of the bronchial semirings

oligomyodian having few syringeal muscles

polymyodian said of a bird having many muscles in the syrinx

Myodocopa an order of ostracod Crustacea which have a well-marked aperture at the anterior portion of the valves through which the antennae may be extruded. They are exclusively marine and distinguished from the Podocopa by the biramose second antenna

Myomorpha an assemblage of rodent mammals containing the families Gliridae, Muridae, Bathyergidae, Spalacidae, Geomyidae, Heteromyidae, Dipodidae and Pedetidae all of which have the mouse-like form indicated by the name

Myophaga a not very widely used suborder of coleopteran insects containing several families of minute fungus-eating beetles

myophrisk a group of myonemes surrounding the base of a radiolarian spine

Myoporaceae a small family of tubuliflorous dicotyledons which may be distinguished from the Scrophulariaceae and Verbinaceae by the pendulous seeds

Myoporineae a suborder of tubulifloral dicotyledonous angiosperms containing the single family Myoporaceae

myosin = ATPase

Myoxidae = Gliridae

Myriapoda a taxon of the animal kingdom, by no means universally accepted, containing the Diplopoda and their allies as well as the Chilopoda

Myricaceae a family of plants containing the single genus Myrica. Characteristics of the family include simple leaves, the absence of a perianth and the indehiscent one-seeded fruit

Myricales an order of dicotyledonous angiosperms containing the family Myricacaceae. These plants are distinguished from other catkin-bearing families by the unilocular bistylar ovary

Myrientomata a class of opisthogoneate arthropoda erected to contain the single order Protura (*q.v.*) by those who do not hold that this order belongs in the class Insecta

Myristicaceae that family of ranalous dicotyledons which contains the nutmeg. The family is distinguished from its rare relatives, the Annonaceae, by the presence of only one set of floral envelopes

Myrmecobiinae a subfamily of dasyurid marsupials containing the numbats (*q.v.*)

Myrmecophagidae that family of xenarthran mammals which contains the true anteaters. They are distinguished by their immensely elongate head

Myrmeleontidae that family of neuropteran insects which contains the forms usually called antlions. They are readily distinguished as the only neuropteran insects to carry their wings extended when at rest and are frequently mistaken for damselflies

Myrmicinae a very large subfamily of formicid Hymenoptera recognizable by the fact that the pedicel of the abdomen is two-segmented. They contain harvester ants, fungus ants, and many others

Myrmosidae a family of scolioidean apocritan hymenopteran insects usually included with the Tiphiidae

myrobalan the combretaceous tree *Terminalia catappa* and its edible fruit

myrobolan the euphorbiaceous shrubby tree *Phyllanthus emblica* (= embal)

Myrothammaceae a small family of South African rosale dicotyledonous angiosperms

myrrh the biblical incense of this name is the gum of the burseraceous tree *Commiphora myrrha.* But the term is nowadays more usually applied to the umbelliferous herb *Myrrhis odorata*

Myrsinaceae a family of primulale dicotyledonous angiosperms differing from the closely allied Primulaceae by the woody habit and from the Theophrastaceae by the nature of the seed

Myrsinales an order of dicotyledonous angiosperms erected to contain the Myrsinaceae by those who do not hold that these belong in the Primulales

Myrtaceae a large family of myrtiflorous dicotyledons containing, *inter alia,* the eucalyptus tree and myrtles, and a number of spice trees including the clove. The pungent oil glands are distinctive of the family

Myrtales = Myrtiflorae

Myrtiflorae that order of dicotyledonous angiosperms which contains *inter alia,* the mousewoods, the buffalo berries, the loosestrifes, the pomegranates, the Brazil nuts, the mangroves, the myrtles, the evening primroses, the water chestnuts and the water milfoils. The most distinguishing feature common to all is the development of a hypanthium and the possession in the stem of interxylary ploem

Myrtineae a suborder of myrtiflorous dicotyledonous angiosperms containing the families Lythraceae, Heteropyxidaceae, Sonneratiaceae, Crypteroniaceae, Punicaceae, Lecythidaceae, Rhizophoraceae, Nyssaceae, Alangiaceae, Combretaceae, Myrtaceae, Melastomaceae, Hydrocaryaceae, Onagraceae and Haloragaceae

myrtle (*see also* grass) any of numerous shrubs, many of the myrtaceous genus Myrtus. Used without qualification, usually refers to *M. communis*

blue myrtle = trailing myrtle

crape myrtle the lythraceous shrub *Lagerstroemia indica*

downy myrtle the myrtaceous shrub *Rhodomyrtus tomentosa*

gum-myrtle any of several myrtaceous trees of the genus Angophora

running myrtle = trailing myrtle

sand-myrtle *U.S.* any of several species of ericaceous shrubs of the genus Leiophyllum, particularly *L. buxifolium*

sea myrtle the compositaceous shrub *Baccharis halimifolia*

trailing myrtle the apocynaceous herb *Vinca minor* (= periwinkle)

wax myrtle any of several myricaceous shrubs of the genus Myrica

-mys- *comb. form* meaning "muscle" often improperly rendered -myo-

Mysidaceae an order of pericarid malacostracous crustaceans in which the carapace extends over the greater part of the thoracic region but does not coalesce dorsally with more than three of the thoracic somites. The pariepods are all biramous and function as swimming organs

mysium muscle

endomysium a connective-tissue sheet, containing capillaries, surrounding an individual muscle fiber

epimysium a connective-tissue sheath surrounding an entire muscle

perimysium white, fibrous connective tissue separating muscle bundles

Mystacoceti a suborder of Cetacea containing all those forms in which the teeth are replaced by whalebone

mystax a tuft of hair-like setae immediately above the mouth in certain Diptera, sometimes referred to as a beard

Mytilidae the family of pelecypod mollusks containing the marine mussels

myurous mouse-tailed

Myxini that order of agnathous craniate Chordata which contains the hag-fishes. They are distinguished from the Petromyzontes or lampreys by the facts that the nasal sac communicates with the alimentary canal and the eyes are greatly reduced

Myxobacterales an order of Schizomycetes, commonly called slime bacteria. They occur either as unicellular rods, or in the form of a slime-like sheet, rapidly spreading over the substrate. Resting cells formed either directly from vegetative cells or in fruiting bodies formed from an aggregate of a large number of vegetable cells

Myxobia = Protista

Myxococcaceae a family of myxobacteriale Schizomycetes distinguished by the spherical or elipsoidal microcysts which contain the resting cells

Myxogastres = Myxomycetes

Myxomycetae a class of Myxomycophyta in which the vegetative stage is a naked amoeboid multinucleate plasmodium

Myxomycetes = Mycetozoa

Myxomycophyta a botanical taxon erected to contain the forms commonly called slime molds, regarded by botanists as related to the fungi, but differing from all other fungi in lacking a cell wall in all stages of vegetative development. These forms are regarded as animals by zoologists who place them among the sarcodine Protozoa in the order Mycetozoa. Some zoologists equate Mycetozoa with the class Myxomycetae

Myxophyceae the only class of the algal phylum Cyanophyta. The characteristics of the class are those of the phylum

Myxospongida an order of tetractinellid Porifera which lack spicules. They are sometimes called the "slime sponges"

Myxosporidia an order of cnidosporidian Sporozoa distinguished as having spores with two valves and from one to four polar capsules

Myzostoma a class of annelids erected to contain the Myzostomidae by those who do not hold them to belong in the Polychaetae

Myzostomidae the only family of the Myzostoma. They are small disc-shaped parasites on many echinoderms

mz a yeast genetic marker indicating a requirement for melezitose

n *see* [naked], [narrow] or [nipple-tipped fruit]

N *see* [Naked] and [Notch]

na *see* [narrow abdomen] or [nana]

na₁ *see* [nana]

Na *see* [Naked neck]

Nabidae a family of predatory bugs with the front femora slightly enlarged and raptorial. They are commonly called damsel-bugs

nacre = mother of pearl

nacreous pearly

NAD = nicotinamide

NAD = nicotinamide-adenine dinucleotide, a frequent receptor in oxido-reductase systems

nadation the implantation of an egg on the uterine wall

NADH = reduced nicotinamide-adenine dinucleotide

NADP = nicotinamide-adenine dinucleotide phosphate. A frequent receptor in oxidoreductase systems

NADPH = reduced nicotinamide-adenine dinucleotide phosphate

naiad the aquatic stage of any metabolous (*see* -bolous) insect

Naiadaceae = Najadaceae

Naididae a family of aquatic, mostly freshwater, oligochaete annelids distinguished by their glasslike transparency. Most have long hair-like chaetae

nail 1 (*see also* nail 2–4) the thin, flattened, homolog of a hoof or claw that covers the upper terminal surface of the fingers and toes in many primate mammals

nail 2 (*see also* nail 1, 3, 4) a horny tip on the upper bill of anseriiform birds

nail 3 (*see also* nail 1, 2, 4) a feather with a waxy spot replacing normal hackles in the grey junglefowl

nail 4 (*see also* nail 1–3) an obsolete measure of length roughly equivalent to half an inch

devil's toenails gum tragacanth

Najadaceae a family of helobe monocotyledonous angiosperms. The family is distinguished by possessing submerged flowers with one-seeded carpels and a reduced perianth. At one time the family also contained the Potamogetonaceae now considered distinct

Najadales = Helobiae

[**naked**] *either* a mutant (gene symbol *n*) on the sex chromosome (linkage group I) of the domestic fowl. The name is descriptive of the phenotypic expression, *or* a mutant (gene symbol *n*) mapped at 0 on linkage group III of the rat. The name is also descriptive of the phenotypic expression

[**Naked**] a mutant (gene symbol *N*) in linkage group VI of the mouse. The phenotypic expression differs

from [hairless] (*q.v.*), in that the hair breaks off rather than sheds

[**Naked neck**] a mutant (gene symbol *Na*) in linkage group IV or linkage group V of the domestic fowl. The phenotypic expression is rarely a completely naked neck but the feathered areas of the neck are much reduced

-nama- *comb. form* meaning "stream"

namatad a plant of streams

name in biology properly a binomial consisting of a generic and a specific designation (*cf.* nym, nomen)

bare name (*nomen nudum*) a published binomial not supported by evidence sufficient to permit its official adoption

gender name a name that of itself indicates sex, *e.g.* bull or cow

group name a term used to designate a group of animals such as a flock of birds or a school of fish

illegitimate name the name of a taxon which does not follow the international rules of nomenclature

juvenile name a name that of itself is indicative of youth (*e.g.* kitten, foal)

trivial name the name by which an organism or chemical compound (*e.g.* "mouse" or "nitrate ester reductase") is known, as distinguished from its scientific name (*e.g. Mus musculus* or *glutathione: polyolnitrate oxidoreductase*)

vernacular name the name by which something is known in the speech of any particular locality

-nan- *comb. form* meaning "dwarf" (*cf.* -nann-)

[**nana**] *either* a mutant (gene symbol *na₁*) mapped at 75 on linkage group III of *Zea mays.* The phenotypic expression is, of course, dwarfism, *or* a mutant (gene symbol *na*) mapped at 28 on linkage group B of the tomato. The phenotypic expression is a super dwarf plant

Nandidae a small family of freshwater acanthopterygian fishes most of which are known as leaffishes

-nann- *comb. form* meaning "dwarf" (*cf.* -nan-)

nannism the condition of being dwarfed

nanny a female goat or other caprid

-nap- *comb. form* meaning "turnip"

nap-at-noon *U.S.* the liliaceous herb *Ornithogalum umbellatum* (= star of Bethlehem)

nape the anterior region of the back of the neck (*cf.* nuchal)

nappy densely hairy

Narcomedusae an order of hydrozoan coelenterates of which the hydroid form is not known and the medusa is distinguished from those of the Trachomedusae by the scalloped umbrella

345

[narrow] *either* a mutant (gene symbol *nw*) mapped at 83.0± on chromosome II of *Drosophila melanogaster*. The term refers to the wings of the phenotype *or* a mutant (gene symbol *n*) in linkage group I of *Habrobracon juglandis*. The name refers to the wing form of the phenotype

[narrow abdomen] a mutant (gene symbol *na*) mapped at 45.2 on the X chromosome of *Drosophila melanogaster*. The phenotypic expression is a cylindrical abdomen. The phenotypic females are sterile

narwhal the delphinapterid cetacean *Monodon monoceros*

nasatus *see under* caste

nascent in the state of being formed

Nassariidae a family of small carnivorous gastropod mollusks having short shells with a pointed spire and a marked columella callus. Commonly called dogwhelk

-nast- *comb. form* meaning "pressed close"

nastic said of a type of curvature produced by the differential growth of the dorsal or ventral side

aitionastic a change in direction or bending caused by external forces

autonastic curvature arising from endogenous forces

econastic said of an ovule that curves towards the horizontal edge of the carpel

apinastic literally, pressed close above, but applied not only, for example, to leaves which lie flat on the ground, but also to ovules which curve in downward direction, because the top grows faster

exonastic said of an ovule with a horizontal curvature in the direction of the upper face of the carpel

geonastic said of curvature towards the ground

geonyctinastic = geonyctitropic

hyponastic said of an organ that curves upwards because the ventral surface grows more rapidly

photohyponastic said of excessive growth caused by the action of light

paranastic said of the longitudinal growth of lateral parts

photonastic said of one-sided growth due to the action of light

seismonastic sensitive to vibration

nasturtium properly the name of a genus of cruciferous plants (*see* cress). The "nasturtium" of the gardener is the tropaeolaceous genus Tropaeolum

-nasty *subs. suffix* indicated of a condition of curvature or differential growth. Compounds are defined under the *adj. suffix* -nastic

nasuti a caste of termites with a long hollow snout secreting a fluid for the repair of the walls of the nest

natality innate ability of a population to increase

natant floating

Natantia a suborder of decapod crustaceans, usually with compressed bodies and rostra, containing the shrimps and prawns which usually swim, as distinct from the Reptantia, with depressed bodies and rostra, containing the lobsters and crabs which usually crawl

Naticidae a family of gastropod mollusca commonly called moon-shells

natriferine a hormone controlling the transport of sodium across the skin of anuran Amphibia

naturalization the enforced adaptation, usually with human aid, of an organism to a foreign environment (*cf.* acclimatization)

nature the sum total of the organic world in its inorganic environment

Naucoridae a family of hemipteran insects commonly called creeping water bugs. They are distinguished from the Belostomatidae and Nepidae both by their creeping habit and absence of venation on the wings

naucum the succulent portions of a drupe

Nautilidae a family of tetrabranchiate cephalopod Mollusca containing the form commonly called Nautilus. Distinguished from all other living cephalopod Mollusca by the possession of an external whorled shell

nautilus either of two zoologically quite distinct cephalopod mollusks

paper nautilus any of several species of argonautid octopod of the genus Argonauta

pearly nautilus the tetrabranchiate nautiloid mollusk *Nautilus pompilius*

pseudonavicella the spore, shaped like the diatom Navicella, of some gregarines

navicular boat-shaped

navigation the process by which an organism determines its position on the surface of the planet, or determines or maintains a course from one position to another

bionavigation the methods of navigation employed by non-human organisms

Navioideae a subfamily of the Bromeliaceae

nawab *I.P.* any of numerous species of nymphalid lepidopteran insect of the genus Eriboea

nc^r a yeast genetic marker indicating growth inhibition by nicotine

Nc *see* [No crescent supernumerary legs]

nd *see* [netted] or [notchoid]

ne *see* [necrotic leaves]

neanic = larval

Nebaliacea the only order of phyllocarid leptostracan malocostracan Crustacea, containing the single genus Nebalia. It is distinguished from all other Malocostraca by having an adductor muscle running between the two valves of the carapace and in possessing an abdomen of seven somites

nebulose cloudy or smokey, either in regard to color or appearance

neck 1 (*see also* neck 2–4, weed) that part of the mammalian body which connects the head with the thorax or analagous structures in other animals

neck 2 (*see also* neck 1, 3, 4) any slender, hollow, tube or rod leading from a cavity or hollow structure to the exterior

septal neck the short tube in the center of each septa of an argonautid shell and through which the siphuncle passes

neck 3 (*see also* neck 1, 2, 4) in *bot.* that part of a plant where the stem joins the root

neck 4 (*see also* neck 1–3) in compound names of organisms

blackneck *Brit.* the noctuid lepidopteran insect *Toxocampa pastinum*

bristle-neck any of several species of pycnonotid birds of the genus Trichophorus

fiddle-neck any of several species of boraginaceous herb of the genus Amsinckia

ring-neck *Austral.* the psittacid bird *Barnardius zonarius* (= twenty-eight parrot) *U.S.* = ringneck pheasant *U.S.*

wryneck either of two species of piciform bird of the family Jyngidae, usually, without qualification *Jynx torquilla*

red-breasted wryneck *J. ruficollis*

necron living material which has died but which has not yet decomposed; applied particularly to plant forms

humus necron partially decomposed vegetable matter not yet humus, but in a condition in which the origin from stems or leaves, etc. can still be determined

[**necrotic leaves**] a mutant (gene symbol *ne*) mapped at 60 on chromosome 2 of the tomato. The phenotypic leaves slowly die from spreading necrotic areas

-nect- *comb. form* meaning "to join." Often confused with -nekt-

nectar a sugary, frequently aromatic, fluid secreted by many flowering plants

nectarine a smooth-skinned horticultural variety of peach

Nectarinidae a large family of passeriform birds commonly called sunbirds. They occur in most regions of the world except America and Europe. They are brilliantly colored in solid patches of red, orange, yellow, purple, etc., and most have tufts of orange feathers at the sides of the breast. The bills are long, decurved and finely serrate

nectary a glandular trichome secreting a sweet substance. Usually, but not always confined to the flowers (*cf.* lepal gland)

circumfloral nectary one on the outside of a flower

nectism the process of swimming by cilia or flagella

necton = nekton

Nectonematoidea a class of the phylum Nematomorpha distinguished from the Gordioidea by the presence of a double row of swimming bristles. The only genus (Nectonema) is marine

needle 1 (*see also* needle 2) any thin pointed object such as a spine

needle 2 (*see also* needle 1) in compound names of organisms

Adam's needle any of several species of the liliaceous genus Yucca, particularly *Y. filamentosa*

devil's darning needle *U.S.* the ranunculaceous climbing plant *Clematis virginiana. Brit.* = dragonfly

devil's needle = devil's darning needle

shepherd's needle *Brit.* the umbelliferous herb *Scandix pecten*

Spanish needle the compositaceous herb *Bidens bipinnata*

Neelidae a family of collembolan insects with globular bodies and no eyes

Neididae a small family of long-legged slender hemipteran insects commonly called stilt bugs

neinei *N.Z.* the epacridaceous tree *Dracophyllum latifolium* (= spiderwood *N.Z.*)

Neisseriaceae a family of gram negative eubacteriale Schizomycetes. Many are pathogenic producing, among other diseases, gonorrhea, meningitis, and scarlet fever

-nekt- *comb. form meaning* "to swim." Often misspelled -nect- (*q.v.*)

nekton swimming organisms sufficiently strong to be independent of the water currents among which they live (*cf.* plankton, tracheron)

edaphonekton that which lives in free water in soil

nelly *N.Z.* the procellariid bird *Macronectes gigantea* (= giant petrel)

Nelumbaceae a family erected to contain the nelumboid members of the family Nymphaeaceae by those who hold them to have separate familial rank

Nelumboidea a subfamily of Nymphaeaceae distinguished by having an indefinite number of sepals usually green but often as large as petals

-nema- *comb. form* meaning "thread." In the form -neme- is frequently confused with -cnem- (*cf.* -neme-)

brochonema when it was thought that chromosomes formed a "spireme" (*q.v.*) in the prophase of mitosis, the looped thread resulting from this was called a brochonema

chromonema coiled intertwined threads of which a chromosome is constituted

diplonema that stage in the prophase of meiosis immediately following pachynema and in which the separation of the paired chromosomes commences

leptonema first stage of the prophase of meiosis in which the elongate thread-like chromosomes show chromomeres

metanema the stage in the development of a moss which immediately follows the protonemata and that is thalloid rather than filamentous

nucleolonema thread-like structures within the nucleolus

pachynema that stage in the process of meiosis immediately following the zygonema. Synapsis is complete and chromosomes are contracting

protonema *see* protonemata

zygonema that stage in the prophase of meiosis in which the synapsis of chromosomes occurs

Nemalionales an order of floridian Rhodophyta distinguished by the lack of a tetrasporophytic generation

nemata *pl.* of nema

paranemata the paraphyses of Algae

protonemata the multicellular branched filaments which develop from the spores of mosses and some liverworts

Nemathelminthes an obsolete taxon of the animal kingdom once regarded as a phylum containing the Nematoda and the Acanthocephala and, from time to time, other thread-like worms

nemathylome a pit containing nematocysts usually occurring in vertical rows on the body wall of certain zoantharian coelenterates

Nematocera a suborder of dipteran insects containing those forms having many-segmented antennae (*cf.* Brachycera)

Nematoda an enormous phylum of pseudocoelomatous animals commonly called the threadworms. They are distinguished by the cylindrical body covered with a cuticle

[**Nematode resistance**] a mutant (gene symbol *Mi*) mapped at 59 on chromosome 6 of the tomato. The phenotypic expression is a resistance to *Meloidogyne incognita*

Nematomorpha a phylum of pseudocoelomatous bilateral animals commonly called the horsehair worms. This name is descriptive of the appearance. This group may be regarded as a class of the phylum Aschelminthes

-neme- anglicized equivalent of -nema-

axoneme literally, "axial thread" and applied to almost anything that could be so described from the central thread of a chromosome to the myoneme of stalked peritrich Protozoa

desmoneme a type of nematocyst in which the tube, closed at one end, is threadlike and coiled into the helix

haploneme a type of nematocyst in which the tube, open at one end, lacks a butt

heteroneme a type of nematocyst in which the tube, open at the tip, has a definite butt

mastigoneme one of the lateral thread-like projections from a flagellum

myoneme a contractile fibril in a protozoan (*cf.* myofibril)

nucleoloneme the filamentous portion of an interphase nucleolus

rhopaloneme a nematocyst, the tube of which is in the form of an elongated sac closed at one end

spasmoneme a bundle of myonemes in the stalk of contractile peritrich Protozoa

spironeme a spasmoneme that contracts into a spiral coil

Nemertea a phylum of acoelomate bilateral animals commonly called boot-lace worms. They are distinguished by the possession of an eversible proboscis and complete digestive and circulatory systems

Nemestrinidae a small family of orthoraphous brachyceran dipteran insects. They are among the larger of the "hover flies" frequently observed around flowers. The complex wing venation has sometimes given them the name tangle-veined flies

heteronemeous said of plants that germinate as a thread (*cf.* protonemata)

Nemichthyidae a small family of abyssal eels distinguished by their extreme length and the curious fact that the upper jaw is bent up and the lower jaw down so that they cannot be closed. Commonly called snipe-eels

-nemo- *comb. form* meaning grassland, pasture, meadow (*cf.* nema-)

Nemocera = Nematocera

nemorose inhabiting woods

Nemouridae a family of plecopteran insecta with an apical cross vein and the second tarsal segment much shorter than the rest

nene the Hawaiian anserine bird *Branta sandvicencis*

-neo- *comb. form* meaning "recent"

Neoceratodontidae = Ceratodontidae

Neognathae an infrequently used taxon of birds distinguished by the fact that the posterior ends of the palantines and the anterior ends of the pterygoids do not articulate with the basisphenoid rostrum and are supported by the hind end of the vomer

Neonychidae = Rhinomaceridae

Neopsyllidae = Hystrichopsyllidae

Neoptera a taxon of insects, in which the wings can be folded and which therefore contrast with the Halioptera. All modern insects except the dragonflies and mayflies lie in this taxon (*cf.* Palaeoptera)

Neornithes a subclass of birds containing all extant forms

Neosittidae a small family of passeriform birds erected to contain the Australian nuthatches. They differ from other nuthatches, which they resemble in habit, by the absence of bristles about the base of the bill

neoteny the condition of remaining indefinitely in an immature state

Neotraginae a subfamily of antelopid artiodactyle mammals containing the forms commonly known as pigmy antelopes. Among these the klipspringers and the dik-diks are the best known

Neotremata an order of ecardine brachiopods in which only the ventral valve is grooved to permit the passage of the pedicel

Nepenthaceae that family of sarraceniale dicotyledons which contains the East Indian and Madagascan pitcher-plants. The peculiar stalked flower differentiates them clearly from the allied Sarraceniaceae

-nephr-, -nephro- *comb. form* meaning "kidney"

nephridium an individual excretory unit as contrasted with a nephron. Used without qualification, the term is often synonymous with metanephridium (*cf.* mixium)

archinephridium the term usually applied to the solenocyte-bearing protonephridium of a trochopore larva

Hatschek's nephridium a tube carrying numerous solenocytes which project forward along the left ventro-lateral side of the notochord from just above the mouth in Amphioxus

meganephridium a large annelid nephridium furnished with both a peritoneal funnel and a duct leading to the exterior

metanephridium a nephridium consisting of a tubule furnished with a ciliated funnel-like end which opens from the coelom

micronephridium an annelid nephridium lacking a coelomic funnel

protonephridium a nephridium consisting of a blind tubule terminating in a flame cell or in a solenocyte

nephron a kidney unit

Nephropidea a superfamily of macruran Crustacea distinguished by the first pair of thoracic limbs being chelate. The lobster and crayfish are typical

nephros kidney

archinephros = holonephros

holonephros a postulated organ ancestral to all vertebrate kidneys

mesonephros a kidney composed of units each of which has both a Bowman's capsule containing a glomerulus and a coelomic funnel

metanephros a kidney composed of units which possess a Bowman's capsule but lack a coelomic funnel

pronephros a kidney composed of units in which a coelomic funnel leads to the tubule but in which the glomerulus is free in the coelomic cavity and does not form part of the nephron

Nephthydidae a family of errant polychaete annelids distinguished by the fact that the prostomium is very small, pentagonal and bearing four small tentacles. The pharynx bears one pair of jaws

nephtyid a modified cirrus which serves as a tango-receptor in polychaete annelids

Nepidae a family of predaceous hemipteran aquatic insects commonly called water scorpions. They are distinguished by their raptorial front legs and caudal breathing tube

nepionic post-embryonic or larval

phylonepionic = acmic

Nepticulidae a family of nepticuloid leaf-mining lepidopterous insects distinguished principally by containing the smallest known members of this order

Neptuneidae a family of large gastropod mollusks closely allied to the Buccinidae with which they are usually confused

nereid a water plant

amphinereid amphibious plant (*cf.* amphiphyte)

autamphinereid an autotrophic amphibious plant

autonereid an autotrophic water plant

halonereid a plant of salt waters, frequently taken to be synonymous with marine alga

limnonereid an association of freshwater algae

Nereidae a family of polychaete annelids distinguished

by their large well developed parapodia and large, transversely biting, jaws. In *U.S.* commonly called clam-worms

nereidion an association of water plants

Nereidiformia an order of phanerocephalous Polychaetae with well-developed tentacles and palps and usually bearing cirri on the peristomium

Neriidae a family of little-known tropical myodarian cycloraphous dipteran insects closely allied to the Micropezidae

nerite popular name of neritid gastropods

neritic a term applied to those organisms that occur in more or less coastal waters in distinction from oceanic organisms

Neritidae a family of gastropod mollusks distinguished by the globular shell with toothed aperture

-nerr- *see* -neur-

nerve 1 (*see also* nerve 2) a fiber, or more usually a bundle of fibers, that conducts impulses through organisms

accelerator nerve a nerve of the sympathetic nervous system which acts on the heart muscle

afferent nerve one which carries impulses to the brain

efferent nerve one which carries impulses from the brain

motor nerve = afferent nerve

nerve 2 (*see also* nerve 1) a specific group of fibers in a common sheath

abducens nerve (= cranial VI). Runs from the ventral surface of the medulla to the lateral rectus muscle of the eye

accessory nerve (= cranial XI)

auditory nerve (= cranial VIII). Runs from the inner ear to the upper surface of the anterior region of the medulla

collector nerve a nerve found in some fish, connecting additional anterior segmental nerves to an appendage

cranial nerve one leaving the central nervous system, within the cranium. They are commonly indicated in serial order by Roman numerals

facial nerve (= cranial VII). Arises from the medulla oblongata. It adheres to the face, the roof of the mouth and the hyoid nerve

glossopharyngeal nerve = cranial IX). Arises from the medulla, close to the root of the tenth nerve

hypoglossal nerve (= cranial XII)

oculomotor nerve = (cranial III). Arises from the ventral surface of the mesencephalon, and supplies the superior medial, and the inferior rectus and inferior oblique muscles of the eye

olfactory nerve (= cranial I.) Conducting impulses from the nasal organ

optic nerve (= cranial II.) Running from the retina to the floor of the diencephalon

scalar nerve the nerve controlling the circular muscles in a rotiferan

somatic nerve a nerve carrying impulses to or from the outer layers of the body

spinal nerve any of numerous nerves leaving the central nervous system posterior to the cranium. They are commonly identified by serial arabic numerals

posttrematic nerve those branches of cranial IX and X which pass behind the gill crest

pretrematic nerve those branches of cranial IX and X which pass in front of the gill crest

trigeminal nerve (= cranial V.) Rises from the dorsal surface of the myelencephalon and divides into the

ophthalmic, maxillary and the mandibulary branches

trochlear nerve (= cranial IV.) Runs from the dorsal surface of the back end of the mesencephalon to the superior oblique muscle of the eye

vagus nerve (= cranial X.) Arises from several roots in a dorso-lateral wall of the medulla. In gilled animals, it sends branches to the gill-slits, as well as to the viscera and the lateral line

visceral nerve a nerve carrying impulses to or from internal organ systems

nerve 3 used synonymously with vein (*q.v.*) in the description of insect wings

brachial nerve in insects, the nerves of the front wing

nerve 4 (*bot.*) completely confused with vein 2 but properly applying to the main vascular bundles in a leaf which cause a prominence on the surface above them. (*see also* leaf)

bast nerves = bast veins

nervose said of a leaf with unusually prominent venation

enervous said of a leaf lacking veins

nest 1 (*see also* nest 2, 3) a structure, usually single and temporary but sometimes semipermanent and colonial, in which eggs are deposited and young are sheltered

nest 2 (*see also* nest 1, 3) term for a group of mice living together, whether or not an actual nest is involved

nest 3 (*see also* nest 1, 2) in compound names of organisms

bird's nest (*see also* fungus) *Brit.* either the orchid *Listera nidusavis* or the pyrolaceous herb *Monotropa hypopitys* (= *U.S.* pinesap)

giant bird's nest the pyrolaceous root-parasitic herb *Pterospora andromedea*

hangnest = hangbird

nestling a bird still in the nest, usually reserved to those having only down feathers

Nestoridae = Psittacidae

net a loose series of anastomizing fibers

branchial net the basket of anastomizing blood vessels that forms the pharynx of ascidians

chromidial net = nuclear net

linin net = chromidial net

nerve net a network of conducting fibrils running over the surface of many invertebrates including the Protozoa. In the latter it consists of the kinetium

nuclear net the network of threads produced as an artifact by the acid fixation of nuclei

[net] a mutant (gene symbol *net*) mapped at 0- on chromosome II of *Drosophila melanogaster*. The phenotypic expression is extreme plexus venation

netrum = initial spindle

[netted] a mutant (gene symbol *nd*) mapped at 72 on chromosome 10 of the tomato. The chlorophyll in the phenotypic leaf is grouped round the veins giving a netted appearance

nettle (*see also* ring, tree) any of several plants that sting, many of the urticaceous genus Urtica. *Brit.*, without qualification, usually *U. dioica*

ball-nettle = horse-nettle

bastard nettle the urticaceous herb *Pilea pumila*

bull-nettle *U.S.* the euphorbiaceous herb *Cnidoscolus stimulosus* (= tread-softly)

burning nettle *U.S. Urtica urens*

dead nettle *Brit.* any of several species of herb of the labiateous genus Lamium

yellow dead nettle *Brit.* the labiateous herb *Galeobdolon luteum*

false nettle *U.S.* any of several urticaceous herbs of the genus Boehmeria

hedge nettle *U.S.* any of several species of the labiateous genus Stachys (= wound wort)

hemp-nettle any of several labiateous herbs of the genus Galeopsis

horse-nettle the solanaceous herb *Solanum carolinense*

spurge-nettle any of several euphorbiaceous herbs of the genus Cnidocolus

stinging nettle *U.S.* the euphorbiaceous herb *Cnidoscolus stimulosus*. *Brit.* any of three species of the genus Urtica but particularly *U. dioica*

wood nettle *U.S.* the urticaceous herb *Laportea canadensis*

-neur- *comb. form* meaning "nerve." Properly, though very rarely -nevr-, frequently, though quite improperly -nerv-

Neuradoideae a subfamily of Rosaceae distinguished by having a superior ovary, indehiscent fruit and a gynoecium of ten or more pistils which are basally connate

neural pertaining to nerves

epineural said of those animals (*i.e.* Chordata) in which the central nervous system is dorsal to the alimentary canal

hyponeural said of those animals in which all save the most anterior ganglia of the central nervous system are ventral to the alimentary canal

myoneural pertaining to muscle and nerve

Neuramphipetalae = Compositae

neurite a general term for axons and dendrons, particularly in the cells of diffuse nervous systems where these are not clearly differentiated

neurium a nerve

endoneurium connective tissue fibers surrounding nerve fibers

epineurium the outer layer of connective tissue around the perineurium

perineurium the outer connective tissue coat of a nerve

neuron a unit of nervous structure consisting of the sum total of a nerve cell and all its processes

adjustor neuron a neuron that is neither sensory nor motor but which correlates the activities of both

internuncial neuron one that connects sensory and motor pathways

motor neuron a neuron that originates stimuli

sensory neuron a neuron that receives stimuli

neuropile the central mass of medullary tissue of an invertebrate ganglion

Neuroptera an order of insects containing, *inter alia,* the antlions, lacewings and dobson flies. All have complete metamorphosis, chewing mouth parts, four membranous wings and lack anal cerci (*cf.* Pseudoneuroptera)

exoneurosis the development of a spine or bristle from the vein of a leaf

neurula that stage in the development of an embryo in which the whole of the neural plate is not yet invaginated

euthyneury the condition of having a symmetrical central nervous system

hyponeustic having less than the normal number of spiracles

neuston organisms floating on the surface of water (*cf.* pleuston)

supraneuston organisms living on or in the surface film of water

neuter without sex

-nevr- *see* -neur-

[New striped] a mutant (gene symbol *S*) mapped at 6.1 on linkage group II of *Bombyx mori*. The phenotypic expression is a dark stripe on the larva

newt the terms newt and salamander are now hopelessly confused. Originally, and still to a great extent in *Brit.* usage, newt is a predominantly aquatic urodele and a salamander is a predominantly terrestrial urodele. However, there is currently an endeavor in *U.S.* to confine the term newt to the family Salamandridae and to use the word salamander for all other urodeles. The term eft is synonymous with newt in both languages, though rarely used. There is currently an attempt in *U.S.* to use the word eft for the terrestrial phrase and newt for the aquatic phase of such forms as the red-spotted newt

alpine newt *Triturus alpestris*

California newt *Taricha torosa*

common newt *Brit. Triturus vulgaris*

crested newt *Brit. Triturus cristatus*

Iberian newt = Spanish newt

Japanese newt *Cynops pyrrhogaster*

marbled newt *Brit. Triturus marmoratus*

palmate newt *Triturus palmata*

Pyrrenean newt *Euproctus asper*

Sardinian newt *Euproctus platycephalus*

smooth newt *Brit. Triturus vulgaris*

Spanish newt *Pleurodeles waltli*

spotted newt *Brit.* = common newt *Brit.*

black-spotted newt *U.S. Diemictylus meridionalis*

red-spotted newt *U.S. Diemictylus viridescens*

striped newt *U.S. Diemictylus perstriatus*

warty newt *Triturus danubialis*. The term is also sometimes applied in England to *T. cristatus*

webbed newt *Brit. Triturus palmatus*

nexus the region of fusion of plasma membranes (*cf.* desmosome)

nfn a bacterial mutant showing nitrofuran resistance or sensitivity

Ng *see* [No glue]

ngaio *N.Z.* the myoporaceous tree *Myoporum laetum*

ngiru-ngiru *N.Z.* the muscicapid bird *Petroica macrocephala*

niacin pyridine-3-carboxylic acid. A water soluble nutrient commonly regarded as a vitamin. Gross deficiency causes pellagra

niacinamide = nicotinamide

nibbler popular name of girellid fish

nic a bacterial mutant and yeast genetic marker indicating a requirement for nicotinic acid

nic-1 to nic -13 *see* [Nicotinic-1] to [Nicotinic 13]

Nicandreae a tribe of the family Solanaceae

niche variously used by ecologists to mean either the position occupied by an individual in an assemblage of individuals or to indicate a microhabitat or biotope

ecological niche a role played by a specific organism in the total ecology of its environment

nicotianous tobacco-colored

nicotinamide niacinamide pyridine-3-carboxamide. Essentially similar to nicotinic acid but lacking some of the side effects when fed to humans

[nicotinic-1] a mutant (gene symbol *nic-1*) in linkage group I of *Neurospora crassa*. The phenotypic expression is a requirement for nicotinamide

[nicotinic-2] *either* a mutant (gene symbol *nic-2*) in linkage group I of *Neurospora crassa*. The phenotypic expression is a requirement of nicotinamide *or* a mutant (gene symbol *nic-2*) in linkage group II of *Chlamydomonas reinhardi*. The phenotypic expression is a requirement of nicotinamide

[**nicotinic-3**] a mutant (gene symbol *nic-3*) on linkage group VII of *Neurospora crassa*. The phenotypic expression is a requirement for nicotinamide

[**nicotinic-7**] a mutant (gene symbol *nic-7*) in linkage group VI of *Chlamydomonas reinhardi*. The phenotypic expression is a requirement for nicotinamide

[**nicotinic-11**] a mutant (gene symbol *nic-11*) on linkage group IV of *Chlamydomonas reinhardi*. The phenotypic expression is a requirement for nicotinamide

[**nicotinic-13**] a mutant (gene symbol *nic-13*) on linkage group X of *Chlamydomonas reinhardi*. The phenotypic expression is a requirement for nicotinamide

[**nicotinic-trytophan**] a mutant (gene symbol *nt*) on linkage group VII of *Neurospora crassa*. The phenotypic expression is a requirement for either nicotinamide or tryptophan

nidosous fetid

nidulant encased in a nest, or by extension, embedded in pulp or other neutral substance

nievitas any of several species of the boraginaceous genus Cryptanthe

nigger *I.P.* the satyrid lepidopteran insect *Orsotriœna medus*

nightingale (*see also* thrush, wren) properly any of several turdid birds of the genus Luscinia, notorious for their song. *Brit.*, without qualification, *L. megarhynchos*. The term is also applied to many other songsters

　Salt Island nightingale *W.I.* the mimid *Mimus gundlachii* (= Bahama mockingbird)

　Jamaica nightingale the mimid *Mimus polyglottos*

　Pekin nightingale the tumaliid *Leiothrix lutea*

　Persian nightingale the turdid *Luscinia megarhynchos*

　Spanish nightingale *W.I.* the mimid *Mimus gundlachii* (= Bahama mockingbird)

　thrush-nightingale the turdid *Luscinia luscinia* (= sprosser)

　Virginian nightingale the fringillid *Richmondena cardinalis* (= cardinal)

nightjar any of several nocturnal birds, particularly Caprimulgids, with raucous cries. *Brit.*, without qualification, *Caprimulgus europaeus*

　dusky nightjar *Caprimulgus binotatus*

　freckled nightjar *Caprimulgus tristigma*

　golden nightjar *Caprimulgus eximius*

　Mozambique nightjar *Scotornis fossii*

　owlet-nightjar *Austral.* any of several aegothelids of the genus Aegotheles. Usually, without qualification, *A. cristatus*

　long-tailed nightjar *Scotornis climacurus*

　white-tailed nightjar *Caprimulgus natalensis*

　pennant-wing nightjar *Semeiophorus vexillarius*

　standard wing nightjar *Macrodipteryx longipennis*

nightshade any of numerous vines, shrubs or herbs usually poisonous

　bastard nightshade any phytolaccaceous shrub of the genus Rivina

　deadly nightshade the solanaceous vine *Atropa belladonna*

　Malabar nightshade the basellaceous herb *Basella rubra*

nikau = nikau palm

nilgaie = nilghai

nilghai the Indian boselaphine bovid artiodactyle *Boselaphus tragocamelus*. The misleading term "antelope" is frequently applied to this form

niltava any of several muscicapid birds of the genus Muscicapa. Usually, without qualification, *M. grandis*

nimble-will any of several species of polygonaceous herb of the genus Polygonum particularly those of the subgenus Tiniaria

ninnuleous the color of a young stag, dark cinnamon-brown

catnip *U.S.* the labiateous herb *Nepeta cataria* (= *Brit.* catmint)

-niph- *comb. form* meaning "snow" but less common than the Latin equivalent niva-

niphic = nival

nipple (*see also* wort) any prominence on a compound gland, particularly the mammary gland, and usually one through which a product is excreted

[**nipple-tipped fruit**] a mutant (gene symbol *n*) mapped at 30 on chromosome 7 of the tomato. The name is descriptive of the phenotypic expression

Nippotaeniidea an order of eucestod cestodes lacking any adhesive organ except an apical sucker

Nissl (properly Nissl's) *epon. adj.* from Franz Nissl (1860–1919)

nit a bacterial mutant showing utilization of nitrate

nit 1 (*see also* nit 2) *W.I.* either of two species of charadriid bird *Charadrius collaris* and *C. wilsonia* (*cf.* plover)

nit 2 (*see also* nit 1) the egg of the head louse

[**n-nit**] = [nit-1]

nit-1–nit 3 *see* [Nitrate-1] to [Nitrate 3]

nitelinous dormouse-colored

nitid lustrous

Nitidulidae a family of coleopteran insects allied to the Cucujidae but differing in that the front coxae are transverse and more or less cylindrical, their cavities usually closed behind. They are sometimes called sap-feeding beetles

nitr = [Nit-3]

Nitrariaceae = Zygophyllaceae

[**nitrate-1**] a mutant (gene symbol *Nit-1*) on linkage group I of *Neurospora crassa*. The phenotypic expression is a failure to reduce nitrate

[**nitrate-2**] a mutant (gene symbol *Nit-2*) in linkage group I of *Neurospora crassa*. The phenotypic expression is an inability to reduce nitrate

[**nitrate-3**] a mutant (gene symbol *Nit-3*) in linkage group IV of *Neurospora crassa*. The phenotypic expression is a failure to reduce nitrate

Nitrobacteraceae a family of pseudomonodale Schizomycetes distinguished by the presence of a polar flagellum and the ability to derive energy from the oxidation of ammonia to nitrite or from the oxidation of nitrite to nitrate

-niva- *comb. form* meaning "snow" (*cf.* -niph-)

nival pertaining to snow, particularly as a habitat for plants

niveal = nival

niveous snow-white

nm a common abbreviation for nothomorph

nme a bacterial mutant showing E-N-methyl-lysine in flagellar protein

NMN = nicotinamide mononucleotide

nname = yam

[**No crescent supernumerary legs**] a mutant (gene symbol *Nc*) mapped at 1.4 in linkage group VI of *Bombyx mori*. The crescents are absent from the phenotypic adult and there is a supernumerary leg in the second abdominal segment of the larva

noctilucous phosphorescent

Noctuidae a large family of noctuoid lepidopteran insects. The family contains an enormous number of (more than 2,500) species many of economic importance. The group is very varied but the presence of ocelli, the relatively narrow antennae, the stout body and the habit of folding the wings tent-wise over the abdomen are usual

Noctuoidea a superfamily of lepidopteran insects containing a very large number of medium-sized moths of which the tiger-moths and gypsy-moths are typical

noctule *Brit.* the vespertilionid chiropteran *Nyctalus noctula*

nocturnal pertaining to the hours of darkness either as a rhythmic activity or in opposition to diurnal and crepuscular

-node- *comb. form* meaning "knot" or "knob"

nodal adjective from node

noodle *Brit. obs.* the human head, particularly the posterior portions

noddy any of several larid birds mostly of the genus Anous. *U.S.*, without qualification, *A. stolidus.* (= brown noddy)

    black noddy = white-capped noddy

    brown noddy = *A. stolidus*

    white-capped noddy *A. minutus*

    blue-grey noddy *Procelsterna cerulea*

    lesser noddy *A. tenuirostris*

node 1 (*see also* node 2, 3, 4) any swelling in an elongate cylindrical structure particularly "joints" on plant stems from which leaves arise

    internode a space between two nodes

    Ranvier's node an interruption in the myelin sheath of axons and neurons

node 2 (*see also* node 1, 3, 4) a transverse septum in the stolon of an ectoproct

    internode the area between two nodes in the stolon of an ectoproct, usually considered to be a modified zooid

node 3 (*see also* node 1, 2, 4) in ants the knob between the thorax and posterior abdomen

node 4 (*see also* node 1–3) a lump or "knot" of tissue in an organ

    heart node a modified lump of muscular tissue exercising nervous functions

    Hensen's node the anterior end of the primitive streak corresponding to the dorsal lip of the blastopore

    primitive node = Hensen's node

    Tawara's node the heart node lying at the junction of the atrium and ventricle

nodosaroid said of the multi-loculate shell of a foraminiferan in which the chambers are arranged in a straight line

nodose knotty or knobby

nodule diminutive of node

    lymphatic nodule an area tightly packed with lymphocytes

    Malpighian nodule = lymphatic nodule

    pseudonodule an area on the frustule of a diatom lacking markings, but not thickened

    root nodule without qualification usually refers to the bacteria containing nodules on the roots of leguminous plants

noggen coarse flax or linen

[No glue] a mutant (gene symbol *Ng*) mapped at 0 on

linkage group XII of *Bombyx mori*. The name indicates a lack of adhesion of the eggs to the surface on which they are laid by a phenotype

noho *Hawaii* the rallid bird *Pennula sandwichensis*

nol a bacterial mutant showing norleucine resistance or sensitivity

noil the knots combed from wool

Nolanaceae a small family of tubuliflorous dicotyledons distinguished from most other families in the order by the peculiarly lobed ovary derived from five carpels

Nolidae a family of noctuoid lepidopteran insects distinguished by the ridges and tufts on their wings

Nolineae a subfamily of amaryllidaceous monocotyledons. Considered by many properly to belong in the Liliaceae

noll the parietal region of the skull

noll-khol = kohl rabi

nolt *obs. Brit.* cattle

-nom- *comb. form* meaning "law"

-noma- *comb. form* meaning "wanderer" but hopelessly confused in *biol.* literature with either or both meanings of -nomo- (*q.v.*)

nomad *zool.* a wandering form *Bot.* a pasture plant

nomadic the adjective is just as confused as the noun. Thus, a nomadic plant could be either a tumbleweed or a pasture plant

Nomadidae a family of apoid apocritan hymenopteran insects usually included with the Apidae. They are usually wasp-like in appearance and because of their parasitic nesting habit are referred to as cuckoo-bees

Nomarthra = Tubulidentata

hyponome the funnel which projects from the mantle of a cephalopod mollusc

nomen name (*cf.* nym)

    nomen conservandum name which is retained in virtue of its long established and wide usage

    nomen dubium a name which cannot be applied with certainty to a specific species

    nomen inquirendum one under investigation by an international commission on nomenclature or other body

    nomen oblitum a senior name which has been ruled unavailable by an international commission, by virtue of not having appeared in primary literature for fifty years or more

    nomen novum a new name applied to a form already having a name which has been ruled invalid

    nomen nudum a name without status since its original publication was not accompanied by an adequate description, definition, or figure

    nomen rejectum an existing name which is no longer considered valid

-nomo- *comb. form* hopelessly confused from two homonymous roots, meaning respectively "custom" and "pasture." Some ecologists have further confused -nomo- meaning "pasture" with -noma- meaning "wanderer"

autonomous used of an organism which is the only expression of its species and not one of an alternation of generations or other polymorphic condition

heteronomous said of things specialized along different lines as the successive segments in some annelid worms

homonomous said of similar things or of those specialized along similar lines as successive identical segments of an annelid. *cf.* heteronomous

-nomy *comb. suffix* properly meaning "the application of laws" but in current usage "the science" or "study of"

non- a *negative prefix*. Such obvious compounds as "non-vascular" are not defined in this dictionary

-non-, -nono- *comb. form* meaning "nine"

[non-agouti] *either* a mutant (gene symbol *a*) at locus 0 on linkage group IV of the rabbit. The phenotypic pelt is black *or* a mutant (gene symbol *a*) in linkage group V of the mouse. This mutation prevents the appearance of yellow bands in what would otherwise be an agouti-type pelt

Noncalcarea a class of the phylum Porifera that is distinguished by the absence of a calcareous skeleton. This taxon is today rarely used

none-so-pretty the caryophyllaceous herb *Silene armeria*

nonesuch the leguminous herb *Medicago lupulina* (= medic)

nonett *obs. Brit.* titmouse (*q.v.*)

nonnant an organism, or said of an organism, that cannot swim

nonpareil the fringillid bird *Passerina ciris* (= painted bunting)

Clifden nonpareil *Brit.* the noctuid lepidopteran insect *Catocala fraxini*

nonsuch = nonesuch

noolbenger (*Austral.*) the tarsipedine phalangerid *Tarsipes rostratus* (= honey-sucker)

noop the shrub *Rubus chamaemorus* and its edible fruit (= cloudberry)

nopal any of several cactus, particularly of the genera Opuntia and Nopalea

nope *Brit. obs.* = bullfinch

nordcaper any of several mystacocetic whales

norril the nostrils of galliform birds

nose 1 (*see also* nose 2) that part of the mammalian head which immediately surrounds the nostrils. Sometimes extended to non-mammalian analogs

afternose = postclypeus

nose 2 in compound names of organisms

arrow-nose the mastacembelid teleost fish *Macrognathus aculeatus*

conenose popular name of reduviid hemipteran insects (*cf.* assassin bug)

blood-sucking conenose *U.S.* the *Triatoma sanguisuga*

elephant-nose popular name of mormyrid fish

red nose the characinid teleost fish *Hemigrammus rhodostomus*

tubenose popular name of aulorhynchid fish

no-see-um *U.S.* popular name of ceratopogonid dipterans

-noso- *comb. form* meaning "disease"

Nosodendridae a small family of coleopteran insects closely allied to the Byrrhidae but distinguished from them in that the head is not bent under the body

nostril one of the paired apertures leading to the nasal cavity of mammals. In insects the term is applied to the rhinarium

nosul *Brit. obs.* = blackbird (*Brit.*)

Notacanthiformes = Heteromi

notal = dorsal

notate marked, or having markings on the surface

Rivinus' notch that dorsal gap between the nearly circular ends of the tympanic bone

[Notch] one of four pseudoalleles (gene symbol *N*) (the others are [notchoid], [split], and [facet])

mapped at 3.0± on the X chromosome of *Drosophila melanogaster*. The name refers to the notched wings of the phenotype

[notchoid] (gene symbol *nd*) one of four pseudoalleles mapped at 3.0± on the X chromosome of *Drosophila melanogaster*. The other pseudoalleles are [split], [facet], and [Notch]

[notchy] a mutant (gene symbol *ny*) mapped at 32.0± on the X chromosome of *Drosophila melanogaster*. The wing tips of the phenotype are notched

note 1 (*see also* note 2) the uropygeal gland of hawks

note 2 (*see also* note 1) *Brit. obs.* = nut

-notero- *comb. form* meaning moist

nothagge *Brit. obs.* = nuthatch

-notho- *comb. form* meaning "bastard"

Notiophilidae = Ephydridae

-noto- *comb. form* meaning "back" or "dorsal"

Notodelphyoidea a heterogenous group of copepod Crustacea resembling each other principally in that all are associated with urochordates either as commensals or parasites. Many have a dorsal brood pouch in which the eggs are incubated and therefore lack the egg masses typical of other copepods

Notodontidae a large family of noctuoid lepidopteran insects, many known in the United States as prominents

notonectal swimming upside down

Notonectidae a family of aquatic hemipteran insects commonly called backswimmers. This name is adequately descriptive

[notopleural] a mutant (gene symbol *np*) mapped at between 58.7 and 60.2 on chromosome II of *Drosophila melanogaster*. The phenotypic expression is broad wings and short bristles

Notopteridae a small family of isospondylous fishes, distinguished by the very long anal fin, commonly called featherbacks

Notoryctidae that family of polyprotodont marsupials which contains *Notoryctes typhlops*, the marsupial mole. The close resemblance to the Talpidae (eutherian moles) is one of the most remarkable known cases of convergent evolution

Notostraca an order of crustacean arthropods commonly called tadpole shrimps. They are distinguished from other branchipods by the fact that they have a broad shield-like carapace on the dorsal surface

notted lacking horns or other protuberances as the beard of wheat

notum the dorsal portion of an arthropod segment (= tergum)

epinotum usually used for the first insect abdominal segment when it is fused with the metathorax as in many Hymenoptera but sometimes applied to the upper surface of the prothorax, more properly called pronotum

eunotum that portion of the insect notum that bears the sclerite to which the wing articulates

metanotum the upper surface of the metathorax in insects

paranotum a lateral outgrowth, precursors of wings, from the dorsal plates of the insect thorax

postnotum an intersegmental plate in the dorsal surface of the insect thorax

pronotum the upper surface of the prothorax in insects

-nov- *comb. form* meaning "new"

-nov-, -novem- *comb. form* meaning "nine"

novice *S.Afr.* the danaid lepidopteran insect *Amauris ochlea*

noxial = nocturnal

Np *see* [Notopleural]

nt *see* [Nicotinic-tryptophan]

nub *see* [nubbin]

nubbin a dwarfed, or imperfect, fruit particularly an ear of corn

[nubbin] a mutant (gene symbol *nub*) mapped at 53.0 on chromosome II of *Drosophila melanogaster*. The phenotypic expression is very small thin wings

nubilous blue-gray

nuble the rhamnaceous shrub *Zizyphus lotus* (*cf.* jujube)

nucal pertaining to a nut

nucament = catkin

nucamentaceous = indehiscent

nucamentum = catkin

nucellus the central portion of a plant ovule and therefore, technically, the megasporangium

nuchal from nape and therefore pertaining to the dorsal area immediately behind the head or back of the neck of mammals; also applied to a similar area of some insects

-nuci- *comb. form* meaning "nut"

micrococcal nuclease an enzyme that hydrolyzes RNA and DNA particularly at the adenine-thymine linkage

ribonuclease an enzyme that catalyzes the formation of a cyclic nucleotide by the transfer of a pyrimidine nucleotide residue

deoxyribonuclease an enzyme that catalyzes the hydrolysis of DNA to yield oligodeoxyribonucleotides

deoxyribonuclease II catalyzes the formation of 3'-nucleotides from DNA

nucleate possessing nuclei

multinucleate possessing many nuclei

uninucleate possessing one nucleus

nucleolid a solid body, other than the nucleolus, apparent in a nucleus

nucleolinus a granule within the nucleolus

nucleolule = nucleolinus

nucleolus a small spherical organelle usually found in the interphase nucleus

amphinucleolus a combination of a nucleolus with some other nuclear granule

lateral nucleolus the second, smaller of two nucleoli

paranucleolus = lateral nucleolus

nucleosidase an enzyme catalyzing the hydrolysis of N-ribosyl-purines into purines and D-ribose

AMP nucleosidase an enzyme catalyzing the hydrolysis of AMP into adenine and D-ribose 5-phosphate

NAD nucleosidase an enzyme catalyzing the hydrolysis of nicotinamide from NAD

uridine nucleosidase an enzyme catalyzing the hydrolysis of uridine into uracil and ribose

nucleotidase an enzyme catalyzing the hydrolysis of nucleotides to yield ribonucleosides and phosphoric acid

phosphoadenylate 3-nucleotidase an enzyme catalyzing the hydrolysis of adenosine 3',5'-diphosphate to yield AMP and orthophosphate

diphosphopyridene nucleotide (DPN) *obs.* = NAD

triphosphophyridine nucleotide (TPN) *obs.* = NADP

nicotinamide adenine dinucleotide (NAD) a frequent receptor in oxidoreductase reactions

nicotinamide-adenine dinucleotide phosphate (NADP) a frequent receptor in oxidoreductase systems

nucleus 1 (*see also* nucleus 2–7) literally the kernel of a nut and thus used for any more or less spherical mass embedded or contained in anything including the central portion of a seed

micronucleus (*see also* nucleus 3) the term is sometimes applied to the centrosome

Pander's nucleus the area of yolk-free cytoplasm immediately under the blastodisc of telolecithal eggs

nucleus 2 (*see also* nucleus 1, 3–7) a clump of associated nerve cells in the brain

habenular nuclei a group of nerve cells in the epithalamus

motor nucleus a group of nerve cells which originates nerve impulses for transmission to a muscle

nucleus 3 (*see also* nucleus 1, 2, 4–7) that organelle with the cell in which almost all of the nucleic acids are concentrated

centronucleus the nucleus of a cell in which centrosomes are active during mitosis

definitive nucleus that which is formed in the plant embryo-sac by the fusion of the two polar nuclei

endonucleus = nucleolus

energic nucleus = resting nucleus

generative nucleus 1 (*see also* generative nucleus 2) a nucleus taking part in reproduction, in contrast to a somatic nucleus. The term is most commonly applied to one of the two nuclei derived from the primary division within a pollen cell (*cf.* tube nucleus)

generative nucleus 2 (*see also* generative nucleus 1) a nucleus in the course of mitosis

germ nucleus a nucleus produced by the fusion of gametes

germinative nucleus = generative nucleus 1

lateral nucleus a second, usually smaller, nucleus as the micronucleus of a ciliate protozoan

macronucleus a large nucleus, without recognizable chromosomal organization, found in many ciliate Protozoa (*cf.* micronucleus)

micronucleus (*see also* nucleus 1) the diploid nucleus that maintains genetic continuity in many ciliate Protozoa (*cf.* macronucleus)

open nucleus the nuclear material of Cyanophyceae

paranucleus the second, usually smaller, of two nuclei in a cell

interphase nucleus a nucleus not actively dividing (= "resting" nucleus)

pronucleus the haploid nucleus of a gamete

rejection nucleus the nucleus of a polar body

resting nucleus one which is not dividing

restitution nucleus one which is restored through the refusion of replicated chromosomes that have failed to separate at mitosis

secondary nucleus a nucleus, which subsequently gives rise to the endosperm, formed by the fusion of two of the eight nuclei derived from the nucleus of a megaspore

somatic nucleus a nucleus which is thought to play no part in sexual reproduction, as the macronucleus of Protozoa or any nucleus of a body cell

trophonucleus (Protozoa) = macronucleus

tube nucleus one of the two nuclei derived from the primary division within a pollen cell (*cf.* generative nucleus)

vegetative nucleus any of those nuclei of the pollen tube which do not take part in fertilization

nucleus 4 (*see also* nucleus 1–3, 5–7) the disc, or shield, containing the sporules on a lichen

nucleus 5 (see also nucleus 1–4, 6, 7) the central part of the perithecium of a fungus

nucleus 6 (see also nucleus 1–5, 7) the first beginnings of a molluscan shell

nucleus 7 (see also nucleus 1–6) the hilum of a starch grain

nuculane a drupe containing more than one seed

Nuculanidae a family of pelecypod mollusks differing from the Nuculidae in that the margins of the shell are not crenulate and that there is a pit for the ligament between the umbones

nuculanium a hard case, containing several seeds, in a fleshy fruit

nucule either any small independent fruit or one of the seeds in a nuculanium

Nuculidae a family of pelecypod mollusks usually having three cornered shells which lack a pit for the ligament between the umbones. The beak is distinctly toothed

nud a verb used to describe the butting action of a young mammal to increase the flow of its mother's milk

Nuda a class of Ctenophora that lack tentacles

Nudibranchiata a suborder of opisthobranchiate gastropods commonly called sea-slugs. The name derives from the dorsal respiratory extensions of the mantle, often brilliantly colored

nugget Austral. = runt Brit.

nuke Brit. obs. spinal cord

nukta I.P. the anatine duck Sarkidiornis melanotus

nukupuu the drepaniid bird Hemignalthus lucidus

nulliplex = bottom recessive

numbat popular name of myrmecobine dasyurids. They resemble anteaters both in structure and habit even to the extent of the long sticky tongue and large front claws

diploid number the number (2N) of chromosomes in a somatic cell

haploid number the number (N) of chromosomes in a gamete

zygotic number = the number of chromosomes in a zygote (= diploid number)

numbles intestines and lungs, particularly of deer

Numididae that family of galliform birds which contains the guineafowls. They differ from the pheasants, apart from the coloration of the plumage, in lacking a protuberance on the second metacarpal of the wing

nun (see also bird) Brit. dial. the parid bird Parus caeruleus (= blue titmouse)

nunlet any of several piciform buceonid birds of the genus Nonnula

nursery a place where young organisms, either plant or animal, are reared

-nut- comb. form meaning "nod"

nut 1 (see also nut 2, cracker, hatch, shell, tree) a hard, dry, single-seeded fruit enclosed in a husk that remains with the fruit as it ripens

apple-nut the seed of the palm Coelococcus amicarum

Barbados nut the euphorbiaceous shrub Jatropha curcas

betel nut the palmaceous tree Areca catechu and its fruit

bitternut the juglandaceous tree Carya cordiformis

bladdernut any of several staphyleaceous herbs of the genus Staphylea and their edible seeds

American bladdernut S. trifolia

Brazil nut the oily three-angled seed of the lethydiaceous tree Bertholletia excelsa

buffalo-nut any of several santalaceous shrubs or trees of the genus Pyrularia, particularly P. pubera

butternut (see also curculio) the juglandaceous tree Juglans cinerea (= white walnut). The term is also occasionally applied to the Brazil nut

candle-nut the fruit of the euphorbiaceous tree Aleurites moluccana

cashew-nut the anacardiaceous tree Anacardium occidentale and its fruit

chestnut (see also nut 2, borer, oak, shell, weevil) any of several trees of the fagaceous genus Castanea or the fruit of these trees. (cf. horsechestnut)

American chestnut C. dentatum

bastard chestnut the hamamelidaceous tree Liquidambar formosana

Cape chestnut the rutaceous tree Calodendrun capensis

Chinese chestnut C. mollisissima

European chestnut C. sativa

horsechestnut (see also nut 2) the hippocastanaceous tree Aesculus hippocastanum or the nutlike fruits of that tree

common horsechestnut A. hippocastanum

Japanese horsechestnut A. turbinata

Japanese chestnut C. crenata

waterchestnut the hydrocaryaceous aquatic herb Trapa natans

Chinese waterchestnut the sedge Eleocharis tuberosa and its edible tuber

Chilean nut the proteaceous shrub Gevuina aveolana

cob nut U.S. any of several species of the betulaceous genus Corylus. Brit., without qualification, C. maxima (cf. hazel nut)

coconut (see also bug, scale, weevil) the palmaceous tree Cocos nucifera and its very large seeds

double coconut the palmaceous tree Lodoicea maldivica and its huge seeds (= maldive nut)

creamnut = Brazil nut

groundnut the leguminous herb Apios tuberosa

hazel nut any of several species of the betulaceous genus Corylus

hedgehog-nut the fruit of the solanaceous herb Datura stramonium (= Jimson weed)

ivory nut the seed of the palm Phytelephas macrocarpa (= vegetable ivory)

Jesuit nut the hydrocaryaceous aquatic herb Trapa natans (= water chestnut)

king-nut the juglandaceous tree Carya laciniosa

macadamia nut = Queensland nut

maldive nut the palmaceous tree Lodoicea maldivica and its huge seed (= double coconut)

marshnut the anacardiaceous tree Semicarpus anacardium

mockernut the juglandaceous tree Carya tomentosa

oilnut the santalaceous shrub Pyrularia pubera

Paranut = Brazil nut

peanut (see also worm) the leguminous plant Archais hypogae and its underground fruit

hog-peanut any of several species of the leguminous genus Amphicarpa, particularly A. bracteata

physic nut any one of three euphorbiaceous shrubs Jatropha multifida, J. podagrica and Croton tiglium

French physic nut Jatropha curcas

pignut the juglandaceous tree Carya glabra, the buxaceous shrub Simmondsia californica and its edible fruit and Brit. the umbelliferous herb Bunium flexuosum and its edible tubers

small pignut C. ovalis

sweet pignut *U.S. C. ovalis*

pond-nut the nymphaeaceous aquatic *Nelumbo lutea*

purging nut = Barbados nut

Queensland nut the proteaceous tree *Macadamia ternifolia* and its fruit

Singhara nut the hydrocaryaceous aquatic herb *Trapa bicornis* and its edible nut

walnut (*see also* aphis, caterpillar, fly, moth, sphinx) any of numerous trees of the juglandaceous genus Juglans

  Arizona walnut *J. major*

  black walnut *J. nigra*

  California walnut *J. californica*

  Circassian walnut = English walnut

  English walnut *J. regia*

  Persian walnut = English walnut

  Texas walnut *J. rupestris*

  Turkish walnut = English walnut

  white walnut *J. cinerea*

nut 2 (*see also* nut 1) any of several lepidopteran insects

chestnut (*see also* nut 1) *Brit.* any of several species, mostly noctuids of the genus Orrhodia, usually without qualification *O. vaccinii*. The term is also applied to some members of the genus Amathes

barred chestnut *Brit.* the noctuid *Noctua dahlii*

horse-chestnut (*see also* nut 1) *Brit.* the geometrid *Pachycnema hippocastanaria*

red chestnut *Brit.* the noctuid *Pachnobia rubricosa*

nutant nodding

nutation the twisting of the growing parts of a plant organ

circumnutation the movement of a growing point of a plant around the axis

revolving nutation = circumnutation

nutmeg 1 (*see also* nutmeg 2, hickory, geranium, shell) the myristicaceous tree *Myristica fragrans* and particularly its seed. The aril surrounding the seed is mace

calabash nutmeg either of two species of annonaceous trees of the genus Monodora

California nutmeg the taxaceous tree *Torreya californica*

nutmeg 2 (*see also* nutmeg 1) *Brit.* the noctuid lepidopteran insect *Mamestra trifolii*

large nutmeg *Brit.* the noctuid *Hama sordida*

nutria the term applied in the fur trade to the pelt of the Coypu and is very occasionally used in biological literature for that animal itself

nutricism the formal association, frequently regarded as a type of symbiosis, in which one partner is almost totally dependent for food and shelter on the other

nutrition the sum total of the processes by which an organism secures and utilizes the materials necessary for the continuation of life

holophytic nutrition a nutrition based on the synthesis of carbohydrates with the aid of chlorophyll

autotrophic nutrition = holophytic nutrition

heterotrophic nutrition the type of nutrition involving the ingestion of organic materials

mixotrophic nutrition the nutrition of an organism that is partly holophytic and partly saprozoic

saprozoic nutrition that type of nutrition that does not involve the ingestion of solids, but only the absorption of nutritive solutes

nw *see* [narrow]

ny *see* [notchy]

nyala any of several species of strepsicerosine bovid artiodactyles of the genus Tragelaphus (*cf.* bushbuck)

-nychium *comb. suffix* from onychium meaning a "claw" or "finger nail"

eponychium the portion of cornified tissue that lies over the base of a nail and that is commonly referred to as the "quick"

hyponychium the cornified epidermal layer immediately under the free edge of the nail of a mammalian finger

paronychium a heavy bristle on the pulvillus

rhizonychium the terminal bone of a toe that bears a claw

-nyct- *comb. form* meaning "night" or "nocturnal"

Nyctaginaceae that family of centrospermous dicotyledons which contains, *inter alia*, the four-o'clocks and the Bougainvillias. The floral bracts and very thin-walled fruit derived from a one-seeded, one-celled ovary are characteristic

Nycteribiidae a family of small, spider-like, wingless pupiparan dipterous insects parasitic on bats and therefore commonly called batflies, or less properly, bat-ticks. In many species the head can be folded back into a groove on the upper side of the thorax

Nycteridae a family of microchiropteran mammals closely allied to the Rhinolophidae from which they are distinguished by the small and cartilaginous premaxilla

Nyctibiidae a small family of caprimulgiform birds commonly called potoos. They lack the rictal bristles of most caprimulgiform birds but the loral feathers have bristle-like tips

epinyctous night-flowering

nye term for a group of pheasants

-nykt- *see* nyct

-nym- *comb. word* meaning "name" (*q.v.*)

basonym the original (valid) name of an animal

caconym a name which is linguistically impossible

chironym a name found in a manuscript

homonym a name rejected on the ground of prior occupancy. Also said of two different names for the same organisms, usually of a species variously assigned to either of two genera

hyponym a name not supported by an actual specimen

metanym a name rejected on the basis that another organism of the same genus already bears this specific designation

tautonym a bionomial in which the genetic and specific names are the same as *Gorilla gorilla*

typonym a name having priority but which is based on the same type as the name currently in use

nymph 1 (*see also* nymph 2, 3) *U.S.* those immature stages of arthropods which inhabit much the same environment as the parent which they more or less resemble in form. *Eur.* the penultimate and sometimes the antepenultimate stage of an arthropod showing metamorphosis

deutonymph the second of three nymphal stages in the development of acarines

protonymph the first of the three nymphal stages of the development of an acarine

pseudonymph = semi-pupa

subnymph an intercalated stage between a nympha and a pupa

tritonymph the third of the three nymphal stages in the development of an acarine

nymph 2 (*see also* nymph 1, 3) various lepidopteran insects

lilac nymph *S.Afr.* the nymphalid *Crenis rosa*

tree-nymph *I.P.* any of several species of danaids of the genus Hestia. *S.Afr.* numerous nymphalids of the genus Crenis

wayward nymph *U.S.* the noctuid *Catocala anti-nympha*

white nymph *Austral.* the nymphalid *Mynes geoffroyi*

woodnymph (*see also* nymph 3) properly the term should be restricted to the satyrid lepidopteran *Cercyonis pegala* but is sometimes used to designate the whole family (*cf.* grayling)

nymph 3 (*see also* nymph 1, 2) in compound names of other organisms

woodnymph any of several apodiform trochilid birds of the genus Thalurania, usually, without qualification. *T. furcata*

Nymphaeaceae that family of ranalous dicotyledons which contains the water lilies. Apart from their habitat, the water lilies are easily distinguished by the peculiar fruit

Nymphaeineae a suborder of ranale dicotyledons containing the families Nymphaeaceae and Ceratophyllaceae

Nymphaeoideae a subfamily of the Nymphaeaceae distinguished by the presence of four to five sepals

Nymphaloidea a very large superfamily of lepidopteran insects distinguished by the great reduction of, or an absence of claws from, the front legs. They are therefore sometimes called the "four-footed butterflies"

Nymphalidae a large, worldwide, family of nymphaloid lepidopteran insects and distinguished by having the discal cell of the hindwing "open," that is not distally closed by crossveins

Nyssaceae a family of myrtiflorous dicotyledonous angiosperms distinguished by the estipulate alternate leaves and solitary flowers. The best-known member of the family is the sourgum tree

Nyssoninae a large subfamily of sphecid hymenopteran insects usually mistaken from their general shape and appearance for vespids but which may be distinguished by the collar-like pronotum. The subfamily is divided into numerous tribes which are not recorded in this work

o *see* [orange]

o₂ *see* [opaque endosperm]

O *see* [Blue egg] or [oval fruit]. Also used as a bacterial genetic marker for a character affecting the activity of somatic antigen

O—symbolic prefix indicating that a serotypic response of a bacterial mutant is somatic (*cf.* H–)

oa *see* [mottled translucent]

oak 1 (*see also* oak 2, bug, girdler, looper, scale, skeletonizer, weevil, worm) any of very many fagaceous trees of the genera Quercus, Lithocarpus, and Pasania

  tanbark oak *Pasania densiflora*
    yellow bark oak *Q. velutina* (= *black oak*)
  basket-oak = cow oak
  bear-oak = scrub oak
  black oak any of several species of Quercus but particularly *Q. rubra, Q. schneckii, P. ellipsoidalis, P. velutina*. If the term black oak is used as a subgenus in distinction from white oak it may be identified by the fact that the wall of the nut is tomentose on the inner surface and requires two years to mature
    California black oak *Q. kelloggii*
  blue oak *Q. douglassii*
  bur oak *O. macrocarpa*
  chestnut-oak *Q. prinus*
    yellow chestnut-oak *Q. muehlenbergii*
  chinquapin oak *Q. prinoides*
  cork-oak *Q. suber*
  cow-oak *Q. michauxii*
  mossy-cup oak = bur oak
    overcup oak = swamp post oak
  English oak *Q. robur*
  holly-oak *Q. ilex*
  holm oak = holly oak
  jack oak *Q. marilandica* or *Q. ellipsoidalis* (= northern pin oak)
  laurel-oak *Q. laurifolia*
  live oak *Q. virginiana*
    California live oak *Q. chrysolepis* (= canyon live oak)
    highland live oak *Q. wislizenii*
    coast live oak *Q. agrifolia*
  marsh oak *Q. palustris*
  maul-oak = California live oak
  Nutall oak *Q. nutallii*
  Oregon oak *Q. garryana*
  pin-oak *Q. palustris*
    northern pin oak *Q. ellipsoidalis*
  post-oak (*see also* oak 2) *Q. stellata*
    swamp post-oak *Q. lyrata*
  red oak *Q. rubra*

  scarlet oak *Q. coccinea*
  scrub oak *Q. ilicifolia*
  shingle oak *Q. imbricaria*
  Shumard oak *Q. shumardii*
  Spanish oak *Q. falcata*
    swamp Spanish oak *Q. pagodaefolia*
  tan oak *Lithocarpus densiflorus*
  Turkey oak *Q. cerris*
  turkey-oak *Q. laevis*
  valonia oak *Q. aegilops*
  water oak *Q. nigra*
  weeping oak *Q. lobata* (= California white oak)
  white oak any of numerous species having the general appearance of *Q. alba*. If used as a subgenus, the white oaks are distinguished from the black oaks by the fact that the walls of the nut are glabrus on the inner surface, and that it matures in one year. In *W.U.S.* the name is confined to *Q. lobata*
    swamp white oak *Q. bicolor*
    Utah white oak *Q. utahensis*
  willow-oak *Q. phellos*
  yellow oak *Q. muehlenbergii* (= yellow chestnut-oak)

oak 2 a number of herbs, trees and shrubs thought to be oak-like

  Jerusalem oak the chenopodiaceous herb *Chenopodium botrys*
  poison oak either of two anacardiaceous shrubby trees *Rhus diversiloba* or *R. quercifolia*. In some parts of the *U.S.* this name is applied to poison ivy (*q.v.*)
  post-oak (*see also* oak 1) the vitaceous vine *V. linsecomii* (= turkey grape)
  she-oak any of several species of tree of the casuarinaceous genus Casuarina
  silk-oak the proteaceous shrub *Grevillea robusta*

oat 1 (*see also* oat 2, 3, grass) any of several species and numerous agricultural varieties of the grass genus Avena, particularly *A. sativa*
  animated oat *A. sterilis*

oat 2 (*see also* oat 1, 3) any of several grasses superficially resembling the agricultural forms
  sea-oats (*see also* oat 3) *U.S. Uniola paniculata*
  swamp-oat *U.S. Trisetum pennsylvanicum*
  water oats *Zizania palustris*
  wild oats (*see also* oat 3) *U.S. Uniola latifolia*
  yellow oats *U.S. Trisetum flavescens*

oat 3 (*see also* oat 1, 2) any organism, other than a grass, thought to resemble oats
  sea-oats (*see also* oat 2) the eggs of gastropod mollusks, particularly of the family Thaisidae
  wild oats (*see also* oat 2) any of several species of liliaceous herb of the genus Uvularia

ob- *comb. form* meaning "opposite" or "reverse"

ob *see* [obese]

[**obese**] a mutant (gene symbol *ob*) on linkage group XI of the mouse. The name is descriptive of the phenotypic expression

obex a barrier, particularly that which divides populations

obices *pl.* of obex

oblate said of a sphere flattened at the poles

obscure apart from its obvious meanings, is used in *bot.*, to indicate a dingy tint

obsitous = squamate

obsolete in *biol.*, sometimes used in the sense of rudimentary

obt *see* [*b*₈-mottled] and [obtuse]

obtect said of an insect pupa having the wings and legs closely pressed to the body and therefore not showing through the surface of the pupal case. (*cf.* exarate)

obtected said of some larval arthropods that have a heavy chitinous coat

obturator anything which closes something

obtuse blunt

[**obtuse**] a mutant (gene symbol *obt*) mapped at 77.5 on chromosome III of *Drosophila melanogaster*. The phenotypic wings are short and blunt

obvalate walled in

obverse one side of anything, the opposite being the "reverse." Which is obverse and which is reverse is a matter of convention. In insects, obverse refers to the "head on" aspect

oc *see* [chinese], [ocelliless]

Oc *see* [Ocellarles]

occipital pertaining to that region of the chordate head that lies in the posterior third, or in the general vicinity of the occipital bone

oticoccipital a general term for the posterior division of the endocranium in fishes

occiput the back part of the crown of the head

ocean a large body of salt water

oceanad an ocean plant

[**Ocellarless**] a mutant (gene symbol *Oc*) mapped at 5.7± on the X chromosome of *Drosophila melanogaster*. The name refers to the lack of ocellar bristles

ocellate eyed, in the sense of having an eye-like marking

[**ocelliless**] a mutant (gene symbol *oc*) mapped at 23.1 on the X chromosome of *Drosophila melanogaster*. The name is descriptive of the phenotype

ocellus 1 (*see also* ocellus 2–4) one of the elements of a compound eye or an eye-like structure.

circular ocellus a closed ocellus with a definite lens

compound ocellus an ocellus with annular divisions

converse ocellus one in whch the distal ends of the retinal cells receive the light

inverse ocellus one in which the free ends of the retinal cells face the light

pigment-cup ocellus an invaginated pigmented ocellus which may contain a lenticular element

pigment-spot ocellus a non-protuberant ocellus containing a pigment and photoreceptors

ocellus 2 (*see also* ocellus 1, 3, 4) a very small, simple eye formed in many invertebrates

ocellus 3 (*see also* ocellus 1, 2, 4) *bot.*, an epidermal, light-sensitive leaf cell

ocellus 4 (*see also* ocellus 1–3) an eye-like pattern (*i.e.* a disc surrounded by a ring) forming part of a pattern, particularly on the wings of insects

ocelot *U.S.* the felid carnivore *Felis pardalis*

Ochnaceae a small family of parietalous dicotyledons distinguished by the many sepals and lobed ovary

Ochtonidae a family of lagomorph mammals properly called pikas, but most commonly known as conies through misapplication of the name of Old World hyracoids which they resemble. They are distinguished from rabbits by the short ears and absence of a tail

[**ochracea**] a mutant (gene symbol *ocr*) mapped at 0 on chromosome II of *Drosophila melanogaster*. The phenotype has a light eye which darkens with age

brindled **ochre** *Brit.* the phalaenid lepidopteran insect *Dasypolia templi*

ochroleucous buff colored

-ochrous *comb. term* probably derived from *Gr. ochros* meaning "pale" rather than *ochra* meaning "yellow-ochre" but which has come to mean "colored"

allochrous said of an organism capable of changing its color

chrysochrous yellow-skinned

galochrous milk-white

myochrous mouse-colored (*cf.* murinous)

Ochteridae a small family of semi-aquatic hemipteran insects

-ochth- *comb. form* meaning "bank" or "ridge"

ochthad a bank plant

pelochthium a mudbank formation of plants

Ochthiphilidae a family of small greyish myodarian cycloraphous dipteran insects the larvae of many of which are predatious on aphids

-ochthon *comb. term* chopped off, without justification, *from the Gr.* autochenon (native inhabitant) and reattached to various prefixes

allochthonous 1 (*see also* allochthonous 2) brought in from outside in contrast to autochthonous, particularly used of materials that enter a body of water by seepage or drainage

allochthonous 2 (*see also* allochthonous 1) pertaining to peat derived from accumulations of drifted vegetation

autochthonous aboriginal. Used to indicate organic materials produced in a body of water

ocotillo the fouquieriaceous shrub *Fouquieria splendens*

ocr *see* [ochracea]

ocreate 1 (*see also* ocreate 2) stipulate

ocreate 2 (*see also* ocreate 1) said of a bird which is booted (*see* boot)

octa *comb. form* meaning "eight"

Octandria that class of the Linnaean classification of plants distinguished by the possession of eight stamens

Octodontidae a family of hystricomorph rodents of rat-like appearance, but with long ears and lacking any trace of a thumb on the forelimb. The group contains, *inter alia*, the spiny rats and porcupine-rats

Octoknemataceae a small family of santandale dicotyledons

Octopoda a suborder of dibranchiate cephalopod Mollusca distinguished from the Decapoda by possessing only eight arms and, in general, a more globose body

Octopodidae a family of octopod dibranchiate cephalopod Mollusca commonly called octopus or devilfish. They are distinguished from the Argonautidae by lacking a shell and in having the mantle united to the head

octopus specifically a genus of dibranchiate cephalopod mollusks but often applied to other shell-less members of the suborder Octopoda

Oculinidae a family of corals, popularly called ivory-corals, distinguished by their slender branching structure

oculus the rudiment of a bud, particularly on a corm or tuber

Ocypodidae a family of brachyuran arthropod Crustacea with a broad, subquadrangular carapace and long eye-stalks situated in a groove-like orbit

od *see* [translucent], [outstreched]

Oddi's *epon. adj.* from Ruggero Oddi

Odinidae a family of brachyceran dipteran insects commonly fused with the Agromyzidae

odk see [mottled]

Odobaenidae a monogeneric family of pinnipede carnivorous mammals containing the walrus, distinguished by the enormous tusks

Odocoileinae a sub-family of cervid artiodactyl mammals containing a large number of deer confined entirely to the western hemisphere. They are generally to be distinguished from the Cervinae by the shape of the antlers which are directed backwards but with forwardly pointing tines

Odonata that order of insects which contains the forms commonly called dragonflies and damselflies. They are distinguished by the four large membranous wings and a freely moveable head bearing chewing mouth-parts

odonate having toothed mouth parts

Odontoceti a suborder of cetacean mammals distinguished from the Mysticocetae by having teeth but no whale bone

odontoid tooth-like

Odontophoridae a family of galliform birds usually included in the Phasianidae

-odos- *comb. form* meaning "way" or "direction"

-oec- *comb. form* meaning "house" (*see also* -ec-)

-oecia plant or animal taxa having this ending are listed alphabetically

-oecious adjectival form derived from oecium. The synonymous -ic, and -icial terminations are very rarely used

androecious pertaining to the housing of male organs

agamoandroecious the condition of having male and neuter flowers in the same capitate flower (*cf.* agamogynaecious, agamogynomonoecious)

autoecious (*see also* parasite) the condition of a moss which has the male and female inflorescence on the same plant

cladautoecious with the male inflorescence of a moss on a true branch

goniautoecious the condition of a moss which has the male inflorescence axillary on a female branch

pseudoautoecious a moss which is sometimes dioecious and sometimes autoecious

rhizautoecious the condition of a moss having the male inflorescence on a short branch and the female on a rhizoid

dioecious having a male and female sex organ in different individuals

androdioecious possessing two forms, one male, the other hermaphrodite

polygamodioecious *either* said of a plant which is functionally dioecious but has scattered through its inflorescence a few perfect flowers *or* of one having perfect and imperfect flowers on separate individuals of the same species

gynodioecious a dioecious plant having hermaphrodite and female flowers on separate plants

euryoecious *biol. jarg.* for ubiquitous

gynoecious pertaining to the housing of female organs

agamogynoecious the condition of having female and neuter flowers in the same capitate flower (*cf.* agamandroecious and agamogynomonoecious)

polygynoecious said of a multiple fruit formed by the combination of the pistils of several flowers

heteroecious *see under* parasite

metoecious *see under* parasite

monoecious said of organisms in which the male and female sex organs occur in the same individual, particularly plants bearing both male and female flowers

andromonoecious having two forms, one with male flowers only, and the other with perfect flowers

dimonoecious the condition of a monoecious plant having not only perfect flowers, but also imperfect flowers

coenomonoecious the condition of a plant which has male, female and hermaphrodite flowers on the same individual

gynomonoecious a monoecious plant, having only female and hermaphrodite flowers

agamogynomonoecious the condition of having neuter, female and perfect flowers in the same capitate flower (*cf.* agamogynaecous and agamandroecious

paraoecious literally, having adjacent dwellings. Said of a moss in which the male and female organs lie in the same inflorescence, or of any cryptogran in which the atheridium or/and archegonium are adjacent, or of ants or other animals, occupying adjacent nests and which sometimes shelter in each others

poloecious the condition of a plant of which some individuals have both fertile and barren flowers and other individuals may have either fertile or barren flowers

synoecious having flowers of different sexes in the same inflorescence

-oecism *comb. suffix* meaning the condition of having a special type of dwelling. The termination -oecy is sometimes used in the same sense

androecism the condition that only male forms of an organism have been observed

gynoecism the condition that only female forms of an organism have been observed

trimonoecism the condition of having male, female and perfect flowers on the same plant

-oecium Latinized form, unknown to the Romans, of *Gr. oikos*, a house. In *biol.* usually, but not invariably (e.g. carcinoecium), used for the housing of reproductive organs. The *adj.* forms -oecic, -oecious and -oecial occur, the first and last only very rarely. The *subst.* form oecism is used in the sense of "pertaining to a type of house" but the *subst.* oekete appears invariably with the "k" spelling. The abbreviation -oeco- is nowadays invariably spelled -eco- (e.g. ecology). **Oecium** (Ectoprocta = ovicell (Ectoprocta)

androecium the male portion of a flower consisting of the stamens and their associated parts

acropetalous androecium = centripetal androecium

basipetalous androecium = centrifugal androecium

centrifugal androecium one in which the inner stamens ripen before the outer ones

centripetal androecium one in which the outer stamens ripen before the inner ones

anthoecium the spikelet of a grass

carcinoecium the replacement of a shell once inhabited by a hermit crab, by the coenechyme of a coral

gamoecium the unit of inflorescence of bryophytes

gynoecium the female portion of a flower

apocarpous gynoecium one in which the carpels are not fused

syncarpous gynoecium one in which the carpels are fused

hydroecium a depression to which the stem is attached in the nectophore of siphonophoran coelenterates

synoecium an association of animals in the dwelling place or places of another

zooecium a case containing an animal particularly applied to the zoids of Ectoprocta

-oeco- *see* -eco-

Oecophoridae a family of small flattened gelechioid lepidopteran insects

-oecy *see* -oecism

Oedemeridae a family of coleopteran insects frequently confused with the Cerambycidae, which they superficially resemble, but from which they can be readily distinguished by the possession of five tarsal joints on each of the two legs and four on the last legs

Oedicnemidae = Burhinidae

Oedogoniales that order of chlorophycean Chlorophyta distinguished by having cylindrical uninucleate cells divided end to end in simple or branched filaments

oekete an inhabitant

synoekete an inhabitant of an ant or termite nest that is neither a guest nor a parasite and, indeed, that has no recognizable relationship to the colony

-oeko *comb. form* meaning dwelling place, but almost universally replaced in *biol.*, literature by -eco- or -oeco-

Oenophilidae = Oinophilidae

Oenotheraceae = Onagraceae

oesophagus in vertebrates that region of the alimentary canal which lies between the mouth and the stomach and, in invertebrates, that portion of the alimentary canal which lies between the pharynx and the stomach

pharyngoesophagus the dorsal of the two tubes into which the pharynx of cyclostomes divides. The ventral is the respiratory passage

Oestridae a family of large, stout-bodied myodarian cycloraphous dipteran insects frequently mistaken for bees from their general appearance. The larvae are parasitic in numerous vertebrates and some, such as the sheep botfly, are serious pests

oestrone *see* estrone

dihydroesterone = estradiol

oestrous pertaining to oestrus

monoestrous said of mammals in which only a single oestrous cycle occurs at one breeding season

polyoestrous said of mammals which have numerous oestrous cycles in the course of one breeding season

oestrus rhythmic changes in the female mammaliam reproductive system which prepare the uterus for the reception of a fertilized egg

metaestrus the period immediately following oestrus

officinal technically, "of the shops" but by convention meaning medicinal herbs of the pharmacopoeia

og *see* [giallo ascoli]

Og *see* [old gold striping]

-ogen a termination once widely used for precurser-compounds, *e.g.* pepsinogen, fibrinogen. The prefix pre- is currently preferred

Ogcocephalidae a family of pediculatous fishes distinguished by their huge arm like pectoral fins and small leg like ventral fins. Commonly called batfishes in virtue of their general appearance

-oicous- = -oecious-

-oicus = -oecious

-oid- *comb. form* meaning "likeness"

-oidea suffix indicating superfamilial, or occasionally subordinal, rank in animal taxonomy

-oideae suffix indicating subfamiliar rank in plant taxonomy. The zoological equivalent is -inae

oidia reproductive cells developed from the fragmentation of fungal hyphae

oi-i *N.Z.* the procellariid bird *Puffinus griseus* (= mutton bird *N.Z.*)

-oikos- *comb. form* meaning "house" usually replaced by oecium (*q.v.*)

ok *see* [kinshiryu]

okapi the African palaeotragine mammal *Okapia johnstoni*

okra the malvaceous herb *Hibiscus esculentus* and its edible fruits (= gumbo)

ol *see* [oligodactyly]

ol *see* [Oleic acid]

Olacaceae a small family of santalalous dicotyledons

[Old gold striping] a mutant (gene symbol *Og*) mapped at 16 on linkage group IX of *Zea mays*. The phenotypic leaves are striped green and yellow-green

old-man-and-woman the crassulaceous herb *Sempervivum tectorum* (= houseleek)

Florida **olea** the oleaceous tree *Osmanthus americanus* (= wild olive, devilwood)

Oleaceae that family of contortous dicotyledons which contains, *inter alia*, forsythia, the ashes, the jasmines, the privets, the olives and the lilacs. The numerical plan of four with a superior ovary is distinctive of the family

oleaginous oily

oleander the apocynaceous shrub *Nerium oleander*

water oleander *U.S.* the lythraceous herb *Decodon verticillatus* (= water willow)

oleaster the elaeagnaceous shrub *Elaeagnus angustifolia*

[Oleic acid] a mutant (gene symbol *ol*) in linkage group IV of *Neurospora crassa*. The phenotypic expression is a requirement for any higher fatty acid

Oleineae a suborder of Contortae containing the single family Oleaceae

oleoid resembling the olive, particularly in regard to the peculiar leaves

Oleoideae a subfamily of the Oleaceae containing all of the genera except those in the Jasminoideae

oleraceous esculent

Olethreutidae a family of tortricoid lepidopteran insects. The distinctive character is a fringe of hairs on the basal portion of the cubitus of the hindwing. The family contains many destructive pests such as the codling moth and the cloverhead caterpillar

olf *Brit. obs.* = finch, olph

-oligo- *comb. form* meaning "few"

Oligochaeta a class of annelid worms containing, *inter alia*, the earthworms. They are distinguished from the Polychaetae by the absence of cephalic appendages and parapodia, and from the Hirudinia by the absence of suckers

[oligodactyly] a mutant (gene symbol *ol*) in linkage

group I of the mouse. The phenotypic expression is a reduction in the number of digits

Oligomera a zoological taxon invented by Butschli to contain the segmented worms or Annelida (*cf.* Polymera and Amera)

[**Oligosyndactylism**] a mutant (gene symbol *Os*) found in linkage group XVIII of the mouse. The phenotypic expression involves stubby, and frequently fused, digits

Olinaceae a small family of myrtiflorous dicotyledonous angiosperms

olivaceous olive colored

olive 1 (*see also* olive 2, 3, 4, fly, scale) any of several species and numerous agricultural varieties of the oleaceous genus Olea particularly *O. europaea*. The term is also applied to a few olive-like trees and shrubs

    wild olive *W.I.* the olacaceous tree *Ximenia americana* and its edible fruit, the euphorbiaceous tree *Putranjiva roxburghii*, and *U.S.* the oleaceous tree *Osmanthus americanus* (= devilwood, Florida olea)

olive 2 (*see also* olive 1, 3, 4) common name of olivid gastropod Mollusca

olive 3 (*see also* olive 1, 2, 4) *Brit.* the noctuid lepidopteran insect *Plastenis subtusa*

olive 4 (*see also* olive 1–3) *Brit. dial.* the haematopodid bird *Haematopus ostralegus* (*cf.* oystercatcher)

Olividae a family of gastropod mollusks commonly called olive shells in virtue of their shape which derives from a large body whorl concealing most of the early volutes

olm the proteid urodele *Proteus anguinus*

-olo- *comb. form* meaning "whole" (= -holo-)

olph *Brit. obs.* = finch

    blood olph = bullfinch

    green olph = greenfinch

olympia *U.S.* the pierid lepidopteran insect *Euchloe olympia*

olynthus *see* larva

Olyreae a tribe of the plant family Gramineae

om *see* [ommatidia]

-omalo- *comb. form* meaning "equal." An initial "h" is frequently appended

omao the turdid bird *Phaeornis obscurus* (= Hawaiian thrush)

omasum the division of the artiodactyl stomach that lies between the abomasum and the reticulum

-ombro- *comb. form* meaning "rain storm" but just as commonly used in the sense of "shade"

omentum that portion of the peritoneum which surrounds and supports the viscera

    greater omentum a membraneous sack of the mesogaster, frequently containing fat bodies

    lesser omentum a ligamentous part of the mesentery extending from the lesser curvature of the mammalian stomach to the liver

omm *see* [ommatoreductum]

-omma- *comb. form* meaning "eye" or "what the eye sees"

Ommastrephidae a family of dibranchiate cephalopod mollusca closely related to the Loliginidae but differing from them in lacking a cornea to the eye

ommatidium one of the units of a compound eye

[**ommatidia**] a mutant (gene symbol *om*) mapped at 0.1± on the X chromosome of *Drosophila melanogaster*. The phenotypic expression is slightly roughened eyes

[**ommatoreductum**] a mutant (gene symbol *omm*) mapped at 12.8 on the X chromosome of *Drosophila melanogaster*. The name refers to the absence of peripheral ommatidia

-omni- *comb. form* meaning "all"

Omophronidae a family of small, oval, heavily convex, coleopteran insects sometimes aquatic but more usually found in wet sand. They are commonly called round sand-beetles

Omosudidae a monospecific family of iniomous fish closely allied to the Alepisauridae but distinguished from them by the short dorsal fin. The single species *Omosudis lowii* (= hammerjaw) is bathypelagic

omphalode with the passage in the hilum leading to the chalaza

Omphralidae = Scenopinidae

omul the *Lake Baikal* teleost fish *Coregonus autumnalis*

-on a terminal symbol used to indicate a group of organisms associated in an aquatic habitat. See *benthon, herpon, neuston, plankton, etc.*

onager an Asiatic wild ass, sometimes held to be a valid species *Equus onager*

Onagraceae that family of myrtiflorous dicotyledons which contains, *inter alia*, the fuchsias, and the evening primroses. The family is clearly distinguished by the perigynous or epigynous flowers with a numerical plan of two or four, and by the inferior ovary containing many ovules

-onco- *see* -onko-

-oncho- *see* -onko-

Oncodidae = Acroceridae

onicous pertaining to a wood-louse either in shape or color

onion (*see also* bug, maggot, thrip) any of numerous liliaceous herbs of the genus Allium but particularly the horticultural varieties of *A. cepa*

    bog-onion *U.S.* the araceous plant *Arisaema stewardsonii*

    sea-onion the liliaceous herb *Urginea scilla* (= squill)

    Welsh onion *Allium fistulosum*

    wild onion any of many species of the genus Allium

Oniscoidea a suborder of isopod Crustacea with free abdominal segments possessing terminal uropods and with the body never very strongly compressed. Distinguished from the closely allied Aselloidea by their usually terrestrial habit

-onko- *comb. form* meaning "hook." Frequently transliterated -onco- or -oncho-

-onych- *see* -onyx- and -nychium-

onychium = pulvillus

Onychopalpida a division of the Acarina distinguished by the presence of ambulacral claws on the pedipalps

Onychophora a group erected to contain the velvetworms as the members of the genus Peripatus are often called. They were at one time a separate phylum but are now usually regarded as a class of Arthropoda. They are caterpillar-like and have from 17 to 40 pairs of ambulatory appendages, papillate skin, and antennae

-onyx- *comb. form* meaning "a claw" almost invariably transliterated -onych-

onyx *I.P.* any of many species of lycaenid lepidopteran insect of the genus Horaga

-oo- *comb. form* meaning "egg"

oo any of several drepaniid birds of the family Moho

Oocystaceae that family of chlorococcale Chlorophyta which contains all those forms that produce autospores

oon egg

ooneion = archegonium

oonyle unfertilized female

ootid the cell which is produced at the end of the meiotic division of oocytes and which metamorphoses directly into an egg

ooze that upper layer of mud which is sufficiently fluid to be subject to slow flow

Op *see* [Opossum]

op *see* [opaca]

[opaca] a mutant (gene symbol *op*) mapped at 13 on chromosome 2 of the tomato. The phenotypic expression is yellow-green patches on the leaves

opah the lamprid teleost fish *Lampris regius* which is the only member of the family Lampridae

Opalinata an order of protociliate Protozoa distinguished by the absence of a mouth

[opaque endosperm] a mutant (gene symbol *o₂*) mapped at 0 in linkage group VII of *Zea mays*. The term is descriptive of the phenotypic seed

Oparoidea a suborder of parasitic isopods distinguished from other parasitic isopods by having a terminal uropod

opercular possessing a lid

operculum literally, a lid and used in *biol.* for almost any structure serving this function, including, *inter alia*, the gill covers in teleost fish and amphibian larvae, the post-pedal plate in many gastropod mollusks, the shell plates of barnacles, the first pair of abdominal appendages in Xiphosura, the plate covering the book lungs in many Arachnida, the lid of a moss capsule, the hinged chitinous lid of an ectoproct zoecium, the protoplasmic covering over the open end of the tube of a nematocyst and the cover of an ascus

operon a group of cistrons that act as a coordinated unit and have a common operator gene

opertous hidden

opesiule one of two holes left in the frontal membrane by a partially developed cryptocyst (*cf.* opesium)

opesium the hole left in the ventral wall of an ectoproct zoecium when the frontal membrane is removed

oph *see* [ophthalmopedia]

Opheliidae a family of sedentary polychaete annelids living in sandy or muddy burrows and with the body not divided into distinct regions

Ophichthidae a family of marine eels commonly called snake-eels. They are distinguished by their slender shape and the fact that the dorsal fin extends for the full length of the body

Ophidia = Serpentes

Ophidiidae a small family of ophiioid acanthopterygian fishes commonly called cusk eels. They are distinguished by the pair of thin narrow barbels under the chin which are actually the remants of the ventral fin otherwise lacking

Ophidioidea a small suborder of acanthopterygian Osteichthyes distinguished by their very elongated shape and reduced pelvic fins sometimes completely lacking. No spines in any fin. The families Brotulidae, Ophidiidae and Caparidae are the principal components

Ophiocephalidae = Channidae

Ophioglossaceae that family of filicine pteridophytes containing the forms commonly called grapeferns and adder's tongues. They are distinguished by the fertile pinna with grape-like, large sporangia

Ophioglossales a small order of primitive pterophyte plants represented today by the Ophioglossaceae

Ophiurae an order of ophiuroid echinoderms distinguished from the Euryalae by the unbranched arms and distinct and regular superficial plates

Ophiuroidea that class of eleuthrozoan Echinodermata which contains the brittle stars. They are distinguished by their long, flexible, sometimes branched arms

ophradium a chemoreceptor gland in the incurrent siphon of many Mollusca

Ophrydoideae the only tribe of orchids in the division Apitonae of the sub-family Monandrae

ophthalm- *comb. form* meaning "eye"

anophthalmic lacking eyes

[ophthalmopedia] a mutant (gene symbol *oph*) mapped at 45.0± on chromosome II of *Drosophila melanogaster*. The phenotypic expression is kidney-shaped eyes, or eyes replaced by an appendage

Opiliaceae a small family of santaeale dicotyledons

Opiliones = Phalangida

-opistho- *comb. form* meaning "posterior"

Opisthobranchiata an order of gastropod Mollusca, mostly marine snails, in which the mantle and shell are either wanting or are greatly reduced

Opisthocoela a suborder of anuran Amphibia distinguished by the possession of opisthocoelous vertebrae and the presence of free ribs in adults of most species

Opisthocomiformes an order of birds erected to contain the hoatzin (*q.v.*)

Opisthoglypha an assemblage of colubrid snakes in which one or more of the posterior maxillary teeth are enlarged and grooved. (*cf.* Aglypha, Proteroglypha)

Opisthognathidae a small group of large-mouthed acanthopterygian fishes commonly called jawfishes

Opisthogoneata a subphylum of Arthropoda, containing the single class Chilopoda (*q.v.*) distinguished from the Progoneata by the more posterior position of the genital apertures

Opisthomi a monofamilial order of Osteichthyes erected to contain the Mastacembelidae or spiny eels which are, however, often included in the Perciformes

Opisthoporia an order of oligochaete annelids in which the male pore is situated on the segment immediately anterior to that which houses the lost pair of testes

Opisthoproctidae a small family of bathypelagic fishes closely allied to the Hiodontidae with which they are often included

opium the dried juice of the opium poppy. The word has been adopted by ecologists to mean a parasitic plant formation

oplarium = scyphus

Opomyzidae an obscure group of seaside dwelling myodarian cycloraphous Diptera

opopanax properly a genus of the Umbelliferae but also applied to a gum obtained from the burseraceous tree *Commiphora erythraea*

opossum (*see also* grape) popular name of didelphiid polyprotodont marsupials. *U.S.*, without qualification, *Didelphis virginiana*. In *Austral.* the term is applied to phalangerids

   four-eyed opossum any of several Central and Southern American didelphids with an eye-like pattern in the fur immediately above each eye

   mouse-opossum any of several species of didelphid of the genus Marmosa

   pigmy opossum (*Austral.*) popular name of phalangerines of the genus Cercaertus (= dormouse-phalanger)

shrew-opossum any of several, long-nosed very small didelphids, mostly of the Monodelphis

water-opossum *S.Amer.* the didelphid *Chironectes minimus* (= Yapok)

woolly opossum any of several species of didelphid, mostly of the genus Philander

[Opossum] a mutant (gene symbol *Op*) found in linkage group V of the mouse. The phenotypic expression is a sparse, ragged coat (*cf.* [Ragged])

Opostegidae a small family of tinioid lepidopteran insects, the larvae of which are all leaf-miners. The hind wings are greatly reduced

opportunism in *biol.*, the adaptation of an organism to the most pressing factors of its environment

Oppositifoliae = Dicotyledoneae

oppositivous the antithesis of alternate

-opse-, *comb. form* meaning "sight"

-opsin a suffix denoting a retinal pigment

iodopsin an analog of rhodopsin found in the cones

lumirhodopsin the first breakdown product of rhodopsin under the influence of light

metarhodopsin the intermediate product between lumirhodopsin and scotopsin and retinene

scotopsin one of the two breakdown products (*cf.* retinene) of metarhodopsin

-opsis *comb. suffix* meaning "having the appearance of"

optic pertaining either to the eye or to sight

holoptic said of those Diptera in which the compound eyes run together in the middle

dermatoptic the condition of responding to light without the use of eyes or other evident photo-sensitive organs

epiopticon the optic segment of the insect brain

Opuntiales that order of dicotyledonous angiosperms which contains the forms, in the single family Cactaceae, usually called cacti. The general appearance is characteristic, though closely paralleled by some *S.Afr.* Euphorbiaceae

Opuntioideae a subfamily of Cactaceae distinguished by the cylindrical leaves and rotate flowers

oquassa the salmonid fish *Salvelinus oquassa* (*cf.* trout)

or *see* [orange]

orach the chenopodiaceous herb *Atriplex hortensis* (= sea purslane)

oral pertaining to the mouth

aboral pertaining to the surface opposite the mouth particularly in echinoderms

orange 1 (*see also* orange 2, 3, dog, melon, root, thrip) any of several species of trees of the rutaceous genus Citrus. Used without qualification usually means *C. sinensis* or *C. aurantium*

calamondin orange *C. mitis*

African cherry orange any of several species of the rutaceous genus Citropsis

common orange = sweet orange

king orange *C. nobilis*

mandarin orange a horticultural variety of *C. nobilis*

navel orange one in which the fruit encloses a small secondary fruit

Seville orange *C. aurantium*

sour orange = Seville orange

sweet orange *C. sinensis*

wild orange (*see also* orange 2) *U.S.* = Seville orange

orange 2 (*see also* orange 1, 3) any of several plants with flowers or fruits resembling those of the true oranges

mock orange any of numerous species of saxifragaceous shrubs of the genus Philadelphus and, in *S.U.S.* the rosaceous tree *Prunus caroliniana*. The term is also applied to the pittosporaceous tree *Pittosporum undulatum* and *U.S.* the styracaceous shrub *Styrax americana*

osage orange the moraceous hedge shrub *Maclura pomifera*

vegetable orange a horticultural variety of musk melon (*q.v.*)

wild orange (*see also* orange 1) the term is sometimes applied to the rosaceous tree *Prunus caroliniana*

orange 3 (*see also* orange 1, 2) any of several lepidopteran insects

frosted orange *Brit.* the noctuid lepidopteran insect *Ochria ochracea*

veined orange *S.Afr.* the pierid lepidopteran insect *Colotis verta.* (*cf.* tip)

[orange] *either* a mutant (gene symbol *o*) in linkage group II of *Habrobracon juglandis*. The name is descriptive of the phenotypic eye color *or* a mutant (gene symbol *or*) mapped at 107.2 in chromosome II of *Drosophila melanogaster*. The name refers to the phenotypic eye color

orang-utan the pongine ape *Simia satyrus*. It is distinguished from the gorilla by its unpigmented skin and from the chimpanzee by its long hair

orania the palmaceous tree *Bentinckia nicobarica*

orbiculate disk-shaped

orbit a cavity housing the eye in the skull of chordates

Orchestiidae = Taltridae

orchid 1 (*see also* orchid 2, orchis, tree, fly) popular name for the monocotyledonous plants of the family Orchidaceae. There is some confusion in popular literature between "orchis", which should properly be confined to members of the genus Orchis, and "orchid" which, as an anglicized form of the familial name, can be applied to any member of the family

baby orchid *Odontoglossum grande*

bee-orchid *Brit. Ophrys apifera*

bog-orchid *Brit. Malaxis paludosa. U.S.* any of several bog-dwelling orchids, particularly *Pogonia ophioglossoides*

dove-orchid *Peristeria elate*

butterfly orchid any of several species of the genus Habenaria. *Brit.* without qualification *H. bifolia U.S. Oncidium papilio*

cranefly orchid *Tipularia discolor*

sawfly orchid *Ophrys tenthrevinisera*

fringed orchid any of several species of the genus Habenaria

green orchid *U.S.* any of several species of the genus Habenaria

holy-ghost orchid *Peristeria elate*

man-orchid *Brit. Aceras anthropophora*

prairie orchid *U.S. Habenaria leucophae*

ragged orchid *Habenaria lacera*

rain-orchid any of several species of genus Habenaria

sweet-scented orchid *Brit. Gymnadenia conopsea*

orchid 2 *adj.* from orchis, meaning "testes"

cryptorchid said of a mammal with concealed (*i.e.* abdominal) testes

Orchidaceae the orchid family of monocotyledons and the sole family in the order Microspermae. The peculiar structure of the flower and the minute seeds are characteristics of the family

Orchidales sometimes considered synonymous with Microspermae but more usually used for the

Orchidaceae when these are regarded as of ordinal rank

orchis 1 (*see also* orchid 1) a genus of orchidaceous monocotyledons. The term should not be used in place of "orchid" which is the proper English name for plants of the family Orchidaceae (*cf.* orchid)

marsh orchis *Orchis latifolia*

monkey-orchis *Brit. O. tephrosanthos*

early purple orchis *Brit. O. mascula*

showy orchis *U.S. O. spectabilis*

sword-orchis *Brit. O. morio*

orchis 2 literally "testes"

cryptorchism anomalous condition in which the testes fail to descend into the scrotum of the embryo

-orchium *comb. suffix* indicating structures connected with, but not part of, the testes

mesorchium that part of the mesentery which supports the vertebrate testes

-orcul- *comb. form* meaning "cask"

order 1 (*see also* order 2) a taxon ranking immediately below a class. *e.g.*, the order Lepidoptera of the class Insecta

suborder a division of an order

superorder a division of a class

order 2 (*see also* order 1) a social stratum

peck-order the social stratum of animals so called from its expression in the behavior of the domestic hen, but applied to other forms

peep-order the equivalent, in anuran Amphibia, of the peck-order of fowls

ordinate in *biol.* used in the sense of "ordered"

inordinate without regular arrangement

ordure mephitic excrement

oread heliophyte

Orectolobidae a family of sharks distinguished by a conspicuous external groove running from the mouth to the nostril

oregano properly the labiateous herbs *Origanum onites* and *O. dictamnus*. Many other species of this genus are, however, sold under this name and much of the dried oregano in American commerce is derived from the verbenaceous herb *Lippia graveolens*. The name is also applied to members of the verbenaceous genus Lantana, and to the labiateous genera Poliomintha and Hyptis

-orga- *comb. form* meaning "meadow" but used by ecologists in the sense of open woodlands

orgadad a plant of open woodland

orgadium an open woodland formation

organ a discrete mass of cells of specific function, or functions

anal organ any organ, frequently glandular, in the vicinity of the anus of any animal. The word is of too varied use to have valid specificity

antennal organ a meaningless term applied to any specialized area in or immediately adjacent to the antenna

apioid organ an accessory sex gland in some Turbellaria

axial organ (Crinoidea) = axial gland (Crinoidea)

bell organ an indented bell-shaped receptor lacking a seta, on the surface of insects

Bidder's organ the remnants of the ovary in some male anuran Amphibia

Bojanus' organ the excretory organ of pelecypod mollusks

cauliflower organ (Enteropneusta) = racemose organ (Enteropneusta)

cerebral organ a hollow organ overlapping the anterior surface of the supraoesophageal ganglion in some sipunculids. It is connected to the exterior by either one or two tubes

retrocerebral organ *either* a complex of one or more epidermal glands in the head region of the Rotifera *or* one of a pair of pit like organs open to the posterior immediately behind the brain of Chaetognatha

chambered organ a five-chambered, aboral division of the coelom in crinoids

chordotonal organ tympanic phonoreceptors on insects

preoral ciliary organ a "U" shaped ciliated depression in the epidermis of Enteropneusta on the ventral surface of the base of the proboscis. It is thought to be chemoreceptor in function

climbing organ a series of thick scales at the root of the tail of anomalurid rodents

conservative organ in *bot.*, those organs that are concerned with nutrition as distinct from reproduction or support

champagne-cork organ insect receptor organs in the form of an indented cylinder with an expanded rounded base

Corti's organ that structure in the cochlea of mammals which is the principal receiver of sound

cupola organ a bell-shaped sensory receptor indented in the surface of an arthropod

dome organ a domed or bell-shaped sensory prominence on an arthropod

dorsal organ (Crinoidea) = axial gland (Crinoidea)

electric organ modified muscular tissue, found in some fishes, capable of generating high potentials and storing considerable charges

enamel organ the layer of ameloblast cells around the outer surface of the developing tooth

end organ the functional termination of either a sensory or motor nerve in tissue

epigonal organ a lymphoid gland of unknown function associated with the gonads of some elasmobranch fishes. In amphibia a similar structure is derived from the genital ridge

epiphyseal organ one or more evaginations from the dorsal anterior end of the forebrain from which the pineal gland subsequently developes

essential organ a botanical term for primary sexual characters

femoral organ a series of epidermal glands lying along the posterior-ventral margin of the thigh in certain lizards

cribriform organ a vertical depression, containing thin sheets of calcium, covered with columnar epithelium, in vertical depressions between adjacent marginal plates on the sides of the arms of some asteroid echinoderms

lyriform organ a presumed chemoreceptor found in spiders

pyriform organ *either* an accessory sex gland in some Turbellaria *or* an invaginated mass of gland cells on the anterior of the oral surface of early ectoproct larvae

frontal organ *either* the remnant of the parietal organ in the anuran amphibian *or* a glandular structure dorsal to, and immediately behind, the cerebral ganglia of some Turbellaria and which is thought to have both chemo- and tangoreceptor functions

fundamental organ plant structures devoted to nutrition as distinct from reproduction

Gene's organ *see under* gland

Graber's organ a complex structure of unknown function found in the larvae of tabanid Diptera

Haller's organ an organ consisting of a sensilla rising

from a pit in the propodosomal area of acarines. It is presumed to be sensory

Hanstrom's organ a neurosecretory gland in the pedicel of many stalk-eyed Crustacea

intromittent organ any male organ, in many forms known as a penis, adapted for insertion into the female for the transfer of sperm

Jacobson's organ an accessory, olfactory organ in many vertebrates, supplied by the first and fifth nerves, usually situated toward the rear of the mouth

Johnstonian organ a sense organ, sometimes thought to be a phonoreceptor, at the base of the antenna in many insects

Julien's organ = corema

Keber's organ clumps of brownish or reddish excretory cells attached to the pericardium in many pelecypod mollusks

Leydig's organ the parietal eye of reptiles

lophophoral organ one of the pair of glandular bodies at the base of the inner ring of tentacles in the lophophore of Phoronida. In some species this organ is associated with neurosensory cells and then called a lophophoral sense organ

Muller's organ an assemblage of phonoreceptors in insects

nuchal organ a lobed, ciliated cushion on the mid-dorsal region of the oral disk of some sipunculids

parietal organ = parietal eye

pearl organ an epidermal sensory organ of actinopterygian fishes

plasmorgan a rudimentary organ

racemose organ a much branched extension of the proboscis coelom into the proboscis stalk in some Enteropneusta

rasping organ the toothed "tongue" of a cyclostome

rudimentary organ one which has not developed to a functional state

sense organ an organ which transmits impulses when stimulated

accessory sex organs = secondary sexual characters (*see* character)

club-shaped organ a tube outside the right wall of the pharynx in the early larva of Amphioxus which disappears when the right gill slits are formed. It is thought to represent a rudimentary gill slit

Siebold's organ a phonoreceptor organ on the tibia of the anterior leg of locustid insects

spongy organ a mass of cellular reticulum, running along one side of the axial gland in certain crinoids

static organ one which assists in preserving equilibrium

pseudostigmatic organ a club-shaped bristle arising from a depression near the hinder margin of the cephalothorax of galumnid mites

stirn organ a small dermal body found in certain anuran amphibia probably equivalent to the pineal

subgenual organ a ganglia in the tibia of some insects

supplementary organ a series of pre-anal, presumably tactile, organs found in some nematodes

intertentacular organ a temporary oviduct formed from the dorsal tentacles of some Ectoprocta

Tomosvary's organ a sensory pit behind the base of the antenna of most chilopods

tympanal organ any arthropod organ presumed to be a phonoreceptor

supratympanal organ = subgenual organ

parauterine organ a secondarily developed fibrous sac which opens into the uterus of cestodes

vegetative organs those plant organs which play no part in sexual reproduction

vestibular organ a transverse row or ridge of papillae immediately behind the teeth of Chaetognatha

Weber's organ the vestigeal uterus in the male mammal. This term is sometimes also applied to lateral glands of the tongue

wheel organ *either* a ciliated tract within the oral hood of Amphioxus associated above with Hatscheks fossa and extending down on either side *or* the corona of rotifers

organelle a discrete portion, with a specific function, of a cell. Cilia and mitochondria are typical of structures termed organelles

organicism the doctrine, closely allied to vitalism, that organisms can only be interpreted by a study of the whole, and that there may be undiscovered biological forces acting on the whole, as important in their way as are the physical and chemical forces which can be understood from examination of the separated parts

organiser an area of a cell of an embryo which influences the organization of other areas

Spemann's organiser the dorsal lip of the blastopore in Amphibia

organism a living entity, acellular, unicellular, or multicellular and capable of producing other organisms

epiorganism = superorganism

microorganism any organism of small size, particularly bacteria

superorganism an aggregate of organisms, or organisms and their environment, which may be studied as though it were a single organism

organogeny the process of the development of organs

organoid any morphological element within a cell. The term is similar to, but somewhat more loosely used than, organelle

orgya a fathom (= six feet)

Oribatei a group of sarcoptiform acarine arachnids commonly called beetle-mites. They are distinguished by their heavily chitinized, dark brown, external covering

oribi a general term for neotragine antelopids of the genera Ourebia and Raphicerus, though the term oribi should properly be confined to Ourebia, and the names steinbok and grysbok applied to Raphicerus

orihau *N.Z.* the araliaceous tree *Nothopanax colensoi* (= mountain ivy tree *N.Z.*)

oriole (*see also* finch, babbler) Old World orioles forming the passerine bird family Oriolidae. Most are of the genus Oriolus. New World orioles are yellowish icterid birds of the genus Icterus

Baltimore oriole *I. galbula*

Bullock's oriole *I. bullocki*

spot-breasted oriole *O. pectoralis*

black-cowled oriole *I. dominicensis*

golden oriole *O. oriolus*

African golden oriole *O. auratus*

black headed oriole *I.P. O. xanthornus, U.S. I. graduacauda, W.Afr. O. brachyrhynchus, E.Afr. O. larvatus*

hooded oriole *I. cucullatus*

Jamaican oriole *I. leucopteryx*

Lichtenstein's oriole *I. gularis*

Martinique oriole *I. bonana*

Montserrat oriole *I. oberi*

black-naped oriole *O. chinensis*

orchard oriole *I. spurius*

St. Lucia oriole *I. laudabilis*

Scott's oriole *I. parisorum*

black-throated oriole = Lichtenstein's oriole

Oriolidae that family of passeriform birds which contains the Old World orioles. They are distinguished by the strong, pointed, and slightly hooked beak and by the long pointed wings. New World "orioles" are in the family Icteridae

ormer *Brit.* the haliotid gastropod mollusk *Haliotis tuberculata* (*cf.* abalone)

Ormyridae a family of chalcidoid apocritan hymenopteran insects closely allied to the Torymidae

Orneodidae = Alucitidae

ornithine α,δ-diaminovaleric acid. $NH_2(CH_2)_3 \cdot CH(NH_2)COOH$. An amino acid not essential to the nutrition of rats. Apparently derived from arginine and forming part of a cyclic reaction with arginine and citrulline

Ornithorhynchidae a monotypic family of monotrematous mammals containing the duck-billed platypus

-oro- *comb. form* meaning a "mountain" but used by ecologists in the sense of "subalpine"

Orobatidae = Galumnidae

oropendola any of several large species of icterid bird of several genera

black oropendola *Gymnostinops guatimozinus*

crested oropendola *Psarocolius decumanus*

chestnut-headed oropendola *Zarhynchus wagleri*

Montezuma oropendola *Gymnostinops montezuma*

dihydroorotase an enzyme catalyzing the hydrolysis of L-4,5-dihydro-orotate to L-3-ureidosuccinate

Orphnephilidae = Thaumaleidae

orpine any of several crassulaceous herbs of the genus Sedum

Ortalidae = Cracidae

Ortheziidae a family of homopteran insects regarded by many as a subfamily of the Coccidae. They are commonly called ensign-coccids in virtue of the white plates of secretion arranged symetrically round the body

Ortheziinae a subfamily of the Coccidae erected by by those who do not hold the ortheziids to have familial rank

-ortho- *comb. form* meaning "upright" though commonly used in the sense of "straight"

Orthonycidae a family of passeriform birds usually included in the Timaliidae

Orthoperidae a small family of fringe-winged beetles distinguished from the Clambidae by the fact that the long hairs cover the whole of the hind wing. Commonly called fringe-winged fungus-beetles

Orthoptera an order of insects at one time including not only the grasshoppers, locusts and crickets but also those forms now more usually placed in the Dictyoptera (cockroaches and mantids), Grylloblattodea and Phasmida (stick insects). The Orthoptera possess eleven apparent abdominal segments, two pairs of wings and have the third pair of legs adapted for jumping

Orthorrhapha an assemblage of those families of brachyceran dipteran insects the adults of which emerge from the pupa through a T-shaped opening. (*cf.* Cylorrhapha)

ortolan the fringillid bird *Emberiza hortulana* (*cf.* ricebird)

Orussidae a small family of symphytan hymenopterous insects very closely allied to the wood-wasp family Xiphydriidae but differing from them in being parasitic. They are commonly called parasitic wood-wasps

orygoma the gemma cups of Marchantia

Oryssidae = Orussidae

oryx any of three species of hippotragine antelope of the genus Oryx (*cf.* gemsbok, beisa)

leucoryx the antelope *Oryx leucoryx*

white oryx *Aegoryx algazel*

os *see* [Osmotic] or [sex-linked]

Os *see* [Oligosyndactylism]

oscar the chichlid fish *Astronotous ocellatus*

Oscillospiraceae a family of caryophanale Schizomycetes occurring in the form of long trichomes partitioned to form narrow cells each with a clear central disc-like nucleus

Oscines a division of passeriform birds, of doubtful taxonomic validity, containing those with the most specialized vocal chords and which are therefore sometimes called the "song birds"

Oscinidae = Chloropidae

inosculating (*bot.*) = anastomosing (*bot.*). In insects two wing veins inosculate when they abruptly join more or less at right angles but anastomose when they join at a sharp angle

osculum literally a "little mouth" and most commonly used in *biol.* for the main exhalent opening of a sponge (*cf.* ostium). In *bot.*, osculum is sometimes used as synonymous with ostiole (*q.v.*)

osh *see* [outshifted]

osier (*see also* wood) properly a detached branch of a willow but used in parts of *Brit.* and *U.S.* as synonymous with willow (*q.v.*)

purple osier *Salix purpurea*

-osis terminal symbol, without specific meaning, indicating "to be in a state of." Words so terminating, as "phragmosis" or "phanatosis" are alphabetized under the principal part of the word

-osm- *see* -osme- and -osmo-

osmazome a culinary, rather than a biological, term referring to that part of red meat which produces the flavor and color of a dish. The O.E.D says that it is "that part of the aqueous extract of meat which is soluble in alcohol."

-osme- *comb. form* meaning "odor"

Osmeridae a family of small fishes commonly called smelts, distinguished by the adipose fin on the dorsal surface of the body

diosmesis the ability of detect odors both in the outside air and separately and distinctly within the oral cavity

monosmesis the ability to detect odor only in outside air

osmeterium an evil-smelling eversible protuberance used as a defence mechanism by some lepidopteran insect larvae

-osmo- *comb. form* meaning to "thrust" from the *Gr.* osmos, but frequently confused with "odor" from *Gr.* osme

osmosis a general term referring to the passage of a solvent through a selectively permeable membrane. The passage is more rapid from a region of low solute concentration to one of high solute concentration than in the reverse direction. This differential flow rate therefore produces "osmotic pressure" against the membrane

endosmosis passage of solvent from a region of low solute concentration to one of high solute concentration

exosmosis the reverse of endosmosis

[Osmotic] a mutant (gene symbol *os*) of linkage group I of *Neurospora crassa*. The phenotypic expression is a sensitivity to high osmotic pressure

Osmundaceae a family of filicale Pterophyta containing, *inter alia*, the forms commonly known as royal ferns and cinnamon ferns. They are distinguished by the fact that all the sporangia on a leaf develop simultaneously

-osous *comb. suffix* indicating unusually large size, as in labiosous, meaning to have a large lip. Terms so ending are not given in this work.

osphradium a chemoreceptor in the incurrent siphon of some pelecypod Mollusca

Osphronemidae = Anabantidae

osprey the pandionid falconiform bird *Pandion haliaetus*

-oss- *comb. form* meaning "bone"

osseous bony

ossicle literally, a little bone but the term is also applied to the small, calcareous particles which replace the exoskeleton in some echinoderms

Meckelian ossicle a small ossification on Meckel's cartilage at the insertion of the adductor muscles of the jaw in antinopterygian fishes

vertebral ossicles ossicles, fancifully resembling vertebrae, forming the endoskeleton of the arm of some ophiuroid echinoderms

ossiculus the pyrene of a fruit

ossification the production of true bones as distinct from calcification

ossified (*bot.*) becoming bone-hard as do some nuts and seeds

-ost- *see* -oste- and -osti-

actinost bones connecting the girdles to the fins in teleost fish

Ostariophysi a very large order of osteichthyes distinguished by possessing a Weberian apparatus

-oste- *comb. form* meaning "bone" frequently confused with -osti-

Osteichthyes a term coined to describe those fishes in which the skeleton consists principally of bone

Osteoglossidae a family of isospondylous fishes that is confined to the southern hemisphere, distinguished by their bony tongues, bony plates covering the head and large scales. The arapaima is the best known example

osteon a unit of bone consisting of a Haversian canal and its surrounding bony lamellae

osteum latinized form of the *Gr.* oste meaning "bone"

coenosteum the calcareous skeleton of a milleporine coelenterate

endosteum the membrane lining the marrow cavity of bones

periosteum the membrane around the outside of bone

-osti- *comb. form* meaning "door" or "aperture"

-ostiol- *comb. form* meaning "little door"

ostiola any small aperture, but specifically applied to the aperture in the stink gland in hemipteran insects. In *bot.* refers to any aperture of a fruiting body, particularly one which liberates spores

ostium literally, a "door" but used in *biol.* for almost any type of aperture, particularly the internal aperture of the oviduct, the inhalent apertures of the arthropod heart, and the inhalent pore in the wall of sponges

Ostomatidae a family of coleopteran insects closely allied to the dermestids and many of which are destructive pests of stored grain. They may

readily be distinguished by the very large mandibles. They are commonly called grain-beetles or bark-gnawing beetles

Ostomidae = Ostomatidae

-ostosis *comb. suffix* used in the sense of the "fusion of bones"

synostosis the fusion of two bones without any connective tissue between them

-ostrac- *comb. form* meaning "shell," particularly a molluscan shell

Ostraciidae = Ostraciontidae

Ostraciontidae a family of plectognathous fishes in which the protective armor forms a solid box-like exoskeleton. Commonly called trunkfishes or boxfishes

Ostracoda a subclass of crustacean Arthropoda. They are distinguished by being totally enclosed in a sometimes calcareous bivalved shell and possessing no more than five pairs of postoral appendages

ostracum a molluscan shell

epiostracum the horny pigmented outer layer of a molluscan shell

hypostracum the inner laminated layer, frequently of mother-of-pearl, of a molluscan shell

periostracum = epiostracum

proostracum the broad anterior portion of the thragmocone of fossil decapod molluscan shells

Ostreidae that family of pelecypod mullusks which contains the edible oysters

ostrich (*see also* fern) the African struthioniform bird *Struthio camelus*

-ot- *comb. form* meaning "ear"

ot *see* [outheld]

Otariidae that family of pinnipede carnivorous mammals which contains the sea-lions. They are distinguished from the seals and walruses by the ability to use their hind feet. Sea-lions are frequently miscalled seals, which term should be confined to the Phocidae

otic pertaining to, or in the vicinity of, the ear

Otidae = Otididae

Otididae a family of gruiform birds containing the bustards. They are distinguished by their broad wings, strong legs, and bristly feathers on the sides of the head and neck

Otitidae a family of myodarian cycloraphous dipteran insects the wings of which are so marked with black, brown, and yellow as to give them the name picture-wings flies

Otocyoninae a monotypic subfamily of canid carnivorous mammals containing the bat-eared fox which differs from all other carnivores, and indeed from most other mammals, in possessing forty-eight teeth

otoporpa used for both a thickened epidermal extension of the lithostyle and for a diminutive peronium

otter any of several species of mustelid carnivorous mammals frequently separated as a separate family Lutridae or subfamily Lutrinae. *U.S.*, without qualification, *Lutra canadensis*. *Brit.*, without qualification, *L. lutra*

Cape otter *Aonyx capensis*

river-otter *U.S. Lutra canadensis*

sea-otter the marine, webbed footed, lutrine *Enhydra lutris*

smooth otter = simung

ou the drepanid bird *Psittirostra psittacea*

ouakari *see* ouakari monkey

ousel properly any of several species of turdid bird of the genus Turdus. In *Brit.* is usually confined to *T. merula* (= *Brit.* blackbird) but is occasionally applied to other thrush-like birds

ring ousel *T. torquotus*

Tickell's ousel *T. unicolor*

water-ousel *U.S.* the cinclid bird *Cinclus mexicanus* (= dipper)

gray-winged ousel *T. boulboul*

[**outheld**] a mutant (gene symbol *ot*) mapped at 65.1± on the X chromosome of *Drosophila melanogaster.* The name refers to the outspread wings of the phenotype

[**outshifted**] a mutant (gene symbol *osh*) mapped at 33.0± on the X chromosome of *Drosophila melanogaster.* The phenotypic expression is a shortening of the wings and a light brown body color

[**outstretched**] a mutant (gene symbol *od*) mapped at 59.2 on the X chromosome of *Drosophila melanogaster.* The wings of the phenotype diverge widely

ouzel = ousel

ov *see* [oval]

oval elliptical, usually, without qualification, broadly elliptical (*cf.* ovate)

lanceoval broadly lanceolate

oboval in the shape of an egg attached by its thick end

[**oval**] a mutation (gene symbol *ov*) mapped at 17.5 on the X chromosome of *Drosophila melanogaster.* The term refers to the eye shape of the phenotype

[**oval fruit**] a mutant (gene symbol *o*) mapped at 29 on chromosome 2 of the tomato. The term is descriptive of the fruit of the phenotype

ovarial pertaining to the ovary

endovarial said of the ovary of a teleost fish formed by the ventral margin curling laterally to meet a fold from the upper margin of the gonad

parovarial said of the ovary of a teleost fish formed by the free margin of the genital fold bending laterally and fusing with the fold from the body wall

ovarium a term related to ovary as orchium (*q.v.*) is to testis

mesovarium that part of the mesentery which supports the ovary in vertebrates

exovation hatching

ovary 1 (*see also* ovary 2) the animal organ from which, or within which, eggs are developed

acrotrophic ovary a closed system in which the reproductive cells remain at the apical region

ovary 2 (*see also* ovary 1) the fertile base of a carpel within which the ovules are developed. After fertilization, the ovary usually gives rise to a fruit

chambered ovary a unilocular plant ovary partially divided by projections from the margins towards the center

compound ovary one formed from more than one carpel

inferior ovary one which appears to be below the calyx

superior ovary one beneath which all the floral parts are inserted

ovate in the shape of a hen's egg (*cf.* oval)

lanceovate narrowly lanceolate

ovellum an immature carpel

overstepping = hypermorphosis

overturn the mixing of zones in a lake, commonly occurring in the fall or spring, through the generation of rotary currents by wind action

-ovi- *comb. form* meaning "egg"

ovi *see* [ovioculus]

Ovibovinae a most misleadingly named subfamily of caprid artiodactyl mammals since the ovine and bovine supposed affinities are purely superficial and the anatomical relationships to the caprids undoubted. The only two representative are the takin and the muskox

ovicapt a muscular thickening at the junction of the ovary and oviduct in some Turbellaria

Ovinae a subfamily of caprid artiodactyl mammals containing the sheep and their immediate allies. There seems little justification for separating these forms into a separate family, the Ovidae. All except the aoudad and the bharal belong in the genus Ovis

[**ovioculus**] a mutant (gene symbol *ovi*) mapped at 0.9± on the X chromosome of *Drosophila melanogaster.* The phenotypic expressions are small egg-shaped eyes and male sterility

ovoid an egg-shaped solid

ovulate possessing ovules

ovulation the production of eggs from an ovary

ovule (*bot.*) ovoid bodies developing from the placenta in the angiosperm ovary. Each contains a nucellus (*q.v.*). After fertilization, the ovule gives rise to the seed

margin ovule an ovule borne on the margin of a carpel

camptotropal ovule an horseshoe shaped orthotropal ovule

lycotropal ovule an orthotropous ovule bent into a horse shoe form

atropous ovule one in which the nucellar apex is at the opposite end to the funiculus

anatropous ovule one in which the nucellar apex is at the same end as the funiculus

campylotropous ovule an ovule which is so recurved as to bring the micropyle near the hilum

orthotropous ovule = atropous ovule

Ovulidae a family of long and slender gastropod mollusks with a straight aperture, notched at each end. They are commonly associated with sea-fans

ovum the L. for egg, (*q.v.*)

deutovum that stage in the development of an acarine which immediately precedes the hatching of the larva

pseudovum an egg destined for parthenogenetic development

owl 1 (*see also* owl 2, parrot) properly any of numerous species of tytonid or strigid bird but applied to a few other birds

barn-owl popular name of tytonid owls, but most commonly applied to *Tyto alba*

bay-owl the tytonid *Phodilus badius*

boobook owl *Austral.* the strigid *Ninox novaeseelandiae* (= *N.Z.* morepork)

boreal owl the strigid *Aegolius funereus*

brown owl *Brit.* = tawny owl *W.I.* = Jamaican owl

burrowing owl the strigid *Speotyto cunicularia*

churn-owl *Brit.* = nightjar

crested owl the strigid *Lophostrix cristata*

death owl *W.I.* = barn owl

eagle-owl any of several species of strigids of the genus Bubo, considered to be of aquiline stature without qualification, *B. bubo*

akun eagle-owl *Afr. B. leucostictus*

barred eagle-owl *Afr. B. shelleyi*

dusky eagle-owl *B. coromandus*

forest eagle-owl *B. nipalensis*

pharoah eagle-owl *Afr. B. bubo*

spotted-eagle owl *B. africanus*

long-eared owl the strigid *Asio otus*

short-eared owl the strigid *Asio flammeus*

elf owl the strigid *Micrathene whitneyi*

fern-owl *Brit.* = nightjar

fish-owl any of several piscivorous strigids of the genus Ketupa, without qualification *K. ketupu*

  brown fish-owl *K. zeylonensis*

  tawny fish-owl *K. flavipes*

fishing owl any of several piscivorous, or supposedly piscivorous owls, but particularly those of the genus Scotopelia, without qualification, *S. peli*

  grey fishing owl *S. bouvieri*

  rufous fishing owl *S. ussheri*

grass owl the tytonid *Tyto capensis*

great grey owl the strigid *Strix nebulosa*

hawk-owl any of several species of strigids of the genera Ninox and Surnia usually, without qualification, *I.P. N. scutulata, U.S.* and *Brit., S. ulula*

  brown hawk-owl *N. scutulata*

horned owl any of several species of strigids of the genus Bubo, without qualification, usually *B. virginianus*

  great horned owl *U.S.* the strigid *Bubo virginianus Brit.* the strigid *B. bubo* (= eagle-owl)

Jamaican owl the strigid *Pseudoscops grammicus*

laughing owl *N.Z.* the strigid *Sceloglaux albifacies*

bare-legged owl the strigid *Gymnoglaux lawrencii*

marsh-owl the strigid *Asio capensis*, occasionally applied to the short eared owl (*A. flammens*)

masked owl the tytonid *Tyto novaehollandiae*

pigmy owl (*see also* owlet) any of several species of strigids of the genus Glaucidium, without qualification, *G. gnoma*

  Cuban pigmy owl *G. siju*

  ferruginous pigmy owl *G. brasilianum*

  least pigmy owl *G. minutissimum*

  northern pigmy owl *G. gnoma*

powerful owl the strigid *Ninox strenua*

scops owl semi-technical name of strigid owls of the genus Otus, without qualification, *Brit. O. scops*

  African scops owl *O. senegalensis*

  collared scops owl *O. bakkamoena*

screech-owl *U.S.* any of several species of strigids of the genus Otus, without qualification usually, *O. asio.* In *Brit.* the term is applied to the barn owl

  flammulated screech-owl *O. flammeolus*

  Puerto Rican screech-owl *O. nudipes*

  spotted screech-owl *O. trichopsis*

  tropical screech-owl *O. choliba*

snowy owl the strigid *Nyctea scandiaca*

sparrow-owl *Brit.* the strigid *Athene noctua* (= little owl)

spectacled owl the strigid *Pulastrix perspicillata*

striped owl the strigid *Rhinoptynx clamator*

tawny owl the strigid *Strix aluco*

saw-whet owl the strigid *Aegolius acadicus*

winking owl the strigid *Ninox connivens*

wood-owl any of several species of strigid owls of the genera Strix and Ciccaba, usually without qualification, *S. seloputo*

  African wood-owl *C. woodfordii*

  Himalayan wood-owl *Strix nivicola*

  Malayan wood-owl *S. leptogrammica*

  mottled wood-owl *U.S., C. virgata, I.P. S. ocellata*

owl 2 (*see also* owl 1, butterfly) *I.P.* and *Austral.* any of several species of satyrid lepidopteran insects of the genus Neorina and also *Taenaris artemia*

owlet (*see also* nightjar, frogmouth) a young owl, or an owl that is unusually small, especially those of the genus Glaucidium

  collared pigmy owlet the strigid *Glaucidium brodiei*

  barred owlet *G. capense* or *G. cuculoides*

  jungle owlet *G. radiatum*

  spotted owlet *Athene brama*

  pearl-spotted owlet *G. perlatum*

ox (*see also* beetle, lip) properly applied to any ruminant artiodactyle of the genus Bos

deer-ox a term sometimes applied to the subfamily Boselaphinae

domestic ox *Bos taurus*

twist-horned ox popular name of Strepsicerosinae

muskox the almost extinct, American-artic, *Ovibos moschatus*. It has from time to time been classified as an ox and as a sheep, but is now regarded as an ovibovine caprid

wild ox a possibly mythical ancestor of the domestic ox called *Bos primigenius*

Oxalidaceae that family of geranialous dicotyledons which contains, *inter alia*, the genus Oxalis. They differ from their close relatives the Geraneaceae principally in possessing a dehiscent capsule or fruit

oxalis properly a genus of oxalidaceous herbs but loosely applied to several other plants

blue oxalis the leguminous herb *Parochetus commumis*

oxbow an almost closed loop in a stream or inlet (*see also* lake)

oxidase the trivial name of a group of oxidoreductase enzymes in which oxygen is used as a hydrogen receptor and hydrogen peroxide or water is therefore a customary by-product of the reaction

acetylindoxyl oxidase synthesizes the production of N-acetylisatin from N-acetylindoxyl

aldehyde oxidase catalyzes the production of the corresponding acids from aldehydes

diamine oxidase catalyzes the production of amino-aldehydes and ammonia from diamines

monoamine oxidase catalyzes the production of aldehydes and ammonia from monoamines

pyridoxamine oxidase catalyzes the production of pyridoxal and ammonia from pyridoxamine

D-aminoacid oxidase catalyzes the production of 2-oxo-acids and ammonia from D-aminoacids

L-aminoacid oxidase catalyzes the production of 2-oxo-acids and ammonia from L-aminoacids

N-methyl-aminoacid oxidase catalyzes the production of L-aminoacids from N-methyl-L-aminoacids

ascorbate oxidase catalyzes the direct oxidation of 2 L-ascorbate to 2-dehydroascorbate

D-asparate oxidase catalyzes the production of oxaloacetate and ammonia from D-aspartate

catechol oxidase catalyzes the oxidation of 2-*o*-diphenol to 2-*o*-quinone

cholesterol oxidase catalyzes the production of Δ^4-cholestene-3-one from cholesterol

cytochrome oxidase catalyzes the oxidation of ferro-cytochrome *c* to ferricytochrome *c*

p- diphenol oxidase catalyzes the oxidation of 2-*p*-diphenol to 2-*p*-quinone

glucose oxidase catalyzes the production of D-glucono-δ-lactone from β-D-glucose (*cf.* glucose dehydrogenase)

glycollate oxidase catalyzes the production of glyoxylate from glycollate

hexose oxidase catalyzes the production of δ-lactones from hexoses

lactate oxidase catalyzes the production of acetate from L-lactate

malate oxidase catalyzes the production of oxaloacetate from malate

nitroethane oxidase catalyzes the oxidation of nitroethane to acetaldehyde, nitrous acid, and hydrogen peroxide

oxalate oxidase catalyzes the production of carbon dioxide from oxalate

pyruvate oxidase catalyzes the production of acetylphosphate from pyruvate and orthophosphate

sarcosine oxidase catalyzes the production of glycine and formaldehyde from sarcosine

sulphite oxidase catalyzes the production of sulphate and hydrogen peroxide from sulphite

urate oxidase catalyzes the oxidation of urate. Terminal products of the reaction are unknown

xanthine oxidase catalyzes the production of urate from xanthine

-oxo- *comb. form* meaning "sour" almost hopelessly confused with -oxy- meaning "sharp"

oxodad a plant of an acid marsh

-oxy- *comb. form* meaning "sharp" (*cf.* -oxo-)

oxygenases a group of enzymes which catalyze the direct transfer of molecular oxygen to a molecule. They therefore differ from the oxidases, in that neither water nor hydrogen peroxide are by-products of the reaction

catechol oxygenase catalyzes the production of *cis-cis*-muconate from catechol

protocatechuate oxygenase catalyzes the production of 3-carboxy-*cis-cis*- muconate from protocatechuate

gentisate oxygenase catalyzes the production of 3-maleylpyruvate from gentisate

homogentisate oxygenase catalyzes the production of 4-maleylacetoacetate from homogentisate

lipoxygenase catalyzes the production of peroxides of unsaturated fats from the unsaturated fats

myo-inositol oxygenase catalyzes the production of D-glucuronate from *myo*-inositol

-oxyl *see* -xyl-

oxyon a heat climax

Oxyrhamphidae = Oxyruncidae

Oxyrhyncha a superfamily of brachyuran Crustacea commonly called spider-crabs because of the small triangular body and elongate appendages

Oxyruncidae a monotypic family of passeriform birds containing the crested sharpbill so called in virtue of its long straight, pointed bill with short bristly feathers at the base

Oxystomata a division of brachyuran decapod Crustacea distinguished by having the mouth-frame triangular and prolonged forward into a groove.

oxytocin a smooth muscle contraction-stimulating hormone found in the neurohypophysis

Oxyuridea a class of parasitic nematodes with a posterior bulb on the pharynx and a caudal bursa

Oxyuriformidae a family of ascaroid nematode worms distinguished from the Ascaridae by the bulbous esophagous. It is to this group that the term pinworm is most frequently applied

Oxyurini a group of anatine birds containing the stiff-tail ducks

oyster 1 (*see also* oyster 2, catcher, crab, leaf, plant, scale) popular name of ostreid pelecypods

coon-oyster one which grows in tidal zones and is presumably available to racoons at low tide

pearly oysters popular name of pteriid pelecypods

rock-oyster popular name of chamiid pelecypods

thorny oyster popular name of spondylid pelecypods

tree oyster any of several species of pteriid mollusks of the genera Pedalion and Pinctada (*cf.* purse shell)

oyster 2 (*see also* oyster 1) in compound names of other organisms

vegetable oyster the compositaceous herb *Tragopogon porrifolius* and its edible roots (= salsify)

P

p *see* [pea comb], [pericarp], [pink], [pink eye], [pink-eyed dilution] [plain.] In yeast genetics this symbol indicates petite

p *see* [Peach]

pa *see* [patulous], [pallid]

paauw *S. Afr.* any of several bustards (*q.v.*)

pab a microbial genetic marker indicating a requirement for paraaminobenzoic acid

pab-1, 2 *see* [para-aminobenzoic acid-1] and [-2]

paca the dasyproctid hystricomorph rodent *Cuniculus paca*

  false paca = pacarana

  mountain paca any of several species of dasyproctid of the genus Stietomys

pacarana the dinomyid hystricomorph rodent *Dinomys branickii*. It resembles the paca but differs from it in having a hairy tail and s-shaped nostrils. Sometimes known as the false paca

pace the term for a group of asses (*cf.* drove)

-pachy- *comb. form* meaning "thick"

Pachycephalidae a family of passeriform birds usually included with the Muscicapidae

Pachyneuridae a family of nematoceran dipteran insects now usually included with the Rhyphidae

pachynosis the increase in thickness of a plant

pachyte the sum total of the secondary phloem and secondary xylem

Pachytroctidae a family of psocopteran insects containing the common booklouse. They are distinguished by the short arched body

Pacini's *epon. adj.* from Filippo Pacini (1812–1883)

pack term for a group of dogs, all engaged in the pursuit of an animal (*cf.* kennel, gang), wolves when hunting (*cf.* rout), weasels and sometimes grouse though the term covey is more common, for these last forms

pad in *biol.*, any cushion like growth. The term is also applied in horticulture to the leaves of water-lilies

  archesporial pad the support below the spore producing tissue in ferns

paddle a limb adapted to swimming (*cf.* nectopod)

paddling a term for a group of ducks moving on the water (*cf.* bed, raft)

pademelons popular name of macropodine marsupials of the genera Setonyx and Thylogale. They are the least kangaroo-like of the macropodines and greatly resemble the rat-kangaroos

-paed- *comb. form* meaning "child" often transliterated **-ped-**

-paedic *comb. suffix* meaning "pertaining to the young"

psilopaedic said of birds which hatch naked from the egg

ptilopaedic said of a bird hatched fully clothed with down

paedium an assemblage of young

  gynopaedium an assemblage consisting of a female and her offspring, or a few females and their direct offspring

  monogynopaedium an assemblage of one female, together with her offspring

    partrogynopaedium a group in which both parents remain with their immediate offspring

    polygynopaedium an assemblage consisting of one female together with her offspring and the parthenogenetic descendents of her offspring

  partopaedium a group consisting of a male and his immediate offspring

  sympaedium an assemblage of young animals which play together

    sysympaedium an association of sympaedia

  synchoropaedium an aggregate of young animals of approximately the same age but different parentage

Paeoniaceae a family of ranale dicotyledonous angiosperms erected to contain the single genus Paeonia by those who do not hold them to belong in the Ranunculaceae

Paeonieae a tribe of the family Ranunculaceae

paeony *obs.* = peony

plover's **page** the scolopacid bird *Erolia alpina*. (= dunlin or red-backed sandpiper)

pagina 1 (*see also* pagina 2) the surface of an insect wing and also the outer, flattened surface of the hind femur of Orthoptera

pagina 2 (*see also* pagina 1) the blade of a leaf

pago- *comb. form* meaning a "peak" but used by ecologists in the restricted sense of "foot hill"

Paguridea a superfamily of anomuran Crustacea in which the first legs are in the form of chelipeds and the abdomen usually soft and noticeably asymmetrical. The group contains not only the hermit-crabs but a few such as the coconut-crab and northern stone-crab which have hardened abdomens

Paguridae a family of thalassinid anomuran Crustacea commonly called hermit-crabs in virtue of their habit of using abandoned gastropod shells to protect their asymmetrical, elongate, soft abdomen

Paictidae = Philepittidae

paigle *Brit.* the primulaceous herb *Primula veris* (= cowslip)

Indian **paint** either the chenopodiaceous herb *Chenopodium capitatum* or the boraginaceous herb *Lithospermum canescens* (= red root)

painted in *biol.*, having irregularly shaped masses of color

pit-**pair** two superimposed pits from adjacent cells

non-homologous **pairing** the condition in which non-homologous chromosomes or parts of chromosomes, pair in meiosis

pakake *N.Z.* the phocid carnivore *Hydrurga leptonyx* (= sea leopard)

-**pala**- *comb. form* meaning "spade"

palaceous when the edge of a thin organ adheres to its support, thus giving a spade-like appearance

Palade's *epon. adj.* from George Emil Palade (1903–)

-**palae**- *comb. form* meaning "ancient"

Palaeacanthocephala an order of Acanthocephala distinguished from the Archiacanthocephala by lacking protonephridia and from the Eoacanthocephala by having the probosis hooks arranged radially

Palaeoptera a taxon of the phylum Insecta erected to contain those insects, mostly fossil, in which the wings cannot be folded. The dragonflies and mayflies are the modern representatives (*cf.* Neoptera)

Palaeotraginae a subfamily of giraffid artiodactyl mammals containing the okapi. This subfamily differs from the Giraffinae in the absence of a long neck and skin-covered horns but agrees in almost all other characteristics

Palaeotremata once regarded as an order of brachiopods but no longer considered a valid taxon

palama the webbing of web-footed birds

Palamedeidae = Anhimidae

palar the condition in which a root continues the stem

palate 1 (*see also* palate 2) the sum total of those structures that separate the roof of the buccal, from the floor of the nasal, cavity of vertebrates

cleft palate a developmental anomaly

palate 2 (*see also* palate 1) a projection or lobe in a gametopetalous corolla

-**pale**- *comb. form* meaning "chaff"

palea the upper bract of a grass floret (*cf.* lemma)

paleaceous chaff-like

intrapalear said of fertilization which takes place within the palea before the opening of a grass flower

[pale green] *either* a mutant (gene symbol pg_{12}) mapped at 66 on linkage group IX of *Zea mays*. The term is descriptive of the appearance of the phenotypic seedling and plant *or* a mutant (gene symbol pg_{11}) mapped at 33 on linkage group VI of *Zea mays*. The name refers to the color of the phenotypic seedlings and plants

[pale-ocelli] a mutant (gene symbol *po*) mapped at 65.2 on chromosome II of *Drosophila melanogaster*. The name is descriptive

paleola = lodicule

Paleonemertea an order of anoplate Nemertea distinguished by a gelatinous dermis

palet = palea

palila the drepaniid bird *Psittirostra bailleui*

-**palin**- *comb. form* confused from two roots and therefore meaning either "backwards" or "again"

Palinura a section of reptant decapod Crustacea including the spiny lobsters. They are distinguished by the extended abdomen with a broad tail-fan and by the reduced size, or absence, of the rostrum

palla the antelope *Aepyceros melampus* (= impalla)

pallescent becoming pale

pallial pertaining to the mantle

[pallid] a mutation (gene symbol *pa*) in linkage group V of the mouse. The phenotypic expression involves

an absence of pigment in the eyes and the reduction of pigment in the pelt

pallium literally, a "mantle." Without qualification, usually applies to the gelatinous envelope of diatoms

archipallium = olfactory cortex

neopallium that portion of the cortex of the brain which covers the dorsal convexity of the cerebral hemispheres

pseudopallium a backwardly directed fold, used as a broodpouch, developed from the anterior margin in some parasitic snails

Pallopteridae a family of medium sized myodarian cycloraphous dipteran insects with maculated wings

palm 1 (*see also* palm 2–4, bug, Civet, creeper, dart, fly, grass, king, swift, weevil) popular name of monocotyledonous trees of the family Palmaceae

assai palm *Euterpe edulis*

barrel-palm = bottle palm

beach-palm *Bactris minor*

blue palm *either Erythea armata or Sabal glabra* (= dwarf palmetto)

bottle-palm *Pritchardia wrightii*

cabbage-palm *Oreodoxa oleracea*

carnauba palm *Copernicia cerifera*

coconut palm *Cocos nucifera*

cohune palm *Attalea cohune*

date palm *Phoenix dactylifera*

fan-palm one with flabellate leaves, or palmately cleft

feather-palm one with pinnately compound leaves

fortune's palm *Trachycarpus excelsa*

grugru palm any of numerous palms with spiny leaves

jaggery palm *Caryota urens*

needle-palm *Rhapidophyllum hystrix*

nikau palm *Rhopalostylis sapida*

oil palm *Elaeis guineensis*

Palmyra palm *Borassus flabellifer*

Para palm *Euterpe edulis*

piassaba palm *Attalea funifera*

royal palm *either Oreodoxa regia or O. borinquena*

sago-palm (*see also* palm 2) any of several species of palm, mostly of genus Metroxylon, from the pith of which an edible substance is extracted

bastard sago-palm *Caryota urens* (= jaggery palm)

silver palm *Coccothrinax garberi*

fish-tail palm any of several species of the genus Caryota

talipot palm *Coryphe umbractulifera*

thatch palm any of many species of the genus Thrinax

toddy palm = wine palm

umbrella palm (*see also* palm 2) *Hedyscepe canterburyana*

wax palm any of several species of the genus Ceroxylon and also *Diplothemium caulescens*

wine palm *Caryota urens*

palm 2 (*see also* palm 1, 3, 4) any of several plants of palm-like appearance

banana palm the musaceous tree-like herb *Musa sapientum*

dracena palm any of numerous species of the liliaceous herb Cordyline

fern palm the cycadaceous tree *Cycas circinalis*

nut palm the cycadaceous tree *Cycas media*

sago palm (*see also* palm 1) the cycadaceous tree *Cycas revoluta*

snake palm the araceous herb *Amorphophallus rivieri*

umbrella palm (*see also* palm 1) the sedge *Papyrus alternifolius*

palm 3 (*see also* palm 1, 3, 4) the plantar surface of the primate hand

palm 4 (*see also* palm 1, 2, 3) an obsolete measure, derived from palm 3, of approximately three inches

Palmae the palm family of monocotyledons. The palms are distinguished by their enormous leaves, frequently born in a terminal center

Palmaceae = Palmae

Palmales = Principes

palmar pertaining to the palm. The term is extended to those plates in the arm of a crinoid echinoderm which lie between the second and third fork (= tertibrach)

antipalmar pertaining to the dorsal surface of the forefoot

postpalmar those plates in the crinoid echinoderm arm which lie beyond the third fork

palmate 1 (*see also* palmate 2) lobed like a hand

palmate 2 said of birds having the toes webbed

fissipalmate said of the feet of a bird in which the individual digits are palmate but the foot is not webbed

semipalmate said of palmate birds in which the webbing is greatly reduced

totipalmate said of birds in which four toes are united by webs

palmella the condition of a flagellate which has lost its locomotory organs and assumed a rounded state, but is not encysted

palmer (*see also* worm) *Brit.* any destructive caterpillar, particularly those of arctiids. The name derives from the depredations of returning, palmleaf-bearing, pilgrims

palmetto any low-growing palm particularly of the genus Sabal

blue palmetto the palmaceous shrub *Rhapidophyllum hystrix*

cabbage-palmetto *S. palmetto*

dwarf palmetto *S. glabra*

saw palmetto *either Serenoa repens, Paurotis wrightii, or Serenoa serrulata*

scrub palmetto *S. megacarpa*

palmus an obsolete measure of length, which may either mean palm (*q.v.*) or span (*q.v.*)

palometa the serranid fish *Trachinotus glaucus* (*cf.* pompano)

palp a sensory appendage

pedipalp the second appendage of the arachnida

palpebral pertaining to the eyelid

palpiger a segment bearing a palp, particularly, the labial palp of insects (*cf.* palpifer)

Palpigrada a very small order of small (less than 2 mm long) arachnid arthropods distinguished by the fact that the thin, elongated, segmented body ends in a many-segmented flagellum

palpon one of the dactylozooids of a siphonophorous coelenterate

gonopalpon a dactylozooid associated with a gonophore

paludal pertaining to marshes (*cf.* palustrine)

palumbinous lead colored

palus one of many small ridges which sometimes occur between the columella and the sclerosepta in corals

palustrine pertaining to bogs (*cf.* paludal)

pam *see* [platinum]

pampa the *S. Amer.* equivalent of steppe

Pamphiliidae a small group of medium-sized symphytan hymenopteran insects distinguished by the fact that the larvae roll leaves as a shelter

-pan- *comb. form* meaning "all"; -pan- is neuter, -pas- is masculine and -pasa- feminine, a fact usually ignored by biological logotechnicians

pan 1 (*see also* pan 2) a bacterial genetic marker indicating a requirement for panothenic acid mapped at 15 mins for *Escherichia coli*

pan 2 (*see also* pan 1) a bare area of a saltmarsh below the level of the surrounding vegetation and in which water stands for much of the year

creek pan a pan formed by blocking off the end of a creek

primary pan a portion of a salt marsh never colonized by vegetation

pan-1 or -2 *see* [Pantothenic-1] or [2]

panchax aquarists' name for almost any cyprinodont teleost fish, though most often applied to *Aplocheilus panchax*

pancreas a gland which discharges digestive enzymes into the intestine and also houses the insulin-secreting islets of Langerhans. The term is also applied to many invertebrate glands primarily concerned with the secretion of digestive enzymes

hepatopancreas a gland found in many invertebrates which combines the food storage functions of the liver of higher forms with the secretion of digestive enzymes

parapancreas secretory cells of unknown function adnexed to the pancreas in many reptiles

panda 1 (*see also* panda 2) two entirely different carnivorous mammals go by this name. The "true" or lesser panda is an ailurine procyonid and therefore closely related to the raccoons. The giant "panda" is a bear and is the black and white animal usually called "panda" in zoological gardens

panda 2 (*see also* panda 1) the araceous climbing shrub *Philodendron panduriformes* that is not related to trees of the genus Panda of the family Pandaceae

Pandaceae a family of pandale dicotyledons containing the single genus Panda which is composed of small *Afr.* and *W.I.* trees

Pandales an order of dicotyledonous angiosperms erected to contain the single family Pandaceae

Pandalidae a family of natant decapod crustacea distinguished from other natant decapods by posessing minute pincers on the first pair of pereiopods

Pandanaceae a family of pandanale angiosperms containing the forms commonly called screwpines. It is distinguished from other pandanale families by its woody stem and palm-like habit

Pandanales an order of monocotyledonous angiosperms containing the cattails, the screw-pines and the bur-reeds. They are distinguished by the linear leaves and the more or less nut-like fruits containing seeds with endosperm

Pander's *epon. adj.* from Heinrich Christian Pander (1794–1865)

Pandionidae a monotypic family of falconiform birds erected to contain the osprey. Readily distinguished by the spiny scales on the clawed feet which adapts them to the capture of birds

pandora *Brit.* the sparid fish *Pagellus erthinus*

Pandoridae a family of pelecypod mollusks commonly called slender clams. They are usually very small with unequal thin and flat valves

windowpane the bothid fish *Scophthlamus aquosus*

Paneth *epon. adj.* from Joseph Paneth (1857–1890)

pangolin any member of the mammalian order Pholidota. Also called scaly anteater

-panic- *comb. form* meaning "tuft"

Paniceae a tribe of the plant family Gramineae

panicle a loose term for a diversely branched cluster of flowers sometimes considered to be synonymous with a racemose type of inflorescence

Panizza's *opon. adj.* from Bartolommeo Panizza (1785–1867)

-pann- *comb. form* meaning "cloth"

Panorpidae that family of mecopteran insects which contain the common scorpion flies: It is the largest family in the order

pansy 1 (*see also* pansy 2, 3) any of numerous species of violaceous herbs of the genus Viola, particularly *V. tricolor* and its many horticultural forms

bedding pansy *V. cornuta*

wild pansy *Brit. V. tricolor U.S. V. rafinesquii*

pansy 2 (*see also* pansy 1, 3) any organism shaped like a pansy flower

sea pansy popular name of renillid alcyonarian coelenterates

pansy 3 (*see also* pansy 1, 2) *S. Afr.* either of several species of nymphalid lepidopteran insect of the genus Precis

panther a term variously applied to several medium sized cats. *Brit.* without qualification, almost invariably *Felis pardus* (= leopard). *U.S.* without qualification *F. concolor* (= mountain lion)

-pantin- *comb. form* meaning "tendril"

Pantodontidae a monotypic family of isospondylous fishes containing the butterfly fish *Pantodon buchholzi*

Pantophthalmidae a small family of orthoraphous brachyoceran insects the larvae of which are woodboring and which are sometimes destructive of tropical timbers

Pantostomatida = Rhizomastigina

[Pantothenic-1] a mutant (gene symbol *pan-1*) on linkage group IV of *Neurospora crassa*. The phenotypic expression is a requirement for pantothenic acid

[Pantothenic-2] a mutant (gene symbol *pan-2*) on linkage group VI of *Neurospora crassa*. The phenotypic expression is a requirement for pantothenic acid

Panuridae a family of passeriform birds usually included in the Timaliidae

moko-papa *N.Z.* the geckonid lacertilian *Hoplodactylus pacificus*

papachi the rubiaceous shrub *Randia thurberi*

papain an enzyme found in the papaya (*q.v.*) which catalyzes the hydrolysis of numerous compounds at basic aminoacid bonds

chymopapain an isoenzyme of papain

papaumu *N.Z.* the cornaceous tree *Griselinia littoralis* (= broadleaf *N.Z.*)

Papaveraceae that family of rheodalous dicotyledons which contains the poppies. Distinguishing characters are the milky juice and the one-celled ovary with parietal placentae and numerous stamens

Papaveroideae a subfamily of the Papaveraceae distinguished by the actinomorphic flowers with numerous stamens

papaya (*see also* fly) the caricaceous tree *Carica papaya* and its edible fruits

Papayineae a suborder of parietale dicotyledons containing the single family Caricaceae

Papilionaceae a family of rosale dicotyledons containing those forms more usually regarded as forming the subfamily Lotoideae of the Leguminosae

papilionaceous said of a flower, such as that found in lotoid Leguminosae, in which the posterior petal is outermost and the two anterior petals form a keel

Papilionate = Lotoideae

Papilionidae a family of papilionoid lepidopteran insects containing the butterflies commonly called swallowtails and parnassians and also known in various parts of the world as bird-wings, bat-wings, windmills, roses, mormons, redbreasts, peacocks, ravens, swordtails, jays, bluebottles and dragontails

Papilionoidea a superfamily of lepidopteran insects containing all those forms commonly called butterflies except for the skippers (*cf.* Hesperioidea) from most of which they are distinguished by having the third, or R vein of the forewing 5-branched, while that of the skippers is 4-branched. Moreover the body of the skippers is thick rather than slender. In contrast, the "moths" are divided among close to a score of superfamilies

papilla a nipple, or any small conical protuberance

adhesive papillae a raised structure, capable of causing adhesion. Usually, without qualification, applies to the protuberant ends of the marginal adhesive glands of triclad Turbellaria

circumvallate papilla one of many large, club-shaped, sensory papillae in the mammalian tongue, sunk deeply into the mucuous membrane and surrounded by a circular furrow

filiform papilla one of the sensory papillae of the mammalian tongue in the form of numerous short, thread-like bodies arising from a thickened base

fixation papilla the anterior attachment mechanism of urochordate larvae (*cf.* fixation disc)

fungiform papilla a sensory papilla of the mammalian tongue, in the form of a pileate lobe, interspersed among the filiform papillae

penis papilla a conical projection caused by the evagination of the sperm duct in some Turbellaria

renal papilla the projecting fused end of numerous collecting tubules projecting into the pelvis of a mammalian kidney

urinary papilla the papilla through which the ureter opens into a cloaca

vateris papilla a papilla round the mouth of the pancreatic duct within the intestine

wheel papilla a wart-like structure containing numerous "wheels" (*q.v.*) found on some holothuria

papillate having nipples

papillose covered with nipples

-papp- *comb. form* meaning "plant down"

pappose downy

papula literally a pimple but applied specifically to a small retractile projection, thought to be respiratory in function, in the fleshy areas of the body wall of asteroids echinoderms. The term is occasionally used in a wider sense for dermal gills in general

papulum a term sometimes mistakenly used as the "singular" of papula, itself a singular noun (*pl.* papulae)

papyraceous resembling paper, either in color or texture

papyrus the sedge *Cyperus papyrus*

-para- *comb. form* meaning "alongside"

[para-aminobenzoic acid-1] a mutant (gene symbol *pab-1*) on linkage group V of *Neurospora crassa*. The phenotypic expression is a requirement for p-aminobenzoic acid

[para-aminobenzoic acid-2] a mutant (gene symbol pab-2) on linkage group V of *Neurospora crassa*. The phenotypic expression is a requirement for p-aminobenzoic acid

Paradisaeidae that family of passeriform birds erected to contain the birds-of-paradise of New Guinea and its adjacent islands. They are distinguished by the fantastic tail wires or plumes and plume fans arising from any or all parts of the body

Paradoxornithidae a family of passeriform birds usually included in the Timaliidae

Paradoxurinae a sub-family of viverrid carnivorous mammals commonly called palm-civets

parakeet 1 (*see also* parakeet 2) any of numerous psittacid birds of a size too small to be regarded as parrots

barred parakeet *Bolborhynchus lineola*
beautiful parakeet *Psephotus pulcherrimus*
red-breasted parakeet *Psittacula alexandri*
Carolina parakeet *Conuropsis carolinensis*
Caribbean parakeet *Aratinga pertinax*
orange-chinned parakeet *Brotogeris jugularis*
many-colored parakeet *Psephotus varius*
Cuban parakeet *Aratinga euops*
orange-fronted parakeet *Aratinga canicularis*
red-fronted parakeet *N.Z. Cyanoramphus novaezelandiae*
yellow-fronted parakeet *N.Z. Cyanoramphus auriceps*
green parakeet *Aratinga holochlora Psittacula*
western blossom-headed parakeet *Psittacula cyanocephala*
slaty-headed parakeet *Psittacula himalayana*
Hispaniolan parakeet *Aratinga chloroptera*
large Indian parakeet *Psittacula eupatria*
West Indian parakeet = Caribbean parakeet
Jamaican parakeet *Aratingahana*
hard-mouthed parakeet *W.I.* = Spanish parakeet *W.I.*
rose-ringed parakeet *Psittacula krameri*
long-tailed parakeet *Psittacula longicauda*
New Zealand parakeet *Cyanoramphus novaezelandiae*
Spanish parakeet *W.I.* = coffee bird
olive throated parakeet *W.I. Arantinga astec*

parakeet 2 (see also parakeet 1) a few birds not of the family Psittacidae

blue-headed parakeet W.I. the thraupid *Tanagra musica*

Paralepididae slender, deep sea, iniomous fishes superficially resembling the barracuda in their possession of numerous teeth and therefore commonly called barracudinas

[Paralyzed-1] a mutant (gene symbol *pf-1*) in linkage group V of *Chlamydomonas reinhardi*. The phenotypic expression is paralyzed flagella

[Paralyzed-2] a mutant (gene symbol *pf-2*) in linkage group XI of *Chlamydomonas reinhardi*. The phenotypic expression is paralyzed flagella

[Paralyzed-3] a mutant (gene symbol *pf-3*) in linkage group VIII of *Chlamydomonas reinhardi*. The phenotypic expression is paralyzed flagella

[Paralyzed-4] a mutant (gene symbol *pf-4*) in linkage group I of *Chlamydomonas reinhardi*. The phenotypic expression is the paralyzed flagella

[Paralyzed-5] a mutant (gene symbol *pf-5*) in linkage group III of *Chlamydomonas reinhardi*. The phenotypic expression is a paralyzed flagella

[Paralyzed-6] a mutant (gene symbol *pf-6*) in linkage group X of *Chlamydomonas reinhardi*. The phenotypic expression is paralyzed flagella

[Paralyzed-6] a mutant (gene symbol *pf-6*) in linkage group of X of *Chlamydomonas reinhardi*. The phenotypic expression is paralyzed flagella

[Paralyzed-12] a mutant (gene symbol *pf-12*) in linkage group II of *Chlamydomonas reinhardi*. The phenotypic expression is paralyzed flagella

[Paralyzed-13] a mutant (gene symbol *pf-13*) in linkage group IX of *Chlamydomonas reinhardi*. The phenotypic expression is paralyzed flagella

[Paralyzed-14] a mutant (gene symbol *pf-14*) in linkage group VI of *Chlamydomonas reinhardi*. The phenotypic expression is paralyzed flagella

[Paralyzed-15] a mutant (gene symbol *pf-15*) in linkage group III of *Chlamydomonas reinhardi*. The phenotypic expression is paralyzed flagella

[Paralyzed-16] a mutant (gene symbol *pf-16*) in linkage group IX of *Chlamydomonas reinhardi*. The phenotypic expression is paralyzed flagella

[Paralyzed-17] a mutant (gene symbol *pf-17*) in linkage group VII of *Chlamydomonas reinhardi*. The phenotypic expression is paralyzed flagella

[Paralyzed-18] a mutant (gene symbol *pf-18*) in linkage group II of *Chlamydomonas reinhardi*. The phenotypic expression is paralyzed flagella

[Paralyzed-19] a mutant (gene symbol *pf-19*) in linkage group IX of *Chlamydomonas reinhardi*. The phenotypic expression is paralyzed flagella

[Paralyzed-20] a mutant (gene symbol *pf-20*) in linkage group IV of *Chlamydomonas reinhardi*. The phenotypic expression is paralyzed flagella

Paramythiidae a family of passeriform birds usually included with the Dicaeidae

paraoa N.Z. the physeterid cetacean *Physeter macrocephalus* (= sperm whale)

parapara N.Z. the nyctaginaceous shrub *Pisonia brunoniana*

parapet a fleshy fold separating the scapus and the capitulum in some anthozoans

parasite one organism which lives on another to the detriment of its host

ecoparasite one adapted to a specific host, or closely allied group of hosts

ectoparasite one the body of which is not lodged within the tissues of the host

endoparasite one the body of which is lodged within the tissues of the host

hemiparasite *either* a form which is partly parasitic (*i.e.* capable of existing in the absence of its host) *or* a parasitic plant which develops from free germinating seeds in the soil

holoparasite a parasite with no other mode of life open to it

hyperparasite one which parasitizes another parasite (= superparasite)

obligate parasite one which cannot exist in any other form

autoecious parasite one which passes its entire life on an individual host

heteroecious parasite *either* one which requires more than one host for its life cycle *or* which is not host specific

metoecious parasite one that is not host-specific

ametoecious parasite one which is host-specific

potential parasite = facultative saprophyte

social parasite an organism which lives on terms of amity with another organism from which it steals food

superparasite a parasite which lives on another parasite (= hyperparasite)

water-parasite an epiphyte, such as the mistletoe, which derives only water from its host

xenoparasite *either* one which infests the host not normal to it, *or* one only capable of invading a host after the latter is injured adventitiously

Parasitidae a family of mesostigmate acarine arthropods occurring in rotting logs and litter. They are of no known medical or economic importance

parasitism the method of life adopted by a parasite

Parastacidae nephrosidean freshwater crustaceans of the southern hemisphere that lack appendages on the first abdominal somite, and with the telson usually lacking, but having the outer ramus of uropods with a transverse suture.

Parazoa a rarely used taxon of the animal kingdom at one time erected to contain the phylum Porifera, in distinction from the Eumetazoa containing all other metazoan animals

parenchyma *see* -chyma

Parenchymata a term at one time used to embrace all the phyla of acoelomate animals

Parenchymula a mythical solid ancestor of the Metazoa, replacing, in theory, the hollow Blastea and Gastraea (*q.v.*)

Pareinae = Amblycephalidae

parera *N.Z.* the anatid bird *Anas superciliosa* (= grey duck *N.Z.*)

pardalote any of several dicaeid birds of the genus Pardalotus

  black-headed pardalote *P. melanocephalus*

  spotted pardalote *P. punctatus*

  striated pardalote *P. substriatus*

  yellow-tipped pardalote *P. striatus*

Paridae that family of passeriform birds which contains the titmice. They are mostly small with short sharp beaks and large heads. They are usually active and gregarious birds

Parideae a subfamily of Liliaceae

-parie- *comb. form* meaning "the wall of a house"

parietal literally pertaining to the wall or periphery. Specifically said of an ovule attached to the wall, rather than to the axis, of a plant ovary, of certain bones (*see* bone) forming the central portion of the roof of the skull of vertebrates and of structures or organs in the vicinity of these bones

Parietales that order of dicotyledonous angiosperms which contains, *inter alia*, the tea plant, the mangosteen, the St. John's worts, the tamerisks, the candlewoods, the violets, the passion flowers, the papayas and the begonias. The order is in general characterized by the biseriate pentamerous perianth with an imbricated calyx and the presence of numerous stamens

yellow **parilla** the menispermaceous vine *Menispermum canadense* (= moonseed vine)

-pario- *comb. form* meaning "produce"

herb-**paris** the liliaceous herb *Paris quadrifolia*

parity in *biol.* usage, the *subst.* form of -parous (*q.v.*)

park *Brit.* an area of cultivated grassland with scattered trees. *Brit.* and *U.S.* a similar environment maintained in large cities and, principally *U.S.* and *S.Afr.*, an area maintained, so far as possible, in the state in which it was before the coming of man

Parkeriaceae a family of eufilicale filicine Pteridophyta containing those water ferns not included in the Hydropteridales. Their aquatic habitat and polymorphic leaves are typical

parnassian popular name of parnassiid lepidopterans

Parnassiidae a family of lepidopteran insects now usually included as a subfamily in the Papilionidae, distinguished by lacking a tail-like elongation of the hindwing

Parnidae = Dryopidae

paroquet *see* parakeet

-parous *comb. suff.* meaning "to produce" usually in the sense of "giving birth to"

  ambiparous said of a bud producing both leaves and flowers

  larviparous an animal from which larvae are born, as in the sarcophagid dipteran insects

  multiparous *bot.* applied to a cyme which has many axes *zool.* bringing forth many young at one birth

  nymphiparous the bearing of young in the nymph stage

  oviparous said of females which produce eggs (*cf.* cenogenous)

  plasmatoparous the condition in which the entire contents of a gonidium germinates as a solid mass

  pupiparous either bringing forth pupa or larvae about to pupate

  ramiparous bearing branches

  scissiparous said of a many-celled animal that reproduced by fission

  tomiparous reproducing by fission

  viviparous *zool.* said of females in which the young are maintained for some time internally before birth (*cf.* cenogenous) *bot.* said of a seed which germinates while attached to the parent plant

    metaviviparous a term suggested as an antithesis of ovoviviparous for those forms in which nourishment is derived from the uterine wall

  ovoviparous said of females which produce eggs that hatch inside the female immediately preceeding birth

[**paroxysm**] a mutant (gene symbol *px*) in the sex chromosome (linkage group I) of the domestic fowl. The phenotypic expression is lethal

parr a term applied to the young of many fishes, but particularly a young salmon that has short vertical dark bars on the side. Male parrs may reach sexual maturity without ever descending to the oceans

parrakeet = parakeet

Parridae = Jacanidae

parrokeet *see* parakeet

parroquet *see* parakeet

parrot 1 (*see also* parrot 2, bill) any of very numerous psittaciform birds

  black-billed parrot *W.I. Amazona agilis*

  yellow-billed parrot *W.I. Amazona collaria*

  brown parrot *Poicephalus meyeri*

  cape parrot *Afr. Poicephalus robustus*

  red-capped parrot *Purpureicephalus spurius*

  lilac-crowned parrot *Amazona finschi*

    red-crowned parrot *Amazona viridigenalis*

    white-crowned parrot *Pionus senilis*

  Cuban parrot *Amazona leucocephala*

  fig-parrot any of several species of the genera Psittaculirostris and Opopsitta

  white-fronted parrot *Amazona albifrons*

  ground parrot *N.Z. Strigops habroptilus* (= kakaps)

  grey parrot *Psittacus erithacus*

  yellow-headed parrot *Amazona ochrocephala*

  brown-hooded parrot *Pionopsitta haematotis*

  Hispaniolan parrot *Amazona ventralis*

  imperial parrot *Amazona imperialis*

  king parrot *Austral. Alisterus scapularis* and, less frequently, *Purpureicephalus spurius* (= red-capped parrot)

  red-lored parrot *Amazona autumnalis*

  yellow-lored parrot *Amazona xantholora*

  Malay parrot *Psittinus cyanurus*

  mealy parrot *Amazona farinosa*

  musk-parrot any of several species of the genus Prosopeia

  night parrot *Austral. Geopsittacus occidentalis* and, less frequently, *Neophema bourkii*

  oak-parrot *Austral. Northiella haematogaster*

owl-parrot *N.Z. Strigops habroptilus* (= kakapo)

Puerto Rican parrot *Amazona vittata*

pigmy parrot any of several species of the genus Micropsitta

red-necked parrot *W.I. Amazona arausiaca*

regent parrot *Austral.* = smoker parrot *Austral.*

St. Lucia parrot *Amazona versicolor*

St. Vincent parrot *Amazona guildingii*

smoker parrot *Austral. Polytelis anthopeplus* (= regent parrot)

superb parrot *Polytelis swainsonii*

swift parrot *Lathamus discolor*

black-tailed parrot *Austral.* = smoker parrot *Austral.*

racket-tailed parrot any of several species of the genus Prioniturus

twenty-eight parrot *Austral. Platycercus zonarius*

plaid-wing parrot *Austral.* = smoker-parrot *Austral.*

parrot 2 (*see also* parrot 1, bill, feather, fish) organisms which resemble a parrot in appearance or behavior

sea-parrot any of numerous alcid birds

Parrotiaceae = Hamamelidaceae

parrotlet any of several species of psittacid bird of the genera Forpus and Touit

blue-fronted parrotlet *T. dilectissima*

blue-rumped parrotlet *F. cyanopygius*

blue-winged parrotlet *F. passerinus*

parsley (*see also* worm) properly the umbelliferous herb *Petroselinum hortense*, but applied to any herb thought to resemble parsley in appearance or flavor

bastard parsley any of several *spps.* of the umbelliferous genus Caucalis

beaked parsley *Brit.* any of several species of umbelliferous herb of the genus Anthriscus

bur-parsley *Brit.* any of several species of umbelliferous herb of the genus Caucalis

corn-parsley *Brit. Petroselinum segetum*

fool's parsley *Brit.* umbelliferous herb *Aethusa cynapium*

hedge-parsley *Brit.* any of several species of umbelliferous herbs of the genus Torilis

hedgehog-parsley the umbelliferous herb *Caucalis daucoides*

hemlock-parsley any of several species of umbelliferous herb of the genus Conioselinum

marsh parsley *Brit.* any of several edible marsh plants including the umbelliferous herbs *Petroselinum galbanum. U.S.* the hydrocharitaceous water plant *Vallisneria spiralis*

mountain-parsley *U.S.* the polypodiaceous fern *Cryptogramma crispa*

stone-parsley *Brit.* the umbelliferous herb *Sison amomum*

parsnip properly the umbelliferous herb *Pastinaca sativa*, but applied to some other forms thought to be parsnip-like

meadow-parsnip any of several species of umbelliferous herb of the genera Thaspium, Sium and Berula

parted in *biol.* cleft, but not to the base

-parthen- *comb. form* meaning "virgin"

parthenita a stage in the development of a trematode platyhelminth (*e.g.* sporocyst, redia) which themselves reproduce parthenogenetically (*cf.* adolescaria and marita)

parthenogenesis *see under* genesis

parthenote a haploid parthenogenetically produced individual

particle in *biol.* practically synonymous with body 1 and 2 (*q.v.*)

attraction particle = centriole

kappa particle a particle self-perpetuating in Paramecium containing a K gene and capable, if this gene is present, of developing a substance poisonous to other Paramecium

palmatipartite palmate, but with the divisions not reaching the base

partridge 1 (*see also* partridge 2, beery, pea, pigeon, shell) any of numerous galliform phasianid birds of the subfamily Phasianinae

bamboo partridge *Bambusicola fytchii*

long-billed partridge *Rhizothera longirostris*

grey partridge *Perdix perdix*

hill partridge = tree partridge

Hungarian partridge = grey partridge

red-legged partridge any of several species of the genus Alectoris. *Brit.*, without qualification, *A. rufa*

rock partridge *Alectoris graeca* (= chukor)

seesee partridge *Ammoperdix griseigularis*

snow-partridge *I.P. Tetraogallus himalayensis*

stone-partridge *Afr. Ptilopachus petrosus*

tree-partridge any of several Asiatic species of the genera Arborophila and Tropicoperdrix

red-wing partridge *S.Afr. Francolinus levaillantii*

bearded wood-partridge *Dendrortyx barbatus*

black wood-partridge *Melanoperdix nigra*

green-crested wood-partridge *Rollulus roulroul*

ferruginous wood-partridge *Caloperdix oculea*

long-tailed wood-partridge *Dendrortyx macroura*

partridge 2 (*see also* partridge 1) *W.I.* columbid birds of the genus Geotrygon (= quail-dove)

blue partridge *W.I. G. versicila* (= crested quail-dove)

red partridge *W.I. G. montana* (= ruddy quail-dove)

partridge 3 in names of other organisms

sea partridge *Brit.* the labrid fish *Crenilabrus melops*

-partur- *comb. form* meaning "to bring forth young"

parturital sexual

parturition the separation of an embryo mammal from its mother (= birth)

Parulidae that family of passeriform birds which contains the American wood-warblers. They are small birds slender in shape and typically arboreal

-parv- *comb. form* meaning "small"

-pas-, -pasa- *see* -pan-

pascual pertaining to pastures having few dicotyledonous inhabitants

pasha *I.P.* the nymphalid lepidopteran insect *Herona marathus*

passage in *biol.*, any open connection

pore passage the connection between the inner and outer surfaces of a stoma

apical pore passage = hydathode

respiratory passage the ventral of the two tubes into which the pharynx of the cyclostome divides. The dorsal is the pharyngoesophagus

Passalidae a family of coleopteran insects closely allied to the Lucanidae but differing from them in having the mentum of the labium deeply notched. They have numerous popular names in the *U.S.* of which bess-beetle is probably the most usual

passalus literally a peg, but applied specifically to a gamosepalous calyx

Passeridae = Ploceidae

Passifloraceae that family of parietalous dicotyledons which contains the passion flowers. The large flower with a crownlike outgrowth of the receptacle is typical

Passeriformes a very large order of birds distinguished as the name indicates, by their habit of perching

pastern that part of the limb of a perissodactyl or artiodactyl mammal which lies immediately adjacent to the coffin joint

Pasteur's *epon. adj.* from Louis Pasteur (1822–1895)

Pasteuriaceae a family of hyphomicrobiale Schizomycetes distinguished by the fact that the buds are sessile

rosy **pastor** the sturnid bird *Sturnus roseus*

pasture "domesticated" grassland used for grazing livestock

pat *see* [Patch]

patagium 1 (*see also* patagium 2, 3, 4) a flap of skin protruding from a mammalian body supported by limbs and used for gliding or flying

　uropatagium that portion of the patagium of chiropterans which lies alongside the tail or beyond which, in Mollossidae, the tail protrudes

patagium 2 (*see also* patagium 1, 3, 4) the fold of skin between the shoulder and manus, or between shoulder region and elbow, in a bird

patagium 3 (*see also* patagium 1, 2, 4) a lobe-like process covering the base of the forewing in some insects

patagium 4 (*see also* patagium 1, 2, 3) one of a pair of lateral expansions of the insect prothorax immediately in front of the base of the wing

patalusas popular name of priacanthid fishes (= bigeye)

[**Patch**] *either* a mutant (gene symbol *Pat*) in linkage group I of *Neurospora crassa*. The phenotypic expression is a circadian rhythm of alternating dense and sparse mycelial growth *or* a mutant (gene symbol *Ph*) found in linkage group III of the mouse. The phenotypic expression resembles [Dominant spotting] but is not accompanied by anemia or sterility

Peyer's **patches** = lymph follicle

patchouli *see* plant

pate 1 (*see also* pate 2) the top of the head

　baldpate *U.S.* the anatine duck *Mareca americana* (*cf.* widgeon). *W.I.* the columbid bird *Columba leucocephala* (= white-crowned pigeon)

pate 2 (*see also* pate 1) *N.Z.* the araliaceous tree *Schefflera digitata*

pateke *N.Z.* the anatid bird *Anas chlorotis*

-patell- *comb. form* meaning a "small dish" or "saucer"

patella a small dish and applied to almost any structure, plant or animal of this shape, including the joint between the femur and the tibia in the leg of chelicerate arthropods, the tibia of acarine arthropods, the adhesive disc on the foreleg of some male dytiscid beetles and a bone (*q.v.*)

patellula = patella of dytiscid beetles

patent used in *bot.*, for spreading

copulation **path** the route followed by the sperm from the end of the penetration path to the female nucleus

　penetration path the route followed by the sperm immediately after penetrating the egg and in some Amphibia delineated by the movement of pigment from the exterior, carried by the male gamete through the egg

antipathetic said of two organisms, the parts of which do not graft or transplant easily to each other, but which are not specifically antagonistic to each other

patoo = potoo

-patr- *comb. form* meaning "father"

marsh **patroller** *S.Afr.* the satyrid lepidopteran insect *Henotesia perspicua*

-patry *comb. suffix* used in *ecol.* in the sense of population

allopatry the condition of two related populations occupying separate geographic areas and which do not interbreed

　contiguous allopatry the condition when the territories of the two populations abut on each other

　disjunct allopatry the condition when the two populations are separated by an area inhabited by neither

sympatry the condition of two populations occupying the same territory

pattern a specific arrangement of things

　prepattern a term used to describe the non-random distribution of chemical and physical factors in an undifferentiated tissue, which accounts for the local initiation of differentiation

patulous spread wide

[**patulous**] a mutant (gene symbol *pa*) mapped at 101.0± on chromosome II of *Drosophila melanogaster*. The phenotypic expression is a wide spreading apart of the wings

-pauc- *comb. form* meaning "few"

paunch any pouch-like or sac-like appendage of the alimentary canal, including the pendulous abdomen of an obese human

pauraque the caprimulgid bird *Nyctidromus albicollis*

Pauropoda an order of minute progoneate Arthropoda with 12 segments and nine pairs of legs. The most typical character is the antennae consisting of four basal joints bearing at the end a pair of styli, the anterior of them with two flagella and the posterior with one flagellum

diapause a condition of arrested larval development shown by many insect larvae over-wintering in temperate climates

　menopause the time of cessation of the menstrual cycle in human females or of the oestral cycle in other mammals (*cf.* menarche)

pausiacous olive-green

pavoninous peacock-blue

paw the foot of a mammal

　cat's paw the spondylid mollusk *Plicatula gibbosa*

paw-paw (*see also* bird) the annonaceous tree *Asimina triloba* and its edible fruit. The term is sometimes applied to papilionaceous tree *Carica papaya* (= papaya)

-paxill- *comb. form* meaning a "tent peg."

paxilla the large, egg-shaped ossicles on the aboral surface of some asteroid echinoderms

pb *see* [proboscipedia]

pd *see* [purpleoid]. This symbol is also used for a bacterial mutant showing a requirement for pyridoxine

Pdr *see* [Purpleoider]

pdx-1, -2 *see* [pyridoxine-1] and [2]

pe *see* [pink-eyed] and [sticky peel]

Pe *see* [Peach]

pea 1 (*see also* pea 2, aphis, cock, curculio, miner, fowl, moth, mouth, thrip, tree, weevil) any of numerous leguminous plants, usually climbing vines, which have a spherical, or subspherical seed, particularly *Pisum sativum,* the garden pea

　Australian pea *Dolichos lignosus*

　beach pea *Lathyrus maritimus*

　black pea *Lathyrus niger* or *L. sylvestris*

　bush-pea *U.S. Thermopsis mollis*

　butterfly-pea *U.S.* any of several species of leguminous herb of the genera Clitoria and Centrosoma, particularly *Centrosoma virginianum*

　chick-pea *Cicer arietinium* (= garbanzo)

　coffee-pea = chick-pea

Congo pea *Cajanus indicus* (= pigeon-pea, dhal)
everlasting pea *Lathyrus grandiflorous* or *L. latifolius*
  Persian everlasting pea *L. rotundifolius*
two-flowered pea *Lathyrus grandiflorus*
garden pea any of very numerous horticultural hybrids of *Pisum sativum*
glory pea *Clianthus dampieri* or *C. puniceus*
green pea (*see also* pea 2) the immature seed of the garden pea
hoary pea *Tephrosia virginiana*
Lord Anson's pea *Lathyrus magellanicus*
milk-pea *U.S.* any of several species of the genus Galactia
partridge-pea *Cassia chamaecrista*
perennial pea *Lathyrus latifolius*
pigeon-pea = Congo pea
scurf pea *U.S.* any of several species of the genus Psoralea
scurfy pea = scurf pea
sea-pea = beach-pea
shamrock-pea *Parochepus commumis*
wing-stemmed pea *Lathyrus palustris*
sweet pea any of numerous horticultural varieties of *Lathyrus odoratus*
  perennial sweet pea *Lathyrus latifolius*
  wild sweet pea *Tephrosia virginiana*
Tangier pea *Lathyrus tingitanus*
wild pea *Lathyrus latifolius*
winged pea *Lotus tetragonolodus*
pea 2 (*see also* pea 1) in compound names of other organisms
green pea (*see also* pea 1) *Brit.* the noctuid lepidopteran insect *Earias chlorana*
heart-pea the sapindaceous vine *Cardiospermum halicacabum* (= balloon vine)
peach 1 (*see also* peach 2, beetle, scale) properly the rosaceous tree *Prunus persica* and its edible fruit. Also applied to a melon
vine peach a horticultural variety of muskmelon (*q.v.*)
peach 2 (*see also* peach 1) in compound names of other animals
sea-peach the pyurid urochordata *Halocynthia pyriformis*
[peach] a mutant (gene symbol *p*) mapped at 9 on chromosome 2 of the tomato. The phenotype has a peach-like (pubescent) fruit
[Peach] a mutant (gene symbol *pe*) on linkage group II of *Neurospora crassa*. The term is descriptive of the phenotypic conidial color
[Pea comb] a mutant (gene symbol *P*) in linkage group IV of the domestic fowl. The name is descriptive of the phenotypic expression (*see* comb 2)
pear 1 (*see also* pear 2, midge) the rosaceous tree *Pyrus communis* and the edible fruit borne by it and its very numerous horticultural varieties
snow pear *P. nivalis*
pear 2 (*see also* pear 1) any tree or fruit thought to resemble the pear, usually in shape
alligator-pear = avocado
balsam-pear the cucurbitaceous vine *Momordica charantia* and its fruit (*cf.* balsam apple)
melon-pear the solanaceous shrub *Solanum muricatum* and its edible fruits
prickly pear *U.S.* any of several species of cactus of the genus Opuntia
pearl 1 (*see also* pearl 2, 3, eye, fish, organ, twist, wort) a more or less spherical mass of small, finely, laminated aragonite crystals produced in several species of pelecypod Mollusca (*cf.* mother-of-pearl)

pearl 2 (*see also* pearl 1, 3) the obtuse lobes on the burr of the horn of a deer
pearl 3 (*see also* pearl 1, 2) *Austral.* any of several species of pierid lepidopteran insect of the genus Elodina
ground pearl popular name of numerous white or cottony coccid homopteran insects
peat partially decomposed vegetable matter
eutrophic peat that which is found in bogs
mesotrophic peat that which is found in moist, but not boggy, environments
oligotrophic peat the peat found in high moors
peb *see* [pebbled]
peba any of several species of dasypodid xenarthran mammals of the genus Tolypeutes. They have three bands of moveable scutes
[pebbled] a mutant (gene symbol *peb*) mapped at 7.3± on the X chromosome of *Drosophila melanogaster*. The phenotype has slightly roughened eyes
rock pebbler *Austral.* the psittacid bird *Polytelis anthopeplus* (= smoker parrot)
pecan (*see also* bearer, bug, mite, moth, weevil) the juglandaceous tree *Carya pecan* and its edible seeds
bitter pecan *C. aquatica*
peccarie = peccary
peccary either of two species of dicotylid suine artiodactyle mammals. *U.S.* without qualification, *Tayassu tajaca* (= javelina)
pecker in compound names of many birds
flowerpecker any of numerous species of dicaeids mostly of the Dicaeum genus
  scarlet backed flowerpecker *D. cruentatum*
  thick-billed flowerpecker *D. agile*
  fire-breasted flowerpecker *D. ignipectus*
oxpecker either of two African sturnids of the genus Buphagus (= tickbird)
  red-billed oxpecker *B. erythrorhynchus*
  yellow-billed oxpecker *B. africanus*
woodpecker any of numerous species of the piciform family Picidae or, more rarely, Jyngidae. In *Austral.* the name is applied to the certhiid *Climacteris rufa*
  acorn woodpecker *Melanerpes formicivorus*
  Arizona woodpecker *Dendrocopos arizonae*
  brown-backed woodpecker *Afr. Dendrocopos obsoletus*
  golden-backed woodpecker *Dinopium benghalense*
  ladder-backed woodpecker *Dendrocopos scalaris*
  red-eared bay woodpecker *Blythipicus pyrrhotis*
  red-bellied woodpecker *Centurus carolinus*
  scaly bellied woodpecker *Picus viridanus*
  ivory-billed woodpecker *Campephilus principalis*
  pale-billed woodpecker *Phloeoceastes guatamalensis*
  black woodpecker *Dryocopus martius*
  smoky-brown woodpecker *Veniliornis fumigatus*
  California woodpecker = acorn woodpecker
  cardinal woodpecker *Dendropicus fuscescens*
  black-cheeked woodpecker *Centurus pucherani*
  red-cockaded woodpecker *Dendrocopos borealis*
  chestnut-colored woodpecker *Celeus castaneus*
  gray-crowned woodpecker *Piculus auricularis*
  downy woodpecker *Dendrocepos pubescens*
  brown-eared woodpecker *Campethera caroli*
  ant-eating woodpecker = acorn woodpecker
  golden-fronted woodpecker *Centurus aurifrons*
  Gila woodpecker *Centurus uropygialis*
  gilded woodpecker *Colaptes chrysoides* (= gilded flicker)
  green woodpecker *Brit. Picus viridis W.I. Centurus radiolatus* (= Jamaican woodpecker)

bamboo green woodpecker *Picus vittatus*
scaly-bellied green woodpecker *Picus squamatus*
Cuban green woodpecker *Xiphidiopicus percussus*
black-naped green woodpecker *Picus canus*
grey woodpecker *Mesopicos goertae*
ground woodpecker *Afr. Geocolaptes olivaceus*. In *U.S.* this term is sometimes applied to the flicker
Guadeloupe woodpecker *Melanerpes herminieri*
hairy woodpecker *Dendrocopos villosus*
pale-headed woodpecker *Gecinulus viridis*
northern pale-headed woodpecker *Gecinulus grantia*
red-headed woodpecker *Melanerpus erythrocephalus*
white-headed woodpecker *Dendrocopos albolarvatus*
blackheart woodpecker *W.I. Colaptes auratus* (= yellow-shafted flicker)
Hispaniolan woodpecker *Centurus striatus*
imperial woodpecker *Campephilus imperialis*
Jamaican woodpecker *Centurus radiolatus*
Lewis' woodpecker *Asyndesmus lewis*
lineated woodpecker *Dryocopus lineatus*
Mexican woodpecker = ladder-backed woodpecker
cuff-necked woodpecker *Meiglyptes tukki*
Nubian woodpecker *Campethera nubica*
golden-olive woodpecker *Piculus rubiginosus*
pied woodpecker *Brit. Dendrocopos major*
Himalayan pied woodpecker *Dendrocopos himalayensis*
pileated woodpecker *Dryocopus pilatus*
pygmy woodpecker any of several species of the genus Dendrocopos
Puerto Rican woodpecker *Melanerpes portoricensis*
rufous woodpecker *Micropternus brachyurus*
great slaty woodpecker *Mulleripicus pulverulentus*
Spanish woodpecker *W.I.* = hairy woodpecker
spotted woodpecker *Brit. Dendrocopos minor*
buff-spotted woodpecker *Afr. Campethera nivosa*
three-toed woodpecker any of several species of the genera Picoides and Dinopium, without qualification, usually *P. tridactylus*
Arctic three-toed woodpecker *P. arcticus*
crimson-winged woodpecker *Picus puniceus*
Pecora an assemblage of ruminant artiodactyle mammals consisting of the families Cervidae, Giraffidae, Antilocapridae and Bovidae, distinguished by the possession of two functional digits on the feet and the presence of two horns
-pect- *comb. form* meaning "coagulation"
pecten any comb-like structure particularly a projection into the vitreous humor from the retina of the eye of birds, a similar structure in teleost fish, and a comb-like structure of unknown function immediately behind the legs in scorpions. The genus Pecten is the pelecypod mollusk known as the scallop
-pectin- *comb. form* meaning "comb-like"
pectinate shaped like a comb (= monopectinate)
bipectinate in the form of a comb having teeth on both sides
pectine a fringed corolla, particularly that of the gentians
pectine used interchangeably with pecten (*q.v.*) particularly as regards the appendages of scorpions
Pectinibranchia an order of streptoneuran Mollusca containing the limpets and their allies. They are distinguished by the usually conical shell, showing few traces of the torsion present in the viscera, and by the presence of a monopectinate ctenidium attached to the mantle throughout its length

Pectinidae that family of mollusks which are commonly called scallops. The lower of the two valves is usually strongly convex and the upper flat. The surface is ribbed and the margin scalloped
pectus the ventral portion of the insect thorax
antepectus the base of the prothorax
postpectus the undersurface of the metathorax
-ped- *comb. form* meaning "foot." Also a very frequent mis-spelling of "-paed-" (*cf.* -pod-)
longiped long-footed
palmiped said of birds having the three front toes fully webbed (*cf.* palmate)
Pedaliaceae a small family of tubiliflorous dicotyledons containing *inter alia, Sesamum indicum* from which sesame oil is extracted. It is distinguished from other allied families by the presence of slime glands
pedalium a tentacle-bearing, gelatinous extension of the umbrella in some Schyphozoa
peddler popular name of the larvae of those chrysomelid beetles which carry their excrement and cast skins on a fork on their tail
-pede- *comb. form* synonymous with -ped-
centipede the common name of the many-legged arthropods belonging to the class Chilopoda. The term is also used for some members of the class Symphyla
garden centipede the symphylan arthropod *Scutigerella immaculata*
house centipede *U.S.* the scutigerid *Scutigera coleoptrata*
millipede popular name of diplopod arthropods
diapedesis the process by which the white cell squeezes through the wall of a capillary
Pedetidae a monotypic family of myomorph rodents containing the Cape jumping hare or spring-haas
pedicel any slender stalk connecting two larger objects. Apart from this general use, which pertains equally to plants and animals, it is specifically applied to that section of a plant stem which is directly connected to a single flower, to the tubefoot of an echinoderm, and to the segment between the node and the abdomen in ants (*cf.* petiole)
anal pedicel a stalk secreted through the anus of certain acarine nymphs for purposes of attachment to the host
optic pedicel a cartilaginous cup in the orbit of chondrichthian fish
post pedicel the third antennae segment in some Diptera (*cf.* post petiole)
pedicellaria one of numerous, minute, usually two or three "fingered" pincer-like organs, found on the surface of many echinoderms
alveolar pedicellaria one which is partly sunk in an endoskeletal depression
excavate pedicellaria a bifurcate alveolar pedicellaria
ophiocephalous pedicellaria a pedunculate pedicellaria with short-toothed jaws on a flexible stalk
crossed pedicellaria a pedunculate pedicellaria in which the "fingers" cross as in a pair of scissors
tridactyl pedicellaria a pedunculate pedicellaria with three long jaws on a flexible stalk
fasciculate pedicellaria a sessile pedicellaria consisting of a number of spines on a single ossicle
trifoliate pedicellaria a sessile pedicellaria with three flattened jaws
gemmiform pedicellaria an aberrant type consisting of a globular head containing a poison sac on the end of a long thin stalk

spiniform pedicellaria sessile pedicellaria consisting of spines borne on adjacent ossicles

furcate pedicellaria a type of straight pedicellaria in which the tips are bifurcate, the whole somewhat resembling the claw of a lobster

peduncular pedicellaria having a short stalk

sessile pedicellaria pedicellaria lacking a stalk

straight pedicellaria a peduncular pedicellaria in which the claws meet at their tips as a pair of forceps

spatulate pedicellaria a form of valvulate pedicellaria of which the tips are expanded

valvulate pedicellaria a sessile pedicellaria formed of two groups of fused spines

Pediculati an order of Osteichthyes distinguished by a lure at the tip of the modified first dorsal fin spine near the tip of the snout. They are also distinguished by the reduced number of gill slits and the footlike pelvic fins

Pediculidae that family of anopluran insects containing the common human lice

pediculus the stalk of a flower, fruit or any reproductive vegetable body (*cf.* pedicel)

pedigerous limb-bearing

Pedilidae a small family of coleopteran insects closely resembling the Anthicidae from which they are most readily distinguished by the smaller size of the facets on the compound eye

-pedio- *comb. form* meaning "flat or country" or "plain" but used by ecologists in the sense of "uplands"

Pedionomidae a monospecific family of gruiform birds erected to contain the collared-hemipodes, readily distinguished by possession of four toes and a broad white black spotted collar

Pedipalpi at one time an order of arachnid arthropods commonly called the whip-scorpions. They are now regarded as two distinct orders the Uropygi, with a posterior whip and the Amblypygi, lacking a whip and with the first legs very long and slender

pediunker *Austral.* the procellariid bird *Procellaria cinerea* (= black-tailed shearwater)

peduncle with great difficulty to be distinguished from pedicel, but usually used in the sense of a stalk bearing one part rather than a connection between two parts and thus applied to the stalk of a barnacle, or the main stem of an inflorescence

anterior peduncle the fiber tract connecting the cerebellum with the mid-brain

caudal peduncle the tapered posterior end of the body of fishes

cerebral peduncle the thickened lower lateral wall of the mesencephalon

middle peduncle the termination of each end of Variolus' bridge in the cerebellum

willie pee *W.I.* the tyrannid bird *Contopus caribaeus* (= greater Antillean peewee)

peel the rind of a fruit

peep (*see also* order) *U.S. dial.* any of numerous small scolopacid birds

marsh peep *Erolia minutilla*

peeper *U.S.* the hylid anuran amphibian *Hyla crucifer*

peewee *Brit.* the charadriid bird *Vanellus vanellus* (= lapwing, pewit). *U.S.* any of several species of tyrannid bird particularly of the genus Contopus. *Austral.* any of several species of grallinid bird of the genus Grallina (= magpie lark)

greater Antillean peewee *C. caribaeus*
lesser Antillean peewee *C. latirostris*

greater peewee *C. pertinax*
eastern wood-peewee *C. virens*
western wood-peewee *C. sordidulus*

peewit = pewit

peg a small, bluntly tapered, body. In *bot.*, without qualification, applied to a peg-shaped body at the lower end of the hypocotyl of some seedlings

auditory peg = scolopale

cuticle peg a portion of cuticle which projects into or between epidermal cells

interpapillary peg a mass of cells which support the neck of a sweat gland as it passes through the papillary layer

Pegasidae a family of hypostomid Osteichthyes erected to contain the "sea moths" a name descriptive of these extraordinary fish with their enormously expanded pectoral fins and concentric rings of armored plates

pekan the mustelid carnivore *Martes pennanti* (= fisher)

pekapeka *N.Z.* the vespertilionid mammal *Chalinolobus morio* (= long-tailed bat)

-pel- *comb. form* meaning "clay" or "mud"

leptopel finely divided dead matter dispersed in water

sapropel leptopel (*q.v.*) which has accumulated on the bottom under anaerobic conditions and is gas-producing

-peleg- *comb. form* meaning "the surface of the sea"

pelagad a plant on the sea surface

pelagic properly, inhabiting the open ocean, as distinct from shores and estuaries; however, frequently in *biol.*, assigned other meanings and sometimes even used as synonymous with "free-swimming"

allopelagic the condition of pelagic organisms occuring at various depths, the cause for the variation in depth being unknown

autopelagic *see* plankton

bathypelagic pertaining to organisms living in the ocean at depths greater than one hundred fathoms

epipelagic pertaining to organisms living in about the top forty fathoms of ocean water

mesopelagic referring to organisms living at moderate depths (sixty to one hundred fathoms) in the ocean

nyctipelagic those pelagic organisms which seek the surface at night

tychopelagic = tycholimnetic

knephopelagile said of that zone of the ocean which extends from about five fathoms to the depth, in that particular area, at which photosynthesis is no longer possible

pelagium a community of marine pelagic animals

pelamid *Brit.* the scombrid fish *Katsuwonus pelamis* (= *U.S.* skipjack tuna)

Pelecanidae that family of pelecaniform birds which contains the pelicans, huge clumsy birds with notoriously large bill-pouches. All have totipalmate feet

Pelecaniformes that order of birds which contains, *inter alia,* the pelicans and cormorants. The throat-pouches are typical

Pelecanoidae = Pelecanoididae

Pelecanoididae that family of procellariiform birds which contains the diving petrels. They are distinguished from other families in the order by their very rapid flight and ability to dive. The tubular nostrils open upwards

Pelecinidae a family of proctotrupoid apocritan hymenopteran insects that includes only the one genus Pelecinus. The sexes are strikingly dimorphic, the female being easily recognized by the long abdomen. Males are rare north of Mexico and reproduction is parthenogenic in male free areas

Pelecypoda a very large class of prorhipidoglossomorphic Mollusca distinguished by their bivalved shells

pelican (*see also* flower, foot) any of several species of pelecanid birds of the genus Pelecanus distinguished by their large beaks and pouches

Australian pelican *P. conspicillatus*

pink-backed pelican *P. rufescens*

spot-billed pelican *P. roseus* or, less frequently, *P. philippensis*

brown pelican *P. occidentalis*

Dalmatian pelican *P. crispus*

rosy pelican *P. onocrotalus*

white pelican *U.S. P. erythrorhynchus Brit. P. onocrotalus* (= rosy pelican)

pelious ivory black

pellicle a thin skin, particularly an outer non-cellular coat of an organism

pellitory *Brit.* and *U.S.* any of several species of urticaceous herbs of the genus Parietaria

bastard pellitory the rosaceous herb *Agrimonia eupatoria*

pellitory-of-the-wall *Brit.* the urticaceous herb *Parietaria officinalis*

pellucid transparent

[pellucid] a mutant (gene symbol *pl*) in linkage group III of *Habrobracon juglandis*. The name is descriptive of the phenotypic eye

Pelmatozoa a subphylum of the Echinodermata distinguished usually by being sessile and always furnished by an aboral attachment often in the form of a stalk. All save one class (the Crinoidea) are extinct

antiopelmous the condition of the feet of piciform birds in which toes 1, 2, and 4 are connected with the flexor hallucis longus muscle and toes 3 with the flexor digitorum longus muscle

Pelobatidae a family of anamocoelous anuran Amphibia in which the upper jaw and vomer are provided with teeth, and the urostyle has a double articular condyle

Pelomedusidae a family of pleurodiran chelonian reptiles in which the neck is completely retractile

Peloplocaceae a family of chlamydobacterial Schizomycetes with unbranched, frequently very long, (up to 0.5 cm.) trichomes

-pelos- *comb. form* meaning "clay." Usually transliterated -peloz- and used by ecologists in the sense of "mud" or "mudbank"

-pelt- *comb. form* meaning a "small round shield" of the type once called a "target" (*cf.* -clyp- and -scut-)

peltate literally, shield shaped, but *biol.*, used to designate flattened structures, such as leaves, expanded tentacles, and the like, in which the stem is attached to the under side rather than to the edge

epipeltate when a leaf is expanded through the growth of its upper surface

hypopeltate when the leaf is expanded through the growth of its lower surface

Peltoperlidae a family of plecopteran insects with short cerci and two ocelli

peludo any of numerous dasypodid xenarthran mammals of the genus Chatophractus and their immediate allies. These armadillos, like the closely allied

weasel-headed armadillos and pigmy armadillos, have six bands of scutes

-pelv- *comb. form* meaning "basin"

pelvis 1 (*see also* pelvis 2) the bony mass formed, in many vertebrates, by the fusion of the pelvic girdle with elements of the vertebral column

pelvis 2 (*see also* pelvis 1) the swollen end of the ureter within the kidney

-pemphi- *comb. form* meaning "bubble" or "blister"

cytopemphis the passage, by a process analogous to pinocytosis, of large molecules through the walls of a capillary

Pemphredoninae a subfamily of sphecid hymenopteran insects commonly called aphid-wasps in virtue of their preferred food

pen a bacterial mutant showing penicillin resistance or sensitivity

pen a female swan

sea-pen popular name of stylatulid alcyonarian coelenterates (*cf.* sea feather) and of arcid pelecypod mollusks)

penacs the leguminous herb *Apios americana* (= ground nut)

-pencill- *comb. form* meaning a "small pointed brush" (*cf.* -penicill-)

Penaeaceae a small family of myrtiflorous dicotyledonous angiosperms

propendent ranging downwards

filipendulous in the shape of a thread with lumps on it

penetrance the extent or regularity with which a gene alteration appears phenotypically (*cf.* expressivity)

complete penetrance said of a gene which always produces an effect

incomplete penetrance = partial penetrance

partial penetrance a condition in which the phenotypic expression overlaps that of the wild type

penetrant = stenotele nematocyst

penguin 1 (*see also* penguin 2) popular name of spheniscid birds

Adelie penguin *Pygoscelis adeliae*

blue penguin *Eudyptula minor*

crested penguin *Eudyptes pachyrhynchus*

big-crested penguin *Eudyptes sclateri*

yellow-eyed penguin *Megadyptes antipodum*

emperor penguin *Aptenodytes forsteri*

fairy penguin = little penguin

white-flippered penguin *Eudyptula albosignata*

Galapagos penguin *Spheniscus mediculus*

grand penguin *N.Z.* = yellow-eyed penguin

gentoo penguin *Pygoscelis papua*

jackass penguin *Spheniscus demersus*

king penguin *Aptenodytes patagonica*

rock-hopper penguin *Eudyptes cristatus*

little penguin = blue penguin

royal penguin *Eudyptes schlegeli*

tufted penguin *Catarrhactes chrysocome*

penguin 2 (*see also* penguin 1) the characinid teleost fish *Thayeria obliqua*

-penicill- *comb. form* meaning "in the shape of a small, sharply pointed brush" of the type known as an artist's pencil (*cf.* -pencill-)

penicillate in the shape of an artist's pencil

penicillinase an enzyme catalyzing the hydrolysis of penicillin to penicilloic acid

penicillium = amastigophore

penicillus a spray of straight arterioles in the red pulp of the spleen

penis (*see also* bulb, papilla) the muscular male intromittent organ of many animals. The insect equivalent is the aedeagus, a flexible membranous organ

hemipenis either of the two halves of paired penes in those forms which possess them

penitent *U.S.* the noctuid lepidopteran insect *Catocala piatrix* (*cf.* underwing)

-penn- *comb. form* meaning "feather" (*cf.* -pinn-)

penna literally a wing, but also used for feather

prepenna a down feather on a nestling bird which precedes the contour feathers

Pennatulacea that order of alcyonarian Anthozoa which contains the sea-pens and sea-pansies. They are distinguished by the presence of a fleshy stalk (axial polyp) itself devoid of polyps, but with lateral secondary polyps at the upper end

Pennatulidae a family of pennatulacean Alcyonarian coelenterates commonly called sea-pens. They are distinguished from the sea-feathers (*q.v.*) by the very small lateral projections and greatly elongated rachis

penniculus the closely adpressed basal granules of the cilia of the cytopharynx in some ciliates or the rows of cilia themselves

water **penny** the flat almost circular stone-clinging larvae of psephenid coleopterans

pennyroyal any of several labiateous herbs *U.S. Hedeoma pulegiodes* or *Monardella villosa. Brit. Mentha pulegium*

American pennyroyal the labiateous herb *Hedoma pulegioides*

bastard pennyroyal any of several species of the genus Trichostema, particularly *T. dichotomum*

false pennyroyal any of several species of the genus Isanthus

mock pennyroyal any of several species of the genus Hedeoma

-penta- *comb. form* meaning "five"

pentact a five-axis sponge spicule that has been derived from a hexactinellid by the loss of half of one axis

Pentacula a postulated common ancestor of the echinoderms (*cf.* Dipleurula)

[**pentagon**] a mutant (gene symbol *ptg*) mapped at 23.2 on the X chromosome of *Drosophila melanogaster.* The phenotype has the thoracid trident and scutellar spot dark

Pentagynia those orders of the Linnaean system of plant classification distinguished by the possession of five pistils

Pentandria that class of the Linnaean classification of plants distinguished by the possession of five stamens

Pentaphyllaceae a small family of sapindale dicotyledons

Pentatomidae a family of hemipteran insects readily recognized by their broad shield-like shape, five-segmented antennae, and large triangular scutellum. They are best known for their bright-colorings and offensive odor, which causes them to commonly be called stink-bugs

Penthoroideae a subfamily of Saxifragaceae considered by some to be of familial rank

peony any of numerous species and very numerous horticultural varieties of the ranunculaceous genus Paeonia

tree pony *P. suffruticosa* and its very numerous horticultural varieties

pepo a type of fleshy fruit in which the hard outer skin derives from the receptacle of the flower. Cucumbers and squashes are examples

Peponiferae = Cucurbitales

Peponiferes = Cucurbitales

pepper 1 (*see also* pepper 2–4, conch, grass, maggot, saxifrage, shrike, tree, weevil) any of several species of the piperaceous genus Piper

Ashanti pepper *P. clusii*

betel pepper *P. betle*

black pepper *P. nigrum*

guinea pepper (*see also* pepper 3) = Ashanti pepper

Japanese pepper *P. futokadsura*

Japanese pepper (*see also* pepper 2) *P. longum*

pepper 2 (*see also* pepper 1, 3, 4) any of several species and numerous horticultural varieties of the solanceous herbs or shrubs of the genus Capsicum, especially *C. frutescens*

bell pepper (*see also* pepper 3) a bell shaped horticultural variety

cayenne pepper *C. frutescens*

chili pepper (*see also* pepper 4) a general term of Capsicum peppers

hot pepper any containing large amounts of capsaicin

long pepper (*see also* pepper 1) a long-fruited, very hot, variety of *Capsicum frutescens*

pepper 3 (*see also* pepper 1, 2, 4) any plant resembling those defined in pepper 1 and 2

African pepper any of several species of the annonaceous genus Xylopia

bell pepper (*see also* pepper 2) the ericaceous shrub *Erica tetralix*

bitter pepper the rutaceous tree *Evodia dantelli*

Guinea pepper (*see also* pepper 1) the annonaceous tree *Xylopia aethiopica*

poor man's pepper *U.S.* the cruciferous herb *Lepidium virginicum*

melegueta pepper the zingiberaceous herb *Elettaria granumparadisi*

wallpepper the crassulaceous herb *Sedum acre*

waterpepper the polygonaceous herb *Persicaria hydropiper*

pepper 4 (*see also* pepper 1–3) in compound names of other organisms

chilipepper (*see also* pepper 2) the scorpaenid fish *Sebastodes goodei*

pepper-and-salt the umbelliferous herb *Erigenia bulbosa*

pepperidge the nyssaceous tree *Nyssa sylvatica*

pepsin an enzyme catalyzing the hydrolysis of peptides

peptidase a group of enzymes catalyzing the hydrolysis of peptides by splitting off residues. The names are descriptive of the function

aspergillopeptidase catalyzes the hydrolysis of peptides and converts trypsinogen into trypsin

clostridiopeptidase catalyzes the hydrolysis of many peptides but particularly collagen and gelatin

streptococcus peptidase a non-specific peptidase derived from various species of Streptococcus

dipeptidase a group of enzymes catalyzing the hydrolysis of dipeptides

glycyl-glycine dipeptidase catalyzes the hydrolysis of both glycyl-glycine and sarcosyl-glycine

aminoacyl-histidine dipeptidase catalyzes the hydrolysis of the dipeptides of L-histidine and their amides

aminoacyl-methylhistidine dipeptidase catalyzes the hydrolysis of anserine into β-alanine and 1-methyl-L-histidine

imidodipeptidase hydrolyzes dipeptides in which the proline residue is bound by the imido group

iminodipeptidase catalyzes the hydrolysis of dipeptides or amides in which the proline residue is bound by the imino group

enteropeptidase an enzyme catalyzing the conversion of trypsinogen into trypsin

subtilopeptidase an enzyme catalyzing the conversion of ovalbumin into plakalbumin

Peramelidae a family of marsupial mammals containing the forms commonly called bandicoots. They are distinguished from the dasyurids in possessing the typical marsupian feature of having the second and third hind toes fused. The common name sometimes causes confusion with Bandicota, a genus of rodents

perch (*see also* bug) any of numerous teleost fish. *U.S.*, without qualification, *Perca flavescens*, *Brit. P. fluviatilis*. In midwestern *U.S.*, where *P. flavescens* does not exist, the term perch is often applied to sunfishes of the genus Lekomis

blue perch the nandid *Badis badis*

climbing perch the anabantid *Anabas testudineus.* It is capable of walking long distances from one body of water to another

dusky perch *Brit.* the serranid *Serranus gigas* (*cf.* bass 1)

glassperch *U.S.* the centropomid *Chanda ranga* (*cf.* glassfish)

Nile perch the huge centropomid *Lates niloticus* which reaches a weight of several hundred pounds

ocean-perch trade name for frozen fillets of either the redfish or rosefish

pikeperch *U.S. local.* the percoid *Stizostedian vitreum* (= wall-eyed pike) *Brit.* the percid fish *Lucioperca lucioperca*

pirate-perch the aphredoderid *Aphredoderus sayanus*

Rio Grande perch the chichlid *Cichlasoma cyanoguttatum*

Sacramento perch the centrarchid *Archoplites interruptus*

sandperch any of several species of serranids of the genus Diplectrum, usually, without qualification *D. formosum*

dwarf sandperch *D. bivittatum*

sea-perch popular name of embiotocids

silver perch the sciaenid *Bairdiella chrysura*

surf-perch popular name of embiotocids

trout-perch popular name of the percopsid *Percopsis omiscomaycus*

western trout-perch *U.S. P. transmontana* (= sandroller)

white perch the serranid *Roccus americanus*

zebra perch *U.S.* the kyphosid *Hermosilla azurea*

Percidae a very large family of freshwater percoid fishes containing the darters and perches. In many the soft-rayed and spiny-rayed dorsal fins are separated

Perciformes = Acanthopterygii

Percoidea a very large suborder of acanthopterygiian Osteichthyes. They are usually regarded as primitive spiny-rayed fishes since there are five soft rays in each of the pelvic fins

Percomorphi = Acanthopterygii

Percophididae = Platycephalidae

Percopsidae a family of salmopercous fishes distinguished by the presence of adipose fins and rough ctenoid scales. Commonly called trout-perch

Percopsiformes = Salmoperci

Perdicidae a family of galliform birds usually included in the Phasianidae

Peregrine = peregrine falcon

-pereio- *comb. form* meaning "to transport." Often confused with -peri-

pereion the thoracic region of isopod Crustacea, and sometimes used as synonymous with prothorax in insects

postpereion = metanotum

perentie the varanid saurian reptile *Varanus giganteus*

Pereskioideae a subfamily of Cactaceae distinguished by the flattened leaves and by bearing flowers in panicles

perfect in *bot.*, applied to a flower which has both pistils and stamens

cotton leaf perforator *U.S.* the lyonetiid lepidopteran insect *Bucculatrix thurberiella*

perfusous completely covered

pergameneous resembling parchment

Pergidae a small family of symphytan Hymenoptera distinguished from other sawflies by the fact that the antenna is six-jointed

-peri- *comb. form* meaning "about" in the sense of "surrounding"

perianth those parts (*e.g.* calyx and corolla) which surround the petals of a flower. Also used for the sheath that, in some lower plants encases the reproductive organs

pseudoperianth the envelope of an archigonium

Pericarida a superorder of malacostracan Crustacea in which the carapace leaves at least four of the thoracic segments distinct but with the first thoracic somite always fused with the head

[Pericarp] a mutant (gene symbol *P*) mapped at 28 on linkage group I of *Zea mays*. The gene controls the color of the pericarp

Pericopidae a family of noctuoid lepidopteran insects usually conspicuously marked in black and white

Pericrocotidae a family of passeriform birds usually included with the Campephagidae

Peridiniales that order of dinophycean pyrrophytes which is distinguished by a cell wall composed of a definite number of plates the arrangement of which is a specific character. It is distinguished from the Gymnodiniales, which it closely resembles, by the fact that the cell wall is vertically differentiated into two opposed halves

peridiole chamber of the gleba containing a nest of spores in puffball fungi

peridium the outer coat of a fungus sporophore

Perilampidae a family of chalcidoid apocritan hymenopteran insects recognizable by the facts that the thorax is larger than the triangular or globose abdomen and that the collar-like prontum is as wide as the mesonotum

Perimelitae plants having honey glands at the bottom of the perianth

perine the outer, sculptured, layers of a pollen grain

perinium the outermost of the three coats of a fern spore

periodism the condition of association with periodic variables

photoperiodism response to periodic variations in the light

Periophthalmidae a family of salmopercous fishes containing the mudskippers, often included in the Gobiidae

Periplomatidae that family of pelecypod mollusks which contains the forms commonly called spoon-shells. The name derives from the typical spoon-shaped chondrophore on each valve

Perisporiales = Erysiphales

-perisso- *comb. form* meaning "uneven"

Perissodactyla an order of placental mammals at one time combined with the Artiodactyla into the group Ungulata. They contain the forms commonly referred to as horses, rhinoceroses, and tapirs, in which there is an odd number of digits on the foot with the axis of symmetry passing through the center digit

peristaltis wave-like contractions passing along the length of a tube

Peristediidae a family of scleroparous fishes closely allied to the Triglidae but distinguished from them by the presence of spiny, bony plates all over the body. They are therefore commonly called armored sea-robins

peristem *see* -istem

Peristeridae a family of columbiform birds now usually included in the Columbidae

peritoneum the lining of the abdominal cavity (*cf.* omentum)

retro**peritoneal** said of an organ covered by peritoneum, but not projecting into the coelom

Peritricha an order of euciliate ciliate Protozoa distinguished by the possession of an adoral row of cilia which passes from left to right round the peristem and by the absence of any other cilia from the body

Peritrichida = Peritricha

perlarious pearly, either in the sense of being shiny or in the sense of being a small round body

Perlidae a family of plecopteran insects recognizable by the remnants of the filamentous gills on the sides of the thorax

Perlodidae = Perlidae

permeant a roving member of a terrestrial community frequently passing from one community to another

permit *U.S.* the carangid fish *Trachinotus falcatus* (*cf.* pompano)

perniciasm the state of host cells killed by a parasite

Perninae that subfamily of accipitrid birds which contains the honey-buzzards and the swallow-tailed kites. They are best distinguished by the very dense feathering on the sides of the head and the lack of a naked superciliary shield

-pero- *comb. form* meaning "maimed"

peroid having defective fruit

peronate densely hairy

peronium a nematocyst-containing extension of the epidermis of the inner face of a tentacle onto the exumbrella of a medusa

Peronosporales an order of biflagellate phycomycete fungi distinguished by the production of zoosporangia

peroxidase a group name for enzymes catalyzing oxidations in which peroxide acts as the oxygen donor and water is therefore a product of the reaction. Peroxidase, without qualification, is used to describe a large group of haemoprotein enzymes oxidizing a great variety of materials by the indicated reaction

  fatty acid peroxidase catalyzes the production of 1-pentadecanal from palmitic acid and corresponding compounds from myristic and stearic acids

  cytochrome peroxidase catalyzes the production of oxidized cytochrome *c* from reduced cytochrome *c* and hydrogen peroxide (*cf.* cytochrome oxidase)

  NAD peroxidase oxidizes NADH and hydrogen peroxide to NAD+

  NADAP peroxidase oxidizes NADPH and hydrogen peroxide to NADP+

  tryptophan peroxidase oxidizes tryptophan to a product not yet established

perpusillous minute

perroquet *W.I.* without qualification = red necked parrot (*cf.* parakeet)

Perseoideae a subfamily of the Lauraceae distinguished from the Lauroidea by the four-celled anthers dehiscing by four valves

persimmon (*see also* borer) any of numerous species of tree of the ebenaceous genus Diospyros; usually, without qualification *D. virginianum*

  American persimmon *D. virginianum*

  Japanese persimmon *D. kaki*

golden **pert** the scrophulariaceous herb *Gratiola aurea*

pertusate perforated

-perul- *comb. form* meaning "a small pouch"

deperulation the shedding of bud-scales

  calyptral deperulation when the scales are thrown as a cap

  tubular deperulation when the scales remain as a collar

perule the scale of a leaf bud

pervious open to passage; the antithesis of impervious

pessimum the worst conditions under which a specific organism can just survive

-petal- *comb. form* meaning "motion towards"

petal 1 (*see also* petal 2, 3) a leaf-like, frequently brightly colored, structure found in one or more cycles immediately surrounding the reproductive portions of a flower

petal 2 (*see also* petal 1, 3) in the sense of movement towards

  acropetal produced in succession towards the apex

  rectipetal produced, or growing in straight lines

petal 3 (*see also* petal 1, 2) in compound names of organisms

  rose petal *U.S.* the tellinid mollusk *Tellina lineata*

petaloid petal-like but used specifically for the petal-like shape assumed by the ambulacra on the aboral surface of many irregular echinoid echinoderms

petalous pertaining to petals

  apetalous without petals

  acropetalous said of those inflorescences in which the flowers first open at the base and later at the tip

  adenopetalous having petals derived from glandular structures

  alternipetalous applied to those flowers in which stamens alternate with petals

  andropetalous a double flower in which the additional petals come from stamens

  anisopetalous with petals of unequal size

  apopetalous having more petals than usual in those cases where these petals are not derived as modified bracts or modified reproductive structures

  basipetalous said of those inflorescences in which the flowers open first at the apex

  catapetalous the union of petals with united stamens

  choripetalous = polypetalous

  cleistopetalous said of a flower of which the petals, but not of necessity the other portions of the perianth, remain permanently closed

  dialypetalous = polypetalous

  dipetalous having two petals

  eleutheropetalous having many petals each free of the other

  epipetalous said of floral parts projecting beyond (*i.e.* "above") the petals

  gamopetalous having the petals united

  haplopetalous with one row of petals

  heteropetalous having more than one type of petal or changing from one type of petal to another

homopetalous having all the petals alike

monopetalous properly possessing only a single petal, but frequently used in place of gamopetalous

olopetalous said of a flower having other parts of the perianth metamorphosed into petals

parapetalous the conditions in which the stamens stand alongside the petals

pleiopetalous the condition of a "double" flower

polypetalous having many petals or, sometimes, more petals than is usual

proteropetalous an obdiplostemonous condition in which the epipetalous stamens are the inner ones

sympetalous = gamopetalous

petalum literally, a "metal plate" but sometimes used in *bot.* for "petal"

mesopetalum (Orchidaceae) = labellum (Orchidaceae)

parapetalum any appendage to a corolla

perapetalum = paraphyllum

Petaluridae a family of anisopteran odonatan insects with well separated eyes and a notched labium

Petauristidae = Trichoceridae

petchary *W.I.* the tyrannid bird *Tyrannus dominicensis* (= grey kingbird)

grey **peter** *Austral.* the meliphagid bird *Meliphaga virescens* (*cf.* honey-eater)

pick-peter *W.I.* the tyrannid bird *Tyrannus dominicensis* (= grey king-bird)

petiolaceous pertaining to a petiole

petiolaneous consisting of a petiole

petiolary pertaining to a petiole

Petiolata = Apocrita

petiolate possessing a petiole

petiole any slender stalk, particularly that which connects the leaf to a stem, or the slender segment or segments connecting the swollen abdomen of a thin-waisted wasp or dipteran to the thorax (= pedicel)

alate petiole one with wing-like extensions at the base

postpetiole the posterior segment when the petiole consists of two segments (*but see* postpedicel)

primary petiole the rhachis of a compound leaf

[**petite**] a group of yeast mutations showing a deficiency in mitochondrial activity controlled by a plasmagene

-peto- *comb. form* meaning "seek"

-petr- *comb. form* meaning "rock" or "stone"

petrel common name of many procellariiform birds

ashy petrel *Oceanodroma homochroa*

Bermuda petrel *Pterodroma cahow*

black petrel *U.S. Loomelania melania. N.Z. Procellaria parkinsoni*

blue petrel *Halobaena caerulea*

Bonin petrel *Pterodroma leucoptera* (= gadfly petrel)

brown petrel *N.Z. Procellaria cinerea* (= black-tailed shearwater)

black-capped petrel *U.Z. Pterodroma hasitata. N.Z. Pterodroma externa*

Chatham Island petrel *N.Z. Pterodroma cookii*

white-chinned petrel *Procellaria aequinoctialis*

Cook's petrel *Pterodroma cookii*

diving-petrel popular name of pelecanoidid birds of the genus Pelicanoides

grey-faced petrel *Pterodroma macroptera*

white-faced petrel *Pelagodroma marina*

blue-footed petrel = Cook's petrel

gadfly petrel *Pterodroma leucoptera*

giant petrel *Macronectes giganteus*

silver-grey petrel *Priocella antarctica*

Guadalupe petrel *Oceanodroma macrodactyla*

white-headed petrel *Pterodroma lessonii*

Leach's petrel *Oceanodroma leucorhoa*

least petrel *Halocyptena microsoma*

scaled petrel *Pterodroma inexpectata*

snow petrel *Pagodroma nivea*

storm-petrel any of many procellariiform birds of the family Hydrobatidae. *Brit.*, without qualification *Hydrobatis pelagicus*

fork-tailed petrel *Oceadroma furcata*

Wilson's petrel *Oceanites oceanicus*

black-winged petrel *N.Z.* = Cook's petrel

hygropetric pertaining to wet zones

Petricolidae a family of pelecypod mollusks often referred to as "rock dwellers." They are burrowing mollusks with a wide gaping shell weakly hinged

petrodad a plant of stony places

Petromyidae a monotypic family erected to contain the *S.Afr.* rockrat, *Petromys typicus*. This form resembles an octodont in possessing four molars in each jaw but differs from all other hystricomorphs in the tuberculation of the teeth

Petromyzontes that order of marsipobranchiiate agnathous craniates which contains the lampreys. They are distinguished from the Myxini or hagfishes by the fact that the nasal sac does not communicate with the mouth and the eyes are normally developed

-petros- *see* -petr-

Petrosavieae a subfamily of Liliaceae

pettichaps *Brit.* the sylviid bird *Sylvia borin* (= garden warbler)

pewee = peewee

pewit *U.S.* the larid bird *Larus ridibundus.* (= black-headed gull) *Brit.* the charadriid bird *Vanellus vanellus* (= lapwing)

Peyer's *epon. adj.* from Johann Conrad Peyer (1653–1712)

peyotl the cactus *Lophophora williamsii*

-pez- *comb. form* meaning "foot" (*cf.* -ped-)

pezizoid in the shape of a pezzizale fungus

Pezizales an order of discomycete Ascomycetae distinguished by a cup-shaped or saucer-like ascocarp. These fungi are often, from this fact, mistaken for Basidiomycetes

pf-1–pf-20 *see* [paralyzed-1] to [20]

Pfd *see* [Pufdi]

pg *see* [pigmy]

pg$_{11-12}$ *see* [pale green 11]–[12]

Ph *see* [patch]

pha a yeast genetic marker indicating a requirement for alanine

–phaen- *comb. form* meaning "to appear" or "appearance." Almost invariably transliterated -pheno-

-phaeo- *comb. form* meaning "brown"

Phaeophyta a subphylum of algae containing the forms commonly known as brown algae. They are distinguished by the yellowish-brown chromatophores

periphaericous circumferential

Phaeosporae = Phaeophyceae

-phaet- *comb. form* meaning "main"

Phaethontidae that family of pelecaniform birds which contains the tropic-birds. They are distinguished by the extreme length of the attenuated middle tail feathers

Phaetontidae = Phaethontidae

-phag- *comb. form* meaning "glutton," though more commonly used in the sense of "to eat"

-phage *comb. suffix* indicating something that devours another

bacteriophage a virus infesting, and usually lysing, bacteria. A phage (as this term is usually abbreviated) can alter the heritable characters of an infected cell

biophage an organism that destroys living things

cholaraphage a bacteriophage specific to vibrios

coliphage a bacteriophage specific to *Escherichia coli*

macrophage = histiocyte

microphage = neutrophil

prophage a factor in a bacterial genome that can, in combination with another factor, cause the production of a bacteriophage

vitellophage a wandering cell in a yolk mass which is alleged to assist in its digestion

-phagous *comb. suffix* used in the sense of "eater of." The suffix -vorous (*q.v.*) is synonymous

acridophagous grasshopper-eating

algophagous alga-eating

biophagous life-eating (*i.e.* parasitic or predatory)

coprophagous dung-eating

creophagous = carnivorous (used mostly of plants)

entomophagous insect-eating

euryphagous eating a wide range of nutrients

heterophagous being able to eat many things, particularly said of a parasite of many varied hosts

isophagous referring to parasites, or predators, which are selective, but not monospecific

meliphagous honey-eating

monophagous said of a parasite which is host specific or a predator of a single organism

myrmecophagous ant-eating

necrophagous an eater of dead bodies

pantophagous = omnivorous

phytophagous plant-eating

pleophagous eating a variety of forms or things

polyphagous said of a parasite infesting many hosts

ryophagous filth-eating

tecnophagous pertaining to an individual that eats its own eggs

adelphophagy the fusion of two gametes of the same sex

Phalacridae = Scaphidiidae

Phalacrocoracidae that family of pelecaniform birds which contains the cormorants. These large, usually black, diving birds are distinguished from other members of the order by the prominent crest, hooked beak and absence of subcutaneous airsacs

Phalaenidae = Noctuidae

phalanger any of several ground-dwelling phalangelid marsupials often separated as a subfamily the Phalangerinae

dormouse-phalanger popular name of phalangerine phalangerids of the genus Cercaetus (= pigmy opossum)

flying phalanger popular name of gliding phalangerine phalangerids mostly of the genus Petaurus

pigmy flying phalangers any of several species of the genus Acrobates

Taguan flying phalanger *Petauroides volans*

long-snouted phalanger any of several species of the genus Tarsipes

squirrel-phalanger *Gymnobelideus leadbeateri*

striped phalanger any of several species of phalangid of the genus Dactylopsila

pen-tailed phalanger any of several species of the genus Distaechurus

ring-tailed phalanger any of several species of the genus Pseudochirus

yellow phalanger *Pseudochirus archeri*

Phalangeridae a large family of marsupial mammals

distributed over the whole Australian region including the Dutch East Indies. They are distinguished by the fact that the big toe is apposed to the other toes and bears a nail in place of a claw

Phalangerinae a sub-family of phalangerid marsupials distinguished by the long tail

phalanges plural of phalanx

Phalangida an order of arachnid arthropods containing those forms known in the *U.S.* as daddy longlegs and in *Brit.* as harvestmen. They are distinguished by the possession of four pairs of walking legs and a segmented abdomen broadly joined to the cephalothorax

phalanx literally, an array of soldiers, but used specifically in *bot.* for rays of stamens and in *zool.* for a bone in the vertebrate skeleton distal to the metacarpal or metatarsal. The term is also occasionally used for a taxon between genus and subfamily

phalarope any of several species of phalaropodid birds

grey phalarope *Brit.* = *U.S.* red phalarope

red phalarope *U.S. Phalaropus fulidarius*

red necked phalarope *Brit.* = *U.S.* northern phalarope

northern phalarope *U.S. Lobipes lobatus*

wilson's phalarope *Steganopus tricolor*

Phalaropodidae a small family of charadriiform birds containing the phalaropes. They are distinguished by the long pointed wing, the longer slender bill and the lobate feet

-phall- *comb. form* meaning "penis"

Phallostethidae a small family of acanthopterygian fishes often placed in the suborder Atherinoidei. They are distinguished by the large muscular and bony copulatory organs under the throat of the male

phallus properly penis but extended to any male intromittent organ

endophallus the interior of a penis or in arthropods, an invagination at the anterior end of the aedeagus

Phaloniidae a small family of tortricoid lepidopteran insects many of which are borers in herbaceous plants

-phan- = -phaner-

nymphophan an inactive stage, corresponding to an insect pupa, in the life history of some acarines

teleiophan the resting stage between nymph and adult in some Acarina

-phaner- *comb. form* meaning "obvious" or "manifest"

Phanerocephala a subclass of polychaete Annelida in which the prestomium, usually bearing eyes, tentacles and palps, is free and exposed

Phaneroglossa an assemblage of anuran Amphibia distinguished from the Aglossa by the presence of a movable secondary tongue

Phanerozonia an order of asteroid echinoderms distinguished by the prominent contiguous marginal plate along the sides of the arms

aphanisis the suppression of a part of an organism

diaphanous transparent

hygrophanous said of a vegetable structure which is diaphanous when moist and opaque when dry

Phareae a tribe of the plant family Gramineae

pharetrone a term used for those calcareous sponges in which the spicules are united into a network

-pharo- *comb. form* confused between four *Gr.* roots and thus variously meaning "a light, or lighthouse," "a garment," "a plough" or "to possess"

-pharyngo- *comb. form* from "pharynx"

pharynx that part of the alimentary canal that lies between the oral cavity and the esophagus. In invertebrates a muscularized portion of the alimentary canal, lying immediately behind the buccal tube or esophagus

ambipharynx a membrane running along the inside edge of the insect mandible

aphelenchoid pharynx similar to the tylenchoid, except that the pharyngeal glands protrude at the posterior region

basipharynx in insects the epigusta and subgusta taken together

cytopharynx a permanently open food-entry passage in some Protozoa

dorylamoid pharynx a nematode pharynx in the shape of an elongate hair with the thin end anterior

epipharynx a supposedly chemoreceptor organ found on the inner surface of the insect labrum and the complex buccal area of certain insect larvae

diplogasteroid pharynx a nematode pharynx with a bulb in the middle

hypopharynx a chemo-receptor organ on the upper surface of the insect labium frequently referred to as the "tongue." Also the median post-oral lobe of the ventral wall of the gnathal region of the head anterior to the labium

mermithoid pharynx a nematode pharynx which is long, tubular and without muscles

nasopharynx that portion of the mammalian pharynx which lies above the plane of the soft palate

oral pharynx that portion of the mammalian pharynx which lies below the plane of the soft palate

parapharynx in insects, the lower part of the prepharynx

prepharynx any undifferentiated portion of the alimentary canal which lies between the pharynx and the buccal cavity

rhabditoid pharynx a nematode pharynx with an anterior wide region, a median pseudobulb, and a narrow posterior region terminating in a bulb

tylenchoid pharynx a nematode pharynx with a very narrow anterior region, a medium bulb, and a wider posterior region

polypharyngy the condition occurring in some planaria of having numerous pharynges with numerous openings

Phascogalinae a subfamily of dasyurid marsupials known in Australia as pocket-mice

Phascolarctidae a family of marsupial mammals containing the koala (q.v.)

-phase- *comb. form* meaning "appearance"

phase a transitory stage, particularly in cell division

anaphase the stage of mitotic division during which the half chromosomes (chromatids) separate and start moving in the direction of the poles

dicaryophase the stage in spore production which terminates in the production of a teleutospore

diplophase = sporophyte generation

halophase = gametophyte generation

interphase = intermitosis or intermeiosis

metaphase the stage of mitotic division at which the chromosomes are aligned on the equatorial plate of the spindle

prometaphase a stage in mitosis between the breakdown of the nuclear envelope and metaphase orientation, during which the chromosomes form a more or less tight clump

prophase the first stage in mitotic division in which the chromosomes are visible within the nucleus

telophase the terminal phase of mitotic division in which the chromosomes reconstitute a nucleus

Phasianellidae a family of gastropod mollusks distinguished by the porcellaneous high-spired shell which lacks a periostracuum. Commonly called pheasant-shells

Phasianidae that family of galliform birds which contains the quails, partridges, pheasants, jungle fowls, etc. They are distinguished from other galliform birds by the very large posterior notch of the sternum and the fact that the hallux is inserted higher than the other toes

Phasianinae that subfamily of phasianid galliform birds which contains the quails, partridges, pheasants, and jungle fowls. They are distinguished from the grouse by their long tails and generally naked legs.

haplophasic = haploid

Phasiidae a family of brachyceran dipteran insects commonly included in the Tachinidae

Phasmatidae the only family of phasmid insects. The characteristics of the family are those of the order

phasmid one of a pair of nematode sense organs, strongly resembling amphids (q.v.) but at the anterior end

Phasmida an order of insects containing those forms commonly called stick insects or walking sticks. They were, together with the Dictyoptera and Grylloblattodea, at one time included in the Orthoptera, a group restricted by this definition to grasshoppers, locusts and crickets. They possess eleven apparent abdominal segments but are distinguished from the Orthoptera proper by the small eyes

Phasmidea a class of Nematoda, distinguished from the Aphasmidea by the presence of phasmids and the fact that the excretory system consists of two lateral canals

polyphasy pertaining to marked differences existing synchronously between individuals of a species living in the same habitat

phe bacterial mutant indicating a requirement for phenylalanine

phe A a bacterial genetic marker indicating a requirement for phenyl alanine, mapped at 49.75 mins. for *Escherichia coli*

phe B a bacterial genetic marker indicating a requirement for phenyl alanine, mapped at 43.1 mins. for *Escherichia coli*

pheasant 1 (*see also* pheasant 2, cuckoo) any of very numerous species of galliform phasianid birds of the subfamily Phasianinae

argus pheasant *Argusianus argus*

fireback pheasant any pheasant with a vivid red back

crested fireback pheasant *Lophura ignita*

crestless fireback pheasant *Houpifer erythrophthalmus*

blood-pheasant *Ithaginis cruentus*

college-pheasant = kaleege

copper pheasant *Syrmaticus soemmeringii*

brown-eared pheasant *Crossoptilon mantchuricum*

white-eared pheasant *Crossoptilon crossoptilon*

golden pheasant *Chrysolophus pictus*

crimson-horned pheasant *Tragopan satyra*

impeyan pheasant properly *Lophophorus impeyanus* though the term is often applied to other species of Lophophorus

kaleege pheasant any of many species of the genus Gennaeus

black-headed kaleege pheasant *G. eumelanos*

koklas pheasant *Puchrasis macrolopha*

ring-neck pheasant *Phasianus colchicus*

ocellated pheasant *Rheinardia ocellata*

peacock-pheasant any of several species of the genera Polyplectron and Chalcurus, without qualification usually *P. malacensis*

pukras pheasant = kiklas pheasant

silver pheasant *Gennaeus nycthemerus*

snow pheasant any of several species of the genus Crossoptilon

wattled pheasant *Lobiophasis bulweri*

pheasant 2 (*see also* pheasant 1) any of several birds not galliform in affinity but more or less resembling pheasants

crow-pheasant any of several species of cuculid bird of the genus Centropus (*cf.* coucal)

　large crow-pheasant *C. sinensis*

　lesser crow-pheasant *C. bengalensis*

　Malaysian crow-pheasant *C. rectunguis*

native pheasant *Austral.* the menurid *Menura superbe* (= lyrebird)

north-west pheasant *Austral.* the cuculid *Centropus phasianinus* (= pheasant coucal)

reed-pheasant *Brit.* the timaliid bird *Panurus biarmicus* (= bearded tit)

sea pheasant the anatine *Anas acuta* (= pintail)

water pheasant the Asiatic jacanid *Hydrophasianus chirurgus* (= pheasant-tailed jacana)

-phell- *see* -phello

phellad a plant of stony soil

phellem cork tissue

-phello- *comb. form* confused between two *Gr.* roots and therefore meaning either "cork" or "stony ground"

phelloid a cell produced by phellogen but which does not contain suberin

　choriphelloid a suberized cell apart from suberized tissue

　pseudophelloid the cork-like tissue found in some ferns

phen *see* [phenylalanine]

ecophene a modification of a genotype through selective breeding, the selection being occasioned by environmental or ecological factors

Phengodidae a small family of coleopteran insects strongly allied to the Lampyridae and, like them luminescent. They may be distinguished by the possession of short pointed elytra which do not cover the hind wings

polyphenism the exhibition of several forms by one individual. It applies equally to larval changes, social castes in insects, or polymorphism and cyclomorphosis

-pheno- comb. form derived from -phaeno- meaning "appearance." Sometimes used as though derived from -phaner- meaning "evident"

phenome the sum total of all the phenotypic characteristics of an organism (*cf.* genome)

isophenous having identical phenotypes

[phenylalanine] a mutant (gene symbol *phen*) in linkage group I of *Neurospora crassa*. The phenotypic expression is a requirement for an aromatic amino acid

[phenylalanine-tyrosine] a mutant (gene symbol *pt*) in linkage group IV of *Neurospora crassa*. The phenotypic expression is a requirement for phenylalanine plus tyrosine

-phero- *comb. form* meaning "to bear" in the sense of "carrying" (*cf.* -phore-)

diaphery the condition of having two flowers in one calyx

pheromene a substance produced by one organism for purposes of chemocommunication with another (*cf.* hormone, from which this word is derived)

-phil-, -philo- *comb. form* meaning "loving" in the sense of being attracted to and hence used in *bot.* in the sense of "fertilized through the agency of" and in *ecol.* in the sense of "preferring to live on or among." The *subs. suffix* -phil is usually confined to leucocytes -phile being used for most other forms, though -phila pertains to associations of organisms. The *adj. suffix* -philic is synonymous with but less usual than -philous

-phil *see* -phil-

　basophil a descendant of a basophilic metamyelocyte (*q.v.*) in which the nucleus is scarcely apparent owing to the dense mass of basophilic granules

　eosinophil a myelocyte (*q.v.*) with a bi- or tri- lobed nucleus and containing eosinophilic granules

　heterophil = neutrophil

　neutrophil a myelocyte (*q.v.*) with a tri-lobed nucleus the granules of which are neither eosinophilic nor basophilic

　polychromatophil the third stage in the development of an erythrocyte

phial- *comb. form* meaning "bowl"

-phila *see* -phil-

　geophila an association of plants growing on bare earth

　rheophila an association of organisms, most frequently algae, in running water

philander = filander

Philanthinae a subfamily of sphecid hymenopterans which share with the closely allied Crabroninae the name digger-wasp. The have a distinct constriction between the first and second abdominal segments and are also variously called bee-killer wasps or beetle-wasps according to their choice of prey for provisioning their nests

-philax- *comb. form* meaning "to guard"

-phile *see* -phil-

　basophile (*cf.* basophil) a plant preferring alkaline soil

　symphile a pet of an ant or termite colony. The word guest, frequently used, does not adequately describe the relationship

Philepittidae a family of passeriform birds containing the Madagascan birds commonly called asities. They are distinguished by having a bluish or greenish caruncle on the orbit

Philesiaceae a family of liliiflorous monocotyledonous angiosperms containing forms usually included within the Liliaceae

-philic *comb. term* synonymous with -philous, the form preferred in this work except for the terms below

　basophilic said of a substance which stains readily with basic dyes

　cyanophilic readily taking a blue stain or color

　thermophilic thriving in high temporatures

Philodendroideae a subfamily of the Araceae

Philopteridae a very large family of mallophagan insects and commonly meant when the word birdlouse is used

Philopotamidae a family of trichopteran insects commonly called finger-net caddisflies. The larvae construct finger-shaped tubular nets

-philous *see* -phil-

　acarophilous attracted to or attractive to mites

　acrophilous pertaining to or dwelling in lofty peaks

　actophilous pertaining to or growing on a coast

　aelophilous said of plants distributed by wind

　agrophilous pertaining to or dwelling in cultivated fields of grain

aigialophilous- (*cf.* psammophilous, ammophilous) beach-dwelling
aithalophilous dwelling in evergreen thickets
alsophilous dwelling in groves
amathophilous dwelling on sandy soils
ammophilous dwelling in sand (*cf.* psammiophileous)
ammochthophilous exhibiting a preference for sandbanks
androphilous loving or dwelling in the neighborhood of man
anemophilous applied to flowers which are fertilized by pollen conveyed by the wind
anthophilous living on or frequenting flowers
anthropophilous = androphilous
autophilous = autogamous
barophilous tolerant of weight, said particularly of bacteria growing at great depths in the ocean
bathyphilous dwelling in lowlands or in the depths of the ocean
biophilous dwelling on living things, though usually confined to plant parasites
chasmophilous loving or dwelling in chinks in rocks
cheradophilous dwelling on sandbars
chimonophilous preferring the winter season
    hemichimonophilous the condition of plants which give rise to aerial growth while freezing temperatures still prevail
chledophilous dwelling in rubbish heaps
chalicodophilous dwelling on gravel slides
coleopterophilous beetle loving
    necrocoleopterophilous the state of being pollinated by carrion beetles
coniophilous dust-loving, particularly of those lichens which appear to benefit from a coat of dust
coprophilous inhabiting dung
    ornithocoprophilous living on the excreta of birds
coryphophilous dwelling in high mountains
cremnophilous cliff-dwelling
crenophilous spring loving
crymophilous dwelling in polar regions
dendrophilous dwelling in orchards
drimyphilous = halophilous
drosophilous dew loving or being fertilized through the agency of dew
entomophilous attracted to insects, attracted by insects, or (flowers) attractive to insects which fertilize them
    dientomophilous the condition of a plant having two kinds of flowers each adapted to a specific, but different insect for purposes of fertilization
    necroentomophilous = necrocoleopterophilous
eremophilous dwelling in deserts
eurotophilous dwelling in leaf mould
geophilous dwelling on land, though the term is sometimes applied to plants bearing underground fruits, or to animals more properly referred to as burrowing
    oxygeophilous humus-dwelling (= oxylophilous)
gypsophilous chalk-loving
halophilous salt-loving
heliophilous sun-loving
helophilous marsh-loving
helogadophilous dwelling in swampy woodlands
hemerophilous agreeable to cultivation
histophilous parasitic
hydrophilous water-dwelling
hygrophilous moisture-loving
hylophilous forest-dwelling
    helohylophilous dwelling in wet forests
    xerohylophilous dwelling in dry forests

hylodophilous living in dry woods
limnodophilous dwelling in marshes
limnophilous lake-dwelling
lithophilous rock-dwelling
lochmophilous dwelling in thickets
    helolochmophilous dwelling in meadow thickets
lochmodophilous dwelling in dry thickets
lophophilous hill-dwelling
melangeophilous dwelling in black loam
mellitophilous attractive to bumble bees
mesophilous said of organisms loving "mediumness" particularly of bacteria thriving at human body temperature and of organisms preferring moderate mositure (*cf.* mesothermophilous)
monophilous dwelling in pastures
myiophilous said of a plant fertilized by dipterous flies
myrmecophilous ant-loving or ant-attracting
namatophilous dwelling in streams
oceanophilous living in oceans
ochothophilous living on banks
    mesochthonophilous dwelling in midlands
    petrochthophilous living on a rock bank
ombrophilous either rain-loving or shade-loving
opophilous sap-loving
orgadophilous dwelling in open woodlands
ornithophilous bird-loving or being pollinated by the action of birds
orophilous dwelling in subalpine regions
oxylophilous humus-loving (= oxygeophilous)
pagophilous dwelling in foothills
pediophilous living in uplands
pelophilous clay-loving
pelochthophilous dwelling on mud banks
petrophilous rock-dwelling
petrodophilous dwelling among stones
phellophilous dwelling in stony fields
photophilous light-loving
pontophilous dwelling in deep waters
poophilous dwelling in meadows
    xeropoophilous loving or dwelling on heaths
potamophilous dwelling in rivers
psamathophilous living on the sea shore
psammophilous dwelling in sand (*cf.* ammophilous, aigialophilous)
psilophilous inhabiting pariries
ptenothalophilous dwelling in deciduous thickets
ptenophyllophilous living in deciduous forests
pyrophilous growing on soil which has been burnt over
rheophilous said of an organism which prefers, or is confined to, running water
rhizophilous growing on roots
rhoophilous dwelling in streams
rhyacophilous dwelling in torrents
saprophilous a plant preferring humus
sciophilous shade loving
scotophilous dwelling in darkness. Used also in the sense of geophilous
siderophilous iron-loving
skotophilous = scotophilous
sphagnophilous dwelling on or among sphagnum
spiladophilous living in clay soils
stasophilous dwelling in stagnant water
sterophilous dwelling on moor lands
syrtidophilous dwelling on dry sandbars
taphrophilous dwelling in ditches
telmatophilous dwelling in wet meadows
termitophilous dwelling with termites
thalassophilous sea-loving or dwelling in the sea
thermophilous heat-loving

eurithermophilous capable of living in a wide range of temperatures

mesothermophilous dwelling in temperate zones (*cf.* mesophilous)

thinophilous dwelling on sand dunes

tiphophilous pond-loving

hydrotribophilous dwelling in "badlands"

tropophilous dwelling in a climate that alternates between sunny drought and torrential rain

xerophilous loving dry situations

helioxerophilous the condition of desert organism which is adapted both to strong sunlight and to drought

subxerophilous preferring a fairly dry situation

xylophilous wood-loving or wood-dwelling in the sense of a parastic fungus or wood-boring organism

zoidophilous said of a plant pollinated through the agency of an animal (*cf.* Zoidogamae)

zoophilous = zooidiophilous

microzoophilous pollinated by small animals

protozoophilous said of water plants that are fertilized by Protozoa or other small animals

symphily the condition of being a symphile (*cf.* metochy, synechthry)

Philydraceae a small family of farinose monocotyledonous angiosperms

-phleb- *comb. form* meaning vein

phlebedesis the process of the suppression of the true coelom by a haemocoel

Phleboptera = Hymenoptera

-phloe- *comb. form* meaning "bark"

phloem the main food-conducting tissue of vascular plants

inactive phloem that part of the phloem in which the sieve elements no longer function

abaxial phloem = external phloem

adaxial phloem = internal phloem

epiphloem = phloem

external phloem that part of the phloem which is on the outside of the xylem

included phloem = interxylary phloem

internal phloem that part of the phloem which is on the inside of the xylem

loptophloem rudimentary phloem

metaphloem phloem that matures after longitudinal growth of the plant has taken place

protophloem phloem elements differentiated before a plant organ completes elongation

interxylary phloem phloem strands included in the secondary xylem

intraxylary phloem = internal phloem

periphloematic pertaining to concentric bundles in ferns

hypophloeodal beneath the bark. Applied particularly to lichens but sometimes also to insects occupying that position

phloeodeous bark-like

Phloeomynae a subfamily of murid rodents containing the Philippine and E. Indian forms known as cloud-rats. They differ from all other murids except the Hydromnae in having three molars in each jaw but differ from this subfamily in their adaption to a terrestrial habitat

Phloeothripidae the only family of tubuliferous thysanopteran insects

-phloic pertaining to bark

amphiphloic said of a plant stem in which the phloem exists on both sides of the xylem

ectophloic said of a plant stem in which the phloem is external to the xylem. The term is also used in the sense of pertaining to the surface of bark

-pho- *comb. form* meaning "a light"

pho a bacterial genetic marker having a character affecting the activity of alkaline phosphatase mapped at 10.25 mins. for *Escherichia coli*

-phob- comb. form meaning "to fear" or "flee from." It is sometimes combined with tropism to indicate negation of the operative word. Thus phobophototropism is negative phototropism

-phobic *adj. suffix* meaning "shunning" or "fleeing from." The meaning is opposite to -philous under which most compounds will be found

anemophobic an organism which shuns wind currents

halophobic salt-shunning

hemerophobic difficult to cultivate

photophobic light avoiding

Phocidae that family of pinnipede carnivorous mammals which contains the true seals (*cf.* Otaridae). Since otarids (properly sea-lions) are frequently called seals the Phocidae are sometimes distinguished as the true seals

Phocinae a subfamily of Phocidae clearly distinguished by having five fingers and toes on each limb

phoebe *U.S.* any of several species of tyrannid bird of the genus Sayornis. Without qualification, usually *S. phoebe*

black phoebe *S. nigricans*

Say's phoebe *S. sayi*

phoeniceous scarlet

Phoenicopteridae that family of ciconiiform birds which contains the flamingoes. They are distinguished by their thick lamellate bill which is sharply bent down in the midpoint giving a "Roman nose" effect

Phoeniculidae that family of coraciiform birds commonly called wood-hoopoes. They are readily distinguished by the long, slender, laterally compressed sickle-shaped beak

phoenix *Brit.* either of two geometrid lepidopteran insects *Eustroma silaceata* and *Lygris prunata*. Without qualification, the latter is usually intended

phoeodium a granule or granular mass in Protozoa consisting of the residual detritus of digestion

-phol- *comb. form* meaning a "horny scale"

Pholadidae a family of rock-boring pelecypod mollusks commonly called piddocks, or angelwings. They are differentiated from the other wood-boring pelecypods (*cf.* Saxicavidae, Gastrochaenidae, and Teredidae) by the thin, elongate, brittle shells

Pholadomyidae that family of pelecypod mollusks which contains the forms commonly called thinwings. They are distinguished by the strongly radiating ribs on the equal-sized gaping valves

Pholidae a family of elongate blennioid acanthopterygian fishes with a completely spiny dorsal fin, an incomplete lateral line, and rudimentary pelvic fins. Commonly called gunnels

pholidosis the arrangement of scales on an organism

Pholidota an order of eutherian mammals containing the pangolins and scaly anteaters, They are distinguished by the peculiar scales—actually plates of fused hair—which cover them

-phon- *comb. form* confused between two *Gr.* roots and therefore meaning either "sound" or "murder"

biophor = cell (*but see* biophore, under -phore)

diaphorase see NADPH₂ diaphorase

NADPH diaphorase an enzyme which catalyzes the dehydrogenation of NADPH to NADP⁺

mesophorbium evergreen meadows

-phore- *comb. form* confused between four *Gr.* roots and therefore meaning "bearer" (and, by extension, the vessel born), "movement," "thief" or "detector." The first of these is the most usual

allophore a red pigment-containing cell in the skin of a fish or amphibian

androphore the peduncle of an androecium or the central core of a column of stamens

gynandrophore a column bearing both stamens and pistils

antherophore an extension of the axis bearing anthers

antheridiophore a gametophore bearing antheridia only

anthophore a stalk between calyx and corolla

archegoniophore a gametophore bearing only archigonia

ascophore the supporting structure of the ascus within an ascocarp

aurophore a constricted off portion, lying among the swimming bells, of the float of some siphonophorans

biophore when genes were called determinants each was supposed to be composed of many biophores

blastophore that part of a spermatid that remains after the formation of a spermatozoan

cadophore a dorsal outgrowth from the posterior end of doliolid urochordates from which new individuals are budded off

calathidiphore the stalk of a capitulum

carpophore the stalk of a sporocarp

cephalophore that part which bears a head, as the stipe of basidiomycetes

chlorophore = chloroplast

chondrophore that portion of the shell of a pelecypod mollusk which supports, either as a depression or a prominence, the internal hinge of cartilage

chromatophore a cell containing pigment

chromophore = chromatophore

cladophore that portion of the stem from which an inflorescence arises

clathrophore the glandular cells at the base of the leaf of the pitcher plant

cnidophore a contractile stalk with a hollow tip in the tentacle of medusae. The tip is filled with nematocysts

collophore a protrusible membranous holdfast organ on the first abdominal segment of collembolan insects

cystophore the stalk bearing the fruiting body of cysts of polyangiancean schizomycetes

colletocystophores a rhopalioid of Stauromedusae in the form of a horseshoe-shaped adhesive organ

embryophore eggshell. The term is usually confined to platyhelminth eggs

gametophore that portion of a lower plant, particularly an alga, which produces gametes

androgametophore the male form of a lower plant having alternation of generations

gonophore any structure bearing reproductive cells including the reproductive polyp of a siphonophoran coelenterate, the reproductive zoid of a hydrozoan coelenterate, and a plant stalk on which a reproductive organ is borne

cryptomedusoid gonophore a gonophore which resembles a medusa, but lacks a velum and radial canals

eumedusoid gonophore a gonophore which resembles a medusa in every way except the production of tentacles

heteromedusoid gonophore a type of cryptomedusoid gonophore in which traces of the entocodon are present

styloid gonophore a type of cryptomedusoid gonophore consisting of a simple protuberance

guanophore a cell containing guanine though often applied to any cell containing a white pigment

gynophore the stipe of a pistil

hymenophore the stalk of the hymenium

conidiophore a aerial hypha that bears conidia

gonidiophore a gonidium on a sporophore

lipophore = xanthophore

lophophore a horseshoe-shaped crown of ciliated tentacles at the anterior end of many invertebrates (*cf.* Lophophoria)

plectolophous lophophore (Brachiopoda) a zygolophous type with a median coiled arm between two uncoiled lateral arms

ptycholophous lophophore (Brachiopoda) a simple quadrilobulate form

scizolophous lophophore (Brachiopoda) a weakly bilobulate form

spirolophous lophophore (Brachiopoda) a zygolophous type with the arms coiled into spirals

trocholophous lophophore (Brachiopoda) a simple disc fringed with tentacles

zygolophous lophophore (Brachiopoda) one having a pair of lateral arms

mastigophore a type of heteroneme rhabdoid nematocyst in which the tube continues beyond the butt

amastigophore a type of rhabdoid heteroneme nematocyst, the tube of which does not continue beyond the butt

melanophore a cell containing melanin

nectophore the gelatinous swimming bells of a siphonophoran coelenterate

nematophore a small dactylozoid, containing numerous nematocysts, found on some hydrozoan coelenterates

odontophore the supporting structure of a radula

phaeophore the chromoplast of a phaeophycean alga

photophore a cell or organ used for the production of light by a living organism

phyllophore the summit, which alone bears leaves, of a stem as that of a palm

pneumatophore *zool.* the floating chamber of a siphonophoran hydrozoan *bot.* air-filled tracheids and air spaces in pleuston

polyphore a torus of many pistils as in a strawberry

pycnidiophore a compound sporophore bearing pycnidia

rhinophore a presumed chemoreceptor on the head of some opisthobranch Mollusca

rhizophore the root of Selaginella

scolopophore a sense organ perceiving continuous vibration as distinct from a tangoreceptor which perceives individual touches. A tuned scolopophore is an organ of hearing (*cf.* phonoreceptor)

sematophore = spermatophore

spermatophore *zool.* a container of spermatozoa external to the body. *bot. either* a structure bearing a spermatium *or* the gynophore of umbelliferous plants

spermophore any portion of a plant bearing either seed-producing structures or gamete producing structures

sporangiophore an aerial hypha of fungus which produces one or more sporangia

sporophore used in the sense of sporophyte in Pteridophyta and Bryophytae, but is also applied to any part of a plant which bears seeds or spores, though more usually used for the later

compound sporophore one formed from the branches of several different hyphae twisted together

microsporophore the organ producing microspores in either meaning of that word

thecaphore the stipe of a carpel

trichophore the structure from which the annelid chaeta is produced (*cf.* trichogen, trichopore)

trochophore *see* larva

xanthophore a cell containing red or yellow pigment

phoresy the condition of being carried about by another. Some pseudoscopions, for example, are phoretic, as distinct from parasitic, on houseflies

Phoridae a family of aschizan cycloraphous brachyceran dipteran insects commonly called hump-backed flies in virtue of the greatly convex thoracic dorsum of most species. The laterally flattened hind tibia are also typical

epophoron an anterior remnant of the mesonephros in adult amniotes

paroophoron a remnant of the embryonic mesonephros remaining in the ovarian region of some mammals

-phorous *adj. suffix* from -phore-. Compounds derived directly from the *subs.* -phore (see above) are not listed

adenophorous bearing glands

anthophorous carrying flowers

eriophorous woolly

hemerodiaphorous changed by cultivation

hormophorous carrying a chain or a necklace

isophorous capable of transformation

mycophorous said of one fungus parasitic on another

parachromatophorous having pigment in the cell wall

rhizophorous producing roots

scyphiphorous bearing cups

thelephorous having the appearance of being covered with nipples

-phos- *comb. form* identical with -phot-

phosphatase group name for enzymes that hydrolyze phosphate ester linkages

acid phosphatase catalyzes the hydrolysis of an orthophosphoric monoester in an acid environment to yield an alcohol and phosphoric acid

acylphosphatase an enzyme that catalyzes the hydrolysis of acylphosphate to acids and orthophosphate

alkaline phosphatase hydrolyzes an orthophosphoric monoester in an alkaline environment to yield an alcohol and phosphoric acid

nucleoside diphosphatase an enzyme that catalyzes the hydrolysis of nucleoside diphosphate to the nucleotide and orthophosphate

glucose 1-phosphatase catalyzes the hydrolysis of D-glucose 1-phosphate to yield D-glucose and orthophosphate

glucose-6-phosphatase catalyzes the hydrolysis of D-glucose 6-phosphate to D-glucose and orthophosphate

diphosphoglycerate phosphatase catalyzes the hydrolysis of D-2,3-diphosphoglycerate to yield D-3-phosphoglycerate and orthophosphate

methylthiophosphoglycerate phosphatase catalyzes the hydrolysis of methylthio-D-3 phosphoglycerate to yield methylthio-D-glycerate and orthophosphate

histidinolphosphatase catalyzes the hydrolysis of histidinol phosphate to yield histidinol and orthophosphate

hexosediphosphatase catalyzes the hydrolysis of D-

fructose 1,6-diphosphate to yield D-fructose 6-phosphate and orthophosphate

polymetaphosphatase general term for enzymes that catalyze the hydrolysis of polyphosphates into pentaphosphates

trimetaphosphatase hydrolyzes inorganic trimetaphosphate to yield triphosphate

phosphatidate phosphatase catalyzes the hydrolysis of an L-α-phosphatidate to yield orthophosphate and a L-1,2-diglyceride or a D-2,3-diglyceride

phosphoserine phosphatase catalyzes the hydrolysis of phosphoserine to yield serine and orthophosphate

phosphorylase phosphatase catalyzes the hydrolysis of phosphorylase *a* to phosphorylase *b* and phosphoric acid

phosphoprotein phosphatase catalyzes the hydrolysis of phosphoproteins to yield proteins and phosphoric acid

pyrophosphatase hydrolyzes inorganic pyrophosphate to yield orthophosphate

nucleotide pyrophosphatase an enzyme which catalyzes the hydrolysis of dinucleotides to two mononucleotides

trehalosephosphatase catalyzes the hydrolysis of trehalose 6-phosphatase to yield trehalose and orthophosphate

phosphorescent said of an organism which emits light

phosphorylase a group of transferase enzymes catalyzing reactions involving orthophosphate

α -glucan phosphorylase catalyzes the production of α-D-glucose phosphate from polysaccharides and orthophosphate

maltose phosphorylase catalyzes the production of β-D-glucose 1-phosphate and D-glucose from maltose and orthophosphate

polynucleotide phosphorylase = polynucleotide nucleotidyltransferase

purine nucleoside phosphorylase catalyzes the production of α-D-ribose 1-phosphate and purine from purine nucleoside and orthophosphate

AMP pyrophosphorylase = adenine phosphoribosyltransferase

FAD pyrophosphorylase = FMN adenyl transferase

IMP pyrophosphorylase = hypoxanthine phosphoribosyltransferase

NAD pyrophosphorylase = NMN adenylyltransferase

NMN pyrophosphorylase = nicotinamide phosphoribosyltransferase

UMP pyrophosphorylase = uracil phosphoribosyltransferase

thymidine phosphorylase catalyzes the production of thymine and 2-deoxy-D-ribose 1-phosphate from thymidine and orthophosphate

uridine phosphorylase catalyzes the production of uracil and D-ribose 1-phosphate from uridine and orthophosphate

-phot- *comb. form* meaning "light"

photeolic pertaining to the "sleep" of plants

photic pertaining to light

aphotic capable of growing in the absence of light

diphotic having two surfaces unequally lighted

euphotic (*see also* zone) literally, well-lighted but applied particularly to plants, such as submerged hydrophytes, which normally are poorly lighted

stenophotic organisms restricted to a narrow range of light intensities

-photo- *see* -phot-

-phragm- *comb. form* meaning "an enclosure"

cephalophragm a v-shaped apodeme which divides the head of some orthopteran insects into two chambers

diaphragm *either* the muscular partition separating the cavity of the thorax from that of the abdomen in mammals *or* a constriction terminating the anterior, in-turned, body wall in some Ectoprocta **nodal diaphragm** transverse sheet across the node in a plant stem

endophragm a partition formed by the thoracic apodeme in Crustacea (*cf.* endothorax)

epiphragm literally, an outer fence; used in *bot.* for the membranous cap of moss thecae and similar structures and in *zool.* for the seasonal, usually dry weather, closure of the aperture in the shell of a terrestrial gastropod mollusk

mesophragm a portion of the dorsal thoracic apodeme of insects derived from the post-scutellum or an analogous structure in other arthropods

periphragm the pericycle of the stem

telophragm = Z line

-phragma = -phragm

phragmigerous divided by partitions

Phragmophora a division of decapod cephalopod mollusks distinguished by the presence of a calcareous shell which is spiral

phragmosis to be in the condition of acting as a barricade, as when an organism uses its body to block the entrance to a burrow

-phret- *comb. form* meaning a "tank" or "cistern"

phretad a tank plant

Phryganeidae a family of large trichopteran insects usually with grey- and brown-mottled wings. The larval cases, which are commonly found in marshes and lakes, are long and slender and composed of narrow strips of vegetable material joined together in a helix

Phrymaceae a family erected to contain the genus Phryma by those who do not hold it to belong in the Verbenaceae

Phrymineae a surborder of tubifloral dicotyledonous angiosperms containing the single family Phrymaceae

Phrynomeridae a monogeneric family of African anuran Amphibia which represents an arboreal offshoot of the Microhylidae

Phryneidae = Rhyphidae

Phthiraptera an order of insects at one time containing all those forms known as lice. They are now divided between the Mallophaga and Anoplura

Phthiriidae a family of anopluran insects, commonly called crablice, parasitic on man

-phthis- *comb. form* meaning "to waste away"

phyad = plant

-phyadic pertaining to a plant

heterophyadic having various types of stem, particularly when some are fertile and some barren

homophyadic applied to a plant, having both fertile and barren stems of identical external appearance

Phycitidae a large family of lepidopteran insects, frequently fused with the Pyralidae. They are distinguished by long, narrow forewings and broad hindwings

-phyco- *comb. form* confused between two *Gr.* roots and therefore meaning either "alga" or "painted"

Phycobryophytes = Characeae

Phycodromidae = Coelopidae

phycoma an alga taken as a whole

Phycomycetae a phylum of Eumycophyta distinguished by the fact that the number of spores formed in any one sporangium is indefinite

Phycophyta = Characeae

-phyko- *comb. form* meaning "alga" or "seaweed" almost invariably transliterated -phyco-

-phyl- *comb. form* meaning "race" or "tribe"

-phylact- *comb. form* meaning "guard" or "guardian"

Phylactolaemata a class or the Ectoprocta containing those forms commonly referred to as freshwater Bryozoa. The are distinguished by the presence of a large horseshoe-shaped lophophore

anaphylaxis excessive susceptibility

-phyletic *adj. term* meaning "pertaining to race" usually in the sense of racial origins

monophyletic descending from a common stock or individual

pleophyletic = polyphyletic

polyphyletic descending from many stocks or indivisuals

-phyll- *comb. form* meaning "leaf." Many other words are given under -phyllous below

carpophyll = carpel

cataphyll a leaf modified as a scale on a bud or underground stem

chlorophyll (*see also* body, corpuscle, granule, vessicle) the photosynthetic pigment of plants. It is a magnesium centered porphyrin containing, as its most distinctive constituents, a hydrophilic 5-membered carbocyclic ring and a lipophilic phytol tail

coleophyll the first true leaf of a monocotyledon, more commonly today called a coleoptile

diplophyll a leaf having palisade tissue on both the upper and lower surfaces

paradiphyll a double leaf deriving from the dichotomous growth of a single blade

epiphyll the blade and petiole of a leaf as distinct from the stipules (*but see* epiphyllous)

hydrophyll the submerged leaf of a hydrophyte

hypophyll that portion of the leaf from which stipules develop

hypsophyll a leaf modified as a floral bract

leptophyll a minute leaf

macrophyll a long leaf

megaphyll a large leaf

megalophyll an enormous leaf

mesophyll all that portion, except the veins, of an angiosperm leaf which is enclosed within the epidermis. In gymosperm leaves, the tissue lying between the central bundle and the epidermis

metaphyll a mature, as compared to a juvenile, leaf

microphyll a small leaf usually with a single unbranched vein

oligophyll a bract

paraphyll a secondary leaflet of mosses

prophyll the first cataphyll on a lateral branch

protophyll a leaf on a protocorm

rhiophyll a basic bract of Hepaticae

sciophyll a leaf adapted to shady condition

spongophyll a shade leaf

sporophyll a leaf, or leaf-like structure, bearing spores

macrosporophyll = carpel

megasporophyll the scales of the female cone of the cycad or, in angiosperms, the carpel

microsporophyll the scales of the male cone of the cycad, a leaf-like microsporangium or, in angiosperms, the stamen

sclerophyll a leaf with an unusually thick layer of sclerenchyma

trophophyll a leaf which does not bear reproductive structures but functions photosynthetically

phyllade = cataphyll

chlorophyllase an enzyme that catalyzes the hydrolysis of chlorophyll to phytol and chlorophyllide, and also the transfer of chlorophyllide

-phyllic = -phyllous

phyllidium the gametophytic equivalent of a leaf

Phyllobothrioidea = Tetraphyllidea

Phyllocarida the only superclass of leptostracan malacostracan Crustacea. This superorder contains the single order Nebaliacea. The characters of the order are those of the superorder

Phyllocladoideae a superfamily of Podocarpaceae

Phyllocnistidae a family of lepidopteran insects now usually included in the Gracilariidae

phyllocolly the production of a leaf from a leaf

phyllode *bot.* a flattened petiole having the appearance of a leaf blade (*cf.* cladode) *zool.* the oral end of the ambulacra surrounding the peristome in echinoid echinoderms

synphyllodium a cone scale

autophyllogeny the production of one leaf by the blade of another

phylloid leaf-like. Applied particularly to leaf-like appendages on the stems of algae

phyllom the sum total of all the leaves of a plant

phyllome the totality of leaves in a bud

postphyllome = leaf

Phyllomyzidae a small group of obscure myodarian cycloraphous dipteran insects

phyllopidium a leaf regarded as an axis

Phyllopoda a previously recognized border of branchiopod Crustacea distinguished from the Cladocera by possessing 11 to more than 60 trunk appendages. The order included the presently recognized Anostraca, Notostraca, and Conchostraca

Phyllornithidae a family of passeriform birds usually included with the Irenidae

Phyllostomatidae a very large family of New World microchiropteran mammals distinguished by the presence of a well-developed nose-leaf and the possession of three phalanges in the middle finger. They are commonly known as leaf-nosed bats though the false vampire bat *Vampyrus spectrum* also belongs to this family (*cf.* Desmodontidae)

-phyllous *adj. suffix* from -phyll (*q.v.*). Forms deriving directly from -phyll words given above are not listed

aphyllous lacking leaves

adenophyllous having glandular leaves

aiophyllous = evergreen

allagophyllous bearing alternate leaves

artiphyllous said of leaf-bearing nodes

brachyphyllous with short leaves

echlorophyllous lacking chlorophyll

choristophyllous having the leaves separated from each other

codiophyllous having fleecy leaves

diphyllous having two leaves

dialyphyllous polyphyllous

eleutherophyllous having the leaves separate, not fused

endophyllous within a leaf

epiphyllous growing on leaves (*but see* epiphyll)

eprophyllous lacking bracteoles

eriophyllous woolly-leaved

exophyllous lacking a sheath for the cotyledon (= exoptile)

gamophyllous having leaves united by their edges

gymnophyllous having leaves borne on stems lacking a cortex

heterophyllous having different kinds of leaves on the same plant

hypophyllous beneath the leaf

inophyllous having leaves with thread-like veins

isophyllous *either* straight-leaved *or* having similar leaves

anisophyllous having pairs of leaves of which one differs in size or shape from the other

lepidophyllous having scaly leaves

leptophyllous slender leaved

malacophyllous with fleshy leaves

melanophyllous having very dark leaves

pachyphyllous thick-leaved

platyphyllous broad leaved

polyphyllous having many, frequently with the connotation more than the usual number of, leaves

rhizophyllous having rooted leaves

sclerophyllous having hard leaves

symphyllous = gamophyllous

therophyllous bearing leaves only in summer (*i.e.* deciduous)

trichophyllous with hair-like leaves

xiphophyllous having sword-shaped leaves

Phylloxera a genus of chermid homopteran insects well known for the devastation caused by *P. vitifoliae* on the European grapevine

grape phylloxera *U.S. P. vitifoliae*

pecan leaf phylloxera *U.S. P. notabilis*

pecan phylloxera *U.S. P. devastatrix*

Phylloxeridae = Chermidae

phyllula = leaf scar

-phylly *subs. suffix* indicating a condition pertaining to a leaf. Only those compounds not directly derived from words ending in -phyll or -phyllous are here given

antiheterophylly = antidimorphism

ovariophylly the metamorphosis of a carpel into a leaf

-phylo- *see* -phyl-

phylum a taxon representing a main division of a kingdom

subphylum a taxon intermediate between phylum and class

Phymatidae a family of hemipteran insects distinguished by their well-developed, broad, raptorial front legs. They are commonly called ambush bugs

phymatodeus warty

-phyom- *comb. form* meaning "to spring from"

-phys- *comb. form* meaning "a growth" or "outgrowth"

physema the frond of an underwater alga, sometimes also applied to a branch of Chara

Physeteridae that family of odontocetan whales which contains the sperm whales and beaked whales

Culver's physic the scrophulariaceous herb *Veronicastrum virginicum* (*cf.* culver's root)

Indian physic *U.S.* any of several species of rosaceous herb of the genus Gillenia

Physidae a family of pulmonate gastropods with a sinistral spiral shell, commonly called pond snails

-physis *subs. suffix* from -phys- (*q.v.*)

ascophysis the hypha which forms the precursor of the ascus

apophysis literally "an outgrowth away from." Specifically an outgrowth of a bone from a vertebra, outgrowths or ingrowths from the exoskeleton of an insect or other arthropod, a basal swelling beneath the capsule of some mosses, and the basal swelling of the cone-scale of some pine

anterior apophysis long slender extensions of the apodeme of the ninth abdominal segment in female Lepidoptera

gonapophysis the sum total of insect genitalia

parapophysis the vertebral prominence with which the lower branch of a two-headed rib articulates

zygapophysis processes which articulate vertebrae

epiphysis literally "an outgrowth above" and specifically applied to the ossicles which join the aboral surface of the pyramids of Aristotle's lantern, a projection from the tibia of the front leg in some Lepidoptera, protuberances around the hilum of a seed, the pineal body and a secondary ossification at the ends of long bones

hypophysis literally "an outgrowth beneath." Specifically applied to the initial of the primary root and rootcap, but, without qualification, almost always refers to an endocrine gland lying between the floor of the brain and the roof of the mouth in vertebrates

adenohypophysis that portion of the hypophysis which develops from the oral epithelium, originally as Rathke's pouch

neurohypophysis that portion of the hypophysis which is derived from the floor of the diencephalon as the infundibulum

metaphysis a segment of a long bone occupied by an epiphyseal disk and by newly formed bone on the diaphyseal side of the disk

nucleophysis septate projection which is a forerunner of an ascophysis

paraphysis literally "an outgrowth alongside" but specifically used to refer to the roof portion of the developing forebrain which lies between cerebral hemispheres, and to sterile filaments in fruiting bodies of cryptogams

periphysis literally "a growth around" but applied in *bot.* to structures similar to paraphyses, but arising at places destitute of asci

symphysis literally "a growth together." Generally refers to a joint in which a bone capped with cartilage is held together with dense fibrous connective tissue and specifically for an amphiarthrosis between the rami of the lower jaw or other bilaterally symmetrical bones

-physo- *comb. term* of confused origin properly meaning a "bubble" or "bladder" but often used in the sense of "swollen"

-phyt- *comb. form* meaning "plant"

hysterophytal fungoid

phytase an enzyme catalyzing the hydrolysis of *myo*-inositol hexaphosphate to yield *myo*-inositol and orthophosphate

-phyte *subs. suffix* meaning "a plant"

achascophyte a plant bearing indehiscent fruit

achyrophyte a plant, the flowers of which are sheathed by a glume

aerophyte a plant living in air without direct contact with either soil or water

aigialophyte a beach plant

aithalophyte a plant of evergreen thickets

aletophyte a plant of the wayside

amphiphyte an amphibious plant (*cf.*. amphinereid)

anchophyte a plant of canyons

androphyte a male plant

anemophyte either a plant growing in windy places or one fertilized by wind-borne pollen

anisophyte an *obs.* term for moss

anthropophyte a plant introduced by cultivation, but not, as in the case of an anthropochore or an ergasiophyte, introduced intentionally

antiphyte that generation of an alternation of generations in which the sex cells are produced

apophyte a native weed which appears after the cultivation of the soil. The term has also occassionally been used for "lichen"

archaeophyte a plant introduced to cultivation before recorded history

arizophyte a plant lacking true roots, *i.e.* bryophytes and thallophytes

aulophyte an epiphyte living in a cavity rather than on the surface

autophyte a plant that does not require dead organic matter for its nourishment

hemiautophyte a parasitic plant, the tissues of which are capable of synthesizing chlorophyll

benthophyte a plant living under water and rooted to the bottom

biophyte a predatory or parasitic plant

carnivorophyte literally, a flesh-eating plant, though a better description is insectivorous plant

carpophyte = phanerogam

syncaryophyte = sporophyte

chalicodophyte a plant of gravel slides

chamaephyte a plant the buds of which rest on or are slightly above the ground

chasmochomophyte a plant living in a rock crevice

cheradophyte a plant of sandbars

chersophyte a plant dwelling in dry waste land

chionophyte a plant dwelling in snow

chledophyte a plant of rubbish heaps

chomapophyte a plant growing on ledges or in the fissures of precipices

chledophyte a plant living on rubbish heaps

coprophyte a plant growing in fecal matter either within or without the animal

cormophyte a plant that produces a stem and a root, as distinct from a thallophyte

coryphophyte a plant of mountain summits

crymophyte a plant of the "polar barrens"

cryophyte a cold-loving plant

cryptophyte *either* a cryptogam *or* any plant of which part is concealed

hemicryptophyte a perennial plant, the buds of which just clear the surface of the ground

protohemicryptophyte a plant which does not bear leaves close to the soil, but only towards the middle tip of the stem; there may be scales substituting the leaves at the base

hydrocryptophyte a plant of which the vegetative parts are always submerged but the flowers may rise above water (*cf.* emophyte)

cteinophyte a parasitic fungus which kills by the exudation of a toxin

cumaphyte a surf plant

cymaphyte = cumaphyte

dermatophyte a plant growing on skin (*e.g.* a parasitic fungus)

diodophyte a vascular plant

dissophyte a plant with xerophytic adaptations above ground and mesophytic adaptations below ground

distrophophyte a plant of firm, moist soil

drymyphyte = halophyte

emophyte a plant all parts of which are completely submerged throughout its whole life history (*cf.* hydrocryptophyte)

endophyte a plant which grows within the interior of another

    holendophyte a parasitic plant which passes its entire life within the host

epiphyte a plant which grows on other plants. There is no connotation of parasitism

    cistern epiphyte an epiphyte lacking roots and which holds a reserve of water between the bases of leaves

    hemiepiphyte *either* a plant which rises from subterranean roots but subsequently produces aerial roots *or* a plant which first produces aerial, and subsequently subterranean, roots

    nest epiphyte a tangled epiphyte which ultimately accumulates humus in the tangle

    pseudoepiphyte a plant which grows from the ground and then, through the withering of its lower parts, becomes an epiphyte

    tank epiphyte an epiphyte which is anchored by its roots to another plant from which it does not draw water and thus functions as a rooted cistern epiphyte

eremophyte a desert plant

ergasiophyte a plant intentionally introduced by man (*cf.* anthrophyte)

escatophyte a plant in a climax

eurotophyte a plant of leaf mould

gametophyte in those plants showing alternation of generations, that generation which reproduces sexually through the production of gametes

female gametophyte (Angiosperm) the embryo sac

male gametophyte (Angiosperm) the germinated polled grain

    megagametophyte that generation in the life history of a plant which produces the megagamete

    microgametophyte that generation in the life history of a plant which produces the microgamete

gamophyte = gametophyte

geophyte a perennial plant, propagated by underground shoots

    eugeophyte a plant having a resting period during which it retains only hypogean parts

    heterogeophyte a parasitic cryptogam

    mat geophyte a plant that does not spread

    melangeophyte a plant of black loam

    rhizome geophyte a plant spreading by underground shoots

    oxygeophyte a humus plant

    saprogeophyte a saprophytic flowering plant

    xerogeophyte a plant which enters a resting stage during a dry period

gynophyte a female plant

gypsophyte a plant living on chalk

halophyte a plant growing in a salty area

heliophyte a sun-loving plant

helophyte a marsh plant

hemerophyte = anthropophyte

heterophyte a term of such diverse use as to be almost meaningless. It has been applied *inter alia* to heterothallic plants, to dioecious plants, to species of plants having a wide range of habitat, to parasitic plants devoid of chlorophyll, to plants bearing leaves and flowers on different stems and to many other diverse forms

histophyte a parastic plant

holophyte a plant which derives its nourishment solely from its own organs, in contrast to a saprophyte or a parasite

hydrophyte a water plant

    mesohydrophyte a plant with a habitat intermediate between that of a mesophyte and a hydrophyte

holophyte a plant of forests or woods

hylodophyte a plant living in dry woodlands

hysterophyte = saprophyte

lepidophyte a fossil plant

leimophyte a plant of wet meadows

limnophyte a lake-plant or marsh-plant (*see* -limno-, -limn-)

lithophyte a plant growing on rocks

    endolithophyte a plant, such as lichen, which penetrates into crevices in rock

lochmodophyte a plant of dry thickets

lochmophyte a plant dwelling in thickets

lophophyte a hill plant

macrophyte an extremely elongate plant (*e.g.* kelp)

megaphyte = spermophyte

mesophyte a plant halfway between a hydrophyte and a xerophyte

metaphyte *either* a many-celled plant (*cf.* metabiont, metazoon) *or* a plant with completely differentiated tissues

myxophyte = myxomycete

namatophyte a plant of streams

neophyte a plant of recent introduction

nomophyte a pasture plant

nosophyte a pathogenic ("nosopoetic") plant

noterophyte = mesophyte

oceanophyte an ocean plant

ochthophyte a plant of banks

    ammochthophyte a plant of sandbanks

    pelochthophyte a plant of mudbanks

    petrochthophyte a plant of rock-banks

oekiophyte a plant cultivated for ornamental pruposes

ombrophyte a shade-loving plant

oophyte = gametophyte

opophyte a parasitic plant

orgadophyte a plant of open woodlands

orophyte a subalpine plant

orthophyte the sum total of a plant's existence from egg to egg through both the sporophyte and gametophyte generation

ostariphyte a plant bearing drupes

oxylophyte a plant of acid soils

pagophyte a plant of foothills

pediophyte an upland plant

pelagophyte a plant of the surface of the sea

pelophyte a plant of muddy soil

petrophyte a plant of rocky places

petrodrophyte plant of stony places

phanerophyte a conspicuous plant

    macrophanerophyte tree

    mesophanerophyte a perennial plant between 25 and 100 feet in height

    microphanerophyte woody plants between six and twenty five feet in height

    nanophanerophyte a woody plant less than six feet in height

phellophyte a plant of gravel

phreatophyte a desert plant with a tap root capable of reaching the water table

phretophyte a plant of water tanks

phyllophyte a leafy plant

    ptenophyllophyte a plant of deciduous forests

planktophyte a member of the phytoplankton

plotophyte a floating plant specifically adapted to that environment

pontophyte a plant of the deep sea

poophyte a meadow plant

xeropoophyte a heath plant
potamophyte a river plant
protophyte *either* a plant, the tissues of which are not particularly well differentiated *or* plant of the sexual generation
psamathophyte a plant of the sea shore
psammophyte a plant inhabiting sand
psilophyte a plant of prairies
psychrophyte a plant of cold places
ptenothalophyte a plant of deciduous thickets
pyrophyte a plant with a forest-fire resisting bark
rhoophyte a plant of streams
rhyacophyte a plant of torrents
saprophyte a plant which derives its nourishment from existing organic matter
  hemisaprophyte = hemiparasite particularly applied to plants
holosaprophyte a plant which is solely dependent on decomposing organic matter for its existence
mesosaprophyte used of a saprophyte which sporulates on the surface of its host, but is otherwise totally enclosed
  parasaprophyte = endosaprophyte
  parasite-saprophyte a parasitic plant which kills its hosts and then lives on the decomposing remains
  potential saprophyte = facultative parasite
  symbiotic saprophyte a higher plant associated with a lower
sathrophyte a plant dwelling on humus. The term is sometimes used as synonymous with saprophyte
sciophyte a shade plant
sclerophyte a plant with hard, stiff leaves
scotophyte a plant of dim places
smathophyte a plant of sandy soils
spermatophyte a seed plant
spermophyte sometimes used in the sense of "flowering plant" or "seed-producing" plant
sphagnophyte a plant growing on sphagnum
spiladophyte a plant of clay soils
sporophyte in plants showing alternation of generations that generation which reproduces asexually by means of spores, or, in higher plants, that generation which produces the seed
  carposporophyte the asexual generation of floridiophycean rhodophytes consisting of a mass of gonimoblast filaments and the carposporangia which they bear, growing parasitically on the female gametophyte
stasophyte a plant of stagnant water
sterrophyte a plant of moor lands
syrtidophyte a plant of dry sandbars
taphrophyte a plant of ditches
telmatophyte a plant of wet meadows
thalassophyte used by ecologists in the sense of a "sea plant" but by many botanists usually taken to mean a marine alga
thallophyte a plant which consists principally of a thallus not organized into root stem, and leaf (*see also* Thallophyta)
mesothermophyte a plant of temperate zones
  microthermophyte an arctic plant
therophyte a plant which grows during the summer and the seeds of which rest over winter, to germinate the next spring
thinophyte a plant of sand dunes
tiphophyte a pool plant
tropophyte a plant dwelling in a climate which alternates between sunny drought and torrential rain
isotrophyte a parasitic plant which derives nourish-

ment from, but does not apparently change the form of, its host
xenophyte the endosperm of an angiosperm if this is regarded as a third generation equivalent to gametophyte and sporophyte
xerophyte a plant specifically adapted to life in dry places
  mesoxerophyte a plant halfway between a mesophyte and a xerophyte
xylophyte a woody plant, but used by some ecologists in the sense of a plant dwelling in woods
zygophyte a plant which has been produced sexually
-phytic 1 (*see also* phytic 2) *adj. suffix* from -phys- (*q.v.*)
  symphytic formed by the fusion of several nuclei
-phytic 2 (*see also* phytic 1) *adj. suffix* from -phyt- (*q.v.*). Most compounds are given under -phyte
  emophytic pertaining to submerged vegetation
  hemiendophytic = hemiendobiotic but specifically of plants
  phalarsiphytic = polyadelphic
-phytium *subs. suffix* indicating a plant community (*cf.* phyton)
  acarophytium a mutualistic association of mites and plants
  cryptophytium an association of hemicryptophytes and geophytes
  drymophytium an association of bushes and small trees
  prodophytium the forerunners of a plant formation
  steganochamaephytium an association of dwarf shrubs under trees
phytodyte a plant living predominantly above ground (*cf.* phytoedaphon)
Phytolaccaceae that family of centrospermous dicotyledons which contains, *inter alia,* the pokeberries and ink-berries. The flowers, with several one-seeded carpels, are distinctive of the family
Phytomastigophora a subclass of mastigophoran Protozoa distinguished by their ability to synthesize chlorophyll and therefore considered by many to be algae. This zoological taxon does not, however, correspond to any acceptable botanical taxon. See, however, the orders Chrysomonadida, Cryptomonadida, Dinoflagellida, Euglenida and Phytomonadida
phytome the sum total of a plant
Phytomonadida a rarely accepted order of phytomastigophoran Protozoa distinguished by a large cup-shaped chloroplast and a cellulose wall. Many biologists hold them to be algae, in which case they are placed in the subphylum Chlorophyta as the order Volvocales of the class Chlorophyceae
phyton = phytomer
-phyton *subs. suffix* equivalent to -phytium (*q.v.*)
  epiphyton a sparsely distributed periphyton (*cf.* lasion)
  periphyton the surface population of submerged objects or substrates (= aufwuchs)
Phytophaga a no longer valid zoological taxon at one time applied to phytophagous Coleoptera. The group is now equivalent to the Chrysomeloidea, provided that this group includes the Cerambycidae
-phytosis *subs. suffix* indicating "to be in the state of a growing plant"
  metanaphytosis formation of the floral envelope
Phytotomidae a family of passeriform birds containing the three plantcutters. They are medium-sized birds with a marked crest and a short, heavy, conical and finely serrated, bill

-phytous *comb. suffix* meaning "pertaining to a plant" (*cf.* phytic 2)

    styridophytous having a cross-shaped flower or petal as in the Crucifereae

    synarmophytous = gynandrous

    systellophytous the condition in which a calyx appears to be fused to and form part of the fruit

-phytum a rarely used *subst. suffix* with the meaning "plant"

    mesophytum the junction of the stem and root of a plant

pi *see* [pied] and [pirouette]

piapiac the corvid bird *Ptilostomus afer* (= black magpie)

Picathartidae a family of passeriform birds usually included in the Timaliidae, but at one time erroneously thought to be corvids (*cf.* bare-headed crow)

piceous pitch black

pichicago = pichiciego

pichiciego either of two species of dasypodid xenarthran mammals of the genus Chlamyphorus that lacks movable bands of scutes between the anterior and posterior shields

Picidae a very large family of piciform birds containing the woodpeckers. They are distinguished by the large strong chisel-like beak and slender but extremely strong neck. The undulating method of flight is a common field-identification character

Piciformes that order of birds which contains the woodpeckers and their allies. They are distinguished by their strong beak and by the rounded wings which gives rise to the peculiar method of flight. The tail is short and stiff and is used as a prop

-pic- *comb. form* meaning "magpie" but now used almost entirely for "woodpecker"

cherry picker *Austral.* any of several species of meliphagid birds of the genus Melithreptus (*cf.* honeyeater)

pickerel (*see also* frog, weed) *U.S.* any of several species of esocid fishes of the genus Esox. *U.S. local* the percid fish *Stizostedion vitrem* (= walleye)

    chain-pickerel *E. niger*

    grass-pickerel *E. americanus vermiculatus*

    mud-pickerel *U.S.* local name for either *E. americanus* or *E. niger*

    redfin pickerel *E. americanus americanus*

picket *Brit. dial.* = tern

    watchy-picket *W.I.* the icterid bird *Ictera leucopteryx* (= Jamaican oriole)

picucule a one-time popular name for dendrocolatid birds now called woodcreepers

piculet popular name of numerous small tropical woodpeckers once separated as the subfamily Picumninae, and now included with the Picidae

    African piculet *Verreauxia africana*

    Hispaniolan piculet *Nesoctites micromegas*

    olivaceous piculet *Picumnus olivaceous*

    rufous piculet any of several species of the genus Sasia, usually without qualification, *S. abnormis*

piculule common misspelling of picucule

Picumninae a group of woodpeckers, popular called piculets, now usually included with Picidae

piddock any boring pelecypod mollusk of the family Pholadidae though by some the name is confined to the genera Zirfaea and Martesia

pie 1 (*see also* pie 2, marker) any of many birds, mostly black and white. The term "magpie," a *Brit.* country colloquialism, has replaced the original term in many cases

French pie *Brit.* either the shrike (*q.v.*) or any black and white woodpecker particularly *Dendrocopos major*

horn-pie = lapwing

magpie 1 (*see also* pie 2, flycatcher, lark, manikin, robin, shrike) any of several species of passeriform birds, mostly of the family Corvidae. *Brit.*, without qualification, *Pica pica* (= black-billed magpie)

    American magpie = black-billed magpie

    bell magpie any of several species of cracticids of the genus Strepera (= currawong)

    black-billed magpie the corvid *Pica pica*

    blue magpie any of several corvids of the genus Urocissa

    red-billed blue magpie *U. flavirostris*

    green magpie the corvid *Cissa chineusis*

    Indian magpie the corvid *Dendrocitta vagabunda* (= fufous tree-pie)

    racquet-tailed magpie the corvid *Crypoirina temia*

    western magpie *Austral.* the cracticid *Gynmorhyna dorsalis*

sea pie *Brit.* any of several curvirostrids. *W.I.* the haematopodid *Haematopus ostralegus* (= oystercatcher)

tree-pie any of several species of the genera Dendrocitta and Crypsirina

    rufous tree-pie *D. vagabunda*

    racket-tailed tree-pie either of two species of the genus Crypsirina

wood pie = woodpecker

pie 2 (*see also* pie 1) any of several lepidopteran insects. *S.Afr.*, without qualification, those of the lycaenid genus Castalius

    magpie (*see also* pie 1) *Brit.* either of two species of the geometrids of the genus Abraxas, without qualification, *A. grossulariata* is usually meant. *U.S.* the saturniid *Pseudohazis hera*

[piebald] a mutant (gene symbol *s*) in linkage group III of the mouse. The name is adequately descriptive of the phenotypic expression

[pied] a mutant (gene symbol *pi*) mapped at 17.0± on chromosome II of *Drosophila melanogaster*. The phenotypic expression is jumbled eye facets

flower-piercer any of several species of thraupid birds of the genus *Diglossa*

    cinnamon-bellied flower-piercer *D. baritula*

    slaty flower-piercer *D. plumbea*

Pieridae that family of papilionoid lepidopteran insects which contains the forms commonly called whites, sulfurs and orange tips

pierrot *I.P.* any of numerous species of lycaenid lepidopteran insects mostly of the genera Talicada, Taraka, Castallius. Without qualification, usually *C. rosimon*

parsley piert *Brit.* the rosaceous herb *Alchemilla arvensis* (= field ladies mantle)

Piesmidae a small family of hemipteran insects closely resembling the Tingididae but far slenderer. They are often called (*U.S.*) ash-gray leaf bugs

Pierzata = Hymenoptera

pig 1 (*see also* pig 2, boar, ear, hog, not) any of numerous suinine artiodactyl ungulates, many of the genus Sus. In *U.S.* the word usually refers to juvenile forms, the adult being a hog

    bush-pig any of several species of the genus Potamochoerus. They differ markedly from wild pigs of the genus Sus in possessing sharply pointed, long-tufted, ears

pig 2 any of several non-suine mammals of pig-like appearance or habit

guinea-pig (*see also* louse) any of numerous caviid rodents of the genus Cavia and their near allies. The domestic guinea-pig is descended from *C. cutleri* of the Andes

pigeon 1 (*see also* pigeon 2, fly, grape, louse, pea, plum) any of numerous columbiform birds of the family Columbidae. There is no real distinction between doves (*q.v.*) and pigeons though the former term is usually applied to smaller birds

    red-billed pigeon *Columba flavirostris*

    short-billed pigeon *C. nigrirostris*

    tooth-billed pigeon *Didunculus strigirostris* (= dodlet)

    black pigeon *W.Afr.* *Streptopelia semitorquata* (= red-eyed dove)

    blue pigeon (*see also* pigeon 2) *W.I.* = red necked pigeon *W.I.*

    common pigeon *Columba livia* (= rock dove)

    crowned pigeon any of several species of the genus Goura

    white-crowned pigeon *Columba leucocephala*

    flock pigeon *Histriophaps histrionica*

    green pigeon any of numerous species of the genera Treron, Sphenus and Butreron, usually, without qualification, *T. calva*

    imperial pigeon any of many species of the genus Ducula

    red-necked pigeon *Columba squamosa*

    New Zealand pigeon *Hemiphaga novaeseelandiae*

    Nicobar pigeon *Caloenas nicobarica*

    olive pigeon *Columba arquatrix*

    partridge-pigeon *Austral.* either of two species of columbid of the genus Geophas

    passenger pigeon the recently extinct *N.Amer. Ectopistes migratorius*

    plain pigeon *Columba inornata*

    plumed pigeon *Lophophaps plumifera*

    box-poison pigeon *Austral. Phaps elegans*

    rock-pigeon either of two species of the genus Petrophassa

    scaled pigeon *U.S. Columba speciosa W.I.* = red-necked pigeon

    speckled pigeon *Columba guinea*

    ring-tail pigeon *Columba caribaea*

    band-tailed pigeon *Columba fasciata*

    wedge-tailed pigeon *Sphenurus sphenurus*

    topknot pigeon *Lopholaimus antarcticus*

    wampoo pigeon *Megaloprepia magnifica*

    wonga pigeon *Leucosarsia melanoleuca*

    West Indian pigeon *Columba inornata* (= plain pigeon)

    wood pigeon *Brit. Columba palumbus* (= ring-dove). *N.Z. Hemiphaga novaeseelandiae*

    purple wood pigeon *C. punicea*

pigeon 2 pigeon-like birds not of the family Columbidae

    blue pigeon *Austral.* (*See also* pigeon 1) the campephagid *Coracina novaehollandiae* (*cf.* cuckoo-shrike)

    Cape pigeon the procellariid *Daption capensis*

pigment a substance which produces the appearance of color

    bile pigment *see* -verdin

    respiratory pigment one that carries oxygen. *See* heme-, hemo, and -cruorin

    visual pigment photosensitive pigments in the eye

[pigmy] *either* a mutant (gene symbol *py*) mapped at 64 on linkage group VI of *Zea mays*. The phenotypic expression is a dwarfing of the plant *or* a mutation (gene symbol *pg*) found in linkage group IV of the mouse the name is descriptive of the phenotypic expression

pignon the seed of the nut-pine (*q.v.*)

piha any of several cotingid birds of the genera Lipangus and Chirocylla

pihoihoi *N.Z.* the motacillid bird *Anthus novaeseelandiae* (= New Zealand pipit)

pika any of several species of ochotonid lagomorphs of the genus Ochotona

pike 1 any of several fresh water teleost fishes of the family Esocidae. *Brit.*, without qualification, *Esox lucius.* (= *U.S.* and *Canad.* great northern pike)

pike 2 (*see also* perch, salmon) any pike-like fish not of the family Esocidae

    adder pike *Brit.* the marine teleost *Trachinus vipera*

    blue pike *U.S.* the percoid fish *Stizostedion vitreum glaucum*

    dwarf pike the poeciliid *Belonesox belizanus* (= pike topminnow)

    walleyed pike *U.S.* the percoid fish *Stizostedion vitreum vitreum*

    garpike *Brit.* the belonid fish *Rhamphistoma belone*

    saury pike *Brit.* the scomberesocid *Scomberesox saurus* (= *U.S.* atlantic saury)

-pil- *comb. form* meaning "hair"

pil a bacterial genetic marker having a character that affects the production of pili. Mapped at 88.0 mins. for *Escherichia coli*

neuropil an obsolete term for the "cell-free" areas of the central nervous system

pilchard any of several small clupeid teleost fishes. The term is so confused with the sardine (*q.v.*) as to be meaningless, but possibly should be confined to *Clupeo philchardus. U.S. Sardinops sagax* (= Pacific sardine)

    false pilchard *Harengula clupeola*

-pile- *comb. form* meaning "cap"

pileate having the form of a cap

pileated said of birds in which the whole top of the head forms a crest

pileum the top of the head, particularly in birds

mesocephalic pillar one of the two beam-like portions of the cephalic apodeme in Hymenoptera that braces the front and back walls of the head

pilose hairy

pilus a fine, hair-like, protuberance from a bacteria

pilwillet usually synonymous with willet, but also applied (*S.W.U.S.*) to the haematopodid bird *Haematopus palliatus* (= oystercatcher)

pimbini either of two caprifoliaceous shrubs *Viburnum edule* or *V. trilobum*

-pimel- *comb. form* meaning "fat"

etepimeletic said of the behavior of a young animal demanding attention from a parent (literally "one who seeks fat")

Pimelodidae a family of naked catfishes distinguished by the four barbels on the chin and the unusually long maxillary barbels (*cf.* Bagridae)

pimento 1 (*see also* pimento 2) the myrtaceous tree *Pimento officinalis* which yields allspice

pimento 2 (*see also* pimento 1) any of several species of small solanaceous bushes of the genus Capsicum and their fruit. More properly spelled pimiento

pimilico *W.I.* the procellariid *Puffinus lherminieri.* (= Audubon's shearwater). *Austral.* = friarbird

pimpernel *Brit.* any of several small-flowered herbs. Without qualification, any of several species of the primulaceous herb Anagallis

    brook pimpernel the scrophulariaceous herb *Veronica anagallis-aquatica*

    common pimpernel *U.S. A. arvensis* (= *Brit.* scarlet pimpernel)

false pimpernel any of several species of herb of the scrophulariaceous genus Lindernia

mountain pimpernel U.S. the umbelliferous herb *Pseudotaenidia montana*

scarlet pimpernel *Brit. A. arvensi*

water pimpernel *U.S.* any of several species of aquatic herb of the genus Samolus

yellow pimpernel *U.S.* the umbelliferous herb *Taenidia integerrima*

picket pin the sciurid rodent *Citellus richardsoni* (*cf.* ground squirrel)

[Pin] a mutant (gene symbol *Pin*) mapped at 107.3± on chromosome II of *Drosophila melanogaster*. The name derives from the pin-like bristles on the thorax of the phenotype

Pin *see* [Pin]

-pinac- *comb. form* meaning "plank" and thus, by extension, "structural"

Pinaceae the pine family of Gymnospermae, containing the cone-bearing trees, except the yews and Gnetales. The dry woody cones, with winged seeds between the scales, are distinctive

pinacoua the annonaceous tree *Fusaea longifolia* and its fruit

pincer a loose term for any organ capable of pinching and applied to such varied structures as the chela of a lobster or the anal forceps of an earwig

pinc-pinc *S.Afr.* the sylviid bird *Cisticola textrix* (= tinc-tinc)

pine 1 (*see also* pine 2, aphis, apple, bird, butterfly, chafer, beetle, midge mite, moth, sawfly, webworm, weevil) properly any of very numerous species of the pinaceous genus Pinus

Aleppo pine *P. halepensis*

Apache pine *P. latifolia* (= Arizona pine)

Arizona pine = apache pine

Austrian pine *P. nigra*

lace-bark pine *P. bungeana*

　white-bark pine = *P. albicaulis*

bastard pine any of numerous species of the genus Pinus, including *P. taeda*, *P. serotina*, *P. elliottia*. Also, rarely, the white fir (*Abies concolor*)

bishop pine *P. muricata*

Japanese black pine *P. thunbergii*

bull-pine = yellow pine

cluster pine *P. pinaster*

bristle-cone pine *P. aristata*

　knob-cone pine *P. attenuata*

　prickle-cone pine *P. muricata*

digger pine *P. sabiniana*

Douglas pine = Douglas fir

frankincense pine = loblolly pine

old field pine = loblolly pine

hickory-pine *P. aristata*

jack pine *P. banksiana*

Jeffrey's pine = *P. jeffreyi*

Jersey pine = scrub pine

long-leaf pine = southern pine

　short-leaf pine *P. echinata*

limber pine *P. flexilis*

loblolly pine *P. taeda*

marsh-pine *P. serotina* (= pond pine)

Monterrey pine *P. radiata*

Table Mountain pine *P. pungens*

　Swiss mountain pine *P. montana*

Norway pine *P. resinosa*

nut pine either of two species *P. parryana* or *P. edulis* of which the seeds are eaten

pitch pine *P. rigida* or *P. coulteri*

lodge pole pine *P. contorta*

pond-pine = marsh pine

ponderosa pine *P. ponderosa*

poverty pine = Table Mountain pine

red pine (*see also* pine 2) = Norway pine

　Japanese red pine *P. densiflora*

sand pine *P. clausa*

Scotch pine *P. sylvestris*

scrub-pine *P. virginiana* or *P. contorta*

slash pine = *P. elliottii*

Soledad pine *P. torreyana*

southern pine *P. palustris*

spruce-pine *P. glabra*

stone-pine *P. pinea* or *P. cembra*

sugar-pine *P. lambertiana*

swamp pine = slash pine

fox-tail pine *P. balfouriana*

white pine (*see also* pine 2) *P. strobus*

　mountain white pine *P. monticola*

yellow pine *P. ponderosa*

pine 2 (*see also* pine 1) any of many other evergreen trees or plants or herbs resembling the true pines in structure or habit

Alaska pine *Tsuga heterophylla*

black pine *N.Z.* the podocarpaceous tree *Podocarpus spicatus* (= matai)

bog-pine *N.Z.* the podocarpaceous tree *Dacrydium bidwillii*

cyprus-pine the pinaceous tree *Callitris robusta*

ground-pine *Brit.* the labiateous herb *Ajuga chamaepitys*

Kauri pine *N.Z.* the araucariaceous tree *Agathis australis*

mountain pine *N.Z.* = bog-pine *N.Z.*

Norfolk Island pine the araucariaceous tree *Araucaria excelsa*

pink pine *N.Z.* the podocarpaceous tree *Dacrydium beforme*

pygmy pine *N.Z.* the podocarpaceous tree *Dacrydium laxifolium*

red pine (*see also* pine 1) *N.Z.* the podocarpaceous tree *Dacrydium cupressinum* (= rimu)

saw-pine any of numerous species of crassulaceous herb of the genus Sedum (= stone crop)

screw-pine any of numerous species of the pandanaceous genus Padanus

silver-pine *N.Z.* the podocarpaceous tree *Dacrydium colensoi*

yellow silver-pine *N.Z.* the podocarpaceous tree *Dacrydium intermedium*

Swiss pine the pinaceous tree *Abies alba*

umbrella pine the pinaceous tree *Sciadopitys verticillata*

white pine (*see also* pine 1) *N.Z.* the podocarpaceous tree *Podocarpus dacrydiodides* (= kahikatea)

pineal an organ of doubtful function arising from the epithalamus and retaining a nervous connection with the left habenular ganglion

parapineal a body similar to the pineal but having a nervous connection to the right posterior commisure. Some cyclostomes have both a pineal and a parapineal

piney *obs.* – peony

pingao *N.Z.* the cyperaceous plant *Desmoschoenus spiralis*

Pinguiculariaceae = Lentibulariaceae

pinha the annonaceous tree *Annona squamosa* and its fruit

pinion 1 (*see also* pinion 2) properly the distal joint of a bird's wing. Frequently used as synonymous with wing

pinion 2 any of several species of noctuid lepidopteran insect of the genus *Xylina* (*U.S.*) and *Calymia* (*Brit.*)

pretty pinion *Brit.* the geometrid lepidopteran insect *Perizoma blandiata*

pinguin the bromeliaceous herb *Bromelia pinguin*

pink 1 (*see also* pink 2, 3) any of numerous species of caryophyllaceous herbs mostly of the genus Dianthus

Cheddar pink *D. caesius*

clove pink any of several species but most usually either *D. caryophyllus* or *D. plumarius*

cushion pink *Silene acaulis*

Deptford pink *D. armeria*

fire-pink *Silene virginica*

garden pink *D. plumarius*

maiden pink *D. deltoides*

Scotch pink = garden pink

pink 2 (*see also* pink 1, 3) in compound names of other plants

election pink *U.S.* the ericaceous shrub *Rhododendron nudiflorum*

French pink the compositaceous herb *Centaurea cyanus* (= cornflower)

grass-pink (*see also* pink 3) *U.S.* any of several species of orchid of the genus *Calopogon*

ground-pink = moss-pink

Indian pink the convolvulaceous vine *Quamoclit pinnata*, the campanulaceous herb *Lobelia cardinalis*, and *U.S.* the loganiaceous herb *Spigelia marilandica*

marsh pink the gentianaceous herb *Sabatia stellaris*

moss-pink the polemoniaceous herb *Phlox subulata*

mullein pink the scrophulariaceous herb *Lychnis coronaria* (= rose-campion)

rose pink *U.S.* the gentianaceous herb *Sabati angularis*

sea pink any of several species of the plumbaginaceous genus Statice (= thrift)

marsh sea pink the gentianaceous herb *Sabatia stellaris*

stud pink = swamp pink

swamp pink *U.S.* the liliaceous herb *Helonias bullata* and any of several species of orchid of the genus Calopogon

pink 3 (*see also* pink 1, 2) any of several birds mostly fringillids. Usually, without qualification, *Fringilla coelebs* (= chaffinch)

grass pink (*see also* pink 2) *W.I.* the fringillid *Ammodramus savannarum* (= grasshopper sparrow)

October pink *W.I.* the icterid *Dolichonyx oryzivorus* (= bobolink)

[pink] *either* a mutant (gene symbol *pk*) in linkage group VII of *Habrobracon juglandis*. The name is descriptive of the eye color of the phenotype *or* a mutant (gene symbol *p*) mapped at 48.0 on chromosome III of *Drosophila melanogaster*. The phenotypic eye-color is dull red

[pink eye] a mutant (gene symbol *p*) found at locus 0 in linkage group I of the rat. The phenotypic expression involves not only a pink eye-color but also a yellow coat

[pink-eyed] a mutant (gene symbol *pe*) mapped at 0 on linkage group V of *Bombyx mori*. The phenotypic expression is a white egg

[pink-eyed dilution] a mutant (gene symbol *p*) in linkage group I of the mouse. The phenotypic expression

is an absence of pigment in the retina but not in the skin

-pinni- *comb. form* meaning "wing" (*cf.* -penna-)

pinna literally "wing" but usually applied to the external lobe that surrounds the orifice of the ear in mammals, sometimes called the outer ear

[pink-wing-c] a mutant (gene symbol *pw-c*) mapped at 70.9± on chromosome II of *Drosophila melanogaster*. The phenotypic expression is a dilution of eye color and short and blunt wings

pinnate (*see also* leaf) said of birds having tufts of feathers on the side of the neck

Pinnidae a family of pelecypod mollusks commonly called pen-shells by reason of their large wedge-shaped shells which gape at the posterior end

Pinnipedia that suborder of Carnivora which contains the seals and their allies in which flippers replace true feet

Pinnotheridae a family of brachyuran Crustacea with membranous cylindrical carapace and very small eyes. These small crabs live in the mantle cavity of pelecypod mollusks or in the tubes of polychaete worms

pinnule literally a "little feather" and applied in *biol.* to many structures of this shape including one of the numerous, tapering, side branches of a crinoid arm, a second rib on the valve of a diatom, and a lateral projection on the tentacle of an alcyonarian coelenterate

-pino- *comb. form* confused between three *Gr.* and one *L.* root and therefore meaning "hunger," "dirt," "drink" or "pine tree"

piñon the seed of the nut-pine

haematopinous blood-drinking (usually of insects)

xylopinous wood-hungry (usually of insects)

cockoo pint the araceous herb *Arum maculatum*

pintado *Austral.* = Cape pigeon

[Pintail] a mutant (gene symbol *Pt*) in linkage group VIII of the mouse. The name of the mutant is descriptive of the phenotypic expression

pinule a hexactine sponge spicule in which one of the axes is greatly elongated and covered with spines

pinyon = piñon

pioneer *I.P.* the pierid lepidopteran insect *Anaphaeis aurota*

Piophilidae a family of small, metallic-black, myodarian cycloraphous dipteran insects commonly called skipper-flies because their larvae are capable of jumping. The larvae are well-known pests of cheese and stored meat

piopio *N.Z.* the turnagrid *Turnagra capensis* (= thrush N.Z.)

pip 1 (*see also* pip 2, 3) a budded rhizome of a Lily of the Valley

pip 2 a popular name for the seed of a pome

pip 3 the action of hatching birds in cutting their way out of the shell

dutchman's pipe the aristolochiaceous herb *Aristolochia macrophylla*

Indian pipe the pyrolaceous parasitic herb *Monotropa uniflora*

pygmy pipe the pyrolaceous parasitic herb *Monotropsis odorata*

swinepipe *Brit. obs.* the turdid bird *Turdus iliacus* (= *Brit.* redwing)

piper 1 (*see also* piper 2, 3) *S. Afr.* either of two species of nymphalid lepidopteran insect of the genus Eurytela. *E. hiarbas* (**pied piper**) or *E. dryope* (**golden piper**)

piper 2 (*see also* piper 1, 3) *Brit.* the triglid fish *Trigla lyra*

piper 3 (*see also* piper 1, 2) sandpiper, any of numerous scolopacid birds

  avocet sandpiper = terek sandpiper

  red-backed sandpiper *Erolia alpina* (= dunlin)

  Baird's sandpiper *Erolia bairdii*

  broad-billed sandpiper *Limicola falcinellus*

    spoon-billed sandpiper *Eurynorhynchus pygmeus*

  buff-breasted sandpiper *Tryngites subruficollis*

  common sandpiper *Actitis hypoleucos*

  curlew-sandpiper *Erolia ferrugisea*

  green sandpiper *Triga ochropus*

  least sandpiper *Erolia minutilla*

  semipalmated sandpiper *Ereunetes pusillus*

  pectoral sandpiper *Erolia melanotos*

  purple sandpiper *Erolia maritima*

  rock-sandpiper *Erolia ptilocnemis*

  white-rumped sandpiper *Erolia fuscicillis*

  solitary sandpiper *Tringa solitaria*

  spotted sandpiper *Actitis macularia*

  stilt sandpiper *Micropalama himantopus*

  sharp-tailed sandpiper *Erolia acuminata*

  terek sandpiper *Xenus cinereus*

  upland sandpiper *Bartramia longicauda* (= upland plover)

  western sandpiper *Ereunetes mauri*

  wood sandpiper *Tringa glareola*

Piperaceae that family of piperalous plants which contains the Old World peppers. Distinctive characteristics are the 1-celled, 1-seeded ovary and spicate inflorescence

Piperales that order of dicotyledenous plants which contains, *inter alia*, the peppers. The order is principally distinguished by having minute flowers, lacking a perianth, in spikes or racines

pipinella the cucurbitaceous vine *Sechium edule* and its edible squash-like fruit

pipiri *W.I.* the tyrannid bird *Tyrannus melancholicus* (= tropical king-bird)

  topknot pipiri *W.I.* the tyrannid *Elaenia flavogaster*

pipistrel = pipistrelle

pipistrelle *Brit.* the vespertilionid bat *Pipistrellus pipistrellus*

  eastern pipistrelle *U.S. P. subflavus*

  western pipistrelle *U.S. P. hesperus*

pipit any of numerous motacilid birds of the genus Anthus except for the ant pipits which are conophagids

  American pipit = water pipit

  antpipit any of several species of conopophagid bird

  Australian pipit = New Zealand pipit

  plain-backed pipit *A. leucophrys*

  long-billed pipit *A. similis*

  yellow-breasted pipit *A. chloris*

  golden pipit *Tmetothylacus tenellus*

  Indian pipit = New Zealand pipit

  meadow-pipit *A. pratensis*

  New Zealand pipit *a. novaeseelandiae*

  petchora pipit *A. gustavi*

  rock-pipit *A. crenatus*

    brown rock-pipit = long-billed pipit

  Spragues pipit *A. spragueii*

  striped pipit *A. lineiventris*

  tawny pipit *A. campestris*

  red-throated pipit *A. cervinus*

  tree-pipit *A. trivialis*

    Indian tree-pipit *A. hodgsoni*

  upland pipit *A. sylvanus*

  water-pipit *A. spinoletta*

pipiwarauroa *N.Z.* the cuculid bird *Chalcites lucidus* (= shining cuckoo)

pippit = pipit

Pipridae a family of passeriform birds commonly called manakins. Many are black birds with brilliantly colored heads

pipsissewa any of several ericaceous herbs of the genus Chimaphila

Pipunculidae a family of minute aschizan cycloraphous brachyceran dipteran insects distinguished by their enormous eyes and therefore frequently called big-headed flies

piramidig *S.U.S.* and *W.I.* the caprimulgid bird *Chordeiles minor* (= nighthawk)

piranha any of several species of aggressive predatory characid fishes of the genus Serrasalmus

pirate *S.Afr.* the nymphalid lepidopteran insect *Catacroptera cloantha*

pirenet the tadornine anatid bird *Tadorna tadorna* (= shelduck)

Pirolaceae = Pyrolaceae

piquero the sulid pelecaniform bird *Sula variegata*

[pirouette] a mutant (gene symbol *pi*) in linkage group III of the mouse. The phenotypic expression is a waltzing syndrome accompanied by deafness

-pis- *comb. form* meaning either "pea" or "meadow"

pisaceous pea-green

Pisauridae a family of large dipneumonomorphous ground spiders. Many hunt on the surface film of bodies of water and frequently dive below the surface, thus earning the name fishing spiders

Pisces a superclass of craniate chordates containing all those forms which are generally called fishes including the sharks

Pisciolidae a family of rhyncobdelloid hirudinean annelids parasitic on fishes. The body is divided into anterior and posterior regions

pismire *Brit. obs.* = ant

Pisoneae a tribe of the family Nyctaginaceae with glabrous ovaries and of shrubby habit

pistachio the anacardiaceous tree *Pistacia vera* and its edible seeds

  Chinese pistachio the anacardiaceous tree *Pistacia chinensis* and its edible seeds

pistil the ovary style and stigma of a mature flower. In flowers having a syncarpous gynoecium, the whole structure is popularly referred to as a pistil

  compound pistil one which is formed by the fusion of two or more carpels

  simple pistil = carpel

pit (=*see also* canal, cavity, chamber, membrane) any open top cavity. Without qualification, usually a cavity in a plant secondary cell wall. The cavities in the primary cell wall are called primary pit fields

  anal pit an invagination precursor to the anus in embryos

  blind pit a plant pit occurring opposite an intercellular space

  bordered pit a plant pit in which the secondary cell wall arches over the pit cavity

  ramiform pit a deep plant pit with two or more cavities and thus apparently branched

  gastric pit a pit in the wall of the stomach forming the common aperture of numerous glands

  subgenital pit a replacement for a peristomial pit in some Schyphozoa

  Hatschek's pit a coelomic pouch lying under the whole length of the notochord in the embryos of cephalochordates

Kollicker's pit a ciliated invagination on the dorsal surface of Amphioxus just in front of the forward end of the neural tube

nasal pit the precursor of the nasal cavity in embryos

primitive pit = a depression in Hensen's node

simple pit a plant pit not partially closed by an arch of secondary wall

peristomial pit one in the interradius of a scyphozoan coelenterate

vestibular pit a glandular depression immediately behind the vestibular organ of Chaetognatha

ventilating pit an analogue of a stoma found in some ferns

vestured pit a plant pit with minute outgrowths from the free surface of the secondary wall

pitanga either of two shrubs of the myrtaceous genus Eugenia but usually *E. pitanga*

Pitcairnioideae a subfamily of bromeliaceous herbs distinguished by their terrestrial habitat

pitcher a term applied to the leaf of a pitcher plant

pites *U.S.* the equisetaceous plant *Equisetum fluviatile* (*cf.* horsetail)

pith 1 (*see also* pith 2) that part of the fundamental supporting system of a plant which lies in the center of the stem or root

ulterior pith piths formed in the root after the separation of the stele

pith 2 (*see also* pith 1) a verb used to describe the method of killing an animal by the destruction of the spinal cord

Pithecidae a family of ceboid primates differing from the Cebidae in lacking prehensile tails.

pitocin = oxytocin

pitressin a hormone found in the neurohypophysis acting on the contraction rate of smooth muscle in the walls of the blood vessels

pitta any of numerous species of passeriform birds of the family Pittidae

antpitta any of several species of formicariid bird mostly of the genus Grallaria

scaled antpitta *G. guatamalensis*

black-breasted pitta *P. iris*

blue-breasted pitta *P. mackloti*

buff-breasted pitta *P. versicolor*

Pittidae a family of passeriform birds commonly called pittas. They are distinguished by the short, rounded, wing and very short tail together with the presence of long and strong legs

pitting the arrangement of pits in a plant stem

alternate pitting with pits arranged in diagonal rows

scalariform pitting with pits arranged in the form of a ladder

opposite pitting pits arranged in parallel rows

sieve pitting with simple pits arranged in clusters

Pittosporaceae a small family of rosalous dicotyledons mostly confined to the Australasian region. This family was for some time fused with the saxifrages

pituitary *see* hypophysis

Pityriasidae a family of passeriform birds usually included with the Sturnidae

pk *see* [pink] and [prickle]

pl *see* [pellucid], [pleated], and [plug]

Pl *see* [Purple plant]

(P1) a bacterial gentic marker indicating lysogenicity for phage P1

-plac- *comb. form* meaning a "tablet," usually, but not of necessity circular in outline

-placent- *comb. form* meaning round, flat or slightly domed "cake"

placenta 1 (*see also* placenta 2) a compound organ derived in part from the wall of a mammalian uterus, and in part from the embryonic chorion and amnion, which provides a mechanism for osmotic exchange between maternal and foetal blood

endotheliochorial placenta one in which both the epithelium and the connective tissue of the uterus are eroded away, so that the embryonic tissues lie in contact with the maternal blood vessels

epitheliochorial placenta one in which the embryo and the maternal tissue lie in close interdigitating contact, with no erosion of cell layers

hemochorial placenta one in which the walls of the maternal blood vessels are penetrated by the embryonic blood vessels, which are therefore directly bathed in maternal blood

syndesmorchorial placenta one in which the maternal uterine epithelium is eroded away, so that the embryonic membranes are in contact with the connective tissue of the uterus

cotyledonary placenta one having patches of villi over the entire surface of the extraembryonic membranes

diffuse placenta one having villi uniformly distributed over the whole surface of the extraembryonic membrane

discoidal placenta one in which the villi are confined to one or two discs on the surface of the extraembryonic membrane

semiplacenta one in which the embryonic and maternal tissue is not fused

zonary placenta one having a band of villi encircling the membranous cover

placenta 2 (*see also* placenta 1) the tissue which carries the ovules in a plant ovary, or which, in cryptogams, bear sporangia

basal placenta with the placenta at the base of the ovary

placentation (*bot.*) the disposition of ovules on the placenta

axile placentation said of the placement of ovules in the axil formed by the fusion of septa of a compound plant ovary

basal placentation that type of placentation in which a single ovule rests on a short stub in the center axis at the base of the ovary

free-central placentation that type of placentation in which the ovules are arranged along a single central stalk in a compound ovary

parietal placentation a type of placentation in which the ovules are arranged on the walls of the ovary

placid a large plate on the second zonite of kinorhynchs

placoid (*see also* scale) having the form of a flat round plate

macroplacoid one of many hardened lumps in the pharynx of Tardigrada

microplacoid a single minute analog of a macroplacoid

pladinium the active larval instar hatched from the egg of some chalcicoid hymenopterans in which hypermetamorphosis is observed

-plagio- *comb. form* meaning "oblique" and by extension "side" or "flank"

plague term for a large group of locusts in the Old-World sense of this word

plaice any of many flatfishes. *Brit.*, without qualification, the pleuronectid *Pleuronectes platessa*

Alaska plaice *U.S.* the pleuronectid *Pleuronectes quadrituberculatus*

American plaice *U.S.* the pleuronectid *Hippoglos-soides platessoides*

[**Plain**] a group of pseudoalleles (gene symbol *P*) mapped at 0 of linkage group II of *Bombyx mori*. The phenotypic larva is white

[**Plain supernumary legs**] a group of pseudoalleles mapped at 0 on linkage group VI of *Bombyx mori*. The supernumerary legs are in the first and second abdominal segments of the larva

plakea a flat colony of flagellated algae

-plan- *comb. form* meaning "motile" or "flat"

complanate flattened

 deplanate flattened down

 explanate flattened out (*cf.* expanded)

plancton = plankton

Planctosphaeroidea a class of the phylum Hemichordata known only from their peculiar transparent pelagic larvae

plane 1 (*see also* plane, tree) a flat surface; in *biol.*, usually a section cut across a solid body

 apical plane the plane joining the apices, but in diatoms specifically the plane at right angles to the valvar plane

 transapical plane in diatoms, the plane at right angles to the valvar and apical planes

 paratransapical plane sections parallel to the apical axis in diatoms

 horizontal plane a plane parallel to the longitudinal axis and at right angles to the sagittal axis. Sections through this plane are often called "frontal" sections

 longitudinal plane a plane parallel to the longitudinal axis. It may be sagittal, horizontal or at any angle between these

 meridian plane that plane of the diatom in which lies the perivalvar axis

 parameridian plane those planes in a diatom frustule, which are parallel to the meridian

 periclinal plane one which conforms to the exterior surface

 sagittal plane a plane parallel to the sagittal axis. A sagittal section is therefore a vertical longitudinal section

 transverse plane a plane parallel to the transverse axis, that is at right angles to the longitudinal axis

 paratransversan plane that which is parallel to the transversan axis of a diatom frustule

 valvar plane that plane which passes through the apical and transapical axes of a diatom

 paravalvar plane that which is parallel to the valvar plane in a diatom

 pervalvar plane in diatoms, the line joining the apices

plane 2 (*see also* plane 1, tree) *Austral. I.P.* the lycaenid lepidopteran insect *Bindahara phocides*

diplanetic the condition of those zoospores which first cluster, sometimes forming a cell wall, and subsequently again escape as free-swimming spores

planetous migratory

planiusculous not quite flat

plankt- *comb. form* meaning "wandering"

plankter an individual of the plankton (*cf.* planktont)

plankton organisms living in the upper part of any body of water and which drift with the current (*cf.* nekton, tracheron)

 diacmic plankton one showing two blooms per season

 monacmic plankton one showing a single bloom per year

 adventitious plankton those organisms which have accidentally become planktonic through the action of waves or currents

aeroplankton organisms floating in air

anthoplankton that which produces algal blooms

bathyplankton plankton living at whatever depth below the mesoplankton the author considers "great depths" (*cf.* hadal)

chaetoplankton plankton in which diatoms with awn-like processes predominate

colaplankton that which consists primarily of organisms with gelatinous sheaths

contortoplankton plankton consisting of a floating mass of diatoms

cryoplankton that of perpetually cold or icy waters, or organisms suspended in snow

desmoplankton that which appears in the form of bands or ribbons

discoplankton plankton in which disc diatoms predominate

elaioplankton that which floats in virtue of contained oil droplets

epiplankton the upper or surface layer of plankton. There is no agreement as to the depth to which the epiplankton extends. A fairly common measure is the limit of light penetration in the waters in question. The term is also used for floating organisms attached to pelagic organisms

euplankton either that which consists entirely of free-floating organisms or open water plankton as distinct from tychoplankton

false plankton that which is composed of fixed organisms which have broken loose

gasoplankton that which consists of organisms floated by air vacuoles

euryhaline plankton that capable of withstanding considerable variations in salinity

haloplankton plankton of salt water

heleoplankton phytoplankton of marshes

heloplankton the plankton of a marsh, usually including relatively large floating plants

hemiplankton phytoplankton of shallow pools, which has come to rest on the tops of submerged plankton

hidroplankton plankton consisting primarily of organisms which float with the aid of secretions

holoplankton plankton of the open sea

hydroplankton occasional misspelling of hidroplankton

kalloplankton that which is supported by gelatinous cases

knephoplankton that which occurs between 15 fathoms and 250 fathoms

kremastoplankton that which consists primarily of organs floating with the aid of hairs, bristles, etc.

eulimnetic plankton that found only in standing water

 tycholimnetic plankton that which is formed of algae which have broken away from the bottom and float in consequence of contained gas bubbles

limnoplankton properly, the plankton of marshes, but widely applied to freshwater plankton

longipes plankton that which is composed mostly of *Ceratium longipes*

macroplankton those planktonic organisms that can be classified without the aid of a microscope

megaplankton that which includes large floating organisms such as the water-hyacinth, jellyfish, etc.

meroplankton that which is seasonal in its appearance or which consists of organisms planktonic for only a portion of their life cycle

mesoplankton variously used to mean the plankton between epiplankton and bathyplankton or alternatively, all plankton below the epiplankton

hyphalmyroplankton marine plankton

nannoplankton planktonic organisms so small that they pass through the meshes of a number 25 plankton net

neidioplankton those planktonic organisms possessing any means of locomotion

neroplankton plankton that is derived from coastal waters

net plankton those constituents of the plankton retained by any particular plankton net

neuroplankton that which is found only at certain seasons of the year

  autopelagic plankton that which lives continuously on the surface without seasonal or diurnal migrations

  bathypelagic plankton that which shows a diurnal vertical migration

  chimopelagic plankton that found on the surface only in winter

  eupelagic plankton that which is found exclusively in oceans

  nyctipelagic plankton that portion of a diurnal migrating plankton which rises at night

  spanipelagic plankton that which only occasionally appears at the surface

phagoplankton that composed of autotrophic algae

phaoplankton that which occurs only in depths to which enough light penetrates to permit photosynthesis

phyktioplankton that part of the plankton which floats by means of cysts or bladders

phytoplankton the plant constituent of the planktonic population (*cf.* phytopleuston)

macrophytoplankton either that which is composed of large plants (which should properly be called megaphytoplankton) or that which is composed of elongate plants such as filamentous algae (*cf.* trichophyton)

triposplankton that consisting chiefly of *Ceratium tripos*

potamoplankton that which occurs in rivers

  eupotamic plankton that which is confined to fresh waters

  tychopotamic plankton the plankton of streams (*cf.* autopotamic)

pseudoplankton that which consists of accidental components not properly part of the plankton

raphidoplankton that which consists of needle-shaped organisms

rheoplankton that which lives in running water

saproplankton that which exists in unusually foul waters or consists primarily of saprophytes

scoticaplankton that composed largely of dinoflagellates

siraplankton that which is composed mostly of Thalassosira

skaphoplankton that which consists largely of boat-shaped organisms

skotoplankton that which occurs below 250 fathoms

soleniaplankton that which is composed mostly of Rhizosolenia

spheroplankton that which is composed mostly of spherical forms

stagnoplankton that of stagnant water

stiliplankton (= styliplankton) that which consists principally of *Rhizosolenia styliformis*

thalassoplankton that which occurs in the open ocean

trichoplankton that which consists mainly of filamentous algae (*cf.* macrophytoplankton)

tychoplankton planktonic forms, particularly algae, which become entangled among mats of vegetation near the shore

ultraplankton = nannoplankton

zooplankton the animal constituent of the planktonic population

planktont an organism of plankton (*cf.* plankter)

Plannipennia that order of insects now usually included in the Neuroptera and containing, *inter alia*, the antlions

-plano- *comb. form* meaning "roaming"

planont an organism which wanders

Planorbidae a family of pulmonate gastropods usually having flattened, or very low-spiraled, shells. Commonly called orb snails

plant 1 (*see also* plant 2, 3, beetle, cutter, hopper, bug) in names describing types, as distinct from species, of organism. For "plant" as the antithesis of "animal" *see* Planta

  ant plant any plant actually, or reputedly, attractive to ants

  ginger-beer plant a symbiotic association of yeasts and bacteria at one time popular in England for the fermentation into alcohol of ginger-flavored sugar solutions

  cellular plant one which does not possess vascular tissue

  compass plant (*see also* plant 2) one in which the edges of the leaves point north and south

  cushion plant any of numerous plants forming a dense cushion-like mass

  mother plant a plant from which others are derived either sexually or asexually

  resurrection plant one of numerous plants, at one time a frequent article of commerce, which can rest for long periods in a dry state, and be restored by soaking in water; several are species of Selaginella. The term is also applied to plants which, by legend or geographical location, are associated with the resurrection of Christ. This category includes the rose of Jericho and the fig-marigold

  rosette plant those having very short internodes and closely set leaves

  ruderal plant one of a rubbish heap

  superplant a plant which lives on another plant either as an epiphyte, a symbiote or a parasite

  vascular plant those which possess vascular bundles

plant 2 (*see also* plant 1, 3) as part of a specific name (*cf.* weed, wort)

  amulet-plant the euphorbiaceous tree *Putranjiva roxburghii*

  artillery-plant the urticaceous herb *Pilea microphylla* the pollen of which is violently dispersed

  bead-plant the rubiaceous herb *Nertera depressa*

  bee-plant the capparidaceous herb *Cleome serrulata*

  Australian bower-plant bignoniaceous climbing shrub *Plandorea jasminoides*

  candle-plant the compositaceous herb *Senecio articulatus*

  century-plant any of several species of the amaryllidaceous genus Agave particularly *A. americana* (= American aloe)

  compass-plant (*see also* plant 1) the compositaceous herb *Silphium laciniatum*

  coral-plant *either* the scrophulariaceous herb *Russelia juncea or* the euphorbiaceous shrub *Jatropha multifida*

  corpse-plant *Brit.* the pyrolaceous herb *Monotropa uniflora* (*cf.* pinesap)

cowplant the asclepiadaceous herb *Oxystelma esculentum*

cruel plant the asclepiadaceous herb *Cynanchum acuminatifolium* (= mosquito-plant)

cup-plant the compositaceous herb *Silphium perfoliatum*

dewplant any of numerous droseraceous herbs of the genus Drosera

egg-plant (*see also* beetle, bug, miner) the solanaceous shrub *Solanum melongena* and its edible fruit
  Ethopian egg-plant = scarlet egg-plant
  scarlet egg-plant *S. integrifolium*

Mexican fire-plant the euphorbiaceous shrub *Euphorbia heterophylla*

frog-plant *U.S.* the crassulaceous herb *Sedum purpureum*

gas-plant any of several species of rutaceous plants of the genus Dictammus

gopher-plant (*W.U.S.*) = mole-plant

harts-horn plant the ranunculaceous herb *Anemone patens* (= pasque flower)

humble-plant the leguminous shrub *Mimosa pudica*

hypocrite plant the euphorbiaceous shrub *Euphorbia heterophylla*

ice-plant the aizoaceous herb *Mesembryanthemum crystallinum*
  New Zealand ice-plant the aizoaceous herb *Tetragonia expansa*

Chinese lantern plant the solanaceous herb *Physalis franchetii*

lead-plant the leguminous shrub *Amorpha canescens*

leopard-plant the compositaceous herb *Ligularia kaempferi*

letter plant the orchid *Grammatophyllum speciosum*

loco-plant = locoweed

love plant any of numerous species of the portulacaceous herb Anacampseros

mole-plant the euphorbiaceous herb *Euphorbia lathyris*

mosquito-plant the asclepiadaceous herb *Cynanchum acuminatifolium* (= cruel plant)

castor oil plant the euphorbiaceous shrub *Ricinus communis*

oyster-plant the compositaceous herb *Tragopogon porrifolius* and its edible roots (= salsify)
  Spanish oyster-plant the compositaceous herb *Scolymus hispanicus* and its edible root (= scorzonera)

paper-plant = papyrus (*q.v.*)
  rice-paper plant the araliaceous shrub *Tetrapanax papyriferus*

patchouli plant the labiateous herb *Pogostemon heyneanus*

peacock-plant the compositaceous herb *Gazania pavonia*

pie-plant *U.S.* = rhubarb

pitcher-plant (*see also* mosquito) *U.S.* any of several species of sarraceniaceous insectivorous herbs of the genus Sarracenia. *Brit.* any of numerous species of insectivorous herbs of the nepenthaceous genus Nepenthes

poker-plant the liliaceous herb *Kniphofia uvaria*

rattlesnake plant the marantaceous plant *Calaphea crotalifera*

rubber plant the moraceous shrub *Ficus elastica*

sensitive plant the leguminous shrub *Mimosa pudica*

sheep-plant any of several species of compositaceous plant of the genus Raoulia (= vegetable sheep)

side-saddle plant any of several species of sarraceniaceous insectivorous herbs of the genus Sarracenia (= pitcher-plant)

slipper-plant any of several species of euphorbiaceous succulent shrub of the genus Pedilanthus

soap-plant the liliaceous herb *Chlorogalum pomeridianum* (*cf.* soapwort)

spider-plant the capparidaceous herb *Cleome spinosa*

tapioca plant the euphorbiaceous herb *Manihot utilissima*

telegraph plant the leguminous herb *Desmodium gyrans*

tortoise-plant any of several species of the dioscoriaceous shrub Testudinaria

umbrella plant *U.S.* any of several species of the polygonaceous genus Eriogonum and the sedge *Papyrus alternifolius*

unicorn-plant the martyniaceous herb *Martynia louisiana*

vanilla plant the compositaceous herb *Trilisa odoratissima* (*but see also* vanilla)

vinegar plant = mother-of-vinegar

Washington plant the nymphaeaceous aquatic plant *Cabomba caroliniana*

wax-plant the ascelepiadaceous woody vine *Hoya carnosa*

weather-plant the leguminous vine *Abrus precatorius*

lampwick plant the labiateous herb *Phlomis lychnitis*

wine plant *U.S.* = rhubarb

wire plant the polygonaceous shrub *Muehlenbeckia complexa*

zebra-plant the marantaceous plant *Calaphea zebrina*

plant 3 (*see also* plant 1, 2) in the sense of planting or implantation

implant any tissue transferred to a new location in other tissues
  interplant an implantation of tissues from one organism into another organism

planta the basal part of the posterior tarsus of pollen-gathering hymenopterous insects

Planta that kindgom of organisms which contains the plants. No single characteristic distinguishes plants from animals though, among plants, only the fungi, and a few parasites among higher forms, lack chlorophyll. In general, the utilization of chlorophyll in photosynthesis, the presence of cellulose cell walls and a relatively slow response to external stimuli distinguish plants from animals

Plantaginaceae that family of plantaginale plants which contains the herbaceous plantains, not to be confused with the banana-like fruits of the same name. The alternate or opposite lanceolate leaves and the peculiarly mucilaginous seeds are typical

Plantaginales that order of dicotyledonous angiosperms which contains the single family Plantaginaceae, the characteristics of which pertain to the order

plantain (*see also* coot, eater, lily, spearwort) a term variously applied to herbs of the family Plantaginaceae, and to numerous other plants including the banana *Musa paradisiaca* and (*Brit.* rare) the tree *Platanus orientalis* (= plane tree)

Indian plantain any of several species of compositaceous herb of the genus Cacalia

mud plantain any of several species of pontederiaceous herb of the genus Heteranthera

rattlesnake plantain *U.S.* any of several species of the orchidaceous genus Goodyera, particularly *G. pubescens*

robin's plantain the compositaceous herb *Erigeron pulchellus*
  poor robin's plantain the compositaceous herb *Hieracium venosum* (= rattlesnake weed)

water plantain the alismaceous water plant *Alisma plantagoaquatica*

wild plantain the musaceous tree-like foliage plant *Helicomia bihai*

plantar pertaining to the sole of the foot or the palm of the hand

acutiplantar said of birds which have the posterior part of the tarsus at an acute angle

antiplantar pertaining to the dorsal surface of the hindfoot

latiplantar said of the foot of a bird in which the hinder part of the tarsus is rounded

scutelliplantar said of birds in which the plantar surface of the tarsus is broken up into small scales

implantation the attachment of the blastocyst of a mammal to the uterine wall

planticle the embryo of a seed

plantling a small, or juvenile plant

plantule = plumule

planula *see* larva

Planuloidea = Mesozoa

plaque = platelet

electroplaque a modified myoblast forming a unit of an electric organ in electric fish

metaplasia the production of one kind of tissue by cells belonging to another kind of tissue

-plasm- *comb. form* properly meaning "molded" but by extension "layer" or "substance" or even "organelle." These meanings shade into each other and are not separated in the following list

alloplasm sum total of organelles in a cell, and also used in the sense of highly specialized protoplasm such as that which forms organelles (*cf.* alloplasmic)

archoplasm *either* specifically the centrosome *or* in the sense of kinoplasm (*q.v.*)

arrhenoplasm male protoplasm

bioplasm = protoplasm

cytoplasm that part of the cell substance which is not the nucleus

deuteroplasm protoplasm in the sol condition

deutoplasm = yolk

ectoplasm the outer, gel layer of the protoplasm of rhizopod Protozoa. The term is also applied to the capsule of bacteria

endoplasm the inner (sol) layer of the protoplasm of rhizopod Protozoa

eoplasm a theoretical intermediate stage between inorganic matter and protoplasm

epiplasm the residual protoplasm of an ascus after spore formation

ergastoplasm basophilic elements in the cytoplasm not otherwise identified. The term was also at one time used for endoplasmic reticulum

germ plasm the cytoplasm of germ cells constituting a germ line from generation to generation

gynoplasm the protoplasm of a female gamete

hyaloplasm that portion of the protoplasm in which granules are not apparent. Since protoplasm was at one time thought to be a reticular structure, this term has also been used in the sense of fibrils or reticulations. It has also been used for the postulated clear substance filling the postulated alveoli in the alveolar theory

cytohyaloplasm the sum total of the protoplasm of cells apart from any inorganic inclusions

nucleo-hyaloplasm = linin

hygroplasm protoplasm in the sol condition

hypnoplasm the protoplasm of a dormant organism, particularly of a seed

hysteroplasm protoplasm in the sol condition

idioplasm an early term for the living portions of protoplasm

karyoplasm = nuclear sap

kinoplasm the cytoplasmic components of a mitotic figure

metaplasm an early term for those parts of the protoplasm which, being granular, were thought to produce the rest of the protoplasm. The term has also been applied to the sum total of nonliving material within a cell

necroplasm dead protoplasm

ooplasm a yolk-free area of cytoplasm immediately surrounding the nucleus in isolecithal eggs

paraplasm protoplasm in a sol condition

periplasm that protoplasm in the sexual organs of plants which does not give rise to gametes

plastidplasm a term derived from an early hypothesis that the protoplasm of plastids differed from that of the rest of the cell

polioplasm that part of the protoplasm of a plant cell which flows

protoplasm the sum total of the contents of a living cell

sarcoplasm the amorphous material in which myofibrils are embedded

somaplasm that part of the protoplasm of an organism which plays no part in the production of germ cells

somatoplasm the sum total of all the tissues of an organism except the gonads

sporeplasm that part of a sporangium which produces spores

sporoplasm an amoeboid mass within a spore, frequently giving rise to a gamete

stereoplasm protoplasm in the gel condition

thelyplasm female protoplasm

plasma blood from which the corpuscles have been removed (*cf.* serum)

plasmatic being in a condition for growth

Plasmatogennylicae term coined to combine angiosperms and gymnosperms

alloplasmic a term applied to motile organelles such as cilia (*cf.* alloplasm)

deutoplasmic *see* deutoplasm and valve

plasmid a term coined to include both intrinsic plasmogenes and extrinsic factors such as viruses the effect of which may be mistaken for that of a true plasmogene

plasmin an enzyme hydrolyzing peptides and esters of arginine and lysine and thus converting fibrin into soluble products

plasmodiation the softening of a spore before germination

Plasmodiophoreae a class of the phylum Myxomycophyta containing those forms in which the thallus is a naked multinucleate plasmodium

plasmodium a multinucleate mass of protoplasm, particularly that formed from the fusion of zoospores in the life history of myxomycetes and sporozoans

aggregate plasmodium a myxomycete plasmodium in which the myxamoebae are congregated without fusion and each cell of which develops separately

fused plasmodium the stage of a myxomycete in which the previously free myxamoebae fuse and subsequently come to fruit

pseudoplasmodium the result of the aggregation, but not fusion, of myxoamoebae

Plasmodroma a sub-phylum of animals containing those Protozoa which lack cilia

plasmon = plasmagene

homo**plasmy** the condition of being homologous, in the sense that this term is antithetic to analogous

plasome = plasmatosome

-plast- *comb. form* indicative of "formation" or "shaping" in the sense, which is the original meaning, of "molding." The *subst. suffix* -plast is often used for plasted. See -blast- for remarks on further confusion

amphiplast a chromosome which has lost a trabant

amyloplast a plastid modified for the storage of starch grains

anaplast = leucoplast

apoplast the continuum of non-living material (*e.g.* cell walls and xylem) in a plant (*cf.* symplast)

autoplast = chloroplast

basiplast a leaf which matures first at the apex

bioplast a minute fragment of living material

blepharoplast *zool.* an organelle located at the base of a cilium or flagellum. *bot.* that portion of the protoplasm from which cilia arise

centroblepharoplast a blepharoplast that functions as a centriole in mitotic division

centroplast a thickened area in the center of some heliozoan Protozoa on which the axopods are implanted

chloroplast an organelle present in photosynthetic plants containing chlorophylls and sometimes other pigments

chromoplast a colored plastid

cytoplast = cytoplasm

hemocytoplast the first stage in the development of a polymorph leucocyte

ectoplast the cell membrane, as distinct from cell wall, of a plant cell. The term is also applied to chromoplasts of Cyanophyceae

elaeoplasts oil droplets or oil containing plastids in plant cells

endoplast the living portion of a cell

genoplast = genotype

gymnoplast a naked cell

idioplast a cell having unusual contents

karyoplast the nucleus

kinetoplast the blepharoplast and parabasal body combined. The term is also applied to a body associated with, and larger than, the blepharoplast of hemoflagellate Protozoa

leucoplast a colorless plastid

libroplast elaeoplasts which lie free on the median line of diatoms

neoplast an organism produced from a few cells (*e.g.* sexually) as distinct from that produced by a bud or shoot

nematoplast a thread-shaped plastid

oikoplast a tunic-secreting cell of urochordates

ooplast = oosphere

periplast that part of the protoplasm immediately surrounding the nucleus

phaeoplast = phaeophore

phragmoplast a fibrous spindle, precursor to the cell plate (*q.v.*) and hence a spindle between two nuclei in the same cell

chlorophylloplast = chloroplast

placoplast an elaioplast of a diatom

polyplast in *bot.* a multicellular, undifferentiated embryonic stage, roughly corresponding to the morula of some animals

proteinoplast a plastid modified for the storage of protein

protoplast the protoplasm contained within one cell

rhizoplast a fiber connecting the centriole of some Mastigophora to the blepharoplast *or* the connection of a blepharoplast to the nucleus

symplast the continuum of living material in a plant (*cf.* apoplast)

gymnosymplast a naked syncytium

tonoplast the limiting layer between a vacuole and a cytoplasm

trophoplast those plastids which were originally assumed to play a role in the nutrition of a cell

-plastic- *comb. form* from Greek having precisely the same meaning as the English word plastic, but used in *biol.* as an *adj. form* from both -plast- and plastid

aplastic that which cannot be shaped or converted. Hence, materials that cannot be used as nutrients

homoplastic 1 (*see also* homoplastic 2) the condition of an abnormal increase in tissue

homoplastic 2 (*see also* homoplastic 1) analogous in the sense that this term is anthetic to homologous

organoplastic capable of producing organs

pleuroplastic said of a leaf with a marginal meristem

polyplastic the condition of a septate spore

plastid 1 (*see also* plastid 2) any discreet organelle particularly those in plant cells. The suffix -plast is often used in this sense—*e.g.* amyloplast

endoplastid a plastid containing a starch granule

erythroplastid a mammalian erythrocyte

leucoplastid any colorless cytoplasmic granule

proplastid a particle in a plant cell from which a plastid is derived

plastid 2 (*obs.*) a cell

aplanoplastid a non-flagellated cell

teleplastid a reproductive cell either sexual or asexual

plastidule the smallest mass of protoplasm that can exist as a living unit

Plastoceridae a family of coleopteran insects closely allied to the Elateridae and often fused with them

plastoid a needle-shaped body found in the stalk-cells of the tentacles of Drosera

plastron the under-shell of a Chelonian reptile

-plasy *comb. suffix* used in *biol.* to denote alteration in structure

callus heteroplasy structures developed either in plants or animals in consequence of a wound

metaplasy a change in the cell content or structure not of necessity caused by a wound

polaplasy the division of an organ into several parts

Platacanthomyidae a family of myomorph rodents closely allied to the Gliridae, but differing from them both in the possession of spines interspersed with the hair and in their habit of boring holes in tree trunks

Platacidae a small family of acanthopterygian fishes commonly called bat fish by reason of their equally long dorsal and anal fins

Plataleidae a family of ciconiiform birds usually included with the Threskiornithidae

Platanaceae that family of rosale dicotyledonous angiosperms containing the plane trees. The dense globose flower head is a characteristic of the family

Platanistidae a family of odontocetan Cetacea closely allied to the Delphinidae but distinguished from them by the long and narrow jaw

platanna *S.Afr.* the pipid anuran amphibian *Xenopus laevis* (= clawed frog)

plate any flat structure of limited area

alar plate the dorsal half of the side of the brain stem or spinal cord

anal plate any plate abutting on or surrounding the

anus. The term is common in arthropods and is used also for the transversely enlarged ventral scale that covers the anus in some reptiles

apophystegal plate thickened plates covering the gonapophyses in orthopteran insects

basal plate 1 (*see also* basal plate 2) the cartilaginous base of the chondrocranium below the foramen magnum

basal plate 2 (*see also* basal plate 1) the ventral half of the lateral side of the brain stem or spinal cord

Bianconi's plate a network of fibers on the sensitive surface of plant tendrils

cardiogenic plate early primordium of the heart of the human embryo

cell plate the first appearance of the future cell wall between two cells at the conclusion of mitosis (*cf.* phragmoplast)

prechordal plate the anterior, undifferentiated mesoderm in the early embryo of some vertebrates, particularly elasmobranch fish

endochrome plate a band of chromatophores in diatoms

cirral plate the skeletal plate in the cirrus of an echinoderm, corresponding to the columnal in the stalk

closing plate a membrane formed by the junction of a pharyngeal pouch with a visceral groove in the vertebrate embryo

columnal plate one of the numerous, skeletal plates, supporting the stalk of crinoid echinoderms

cribriform plate that part of the ethmoid bone of the mammal which separates the cranial from the nasal cavity and contains several foramina for the divisions of the olfactory nerve. Sometimes preformed from cartilage in the embryo

coxal plate a thickening on the sternite of an arthropod supporting the base of the coxa and, in certain insects, a plate-like dilation of the coxa

dermal plate a dermal bone forming much of the side of the cranium of monotremes, and which is either a development of, or at least fused with, the petrosal

dental plate a projection from the edge of the palintrope of a brachiopod valve

end plate a terminal ossicle on the podium of an echinoderm

epigynal plate a median thickening on the lower surface of acarines lying immediately posterior to the sternal plate and immediately anterior to the ventral plate

equatorial plate the wide central portion of the mitotic figure on which the chromosomes lie at metaphase

filament plate a lateral plate of cells in the insect embryo connecting the genital and cardiac rudiments

floor plate a narrow plate along the ventral surface of the brain stem or spinal cord separating the basal plates

genital plate 1 (*see also* genital plate 2) a plate covering the aperture of the sex organs in some arachnids

genital plate 2 (*see also* genital plate 1) one of those plates which lie in the interambulacral position on the aboral surface of echinoid echinoderms and each of which bears a gonopore

gulamental plate the basal plate of the insect labium, particularly the fused gula and submentum of termites

jugular plate thickenings on the undersurface of acarines lying immediately anterior to the sternal plate on each side of the middle line and usually bearing setae

lateral plate the non-segmented, lateral part of the mesodermal mantle of an embryo

madroporic plate the sieve plate which connects the internal water-vascular system of echinoderms with the exterior

medullary plate = neural plate

ventral nerve plate a thick, nervous layer, replacing the ventral nerve chord in some terrestrial Turbellaria

neural plate those thickened cells on the mid-dorsal surface of an early embryo which will later sink in to form the central nervous system

ocular plate 1 (*see also* ocular plate 2) echinoderm = terminal plate (echinoderm)

ocular plate 2 (*see also* ocular plate 1) a thickening on the dorsal surface of the gnathosoma of acarines which surrounds and supports the eyes

metapodal plate lateral thickenings on the ventral surface of acarines posterior to the parapodal and endopodal plates and further from the medial line than the lateral plates

parapodal plate a thickening on the undersurface of acarines lying immediately outside the attachment of the legs

parietal plate a pair of cartilages in a chondrocranium immediately above the otic capsules. In most embryonic forms they are connected by the synotic tectum

perforation-plate the area of a xylem vessel member (*q.v.*) which bears one or more perforations

foraminate perforation-plate one irregularly covered with circular holes

reticulate perforation-plate one having the appearance of a net

scalariform perforation-plate one in which the perforations are arranged in parallel rows

peritremal plate a thickening on the undersurface of acarines corresponding in position to, but outside of, the parapodal plates

perpendicular plate a projection of the ethmoid bone which separates the right and left nasal passages

hypophyseal plate a cartilage formed in the embryonic chondrocranium from the fusion of the hypophyseal cartilages

endopodal plate a thickened area of the undersurface of acarines lying between the attachments of the walking appendages on the inner (*i.e.* medial) side

preendopodal plate a small thickening on the undersurface of acarines· immediately anterior to the jugular plates

podical plate the paraproct, but also used for latero-ventral plates on the tenth abdominal segment of an orthopteran insect

pore-plate a porous plate lying between contiguous zoecia in some Ectoprocta

roof-plate a narrow area along the dorsal side of the brain stem separating the alar plates

rosette plate (Ectoprocta) = pore plate (Ectoprocta)

sclerotic plate one of a ring of bones in the outer part of the eyeball surrounding the area exposed to the exterior

sieve-plate 1 (*see also* sieve plate 2, 3) a sieve area with larger holes and much larger callose cylinders than most, and usually placed on the terminal walls of elongated phloem sieve elements

sieve plate 2 (*see also* sieve plate 1, 3) a transverse porous partition below the osculum of some hexactinellid sponges

sieve plate 3 (*see also* sieve plate 1, 2) = madreporic plate

skeletal plate any flat skeletal structure but specifically the thickened basal membrane in the anterior region of Chaetognatha

skeletogenous plate a small plate lying under the stomochord in hemichordates

hysterosomal plate a thickening on the base of the ventral surface of the hysterosome of acarina

propodosomal plate a thickened area on the underside of the propodosoma of acarines

perisomatic plate the skeletal plates of the vault of a crinoid echinoderm

sternal plate a thickened area, often ornamented, on the ventro-posterior surface of acarines

metasternal plate a thickening on the undersurface of acarines lying immediately anterior to the lateral plate and posterior to the endopodal plates, and therefore placed on the medial side of the gap between the last two legs

stigmal plate a thickened area surrounding the stigma of arthropods

tarsal plate a dense connective tissue plate, supporting the edge of the eyelid

terminal plates those plates which lie intermediate tween the genital plates on the aboral surface of an echinoid echinoderm

basitrabecular plate the anterior region of the base of the chondrocranium of cyclostomes

ventral plate 1 (*see also* ventral plate 2) a large thickening on the undersurface of acarines lying between the anal plate and the epigyneal plate

ventral plate 2 (*see also* ventral plate 1) a layer of cells of the insect blastoderm on the ventral side of the egg and which later becomes the germ band

platelet detached portions of megakaryocyte cytoplasm found in blood

platinum] a mutant (gene symbol *pam*) mapped at 23.1 on the X chromosome of *Drosophila melanogaster*. The male body and bristles of the phenotype are colorless with a trace of dark shading at the base of the bristles. The males are sterile

-platy- comb. form meaning "broad" but widely used in *biol.* as though it meant "flat"

platy aquarists' name for several poeciliid fishes of the genus Xiphophorus (*cf.* swordtail)

amphiplatyan said of a vertebra in which both articular surfaces of the centrum are flat (*cf.* -coelous)

Platycephalidae a family of acanthopterygian fishes containing the flatheads

Platycopa an order of ostracod Crustacea containing a few marine species distinguished by the absence of an opening between the valves through which the antennae may extend and in possessing only four pairs of postoral appendages

Platyctenea a curious order of aberrant tentaculate ctenophores which is thought by many to be intermediate between this group and the lower Platyhelminthes. They are flattened creeping forms and in some the ciliated plates may be present only in the larva

Platygastridae = Platygasteridae

Platygasteridae a family of proctotrupoid hymenopteran insects

Platyhelminthes that phylum of the animal kingdom which contains the flatworms. These are usually regarded as the most primitive phylum in the group Bilateria and they are distinguished by the presence of a mouth, the absence of an anus, and by the fact that the space between the endoderm and ectoderm is filled with parenchymatous tissue

Platypezidae a family of aschizan cycloraphous brachyceran dipteran insects distinguished by the flattening of the basal segment of the hind tarsus. Sometimes called flat-footed flies

Platypsyllidae a monotypic family of coleopteran insects erected to contain the unique form *Platypsyllus castoris* which is an ectoparasite on beavers. They lack both eyes and hindwings. The order Achreioptera has been proposed for this insect and some authors include it in the family Leptinidae

Platypodidae a family of scolytoid coleopterans commonly called pin-hole borers

duck-billed platypus the monotreme *Ornithorhynchus anatinus*

Platyrrhina a division of the primates containing the New World as distinct from the Old World forms. They derive their name from the fact that the nostrils point either upwards or forwards. This division comprises the sub-orders Hapaloidia and Ceboidia

Platyrrhinidae = Anthribidae

Platysternidae a family of cryptodiran chelonian reptiles erected for the Asiatic water turtle *Platysternon megacephalum*. It is distinguished from other chelonians by the peculiar arrangement of the bones in the skull. This family is treated by some as a subfamily of the Testudinidae

Platystomatidae a family of brachyceran dipteran insects usually included with the Otitidae

Platystomidae = Anthribidae

Platystomoidea = Anthriboidea

Platyuridae a family of nematoceran dipteran insects now usually included with the Mycetophilidae

[pleated] a mutant (gene symbol *pl*) mapped at 47.9 on the X chromosome of *Drosophila melanogaster*. The name refers to the longitudinal pleating of the wings of the phenotype

Plecoptera that order of insects containing the forms commonly called stoneflies. They are distinguished by four membranous wings lying flat on the body and by the presence of segmented cerci on the last abdominal segment

-plect- comb. form meaning "woven," "twisted" or, by extension "folded" (*cf.* -ploc-)

plectane a striated, cuticular plate supporting the posterior papillae of some nematodes

amplectant embracing

Plectascales = Aspergillales

amplected said of an insect in which the head is set in a cavity of the thorax

Plectognathi an order of Osteichthyes usually posessing small mouths and small opening from the gill cavities. Many have bony plates and spines over the body. Some are poisonous and a few venomous. The best known members are the filefishes, triggerfishes and trunk-fishes

Plectomycetes a subclass of euascomycete Ascomycetae in which the liberation of ascospores is produced by weathering of the peridium

plectrum a thick, curved bristle on an insect, particularly a dipteran

Plegadidae = Threskiornithidae

-pleio- comb. form meaning "more." Frequently transliterated -pleo- and thus confused with this form

-pleo- comb. form confused between two *Gr.* roots and therefore meaning either "full" or "to sail." *See* also -pler- and -pleo-

pleon the abdominal region of those Crustacea (*e.g.* crayfish, lobster) in which the whole abdomen is used for swimming. The term is also used as synonymous with abdomen in some other Crustacea

posterior pleon the terminal segment of the crustacean abdomen

pleonasm reduncancy

-pler- *comb. form* meaning "full"

plerome the growing tip of a root encased in periblem

plesiasmy the condition of having an abnormally shortened stem

-plesio- *comb. form* meaning "near," "approximate" or "recent"

Plesiopora an order of oligochaete annelids distinguished by the fact that the male aperture is found on the segment behind that in which the testis is located

-pleth- *comb. form* meaning to "swell" or "full up"

Plethodontidae a family of salamandrid urodele Amphibia distinguished by the presence of a nasolabial groove and absence of lungs

-plethysmo- *comb. form* meaning "swelling" or "enlargement"

-pleur- *comb. form* meaning "side" or "rib" (*see also* -pleura-)

-pleura- *comb. form* meaning "rib cage" or "side" in the sense of a "side of beef"

pleura the lining of the thoracic cavity of mammals

  endopleura the inner seed coat

  epipleura the outer half of a diatom girdle

  exopleura = testa

  hyopleura the inner half-girdle of the frustule of a diatom

pleural pertaining to the pleuron

metapleure one of a pair of integumentary folds on the latero-ventral surface of cephalochordates

pleurion one of a pair of lateral scaly plates on the head of a gastrotrich (*cf.* cephalion, hypostomium)

pleurite used both in the sense of sclerite 2 (*q.v.*) and for flexible, intersegmental, membranes of arthropods

  anapleurite a thickened plate in the insect pleuron immediately above the attachment of the coxa

  endopleurite a groove or flap between pleurites

  epipleurite the upper half of a pleuron when it is separated by a horizontal suture

Pleuroceratidae a family of ctenobranchiate Mollusca having a paucispiral operculum; commonly called river snails

Pleurodira an assemblage of chelonian reptiles distinguished by the fact that the neck bends laterally when retracted in distinction from the Cryptodira in which it bends in the vertical plane. The two families Pelomedusidae and Chalididae are distinguished by the extent to which the neck can be retracted

pleurodire a chelonian reptile that pulls its head under the carapace by bending the neck horizontally

pleuron the lateral region of the segment of an arthropod body

  antepleuron the episternum of the last two thoracic segments of insects

  epipleuron 1 (*see also* epipleuron 2) a lateral projection from the rib of the teleost fish

  epipleuron 2 (*see also* epipleuron 1) that portion of the elytrum of the coleopteran insects which is turned back from the edge

  eupleuron the dorsal arch of the pleuron

  mesopleuron the lateral wall of the mesothorax in insects

  metapleuron the lateral wall of the metathorax in insects

postpleuron = epimeron

propleuron the lateral wall of the prothorax in insects

Pleuronectidae a family of heterostomate Osteichthyes in which both eyes are on the right side of the head. Commonly, therefore, called right-eye flounders

Pleuronectiformes = Heterostomata

Pleurophoridae a family of pelecypod mollusks commonly called the black clams. They are distinguished by the circular shell lacking a lunule and with a very thick and heavy periostracum

pleuston organisms that float because of their low specific gravity (*cf.* neuston)

[Plexate] a mutant (gene symbol *Px*) mapped between 107.0 and 107.4 in chromosome II of *Drosophila melanogaster*. The venation of the phenotype is as in [blistered] but the veins are thickened and broken

implexous interlaced

-plexus *see also* amplexus

plexus a network of nerves or blood vessels

  Auerbach's plexus an autonomic plexus in the muscle layers of the intestine

  labial plexus (crinoid) = spongy organ (crinoid)

  Meissner's plexus a parasympathetic ganglionated plexus found in the intestinal submucosa

  myenteric plexus = Auerbach's Plexus

  nodal plexus a tangle of vessels at a plant node

  solar plexus = coeliac ganglia

  chorioid plexus an extension of the vascularized roof into the inner cavity of the brain

[plexus] a mutant (gene symbol *px*) mapped at 100.5 on chromosome II of *Drosophila melanogaster*. The phenotypic expression is a plexus of extra veins in the wings

-plic- *comb. form* meaning "fold" or "to fold" or occasionally "to plait"

plicate folded

  complicate folded together

  conduplicate folded together lengtwise

  reduplicate apart from its usual meaning, is applied in botany to a flower bud the tips of which are bent back before opening

  contortuplicate twisted upon itself and then folded back on itself

  eplicate not folded

  implicate woven

  multiplicate repeatedly folded

  replicate bent back on itself or several times repeated. In *bot.* used of stamens united by their filaments

plication the condition of being folded

duplication literally doubled but used in genetics specifically for the gain of a section of chromosome in excess of the normal diploid number

  conduplication the condition in a folded flower in which organs are pressed against each other, face to face

  endoreduplication a doubling of chromosomes within the interphase nucleus without the separation of the chromatids or the appearance of any of the phenomena of mitosis (*cf.* endomitosis)

plicatulate slightly folded

plicature a fold

-ploc- *comb. form* meaning "to weave" or "twist together" (*cf.* -plect-)

Ploceidae a very large family of passeriform birds commonly called weaverbirds from the elaborate solitary or colonial nest made by many

ortho**ploceous** the condition of having the cotyledons folded round the radicle

sym**plocium** = sporangium

Ploiariidae a family of hemipteran insects commonly called thread-legged bugs, in virtue of the greatly elongate second and third pairs of legs. The first pair of legs are reflected predatory organs resembling those of the preying mantis

-ploid *comb. suffix* which has come to mean "replicate" by extension from the *Gr.* root meaning "doubled"

amphiploid a type of polyploidy characterized by the addition of two sets of chromosomes from each of two species

aneuploid an irregular polyploid in that the nuclei do not contain a multiple of the haploid number of chromosomes. Also used in the sense of "lacking chromosomes"

artioploid having chromosomes multiplied by even numbers

autoploid a type of polyploid in which each of the chromosome sets has been derived from the same species

diploid having twice the number (2N) of chromosomes found in a gamete, or in the other of two alternate generations. The diploid number is usually given as the chromosome number of an organism (*cf.* haploid)

amphidiploid an allopolyploid having two complete diploid sets from each of the two species from which it was hybridized

didiploid the condition of a nucleus formed by the fusion of diploid nuclei

hyperdiploid the condition in which the normal complement of chromosomes is augmented by a translocated portion of another chromosome

syndiploid a diploid produced through failure of chromosomes to separate in meiosis

dysploid having a number of chromosomes which is not a multiple of the haploid number (= aneuploid)

euploid a polyploid, the chromosome number of which is an exact multiple of the chromosome number of its ancestral species, though also used in the sense of having a chromosome set that is a multiple of a monoploid

haploid having half the number of chromosomes found in a somatic cell or half the number found in the other of two alternating generations. The haploid is usually said to have N chromosomes in contrast to the diploid (*cf.* diploid)

double haploid one having a complete genome from each of two species

synhaploid the condition arising from the fusion of two or more haploid nuclei

hypoploid having less than the normal number of chromosomes

monoploid having a chromosome set that is a complete genome

orthoploid a polyploid the chromosomes in which are an exact multiple of the haploid number of chromosomes of the organism from which it was derived

perissoploid having chromosomes in uneven multiples

polyploid having more than the usual number of chromosomes

allopolyploid the condition in which the replicated diploid sets come from genetically different strains

amphipolyploid = amphiploid

autopolyploid one in which all the diploid sets have come from the same parent species

duplicational polyploid = autopolyploid

endopolyploid a polyploid resulting from the replication of chromosomes within an undivided cell

partial polyploid one in which segments of chromosomes but not entire genomes are replicated

secondary polyploid an allopolyploid with a predominance of genomes from one ancestor

tetraploid having twice (*i.e.* 4N) times the usual number of chromosomes

allotetraploid a tetraploid in which one diploid set has been derived from a genetically different parent

autotetraploid one in which the four diploid sets are genetically identical

double tetraploid one which carries four genomes from each of two distinct species

triploid applied to a nucleus having one and a half times the diploid number of chromosomes (*i.e.* 3N)

allotriploid a triploid with two similar and one dissimilar genomes

autotriploid one in which the three diploid sets are identical

ploidy the condition of having the chromosome sets replicated. Compound terms are given under -ploid above

Ploima an order of monogonate Rotifera distinguished from the Flosculariacae and Collothecacae by the fact that the corona, though variable, never resembles that of these two groups. The foot, when present, has two toes and two pedal glands

epiploon (Insecta) = caul (Insecta)

diplostic said of rootlets having only two vascular bundles

Plotidae = Anhingidae

Plotosidae a family of scaleless catfishes with long dorsal fins and long anal fins confluent with the caudal fin. Many have venemous spines

plough-girl *U.S.* the noctuid lepidopteran insect *Richia aratrix*

plover properly any of numerous species of charadriiform birds of the family Charadriidae, but also applied to many other plover-like birds

banded plover *Zonifer tricolor*

bastard plover *Vanellus vanellus* (= lapwing)

black-bellied plover *U.S. Squatarola squatarola* (= grey plover *Brit.*)

wrybill plover *Anarhynchus frontalis*

blacksmith plover *Hoplopterus armatus*

Caspian plover *Eupoda yeredus*

collared plover *Charadrius collaris*

crab-plover the dromadid *Dromas ardeola*, a crepuscular shore bird of Africa and parts of Asia

Egyptian plover the glariolid *Pluvianus aegyptuis* (= crocodile-bird)

golden plover *U.S. Pluvialis dominica Brit. P. apricaria*

grey plover *Brit.* = black-bellied plover *U.S.*

blackhead plover *Sarciophorus tectus*

white-headed plover *Xiphidiopterus albiceps*

Kentish plover *Brit. Charadrius alexandrinus*

marsh plover *U.S.* the scolopacid *Erolia melanotos* (= pectoral sandpiper) *Brit.* (rare) the scolopacid *Scolopax rusticola* (= woodcock)

mountain plover *Eupoda montana*

Norfolk plover the burhinid *Burhinus oedicnemus* (= stone-curlew)

piping plover *Charadrius melodus*

quail-plover the turnicid *Ortyxelos meiffrenii*

ringed plover *Brit. Charadrius hiaticula*

little ringed plover *Charadrius dubius*

sand-plover *N.Z. Thinornis novaeseelandiae.* Also many of several Asiatic and African species of the genus Charadrius
    African sand-plover *C. pecuarius*
    chestnut-banded sand-plover *C. venustus*
    large sand-plover *C. leschenaultii*
    Malay sand-plover *C. peronii*
semipalmated plover *Charadrius semipalmatus*
shore plover *Thinornis novaeseelandiae*
snowy plover *U.S. Charadrius alexandrinus* (= *Brit.* Kentish plover)
upland plover the scolopacid *Bartramia longicauda* (= upland sandpiper)
wattled plover *Afribyx senegallus*
stone-plover = thickknee
swallow-plover = pratincole
black-winged plover *Stephanibyx melanopterus*
spur-winged plover *Afr. Hoplopterus spinosus Austral. Lobibyx novaehollandiae I.P. Hoplotpterus duvaudcelii*
Wilson's plover *Charadrius wilsonia*
ploward *W.I.* = killdeer
    little ploward *W.I.* either of two species of charadriid bird *Charadrius collaris* and *C. wilsonia* (*cf.* plover)
plug anything that blocks a hole
    copulatory plug a plug of hardened secretion that, in some animals, blocks the female genital aperture after mating
    pharyngeal plug a projection of the pharyngeal lining into the midgut in some gastrotrichs
    slime plug the accumulation of slime on or around sieve plates
    yolk plug the heavily yolked unpigmented endodermal cells which are visible in the blastopore of the amphibian gastrula
[Plug] a mutant (gene symbol pl) on linkage group V of *Neurospora crassa.* The phenotypic expression is a dense growth of aerial hyphae
plum 1 (*see also* plum 2, borer, boy, curculio, gouger, hopper, weevil, yew) any of numerous species and hybrids of rosaceous trees of the genus Prunus and their edible fruits
    Allegheny plum *P. allegheniensis*
    alpine plum *P. brigantiaca*
    apricot plum *P. simonii*
    beach plum *P. maritima*
    Canada plum *P. nigra*
    cherry plum *P. cerasifera*
    Chickasaw plum *P. angustifolia*
    creek plum *P. rivularis*
    garden plum the tree *P. domestica*
    wild goose plum *P. munsoniana*
    hog plum (*see also* plum 2) *P. reverchonii*
    hortulana plum *P. hortulana*
    Italian plum *P. cocomilia*
    Japanese plum *P. salicina*
    Oklahoma plum *P. gracilis*
    Pacific plum *P. subcordata*
    figtree plum *P. mexicana*
    wild plum *U.S. P. americana*
plum 2 (*see also* plum 1) any of numerous trees and shrubs having plum-like fruit
    cocoa-plum the rosaceous shrub *Chrysobalanus icaco* and its fruit
    darling plum the rhamnaceous tree *Reynosia latifolia*
    governor-plum the flacourtiaceous shrub *Flacourtia ramontchi* and its fruit
    ground-plum any of several species of leguminous herbs of the genus Astragalus (= milk-vetch) par-

ticularly *A. caryocarpus, A. mexicanis, A. plattensis* and *A. tennesseensis*
Guiana plum the euphorbiaceous shrub *Drypetes lateriflora*
hog-plum (*see also* plum 1) either of two anacardiaceous trees *Spondias lutea* (= mombin) or *S. pinnata* (= amra)
jambolan plum the myrtaceous shrub *Eugenia jambolana* and its fruit
Kafir plum either of two species of the anacardiaceous genus Harpephyllum and their fruits
moxie plum the ericaceous shrub *Gaultheria hispidula*
Natal plum the apocynaceous shrub *Carissa grandiflora*
pigeon-plum the polygonaceous tree *Coccoloba floridana*
Spanish plum the anacardiaceous tree *Spondias mombin* and its edible fruit
sugar-plum *U.S.* any of several species of shrub or tree of the rosaceous genus Amelanchier
plumage the feathered covering of birds
    eclipse plumage colorless plumage replacing colored plumage, particularly flight feathers, after a molt
plumatous = plumose
plumbaceous lead colored
Plumbaginaceae a family of primulalous dicotyledons containing, *inter alia,* the sea-pinks and thrifts. The family may easily be distinguished from the colsely related Primulaceae by the one-seeded fruits
Plumbaginales a monofamilal order erected to contain the family Plumbaginaceae. The characteristics of the family are those of the order
plumbeous lead-colored
plume 1 (*see also* plume 2, 3, 4) a pinnate feather
    corniplume a hornlike tuft of feathers on the head of a bird
    filoplume a "feather" in the form of a simple bristle
    semiplume soft and fluffy feathers that have a rachis but lack hooks on the barbs
plume 2 (*see also* plume 1, 3, 4) (*bot.*) = plumule *bot.*)
plume 3 (*see also* plume 1, 2, 4) any of several plants with feathery flowers or foliage
    Apache plume the rosaceous shrub *Fallagia paradoxa*
    orange plume *U.S.* the orchid *Habenaria ciliaris*
    scarlet plume the euphorbiaceous shrub *Euphorbia fulgens*
plume 4 (*see also* plume 1–3) any feathery structure
    vibratile plume a name sometimes applied to long tufts of cilia particularly on invertebrate larvae
plumleteer any of several trochilid apodiform birds of the genus Chalybura
-plumil- *comb. form* meaning "plume"
plumose having a feathered appearance
preplumula a nestling down feather which precedes adult down feathers
plumule 1 (*bot.*) (*see also* plumule 2, bulb) the rudimentary shoot containing a leaf primordium lying immediately above the cotyledons (*cf.* epicotyl, hypocotyl)
plumule 2 (*see also* plumule 1) a feather structure of other than pinnate form
-plur- *comb. form* meaning "many" or "several"
Plutellidae a family of yponomeutoid lepidopteran insects commonly called diamondback moths in virtue of the fact that the males show a series of three diamond-shaped yellow marks when the wings are folded
pmx a bacterial mutant showing polymyxin sensitivity resistance

pn see [prune] and [pugnose]

-pneum- comb. form meaning "wind" and by extension "breathing"

pneumatode a hole for the passage of air in the surface of a plant (i.e. the stomata or lenticels)

pneustic pertaining to breathing

  apneustic having spiracles that do not open to the air

  amphipneustic having two methods of respiration, as amphibia having both gills and lungs or aquatic insect larvae having functional spiracles at both the anterior and posterior ends

  branchiopneustic the condition of an aquatic insect larva in which the spiracles are replaced by gills

  holopneustic having spiracles that open to the air

  metapneustic having functional spiracles only at the posterior end

  peripneustic having spiracles along the whole body

  propneustic with functional spiracles only on the prothorax

po see [pale-ocelli] and [polymitotic]

Po see [Polydactyly]

-poa- comb. form meaning "grass" or "meadow"

Poaceae = Graminaea

sea **poacher** popular name of agonid fish

poad a meadow plant

  xeropoad a heath plant

poaka N.Z. any of several species of recurvirostrid bird of the genus Himantopus (= stilt)

Poales = Glumiflorae

poataniwha N.Z. the rutaceous shrub Melicope simplex

pochard any of several aythyine anatid birds. Usually, without qualification, Aythya ferina or, more properly, the male of this species (cf. dunbird)

  African pochard Aythya erythrophthalma

  common pochard A. ferina

  red-crested pochard Netta rufina

  white-eyed pochard A. nyroca

pochote the bombaceous tree Ceiba casearia

pockard = pochard

pocket (see also pouch) a flat pouch

  Sessel's pocket a diverticulum of the embryonic pharynx anterior to the cranial attachment of the bucco-pharyngeal membrane

  subcephalic pocket the area underlying the head fold in an embryo

-pocul- comb. form meaning a "cup" deeper than is indicated by -calyc- but shallower than is indicated by -cyath-

-pod- comb. form meaning "foot" or, by analogy, "petiole"

pod 1 (see also pod 2–6) a popular name of a seed or egg containing structure, particularly a legume or silique

  egg pod a pouch of eggs such as that deposited by orthopteran insects

pod 2 (see also pod 1, 3–6) term for a group of whales, or porpoises, though school is more frequently used (cf. gam). Also, less usually, a group of coots, seals or walruses

pod 3 (see also pod 1, 2, 4–6) in compound names of plants with distinctive pods or of animals resembling pods

  anglepod any of several species of asclepiadaceous herb of the genus Gonolobus

  peapod see peapod shell

  ribbed pod the solenid mollusk Siliqua costata

  right pod the solenid pelecypod mollusk Siliqua lucida

  sheep-pod any of several leguminous herbs of the genus Astragalus (cf. loco weed)

sicklepod U.S. the cruciferous herb Arabis canadensis or the leguminous shrub Cassia tora

pod 4 (see also pod 1–3, 5, 6, podite, podium) in the sense of "foot" or "limb"

  baenopod a thoracic leg on an arthropod

  cercopod posterior, segmented projections from an arthropod abdomen. Includes the anal cerci of insects as well as analogous structures in some Crustacea

  endopod = endopodite

  gonopod an arthropod appendage modified for copulation

  gnathopod variously used as synonymous with maxillipede, or for the prothoracic leg of insects, or for any mouth part

  hexapod (see also larva) six-footed

  nectopod a limb adapted for swimming

  periopod in insects the second or third pair of thoracic legs of the larva or the second pair of the adult. In Crustacea, any thoracic limb

  pleopod a crustacean abdominal leg usually one adapted for swimming

    posterior pleopod the anal claspers or clasping legs of lepidopterous larvae, or the posterior legs of adult insects

  polypod that stage in the embryonic development of an insect in which there are appendage buds on all segments including the abdominal segments

  pygopod the appendages on the tenth abdominal segment of insects

  telepod a modified leg serving as an intromittent organ in Diplopods

  tetrapod having four limbs or feet

pod 5 (see also pod 1–4, 6) as a synonym of pseudopodium

  axopod a permanent pseudopodium, stiffened by an axial filament

  filopod a pseudopodium composed entirely of ectoplasm usually thin and with a pointed tip

  lobopod a bluntly ending pseudopodium as that of most amebas

  pseudopod see pseudopodium

  reticulopod thread-like pseudopodia which branch and anastomose into networks

  rhizopod (see also rhizopodous) a narrow lobopodium

pod 6 (see also pod 1–5) in various botanical terms, mostly analagous to pod 4

  glosopod the leaves at the base of Isoetes

  hymenopod = hypothecium

  mastigopod the swarm spores of a myxomycete

  myxopod the amoeboid stage of a myxomycete

  stylopod the swollen base of the styles in Umbelliferae

  undulipod any locomotor organelle of a protozoan such as cilium, a compound cilium or a flagellum

Podagrionidae a small family of chalcidoid hymenopteran apocritan insects

-podal- see -podous, -podial

Podargidae that family of caprimulgiform birds which contains the frogmouth. They are distinguished by the broad, flat, strongly hooked, and wide-gaped mouth from which the name is derived

podeon the petiole of hymenopteran insects

  metapodeon that part of the abdomen in hymenopteran insects which lies behind the petiole

-podet- comb. form meaning "foot" sometimes in the sense of "pedicel"

podetium a star shaped support of an apothecium

-podial adj. suffix meaning "footed" and therefore

synonymous with -podous (*q.v.*). The alternate form -podal is rarely used

autopodial said of all those elements of a vertebrate limb which lie distal to the zeugopodial elements

monopodial 1 (*see also* monopodial 2) said of the type of growth in the hydrozoan hydranth, in which the branch is topped by a terminal hydranth, and elongates by reason of a growth zone below the hydranth

monopodial 2 (*see also* monopodial 1) said of a type of branching in plant stems in which the main axis dominates the laterals

stylopodial descriptive of the basal element (femur or humerus) of a vertebrate limb

sympodial 1 (*see also* sympodial 2) said of a method of branching in plant stems in which the main axis ceases growth and auxilliary buds near the tip assume the major role in shoot development

sympodial 2 (*see also* sympodial 1) said of the type of growth in a coelenterate colony in which temporary, terminal hydranths are produced alternately on each side

zeugopodial said of the middle segment (radius and ulna) or (tibula and fibula) of the vertebrate limb

podicellate stalked

Podicepidae = Podicipedidae

Podicipedidae the only family of podicipediform birds. The characteristics of the order apply equally to the family

Podicipidae = Podicipedidae

Podicipitidae = Podicipedidae

Podicipediformes that order of birds which contains the grebes. They are characterized by their rudimentary tails, flattened nails, and lobate feet

-podite (*see also* pod 4) an arthropod limb or any part of it

basipodite that joint of a crustacean appendage which lies between the ischiopodite and the coxopodite

carpopodite that segment of a crustacean appendage which lies between the propodite and the meriopodite

coxopodite the basal segment of an arthropod appendage

dactylopodite 1 (*see also* dactylopodite 2) a thin second tarsal segment rising from an enlarged first tarsal segment in some insects

dactylopodite 2 (*see also* dactylopodite 1) the terminal segment of a crustacean or insect appendage

endopodite the adaxial branch of a biramous crustacean appendage

epipodite a structure, usually a gill, but sometimes a gill-separator, arising from the base of a crustacean appendage

exopodite 1 (*see also* exopodite 2) the third segment of the insect maxillary palp

exopodite 2 (*see also* exopodite 1) the abaxial branch of a biramous crustacean appendage

ischiopodite 1 (*see also* ischiopodite 2) in insects, variously the second segment of the telopodite, the second trochanter or the prefemur

ischiopodite 2 (*see also* ischiopodite 1) the segment of a crustacean appendage that lies between the meriopodite and the basipodite

meriopodite the segment of the crustacean appendage that lies between the carpopodite and ischiopodite

meropodite the femur of Chelicerata

propodite the segment of a crustacean appendage which lies between the dactylopodite and the carpopodite

protopodite in insects, the basal portion of the maxilla or, in some other arthropods, of any appendage

telopodite in arthropods, all those segments distal to the coxopodite

podium literally, foot. Specifically, applied to the tube feet of echinoderms

acropodium 1 (*see also* acropodium 2) the tegumentary covering of the dorsal surface of the toes of birds

acropodium 2 (*see also* acropodium 1) the division of the autopodial segment containing the phalanges

axopodium *see* -pod

basopodium a division of the autopodial segment consisting of the tarsals or carpals

buccal podium one of five pairs of modified podia, probably sensory and possibly chemoreceptor, surrounding the mouth of some echinoid echinoderms

carpopodium the fruit-stalk

penicillate podium an echinoid tube foot of which the branched expanded end has skeletal support in each branch

empodium a median appendage found between the terminal claws on the walking legs of many insects and arachnids

epipodium 1 (*see also* epipodium 2) something that rises above the foot particularly lobes on the side of the feet of gastropod Mollusca

epipodium 2 (*see also* epipodium 1) the apical portion of a leaf axis

hypopodium the stalk of a carpel

mesopodium = petiole

metapodium the middle region of the foot, either in vertebrates or invertebrates

myxopodium = pseudopodium

neuropodium the lower of the two lobes of a parapodium

notopodium the upper of the two lobes of a parapodium

parapodium a lateral extension from each segment of a polychaete annelid

papillate podium any of numerous, non-locomotory protrusions of the body wall of holothurian echinoderms

phyllopodium a broad leaflike limb

propodium the anterior foot, particularly when differentiated in Mollusca (*cf.* propodeum)

pseudopodium 1 (*see also* pseudopodium 2, pod 5) a protuberance, either temporary or permanent, from the body of a sarcodine protozoan

pseudopodium 2 (*see also* pseudopodium 1) the stalk of an oophyte of a moss

stenopodium a slender, elongate limb

sympodium an apparent plant stem actually made up of many branches closely pressed together

xylopodium a fruit in the form of a nucule borne on a fleshy support

Podocarpaceae a family of coniferous gymnosperms distinguished by the solitary ovulate flower

Podocarpoideae a subfamily of Podocarpaceae

Podocopa an order of ostracod Crustacea, distinguished from the Myocopa by valves which lack an anterior aperture, by possessing five pairs of post oral appendages, and by the uniramose second antenna. Most Ostracoda belong to this order

Podogona = Ricinulida

Podopidae a family of hemipteran insects resembling the Scutelleridae but differentiated from them by the strong lobe or tooth in front of the humeral angle of the pronotum. Commonly called turtle-bugs

Podostemaceae a family of podostemonale dicotyledons containing the forms usually called river-weeds.

They are distinguished by living in running streams and their moss-like appearance

Podostemonales a monotypic order of dicotyledonous angiosperms. The characteristics of the order are those of the family

-podous pertaining to -pod, or -podium. The form -podial (*q.v.*) is synonymous

apodous lacking a foot or, in fishes, a fin

acanthopodous 1 (*see also* acanthopous 2) with spines on the leg

acanthopodous 2 (*see also* acanthopodous 1) said of a leaf having a spiny petiole

adenopodous said of a leaf with a glandular petiole

brachypodous with a short stalk

coinopodous having a base in the form of an inverted cone

dichopodous said of a multiple dichotomous structure, particularly an inflorescence

macropodous used of an embryo with unusually large or elongate hypocotyl

pteropodous having wing-like structures at the base of the petiole

pterygopodous having a winged peduncle

pseudophyllopodous when the lower leaves of an aphyllopodous species are closely pressed to the ground

ptilopodous said of a bird in which the legs and toes are feathered

rhizopodous (*see also* rhizopod *under* pod 5) said of those algae in which there is an amoeboid tendency in temporary stages of gametes and zoospores. The term is also used in the sense of amoeboid

Poduridae a family of collembolan insects distinguished by the elongate body and well-developed prothorax furnished with bristles

-pody *subs. suffix* meaning to be in the condition of -podial, or -podous

-poecil- *see* -poikel-

Poeciliidae a large family of microcyprinous fishes closely allied to the Cyprinodontidae but distinguished from them by the ovoviviparous habit and the large intromittent organ (gonopodium) of the male. The well known guppy is typical of the group

Poecilosclerina an order of monaxoid demospongious Porifera distinguished by possessing microscleres which are usually chelas and toxas

pogge *Brit.* popular name of agonid fish (= *U.S.* poacher or alligator-fish), without qualification usually *Agonus cataphractus*

pogon literally a beard and applied to numerous biological structures of this appearance

pohuehue *N.Z.* the polygonaceous herb *Muehlenbeckia complexa*

pohutukawa *N.Z.* the myrtaceous tree *Metrosideros tomentosa*. (= Christmas tree *N.Z.*)

poic pertaining to grasslands

-poes- *see* -poies-

-poies- *comb. form* meaning "to produce"

biopoiesis the production of life. Often used specifically for the hypothetical evolution of life from non-living molecules

erythropoiesis the formation of red blood cells

galactopoiesis the maintenance of lactation in a mammal

haemopoiesis blood-formation

prehepatic haemopoiesis the formation of blood cells from embryonic blood islands

nosopoiesis disease production

erythropoietin a substance supposed to be created by the kidneys and which controls the rate of erythropoiesis

-poikel- *comb. form* meaning "variegated" or "various" frequently transliterated -poicel-

poinsettia the florist's name for the euphorbiaceous herb *Euphorbia pulcherrima*

point a typographic measurement, once common in biological measurements, equaling one-sixth of a line, that is one seventy-second of an inch

white-point *Brit.* the noctuid lepidopteran insect *Leucania albipuncta*

[Pointed wing] a mutant (gene symbol *Pw*) mapped at 94.1 on chromosome III of *Drosophila melanogaster*. The name is descriptive of the phenotype

poion a meadow association

poirei a variety of Swiss chard (*q.v.*)

poiser = halter

poison 1 (*see also* poison 2) a reagent capable of removing the property of life from an organic compound of otherwise known structure

poison 2 (*see also* tree) numerous plants either actually, or supposedly, poisonous. The term bane (*q.v.*) is also used in this sense

bushman's poison the apocynaceous shrub *Acocanthera venenata*

crow-poison *U.S.* the liliaceous herb *Zigadenus densum*

fever-poison the umbelliferous herb *Cictua maculata*

fly-poison *U.S.* the liliaceous herb *Amianthium muscaetoxicum*

hyena-poison the euphorbiaceous tree *Hyaenanche capensis*

river-poison the euphorbiaceous tree *Excoecaria bicolor*

sheep-poison the leguminous herb *Lupinus densiflorus*

pokaka *N.Z.* the elaeocarpaceous tree *Elaeocarpus hookerianus*

poke (*see also* weed) a word of numerous meanings and origins

feather poke *Brit.* the parid bird *Aegithalos caudatus* (= long tailed tit)

Indian poke the liliaceous herb *Veratrum viride*

poker = pochard

red-hot poker the liliaceous herb *Kniphofia uvaria*

polar pertaining to the poles either of the Earth or of a spherical egg

heteropolar a term applied to a radially symmetrical object, the two ends of which are different

polarity in *biol.,* the existence of a difference between the two ends of an organism

redpole = redpoll

root pole that end of a plant embryo from which the root develops

Polemoniaceae a family of tubiflorous dicotyledons distinguished with difficulty from the Convolvulaceae by the three many-ovuled cells of the ovary

polex the terminal segment of the insect abdomen (*cf.* pollex)

Polioptilidae a family of passeriform birds usually included with the Sylviidae

Polistinae a subfamily of elongate, reddish or brownish vespid hymenopteran insects. They make a comb of paper attached to a support by a slender stalk

poll an old word for head, specifically applied to the area of a horse's, and occasionally other ungulate's, head between and immediately posterior to the ears

redpoll *U.S.* any of several fringillid birds of the genus Acanthus. *Brit.*, without qualification, *A. linaria*

-pollac- *comb. form* meaning "frequent" or "numerous"

pollack alternative spelling, preferred in parts of U.S. for pollock

pollan *Brit.* any of several species of salmonid fish of the genus Coregonus

pollard a tree, the trunk of which is continually cut back to the same level, usually from five to ten feet above ground, so that numerous branches grow out from just below the cut area. The term pollard is also sometimes applied to branches so produced (*cf.* ratoon)

polled de-horned or, of trees, with the branches cut short

pollen (*see also* basket, comb) the microspore of seed plants

pollenine the contents of a pollen grain

pollex (*cf.* polex) the first digit of the forelimb. (= thumb in primates). Also any finger-like spine on an arthropod limb, particularly a stout spur on the inside of the tip of the tibia or a spiney process on the male genitalia of Lepidoptera. The term was once synonymous with inch

[pollex] a mutant (gene symbol px) in linkage group I of the guinea pig. The phenotypic expression is the production of a pollex

pollinarious floury (*but see* pollinosous)

pollinarium = androecium

pollination the act, or method, of transferring pollen to the female portion of the flower

pollinodium the male sex organ of an ascomycete

pollinosous covered in pollen (*but see* pollinarious)

pollock any of several gadid fishes. Usually, without qualification, *Pollachius virens* The spelling pollack is preferred in parts of *U.S.*

Alaska pollock = walleye pollock

walleye pollock the gadid fish *Theragra chalcogrammus*

pollywog *U.S. obs.* tadpole

-poly- *comb. form* meaning "many"

Polyadelphia that class in the Linnaean system of plant classification distinguished by the possession of stamens combined by their filaments into three or more sets

polyam a transitional form in the development of a new type

Polyandria that class of Linnaean classification of plants distinguished by the possession of many stamens inserted in the receptacle

Polyangiaceae a family of myxobacteriale Schizomycetes distinguished by the fact that the resting cells, in the form of thickened rods, are always enclosed in cysts which may be sessile either singly or as a group or raised on stalks

Polybiinae a subfamily of mostly tropical vespid hymenopterans strongly resembling the Polistinae

Polyboridae a family of falconiform birds usually treated as a subfamily of the Falconidae

Polyborinae that subfamily of falconid birds which contains the caracaras. They are distinguished by their more or less vulturine habit, long legs, and short toes.

Polycentridae a small family of teleost fishes usually included with the Nandidae

Polychaeta that class of annelid worms which are commonly called bristle worms. They are distinguished by the presence of parapodia

[polychaetoid] a mutant (gene symbol *pyd*) mapped at

39.0± on chromosome III of *Drosophila melanogaster*. As the name indicates the phenotypic expression is extra bristles

[polychaetous] a mutant (gene symbol *pys*) mapped at 52.0± on chromosome II of *Drosophila melanogaster*. The phenotypic expression is numerous extra bristles

Polychondreae a tribe of orchids of the division Acrotonae of the subfamily Monandreae. They are distinguished from the Kerosphaereae by the possession of granular pollen

Polycladida an order of turbellarian Platyhelminthes distinguished by having numerous radially disposed branches from the intestine

Polyctenidae a small family of wingless hemipteran insects parasitic on chiropterans. Commonly called batbugs

[polydactyly] a mutant (gene symbol *py*) found in linkage group XIII of the mouse. The name is descriptive of the phenotypic expression

[Polydactyly] a mutant (gene symbol *Po*) in linkage group VI of the domestic fowl. The name is descriptive of the phenotypic expression

Polygalaceae that family of geranialous dicotyledons which contain the milkworts. The peculiarly irregular perianth and stamens and the two-celled ovary are typical of the family

Polygalineae a suborder of geraniale dicotyledonous angiosperms containing the families Tremandraceae and Polygalaceae

Polygamia that class in the Linnaean system of plant classification containing those plants possessing three kinds of flowers, some bearing both stamens and pistils, some stamens only and some pistils only

polygamist *U.S.* the noctuid lepidopteran insect *Catocala polygama* (*cf.* underwing)

Polygonaceae the family of plants which contains, *inter alia*, the docks, buckwheats, rhubarbs and sorrels. The very simple floral structure and the one-celled fruit with its single orthotropous seed are characteristic

Polygonales an order of dicotylodenous spermatophytes which contains the single family Polygonaceae. The order is characterized by bisexual flowers with a superior unilocular uniovular two-four-carpellate ovary

Polygynia those orders of the Linnaean system of plant classification distinguished by the possession of many pistils

Polymastigida an order of zoomastigophorous Protozoa with numerous flagella often trailing or forming undulating membranes. This group contain, *inter alia*, the well-known intestinal commensals of termites

polymely the abnormal condition of a tetrapod having more than four limbs

Polymera the zoological taxon erected by Butschli to contain those worm-like animals with two or three coelomic divisions (*i.e.* the Phoronida, Bryozoa, Brachiopoda and Chaetognatha) (*cf.* Oligomera and Amera)

[polymitotic] a mutant (gene symbol *po*) mapped at 0 on linkage group VI of *Zea mays*. The spores of the phenotype show more than the usual number of meiotic divisions

Polymyxiidae a monogeneric family of berycoidous fishes distinguished by the large number of scales on the lateral line and barbels on the chin

Polynemidae the only family of polynemoid acanthopterygians. The characteristics of the family are those of the suborder

Polynemoidea a monofamilial suborder of acanthopterygian fishes commonly known as thread-fins or tassel-fishes by reason of the peculiar division of the pectoral fin into a normal dorsal portion and a ventral section of filamentous rays

Polyodontidae a family of chondrostean fishes distinguished by the monstrous paddle projecting from the snout

Polyopisthocotylea an order of monogeneous trematode Platyhelminthes distinguished by having many posterior adhesive suckers

polyp literally an aquatic animal but specifically used to distinguish the sessile form of a polymorphic coelenterate in contrast to the medusa

machopolyp = machozooid

Polyphaga a suborder of coleopteran insects containing the great majority of extant Coleoptera distinguished from the Adephaga by the fact that the hind coxae are not fused to the metasternum and from the Archostemata in lacking notopleural sutures

polypide the living portion of an ectoproct zooid (*cf.* cystid)

Polyplacophora an order of amphineuran Mollusca commonly referred to as chitons. They are distinguished by having a shell of eight transverse separated plates and in the fact that the mouth and anus are at opposite ends of the elongate body

Polypodiaceae that family of filicale filicine pteridophytes containing the great majority of ferns. They are distinguished by the possession of a vertical incomplete annulus associated with a lenticular long stalked sporangium

Polyporaceae that family of basidomycete fungi which contains the forms called polypores and shelf-fungi. They have the distinctive habit of growing as a plate or shelf-like prominence

Polyprotodontia a main division of marsupials containing all forms except the kangaroos and wombats (Diprotodontia). They are distinguished by having four or five incisors on each side of the upper jaw

Polypteridae the only family of the fish order, Cladistia. These lobe-finned fishes have rhomboidal ganoid scales, spiracles and non-alveolated lungs

Polystoechotidae a small family of very large neuropteran insects distinguished from all neuropterans except the dobson fly by their size and from the dobson fly by the tent-shaped wings

anapolytic a term used to describe those tapeworms which retain their ripe proglottids

apolytic a term used to describe those tapeworms that shed their ripe proglottids

-pom- *comb. form* meaning "apple"

Pomacentridae a family of acanthopterygian marine fishes distinguished from other acanthopterygians, except the freshwater cichlids, by the presence of a single nostril, rather than two nostrils, on each side of the snout

pomaceous pertaining to apples

Pomadasyidae a family of acanthopterygian osteichthes commonly called grunts from the sound which they produce by grinding their large pharyngeal teeth which are typical of the family

Pomatomidae a monospecific family of percoid fishes containing the bluefish *Pomatomus salatrix*

pome a fleshy fruit consisting of a thin skin and an outer

zone of edible flesh. The apple and pear are typical examples

pomegranate the large, pink to red, fruit of the punicaceous tree *Punica granatum*. This fruit, which contains numerous seeds in a dark red pulp, is actually a multilocular berry. The name is also applied in Canada to the fruit of *Prunus nigra* (= Canada plum)

pomelo the rutaceous tree *Citrus grandis* (= grapefruit)

Pomoideae a subfamily of Rosaceae distinguished by having an indehiscent fruit and an inferior ovary

pompadour the cotingid bird *Xipholena pompadora*

pompano (*see also* shell) any of several species of carangid fishes but usually, without qualification, *Trachinotus carolinus*

Irish pompano the gerrid fish *Diapterus olisthostomus*

Pacific pompano the stromateid fish *Palometa simillima*

Pompilidae a family of vespoid hymenopterans commonly called spider-wasps by reason of the fact that most of them feed on spiders. Some reach a length of more than two inches and are capable of stinging severely

pond a standing body of fresh, brackish, or rarely sea, water too small to be called a lake

pone posterior

Ponerinae a subfamily of Formicidae containing those ants in which the pedicel of the abdomen is one-segmented. They form small colonies and are carnivorous

Pongidae a family of anthropoids containing the subfamilies Poninae and Hylobatinae and forming, with the family Hominidae, the superfamily Hominoidea. Some workers consider the Hominidae to be a sub-family (Homininae) of the Pongidae or else place man in the Ponginae

Ponginae that subfamily of apes which contains the chimpanzee, gorilla, orangutan, and, in the opinion of many, man. These are often called the great apes leaving the term lesser apes to apply to the Hylobatinae

pons *see* bridge

-pont- *comb. form* meaning "sea" and more particularly the Black Sea, but used by some ecologists in the sense of "deep sea"

Pontederiaceae that family of farinosean monocotyledons which contains the pickerel-weeds and water hyacinths. The pickerel-weeds can be distinguished from the few marshy lilies by the mealy endosperm and the irregular flowers

pony (*see also* grass) a small domesticated race of horse

poor-man's-weather-glass the primulaceous herb *anagallis arvensis*

poorwill *U.S.* any of several caprimulgid bird usually, without qualification, *Phaeaenoptilus nuttallii*

eared poorwill *Otophanes mcleodii*

ocellated poorwill *Nyctiphrynus ocellatus*

Yucatan poorwill *Otophanes yucatanicus*

whip poorwill *Caprimulgus vociferus*

pope *Brit. dial.* = puffin or bullfinch

popikatea *N.Z.* the malurine bird *Mohoua ochrocephala* (= whitehead)

popinac the leguminous shrub *Acacia farnesiana*

poplar 1 (*see also* poplar 2, aphis, borer, grey, moth, sawfly) properly any of numerous species of the salicaceous genus Populus but also used of a few other plants. (*see also* aspen and cottonwood)

balsam poplar *P. balsamifera*

black poplar *P. nigra*
Carolina poplar *P. eugenii*
downy poplar *P. heterophylla* (= black cottonwood)
grey poplar *P. canescens*
Japanese poplar *P. maximowiczii*
Lombardy poplar *hort. var.* of black poplar
native poplar *Austral.* the phytolaccaceous tree *Co-donocarpus cotinifolius*
Ontario poplar *P. candicans*
Queensland poplar *Austral.* the euphorbiaceous tree *Homalanthus populifolius*
white poplar *P. alba*
  Chinese white poplar *P. tomentosa*
yellow poplar the magnoliaceous tree *Liriodendron tulipifera* (= tulip-tree)
poppy any of numerous flowers of the family Papaveraceae. Without qualification, usually refers to the genus Papaver
alpine poppy *P. alpinum*
California poppy *Eschscholtzia californica*
celandine poppy any of several species of the genus Chelidonium and also *Stylophorum diphyllum*
yellow Chinese poppy *Meconopsis integrifolia*
corn-poppy *P. rhoeas*
flamingo-poppy = wind poppy
horned poppy any of several of the genus Glaucium and particularly *G. flavum*
Iceland poppy *P. nudicaule*
matilija poppy *Romneya coulteri*
opium poppy *P. somniferum*
oriental poppy *P. orientale*
peacock-poppy *P. pavonium*
plume-poppy any of several species and numerous horticultural varieties of the genus Bocconia
prickly poppy any of several species of the genus Argemone, particularly *A. mexicana*
satin-poppy the papaveraceous herb *Meconopsis wallichii*
sea poppy *Glaucium flavum*
tulip-poppy *P. glauca*
Welsh poppy *Meconopsis cambrica*
wind-poppy *Meconopsis heterophylla*
wood-poppy *Stylophorum diphyllum*
population an assemblage of organisms
Mendelian population a group of individuals who interbreed and form a community by themselves
amphimictic population = panmictic population
panmictic population one which results from prolonged, random, interbreeding
populeous blackish green
-por- *comb. form* confused between four *Gr.* roots and therefore meaning "blind," "soft" (or by extension "porous"), "shore" and "a hole"
cribriporal that condition of a sponge in which clusters of pores open into a chone
porbeagle the lamnid shark *Lamna nasus*
porcatous ridged
porcupine (*see also* fish, grass) the name is applied to members of two families of hystricomorph rodents, the Hystricidae (Old World porcupines) and the Erethizontidae (New World porcupines). These resemble each other in practically no characteristic save the possession of spines. The large common porcupine (Hystrix) of the Mediterranean basin and parts of Asia is most commonly called the crested porcupine. *U.S.*, without qualification, *Erethizon dorsatum*. *Brit.*, without qualification, *Hystrix cristata*
rat-porcupine the hystricid *Trichys lipura*. It is dis-

tinguished by its long tail, small size and the fact that the spines are flattened from side to side
Sumatran porcupine any of several hystricids of the genus Thecurus distinguished by having a spiny, instead of a hairy, head
brush-tailed porcupine African and Asiatic hystricids of the genus Atherura distinguished by a long naked tail terminating in a mass of hollow spines
tree porcupine any of several cercolabid rodents but particularly, without qualification, *Sphingurus prehensilis*
pore (*see also* plate) a small hole
acanthopore (Ectoprocta) = acanthozoid
ascopore the opening of the ascus in Ectoprocta
atriopore the opening of the atrium particularly in Amphioxus
blastopore the aperture leading to the archenteron in a gastrula
bordered pore = bordered pit
breathing pore = stoma
cortex pore = lenticel
cortical pore = lenticel
cryptopore a stoma which lies under the epidermis
dactylopore the small pore in a coenosteum through which the dactylozoids protrude
gastropore the large pore in a coenosteum through which the gastrozoid protrudes
germ pore the hole through which a germ-tube leaves a spore
gonopore the opening of a gonoduct
hydropore the connection of the echinoderm water-vascular system to the exterior
madrepore any of numerous stoney corals. *See also* madreporite
mesopore (Ectoprocta) = mesozoid
nephridiopore the opening to the surface from either a protonephridium or a metanephridium
supraneural pore a temporary ovopore formed on the lophophore of some Ectoprocta
neuropore the connection between the neural canal and the exterior in the neurula
nuclear pore a hole in the nuclear envelope
nullipore *obs.* calciferous algae
opisthelial pore the posterior border of a stoma
ovipore the external opening of the female genital duct in many invertebrates
ovopore a hole providing a passage for eggs to the exterior
pharyngeal pore one of a pair of connections between the central portion of the pharynx and the exterior in some gastrotrichs
pseudopore a hole which is not actually an aperture. Used particularly of the pores in the calcified zoecia of some Ectoprocta
taste pore a passageway through the epithelium reaching to a taste bud
trichopore a pore under the trichogen
uropore the opening of an excretory organ particularly that of an arthropod
interzoidal pore pores connecting zoecia of Ectoprocta
porgy popular name of sparid fish
poricidal said of anthers which open by pores
Porifera that phylum of the animal kingdom which contains the sponges. In general they consist of aggregates of cells, mostly choanocytes, which often line canals and chambers usually supported by a skeleton of fibers or spicules
porite diminutive of pore
madreporite the calcareous sieve plate through which

the hydropore (*q.v.*) of some echinoderms is connected to the exterior

more**pork** *N.Z.* the strigid bird *Ninox novaeseelandiae* *Tasmania* the podargid *Podargus strigoides* (= tawny frogmouth)

sea-**pork** the krikobranchiate colonial ascidian *Amaroucium stellatum*

poroid a dot on a diatom frustule resembling a pore

porokaiwhiri *N.Z.* the monimiaceous tree *Hedycarya arborea* (= pigeonwood *N.Z.*)

Poromyidae a family of pelecypod mollusks commonly called granular shells. The name derives from the granulated appearance of the unequal, thin valves

porose having pores

porous pertaining to pores

oligoporous said of those ambulacral plates of echinoid echinoderms which bear two or three pores for the passage of podia

phaneroporous the condition of stomata which lie in the plane of the epidermis

polyporous said of ambulacral plates in echinoid echinoderms, which bear more than three pair of pores for the passage of podia

porphyreus light purple

porpoise (*see also* whale) properly any delphinid cetacean of the genus Phocaena. *Brit.*, without qualification, *P. communis*. *N.Z.* the delphinid cetacean *Cephalorhynchus hectori*

porraceous leek-green

porrect stretched outwards and forwards, though sometimes used in the sense of arrect (*q.v.*)

subgenital **porticus** the cavity resulting from the fusion of the four subgenital pits of some Schyphozoa

Portuguese-man-of-war the large siphonophoran coelenterate Physalia

Portulacaceae that family of centrospermous dicotyledons which contains, *inter alia*, the purslanes. Distinctive characteristics are the two sepals and the one-celled ovary

Portunidea a family of brachyuran decapod Crustacea commonly called swimming crabs. They are distinguished by the broad, flattened, fifth pereipods

Posidonieae a tribe of the family Najadaceae when this family includes the Potamogetonaceae

[**positional-sterile**] a mutant (gene symbol *ps*) mapped at 20 on chromosome 2 of the tomato. The position of the phenotypic flower is such that the corolla cannot open

-**positor**- comb. form meaning "to place"

larvipositor an ovipositor modified for a larviparous organism

ovipositor a more or less elongate projection from the posterior end of many insects which is adapted to placing eggs

possum (*see also* haw) *U.S. dial.* for opposum

-**post**- *comb. form* meaning "behind" either in time or place

[**postaxial hemimelia**] a mutant (gene symbol *px*) found in linkage group XI of the mouse. Phenotypic expression is a defective postaxial portion of the limbs

postical = posterior

blue **posy** *I.P* the lycaenid lepidopteran insect *Biduanda melisa*

pot a rotund container

honeypot = replete

skill-pot *U.S.* = stinkpot *U.S.*

stinkpot *U.S.* popular name of cinosternine chelonian reptiles usually, without qualification, *Sternothaerus odoratus* (*cf.* musk-turtle)

-**potam**- *comb. form* meaning a "river"

potamad a river plant

autopotamic a term applied to the algae component of tychopotamic plankton

potamium a community of river dwellers

-**potamo**- *see* -potam-

Potamogalidae a small family of West African insectivores strongly resembling otters in appearance

Potamogetonaceae that family of heliobe monocotyledonous angiosperms which contains the forms commonly called pond weeds, They are distinguished by their submerged aquatic habit and the rudimentary perianth on the flowers

Potamogeoneae a tribe of the family Najadaceae when this family includes the Potamogetonaceae

potato 1 (*see also* potato 2, aphis, beetle, borer, bug, fly, dandelion, gnat, hopper, vine, weevil, worm) the solanaceous herb *Solanum tuberosum* and its very numerous horticultural varieties and hybrids. Also a few other species of the same genus

Darwin potato *S. maglia*, thought by some to be an ancestor of some agricultural potatoes

wild potato *S. jamesii*

potato 2 (*see also* potato 1) numerous other plants having edible tubers

air potato the dioscoreaceous herb *Dioscorea bulbifera*

Chinese potato the dioscoraceous herb *Dioscorea batatas* and its edible tubers (= yam)

duck-potato the alismataceous herb *Sagittaria latifolia* and its edible tubers

Indian potato the compositaceous herb *Helianthus giganteus* and its edible tubers

swamp potato any of several species of the alismataceous genus Sagittaria and its edible tubers (= arrow head)

sweet potato (*see also* worm) the convolvulaceous vine *Ipomoea batatas* and its edible tubers. The sweet potato is frequently miscalled a yam

prepotence a term of animal breeding reflecting genetic dominance of a male or less usually, female parent

potency 1 (*see also* potency 2) the potential of a part to deviate from its fate (*q.v.*) (*e.g.* the production of a whole organism from either of two separated blastomeres)

potency 2 (*see also* potency 1) the genetically controlled ability of an egg to develop

prepotency the condition of a pollen which enables it to fertilize a specific pistil which other pollen cannot fertilize

totipotency the ability of an egg to reproduce a whole organism and, particularly the retention of this power by some early blastomeres

biotic **potential** either the possibility of a specific organism surviving in a specific environment, particularly an unfavorable one or the specific growth rate of a population under conditions of a stable age distribution

Pothoideae a subfamily of the Araceae

potoo any of several species of nyctibiid bird of the genus Nyctibus, particularly *N. griseus*. In some parts of the *W.I.* applied to *Pseudoscops grammicus* (= Jamaican owl)

common potoo *N. griseus*

great potoo *N. grandis*

Potoroinae a subfamily of phalangerid marsupials containing the forms commonly called rat-kangaroos. They differ from the Macropodinae (kangaroos and wallabies) by the possession of well-developed canine teeth

potto the lemuroid primate *Perodicticus potto.* This term is also most misleadingly applied to the kinkajou

pou = paau

pouch (*see also* pocket) a hollow protuberance, too small to be called a lobe, or an inwardly directed sac

amniotic pouch the sac which contains the amniotic fluid

branchial pouch a series of respiratory pouches in cyclostomes corresponding to the gills of higher forms

egg pouch = ootheca

Rathke's pouch a dorsal evagination from the roof of the embryonic buccal cavity which develops into the adenohypophysis

Saefftigen's pouch a muscular pouch in the genital sheath of acanthocephalans serving to assist in the eversion of the bursa

metasome pouch an eversible ectodermal invagination in the midventral line of an actinotroch larva

poult a young game bird, most commonly used of the grouse and turkey

moor poult = red grouse

pout any fish with a large head. *Brit.,* without qualification, the gadid fish *Gadus bismarckii*

eelpout popular name of zoarcid fish

horned pout *U.S.* the ictalurid *Ictalurus metas*

ocean pout the zoarcid fish *Macrozoarces americanus*

powan *Brit.* the salmonid fish *Coregonus clupeaformis* (= *U.S.* lake whitefish)

powee *W.I.* the cracid bird *Crax nigra* (*cf.* curassow)

pr *see* [purple] or [red aleurone color]. This symbol is also used for a yeast genetic marker indicating a requirement for protein

Pr *see* [Prickly]

ethmopract said of those echinoid echinoderms in which the terminal plates encircle the genital plates (*cf.* ethymolytic)

prae- *adj. prefix* meaning "before" either in time or place (= -pre)

praeustous scorched

prairie (*see also* dock) *U.S.* loosely applied to almost any type of grassland habitat, usually that in which tall grasses predominate. The term has also been applied to open marshland in swamps

prasinous leek green

Prasiolales = Schizogoneales

pratal pertaining to meadows containing large numbers of dicotyledonous herbs

pratincole any of numerous species of glareolid birds. *Brit.,* without qualification, usually *Glareola pratincola*

East African pratincole = collared pratincole

Australian pratincole *Stiltia isabella*

collared pratincole *Glareola pratincola*

white-collared pratincole *G. nuchalis*

Indian pratincole *G. lactea*

Madagascan pratincole *G. ocularis*

Sudan pratincole = collared pratincole

black-winged pratincole *Glareola nordmanni*

prawn properly a palaemonid natant decapod crustacean but often applied to other forms. *Brit.* without qualification, *Palaemonetes vulgaris.* In *U.S.* the terms shrimp and prawn are largely interchangeable, but in *Brit.* and *W.I.* the distinction is one of size

pre- *comb. prefix* meaning "before," either in time or place (= prae-)

precocial said of birds hatched in a relatively advanced state

predation the act of an organism which catches and kills its food

predator an organism which lives by predation

preen the actions of a bird in arranging its feathers

allopreening the mutual preening of feathers by two birds

pregnancy the condition of a female carrying unborn young

compressed flattened laterally

depressed flattened dorso-ventrally

vasopressin a hormone secreted by the neurohypophysis which acts on the capillaries and arterioles

osmotic pressure *see* osmosis

prey the victim of a predator

Priacanthidae a small family of percoid fishes distinguished by their small scales and very large eyes from which the popular name bigeye derives

Priapulida a small group of marine worms that has been variously regarded as a separate phylum, an order of the phylum Gephyrea or a class of the Annelida. They are now usually regarded either as a separate phylum, or as a class of the pseudocoelomate phylum Aschelminthes. They are distinguished by having the appearance of a warty cylinder with an introversible presoma

pricket a two year old fallow deer

[prickle] a mutant (gene symbol *pk*) mapped at 55.3 on chromosome II of *Drosophila melanogaster.* The phenotypic expression is an irregularity in the bristles and hairs

[Prickly] a mutant (gene symbol *Pr*) mapped at 90.0 on chromosome III of *Drosophila melanogaster.* The phenotypic expression is short, thin, twisted bristles

pride 1 (*see also* pride 2) term for a group of lions, or peacocks though muster is preferable for the latter

pride 2 (*see also* pride 1) in compound names of organisms

Barbados pride the leguminous tree *Caesalpinia pulcherrima*

London pride the saxifragaceous herb *Saxifraga umbrosa*

pride-of-Barbados = Barbados pride

pride-of-California the leguminous vine *Lathyrus splendens*

pride-of-China the sapindaceous herb *Sapindus saponaria*

pride-of-India the meliaceous tree *Melia azedarach* (= China tree)

pride-of-the-Congo the araceous plant *Zantedeschia oculata* (= yellow calla)

pride-of-the-peak the orchid *Habenaria peramoena*

prim = privet

Primates that order of placental mammals which contains, *inter alia,* the monkeys, baboons and apes. They are distinguished by a pentadactyl pattern in the limb and an orbit completely encircled by bone. Most have pectoral mammary glands

primine the outermost coat of an ovule

primitive properly, the first of its kind, and thus the ancestral form. It is also used to mean un- or underdeveloped, and also occasionally for "wild type" in genetics

primordium the earliest stage at which the differentiation of an organ can be perceived. Probably the best translation of the widely used *Ger.* word *anlage*

primrose *see* -rose

Primulaceae that family of primulale dicotyledonous angiosperms which contains, *inter alia,* the primroses and cyclamen. They are distinguished from most other members of the order by the herbaceous habit and the single whorled stamens in the gamopetalous corolla

Primulales that order of dicotyledonous angiosperms which contains, *inter alia,* the primroses, the pimpernels, sea lavender, and thrift. The order is characterized by the gamopetalous pentamerous flowers

brown **prince** *I.P.* the nymphalid insect *Apatura parvata.* (*cf.* emperor)

Principes that order of monocotyledons which contains the palms and allied forms. The habit of growth is typical

principle a term frequently used, as is also factor, to indicate an unidentified compound

posterior-lobe principle = oxytocin

vasopressor principle = vasopressin

vasopressor-antidiuretic principle = vasopressin

butter **print** *U.S.* the malvaceous herb *Abutilon theophrasti*

-prion- *comb. form* meaning "saw"

prion *Austral.* any of several species of the procellariid bird of the genus Pachyptila (*cf.* whalebird)

broad-billed prion *P. forsteri*

thin-billed prion *P. belcheri*

dove-prion *P. desolata*

fairy prion *P. turtur*

Prionodesmata a subclass of pelecypod Mollusca having the right and left lobes of the mantle separated on their ventral and hinder margins

Prionopidae that family of passeriform birds which contains the wood-shrikes. They are distinguished by the brightly colored, fleshy, orbital ring

Pristidae a family of rajiform Chondrichthyes bearing a blade-like snout with teeth on the side in the form of a "saw" (*cf.* Pristiophoridae)

Pristiophoridae a family of sharks carrying a "saw." May be distinguished from the sawfishes (Pristidae) by the fact that the gill openings are anterior to the pectoral fins

privet (*see also* aphis, mite, moth, thrip) any of several species, and numerous horticultural varieties of oleaceous shrub of the genus Ligustrum

amur privet *L. amurense*

border privet *L. obtusifolium*

California privet *L. ovalifolium*

common privet = European privet

delavay privet *L. ionandrium*

European privet *L. vulgare*

Formosa privet *L. formosanum*

glossy privet *L. lucidum*

Himalayan privet *L. massalongianum*

Indian privet *L. nepalense*

sharpleaf privet *L. acuminatum*

slender privet *L. gracile*

Yunnan privet *L. compactum*

-pro- *comb. form* meaning "before" and "early"

pro a bacterial mutant indicating a requirement for protein

proA a bacterial marker indicating a requirement for proline, mapped at 6.4 mins. for *Escherichia coli*

proB a bacterial genetic marker having a character which supports syntrophic growth of *proA* mutants, mapped at 8.25 mins. for *Escherichia coli*

proC a bacterial genetic marker having a character which supports syntrophic growth of *proA* mutants, mapped at 11.5 mins. for *Escherichia coli*

Proboscidea 1 (*see also* Proboscidea 2) an order proposed for those insects now in the homopteran family Coccidae

Proboscidea 2 (*see also* Proboscidea 1) that order of placental mammals which contains forms commonly known as elephants. The elongation of the nose into a trunk or proboscis is typical of the order

proboscis (*see also* flower) a protuberance from the head or face

[Proboscipedia] a mutant (gene symbol *pb*) mapped at 47.7 on chromosome III of *Drosophila melanogaster.* The phenotype has oral lobes changed to tarsus-like or arista-like processes

Procellariidae that family of procellariiform birds which contains the shearwaters. They are pelagic birds with short heavy hooked bills protected with bony plates and short thick legs

Procellariiformes that order of birds which contains *inter alia* the albatrosses, shearwaters and petrels. They are distinguished from other gull-like birds by the fact that the nostrils are in raised tubes

process anything which projects from something

angular process a process on the dentary bone posterior (or ventral) to the articular process. In the marsupials, the angular process is directed medially

antennal process the reverse of antacava, that is, prominences from which antenna arise

articular process in mammals, a process of the dentary bone, posterior to the coronoid process which articulates the dentary with the squamosal

basal process a nonarticulating bony projection on the caudal vertebrae of actinopterygian fishes

ciliary process the free, anterior edge of the choroid in the eye

clinoid process a process on the orbitosphenoid bone

coronoid process in mammals a posterior process that rises from the dentary bone inside the zygomatic arch. In reptiles, a process of the lower jaw associated with the adductor muscles

postglenoid process a downwardly directed process from the base of the zygomatic arch

iliac process one of a pair of lateral projections from the cartilaginous pelvic girdle of primitive fish

ischial process an unpaired, posterior projection from the cartilaginous pectoral girdle of primitive fishes

lingual process a forward extension of the copula

mastoid process in mammals a bony extension of the petrosal bone just behind the tympanic bulla. It contains air spaces communicating with the middle ear

paroccipital process one of a pair of processes projecting around the dorsal and posterior margins of the otic capsule in chondrocrania. In reptiles a process of the opistholic bone

otic process a process of the palato-quadrate extending up toward, or continuing through the otic capsule in many vertebrates

palatal process those processes from the palatine bone which meet along the midline between the anterior palatine fenestrae

pisiform process a bony bump on the under surface of the carpometacarpus of birds

prenasal process a process of the premaxilla extending upward, medial to the external nares and contacting the nasal bone

pterygoid process the anterior end of the palatoquadrate in developmental stages of the tetrapods

pubic process an anterior process of the cartilaginous pectoral girdle of primitive fishes

spinous process = neural spine

Tomes' processes (*cf.* fibril) processes from ameloblast cells

uncinate process a thin plate, found in all birds except Anhimidae, projecting upwards from a rib over the rib above it and a dorso-lateral projection from the central portion of the rib in Sphenodon

zygomatic process either of the processes from the squamosal or maxilla, which articulates with the jugal bone to form the zygomatic arch

processus alaris an antero-lateral process of the polar cartilage

Procniatidae = Tersinidae

Procoela a suborder of anuran Amphibia containing the great majority of living frogs and toads. They are distinguished by the procoelous vertebrae and the absence of ribs

-proct- *comb. form* meaning "hindgut" or "anus"

epiproct anything which lies immediately above the anus and therefore applicable to the pygidium and to the supra-anal plates of many insects

paraproct anything alongside the anus and particularly the sclerite in this position in many arthropods

periproct the areas surrounding the anus

Proctotrupidae a small family of proctotrupoid apocritan hymenopteran insects

Proctotrupoidea a superfamily of apocritan hymenopteran insects much better known under their obsolete name of Serphoidea. Almost all are parasites of other insects and they are therefore of great economic importance. The proctotrupoids may be distinguished from the chalcidoids by the fact that the pronotum extends back to the tegulum and from other parasitic Hymenoptera by the fact that the antennae never have more than 15 segments

Procyonidae a family of fissipede carnivorous mammals closely allied to the Viverridae and containing those forms commonly called raccoons, coatimundis and kinkajous. They are distinguished by having two molars in each half of both jaws

-prodo- *comb. form* meaning "pioneer"

Prodoxidae a family of incurvarioid lepidopteran insects containing *inter alia*, the yucca moths, the remarkable adaptations of the mouth parts of which to the fertilization of plants of the genus Yucca has been the subject of much comment. This family is closely allied to the Incurvariidae

production the sum total of the process of producing

[Production of atropinesterase] a mutant (gene symbol *At*) at locus 26.2 on the linkage group VI of the rabbit. The name describes the phenotypic expression

productivity production rate in populations, communities or ecosystems

biological productivity usually measured in protein-time units

profundal pertaining to deep waters. In limnology confined to the area free of rooted aquatic plants

progesterone a hormone secreted by the ovary when stimulated by luteinizing hormones. Active in the preparation of the uterine wall for the implantation of the embryo

progestin = progesterone

Progoneata a subphylum of the phylum Arthropoda erected to contain those "myriapod" forms in which the genital aperture is placed toward the anterior end of the body in contrast to those (Opisthogoneata) in which it is placed towards the middle or posterior end

progredient said of a plant which changes its position by growing in one direction and dying behind

Prohepatica a postulated ancestral form intermediate between a thallophyte and a higher plant

Projapygidae a family of thysanuran insects with short cerci and a stylus on the first abdominal segment

projectura a small, lateral leaf-bearing branch or projection

-prol- *comb. form* meaning "offspring"

prol-1 *see* [Proline-1]

prolan = chorionic gonadotropin

prolate drawn out, particularly of a sphere which is changed to an ovoid

prole offspring, progeny, the result of breeding. Used frequently in German, and sometimes in English, for "subspecies"

prolidase = imidodipeptidase

proliferation in *bot.*, refers specifically to the production of leaf buds within a flower

prolific producing numerous offspring

prolified in *bot.*, refers to a tuft of leaves on the end of a cone as in a pineapple

[proline-1] a mutant (gene symbol *prol-1*) in linkage group III of *Neurospora crassa*. The phenotypic expression is a requirement for proline and an inability to use ornithine, citrulline or arginine

L-**proline** 2-pyrrolidinecarboxylic acid. An amino acid not essential to the nutrition of rats. It is a constituent of most plant proteins

L-γ-**hydroxyproline** an amino acid, 4-hydroxy-2-pyrrolidine-carboxylic acid. Found, together with choline, in many collagens

Prolycopod a postulated intermediate stage between Prohepatica and a higher plant

Promeropidae a family of passeriform birds commonly included with the Meliphagidae

prominent *U.S.* popular name of notodontid noctuoid lepidopteran insects

saddled prominent *U.S. Heterocampa guttioitta*

-pron-, -prono- *comb. form* confused from two *Gr.* roots and therefore meaning "to bend down" or "a prominence" or "headland"

prone (*cf.* supine) lying flat on the face

propago that part of a plant which is used for vegetative reproduction

propagule that which can be, or has been, propagated

Propionibacteriaceae a family of eubacteriale schizomycetes occuring in the form of gram-positive, irregularly shaped, rods. They typically ferment carbohydrates and lactic acids to propionic, butyric, and acetic acids, though one genus (Zymobacterium) produces ethanol from glucose. The propionic acid producers commonly occur in dairy products

propodeum that portion of the hymenopteran metathorax which immediately surrounds the insertion of the abdominal stalk (*cf.* propodium)

-proprio- *comb. form* meaning "proper" in the sense of "one's own"

3-ureidopropionase an enzyme catalyzing the hydrolysis of N-carbamoly-β-alanine to β-alanine, carbon dioxide, and ammonia

proprious partial

Prorhipidoglossomorpha a subphylum of Mollusca containing, *inter alia*, the snails, slugs, elephant-tooths, and clams. They are distinguished from the Isopleura by the fact that the head is anterior to the foot and that the gonads do not connect directly with the pericardium. They are distinguished from the Siphonopoda in the shape of the foot

-pros- *comb. form* meaning "towards"

proscolla the sticky gland on the stigma of orchids to which pollen masses become attached (= retinaculum)

Prosobranchiata an order of gastropod Mollusca, mostly marine snails, in which the ctenidium lies in the mantle cavity at the forward end of the body in front of the heart

prosodus a narrow tube between the incurrent channel and the flagellated chamber in some sponges

Prosoporia an order of oligochaete Annelida in which the male pore is situated on the segment which houses the testis

Prosostomata a suborder, or if Digenea is regarded as a subclass an order, of trematode Platyhelminthes. They are distinguished by the presence of the mouth at the anterior end

prostal a siliceous sponge spicule which projects from the surface of the sponge

-prostho- *comb. form* meaning "before" in the sense of position

prostrate lying flat

-prot-, -proto- *comb. form* meaning "first in time" and by extension "primitive"

Protaxonia a zoological taxon erected by Hatschek to contain the sponges, coelenterates and Ctenophora

-prote- *comb. form* meaning "polymorph"

protea *S.Afr.* any of numerous species of lycaenid lepidopteran insect of the genus Capys

Proteaceae that family of proteale dicotyledons which contains, *inter alia,* the silk oaks. Distinguishing characteristics are the four valvate perianth parts, the four stamens and the small flowers aggregated in spikes or heads surrounded by bracts

Proteales an order of dicotyledonous angiosperms containing the single family Proteaceae. The characteristics of the only family apply to the order

Proteidae a family of perennibranchiate urodele Amphibia distinguished by the absence of maxillary bones and eyelids

protein any of a vast number of organic compounds basically consisting of peptide-linked amino-acids

  conjugated protein = compound protein

  chromoprotein a compound protein which contains a pigment, usually metal-containing, but sometimes a carotenoid. Many (*e.g.* hemoglobin) function as oxygen carriers

  compound protein one which has one or more prosthetic groups in addition to the basic amino acid structure

  fibrous protein one in which the molecules are arranged in compound spirals with hydrogen bonds in many planes

  glycoprotein a compound protein having a sugar-like prosthetic group

  lecithoprotein a compound protein with lecithin as the prosthetic group

  nucleoprotein a compound protein, mostly found in the nucleus, combined with nucleic acids

  phosphoprotein a protein which yields phosphorus on hydrolysis

  lipoprotein a compound protein with a fatty acid prosthetic group

  scleroprotein any protein which forms a hard skeletal or protective structure

  simple protein one consisting of amino acids only and lacking prosthetic groups

  soluble protein one in which chains of molecules are cross linked with weak internal hydrogen bonds and which therefore readily disperse in water

Protelinae a monotypic subfamily of hyaenid carnivorous mammals erected to contain the aard-wolf *Proteles cristata*. It differs from the true hyaenas in having five toes on the fore feet

Proteocephaloidea an order of eucestode cestodes distinguished by the presence of four acetabula and an apical sucker on the scolex

-protero- *comb. form* meaning "before," either in time or space

Proteroglypha an assemblage of colubrid snakes in which the anterior maxillary teeth are grooved (*cf.* Aglypha, Opisthoglypha)

Protista a taxon erected to contain those acellular organisms that in contrast to the Monera (*q.v.*) have a diploid nucleus bounded by a nuclear membrane

-proto- *comb. form* meaning "first"

Protobranchia an order of pelecypod Mollusca containing, *inter alia,* the nut-shells. They are distinguished by the possession of gills with flat and nonreflected filaments disposed in two rows on opposite sides of the branchial axis

Protochula a postulated ancestor of the Protostomia resembling a primitive ctenophore

Protociliata a subclass of ciliate Protozoa distinguished by the absence of a mouth

Protococcaceae that family of ulotrichale Chlorophyceae containing the common protococcus which occurs either single or as double or four-celled thalli

Protocoelomata a no longer acceptable zoological taxon at one time containing the Ctenophora and thought to represent a bridge between the coelenterate and coelomate animals

protocol *biol. jarg.* for a detailed account of test methods and results

protodoche = prisere

Protoflorideae = Bangioideae

Protolycopodiasea the only extant family of lycopodiale Lepidophyta. They are commonly called clubmosses and are distinguished by having numerous small leaves without ligules on the herbaceous sporophytes

Protomastigida an order of zoomastigophorous Protozoa containing *inter alia* the blood parasitic trypansomes. They are distinguished by the large polybasal body and are usually possessed of an undulating membrane

Protophyta a little used botanical taxon originally erected to contain the unicellular plants, but from time to time expanded to contain variously the fungi, the algae, the myxomycetes, and the lichens

Protopteridae a family of dipneustian fishes containing the South African lung fish of the genus Protopterus. They are distinguished from the Lepidosirenidae by the possession of six gill arches and five gill clefts

Protosiphonaceae that family of chlorococcale Chlorophyceae which consists of solitary, usually spherical, multinucleate cells in which one side may be prolonged into a rhizoidal process

Protospondyli an order of Osteichthyes containing as extant examples only the monotypic family Amiidae the single species of which *Amia calva,* is distinguished by the long spineless dorsal fin and by the amphicoelous vertebrae

Protostomia a term coined to define those phyla of bilaterally symmetrical animals in which the embryonic blastopore becomes a mouth. This division therefore includes all phyla of the Bilateria except Chaetognatha, Echinodermata, Hemichordata and Chordata

Prototheria a subclass of the Mammalia containing the single order Monotremata. The characters of the order are typical of the subclass

Protozoa a phylum of animals which are not obviously divided into cells. Many are solitary uninucleate organisms but they may also exist as multi-nucleate individuals or as large colonies

Protremata once considered an order of brachipods but no longer regarded as a valid taxon

Protura a small, anomalous group of hexapod Arthropoda excluded by many from the Insecta on the ground that they have fifteen postcephalic segments, some being added during postembryonic development

biotic **province** a biogeographical division of a land mass

prowl the careful walk of a predator in search of prey

proximal that which lies nearer to. In general biological usage proximity starts at the anterior end, or at the main axis; thus the esophagus is proximal to the stomach and the shoulder is proximal to the arm (*cf.* distal)

pruinose having a bloom, as a ripe grape

   epruinoise lacking a surface bloom

[**prune**] a mutant (gene symbol *pn*) mapped at 0.8 on the X chromosome of *Drosophila melanogaster.* The name refers to eye color

Prunellidae that family of passeriform birds which contains the accentors and hedge-sparrows. They are distinguished from other sparrows by the thrush-like bill and rounded wings

twig **pruner** *U.S.* the cerambycid coleopteran insect *Elaphidion villosum* (*cf.* twig girdler)

Prunoideae a subfamily of Rosaceae distinguished by having an indehiscent fruit, a superior ovary, and actinomorphic flowers

ps see [positional-sterile]

psalterium the third of the four divisions of the stomach of artiodactyls (*cf.* abomasum, reticulum, rumen)

-psamm- *comb. form* meaning "sand"

psamma *U.S.* any of several species of the grass of the genus *Ammophila* (= sand reed), particularly *A. arenaria*

psammic pertaining to sand

   perpsammic pertaining to sandy detritus

psammath- *comb. form* meaning "sea sand"

psammathad a plant formation of the sea shore

Psammocharidae = Pompilidae

psammon *ecol. jarg.* sandy beach

   mesopsammon = psammon

Pselaphidae a small family of coleopteran insects closely allied to the Staphylinidae with which they are sometimes fused. Some of these have been domesticated by ants which feed from a secretion of the beetle

-pseud- *comb. form* meaning "false"

Pseudidae a small family of highly aquatic anuran Amphibia, related to the true toads but differing from them by possessing an extra phalanx in each digit and maxillary teeth

-pseudo- *see* -pseud-

Pseudococcidae a family of homopteran insects regarded by many as a sub-family of the Coccidae. They are commonly called mealy bugs by reason of the fine granular secretion which covers their body

Pseudocoelomata a term coined to describe those phyla of the animal kingdom which are neither acoelomate nor coelomate. This taxon therefore embraces Acanthocephala, Rotifera, Gastrotricha, Kinorhyncha, Nematoda, Nematomorpha and Entoprocta and possibly the Priapulida. All these phyla except the Acanthocephala and Priapulida are sometimes regarded as classes of the phylum Aschelminthes

pseudoidia cells broken from a hypha which can reproduce further hyphae

Pseudomonadaceae a very large family of pseudomonodale Schizomycetes most of which are gram-negative, elongate rods. They produce water soluble pigments, some are phosphorescent, and many are pathogens. A few are capable of existing in strong brine or even saturated solutions

Pseudomonadales an order of Schizomycetes which may be coccoid, straight curved, or spiral, sometimes occuring in chains, but never in trichomes and usually motile by means of polar flagella

Pseudomonadineae a suborder of pseudomonodale schizomycetes differing from the Rhodobacteriinae in lacking photosynthetic pigment. Some produce other pigments

Pseudoneuroptera all those orders of insects (Ephemera, Odonata, Plecoptera, Isoptera, etc.) which were at one time included in the Neuroptera

Pseudophyllidea an order of eucestode Cestoda distinguished by the presence of from two to six bothria on the scolex. Some unsegmented forms are known

Pseudoptera an obsolete order at one time containing the scale-insects

Pseudoscorpionida an order of arachnid arthropods containing the animals commonly called pseudoscorpions. Save for the very much smaller size they resemble scorpions in external appearance though they have a comparatively broad abdomen and lack a sting on the last abdominal segment

Pseudosolaneae a subfamily of the Scrophulariaceae

-psil- *comb. form* meaning "nude" but commonly used in *biol.* for "slender." It is used by ecologists in the sense of "prairie"

psilad a prairie plant

Psilidae a family of medium sized slender myodarian cycloraphous dipteran insects some of which are pests of crops commonly called rust-flies

-psilo- *see* -psil-

Psilophyta a phylum of pteridophyte plants of which only one order containing two genera is extant. They are distinguished from all other vascular plants in having a rootless sporophyte

Psilotaceae the only family of the class Psilotinae and the order Psilotales of Pteridophyta. They are distinguished from other pteridophytes by the absence of leaves

Psilotales the only living order of psilophyte Pteridophyta. They are distinguished by having a rootless dicotomously branched sporophyte either leafless or with a few very small leaves. They occur only in the Australasian region

Psittacidae a very large family of psittaciform birds containing the parrots, cockatoos and their allies. They are distinguished by hard, glossy, and often brightly colored, plumage and their short, stout, strongly hooked bills

Psittaciformes that order of birds which contains the parrots and their allies. They are distinguished by the short hooked bill with a cere and imperforate nares and by their harsh and dense plumage, frequently brilliantly colored

-psoc- *comb. form* meaning to "grind away"

psocid a term properly applied to insects of the family Psocidae but usually extended to cover the whole order

Psocoidia an order of small insects containing *inter alia,* the forms commonly called booklice for the reason that several wingless members of the family Atropidae are common in old books. The majority of psocids are of the family Psocidae and possess four functional wings. The order is distinguished by the possession of chewing mouthparts which include an unusually elongate and chisel-like lacinia of the maxilla. The name Corrodentia is preferred by many

Psoidae a family of coleopteran wood-boring insects distinguished from the closely allied Bostrichidae by having the head clearly visible from above. They are commonly called twig-beetles

Psophiidae a small family of gruiform birds containing those forms commonly called trumpeters. They are distinguished by the loose-webbed feathers, in many cases reduced to hair-like strands

Psophiidae at one time synonymous with Gruidae

psyche *I.P.* the pierid lepidopteran insect *Leptosia nina*

Psychidae a family of tineoid lepidopteran insects commonly called bag-worms because of the characteristic cases which are made and carried about by the larvae. They are extremely destructive pests

Psychodidae a family of minute nematoceran dipteran insects distinguished by having very hairy wings. Most are called moth-flies but the sand-flies, notorious blood sucking vectors of many diseases, are also included

Psychoidea = Zygaenoidea

Psychomyiidae a family of trichopteran insects called the trumpet-net caddisflies from the shape of their larval cases

-psychro- *comb. form* meaning "cold"

psyllid popular name of psyllid insects
    alder psyllid *Psylla floccosa*
    boxelder psyllid *U.S. P. negundinis*
    pear psyllid *U.S. P. pyricola*
    persimmon psyllid *U.S. Trioza diospyri*
    potato psyllid *U.S. Paratrioza cockerelli* (= **tomato psyllid**)
    boxwood psyllid *U.S. P. boxi*

Psyllidae a family of homopteran insects commonly called jumping plant-lice. Their general appearance is that of miniature cicadas though they have strong jumping legs and relatively long antennae

Psylliidae = Psyllidae

pt *see* [Phenylalanine-tyrosine]

Pt *see* [Pintail]

ptarmigan any of several species of galliform birds of the tetraonid genus Lagopus. In *Brit.* usage the term ptarmigan is reserved for those grouse which develop white plumage in winter
    rock ptarmigan *L. mutus*
    white-tailed ptarmigan *L. leucurus*
    willow ptarmigan *L. lagopus*

Pteleaceae = Rutaceae

-pter- *comb. form* meaning "wing," frequently confused with -pteris-

hypoptera = tegula

pteralium one of the articular sclerites surrounding the attachment of the insect wing

pterampelid a climbing fern

pteratous winged

microptere a furrow along the length of a plant stem

Pteridaceae a taxon of Pteridophyta not usually accepted but which contained all the Dicksoniaceae and many other genera derived from the Polypodiaceae

pteridium = samara

Pteridophytea a subkingdom of plants containing, *inter alia,* the clubmosses, horsetails and ferns. They differ from the bryophytes and spermatophytes in the fact that both the gametophytes and the sporophytes are independent plants at maturity. Many taxa of pteridophytes are known only as fossils and are therefore not defined in this dictionary

Pteriidae the family of pelecypod mollusks containing, *inter alia,* the pearl oysters

pterin a general term for a group of white, yellow and red biochromes closely allied to the purines. They are responsible for the body and wing colors of many insects

-pteris- literally "fern," but frequently abbreviated to -pter- and thus confused with wing

apterium the space of bare skin between the tracts of contour feathers in birds

-ptero- *see* -pter-

Pterobranchia a class of the phylum Hemichordata distinguished by the presence of a U-shaped digestive tract and the possession of secreted encasements resembling those of the Ectoprocta

Pterocallidae a family of brachyceran dipteran insects usually included with the Otitidae

Pteroclidae a family of columbiform birds containing the sandgrouse. They are distinguished from the pigeons by having the tarsae, and in some cases also even the toes, completely feathered over

pteroid either wing-like or fern-like according to the intent of the writer

Pteromalidae an extensive and poorly delimited family of chalcidoid apocritan hymenopteran insects. Most included genera possess a 13-segmented antenna with 2–3 ring-segments, a transverse pronotum, incomplete parapsidal furrows, and a non-exserted ovipositor. The abdomen is variable in shape

parapteron a sclerite immediately below the wing of insects

Pteronarcidae a family of plecopteran insects distinguished by the presence of two or more rows of cross veins in the anal area

Pterophoridae a family of pyraloid lepidopteran insects commonly called plume-moths because the wings are split into feather-like processes

Pterophyta a phylum of plants containing those forms commonly referred to as ferns. They are distinguished from all other cryptogamic plants by the fact that the sporophyte is differentiated into stems, leaves and roots

Pteropodidae a family of megachiropteran mammals including many of the fruit-bats and flying foxes

Pteropsida a subphylum erected to contain the filicine Pteridophyta and the Spermatophyta in contrast to the Tracheophyta (*q.v.*)

Pteroptochidae = Rhinocryptidae

-pterous pertaining to the wing

epipterous with a winged tip

heteropterous either having different kinds of wings or, as in hemipterous insects, having two textures on one wing

peripterous having a winged border

podopterous having a winged peduncle

-pterygium literally a wing but used in *biol.* for fin

mesopterygium the middle cartilaginous element of the base of the pectoral or pelvic fin in elasmobranch fishes

metapterygium the posterior cartilaginous element of the base of the pectoral or pelvic fin in elasmobranch fishes

protopterygium the anterior element of the base of the pectoral or pelvic fin of elasmobranch fishes

pterygoid (*see also* bone) wing-shaped

pterygophore *see* bone

pteryla (*see also* tract) an area of bird skin from which contour feathers spring (*cf.* apterium)

pterylosis the state of distribution of pterylae on a bird's skin

ptg *see* [pentagon]

PTH = parathyroid hormone

-ptil- *comb. form* meaning "wing" or "wing-like" but frequently used in *biol.* in the sense of "sheath"

coleoptile the first true leaf of a monocotyledon (= coleophyllum)

endoptile a plant embryo in which the plumule is sheathed in the cotyledon

exoptile lacking a sheath for the cotyledon

hemiptile the juvenile plumage of a bird

mesoptile a feather which chronologically follows a protoptile and precedes the regular plumage

neossoptile downy plumage of young birds in general

protoptile a nestling down feather, pertaining to the first series in birds which have two

teleoptile an adult feather

trichoptile a hair like prolongation of a developing feather

Ptilidae = Ptiliidae

Ptiliidae a family of minute coleopteran insects commonly called feather-winged beetles the term being descriptive of the hind wings

ptilinum an inflatable organ just above the attachment of the antenna possessed by cyclorrhapid Diptera when they emerge from the pupa

Ptilogonatidae a small family of passeriform birds containing the four species of silky-flycatchers of North and Central America. They are distinguished by their short legs, marked crests, and silky plumage

Ptilonorhynchidae a family of passeriform birds commonly called the bower-builders from their habit of building elaborate bowers decorated with brightly colored ornaments of seeds, grass, flowers, etc.

ptilosis the sum total of the plumage of a bird

hypoptilum = hyporachis

-ptin- *comb. form* meaning "feathered"

Ptinidae a small family of coleopteran insects closely allied to the Tenebrionidae but readily distinguished from them by their broader shape and long narrow legs from which the name spider-beetle derives

Ptinoidea = Bostrichoidea

-ptos- *comb. form* meaning "fallen" or "shed"

anthoptosis the shedding of flowers

-ptych- *comb. form* meaning a "fold" or "layer"

Ptychobranchia an order of ascidiacean urochordates distinguished by their opaque tunic and branched

tentacles. They are almost without exception compound forms living in large colonies

Ptychopteridae a family of slender long-legged nematoceran brachyceran insects commonly called phantom craneflies and distinguished from other craneflies by the conspicuously black and white banded legs

-ptygm- *comb. form* meaning "folded" or "layered"

ectoptygma the outer membranes round an insect embryo

entoptygma that layer immediately surrounding the insect embryo

-ptyl- *see* -ptil-

pu *see* [pupal] or [pudgy]

puaiohi the Hawaiian turdid bird *Phaeornis palmeri*

pubera the stage in a fruit between fertilization and the production of a recognizable fruit

pubic pertaining to the area of the genitalia (*cf.* pudendic)

puccoon the boraginaceous herb *Lithospermum canescens* (= red root)

red puccoon the papaveraceous herb *Sanguinaria canadensis* (= blood root)

yellow puccoon the ranunculaceous herb *Hydrastis canadensis*

puckerige *Brit. dial.* = nightjar

pudendic pertaining to the external genitalia (*cf.* pubic)

[pudgy] a mutant (gene symbol *pu*) found in linkage group I of the mouse. The phenotypic expression is adequately described by the name

pudu any of several species, or geographic races, of odocoiline cervid artiodactyles of the genus Pudu. They are distinguished from all other cervids by their small size which rarely exceeds one foot in height at the shoulder

puf *see* [puff]

[Pufdi] a mutant (gene symbol *Pfd*) mapped at 70.8 on chromosome II of *Drosophila melanogaster*. The phenotypic expression is puffed, divergent wings

[puff] a mutant (gene symbol *Pufd*) mapped at 58.0± on chromosome II of *Drosophila melanogaster*. The phenotypic expression is blistered wings

puffer popular name of tetraodontid fish

sharpnose puffer popular name of canthigasterid fish

puffin 1 (*see also* puffin 2, neb) any of several charadriiform birds of the family Alcidae usually of the genera Lunda and Fratercula. Much confusion is occasioned by the fact that the genus Puffinus contains the procellariiform birds known in English as shearwaters

Atlantic puffin = common puffin

common puffin *F. arctica*

horned puffin *F. corniculata*

tufted puffin *L. cirrhata*

puffin 2 (*see also* puffin 1) *I.P.* any of several species of pierid lepidopteran insect of the genus Appias. (*cf.* albatross)

Puffinidae = Procellariidae

pufley portulacaceous herb *Portulaca oleracea* (= purslane)

pug *Brit.* any of numerous species of geometrid lepidopteran insect mostly of the genus Eupithecia

-pugio- *comb. form* meaning "dagger"

[pugnose] a mutant (gene symbol *pn*) in linkage group III of the mouse. The name is adequately descriptive of the phenotypic expression

puka either the Hawaiian cornaceous tree *Griselinia lucida* or *N.Z.* the araliacean tree *Meryta sinclairii*

pukatea *N.Z.* the large monimiaceous tree *Laurelia novaezealandiae*

pukeko *N.Z.* the rallid bird *Porphyrio porphyrio* (*cf.* swamp hen)

pukras *Ind.* the pheasant *Pucrasia macrolopha* (= koklas pheasant)

puku the reduncine antelopid *Adenota kob* (= kob)

Pulicidae the largest family of siphonapteran insects including most of the fleas associated with man and domestic animals

pullatous black-cloaked

pullous black with a tinge of gray

pullulate to bud

pullus a juvenile bird showing juvenile plumage

-pulmo- *comb. form* meaning "lung"

[pulmonary tumors] a mutant (gene symbol *tm*) in linkage group VII of the mouse. Phenotype is susceptible to both induced and spontaneous tumors

Pulmonata an order euthyneuran gastropod Mollusca containing the land and fresh-water snails and slugs. The respiration is by "lung"

pulsellum a flagellum as distinct from tractellum

pulveraceous powdery

pulvillus a pad, usually hairy, between or under the tarsal joints of some arthropods

-pulvin- *comb. form* meaning "cushion"

pulvinar a bump on the posterior side of the thalamus

pulvinus the swollen "cushion-like" base of a petiole

puma *U.S.* the felid carnivore *Felis concolor* (= panther *U.S.*)

pumilous low growing

salivary **pump** the organ at the base of the stylet in piercing insects through which the products of the salivary gland are pushed into the victim

    subclypeal pump in some insects a bulbous thickening at the anterior end of the oesophagus

pumpkin (*see also* seed 3) a large, orange, horticultural variety of the cucurbitaceous vine *Cucurbita pepo*. The term is loosely applied to other cucurbits

    ash pumpkin the cucurbit plant *Benincasa hispida* and also its fruit

pun *see* [puny]

punch *I.P.* any of several species of riodinid lepidopteran insect of the genus Dodona

punchinello *I.P.* the riodinid lepitopteran insect *Zemeros flegyas*

-punct- *comb. form* meaning "puncture" or "sting"

Punctariales an order of heterogenerate Phaeophyta distinguished by their habit of growth through intercalary cell divisions and which therefore lacks a definite meristem

punctate dotted

pungent properly, having a piercing, sharp point though by extension, in popular speech, extended to a piercing aroma

Punicaceae that family of myrtiflorous dicotyledonous angiosperms which contains the pomegranate. The appearance of the fruit is distinctive of the family

puniceous crimson

punkie any minute dipteran but particularly *U.S.* popular name of ceratopogonid midges

punui *N.Z.* the araliaceous herb *Kirkophytum lyallii*

[puny] a mutant (gene symbol *pun*) mapped at 41.1± on the X chromosome of *Drosophila melanogaster*. The name is descriptive of the phenotype

pup (*see also* puppy) a young canid, though the term is also applied to many young carnivores, such as sea-lions and seals

pupa the dormant pre-imagal stage of holometabolous organisms. The term was at one time applied to the doliolaria larva (*q.v.*) of Holothuria

    coarctate pupa one which lies within the "skin" of the last pre-pupal molt (*cf.* pseudopupa)

    follicular pupa one which lies in a case

    masked pupa one in which the segmentation and parts of the contained insects are clearly visible on the surface

    prepupa a quiescent larval instar that has not yet pupated

    pseudopupa a resting stage between insect instars which is not covered with a pupal case and in which histolysis and histogenesis do not take place. The term is sometimes used as synonymous with coarctate pupa

    semi-pupa a quiescent, torpid instar of an insect which is about to pupate

[pupal] a mutant (gene symbol *pu*) mapped at 51.0± on chromosome II of *Drosophila melanogaster*. The wings of the phenotype remain in the pupal condition

pupango *N.Z.* the aythyine bird *Aythya novaeseelandiae* (= New Zealand scaup)

puparium the thickened larval case within which the pupa of many Diptera is formed

pupil the contractile aperture in the iris of the eye

Pupipara an assemblage of those schizophoran cyclorrhaphous brachyceran dipteran insect which are permanent ectoparasites on other animals. The majority either lack wings or have wings so poorly developed that they cannot pass from one host to another. Some, such as the Hippoboscidae, fly well but are rarely found far from the host (*cf.* Myodaria)

puppy any immature canine (*cf.* pup)

    mudpuppy *U.S.* the proteid urodele amphibian *Necturus maculosus*. The term is also used in various parts of *U.S.* for any large urodele amphibian, especially neotonic forms

    sand-puppy either of two species of bathyergid rodent of the genus Heterocephalus. They are distinguished from all other rodents by the complete absence of hair and are confined to the deserts of Somaliland

pur a bacterial mutant indicating a requirement for purine

purA a bacterial genetic marker having a character which affects the activity of adenylosuccinic synthetase, mapped at between 83.25 mins and 88.0 mins for *Escherichia coli*

purB a bacterial genetic marker having a character which affects the activity of adenylosuccinase, mapped at 23.5 mins. for *Escherichia coli*

purC a bacterial genetic marker having a character which affects the change of 5-aminoimidazole ribotide (AIR) to 5-aminoimidazole 4-(N succinocarboxamide) ribotide, mapped at 44.25 mins. for *Escherichia coli*

purD a bacterial genetic marker having a character which affects the biosynthesis of 5-aminoimidazole ribotide (AIR) mapped at 77.25 mins. for *Escherichia coli*

purE a bacterial genetic marker having a character which affects the change of 5-aminoimidazole ribotide (AIR) to 5-aminoimidazole 4-(N succinocarboxamide) ribotide, mapped at 11.0 mins. for *Escherichia coli*

puriri *N.Z.* the large verbenaceous tree *Vitex lucens*

Purkinje *epon. adj.* from Johannes Evangelista Purkinje (1787–1869)

purple 1 (*see also* purple 2, 3) popular name of several species of nymphalid lepidopteran insects, many of the genus Basilarchia (*cf.* viceroy)

    banded purple *U.S.* *Limenitis arthemis* (= white admiral)

    dusky purple *S.Afr.* *Aslauga purpurascens*

    red-spotted purple *Limenitis astyanax*

purple 2 (*see also* purple 1, 3) in the sense of color

    Tyrian purple an indigoid biochrome derived from several species of mollusk particularly the gastropod genera Murex, Mitra and Purpura

    visual purple = rhodopsin

rock purple 3 (*see also* purple 1, 2) the thaisid gastropod mollusk *Thais lapillus*

[purple] a mutant (gene symbol *pr*) mapped at 54.5 on chromosome II of *Drosophila melanogaster*. The description refers to the eye color of the phenotype

[purpleoid] a mutant (gene symbol *pd*) mapped at 106.4 on chromosome II of *Drosophila melanogaster*. The phenotypic eye is dark pink

[Purpleoider] a mutant (gene symbol *Pdr*) mapped at 46.0± on chromosome II of *Drosophila melanogaster*. The phenotypic expressions is an intensification of purpleoid

[Purple plant] a mutant (gene symbol *Pl*) mapped at 44 on linkage group VI of *Zea mays*. The term is descriptive of the appearance of the phenotype

atropurpureous purple-black

shepherd's purse *Brit.* the cruciferous herb *Capsella bursapastoris*

purslane any of several species of herb of the portulaceous genus Portulaca but particularly *P. oleracea*. The term is also applied to a few other plants

    marsh purslane the onagraceous herb *Ludwigia palustris*

    milk-purslane *U.S.* the euphorbiaceous herb *Euphorbia supina*

    sea purslane *U.S.* any of many herbs of the family Iazoaceae particularly those of the genera Sesuvium and Prianthema. *Brit.* the caryophyllaceous herb *Arenaria peploides* and the chenopodiaceous herb *Atriplex hortensis*

    water purslane the onagraceous herb *Ludwigia palustris* and the lythraceous aquatic herb *Peplis diandra*

    winter purslane the portulacaceous herb *Montia perfoliata*

pusillous weak, or fragile

puss a female hare, also a term of endearment applied to the domestic cat

pustular with blisters

pusule one of a series of non-contractile vacuoles found in dinoflagellates

putamen a hardened covering of an edible substance such as the shell of a nut, or the endocarp of a peach seed

putaminaceous as hard as a peach or plumb stone

putangitangi *N.Z.* the tadornine bird *Casarca variegata* (= paradise duck)

putaputaweta *N.Z.* the saxifragaceous shrub, or small tree *Carpodetus serratus*

putoto *N.Z.* the rallid bird *Porzana tabuensis* (= spotless crake)

puttock *Brit. dial.* almost any diurnal bird of prey

monkey puzzle *I.P.* the lcaenid lepidopteran insect *Rathinda amor*

monkey puzzler the araucariaceous tree *Araucaria imbricata*

Pw *see* [Pointed-wing]

pw-c *see* [pink-wing-c]

px *see* [paroxysm], [plexus] or [post axial hemimelia]

Px *see* [Plexate] or [Pollex]

py *see* [polydactyly] or [pigmy]

pyarg a biblical beast (Hebrew *dishon*) usually supposed to be the addax

-pycn- *comb. form* meaning "dense"

macropycnid = stylospore

    micropycnid = pycnoconidium

pycnidium a cavity containing gonidia in a lichen

pycnium a dense mass of cells giving rise to the sorus of the Uredineales

Pycnogonida an aberrant class of marine arthropods commonly called sea spiders. They are distinguished by a greatly reduced slender body to which are attached four, five, six or twelve pairs of frequently very long legs

Pycnonotidae a large family of passeriform birds commonly called bulbuls. They mostly have curved heavy bills and numerous hairlike feathers on the nape of the neck. The wings are rounded and short

pycnosis 1 (*see also* pycnosis 2) the condition of having a contracted nucleus, usually in moribund cells

    heteropycnosis the condition of darkly staining parts of a chromosome

pycnosis 2 (*see also* pycnosis 1) those cases of insect larval histolysis in which the chromatin becomes distributed in the nodules of histolyzing tissue

pyd *see* [polychaetoid]

-pyg- *comb. form* indicating "rump" or "buttocks"

cytopyge the more or less definite point from which flagellates and ciliates eject the undigested remnants of their ingested food

Pygidiidae = Trichomycteridae

pygidium any terminal region of the body which cannot properly be called a tail for example, the terminal tergite of the insect's abdomen, the tail shield of the trilobite, and the terminate segments of an adult polychaete worm

hypopygium literally, beneath the tail and thus applying to the anus particularly in insects in which it is also used for some intromittent organs

Pygopodidae a family of snake-like lacertilian reptiles lacking forelimbs but with the hindlimbs represented by a pair of paddle-shaped structures or scaly flaps

pyknosis = pycnosis

-pyle- *comb. form* properly meaning "entrance" but also misused in the sense of "hole" or "pore"

    aeropyle a small hole at the base of a leguminous pod

    apopyle the opening of the flagellated chamber of the sponge into an excurrent canal

    astropyle the basal hole penetrating the central capsule of phaeodarian radiolaria (*cf.* parapyle)

    micropyle a perforation in the envelope of an egg; specifically, in insects the aperture in the egg membrane through which the sperm enters. In *bot.* an opening from the integument from the nucellus to the exterior of an ovule

    parapyle two lateral holes penetrating the central capsule of phaeodarian Radiolaria (*cf.* astropyle)

    prosopyle the pore between the incurrent and the radial canals in a syconoid type sponge

pylome the aperture of the shell of a testacean rhizopod

pylorus the opening between the stomach and the intestine

pylstaart *S.Afr.* = tropicbird

-pyr- *comb. form* confused from three *Gr.* roots and therefore meaning either "pear," "fire" or "wheat"

pyr a bacterial mutant indicating a requirement for pyrimidine

pyrA a bacterial genetic marker having a character affecting the activity of carbamate kinase, mapped at 0.2 mins. for *Escherichia coli*

pyrB a bacterial genetic marker having a character affecting the activity of aspartate transcarbamylase, mapped at 83.25 mins. for *Escherichia coli*

pyrC a bacterial genetic marker having a character affecting the activity of dihydroorotase, mapped at 20.0 mins. for *Escherichia coli*

pyrD a bacterial genetic marker having a character which affects the activity of dihydroorotic acid dehydrogenase, mapped at 20.0 mins for *Escherichia coli*

pyrE a bacterial genetic marker having a character which affects the activity of orotidylic acid pyrophosphorylase, mapped at 71.4 mins. for *Escherichia coli*

pyrF a bacterial genetic marker having a character which affects the activity of orotidylic acid decarboxylase, mapped at 20.0 mins. for *Escherichia coli*

pyr-1 -3 *see* [Pyrimidine-1, 2, 3]

Pyralidae a very large family of pyralidoid lepidopteran insects. Many are pests of both growing and stored crops

Pyralididae = Pyralidae

Pyralidoidea a superfamily of lepidopteran insects containing a number of small moths of which the plume-moths are typical

pyramid a solid body the sides of which are triangles coming to a common point. Specifically applied, without qualification, to the ventrolateral portion of the cerebellum and the expanded floor and walls of the medulla oblongata. The term is also used for the five vertical interradial supports of Aristotle's lantern

anal pyramid a ring of calcareous plates surrounding the anus in some echinoderms

Eltonian pyramid the pyramidal arrangement of numbers of organisms, of biomass, or of the trophic levels in an ecosystem

renal pyramid a compact pyramid-shaped portion of the medulla in the kidneys of birds and mammals

Pyramidellidae a family of gastropod Mollusca distinguished by their many-whorled pyramidal shells. Most are small and some are minute

Pyraustidae a large family of lepidopteran insects often fused with the Pyralidae

-pyre- *comb. form* meaning the "stone" of a fruit or "kernel" of a nut

Pyrenaceae = Verbenaceae

pyrenarious said of a pome in the form of a drupe

pyrenarium a pome in the shape of a pear

-pyrene- *comb. form* meaning "fruit stone" but once widely used for "nucleus"

pyrene a nucleus or the stone of a drupe

eupyrene having a normal nucleus, said particularly of spermatozoa

oligopyrene having fewer than the normal number of chromosomes

micropyrenic having unusually small nuclei

Pyrenidae a family of gastropod mollusks having small fusiform shells with an outer lip thickened in the middle and with an inner lip bearing a small tubercule. Commonly called dove-shells

pyrenodeous wart-like

pyrenoid a dense proteinaceous mass in the center of a chloroplast

Pyrenomycetes a subclass of euascomycete Ascomycete distinguished by the presence of a hole in the pteridium

apyrenous said of a fruit lacking seeds

monopyrenous containing a single seed

dipyrenous a fruit containing two seeds, or, more usually, a fleshy fruit containing two stones

pyridion = pyrenarium

pyridoxine *see* vitamin B_6

[Pyridoxine-1] a mutant (gene symbol *pdx-1*) in linkage group IV of *Neurospora crassa.* The phenotypic expression is a requirement for pyridoxine

[Pyridoxine-2] a mutant (gene symbol *pdx-2*) in linkage group IV of *Neurospora crassa.* The phenotypic expression is a requirement for pyridoxine

dihydropyrimidinase an enzyme catalyzing the hydrolysis of dihydrouracil to ureidopropionic acid

[Pyrimidine-1] a mutant (gene symbol *pyr-1*) in linkage group IV of *Neurospora crassa.* The phenotypic expression is a requirement for pyrimidine

[Pyrimidine-2] a mutant (gene symbol *pyr-2*) in linkage group IV of *Neurospora crassa.* The phenotypic expression is a requirement for pyrimidine

[Pyrimidine-3] a mutant (gene symbol *pyr-3*) in linkage group IV of *Neurospora crassa.* The phenotypic expression is a requirement for pyrimidine

Pyrochroidae a family of medium to large beetles so usually reddish or yellowish as to be distinguished as the fire-colored beetles. The antennae of the males are so deeply pectinate as to be almost plumose

Pyrolaceae a family of ericale dicotyledonous angiosperms closely related to the Ericaceae from which it may be distinguished by the herbaceous habitat and polypetalous flowers with five carpels

Pyrolideae a subfamily of Pyrolaceae containing those genera which have chlorophyll in distinction from the Monotropoideae

Pyromorphidae a family of lepidopteran insects, commonly called smoky moths, closely allied to, and usually fused with, the Zygaenidae

Pyrosomida an order of thaliacious tunicates, thought by some to belong with the ascidian tunicates. They are distinguished by existing in regular colonies contained in a common tunic and by their pelagic mode of life

Pyrrhocoridae a family of elongate oval hemipteran insects almost invariably brightly marked with red and black and thus commonly called redbugs

Pyrrophyta a phylum of algae distinguished by the fact that the pigments in the chromatophores are brownish-green through the presence, in addition to chlorophyll, of carotenes and xanthophylls

pys *see* [polychaetous]

pythmic pertaining to lake bottoms (*cf.* bathile, chilile)

python any pythonine boid snake. The English and Latin words python are interchangeable

African python *P. sebae*

Angolan python *P. anchietae*

Australian python *Morelia argus*

ball python *P. regius*

burrowing python *Calabaria reinhardti*

carpet-python *Austral.* = carpet-snake *Austral.*

Indian python *P. molurus*

reticulated python *P. reticulatus*

rock-python *Afr.* = African python. *Austral.* any of several species of the genus Liasis

short-tailed python *P. curtus*

Timor python *P. timorensis*

Pythoninae a subfamily of boid snakes distinguished from the Boinae by the presence of supraorbital bones and an oviparous mode of reproduction

pytilia any of several African ploceid birds of the genus Ptylia (*cf.* melba finch)

Pyuridae a family of ptychobranchiate ascidian urochordates having four lobed siphons, branched tentacles, and an opaque test

-pyxid- *comb. form* meaning "chest" in the sense of a box with a hinged lid

pyxidate having a lid

pyxie the diapensiaceous creeping herb *Pyxidanthera barbulata*

pyxis a capsule with a circumscissile dehiscence

q *see* [quail]

-quadr- *comb. form* meaning "four"

quadrat a square area in which the organisms are intensively examined and one or several of which form the basis for assessing the entire population of the area

quadrate (*see also* bone) more or less square

quadrieremus a colony or coenobium of four independent organisms

quadrulus four rows of cilia opposite to and parallel to the penniculus (*q.v.*)

quagga the zebra-like wild ass *Equus quagga*

quaha the jackal *Canis lateralis*

quahog the venerid pelecypod mollusk *Venus mercenaria* (*cf.* littleneck clam)

quail (*see also* dove, finch, hawk, plover, snipe, thrush) properly any of numerous phasianid galliform birds but occasionally used for other quail-like forms. *U.S.* the usage of "quail" corresponds closely to *Brit.* usage of "partridge." *Brit.* any of several species of the genus Coturnix, particularly *C. coturnix*

red-backed quail the turnicid *Turnix maculosa*

banded quail *Philortyx fasciatus*

blue quail *Excalfactoria adansbnii*

black-breasted quail the turnicid *Turnix melanogaster*

blue-breasted quail *Excalfactoria chinensis*

brown quail *Synoicus ypsilophorus*

bush-quail any of several species of the genera Perdicula and Cryptoplectron

bustard-quail = button-quail

button-quail properly, any of several turnicids of the genus Turnix, without qualification, *T. sylvatica*; *I.P.*, used for *Excalfactoria chinensis* (= blue-breasted quail)

California quail *Lophortyx californica*

red-chested quail the turnicid *T. pyrrhothorax*

common quail *Coturnix coturnix*

desert quail *Lophortyx gambelii*

elegant quail *Lophortyx douglasii*

tawny-faced quail *Rhynchortyx cinctus*

Gambel's quail = desert quail

harlequin quail *U.S. Cyrtonyx montezumae Afr. Coturnix delegorguei*

hemipode-quail popular name of turnicid birds (*cf.* button quail)

little quail the turnicid *Turnix velox*

marsh quail *U.S.* rare the icterid bird *Sturnella magna* (= Eastern meadow-lark)

migratory quail = common quail

mountain quail *Oreortyx picta*

New Zealand quail *Coturnix novaezeelandiae*

ocellated quail *Cyrtonyx ocellatus*

painted quail the turnicid *Turnix varia*

plover-quail *Austral.* the pedeonemid gruiform bird *Pediomomus torquatus* (= collared hemipode)

rain quail *Coturnix coromandelica. W.I.* = rain bird

scaled quail *Callipepla squamata*

singing quail *Dactylortyx thoracicus*

Spanish quail *B.W.I.* the thraupid bird *Spindalis zena* (= stripe-headed tanager)

stubble quail *Austral. Coturnix pectoralis*

valley quail = California quail

Virginia quail *Cotinus virginiarus* (= common bobwhite)

bobwhite quail = Virginia quail

wood-quail any of several species of the genus Odontophorus

spotted wood-quail *Odontophorus guttatus*

[quail] a mutant (gene symbol *q*) mapped at 0 on linkage group VII of *Bombyx mori*. The phenotypic expression is a larval body tinted reddish-purple

quaker any of many organisms somber in color and mild in habits, mostly lepidopteran insects. Specifically the procellariid bird *Phoedetria fuliginosa*. (= dusky albatross). Also *U.S.* any of several species of noctuid lepidopterans of the genus Orthodes. *Brit.* any of several species of noctuid lepidopteran insects of the genus Taeniocampa and Amathes. *Austral.* and *I.P.* the lycaenid lepidopteran insect *Neopithecops zalmora*

quan = guan

Brisbane quandary = quandong

quandong the elaeocarpaceous tree *Elaeocarpa grandis* and its edible fruit

qua qua *W.I.* the formicariid bird *Thamnophilus doliatus* (= barred antshrike)

quarrion *Austral.* the psittacid bird *Leptolophus hollandicus* (= cockatiel)

quarterdeck the crepidulid mollusk *Crepidula fornicata* (*cf.* boat-shell, slipper-limpet)

quaternary arranged in fours

queen a regnant female, particularly the fertilized egg-laying female of an insect colony. Also *U.S.* the danaid lepidopteran *Danaus plexipus* (*cf.* monarch)

forest queen *S.Afr.* the nymphalid lepidopteran insect *Euxanthe wakefieldii*

junglequeen *I.P* any of numerous species of amathusiid insects of the genus Sticopthalma

trailing queen the onagraceous trailing shrub *Fuchsia procumbens*

queen-of-lilies = golden-banded lily

queen-of-Spain *Brit. Argynnis lathonia*

queen-of-the-meadow the rosaceous herbs *Spiraea alba* (= meadow sweet) and *Filipendula rubra*

queen-of-the-prairie the rosaceous herb *Filipendula rubra*

queest *Brit.* columbid bird *Columba palumbus* (= wood pigeon)

quetzal any of several trogonid birds of the genus Pharomacrus, usually *P. mocinno*

quick (of nails) = eponychium

quicken *Brit.* the rosaceous shrub *Pyrus aucuparia*

Quiinaceae a small family of parietale dicotyledons

quill popular name of the shaft of a feather

hand quill the quill of a primary feather

quilled said of a ligulate floret which is not tubular

quinary arranged in fives

triquinate divided first into three and then into five

quince (*see also* curculis, hopper) the rosaceous tree *Cydonia oblonga* or any of its numerous horticultural varieties. Also the fruit of this tree

bastard quince the rosaceous shrub *Sorbus chamaemespilus*

Japanese quince the rosaceous shrub *Chaenomeles japonica*

quincuncial *see* quincunx

quincunx a pattern of five objects, four placed at the corners of a square, and the fifth at its center. Serial repetition of quincunces produces a pattern of alternating rows to which the term quincuncial is commonly applied

quinine the alkaloid obtained from the bark of the rubiaceous tree *Cinchona ledgeriana* and some other species of the same genus

wild quinine *U.S.* the compositaceous herb *Parthenium integrifolium*

quink = brant goose

quinoa the chenopodiaceous herb *Chenopodium quinoa* (*cf.* pigweed)

-quinq- *comb. form* meaning "five"

quit *W.I.* a name applied to several birds unrelated save as to the cry they utter

bananaquit the parulid bird *Coereba flaveola*

blue quit the thraupid *Pyrrhuphonia jamaica*

cho-cho quit = bluequit

grassquit any of several species of fringillids of the genera Loxipasser, Tiaris and Volatinia

Cuban grassquit *T. canora*

black-faced grassquit *T. bicolor*

yellow-faced grassquit *T. olivacea*

yellow-shouldered grassquit *L. anoxanthus*

marley quit *W.I.* = bananaquit

short-mouthed quit = bluequit

orangequit the thraupid *Euneornis campestris*

[quivering] a mutant (gene symbol *qv*) in linkage group I of the mouse. The phenotypic expression is adequately described by the name

qv *see* [quivering]

R

r *see* [reduced], [rodless retina], [red-eyed yellow] or [yellow fruit flesh]

r a symbol used for biotic potential (*q.v.*) in the sense of population growth rate

-r *symb. suffix* appended to bacterial mutant symbols to indicate resistance to the compound or organism indicated (*cf.* -d, -s)

R a bacterial genetic marker indicating that the mutant is a repressor gene

R *see* [Aleurone], [Plant color], [Rough fur], [Roughened] or [Rose comb]

r₁, r₂ *see* [rex₁], [rex₂]

r⁹ *see* [rudimentary ⁹]

ra *see* [rase]

Ra *see* [Ragged]

ra₁ *see* [ramosa]

rabbit (*see also* bandicoot, louse, tick) the confusion between rabbit and hare is similar to that between rat and mouse: that is, the distinction, in popular parlance, is one of size. All are, of course, leporid lagomorphs. The European rabbit is Oryctolagus (usually miscalled Lepus). European hares belong to the genus Lepus, as do most American hares, and many large American rabbits. The majority of small American rabbits belong to the genus Sylvilagus. The laboratory rabbit is *Oryctolagus cuniculus*

brush rabbit *U.S. S. bachmani*

jackrabbit any of several *U.S.* species of Lepus. Without qualification, usually *L. americanus*

antelope jackrabbit *L. alleni*

blacktail jackrabbit *L californicus*

whitetail jackrabbit *L. townsendi*

jumping rabbit the dipodid rodent *Alactaga jaculus* (*cf.* jerboa)

marsh rabbit *U.S. S. palustris*

pygmy rabbit *U.S. S. idahoensis*

swamp rabbit *U.S. S. aquaticus* (= cane-cutter)

raccoon (*see also* dog, grape) any of numerous fissipede canoid Carnivora of the family Procyonidae. *U.S.*, without qualification, *Procyon lotor*

South American raccoon *Procyon cancrivorus*

race 1 (*see also* race 2) a geographic enclave of a species the gene pool of which differs from that of other similar enclaves of the same species: in this sense the word "subspecies" is preferable. The term is also loosely used in the sense of a variety that breeds true

adaptive race one which is physiologically, not morphologically, distinguished

between race the offspring of a cross that shows some new types, but that mostly resemble the parent

biological race = adaptive race

climatic race one which is adapted to a different climatic environment from that in which its nearest relatives live

ecological race variously used to mean geographic race or subspecies

mid race one which can be improved by artificial selection and breeding

race 2 (*see also* race 1) a *bot.* taxonomic category ranking below forma

-racem- *comb. form* meaning a "bunch of berries"

racemase a group name for those enzymes which change L to D forms. The names are all completely descriptive (*e.g.* glutamate racemase catalyzes the production of D- glutamate from L-glutamate) and are not listed separately

raceme an indeterminate centripetal inflorescence with a long axis; the term derives from the appearance of a bunch of grapes

botrycymose raceme one composed of botryoid clusters gathered together in cymes

compound raceme = panicle

false raceme = helicoid cyme

racemose a type of branching in which the main stem coninues to grow throwing off lateral branches as it does, so that the lowest branches are the oldest and largest. This results in a conical or bushy formation

racer *U.S.* any of numerous species of colubrine snake mostly of the genus Coluber

black racer *C. constrictor* many races and subspecies of this form are recorded under a variety of names

speckled racer *Drymobius margaritiferus*

-rach- *comb. form* meaning "cliff" or "crag" but almost universally used as a misspelling of -rhach- (*q.v.*)

aporachial in a direction away from the rachis

Rachiceridae a family of dipteran insects variously placed in the Brachycera because of ther wing venation and in the Nematocera because of thir many segmented antennae. They strongly resemble the sawflies in appearance

rachilla diminutive of rachis. Applied particularly to the short, lateral axes in a compound inflorescence

rachion the area of maximum wave turbulence on a shore

rachis 1 (*see also* rachis 2, 3, 4) the axis to which leaflets are attached

rachis 2 (*see also* rachis 1, 3, 4) that portion of a pennatualacean which bears polyps

rachis 3 (*see also* rachis 1, 2, 4) the central shaft of a feather

hyporachis the very small counterpart of a feather springing from the inner surface of a quill common to both or the aftershaft of a feather

rachis 4 (*see also* rachis 1–3) the vertebrate axial skeleton

endorachis the connective tissue lining of the skull and vertebral column

Rachycentridae a monospecific family of percid fish containing the cobia (*Rachycentron canadum*)

rad a bacterial mutant showing radiation resistance or sensitivity

radial pertaining to the radius of the circle or combinations of radii of a circle or to radius 1 or 2 below. It is also used as the adjective (in *bot.*) from a ray, as in the ray florets of compositaceous flowers, for spines of echinoderms and for the skeletal support of the fins of fishes

radiale = fin ray of fish (*see also* bone)

radiant either progressing outwards from a common center or pertaining to radiant energy

Radiata a zoological taxon erected by Lamarck to contain the coelenterates, ctenophores, and echinoderms. In Cuvier's system, the Radiata were extended to contain all the invertebrates except the Mollusca and the Articulata

radiate arranged round the common center

adaptive **radiation** the development within a taxon, of forms adapted to specific ecological niches

-radic- *comb. form* meaning "root"

radicant rooting, that is the production of roots, particularly from unusual organs

radicle 1 (*see also* radicle 2) arising from the root or the primordium of the root

radicle 2 (*see also* radicle 1) in insects, that joint of the antenna which is attached to the head

radicose having unusually abundant roots

eradiculose lacking roots

Radiolaria a large order of actinopodous Protozoa usually producing a silicious skeleton

radish (*see also* weevil) the cruciferous plant *Raphanus sativus* and particularly its edible roots

horseradish (*see also* beetle, tree) the cruciferous herb *Armoracia lapathifolia* and its allegedly edible roots

water radish *Brit.* the cruciferous herb *Armoracia amphibia*

radius 1 (*see also* radius 2, bone) a straight line running from the center of a circle to the circumference or a structure lying in the position of such a line

radius 2 (*see also* radius 1) in insects the third primary longitudinal wing vein counting from the anterior

adradius the midradius of an interradius

interradius a radius of a radially symmetrical organism which lies between the perradii

perradius that radius of a radially symmetrical organism on which organs are grouped

[**radius incompletus**] a mutant (gene symbol *ri*) mapped at 47.1 on chromosome III of *Drosophila melanogaster*. The vein L2 of the phenotype shows a gap

raduculose bearing rootlets

radula a ribbon-like band bearing transverse rows of teeth found about the mouth in many mollusks

raf a yeast genetic marker indicating a multiple requirement for sucrose and raffinose

Rafflesiaceae a family of aristolochiale angiosperms distinguished by their parasitic habit and the fact that the vegetative is reduced to a thalloid, or even mycelium-like form. The leaves are scale-like and

flowers vary from minute to that of *Rafflesia arnoldii* which is the largest known flower sometimes reaching a diameter of 36 inches

raft a term for a group of ducks stationary on the water (*cf.* paddling, bed)

rafter term for a group of wild turkeys

rag 1 (*see also* rag 2) term for a group of colts (*cf.* rake)

rag 2 (*see also* rag 1) in compound names of several animals

blackrag the stromateid fish *Icticus pellucidus*

silver rag the stomateid fish *Cubiceps nigriargenteus*

whiterag *Brit.* popular name of nereid worms particularly when used as fish bait

[**Ragged**] *either* a mutant (gene symbol *Ra*) found in linkage group V of the mouse. The phenotypic expression is a sparse coat *or* a mutant (gene symbol *Rg*) in linkage group I of *Neurospora crassa*. The term is descriptive of the phenotypic colony

[**Ragged leaves**] a mutant (gene symbol *Rg*) mapped at 40 in linkage group III of *Zea mays*. The ragged appearance of the phenotypic leaf is due to irregular dead patches

ra₁ *see* [raisin]

rail (*see also* babbler) any of numerous gruiform birds of the family Rallidae, but occasionally applied to other rail-like buds

Bensch's rail the mesoenatid *Monias benschi*

black rail *Laterallus jamaicensis*

slate-breasted rail *Rallus pectoralis*

cape rail *Rallus caerulescens*

clapper rail *Rallus longirostris*

king rail *Rallus elegans*

land rail *Brit. Crex pratensis* (= corncrake)

marsh rail *N.Z. Porzana pusilla* (*cf.* crake)

pectoral rail *Rallus philippensis*

pygmy rail any of numerous African rallid birds of the genus Sarothura (*cf.* crake)

spotted rail *Pardirallus maculatus*

swamp rail *N.Z. Porsana tabuensis*

grey-throated rail *Canirallus oculeus*

Virginia rail *Rallus limicola*

water rail *Brit. Rallus aquaticus*

wood-rail any of several species of the genus Aramides

grey-necked wood-rail *A. cajanea*

rufous-necked wood-rail *A. axillaris*

yellow rail *Coturnicops noveboracensis*

Zapata rail *Cyanolimnas cerverai*

[**raised**] a mutant (gene symbol *rsd*) mapped at 95.4 on chromosome III of *Drosophila melanogaster*. The phenotype folds its wings vertically

raisin (*see also* moth, tree) properly a dried grape but applied to two species of caprifoliaceous shrubs of the genus Viburnum which bear raisin like fruit

wild raisin *U.S. V. cassinoides or V. lentago*

[**raisin**] a mutant (gene symbol *rai*) mapped at 17.0± on chromosome III of *Drosophila melanogaster*. The phenotype has raisin-colored eyes

raj a bacterial mutant showing a utilization of raffinose

rajah *I.P.* any of several species of nymphalid insect of the genus Charaxes

Rajidae a family of the chondrichthyan order Rajiformes commonly called skates or rays

rake 1 (*see also* rake 2) term for a group of colts (*cf.* rag)

rake 2 (*see also* rake 1) a structure resembling a gardener's rake

diatom rake the series of spines on the galea of trichopteran insect larvae used for detaching diatoms from rocks

choanal **raker** a fimbriate flap, partially closing the choana (*q.v.*) of some marine turtles

    gill raker a series of protuberances varying in shape from nobs to filaments on the inner edge of the gills of fish

Rallidae a large family of gruiform birds containing the forms commonly called rails, gallinules and coots. They are distinguished by the laterally compressed body and relatively long legs and toes. The tibia are usually naked

-ram- *comb. form* meaning "branch"

ram (*see also* horn) a male sheep, or other ovid, though the term is occasionally applied to other artiodactyls such as the impala

ramal pertaining to a branch

ramarama *N.Z.* the myrtaceous shrub *Myrtus bullata*

ramate branched

rameal = ramal

ramentum a thin epidermal scale

rameous pertaining to a branch

ramet one individual of a clone

-rami- *comb. form* meaning a branch

[ramosa] a mutant (gene symbol *ra₁*) mapped at 22 on linkage group VII of *Zea mays*. The phenotypic expression is a branching of the ear and tassel

ramose branched

-ramous *comb. suffix* meaning "pertaining to a branch"
    biramous two branched
    breviramous having short branches
    eramous unbranched

ramp *U.S.* the liliaceous herb *Allium tricoccum* (= wild leek)

Ramphastidae that family of piciform birds which contains the toucans. The monstrous beak is typical

rampion either of two campanulaceous herbs *Campanula rapunculus* or *Phyteuma orbiculare*
    horned rampion any of numerous species of campanulaceous herb of the genus Phyteuma

ramson *Brit.* the liliaceous herb *Allium ursinum*

ramule the branch of a moss

ramuline appended to the branch of a moss

ramulus a little or subordinate branch
    pseudoramulus a branch which adheres to another, but is not derived from it

ramus literally, an awl. Particularly one of the paired trophi in the mastax of Rotifera

Ranales a very large order of dicotyledonous angiosperms which contains, *inter alia*, the water lilies, the buttercups, the peonies, the barberries, the moonseeds, the magnolias, the pawpaws, the nutmegs, and the laurels. The order is distinguished by the spirally arranged floral parts in which the perianth is frequently not differentiated into calyx and corolla and the gynoecium is apocarpous

ranette = ventral gland (*q.v.*) of nematodes

ranger *S.Afr.* any of numerous species of hesperid lepidopteran insect of the genus Kedestes

Rangiferinae a subfamily of cervid artiodactyl mammals containing the reindeer and caribou

rangiora *N.Z.* the compositaceous shrubby tree *Brachyglottis repanda*

Ranidae a family of anuran Amphibia distinguished by having a fused sternum but with cylindrical sacral diapophyses. Most of the common frogs of Europe and many of those in the *U.S.*, belong to this family which is occasionally referred to as the "true frogs"

rank a vertical column of something in contrast to a horizontal row

ranny the soricid insectivorous mammal *Sorex araneus* (= shrew)

Ranunculaceae a very large family of ranalous dicotyledons that contains, *inter alia*, the aconites, the anemones, the columbines, the peonies and the buttercups. Distinctive characters are the numerous hypogynous stamens and the spiral floral structure

Ranunculineae a suborder of ranale dicotyledonous angiosperms containing the Ranunculaceae, Lardizabalaceae, Berberidaceae and Menispermaceae

ranunculus 1 (*see also* ranunculus 2) any of very many ranunculaceous herbs. In *hort.* practice not confined to the genus Ranunculus

ranunculus 2 (*see also* ranunculus 1) any of several species of noctuid lepidopteran moth
    feathered ranunculus *Brit. Epunda lichenea*
    large ranunculus *Brit. Polia flavicincta*
    small ranunculus *Brit. Hecatera chrysozona*

Ranvier's *epon. adj.* from Louis Antoine Ranvier (1835–1922)

rapaceous turnip shaped

Rapataceae a small family of farinose monoctyledonous angiosperms

rape a cultigen of the cruciferous plant *Brassica napus*
    bird's rape the cruciferous herb *Brassica rapa*
    broom-rape any of several species of parasitic herb of the orobanchaceous genus Orobanche
    fir-rape *Brit.* the pyrolaceous herb *Monotropa hypopitys* (= *U.S.* pinesap)

-raph- *comb. form* meaning "needle" or "seam"

Raphanaceae = Cruciferae

raphe properly a seam. Specifically the median line along the length of a diatom frustule, the suture between carpels, the ridge on a plant placenta to which ovules are attached, or the line of division between the two sides of a bilaterally symmetrical organ
    canal raphe a diatomaceous raphe in the form of a longitudinal fissure
    pseudoraphe a marking on the frustule of a diatom that appears to be, but is not, a raphe

raphid a bundle of styloids in a plant cell

exraphidian lacking raphides

Raphidae a family of birds erected to contain the dodo and two solitaires of the Mascarene Islands. The poor tail and the reduction of wings distinguishes this family from the Columbidae in which the dodo was once placed

Raphidiidae a family of neuropteran insects occasionally advanced to the rank of order. They are the forms popularly called snakeflies because the head is borne on a greatly elongate prothorax

Raphidioidea an order of insects erected to contain the Raphidiidae by those who do not hold them to belong in the Neuroptera

R arg a bacterial genetic marker having a character which affects the activity of an arginine repressor, mapped at between 61.0 mins and 64.0 mins for *Escherichia coli*

rarous sparse

ras *see* [raspberry]

[rase] a mutant (gene symbol *ra*) mapped at 97.3 on chromosome III of *Drosophila melanogaster*. The phenotypic expression is a reduction in bristles and hairs

rasorial having the custom of scratching for food, as do many birds

[raspberry] a mutant (gene symbol *ras*) mapped at 32.8 on the X chromosome of *Drosophila melanogaster*. The name refers to the eye color of the phenotype

rasse the Far Eastern viverrine carnivore *Viverricula malaccensis*

rastellum a row of short thick spines, used for digging, on the basal joint of the chelicera of ctenizid spiders

rat (*see also* fish, kangaroo, mite, tail) the term mouse and rat are indiscriminately applied in all parts of the world to any long-tailed rodent. The distinction between rat and mouse is one of size only and is not uniform in various parts of the world. Old World rats and mice belong to the family Muridae and three of them (the roof rat, the Norway rat and the house mouse) are now to be found in almost every inhabited part of the planet. There are no indigenous murid rodents in the New World

bamboo rat any of several species of spalacid rodents of the genus Rhizomys

black rat *Rattus rattus* (= roof rat)

brown rat *Brit.* *Rattus norvegicus* (*cf.* white rat)

cane-rat = cutting-grass

cloud-rat popular name of phloeomyine rodents

cotton-rat *U.S.* any of several species of cricetid of the genus Sigmodon

ground-rat *S.Afr.* the octodontid hystricomorph *Thrynomys swindernianus*

Hanoverian rat *Brit.* = brown rat *Brit.*

hedgehog-rat any of several octodontid rodents

hooded rat a race of white rats but with a black head

jerboa-rat any of several species of the genus Conilurus

kangaroo-rat = rat-kangaroo

maned rat = crested hamster

marsupial-rat any large dasyurid marsupial (*cf.* marsupial mouse)

mole-rat any of several species of spalacid myomorph rodents

pack rat *U.S.* = woodrat *U.S.*

procupine-rat any of many species of echimyid hystricomorph rodent, particularly *Loncheres quinae*

rice rat *U.S.* and *S.Amer.* any of several species of cricetid of the genus Oryzomys

African rock-rat the petromyid *Petromys tipicus*

roof-rat *Rattus rattus* (= black rat)

root-rat = bamboo rat

shrew-rat a short-legged, long-nosed Phillipine rodent *Rhynchomys soricioides* of doubtful affinities but usually included with the murids

spiny-rat any of many species of heteromyid sciuromorph rodent distinguished by the fact that the hairs are fused into soft bristles or spines

strand-rat any of several species of bathyergid rodent mostly of the genus Bathyergus (*cf.* blesmol)

trade-rat *U.S.* = woodrat *U.S.*

trumpet-rat = elephant shrew

wading rat the dendromyine murid rodent *Deomys ferrugineus*

water-rat *Brit.* *Microtus amphibius*

giant water-rat any of several species of hystricomorph octodontid rodents of the genus Myocastor

white rat a domesticated race of *Rattus norvegicus* (*cf.* brown rat)

woodrat *U.S.* any of several species of cricetid of the genus Neotoma

rata *N.Z.* either of two myrtaceous trees of the genus Metrosideros

ratahuihui *N.Z.* the balaenid cetacean *Balaenoptera musculus* (= rorqual)

ratel any of several species of thick-set, pugnacious, mellivorine carnivores of the genus Mellivora

Rathke's *epon. adj.* from Martin Heinrich Rathke (1793–1860)

karyoplasmic ratio the ratio of nuclear to cytoplasmic mass

Ratitae a taxon of birds erected to include the flightless ostriches, emus, cassowaries, and kiwis

ratoon the initial secondary growth which develops around the base of a felled tree

rattle any of several scrophulariaceous plants

red rattle *Brit.* any of several species of the genus Pedicularis

yellow rattle *U.S.* any of several species of the genus Rhinanthus. *Brit.*, without qualification, *R. cristagalli*

rattler *U.S.* = rattlesnake *U.S.*

swamp rattler *U.S.* *Sistrurus catenatus* (= massasauga or black snapper)

ratufa any of several species of large oriental sciurid rodents, distinguished by the long plumes rising from the ears

raukawa *N.Z.* the araliaceous tree *Nothopanax edgerleyi*

raurekau *N.Z.* the small rubiaceous tree or shrub *Coprosma grandifolia*

raven 1 (*see also* raven 2, 3) any of several species of corvid bird of the genus Corvus. *Brit.* and *U.S.* without qualification *C. corax.*

Australian raven *C. coronoides*

thick-billed raven *C. crassirostris*

common raven *C. corax*

white-necked raven *U.S.* *C. cryptoleucus, Afr.* *C. albicollis*

fan-tailed raven *C. rhipidurus*

raven 2 (*see also* raven 1, 3) *I.P.* either of two species of papilionid lepidopteran insect of the genus Papilio but usually, without qualification, *P. castor.* (*cf.* peacock, redbreast, mormon, swallowtail, spangle helen)

raven 3 (*see also* raven 1, 2) the cottid fish *Hemitricterus americanus* (= sea raven)

ravidous any medium dark tint for which the user cannot find a specific name

ray 1 (*see also* ray 2–8) any of numerous flattened elasmobranch fishes. The terms ray and skate (*q.v.*) are subject to much local interchange

bat-ray any of several species of myliobatids

butterfly-ray any of several species of dasyatids

devil-ray any of numerous large batoids particularly those of the genera Manta and Mobula. Without qualification, usually *Mobula hypostoma*

eagle-ray any of several species of myliobatids

electric ray popular form of torpedinids

cownose ray any of several myliobatids of the genus Rhinocera

stingray any of several species of dasyatids carrying a venomous spike at the root of the tail

round stingray any of several species of the genus Urolophus

whip-ray any of several species of dasyatids with unusually long or thin tails

ray 2 (*see also* ray 1, 3–8) woody tissue growing at right angles to the axis of the organ and therefore appearing as "rays" in transverse section

parenchyma ray parenchyma cells in horizontal phloem

heterocellular ray = heterogeneous ray

homocellular ray = homogeneous ray

heterogeneous ray a ray containing both upright and procumbent cells

homogeneous ray a ray consisting entirely of one type (procumbent or upright) of parenchyma cells

medullary ray (*see also* ray 3) a plate of parenchyma tissue radiating from the center of a tree

vascular ray the phloem and xylem components of a ray taken together

ray 3 (*see also* ray 1, 2, 4–8) in the sense of radius

astral ray one of the fibers in the cytoplasm which radiate from each centriole and are thus the basis of the cytoplasmic spindle

medullary ray (*see also* ray 2) rays running from the cortex through the medulla of some mammalian kidneys

ray 4 (*see also* ray 1–3, 5–8) thin bony structures stiffening the fins, tail and gill chambers of fishes

branchiostegal ray one of a series of bony rays projecting backwards from the hyoid arch of fish. They support the floor and sides of the gill chamber

dorsal ray the skeletal structure of a dorsal fin

fin ray any fine bony structure supporting the fin of a fish

hyoid rays a series of cartilaginous rays, rising from the ceratohyal in some chondrichthian fishes and serving the same function as the branchiostegal of bony fishes

ray 5 (*see also* ray 1–4, 6–8) in the sense of radiation

mitogenetic ray a form of radiation, for the existence of which no experimental evidence has been produced, supposed to be emitted by mitotic figures and to inagurate mitosis in neighboring cells

ray 6 (*see also* ray 1–5, 7, 8) the marginal portion of a disk of a compositaceous flower

ray 7 (*see also* ray 1–6, 8) the branch of an umbel

ray 8 (*see also* ray 1–7) a term preferred by some to "arm" for the appendages of sea stars

thorn-tailed rayadito the furnariid bird *Aphrastus spinicauda*

razor (see also bill, shell, clam) any of several species of pelecypod mollusks mostly of the family Sanguinolariilae and Solenidae

rb *see* [ruby]

rbs a bacterial mutant showing a utilization of ribose

Rc *see* [Rusty]

RC a bacterial genetic marker having a character affecting the activity or the regulation of RNA synthesis, mapped at between 56.4 mins and 60.5 mins. for *Escherichia coli*

rd *see* [clumpy], [red] or [reduced]

rdo *see* [reduced ocelli]

re *see* [red egg]

Re *see* [Rex]

cortical **reaction** changes in the surface of an egg at fertilization or activation

Reaumurieae a tribe of Tamaricaceae distinguished from the Tamariceae by the solitary flowers

λrec-malA a bacterial genetic marker having a character affecting both the activity of maltose permease and resistance to phage λ mapped at 64.9 mins. for *Escherichia coli*

receptacle that which receives or bears anything. Its biological usages vary from the placenta of a plant ovary which bears ovules to cup-like hollows in fungi which bears spores

inflorescence receptacle the rachis or principal axis

flower receptacle the swollen end of the axis on which the components of the flower are placed

receptaculum seminalis a vessel used in hermaphrodite animals for the storage of sperm derived from another animal

receptor 1 (*see also* receptor 2, 3 -ceptor-) an organ or cell designed to perceive stimuli.

chemoreceptor a receptor capable of perceiving dissolved or dispersed substances foreign to the normal environment

equilibrium receptor an organ (*e.g.* statoblast) which orients an organism in relation to gravity

gravity receptor = equilibrium receptor

interoreceptor one which responds to internal stimuli

mechanoreceptor a general term for receptors responding to physical forces

olfactory receptor a chemoreceptor activated by gasses

phonoreceptor a receptor distinguishing variable vibrations and therefore an organ of hearing (*cf.* scolopophore)

photoreceptor a receptor capable of perceiving photon energy

pressure receptor a mechanoreceptor stimulated by compression forces and thus one of the organs of touch. There are also pressure receptors in the arterial system

rheoreceptor a receptor perceiving the direction of flow, usually that of water

statoreceptor a receptor for perceiving the direction of gravitational pull

stretch receptor a mechanoreceptor stimulated by the stretching of peripheral nerve endings and thus an organ of touch

tangoreceptor a receptor perceiving contact stimuli

taste receptor a chemoreceptor activated by solutions

thermoreceptor a receptor sensitive to temperature change

touch receptor usually a combination of tango- and stretch receptors

receptor 2 (*see also* receptor 1, 3) a molecule to which atoms are transferred in the course of an enzyme-catalyzed reaction

receptor 3 (= acceptor) (*see also* receptor 1, 2) the individual to which a tissue transplant is transferred from a donor

recess in *biol.* commonly used in the sense of cavity

bullar recess a cavity in the tympanic bullae of some mammals, separated from the ossicular cavity by a septum derived from the tympanic bone

mamillary recess an evagination from the floor of the embryonic forebrain where the latter joins the midbrain

preoptic recess that slightly swollen portion of the embryonic forebrain from which the eye will subsequently develop

recessive (*see also* character) a term applied to organisms having recessive characters and to the gene (*q.v.*)

bottom recessive an individual possessing all the recessive alleles of a group

double recessive having homozygous recessive alleles at different loci

recombination the result of the exchange of linked genes through crossing over. In bacterial genetics the result of pairing opposite mating types but with a unidirectional transmission from donor to recipient with the recombinant then developing within the recipient cell

recon the smallest unit of recombination within a cistron (*q.v.*)

-rect- *comb. form* meaning "straight"

rectrix a tail feather of birds

rectum that portion of the large intestine immediately adjacent to the anus

Recurvirostridae that family of charadriiform birds which contains the avocets and stilts. They are distinguished by their long, slender, neck and very long, very slender, bill which in some species is recurved

[red] *either* a mutant (gene symbol *rd*) in linkage group I of *Habrobracon juglandis*. The name refers to the phenotypic eye color *or* a mutant (gene symbol *red*) mapped at 55.5± on chromosome III of *Drosophila melanogaster*. The color refers to the malpighian tubules of the phenotype

[red aleurone color] a mutant (gene symbol *pr*) mapped at 32 in linkage group V of *Zea mays*. The phenotypic plant has a red aleurone layer in the seed

redd the depression in which salmonid fishes deposit their eggs

[red egg] a mutant (gene symbol *re*) mapped at 31.7 on linkage group V of *Bombyx mori*. The name is descriptive

[red-eyed yellow] a mutant (gene symbol *r*) at locus 20.5 in linkage group I of the rat. In addition to red eyes the phenotype has a yellow coat

[reduced] *either* a mutant (gene symbol *r*) in linkage group I of *Habrobracon juglandis*. The phenotypic expression is small wings *or* a mutant (gene symbol *rd*) mapped at 51.0 on chromosome II of *Drosophila melanogaster*. The reduction is in the bristles of the phenotype

[reduced ocelli] a mutant (gene symbol *rdo*) mapped at 53.0± on chromosome II of *Drosophila melanoogaster*. The name is descriptive

[reduced wings] a mutant (gene symbol *rwg*) mapped at 59.5± on the X chromosome of *Drosophila melanogaster*. The reduced wings of the phenotype are held upright

reductase a group name for enzymes that catalyze a reduction reaction. In the great majority of cases, and unless specifically noted below, NDAP is the hydrogen receptor becoming reduced NADP

aldose reductase catalyzes the reduction of polyol to aldose

oxidized ascorbate reductase catalyzes the reduction of oxidized ascorbate using either reduced NADP or NAD as the hydrogen donor

azobenzene reductase catalyzes the production of dimethyl-*p*-phenylene-diamine and aniline from dimethylaminoazobenzene, using reduced NADP as the hydrogen donor

pyrroline-2-carboxylate reductase catalyzes the production of Δ¹-pyrroline-2-carboxylate from L-proline

pyrroline-5-carboxylate reductase catalyzes the production of Δ¹-pyrroline-5-carboxylate from L-proline

acetoacetyl-CoA reductase catalyzes the reduction of D-3-hydroxyacyl-CoA to 3-oxo-acyl-CoA

cortisone reductase catalyzes the reduction of dihydrocortisone to cortisone

cystine reductase catalyzes the production of cysteine from cystine using reduced NAD as a hydrogen donor

nitrate ester reductase catalyzes the production of oxidized glutathione and nitrite from 2-reduced glutathione and polyolnitrate

glucuronate reductase catalyzes the reduction of L-gulonate to D-glucuronate

hydroxymethylglutaryl-CoA reductase catalyzes the production of 3-hydroxy-3-methyl-glutaryl-CoA from mevalonate + CoA

glutathione reductase catalyzes the production of reduced glutathione from oxidized glutathione using either reduced NAD or NADP as the hydrogen donor

glucuronolactone reductase catalyzes the reduction of L-gulono-γ-lactone to D-glucurono-γ-lactone

glyoxylate reductase reduces glyoxylate to gycollate using reduced NAD as the hydrogen donor

hyponitrite reductase catalyzes the reduction of hyponitrite to hydroxylamine using reduced NAD as the hydrogen donor

hydroxylamine reductase catalyzes the production of ammonia from hydroxylamine. Reduced pyocyanines and flavins can act as the hydrogen donors

menadione reductase catalyzes the production of 2-methyl-naphthohydroquinone from 2-methyl-1, 4-naphthoquinone, using either reduced NAD or NADP as the hydrogen donor

mevaldate reductase catalyzes the reduction mevaldate to mevalonate using either reduced NAD or NADP as the hydrogen donor

NAD reductase any of several enzymes catalyzing the production of NAD from reduced NAD using cytochromes as the hydrogen receptor

nitrate reductase catalyzes the production of an oxidized cytochrome and nitrite from a reduced cytochrome and nitrate

NAD nitrate reductase catalyzes the reduction of nitrate to nitrite using reduced NAD as the hydrogen donor

NADP nitrate reductase catalyzes the production of nitrite from nitrate using reduced NADP as the hydrogen donor

nitrite reductase catalyzes the reduction of nitrite to either hydroxylamine or ammonia, using either reduced NADP or NAD as the hydrogen donor

nitric oxide reductase catalyzes the direct production of nitric oxide from nitrogen. Pyocyanine acts as the acceptor and reduced pyocyanine as the donor

oxidoreductases the preferred term for all enzymes entering into oxidation-reduction reactions and therefore including all those enzymes previously known as oxidases, reductases and dehydrogenases. These three terms are, however, retained in the trivial names of enzymes and are used in this dictionary

D-proline reductase catalyzes the production of D-proline from 5-amino-valerate using NAD as the hydrogen receptor

quinone reductase catalyzes the oxidation of reduced NAD or NADP using quinone as the hydrogen acceptor

sulphite reductase catalyzes the production of sulphite from hydrogen sulfide

ubiquinone reductase catalyzes the production of dihydro-ubiquinone from ubiquinone using either reduced NAD or NADP as the hyrogen donor

D-xylulose reductase catalyzes the production of D-xylulose from xylitol using NAD as the hydrogen receptor

L-xylulose reductase similarly catalyzes the production of L-xylulose but uses NADP as the receptor

Reduncinae a subfamily of antelopid artiodactyl mammals containing the forms frequently called marsh antelopes. They are the most deer-like of the antelopes in shape. The best known are the waterbucks

[reduplicated sex combs] a mutant (gene symbol *rsc*) plotted at 0.9± on the X chromosome of *Drosophila melanogaster*. The phenotypic male has six sex combs on all six legs

Reduviidae a large family of predatory hemipteran insects distinguished by the neck-like constriction behind the eyes and greatly broadened abdomen often extending beyond the wing. They are commonly called assassin-bugs

reed (*see also* fish, grass, haunter, hen, mace, pheasant) any of numerous tall thin plants, usually monocotyledonous and almost invariably of a marshy or aquatic habitat

burr-reed any of numerous submerged aquatic plants of the pandale family Sparganiaceae, especially any of several sparganiaceous herbs of the genus Sparganium

common reed the grass *Phragmites communis*

giant reed the grass *Arundo donax*

sand-reed *U.S.* any of several species of grass of the genus Ammophila

seasand-reed the grass *Ammophila arenaria*

reef a rocky ridge, frequently of coral origin, projecting near or slightly above the surface of oceans adjacent to the shore

barrier reef a fringing coral reef, which is separated from the land by a deep and wide lagoon

fringing reef a coral reef parallel to the shore and separated from it by a relatively shallow, usually narrow, lagoon

reeler *Brit.* = reel-bird *Brit.*

[reeler] a mutant (gene symbol *rl*) found in linkage group III of the mouse. The phenotypic expression is an unsteady gait leading to early death

reeve the scolopacid bird *Philomachus pugnax* (= ruff). The term is usually confined to the female ruff and is also used locally for female sand-pipers

reflex an automatic response

ascaphus reflex = thanatosis

refracted in *biol.* has the meaning of bent sharply backwards from the base

refugium an area that remains unaffected by a general climatic change and which therefore contains a population of organisms typical of those once spread over the whole region. The English word refuge is preferred by some

Regalecidae a monotypic family of allotriognathous fishes containing the giant oarfish (*Regalecus glesne*)

regeneration (*see also* -generation) the recreation of cells, tissues, or organs which have been lost by degeneration, dedifferentiation, or removal

epimorphic regeneration regeneration which takes place from a blastema of undifferentiated cells, in contrast to morphallaxis (*q.v.*)

regma a schizocarp with elastically opening segments

Regulidae a family of passeriform birds usually included with the Sylviidae

Reil's *epon. adj.* from Johann Christian Reil (1759–1813)

rejuvenescence the recovery of youthful characteristics

relaxin a hormone secreted by the ovary and producing relaxation of the pubic ligament and softening of the cervix in pregnancy

releasin = relaxin

relict an organism or group of organisms belonging to an earlier epoch than the other members of the biopopulation in the area in which the relict occurs

setireme the swimming leg of an aquatic insect

remex a primary flight feather attached to the ulna of birds

remicle diminutive of remex (*q.v.*)

remiges *pl. of* remex

remora popular name of discocephalous fishes. *U.S.*, without qualification, usually *Remora remora*

remote in *biol.*, has the sense of scattered

remuda term for a group of work horses in *U.S.* cow country

-ren- *comb. form* meaning "kidney"

renal pertaining to the kidney

renette excretory cell of nematodes

reniculus an individual lobule of a polylobular kidney when the lobes lie free and separate

renierine a term applied to a type of sponge skeleton consisting of a net-work of uniformly sized meshes, each of the length of a spicule

Renillidae a family of pennatulacean alcyonarian coelenterates commonly called sea-pansies. They are distinguished by the broad circular or kidney shaped rachis on a short stalk. They are frequently violet or pansy-colored

renin an enzyme catalyzing the conversion of hypertensinogen to hypertensin

suprarenine = epinephrine

rennin an enzyme catalyzing the hydrolysis of peptides (*see also* renin)

repand said of a bacterial colony, on agar, which has a wavy surface

repent creeping (= reptant)

replete a caste of myrmecine ants of the genus Myrmecocystus so called because they serve as reservoirs for the honey collected by other workers. Sometimes called honeypots or perergates

repletum a fruit in which the valves remain connected together after dehiscence

replum the framework of the placenta which remains after the valves of a silique fall away in dehiscence

reproduction the replication of an organism

asexual reproduction the replication of an organism of and by itself

sexual reproduction the replication of an organism from the fusion of two, or parts of two, parents

vegetative reproduction = asexual reproduction

reptant creeping (= repent)

Reptantia a suborder of decapod Crustacea with a dorso-ventrally flattened body and the first abdominal segment significantly smaller than the posterior segments. The pleopods are frequently reduced or absent and not usually used for swimming (*cf.* Natantia)

Reptilia a class of gnathostomatous craniate chordates containing the reptiles. They are commonly distinguished by the presence of epidermal scales on the skin but are better characterized by the 3-chambered heart and their poikilothermy

repulsion the condition of two loci on the same chromosome one having a dominant allele and the other a recessive allele

Resedaceae that family of rheodalous dicotyledons which contains the mignonettes and dyer's weed. The family is with great difficulty separated from the Cruciferae though there is often a gaping aperture in the summit of the ovary which is distinctive

Resedineae a suborder of Rhoedales containing the family Resedaceae

food reservoir any pouch or diverticulum from the alimentary canal anterior to the digestive regions

resin any water insoluble exudate of a tree. Water soluble exudates are properly gums (*q.v.*). Used without qualification, resin is sometimes improperly used for rosin

[Resistance to Puccinia] a mutant (gene symbol *Rp*) mapped at 0 on linkage group X of *Zea mays*. The term refers to a characteristic of the phenotype

respiration the sum total of the method or methods by which an organism utilizes gases

aerobic respiration that which requires oxygen

anaerobic respiration that which does not require oxygen

fermentative respiration = fermentation

restant persistent

restibilic = perennial

Restionaceae a small family of farinose monocotyledonous angiosperms

restitution the rejoining of a severed part of a chromosome in such a manner as to cause no genetic effect (*cf.* recombination)

-ret- *comb. form* meaning "net"

ret a verb used to describe the maceration of plants to liberate, for commercial purposes, fiber. The word is actually an obsolete form of "rot"

-rete- *see* -ret-

rete mirabile the internal network of collecting ducts within the testes

reticular net-like

reticulate netted or in the form of a net

[reticulate virescent] a mutant (gene symbol *rv*) mapped at 0 on linkage group A of the tomato. The greenish phenotypic leaves have prominent veins

reticulum 1 (*see also* reticulum 2) any net-like structure

endoplasmic reticulum a network or organization of double membranous structures spreading through the cytoplasm. It has various times been reported to open both to the exterior of the cell and from the nucleus

reticulum 2 (*see also* reticulum 1) the second of the four divisions of the stomach of artiodactyls (*cf.* abomasum, psalterium, rumen)

retina the photosensitive portion of an eye

retinacula a rope or chain particularly one which restrains or tethers something. This word is not the plural of retinaculum a little net. The word has been given so many meanings in biology as to be almost worthless particularly as it is frequently confused with retinaculum. It is also misused to mean a retractile muscle in many invertebrates

retinaculatous hooked

retinaculum a little net (*pl.* retinaculi). Retinacula (*pl.* retinaculae) derives from a different root. There are many meanings in *biol.* but the word usually refers to one body which supports or contains another. Best known as applying to the gland by which pollen-masses are retained in orchids though applying to a structure of similar function but different shape in asclepiads. It is also used, probably by confusion with retinacula, of the hooked funicle which retains the seeds of Acanthaceae until they mature. The term is, in insects, used mostly for restraining parts, such as the ring which prevents the hymenopteran sting being pushed out too far. There seems little justification for its application to the muscular sheath of a nerve in acanthocephalans

retinene one of the two breakdown products of metarhodopsin (the other is scotopsis) in the presence of ample supplies of vitamin A. Retinene and scotopsis recombine into rhodopsin

retinula a group of photoreceptor cells at the base of an ommatidium

retraction the drawing back

Retzius' *epon. adj.* from Gustav Magnus Retzius (1842–1919) [Retzius' lines] and Andreas Adolf Retzius (1796–1860)

reversion a term of plant and animal breeding used to explain the appearance of a phenotype which has not occurred for several generations

revolute rolled backwards

revolutivous said of a flower bud in which the outer portions are rolled back spirally

rewarewa *N.Z.* the large proteaceous tree *Knightia excelsa* (= honeysuckle *N.Z.*)

[Rex] a mutant (gene symbol *Re*) found in linkage group VII of the mouse. The phenotypic expression is almost identical with that of [caracul]

[rex₁] a mutant (gene symbol *r₁*) at locus 0 on linkage group III of the rabbit. The phenotypic expression is a velvety pelt

[rex₂] a mutant (gene symbol *r₂*) at locus 17.2 in linkage group III of the rabbit. The phenotype is identical with [rex₁]

rey *see* [rough eye]

reynard a male fox

rf *see* [roof wings]

Rf *see* [Roof]

rg *see* [Ragged] or [rugose]

Rg *see* [Ragged leaves]

R gal a bacterial genetic marker having a character which acts as a galactose repressor, mapped at 54.1 mins for *Escherichia coli*

rh *see* [roughish]

rha a bacterial genetic marker having a character affecting the utilization of D-rhamnose, mapped at 73.7 mins for *Escherichia coli*

statorhab a short thick tentacle carrying a statolith in some coelenterate medusae

-rhabd- *comb. form* meaning "rod"

rhabd a monaxon sponge spicule developed by growth in both directions from a central point in distinction from a style

Rhabdiasidae a class of nematodes with smooth cuticles, and no pharyngeal bulb

rhabdion a sclerotized concretion found in the protostom of some nematodes

rhabdite 1 (*see also* rhabdite 2) a rhabdoid in the form of a straight or slightly curved rod

adenal rhabdite a rhabdite occuring in the mesochyme instead of the epidermis

pseudorhabdites rhabdite shaped masses of mucous in the epidermis of some turbellaria

rhabdite 2 (*see also* rhabdite 1) in insects, the sharp-edged components of stings and ovipositors

Rhabditida an order of phasmid nematodes, distinguished from the Tylenchida by the presence of a mouth stylet and from the Spirurida by the tripartite esophagus

Rhabditidea a class of nematodes with a smooth cuticle small pocket-like amphids and a posterior bulb on the pharynx

rhabditoid *see* pharynx

-rhabdo-*see* -rhabd-

Rhabdocoela an order of turbellarian Platyhelminthes distinguished by having a simple sac-like digestive tract with no lobes or branches

rhabdoid 1 (*see also* rhabdoid 2, 3) a solid rod or spicule in the epidermis of Turbellaria (*cf.* rhabdite, rhamnite, chrondrocyst, sagittocyst)

rhabdoid 2 (*see also* rhabdoid 1, 3) a type of heteroneme nematocyst in which the butt is cylindrical (*cf.* eurytele, mastigophore, stenotele)

rhabdoid 3 (*see also* rhabdoid 1, 2) a rod shaped body found in the cells of the tentacles of Drosera

rhabdom a group of four rhabdomeres in the center of the retinula

Rhabdopleurida an order of pterobranchiate Hemichordata distinguished by the fact that the members of the colony are in organic continuity with each other

-rhach- *comb. form* literally meaning "spine" or "backbone" but used in *biol.* for any axial support. The almost universal misspelling -rach- (*q.v.*) causes much confusion

Rhachiceridae = Rachiceridae

Rhachiodontinae = Elachistodontinae

Rhacophoridae a family of anuran amphibia once considered part of the Hylidae but now thought to represent an arboreally adapted branch of the Ranidae

rhagadiose cracked

Rhagionidae a family of medium sized to large orthoraphous brachyceran dipteran insects mostly brownish or grey with spotted wings. The popular name snipe-fly derives from the woody habitat and long proboscis. Most are predatory, though a few bite humans

-rhag- *comb. form* confused from five *Gr.* roots and therefore variously meaning a "break," "berry," "violence" and "poisonous spider." An extension of this last provides the justification for the next entry

rhagus a nematocyst-filled protrusion from a coelenterate

acrorhagus nematocyst-bearing protrusions immediately outside the ring of tentacles in some Zoantharia.

cnidorhagus a rounded, nematocyst-filled projection from the septal filament of some Zoantharia

pseudorhagus an acrorhagus lacking nematocysts

Rhamnaceae that family of rhamnalous dicotyledons which contains *inter alia*, the buckthorns and the jujubes. It is distinguished by the strongly perigynous flowers and the fact that the stamens are opposite the petals

rhammite a long, slender, sinuous rhabdoid found in some Turbellaria

Rhamnales that order of dicotyledonous angiosperms which contains, *inter alia*, the buckthorns and the grapes. The order is distinguished by the small number of stamens and the presence of a disc surrounding or subtending the ovary

Rhamphastidae = Ramphastidae

Rhamphocottidae an occasionally recognised family of scleroparous fishes usually placed with the Cottidae.

rhape = raphe

rhaphid = raphid

Rhaphidiodea = Raphidioidea

Rhapidae = Raphidae

rhea either of two species of rheid birds of South America

greater rhea

lesser rhea *Pterocnemia pennata*

rhebok the *S.Afr.* reduncine antelopid *Pelea capreolus*

Rheidae the only family of rheiform birds

Rheiformes an order of birds containing the *S.Amer.* rheas. They are distinguished from the Struthioniformes (ostriches) by the three-toed feet and the fact that the head and neck are partially feathered

-rheo- *comb. form* meaning "to flow" or pertaining to that which flows *e.g.* a "stream" (*cf.* -rhya-)

-rhexi- *comb. form* meaning "to rend apart"

ribazuba *Brit. obs.* walrus ivory

-rhin- (*see also* -rhynch- and -rostr-) *comb. form* confused between two *Gr.* roots and therefore meaning either "nose" or "file"

amphirhinal said of birds having both an anterior and posterior bony nostril opening on each side of the bill

holorhinal said of birds having a single, rounded, bony nostril opening and in which the anterior border of the nasal bones is only slightly notched

schizorhinal said of birds having a single, slit-like, bony nostril opening and in which the anterior border of the nasal bones is deeply cleft

Rhinanthoideae a subfamily of the Scrophulariaceae

rhinarium 1 (*see also* rhinarium 2) a naked and moist glandular skin surrounding the nostrils in many mammals

rhinarium 2 (*see also* rhinarium 1) that portion of an insect which would be the nose did it have one

Rhincodontidae a monotypic family containing the whale shark *Rhincodon typus*

mesorhinium that portion of a bird's bill which lies between the nostrils

Rhinobatidae a family of rajiform Chondrichthyes distinguished by the elongate body flattened along the sides of the head. Commonly called guitarfishes

rhinoceros (*see also* anklet, hornbill) popular name of rhinocerotid mammals

black rhinoceros the African, two-horned *Diceros bicornis*

hairy-eared rhinoceros the Asiatic two-horned *Didimoceros lasiotis*

one-horned rhinoceros *Rhinoceros indicus* (= great Indian rhinoceros)

Javan rhinoceros *R. sondanicus*

Sumatran rhinoceros the two-horned *Didimoceros sumatrensis*

white rhinoceros the African two-horned *Ceratotherium simus.* The term "white" is a corruption of the Dutch "weit" meaning wide

Rhinocerotinidae that family of perissodactyle mammals which contains the rhinoceroses. They are distinguished by their heavy build and horned snout

Rhinochetidae = Rhynochetidae

Rhinochimaeridae a family of abyssal holocephalous fishes distinguished by the extremely elongate nose

Rhinocryptidae a family of passeriform birds commonly called tapaculos. They are distinguished by the large and strong claws and feet, and their habit of creeping about like mice rather than hopping

Rhinodermatidae a monotypic family of anuran Amphibia erected to contain *Rhinoderma darwini* frequently called the mouth-breeding frog

Rhinolophidae a family of microchiropteran mammals possessing a large leafy outgrowth around the nostrils. They are popularly called horseshoe, or horseshoe-nosed bats though the term leaf-nosed bat, properly applied to the Phyllostomatidae, is sometimes used

Rhinomaceridae a family of anthriboid coleopteran insects commonly called pine-flower snout-beetles. The snout is about as long as the prothorax

Rhinophoridae a family of brachyceran dipteran insects commonly included with the Tachinidae

Rhinophrynidae a monotypic family of anuran Amphibia erected to contain the Mexican burrowing toad *Rhinophrynus dorsalis*

Rhinotermitidae a family of isopteran insects distinguished by the presence of a fontanelle and the absence of an empodium

-rhinous *comb. term* meaning pertaining to the nose (*cf.* rhinal)

Rhipiceratidae = Rhipiceridae

rhipidium a fan shaped cyme

Rhipiduridae a family of birds usually included with the Muscicapidae

Rhipiptera = Strepsiptera

-rhiz- *comb. form* meaning "root"

  coleorhiza the root sheath of a monocotyledonous embryo

  hydrorhiza the root-like stolons of some hydrozoan coelenterates

  isorhiza a haploneme nematocyst with a tube of the same diameter throughout

    anisorhiza a haploneme nematocyst with the tube slightly dilated toward the base

  mycorrhiza fungus hyphae symbiotic with the root of a higher plant

    pseudomycorrhiza an association of a fungus with a root which is purely parasitic

    ectotrophic mycorrhiza an association in which the hyphae invest the root

    endotrophic mycorrhiza an association in which the fungal hyphae penetrate the root tissue

  phyllorhiza a leaf functioning as a root, as in some water plants

  pseudorhiza a swollen monocotyledonous root

rhizal pertaining to a root, rhizome or rhizoiod

  arrhizal lacking roots

  chordorrhizal pertaining to those plants in which the root stalk produces numerous flowering stems

  epirhizal growing on roots

  heterorhizal the production of roots rather at random

  hydrorhizal said of a hydrozoan coelenterate the polyps of which spring directly from the stolon

  monacrohizal the condition of having the entire root derived from a single initial

  eligorhizal having few roots

  polyrhizal with many roots or rootlets

rhizina the equivalent of a root hair in a moss

Rhizobiaceae a family of eubacterial schizomycetes, gram negative and sparsely flagellated. They are best known from the genus Rhizobium, though other genera have free-living forms and pathogenic forms are known

Rhizocephala an order of cirripede Crustacea thought by some to be a separate subclass. Most are parasitic on decapod crustaceans in which the female has degenerated to a tumor-like sac protruding from the abdomen of the host with root-like extensions ramifying to most parts of the body of the host

Rhizochloridales an order of xanthophycean Chrysophyta containing amoeboid forms with pseudopodia which sometimes asastomose with each other, and with the pseudopodia of other cells

Rhizochrysidales an order of chrysophycean Chrysophyta containing those forms in which the amoeboid phase is dominant

Rhizodiniales a small order of aberrant dinophysian Pyrrophyta in which the vegetative cell is naked and amoeboid

Rhizogenum the thickened base of a marine alga which serves as, but has not the structure of, a holdfast

rhizoid a root-like filament of a non-vascular plant. The term is also applied to anchoring extensions of the stolon in Ectoprocta

Rhizomastigina (= Pantostomatida) a small order of

zoomastigophorous Protozoa showing permanent amoeboid motion but possessing a single flagellum

rhizome a swollen underground stem commonly used for reproduction and food storage. The term is also applied to the stolon of hydroid coelenterates

Rhizomyidae a family of myomorph rodents frequently fused with the Spalacidae but differing from them in that the eyes are not covered with skin and in possessing ears. They are variously known as root-rats and bamboo-rats

Rhizophagidae a small family of minute coleopteran insects

Rhixophoraceae a family of myrtiflorous dicotyledons containing the mangrove trees. The habit of growth, and habitat, of the trees is typical

Rhizopoda the term is used either as synonymous with Sarcodina (*q.v.*) or as a subclass of Sarcodina. In the latter case, it comprises all of the orders of Sarcodina except the Heliozoa and Radiolaria which are placed in the subclass Actinopoda

Rhizostomeae an order of scyphozoan coelenterates in which the mouth is replaced by numerous small openings, leading to a canal, in the oral lobes

-rhizous *adj. term.* synonymous with -rhizal (*q.v.*) which is preferred in this work

-rhod- *comb. form* meaning "red"

rhodanase = thiosulfate sulfurtransferase

rhodellous rosy pink

cynarrhodion a hip that is a fleshy, hollow fruit, including achenes

Rhodobacteriinae a suborder pseudomonodale Schizomycetes that differs from the Pseudomonadineae in having photosynthetic pigments

rhodochrous rose colored

Rhododendroideae a subfamily of the Ericaceae distinguished by the septicidal capsule and ribbed seed

rhododendron (*see also* bug) a very large genus of ericaceous shrubs. The distinction between a rhododendron and an azalea is horticultural rather than botanical

Rhodophyta a subphylum of algae commonly called the red algae. They are the only algae to show a form of sexual reproduction in which non-flagellated male gametes are transported to the female sex organ

rhodopsin primary reagent which converts photon energy into nerve impluses in the rods of the retina

rhodora the ericaceous shrub *Rhododendron canadense*

Rhodoraceae = Ericaceae

Rhodymeniales an order of tetrasporophytic floridian Rhodophyta in which the auxiliary cell is a special cell of the procarp

Rhoeadales that order of dicotyledonous angiosperms which contains *inter alia*, the poppies, the bleeding hearts, the cabbages and their allies, the capers and the spiderplants, dyer's weed and the so-called horseradish tree

Rhoeadineae a suborder of Rhoeadales containing the family Papaveraceae

-rhomb- *comb. form* meaning either "lozenge-shaped" or "whirling"

rhopalioid a modified tentacle that differs from a rhopalium in lacking a sensory function

rhopalium a hollow clavate organ functioning as a statocyst in Schyphozoa

Rhopalocera a term at one time used to combine the "butterflies" (Papilionoidea and Hesperioidea) as distinct from the "moths" (*cf.* Heterocera)

Rhopalosomidae = Rhopalosomatidae

Rhophoteira = Siphonoptera

rhubarb (*see also* curculio) the polygonaceous edible herb *Rheum rhaponticum*. The term is sometimes extended to other members of the genus

    bastard rhubarb the polygonacean herb *Rumex alpinus*

    Guatemala rhubarb the euphorbiaceous shrub *Jatropha podagrica*

-rhya- *comb. form* meaning "torrent" (*cf.* -rheo-)

rhyacad a plant of torrents

Rhyacophilidae a family of primitive trichopteran insects with ocelli and five-segmented maxillary palpi

-rhynch- *comb. form* meaning "beak" (*cf.* -rhin- and -rostr-)

Rhynchobdellida an order of hirudinian annelids with a protrusible proboscis and which lack the jaws typical of most leeches

Rhynchocephalia an order of diapsid reptiles containing the single extant representative Sphenodon, the tuatara of New Zealand. The order differs from other diapsids in that the quadrate bone is immovable and there are upper and lower temporal arcades

Rhynchocoela = Nemertea

Rhynchophora a no longer valid entomological taxon at one time containing all the snout-headed weevils now distributed among several superfamilies

Rhynchopidae = Rynchopidae

Rhynchosporaceae the name applied to a subfamily of Cyperaceae sometimes elevated to familial rank

myzorhynchus a protrusible muscular mass in the scolex of some tapeworms

rhynchous pertaining to the "snout" particularly in insects

    auchenorrhynchous with the beak coming from the under surface of the head as in many biting insects

    protorhynchous with the beak protruding forwards

-rhyno- *see* -rhino-

Rhynochetidae a monospecific family of gruiform birds containing the kagu distinguished by its loosely webbed plumage, long sharp bill and conspicuous bushy crest

Rhyphidae a small family of nematoceran dipteran insects commonly called wood-gnats

Rhysodidae a small family of coleopteran insects commonly called wrinkled bark-beetles. They are distinguished by the three deep grooves on the pronotum and the moniliform antennae

Rhyssodidae = Rhysodidae

rhythm a repeated cyclic change

    circadian rhythm a rhythm or cycle of approximately 24 hours. Now commonly used in place of diurnal

    endogenous rhythm a cyclic change in an organism induced by causes presumed to be within the organism itself

    exogenous rhythm a rhythmic behavior pattern in an organism presumed to be caused by factors external to the organism, such as the phases of the moon or geomagnetism

    neurogenic rhythm a rhythm of neural origin, particularly the rhythm of the alary muscles of insects

    seasonal rhythm one dependent on the seasons of the year but specifically an endogenous rhythm of hibernators which makes it difficult to induce artificial hibernation by lowering the temperature out of season

ri *see* [radius incompletus]

rib a microbial genetic marker indicating a requirement for riboflavin

rib 1 (*see also* rib 2, 3) a cartilaginous or bony rod articulating with, or pendant from, a vertebra

    abdominal rib a sternal rib-like bone found in the ventral abdominal wall between the last true rib of the pelvis in Crocodilia, Sphenodon and some fossil reptiles (= parasternalia, gastralia)

    false rib a pleural rib which reaches the sternum indirectly through a crest or cartilage

    floating rib a pleural rib that does not reach the sternum

    hemal rib a rib which lies inside the muscles forming the body wall

    pleural rib a rib lying in the horizontal septum which separates the body musculature into dorsal and ventral divisions

    sacral rib a rib articulating with the pelvis

    true rib a pleural rib which articulates directly with the sternum

rib 2 (*see also* rib 1, 3) a leaf vein that protrudes above the surface

    midrib the central nerve in a leaf

rib 3 (*see also* rib 1, 2) any of several rod-like or tube-like structures

    Semper's rib a functionless trachea alongside a functional trachea in some insects

rib-1, -2 *see* [riboflavin-1, -2]

[riboflavin-1] a mutant (gene symbol *rib-1*) on linkage group VI of *Neurospora crassa*. The phenotypic expression is a requirement for riboflavin

[riboflavin-2] a mutant (gene symbol *rib-2*) in linkage group IV of *Neurospora crassa*. The phenotypic expression is a riboflavin requirement

Ricciaceae a family of marchantiale Hepaticae in which the sex organs are borne in a sagittal strip extending the length of the thallus. The common Riccia of aquarists is typical

rice (*see also* bird, bug, flower, shell, weevil) properly the annual grass *Oryza sativa* and its seeds, but the term is extended to many other grasses

    ant-rice *Aristida oligantha*

    Indian rice *U.S. Zizania palustris*

    mountain rice *U.S.* any of several species of the genus Oryzopsis

    water rice *U.S.* = Indian rice *U.S.*

    wild rice *U.S.* = Indian rice *U.S.*

Richardiidae a family of brachyceran dipteran insects usually included with the Otitidae

richness term for a group of martens

Ricinidae a small family of mallophagan insects parasitic on passerine birds and distinguished from the closely allied Laemobothriidae by the absence of a swelling in front of the eye

Ricinulida a very small order of arachnid arthropods distinguished by possessing an opisthosoma with nine somites, two-jointed chelate chelicerae, and small forcipate pedipalpi

[rickets] a mutant (gene symbol *rk*) mapped at 46.0± on chromosome II of *Drosophila melanogaster*. The phenotype has flattened and bent hind legs

Rickettsiaceae a family of Rickettsiales distinguished as vertebrate parasites transmitted by arthropod vectors. They are the cause of, *inter alia*, epidemic typhus, Rocky Mountain spotted fever, and "Q" fever

Rickettsiales an order of microtatobiotes containing the forms commonly called rickettsias. Structurally they resemble minute bacteria and are mostly below 0.1 microns in size

rictous open mouthed

rictus the gape of the mouth

ridge an elongate prominence rising from a flat surface

apical ridge the thickened ectodermal ridge on the developing limb bud (= Saunder's ridge)

hypobranchial ridge a strip dividing the pharyngeal and digestive portions of the alimentary canal in some Hemichordata

parabranchial ridge an unusually large hypobranchial ridge

canthal ridge the border between the snout and the top of the head in snakes

genital ridge a ridge along the dorsal-lateral portion of the embryonic coelum to which primary sex cells migrate. There is an analagous structure in some invertebrates

neural ridge the lateral margins of the neural plate as it begins to invaginate

Saunder's ridge = apical ridge

ridley *U.S.* the cheloniid chelonian reptile *Lepidochelys kempi*

Riel's common misspelling of Reil's

Riellaceae a family of spherocarpale Hepaticae distinguished by the lack of bilateral symmetry

rigidity stiffness

winter rigidity a term used to describe the hibernation of poikilothermic animals in distinction to the dormancy of homoiothermic animals

Sanio's rim a ridge on the radial wall of the tracheids of conifers between two pits

rim *see* [rimy]

-rim- *comb. form* meaning "fissure"

rima literally a cleft and applied particularly to the ostioles of some fungi

birimose having two slits

rimous covered with cracks or chinks

rimu *N.Z.* either the podocarpaceous tree *Dacrydium cupressinum* (= red pine *N.Z.*) or the rhodophyte alga *Caulerpa brownii*

rimule a median indentation or groove in the posterior rim of the orificial collar of some Ectoprocta

[rimy] a mutant (gene symbol *rim*) mapped at 48.1± on the X chromosome of *Drosophila melanogaster.* The wings of the phenotype resemble [pleated] but the eyes are brownish with white hairs

rind a popular term for the outer surface of any plant structure

ring 1 (*see also* ring 2) an annular structure

annual ring a growth ring in forms having annual periodicity

false annual ring a second growth ring produced in the same year, due to temporary interruption of growth

circumenteric ring the nerve ring which encircles the pharynx in nematodes

circumoral ring the ring from which nerves arise to the outer tentacles of the lophophore of some Ectoprocta

epistomial ring the ring from which the radial nerves to the median tentacles of some Ectoprocta arise

fairy-ring either fungi growing in a circle or the pattern left after such growth

growth ring an annular growth layer such as that seen on the scale of a fish or on the stem of a plant in transverse section

milled ring the groved ring round the base of radioles (*q.v.*) when these are constricted above the extension for the ball-and-socket type joint with the tubercle

nettle-ring a thick band of nematocysts round the margin of some medusae

pneumatic ring the air cells surrounding the statoblast of some Ectoprocta

sclerotic ring = sclerotic plate

opisthospondylous ring a ring of cartilage around the notochord marking the posterior end of a vertebra

prospondylous ring a ring of cartilage around the notochord marking the anterior end of a vertebra

vertebral ring one of those cartilaginous elements of a vertebra which lie between the prospondylous and opisthospondylous rings

ring 2 (*see also* ring 1) in compound names of organisms

dingy ring *Austral.* the satyrid lepidopteran insect *Yphthima arctous*

ringent either stiff or gaping

ringhal the elapid snake *Haemachatus haemachatus* (= spitting cobra)

ringlet any of many satyrid lepidopteran insects. *U.S.* any of several species and subspecies of the genus Coenonympha. *Brit. Aphantopus hyperanthus I.P. Ragadia crisilda S.Afr.* any of several species of the genus Physcaeneura. *Austral.* any of numerous species of the genus Hypocysta

marsh ringlet *Brit. Caenonympha typhon*

Riodinidae a family of lycaenoid lepidopteran insects, mostly tropical, commonly called metal-marks. They are closely allied to the Lycaenidae

riparian pertaining to the shores or banks of rivers

rippock *Brit. dial.* = tern

riro-riro *N.Z.* the sylviid bird *Gerygone igata* (= grey warbler)

rivose having sinuate vessels or channels

rivulet *Brit.* any of several species of geometrid lepidopteran insect of the genus Perizoma. Without qualification *P. affinitata* is usually meant

rivulose the diminutive of rivose

rk *see* [rickets]

rl *see* [rolled] or [reeler]

rmp *see* [rumpled]

rn *see* [rotund]

ro *see* [rough] or [rosette]

ro-1, 2 *see* [Ropy-1] or [-2]

roach 1 (*see also* roach 2, 3) common *U.S.* name of dictyoperous blattid insects (= cockroach)

brown-banded cockroach *U.S. Supella supellectilium*

American cockroach *U.S. Periplaneta americana*

Australian cockroach *U.S. Periplaneta australasiae*

Cuban cockroach *U.S. Panchlora nivea*

German cockroach *U.S. Blattella germanica*

oriental cockroach *U.S. Blatta orientalis*

Madeira cockroach *U.S. Leucophaea maderae*

spotted Mediterranean cockroach *U.S. Ectobius pallidus*

Surinam cockroach *U.S. Pycnoscelus surinamensis*

roach 2 (*see also* roach 1, 3) *Brit.* the cyprinid fish *Rutilus rutilus. U.S.* any of several species of centrarchid fishes

California roach the cyprinid fish *Hesperoleucus symmetricus*

roach 3 (*see also* roach 1, 2) a short crest or mane

vinegar roan = vinegaroon

robalo popularn name of centropomid fish (= snook)

nest-robber *W.Afr.* the accipitrid bird *Gymnogenys typicus* (= harrier hawk)

herb robert the geraniaceous herb *Geranium robertianum*

robin (*see also* robin 1, 2) this word originally meant "poorman" or "peasant." Since these persons were

often ragged, the name was applied to many strag-
gling plants (*see* robin 2). The name was also applied to urchins, as the children of robins were
called since, like urchins (now called hedgehogs),
they frequented hedgerows. Many of these robin
urchins were pert in their manner, friendly to all
men and amazingly inquisitive. These characteristics so exactly coincided with those of the delightful
sparrow-sized turdid bird (*see* robin 1), then known
in England as the redbreast, that he became robin
redbreast and so he has remained. Early settlers in
North America fround a thrush-sized turdid bird
which they, quite correctly, called robin, marveling
only that their old friend should have grown so
monstrous large in his new land. Today there is
scarcely an English speaking country that does not
have a red-breasted bird nostagically called "robin,"
without regard to its taxonomic affinities

robin 1 (*see also* chat, run-away robin, robin 2, 3) any of
many redbreasted birds, mostly turdids of the genus
Turdus. Without qualification:—*U.S. T. migratorius. Brit. Erithacus rubitola. W.I. Mimocichla
plumbea* (= red-legged thrush). *W.I.* the thraupid
Spindalis zena (= stripe-headed tanager). *W.I.* the
fringillid *Loxigilla noctis* (= Lesser Antillean bullfinch). *N.Z.* the muscicapid *Petroica australis*
  American robin *T. migratorius*
  rufous-backed robin *T. rufopalliatus*
  black-billed robin *T. ignobilis*
  black robin *T. infuscatus*
  blue robin *U.S.* any of several turdid birds of the
  genus Sialia (= bluebird)
    Siberian blue robin *Luscinia cyane*
    white-tailed blue robin *Cinclidium leucurum*
  black bush-robin *Cercotrichas podobe*
    white-starred bush robin *Pogonocichla stellata*
  red-capped robin the muscicapid *Petroica goodenovii*
  rufous-collared robin *T. rufitorques*
  clay-colored robin *T. grayi*
  flame robin the muscicapid *Petroica phoenicea*
  forest-robin *Stiphrornis erythrothorax*
  hill robin = Peking robin
  hooded robin the muscicapid *Melanodryas cucullata*
  magpie-robin *Copsychus saularis* (= dyal)
  marsh robin *U.S.* (rare) the fringillid *Dipilo
  erythrophalmus* (= towhee)
  mountain robin *T. phebejus*
  Peking robin the timaliid *Leiothrix lutea*
  pink robin the muscicapid *Petroica rodinogaster*
  rooster robin *W.I.* the thraupid *Spindalis zena* (=
  stripe-headed tanager)
  rose robin the muscicapid *Petroica rosea*
  scarlet robin the muscicapid *Petroica multicolor*
  scrub-robin *Afr.* any of several species of the genus
  Erythropygia; *Austral.* either of two species of
  the genus Drymodes
    bearded scrub-robin any of several species of the
    genus Erythropygia, without qualification, *E. bar-
    bata*
    brown-backed scrub-robin *E. hartlaubi*
    red-backed scrub-robin *E. zambesiana*
    white-browed scrub robin *E. leucophrys*
    northern scrub-robin *D. superciliaris*
    southern scrub-robin *Austral. D. brunneopygia*
    white-winged scrub-robin *E. leucoptera*
  starred robin *S.Afr.* = white-starred bush robin
  white-throated robin *T. assimilis*
  tropical robin *Afr.* the ploceid *Lagonosticta rhodopareia* (= ruddy waxbill)

  wood robin *N.Z.* any of several species of muscicapids
  of the genus Petroica
  yellow robin any of several muscicapids of the genus
  Eopsaltria
    southern yellow robin *E. australis*
robin 2 (*see also* plantain, pincushion, robin, robin 1,
robin 3) any of several ragged plants
  ragged robin the scrophulariaceous herb *Lychnis
  floscuculi*
  red robin *Brit.* the caryophyllaceous herb *Lychnis
  diurna* and the geraniaceous herb *Geranium
  robertianum*
  wake robin the araceous herb *Arum maculatum* and
  any of many species of liliaceous herbs of the genus
  Trillium
robin 3 (*see also* robin, robin 1, robin 2) in compound
names of other ragged or red organisms
  sea-robin any of many species of triglid fishes
    armored sea-robin popular name of peristediid
    fishes
robin-run-away the rosaceous herb *Dalibarda repens*
roborinous silvery gray
rock in compound names of many organisms
  jumprock any of several species of catostomid fishes
  of the genus Moxostoma (*cf.* redhorse)
    bigeye jumprock *M. ariommum*
    black jumprock *M. cervinum*
    greater jumprock *M. lachneri*
    striped jumprock *M. rupiscartes*
  living rock the cactus *Ariocarpus fissuratus*
rocket *Brit.* the cruciferous herb *Eruca sativa. U.S.* any
of several species of the cruciferous genus Hesperis
but particularly *H. matronalis*
  dyer's rocket *Brit. Reseda luteola* (= dyer's weed)
  garden rocket *U.S.* the cruciferous herb *Eruca sativa*
  London rocket *Brit.* the cruciferous herb *Sisymbrium
  irio*
  prairie rocket *U.S.* the cruciferous herb *Erysimum
  asperum*
  purple rocket *U.S.* any of several species of cruciferous herbs of the genus Iodanthus
  sea rocket *Brit.* the cruciferous herb *Cakile maritima*
  sweet rocket = rocket
  yellow rocket the cruciferous herb *Barbarea vulgaris*
  (= wintercress)
rockling any of several species of gadid fishes. In *Brit.*
confined to those of the genus Motella
  fourbeard rockling the gadid fish *Enchelyopus
  cimbrius*
rod 1 (*see also* cell, rod 2) a short cylindrical object particularly the light-sensitive rods in the retina
  callous rod thread-like thickenings in the walls of
  sieve-tubes
  chromatic rod a rod attached to a blepharoplast
  pharyngeal rod = trichite
  retinal rod = rhabdomere
rod 2 (*see also* rod 1) in compound manes of organisms
  Aaron's rod the scrophulariaceous herb *Verbascum
  thapsus* (= *Brit.* great mullein)
  goldenrod (*see also* beetle) any of several compositaceous herbs of the genus Solidago
    European goldenrod *S. virgaurea*
    sweet goldenrod *S. odora*
    wreath goldenrod *S. caesia*
Rodentia a large order of placental mammals containing,
inter alia, the rats, mice, squirrels, and porcupines.
They are distinguished by the large single pair of
upper and lower incisors adapted to gnawing, from
which the name is derived

[rodless retina] a mutant (gene symbol *r*) found in linkage group IV of the mouse. The name is descriptive of the phenotypic expression

roe (*see also* deer) the gonads of fishes

hard roe ovary

soft roe testis

rohutu *N.Z.* the myrtaceous shrub *Myrtus obcordata*

[rolled] a mutant (gene sumbol *rl*) mapped at 55.1— on chromosome II of *Drosophila melanogaster*. The rolling refers to the wing of the phenotype

roller 1 (*see also* roller 2, 3) any of many species of bird of the family Coraciidae, and occasionally used for birds in related families. *Brit.*, without qualification, *Coracias garrulus*

Abyssinian roller *Coracias abyssinica*

broad-billed roller any of several species of the genus Eurystomus. Without qualification *Afr. E. glaurcurus I.P.* and *Austral. E. orientais* (= dollar bird)

blue-breasted roller *Coracias cyanogaster*

lilac-breasted roller *Coracias caudata*

rufous-crowned roller *Coracias noevia*

cuckoo-roller the leptosomatid bird *Leptosomus discolor*

European roller *Coracias garrulus*

ground-roller any of several Madagascan species of the genera Brachypteracias, Atelornis, and Uratelornis

Indian roller *Coracias benghalensis*

racket-tailed roller *Coracias spatulata*

blue-throated roller *Eurystomus gularis*

roller 2 (*see also* roller 1, 3) any of many lepidopteran insect larvae that roll plant leaves into a protective cylinder

gray-banded leaf-roller *U.S.* the larva of the tortricid *Argyrotaenia mariana*

oblique-banded leaf-roller *U.S.* the larva of the tortricid *Archips rosaceanus*

red-banded leaf-roller *U.S.* the larva of the tortricid *Argyrotaenia velutinana*

basswood leaf-roller *U.S.* the larva of the pyraustid *Pantographa limata*

bean leaf-roller the larva of the hesperiid *Urbanus proteus*

boxelder leaf-roller *U.S.* the larva of the gracilariid *Gracilaria negundella*

larger canna leaf-roller *U.S.* the larva of the hesperid *Calpodes ethlius*

lesser canna leaf-roller *U.S.* the larva of the pyraustid *Geshna cannalis*

coconut leaf-roller *U.S.* the larva of the pyraustid *Hedylepta blackburni*

European honeysuckle leaf-roller *U.S.* the larva of the yponomeutid *Harpipteryx xylostella*

hickory leaf-roller *U.S.* the larva of the tortricid *Argyrotaenia juglandana*

locust leaf-roller *U.S.* the larva of the phycitid *Nephopteryx subcaesiella*

Mexican leaf-roller *U.S.* the larva of the tortricid *Amorbia emigratella*

sweetpotato leaf-roller *U.S.* the larva of the pyraustid *Pilocrocis tripunctata*

raspberry leaf-roller *U.S.* the larva of the olethreutid *Exartema permundanum*

strawberry leaf-roller *U.S.* the larva of the tortricid *Ancylis comptana* (*cf.* cloverleaf dyer)

sugarcane leaf-roller *U.S.* the larva of the pyraustid *Hedylepta accepta*

fruit-tree leaf-roller *U.S.* the larva of the tortricid *Archips argyrospilus* (*cf.* spruce budworm)

roller 3 (*see also* roller 1, 2) in compound names of other organisms

sandroller the percopsid fish *Percopsis omisomaycus* (*cf.* troutperch)

stoneroller either of two species of cyprinid fishes of the genus Campostoma

ronquil popular name of bathymasterid fishes

[Roof] a mutant (gene symbol *Rf*) mapped at 59.0± on chromosome III of *Drosophila melanogaster*. The phenotypic wings droop alongside the body

[roof wings] a mutant (gene symbol *rf*) mapped at 81.0± on chromosome II of *Drosophila melanogaster*. The phenotypic expression is a drooping of the wings along the side of the body

rook the corvid bird *Corvus frugilegus*

rookery properly the breeding place of a colony of rooks but now often used of any colonies of breeding birds or even of some mammals (*e.g.* seals)

rooster (*see also* fish) a male domestic chicken

root 1 (*see also* root 1, 3, aphis, curculio, hair, worm) that portion, usually underground, of a vascular plant that is adapted both to absorb water and minerals and to anchor the aerial portions. (*cf.* rhizoid, rhizome)

adventitious root a root developed out of the normal sequence or from some unusual part

cladogenous root roots or root-like structures developed from stems and not functioning as roots

nest root the lower part of a nest-epiphyte

primary root that which develops directly from the radicle

pull root one which by its contraction draws a plant deeper into the ground

stilt root large adventitious roots that support the trunk of a tree, such as in the mangrove

tabular root = stilt root

tap root the straight, long root which, when it occurs, derives directly from the continued growth of the radicle

root 2 (*see also* root 1, 3) in compound names of plants with distinctive roots

alum-root any of several saxifragaceous herbs of the genus Heuchera

false alum-root the saxifragaceous herb *Tellima grandiflora*

apple-root any of several shrubby plants of the genus Euphorbia

arrowroot any of numerous herbs of the genus Maranta. Also a commercial edible starch derived either from Maranta, or from Tacca, or, most commonly, from *Canna edulis*

beetroot *Brit* = *U.S.* beet

birth root any of many species of liliaceous herbs of the genus Trillium (= wake-robin)

bitter root the portulacaceous herb *Lewisia rediviva*

bloodroot popular name of plants of the monocotyledonous family Haemodoraceae. The term is also applied to the papaveraceous herb *Sanguinaria canadensis*

bowman's root *U.S.* the rosaceous herb *Gillenia trifoliata*

briar root the root of the Mediterranean ericaceous shrub *Erica arborea*

buckroot *U.S.* the leguminous herb *Psoralea canescens*

cancer root = panther root

canker-root *either* the ranunculaceous herb *Coptis groenlandica or* the scrophulariaceous herb *Kickxia elatine*

chinaroot the climbing lily *Smilax china*
    bastard chinaroot *S. hispida*
colic-root *U.S.* any of several species of liliaceous herb of the genus Alectris
convulsion-root the pyrolaceous parasitic herb *Monotropa uniflora* (*cf.* pinesap)
coral-root any of several species of orchidaceous saprophytes of the genus Corallorhiza
Indian cucumber-root the liliaceous herb *Medeola virginiana* and particularly its edible root
Culver's root the scrophulariaceous herb *Veronicastrum virginicum*
devil's root any of several parasitic herbs most particularly the orobanchaceous *Orobanche minor* (*cf.* broomrape)
dragon root *U.S.* the araceous plant *Arisaema dracontium*
feverroot the caprifoleaceous herb *Triosteum perfoliatum* (= horse-gentian)
fits root *U.S.* the leguminous herb *Astragalus glycyphyllos*. *Brit.* the pyrolaceous herb *Monotropa uniflora* (*cf.* pinesap)
goat-root the leguminous herb *Ononis natrix* (= restharrow)
iris-root = orris root
knot-root the labiateous herb *Stachys sieboldii* and its edible root (= Japanese artichoke)
longroot *U.S.* the caryophyllaceous herb *Arenaria caroliniana* (= pine barren sandwort)
musquash-root the umbelliferous herb *Cicuta maculata*
orange-root the ranunculaceous herb *Hydrastis canadensis*
orris root the iridaceous herb *Iris pallida* and the fragrant rhizome obtained from it
panther-root the orobanchaceous herb *Conopholis americana*
pepper-root the cruciferous herb *Dentaria diphylla*
pinkroot any of several species of loganiaceous herb of the genus Spigelia, particularly *S. marilandica*
pleurisy-root the asclepiadaceous herb *Asclepias tuberosa*
putty-root *U.S.* the orchid *Aplectrum hyemale*
queen's root *U.S.* the euphorbiaceous herb *Stillingia sylvatica*
redroot *U.S.* any of numerous plants including several species of shrub of the rhamnaceous genus Ceanothus, the haemodoraceous herb *Lachnanthes tinctoria*, the leguminous herb *Psoralea esculenta* and the boraginaceous herb *Lithospermum canescens*
rose-root *Brit.* the crassulaceous herb *Sedum rhodiola*
snake-root any of numerous plants with serpentine roots
    black snake-root *U.S.* the liliaceous herb *Zigadenus densum*, the ranunculaceous herb *Cimicifuga racemosa* and any of several umbelliferous herbs of the genus Sanicula
    broom snake-root any of several species of compositaceous herb of the genus Gutierrezia
    button snake-root the umbelliferous herb *Eryngium aquaticum* and any of numerous species of compositaceous herb of the genus Liatris
    Canada snake-root the aristolaceous herb *Asarum canadense*
    rattlesnake-root any of several species of herb of the compositaceous genus Prenanthes
    Sampson's snake-root *U.S.* the gentianaceous herb *Gentiana villosa*
    seneca snake-root the polygalaceous herb *Polygala seneca*

    Virginia snake-root the aristolaceous herb *Aristolochia serpentaria*
    white snake-root the compositaceous herb *Eupatorium rugosum*
squaw-root the orobanchaceous herb *Conopholis americana*
stone-root any of several species of the labiateous herbs of the genera Collinsonia and Micheliella, particularly *C. canadensis*
whiteroot *Brit.* the umbelliferous herb *Hydrocotyle vulgaris* (= *U.S.* water pennywort)
yellowroot the ranunculaceous herb *Hydrastis canadensis*
    shrub yellowroot *U.S.* the ranunculaceous shrub *Xanthorhiza simplicissima*
root 3 (*see also* root 1, 2) the basal portion of an organ, or part of an organ
aortic root one of a pair of dorsal blood vessels into which the aortic arches open
dorsal root the metameric sensory ganglia, external to the cord, forming the root of spinal nerves
ventral root the metameric motor ganglia, inside the chord, forming the ventral roots of spinal nerves
[rootless] a mutant (gene symbol *rt*) mapped at 32 on linkage group III of *Zea mays*. The name is descriptive of the phenotype
rootlet diminutive of root
Roproniidae a small family of proctotrupoid apocritan hymenopteran insects
[Ropy-1] a mutant (gene symbol *Ro-1*) on linkage group IV of *Neurospora crassa*. The name derives from the rope-like aggregations of the phenotypic hyphae
[Ropy-2] a mutant (gene symbol *Ro-2*) found in linkage group III of *Neurospora crassa*. The phenotype has rope-like aggregations of hyphae
roridous covered with, or apparently covered with, dewdrops
Roridulaceae a small family of South African rosale dicotyledonous angiosperms
rorqual popular names of balaenopterid whales of the genus Balaenoptera
Rosaceae the largest family of rosalous plants and one of the largest families in the plant kingdom. Apart from the roses and their flowering allies, most temperate zone edible fruits and berries belong in this family. The family is clearly distinguished by the perigynous flower, the numerous cyclic stamens and the numerous cyclic carpels. Many of the flowers superficially resemble those of the Ranunculaceae but these have hypogynous flowers
Rosales the largest order of dicotyledonous angiosperms. It contains, *inter alia*, the stone crops, the saxifrages, the witch hazels, the plane trees, the roses, the strawberries and legumes. The order is distinguished by the pentamerous cyclic flowers with an androecium of many whorls, the gynoecium with the styles distinct
rose 1 (*see also* rose 2, 3, aphis, apple, bay, beetle, calla, caterpillar, chafer, curculio, finch, girdler, hill, moss, petal, pink, root, scale, slug, wasp) any of numerous flowering shrubs of the family Rosaceae, almost all of the genus Rosa. The "rose" of classical literature was the rhododendron
bourbon rose *Rosa borbonica*
burnet rose = scotch rose
cabbage-rose any of numerous horticultural varieties and hybrids of *R. centifolia*
champney-rose *R. noisettiana*
Cherokee rose *R. laevigata*
cinnamon rose *R. cinnamomea*

damask rose *R. damascena*
dog-rose *R. canina*
    trailing dog-rose *Brit.* *R. arvensis*
French rose *U.S.* *R. gallica*
Japanese rose *Kerria japonica*
burnet-leaved rose *Brit.* = Scotch rose *U.S.*
downy-leaved rose *R. tomentosa*
Macartney rose *U.S.* *R. bracceata*
multiflora rose *R. multiflora*
musk-rose *R. moschata*
    Himalayan musk-rose *R. brunonii*
prairie rose *R. setigera*
Scotch rose *R. spinosissima*
tea-rose any of numerous horticultural varieties and hybrids of *R. odorata*

rose 2 (*see also* rose 1, 3) in compound names of numerous other plants
    alpine rose *Rhododendron ferrugineum*
    Andes rose any of several species of the ericaceous genus Befaria
    California rose the convolvulaceous vine *Convolvulus japonicus*
    Christmas rose the ranunculaceous herb *Helleborus niger*
    cotton-rose the compositaceous herb *Filago germanica*
    guelder-rose the caprifoliaceous shrub *Viburnum opulus*
        mealy guelder-rose *V. lantana*
    Jamaica rose the melastomaceous plant *Blakea trinervia*
    mallow-rose = rose mallow
    mountain rose polygonaceous climbing vine *Antigonon leptopus*
    primrose any of very numerous species of herb of the primulaceous genus Primula (*Brit.*, without qualification, *P. vera*) and many more or less similar flowers
        Arabian primrose the boraginaceous herb *Arnedia cornuta*
        baby primrose *P. forbesii*
        Cape primrose any of numerous gesneriaceous herbs of the genus Streptocarpus
        evening primrose any of numerous species and horticultural varieties of the onagraceous genus Oenothera but particularly *O. biennis*
        fairy primrose *P. malacoides*
        sea primrose *N.Z.* the primulaceous herb *Samolus repens*
        sun-rose any of several species of the cistaceous genus Helianthemum

rose 3 (*see also* rose 1, 2) *I.P.* any of several species of papilionid lepidopteran insect of the genus Tros (*cf.* batwing, clubtail, windmill)

[rose] a mutant (gene symbol *rs*) mapped at 35.0 on chromosome III of *Drosophila melanogaster*. The name refers to the phenotypic eye color

[Rose comb] a mutant (gene symbol *R*) in linkage group II of the domestic fowl. The name is descriptive of the phenotypic expression

rosella 1 (*see also* rosella 2) = rosette
rosella 2 (*see also* rosella 1) any of several psittacid birds of the genus Platycercus. Without qualification usually *P. eximius*
    crimson rosella *P. elegans*
    eastern rosella *P. eximius*
    pale-headed rosella *P. adscitus*
    yellow rosella *P. flaveolus*
roselle the malvaceous herb *Hibiscus sabdarissa*
rosemary the labiateous herb *Rosmarinus officinalis*

dog-rosemary the ericaceous dwarf shrub *Andromeda glaucophylla*
marsh rosemary a term variously applied to either of two ericaceous shrubs *Andromida polifolia* and *Ledum palustris*. Also at times to several members of the plumbaginaceous genus Limonium
wild rosemary the ericaceous shrub *Ledum plaustris*
rose-of-heaven the caryophyllaceous plant *Lychnis coelirosa*
rose-of-Jericho the cruciferous herb *Anastatica hierochuntina*
rose-of-Sharon the malvaceous tree *Hibiscus syriacus*. The name is also sometimes applied to the hypericaceous herb *Hypericum calycinum*
roseolous pinkish
roseous pale pink
rosette any structure the parts of which are arranged radially. Specifically applied to a central disk of fused plates in the vault of a crinoid calyx
    double rosette the prophase of mitosis or meiosis
    pulsatile rosette a rounded, slowly pulsating, projection of the coelomic wall, found near the calcareous ring of Holothuria
    umbilical rosette the star-shaped structure in the center of the frustule of some diatoms
[rosette] a mutant (gene symbol *ro*) on chromosome 2 of the tomato. The phenotypic plant has very short internodes and does not flower
rosin (*see also* weed) a resin derived from numerous pine trees
Rosineae a suborder of Rosales containing the Platanaceae, Crossosomaceae, Rosaceae, Connaraceae and Leguminosae
Rosoideae a subfamily of Rosaceae distinguished by having a superior ovary, indehiscent fruit, and a gynoecium of ten or more pistils which are basally connate, pistils distinct and free from hypanthium walls
-rostell- *comb. form* meaning "a little beak" (diminutive of rostrum)
rostellar *adj.* from rostellum but frequently used in the sense of rostral
rostellum diminutive of rostrum and applied to numerous plant and animal structures in the form of a short or small beak and specifically to the mobile, cone-shaped tip of the scolex of some tapeworms
-rostr- (*see also* -rhynch- and -rhin-) *comb. form* meaning "beak" or "rostrum"
rostral pertaining to the rostrum (*cf.* rostellar) but commonly used as pertaining to the beak
    conirostral said of a bird having a conical beak
    dentirostral said of a bird in which the commissure of the beak is dentate or serrate
    fissirostral said of a bird having a broad widely gaped beak
    pressirostral said of a bird having a beak like a plover
    tenuirostral said of a bird having a long slender bill with a small gape
Rostratulidae a family of charadriiform birds containing the two species of painted snipe. They are distinguished by the swollen tip of the bill
rostrate with a beak, particularly a pointed beak
    erostrate lacking a beak
rostrum literally a beak but applied to almost any pointed non-retractile prominence in organisms, including the solid conical terminal portion of the shell of fossil decapod mollusca such as the belemnitidae
rosulate having the form of rosettes. The word is fre-

quently, particularly in anatomical descriptions of turbellarian pharynges, misused for spherical

[**rosy**] a mutant (gene symbol *ry*) mapped at 51.0± on chromosome III of *Drosophila melanogaster*. The phenotypic eye color is not rose but deep ruby

-**rota**- *comb. form* meaning "wheel"

rotaceous wheel-shaped

[**rotated**] a mutation (gene symbol *rt*) mapped at 37.0± on chromosome III of *Drosophila melanogaster*. The phenotypic expression is a counterclockwise twisting of the abdomen

[**rotated penis**] a mutant (gene symbol *rp*) mapped at 41.7± on chromosome III of *Drosophila melanogaster*. The name is descriptive

Rotatoria = Rotifera

rotche the alcid bird *Plautus alle* (= dovekie)

Rotifera a taxon of the animal kingdom which may either be regarded as a phylum of pseudocoelomate Bilateria or as a class of the Aschelminthes. They are distinguished by the presence of an anterior ciliated corona from which the popular name "wheel animalcule" is derived

rotula a small, intercalary subsegment found in some insect antennae

rotulae one of five heavy bars lying beneath the compasses of Aristotle's lantern

[**rotund**] a mutant (gene symbol *rn*) mapped at 54.5± on chromosome II of *Drosophila melanogaster*. The wings of the phenotype are rounded and the tarsae are 3-jointed

rou a bacterial genetic marker indicating that the mutant makes a rough colony on agar

[**rough**] *either* a mutant (gene symbol *ro*) mapped at 91.1 on chromosome III of *Drosophila melanogaster*. The phenotypic eyes are rough and small *or* a mutant (gene symbol *ro*) found in linkage group I of *Habrobracon juglandis*. The phenotypic expression involves the absence of the 4th. radius vein and the roughening of other veins *or* a mutation (gene symbol *ro*) found in linkage group V of the mouse. The phenotypic expression is coarse hair and wavy whiskers

[**Roughened**] a mutant (gene symbol *R*) mapped at 1.4 on chromosome III of *Drosophila melanogaster*. The name is descriptive of the phenotypic eye

[**roughest²**] a mutant (gene symbol *rst²*) mapped at 1.7 on the X chromosome of *Drosophila melanogaster*. The phenotypic expression is rough eyes and a dwarfed body

[**roughex**] a mutant (gene symbol *rux*) mapped at 15.0 on the X chromosome of *Drosophila melanogaster*. The phenotype has small rough eyes

[**rough eye**] a mutant (gene symbol *rux*) mapped at 0.7± on the X chromosome of the *Drosophila melanogaster*. The phenotypic expression is small rough eyes

[**rough fur**] a mutant (gene symbol *R*) in linkage group I of the guinea pig. The phenotypic expression is rough fur, particularly on the hind toes

[**roughish**] a mutant (gene symbol *rh*) mapped at 54.7± on chromosome II of *Drosophila melanogaster*. The phenotypic expression is moderately rough eyes

[**roughoid**] a mutant (gene symbol *ru*) mapped at 0 on chromosome II of *Drosophila melanogaster*. The phenotypic expression is small rough eyes with erupted facets

rout term for a group of wolves, when not hunting (*cf.* pack)

row a number of objects placed side by side in a hori-

zontal row, particularly in contrast to a vertical rank though this distinction is not universal

axial row the egg ventral canal cell and neck canal cells in the archegonium of the bryophyte

comb row one of the bands of ctenes on a ctenophore

royal (*see also* pennyroyal) *I.P.* any of numerous lycaenid lepidopteran insects of the genus Tajuria

tufted royal *I.P.* any of numerous lycaenid lepitopteran of the genus Pratapa

rp *see* [rotated penis]

Rp *see* [Resistance to *Puccinia*]

R1 pho a bacterial genetic marker having a character which acts as an alkaline phosphatase repressor, mapped at 10.5 mins. for *Escherichia coli*

R2 pho a bacterial genetic marker having a character which acts as an alkaline phosphatase repressor, mapped at 73.25 mins. for *Escherichia coli*

karyo**rrhexis** the condition of a necrotic nucleus which breaks into fragments

rs *see* [rose]. The term is also in yeast genetics to indicate radiation sensitivity

rsc *see* [reduplicated sex combs]

rsd *see* [raised]

rst² *see* [roughest²]

rt *see* [rotated] or [rootless]

R try a bacterial genetic marker having a character which acts as a tryptophan repressor, mapped at between 88.0 mins. and 90.0 mins. for *Escherichia coli*

ru *see* [roughoid] or [ruby eye]

rubber coagulated latex, usually commercially derived from euphorbiaceous trees of the genus Hevea

virgin rubber the euphorbiaceous tree *Sapium verum*

rubellous reddish

rubeolous ruddy

e**rubescent** suffused with red

Rubiaceae that family of rubialous dicotyledons which contains *inter alia* the coffee shrub, the cinchona, the cape jasmine and the partridge berry. The family is closely related to the Caprifoliaceae but is distinguished from them by having stipules or whorled leaves

Rubiales that order of dicotyledonous angiosperms which contains *inter alia*, coffee, cinchona, the honeysuckles, the valerians and the teasels. The order is characterized by the possession of gamopetalous flowers with an inferior ovary and distinct anthers borne in cymose inflorescences

rubicundous deep blush-pink

rubidous reddish

rubiginous rust colored

bili**rubin** the breakdown product of biliverdin (*q.v.*)

-**rubu**- *comb. form* meaning "bramble"

[**ruby**] a mutant (gene symbol *rb*) mapped at 7.5 on the X chromosome of *Drosophila melanogaster*. The name applies to the eye color

[**ruby eye**] a mutant (gene symbol *ru*) found on linkage group XII of the mouse. The phenotypic expression involves an overall reduction of pigment

black **ruby** the cyprinid teleost fish *Puntius nigrofasciatus* (= nigger barb)

Brazilian ruby the trochilid bird *Clytolaema rubricauda*

ruched pleated or gathered in folds. Sometimes, quite incorrectly, used as synonymous with goffered

-**rud**- *comb. form* meaning waste heap or rubbish heap

rud *see* [ruddle]

rudd *Brit.* the cyprinid fish *Scardinius erythrophthalmus*

[ruddle] a mutant (gene symbol *rud*) mapped at 3.3± on the X chromosome of *Drosophila melanogaster*. The phenotypic expression is reddish-brown eyes

ruddock *Brit.* = redbreast *Brit.*

-ruderal- *comb. form* meaning "rubbish heap" (*cf.* -chled-)

ruderal pertaining to rubbish heaps

rudiment the first vestige of a developing organ or a functionally useless organ

rudimentary arrested in a early stage of development

[rudimentary⁹] a mutant (gene symbol r^9) mapped at 54.5 on the X chromosome of *Drosophila melanogaster*. The phenotypic expression includes truncated wings and sterile females

rue (*see also* anemone) the rutaceous herb *Ruta graveolens*

castle-rue the ranunculaceous herb *Trautvetteria caroliniensis*

goat's rue *Brit.* the leguminous herb *Galega officinales U.S. Tephrosia virginiana*

lady rue *U.S.* the ranunculaceous herb *Thalictrum clavatum*

meadow rue any of several species of herb of the ranunculaceous genus Thalictrum

tall meadow rue *T. polygamum*

wall rue *U.S.* the polypodiaceous fern *Asplenium cryptolepis*

ruff 1 (*see also* ruff 2–4) *U.S.* any of several species of stromateid fishes of the genus Centrolophus

black ruff *C. niger*

ruff 2 (*see also* ruff 1, 3, 4) a male sandpiper

ruff 3 (*see also* ruff 1, 2, 4) the scolopacid bird *Philomachus pugnax* (*cf.* reeve)

ruff 4 (*see also* ruff 1–3) a collar of protruding feathers, or feathers which may be protruded or elevated, around the neck of a bird

Ruffini's *epon. adj.* from Angelo Ruffini (1874–1925)

ruffe *Brit.* the percid fish *Acerina vulgaris*

rufidulous pinkish

rufous reddish

-rugo- *comb. form* meaning "wrinkle"

rugose wrinkled

[rugose] a mutant (gene symbol *rg*) mapped at 11.0 on the X chromosome of *Drosophila melanogaster*. The phenotypic expression is rough eyes with thin friable wings

rule a term preferred to "law" by many contemporary scientists. "So-and-so's rule" is, to avoid repetition, given in this work as "so-and-so's law." The various international bodies controlling nomenclature and the like, promulgate "rules" having the force of laws

rumen the first of the four divisions of the stomach of artiodactyles (*cf.* abomasum, psalterium, reticulum)

Ruminantia a group of artiodactyle mammals containing the families Tragulidae, Camelidae, Cervidae, Giraffidae, Antilocapridae and Bovidae. They differ from the Suinae in having a relatively complex stomach

ruminate having the appearance of having been chewed

[rumpled] a mutant (gene symbol *rmp*) mapped at 14.4± on the X chromosome of *Drosophila melanogaster*. The phenotypic expression is rumpled (unexpanded) wings

rumule a teat-like prominence

run 1 (*see also* run 2) to progress in the manner of a walk but more rapidly

run 2 (*see also* run 1) term for a group of fish, particularly those ascending a river from the sea

run-away-robin *U.S.* the labiateous creeping herb *Glechoma hederacea* (= ground ivy)

runcinate with incised retrorse teeth

runner 1 (*see also* runner 2, 3, bulb) a popular term for stolon

runner 2 (*see also* runner 1, 3) in the sense of that which runs

lark-like brush-runner the furnariid bird *Coryphistera alaudina*

logrunner either of two timaliid birds of the genus Orthonyx

black-headed logrunner *O. spaldingi*

spine-tailed logrunner *O. temmincki*

racerunner any of several species of tejid saurian reptiles of the genus Cnemidophorus, particularly *C. sexlineatus*

road runner *U.S.* either of two species of cuculid birds of the genus Geococyx

greater roadrunner *G. californianus*

lesser roadrunner *G. velox*

stone runner the charadriid bird *Charadrius hiaticula* (= ringed plover)

tree-runner properly, any of several furnariid birds of the genera Margarornis and Pygarrhicus *Austral.* = sittella

runner 3 (*see also* runner 1, 2) pertaining to a run in the sense of run 2 (above)

blue runner the carangid fish *Caranx crysos*

rainbow runner *U.S.* the carangid fish *Elagatis bipinnulatus*

runt the smallest of a litter or clutch

-rupes- *comb. form* meaning "rock"

rupestral pertaining to rocks

Rupicaprinae a subfamily of caprid artiodactyle mammals containing the forms commonly called rock-goats

Rupicolidae a family of passeriform birds usually included with the Cotingidae

-rupt- *comb. form* meaning "broken"

ruptile dehiscing through irregular fracture

Ruralidae = Lycaenidae

rusa a cervine artiodactyle (*Cervus timoriensis*) of the East Indies

Ruscaceae a family of liliiflorous monocotyledonous angiosperms containing forms usually included within the Liliaceae

rush 1 (*see also* rush 2, bird) properly any of numerous plants of the monocotyledonous family Juncaceae and particularly those of the genus Juncus

bog rush *U.S.* almost any member of the genus Juncus

bulrush (*see also* rush 2) *U.S. J. effusus*

hard rush *Brit. J. glaucus*

soft rush *Brit. J. effusus*

wood rush *U.S.* any of several species of the genus Luzula

rush 2 (*see also* rush 1, broom, lily) any of numerous plants resembling the true rushes either in appearance or habitat

bald rush *U.S.* any sedge of the genus Psilocarya

beak rush *U.S.* any of several species of sedge of the genus Rhynchospora

bullrush (*see also* rush 1) any of several sedges of the genus Scirpus but especially *S. lacustris* also known as the great bullrush or giant bullrush

hedgehog club-rush either of the sedges *Cyprerus ovularis* or *C. globulosus*

chair maker's rush *U.S.* the sedge *Scirpus americanus*

horned rush *U.S.* any of several species of sedge of the genus Rhynchospora particularly those of the subgenus Calyptrostylis

nut-rush *U.S.* any of several species of sedge of the genus Scleria

scouring rush the equisetaceous plant *Equisetum hiemale*

spike rush *U.S.* any sedge of the genus Eleocharis

white rush *U.S.* the grass *Spartina monogyna*

russous russet

Russulaceae a family of agaricale basidomycete fungi distinguished by lacking both a volva and an annulus

rustic *U.S.* any of several species of noctuid lepidopteran insects of the genus Caradrina, usually, without qualification, *C. tarascaci. Brit.* any of several species of noctuid moth of the genera Agrotis, Luberina, Apamea, Aporophyla, and Tholera. *Austral.* and *I.P.* any of several species of nymphalid lepidopteran insects of the genus Cupha. Usually, without qualification, *C. prosope* (*Austral.*) or *C. erymanthis* (*I.P.*)

hedge rustic *Brit.* the noctuid lepidopteran insect *Tholera cespitis*

[Rusty] a mutant (gene symbol *Rc*) mapped at 31.8 on linkage group II of *Bombyx mori.* The phenotypic expression is a yellowish brown cocoon

rut the condition of a male animal ripe to breed

rutabaga any of numerous horticultural varieties of the cruciferous plant *B. campestris var. napobrassica* (= swedish turnip)

Rutaceae that family of geranialeous dicotyledons which contains *inter alia,* the citrus fruits and their less known allies. The presence of berry-like fruits with fragrant oils in both the fruit and the leaves is typical of most of the family, as is also the lobed ovary from which the fruit develops

Rutelidae a family of coleopteran insects closely allied to the Melolonthidae. Both families are frequently treated as sub-families of the Scarabeidae

rutilant glowing, in the sense that this term modifies colors

rux *see* [roughex]

rv *see* [reticulate virescent]

rwg *see* [reduced wings]

ry *see* [rosy]

rye (*see also* grass) any of several species of grass of the genus Secale and numerous horticultural hybrids mostly derived from *S. cereale*

wild rye any of several species of the genus Elymus

Rynchopidae a small family of charadriiform birds erected to contain the three species of skimmers. These are gull-like birds with long bills laterally compressed to thin blades with the lower mandible much longer than the upper

S

s *see* [piebald], [sable], [silvered] or [compound infloresence]. This symbol is also used as a yeast genetic marker for an active recessive suppressor

S *see* [New striped], [Silver] or [Star]. This symbol is also used as a yeast genetic marker for an active dominant suppressor

S+ a yeast genetic marker for a non-active, wild type, suppressor allele

s+ a yeast genetic marker for a non-active, wild type suppressor allele

-s *symb. suffix* appended to bacterial mutant symbols to indicate sensitivity to the compound indicated (*cf*. -d, -r)

sa *see* [satin]. In yeast genetics indicates a site of affinity

Sabellariidae a family of sedentary polychaete annelids living in calcareous tubes. They are distinguished from the Serpulidae by having setae on the peristomial collar

Sabellidae a family of sedentary polychaete worms living in membranous tubes in mud and sand. They are distinguished by having a peristomial collar which does not bear setae

Sabelliformia an order of cryptocephalic polychaete annelids with very small tentacles. The tubes are formed either of a mucilaginous material covered with sand or pieces of shells or calcareous matter

Sabiaceae a small family of sapindale dicotyledonous angiosperms distinguished by the pentamerous flowers the stamens of which are opposite the petals

Sabiineae a suborder of Sapindales containing the single family Sabiaceae

sable the musteline carnivore *Martes zibellina*. The term is also, particularly in *U.S.*, applied to the pine marten *M. americana*

[sable] a mutant (gene symbol *s*) mapped at 43.0 on the X chromosome of *Drosophila melanogaster*. The name refers to the dark body color of the phenotype

-sabul- *comb. term* meaning "sand"

sabulose pertaining to sand

sac a bag or pouch

  air sac 1 (*see also* air sac 2, 3) one of five pairs of outgrowths, extending through the body cavity of birds, derived from the lungs. These are called according to their location: cervical, anterior, posterior, clavicular, thoracic and abdominal air sacs

  air sac 2 (*see also* air sac 1, 3) the gas-containing portion of the pneumatophore of a siphonophoran coelenterate

  air sac 3 (*see also* air sac 1, 2) a pouch-like expansion of a tracheal tube in winged insects (= vesicle)

pericardial sac that division of the coelom of vertebrates which contains the heart

cement sac that portion of an invertebrate oviduct in which eggs are coated with a capsule or shell

compensation sac a thin-walled sac permitting volume changes in Ectoprocta with calcified zoecia (= ascus)

digestive sac walls covering secondary rootlets within the main root before the former reach the exterior

dorsal sac (Echinodermata) = madreporic vesicle

egg sac the term is particularly applied to the egg masses of copepod Crustacea. It is also used, in *bot*., to describe membranes surrounding eggs, ovules and oospheres (*cf*. ovisac)

embryo sac specifically the female gametophyte which develops from the megaspore within the nucellus

enteric sac that remnant of the archenteron of the crinoid larva which becomes the future intestine

germinal sac (Platyhelminthes) = sporocyst larva

ink sac a gland found in some cephalopod mollusks that produces a black ink-like fluid used to deceive predators

membranous sac a subdivision of the proximal coelom in some Ectoprocta

metameric sac = osmeterium

nectosac that portion of the nectophore of siphonophoran medusans which contains the radial canals

ovisac *either* a membrane within which many eggs are extruded from the body, *or* that portion of a female reproductive system filled with ripe eggs (= egg sac)

plasm-sac the protoplasm lining the frustule of diatoms

pleural sac the fold of peritoneum containing the lung of a mammal

pollen sac the cavity in the stamen in which the microspores are formed

spore sac = moss capsule

sporosac an area on the blastostyle of some hydroid coelenterates in which sex cells develop directly, without the production of a medusoid form

sacaline the polygonaceous herb *Polygonum sachalinense*

sacaton the grass *Sporodolus wrightianus*

-sacc- *comb. form* meaning "sack"

saccate in the form of or possessing a pouch or sack. Used specifically by microbiologists to describe the appearance of the liquified area in a stab culture when it is a blunt-nosed cone not reaching to the bottom of the tube

-sacchar- *comb. form* meaning "sugar"

saccharatous sugary

saccule one of numerous small spherical bodies, embedded in the surface tissue, between or alongside the podia of some echinoderms

sacculus 1 (*see also* sacculus 2) the ventral of the two divisions into which the auditory vesicle becomes first divided (*cf.* utriculus)

sacculus 2 (*see also* sacculus 1) the ventral, infolded part of the harpe

saccus in the genitalia of female Lepidoptera, a sclerotized pocket formed by invagination of the ninth abdominal segment. In males it is a midventral cephalad projection of the ninth abdominal segment serving for the attachment of penis muscles

sacellus a fruit consisting of a one-seeded indehiscent pericarp inclosed in a papery calyx

Sach's *epon. adj.* from Julius von Sachs (1832–1897)

saiga a saigine caprid mammal, *Saiga tartarica,* the hugely arched snout of which gives it an air of supercilious superiority unequalled by any other animal. It is today confined to the arid steppes of the south of Kazakh (*cf.* chiru)

Saiginae a sub-family of caprid artiodactyle mammals commonly called the gazelle-goats. The only two members of the group are the chiru and the saiga

sacrifice *biol. jarg.* for kill

sacrum that part of the vertebral column which articulates with the pelvis

synsacrum that part of the pelvic girdle which is derived from modified vertebrae

saffron 1 (*see also* saffron 2, crocus) the bulbous liliaceous plant *Crocus sativus* and an article of commerce consisting of the dried stamens of this plant

bastard saffron the compositaceous herb *Carthamus tinctorius* (= safflower) and also the euphorbiaceous tree *Mallotus phillippinensis*

horse-saffron the compositaceous herb *Carthamus tinctorius* (= safflower)

meadow saffron the liliaceous herb *Colchicum autumnale* and occasionally some other species of the same genus

saffron 2 (*see also* saffron 1) *I.P.* the lycaenid lepidopteran insect *Mota massyla*

[safranin] a mutant (gene symbol *sf*) mapped at 71.5± on chromosome II of *Drosophila melanogaster.* The phenotypic expression is a dark chocolate, not safranin, eye color

sage (*see also* defoliator, wort) any of numerous species of the labiateous genus Salvia, but particularly the culinary herb *S. officinalis*

Bethlehem sage the boraginaceous herb *Pulmonaria saccharata*

germander sage = wood sage

Jerusalem sage the labiateous herb *Phlomis fruticosa*

scarlet sage the labiateous herb *Salvia splendens* and its numerous horticultural varieties

white sage *U.S.* the compositaceous herb *Artemisia ludoviciana*

wood sage either of two species of herb of the labiateous genus Teucrium (*T. canadense* and *T. scorodonia*)

-sagitta- *comb. form* meaning "arrow head"

Sagittariidae a monotypic family of falconiform birds erected to contain the snake-eating secretary-birds readily distinguished by the conspicuous crest of long paddle-like feathers on the nape of the neck

sagittate in the form of an arrow head

sago *see* palm

cascara sagrada the rhamnaceous shrub *Rhamnus purshiana* and particularly the medical bark collected from it

saguaro the tall, tree-like cactus *Carnegiea gigantea*

sailor *I.P.* and *S.Afr.* any of several species of nymphalid lepidopteran insects of the genus Neptis (*cf.* lascar)

ragged sailor the compositaceous herb *Centaurea cyanus* (= cornflower)

sailor's-choice the pomadasyid fish *Haemulon parrai*

St. John's *epon. adj.* from numerous saints. For specific derivations see compound (*e.g.* St. John's wort)

saithe *Brit.* the gadid fish *Gadus virens* (= coalfish)

saker the falconid bird *Falco cherrug* (*cf.* falcon)

saki pithecinid primates of the genus Chiropotes. They are principally distinguished by the hair on the head and the huge beard

sakiwinki pithecinid primates of the genus Pithecia. They are distinguished by their long coarse hair and bushy tails

corn salad any of several species of the valerianaceous herb Valerianella but particularly *V. olitoria*

salamander the terms newt and salamander are now hopelessly confused. Originally, and still to a great extent in *Brit.* usage, newt is a predominantly aquatic urodele and a salamander is a predominantly terrestrial urodele. However, there is currently an endeavor in *U.S.* to confine the term newt to the family Salamandridae and to use the word salamander for all other urodeles. The term eft is synonymous with newt in both languages, though rarely used. There is currently an attempt in *U.S.* to use the word eft for the terrestrial phrase and newt for the aquatic phase of such forms as the red-spotted newt

alpine salamander the salamandrine salamandrid *Salamandra atra*

black-bellied salamander *U.S.* the plethodontid *Desmognathus quadramaculatus*

blind salamander a term generally applied in *U.S.* to troglodytic plethodonts of the genera Typhlotriton, Typhlomolge and Haideotriton

brook salamander *U.S.* general term for plethodontids of the genus Eurycea

cave salamander *U.S.* either of two plethodontids, *Gyrinophilus palleucus* or *Eurycea lucifuga*

dusky salamander *U.S.* any of numerous species of plethodontids of the genus Desmognathus, particularly *D. fuscus*

flatwoods salamander the ambystomid *U.S. Ambystoma cingulatum*

fire salamander *Salamandra salamandra*

giant salamander the Japanese cryptobranchid *Megalobatrachus japonicus,* which attains a length of over five feet

Chinese giant salamander the cryptobranchid *M. davidianus* which attains a length of about 3 feet 6 inches

Pacific giant salamander the ambystomid *Dicamptodon ensatus* which attains a length of one foot

green salamander *U.S.* the plethodontid *Aneides aeneus*

grotto salamander *U.S.* the plethodontid *Typhlotriton spelaeus*

lungless salamander a term generally applied in *U.S.* to plethodontids

marbled salamander the ambystomid *Ambystoma opacum*

mole salamander *U.S.* any urodele of the family

Ambystomidae. Usually, without qualification *Ambystoma talpoideum*

Monterey salamander *U.S.* the plethodontid *Ensatina eschscholtzii*

mud salamander *U.S.* the plethodontid *Pseudotriton montanus*

shovel-nosed salamander *U.S.* the plethodontid *Leurognathus marmoratus*

pigmy salamander *U.S.* the plethodontid *Desmognathus wrighti*

purple salamander *U.S.* = spring salamander *U.S.*

ravine salamander *U.S.* the plethodontid *Plethodon richmondi*

seal-salamander the plethodontid *Desmognathus monticola*

slender salamander *U.S.* any of several species of plethodontid of the genus Batrachoseps

slimy salamander *U.S.* the plethodontid *Plethodon glutinosus*

spotted salamander the salamandrid *Salamandra salamandra maculosa*

spring salamander *U.S.* either of two species of plethodontid of the genus Gyrinophilus, particularly *G. porphyriticus*

four-toed salamander *U.S.* the plethodontid *Hemidactylium scutatum*

tree salamander *U.S.* the plethodontid *Aneides lugubris*

woodland salamander *U.S.* a general term for plethodontids of the genus Plethodon

zigzag salamander *U.S.* the plethodontid *Plethodon dorsalis*

Salamandridae a family of urodele Amphibia distinguished by the presence of maxillary bones and moveable eyelids

Salamandrinae a subfamily of salamandrid urodele Amphibia distinguished by a series of palatal teeth in two longitudinal series inserted on the inner margin of the long palatine processes

Saldidae small brownish bugs commonly called shore bugs by virtue of their freshwater or marine habitats. They are excellent fliers and are distinguished by the possession of four or five long closed cells in the membrane of the hemelytra

salema the pomadasyid fish *Xenistius californiensis*

Salicaceae the family of plants which contains the willows and poplars. Distinctive characteristics are that the flowers of both sexes are borne in catkins and that they have a dehiscent many-seeded capsule

Salicales an order of dicotyledenous plants containing the single family Salicaceae. The characteristics of the family are those of the order

Salicariaceae = Lythraceae

salient projecting forward

Salientia a large order, or sub-class of Amphibia distinguished by the absence of tail in the adult, the development of the hind legs for jumping and the reduction or absence of ribs. Commonly called frogs and toads (= Anura)

saliva a secretion, accumulated in the mouths of many mammals, from the salivary glands

salivarium the chamber into which the salivary glands of insects open

sallow any of numerous species of noctuid lepidopteran insects. *Brit.* without qualification, usually *Xanthia fulvago*

dusky sallow *Brit. Eremobia ochroleuca*

red-winged sallow *U.S. Jodia rufago*

salmon (*see also* king-of) *U.S.* any of several species of salmonid fishes of the genera Oncorhynchus and

also used for a few other forms. *Brit.* the salmonid *Salmo salar* (= *U.S.* Atlantic salmon)

Atlantic salmon the salmonid *Salmo salar*

beaked salmon the gonorhynchid *Gonorhynchus gonorhynchus*

Chinook salmon *Oncorhynchus tshawytscha*

Chun salmon *O. keta*

coho salmon *O. kisutch*

sockeye salmon *O. nerka*

pike-salmon the characid *Acestrorhynchus microlepis*

pink salmon *O. gorbuscha*

quinnat salmon *U.S.* = Chinook salmon *U.S.*

salmoneous salmon colored

Salmonidae a very large family of isospondylous fishes containing, *inter alia,* the salmon, trout, white fish (sometimes placed in a special family—Coregonidae) and grayling (sometimes placed in a special family—Thymallidae). They are easily identified by the fleshy adipose fins on the posterior dorsal surface of the body

[salmon silk] a mutant (gene symbol *sm*) mapped at 54 in linkage group VI of *Zea mays.* The term is descriptive of the "silk" of the phenotype

Salmopercae an order of Osteichthyes intermediate between the soft-boned and spiny-finned forms. All have soft fin rays but with one or more spines preceding the dorsal, anal, and pelvic fins

-salp- *comb. form* meaning "trumpet"

salp general name for free-swimming thalliacean tunicates particularly those of the genus Salpa

Salpida an order of thaliaceous tunicates, either solitary or aggregated in chains, distinguished by a single pair of large branchial gills

Salpiglossideae tribe of the family Solanaceae

salpinges plural of salpinx

Salpingidae = Pythidae

salpinx literally, a trumpet, but used for any tube, such as the oviduct or eustachian tube, which terminate in a trumpet shaped orifice

mesosalpinx that part of the mesentery which supports the oviduct

salsify the compositaceous herb *Tragopogon porrifolius* and its edible roots

black salsify the compositaceous plant *Scorzonera hispanica* and its edible root

salsuginous growing in brackish places

saltant literally, jumping or leaping, but used primarily in *biol.* for a major mutational change

saltation a type of evolution which proceeds by leaps and bounds through the production of mutants which differ grossly from the parent (*cf.* anamorphosis)

saltator any of several fringillid birds of the genus Saltator

greyish saltator *S. coerulescens*

black-headed saltator *S. atriceps*

buff-throated saltator *S. maximus*

Saltatoria a group of insects, once considered to be a suborder of Orthoptera, containing the crickets and grasshoppers

saltatory with the habit of leaping

Salvadoraceae a small family of sapindale dicotyledons

Salvia the genus of labiateous herbs commonly called sages. However, *S. fulgens* is known as Mexican red salvia or cardinal salvia

Salvineaceae the only family of the order Salviniales. The characters of the family are those of the order

Salviniales a small order of leptosporangiate Pterophyta distinguished by being heterosporous and producing their sporangia with sporocarps that contain either macrosporangia or microsporangia

saman = rain-tree

samara a single-seeded winged fruit like that of the maple or ash

sambar the asiatic cervine artiodactyl *Cervus unicolor* (*cf.* deer)

samphire any of numerous seaside plants particularly *U.S.* the amaranthaceous herb *Thiloxerus vermicularis* and *Brit.* the umbelliferous plant *Crithmum maritimum*

   golden samphire *Brit.* the compositaceous herb *Inula crithmoides*

   marsh samphire any of several species of the chenopodiaceous genus *Salicornia*

sand (*see also* flea, martin, piper) rock particles, frequently silicious, from 1/20 of a millimeter to 1 millimeter in size. The term is also used for other small particles (*cf.* gravel, silt)

   oolitic sand that composed entirely of granules of calcium carbonate

   belisand sand from which nutrients have been washed out

   hydatid sand small secondary cysticerci found in hydatid cysts (*q.v.*)

   brain sand concentrically laminated concretions, thought to consist principally of alkaline carbonates and phosphates, found in the pineal gland

Sandalidae a small family of coleopteran insects distinguished by the large mandibles and curiously lobed tarsae. Commonly called cedar beetles

sandarach the pinaceous tree *Callitris quadrivalvis* and the gum secured from it

sanderling the scolopacid bird *Crocethia alba*

sanguine bloody, with regard to color

consanguineous an obsolete term once used to designate individuals related by descent from a common ancestor

Sanguinolariidae a family of pelecypod mollusks commonly called the long-siphon clams. They have elongate equivalved shells

sanicle *Brit.* anglicized form of the umbelliferous genus Sanicula

Santalaceae that family of santanalous dicotyledons which contains the sandalwoods. This family is distinguished from the closely allied Loranthaceae by the more numerous ovules and the fact that they are not aerial parasites

Santalales that order of dicotyledonous angiosperms which contains, *inter alia* the mistletoes and sandal woods. They may be distinguished from similar organisms by the unisexual flowers with small perianth and opposite adnate stamens

Santalineae a suborder of Santalales containing all families except the Loranthaceae

Santorini's *epon. adj.* from Giandominico Santorini (1681–1737)

sap 1 (*see also* sap 2, beetle, sucker) properly, the fluid which circulates in plants, but applied to other materials regarded as similar

   nuclear sap the clear portions of the nucleus. Sometimes restricted to those portions of a prophase mitotic nucleus which are not discretely chromosomal

sap 2 (*see also* sap 1) in compound names of organisms

   pinesap any of several species of parasitic herb of the genus Monotropis

Sapindaceae a large family of sapindalous dicotyledonous trees. The small flowers with ten stamens, and the three-celled fruits are distinctive

Sapindales that order of dicotyledonous angiosperms which contains *inter alia*, the boxwoods, the crow berries, the cashew nuts, the hollies, the bladder nuts, the maples, the horsechestnuts, the soap berries and the balsams. The order is distinguished from the Geraniales by possessing pendulous ovules

-sapo- *comb. form* meaning "soap" and by extension "slippery"

saponaceous soapy

sapote the sapotaceous tree *Lucuma mammosa*

Sapotineae a suborder of the Ebenales containing the families Sapotaceae and Hoplestigmataceae

sapphire 1 (*see also* sapphire 2) *I.P.* any of numerous species of lycaenid lepidopteran insects of the genus Heliophorus. *S.Afr.* any of numerous species of lycaenid lepidopteran insect of the genus Iolaus, usually, without qualification, *I. silas*

sapphire 2 (*see also* sapphire 1) any of several trochilid birds of the genus Hylocharis

-sapro- *comb. form* meaning "decomposed" in the sense of "rotten" or "putrid"

saprobe = saprophyte

   oligosaprobe an organism that flourishes in oligodynamic waters

   mesosaprobe an organism that lives in an aqueous habitat with a relatively low oxygen, and relative high decomposing organic matter, content

   polysaprobe an organism that is adapted to live in heavily contaminated water

Saprolegniales an order of biflagellate phycomycete Fungi found so frequently in water on decaying fish that they are sometimes called "fish molds." They are distinguished by the extensive mycelial thallus that lacks a hold fast

Sapromyzidae a family of brachyceran dipteran insects now distributed between the Lonchaeidae, Pallopteridae and Lauxaniidae, the first two of these sometimes being included in the last

Sapromyiophilae plants having the smell of decomposing flesh and therefore fertilized by carrion-flies

-sarc- *comb. form* meaning "flesh"

   amphisarc = pepo

   coenosarc the living portion of a hydrozoan hydroid colony

   cytosarc = cytoplasm

   perisarc the sheath of hydroid coelenterates. Sometimes used as synonymous with periderm in the sense of -derm 1 (*q.v.*)

Sarcodina a class of plasmodromous Protozoa containing, *inter alia*, the amebas, Foraminifera and the Radiolaria. They are distinguished from the Mastigophora and Ciliata by lacking locomotor organelles and from the Sporozoa by their methods of reproduction

Sarcopsyllidae = Tungidae

Sarcophagidae a family of myodarian cycloraphous dipterans closely allied to the Calliphoridae but commonly called flesh-flies. Many are larviparous

Sarcoptidae a family of acarine arthropods distinguished by the minute globular body lacking a suture between the propodosoma and hysterosoma and with very short legs that may, or may not, have caruncles or claws. They burrow in the skin of mammals which gives rise to their popular name of itch mites

Sarcoptiformes a division of acarine Arachnida which lack stigmata. This group includes the cheese mites, itch mites, feather mites and beetle mites

Sarcoramphidae = Cathartidae

sarcosine methyl-glycine. A decomposition product

of the alkaline hydrolysis of creatine sometimes, for this reason, classified among the amino acids

Sarcosporidia a subclass of sporozoan Protozoa distinguished by possessing large capsules filled with crescentic spores that lack polar caps

sardine any of numerous small clupeid fishes. The young of many clupeids are also caught and marketed as sardines. It is very doubtful whether there is any difference except in size between the so-called "European sardine" ("*Clupea sardinia*") and the *Eur.* pilchard (*q.v.*)

Pacific sardine *Sardinops sagax* (= pilchard *U.S.*)

red-ear sardine *Harengula humeralis*

scaled sardine *Harengula pensacolae*

Spanish sardine *Sardinella anchovia*

Sargentodoxaceae a family of ranale dicotyledonous angiosperms erected to contain the genus Sargentodoxa, more usually placed in the Lardizabalaceae

sargo the pomadasyid fish *Anistremus davidsoni*

-sarment- *comb. form* meaning "twig"

sarment a slender long stolon like that of a strawberry

sarmentitious pertaining to twigs

saro the Amazonian lutrine mammal *Pteroneura brasiliensis*

Sarraceniaceae that family of dicotyledons which contains the pitcher plants. The unmistakable shape of this insectivorous plant is sufficient identification

Sarraceniales that order of dicotyledonous angiosperms which contains the pitcher plants and sundews both of which are insectivorous. The presence of the viscid sensitive hairs necessary to this habit distinguishes the Droseraceae from the Sarraceniaceae

sarsaparilla a flavoring agent extracted from various species of liliaceous climbing vines of the genus Smilax. The term is also applied to other plants having more or less the same flavor

bristly sarsaparilla the araliaceous herb *Aralia hispida*

wild sarsaparilla *A. nudicaulis*. In *U.S.* the term is also applied to the liliaceous herb *Smilax glauca*

sassaby a *S.E. Afr.* alcelaphine antelopid *Damaliscus lunatus* (*cf.* korrigum, topi, blesbok and bontebok)

sassafras the lauraceous tree *Sassafras albidum*

satellite 1 (*see also* satellite 2, 3) a smaller body adnexed to a larger one

perineural satellite oligondendroglia closely pressed to the cell body

perivascular satellite oligodendroglia surrounding capillaries in the brain

satellite 2 (*see also* satellite 1, 3) *Brit.* the noctuid lepidopteran insect *Eupsilia satellitia*

satellite 3 (*see also* satellite 1, 2) = trabant

-sathr- an occasional variant of -saphr- (*q.v.*) and sometimes applied to humus in distinction from other forms of decaying matter

-sati- *comb. form* meaning "to plant"

[satin] a mutant (gene symbol *sa*) found in linkage group XVII of the mouse. The name describes the appearance of the pelt of a phenotype

Saturniidae a family of saturnioid lepidopteran insects. The largest known Lepidoptera belong to this family which are often called giant silkworm moths. Many tropical saturniids have clear areas on the wing

Saturnioidea a superfamily of lepidopteran insects containing for the most part very large moths such as the cecropia, luna and the like

satyr *U.S.* popular name of satyrid lepidopteran insects, though also applied to the noctuid *Ufeus*

satyricus. I.P., without qualification, usually *Aulocera swaha*

banded satyr *I.P.* any of numerous satyrid lepidopteran insects of the genus Aulocera

wood satyr *U.S.* any of numerous species of satyrid lepidopteran of the genus Euptychia

Satyridae a family of nymphaloid lepidopteran insects containing, *inter alia*, the forms known as satyrs and wood nymphs in the *U.S.* and meadow browns in *Brit.* Almost all have one or more eye-like spots on the front wings

sauce-alone *Brit.* the cruciferous herb *Alliaria officinalis* (= garlic-mustard)

sauger the percid fish *Stizostedion canadense*

Saunder's *epon. adj.* from John Warren Saunders, Jr. (1919–)

-saur- *comb. form* meaning "lizard" but widely used in the sense of "reptile"

saurel the carangid fish *Trachurus trachurus* (= horse-mackerel)

saurian pertaining to lizards

Sauropsidia a general term, not usually accepted as a taxon, for birds and reptiles taken together

Saururaceae that family of piperalous plants which contains the lizard's tails. This small family is distinguished from its close relative the peppers, by having several carpels in each flower

saury common name of scomberosocid fish

savannah a type of grassland habitat found in the tropics in which forbes usually predominate over grasses and on which there are scattered clumps of shrubbery and small trees

savin the pinaceous tree *Juniperus virginiana* and occasionally other species of the same genus

savory any of several species of labiateous herb of the genus Satureja

summery savory *S. hortensis*

winter savory *S. montana*

saw *see* [sawtooth]

[sawtooth] a mutant (gene symbol *saw*) mapped at 0+ on the X chromosome of *Drosophila melanogaster*. The phenotypic expression is serrated wing hairs

sawwhet *U.S.* the strigid owl *Aegolius acadicus* (= saw-whet owl)

sawyer *U.S.* any of many species of cerambycid coleopteran insect of the genus Monochamus

balsam-fir sawyer *U.S. M. marmorator*

Oregon fir sawyer *U.S. M. oregonensis*

northeastern sawyer *U.S. M. notatus*

southern pine sawyer *U.S. M. titillator*

spotted pine sawyer *U.S. M. maculosus*

white-spotted sawyer *U.S. M. scutellatus*

saxatile = saxicolous

Saxicavidae a family of pelecypod mollusks commonly called rock-borers by reason of their habit of so acting. The valves are elongate and unequal

Saxifragaceae that family of rosalous dicotyledons which contains, *inter alia*, the saxifrages and mock oranges. They are distinguished from the closely allied Crassulaceae by the nonfleshy leaves and from the Rosaceae by the small number of carpels and stamens

saxifrage any of numerous species of the saxifragaceous genus Saxifraga and some closely allied genera

burnet saxifrage *Brit.* any of several species of umbelliferous herbs of the genus Pimpinella. *U.S.* without qualification *Pimpinella saxifraga*

golden saxifrage *Brit.* any of several species of the

saxifragaceous genus Chrysosplenium, in *U.S.* usually *C. americana*

meadow saxifrage any of several species of umbelliferous herbs of the genus Seseli

pepper saxifrage *Brit.* the umbelliferous herb *Silaus pratensis*

purple saxifrage the saxifragaceous herb *Saxifraga oppositifolia*

swamp saxifrage *S. pennsylvanica*

Saxifraginae a suborder of Rosales containing all the families not placed in Rosineae (*q.v.*)

saxifragous dwelling in cracks in rocks

saxon *Brit.* the noctuid lepidopteran insect *Lithomoea rectilinea*

saysie *S.Afr.* any of several species of fringillid bird of the genus Serinus (*cf.* seed eater)

sb *see* [stubby]

Sb *see* [Stubble]

sbr *see* [small bristle]

sc *see* [scute] or [scumbo]

Sc *see* [Scotched eye]

sca *see* [scabrous]

scabious *Brit.* anglicized form of the dipsaceous genus Scabiosa. (= *U.S.* mourning bride)

field scabious *Brit.* any of several species of dipsaceous herb of the genus Knautia

premorse scabious *Brit. Scabiosa succisa*

sheep scabious = shepherd's scabious

sheep's scabious *Brit.* the campanulaceous herb *Jasione montana*

shepherd's scabious the campanulaceous herb *Jasione perennis*

sweet scabious *S. atropurpurea* and also the compositaceous herb *Erigeron annuus*

scabrate roughened

scabrous feeling rough

[scabrous] a mutant (gene symbol *sca*) mapped at 66.7 on chromosome II of *Drosophila melanogaster*. The phenotypic expression is rough eyes

scad *U.S.* any of several carangid fish mostly of the genus Decapterus

-scalar- *comb. form* meaning "ladder"

scalare aquarists' name for the freshwater angelfish

scale 1 (*see also* scale 2–6) a dermal (fish) or epidermal (Sauropsidia) protective platelet

cosmoid scale a term used to refer to the "decorative" scale of crossopterygian fishes

ctenoid scale a teleost scale one side of which has finger-like or comb-like processes

cycloid scale a large smooth edged teleost scale showing concentric growth rings

dermal scale any scale, such as those of chondrichthine and teleost fishes, derived from dermal structure

ganoid scale scales, found in many actinopterygian fishes, in which the basal part is bony and the surface covered with a peculiar enamel called ganoin

placoid scale a scale found in elasmobranch fishes having a plate-like bony foot from which rises, above the surface of the skin, a tooth-like surface plate

scale 2 (*see also* scale 1, 3–6) a modified, flattened, seta found on many arthropods, particularly insects

anal scale the lateral process from the ovipositor in some hymenopteran insects

scale 3 (*see also* scale 1, 2, 4–6) a modified, flattened, trichome (= peltate plant hair)

scale 4 (*see also* scale 1–3, 5, 6) hardened, protective, leaves on the surface of tree buds, cones and catkins

bract scale a cone scale supporting or enveloping a seed bearing scale

bud scale the outer covering of a bud

bulb scale one of the modified leaves forming a bulb

intervaginal scale a scale occurring between the leaves of aquatic monocotyledons

scale 5 (*see also* scale 1–4, 6) scaly plates of other types

tentacle scale a calcareous plate, projecting over each podium in the arm of ophiuroid echinoderms

scale 6 (*see also* scale 1–5) as an abbreviation of coccid homopterans of several sub-families. Further names are given under scale insect (*q.v.*)

azalea bark scale the eriococcine *Eriococcus azaleae*

barnacle scale the lecaniine *Ceroplastes cirripediformis*

beech scale the dactylopiine *Cryptococcus fagi*

black scale the lecaniine *Saissettia oleae*

spruce-bud scale the diaspidine *Physokermes piceae*

cactus scale the diaspidine *Diaspis echinocacti*

calico scale the lecaniine *Lecanium cerasorum*

camellia scale the diaspidine *Lepidosaphes camelliae*

camphor scale the diaspidine *Pseudaonidia duplex*

chaff scale the diaspidine *Parlatoria pergandii*

citricola scale the coccine *Coccus pseudomagnoliarum*

coconut scale *U.S.* the diaspidine *Aspidiotus destructor*

cottony-cushion scale any of several monophlebines but particularly (*U.S.*) *Icerya purchasi*

red date scale the dactylopiine *Phoenicococcus marlatti*

dogwood scale the diaspine *Chionaspis corni*

European elm scale the dactylopiine *Gossyparia spuria*

euonymus scale the diaspidine *Unaspis euonymi*

greenhouse ensign scale *U.S.* the orthexiine *Orthezia insignis*

fern scale the diaspidine *Pinnaspis aspidistrae*

fig scale the diaspidine *Lepidosapkes ficus*

Fletcher scale the lecaniine *Lecanium fletcheri*

Forbes scale the diaspidine *Aspidiotus forbesi*

European fruit scale the diaspidine *Aspidiotus ostreaeformis*

globose scale the lecaniine *Lecanium prunastri*

gloomy scale the diaspidine *Chrysomphalus tenebricosus*

Glover's scale the diaspidine *Lepidosaphes gloverii*

grape scale the diaspidine *Aspidiotus uvae*

Rhodes grass scale the eriococcine *Antonina graminis*

greedy scale the diaspidine *Aspidiotus camelliae*

green scale the coccine *Coccus viridis*

Hall's scale the diaspidine *Nilotaspis halli*

hemlock scale the diaspidine *Aspidiotus ithacae*

hemispherical scale *U.S.* the lecaniine *Saissetia hemisphaerica*

holly scale the diaspidine *Aspidiotus britannicus*

Howard scale the diaspidine *Aspidiotus howardi*

juniper scale the diaspidine *Diaspis carueli*

blackpine leaf scale the diaspidine *Aspidiotus californicus*

magnolia scale the lecaniine coccid homopteran *Neolecanium cornuparvum*

cottony maple scale the lecaniine *Pulvinaria amygdali*

mining scale the diaspidine *Hoevardia biclavis*

pine needle scale the diaspidine *Phenacaspis pinifoliae*

golden oak scale the lecaniine *Asterolecanium variolosum*

obscure scale the diaspidine *Chrysomphalus obscurus*

Chinese obscure scale *Parlatoria chinensis*

oleander scale the diaspidine *Aspidiotus hederae*

olive scale the diaspidine *Parlatoria oleae*

oysterskill scale the diaspidine *Lepidosaphes ulmi*

peach scale the lecaniine *Lecanium oleae*

cottony peach scale *Pulvinaria amygdali*

European peach scale *Lecanium persicae*

white peach scale *U.S.* the diaspidine *Pseudo-daulacaspis pentagona*

Italian pear scale the diaspidine *Epidiaspis piricola*

pineapple scale the diaspidine *Diaspis bromeliae*

purple scale the diaspidine *Lepidosaphes beckii*

Putnam scale the diaspidine *Aspidiotus ancylus*

pyriform scale the lecaniine *Protopulvinaria pyriformis*

California red scale the diaspine *Aonidiella aurantii*

Florida red scale *Chrysomphalus aonidum*

rose scale *U.S.* the diaspidine *Aulacaspis rosae*

San Jose scale the diaspidine *Aspidiotus perniciosus*

scurfy scale the diaspidine *Chionaspis furfura*

elm scurfy scale *C. americana*

green shield scale *U.S.* the lecaniine *Pulvinaria psidii*

brown soft scale the coccine *Coccus hesperidum*

terrapin scale the lecaniine homopteran *Lecanium nigrofasciatum*

black thread scale the diaspidine *Ischnaspis longirostris*

pine tortoise scale the lecaniine *Toumeyella numismaticum*

walnut scale the diaspidine *Aspidiotus juglansregiae*

Chinese wax scale the lecaniine *Ericerus pela*

Florida wax scale *U.S. Ceroplastes floridensis*

Indian wax scale *C. ceriferus*

red wax scale *C. ruben*

yellow scale the diaspidine *Aonidiella citrina*

scalid one of several circlets of spines round the head of a kinorhynch

trichoscalid those posterior scalids which are covered in bristles

scallop 1 (*see also* scallop 2) popular name of pectinid pelecypods

giant scallop *U.S. P. irradians*

scallop 2 (*see also* scallop 1) *Brit.* either of two geometrid lepidopteran insects, *Ania emarginata* or *Eucosmis undulata*

scalloped crenate

[scalloped] a mutant (gene symbol *sd*) mapped at 51.5 on the X chromosome of *Drosophila melanogaster*. The name refers to the condition of the wing margins in the phenotype

-scalpr- *comb. form* meaning "chisel"

scamp *U.S.* the serranid fish *Mycteroperca phenax*

-scand- *comb. form* meaning "climb"

scandent climbing

scansorial adapted to climbing or having the habit of climbing

-scap- *comb. form* meaning "stem"

scape 1 (*see also* scape 2) a stem rising from the ground which bears a flower at its tip, and which is devoid of leaves. The term is also sometimes used as synonymous with stipe

scape 2 (*see also* scape 1) in insects, the long basal joint of a geniculate antenna

-scaph- *comb. form* meaning a "boat" ("skiff") or boat-shaped container

Scaphandridae a family of gastropod mollusks practically lacking a cone and with the greatly expanded body volute wrapped like a scroll round the central portion

Scaphidiidae a small group of coleopteran insects commonly called shining fungus beetles

scaphidium the sporangium of some algae

scaphium 1 (*see also* scaphium 2) the keel of a leguminous flower

scaphium 2 (*see also* scaphium 1) a ventral rod, formed from the fusion of the gnathos which lies immediately below the uncus in the genitalia of the male Lepidoptera

scaphoid boat shaped

Scaphopoda a class of prorhipidoglossomorphic Mollusca commonly called the elephant-tooth shells. They are distinguished by the bilaterally symmetrical tusk-like shell from the anterior of which projects a cylindrical foot and numerous tentacles

scapula (*see also* bone) in insects, the term is applied to numerous structures arising from the latero-dorsal aspect of the thorax

scapulet a mouth-bearing projection from the brachial arms of some Schyphozoa

scapulus a region between the capitulum and scapus in some Zoantharia

scapus the lower, thick-walled, portion of the "body" of an anthozoan (*cf.* capitulum)

leaf scar that portion of the abscission layer which remains attached to the branch

Scaridae a family of marine acanthopterygian fishes commonly called parrot fishes by reason of their sharp chisel-like beaks

Scarabaeidae an enormous family of coleopteran insects containing the scarabs, june beetles, chafers, tumblebugs and the like. They are a varied group, but all are heavily bodied, oval, convex beetles with lamellate antennae

Scarabaeinae a subfamily of scarabaeid coleopterans containing those that are coprophagous, as distinct from those that are phytophagous. This habit caused them, at one time, to be placed in the family Copridae

Scarabaeoidea a superfamily of polyphagous coleopteran insects mostly distinguished by the possession of lamellicorn antennae

scarf *Brit. dial.* = cormorant

scarious like a thin, dry membrane

scarlet *S. Afr.* any of several species of lycaenid lepidopteran insect of the genus *Axiocerses*

[scarlet] a mutant (gene symbol *st*) mapped at 44.0 on chromosome III of *Drosophila melanogaster*. The name refers to the eye color of the phenotype

scat popular name of scatophagid fish

scat feces

Scatomyzidae = Scopeumatidae

Scatophagidae a small family of Indo-Pacific brightly colored acanthopterygian fishes popular with aquarists under the name of scat

Scatophagidae (insecta) = Scopeumatidae

Scatopsidae a small family of minute nematoceran dipteran insects sometimes called minute black scavenger flies

scaup any of several aythyine ducks of the genus *Aythya*. *Brit.*, without qualification, *A. marila*. *N.Z.* without qualification *A. novaeseelandiae*

greater scaup *A. marila*

lesser scaup *A. affinis*

scaurie *N. Brit.* immature larine birds, particularly fledging *Larus argentatus* (= herring gull)

scavenger the lethrinid fish *Lethrinus nebulosus*

Scenedesmaceae that family of chlorococcale Chlorophyceae which reproduce by autospores which themselves become opposed to one another and form a coenobium

Scenopinidae a family of orthoraphous brachyceran dipteran insects resembling small black houseflies but differing from them in having a cylindrical rather elongate abdomen. Sometimes called window-flies from the fact that the commonest species (*Scenopinus fenestralis*) was observed by Linnaeus in some quantity on a window pane

scepter a monactine derivative of a hexactinellid sponge spicule with short spines along one end

Schisandraceae a small family of ranale dicotyledonous angiosperms distinguished from the closely allied Illiciaceae by its climbing habit and unisexual flowers

-schist- *comb. form* meaning "split" and by extension, an easily split stone (slate) (*cf* -schizo-)

anaschist said of a zygotene which separates longitudinally

aposchist a cell which metamorphoses into a gamete without further division

bradyschist a cell which shows several nuclear divisions before cytoplasmic divisions appear

euthyschist said of cell masses in which each nuclear division is accompanied by a cytoplasmic division in cases such as pollen formation, where this is not universal

hemischist the condition of a brood cell which divides into a syncytium

isochist one of a brood of identical cells developed from one parent cell

anisochist a gamete which differs from other gametes of the same species and the same sex

schistaceous slate colored

dia**schistic** dividing transversely

-schisto- *see* -schist-

Schistogamae = Characeae

Schistotsomatidae that family of digenetic trematodes which are blood flukes of birds and mammals and which are distinguished by sexual dimorphism

Schizaeaceae that family of eufilicale filicine pteridophyta containing the forms commonly called the grassy ferns (curly grasses) and the climbing ferns. They are distinguished by the large sporangia borne on the naked leaves

heteroschizis the division of a nucleus into many parts

-schizo- *comb. form* meaning "split" (*cf.* -schist-)

Schizogoniales that order of chlorophycean Chlorophyta which are distinguished by their uninucleate cells each containing a single stellate chloroplast

Schizomeridaceae that family of ulvale Chlorophyceae containing those forms in which the thallus is a solid cylinder several cells in diameter

schizont an organism derived by splitting or which will give rise to other organisms by splitting

Schizophora an assemblage of those families of cyclorrhaphous brachyceran dipteran insects which possess a frontal suture

Schizophyceae = Cyanophyceae

Schneiderian *epon. adj.* from Conrad Viktor Schneider (1614–1680)

schelly *Brit.* the salmonid fish *Coregonus stigmaticus*

Scheuchzeriaceae that family of helobe monocotyledonous angiosperms which contains the forms commonly called arrow-grasses. They are rush-like marsh herbs with tuberous roots

Schilbeidae a small family of naked catfishes distinguished by a forked tail, a long anal fin, and a very small adipose fin. The huge trey-reach (*Pangasianodon gigas*) is a member of this family

Schindleriidae = Hemirhamphidae

Schizomycetes unicellular plants lacking chlorophyll and reproducing by fission. Some forms may occur in chains, in sheets held together by noncellular material, and occasionally as filaments. They are popularly referred to as "bacteria" though the term Bacterium properly pertains to a one-time genus, the species of which are now distributed among many genera and families

school a term for a group of many kinds of fishes, also improperly applied to porpoises and whales for which gam or pod is better

Schreger's *epon. adj.* from Christian Heinrich Theodor Schreger (1768–1883)

Schwann's *epon. adj.* from Theodor Schwann (1802–1884)

sciad a shade plant

Sciadopityoideae a subfamily of Taxodiaceae

Sciaenidae a family of acanthopterygian Osteichthyes distinguished by the complete separation of the two dorsal fins. They are commonly called croakers from the very loud noise that they can make with the aid of their air bladder. It is recorded that some large schools synchronize their sounds

Sciaridae a family of minute dark winged nematoceran dipteran insects closely allied to the Mycetophilidae both in structure and habit. They are sometimes called the dark-winged fungus-gnats and are a well known pest in commerical mushroom beds

Scincidae a family of lacertilian reptiles with pleurodont dentition, a scaly tongue and osteoderms. Most have short, stubby tails and are called skinks

Sciomyzidae a family of myodarian cycloraphous dipteran insects usually yellow or brownish with spotted wings. All commonly live in marshes and have aquatic larvae

scion that part of a plant which is inserted onto or into the stalk when grafting

Sciophilidae a family of nematoceran dipteran insects now usually included with the Mycetophilidae

scissile splitting or separating

Scitaminales an order of monocotyledonous angiosperms containing, *inter alia*, the bananas, the gingers, the cannas, and the arrowroots; characterized by having from one to five functional stamens in zygomorphic flowers

-sciur- *comb. form* meaning "squirrel"

Sciuridae a large family of sciuromorph rodents containing the squirrels. They are distinguished by their large eyes and ears and thick bushy tails

sciuroid squirrel-like, but in *bot.* particularly applying to a bushy plant like a squirrel tail

Sciuromorpha an assemblage of rodents containing the families Anomaluridae, Sciuridae, Castoridae and Haplodontidae

-scler- *comb. form* meaning "hard"

sclera the outer coat of the back of the eye

sclere = sclerite 1

sclereid a short sclerenchyma cell (*cf.* fiber)

astrosclereid a more or less star-shaped sclereid

brachysclereid an isodiametric sclereid (= stone cell)

macrosclereid a long sclereid with blunt ends

osteosclereid a sclereid with a bump at each end thus having the form of a stylized "bone"

mega**scleric** said of a foraminiferan in which the proloculus is large

microscleric said of a foraminiferan in which the proloculus is small

sclerite 1 (*see also* sclerite 2) a general term for sponge spicules

megasclerite a large spicule frequently uniting with others to form a definite skeleton

microsclerite a small sponge spicule that does not unite to form a definite skeleton

sclerite 2 (see also sclerite 1) rigid portions of the arthropodan exoskeleton, usually bounded by sutures

articular sclerite a thickened area between the base of an arthropod limb and an arthropod body, but not forming one of the segments of the appendage

Scleroparei an order of Osteichthyes frequently regarded as a suborder of Acanthopterygii containing the families Scorpaenidae, Triglidae, Peristediidae, Cottidae, Hexagrammidae, Agonidae, Litaridae, Cyclopteridae and several other small families. They are commonly referred to as the mailcheeked fishes because of the presence of a bony plate extending across the cheek from the eye to the gill cover. They are distinguished by their rugged appearance. Many are called scorpion fishes by reason of the venomous nature of the dorsal spines in some genera. Some are locally called rockfishes

sclerotic hardened

sclerotium a compact mass of dormant hyphae in a fungus corresponding more or less to gemmae in higher forms

sclerotize used specifically to describe the hardening of arthropod appendages

scobby N. Brit. = chaffinch

scobicular in irregularly sized polyhedral grains

-scobis- comb. form meaning "sawdust"

scobinate roughened as though by a file

-scol- comb. form confused from three Gr. roots and therefore variously meaning "a thorn," a "worm" or "crooked"

scolder any of several ducks or other birds. N. Brit. either the merganser or the oystercatcher. E.U.S. the aythyine Clangula hyemalis (= old squaw)

-scolec- see "-skolex-"

Scolecida a taxon erected by Hatschek to contain the Platyhelminthes, the Rotifera, the Nematoda, the Acanthocephali and the Nemertini

Scoleciformia an order of phanerocephalous polychaete annelids with greatly reduced or absent parapodia and with a prestomium that lacks sensory processes

scolex the anterior strobilating proglottid of a cestode platyhelminth worm, often furnished with hooks, suckers or both; or the same structure in a prestrobilating condition, in the cysticercus larva

metascolex a collar of thin, flexible leaf-like structures, arising from the posterior region of the scolex in some tapeworms

pseudoscolex a crest-like prominence, which arises on the first proglottid of some tapeworms after the scolex has been lost

scolidium = bothrium

Scoliidae a family of scolioid apocritan hymenopteran insects resembling the Mutiliidae in being hairy but differing from them in frequently having black and yellow bands on the abdomen

-scolio- see -skolio-

Scolopacidae a large family of charadriiform birds containing those forms commonly called sandpipers. They are distinguished by their long slender bill and relatively long neck

scolopale a peg-like structure in the scolophore

scolus a spiny projection from a caterpillar

Scolytidae a family of scolytoid coleopteran insects commonly called engraver, ambrosia, or timber beetles. They are small cylindrical beetles either engraving the underside of the bark of trees or (ambrosia beetles) burrowing in the hardwood

Scolytoidea a superfamily of polyphagous coleopteran insects containing the short-snouted wood- or bark-eating weevils

Scomberosocidae a family of synentognathous fishes strongly resembling the Belonidae but differing from them in the shortness of the jaws and in the series of small finlets lying behind the dorsal fin. The common name is saury

Scombridae a family of scombroid fishes typically having a heavy streamlined body with a narrow caudal peduncle and a large deeply forked tail. The family contains the tunas, mackerels and their allies

Scombroidea a subfamily of acanthopterygian fishes distinguished from the Percoidea by the deeply forked tail. The order contains the families Scombridae, Istiophoridae, Xiphiidae and Luvaridae

[scooped] a mutant (gene symbol scp) mapped at 19.3 on the X chromosome of Drosophila melanogaster. The name derives from the upturned scoop-shaped wings of the phenotype

scooper Brit. dial. the recurvirostrid bird Recurvirostra avosetta (= avocet Brit.)

-scop- comb. form meaning "watchman" and thus, by extension, "one who examines things" or to the actual "process of examination"

scopa a comb-like structure commonly called a pollen-comb on the tibia of the hind leg of a bee; the scopa of one leg selects pollen which is transferred to the corbiculum (q.v.) of the other leg (cf. scopula)

Scopelarchide a family of bathypelagic iniomous fishes distinguished by the hooked teeth on the tongue and telescopic eyes commonly called pearl-eyes

Scopeumatidae as formerly considered a family of small yellowish or brownish hairy myodarian cycloraphous dipterans most of which breed in cow dung. They are commonly called dung-flies which sometimes causes confusion with the Sphaeroceridae. The family is now usually included with the Scatophaginae, a subfamily of the Anthomyiidae

Scopeumidae = Scopeumatidae

-scopic pertaining to the appearance

endoscopic the condition of a plant embryo when the apical pole is turned toward the base of the archegonium

exoscopic the condition of a plant embryo, the apical pole of which is turned toward the neck of the archegonium

macroscopic visible to the naked eye

microscopic invisible to the naked eye, but visible under an optical microscope

ultra microscopic invisible under an optical microscope

phytoscopic pertaining to an organism, particularly an insect larva, which has the same color as a leaf in virtue of its own pigment and not because of ingested chlorophyll

Scopidae a monotypic family of ciconiiform birds which contains the hammerhead. It is distinguished from other families in the order by the greatly laterally compressed bill and elongate head

scops properly owls of the genus Otus but often loosely used for any medium-sized owl (cf. scops owl)

scopula literally, a "small broom" and used for any small brush-like structure particularly on the legs of arthropods, for compound cilia in some Protozoa, particularly for the adhesive disc on the posterior end of some peritrichous ciliates and for a monactine derivative of a hexactine sponge spicule with branched ends. The term is also used as a synonym for scopa (*q.v.*)

scopulus a "cliff," "rock" or "crag" frequently confused with scopula (*q.v.*)

scorey = scaurie

Scorpaenidae a small family of scleropareid fish commonly called scorpion-fishes or rock fishes. Many have dangerously venomous spines.

scorpion (*see also* fish, weed) popular name of scorpionid arthropods

  book-scorpion *Brit.* the cheirid pseudoscorpion *Cheiridium museorum*

  house-scorpion *U.S.* the pseudoscorpion *Chelifer cancroides*

  pseudoscorpion popular name of members of the order Pseudoscorpionida

  sea scorpion *Brit.* any of several species of cottid fish of the genus Cottus

  water scorpion popular name of nepid hemipterous insects

  whip-scorpion common name of pedipalp arthropods of the family Thelyphonidae

  wind-scorpion popular name of members of the group Solifuga

Scorpionida an order of arachnid Arthropoda commonly called scorpions. They are clearly distinguished by the modification of the pedipalps into chelate appendages, the presence of a comb-like appendage on the second mesosomatic segment and the segmented tail terminating in a sting

scorzonera the compositaceous plant *Scorzonera hispanica* and its edible root (= black salsify)

[**scotched eye**] a mutant (gene symbol *Sc*) mapped at 4.0± on the X chromosome of *Drosophila melanogaster*. The phenotypic expression is a disarrangement of the ommatidia in the eye

scoter any of several species of aythyine ducks of the genus Melanitta and Oidemia

  American scoter = common scoter

  black scoter = common scoter

  common scoter *O. nigra*

  surf scoter *M. perspicillata*

  velvet scoter *M. fusca*

  white-winged scoter *M. deglandi*

-scoto- *see* -skoto-

scp *see* [scooped]

scr a bacterial mutant showing a utilization of sucrose

scraber any of several alcid birds but particularly *Plautus alle* (= dovekie)

leaf scraper any of several species of furnariid bird of the genus Sclerurus

  scaly-throated leaf scraper *S. guatamalensis*

  tawny-throated leaf scraper *S. Mexicanus*

screamer any of several species of anhimid birds

  crested screamer *Chauna chavaria*

  horned screamer *Anhima cornuta*

  ringed screamer *C. torquata*

scribbler *U.S.* geometrid lepidopteran insect *Cladora atroliturata*

scrobe a groove on the surface of an insect into which an appendage can be laid and thus concealed; the best known example is a groove on the side of a weevil snout into which the antenna fits

scrobiculate minutely pitted

scrod a young gadoid fish, particularly the cod (*cf.* dorse)

Scrophulariaceae an enormous family of tubuliflorous dicotyledons containing, *inter alia*, the snapdragons, toadflaxes, and foxgloves. The peculiar irregular

corolla and the reduced number of stamens are distinctive of this easily recognizable family

-scrot- *comb. form* meaning "pouch"

scrotum 1 (*see also* scrotum 2) the sac in which the descended testes lie in those mammals in which they leave the body cavity

scrotum 2 (*see also* scrotum 1) the vulva of some fungi

scrub a habitat consisting principally of low growing bushes and stunted trees

  arctic scrub a low-growing scrub consisting principally of birches and willows

  mulga scrub a thicket of spiny acacias

sculpin 1 (*see also* sculpin 2) any of very many species of fishes of the family Cottidae

  grunt sculpin common name of *Rhamphocottus richardsoni*

  mottled sculpin *Cottus bairdi*

sculpin 2 (*see also* sculpin 1) an extremely venomous scorpaenid fish *Scorpaena guttata*

scum a surface film

  pond scum a common name for floating algae

[**scumbo**] a mutant (gene symbol *sc*) found in linkage group III of *Neurospora crassa*. The phenotypic colony is flat and irregularly concentric

scup *U.S.* the sparid fish *Stenotomus chrysops*

[**scurfy**] a mutant (gene symbol *sf*) found in the sex chromosome (linkage group XX) of the mouse. The genotype is scaly and dies within the first month

-scut- *comb. form* meaning an "elongate quadrangular shield" or "buckler" (*cf.* -clyp- and -pelt-)

scutate shield shaped

scute 1 (*see also* scute 2) a scale or shield, specifically the large ventral scales on the lower surface of a snake, a tergal plate of a chilopod arthropod, the chitinous plates on many larval insects and the scale-like notopodial cirri on some polychaete annelids

  corneoscute epidermal scales of the type that cover reptiles

scute 2 (*see also* scute 1) *Brit. obs.* = coot

[**scute**] a mutant (gene symbol *sc*) mapped at 0+ on the X chromosome of *Drosophila melanogaster*. The phenotypic expression is a lack of scutellar bristles

-scutel- *comb. form* meaning a "flat round dish with a raised edge"

scutellate like a round dish with a raised edge (*cf.* scutelliform)

Scutelleridae a family of hemipteran insects strongly resembling the Pentatomidae but differing from them in the greatly enlarged scutellum. Commonly called shield-backed bugs

Scutellidae a family of exocycloid echinoids distinguished by their extremely flattened shape. The sand-dollar of the *U.S.* coast is typical

scutellum 1 (*see also* scutellum 2, 3) the dorsal portion of the thoracic skeleton of an insect

  laminiplantar scutellum one that is fused on the antero-posterior aspect

  mesoscutellum the scutellum of the mesothorax

  metascutellum the scutellum of the metathorax

  postscutellum in various insects, any thoracic sclerite immediately behind the scutellum

scutellum 2 (*see also* scutellum 1, 3) a scale on the leg of a bird

scutellum 3 (*see also* scutellum 1, 2) the cotyledon, or sometimes a secondary cotyledon, in some grasses and also the cap over the growing tip of some other monocotyledons

scuttle the rapid movement of a many-legged animal

scutum (*see also* scutum 1–4) literally an oblong-shaped shield. The plural is scuti, *not* scuta as usually given (*cf.* scutellum)

scutum 1 (*see also* scutum, scutum 2-4) the dorso-central sclerite of an insect notum

mesoscutum the scutum of the mesothorax of insects which is frequently concealed under the scutum of the prothorax

metascutum the scutum of the insect metathorax

prescutum the anterior, dorsal surface of the second and third thoracic segments in insects

postscutum = postscutellum

scutum 2 (*see also* scutum, scutum 1, 3, 4) the propodosomal plate in several groups of acarines

alloscutum a plate lying immediately behind the scutum in some acarines

conscutum a plate produced by the fusion of the scutum and alloscutum

scutum 3 (*see also* scutum 1, 2, 4) an irregularly shaped shield or spine projecting over the frontal membrane of ectoprocts

scutum 4 (*see also* scutum, scutum 1-3) the rectangular basal plates of a cirriped crustacean exoskeleton

Scyllaridea a superfamily of palinuran Crustacea distinguished by the absence of chelae on the first four pairs of legs. The spiny rock lobster is typical

Scylliorhinidae a family of small coastal sharks known as cat sharks or dogfish

Scydmaenidae a family of small beetles closely allied to the Staphylinidae but in which the elytra completely cover the abdomen

-scyph- *comb. form* meaning "cup" (*cf.* -calyc-, -cyath-, -pocul-)

scyphus a cup-like, peltate-edged podetium in lichens. Frequently mis-spelled scypha

ascyphous lacking scyphi

Scyphozoa that class of Coelenterata which contains the jellyfish. The hydroid generation is greatly reduced or even entirely wanting. The hydroid generation is the scyphistoma (*q.v.*)

Scythrididae = Scythridae

Scytopetalaceae a small family of malvale dicotyledonous angiosperms

Scytopetalinae a suborder of malvale dicotyledons containing the single family Scytopetalaceae

sd *see* [scalloped]

Sd *see* [Danforth's short tail] or [Dilution]

se *see* [sepia] or [short ear]

Se *see* [White side egg]

sea an area of saltwater too small to be called an ocean

seal 1 (*see also* seal 2, salamander, louse, elephant) any of numerous pinnipede carnivores, properly those of the family Phocidae though many sea-lions (Otariidae) are also called seals

Atlantic seal *Brit.* = grey seal *Brit.*

saddleback seal = harp seal *U.S.*

bearded seal *Erignathus barbatus*

crab-eating seal the lobodotine phocid *Lobodon carcinophaga*

common seal *Brit. Phoca vitulina*

crested seal the cystophornine phocid *Cystophorinae cristata* (= *Brit.* hooded seal)

elephant-seal = seal-elephant

fur-seal any otarid the fur of which is commercially valuable. The term is sometimes confined to *Otaria nigrescens* and in *N.Z.* to *Arctocephalus forsteri*

Alaska fur-seal *Callorhinus ursinus*

Guadalupe fur-seal *Arctocephalus philippi*

northern fur-seal *Otaria ursina*

gray seal *Halichoerus grypus*

hair-seal any otarid the hair of which is too coarse to be called fur

harbor seal *U.S. Phoca vitulina. Brit.* = common seal (*Brit.*)

harp-seal *Pagophilus groenlandicus*

hooded seal *Cystophora cristata*

Jamaica seal *Monachus tropicalis*

leopard-seal the lobondontine pinniped carnivore *Hydrurga leptonyx* (*cf.* sea-leopard)

monk-seal *Monachus albiventer*

bladdernosed seal = hooded seal

ribbon-seal *U.S. Histriophoca fasciata*

ringed seal *U.S. Phoca hispida*

seal 2 (*see also* seal 1) in compound names of plants

goldenseal the ranunculaceous herb *Hydrastis canadensis* (= yellow root)

Solomon's seal any of several liliaceous herbs of the genus Polygonatum

false Solomon's seal any of several liliaceous herbs of the genus Smilacina

two-leaved Solomon's seal *U.S.* the liliaceous herb *Maianthemum canadense* (= wild lily-of-the-valley)

searcher the bathymasterid fish *Bathymaster signatus*

sebaceous like tallow

sebum the oily material secreted by sebaceous glands

-secret- *comb. form* meaning "to part from" and, by *biol.* extension "to secrete"

secretagogue a material, not in itself endocrine, which initiates endocrine secretion

secretin a hormone secreted in the upper intestinal mucosa. It acts in controlling the volume rate of the secretion of pancreatic enzyme

secretion the production by an organism of a useful product (*cf.* excretion)

secretory that which secretes

neurosecretory said of a nerve cell which also produces a hormone

sectile cut up

section a thin slice (*for* section plane, *see* plane)

-secund- *comb. form* properly meaning "second" or "following" but widely used in *biol.* to mean "directed in one sense"

-secur- *comb. form* meaning "axe," particularly the form having a broad, almost semicircular, edge

sed *see* [sepiaoid]

Sedaceae = Crassulaceae

Sedentaria an assemblage, sometimes regarded as an order, of those polychaete annelids which live in tubes or burrows (*cf.* Errantia)

sedge 1 (*see also* sedge 2, wren) any of numerous glumifloral monocotyledonous plants of the family Cyperaceae particularly in *U.S.* those of the genus Carex. They are readily distinguished from the Gramineae, though some grasses are called sedges, by the fact that the stem, usually triangular in section, is not hollow

broom sedge the grass *Andropogon virginicus*

hopper sedge *C. subspathacea*

cypress-knee sedge *U.S. C. decomposita*

salt marsh sedge *C. salina*

umbrella sedge *U.S.* any sedge of the genus Cyperus

sedge 2 (*see also* sedge 1) term for a group of bitterns (*cf.* siege)

see-see *W.I.* any of several small birds. *Afr.* and *Asia* either of two phasianid birds of the genus Ammoperdix (*cf.* partridge)

white-beaked see-see *W.I.* the fringillid *Sporophila nigricollis* (= yellow-bellied seedeater) blue-black see-see *W.I.* the fringillid *Volatinia jacarina* (= blue-black grassquit)

red-throated see-see *W.I.* the fringillid *Loxigilla noctos* (= Lesser Antillean bullfinch)

white see-see *W.I.* the fringillid *Tiaris bicolor* (= black-faced grassquit)

yellow see-see *W.I.* the parulid *Coereba flaveola* (= banana-quit)

seed 1 (*see also* seed 2, 3, beetle, box, chalcid, coat, cracker, eater, finch, snipe) the resting stage of a higher plant, containing a partially developed sporophyte and a food reserve called endosperm, both enclosed in a seed coat or testa

albuminous seed one in which some part of the endosperm is retained until germination

exalbuminous seed one in which the endosperm is completely absorbed by the time the seed is ready for dispersion

perispermous seed one in which part of the nucellus is used for food storage

seed 2 (*see also* seed 1, 3) in compound names of plants with distinctive seeds

allseed *U.S.* the linaceous herb *Millegrana radiola. Brit.* any of several species of the caryophyllaceous genus Polycarpon

cornish bladderseed *Brit.* the umbelliferous herb *Physospermum cornubiense*

bugseed any of several species of chenopodiaceous herb of the genus Corispermum

chaffseed *U.S.* the scrophulariaceous herb *Schwalbea americana*

coleseed *Brit.* the cruciferous herb *Brassica napus* (= rape)

cutseed the menispermaceous herb *Calycocarpum lyoni*

dropseed the grass *Sporobolus cryptandrus.* The term is sometimes applied to allied species

northern dropseed *U.S.* the grass *Sporobolus heterolepis*

flaxseed *Brit.* = all-seed *U.S.*

water flaxseed *U.S.* the lemnaceous aquatic *Spirodela polyrhiza*

heartseed the sapindaceous vine *Cardiospermum halicacabum* = balloon vine

jumpseed *U.S.* any of several species of polygonaceous herb of the genus Tovara

lopseed the phrymaceous herb *Phryma leptostaehya*

marbleseed any of several species of boraginaceous herb of the genus Onosmodium (= false gromwell)

moonseed a few species of the menispermaceous genus Menispermum

Carolina moonseed the menispermaceous shrub *Cocculus carolinus*

snailseed *U.S.* the menispermaceous herb *Cocculus carolinus*

stickseed any of several species of boraginaceous herb of the genus Lappula

tickseed any of numerous species of compositaceous herbs with seeds resembling ticks, particularly those of the genera Bidens and Coreopsis

seed 3 (*see also* seed 1, 2) in compound names of other organisms

apple seed the eratid gastropod *Erato vitellina*

Kefir seed dried masses of microorganisms in a granular matrix of coagulated milk protein

pumpkin seed *U.S.* centrarchid fish *Lepomis gibbosus*

tobacco seed *W.I.* the fringillid bird *Tiaris bicolor*

seedling a young plant produced from a seed

Seessel's (often mispelled Seesel) *epon. adj* from Albert Seessel (1850–1910)

segetalile growing in grain fields

segge *Brit. obs.* any small bird but particularly the hedge sparrow

segment 1 (*see also* segment 2) in the sense of a discrete portion of an organism or appendage. In arthropods, the term has become hopelessly confused with "joint" which is properly the connection between segments, or segments of limbs and appendages. The term, for example, "five-jointed antenna" is both inaccurate and misleading

antennal segment that segment of the arthropod head from which the antennae arise, usually considered to be the second segment

intercalary segment one which is inserted between two regular segments, though the term is also applied to the premandibular segment of the head of insects

Latreille's segment that abdominal segment which is fused to the thorax in some Hymenoptera

subsegment a subdivided arthropod segment distinguished from a true segment by the lack of specific musculature

superlingual segment in insects, the fifth segment of the head

supernumary segment one which is intercalated between segments

segment 2 (*see also* segment 1) in the sense of a portion of a chromosome

differential segment that portion of a chromosome which has no corresponding part in its bivalent or any portion of a chromosome showing other than typical staining

dislocated segment a segment of a chromosome homologous with, but in a different position from, a similar segment in another chromosome

pairing-segment that portion which has a exact equivalent in the other sex chromosome

segmentation the division of an elongate organism into a number of similar parts, but, inaccurately, usually considered to be synonymous with metamerism. (for the division of a fertilized egg, sometimes called segmentation, *see* cleavage)

heteronomous segmentation the condition of an animal in which successive segments are markedly different

homonomous segmentation the condition in which successive segments are for all practical purposes identical

metameric segmentation (*see also* theory) the form of segmentation involving the replication of similar parts throughout the length of one organism

superficial segmentation a form of metameric segmentation which involves only the outer layers of an organism

segregation the maintaining of closely connected things separate from each other. In genetics, the separation of pairs of alleles at meiosis

autosegregation gene rearrangement in diplospores resulting from the failure of meiosis through the formation of a restitution nucleus

seiche a rhythmic rise and fall at the surface level of a lake following strong winds or unusual barometric changes or any rhythmic oscillation of water having a specific periodicity in any given lake basin

-seio- *comb. form* meaning "shape" (*cf.* -seism-) frequently transliterated -sio-

-seiro- *comb. form* meaning a "rope"

-seismo- *comb. form* meaning "vibration"

Seisonidea a class of the phylum Rotifera distinguished by the elongated form and extremely elongated neck region

Selachoidea that subclass of elasmobranch fishes which contains the sharks. Most may be superficially distinguished from the rays (Batoidea) by the fact that the body is not significantly flattened

Selaginallales an order of lepidophyte plants distinguished from the Lycopodiales by the presence of ligules on the leaves and from the Isoetales by the absence of any secondary thickening on the stem

Selaginellaceae that family of lycopodine Pteridophyta which contains the small club-mosses of the genus Selaginella. They are distinguished from the club-mosses (Lycopodiaceae) by the presence of ligules

natural **selection** the selection, either of individuals or breeding rates, by environmental conditions

    orthogenetic selection a term used before the discovery of mutations, to account for the continued production of characters not apparently specifically adapted to a given environment

-selen- *comb. form* meaning "moon"

Selenodontia = Ruminantia

Seleviniidae a monotypic family of myomorph rodents containing the Russian *Selevinia paradoxa*

[self-pruning] a mutant (gene symbol *sp*) mapped at 2 on chromosome 6 of the tomato. The phenotypic expression is a determinate stem

-sell- *comb. form* meaning "saddle"

sella turcica literally, "Turkish saddle." Usually applied to the shallow depression in the basisphenoid bone into which the pituitary gland fits, but also, in some decapod Crustacea, to a saddle-shaped portion of the abdominal apodeme

-sem- *comb. form* meaning "sign" or "signal"

Semanturidae = Manidiidae

Semastomeae an order of scyphozoan coelenterates distinguished by having the corners of the mouth prolonged into four long lobes

sematic pertaining to methods of signalling

    aposematic possessing warning coloration or giving warning signals

    pseudoposematic a protective or warning coloration borne by a defenseless species in imitation of one well able to defend itself

    allosematic having coloration mimicking that of another species supposed to be protected by its own coloration

    episematic pertaining to warning or recognition colors or patterns

    eposematic said of coloration, such as the white tuft at the base of a deer tail, which is supposed to be a recognition signal

aposeme a population, all the individuals of which, though taxonomically distinct, share the same aposematic coloration

semen seed or spermatic fluid

semet = anther

-semi- properly meaning half, but widely used in *biol.* in the sense of "partly"

[semilong] a mutant (gene symbol *sl*) in linkage group I of *Habrobracon juglandis*. The segments of both antenna and leg are lengthened in the phenotype

semine the genitalia of plants

Semiontiformes an order of osteichthyes more usually divided into the Ginglymodi and the Protospondyli

-semper- *comb. form* meaning "always"

Semper's *epon. adj.* from Karl Semper (1832–1893)

senary pertaining to six

Senecioneae a tribe of tubiflorous Compositae

senegali any of numerous species of plocoid birds

senescence aging

senna any of several leguminous trees and shrubs of the genus Cassia and their pods used in pharmacy, and other plants of similar taxonomic affinities or pharmaceutical properties

    bastard senna any of several species of the leguminous shrubs of the genus Colutea (= bladder senna)

    bladder-senna = bastard senna

    coffee-senna *Cassia occidentalis*

    prairie senna *C. fasciculata*

    scorpion-senna the leguminous shrub *Coronilla emerus*

    wild senna *C. marilandica*

sennet any of several species of sphyraenid fish of the genus Sphyraena (*cf.* barracuda)

senorita the labrid fish *Oxyjulis californica*

sense perception (*see also* -ceptor and -receptor)

    time-sense the intuitive perception of chronological periods

sensible sometimes used in old fashioned writing in its proper meaning of sensitive

sensilla a sense organ, or a collection of neurosensory cells, not always of known function, in invertebrates

    slit sensilla a phonoreceptor found in spiders

sensorium a hemispherical or flask-shaped cavity covered by a membrane on the surface of some insects. The use of the term usually denotes a suspected, but unidentified sensory function

sensory pertaining to the detection of stimuli

sensu in the sense of

    sensu lato in a broad sense, as when a species name is used in the sense of including subspecies

    sensu stricto in a restricted sense, in contrast to sensu lato

heterosepalody the change of one kind of sepal into another

sepalous pertaining to sepals

    alternisepalous applied to those flowers in which petals alternate with sepals

    anisosepalous with unequal sepals

    aposepalous having supernumerary sepals

    chorisepalous = polysepalous

    dialysepalous having separate sepals

    eleutherosepalous having many sepals free of each other

    gamosepalous having the sepals united

    monosepalous properly possessing a single sepal, but almost invariably used for gamosepalous

    synsepalous said of those flowers in which the sepals are fused

gill **separator** epipodite

sepia a genus of sepid debranchiate Mollusca which has given its name to a color derived from the ink glands

[sepia] a mutant (gene symbol *se*) mapped at 26.0 on chromosome III of *Drosophila melanogaster*. The phenotypic eye color is brownish red

sepiaceous sepia colored

[sepiaoid] a mutant (gene symbol *sed*) mapped at 64.5± on chromosome III of *Drosophila melanogaster*. The phenotypic eye color is chocolate

-sepes- *comb. form* meaning "a hedge"

Sepiidae a family of dibranchiate decapod cephalod Mollusca with an oval body and long narrow fins running the length of the body. Commonly called cuttlefish

Sepiolidae a family of dibranchiate cephalopod Mollusca distinguished from Sepiidae by having the fins only in the central region of the body

Sepiophora a division of decapod cephalod mollusks distinguished by the presence of a straight calcareous shell. Cuttlefish belong in this group

Sepsidae a family of small black myodarian cycloraphous dipteran insects distinguished by having an almost spherical head and an abdomen narrowed at the base. Commonly called black scavenger flies

septate having a septum or septa

aseptate lacking septa

eseptate lacking septa

latiseptate broadly partitioned

Septibranchia a small order of pelecypod Mollusca containing, *inter alia*, the shells commonly called dippers. They differ from all other pelecypod mollusks in lacking gills

Septobasidiales a small order of heterobasidiomycete Basidiomycetes parasitic on scale insects

septulate indistinctly, or spuriously, septate

septum a division or wall. In Coelenterata, specifically one of the radial partitions projecting into the coelenteron

myoseptum the interval between, or the collagen filling the interval between, successive myotomes

nasal septum a vertical plate of the ethmoid bone dividing the right and left nares

scleroseptum one of the ridges surrounding or separating corallites

transverse septum the posterior wall of the pericardial cavity in vertebrates

ser a microbial genetic marker indicating a requirement for serine

-ser- *comb. form* confused from four Latin roots and therefore variously meaning "watery," "to plant," "to fasten or interweave" or "late" in the sense of a late season crop

serA a bacterial genetic marker having a character affecting the activity of 3-phosphoglycerate dehydrogenase, mapped at 61.5 mins. for *Escherichia coli*

serB a bacterial genetic marker having a character affecting the activity of phosphoserine phosphatase, mapped at between 88.0 mins and 0.0 mins for *Escherichia coli*

ser-1, 2 *see* [serine-1], [-2]

seral pertaining to seres and therefore carrying the connotation "developmental"

seraphim *Brit.* either of two species of geometrid lepidopteran insect of the genus Lobophora. Without qualification *L. halterata* is usually meant

-sere- *see* -ser-

sere one of a chain of successional ecological stages leading to a climax

adsere a sere which is going to turn into another sere but not into a climax

angeosere a climax of angiosperms

aquatosere one which commences in a wet area and develops to an aquatic climax

clisere a condition intermediate between two seres

postclisere proceeding from a lower to a higher climax

preclisere one which proceeds from a higher to a lower climax

cosere a series of unit successions, usually of vegetation in the same place

consere = cosere

eosere a climax occurring in a specific era

gymnosere a sere in which gymnosperms predominate

hydrosere an association in a moist environment (= hydrarch)

lithosere a stage of vegetation inhabiting relatively bare rocks

mesosere an intermediate between two other stages

oxysere a hydrosere of acid land

prisere = primary sere

prosere a migratory community playing a temporary role in seral development

psammosere an adsere of loose sand

subsere either a secondary sere or a partially developed climax

thallosere = proteosere

xerosere a succession developing because of a scanty supply of water

sergeant any organism having the color, or behavior pattern, popularly attributed to noncommissioned officers. *I.P.* any of numerous nymphalid lepidopteran insects of the genus Pantoporia. Without qualification, usually *P. perius*

night sergeant the pomacentrid fish *Abudefduf taurus*

wildpine sergeant *W.I.* the icterid bird *Nesopsar nigerrimus*

Sergeant Baker *Austral.* the aulopid fish *Aulopus purpurissatus*

sergeant-major the pomacentrid fish *Abudefduf saxatilis*

serial disposed in rows or ranks. The term is synonymous, but not interchangeable with, seriate

biserial said of those arms of a crinoid echinoderm in which brachials alternate on each side of the arm

curviserial in curved ranks

retiserial in straight lines or ranks

uniserial said of those echinoderm arms in which brachials occur only on one side

seriate pertaining to rows (*cf.* serial)

multiseriate in many rows

uniseriate with one row

sericeous silky

holosericeous completely covered with silky down

seriema either of two cariamid birds (*cf.* cariama)

black-legged seriema *Chunga burmeisteri*

red-legged seriema *Cariama cristata*

floral series the successive whorls or spirals of members such as petals, sepals, etc., which make up a flower

serin any of several fringillid birds of the genus Serinus. Brit, without qualification, *S. serinus*

serine 2-amino-3-hydroxy-propanoic acid (= β-hydroxyalanine), $HOCH_2CH(NH_2)COOH$. A widely distributed amino acid not necessary for the nutrition of rats

[serine-1] a mutant (gene symbol *ser-1*) found in linkage group III of *Neurospora crassa*. The phenotypic expression is a requirement for serine but the phenotype can use glycine

[serine-2] a mutant (gene symbol *ser-2*) in linkage group V of *Neurospora crassa*. The phenotypic expression is a requirement for serine

seringa the euphorbiaceous tree *Hevea guianensis* (*cf.* syringa)

seringueira the euphorbiaceous tree *Hevea brasiliensis*

serotinal pertaining to the late summer and the early fall

serotine 1 (*see also* serotine 2) the vespertilionid cheiropteran *Eptesicus serotinus*

serotine 2 (*see also* serotine 1) late flowering

serotinous produced late in the season

serous watery, usually in distinction from mucous (*q.v.*)

serow any of several species of rupicaprine caprids of the genus Capricornis. The Japanese *C. crispus* is sometimes known as the Japanese goat-antelope (*cf.* goral)

Serpentariidae = Sagittariidae

Serpentes a suborder containing the snakes of the reptilian order Squamata

serpentine bent into abrupt zigzags, with sharper curves than is indicated by the term flexuose

Serphidae = Proctotrupidae

Serphoidea = Proctotrupoidea

Serpulidae a family of sedentary polychaete annelids that dwell in calcareous tubes

-serr- *comb. form* meaning "saw"

Serranidae a large family of percoid acanthopterygians, commonly called groupers and sea basses. Many have heavy coarse fins and massive underslung jaws giving an upwardly directed mouth

serrate toothed, particularly if the teeth, as in many leaves, point towards the apex

biserrate with serrated serrations

serried in ranks which are close together

Serropalpidae = Melandryidae

serrulate minutely serrate

insert said of those echinoderm terminal plates which touch the periproct

exsert said of those echinoderm terminal plates which do not touch the periproct

Sertoli's *epon. adj.* from Enrico Sertoli (1842–1910)

serum defibrinated blood plasma

antiserum a serum containing antibodies

fowlers' **service** *Brit.* the rosaceous shrub *Pyrus aucuparia*

sesame (*see also* grass) the pedaliaceous herb *Sesamum indicum* and its oily seeds.

Sesiidae = Aegeriidae

-sesqui- *prefix* meaning "one and one half"

sessile lacking a stalk

pseudosessile used of those petiolate hymenopteran insects in which the petiole is so short as to be perceived with great difficulty

-sessor- *comb. form* meaning "seated," originally in the sense of having a "seat" or "house" in a community

insessorial perching

-sesto- *comb. form* meaning "sieve"

seston the sum total of suspended matter in a given body of water

bioseston the living component of the seston

abioseston non-living matter suspended in water (= tripton)

-set- *comb. form* meaning "bristle" (*cf.* setul)

bone **set** the compositaceous herb *Eupatorium perfoliatum*

false bone set any of several species of compositaceous herb of the genus Kuhnia

offset a small plant developing alongside the parent from a lateral shoot

seta literally a bristle. Applied in *bot.* to any bristle-shaped body, including the stalk of a moss sporangium. Confined in *entomol.* to any monocellular cuticular outgrowth. (*cf.* chaeta). Sometimes used in the sense of filoplume (*q.v.*)

glandular seta a tubular seta which serves as the outlet of a hypodermal gland in insects (*cf.* glandular chaeta)

poison seta a hollow seta through which many insects secrete a repellant toxin

setaceous bristle like

setigerous bristly

setireme the swimming leg of an aquatic insect

setoceous = setaceous

Setomorphidae a family of lepidopteran insects now usually included in the Tineidae

setose very bristly

-setul- *comb. form* meaning "small bristle" (*cf.* -set-) but extended to any fine, thread-like structure

Sewall Wright *see* Wright's

sewellel the tailless fossorial aplodontid sciuromorph rodent *Aplodonta rufa* (= mountain beaver)

sewi sewi *W.I.* the vireonid bird *Vireo modestus* (= Jamaican vireo)

-sex- *comb. form* meaning "six"

sex a necessary attribute on a gamete producing organism. The term is definable only when anisogametes are produced in which case the parent of the larger gamete is usually referred to as having female sex and that of the smaller gamete as having male sex

heterogametic sex that sex which is characterized by the XY or XO arrangement of the sex chromosomes

homogametic sex that sex which is characterized by the XX arrangement of chromosomes

intersex an organism, not normally hermaphroditic, which shows characters intermediate between those of the two sexes

phenotypic intersex = freemartin

super sex an individual with an abnormal ratio of sex chromosomes

[**Sex**] *either* a group of pseudo alleles (gene symbol X) in linkage group I of *Habrobracon juglandis or* a mutant (gene symbol A/a) in linkage group I of *Neurospora crassa*. The phenotypic expression is in the mating type

[**sex-linked**] a mutant (gene symbol *os*) mapped at 0 in linkage group I of *Bombyx mori*. The phenotypic expression is a low translucency of the larva

sexual pertaining to sex

asexual lacking sex

parasexual said of the activities of those organisms, such as the fungi imperfecti, in which protoplasmic fusion, nuclear fusion and chromosome reduction occur hapazardly and unpredictably rather than in an orderly sequence

sf *see* [safranin], [scurfy] or [solanifolium]

sfo *see* [sulfonamide]

sge *see* [shifted genitals]

sh *see* [shaker] or [shallow]

sh-1, -2 *see* [shaker-1], [-2]

sh$_1$, $_2$ *see* [shrunken endosperm$_1$], [-$_2$]

Sh *see* [Shaggy] or [Shaker]

shad any of several species of clupeid fishes particularly of the genera Alosa and Dorosoma

Alabama shad *A. alabamae*

allis shad *Brit. A. alosa*

American shad *A. sapidissima*

threadfin shad *D. petenense*

yellowfin shad *Brevoortia smithi*

gizzard shad any of several species of the genus Dorosoma, usually, without qualification, *D. cepedianum*

hickory shad *A. mediocris*

Ohio shad *A. ohiensis*

shaddock the rutaceous tree *Citrus grandis* (= grapefruit)

angle **shade** *Brit.* any of several species of noctuid lepidopteran insect of the genera Euplexia and Phlogophora

night shade any of numerous shrubby plants many of the family Solanaceae

black nightshade *Solanum nigrum*

deadly nightshade *Atropa belladonna*

enchanter's nightshade any of several species of onagraceous herbs of the genus Circaea

woody nightshade *Brit. Solanum dulcamara*

shaft the central part of a feather, consisting technically of the basal calamus and distal rachis. Applied to other elongate structures

aftershaft the vestigial feather that rises behind the shaft of the main feather

antenna shaft those portions of an antenna which are neither pedicel nor scape

shag *N.Z.* any of many species of phalacrocoracid bird of the genus Phalacrocorax. *Brit.*, without qualification, *P. aristotelis*. In *U.S.* the term cormorant (*q.v.*) is preferred

black shag *N.Z. Phalacrocorax carbo* (= great cormorant)

little black shag *N.Z. P. sulcirostris*

rough-faced shag *N.Z.* = king shag

frilled shag *N.Z.* = white-throated shag

pink-footed shag *N.Z.* = Stewart Island shag

Auckland Island shag *N.Z. P. colensoi*

Bounty Island shag *N.Z. P. ranfurlyi*

Campbell Island Shag *N.Z. P. campbelli*

Chatham Island shag *N.Z. P. onslowii*

Macquarie Island shag *N.Z. P. purpurascens*

Pitt Island shag *N.Z.* = Chatham Island shag

Stewart Island shag *N.Z. P. chalconotus*

king shag *N.Z. Phalacrocorax carunculatus*

pide shag *N.Z. P. varius*

spotted shag *N.Z. P. punctatus*

white-throated shag *N.Z. P. melanoleucos*

[Shaggy] a mutant (gene symbol *Sh*) mapped at 0 on linkage group II of the rat. The hair and whiskers of the phenotype are curved

shagreen skin covered in small bony denticles, particularly when tanned, smoothed and polished for decorative surfaces

[shaker] a mutant (gene symbol *sh*) on the sex chromosome (linkage group I) of the domestic fowl. The name is descriptive of the phenotypic expression which is lethal

[shaker-1] a mutant (gene symbol *sh*-1) in linkage group I of the mouse. The phenotypic expression is deafness, circling motions and head waving

[shaker-2] a mutant (gene symbol *sh*-2) in linkage group VII of the mouse. The phenotypic expression is identical with [shaker-1]

[Shaker] a mutant (gene symbol *Sh*) mapped at 58.0 on the X chromosome of *Drosophila melanogaster*. The shaking under anesthesia is the same as that of [Shaker-downheld] but the wings do not droop

[Shaker-downheld] a mutant (gene symbol *Shw*) mapped at 53.3 on the X chromosome of *Drosophila melanogaster*. The droopy winged phenotype shakes under ether anesthetization

shallot the liliaceous plant *Allium ascalonicum*

[shallow] a mutant (gene symbol *sh*) on linkage group V of *Neurospora crassa*. The term is descriptive of the thin spreading layer of hyphae of the phenotypic colony

shama any of several species of asiatic turdid birds of the genus Copsychus, usually, without qualification, *C. malabaricus*

shamrock either the oxalidaceous herb *Oxalis acetosella* or the leguminous *Trifolium repens*. Neither the Irish, nor anyone else, has ever agreed which of these two similarly leaved plants is the true shamrock

Indian shamrock *Brit.* any of several species of liliaceous herb of the genus Trillium (= *U.S.* wakerobin)

wood shamrock *U.S.* the oxalidaceous herb *Oxalis montana* (= wood sorrel)

shank 1 (*see also* shank 2) in birds the tarsometatarsus setion of the leg. In other animals loosely applied to the tibial section

shank 2 (*see also* shank 1) in compound names of long-legged birds

greenshank the scolopacid bird *Tringa nebularia* (*cf.* yellow-legs)

redshank either of two scolopacid birds of the genus Totanus. *Brit. T. totanus. W.I.* the recurvirostrid *Himantopus mexicanus* (*cf.* stilt)

common redshank *T. totanus*

spotted redshank *T. erythropus*

yellowshanks = yellow legs

shanny *U.S.* the sticheid fish *Lumpenus maculatus Brit.*

shark 1 (*see also* shark 2, 3) any of numerous selachoid elasmobranch fishes

angel-shark any of several species of the family Squatinidae

basking shark the lamnoid *Cetorhinus maximus*

blue shark the carcharninid *Prionace glauca*

bramble shark the squalid *Echinorhinus brucus*

carpet shark = nurse shark

cat-shark popular name of scyliorhinids

cow-shark popular name of hexanchids

fox-shark *Brit.* = thresher shark *U.S.*

frill shark popular name of chlamydoselachid sharks

hammerhead shark any of numerous sharks of the family Sphyrnidae

horn-shark popular name of heterodontids

leopard-shark the carcharhinid *Triakis semifascita*. Sometimes locally used for other spotted sharks

mackerel-shark popular name of lamnids

nurse shark any of several species of orectolobids

salmon-shark the lamnid *Lamna ditropis*

sand-shark any of several species of the family Carchariidae

saw-shark any of several species of pristids (*cf.* saw fish)

spinous shark *Brit.* the squalid *Echinorhinus spinosus*

thresher shark *U.S.* the alopiid *Alopias vulpinus*

tiger shark the large carcharinid *Galeocerdo cuvieri*. The term has also been applied to *Carcharias taurus*

whale-shark the rhincodontid *Rhincodon typus*

white shark the lamnid *Carcharodon carcharias*

shark 2 (*see also* shark 1, 3) improperly applied to some teleost fishes

black shark aquarist's name for the cyprinid fish *Morulius chrysophekadion*

shark 3 (*see also* shark 1, 2) *Brit.* either of two species of noctuid lepidopteran insect of the genus Cucullia but usually, without qualification *C. umbratica*

saw sharpener *Brit.* the parid bird *Parus major* (= great titmouse)

[shaven] a mutant (gene symbol *sv*) mapped at 3.0 on chromosome IV of *Drosophila melanogaster*. The phenotypic expression is a reduction in the number of abdominal bristles

shears *Brit.* any of several species of noctuid lepidopteran insects

dingy shears *Dyschorista fissipuncta*

tawny shears *Dianthoecia carpophaga*

sheath (*see also* extension) anything which closely envelopes something else

bundle sheath cells surrounding the vascular bundles in a leaf

genital sheath a muscular sheath, enclosing the sperm ducts, cement ducts, and protonephridial canals in Acanthocephala

hydrome sheath a layer between hadrome and leptome

leaf sheath that part of the petiole which invests the stem

medullary sheath xylem vessels immediately surrounding, and sometimes projecting into, the pith

mestom sheath a bundle sheath of thick-walled cells lacking chloroplasts

plerome sheath = bundle sheath

external root sheath the outer walls of a hair follicle

internal root sheath a cellular tube around the lower portion of the hair separating it from the external root sheath

Schweiger-Seidel sheath a sheath consisting of densely packed reticular cells round the arterioles in the red pulp of the spleen

starch sheath a sheath of starch-containing cells occupying the same position as the endodermis on the outside of the pericycle

sheep 1 (*see also* sheep 2, bit, grass, louse, mite, moth, plant, scabious and sorrel) any of numerous ovid artiodactyl mammals of the genus Ovis, without qualification usually *O. aries*, the domestic sheep. Whether an animal is called a sheep or a goat is, with the exception of the domestic forms, dependent on size, habitat and local custom

barbary sheep aoudad

blue sheep = bharal

Cyprus sheep *O. ophion* (= cyprus mouflon)

Dall's sheep a geographic white race of the bighorn sheep sometimes distinguished as a separate species

domestic sheep *O. aries*

bighorn sheep *O. canadensis,* which is the common wild sheep of the Rocky Mountains

mountain sheep *O. nidicola*

Pamir sheep *O. poli*

white sheep *U.S.* = Dall's sheep

sheep 2 (*see also* sheep 1) in compound names of other organisms

Cape sheep = albatross

vegetable sheep *N.Z.* any of several species of compositaceous herbs of the genus Raoulia (= sheep-plant)

shelder either the shelduck or the oystercatcher

lingual shelf one on the inside of the dentary bone in some reptiles

sulphur shelf *U.S.* the polyporaceous basidomycete fungus *Lateiporus sulphureus*

shell 1 (*see also* shell 2–4 apple, flower) any hard covering over an organism or part of an organism, most particularly the calcareous covering of the Mollusca and the hardened plates of chelonian reptiles

tortoise shell (*see also* shell 3) the translucent, mottled brown, horny plates covering the carapace of the hawksbill and some other marine turtles

shell 2 (*see also* shell 1, 3, 4) in compound names of Mollusca, based on the appearance of the shell

apple shell any of several species of ampullariid of the genus Ampullaria

ark shell popular name of arcid pelecypods

auger shell popular name of terebrid gastropods (= screw shell) but particularly *Terebra dislocata*

awning shell the solemyied pelecypod *Solemya velum*

band shell any of several species of fasciolarid gastropods (*cf.* tulip shell)

cross-barred shell popular name of cancellariid gastropods

basket shell either the nassariid gastropod *Nassarius obsoleta* or the corbulid pelecypod *Aloides luteola*

coffee-bean shell the ellobiid gastropod *Melampus coffeus*

bird shell the carditid pelecypod *Cardita floridana*

blind shell the caecid gastropod *Caecum pulchellum*

boat shell any of several species of crepidulid gastropods of the genus Crepidula but particularly *C. fornicata* (*cf.* slipper limpet) and *C. convexa*

bonnet shell any of numerous species of cassidid gastropods of the genus Cypraecassis (*cf.* helmet shell, cameo shell)

bubble shell any of numerous gastropods with a reduced or lacking cone. The family Bullidae are often referred to as the **true bubbles,** the family Akeridae as the **glassy bubbles,** the family Scaphandridae as the bubbles and the family Hydatinidae as the **painted bubbles**

small bubble shell popular name of acteocinid gastropods

tiny bubble shell popular name of philinid gastropods

true bubble shell popular name of bullid gastropods

buttercup shell the lucinid pelecypod *Loripinus chrysostoma*

butterfly shell any of several species of donacid pelecypod of the genus Donax but usually *D. variabilis* (= coquina)

button shell the arcid pelecypod *Glycymeris sub-obsoleta*

cameo shell popular name of cassidid gastropods (*cf.* bonnet shell, helmet shell)

canoe shell popular name of scaphandrid gastropods. The name is sometimes reserved for *Scaphander punctostriatus*

cap shell popular name of capulid gastropods

carrier shell popular name of xenophorid gastropods but particularly of *Xenophora trochiformis*

cask shell popular name of tonnid gastropods particularly those of the genus Tonna

chank shell popular name of xancid gastropods

chink shell popular name of lacumid gastropods

conch shell popular name of strombid gastropods

cone shell popular name of conid gastropds

crooked shell the colubrariid gastropod *Colubraria tortuosa*

cup-and-saucer shell the calyptraeid gastropod *Crucibulum spinosum* (*cf.* cup-and-saucer limpet)

dipper shell popular name of cuspidariid pelecypods

disk shell any of several species of venerid pelecypods of the genus Dosinia

dove shell popular name of columbellid gastropods

dye shell popular name of thaisid gastropods

ear shell any haliotid gastropod (*cf.* abalone, ormer) The term is also applied to the naticids *Sinum scopulosum* and *S. perspectivum*

egg shell popular name of amphiperatid gastropods *U.S.* particularly those of the genus Neosimnia. The term is also applied to ovulids

fig shell any of several species of tonnid gastropods of the genus Ficus

file shell any of several species of nuculanid pelecypods of the genus Youldia and also limids in general

flood shell popular name of rissoid gastropods

frog shell popular name of bursid gastropods but in

U.S., without qualification, usually *Bursa californica*. Also applied to cymatids in general

gem shell the venerid pelecypod *Gemma gemma*

granular shell popular name of poromyid pelecypods

hedgehog shell any of numerous species of murid gastropods

helmet shell popular name of cassidid gastropods of the genus Cassis

hieroglyphic shell the venerid pelecypod *Lioconcha hiroglyphica*

hoof shell popular name of hipponicid gastropods but particularly of *Hipponix antiquata*

horn shell popular name of cerithiid gastropods, particularly those of the genus Zerithium

ivory shell any of several large buccinid gastropods

jingle shell popular name of anomiid pelecypods

kelly shell commonly applied to any pelecypod of the family Leptonidae but, *U.S.* without qualification, usually *Kellia laperousii*

kneecap shell the patellid gastropod *Helcionicus argentatus*

laddershell any of several species of epitoniids but particularly *Epitonium groenlandicum*

lantern shell any of several species of periplomatid pelecypods of the genus Periploma (*cf.* spoon shell)

magpie shell the trochid gastropod *Livona pica*

margin shell popular name of marginellid gastropods

salt-marsh shell popular name of ellobiid gastropods

miter shell popular name of mitrid gastropods

money shell = money cowry

moon shell popular name of naticid gastropods

nut shell popular name of nuculid pelecypods

    chestnut shell the astartid pelecypod *Astarte castanea*

    sculptured nut shell the solemyid pelecypod *Acila castrensis*

nutmeg shell the cancellariid gastropod *Cancellaria reticulata*

obelis shell popular name of melanellid gastropods but is also specifically applied to the pyramidellid *Pyramidella dolabrata*

olive shell popular name of olivid gastropods

paper shell popular name of lyonsid pelecypods

partridge shell the tonnid gastropod *Tonna perdix*

deepwater pearly shell popular name of verticordiid pelecypods

pheasant shell any gastropod mollusk of the family Phasianellidae and particularly of the genus Phasianella

peapod shell any of several species of mytilid pelecypod of the genus Botula

pompano shell any of several species of donacid pelecypods of the genus Donax but usually *D. variabilis* (= coquina)

pyramid shell popular name of pyramidellid gastropods

rattle shell the thaid gastropod *Sistrum nodulosum*

sunray shell the venerid pelecypod *Macrocallista nimbosa*

broad razor shell (*see also* razor clam) the garid pelecypod *Tagelus californianus*

rice shell popular name of marginellid gastropods

rim shell common name of marginellid gastropods

rock shell popular name of muricid gastropods, particularly those of the genus Murex

blacksand shell the unionid pelecypod *Lampsilis recta*

screw shell popular name of terebrid and turritellid gastropods

slipper shell popular name of crepidulid gastropods (= slipper limpets)

slit shell popular name of turrid gastropods

spindle shell any of numerous species of fastiolariid gastropods of the genus Fusinus and also buccinid gastropods of the genus Cantharus

spoon shell popular name of periplomatid pelecypods

star shell any of several species of turbinid gastropods of the genus Astraea

sundial shell popular name of architectonicid gastropods

sunset shell the garid pelecypod *Gari californica*

toad shell the cymatiid gastropod *Bursa buffonia*

tooth shell general name for members of the order Scaphopoda

    elephant-tooth shell = tooth shell

top shell any of several species of trochid gastropods of the genera Calliostoma and Margarites

    knobby top shell the littorinid gastropod *Tectarius muricatus*

triangular shell popular name of gouldiid pelecypods

trumpet shell any of several species of cymatid gastropods of the genus Charonia but particularly *C. tritonis*

turret shell popular name of turritellid gastropods

unicorn shell any of several species of thaisid gastropods of the genus Acanthina

vase shell the xancid gastropod *Vasum muricatum*

velvet shell popular name of gastropods of the family Velutinidae but, *U.S.*, without qualification, usually *Velutina laevigata*

Venus shell popular name of venerid pelecypods

volcano shell popular name of fissurellid gastropods

warped shell the cymatiid gastropod *Distorsio anus*

wedge shell popular name of donacid pelecypods

wing shell the pholadid pelecypod *Pholas campechiensis*

worm shell popular name of vermetid gastropods

writhing shell the cymatiid gastropod *Distorsio clathrata*

zebra shell the neritid gastropod *Purperita pupa*

shell 3 (*see also* shell 1, 2, 4) in compound name of other organisms

acorn shell a sessile cirripede crustacean (= barnacle)

dingy shell *Brit.* the geometrid lepidopteran insect *Euchoeca obliterata*

lamp shell popular name for members of the phylum Brachiopoda

softshell *U.S.* any of numerous species of trionychid chelonian reptile of the genus Trionyx

tortoise shell (*see also* shell 1) any of numerous nymphalid lepidopteran insects once placed in the genus Vanessa but now usually placed in the genus Nymphalis

    California tortoise-shell *U.S. N. californica*

    large tortoiseshell *Brit. N. polychloros. I.P. N. xanthomelas*

    mountain tortoiseshell *I.P.* = small tortoiseshell *Brit.*

    small tortoiseshell *Brit. N. urticae*

shell 4 (*see also* shell 1-4) the hardened case of a nut

shellac *see* lac

shelly *Brit. dial.* the fringillid bird *Fringilla coelebs* (= chaffinch)

shepster *Brit. dial.* = starling

shield 1 (*see also* shield 2) a term used in *biol.* in the sense of plate, not of necessity shield-shaped. Specifically a calcarous, or horn-like plate found on the anterior end of the trunk of some Sipunculids

cervical shield that part of the exoskeleton which lies immediately above the upper side of the head in many insect larvae

foot shield the lateral sclerite on the abdominal feet of caterpillars

gastric shield a thick plate in the stomach of many pelecypod mollusks

head shield the anterior region of the lorica of certain Rotifera, demarcated by a clear groove in some species

radial shield one of a pair of conspicuous plates at the base of each arm on the aboral surface of ophiuroid echinoderms

shield 2 (*see also* shield 1) in compound names of organisms

water shield popular name of nymphaeaceous aquatics of the genus Drasenia

[shifted genitals] a mutant (gene symbol *sge*) mapped at 48.4± on the X chromosome of *Drosophila melanogaster*. The name is descriptive of the condition of the phenotype

shf² *see* [shifted²]

[shifted²] a mutant (gene symbol *shf²*) mapped at 17.9 on the X chromosome of *Drosophila melanogaster*. The phenotypic expression is a shifting of the wing veins together

shikim = Japanese star anise (*cf.* shikimic acid)

shikra the acciptrid bird *Accipiter badius*

shin = tibia

shiner 1 (*see also* shiner 2) *U.S.* any of very many small freshwater cyprinid fishes mostly of the genus Notropis

Alabama shiner *N. callistius*
altamaha shiner *N. xaenurus*
Arkansas River shiner *N. girardi*
silverband shiner *N. illecebrosus*
sandbar shiner *N. scepticus*
fiery-black shiner *N. pyrrhomelas*
bleeding shiner *N. zonatus*
blue shiner *N. caeruleus*
Brazos shiner *N. brazosensis*
bridle shiner *N. bifrenatus*
chihuahua shiner *N. chihuahua*
blackchin shiner *N. heterodon*
chub shiner *N. potteri*
coastal shiner *N. petersoni*
ironcolor shiner *N. chalybaeus*
steelcolor shiner *N. whipplei*
tricolor shiner *N. trichroistius*
colorless shiner *N. perpallidus*
common shiner *N. cornutus*
comely shiner *N. amoenus*
coosa shiner *N. xaenocephalus*
crescent shiner *N. cerasinus*
dusky shiner *N. cummingsae*
emerald shiner *N. atherionoides*
bigeye shiner *N. boops*
popeye shiner *N. ariommus*
smalleye shiner *N. buccula*
bluntface shiner *N. camurus*
rosyface shiner *N. rubellus*
bandfin shiner *N. zonistius*
cherryfin shiner *N. roseipinnis*
flagfin shiner *N. signipinnis*
greenfin shiner *N. chloristius*
highfin shiner *N. altipinnis*
redfin shiner *N. umbratilis*

sailfin shiner *N. hypselopterus*
satinfin shiner *N. analostanus*
spotfin shiner *N. spilopterus*
whitefin shiner *N. niveus*
yellowfin shiner *N. lutipinnis*
stargazing shiner *N. uranoscopus*
ghost shiner *N. buchanani*
golden shiner *Notemigonus crysoleucas*
burrhead shiner *N. asperifrons*
greenhead shiner *N. chlorocephalus*
kiamichi shiner *N. ortenburgeri*
redlip shiner *N. chiliticus*
taillight shiner *N. maculatus*
mimic shiner *N. volucellus*
mirror shiner *N. spectrunculus*
mountain shiner *N. lirus*
bigmouth shiner *N. dorsalis*
whitemouth shiner *N. alborus*
New River shiner *N. scabriceps*
blacknose shiner *N. heterolepsis*
bluenose shiner *N. welaka*
bluntnose shiner *N. simus*
longnose shiner *N. longirostris*
pugnose shiner *N. anogenus*
sharpnose shiner *N. oxyrhynchus*
ocmulgee shiner *N. callisema*
ohoopee shiner *N. leedsi*
Ozark shiner *N. ozarcanus*
warpaint shiner *N. coccogenis*
pallid shiner *N. amnis*
plains shiner *N. percobromus*
plateau shiner *N. lepidus*
pretty shiner *N. bellus*
proserpine shiner *N. proserpinus*
rainbow shiner *N. chrosomus*
red shiner *N. lutrensis*
river shiner *N. blennius*
Red River shiner *N. bairdi*
ribbon shiner *N. fumeus*
Rio Grande shiner *N. jemezanus*
rough shiner *N. baileyi*
sabine shiner *N. sabinae*
saffron shiner *N. rubricroceus*
sand shiner *N. stramineus*
highscale shiner *N. hypsilepis*
silver shiner *N. photogenis*
blackspot shiner *N. atrocaudalis*
wedgespot shiner *N. greenei*
bluestripe shiner *N. callitaenia*
broadstripe shiner *N. euryzonus*
silverstripe shiner *N. stilbius*
blacktail shiner *N. venustus*
spottail shiner *N. hudsonius*
swallowtail shiner *N. procne*
whitetail shiner *N. galacturus*
tamaulipas shiner *N. braytoni*
Tennessee shiner *N. leuciodus*
Texas shiner *N. amabilis*
Topeka shiner *N. topeka*
weed shiner *N. texanus*
white shiner *N. albeolus*
pinewoods shiner *N. matutinus*

shiner 2 (*see also* shiner 1) *Brit.* the gyrinid coleopteran insect *Gyrinus natator* (= whirligig beetle)

Shinisauridae a monotypic family of lacertilian reptiles usually fused with Xenosauridae

mother Shipton *Brit.* the noctuid lepidopteran insect *Euclidia mi*

shirle the turdid bird *Turdus viscivorus* (= mistle thrush)

shoal a term for a group of fish, applied to many species (*cf.* school)

shoat a young pig

shoot (*see also* moth) a plant stem together with the leaves and buds that it bears

    primary shoot that which develops directly from the plumule

    rosette shoot a shoot having a cluster of leaves at its base

        double rosette shoot = dynaster

shore that portion of land which is contiguous to a body of still water (*cf.* bank)

[short ear] a mutation (gene symbol *se*) in linkage group II of the mouse. The phenotypic expression involves not only the shortening of the ear but also a general reduction of the cartilaginous skeleton

[short vein] a mutant (gene symbol *shv*) mapped at 3.8± on chromosome II of *Drosophila melanogaster*. The phenotype has terminal gaps in veins L2 and L4

[short-wing] a mutant (gene symbol *sw*) mapped at 64.0 on the X chromosome of *Drosophila melanogaster*. The phenotypic expression involves not only warped, short, wings but also reduced, rough eyes

Indian shot any of several species of the cannaceous herb Canna, particularly *C. indica*

shote = shoat

[shot-veins] a mutant (gene symbol *sv*) in linkage group IV of *Habrobracon juglandis*. The phenotypic expression is in broken and distorted wing veins

shou a Tibetan race of the cervine artiodactyl *Cervus elaphas* (= red deer)

flame shoulder *Brit.* the noctuid lepidopteran insect *Noctua plecta* (*cf.* flame)

shoveller any of several species of anatine ducks of the genus Spatula. Without qualification *U.S.* and *Brit. S. clypeata. N.Z.* and *Austral. S. rhyncotis*

    Australian shoveller *S. rhyncotis*

    Cape shoveller *S. capensis*

    common shoveller *S. clypeata*

golden shower the leguminous tree *Cassia fistula*

    pink shower *Cassia grandis*

shp *see* [shrimp]

shr *see* [shrunken]

shrew (*see also* note) any of numerous small insectivores mostly of the family Soricidae, except for the "tree shrew" which is a primate

    armored shrew = girder-backed shrew

    ashland shrew *Sorex trigonirostris*

    girder-backed shrew the Congolese soricid Scutisorex, so called because of the massive overdevelopment of its vertebral column

    common shrew *Brit.* without qualification, *Sorex araneus*

    desert shrew *U.S. Notiosorex crawfordi*

    dwarf shrew *U.S. Sorex nanus*

    elephant shrew popular name of macroscelid insectivores

    web-footed shrew any of several aquatic Tibetan soricids of the genus Nectogale

    gray shrew *U.S.* = desert shrew *U.S.*

    inyo shrew *Sorex tenellus*

    least shrew *U.S. Cryptotis parva*

    lesser shrew *Brit. Sorex minutus*

    malheur shrew *Sorex preblei*

    masked shrew *Sorex cinereus*

    mole shrew popular name of asiatic soricids of the genus Anourosorex

    musk-shrew popular name of soricids of the genus Crocidura, usually, without qualification, *C. russula*

    pygmy shrew *U.S. Microsorex hoyi Brit. Sorex minutus*

    Santa Catalina shrew *Sorex willeti*

    smoky shrew *Sorex fumeus*

    southeastern shrew *U.S. Sorex longirostris*

    suisun shrew *Sorex sinuosus*

    longtail shrew *Sorex dispar*

        shorttail shrew *U.S. Blarina brevicauda*

    fat-tailed shrew popular name of soricids of the genus Suncus

    white-toothed shrew *Brit.* either of two species of the genus Crocidura

    tree-shrew popular name of tupaiad primates

    Unalaska shrew *Sorex hydrodromus*

    water shrew popular name of soricid insectivores of the genus Neomys. *U.S.* without qualification, either *N. palustris* or *N. benderei*

        asiatic water-shrew popular name of soricids of the genus Chimarrogale

        giant water-shrew the African potamogalid *Potamogale velox*

shrewdness the term for a group of apes

shrieker *Brit.* either of two species of the scolopacid genus Scolopax (= godwit)

shrike (*see also* babbler, starling, tanager, thrush, tit, vireo) any of numerous species of birds of the families Vangidae, Prionopidae, and Laniidae and occasionally for other shrike-like birds. Without qualification, the laniids are usually meant. *Brit.*, without qualification, *Lanius collurio* (= red-backed shrike)

    antshrike any of numerous species of formicariid birds mostly of the genus Thannophilus

        barred antshrike *T. doliatus*

        great antshrike *Taraba major*

        russet antshrike *Thamnistes anabatinus*

        slaty antshrike *T. punctatus*

    puffback shrike *W.Afr.* any of several species of the laniid genus Dryoscopus

    red-backed shrike *Lanius collurio*

    bill shrike *Laniarius ferrugineus* (= boubou)

    red-billed shrike any of several prionopids of the genus Sigmodus; without qualification, *S. caniceps*

        Retz's red-billed shrike *S. retzii*

        yellow-billed shrike *Corvinella corvina*

    brown shrike *Lanius cristatus*

    bush-shrike any of numerous, mostly African, laniid birds many of the genera Laniarius, Tchagra, and Chlorophoneus

        sulphur-breasted bush-shrike *C. sulfureopectus*

        blackcap bush-shrike *Bocagia minuta*

        four-colored bush-shrike *Telophorus quadricolor*

        many-colored bush-shrike *C. multicolor*

        chestnut-crowned bush-shrike *L. luetideri*

        black-fronted bush-shrike *C. nigrifrons*

        grey-green bush-shrike *S. bocagei*

        black-headed bush-shrike *T. senegala*

        brown-headed bush-shrike *T. australis*

        grey-headed bush-shrike *Malaconotus blanchoti*

        red-naped bush-shrike *L. ruficeps*

        three-streaked bush-shrike *T. jamesi*

    caterpillar-shrike = cuckooshrike

    chat-shrike *Lanioturdus torquatus*

    white-crowned shrike the prionopid *Eurocephalus anguitimens*

    cuckooshrike any of very numerous campiphagid birds mostly of the genus Coracina barred cuckooshrike *C. lineata*

black cuckooshrike *Campephaga flava*

white-breasted cuckooshrike *C. pectoralis*

black-faced cuckooshrike *Coracina novaehollandiae*

grey cuckooshrike *C. caesia*

black-headed cuckooshrike *C. melanoptera*

little cuckooshrike *C. robusta*

purple-throated cuckooshrike *Campephaga quiscalina*

red-throated cuckooshrike *Campephaga phoenicea*

fiscal shrike any of several species of the genus Lanius

flycatcher shrike any of several species of muscicapid bird of the genus Pachycephala (= whistler) or any of several campephagids of the genus Tephrodornis (= pygmy triller)

chestnut-fronted shrike the prionopid *Sigmodus scopifrons*

grey shrike *U.S.* and *Brit. Lanius excubitor; Afr. Lanius elegans*

lesser grey shrike *Lanius minor*

helmet-shrike any of several prionopids, mostly of the genus Prionops

curly-crested helmet-shrike *P. cristata*

grey-crested helmet-shrike *P. poliolopha*

straight-crested helmet-shrike *P. plumata*

loggerhead shrike *Lanius ludovicianus*

magpie-shrike *Urolestes melanoleucus*

masked shrike *Lanius nubicus*

northern shrike *Lanius excubitor* (= grey shrike *Brit.*)

rosy-patched shrike *Rhodophoneus cruentus*

pepper-shrike either of two species of cyclarhid bird of the genus Cyclarhis

rufous-browed pepper-shrike *C. gujanensis*

swallow-shrike = woodswallow

yellow-spotted shrike the pycnonotid *Nicator chloris*

vanga shrike any of several species of bird of the family Vangidae

wood-shrike properly, any of several campephagids of the genus Tephrodornis; without qualification, *T. pondicerianus*. Sometimes used for prionopids of the genus Prionops (= helmet-shrike)

woodchat shrike *Lanius senator*

shrimp (*see also* fish) Properly a cragonid natant decapod crustacean but the term is commonly applied to almost any small crustacean. *Brit.*, without qualification, the cragonid *Crago septemspinosus*. (commonly miscalled *Crangon vulgaris*). In *U.S.* the term shrimp and prawn (*q.v.*) are largely interchangeable

brine shrimp any of several species of branchiopod crustacean of the genus Artemia; usually, without qualification *A. salina*

burrowing shrimp popular name of thalassinid macruran Crustacea

California shrimp *Crago franciscorum*

clam-shrimp any of numerous conchostracan branchiopod crustaceans

claw-shrimp = clam shrimp

devil-shrimp any of numerous marine amphipod Crustacea of the suborder Caprellidea (= skeleton shrimp)

fairy-shrimp any of several anostracan branchiopodan Crustacea

ghost-shrimp *Brit.*, popular name of palinurid decapods. *W. U.S.* the calianassid *Callianassa californiensis*

skeleton shrimp = devil shrimp

snapping shrimp popular name of crangonid decapods (*see* Cragonidae)

coon-spiked shrimp the pandalid *Pandalus danae*

tadpole-shrimp any of several notostracan branchiopod crustaceans

blacktailed shrimp *Crago nigricauda*

[shrimp] a mutant (gene symbol *shp*) mapped at 47.5± on the X chromosome of *Drosophila melanogaster*. The phenotypic expression is a general reduction of size

shrub any low growing woody plant with many stems and no main trunk

melon shrub the solanaceous shrub *Solanum muricatum* and its edible fruits

subshrub a shrub only some of the stems of which are lignified

undershrub one which does not rise to the level of surrounding vegetation

[shrunken] a mutant (gene symbol *shr*) mapped at 2.3± on chromosome II of *Drosophila melanogaster*. The name is descriptive of the body of the phenotype

[shrunken endosperm] *either* a mutant (gene symbol *sh₂*) mapped at 103.0± on linkage group III. The name is descriptive of the phenotypic seed *or* a mutant (gene symbol *sh₁*) mapped at 29 on linkage group IX of *Zea mays*. The term is descriptive

honey shuck *U.S.* the leguminous tree *Gleditsia triacanthos* (= honey locust)

longshucks the tree *Pinus taeda* (= loblolly)

shv *see* [short vein]

Shw *see* [Shaker-downheld]

si *see* [silvered]

Sialidae a family of neuropteran insects containing the forms commonly called alderflies. They are distinguished from all other Neuroptera except the Corydalidae by the fact that the hindwings posess a large anal area that is folded fanwise and from the Corydalidae by the absence of ocelli

siamang the hylobatid primate *Symphalangus syndactylus* found only on the island of Sumatra. It differs from the true gibbon in possessing a throat pouch and having the third and second toes joined to the second joint by a web

sib *syn.* of sibling (*q.v.*) but often misused as *adj.* form

sibia any of several timaliid birds of the genus Heterophasia

sibling properly "kindred" but used in *biol.* to indicate animals derived from a single birth, a single clutch of eggs or, in insects, a single mating and in *bot.* for plants derived from the ovaries or the pollen of the same flower

-sicc- *comb. form* meaning "dry"

siccitate dried

siccous dryish

Sicyoideae a subfamily of the Cucurbitaceae

redside *U.S.* any of several cyprinid fish of the genus Richardsonius

silversides popular name of atherinid fishes

-sidero- *comb. form* meaning "magnet" but widely used for "iron"

Siderocapsaceae a family of pseudomonodale schizomycetes forming a thick mucilagenous cap rich in iron and manganese compounds. These bacteria usually form deposits of iron and manganese in fresh water pools

[side wing] a mutant (gene symbol *siw*) mapped at 58.5± on the X chromosome of *Drosophila melanogaster*. The phenotypic expression is the holding of wings closely parallel to the side of the abdomen

siege term for a group of bitterns (*cf.* sedge), cranes (*cf.* -herd) or herons

Sierolomorphidae a family of scolioid apocritan hymenopteran insects of a shiny black color

sieve *see* pitting, plate

sifaka any of several species of indriid lemuroid primates of the genus Propithecus

Siganidae a small family of acanthopterygian fishes distinguished by the rabbit-like appearance of the mouth and two spines in each pelvic fin. Commonly called rabbit fishes

Siganoidea a suborder of acanthopterygian fishes containing the single family Siganidae

sigma an s-shaped microscleric, monaxonic sponge spicule

sigmoid s-shaped

signal any communication between animals, or parts of animals

    alarm signal one made by animals denoting the presence of a predator or potential predator. Alarm signals are often shared between species or even phyla. Thus the alarm signals of a mammal may be comprehensible to birds

    territorial signal one that establishes a breeding, or more rarely, hunting, territory. Such signals are usually olfactory in mammals and auditory in birds

Signiphoridae = Thysanidae

sika the Japanese cervine artiodactyl *Cervus nippon*

silicle a little silique

silique the pod of a cruciferous plant which dehisces by the two valves falling away from the replum

multisiliquous having many seed parts

silk 1 (*see also* grass, oak, tree, worm, vine) a thin fiber of solidified brin (*q.v.*)

silk 2 (*see also* silk 1) the style and stigma of corn (*Zea mays*)

[silkie] a mutant (gene symbol *h*) in linkage group V of the domestic fowl. The name refers to the appearance of the feathers

Silphidae a very large family of coleopteran insects containing the forms commonly called carrion beetles and burying beetles. They are usually large and brightly colored and are well known for their habit of burying carrion with a view to laying their eggs on it

[silkless] a mutant (gene symbol *sk*) mapped at 56 on linkage group II of *Zea mays*. The phenotypic ear lacks silk

silt rock particles less than 0.625 mm size

Siluridae a family of naked catfishes distinguished by lacking both the spine in front of the dorsal fin and the adipose fin. The anal fin is unusually long

Siluroidea a suborder of osteriophysous Osteichthyes erected to contain the various families of catfishes

Silvanidae a family of small coleopteran insects frequently fused with the Cucujidae but separated from them by the possession of a clavate antenna

[silver] a mutant (gene symbol *svr*) mapped at 0+ on the X chromosome of *Drosophila melanogaster*. The phenotypic expression is dark bristles on a silvery body

[Silver] a mutant (gene symbol *S*) on the sex chromosome (linkage group I) of the domestic fowl. The name refers to the color of the plumage

[silvered] *either* a mutant (gene symbol *si*) found in linkage group IV of the mouse. The phenotypic expression is the reduction of the pigment of the hair *or* a mutant (gene symbol X) in linkage group II of the guinea pig. The phenotypic expression is

silvery fur *or* a mutant (gene symbol *s*) mapped at 47 on linkage group II of the rat. The name refers to the appearance of the pelt

silverling *U.S.* the caryophyllaceous herb *Paronychia argyrocoma*

Silvicolidae = Rhyphidae

Simarubaceae that family of geraniale dicotyledons which contains *inter alia*, the quassia tree and the tree-of-heaven. The family is most readily distinguished from the closely allied Rutaceae by the absence of oil glands in the leaves

Simenchelidae a monotypic family of apodus fishes erected to contain the parasitic abyssal *Simenchelys parasiticus* sometimes called the snub-nosed eel

Simiidae a family of primates at one time combining the Hylobatidae and Pongidae, but regarded as not containing the genus Homo which was placed in a separate family the Hominidae

consimilar said of a diatom valve in which both sides are alike

Simioidea a suborder of primate mammals containing the Old World monkeys as distinct from the apes (Anthropodea). They differ from the New World monkeys in having downwardly directed nostrils and from the apes in possessing tails, which are never, however, prehensile

subsimple with few divisions

Simuliidae a family of viciously predatory nematoceran dipteran insects with a hump-backed appearance, short broad wings, and short legs. They are commonly called black flies or buffalo-gnats

simung the Indian lutrine carnivore Lutrogale

sinciput the forepart of the crown of the head in insects (= frons)

[singed] a pseudo-allelic locus mapped at 21.0 on the X chromosome of *Drosophila melanogaster*. The bristles have a singed appearance in the phenotype

sing-sing the antelope *Cobus unctuosus* (*cf.* waterbuck)

goldsinny *Brit.* the labrid fish *Ctenolabrus rupestris*

sinuous bent into a very shallow zigzag, not as abrupt as flexuose, but more abrupt than serpentine

sinus literally a "lake," "bay" or "fold of a garment," used mostly in the first two senses in *zool.* and the last sense in *bot.*

sinus 1 (*see also* sinus, sinus 2) in the sense of a "lake"

    axial sinus that part of the crinoid coelom that runs from the chambered organ to a region surrounding the esophagus. A similar structure, though less well developed, is found in other echinoderms

    cardinal sinus the principal blood collecting vessel of vertebrate embryos running laterally along the body from front to back. The anterior and posterior cardinal sini unite to form the Cuverian sinus

    cervical sinus a depression in the pharyngeal region of the embryo. The second arch overgrows the next three arches

    Cuverian sinus a blood collecting vessel of vertebrate embryos running from the junction of the cardinal sini to the sinus venosus

    marginal sinus (Elasmobranchs) = sinus terminalis

    sinus oceanicus = sinus terminalis

    sinus terminalis the circular, peripheral blood vessel, which delimits the outer edge of the area vasculosa of the blastoderm in the development of a telolecithal egg

    sinus venosus the large sinus into which venous blood drains before reaching the atrium

sinus 2 (*see also* sinus, sinus 1) in the sense of a fold,

specifically the pore of some fungi, the recess between the half-cells of desmids, the gap in the micropyle of some seeds, and the reentrant space between the lobes of a leaf

sion *see* scion

-siph-, *comb. form* meaning tube

-sipho- *see* -siph-

siphon almost any type of tubular structure, plant or animal, through which a fluid flows, specifically the inhalant and exhalant tubes of arenicolous pelecypod Mollusca, similar tubes of Ascidia, the sucking structures of certain arthropods, the gastrozooid of siphonophoran Hydrozoa, a slender tube constructed from the inner border of the small intestine of echinoid echinoderms and which conducts to the anterior water removed from the food, and the elongated tube of Polysiphonia and similar algae

 anal siphon the breathing tube of certain aquatic dipterous larvae

Siphonales an order of chlorophycean Chlorophyta in which the thallus consists of a single multinucleate cell, often having the form of a branched tube

Siphonaptera that order of insects containing the forms commonly called fleas. They are characterized by the lack of wings, the mouth parts modified for biting, the laterally compressed shape of the body and the legs modified for jumping

siphoneous pertaining to algae having a tubular structure

siphonic pertaining to a siphon in any of its meanings. The alternative form siphonous is sometimes used

 monosiphonic consisting of a single tube or single row of cells

 polysiphonic said of the hydrocaulus of a hydroid coelenterate when it is covered by rhizocaulomes (*cf.* polysiphonous)

Siphonocladales an order of chlorophycean Chlorophyta containing those forms that have a multicellular thallus from the base of which rhizoids project

Siphonophora an order of hydrozoan coelenterates which exists as free swimming colonies of polymorphic individuals. The Portuguese-man-of-war is a well known example

Siphonopoda a subphylum of the Mollusca containing the single class Cephalopoda. The characteristics of which are those of the subphylum

polysiphonous a filament composed of several parallel filaments joined together (*cf.* polysiphonic)

Siphunculata = Anaplura

siphuncle 1 (*see also* siphuncle 2, 3) in aphids, a hollow projection through which a defensive, sticky, liquid is squirted

siphuncle 2 (*see also* siphuncle 1, 3) the connective tissue cord which attaches the base of the body of argonautid Mollusca to the shell in the first chamber (*cf.* septal neck)

siphuncle 3 (*see also* siphuncle 1, 2) used as a synonym of cornicle (*q.v.*)

Sipunculoidea a small phylum of the animal kingdom containing a group of unsegmented worms without bristles. The main characteristic is a crown of short hollow tentacles on the anterior end. This group has variously been regarded as a class of the Annelida, a class of the Gephyrea and as a separate phylum

sire male parent (*cf.* dam)

siren 1 (*see also* siren 2) *U.S.* a genus of sirenid urodele Amphibia. In the *U.S.*, the Latin name is used as an English word

 dwarf siren the sirenid urodele *Pseudobranchus striatus*

 greater siren *S. lacertina*

 lesser siren *S. intermedia*

siren 2 (*see also* siren 1) *I.P.* any of several species of nymphalid lepidopteran insect of the genus Diagora

Sirenia that order of placental mammals containing the forms commonly called sea cows, dugongs and manatees. They are distinguished by their petal-like fins and forked tails

Sirenidae a family of urodele Amphibia distinguished by the absence of hind limbs, maxillary bones and eyelids

Siricidae a family of large symphytan hymenopteran insects commonly called horntails in virtue of the horny plate on the last abdominal segment of both sexes and the heavy long ovipositor of the female

sirkeer the cuculid bird *Taccocus leschenaultii*

sirystes the tyrannid bird *Sirystes sibilator*

siskin *U.S.* any of several species of fringillid birds mostly of the genus Spinus

 antillean siskin *W.I. Loximitris dominicensis*

 black-capped siskin *S. atriceps*

 black-headed siskin *S. notatus*

 pine siskin *S. pinus*

sisserou *W.I.* the psittacid bird *Amazona pimperialis* (= imperial parrot

little sister *U.S.* the noctuid lepidopteran insect *Catocala fratercula* (*cf.* underwing)

Sisyridae a family of neuropteran insects resembling brown lacewings but far smaller in size. Sometimes called spongilla flies

sitatunga the strepsicerosine bovid artiodactyle *Tragelaphus spekei* (*cf.* bushbuck)

sittella any of several sittid birds of the genus Neositta

 black-capped sittella *N. pileata*

 white-headed sittella *N. leucocephala*

 orange-winged sittella *N. chrysoptera*

rocksitter *S.Afr.* any of numerous species of lycaenid lepidopteran insect of the genus Durbania

Sittidae that family of passeriform birds which contains the nuthatches. They are usually small birds with long pointed wings and short truncated tails which enable them to climb about on trees with a jerky motion

siva any of several timaliid birds of the genus Minla

siw *see* [side wing]

sk *see* [skin], [stick] or [silkless]

Sk *see* [Speckled] or [Streak]

skate any of numerous chondrichthyean fishes of the suborder Batoidei, particularly those of the family Rajidae and the genus Raja. The terms skate and ray are subject to much local interchange though in *U.S.* usage the term ray is preferred for those forms which have a venomous spine in the tail

 barndoor skate *R. laevis*

 big skate *R. binoculata*

 black skate *R. kincaidi*

 tobacco-box skate = little skate

 California skate *R. inornata*

 hedgehog skate *Brit.* = little skate *U.S.*

 little skate *U.S. R. erinacea* (= *Brit.* hedgehog skate)

 clearnose skate *R. eglanteria*

 smooth skate *R. senta*

 thorny skate *R. radiata*

 winter skate *R. ocellata*

skein term for a group of geese in flight (*cf.* gaggle)

skeleton (*see also* weed) those supporting structures of animals to which muscles are attached and, more loosely, supporting and protective structures such as the "skeleton" of some Protozoa, and even such supporting structures as the "skeleton" of leaves or insect wings

autoskeleton a skeleton produced by the organism

axial skeleton that part of the vertebrate skeleton which runs down the median line

cytoskeleton = hyaloplasm

endoskeleton a skeleton under the surface, as that of vertebrates, in contrast to a surface skeleton as that of arthropods. Also any part of the skeleton which lies within another part, as the endoskeleton of the vertebrate skull

exoskeleton skeletal elements forming the surface of an animal as in arthropods

lithistid skeleton a term applied to that type of sponge skeleton which consists of desmas cemented into a framework

pseudoskeleton a supporting structure derived through the accretion of foreign substances

visceral skeleton the visceral arches of fish and their derivatives or equivalents in other groups

skeletonizer any of many lepidopteran insect larva that reduce leaves to skeletons

apple-and-thorn skeletonizer *U.S.* the larva of the glyphipterygid *Anthophila pariana*

birch skeletonizer *U.S.* the larva of the lyonetiid *Bucculatrix canadensisella*

apple-leaf skeletonizer *U.S.* the larva of the *Psorosina hammondi*

bean-leaf skeletonizer *U.S.* the larva of the noctuid *Autoplusia egena*

grape-leaf skeletonizer *U.S.* the larva of the pyromorphid *Harrisina americana*

western-grape leaf skeletonizer *U.S.* the larva of *H. brillians*

palm-leaf skeletonizer the larva of the cosmopterygid *Homaledra sabalella*

oak skeletonizer *U.S.* the larva of lyonetiid *Bucculatrix ainsliella*

maple trumpet-skeletonizer the larva of the olethreutid *Epinotia aceriella*

-skia- *comb. form* meaning "shade"

skiddy the rallid bird *Rallus aquaticus* (= water rail)

skimmer 1 (*see also* skimmer 2) *U.S.* popular name of dragonflies of the family Libellulidae

ten spot skimmer *Libellula pulchella*

white tailed skimmer *Libellula lydia*

amber-wing skimmer *Perithemis tenera*

skimmer 2 (*see also* skimmer 1) popular name of charadriiform rhyncopid birds of the genus Rhyncops

African skimmer *R. flavirostris*

black skimmer *R. nigra*

Indian skimmer *R. albicollis*

skin loosely the outer coat of an organism, and specifically a covering of many animals consisting of both ectodermal and mesodermal elements

[Skin] a mutant (gene symbol *sk*) on linkage group VII of *Neurospora crassa*. The phenotypic expression is a flat nonconidiating growth

skink popular name of scincid lacertan reptiles

blind skink any of several species of the genus Typhlosaurus

Cape Verde skink *Macroscincus coctaei*

dart skink any of several species of the genus Acontias

keeled skink any of several species of the genus Tropidophorus

lidless skink any of several species of the genus Ablepharus

sand skink *Neoseps reynoldsii*

slender skink any of several species of the genus Lygosoma

snake skink any of several species of the genus Ophioscincus

blue-tongued skink *Austral.* *Tiliqua scincoides*

skipper 1 (*see also* skipper 2, fly, hopper 3) a term applied to hesperioid lepidopteran insects to distinguish them from the true butterflies (Papilionoidea). The skippers are relatively thick-bodied forms and in many cases the tip of the antenna is recurved. The technical difference lies in the venation of the wing (*cf.* butterfly, moth)

golden-banded skipper *U.S. Autochton cellus*

checkered skipper *U.S.* popular name of members of the genus Pyrgus. *Brit. Caetoerocephalus palaemon*

dingy skipper *Brit. Nisoniades tagos*

giant skipper popular name of megathymid hesperioid insects

grizzled skipper *Brit. Syrichthus malvae*

hammock skipper *U.S. Polygonus lividus*

large skipper *Brit. Hesperia sylvanus*

lesser skipper any of numerous hesperiids of slightly smaller than average size

Lulworth skipper *Brit. Hesperia actaeon*

mangrove skipper *U.S. Phocides batabano*

pearl skipper *Brit. Hesperia comma*

small skipper *Brit. Hesperia lineola*

silver-spotted skipper *U.S. Epargyreus clarus*

long-tailed skipper *U.S.* any of several species of the genera Urbanus and Chioides. Usually, without qualification *Urbanus proteus*

white skipper popular name of hesperiid lepidopteran insect of the genus Heliopetes

skipper 2 in compound names of other organisms *Brit.* the scomberesocid fish *Scomberesox saurus* (= *U.S.* Atlantic saury)

cheese skipper the larva of any of several piophilid dipterans, particularly *Piophila casei*

mudskipper any of several species of periophthalmid fishes of the genus Periophthalmus

skipperling any of numerous hesperiid lepidopteran insects of the genus Copaeodes

skirret the umbelliferous herb *Sium sisarum*

skitter to scuttle on the surface of the water

-skler- *comb. form* meaning "hard" frequently transliterated -scler-

skoke the phytolactaceous herb *Phytolacca americana* (= poke weed)

-skolex- *comb. form* meaning "worm" almost invariably transliterated -scolec-

-skolio- *comb. form* meaning "bent" or "curved," frequently transliterated -scolio-

skolly *S.Afr.* any of numerous species of lycaenid lepidopteran insect of the genus Thestor

-skoto- *comb. form* meaning "dark" or "dim" commonly transliterated -scoto-

skua any of four species of predatory birds of the family Stercorariidae, which steal the prey of other shore birds. Without qualification, *Catharacta skua* (*cf.* jaeger)

Arctic skua *Brit.* = parasitic jaeger *U.S.*

great skua *Catharacta skua*

pomatorhine skua *Brit.* = pomarine jaeger *U.S.*

long-tailed skua *Brit.* = long-tailed jaeger *U.S.*

skulk term for a group of foxes in cover (*cf.* cloud, troop) skull (*see also* cap) that skeletal element of chordates which encloses the brain and articulates with the jaw or jaws

platybasic skull said of a skull the base of which is formed from a cartilagenous plate to the sides of which the trabeculae fuse

tropibasic skull said of a skull the base of which is derived from the fused trabeculae

skunk (*see also* grape, weed) any of several species of mephitic mustelid carnivores. Usually, *U.S.* without qualification *Mephitis mephitis*

Cape skunk = zorilla

hooded skunk *Mephitis macoura*

Japanese skunk the badger *Mydaus meliceps*

hog-nose skunk *Conepatus leuconotus* (= rooter skunk)

rooter skunk = hog-nose skunk

spotted skunk *Spilogale putorius*

striped skunk *Mephitis mephitis*

sl *see* [semilong], [small wing] or [slashed leaves]

s.l. (or, rarely, S.L.) abbreviation for *sensu lato* (*q.v.*)

sla *see* [slimma]

[slashed leaves] a mutant (gene symbol *sl*) mapped at 40 in linkage group VII of *Zea mays*. The name is descriptive

slc *see* [slim chaete]

sleeper common name of eleotrid fish

sleuth a term for a group of bears (*cf.* sloth)

slider *U.S.* any of several species of testudinid chelonian reptile of the genus Pseudemys (*cf.* cooter). Without qualification frequently *P. concinna*

[slight] a mutant (gene symbol *slt*) mapped at 106.3 on chromosome II of *Drosophila melanogaster*. The body of the phenotype is small with reduced bristles

[slim chaete] a mutant (gene symbol *slc*) mapped at 3.6± on the X chromosome of *Drosophila melanogaster*. As the name indicates the phenotype has fine short bristles

slime in *zool.*, any smooth slippery substance, usually an exhudate. In *bot.* a proteinaceous material produced by slime bodies, mostly in sieve-tube members

[slimma] a mutant (gene symbol *sla*) mapped at 50.0± on the X chromosome of *Drosophila melanogaster*. The body of the phenotype is unusually narrow

slipper (*see also* shell, limpet) a slipper shaped organism; without qualification, any of several species of gastropod mollusk of the genus Crepidula but particularly *C. aculeata* and *C. plana*

baby's slippers the leguminous herb *Lotus corniculatus*

lady's slipper 1 (*see also* lady's slipper 2) any of several orchids of the genus Cypripedium. *Brit.* without qualification, *C. calceolus*

lady's slipper 2 (*see also* lady's slipper 1) the large flowered horticultural varieties of herbs of the scrophulariaceous genus Calceolaria

slit any elongate aperture

cephalic slit one of the grooves along the side of the cephalic lobe of Nemertea

gill slit the aperture between the pharynx and the exterior which, in the adult animal, is kept open by gill arches bearing gills

slo *see* [Slow]

sloe a small, black plum *U.S. Prunus alleghaniensis. Brit. P. communis*

black sloe *P. umbellata*

sloth (*see also* sloth 2) a term for a group of bears (*cf.* sleuth)

sloth 2 (*see also* sloth 1) any of several mammals forming the mammalian order Xenarthra

three-toed sloth *Bradypus tridactylus* which actually possesses five toes but only three fingers

two-toed sloth *Choloepus didactylus*. It actually possesses five toes but only two fingers

[Slow] a mutant (gene symbol *slo*) in linkage group 1 of *Neurospora crassa*. The term is descriptive of the phenotypic growth pattern

[Slow feathering] a mutant (gene symbol *K*) on the sex chromosomes (linkage group I) of the domestic fowl. The name is descriptive of the phenotypic expression

slt *see* [slight]

slug 1 (*see also* slug 2) a general term for those terrestrial gastropod mollusks having no visible shell. They are distributed among several families in the suborder Stylommatophora

gray field-slug *U.S.* the limacid *Deroceras laeve*

gray garden-slug *U.S.* the limacid *Deroceras reticulatum*

spotted garden-slug *U.S.* the limacid *Limax maximus*

tawny garden-slug *U.S. L. flavus*

greenhouse slug *U.S.* the limacid *Milax gagetes*

sea slug slug-like, often brilliantly colored, marine gastropods with external gills, distributed among several families in the suborder Nudibranchiata of the gastropod order Opisthobranchiata

slug 2 (*see also* slug 1) *U.S.* the larvae of any of many species of cochlidiid lepidopteran insect and of tenthredinid hymenopteran insects

spiny oak-slug *U.S.* the larva of lepidopteran *Euclea delphinii*

pear-slug *Brit.* the larva of the tenthredinid hymenopteran *Eriocampoides limacina*

California pearslug *U.S. Pristiphora californica*

rose-slug *U.S.* the larva of the tenthredinid hymenopteran *Endelomyia aethiops*

bristly rose-slug *U.S.* the larva of *Cladius isomerus*

sm *see* [smooth] or [salmon silk]

sma *see* [smaller]

smallage *Brit.* the wild form of the umbelliferous herb *Apium graveolens* (= celery)

[small bristle] a mutant (gene symbol *sbr*) mapped at 33.4 on the X chromosome of *Drosophila melanogaster*. The name is descriptive of the phenotypic expression

[smaller] a mutant (gene symbol *sma*) mapped at 29.9± on the X chromosome of *Drosophila melanogaster*. The name is descriptive of the body size of the phenotype

[small-eye] a mutant (gene symbol *sy*) mapped at 59.2 on the X chromosome of *Drosophila melanogaster*. The name is descriptive of the phenotype

[smalloid] a mutant (gene symbol *smd*) mapped at 60.1± on the X chromosome of *Drosophila melanogaster*. The name refers to the small body size of the phenotype

[small pollen] a mutant (gene symbol sp_1) mapped at 66 in linkage group IV of *Zea mays*. The name is descriptive of the phenotypic expression

[small thorax] a mutant (gene symbol *smt*) mapped at 51.9± on the X chromosome of *Drosophila melanogaster*. The name is descriptive of the phenotype

[small wing] a mutant (gene symbol *sl*) mapped at 53.5

on the X chromosome of *Drosophila melanogaster*. In addition to the short wing the phenotypic expression involves large eyes

smaragdine emerald green

smd *see* [smalloid]

smelt popular name of osmerid and atherine fishes

　American smelt *Osmerus mordax*

　European smelt *O. eperlanus*

　jacksmelt the atherinid fish *Atherinops californiensis*

　sandsmelt *Brit.* popular name of atherinid fishes in general (= *U.S.* silver-sides)

　topsmelt the atherinid fish *Atherinops affinis*

Smerinthidae = Sphingidae

smew the mergine duck *Mergellus albellus* (*cf.* merganser)

Smilacaceae a family of liliiflorous monocotyledonous angiosperms containing forms usually included within the Liliaceae

smilax properly the name of the genus of liliaceous climbers containing greenbriars (*q.v.*). Improperly applied in the florist's trade to the liliaceous climber *Asparagus asparagoides*

Sminthuridae a family of collembolan insects distinguished from the Entomobryidae by the possession of globular bodies

blacksmith the pomacentrid fish *Chromis punctipinnis*

　mountain blacksmith *W.I.* the fringillid bird *Loxigilla portoricensis* (= Puerto Rican bullfinch)

　coppersmith the capitonid piciform bird *Megalaima haemacephala* (*cf.* barbet)

smk *see* [smoky]

lady's smock the cruciferous herb *Cardamine pratense*

earth smoke any of several species of herbs of the papaveraceous genus Fumaria (= fumitory)

　prairie smoke the ranunculaceous herb *Anemone patens* (= pasque flower)

[smoky] a mutant (gene symbol *smk*) mapped at 58.6± on chromosome II of *Drosophila melanogaster*. The phenotypic body is dark in color

smolt the migratory stage of salmon

[smooth] a mutant (gene symbol *sm*) mapped at 91.5 on chromosome II of *Drosophila melanogaster*. The abdomen of the phenotype is hairless

smt *see* [small thorax]

smuck term for a group of jellyfish

smut 1 (*see also* smut 2, beetle) popular name of black ustinaginale basidomycete fungi, parasitic on many plants

smut 2 (*see also* smut 1) *Brit.* a flying swarm of any minute black flies

sn *see* [singed]

Sn *see* [Snowflake]

snag 1 (*see also* snag 2) that part of a dicotyledonous pruned twig which lies distal to the last bud and therefore dies

snag (*Brit. obs.*) **2** (*see also* snag 1) a slug or snail

snagrel the elaeocarpaceous herb *Aristolochia serpentaria* (= Virginia snakeroot)

snail 1 (*see also* snail 2, kite) any of numerous gastropod mollusks with a coiled shell

　apple snail popular name of ampullariids

　edible snail the helicid *Helix pomatia*

　diminutive flat snail popular name of skeneids

　French snail *U.S.* = edible snail

　brown garden-snail *U.S.* the helicid *Helix aspersa*

　white garden-snail *U.S.* the helicid *Theba pisana*

　green snail *U.S.* the turbinid *Turbo marmoratus*

　grooved snail popular name of planaxids

　left-handed snail popular name of triphorids

　hairy-keeled snail popular name of trichotropids

　knobby snail popular name of modulids

　moon snail any of several species of natacid of the genus Polynices

　wide-mouth snail popular name of lamellariids

　orb-snail popular name of freshwater planorbids

　pearly snail popular name of trochids

　pond snail popular name of lymnaeids and physids

　two-ridged snail popular name of colubrariids

　high-spired snail popular name of cerithiopsids

　tubular snail popular name of caecids

　violet snail popular name of janthinids

　wing snail popular name of aporrhaids

　banded wood snail *U.S.* the helicid *Cepaea nemoralis*

snail 2 (*see also* snail 1) in the names of other organisms. The leguminous herb *Medicago scutellata* is called "snails"

sea snail *U.S.* the liparid fish *Liparis atlanticus*

snake 1 (*see also* snake 2, bird, blenny, eel, feeder, fly, gourd, grass, head, iris, mackerel, root, skink) popular name of members of the Serpentes

　Aesculapian snake the colubrid *Elaphe longissima*

　beaked snake the colubrid *Scaphiophis albopunctata*

　red-bellied snake *U.S.* the colubrid *Storeria occipitomaculata* (*cf.* brown snake)

　black snake *U.S.* without qualification, *C. constrictor* (= black racer) *Austral.* the elapid *Pseudechis porphyriaceus*

　blind snake *U.S.* the leptotyphlopid *Leptotyphlops dulcis*

　brown snake *U.S.* any of several species of colubrid of the genus Storeria

　bullsnake *U.S.* the colubrid *Pituophis melanoleucus sayi*

　salt-bush snake *Austral.* any of several species of pygopodid saurians of the genus Pygopus (= scaly foot)

　carpet snake *U.S.* the pythonine *Python spilotes*. *Austral.* the pythonid *Morelia argus* (= Australian python)

　chicken snake *U.S.* = rat snake *U.S.*

　coral snake any of numerous snakes with red and black rings. *U.S.*, without qualification, the elapid *Micrurus fulvius*

　"false" coral snake any of several colubrids of the genera Erythrolamprus or Pliocercus but usually, without qualification, the aniliid *Anilius scytale*

　"true" coral snake any red, yellow, and black banded elapid of the genera Micrurus and Micruroides

　corn snake *U.S.* the colubrine *Elaphe guttata* (*cf.* rat snake)

　crowned snake *U.S.* any of several species of colubrid of the genus Tantilla

　earth snake *U.S.* either of two species of colubrid of the genus Haldea

　egg-eating snake popular name of rhachiodontine and dasypeltine colubrids

　file snake any of several colubrids of the genus Mehelya

　fox snake *U.S.* the colubrine *Elaphe vulpina* (*cf.* rat snake)

　garter snake *U.S.* any of very numerous colubrids of the genus Thamnophis

　glossy snake the colubrid genus *Arizona elegans*

　aglyph snake one having solid conical teeth

　opisthoglyph snake one having enlarged grooved teeth at the rear of the maxillary series

proteroglyph snake one having permanently erected hollow fangs in front of the maxillary series

solenoglyph snake one having canuled fangs attached to rotatable maxillaries

grass-snake *Brit.* the natricine colubrid *Natrix natrix*

green snake *U.S.* either of two species of colubrine of the genus Opheodrys

ground snake *U.S.* the colubrid *Sonora episcopa*

black-headed snake any of several species of colubrid of the genus Tantilla

flat-headed snake *U.S.* the colubrid *Tantilla gracilis*

house snake *E.Afr.* the colubrid *Boaedon lineatum*

indigo snake *U.S.* the colubrine *Drymarchon corais*

jumping snake *Austral.* = salt-bush snake *Austral.*

kingsnake *U.S.* any of several species of colubrid of the genus Lampropeltis

lined snake *U.S.* the colubrid *Tropidoclonion lineatum*

lyre snake any of several species of colubrid of the genus Trimorphodon

mangrove snake the colubrid *Boiga dendrophila*

milk snake *U.S.* the colubrid *Lampropeltis doliata*

buttermilk snake *U.S.* the colubrine *Coluber constrictor anthicus*

mole snake *U.S.* the colubrid *Lampropeltis calligaster*

ringneck snake *U.S.* colubrids of the genus Diadophis

night snake either of two species of colubrid of the genus Hypsiglena, particularly *H. torquata*

hognose snake *U.S.* any of several species of colubrid of the genus Heterodon

hook-nosed snake *U.S.* either of two species of colubrid of the genus Ficimia

leaf-nosed snake any of several colubrids of the genus Phyllorhynchus

long-nosed snake any of several species of colubrid of the genus Rhinocheilus, particularly *R. lecontei*

patch-nosed snake any of several species of colubrid of the genus Salvadora

sharp-nosed snake any of several species of colubrid of the genus Rhamphiophis

shovel-nosed snake any of several species of colubrid of the genus Chionactis

pine snake *U.S.* the colubrid *Pituophis melanoleucus*

pipe snake *U.S.* the anilid *Anilius scytale* (= false coral snake). *I.P.* any of several species of anilid of the genus Cylindrophis

queen snake *U.S.* the colubrine *Regina septemvittata*

rat snake popular name of the colubrine genus Elaphe

Indian rat snake *I.P. Ptyas mucosus*

rattlesnake (*see also* grass, master, plant, plantain, root, weed) *U.S.* properly any of several species of crotaline of the genus Crotalus without qualification usually *C. horridus*, but the term is also extended to other crotalines

diamondback rattlesnake *C. adamanteus*

western diamondback rattlesnake *C. atrox*

pigmy rattlesnake *U.S. Sistrurus miliarius*

prairie rattlesnake *C. viridis*

rock rattlesnake *C. lepidus*

black-tailed rattlesnake *C. molossus*

reed snake any of numerous colubrids of the genus Calamaria

ribbon snake *U.S.* the colubrid *Thamnophis sauritus* (*cf.* garter snake)

Round-Island snake popular name of boine snakes of the genera Casarea and Bolyeria

sand snake any of numerous African colubrids of the genus Psammophis

banded sand snake any of several species of colubrid of the genus Chilomeniscus

scarlet snake *U.S.* the colubrid *Cemophora coccinea*

sea-snake any of several species of the family Hydrophiidae

smooth snake *Brit.* the colubrid *Coronella austriaca*

spindle snake any of several colubrid snakes of the genus Atractus

sunbeam snake the xenopeltid *Xenopeltis unicolor*

swamp snake *U.S.* the colubrid *Regina alleni*

short-tailed snake *U.S.* the colubrid *Stilosoma extenuatum*

tiger snake *Austral.* the elapid *Notechis scutatus*. *Aft.* the colubrid *Tarbophis semiannulatus*

tree snake any of numerous species of elongate tree-dwelling snakes, usually, without qualification, colubrines of the subgenus Letophis. Many of the dipsadomorphine colubrids are also called tree snakes

green tree snake any of numerous African colubrids of the genus Chlorophis

elephant's trunk snake = oriental water snake

vine snake any of several species of colubrid of the genus Oxybelis

Java wart snake = oriental water snake

water snake *U.S.* any of several natricine colubrids of the genus Natrix. *Brit.* the natricine *Natrix natrix*

black-bellied water snake *E.Afr.* the colubrid *Hydraethiops melanogaster*

oriental water snake popular name of acrochordids

whipsnake *U.S.* the colubrid *Masticophis flagellum*

wolf snake any of several species of colubrid of the genus Lycophidion

wood snake any of numerous colubrid snakes of the genus Coluber

snake 2 (*see also* snake 1) a few organisms of snake-like appearance, though in this connection the word is more frequently used as an adjective (*e.g.* snake blenny)

glass-snake the anguid lacertilian *Ophisaurus gracilis*

snapper 1 any of several fish particularly (*U.S.*) lutjanids mostly of the genus Lutjanus. The term is also locally applied to juvenile bluefish. *Austral.* any of several sparid fishes but particularly *Pagrosomus auritus*

black snapper the lutjanid *Apsilus dentatus*

Cubera snapper the *L. cyanopterus*

dog snapper the *L. jocu*

glasseye snapper the priacanthid *Priacanthus cruentatus*

blackfin snapper *L. buccanella*

gray snapper *L. griseus*

lane snapper *L. synagris*

mahogany snapper *L. mahogoni*

mutton snapper *L. enalis*

queen snapper the lutjanid *Etelis oculatus*

red snapper *L. blackfordi*

Caribbean red snapper *L. campechanus*

silk snapper *L. vivanus*

vermilion snapper the lutjanid *Rhomboplites aurorubens*

yellowtail snapper the lutjanid *Ocyurus chrysurus*

snapper 2 any of several reptiles but usually, without qualification, the chelydrid turtle *Chelydra serpentina* (= snapping turtle) black **snapper** *U.S.* the crotaline snake *Sistrurus catenatus* (= swamp rattler)

alligator **snapper** the chelydrid turtle *Macroclemys temmincki*

snare any device by which an organism, including humans, entangles its prey. *Brit.* specifically, a wire loop used to catch birds, and more particularly rabbits

[**Snell's waltzer**] a mutation (gene symbol *sv*) in linkage group II of the mouse. The phenotypic expression is the waltzing mouse syndrome

snipe (*see also* eel) any of several charadriiform birds mostly of the scolopacid genus Gallinago and its near allies, usually, without qualification, *G. gallinago*. *W.I.* either of two charadriid birds *Charadrius collaris* and *Charadrius wilsoni* (*cf.* plover) *N.Z.* the scolopacid *Coenocorypha aucklandica*

common **snipe** *G. gallinago*

Ethiopian **snipe** *G. nigripennis*

grass **snipe** *Brit.* (rare) = jacksnipe

great **snipe** *G. media*

jacksnipe *U.S. Lymnocryptes minimus*. The term is also applied to the male of the common snipe

painted **snipe** either of two species of rostratulid bird

pool **snipe** the scolopacid *Totanus totanus* (= common redshank)

quail-**snipe** = seedsnipe

seedsnipe any of four species of thinocorid birds

Swinhoe's **snipe** *G. megala*

fantail **snipe** = common snipe

Wilson's **snipe** = common snipe

snipper *U.S. dial.* = hellgrammite

snook any of several centropomid fishes of the genus Centropomus, *U.S.*, without qualification, usually *C. undecimalis*

little **snook** the *C. parallelus*

swordspine **snook** *C. ensiferus*

tarpon **snook** *C. pectinatus*

snout 1 (*see also* snout 2–4) that portion of a muzzle (*q.v.*) immediately surrounding the nasal apertures

snout 2 (*see also* snout 1, 3, 4) any of several species of lepidopteran insect. *Brit.* any of many species of noctuids of the genera Bomolocha, Hypena and Hypenodes. Usually without qualification, *Hypena proboscidalis. S.Afr.* the libytheid *Libythea lebdaca*

snout 3 (*see also* snout 1, 2, 4) any of several species of agonid fishes (star-snouts) of the genus Asterotheca

snout 4 (*see also* snout 1–3) *Brit.* the labiateous herb *Galeobdolon luteum* (weasel snout)

red **snow** snow colored by colonies of cryophilic, red-colored, chlorophyaceans, most commonly *Chlamydomonas nivalis*

[**Snowflake**] a mutant (gene symbol *sn*) in linkage group I of *Neurospora crassa*. The term refers to the appearance of the conidiated colonial growth

snow-on-the-mountain *U.S.* the euphorbiaceous herb *Euphorbia marginata*

snowy used by *biol.* as *syn.* with "white"

so *see* [Soft]

soboIe a shoot rising directly from the ground

-sobrin- *comb. form* meaning "cousin"

social pertaining to the interrelationship of organisms

sociation a stable plant community with one or more dominants at each level

association 1 (*see also* association 2) so many ecologists have applied so many meanings to this word that it is now valueless as a specific term; in general, it usually indicates a large assemblage of organisms in a specific area with one or two dominant species (*cf.* faciation, lociation)

chief **association** = stable association

closed **association** one in which there is no room for further growth

complementary **association** an association of plants which do not compete either because they root at different depths or have their principal above-ground forms at different seasons

homotypical **association** an assemblage of organisms of the same species occurring together because they are descendents of the same parent (**primary homotypical association**) or parents (**secondary homotypical association**)

intermediate **association** one in which a little room for growth remains

mixed **association** one in which several species compete for dominance

Mullerian **association** an assemblage of different organisms in one geographical locality all showing similar aposematic colors

open **association** one with ample room for further growth

passage **association** an association in the course of changing from one type to another

progressive **association** any association which is not stable

pure **association** an association completely dominated by a single species

retrogressive **association** an association which has passed the stable phase

stable **association** an association in a state of equilibrium

unstable **association** one which is just beginning

subassociation (*see also* subsociation) an assemblage of organisms which do not, according to the particular usage of the individual writer, agree completely with an association

subordinate **association** an association which is either progressive or regressive

substitute **association** a secondary formation which has replaced a stable association

transitional **association** one in the course of development

association 2 (*see also* association1) pertaining to chromosomes

heterogenetic **association** the pairing of chromosomes derived from different ancestors in an allotetraploid

consociation an association having a single dominant and distinguished from other types of associations by this single characteristic

bacterial **dissociation** the appearance of types differing from the original

subsociation a portion of an association which is distinguished by the presence of a subdominant which is, however, itself clearly under the influence of the dominant

socies a sere, or seral community, with one or more dominants

associes a term only slightly more specific than association (*q.v.*). It refers in general to a transitory or intermediate stage in the development of an association taken as a whole. It has also been defined as a developmental unit of consocies

consocies a portion of an association lacking one or more of its dominant species, or a portion of an association characterized by one or more of the dominants of the association; the term is also used of a seral community with a single dominant

isocies a group of organisms associated together but of different taxonomic affinities, sometimes used merely in the sense of habitat group

subsocies this term is to associes as subsociation is to association

society a term used in so many various ways that it no longer has any specific meaning, as distinct from the generalized meaning of an assemblage of organisms

adoption society an assemblage of one or more organisms living together though free to dissociate should they wish and to none of which does the continued association bring any apparent advantage

closed society a group of animals, human or not, which will not admit any addition to their ranks

complementary society an association of plants which can exist together in virtue of different rooting depths (= complementary association)

[soft] a mutant (gene symbol *so*) in the linkage group I of *Neurospora crassa*. The phenotypic expression is a densely pigmented growth in the lower part of the slant

sokhor any of several species of small fossorial myomorph rodent of the genus Myospalax

Solanaceae a large family of tubuliflorous dicotyledons containing in addition to the potato, eggplant, and tobacco the extremely poisonous nightshades and daturas. The numerous seeds and pleated corolla is typical of the family

[solanifolium] a mutant (gene symbol *sf*) mapped at 34 on linkage group A of the tomato. As the name indicates the phenotypic leaf is potato-like

Solanineae a suborder of tubifloral dicotyledonous angiosperms containing the families Nolanaceae, Solanaceae, Scrophulariaceae, Bignoniaceae, Pedaliaceae, Martyniaceae, Orobanchaceae, Gesneriaceae, Columelliaceae, Lentibulariaceae, and Globulariaceae

solation the passage of colloid from the gel to the sol condition

soldier (*see also* beetle, bug, ergate, fly) an organism with the red color, haughty bearing or predatory habits attributed to the soldiery of past ages. *W.I* the recurvirostrid bird *Himantopus himantopus*

British soldier the lichen *Cladonia cristatella*

brown soldier *Austral.* the nymphalid lepidopteran insect *Precis hedonia*

poor soldier *Austral* = fire-bird *Austral.*

water soldier the hydrocharitaceous aquatic herb *Stratiotes aloides*

-sole- *comb. form* meaning "sandal"

sole 1 (*see also* sole 2, 3) in the sense of footgear, specifically applied to the distal end of a carpel

creeping sole a median ridge, bearing unusually large cilia on the underside of the surface of many terrestrial Turbellaria

sole 2 (*see also* sole 1, 3) in the sense of a sandal-shaped fish. Properly any heterostomate fish of the family Soleidae but often applied to other flatfishes of the families Bothidae and Pleuronectidae. *Brit.*, without qualification, the soleid fish *Solea solea*

butter sole *U.S.* the pleuronectid *Isopsetta isolepis*

C-O sole the pleuronectid *Pleuronichthys coenosus*

Dover sole *U.S.* the pleuronectid *Microstomus pacificus. Brit.* the soleid *Solea solea*

English sole *U.S.* the pleuronectid *Parophrys vetulus. Brit.* the term is not used

curlfin sole *U.S.* the pleuronectid fish *Pleuronichthys decurrens*

yellowfin sole *U.S.* the pleuronectid fish *Limanda aspera*

flathead sole *U.S.* the pleuronectid *Hippoglossoide elassodon*

lemon sole *Brit.* the pleuronectid fish *Pleuronectes limanda*

lined sole *U.S.* the soleid fish *Achirus lineatus*

bigmouth sole the bothid *Hippoglossina stomata*

naked sole *U.S.* the soleid *Gymnachirus nudus*

petrale sole *U.S.* the pleuronectid *Eopsetta jordani*

rex sole *U.S.* the pleuronectid *Glypocephalus zachirus*

rock sole *U.S.* the leuronectid *Lepidopsetta bilineata*

sand sole *U.S.* the pleuronectid *Psettichthys melanostictus*

scrawled sole *U.S.* the soleid *Trinectes inscriptus*

deep-sea sole the pleuronectid *Embassichthys bathybius*

slender sole *U.S.* the pleuronectid *Lyopsetta exilis*

fantail sole *U.S.* the bothid *Xystreurys liolepsis*

tongue sole popular name of cynoglossid fishes

turn **sole 3** (*see also* sole 1, 2) any of several species of the boraginaeous family Heliotropium

Soleidae a family of heterostomate fishes properly called "true soles." They have both eyes on the right side of the head but are distinguished from the Pleuronectidae by having teeth only on the blind side, barbels on the lower side of the head and the preopercular margin of the gill cover hidden by the skin and scales. Most of the true soles are European and most American fishes called soles are pleuronectids

Solemyidae a family of pelecypod mollusks distinguished by the elongate shells with a tough and glossy periostracum extending beyond the valves

-solen- *comb. form* meaning a "tube"

Solenichthyidae = Solenostomidae

Solenidae a family of pelecypod mollusks commonly called razor-clams in the *U.S.* and razor shells in *Brit.* They have elongate flattened shells much the size and shape of the hand razor of the past

soienidion a striated sensory seta found on some mites

solenodon popular name of solenodontid insectivores

Solenodontidae a family of recently extinct, or thought to be extinct, insectivores confined to the West Indies

Solenogastres = Aplacophora

Solenostomidae a family of gasteriform fishes closely allied to the Syngathidae and commonly called ghost pipefishes. They have a long tube-like snout and many species have armor plates

Solifugae = Solpugida

solitaire any of several, mostly passeriform, turdids of the genera Entomodestes and Myadestes. Without qualification frequently refers to the extinct raphid bird *Pezophaps solitaria*

brown-backed solitaire *M. obscurus*

slate-colored solitaire *M. unicolor*

Cuban solitaire *M. elisabeth*

green solitaire the drepaniid *Loxops sagittirostris*

rufous throated solitaire *M. genibarbis*

Townsend's solitaire *U.S. M. townsendi*

Solpugida a small order of arachnid arthropods commonly called the sun-spiders. They are distinguished by the possession of very large chelicerae, a prosome with the posterior three somites free, an opisthosoma of ten somites and the presence of peculiar stalked leaf-like organs (malleoli) on the fourth pair of legs

solute apart from its regular physical and chemical meaning, this word is used in *biol.* in the sense of separate or non-adherent

som *see* [sombre]

-soma- *comb. form* meaning "body." The forms -soma, -some and somite all enter into compounds, the last usually being used as synonymous with metamere

soma a body division (*cf.* -some-, somite)

   gnathosoma that division of the acarine body which bears a mouth opening and mouth parts

   hysterosoma that region of the acarine body that is composed of the metapodosoma and opisthosoma

   idiosoma that region of the acarine body which is posterior to the gnathosoma (*cf.* idiosome)

   opisthosoma that division of the acarine body which is posterior to those regions bearing legs

   podosoma that region of the acarine body which carries legs

   metapodosoma that division of the acarine body which carries the third and fourth pair of legs

   propodosoma that region of the acarine body that carries the first and second pair of legs

   polysoma = polyploid

   presoma 1 (*see also* presoma 2) the short, anterior division of the body of priapulids and acanthocephalans

   presoma 2 (*see also presoma* 1) the larval rudiment of the proboscis in the nematomorph embryo

   proterosoma that region of the acarine body that consists of the gnathosoma and propodosoma

-somal pertaining to -soma- or -some- in any of their varied meanings (*cf.* -somic)

Somateriini a group of Anatinae containing the eider ducks. They closely resemble the Aythyini but are marine, non-vegetarian, forms

somatic properly pertaining to the body but when used alone specifically means pertaining to the body wall as distinct from the splanchnic mass

   asomatic the condition of containing only embryonic tissues

   perisomatic said of those echinoderm skeletal elements which are secondarily developed and to which the terms "actinal" and "abactinal" (*q.v.*) cannot therefore be applied

   polysomatic pertaining to somatic nuclei containing replicated chromosomes

[sombre] a mutant (gene symbol *som*) mapped at 40.8 on the X chromosome of *Drosophila melanogaster*. The name refers to the dark body color of the phenotype

-some- *see* -soma the terminations -some and -soma (*q.v.*) are used interchangeably

-some 1 (*see also* -some 2) in the sense of a division of a body

   acrosome a cytoplasmic cap-like structure on the front of a spermatozoan

   beanosome = thorax

   choanosome the sum total of all sponge tissues lying within the outer cover or ectosome

   cytosome = cytoplasm

   ectosome the outer layer of a leuconoid sponge which is devoid of flagellated chambers

   elaiosome that portion of a plant which produces an oily secretion

   gonosome the sum total of the gonophores, gonothe-

cae and related structures in a hydrozoan coelenterate

   mesosome the "collar" of hemichordates

   metasome in hemichordates, that part of the body which lies posterior to the collar

   nectosome that portion of the stem of a siphonophoran hydrozoan which bears swimming cells

   protosome the proboscis of hemichordates

   siphonosome that portion of the stem of a siphonophoran hydrozoan which lies below the nectosome

   stichosome that part of a trichuroid nematode pharynx which is covered with stichocytes

   trophosome an organ concerned in the production or securing of nourishment as distinct from reproduction

   urosome in insects, the abdomen

-some 2 (*see also* -some 1) in the sense of a discrete body, usually a cell inclusion or organelle

   acrosome *see* some 1

   allosome a chromosome which is different from the rest, usually the sex chromosome

   androsome a male chromosome

   autosome any chromosome except a sex chromosome

   centrosome a minute area of differentiated cytoplasm which gives rise to two centrioles at the beginning of mitosis

   chondriosome = mitochondrion

   chromosome (*see also* segment) thread-like bodies into which the hereditary material of the nucleus is organized. They usually become apparent only in the course of cell division

     Balbiani chromosome = giant chromosome

     lamp-brush chromosomes chromosomes from which, as the name indicates, lateral loops stick out so as to provide a lamp-brush or bottle-brush appearance. They are principally found in the mid prophase of the first meiotic division particularly in the oocyte nuclei of those vertebrates having heavily yolked eggs

     acrocentric chromosome one with a subterminal centromere

     metacentric chromosome one in which the centromere is medially located

     paracentric chromosome an inverted chromosome in which the centromere lies outside the area of inversion

     telocentric chromosome one in which the centromere is terminal

     daughter chromosome those chromosomes developed through the separation of the chromatids in mitosis

     euchromosome = autosome

     giant chromosomes unusually large chromosomes, particularly those found in the salivary glands of certain diptera

     heterochromosome one of those, such as the sex chromosome, which is morphologically distinguishable but pair in meiosis

     heterobrachial chromosome one in which the centromere is not in the middle

     heteropycnotic chromosome one which is shortened and thickened more than the other chromosomes of the set

     homologous chromosome one which is identical to another with respect to the position of the gene loci

     idiochromosome an early name for a sex chromosome

isochromosome a metacentric chromosome with two identical arms

polytene chromosome a giant chromosome resulting from the endoreduplication of numerous parallel chromatids

prochromosome a mass of heterochromatin found in a resting nucleus and once thought to be the precursor of a chromosome

ring chromosome one fused into a ring

sex chromosome that chromosome which has no exact homologue among the chromosomes of the opposite sex (*see* W, X, Y and Z chromosomes)

telomitic chromosome one having a terminal centromere

W chromosome a term used by some biologists for a sex chromosome which is heterozygous in the female (ZW) and absent in the male (ZZ)

X chromosome that sex chromosome which is commonly homozygous (XX) in the female and heterozygous (XY) in the male

Y chromosome that sex chromosome which is commonly heterozygous (XY) in the male and absent from the female

Z chromosome a term for the sex chromosome of those forms in which the male is homozygous (ZZ) and the female heterozygous (ZW)

cytosome = cytomicrosome

desmosome 1 (*see also* desmosome 2) thickened areas of closely opposed, but not fused, plasma membranes (*cf.* nexus)

desmosome 2 (*see also* desmosome 1) intercellular connections between epithelial cells

dictyosome any portion of the Golgi apparatus visible by light microscopy. The term is also used for laminations disclosed by electron microscopy

diplosome one of a pair of centrosomes

ectosome (*see also* -some 2) one of numerous granules found in a primordial germ cell and which distinguishes this cell from other blastomeres

electosome = mitochondrion

endosome a body, staining deeply with nuclear dyes, within the nucleus

episome a genetic factor that can exist either in the chromosome or in the cytoplasm, particularly elements in the genetic makeup of bacteria which can exist either integrated in the chromosome or as autonomous, virus-like, particles (*cf.* plasmagene)

genosome an early term for what is now called locus

heterosome = heterochromosome

hyalosome a granule in protoplasm which does not readily stain

deuthyalosome the nucleus remaining in the ovum after the second reduction division

karyosome a nucleolus-like body, exhibiting the reactions of chromatin

kinetosome the basal body of a cilium or flagellum

lysosome a membrane-bounded particle (*circ.* 0.1 microns in diameter) in the cytoplasm (*cf.* ribosome) showing acid phosphotase activity

microsome 1 (*see also* microsome 2) any cell granule, not otherwise identified as to structure or function. This usage is nowadays largely obsolete

cytomicrosome a microsome in the cytoplasm

microsome 2 (*see also* microsome 1) small ribonucleic acid-containing granules within the cytoplasm. Some are associated with endoplasmic reticulum

monosome = haploid

mycetosome a fungus-infected polar body which apparently plays some role in the reproduction of coccid Insecta

oleosome a fatty inclusion within a cell

pangenosome a complex of pangens (*see under* gen 1)

phagosome = lysosome (particularly one which has engulfed another organelle)

phragmosome a cytoplasmic precursor of the phragmoplast appearing before the commencement of cytokinesis

plasmasome 1 (*see also* plasmosome 2) any protoplasmic granule

plasmosome 2 (*see also* plasmosome 1) = nucleous

plastosome = mitochondrion

ribosome a discrete particle (*circ.* 200 Å in diameter) not membrane-bounded, on the walls of the endoplasmic reticulum

spherosome any spherical cytoplasmic granule

zygosome a pair of fused chromosomes

somic *adj. suffix* pertaining to -soma- in any of its forms or meanings; -somal is synonymous but is used also as an *adj. suffix* from -soma- in all its forms

disomic having an additional chromosome, usually in virtue of nondisjunction

monosomic the condition of lacking one chromosome from the genone

nullisomic lacking one or more pairs of chromosomes, usually by virtue of nondisjunction

somite literally a little body, but usually used for one segment of a metamerically segmented organism (*cf.* metamere *under* mere) particularly used of vertebrates in the embryonic condition

soncoya the annonaceous tree *Annona purpurea* and its fruit

song any rhythmic, repetitive noise produced by an animal

sonification = noise making

Sonneratiaceae a small family of myrtiflorous dicotyledonous angiosperms

sook *W.I.* the picid bird *Dendrocopos villosus* (= hairy woodpecker)

soursop (*see also* bird) the fruit of the annonaceous tree *Annona muricata*

mountain soursop *A. montana*

sweet sop the annonaceous tree *Annona squamosa* and its fruit

-sor- *comb. form* meaning "a heap of anything"

sor a shallow lagoon at the mouth of a large river

sor a bacterial mutant showing a utilization of sorbose

sora 1 (*see also* sora 2) the rallid bird *Porzana carolina* (*cf.* crake)

sora 2 (*see also* sora 1) pl. of sorus

Sorangiaceae a family of myxobacteriale Schizomycetes distinguished by the polygonal, angular cysts containing the fruiting body rods. Many cysts are often included within a single membrane

sord term for a group of mallards, though flock and flight are frequently used

sordid dirty white

sore- *comb. prefix* predominantly used in falconry indicating a bird of the first year that has not molted

e **sorediate** lacking soredia

soredium an algal cell surrounded by hyphal tissues and which can, when detached, produce a lichen thallus (*cf.* sorus)

sorel a three year old fallow deer

sorema a heap of anything, but particularly a mass of reproductive structures as in a flower

sorghum (*see also* midge, worm) *hort. var.* of the grass *Holcus halepensis*

sorgo = sorghum

Soricidae a family of insectivorous mammals containing the true shrews. They are distinguished from other insectivores by the forwardly directed lower front teeth. Most have long tails and snouts.

-soro- *see* -sor-

sorose a fleshy multiple furit

sorrel any of several polygonaceous herbs of the genus Rumex, usually, without qualification *R. acetosa*

field sorrel *R. acetosella*

French sorrel *R. scutatus*

garden sorrel *R. acetosa*

Jamaica sorrel the malvaceous herb *Hibiscus sabdariffa*

lady's sorrel any of several species of herb of the oxalidaceous genus Oxalis

mountain sorrel *U.S.* the polygonaceous herb *Oxyria digyna. Brit. O. reniformis*

sheep sorrel = field sorrel

wood-sorrel any of several oxalidaceous herbs of the genus Oxalis

sorus a cluster of sporangia in ferns or fungi or a mass of soredia on the surface of a lichen.

sounder term for a group of boars, or pigs in general

souslik any of very numerous squirrels of the genus Spermophilus

sow (*see also* bread, bug, thistle) without qualification, a female domestic pig, but applied also to the females of several other animals such as the bear, mink

sp *see* [self-pruning], [speck] or [spray]

sp. nov. abbreviation for *species nova* (new species)

sp., spps. abbreviation for species (*sing.*) and species (*pl.*)

sp₁ *see* [small pollen]

Sp *see* [Splotch] or [Sternopleural]

peridural space the perimeningeal space in those forms having a dura spinalis

subdural space the space between the pia mater and the dura spinalis

epimeningeal space the space between the single menix and the endorhachis in primitive vertebrates

perivitelline space the space between the vitelline and perivitelline membrane

[spade] a mutant (gene symbol *spd*) mapped at 22.3 on chromosome II of *Drosophila melanogaster*. The phenotype has shortened spade-shaped wings

-spadic- *comb. form* confused from two roots meaning "palm frond" and "brown"

spadiaceous either pertaining to a spadix or to a date palm, according to context

spadix 1 (*see also* spadix 2, 3) a spike with an unusually fleshy axis in the Araceae

spadix 2 (*see also* spadix 1, 3) that portion of a blastostyle on which sporosacs develop

spadix 3 (*see also* spadix 1, 2) the copulatory organ of the cephalopod mollusk Nautilus

spain-spain *W.I.* the sylviid bird *Polioptila caerulea* (= blue-grey gnat-catcher *U.S.*)

Spalacidae a family of myomorph rodents resembling moles in their adaptation to a permanently fossorial existence. They are also distinguished from other rodents by the enormous lower incisor teeth used for digging

span an obsolete measure of length of approximately nine inches

spangle I.P. the papilionid lepidopteran insect *Papilio protenor*

gold spangle *Brit.* the noctuid lepidopteran insect *Pusia bractea*

spaniard *N.Z.* the umbelliferous herb *Aciphylla squarrosa* (= spear grass *N.Z.*)

jackspaniard *W.I.* the fringillid bird *Loxigilla violacea* (= Greater Antillean Bullfinch)

Sparassidae a family of Aranaea distinguished by eight similar eyes arranged in two rows, the presence of six spinerettes together with the presence of scopulae on the tarsi

Sparganiaceae that family of pandanale angiosperms which contains the bur-reeds. The plants forming the family are readily distinguished from the cat-tails by the globose heads

Sparidae a large family of acanthopterygian fishes distinguished by the very powerful cutting and grinding teeth in the jaws. Members of the family are known as porgies, sea breams, sheepshead and scup

spark *i.p.* any of numerous lycaenid lepidopteran insects of the genus Sinthusa

sparling the osmerid teleost fish *Osmerus eperlanus*

sparrow (*see also* hawk, lark, owl) any of very numerous passeriform birds of the families Prunellidae, Ploceidae, and Fringillidae

green-backed sparrow the fringillid *Arremonops chloronotus*

Bachman's sparrow the fringillid *Aimophila aestivalis*

Baird's sparrow the fringillid *Ammodramus bairdii*

orange-billed sparrow the fringillid *Arremon aurantiirostris*

black sparrow *W.I.* the fringillid *Melopyrrha nigra* (= Cuban fullfinch), *Loxigilla violacea* (= Greater Antillean bullfinch) and *Tiaris bicolor* (= black-faced grassquit) are all called by this name

Botteri's sparrow the fringillid *Aimophila botterii*

Brewer's sparrow the fringillid *Spizella breweri*

bridle-sparrow the fringillid *Aimophila mystacalis*

cane sparrow *W.I.* = grass sparrow *W.I.*

Cape sparrow the ploceid *Passer melanurus*

Cape Sable sparrow the fringillid *Ammospiza mirabilis*

Cassin's sparrow the fringillid *Aimophila cassinii*

black-chested sparrow the fringillid *Aimophila humeralis*

chestnut sparrow the ploceid *Sorella eminibey*

black-chinned sparrow the fringillid *Spizella atrogularis*

chipping sparrow the fringillid *Spizella passerina*

cinnamon sparrow the ploceid *Passer rutilans*

rufous-collared sparrow the fringillid *Zonotrichia capensis*

clay-colored sparrow the fringillid *Spizella pallida*

golden-crowned sparrow the fringillid *Zonotrichia atricapilla*

rufous-crowned sparrow the fringillid *Aimophila ruficeps*

white-crowned sparrow the fringillid *Domotrichia leucophyrs*

desert sparrow *U.S.* = black-throated sparrow; *Afr.*, the ploceid *Passer simplex*

English sparrow *U.S.* = house sparrow *Brit.*

field-sparrow the fringillid *Spizella pusilla*

fox-sparrow the fringillid *Passerella iliaca*

golden sparrow any of several ploceids of the genus Auripasser. Usually, without qualification, *A. luteus*

grass sparrow *W.I.* the fringillid *Sicalis luteola*

grasshopper-sparrow the fringillid *Ammodramus savannarum*

ground-sparrow any of several fringillid birds of the genus Melozone. *W.I.* = yellow grass-finch

white-eared groundsparrow *M. leucotis*

rusty-crowned ground-sparrow *M. kieneri*

Prevost's ground-sparrow *M. biarcuatum*

grey-headed sparrow the ploceid *Passer griseus*

stripe-headed sparrow the fringillid *Aimophila ruficauda*

hedge sparrow any of numerous passeriform birds of the family Prunellidae. *Brit.*, without qualification *Prunella modularis* (= dunnock)

Henslow's sparrow the fringillid *Passerherbulus henslowii*

house sparrow *Brit.* the ploceid *Passer domesticus*

Ipswich sparrow the fringillid *Passerculus princeps*

jacksparrow *W.I.* = Cuban bullfinch

Java sparrow the ploceid *Munia oryzivora* (*cf.* rice bird)

lark-sparrow the fringillid *Chondestes grammacus*

Le Conte's sparrow the fringillid *Passerherbulus caudacutus*

Lincoln's sparrow the fringillid *Melospiza lincolnii*

Oaxaca sparrow the fringillid *Aimophila notosticta*

olive sparrow the fringillid *Arremonops rufivirgatus*

parson-sparrow *W.I.* the fringillid *Tiaris bicolor* (= black-faced grassquit)

reed-sparrow *Brit.* = black-headed bunting

lesser rock-sparrow the ploceid *Petronia dentata*

pale rock-sparrow the ploceid *Carpospiza brachydactyla*

rufous sparrow the ploceid *Passer motitensis*

rusty sparrow the fringillid *Aimophila rufescens*

sage-sparrow the fringillid *Amphisziza belli*

savannah sparrow the fringillid *Passerculus sandwichensis*

seaside sparrow the fringillid *Ammospiza maritima*

dusky seaside sparrow the fringillid *Ammospiza baileyi*

Sierra Madre sparrow the fringillid *Xenospiza baileyi*

song sparrow the fringillid *Melospiza melodia*

Spanish sparrow the ploceid *Passer hispaniolensis*

striped sparrow the fringillid *Oriturus superciliosus*

black-striped sparrow the fringillid *Aimophila conirostris*

five-striped sparrow the fringillid *Aimophila quinquestriata*

swamp sparrow the fringillid *Melospiza georgiana*

cinnamon-tailed sparrow the fringillid *Aimophila sumichrasti*

sharp-tailed sparrow the fringillid *Ammospiza caudacuta*

black-throated sparrow the fringillid *Amphispiza bilineata*

white-throated sparrow the fringillid *Zonotrichia albicollis*

tree-sparrow *Brit.* the ploceid *Passer montanus; U.S.* the fringillid *Spizella arborea*

cinnamon tree-sparrow the ploceid *Passer cinnmomeus*

vesper sparrow the fringillid *Pooecetes gramineus*

rufous-winged sparrow the fringillid *Aimophila carpalis*

zapata sparrow the fringillid *Torreornis inexpectata*

sparse in *bot.*, frequently used in the sense of scattered or widely spaced

spat post-larval young oyster

-spat- *see* -spath-

-spath- *comb. form* meaning "a blade"

espathaceous lacking a spathe

spathe a large bract enclosing a spadix

bispathelulate having two glumes

Spathiflorae that order of monocotyledenous angiosperms which contains the Araceae and Lemmaceae. The order is distinguished by the minute, non-petaloid flowers

spathulate = spatulate

spatulate in the form of a spatula

spawn the eggs of fishes and amphibia. The term is sometimes applied to fish fry still having the yolk sac attached

Spc *see* [Speckle]

spd *see* [spade]

spear 1 (*see also* spear 2) any large pointed shoot, especially that of asparagus, rising directly from the soil

spear 2 (*see also* spear 1) of or pertaining to relationship on the male side (*cf.* distaff)

speciation the process by which new species are formed

allochronic speciation the production of morphological discontinuity between species based solely on the passage of time

allopatric speciation the production of morphological discontinuity arising from geographic fragmentation combined with the passage of time

sympatric speciation the production of morphological discontinuity without geographical separation

ring speciation the condition that occurs when an extensive cline curves round on itself so that the extremes overlap but interbreeding does not occur

species an organism which is, and remains, distinct because it does not normally interbreed with other organisms. Interspecific hybrids produced by domestication (*e.g.* mule) or captivity (*e.g.* tigrlion) are rarely fertile. (*cf.* cenospecies)

agamospecies a species in which sexual reproduction is unknown

allochronic species species separated in time

"biological" species an ill-defined term often used in popular writings in the sense of "race"

buffer species a form eaten by a predator, but which is not the natural prey of that predator and, therefore, "buffers" the effect of the predator on its normal prey

cenospecies species that can and do interbreed with each other. Cenospecies of the same comparium can produce fertile offspring (*cf.* plastospecies)

coenospecies = cenospecies

cryptic species = sibling species

directive species one which attracts a predator, of which it is not the customary prey, to an area rich in the customary prey

condominant species two or more, not necessarily mutually antagonistic, species that dominate a community (*cf.* cenosis)

ecospecies one of a group of populations associated with a specific environmental niche but capable of interbreeding with other neighboring ecospecies. The term is cometimes used as a synonym of ecotype (*q.v.*)

exclusive species a species of animal, the distribution of which is rigidly limited to environments showing specific types of plants or *vice versa*

holocyclic species those species of aphids in which reproduction is always sexual

incipient species a variation which, if stabilized, would become a subspecies

index species one which is typical of, and unique to, a particular habitat or geological stratum

indicator species = index species

indifferent species one which occurs in many types of habitats or communities

infraspecies a general term used for all categories (subspecies, variety, etc.) below specific rank

microspecies a genetically isolated species, usually one reproducing only parthenogenetically

nominal species one that has a name but lacks a taxonomically acceptable definition

allopatric species one that has been developed through geographic isolation

  sympatric species one which has developed in contact with another population with which it potentially could have, but did not, interbreed

plastospecies a population of plastospecies can, but usually does not, interbreed with another population of plastospecies (*cf.* cenospecies)

semispecies the species which are grouped into a superspecies

sibling species one which is morphologically identical with another species or group of species (themselves the other sibling species) but from which it is geographically isolated

subspecies a geographically restricted population which differs so consistently from other populations of the same species that it can be taxonomically defined

  infrasubspecies a term ocasionally applied to forms resulting from seasonal or other temporary polymorphism

superspecies a term used to designate a group of closely related allopatric species

type species the species from which the characteristics of a genus are derived

verspecies a term applied to a species, as such, when it is desired to indicate that it is neither a superspecies nor a subspecies

specific pertaining to species. Compounds derived directly from the species above are not given

conspecific being of the same species as

monospecific having a single species

[**specific dilutor**] a mutant (gene symbol *dil*) mapped at 57.0± on chromosome II of *Drosophila melanogaster*. It is a dilutor of [brown] and [white] alleles

[**speck**] a mutant (gene symbol *sp*) mapped at 107.0 on chromosome II of *Drosophila melanogaster*. The name comes from the black specks on the wings of the phenotype, the body of which is olive

[**Speckle**] a mutant (gene symbol *Spc*) mapped at 33.1 on linkage group IV of *Bombyx mori*. The phenotypic larva is speckled

[**Speckled**] a mutant (gene symbol *Sk*) in linkage group I of *Habrobracon juglandis*. The phenotypic eye is white with bright red flecks

crimson speckled *Brit.* the arctiid lepidopteran insect *Deiopeia pulchella*

spectacle *Brit.* either of two species of noctuid lepidopteran insect of the genus Abrostola but usually without qualification, *A. tripartita*

speculum the brightly metallic-colored area in the wing of some birds (*e.g.* a drake mallard)

speedwell any of very many species of scrophulariaceous herbs or shrubs of the genus Veronica, particularly *V. officinalis*

  bastard speedwell *V. spuria*

  common speedwell *V. officinalis*

  germander speedwell *V. chamaedrys*

  marsh speedwell *V. scutellata*

  mountain speedwell *V. montana*

  St. Paul's speedwell *V. serpyllifolia*

-speir- *comb. form* meaning "twist" or "twisted" usually transliterated -spir-

speirema a gonidium of a lichen

spekvreter *S. Afr.* the turdid bird *Cercomela familiaris* (= red-tailed chat)

spelt the grass *Triticum spelta*, a wheat of little commercial value

Spemann's *epon. adj.* from Hans Spemann (1869–1941)

-sperm- *comb. form* meaning "seed"

  allosperm an ovule produced by allogamy (*q.v.*)

  archesperm fertilized content of an archegonium

  carposperm the female gamete of algae after fertilization (*cf.* carposphere)

  cochlidiosperm a seed in a form of a dish, or of the bowl of a spoon

  bastard embryosperm an embryo produced parthenogenetically by a plant the parthenogenesis being induced by a foreign pollen

  geitonembryosperm a plant with a parthenogenetic embryo, produced after fertilization with pollen from another flower of the same plant

  xenembryosperm = parthenembryosperm

  endosperm the food reserve of a seed

    autendosperm an endosperm resulting from the fertilization of a flower with its own pollen

    geitonenosperm a plant with parthenogenetic endosperm produced after fertilization with pollen from another flower on the same plant

    parthendosperm a seed, the endosperm of which is parthenogenetic, and the embryo the result of fertilization

    ruminate endosperm one mottled in appearance through the penetrating of the seed coat into the endosperm

    xenoendosperm a plant embryo produced by fertilization in which the endosperm is parthenogenetic and the pollen derived from another individual

  gamosperm a plant which does not reproduce parthenogenetically

  geitonosperm a plant with sexually produced embryos resulting from fertilization by another flower on the same plant

  pseudogymnosperm = cycad

  hyposperm that part of an ovule in which the nucellus is not connected to the integument

  mesosperm the middle coat of a seed

  metasperm = angiosperm

  nomosperm the type of seed customary in the group

  oosperm = zygote

  opseosperm a tuberculous oophore on the surface of some algae

  orthosperm a seed with a ventral groove

  parthenosperm that which is involved in etheogenesis

    euparthenosperm used of plants the entire seed of which is produced parthenogenetically

    hemiparthenosperm a plant in which either the embryo or endosperm is parthenogenetic, but not both

  perisperm storage tissue in the seed derived from the remnants of the nucellus

  podosperm = funicle

  phyllosperm a seed, or other reproductive organ, borne on a leaf

  pseudosperm a fruit which is indehiscent

  pterosperm a winged seed

  stachyosperm an organism bearing seeds on the stem

  trachysperm a seed with a rough coat

  trophosperm = placenta (*Bot.*)

  zoosperm = zoospore

spermarium = antheridium

spermary = pollen tube

spermatid the mother cell of an antherozoid or of a spermatozoan

spermatium a non-motile male gamete

paraspermatium reproductive bodies of algae

Spermatophyta that great division of the plant kingdom which contains the "flowering" or more properly, "seed bearing" plants

spermatozoa motile male gametes

-spermia plant or animal taxa having this ending are listed alphabetically

spermous pertaining to sperm. The alternative form spermic is rarely used

aspermous lacking seeds

amphispermous the condition in which the pericarp functions as the seed coat

cyclospermous a plant embryo which has coiled round its own food reserve

exendospermous said of a seed in which the food reserve is in the embryo

agamospermous said of a plant which produces seed without a sexual process through the direct outgrowth of an embryo from the cell of the parent sporophyte

epiphyllospermous bearing reproductive organs on fronds or leaves

lithospermous with a hard seed

angiomonospermous said of a plant in which each carpel produces a single ovule

epiperispermous said of a seed lacking perisperm or food reserves

heterospermous possessing more than one kind of seed or sperm

pleiospermous having an unusually large number of seeds, as for example, in many orchids

polyspermous 1 (*see also* polyspermous 2) the condition of an egg fertilized by more than one sperm

polyspermous 2 (*see also* polyspermous 1) said of a pericarp containing numerous seeds

synspermous the condition of having several seeds fused together

synaptospermous the condition of a plant the seeds of which are not dispersed but germinate alongside the parent

spet = sennet

-sphac- *see* -sphak-

Sphacelariales an order of isogenerate Phaeophyta distinguished by the fact that growth is derived from a single apical cell

sphacelate shrunken and withered

-sphaer- *comb. form* meaning "sphere" usually transliterated -spher-

Sphaeriales an order of pyrenomycete Ascomycetae distinguished by the presence of discrete, sessile, globose perithecia. The genetically well-known genus Neurospora belongs in this group

sphaeridium a minute, glassy stalked, ovoid or spherical body, found in the ambulacral areas of echinoid echinoderms

Sphaertiidae a family of coleopteran insects with markedly club-shaped antennae closely allied to the Histeridae

Sphaerocarpaceae a family of sphaerocarpale Hepaticae distinguished by their bilateral symmetry

Sphaerocarpales an order of hepaticate Bryophyta easily distinguished by the flask-shaped envelope surrounding the sex organs

Sphaeroceratidae = Sphaeroceridae

Sphaeroceridae a family of small myodarian cycloraphous dipteran insects which are commonly called samll dung-flies to distinguish them from the Scatophaginae or true dung-flies. They are distinguished by the short, and usually dilated, basal segment of the hind tarsus. The family is widely known as the Borboridae

sphaerome a hollow, ovoid body, replacing a normal tentacle in some Antozoa

Sphagnobrya a class of muscose bryophytes commonly called the bogmosses or sphagnums. They are distinguished from all other mosses by the fact that the broadly thallose protonema produces a single gametophore

-sphak- *comb. form* meaning "gangrene" frequently transliterated -sphac-

Sphargidae = Dermochelyidae

Sphecidae a huge family of sphecoid hymenopteran insects commonly called the solitary wasps in contrast to the vespids which are social and colonial wasps. They are divided into very numerous subfamilies the more important of which are noticed in this dictionary

Sphecinae a subfamily of sphecid wasps popularly called thread-waisted wasps in view of the extremely long and slender petiole of the abdomen

Sphecoidea a superfamily of apocritan Hymenoptera. They resemble the Vespoidea in having an anal lobe on the hindwing but differ from them in that the lateral extension of the pronotum is in the form of a distinctly differentiated rounded lobe which does not reach the tegula. They are differentiated from the Apoidea, a superfamily by some included in the Sphecoidea, by the fact that the posterior tarsal cylinder of the sphecoids is naked and that the hairs on the thorax are not plumose. Most are solitary wasps

Sphecotheridae a family of passeriform birds commonly included with the Oriolidae

-sphen- *comb. form* meaning "wedge"

hemosphene articulating processes between the vertebrae of Actinopterygian fishes and which are derived from the hemal arches

zygosphene an articular process on the vertebrae of snakes

Spheniscidae that family of sphenisciform birds containing the forms commonly called penguins. The paddle-like wings covered with scale-like feathers, and upright carriage, are sufficient identification

Sphenisciformes an order of birds containing the family Spheniscidae

sphenoid (*see also* bone) wedge shaped

sphere a body of such shape that any cut surface appears circular

archisphere contents of an archegonium before fertilization (*cf.* archisperm)

biosphere that portion of the Earth in which ecosystems operate

carposphere the female gamete of an alga before fertilization (*cf.* carposperm)

centrosphere a hyaline area, functioning as a centrosome

chondriosphere 1 (*see also* chondriosphere 2) a spherical mitochondrion

chondriosphere 2 (*see also* chondriosphere 1) a group of mitochondria aggregated into a roundish body (*cf.* chondriomite)

chromatic sphere an old term for the telophase nucleus

coenosphere = coenocentrum

directive sphere = centrosome

ectosphere the outer part of the centrosphere

episphere that part of a trochophore larva which lies above the equatorial ciliated band

gonosphere the zoogonidium of some algae

hydrosphere the sum total of all the waters on the planet Earth

limosphere a spherical vacuole in the spermatid of a bryophyte

nematosphere a swollen, nematocyst bearing, terminal portion of the tentacles of some Anthozoa

oncosphere = hexacanth larva

oosphere an unfertilized female gamete

archoplasmic sphere = acromatic spindle

sorosphere a hollow sphere of cells each of whose members ultimately becomes a spore

zoosphere a ciliated swarm cell of an alga which subsequently forms an oosphere

zoozygosphere = planogamete

sphincter a circular muscle capable of closing, by contraction, a tube or aperture

Boyden's sphincter that part of Oddi's sphincter which lies around the preampullary portion of the bile duct

coronal sphincter the network of muscles closing the neck over the retracted corona of Rotifera

Oddi's sphincter at the aperture of the bile duct at its entrance to the duodenum

Sphindidae a small family of coleopteran fungus-eating beetles closely allied to the Bostrichidae

Sphingidae that family of sphingoid lepidopteran insects which contains the forms commonly called hawk-moths. They are distinguished by their long narrow front wings and elongate proboscis

Sphingophilae plants fertilized by nocturnal moths, particularly by hawk-moths

sphinx U.S. popular name of sphingid moths (= hawk-moth)

achemon sphinx U.S. Pholus achemon

great ash-sphinx U.S. Sphinx chersis

catalpa sphinx U.S. Ceratomia catalpae

Virginia-creeper sphinx U.S. Ampeloeca myron

day-sphinx U.S. any of several species of the genera Pogocolon and Proserpinus

elm-sphinx U.S. Ceratomia amyntor

giant sphinx U.S. Cocytius antaeus

white-lined sphinx U.S. Celerio lineata

twin-spot sphinx U.S. Smerinthus jamaicensis

walnut-sphinx U.S. Cressonia juglandis

clear-winged sphinx U.S. any of several species lacking scales from most of the wing

Sphyraenidae a family of perciform fishes, commonly called barracudas, distinguished by the protuberant lower jaw, fang-like teeth, and wide separation of the two dorsal fins

Sphyrinidae a family of sharks containing the hammerheads, a term in itself sufficiently descriptive

-spic- comb. from meaning a "point" or, by extension, any pointed object

-spicat- comb. form meaning "spike"

spice a vegetable product used in seasoning food

allspice the myrtaceous tree Pimenta officinalis

Carolina allspice any of several species of the calycanthaceous genus Calycanthus

wild allspice U.S. any of several species of lauraceous trees or shrubs of the genus Lindera

pond spice the lauraceous shrub Litsea aestivalis

spicigerous bearing spikes, but used particularly of flowering plants having an elongate, spike-shaped inflorescence

spicule a small, inorganic body, commonly calcareous or silicious, found embedded in the tissues of, or forming the skeleton of, certain invertebrates particularly sponges

diaene spicule a sponge spicule in the form of a cladome that has lost one ray

triene spicule a sponge spicule in the form of three short rhabdomes at the end of a very long cladome

hexactine spicule a sponge spicule consisting of three axes crossing at right angles to produce six rays extending from a single point

stauractine spicule a hexactinellid sponge spicule which has lost one complete axis and therefore appears to belong to a tetractinellid sponge

monaxon spicule a spicule with only one axis

tetraxon spicule a spicule, particularly a sponge spicule with four axes

birotulate spicule = amphidisk

triradiate spicules a sponge spicule of three rays

spiculin the organic component of siliceous sponge spicules

spider (see also beetle, catcher, crab, fish, hunter, lily, mite, plant, milkweed, wort) popular name of members of the aranean arachnids. Sometimes extended to include other arachnids (e.g. red "spider.") The biblical spider (Hebrew akkabish) is usually that, though sometimes (Hebrew semamith) a gecko

ant-like spider the attid Synemosyna formica

banana-spider any large neotropical spider imported accidentally with bunches of bananas; most are sparassids and so many belong to the genus Heteropoda that the term is sometimes confined to this genus

bird-spider the theraphosid Avicularia avicularia

crab spider popular name of thomisids

trapdoor spider popular name of mygalomorph spiders of the family Ctenizidae which build a moveable lid to their burrows

fishing spider popular name of those pisaurids that hunt under water

garden spider = orb-web spider

brass spider U.S. the agelenid Agelena naevia

ground spider = wolf spider

jumping spider popular name of attids. They are distinguished by their habit of running and jumping sideways, forwards and backwards on their prey and of not spinning webs except for occasional bag-like lairs. The cephalothorax is unusually large and wide

red "spider" popular name of tetranychid mites (cf. spider-mite)

funnel-web spiders popular name of agelenids

orb-web spiders popular name of argiopids

purse-web spider the atypid mygalomorph spider Atypus abboti

wolf-spider popular name of lycosids

Spigeliaceae = Loganiaceae

spignel the umbelliferous herb Geum athamanticum

spike 1 (see also spike 2, 3) a young mackerel

spike 2 (see also spike 1, 3) an inflorescence with sessile, or nearly sessile flowers, on an elongate axis

compound spike a spike inflorescence made up of other spikes

spike 3 (see also spike 1, 2) in compound names of organisms

marlin-spike the macrurid fish Nezumia bairdi

spikenard the araliaceous herb Aralia racemosa

false spikenard the liliaceous herb *Smilacina race-mosa*

ploughman's spikenard *Brit.* the compositaceous herb *Inula helenium* (= elecampane) and *I. canyza*

small spikenard *Aralia nudicaulis*

-spila- *comb. form* properly meaning "crag," but used by ecologists in the sense of "clay"

spinach 1 (*see also* spinach 2, beetle, dock, minor) properly either of two species of the chenopodiaceous genus Spinacia and their numerous horticultural hybrids and varieties but also applied to numerous other edible-leaved herbs

New Zealand spinach the aizoaceous herb *Tetragonia expansa*

spinach 2 (*see also* spinach 1) *Brit.* either of two species of geometrid lepidopteran insect of the genus Lygris. Without qualification, *L. associata* is usually meant

spindle 1 (*see also* spindle 2, shell, tree) an active axis, particularly the "fibrillar" structure running from one cytocentrum to the other in sections of cells in mytosis

initial spindle the spindle of a dividing centriole

mitotic spindle the figure formed in a dividing cell by the actual rays spreading out from chromosomes to poles

neuromuscular spindle thickened neuromuscular structures found in striped muscle at the junction of muscle and tendon of higher vertebrates

primitive spindle 2 (*see also* spindle 1) the young filamentous plant embryo

spine 1 (*see also* spine 2) popular term for the vertebral column

spine 2 (*see also* bill, spine 1, tail) any stiff process not of necessity sharp, projecting from a plant or animal, particularly the calcareous, articulated projections, from the test of echinoid echinoderms and the sharp pointed, woody spikes protruding from plant stems

miliary spine the smallest size of spine found on echinoderms

pungent spine one that is stiff and pierces readily. Particularly used to distinguish spines from fin rays in fishes

[spineless] a mutant (gene symbol *ss*) mapped at 58.5 on chromosome III of *Drosophila melanogaster*. The phenotype is not spineless but the bristles are greatly reduced

spinescent terminating in a spine

spink 1 (*see also* spink 2) the fringillid bird *Fringilla coelebs* (= chaffinch)

spink 2 (*see also* spink 1) *Brit. obs.* the plumulaceous herb *Primula vera* (= primrose)

silken-tube spinner *U.S.* trichopteran insects of the family Philopotamidae

spinneret 1 (*see also* spinneret 2) the adhesive tube at the posterior end of some nematodes

spinneret 2 (*see also* spinneret 1) the fine tubular processes through which silk is spun

spinney a group of trees distinguished from a coppice by the fact that it is not regularly cut

spinose spiny

spinule diminuitive of spine

Spinulosa an order of asteriod Echinodermata lacking the marginal plates of the Thaneroconia and distinguished from the Forciculata by the relative absence of pedicellariae

Sponionidae a family of burrowing sedentary polychaete

annelids distinguished by a pair of very long peristomial cirri. The dorsal cirri act as gills

Spioniformia an order of phanerocephalous polychaete annelids distinguished by the absence of tentacles and palps

-spir- *see* -speir-

spiracle 1 a perforation in the arthropod exoskeleton through which air enters the tracheae (*cf.* spricle)

spiracle 2 the remnants of the hyoid gill slits in adult chondrichthyean fishes

Spiraeoideae a subfamily of Rosaceae distinguished by the possession of a follicular or capsular fruit

spiral popular term for helix

conjugate spiral a phyllotactic arrangement of two parallel spirals

fundamental spiral = genetic spiral

generating spiral = genetic spiral

genetic spiral an imaginary line drawn down a plant axis and which passes successively through the origins of lateral bodies in order of their age

multispiral a spiral having many turns

orthospiral a helix made of a material which is not twisted as it is formed into the helix

anorthospiral a helix made of a substance which is itself twisted along its own axis

paucispiral a spiral having only a few turns

spiramen a hole between the orifice and the ascus in some Ectoprocta

acrospire = radicle primordium

chromospire the folds of the spireme in nuclear division

sigmaspire a sponge spicule in the form of the spirally twisted sigma

sweet-spire any of several species of saxifragaceous shrub of the genus Itea

spirea any of several rosaceous shrubs of the genus Spiraea

false spirea the rosaceous herb *Sorbaria sorbifolia*

spireme a long looped or coiled thread into which chromosomes were at one time supposed to join in the prophase of mitosis or meiosis

spiricle a water sensitive, coiled thread in the coats of some seeds (*cf.* spiracle)

Spirillaceae a family of pseudomonodale Schizomycetes most of which have the form of a spirally twisted rod, and some occurring in chains of spirally twisted rods. Many are dangerous pathogens, including the genus Vibrio the species of which produce cholera in man and many diseases of domestic animals

Spirochaetaceae a family of spirochaetale Schizomycetes usually of large size and showing well developed protoplasmic structures. The majority are free living though some are found in pelecypod Mollusca

Spirochaetales an order of Schizomycetes consisting of spiral forms swimming freely by flexure of the body. Some reach 500 microns in length and were at one time classed among the Protozoa

Spirolobeae a subfamily of Chenopodiaceae distinguished from the Cyclobeae by having a spirally coiled embryo

spirolobous with spirally twisted cotyledons

Spirotrichida an order of euciliate ciliate Protozoa distinguished by an adoral row of membranelles, twisting spirally from the peristome round the body

spiruleum = basitrichous isorhizum

Spirurida an order of phasmid nematodes, distinguished from the Tylenchida and Rhabditida by the fact that the esophagus is bi-partite, not tri-partite

Spiruridea a class of nematodes with two lateral lips, no bulb on the pharynx and lacking a bursa in the male

spit 1 (*see also* spit 2–4) spittle

spit 2 (*see also* spit 1, 3, 4) to eject a fluid from the mouth or, in the case of some snakes, directly from the poison fang

spit 3 (*see also* spit 1, 2, 4) any long, narrow object such as a point of land extending into a body of water

spit 4 (*see also* spit 1–3) an organism deriving its name from any of the above

adder-spit the fern *Pteridium aquilinum*

cuckoo-spit *Brit.* the cercopid homopteran insect *Philaenus spumarius*. (*cf.* spittle-bug, frog-hopper)

spithema an obsolete measure of length corresponding to a span of seven inches

spittle expectorated saliva or anything resembling it (*cf.* spit)

frog-spittle any surface aggregation of fresh water algae or of the lemnacean plant Wolfia

spl *see* [split]

-splachn- *comb. form* meaning "moss," frequently confused with -splanch-

splake a fish alleged to be a hybrid between *Salvelinus namaycush* (lake trout) and *S. fontinalis* (brook trout)

-splanch- *comb. form* meaning viscera (*cf.* -splachn-)

splanchnic pertaining to the visceral mass as distinct from the body wall (*cf.* somatic)

spleen (*see also* fern, wort)

splendent shining

[split] one of four pseudo-alleles with the gene symbol *spl* (the others are [notchoid], [facet] and [notch]) mapped at 3.0± on the X chromosome of *Drosophila melanogaster*. The phenotypic expression includes not only split bristles but also small rough eyes

splitter a derogatory term applied by taxonomists to those of their number who prefer to distribute organisms into very small groups (*cf.* lumper)

[Splotch] a mutant (gene symbol *Sp*) in linkage group XIII of the mouse. The phenotypic expression resembles patch but the white spotting is usually confined to the belly and feet

-spodo- *comb. form* meaning "ash"

spodochrous ash grey

spoile the cast skin of an insect instar or snake

Spondiaceae = Anacardiaceae

spondyl a spool or vertebra, usually used in *biol.* in the adjectival form spondylous (*q.v.*)

Spondylidae a family of pelecypod mollusks commonly called spiny oysters, or thorny oysters

spondylous (*see also* vertebra) pertaining to a spool and, by extension, a spool-shaped articulation

streptospondylous a type of articulation found in the "arm" of ophiuroid echinoderms in which a distally directed hourglass-shaped structure on the proximal tubule, articulates with a transversely directed hourglass-shape structure on the proximal tubule. This permits motion in every direction. (*cf.* zygospondylous)

zygospondylous a form of articulation between the "joint" of the ophiuroid echinoderm arm in which a median projection of the proximal plate fits into median depression of the distal plate, thus permitting only lateral motion (*cf.* streptospondylous)

sponge 1 (*see also* sponge 2, mushroom) popular name of members of the phylum Porifera

bath sponge any of numerous keratose sponges of the family Spongiidae, particularly in *Eur. Euspongia officinalis*

grass-sponge *U.S.* = horse-sponge *U.S.*

horse-sponge any of several species of keratose sponges of the genus Hippospongia. Particularly *H. equina*

velvet-sponge = horse-sponge

sheep's wool sponge the keratose sponge *Hippospongia gossytina*

sponge 2 other spongy organisms

vetetable sponge the fruit of the cucurbitaceous vine *Lussa cylindrica*

spoon (*see also* bill) the spoon-shaped organ which terminates the tongue of the honey bee

-spor- *comb. form* properly meaning "seed" but used today in compounds to mean either seed or spore, through -sperm- is more common for the former

-sporadi- *comb. form* meaning "scattered"

sporadic used in *biol.* in the sense of widely dispersed

ambisporangiate said of flowers having functional male and female parts

amphisporangiate = ambisporangiate

bisporangiate = ambisporangiate

-sporangic *adj. suff.* from sporangium

heterosporangic said of a plant which produces male and female spores in different sporangia

homosporangic yielding only one kind of a spore

sporangium the spore-bearing organ of ferns and spermatophytes

androsporangium = microsporangium

antherosporangium = microsporangium

carposporangium a sporangium in the cystocarp of Rhodophyta

trichosporangium the multi-locular sporangium of Phaeophyta

clinosporangium = pycnidium

episporangium the indusium of ferns

leptosporangium a sporangium derived from a single cell

macrosporangium (Angiosperm) = nucellus

megasporanguim that part of the ovule lying inside the integument of spermatophyte

microsporangium that part of the stamen which contains the pollen

perisporangium = indusium

tetradsporangium a tetrad mother-cell

spore used both in the sense of a resting stage or more rarely for any cell or group of cells capable of producing a new organism. The meanings of seed, sperm and spore are very confused

acrospore a spore borne at the terminal end of some structure

aecidiospore a teleutospore which always produces a dikaryotic mycelium on germination

agamospore a spore produced asexually

allospore a spore from which a gametophyte is produced

androspore a swarmspore which gives rise to dwarf-males in algae

zoandrospore a motile androspore

aplanospore 1 (*see also* aplanospore 2) a non-motile spore

aplanospore 2 (*see also* aplanospore 1) a spore having a relatively thick wall

microplanospore a small, non-motile spore

archespore a cell which gives rise to a spore mother cell

arthrogenous spore one developed by cutting off a portion of a bacterium

ascospore one of usually eight spores produced in the ascus of an ascomycete fungus

auxospore a diatomaceous spore produced by the union of two individuals

basidiospore a spore produced from the basidium of a basidiomycete fungus

bispore one of two spores formed in a bisporangium of Rhodophyta

blastospore an ascomycete fungal spore produced by budding and which can itself reproduce by budding

carpospore 1 (*see also* carpospore 2, 3) a non-motile, unicellular spore of Rhodophyta

carpospore 2 (*see also* carpospore 1, 3) a spore produced from a sporocarp

carpospore 3 (*see also* carpospore 1, 2) a haploid spore produced by meiosis from a zygote. Diploid carpospores are produced by direct proliferation of the zygote

cellular spore = sporidesm

chlamydospore a spore with an unusually thick wall

chronispore = resting spore

ciliospore a ciliated, as distinct from a flagellated, spore

clinospore = stylospore

comospore a tufted seed

compound spore = sporidesma (*see under* desma 2)

cystospore 1 (*see also* cystospore 2) = carpospore

cystospore 2 (*see also* cystospore 1) a group of intracellular spores

daughter spore one spore produced from another spore

diplospore = teleutospore

ectospore = basidiospore

egg spore = oospore

endospore 1 (*see also* endospore 2) the innermost of two coats of a spore or pollen grain

endospore 2 (*see also* endospore 1) the result of internal fragmentation of the cell, which is liberated without the formation of a wall in contrast to aplanospore

endogenous spore one formed within the cell

epiteospore one in a sorus surrounded by prominent paraphyses

epispore the outer layer of some spores but distinct from, and surrounding, the exospore

exospore = epispore

flagellispore a flagellated spore

agamospore a spore produced asexually

agametospore = agamospore

glacospore a seed which is distributed in virtue of its sticky exterior

gynospore = megaspore

gymnospore a naked spore or one produced in a "naked" sporangium as in gymnosperms

haplospore = asexual spore

heterospore a spore containing a random mixture of male and female components

hypnospore a resting spore

    agamohypnospore an unusually large resting spore produced asexually

conidiospore the sexual spore of Peronosporeae

preaecidiospore = trichogyne

involution spore = resting spore

isospore a spore produced in only one kind in contrast to anisospore (*q.v.*)

    anisospore a spore which is produced in two types and is presumably therefore a form of sexual reproduction, commonly contrasted with isospores produced by the same form as in the "sexual" and "asexual" reproduction of radiolarian Protozoa

lophospore a seed with a plumose pappus

macrospore a large spore, or an elongate spore, according to the meaning of the writer

mastigospore a plant with a flagellated spore

megaspore the larger of two spores or the spore containing the ovule of spermatophytes

    primary megaspore = megaspore mother-cell

merispore the segment of a sporidesma

mesospore a uredospore that will only germinate after a resting period

microspore the pollen grain

monospore 1 (*see also* monospore 2) a single spore formed in a monosporangium of Rhodophyta

monospore 2 (*see also* monospore 1) a spore functioning as a gemma, particularly in Ectocarpus

multilateral spore = sporidesm

neutral spore a spore of a rhodophyte alga that is not produced within a sporangium and forms by direct metamorphosis of vegetative cells

octospore a porphyraceaen carpospore

oncospore a seed having hooks appended to the seed coat

oospore 1 (*see also* oospore 2) a durable resting zygote resulting from the "fertilization" of an oogonium by a hyphal ingrowth

oospore 2 (*see also* oospore 1) the product of the fusion of male and female gametes

osmospore an odiferous spore or seed

paraspore one of a clustered mass of spores in Rhodophyta

paulospore = chlamydospore

perispore the outer case of a spore or, rarely, the mother cell of algal spores

petasospore a seed furnished with a parachute

phragmultispore a multicellular spore

planospore a mobile spore

pleurospore one formed on the side of a basidium in a Basidiomycete

pollen spore = pollen grain

polyspore a spore occurring in multiples of four in a tetrasporangium

protospore a spore which produces a promycelium

pseudospore a gemma

pseudopodiospore a non-flagellated spore resulting from the multiple fission of a protozoan

quintospore one which has reached sexual potentiality

resting spore one with an unusually thick wall designed to survive adverse climatic conditions

rhynchospore a beaked seed

scolecospore a spore having a worm-like shape

secundospore a spore which can also act as a gamete

seirospore a spore produced in a row, as a result of the breaking off of terminal cells

septate spore = sporidesm

spindle spore one that is fusiform

sporangiaspore a term occasionally applied to the spores of Myxomycetes

statospore = resting spore

stylospore a spore borne on a style-like body

summer spore one which germinates immediately without a resting period

tachyspore a plant the seeds of which are quickly dispersed

teknospore a spore directly produced from male or female oragns

teleutospore 1 (*see also* teleutospore 2) a spore produced by the bikaryotic mycelium of a urdiniale basidiomycete fungus

teleutospore 2 (*see also* teleutospore 1) an elongate resting spore of Uredineae

teliospore = teleutospore

tetraspore a spore produced by meiosis from a diploid generation descended from a diploid carpospore

thecaspore = ascospore

uredospore a spore developed from the end of a sterigma which germinates immediately to produce a mycelium which itself splits into other uredospores or, sometimes, teleutospores

winter spore = resting spore

zoospore a motile spore

  azoospore a nonmotile reproductive plant cell

  chronizoospore a resting microzoospore

  gametozoospore the zoospore of Ulothrix

  macrozoospore the larger of two comparable zoospores

  megazoospore the larger of two anisotropic zoospores

  microzoospore a small, motile spore

zygospore 1 (*see also* zygospore 2) a thick-walled spore resulting in phycomycete fungi from the fusion of the tips of two hyphae

zygospore 2 (*see also* zygospore 1) the product of the fusion of two isogametes

  azygospore = parthenospore

-sporic *adj. term* synonymous with -sporous, the form preferred in this work

sporidium properly, a spore produced from a mycelium as in the Ascospores, but sometimes used for any spore

-sporium *comb. suffix* indicating "spore coat"

  ectosporium the outer layer of a bacterial spore

  mesosporium the middle coat of a pollen grain

  perisporium the outermost layer of a pollen grain

Sporochnales an order of heterogenerate Phaeophyta distinguished by the presence of a tuft of hairs on the terminal end of each branch of the sporophyte

sporont an individual that gives rise to spores

-sporous *adj. suffix* indicating "pertaining to a spore" in any sense of that word. Adjectives derived directly from -spore substantives given above are not given

  acosporous said of a plant bearing awned seeds

  angiosporous having spores enclosed within a receptacle

  carphosporous said of a seed which is disseminated with the aid of scales on the pappus

  centrosporous said of a plant having a spurred fruit

  ectosporous pertaining to spores of exogenous origin

  endosporous the condition of producing spores within something

  gynandrosporous used of a female plant which produces androspores

  homosporous 1 (*see also* homosporous 2) having identical seeds

  homosporous 2 (*see also* homosporous 1) developed from only one kind of spore

  heterosporous producing two kinds of spores, one of which gives rise to male and the other to female gametophytes

  isosporous producing only one kind of spore

  sarcosporous said of a plant having a fleshy fruit

  synsporous said of the propagation of lower plants by the fusion of cells

Sporozoa a class of plasmodromous Protozoa distinguished by their universally parasitic habit, their usual reproduction by encysted zygotes, and lack of organelles for locomotion

sport a term used in animal and plant breeding for a mutation

the sport *U.S.* the noctuid lepidopteran insect *Charadra illudens*

sporulation the production of spores

-spory *comb. suffix* descriptive of the production of spores. Terms duplicating meaning given under spore are not listed

  apospory 1 (*see also* apospory 2) the condition of lacking spore-formation so that the prothallus develops directly from the diploid generation

  apospory 2 (*see also* apospory 1) the production of an adventitious diploid embryo sac produced by other than the archesporial cell

    euapospory when sex plays no part in reproduction

    panapospory the condition of those ferns in which prothalli develop all over the surface of the frond, without the intervention of a spore

  thecaspory the condition of bearing asci

  prospory the precocious development of spores

spot *U.S.* the sciaenid fish *Leiostonus xanthurus*

  checkerspot *U.S.* any of numerous species of nymphalid butterfly of the genus Melitaea

  crescent-spot *U.S.* any of numerous nymphaline nymphalid butterflies of the genus Phycioides

  finspot the clinid fish *Paraclinus integrepinnis*

  forty-spot *Austral.* = diamond-bird *Austral.*

  gold-spot *Brit.* the noctuid lepidopteran insect *Plusia festucae*

  receptive spot the point of entry of antherozoids into the oospheres of ferns

  redspot *I.P.* the lycaenid lepidopteran insect *Zesius chrysomallus*

  ruby spot *U.S.* the agrionid zygopteran odonate insect *Hetaerina americana*

  silverspot *U.S.* any of several species of nymphalid lepidopteran of the genus Speyeria (= greater fritillary). *I.P.* the nymphalid lepidopteran insect *Argynnis jerdoni* (*cf.* fritillary, silverstripe)

  square-spot *Brit.* either of several species of noctuid moth of the genus Noctua

  ten-spot *U.S.* the libellulid odonaton *Libellula pulchella*

  twin-spot any of several ploceid birds of the genera Cryptospiza, Hypargos and Mandingea. Without qualification usually *M. nitidula*

  Peter's twin-spot *H. niveoguttalus*

  white spot *Brit.* either of two noctuid insects *Dianthoecia albimacula* or *Lithacodia fasciana*

spouting said of the action of whales in ejecting their breath at the surface

spowe the scolopacid bird *Numenius phaeopus* (= whimbrel)

sprag a young cod (*cf.* dorse)

sprat 1 (*see also* sprat 2) *Brit.* without qualification the clupeid fish *Clupea sprattus*. Loosely applied to many small fishes in various parts of the world

sprat 2 (*see also* sprat 1) *Brit.* any of several rushes

sprawler *Brit.* any of several species of noctuid moth of the genus Brachionycha

[Spray] a mutant (gene symbol *sp*) on linkage group V *Neurospora crassa*. The phenotypic expression is an outwardly directed fanning of aerial mycelia

spring 1 (*see also* spring 2) term for a group of teal

spring 2 (*see also* spring 1) a method of locomotion in which a pair of limbs or occasionally (*e.g.* Collembola) some other part of the body, is used to throw the organism from one place to another

spring-haas the *S.Afr.* jumping hare *Pedetes caffer*. It

is usually, but not always, included in the sciuro-morph rodents

sprosser the turdid bird *Luscinia luscinia* (= thrush-nightingale)

sprout the young root or shoot of a freshly germinated plant

Brussels sprouts a stem-budding variety of cabbage

spruce (*see also* aphis, beetle, grouse, mite, sawfly, scale, squirrel, weevil) properly, any of many pinaceous trees of the genus Picea, but occasionally used of other evergreen grees

bastard spruce the pinaceous tree *Pseudotsuga taxifolia* (= Douglas fir)

black spruce *Picea mariana*

blue spruce = Colorado spruce

Colorado spruce *P. pungens*

douglas spruce the pinaceous tree *Pseudotsuga taxifolia*

hemlock spruce = hemlock

Himalayan spruce *Picea morinda*

Norway spruce *P. excelsa*

red spruce *P. ruba*

Sitka spruce = tide-land spruce

tide-land spruce *P. sitchensis*

white spruce *P. canadensis*

spumose frothy

spur 1 (*see also* spur 2–4, fowl) a bony projection covered with horny skin particularly such projections on the tarsometatarsus of galliform birds or the carpometacarpus of columbiiform birds. The term is applied also to short, stout spines on insect legs

apical spur a term usually applied to the spur on the legs of dipterous insects

spur 2 (*see also* spur 1, 3, 4, 5) a short branch of a tree, sometimes leaf-bearing and sometimes fruit-bearing

spur 3 (*see also* spur 1, 2, 4, 5) the "knee" or "elbow" of a tree trunk

spur 4 (*see also* spur 1, 2, 3, 5) a spur-shaped extension of the calyx and in compound names of herbs having spurred flowers

larkspur (*see also* miner) any of numerous species of the ranunculaceous genus Delphinium

dwarf larkspur *D. tricorne*

field larkspur *Brit. D. consolida*

musk larkspur *D. brunoianum*

rocket larkspur *D. ajacis*

sand spur *U.S.* the grass *Cenchrus tribuloides*

spur 5 (*see also* spur 1–4) in compound names of birds

longspur *U.S.* any of several fringillid birds of the genera Rhynchophanes and Calcarius

chestnut-collared longspur *C. ornatus*

Lapland longspur *C. lapponicus*

McCown's longspur *R. mccowni*

Smith's longspur *C. pictus*

spurge (*see also* moth) any of several species of herbs of the euphorbiacous genus Euphorbia and a number of similar plants

Allegheny spurge = mountain spurge

caper-spurge *Euphorbia lathyris*

cypress-spurge *E. cyparissias*

flowering spurge *E. corollata*

hyssop-spurge the euphorbiaceous herb *Tithymalus peplus*

ipecac-spurge *E. ipecacuanhae*

melon-spurge the cactus-like *E. meloformis*

mountain spurge the buxaceous shrubby herb *Pachysandra procumbens*

slipper-spurge any of several species of euphorbiaceous succulent shrubs of the genus Pedilanthus

spurrey = spurry

spurry the caryophyllaceous herb *Spergula sativa*

corn spurry *Bri.* the caryophyllaceous herb *Spergula arvensis*

sandwort spurry *Brit.* any of several species of the caryophyllaceous genus Spergularia

sea spurry *N.Z. Spergularia marginata*

squab a fledgling pigeon

Squalidae a family of sharks distinguished by the presence of a spine in front of each of the dorsal fins. Commonly called spiny dogfish

Squaliformes = Selachoidea

-squam- *comb. form* meaning "scale"

Squamata an order of diapsid reptiles containing the snakes and lizards. They are distinguished from the Crocodilia by the fact that the cloaca is a transverse slit

squamate scaly

squamody the condition of having leaves change into scales or scale levaes

squamosal *see* bone

carpenter's square *U.S.* the scrophulariaceous herb *Scrophularis marilandica*

three-square *U.S.* either of the sedges *Scirpus americanus* or *Dulichium arundinaceum*

squarose scurfy

squash (*see also* beetle, borer, bug) any of several species and very numerous horticultural hybrids of the cucurbitaceous genus Cucurbita. It is doubtful whether specific ancestry can be at present assigned to any horticultural squash

Guinea squash the solanaceous vine *Solanum melongena* (= eggplant)

Squatinidae a family of flattened sharks, known as angel sharks intermediate in appearance between the sharks and skates

oldsquaw *U.S.* the aythyine duck *Clangula hyemalis*

squeaker *U.S.* any of several mochocid catfish. *Austral.* popular name of rat-kangaroos of the genus Bettongia

squealer a young grouse, partridge or quail (*cf.* poult)

squid popular name of decapod dibranchiate cephalopod Mollusca, though by some confined to the family Loliginidae

flying squid any of several species of ommastrephid mollusks of the genus Ommastrephes (= sea arrow)

giant squid any of several species of ommastrephid cephalopod mollusks of the genus Architeuthis. *A. princeps* is the largest living mollusk having been recorded up to seventy-five feet long

peacock-squid the cranchiid mollusk *Loligopsis pavo*

squill any of numerous liliaceous herbs of the genus Scilla. (*cf.* hyacinth). The term is also applied to *Urginea scilla* (= sea onion)

autumn squill *S. autumnalis*

bellflower-squill *S. hispanica*

common blue squill *S. nonscripta*

dwarf squill *S. monophyllos*

hyacinth-squill *S. hyacinthoides*

spring squill *S. verna*

squinancy *Brit.* the rubiaceous herb *Asperula cynanchica*

squirrel (*see also* corn, cuckoo, grass, monkey, phalanger) any of numerous simplicidentate sciuromorph rodents of the family Sciuridae. *Brit.*, without qualification, *Sciurus vulgaris*

abert squirrel = kaibab squirrel

antelope-squirrel any of numerous members of the genus Ammospermophilus

barking squirrel *U.S.* = prairie dog *U.S.*

brown squirrel *Brit.* = red squirrel

bush squirrel any of several species of African squirrels of the genus Paraxerus

tassel-eared squirrel = kaibab squirrel

flying squirrel any of numerous sciurid rodents or African anomalurid rodents provided with a petagium and therefore capable of gliding. In *U.S.* the term is confined to scruids of the genus Glaucomys. In *Austral.* the term is applied to the flying phalangers (*q.v.*)

fox-squirrel *U.S.* either of two large sciurids *Sciurus niger* or *S. apache.* Any large squirrel, from time to time, is so designated

eastern fox-squirrel *U.S. S. niger*

grey squirrel *Brit.*, without qualification *Sciurus cinereus. U.S.* either *S. griseus* (**western grey squirrel**) or *S. carolinensis* (**eastern grey squirrel**)

ground-squirrel *U.S.* any of numerous burrowing sciurid rodents, particularly the chipmunks, though also applied to paririe dog. *S.Afr.* any of numerous sciurids of the genus Xerus

Arctic ground-squirrel *Citellus ungulatus* (= parka-squirrel)

California ground-squirrel *C. beecheyi*

Ethiopian ground-squirrel any of several species of the genus Xerus

Franklin's ground-squirrel *C. franklini* (= grey gopher)

Idaho ground-squirrel *C. brunneus*

thirteen-lined ground-squirrel *C. tridecemlineatus* (= gopher)

Uinta ground-squirrel *C. armatus*

kaibab squirrel *Sciurus alberti* (= tassel-eared squirrel, albert squirrel)

Malabar squirrel *S. maximus*

golden-mount mantled squirrel *C. lateralis*

palm-squirrel a term loosely applied to many Oriental and African sciurid rodents associated with palm trees. Without qualification, the term often means *Funambulus rufa*

parka squirrel = Arctic ground-squirrel

pygmy squirrel any of several species of the genus Nannosciurus

red squirrel *Brit. Sciurus vulgaris. U.S. Tamiasciurus hudsonicus*

rock-squirrel *Citellus variegatus*

spruce-squirrel *U.S.* = red squirrel *U.S.*

sugar-squirrel *Austral.* = flying phalanger

groove-toothed squirrel *Rheithrosciurus macrotis*

sea **squirt** popular name of sessile ascidians

sr *see* [stripe] or [striate]

srl a bacterial mutant showing a utilization of sorbitol

sr-la *see* [Streptomycin-la]

ss *see* [spineless]

s.s. (or, rarely, S.S.) abbreviation for *sensu stricto* (*q.v.*)

st *see* [scarlet], [stub] or [stumpy]

ST a yeast genetic marker indicating a requirement for starch

sta *see* [stubarista]

stable term for a group of horses kept by one individual

mono**stachous** having a single spike

pycnostachous dense tufts of spikes

-stachy- *comb. form* meaning "spike" in the sense of the seed head of a grass

Stachyuraceae a small family of parietalous shrubs

Stackhousiaceae a small family of Australasian sapindalous dicotyledons

plo**adostadion** a community of plants with their lower parts in water and their upper parts in the atmosphere

stadium = instar

Jacob's **staff** the fouquieriaceous tree *Fouquieria splendens* (= ocotillo)

stag (*see also* beetle) a male deer, in *Brit.* usage properly confined to a red deer in its fifth year

stage one of a series of specific steps in a continuing process

aeciostage the spore-bearing stage in the life history of a uredinale fungus

bouquet stage early leptotene stage of meiosis in which chromosomes are arranged in a bouquet pattern

dictyotic stage the terminal stage in nuclear division in which chromosomes are no longer apparent by normal techniques of light microscopy

diplotene stage that stage in meiotic division which immediately follows the pachytene stage

involution stage = resting stage

stainer any of numerous red and black pyrrhocoid hemipteran insects. Also applied to a few other insects

apple-wood stainer *U.S.* the scolytid coleopteran insect *Monarthrum mali*

cotton-stainer any of several species of pyrrhocorids insect of the genus Dysdercus

Arizona cotton-stainer *U.S. D. mimulus*

stalk 1 (*see also* stalk 2-4) popular name for any above-ground axillary structure of a plant

stalk 2 (*see also* stalk 1, 3, 4) the narrow attachment of an organ or organism to something else

leaf stalk = petiole

optic stalk the connection in the embryo between the optic vesscile and the prosencephalon

seed stalk = funicle

stalk 3 (*see also* stalk 1, 2, 4) in compound names of organisms

twisted stalk any of several liliaceous herbs of the genus Streptopus

stalk 4 (*see also* stalk 1-3) the action of a predator which has seen its prey and is endeavoring to approach it unseen

stallion a male horse

peristalsis a contraction that moves along the length of a tube

stamen the microsporophyll of seed plants

anantherous stamen one which consists of a filament that lacks an anther

parastamen an aborted stamen

aniso**stamenous** with unequal stamens

staminoid an abortive stamen

stand term for a group of plovers

head**stander** the characid fish *Abramites microcephalus*

tailstander the characid fish *Poecilobrycon unifasciatus* (*cf.* pencil fish)

Strangerioideae a subfamily of Cycadiacea containirg the single genus from which the name is derived

Stannius's *epon. adj.* from Freidrich Hermann Stannius (1808–1883)

stapes *see under* bone

Staphylinidae a huge family of elongate beetles commonly called rove beetles and readily distinguished by the short elytra which cover the intricately folded-back wings

Staphylinoidea a very large superfamily of polyphagous coleopteran insects containing the elongate beetles with short elytra like the staphylinids and their immediate allies

Staphyleaceae that family of sapindalous dicotyledons which contains, *inter alia* the bladder-nuts. It is distinguished from the closely related sapindaceae by the abundant endosperm of the seeds

star (*see also* frontlet, throat) any radially symmetrical or apparently radially symmetrical animal or plant. Often applied without qualification to turbinid gastropod mollusks of the genus Astraea (*cf.* star-shell)

blazing star any of numerous species of compositaceous herb of the genus Liatris. The term is also applied to the liliaceous herb *Chamaelirium luteum* and any of several species of the iridaceous genus Tritonia

bog-star *U.S.* any of several species of the saxifragaceous genus Parnassia

brittle star popular name of ophiuroid echinoderms (= serpent star)

daisy brittle star *U.S. Ophiopholis aculeata*

evening star the amaryllidaceous herb *Cooperia drummondii*

feather-star the popular name of stalkless crinoid echinoderms

floating star the liliaceous bulbous herb *Milla diflora*

hill-star any of several trochilid birds of the genera Aureotrochilis and Urochroa

mother-star = monaster

purple star *Asterias vulgaris*

sea star popular name of asteroid echinoderms, also occasionally applied to ophiruoids

serpent-star = brittle star

shooting-star the primulaceous herb *Dodecatheon meadia* and some similar species

sun-star popular name of stimulose asteroid echinoderms of the family Solasteridae

wood-star any of numerous trochilid birds, mostly of the genera Myrtis and Ancestrura

[Star] a pseudo allele (gene symbol S) mapped at 1.3 on chromosome II of *Drosophila melanogaster*. The phenotypic expression is small rough eyes

starch a complex polysaccharide, usually a mixture of D-amylose and amylopectin, that is the principle food storage material of plants

assimilation starch starch produced in a chloroplast for use, not storage

storage starch starch produced in a leucoplast and usually stored for future use

stare any of several sturnid birds, mostly of the genus Sturnus

starlet any of several small sturnid birds of the genus Sturnus

starling any of numerous passerine birds, mostly of the family Sturnidae and many of the genus Sturnus. Without qualification usually *S. vulgaris*. The name is also used for several icterids of the genera Pezites and Leistes

amethyst starling = violet-backed starling

white-winged babbling-starling *Neocichla gutturalis*

violet-backed starling *Cinnyricinclus leucogaster*

golden-breasted starling *Cosmopsarus regius*

red bellied starling *Afr. Spreo pulcher*

Chinese starling *S. sinensis*

common starling *S. vulgaris*

bristle-crowned starling *Galeopsar salvadorii*

daurian starling *S. spurninus*

emerald starling *Afr. Coccycolius iris*

European starling = common starling

glossy starling properly any sturnid of the genera Aplonis, Lamprocolius, and Lamprotornis but occasionally used for other genera

fringed winged glossy starling *Afr. Onychognathus fulgidus* (= chestnut-winged starling)

black-breasted glossy starling *Lamprocolius corruscus*

blue-eared glossy starling *L. chalybaeus*

lesser blue-eared glossy starling *L. chloropterus*

long-tailed glossy starling any of several species of the genus Lamprotornis; without qualification, *L. purpuropterus*

Himalayan starling = common starling

magpie-starling *Speculipastor bicolor*

mountain starling *Onychognathus walleri*

Sharpe's starling *Pholia sharpii*

superb starling *Spreo superbus*

narrow-tailed starling *Poeoptera lugubris*

tree starling any of several species of sturnids of the genus Aplonis

wattled starling *Creatophora carunculata*

chestnut-winged starling any of several sturnids of the genus Onychognathus; without qualification, *O. fulgidus*

red-winged starling *Afr. Onychognatus morio*

[staroid] a mutant (gene symbol *std*) mapped at 56.5± on chromosome II of *Drosophila melanogaster*. The phenotypic expression is very rough small eyes. The males of the population are sterile

star-of-Bethlehem the amaryllidaceous herb *Eucharis grandiflora*

redstart *Brit.* and *Afr.* any of several turdid birds, mostly of the genera Phoenicurus and Rhyacornis usually, without qualification, *P. phoenicurus*. Also, in *U.S.* any of numerous parulid birds of the genera Setophaga and Myioborus usually, without qualification, *S. ruticilla*

American redstart *S. ruticilla*

black redstart *P. ochrurus*

common redstart *P. phoenicurus*

Daurian redstart *P. auroreus*

eastern redstart = Daurian redstart

blue-fronted redstart *P. frontalis*

painted redstart *M. pictus*

plumbeous redstart *R. fluiginosus*

slate-throated redstart *M. miniatus*

white-capped redstart the turdid *Chaimorrornis leucocephalus*

-stas- *comb. form* meaning "to be without movement" (*cf.* stat-)

stasad a plant of stagnant water

hypostase a disk of wood at the base of an ovule

stasis stability or cessation of movement or growth

anastasis the condition of recovering from anabiosis

epistasis the condition of having one gene which masks the effect of another to which it is non-allelic

genepistasis the condition of a form which has failed to evolve over a long period of time

homeostasis the constancy of the internal environment of an organism brought about by dynamic biological processes

hypostasis the condition of having one gene which is masked by another to which it is non-allelic

stasium a community of organisms in a stagnant pool

-stat- *comb. form* meaning "stand," usually in the sense of "stand still" or "stablize" (*cf.* -stas-)

aerostat one of a pair of air sacs in the abdomen of dipterous insects

-static *adj. suffix* from either -stas- or -stat-

epistatic a term once applied to genotypic, as distinct from phenotypic, characters

hydrostatic used by biologists for an assemblage of

organisms which is not attracted to a region of higher moisture

mesostatic a medium succession produced through the action of water

xerostatic the condition of a plant succession which has survived drought

-staur- *comb. form* meaning "cross"

Stauromedusae an order of sessile scyphozoan Coelenterata developing directly from a scyphostoma and remaining attached by an adoral stalk

stauros a smooth, transparent central area, occasionally in the form of a band or cross, on the frustule of a diatom

std *see* [staroid]

Steatornithidae a monotypic family of caprimulgiform birds erected to contain the oilbird. Distinguished from most other caprimulgids by the very long rictal bristles coupled to the wide-gaped hooked bill

steenbok = steinbok

steer properly a castrated bull though in *U.S.* usage frequently applied to any bovine raised for beef

-steg- *comb. form* meaning "roof"

gastrostege a ventral scale immediately anterior to the anus in snakes

hypostege a coelomic cavity lying between the frontal membrane and the cryptocyst in some Ectoprocta

urostege a scale immediately posterior to the anus in snakes

oostegite one of the plates derived from thoracic appendages which form the brood pouch in pericarid Crustacea

gynostegium the central columnar portion of an orchid flower once considered to be an extension of the floral axis but now considered to be a portion of the sex organs

steinbok any of several species of neotragine antelopid of the genus Raphicerus. Used without qualification, frequently refers to *R. campestris*

--stel- *comb. form* meaning "pillar" or "column"

stele 1 (*see also* stele 2) a unit consisting of the vascular system and any tissue enclosed within it which forms a cylinder running through the stem of vascular plants

atactostele the vascular system of monocotyledonous plants in which the strands are scattered through the stem rather than arranged in a circle

homodesmic atactostele one in which all the vascular bundles are of the same type

gamodesmic stele one in which the vascular bundles are fused to each other

dictyostele a siphonostele so cut up by leaf gaps that it appears as a network

dissected dictyostele one in which the vessels are reduced to thin strands

perforated dictyostele one which has gaps other than those occasioned by leaf gaps

siphonic dictyostele one which is simple and tubular

adelosiphonic dictyostele one which is no longer tubular

eustele a siphonostele resembling a dictyostele but with the leaf gaps forming no regular pattern

gamostele one in which the vascular bundles are fused at intervals

haplostele one which consists of xylem surrounded by phloem

hydrome stele = hydrome cylinder

hysterostele one which has lost part of the normal structure of the stele

meristele the portion of a monostelic stem which leads to each leaf

dimeristele having two vascular bundles

monomeristele one with a single leaf trace leaving it

gamomeristele one in which individual bundles are fused

haplomeristele one which consists of axial vessels surrounded by a ring of phloem

trimeristele a stele of three vascular bundles

monarch stele a xylem bundle with a single protoxylem group

oligarch stele one possessing few protoxylem groups

pentarch stele one with five protoxylem groups

polyarch stele one possessing many protoxylem groups

protostele a primitive type of stele consisting of a solid column of vascular tissue with xylem occupying the center and pholem forming the surrounding tissue

peripheral stele one from which the vessels of adventitious roots take their origin

pseudostele a stele-like structure running up a petiole

reparative stele one of four bands corresponding to four orthostichies

siphonostele a tubular stele, which therefore contains pith

phyllosiphonostele a siphonic stele with permanent leaf gaps

amphiphloic siphonostele one with phloem both on the inside and outside of the xylem

ectopholic siphonostele one with the phloem on the outer surface of the xylem cylinder

solenostele an amphiphloic stele with widely separated leaf gaps

perforated solenostele one in which gaps other than leaf gaps occur

tetrarch stele one with four protoxylem groups

hypotetrarch stele the division of the median protoxylem in a triach stele

triarch stele one possessing three protoxylem groups

hypotriarch stele one in which the median protoxylem group is the lowermost

mesotriarch stele a triarch stele in which two xylem bundles are fused

stele 2 (*see also* stele 1) columnar structures in animals

anthostele the thickened aboral portion of the polyp into which the anthocodium withdraws in those alcyonarian coelenterates that do not form colonies

-stelic pertaining to stele, usually in the sense of stele 1

dialystelic said of a stem having two steles

monostelic said of a stem with one central cylinder of vascular tissue

polystelic said of a stem with more than one stele

prostelic said of a stem having a single concentric bundle

schizostelic said of a stele which has broken up into individual vascular bundles

-stelli- *comb. form* meaning "star"

stellate star shaped

Steller's *epon. adj.* from George Wilhelm Steller (1709–1746)

-stem- *comb. form* meaning "garland" or "stamen"

stem 1 (*see also* stem 2, 3, -istem, girdler) that part of the plant axis that bears leaves and buds

compound stem one which is branched

determinate stem one in which there are no axillary shoots hence, in tomatoes, called self-pruning

endistem = pith

herbaceous stem one that lacks woody tissue and that has relatively small increments

meristem *see* -istem

simple stem one which is unbranched

woody stem a stem producing wood and which therefore increases radially in diameter

stem 2 (*see also* stem 1, 3) in compound names of plants

blue stem *U.S.* the grass *Andropogon scoparius*

pipestem *U.S.* the rosaceous herb *Spiraea alba* (= meadow sweet)

screw stem the gentianaceous herb *Bartonia paniculata*

wing stem the compositaceous herb *Actinomeris alternifolia*

stem 3 (*see also* stem 1, 2) the main, or central part of an organ

brain stem that portion of the brain which remains after the various evaginations forming the cerebrum, cerebellum, etc., have been produced

gynostemium the sum total of the sex organs in an orchid

Stemonaceae a small family of liliiflorous monocotyledonous angiosperms

-stemone a *subs. suffix* sometimes used as a derivative of stemonous. Thus a petalostemone is a plant with petalostemonous flowers

-stemonous *adj. suffix* meaning "pertaining to stamens"

adenostemonous having glands on the stamen

allagostemonous having alternate stamens attached to petal and torus

brachybiostemonous having short-lived stamens

macrobiostemonous having persistent stamens

cheirostemonous possessing five stamens united at the base

diplostemonous a bicyclic androecium with the stamens of the outer whorl alternating with the petals or corolla lobes

obdiplostemonous having twice as many stamens as petals, the outer series of the former being opposite the petals

haplostemonous with a single whorl of stamens

isostemonous having equal numbers of stamens and petals or sepals

meiostemonous with fewer stamens than petals

petalostemonous with the stamens fused to some part of the corolla

stylostemonous = hermaphrodite

Steno's *epon. adj.* from Nicolaus Steno better known as Stensen (*q.v.*)

stenoky the condition of an organism of ecologically restricted habitat

Stenomatidae = Stenomidae

Stenomerideae a subfamily of Dioscoreaceae distinguished from the Dioscoreae by the possession of bisexual flowers

Stenomidae a small family of gelechioid lepidopteran insects mostly distinguished for their mimicry

Stenostomata = Cyclostomata (Ectoprocta)

stenotele a type of heteroneme nematocyst in which the butt is dilated at the base (*cf.* rhabdoid and eurytele)

Stensen's (*see also* Steno's) *epon. adj.* from Niels Stensen (1638–1686)

Stephanidae a very small family of ichneumonoid apocritan hymenopteran insects. They are quite unmistakable since the head is spherical and set on the end of a long neck

-stephanos- *comb. form* meaning "wreath"

Stephanosomatidae = Sarcophagidae

steppe grasslands which lie in the rainshadows of mountains in temperate latitudes with desert on the drier side and forest on the wetter side

Stercorariidae a small family of charadriiform birds containing the skuas and jaegers, readily distinguished from other charadriiforms by the webbed feet and complex rhamphotheca

Sterculiaceae that family of malvalous dicotyledons which contains, *inter alia,* the trees which yield the cocoa and cola of commerce. The family may be distinguished from the closely allied Malvaceae by the two-celled anthers, the monodelphous stamens and the copious endoderm of the seeds

Stereogenylae = Bryophyta

stereognesis the ability to recognize objects with senses other than vision

stereome the sum total of the collenchyma and sclerenchyma in a plant

sterigma a hypha from which a spore is split off

sterile 1 (*see also* sterile 2) in the sense of lacking the ability to breed or interbreed

chromosomal sterility a form of sterility often present in hybrids, resulting from structural differences between chromosomes

cytoplasmic sterility that which is due to discordance between the chromosomal gene complement and the cytoplasm

diplontic sterility = zygotic sterility

genic sterility a variety of hybrid sterility, resulting from failure to produce functional gametes

hybrid sterility inability of an F_1 generation to reproduce

self-sterility when an organism cannot fertilize itself even though male and female gametes are mature at the same time

zygotic sterility the condition in which gametes can fuse but the zygote does not develop further

sterile 2 (*see also* sterile 1) said of anything which is devoid of viable organisms

stern = tern

sternal pertaining to the sternum

firmisternal said of those animals in which the procoracoid cartilage is fused in the midline (*cf.* arciferal)

parasternalia sternal rib-like bones found in the ventral abdominal wall between the last true rib of the pelvis in Crocodilia, Sphenodon and some fossil reptiles (= gastralia)

sternellum a free sclerite in the ventral portion of the thoracic segment of an insect

Sternidae a family of charadriiform birds containing the terns, now usually included with the Laridae

sternite the lower, or ventral plate of the exoskeleton of a segmented animal, particularly an arthropod. The term is sometimes considered synonymous with sternum (*q.v.*) though most writers confine this term to chordates

acrosternite a marginal flange on the anterior edge of the abdomen of insects

episternite that part of the sternite that lies anterior to the legs

anepisternite the upper of the two divisions into which the episternite is divided in some insects

mesepisternite the sternite of the mesothorax

notepisternite the dorsal portion of the episternite

eusternite an intrasegmental plate at the base of a thoracic segment

laterosternite the sides of the sternite

mesosternite the underside of an insect mesothorax

metasternite the underside of the insect metathorax

spinasternite the portion of the sternite that covers the segmental boundaries

prosternite that part of the insect thorax that lies between the first pair of legs

tritosternite an unpaired anterior median appendage found on the under-surface of acarines that consists of a basal portion and two or three setiform processes

[**Sternopleural**] a mutant (gene symbol *Sp*) mapped at 22.0 on chromosome II of *Drosophila melanogaster*. The phenotype has an unusually large number of sternopleural bristles

Sternoptychidae a small family of isospondylous fishes distinguished by a dorsal blade anterior to the dorsal fin. Many have ventral phosphorescent organs. Called hatchet fish in reference to their shape

Sternorrhyncha a suborder of homopteran insects containing those forms in which the antennae are usually long and filiform (*cf.* Auchenorrhyncha)

Sternoxia = Elateroidea

sternum a cartilagenous or bony structure with which usually the ribs, but sometimes the pectoral girdle, are articulated in the ventral median plane. This term is sometimes used in the sense of sternite

-stero- *comb. form* meaning "solid"

-sterone *comb. suffix* used in the name of many steroid hormones

androsterone a hormone secreted by the testes having the same activity as testosterone

dehydroepiandrosterone a hormone secreted by the testes having the same activity as testosterone

dehydroisoandrosterone = dehydroepiandrosterone

luteosterone = progesterone

testosterone the hormone secreted in the testes necessary for the development and the maintenance of male tertiary sexual characters

-sterr- *comb. form* properly meaning "rugged" or "rocky" terrain, but used by ecologists in the sense of "moor land"

sterrhad a plant of moor lands

stethidium the trunk (in the sense of trunk 1) of a mammal. In insects, the thorax with all of its appendages

Stichaeidae a family of blennioid acanthopterygian fishes distinguished by the elongate bodies with overlapping scales and the large number of spines in the dorsal fin. Often called pricklebacks

-sticho- *comb. form* meaning a "row"

stichous *comb. form* meaning "rank" or "ranked"

astichous arranged at random, or at least, not in definite rows

distichous disposed in two rows or ranks

diplostichous arranged in two rows

haplostichous having a single row

monostichous arranged in a single column

orthostichous arranged in straight ranks

parastichy a secondary spiral in phyllotaxis

contact parastichy a helix that passes through leaves in contact with each other when they emerge at the apex

stick in compound names of many organisms

chawstick the rhamnaceous shrub *Gouania lapuloides*

walking stick *U.S.* any of numerous phasmid insects.

Usually, *U.S.* without qualification *Diapheromera femorata*

devil's walking stick the araliaceous tree *Aralia spinosa*

two-striped walking stick *U.S.* the phasmatid insect *Anisomorpha buprestoides*

stick-tights any of several compositaceous herbs of the genus Bidens

[**stick**] a mutant (gene symbol *sk*) mapped at 25.8 on linkage group IV of *Bombyx mori*. The phenotypic larva is slender

[**sticky peel**] a mutant (gene symbol *pe*) mapped at 0 on chromosome 10 of the tomato. The ripe phenotypic fruit is sticky

stifle the lower portion of the thigh of quadruped mammals

-stigm- *comb. form* meaning a "sharp point" or a "circular mark" or "hole made with a sharp point"

stigma 1 (*see also* stigma 2–6) the upper end of the style modified for the reception of pollen

stigma 2 (*see also* stigma 1, 3–6) the eye spot of Protozoa

stigma 3 (*see also* stigma 1, 2, 4–6) the spiracle of Arthropoda

stigma 4 (*see also* stigma 1–3, 5, 6) the interstices of a branchial net

stigma 5 (*see also* stigma 1–4, 6) the androconium of lepidopteran insects and also a dense, often discolored, part of the costal margin of an insect wing

stigma 6 (*see also* stigma 1–5) the sac of a mature egg follicle on a mammalian ovary

pseudostigma a pit from the base of which rise sensillae, found on the propodosoma of acarines

stigmata *pl.* of stigma

stigmatic pertaining to stigma in any of its meanings

brachybiostigmatic having short-lived stigmas

macrobiostigmatic a plant in which the stigmas remain capable of fertilization for a long time, specifically until the plant's own anthers are ripe

Stilbaceae a family erected to contain the genera Eurylobium, Euphystachys and Stilbe by those who do not hold them to belong in the Verbenaceae

stilidium the elongate portion of the archegonium of a moss

stilt (*see also* sandpiper) any of several species of recurvirostrid bird of the genera Himantopus and Cladorhynchus. *Brit.,* without qualification *H. himantopus*

banded stilt *C. leucocephalus*

black stilt *H. novaezelandiae*

black-necked stilt *H. mexicanus*

pied stilt *H. himantopus*

black-winged stilt *Brit.* = pied stilt

sting an ovipositor, modified for the ejection of venom, found in some Hymenoptera. Also applied to other injectors of venom such as the calcareous projection at the base of a stingray's tail or the end of a scorpion's tail

horse stinger *Brit.* = dragonfly

stint any of several scolopacid birds of the genus Erolia. Also, *Brit.* a dunlin in breeding plumage

little stint *E. minuta*

red-necked stint *E. ruficollis* (= rufous-necked sandpiper)

Temminck's stint *E. temmincki*

long-toed stint *E. subminuta*

stipate crowded together

stipe the term can be applied to any stalk-like object but is most usually applied to the stalk of a capped basidiomycete fungus

stipes a segment of the insect maxilla next distal of the cardo and on which the galea and lacinia articulate when they are present

parastipes a sclerite bordering the inner edge of the stipes

stipites *pl.* of stipes

-stipul- *comb. form* meaning a "branch" or "long rod"

stipula a newly sprouted feather particularly one which has not yet "feathered out"

stipulaceous pertaining to a stipule or with unusually large stipules

exstipulate lacking stipules

stipule the basal appendage of a leaf

sto *see* [stocky]

stoat any of several species of musteline carnivorous mammals particularly *Mustela erminea* (= *U.S.* short-tailed weasel)

stock 1 (*see also* stock 2) any of numerous cruciferous herbs more particularly those of the genus Matthiola

common stock *M. incana*

Grecian stock *M. bicornis*

mahon stock = Virginia stock

sea stock *M. sinuata*

shrubby stock *Brit.* = common stock

Virginia stock the cruciferous plant *Malcomia martima*

stock 2 (*see also* stock 1) in the *obs. Brit.* sense of log

rootstock = rhizome

[stocky] a mutant (gene symbol *sto*) mapped at 32.5 on the X chromosome of *Drosophila melanogaster*. The name is descriptive of the phenotype which also has large pear-shaped eyes

-stol- *comb. form* meaning "shoot"

stole = stolon

stolon a lateral branch, frequently reproductive in function, produced either from the base of a stem of a plant or from the base of many sessile animals

black stolon the basal stolon joining all the individuals in a colony of pterobranchs

Stolonifera an order of alcyonarian Anthozoa distinguished by the fact that the single polyps are interconnected by basal stolons

-stom- *comb. form* meaning "mouth"

cheilostom the anterior of the three regions into which the buccal capsule of a nematode may be divided

protostom the middle of the three cavities into which the buccal capsule of a nematode may be divided

telostom the terminal of the three cavities into which the buccal capsule of a nematode may be divided

stoma 1 (*see also* stoma 2, 3, stome, stomium) an organ consisting of two guard cells, with a pore between them, on the surface of the leaf or stem of a higher plant

false stoma the epidermal pore of Equisetum

stoma 2 (*see also* stoma 1, 3) in the sense of an animal mouth. Used also for openings through the septa of some anthozoan coelenterates

stoma 3 (*see also* stoma 1, 2) used in *entomol.* in the senses of spiracle 1 and stigma 3

hypostoma in insects, the lower anterior part of the face above the mouth

metastoma the chitinous lobe immediately below the mouth in Crustacea. If divided into two lobes, each is called a paragnath (*q.v.*)

stomach in vertebrates, that portion of the alimentary canal into which the esophagus leads and also any analogous structure in other animals

cardiac stomach the region next to the esophagus

coronal stomach a general term applied to the four gastric pouches of Scyphozoa

honey-stomach the crop of nectar-collecting Hymenoptera

prestomach = proventriculus

pyloric stomach the region next to the intestine

stomata *pl.* of stoma

Stomatopoda the only order of haplocarid malacostracan Crustacea. In addition to the characters of the superorder they are easily recognized by the fact that the second maxillipede is modified as a grasping organ resembling that of the preying mantis

stomatous pertaining to stoma in any of its meanings

astomatous lacking stomata

cryptostomatous said of algae which possess pits of no known function

epistomatous having the shape of a spigot

gymnostomatous having a naked mouth (*i.e.,* one devoid of teeth)

hypostomatous said of a leaf with stomata only on the under surface

hyperstomatous said of a leaf with stomata only on the upper surface

lophiostomatous having tufts of hairs around an aperture

isostomatous said of a flower in which the calyx and corolla are of the same size

haploperistomatous having a single row of teeth on the peristome

-stome (*see also* stoma, stomium) *adj. suffix* meaning pertaining to mouths or to any other aperture or cavity which might conceivably be called a mouth (*cf.* stoma, stomium)

amphistome a term used to describe those trematode Platyhelminthes in which the acetabulum is at the posterior end

camerostome a cavity at the anterior end of some acarines formed by the extension of the propodosoma over the gnathosoma

carpostome the aperture of a cystocarp

coelomostome a primary coelomic funnel or nephrostome derived directly from an expansion of the coelom, in distinction to other metanephridia in which the connection to the coelom is secondary

cytostome the "mouth" of ciliate Protozoa

distome a term used to describe those trematode Platyhelminthes in which the acetabulum is near the oral sucker

diplostome (trematode) = hemistome

echinostome a term used to describe those trematode Platyhelminthes in which the acetabulum is edged with a row of spikes

endostome that portion of the micropyle which pierces the inner coat of an ovule

epistome any flap which covers the mouth. It has been applied to any structure answering this definition, from a portion of the rostrum in many insects to a projection separating the mouth and anus in Phoronidea

exostome that part of the foramen of a seed which penetrates the outer coat

hemistome a term applied in distinction from holostome to those trematodes in which the spoon-like forebody carries an undivided holdfast

holostome a term used to describe those trematode Platyhelminthes in which the forebody has an additional large adhesive organ (the holdfast)

hypostome 1 (*see also* hypostome 2, 3) in insects = hypostoma

hypostome 2 (*see also* hypostome 1, 3) a raised area bearing the mouth in coelenterates

hypostome 3 (*see also* hypostome 1, 2) the piercing organ of ixode acarines

mestome *see under* -tome

monostome a term used to describe those trematode platyhelminthes in which the acetabulum has been lost

nephrostome a peritoneal funnel forming part of, or directly derived from, a nephron

nephridiostome = nephrostome

notostome the postulated "dorsal mouth" of postulated "ancestral vertebrates"

oeciostome the orifice of the gonozooid of an ectoproct

peristome 1 (*see also* peristome 1–4) the orificial ring of Ectoprocta

peristome 2 (*see also* peristome 1, 3–5) the region surrounding the entrance to the cytopharynx in Protozoa

peristome 3 (*see also* peristome 1, 2, 4, 5) the area surrounding the aperture of a gastropod shell

peristome 4 (*see also* peristome 1–3, 5) that segment of an annelid which lies immediately behind the prostome

peristome 5 (*see also* peristome 1–4) the structures immediately surrounding the orifice of a moss capsule

aplolepideous peristome the peristome of a moss, having a single row of teeth

prostome that which lies in front, usually immediately in front, of the mouth. Specifically, that segment of an annelid which lies immediately in front of the mouth

stomium (*see also* stoma, stome) any aperture that can conceivably be called a mouth. Specifically, the opening through which pollen escapes from an anther (*cf.* stome)

hypostomium an unpaired scaly plate, behind the mouth of gastrotrichs (*cf.* cephalion, pleurion)

peristomium = peristome

prostomium = prostome

Stomochorda a zoological taxon erected to contain the Pterobranchia and the Enteropneusta

-stomous *adj. term* meaning "pertaining to a single mouth or opening." The plural form -stomatous is more usual and only the compounds of this form are given in this work

brimstone *Brit.* the pierid lepidopteran insects *Gonepteryx rhamni* and the geometrid *Opisthograptis luteolata*

turnstone either of two species of charadriid birds of the genus Arenaria. *Brit.*, without qualification, usually *A. interpres*

black turnstone *A. melanocephala*

ruddy turnstone *A. interpres*

stool the crown of a plant from which several branches arise, as a strawberry plant which puts out runners

stoop the action of a hawk in diving on its prey

honey **stopper** the valve between the crop and the stomach of a nectar-collecting hymenopteran

stopple the lid of a pollen grain

storax 1 (*see also* storax 2) a gum obtained from the hamamelidaceous tree *Liquidambar styraciflua* (= sweet gum)

storax 2 (*see also* storax 1) anglicized name of the styracaceous tree *Styrax officinalis* which yields gum

benzoin. A gum resin called storax was at one time obtained from *S. officinalis* but is no longer in commerce

stork any of numerous long-legged birds mostly of the family Ciconiidae. *Eur.*, without qualification, *Ciconia ciconia*

Abdim's stork *Sphenorhynchus abdimii*

adjutant stork either of two species of the genus *Leptoptilos*. The term is sometimes applied to the marabou stork (*q.v.*)

white-bellied stork = Abdim's stork

openbill stork either of two species of the genus Anastomus (= openbill)

saddle-billed stork *Ephippiorhynchus senegalensis* (= saddle bill)

black stork *Ciconia nigra*

whale-headed stork = Abdim's stork

marabou stork *Leptoptilos crumenifer* (= U.S. marabou, *cf.* adjutant stork)

milky stork *Ibis cinereus*

black-necked stork *Xenorhynchus asiaticus*

white-necked stork = wooly-necked stork

wooly-necked stork *Dissoura episcopus*

painted stork *Ibis leucocephalus*

white stork *Ciconia ciconia*

str a bacterial mutant showing streptomycin resistance, sensitivity or dependence

str a bacterial genetic marker having a character affecting the resistance sensitivity or dependence on streptomycin, mapped at 64.0 mins. for *Escherichia coli*

-stracum- *comb. form* meaning "shell"

ectostracum that layer of the acarine integument which lies between the epiostracum and the hypostracum

epiostracum that layer of the acarine integument that lies immediately under the tectostracum

hypostracum the innermost of the four layers of the acarine integument

periostracum the outermost portion of a molluscan shell

tectostracum the thin outermost covering of the acarine integument

strain a relatively meaningless term occasionally used in the sense of variety or subvariety

-stramen- *comb. form* meaning "straw"

stramenieous resembling straw either in shape, texture or color

strand 1 (*see also* strand 2) one of the fibrous components of a string, rope or bundle

connecting strand the cytoplasmic connections which pass through sieve plates

transfusion strand the parenchymatous cells which lie between the xylem and phloem bundles

hydrome strand a strand of water-carrying vessels

strand 2 (*see also* strand 1) *obs.* beach

stranger *Brit.* the noctuid lepidopteran insect *Mamestra peregrina*

Strasburgeriaceae a small family of parietale dicotyledons

-strat- *comb. form* originally meaning a "paved highway" but now used almost entirely in the sense of "layer"

Stratiomyidae a family of orthoraphous brachyceran dipteran insects mostly medium-sized and many brightly colored or wasp-like. The wing venation is distinctive. The more brilliant colored forms are commonly called soldierflies

Stratiomyiidae = Stratiomyidae

stratiote any individual belonging to a particular social stratum or caste

mermithostratiote a worker ant, structurally altered in consequence of being parasitized by mermithid nematodes (*cf.* aner)

stratous arranged in layers

bistratous disposed in two layers

humistratous cast flat against, or hugging the surface of, the soil

straw 1 (*see also* straw 2, 3) the dried, hollow stem of grains

straw 2 (*see also* straw 1, 3) in compound names of plants usually with wiry stems

bedstraw any of numerous species of herbs of the rubiaceous genus Galium

cross-leaved bedstraw (*Brit.* = northern bedstraw)

heath bedstraw *G. saxatile*

hedge bedstraw *G. mollugo*

lady's bedstraw any of several species of the genus Galium, usually without qualification, *G. verum*

marsh bedstraw *G. uliginosum*

northern bedstraw *G. boreale*

water bedstraw *G. palustre*

yellow bedstraw *G. verum*

straw 3 (*see also* straw 1, 2) in compound names of animals, usually straw-colored

barred straw *Brit.* the geometrid lepidopteran insect *Cidaria pyraliata*

bordered straw *Brit.* either of two species of noctuid lepidopteran insect of the genus Heliothis but usually, without qualification *H. peltigera*

scarce bordered straw *Brit. H. armiger* (= *U.S.* bollworm)

[straw] a mutant (gene symbol *stw*) mapped at 55.1 on chromosome II of *Drosophila melanogaster*. The phenotype is yellow

streak 1 (*see also* streak 2) in the sense of a linear mark

primitive streak an area of concreted cells, corresponding to the blastopore, at the posterior end of the blastodisc of a developing telolecithal egg

streak 2 (*see also* streak 1) in the names of numerous lepidopteran insects with streaky wings. *Brit.*, without qualification, the geometrid *Chesias spartiata*

hairstreak any of the members of the family Lycaenidae (*cf.* copper, blue)

black hairstreak *Thecla pruni*

brown hairstreak *Brit. Zephyrus betulae*

common hairstreak *U.S. Strymon melinus*

green hairstreak *Brit. Thecla rubi*

purple hairstreak *U.S. Atlides halesus. Brit. Zephyrus quercus*

white letter hairstreak *Brit. Thecla w-album*

honey-streak *U.S.* the geometrid lepidopteran *Sciagraphia mellistrigata*

[streak] a mutant (gene symbol *Sk*) mapped at 16.0 on chromosome II of *Drosophila melanogaster*. The name derives from a central streak on the thorax of the phenotype

streamer *Brit.* the geometrid lepidopteran insect *Anticlea nigrofasciaria*

Streblidae a family of pupiparous dipteran insects parasitic on bats and closely allied to the Nycteribiidae from which they may be distinguished by the fact that some species are winged and that the head cannot be bent back into a groove on the thorax

Strelitzioideae a subfamily of musaceous scitaminale monocotyledons containing the bird-of-paradise flowers

Streperidae = Cracticidae

-streph- *comb. form* meaning "twisted" (*cf.* -strept-)

Strepsicerosinae a subfamily of bovid artiodactyle mammals commonly called the twist-horned oxen. The elands and bushbucks are the best known members of this subfamily, which is, of course, distinguished by the spiral twist on the horns

Strepsiptera a small order of minute parasitic insects distinguished by the fact that the hindwings of the male are functional but the forewings are reduced to slender club-like appendages. The females are wingless. Sometimes regarded as a family (Stylopidae) of Coleoptera

biastrepsis = torsion

-streph- *comb. form* meaning "twisted" (*cf.* = strept-)

Streptococceie *see* Lactobacillaceae

Streptomycetaceae a family of actinomycetale Schizomycetes in which the mycelium, unlike that of the Actinomycetaceae, never fragments into rods. Conidia are borne on sporophores. Typically soil organisms

[Streptomycin-la] a mutant (gene symbol *sr-la*) in linkage group IX of *Chlamydomonas reinhardi*. The phenotypic expression is a resistance to streptomycin

Streptoneura a subclass of gastropod Mollusca containing the abalone, conchs, whelks, cowries, limpets and slipper limpets. They are distinguished by the extreme torsion of the visceral mass which results in the visceral commissures being twisted into a figure eight

stria literally a furrow, but applied particularly to markings on the frustules of diatoms which have the appearance of lines until they are resolved into dots

striate laying in parallel lines

[striate] a mutant (gene symbol *sr*) mapped at 0 on linkage group I of *Zea mays*. The name refers to the appearance of the phenotypic leaf

striatum a pseudo-Latin word derived from *stria*, a groove

neostriatum that part of the corpus striatum which first appeared in Amphibia and is probably homologous to the putamen of mammals

paleostriatum that portion of the corpus striatum which is found in primitive forms and which is probably homologous to the globus pallidus of mammals

water **strider** popular name of gerrid hemipteran insects

stridulate to make a noise by rubbing two toothed structures together

-strife *see* loosestrife or willowstrife. The suffix is of itself meaningless

strig *Brit. obs.* stem, in the broadest possible sense

Strigidae a large family of strigiform birds containing the true, or typical owls. They are distinguished by the peculiar eyes and facial feathering and by short strong and hooked bills which have a cere at the base

Strigiformes that order of birds which contains the nocturnal birds of prey or owls. They are distinguished by their short hooked bill and forwardly directed eyes surrounded by a ruff of feathers

strigil literally, a stiff comb-like scraper and specifically applied to an organ of this function on the tibia of many insects

Strigopidae = Psittacidae

strigose bristly

string term for a small group of horses maintained for racing

lutestring *Brit.* either of two thyatirid lepidopteran

insects *Palimpsestis or* and *Asphalia diluta*

strip a ribbon-like mass

Casparian strip a strip of lignified, and often suberized, material in the radial and transverse walls of plant endoderm cells

stripe a strip of contrasting color or shade

silverstripe *I.P.* any of several nymphalid lepidopteran insects of the genus Argynnis (*cf* fritillary)

tooth-stripe *Brit.* any of several species of geometrid lepidopteran insect of the genus Lobophora

[stripe] a mutant (gene symbol *sr*) mapped at 62.0 on chromosome III of *Drosophila melanogaster.* The phenotypic expression is a dark dorsal stripe

-strobil- *comb. form* meaning a "cone" in the sense of a pine cone

strobila (*see also* larva) an organism, or stage in the life history of an organism, which buds off successive parts from one end. In this meaning, a tapeworm, as well as a larval stage of a jellyfish, may be called a strobila

strobilate having the form of a pine cone

Strobiloideae a term used to describe those plant families, either monocot or dicot, in which the flower is thought to have originated from spirally arranged phyllomes on a vegetative shoot

strobilus a pine cone or anything resembling it

anthostrobilus a postulated, theoretical ancestor of the angiosperm flower

megastrobilus the female cone of a cycad

microstrobilus the male cone of cycads

-strom- *comb. form* meaning "mattress"

stroma a sponge-like framework usually produced by reticular cells. A section of a stroma has the appearance of a net

hypostroma = mycelium

Stromateidae a family of stromateoid acanthopterygians distinguished by the lack of pelvic fins. Commonly called butterfishes

Stromateoidea a small suborder of acanthopterygian Osteichthyes distinguished by their expanded and muscular esophagus often armed with teeth. The suborder includes the families Stromateidae and Nomeidae

distromatic said of a thallus having two distinct layers

-stromb- *comb. form* meaning "spirally twisted" like a whelk shell

Strombidae a family of active carnivorous gastropod Mollusca with thick heavy solid shells with a greatly enlarged body whorl. The long and narrow aperture is thickened and expanded and is partially closed by a claw-like operculum. This family is commonly called conches

strombus a spirally coiled leguminous seed pod

strongyle a rhabd rounded at each end

Strongylidae a family of parasitic nematode worms distinguished by the large teeth on the mouth, but lacking the ventral cutting plates of the Ancylostomidae. The head is surrounded by a fringe of narrow papillae known as the leaf crown

Strongyloidea a suborder of telogonian nematodes distinguished from the Filarioidea by the large mouth and from the Ascaroidea by lack of the three prominant lips of this group. The large copulatory expansion, supported by rays in the male is also characteristic

-stroph- *comb. form* meaning "turn" (*cf.* -trop-)

Strophariaceae a family of agaricale basidiomycete fungi closely allied to Coprinaceae but differing in that the spores are purple rather than black

apastrophe the removal of chloroplasts to the periphery of the cell under the influence of intense light

epistrophe the arrangement of chloroplasts on the upper and lower phases of cells under conditions of poor illumination

parastrophe = apostrophe

protostrophe a secondary leaf spiral

systrophe the aggregation of chloroplasts into masses under the influence of light

epistropheus = axis

strophiole a swollen portion of the hilum of a seed (*cf.* epiphysis)

strophism a tendency to, or act of, twisting

-strot- *comb. form* meaning "to spread"

-strote *adj. suffix* indicating an organism defined by its method of distribution

carpostrote a plant which is dispersed solely by fruits

phytostrote 1 *see also* phytostrote 2) a plant which is dispersed by other plants

phytostrote 2 (*see also* phytostrote 1) the distribution of plankton over a surface

spermatostrote a plant which is dispersed by seeds

sporostrote an organism which is dispersed spores

thallostrote a plant which spreads by the production of offshoots or runners

zoostrote an organism which is dispersed by animals

ultrastructure those features of cellular structure which are disclosed by electron microscopy

struma a surface swelling like a goiter

strumous having the appearance of being locally swollen

Struthidiidae a family of passeriform birds commonly included with the Grallinidae

Struthionidae a monotypic family of struthioniform birds which contains the ostrich, distinguished by the naked thighs and neck and by the possession of only two toes

Struthioniformes an order of birds containing the ostriches and their extinct allies. They are distinguished by the naked neck and head as well as the lack of ability to fly

[stub] a mutant (gene symbol *st*) mapped at 34.1 on linkage group IV of the rat. The name refers to the condition of the tail

[stubarista] a mutant (gene symbol *sta*) mapped at 0.3± on the X chromosome of *Drosophila melanogaster.* The phenotypic expression involves not only shortened aristae, but also generally reduced bristles and short antennae

[Stubble] a mutant (gene symbol *Sb*) mapped at 58.2 on chromosome III of *Drosophila melanogaster.* The phenotypic expression is short thick bristles

[stubby] a mutant (gene symbol *sb*) in linkage group I of *Habrobracon juglandis.* The phenotypic expression is a shortened antenna

stud term for a group of horses maintained for breeding purposes

[stumpy] a mutant (gene symbol *st*) in linkage group III of *Habrobracon juglandis.* The phenotypic expression involves a reduction of the tarsae

-stup- *comb. term* meaning "tow" in the sense of the fiber and, by extension, "hairy"

stupa = stuppa

stupeous felted, but with some loose surface hairs

stuppa a mass of matted fibers

sturgeon any of several species of chondrostean fishes of the family Acipenseridae

Atlantic sturgeon *Acipenser oxyrhynchus*

green sturgeon *Acipenser medirostris*

lake sturgeon *Acipenser fulvescens*

shortnose sturgeon *Acipenser brevirostrum*
shovelnose sturgeon *Scaphirhynchus platorynchus*
pallid sturgeon *Scaphirhynchus albus*
white sturgeon *Acipenser transmontanus*
Sturnidae a large family of passeriform birds containing, *inter alia*, the starlings. They are distinguished by their silky plumage and dark metallic colors. The wings are variable but the tail is usually short and square
stw *see* [straw]
Styelidae a family of ptychobranchiate ascidians occurring both as compound and simple forms. They are distinguished from other ptychobranchiates by having simple unbranched tentacles
stygious pertaining to foul waters
-styl- *comb. form* meaning "column"
Stylasterina an order of hydrozoan coelenterates forming a calcareous skeleton like the Milliporina but lacking free-swimming medusae
style literally a "column" but used in *biol.* in a wide variety of senses
style 1 (*see also* style 2–5) the part of a carpel lying above the ovary. Ovary and style, capped by the stigma, make up the pistil
    parastyle an aborted style
style 2 (*see also* style, style 1, 3–6) in the sense of a cusp on a mammalian tooth
    hypostyle a small cusp between the hypocone and the metacone
    mesostyle a small cusp situated between the metacone and the paracone
    metastyle a small cusp posterior to the metacone
    parastyle (*see also* style 6) a small cusp just anterior to the paracone
style 3 (*see also* style, style 1, 2, 4–6) in the sense of a sponge spicule that grows from one end in contrast to a rhabd
    acanthostyle a style covered with spines
    tylostyle a style in which the broad end is knobbed
style 4 (*see also* style, style 1–3, 5, 6) in the sense of a rod or column-like organ or skeletal element in a chordate
    endostyle a flattened or trough-like structure, grooved on its dorsal surface by the hypo-branchial groove, which runs along the ventral margin of the pharynx in prochordates
    pygostyle a short squat "tail" of fused vertebrae found in birds
    urostyle an unsegmented, posterior prolongation of the vertebral column. It is a thin, rod-like structure in anuran Amphibia, a group of fused vertebrae in certain Chelonia and a terminal group of hypural bones in actinopterygian fishes
style 5 (*see also* style, style 1–4, 6) in the sense of a projection from or rodlike structure in, an invertebrate
    anal style = anal cercus
    blastostyle (Coelenterata) = gonozoid (Coelenterata)
    crystalline style a transparent body, supposed to liberate enzymes, found in the alimentary canal of some pelecypod mollusks
    lithostyle a club-shaped group of lithocytes hanging from the bell of a medusa
    odontostyle an enlarged tooth or stylet found in some nematodes
    sarcostyle = nematophore
style 6 (*see also* style, style 1–5) in the sense of a rod-like organelle
    axostyle a rod-shaped body attached to the blepharoplast

parastyle (*see also* style 2) an oval body attached to a blepharoplast
Stylephoridae a small family of deep water allotriognathous fish with protruding eyes
Stylommatophora a sub-order of pulmonate gastropod mollusca distinguished by bearing two pairs of contractile tentacles, of which the posterior bear eyes
stylet diminutive of style in any of its meanings
    buccal stylet a spear shaped style projecting from the buccal organ of nematodes
    penis stylet a hard spine which in some invertebrates substitutes for a penis
stylic pertaining to a style in any of its meanings except style 1 (*q.v.*) for which stylous (*q.v.*) is more usual
Stylidiaceae a small family of campanulate dicotyledonous angiosperms
styloid an elongate, usually single, crystal in a plant cell
Stylopidae a family of coleopteran insects now regarded as the separate order Strepsiptera
-stylous pertaining to the style of a flower (*see* style 1, stylic)
    aristostylous pertaining to a flower with an exerted style with a sinistral bend or twist
    brachystylous = microstylous
    heterodistylous a form of dimorphism found in some plants which are distinguished by the possession of either long or short styles
    dolichostylous having a long style
    enantiostylous the condition of having styles protrude from one side of a flower and anthers from the other
    heterostylous = heterogamous
        homoheterostylous the condition of a species of flower of which some have similar, and others dissimilar, styles in the flower
    macrostylous having a long style
    polexostylous = carcerule
    peristylous the condition of having stamens inserted between the styles and the limb of the calyx
    systylous the condition of having the lid attached to the capsule in a moss
-styly *comb. suffix* indicating the manner in which the lower jaw of vertebrates articulates with the cranium
    amphistyly the condition of having the upper jaw supported from the cranium by both the palatoquadrate and hyomandibular cartilages
    cranio-amphistyly = craniostyly
    autostyly the condition of having the palatoquadrate either sutured or fused to the cranium
    euautostyly = quadratostyly
    holoautosystyly = holostyly
    parautostyly the peculiar form of autostyly found in cyclostomes
    craniostyly the condition, as in mammals, in which neither the palatoquadrate or hyoid arch are involved and the dentary bone articulates with the squamosal of the cranium
    autodiastyly autostyly in which the palatoquadrate articulates with the cranium both at the front and rear
    holostyly the condition of having the palatoquadrate fused with the cranium without involvement of the hyoid arch. This is typical of holocephalan fishes
    hyostyly the condition of having the lower jaw supported from the cranium largely by means of the hyoid
    monimostlyly the condition in which the quadrate is fused to the cranium
    methyostyly a condition similar to hyostyly but in

which the symplectic is combined with the hyo-mandibular

palaeostyly the postulated primitive state in which the mandibular arch is identical with the other visceral arches

quadratostyly auostyly involving only the quadrate

streptostyly the condition in which the quadrate bone moves freely against the cranium

prostypus = raphe

Styracaceae a small family of ebenalous dicotyledons containing, *inter alia*, the styrax tree and silver bill. The several-celled ovary, bisexual flowers and absence of milky juice distinguish it from related families

su a yeast genetic marker indicating a requirement for sucrose. The symbol is also used in conjunction with another symbol to indicate that a mutant is a supressor

su₁ *see* [sugary endosperm]

su-b *see* [suppressor of black]

suB-pr *see* [suppressor of purple]

su-Cbx *see* [suppressor of Contrabithorax]

Su-dx *see* [suppressor of deltex]

su-f *see* [suppressor of forked]

Su-H *see* [Suppressor of Hairless]

su-Hw *see* [Suppressor of Hairywing]

su-l-me *see* [Suppressor-l-methionine]

su-mel-3 *see* [Suppressor-melon-3]

Su-S *see* [Suppressor of Star]

su-pd *see* [suppressor of purpleoid]

su-s *see* [suppressor of sable]

su-t *see* [suppressor of tan]

su ^{s2}-vpr *see* [suppressor of vermilion and purple]

su-ve *see* [suppressor of veinlet]

su-wᵃ *see* [suppressor of apricot]

suˣ-dx see [suppressor of first chromosome of deltex]

suaveolent fragrant

-sub- *comb. form* meaning "below" though used in *biol.* compounds to mean "almost" as in the word "subacute" meaning not quite pointed. Most compounds of this type are not given in this dictionary, since the meaning is self-evident

suber = cork

suberin a fatty substance, the presence of which in cork cells causes the the consistency peculiar to the tissue cork

suberization the conversion of other tissues into cork

suberose corky

subex that part of the axis on which the cataphyllary leaves are found

subiculum a mass of hyphae bearing perithecia

-subit- *comb. form* meaning "sudden"

substance matter or material. Usually used in *biol.* for acellular matter of varied or ill-defined composition

intercellular substance the material between cells (often called ground substance)

cement substance a hard, amorphous, intercellular material

chromidial substance = ergastoplasm

ground substance 1 (*see also* ground substance 2) a soft, amorphous intercellular material (often called intercellular substance)

ground substance 2 (*see also* ground substance 1) the postulated basic material of "protoplasm" in the days before electron microscopy

Nissl substance a granular endoplasmic reticulum in nerve cells

gene substitution the replacement of one allele by another

subtend properly to extend underneath, but almost invariably used to mean opposite to

subtilisin = subtilopeptidase

-subul- *comb. form* meaning "awl"

subulate awl shaped

suc a bacterial genetic marker indicating a requirement for succinic acid, mapped at between 13.25 mins. and 16.0 mins. for *Escherichia coli*

Suc *see* [Succinic]

succession the event of one thing following another chronologically. Specifically, in ecology, the sequence of changes in the community of a given area

halarch succession one occuring under saline conditions

autogenic succession one resulting from biotic, as distinct from climatic, causes

ecological succession the gradual change of uninhabited land to a climax community

hydrostatic succession one which does not change with variations of moisture

subsuccession a sere beginning on a rock surface and ending in a matgrowth

hydrotopic succession one which changes in virtue of an increased water content

-succin- *comb. term* meaning "amber"

succinase general term for a group of enzymes

argininosuccinase = argininosuccinate lyase

ureidosuccinase an enzyme hydrolyzing N-carbamoyl-L-aspartate to L-aspartate, carbon dioxide and ammonia

succineous amber colored

[Succinic] a mutant (gene symbol *Suc*) in linkage group I of *Neurospora crassa*. The phenotypic expression is a requirement for succinic acid

succory the compositaceous herb *Cichorium inthybus* (= chicory)

gum-succory any of several species of compositaceous herbs of the genus Chondrilla

succose sappy

exsuccous lacking juice

succulent literally juicy but usually used to describe plants with thick fleshy leaves

sucker 1 (*see also* sucker 2, 3, 4, 5) a shoot which derives from an underground lateral root

sucker 2 (*see also* sucker 1, 3, 4, 5) any of numerous teleost fishes. In *U.S.* confined to catostomids but in *Brit.* extended to gobiesocids

blue sucker *Cycleptus elongatus*

blackfin sucker *Moxostoma atripinne* (*cf.*, redhorse)

carpsucker *U.S.* any of several species of catostomids of the genus Carpiodes

highfin carpsucker *C. velifer*

plains carpsucker *C. forebesi*

river carpsucker *C. carpio*

chubsucker any of several species of catastomids of the genus Erimyzon

creek chubsucker *E. oblongus*

lake chubsucker *E. sucetta*

sharpfin chubsucker *E. tenuis*

humpback sucker *Xyrauchen texanus*

Gila sucker *Pantosteus clarki*

green sucker *Pantosteus virescens*

bluehead sucker *Pantosteus delphinus*

hogsucker any of several species of catastomids of the genus Hypentilium

Alabama hogsucker *H. etowanum*

northern hogsucker *H. nigricans*
Roanoke hogsucker *H. roanokense*
June sucker *Chasmistes liorus*
lumpsucker popular name of cyclopterid fishes, particularly *Liparis atlanticus*
lahontan sucker *Pantosteus lahontan*
bridgelip sucker *Catostomus columbianus*
harelip sucker *Lagochila lacera*
Lost River sucker *Catostomus luxatus*
Modoc sucker *Catostomus microps*
mountain sucker *Pantosteus platyrhynchus*
flannelmouth sucker *Catostomus latipinnis*
mudsucker the gobiid fish *Gillichthys mirabilis*
longnose sucker *Catostomus catostomus*
shortnose sucker *Chasmistes brevirostris*
Rio Grande sucker *Pantosteus plebeius*
Sacramento sucker *Catostomus occidentalis*
Santa Ana sucker *Pantosteus santaanae*
largescale sucker *Catostomus macrocheilus*
Klamath largescale sucker *Catostomus snyderi*
Klamath smallscale sucker *Catostomus rimiculus*
rustyside sucker *Moxostoma hamiltoni*
Sonora sucker *Catostomus ingignis*
spotted sucker *Minytrema melanops*
Tahoe sucker *Catostomus tahoensis*
torrent sucker *Moxostoma rhothoecum*
Utah sucker *Catastomus ardens*
warner sucker *Catostomus warnerensis*
Webug sucker *Catostomus fecundus*
white sucker *Catostomus commersoni*
White River sucker *Pantosteus intermedius*
sucker 3 (*see also* sucker 1, 2, 4, 5) any of several psyllid homopteran insects of the genus Psylla
apple sucker *U.S. P. mali*
sucker 4 (*see also* sucker 1, 2, 3, 5) any of numerous animals reputed in folklore to suck, or which actually do suck, fluids from other organisms
bloodsucker 1 (*see also* blood sucker 2) any of numerous species of agamid lizards with red gular pouches. Without qualification, the Asiatic *Calotes versicolor* is usually meant. *Austral. Amphibolurus muricatus*
bloodsucker 2 (*see also* bloodsucker 1) general term for hirudinean annelids (= leech)
cherry-sucker *Brit.* the muscicapid bird *Muscicapa striata* (= spotted flycatcher)
cowsucker any of numerous animals reputed to suck cow's milk. In the *U.S.* this legend pertains mostly to snakes and in Europe mostly to the hedgehog
goatsucker *U.S.* popular name of caprimulgid birds (= *Brit.* nightjar)
honey-sucker 1 (*see also* honey-sucker 2) popular name of tarsipedine phalangerid marsupials, particularly *Tarsipes rostratus* (= noolbenger)
honey-sucker 2 (*see also* honey-sucker 1) *W.I.* the parulid bird *Coereba flaveola* (= bananaquit)
sapsucker *U.S.* either of two species of picid bird of the genus Sphyrapicus
yellow-bellied sapsucker *S. varus*
red-breasted sap sucker = yellow-bellied sapsucker
Williamson's sap sucker *S. thyroideus*
sucker 5 (*see also* sucker 1, 2, 3, 4) a saucer shaped adhesive disc or organisms distinguished by possessing such an organ
marlin-sucker the echeneid fish *Remora osteochir*
shark-sucker any of several species of echeneid fishes, particularly *Echeneis naucrates*
whale-sucker the echeneid fish *Remora australis*

honeysuckle *see* honeysuckle
suckling an unweened mammal, particularly of a primate
amylosucrase = sucrose glucosyltransferase
dextransucrase = sucrose 6-glucosyltransferase
inulosucrase = sucrose l-fructosyltansferase
levansucrase = sucrose 6-fructosyltransferase
Suctoria 1 (*see also* Sictoria 2) = Siphonaptera
Suctoria 2 (*see also* Suctoria 1) a group of Protozoa possessing tentacles and reproducing by ciliated young. They are variously regarded as a class of the subphylum Ciliophora or an order of the class Ciliata
suf *see* [sufflava]
[sufflava] a mutant (gene symbol *suf*) mapped at 33 on chromosome 2 of the tomato. The phenotypic expression is light green leaves
suffultous being propped up
horse sugar the styracaceous tree *Symplocus tinctoria*
[sugary endosperm] a mutant (gene symbol *su₁*) mapped at 71 in linkage group IV of *Zea mays*. The name is descriptive of the phenotypic endosperm
Suidae that family of suine artiodactyles which contains the pigs. They differ from the Hippopotamidae in having terminal nostrils and a mobile snout and in the fact that only two of the four toes reach the ground in walking
Suina a group of artiodactyle mammals containing the families Hippopotamidae, Suidae and Dicotylidae. They differ from the Ruminantia in the simple nature of the stomach
sul a bacterial mutant showing sulfonamide sensitivity or resistance
-sulca- *comb. form* meaning a "groove"
sulcal an adjective applied to that side of an alcyonarian coelenterate on which the siphonoglyph (sulcus) lies. This is frequently referred to as the ventral side
asulcal an adjective applied to that side of an alcyonarian which is opposite to the siphonoglyph (sulcus). This is often referred to as the dorsal side
sulcate grooved
sulculus a term applied to the smaller of two siphonoglyphs in those anthozoan coelenterates which have two (*cf.* sulcus)
sulcus 1 (*see also* sulcus 2) the siphonoglyph of an alcyonarian or the larger of two siphonoglyphs in some anthozoan coelenterates (*cf.* sulculus)
sulcus 2 (*see also* sulcus 1) a groove (as distinct from a fissure) in the cerebral hemisphere
Monro's sulcus the longitudinal groove that divides the embryonic central nervous system into dorsal and ventral regions
[sulfonamide] a mutant (gene symbol *sfo*) on linkage group VII of *Neurospora crassa*. The phenotypic expression is a requirement for sulfonamide
Sulidae that family of pelecaniform birds which contains the boobies and gannets. They are distinguished from the Phaethontidae by the absence of an attenuated central tail feather, from the frigatebirds by their smaller tails and from other families of the order by their short necks. They have the general appearance of a short, squat, clumsy gull
sulfatase any of several enzymes that catalyze the hydrolysis of organic sulfates with the liberation of sulfuric acid
arylsulfatase hydrolyzes phenol sulfate to phenols and sulfuric acid

glycosulfatase hydrolyzes D-glucose 6-sulfate to glucose and sulfuric acid

myrosulfatase hydrolyzes sinigrin to merosinigrin and sulfuric acid

sterol sulfatase hydrolyzes dehydroepiandrosterone 3-sulfate, and some related sterol sulfates, to dehydroepiandrosterone and sulfuric acid

cysteine desulfhydrase catalyzes the production of pyruvate, ammonia and hydrogen sulfide from cysteine

homocysteine desulfhydrase catalyzes the production of oxobutyrate, ammonia and hydrogen sulfide from homocysteine

sulfur (*see also* bottom) popular name of those lepidopteran insects which are uniformly yellow or orange (*cf.* orange-tip), usually, without qualification, pierids of the genus Colias

angled sulfur popular name of pierid lepidopteran insect of the genus Anteos

little sulfur popular name of tropical pierid lepidopteran insects of the genus Eurema

orange sulfur *Colias philodice*

spotted sulfur *Brit.* the noctuid lepidopteran insect *Emmelia trabealis*

tropical sulfur popular name of pierid lepidopteran insects of the genus Phoebis

sulfureous the color of sulfur

sweet **sultan** the compositaceous herb *Centaurea moschata*

sumac (*see also* beetle) any of several species of anacardiaceous shrub or tree of the genus Rhus

black sumac *R. copallina*

dwarf sumac = black sumac

fragrant sumac *R. aromatica*

ill-scented sumac *R. trilobata*

lemon-sumac = fragrant sumac

mountain sumac = black sumac

poison sumac *R. vernix*

shining sumac = black sumac or wingrib sumac

smooth sumac *R. glabra*

staghorn sumac *R. typhina*

velvet sumac = staghorn sumac

wingrib sumac = black sumac

sumach = sumac

summing the complete anter of a deer

rising **sun** *U.S.* the tellinid mollusk *Tellina radiata*

sundew *see* dew

suni any of several species of neotragine antelopid of the genus Nesotragus

super literally "above" but also used in *biol.* in the sense of "greater than" or "superior to" (*cf.* supra)

superposed placed vertically above

supine lying flat on the back (*cf.* prone)

resupinate upside down

cross-over suppressor a gene supposed to inhibit crossing over

[**suppressor -1-methionine**] a mutant (gene symbol *su -1-me*) in linkage group 1 of *Neurospora crassa*. It suppresses both [methionine 2] and [methionine 7]

[**suppressor in first chromosome of deltex**] a mutant (gene symbol *su^x-dx*) which changes [deltex] almost to the white wild type. Mapped at 5.0± on the X chromosome of *Drosophila melanogaster*

[**suppressor-melon-3**] a mutant (gene symbol *su-mel-3*) found in linkage group III of *Neurospora crassa*. The name is descriptive

[**suppressor of apricot**] a mutant (gene symbol *su-w^a*) mapped at 0 on the X chromosome of *Drosophila melanogaster*. It suppresses the apricot eye color of pseudo allele *w^a*

[**suppressor of black**] a mutant (gene symbol *su-b*) mapped at 0.1 on the X chromosome of *Drosophila melanogaster*. It suppresses [black]

[**suppressor of Contrabithorax**] a mutant (gene symbol *su-Cbx*) mapped at 30.0± on the X chromosome of *Drosophila melanogaster*. The phenotypic expression is the suppression of [Contrabithorax]

[**Suppressor of deltex**] a mutant (gene symbol *Su-dx*) located on Chromosome II of *Drosophila melanogaster*. The phenotypic expression is the suppresion of [deltex]

[**suppressor of forked**] a mutant (gene symbol *su-f*) mapped at 64.0± on the X chromosome of *Drosophila melanogaster*. It changes some of the alleles of [forked] to wild type

[**Suppressor of Hairless**] a mutant (gene symbol *Su-H*) mapped at 50.5 on chromosome II of *Drosophila melanogaster*. The name is descriptive

[**suppressor of Hairy-wing**] a mutant (gene symbol *su-Hw*) mapped at 54.8 on chromosome III of *Drosophila melanogaster*. The phenotypic expression is suppression of [Hairy-wing]

[**suppressor of purple**] a mutant (gene symbol *su^B-pr*) mapped at 95.5 on chromosome III of *Drosophila melanogaster*. The name is descriptive

[**suppressor of purpleoid**] a mutant (gene symbol *su-pd*) located on chromosome III of *Drosophila melanogaster*. This mutant turns the eye color of [purploid] to wild type.

[**suppressor of sable**] a mutant (gene symbol *su-s*) mapped at 0+ on the X chromosome of *Drosophila melanogaster*. It suppresses the pseudo-allele [vermilion] and the allele [sable]

[**Suppressor of Star**] a mutant (gene symbol *Su-S*) mapped at 1.3± on chromosome II of *Drosophila melanogaster*. The name is descriptive

[**suppressor of tan**] a mutant (gene symbol *su-t*) mapped at 26.0± on chromosome III of *Drosophila melanogaster*. The mutant converts [tan] to wild type

[**suppressor of vermilion and purple**] a mutant (gene symbol *Au^S2-v pv*) located at 0+ of the X chromosome of *Drosophila melanogaster*. It suppresses the mutant genes [vermilion] and [purple].

[**suppressor of veinlet**] a mutant (gene symbol *su-ve*) mapped at 0- on chromosome III of *Drosophila melanogaster*. The phenotypic expression is the suppression of [veinlet].

-supra- *comb. form* from "super" (q.v.)

surculose producing suckers

surcurrent the condition of having winged extensions from the base of the leaf prolonged up the stem

surficial a term used to describe the flatter of the two surfaces of a tapeworm proglottis, when dorsal and ventral are not clearly defined

suricat the viverrid carnivore *Suricata tetradactyla* (= meerkat)

surmullet the popular name of mullid fishes (*cf.* goatfish)

black-eyed **susan** any of several species of compositaceous herb of the genus Rudbeckia most commonly either *R. hirta* or *R. serotina*. The term is also applied to the acanthaceous herb *Thunbergia alata*

intus**susception** the increase in thickness of a plant cell wall, or other similar structure, through the deposition of materials in its interior (*cf.* apposition)

suspensor any thing which suspends anything

susu the platanistid cetacean *Platanista gangeticae* (= Ganges dolphin)

sute term for a group of bloodhounds

suture 1 (*see also* suture 2, 3) in insects, the line of junction of sclerites

basal suture in termites, the abscission layer along the base of the wing

frontal suture a curved, or horseshoe-shaped, groove immediately behind the antennae in some dipteran flies (*cf.* Schizophora)

suture 2 (*see also* suture 1, 3) a tight fibrous joint between bones

false suture one in which the bones do not interlock

squamose suture one in which the bones overlap

true suture a synarthrosis formed by interlocking finger-like or tooth-like projections of the bone margin

suture 3 (*see also* suture 1, 2) the line of junction between other structures

dorsal suture that which is exterior to the axis of a seed, follicle or fruit

suwarro the very large, tree-like, cactus *Carnegiea gigantea* (= saguaro)

sv see [shaven], [shot veins] or [Snell's walzer]

svr *see* [sliver]

sw *see* [short-wing]

swallow 1 (*see also* swallow 2, bug, plover, shrike, tanager, wing, wort) any of numerous passeriform birds of the family Hirundinidae. *Brit.* without qualification, *Hirundo rustica*

Bahama swallow *Callichelidon cyaneoviridia*

bank swallow *Riparia riparia*

barn swallow *Hirundo rustica*

blue swallow *Hirundo atrocaerulea*

pearl-breasted swallow *Hirundo dimidiata*

stripe-breasted swallow = striped swallow

black-capped swallow *Notiochelidon pileata*

cave swallow *Petrochelidon fulva*

red-chested swallow *Hirundo semirufa*

cliff swallow any of several species of the genus Petrochelidon, without qualification *U.S. P. pyrrhonota, S.Afr. P. spilodera*

Angolan cliff swallow *P. rufigula*

Indian cliff swallow *P. fluvicola*

West African cliff swallow *P. preussi*

eastern swallow = barn swallow

Ethiopian swallow *Hirundo aethiopica*

golden swallow *Kalochelidon euchrysea*

white-gorgeted swallow *Hirundo nigrita*

violet-green swallow *Tachycineta thalassina*

house swallow = barn swallow or Pacific swallow

mangrove swallow *Iridoprocne albilinea*

mosque swallow *Hirundo senegalensis*

Pacific swallow *Hirundo tahitica*

palm swallow = palm swift

resident swallow Malay = Pacific swallow

red-rumped swallow *Hirundo daurica*

striated swallow = red-rumped swallow

Sykee's striated swallow *Hirundo erythropygia*

striped swallow *Hirundo abyssinica*

summer swallow *W.I.* = Bahama swallow

wire-tailed swallow *Hirundo smithii*

white-throated swallow *Hirundo albigularis*

tree swallow *Iridoprocne bicolor*

welcome swallow *Austral. Hirundo neoxena*

saw-wing swallow any of several species of the genus Psalidoprocne

black saw-wing swallow *P. holomelaena*

eastern saw-wing swallow *P. orientalis*

white-headed saw-wing swallow *P. albiceps*

pied-winged swallow *Hirundo Leucosoma*

rough-winged swallow *U.S. Stelgidopteryx ruficollis Afr.* = saw-wing swallow

swallow 2 (see also swallow 1) any of several birds not of the family Hirundidae

bee swallow *W.I.* the sternine bird *Sterna albifrons* (= least tern)

carr swallow *Brit. obs.* for tern

golden-swallow *Austral.* the meropid *Merops ornatus* (*cf.* bea-eater)

sea swallow any of several sternine birds but particularly *Sterna paradisaea* (= Arctic tern)

woodswallow any of many species of artamid birds of the genus Artamus

white-breasted woodswallow *A. leucorhynchus*

white-browed woodswallow *A. superciliosus*

black-faced woodswallow *A. cinereus*

little woodswallow *A. minor*

masked woodswallow *A. personatus*

swan any of several species of very large swimming birds of the genus Cygnus distinguished from the geese principally by their longer necks, and placed in the subfamily Cygninae. Used without qualification, usually means *C. olor*.

Bewick's swan *C. bewickii*

black swan *C. atratus*

mute swan *C. olor*

black-necked swan *C. melancoriphus*

trumpeter swan *C. buccinator*

whistling swan *U.S. C. columbianus Brit.* (= whooper swan)

whooper swan *C. cygnus*

swarm term for a large number of insects, particularly bees, though the term is also applied to a few fish such as the eel

hybrid swarm a large number of interspecific hybrids occurring at the boundaries of two species population

swarmer one of numerous motile spores liberated at the same time

chimney-sweeper the geometrid lepidopteran insect *Odezia atrata*

sweet in compound names of several organisms

bittersweet 1 (*see also* bittersweet 2) the solanaceous climbing shrub *Solanum dulcamara* (*cf.* nightshade)

false bittersweet = shrubby bittersweet

shrubby bittersweet the celastraceous climbing shrub *Celastrus scandens*

bittersweet 2 (*see also* bittersweet 1) any of several species of arcid mollusks of the genus Glycymeris

meadow sweet either the rosaceous herb *Spiraea alba* (= pikestem) or any of several rosaceous herbs of the genus Filipendula but particularly *F. hexapetala*

summer sweet *U.S.* any of several species of the clethraceous genus Clethra (= white alder)

sweet-joe-clear *W.I.* the vireonid bird *Vireo crassirostris* (= thick-billed vireo)

swift 1 (*see also* swift 2, moth) popular name of many lepidopteran insects and particularly (*U.S.*) of those hepialid forms superficially resembling sphingids. *Brit.* any of several species of hepialids of the genus Hepialus. *S. Afr.* any of several species of hesperids of the genera Baoris and Pelopidas. *Austral.* any of several species of hesperids usually without qualification *Baoris mathias I.P.* any of several species of hesperids of the genus Gegenes

lesser horned swift *S.Afr. B. lugens*

longhorned swift *S.Afr. Baoris fatuellus*

swift 2 (*see also* swift 1) properly apodiform birds, mostly of the family Apodidae. In popular terminology they are frequently confused with swallows which are passerine birds of the family Hirundinidae. *Brit.*, without qualification, *Apus apus*

Alpine swift *Apus melba*

black swift *U.S. Cypseloides niger; Afr. Apus barbatus*

chimney swift *Chaetura pelagica*

chestnut-collared swift *Cypseloides rutilus*

mouse-colored swift *Apus pallidus*

common swift *Apus apus*

crested swift = tree swift

horus swift *Apus horus*

house swift *Apus affinis*

mottled swift *Apus aequatorialis*

white-naped swift *Streptoprocne semicollaris*

palm swift either of two species of the genera Tachornis and Cypsiurus. Without qualification, *W.I. T. phoenicodia Afr.* and *I.P. C. parvus*

white-rumped swift *Afr. Apus caffer; Austral. Apus pacificus*

scarce swift *Apus myoptilus*

spinetail swift any of numerous species of apodid bird of the genera Mearnsia, Hirundapus and Chaetura

swallow-tail swift either of two species of the genus Panyptila

great swallow-tailed swift *P. sanctihieronymi*

lesser swallow-tailed swift *P. cayennensis*

tree-swift any of three hemiprocnids of the genus Hemiprocne

crested tree-swift *H. longipennis*

white-throated swift *Aeronautes saxatalis*

swiftlet any of numerous species of apodid birds of the genus Collocalia

backswimmer popular name of hemipteran notonectid insects

sideswimmer *U.S.* popular name of amphipod Crustacea

sy *see* [small-eye]

sycamore (*see also* bug, fig) *Brit.* usually the aceraceous tree *Acer pseudoplatanus* but the term is frequently applied in *U.S.* to *Platanus occidentalis* (= American plane tree)

Arizona sycamore *Platanus wrightii*

California sycamore *P. racemosa*

-sychno- *comb. form* meaning "frequent"

sycon a multiple, hollow fruit

syconoid a level of sponge structure in which the choanocytes are in simple pouch-like outfoldings from the central chamber (*cf.* asconoid, leuconoid, sylleibid)

Sygnathidae an order of gasterosteiform fishes containing the pipefishes and seahorses. The elongate snout and bony plates are typical

Syllidae a family of errant polychaete annelids distinguished by their long and slender dorsal cirri

sylleibid a grade of sponge structure intermediate between the syconoid and the leuconoid, in which each radial canal is subdivided into elongated, flagellated chambers grouped around a common excurrent canal (*cf.* asconoid, leuconoid, synconoid)

sylph 1 (*see also* sylph 2) *S. Afr.* any of numerous species of hesperid lepidopteran insect of the genus Metisella

sylph 2 (*see also* sylph 1) either of two trochilid birds of the genus Aglaiocercus

-sylv- *comb. form* meaning a "wood"

sylvestral pertaining to woods and woody places

Sylviidae a very large family of passeriform birds containing the Old World warblers. They are distinguished by their very small bills and short, often broad and fan-shaped, tails

Sylvius's *epon. adj.* from Francois de la Boe Sylvius (1614–1672) [aqueduct fissure] or Jacques duBois Sylvius (1478–1555) [bone]

-sym- = -syn-

symbiosis *see* -biosis

symbol (*see also* marker) a letter, or group of letters, with or without numerals, that is used to convey a meaning not immediately apparent and, less usually, a mark not part of an alphabet or numeral system, used for the same purpose. In this work letter combinations are entered alphabetically using only the roman letters of the symbol. Symbols containing letters of the roman alphabet are entered in this dictionary under the first such letter without reference to letters from other alphabets or numerals. Thus "λ co-mi" is entered under "c" not lambda. The symbols given below, many of alchemical or astrological origin, cannot be so entered

♂ the astrological symbol for Mars used generally in *biol.* to indicate the male sex and specifically in *bot.* to denote antheridia

♀ the astrological symbol for Venus used generally in *biol.* to indicate the female sex and specifically in *bot.* to denote oogonia

☿ the astrological symbol for Mercury used generally in *biol.* to indicate a hermaphrodite and specifically when that condition is normal to the organism

⚥ a symbol without classical origin used to indicate a hermaphrodite when that condition is not normal to the organism

α the *Gr.* letter alpha occasionally used to designate the female sex or a female gamete

β the *Gr.* letter beta occasionally used to designate the male sex or a male gamete

ø the *Gr.* letter phi. When appended to the name of an organism it indicates a polymorph, or, in the case of an alga, it may indicate the possession of sporophores

+ the plus sign. It is used in *bot.* to designate a strain of lower plants, the spores of which are characterized as "male." In microbial genetics it is used in two senses. When appended to a bacterial mutant symbol derived from the name of an energy source, it indicates a utilization of that compound. If applied to a bacterial mutant symbol for an auxautotropic form, it indicates that the substance denoted by the symbol is not required. Thus *arg*+ indicates a form autoauxotrophic for amino acids but that does not utilize arginine

− the minus sign. When appended to a bacterial mutant symbol derived from the name of an energy source, it indicates that the indicated compound is not utilized.

± the plus or minus sign. It indicates uncertainty in quantitative measurements and particularly those referring to the position of chromosome loci determined by cross-over or, in *Escherichia coli*, by timing the transfer of genes

: the colon. In yeast genetics it indicated a separation of chromosomes

✕ the multiplication sign. In *biol.* this sign placed

between the names of two organisms, or symbols for organisms, indicates a cross between, or hybrid of, the organisms in question.

(✱) when appended to the name of a plant, this symbol indicates an annual habit

(✱✱) when appended to the name of a plant this symbol indicates a biennial habit

♄ the astrological sign for Saturn. When appended to the name of a plant it indicates the existence of a woody stem

⊕ when appended to the name of a lower plant, this symbol indicates the existence of tetrasporangia

o═ when appended to the name of a plant, this symbol indicates that the cotyledons within the seed are accumbent

o‖ when appended to the name of a plant, this symbol indicates that the cotyledons within the seed are procumbent

o‖‖ when appended to the name of a plant, this symbol indicates that the cotyledons within the seed are spirobolous

≪o when appended to the name of a plant. this symbol indicates that the cotyledons within the seed are conduplicate

═ the equals sign. When placed between the names of organisms, it indicates synonymy

✱ the asterisk. In addition to its common usage as a reference mark, when prefixed to an apparent prefixed to second part of a binomial designation, indicating that the specific name has been omitted

§ the section. In addition to its common usage as a reference mark, when prefixed to an apparent genus name it indicates that the name is actually that of a subgénus

† the dagger. In addition to its common usage as a reference mark, when either prefixed or suffixed to a species name it indicates that the author considers the species to be of doubtful validity

! the exclamation mark. In addition to its common usage, when suffixed to a species name, it indicates that the author has verified the identification by comparison with the type.

gene symbol a symbol, usually composed of letters of the roman alphabet and frequently an abbreviated form of the mutant name. Gene symbols capitalized indicate a dominant allele

symmetry the arrangement of similar parts on each side of a common axis or radially around a point

asymmetry a pattern within which no axis or repeating planes can be perceived

anaxial symmetry = asymmetry

monaxial symmetry a type of symmetry, such as radial symmetry, in which only one axis can be cut, to produce identical halves

triaxial symmetry a type of symmetry such as biradial symmetry or bilateral symmetry, in which there are three axes commonly known as the longitudinal, sagittal and transverse

bilateral symmetry a type of triaxial symmetry in which all three axes differ from each other, *i.e.*, in which the dorsal and ventral surface are distinguished

radiobilateral symmetry a variety of radial symmetry which is strongly modified, as in the Anthozoa, in the direction of bilateral symmetry

multilateral symmetry = radial symmetry

dissymmetry = biradial symmetry

polysymmetry = radial symmetry

tetramerous symmetry a variety of radial symmetry in which four major axes can be distinguished

monosymmetry = bilateral symmetry

radial symmetry a pattern in which the parts are so arranged around a central point or shaft, that any vertical cut through the center would divide the whole into two identical halves. A radially symmetrical organism may be heteropolar (*q.v.*)

biradial symmetry a type of triaxial symmetry in which only two axes differ from each other so that, in general terms, there is no difference between the ventral and dorsal surface. It differs in this from bilateral symmetry

ray symmetry = radial symmetry

spherical symmetry a pattern in which parts are arranged around the center of a sphere on which no definite poles are apparent

biverted symmetry an inverted diagonal symmetry, particularly in diatoms

triverted symmetry asymmetry

univerted symmetry bilateral, mirror-image symmetry

Symphoremaceae a family erected to contain the genera Congea, Sphenodesma and Symphorema by those who do not hold them to belong in the Verbenaceae

Symphyla a small class of progoneate Arthropoda with a centipede-like body of fourteen segments of which the first twelve bear legs. They are easily distinguished from the true centipedes by their white translucent bodies

Symphypleona a suborder of collembolan insects distinguished from the Arthropleona by the globose body. The lucerne, and turnip, fleas are representative

-symphyt- *comb. form* meaning "innate"

Symphyta a suborder of hymenopteran insects containing those forms, such as sawflies and horntails, in which the abdomen is broadly joined to the thorax and therefore lack the typical wasp-waisted appearance of the Apocrita

Symplocaceae a small family of ebenalous dicotyledons closely related to the Styracaceae from which it can be distinguished by the several-celled ovary and the numerous stamens

-syn- *comb. form* meaning "with" or "combined with" very frequently mistranslated as -sym-

Syngamidae a family of strongyloid nematodes with a large buccal cavity lacking teeth

Synanthae that order of monocotyledons which contains the Cyclanthaceae. Characteristics of the family apply to the order

synapse the surface of contact between nerve endings derived from separate cells

synapsis the pairing of homologous chromosomes in meiosis

allosynapsis pairing of meiosis of chromosomes derived from different ancestors in an amphipolyploid

autosynapsis the meiotic pairing, within an allopolyploid, of homologous chromosomes from similar sets

desynapsis the precocious separation of synapsing chromosomes in meiosis

homosynapsis that which occurs between homologous chromosomes

intermediate synapsis one which does not occur in any recognizable pattern or order

procentric synapsis that which first occurs at the centromere and continues outwards

proterminal synapsis that which begins at the tips of the chromosomes and continues toward the center

somatic synapsis the pairing of homologous chromosomes in mitosis

synaptene = zygonema

Synaptera = Thysanura

synapticulum a transverse rod joining adjacent pharygeal bars in prochordates

Synbranchidae a family of the order Synbranchii containing the so-called "swamp eels." They bear no relation to the true eels. Some genera are air breathing forms while others possess three or four rudimentary gills. The pelvic and pectoral fins are lacking and the dorsal and anal fins are reduced to a ridge

Synbranchii a curious order of Osteichthyes containing the single families Synbranchidae, Alabetidae and Amphipnoidea

Syncarida a superorder of eumalacostracan Crustacea containing the single order Anaspidacea. The characters of the order are those of the superorder

syndesis a connection of two parts in such a manner as to permit their mutual motion

asyndesis the non-fusion of chromosomes at meiosis

syndrome the sum total of a set of concurrent events, in *biol.* usually disease symptoms or behavior patterns

Kleinfelter's **syndrome** testicular dysfunction in humans, and failure to develope secondary and tertiary characters in other forms, caused by an XXY sex chromosome

waltzing syndrome a mouse behavior pattern, controlled by many genes at different locations, which involves a rambling, circling movement with head shaking

synechthry the condition of a visitor to an ant's nest, who, though not *persona grata* with his host, successfully resists efforts to expel him (*cf.* symphily, metochy)

synema a column of stamens

Synentognathi a small order of osteichthyes containing the families Scomberosocidae, Belonidae, Hemiramphidae and Exocoetidae. They are distinguished by the possession of soft-rayed fins of which the dorsal, anal and ventral are set very far back on the body and by the fact that the lateral line is set very low on the body

-synergo- *comb. form* meaning "assistant" or "assist"

Syngenesia in the Linnaean system of plant classification that class distinguished by possessing stamens united by their anthers into a tube

synizesis the clumping of chromosomes to one side of the cell in the leptotene nucleus

synocreate the condition of a stem which is sheathed by the fusion of the stipules of opposite leaves

Synodontidae a family of imiomious fishes commonly called lizard fishes. They are characterized by a cylindrical shape, a reptile-like head furnished with very sharp teeth, and an adipose fin

synovial said of a joint in which an albuminous fluid separates the cartilage-capped surfaces

photosyntax = photosynthesis

syntechny the condition of two organisms which resemble each other in virtue of adaptation to the same environment

Syntectidae a small family of symphytan hymenopteran insects

Syntexidae = Syntectidae

synthase a term used to distinguish compound-building lyase enzymes from compound-building ligase enzymes which are properly called synthetases

citrate synthase catalyzes the production of acetyl-CoA and oxaloacetate from citrate and CoA

cysteine synthase catalyzes the production of cysteine from serine and hydrogen sulfide

methylcysteine synthase catalyzes the production of S-methyl-L-cysteine from serine and methanethiol

porphobilinogen synthase catalyzes the production of porphobilinogen from 2, 5-aminolevulinate

threonine synthase catalyzes the production of threonine and phosphate from O-phosphohomoserine

cystathionine synthase catalyzes the production of cystathionine from serine and homocysteine

tryptophan synthase catalyzes the production of tryptophan from serine and indole

synthesis *see* thesis

synthetase a term used to distinguish compound-building ligase enzymes from compound-building lyase enzymes which are properly called synthases. Unless otherwise stated, synthetases utilize energy derived from the breakdown of ATP

D-alanylalanine synthetase catalyzes the production of D-alanyl-D-alanine from D-alanine

phosphoribosyl - aminoimidazole - succinocarboxamide synthetase catalyzes the production of 5'-phosphoribosyl-4-(N-succinocarboxamide)-5-aminoimidazole from 5'-phosphoribosyl-4-carboxy-5-aminoimidazole and L-aspartate

phosphoribosyl-glycineamide synthetase catalyzes the production of 5'-phosphoribosylglycineamide from 5'-phosphoribosylamine and glycine

phosphoriboxyl-formyl-glycineamidine synthetase catalyzes the production of 5'-phosphoribosyl-formylglycineamidine and glutamate from 5'-phosphoribosylformylglycineamide and glutamine

asparagine synthetase catalyzes the production of L-asparagine from L-aspartate and ammonia

CTP synthetase catalyzes the production of CTP from UTP and ammonia

γ-glutamyl-cysteine synthetase catalyzes the production of γ-L-glutamyl-L-cysteine from L-glutamate and L-cysteine

phosphopantothenoylcysteine synthetase catalyzes the production of 4'-phospho-L-pantothenoyl-L-cysteine from 4'-phospho-L-pantothenate and cysteine using energy derived from the breakdown of CTP

formyltetrahydrofolate synthetase catalyzes the production of 10-formyltetrahydrofolate from formate and tetrahydrofolate

glutamine synthetase catalyzes the production of glutamine from glutamate and ammonia

glutathione synthetase catalyzes the production of reduced glutathione from γ-L-glutamyl-L-cysteine and glycine

GMP synthetase catalyzes the production of GMP and L-glutamate from xanthosine 5'-phosphate and L-glutamine

phosphoribosyl-amino-imidazole synthetase catalyzes the production of 5'-phosphoribosyl-5-aminoimidazole from 5'-phosphoribosyl-formyl-glycineamidine

UDP-muramyl-alanyl-D-glutamyl-lysine synthetase catalyzes the production of UDP-muramyl-L-alanyl-D-glutamyl-L-lysine from UDP-muramyl-L-alanyl-D-glutamate and lysine

NAD synthetase catalyzes the production of NAD and L-glutamate from deamido-NAD and L-glutamine

farnesylpyrophosphate synthetase = dimethyllallyl-transferase

argininosuccinate synthetase catalyzes the production of L-argininosuccinate from L-citrulline and L-asparate

adenylosuccinate synthetase catalyzes the production of adenylosuccinate from IMP and L-aspartate using energy derived from the breakdown of GTP

panothenate synthetase catalyzes the production of L-pantothenate from L-pantoate and β-alanine

Syntomidae = Amatidae

stenosynusic restricted in distribution

synusium a group of organisms having the same ecological requirements, and reacting in much the same way to, their environment but which are not taxonomically related

syphon = siphon

syringa properly a genus of oleaceous shrubs commonly called lilacs. The term is often misapplied to Philadelphus (see mock orange, seringa)

syrinx the "voice box" of birds at the bifurcation of the trachea into the bronchi

Syrphidae a very large and very abundant family of aschizan cycloraphous brachyceran dipteran insects commonly called flower flies. They are one of the many groups also to which the name hover fly is applied. Many are brightly colored and mimic Hymenoptera

-syrt- comb. form meaning a "sandbank" but used by ecologists in the sense of "dry sandbars"

syrtidad a plant of dry sandbars

-syst- comb. form meaning "contract"

system 1 (see also system 2, 3) a group of related organs or structures

apical system the combination of the genital and terminal plates at the aboral surface of an echinoid echinoderm (cf. corona)

arterial system the sum total of those blood vessels which carry blood under high pressure from the heart to all parts of the body

autonomic system = autonomic nervous system

exteroceptive system the sum total of those sensory receptors which perceive external stimuli

interoceptive system the sum total of those receptors which perceive internal stimuli

proprioceptive system the sum total of those receptors which receive stimuli from muscles, tendons and joints

circulatory system the sum total of the hollow structures through which blood flows

closed circulatory system one in which arteries join veins through arterioles, venules and capillaries

lacunar circulatory system a circulatory system in which there is no close connection between the end of the artery and the beginning of a vein

open circulatory system = lacunar circulatory system

Demanian system a glandular complex found in the posterior region of some female marine Nematodes

digestive system the sum total of those hollow structures in which food is digested and the associated glands that assist in this process

endocrine system a very loose term descriptive of the sum total of the ductless glands

Haversian system = osteon

nervous system the sum total of all those structures in an organism which carry impulses

autonomic nervous system that part of the vertebrate nervous system that controls involuntary muscles, and is therefore the sum of the sympathetic and parasymphathetic nervous system

central nervous system the brain and spinal cord

concentrated nervous system one in which at least some of the nerve cells are concentrated in the ganglia

diffuse nervous system one in which nerve cells are spread over the surface of the body, and not concentrated in the ganglia

enteric nervous system that part of the nervous system of invertebrates concentrated in ganglia supplying the visceral organs. It is thought by some to be ancestral to the sympathetic nervous system

peripheral nervous system the sum total of all those parts of the nervous system which are not included under the term central nervous system

sympathetic nervous system a series of ganglia, the principal of which are the chain ganglia and the pre-vertebral ganglia which are a part of the autonomic nervous system (q.v.)

sympathetic nervous system (Arthropoda) the functional equivalent of the vertebrate system, composed principally of the frontal (rostral), hypocerebral and ventricular ganglia

parasympathetic nervous system that portion of the autonomic nervous system which controls the involuntary responses of the head, thorax and upper abdomen through its cranial parts and of the lower abdomen and reproductive system through its sacral portion

parasympathetic nervous system nerves that bring about opposite effects from those in the sympathetic nervous system. The principal parasympathetic nerve is the visceral branch of the vagus nerve

ectoneural system that portion of the nervous system of an echinoderm which consists of a ring round the mouth from which ganglionated fibers go to the ambulacra (cf. entoneural system)

entoneural system that portion of the central nervous system of echinoderms which consists of a ring surrounding the anus (cf. ectoneural system)

hyponeural system = entoneural system

silver-line system those striations, fibrils, etc. in Protozoa that are demonstrated by silver staining methods

sympathetic system = sympathetic nervous system

vascular system (see also system 2) the sum total of fluid conducting vessels in an organism and therefore includes arteries, veins and lymph vessels

venous system the sum total of those blood vessels which return blood under low pressure to the heart

system 2 (see also system 1, 3) term used in bot. to describe tissue types and assemblages of tissue types

dermal system the outer protective covering of a plant

fundamental system all plant tissues other than dermal and vascular

horizontal system the ground, as distinct from the vascular, substance of a plant

medullary system the pith and medullary rays

micellar system that portion of a plant cell wall, which consists of interconnecting chain molecules of cellulose

intermicellar system those substances, probably not cellulose, which interpenetrate the micellar system

vascular system (see also system 1) the conducting tissues of a plant

ecosystem an ecological system composed of both biotic and abiotic components. This basic ecological unit the structure and function of which is considered to be at least partially self-sustaining and self-regulating usually requires only a continuing supply of solar radiation

systole the contraction of a hollow body such as a contractile vacuole or heart

syzygy the apposition of two bodies particularly the apposition, but not union, of two sporozoan Protozoa and the apposition, and fusion, of two joints in the arm of a crinoid echinoderm

t *see* [tan] and [tangerine fruit color]

T *see* [Brachyury], [Translocation] and [Tyrosinase thermo-stability]

T4 = thyroxine

ta *see* [tapered] or [tapering]

Ta *see* [Tabby]

Tabanidae a very large family of viciously biting orthoraphous brachyceran dipteran insects commonly called horse-flies and deer-flies. Most have brightly colored eyes, the color usually divided into two distinct regions

[Tabby] a mutant (gene symbol *Ta*) found in the sex chromosome (linkage group XX) of the mouse. The phenotype is transversely striped

tabby *I.P.* the nymphalid lepidopteran insect *Pseudergolis wedah*

tabescent shrivelling

table that type of holothurian ossicle which resembles a button (*q.v.*) from which rises a perforated tower of calcium

tablet a rectangular plant, such as a rectangular diatom or desmid colony

tabular horizontally flattened

tabulum a horizontal calcareous plate running across the cavities in a coenosteum

tacamahac = hackmatac

Taccaceae a family of liliflorous monocotyledons containing the single genus Tacca which is distinguished by the one-celled, many-ovuled ovary and the six deformed stamens

taccamahac = tacamahac

Tachinidae an extremely large family of myodarian cycloraphous dipteran insects easily recognized by the thick coating of bristles on the body and the prominent postscutellum. They are almost all parasitic on insects, though a few attack other arthropods, and they are of the utmost economic value

-tachy- *comb. form* meaning "quick"

Tachypetidae = Fregatidae

-tact- *comb. form* properly meaning "to touch" but frequently used as an adjectival form from -tax- (*q.v.*)

tactic 1 (*see also* tactic 2) pertaining to touch

  thigmotactic the change in shape or direction induced by contact

tactic 2 (*see also* tactic 1) *adj. suffix* derived from taxis 2 (*q.v.*)

Tadornini a group of the subfamily Anatinae containing the shellducks, which are large grazing ducks resembling geese. Sometimes called whistling ducks

tadpole (*see also* larva) the larva of an anuran amphibian

-taen- *comb. form* meaning a "band" or "ribbon"

Taenioidea an order of eucestod cestods distinguished by the possession of a scolex bearing four acetabuli. Hooks are also sometimes present

Taenioididae a family of gobioid acanthopterygian fish possessing a sucker but distinguished from the true gobies by the single long dorsal fin. Commonly called eel gobies

Taeniopterygidae a family of plecopteran insects distinguished by a second tarsal segment equal in length to the other segments and with short 1-6 segmented cerci

tagma literally, that which has been ordered or arranged and used of a section of an arthropod body, such as the abdomen, in which a number of segments are combined in one main division

tagmata *pl.* of tagma

tagmosis the condition of having tagmata or, more frequently but incorrectly of having tagmata fused as when a cephalothorax replaces a separate head and thorax

taguan the East-Indian flying squirrel *Petaurista petaurista*

tahr any of three species and several geographical races of caprine caprid of the genus Hemitragus

taiga a zone of intermittent scattered trees between the tundra (*q.v.*) and the forest-tundra (*q.v.*). It is equivalent to "boreal coniferous forest"

tail 1 (*see also* tail 2, 3) a posterior appendage to the body

tail 2 (*see also* tail 1-3) in compound names of animals with distinctive tails

  barbtail any of several furnariid birds of the genera Roraimia and Premnoplex

  red-flanked bluetail the turdid bird *Erithacus cyanurus*

  boat-tail = boat-tailed grackle (*q.v.*)

  bobtail the polycentrid teleost fish *Polycentrus schomburghii* (= leaf-fish)

  bonytail the cyprinid fish *Gila robusta*

  jumping bristletail popular name of machilid insects

  browntail *Brit.* the liparid lepidopteran insect *Euproctis chrysorrhoea*

  clubtail *I.P.* the papilionid lepidopteran insect *Tros coon* (*cf.* clubtail, rose, windmill)

  combtail the anabantid teleost fish *Belontia signata*

  cottontail any of several leporid lagomorphs of the genus Sylvilagus

    desert cottontail *S. audoboni*

    eastern cottontail *S. floridanus*

    mountain cottontail *S. nuttalli*

    New England cottontail *S. transitionalis*

dragontail *I.P.* any of several species of the papilionid lepidopteran insect of the genus Leptocircus

fantail any of several birds of the family Muscicapidae, particularly those of the genus Rhipidura

black-and-white fantail *R. leucophrys* (= willy-wagtail)

grey fantail *R. fuliginosa*

rufous fantail *R. rufifrons*

feather-tail (*Austral.*) any of several species of phalangerine phalangerids of the genera Distoechurus and Acrobates

firetail = redstart *Brit. Austral.* the fringillid bird *Zonaeginthus bellus* (= firetail finch)

flickertail *U.S.* the sciurid mammal *Citellus richardsoni*

fluff-tail any of several species of rallid bird of the genus Sarothrura (*cf.* crake)

forktail any of several species of turdid bird of the genus Enicurus

eastern forktail = spotted forktail

little forktail *E. scouleri*

spotted forktail *E. maculatus*

western forktail = spotted forktail

girdle-tail the cordylid saurian reptile *Cordylus cordylus*

goldentail any of several species of trochilid bird of the genus Hylocharis

blue-throated goldentail *H. eliciae*

greytail either of two furnariid birds of the genus Xenerpestes

hairtail 1 (*see also* hairtail 2) *Brit.* the trichiurid teleost *Trichiurus lepturus* (= *U.S.* Atlantic cutlasfish)

hairtail 2 (*see also* hairtail 1) *S.Afr.* either of two lycaenid lepidopteran insects of the genus Athene

common hairtail *S.Afr. A. definita*

large hairtail *A. lemnos*

hawktail *Brit.* = kite. *I.P.* a group of small birds, by many placed in a separate family the Enicuridae, which is by others fused with the Turdidae (= forktail)

horntail popular name of siricid symphytan hymenopteran insects

blue horntail *U.S. Sirex cyaneus*

long-tail *W.I.* any of several species of phaethontid birds

lyretail the cyprinodont teleost fish *Aphysemion australe*

metaltail any of several trochilid birds of the genus Metallura

piedtail either of two trochilid birds of the genus Phlogophilus

pintail any of several species of anatine birds of the genus Anas. *Brit.* and *U.S.*, without qualification *A. acuta*

Bahama pintail *A. bahamensis*

brown pintail *A. georgica* (*cf.* teal)

common pintail *A. acuta*

red-billed pintail *A. erythroryncha* (= red-billed duck)

spreckled prickletail the furnariid bird *Siptornis striaticollis*

rackettail = racquettail

racquettail properly the trochilid bird *Ocreatus underwoodii,* but also applied to birds of the family Momotidae (= motmot)

rattail common name of macrurid fishes

red-tail *U.S.* = red-tailed hawk *U.S.*

ringtail *U.S.* the bassariscid carnivore *Bassariscus*

astutus (= ring tail cat), though the term is also applied to the raccoon. *Brit. obs.* the hen harrier. *Austral.* any of numerous phalangerine phalangerids with ringed tails

scale-tail general name for anomalurid sciuromorph rodents

scissorstail 1 (*see also* scissorstail 2, 3) *W.I.* the fregatid bird *Fregata magnificens* (= magnificent frigatebird)

scissorstail 2 (*see also* scissorstail 1, 3) the cyprinid teleost fish *Rasbora trilineata*

scissorstail 3 (*see also* scissorstail 1, 2) the tyrannid bird *Muscivora forficata* (= scissor-tailed flycatcher)

sheartail either of two species of trochilid bird of the genus Doricha

Mexican sheartail *D. eliza*

slender sheartail *D. enicura*

softtail any of several furnariid birds of the genus Thripophaga

marvelous spatuletail the trochilid bird *Loddigesia mirabilis*

spinetail 1 (*see also* spinetail 2, duck) any of several species of apodid bird of the genera Chaetura, Hirundapus, and Mearnisia (*cf.* swift)

rufous-breasted spinetail *Synallaxis erythrothorax*

white-necked spinetail *H. caudacutus*

long-tailed spinetail *C. sabini*

stumpy-tailed spinetail *M. cassini*

mottle-throated spinetail *C. ussheri*

spinetail 2 (*see also* spinetail 1) any of numerous furnariid birds (*cf.* ovenbird, foliage gleaner)

tit-spinetail any of several species of the genus Leptasthenura

splittail *U.S.* the cyprinid fish *Pogonichthys macrolepidotus*

springtail 1 (*see also* springtail 2) popular name of collembolan insects

armored springtail the podurid *Achorutes armatus*

garden springtail *U.S.* the sminthurid *Bourletiella hortensis*

marsh springtail the entomobryid *Isotomurus palustris*

seashore springtail the podurid *Anurida maritima*

springtail 2 (*see also* spring tail 1) *Brit. dial.* for oxyurine ducks

square tail popular name of tetragonurid fish

stiff-tail any of several species of oxyurine birds

streamertail the trochilid bird *Trochilus polytmus*

swallowtail (*see also* moth) popular name of those papilionoid butterflies that have a tail-like protuberance on the hind wing (*cf.* parnassian) Most are of the genus Papilio. *Brit.* without qualification, *P. machaon*

black swallowtail *P. polyxenes* or *P. ajax*

chequered swallowtail *Austral. P. demoleus*

dingy swallowtail *Austral. P. anactus*

eastern swallowtail *U.S.* = black swallowtail *U.S.*

giant swallowtail *U.S. P. cresphontes*

mimicking swallowtail *Austral. P. laglaizei*

orchard swallowtail *Austral. P. aegeus*

pipe-vine swallowtail *U.S. P. philenor*

purple-spotted swallowtail *Austral. P. weiskei*

red-bodied swallowtail *Austral. P. polydorus*

silky swallowtail *Austral. P. codrus*

spicebush swallowtail *P. troilus*

tiger swallowtail *U.S. P. glaucus*

zebra swallowtail *U.S. P. marcellus* (*cf.* zebra butterfly)

swordtail 1 (*see also* swordtail 2, 3) any of several papilionid lepidopteran insects. *S.Afr. Papilio colonna I.P.* any of several species of the genus Graphium

fivebar swordtail *Austral. Papilio aristeus*

fourbar swordtail *Austral. P. leosthenes*

swordtail 2 (*see also* swordtail 1, 3) any of several species of poeciliid fishes of the genus Xiphophorus

swordtail 3 (*see also* swordtail 1, 2) any of several insects with prominent ovipositors or anal cerci including those of the hemipteran genus Xiphophorus and the orthopteran tethgonid genus Chonocephalus

tripletail popular name of lobotid fish. *U.S.* without qualification, *Lobotes surinamensis*

thistletail any of several furnariid birds of the genus Schizoeca

thorntail any of several trochilid birds of the genus Popelairia

wagtail 1 (*see also* wagtail 2) properly, any of numerous species of motacillid birds, mostly of the genus Motacilla, distinguished by brisk flicks of the tail

blue-headed wagtail (= yellow wagtail)

Cape wagtail *M. capensis*

forest wagtail *Dendronanthus indicus*

grey wagtail *M. cinerea*

mountain wagtail *M. clara*

pied wagtail *M. alba*

African pied wagtail *M. aguimp*

white wagtail (= pied wagtail)

yellow wagtail *M. flava*

wagtail 2 (*see also* wagtail 1) birds other than motacillids

willy wagtail *Austral.* the muscicapid bird *Phipidura leucophrys* (= black-and white fantail)

washtail = wagtail

weapontail any of several species of iguanid saurian reptiles of the genus Hoplocercus

whiptail 1 (*see also* whiptail 2) *U.S.* popular name of teiid reptiles

whiptail 2 (*see also* whiptail 1) the loricariid teleost fish *Loricaria parva*

whitetail = whitetail deer

wiretail the furnariid bird *Sylvio desmursii*

yellow-tail *Brit.* the liparid lepidopteran insect *Porthesia similis* and the carangid fish *Seriola borsalis* (*cf.* amberjack)

tail 3 (*see also* tail 1, 2, grass, palm) plants resembling animal tails

cattail (*see also* moth) any of several species of pandanale typhaceous herbs of the genus Typha

redheart cattail the euphorbiaceous shrubby herb *Acalypha hispida*

fintail *U.S.* the grass *Pholiurus incurvus* (= hard grass)

foxtail *U.S.* any of several species of grasses

bristly foxtail *Setaria glauca*

marsh foxtail *Alopecurus geniculatus*

horsetail any equisetaceous plant, particularly *Equisetum arvense*

lion's tail the labiateous shrub *Leonotid leonurus* (= lion's ear)

lizard's tail popular name for herbs of the family Saururaceae, particularly *Saururus cernuus*

mare's tail *either* any equisetale plant, more properly called horsetail *or* the haloragidaceous aquatic herb *Hippuris vulgaris*

mousetail *Brit.* the ranunculaceous herb *Myosurus minimus*

squirreltail *U.S.* the grass *Sitanion hystrix*

[tail-kinks] a mutant (gene symbol *tk*) in linkage group II of the mouse. The phenotypic expression is adequately described by the name

tailor (*see also* bird) *U.S.* juvenile bluefish (*q.v.*)

proud tailor *Brit. dial.* the fringillid bird *Carduelis elegans* (= gold finch)

-taine- *comb. form* meaning "a ribbon" frequently transliterated -tene-

taipan *Austral.* the elapid snake *Oxyuranus scutellatus*

takahe *N.Z.* the flightless rallid bird *Notornis mantelli*

takapu *N.Z.* the sulid bird *Morus serrator* (= gannet *N.Z.*)

takin the Himalayan ovibovine caprid *Budorcas taxicolor*. There are numerous geographic races varying markedly in size and color, that from Szechwan being brilliant metallic gold and, at one time, known as the golden goat antelope

Talaeporiidae a family of lepidopteran insects now usually included in the Psychidae

talapoin the cercopithecid primate *Cercopithecus talapoin* (*cf.* guenon)

tell-tale *U.S.* either of two species of scolopacid birds of the genus Totanus (= yellowlegs). *W.I.* the recurvirostrid bird *Himantopus mexicanus* (= black-necked stilt)

Talitridae very large family of gammaridean amphipod Crustacea containing, *inter alia,* the well-known beach-fleas and beach-hoppers. They are distinguished by the short simple first antenna, the uniramous third uropod and by the absence of a mandibular palp

tallow (*see also* gum) animal fat

vegetable tallow the euphorbiaceous tree *Sapium sebiferum*

talon 1 (*see also* talon 2) a heavy recurved claw of a predator particularly of a bird

talon 2 (*see also* talon 1) an upper molar

talonid the basin-like heel on a lower molar

Talpidae that family of insectivorous mammals which contains the moles

talus = astragalus

tamandua the xenarthran mammal *Tamandua tetradactyla* (= lesser anteater)

tamarack the pinaceous tree *Larix laricina* (*cf.* larch)

Tamaricaceae that family of parietalous dicotyledons which contains the tamarisks. Most genera are clearly distinguished by their fluffy foliage

Tamariceae a tribe of the family Tamaricaceae distinguished from the Reaumurieae by bearing flowers in spike-like racines

Tamaricineae a suborder of parietale dicotyledons containing the families Elatinaceae, Frankeniaceae and Tamaricaceae

tamarin popular name of those callithricid hapaloid mammals that have larger tusks than marmosets. The distinction, however, between a marmoset and a tamarin is in no way clear

tamarind the leguminous tree *Tamarindus officinalis* and its edible fruit

bastard tamarind the leguminous tree *Albizzia julibrissin* (= silk tree)

manilla tamarind the leguminous tree *Pithecolobium dulce* and its edible fruit (= guaymochil)

tamarisk anglicized form of Tamarix, a tamaricaceous genus of trees and shrubs

tamarou the Phillipine buffalo *Anoa mindorensis* (*cf.* anoa)

tambourine *S.Afr.* the columbiform bird *Tympanistria tympanistria* (*cf.* dove)

tame said of animals which are tolerant of man

[**tan**] a mutant (gene symbol *t*) mapped at 27.5 on the X chromosome of *Drosophila melanogaster*. The name is descriptive of the phenotypic color

tanager popular name of passerine birds, usually of the family Thraupidae, though extended to other brilliantly colored birds

  ant-tanager any of several species of the genus Habia

    red-crowned ant-tanager *H. rubica*

    red-throated ant-tanager *H. fuscicauda*

  bishop tanager *W.I.* the fringillid *Passerina cyanea* (= indigo bunting)

  blue-grey tanager *Thraupis virens*

  bush-tanager any of several species of the genus Chlorospingus, particularly, *C. ophthalmicus*

  chat-tanager *Calyptophilus frugivorus*

  crimson-collared tanager *Phogothraupis sanguinolenta*

  flame-colored tanager *Piranga bidentata*

  grey-headed tanager *Eucometis penicillata*

    red-headed tanager *Piranga erythrocephala*

    stripe-headed tanager *Spindalis zena*

  hepatic tanager *Piranga flava*

  hooded tanager *Tangara cucullata*

  golden-masked tanager *Tangara larvata*

  mountain-tanager any of several species of the genera Anisognathus and Buthraupis

  palm-tanager either of two species of the genus Phaeniciphilus

    black-crowned palm-tanager *P. palmarum*

    grey-crowned palm-tanager *P. poliocephalus*

  Puerto Rican tanager *Nesospingus speculiferus*

  scarlet-rumped tanager *Ramphocelus passerinii*

  scarlet tanager *Piranga olivacea*

  shrike-tanager any of several species of the genus Lanio

    black-throated shrike-tanager *L. aurantius*

  summer tanager *Piranga rubra*

  swallow tanager the tersinid *Tersina viridis*

  rose-throated tanager *Piranga roseogularis*

  rose-breasted thrush-tanager *Rhodinocicha rosea*

  western tanager *Piranga ludoviciana*

  white-winged tanager *Piranga leucoptera*

  yellow-winged tanager *Thraupis abbas*

Tanagridae = Thraupidae

Tanaidaceae an order of pericarid malacostracous crustaceans in which the carapace coalesces dorsally with the first two thoracic somites and overhangs on each side to enclose a branchial cavity. The body is generally narrow in these minute marine forms which have a curious resemblance to the Isopoda

Tanaioidea a suborder to which the Tanaidaceae are assigned if treated as isopods

Tanaostigmatidae a small family of chalcidoid apocritan hymenopterans

tanekaha *N.Z.* the podocarpaceous tree *Phyllocladus trichomanoides*

blue **tang** the acanthurid fish *Acanthurus coeruleus*

Tangaridae = Thraupidae

tangerine any of numerous citrus fruits, usually flattened and with a loose skin (*cf.* mandarin)

[**tangerine fruit color**] a mutant (gene symbol *t*) mapped at 92 on chromosome 10 of the tomato. The term is descriptive of the phenotypic expression

tannase an enzyme that catalyzes the hydrolysis of ester links in tannins

tanrec = tenrec

tansy any of several species of compositaceous herbs of the genus Tanacetum but particularly *T. vulgare*

Tantalidae = Threskiornithidae

Tanyderidae a family of slender long-legged hematoceran dipteran insects commonly called primitive crane-flies and distinguished from other crane-flies by having banded wings

Tanypezidae an obscure family of sylvicolous myodarian cycloraphous dipteran insects

tapaculo popular name of birds of the family Rhinocryptidae

[**tapered**] a mutant (gene symbol *ta*) mapped at 56.6± on chromosome II of *Drosophila melanogaster*. The phenotypic expression is pointed and narrowed wings

[**tapered wing**] a mutant (gene symbol *tpw*) mapped at 30.8± on the X chromosome of *Drosophila melanogaster*. The name is descriptive of the phenotype

[**tapering**] a mutant (gene symbol *ta*) in linkage group VI of *Habrobracon juglandis*. The name is descriptive of the abnormal antennae of the phenotype

tapesium the layer of felted mycelium on which ascophores are borne

-tapet- *comb. form* meaning a "carpet"

tapetum 1 (*see also* tapetum 2) a temporary layer of cells around sporanginous cells in the archesporium

tapetum 2 (*see also* tapetum 1) that layer of the anther which lies immediately beneath the developing mother cell

-taphr- *comb. form* meaning "ditch"

taphrad a ditch plant

Taphrinales an order of hemiascomycete Ascomycetae distinguished by the lack of a peridinium around the asci

tapioca a farinaceous foodstuff derived from the tapioca plant

tapir any of several species of tapidid perissodactyle mammals of the genus Tapirus

tapira any of several anacardiaceous trees of the genus Cyrtocarpa

Tapiridae that family of perissodactyle mammals which contains the forms commonly called tapirs. They are readily distinguished from the Equidae by the separate toes, elongated snout and absence of a tail

tar *see* [tarry]

tara *N.Z.* the sternid birds *Sterna striata* (= white-fronted tern) and *Chilodonias albistriata* (= black-fronted tern)

taraire *N.Z.* the lauraceous tree *Beilschmiedia tarairi* (*cf.* tawa)

tarantula *W.U.S.* popular name of tarantulid Pedipalpi, which are not spiders *Brit.* the lycosid spider *Lycosa tarentula*. *U.S.* any large spider, particularly the Theraphosidae

Tarantulidae a family of Pedipalpi (not Aranaea) distinguished from the Thelyphonidae (whip scorpions) by the fact that the cephalothorax is much broader than the abdomen

tara-nul *N.Z.* the sternid bird *Hydroprogne caspia* (= Caspian tern)

tarapunga *N.Z.* the larid bird *Larus scopulinus* (= red-billed gull *N.Z.*)

tarata *N.Z.* the pittosporaceous tree *Pittosporum eugenioides*

tarcel = tercel

Tardigrada a small group of apparent arthropods commonly called the water bears. They have variously been regarded as a separate phylum, as part of the Arthropoda or as an order of the Arachnida

tare (*see also* thrip) almost meaningless as a botanical term though it is possibly most frequently used of small leguminous herbs and particularly of the vetches (*q.v.*). The biblical tare was probably a grass. *Brit. V. hirsuta, Vicia tetrasperma* or *V. gracilis*

water **target** any of several aquatic plants of the nymphaceous genus Brasenia

Targioniaceae a family of marchantiale Hepaticae distinguished by having the female receptacle enclosed by a sheath

tarn a large pond or small lake, with steeply sloping mountainous sides

taro the araceous herb *Colocasia esculenta* and its edible root

tarpon (*see also* snook) the elopid fish *Melgalops atlantica*

tarragon the compositaceous herb *Artemsia dracunculus*

tarrock juvenile larine birds of the genus Rissa (= kittiwake)

[**tarry**] a mutant (gene symbol *tar*) mapped at 27.7± on the X chromosome of *Drosophila melanogaster*. The phenotypic expression is blackened femurs and tibia

tarsal 1 (*see also* tarsal 2, tarsal bone) pertaining to the terminal joints of a limb

tarsal 2 (*see also* tarsal 1) = tercel

tarsier anglicized form of Tarsius a genus of tarsiid lemurs

Tarsiidae a family of lemuroid primates containing the single genus Tarsius commonly called tarsiers

[**tarsi irregular**] a mutant (gene symbol *ti*) mapped at 55.9 on chromosome II of *Drosophila melanogaster*. The name is descriptive

Tarsioidea a suborder of lemuroid mammals principally characterized by the enormous eyes

Tarsipedinae a subfamily of phalangerid marsupials containing the forms commonly called honeysuckers

acro**tarsium** the integumentary covering of the dorsal surface of the tarsus of birds

tarsus one of the terminal segments of a limb

    hypo**tarsus** a bony structure on the proximal end of the posterior surface of the tarsometatarsus bone of birds; it supports or encloses the flexor tendons

    meta**tarsus** in insects, the proximal tarsal segment, particularly when it is sufficiently conspicuous to be differentiated easily (*cf.* sarothrum)

    pre**tarsus** a segment distal to the tarsus on the appendages of some arachnids

tartago the euphorbiaceous shrub *Jatropha podagrica*

tartareous having a rough crumbling surface like freshly dried mud

tassel = tercel

[**tassel seed**] *either* a mutant (gene symbol s_2) mapped at 27 on linkage group I of *Zea mays*. The phenotypic expression is the occurrence of female flowers in the male inflorescence *or* a mutant (gene symbol ts_4) mapped at 47 on linkage group III of *Zea mays*. The phenotypic expression is the presence of female flowers in the terminal inflorescence *or* a mutant (gene symbol ts_1) mapped at 74 on linkage group II of *Zea mays*. The phenotypic expression is the presence of female flowers in the terminal inflorescence

or a mutant (gene symbol Ts_6) mapped at 157 on linkage grup of *Zea mays*. The phenotypic expression is the occurrence of female flowers in the male inflorescence

tattler *U.S.* the serranid fish *Prionodes phoebe*

    grey-rumped **tattler** the scolopacid bird *Heteroscelus brevipes*

    wandering **tattler** *H. incanum*

tatu a general name for nine-banded armadillos

tauhou *N.Z.* the zosteropid bird *Zosterops caerulescens* (*cf.* white-eye)

taupata *N.Z.* the rubiaceous tree *Coprosma retusa*

[**taupe**] a mutation (gene symbol *tp*) in linkage group I of the mouse. The phenotypic expression is a yellowish brown-gray coat

tautog *U.S.* the labrid fish *Tautoga onitis*

phenylpyruvate **tautomerase** an enzyme that catalyses the change of ketophenylpyruvate to enolphenyl pyruvate

taw see [tawny]

tawa *N.Z.* the lauraceous tree *Beilschmiedia tawa* (*cf.* taraire)

maire **tawake** *N.Z.* the myrtaceous tree *Eugenia maire*

tawaki *N.Z.* the spheniscid bird *Eudyptes pachyrhynchus* (= crested penguin)

tawapou *N.Z.* the sapotaceous tree *Sideroxylon novozelandicum*

Tawara's *epon adj.* from Sunao Tawara (1873–)

tawari *N.Z.* the saxifragaceous tree *Ixerba brexioides*

tawherowhero *N.Z.* the saxifragaceous tree *Quintinia serrata*

tawhiwhi the pittosporaceous shrubby tree *Pittosporum tenuifolium*

[**tawny**] a mutant (gene symbol *taw*) mapped at 41.1± on the X chromosome of *Drosophila melanogaster*. The body of the phenotype is bicolored with a dark heavy thorax and a light-colored light abdomen

-tax- *comb. form* properly meaning "to arrange" but which has come to mean "to arrange, or move, in the direction of" hence "to be attracted to" or "move towards"

Taxaceae the yew family of Coniferales. The reduction of the female cone to a single ovule is characteristic

[**taxi**] a mutant (gene symbol *tx*) mapped at 91.0± on chromosome III of *Drosophila melanogaster*. The phenotypic wings are divergent

-taxia *subst. suffix* meaning "the result of -taxis"; *e.g.* heterotaxia is the condition of a chicken embryo in which the head is twisted to the left. *Comb. forms* are entered under -taxis

-taxic *comb. suffix* from -tax- (*q.v.*) in the sense of arrangement (*cf.* taxis 2)

    diastataxic said of a bird's wing with a gap between the fourth and fifth secondaries

    eutaxic said of a bird's wing that lacks a gap between the fourth and fifth secondaries

taxis 1 (*see also* taxis 2) a movement of an organism oriented with respect to a source of stimulation. The adjective "negative" indicates movement away from, and the adjective "positive" indicates movement toward

    anemotaxis movement toward wind

    argotaxis movements due to surface tension

    biotaxis movement in consequence of biological stimuli

        neurobiotaxis the migration of embryonic nerves in the direction of stimuli

chemotaxis movement in consequence of chemical stimuli

cytotaxis (*see also* taxis 2) the movement of cells in relation to each other

  negative cytotaxis the tendency of cells to separate

  positive cytotaxis the tendency of cells to aggregate

electrotaxis = galvanotaxis

galvanotaxis orientation or movement in relation to direct current

  apogalvanotaxis negative galvanotaxis

geotaxis orientation or movement in relation to gravitational stress

  apogeotaxis = negative geotaxis

haptotaxis curvature induced by contact stimulus

heterotaxis movement of organs to an unusual position or twisted in an unusual direction, as when the anterior end of the chicken embryo twists to the left instead of to the right

hydrotaxis movement in the direction of moisture

  aphydrotaxis being repelled by water

klinotaxis the response of an organism that moves its head from side to side symmetrically in moving towards the stimulus (*cf.* tropotaxis)

menotaxis movement at a fixed angle in relation to a source of stimulation

pharotaxis the recognition and use of landmarks in bionavigation

phobotaxis a movement away from a concentration or intensity of something without regard to location

phototaxis movement in response to light

  aphototaxis the condition of being unaffected by light

  apophototaxis negative phototaxis

  diaphototaxis the arrangement of an organism or part of an organism at right angles to the incident light

  orthophototaxis standing up under the influence of light

  photohorotaxis response to a color or light pattern. The term should properly be photohoramotaxis. As here spelled, it should, but is rarely taken to mean "response to the time of day"

polytaxis discontinuous change

rheotaxis a movement in response to flow—either of air or water

telotaxis the responses of an animal capable of moving towards a source using only one remaining, of previously paired, receptors (*cf.* tropotaxis)

thermotaxis orientation or movement in relation to temperature

  apothermotaxis the condition of not responding to temperature changes

thigmotaxis movement in response to tactile stimuli

  apothigmotaxis a repulsion or discomfort induced by contact with a surface

topotaxis a movement in regard to the location of something

tropotaxis the response of an animal using paired receptors without moving its head from side to side (*cf.* klinotaxis)

taxis 2 (*see also* taxis 1) in the sense of arrangement rather than movement

  anthotaxis the arrangment of parts of a flower (*cf.* anthesmotaxis)

  anthesmotaxis the organization of the parts of the flower (*cf.* anthotaxis)

  carpelotaxis the arranging of carpels in a flower or fruit

chaetotaxis the arrangement or system of arrangement of bristles

cytotaxis (*see also* taxis 1) the arrangement of cells in an organ

epitaxis the state of being placed in a rank, as distinct from a row

homotaxis the arrangement of an inflorescence in which all the parts have the same arrangement that is, all are spirals or all are whorls

meiotaxis the suppression of an entire whorl in phyllotaxis (*cf.* oligotaxis)

oligotaxis the reduction of the number of whorls in a flower (*cf.* meiotaxis)

pantotaxis the condition of a fern in which a sorus can arise from any vein

parataxis the arrangement of sori on an aborted vein

phyllotaxis (*see also* fraction, helix) the arrangement or the laws covering the arrangement of the spatial distribution of leaves along a stem

  alternate phyllotaxis a condition in which each node bears a single leaf

  decussate phyllotaxis = opposite phyllotaxis

  discontinuous phyllotaxis when the same ratio does not pertain along the stem

  falling phyllotaxis the condition when the diminution in the rate of spiraling progresses down the course of the stem

  rising phyllotaxis the opposite of falling phyllotaxis

  opposite phyllotaxis the condition in which each node bears two leaves

  whorled phyllotaxis the condition in which each node bears more than two leaves

rhizotaxis arrangement of roots

xerotaxis the condition of a plant succession that is not affected by drought

Taxodiaceae a family of coniferous spermatophyta containing, *inter alia*, the sequoia. They are both distinguished by the flat, or peltate, cone scales each producing two to nine seeds

Taxodioideae a subfamily of Taxodiaceae

taxon a category in the taxonomic hierarchy

taxonomy the science of arranging organisms in logical and natural groups, in such a manner as to cast light on their evolution and affinities

  cytotaxonomy that branch of taxonomy which is based on a study of chromosomes and therefore deals with the genotypic characters rather than the phenotypic

Taxotidae a small family of acanthopterygian fishes called archerfishes by reason of their habit of projecting pellets of water at their prey

-taxy *subst. term.* synonymous with -taxis which is preferred in the present work

Tayassuidae a family of suine mammals containing the forms commonly called peccaries

tayra any of several species of *S.Amer.* musteline carnivores, usually without qualification, *Galictis barbara* (*cf.* grison)

tbd *see* [tiny-bristloid]

tc *see* [tinychaete] and [truncate]

td *see* [truncated] this symbol has also been used for *tryp-3* (*q.v.*)

te *see* [tenuchaete]

tea properly, any of several species of the ternstroemiaceous genus Thea and their dried leaves forming the "tea" of commerce. The term is also applied to many other plants from the leaves of which potable infusions can be prepared

  khat tea the celasteraceous shrub *Catha edulis*

Labrador tea the ericaceous shrub *Ledum groen-landicum*
   narrow leaved Labrador tea *L. palustre*
marsh tea the ericaceous shrub *Ledum palustre*
Mexican tea the chenopodiaceous herb *Chenopodium ambrosioides*
mountain tea the ericaceous shrub *Gaultheria procumbens*
New Jersey tea the rhamnaceous shrub *Ceanothus americanus*
oswego tea the labiateous herb *Monarda didyma*
prairie tea the euphorbiaceous herb *Croton monanthogynus*
teak properly any of several species of trees of the verbenaceous genus Tectona and usually, without qualification, *T. grandis*, but extended to numerous timber-trees the wood of which resembles teak
   African teak the moraceous tree *Chlorophora excelsa*. The term is also applied to the euphorbiaceous tree *Dissiliaria baloghioides* (**Australian teak**)
   Philippine teak *T. philippinensis*
teal any of numerous anatid anseriform birds, mostly anatines of the genus Anas. *Brit.*, without qualification, *A. crecca*. *N.Z.*, without qualification, the aythyine *Aythyna novaeseelandiae* (= New Zealand scaup) *Austral.*, without qualification *Stictonetta naevosa* (= freckled duck)
   Baikal teal *A. formosa*
   red-billed teal *A. erythrorhyncha* (= red-billed duck)
   black teal *N.Z.* the aythyine *Aythya novaeseelandiae* (= New Zealand scaup)
   brown teal *A. chlorotis*
   chestnut teal *A. castanea*
   cinnamon teal *A. cyanoptera*
   common teal *A. crecca*
   diving teal *W.I.* = ruddy duck
   European teal (= common teal)
   falcated teal *A. falcata*
   garganey teal *A. querquedula*
   Georgian teal *A. georgica*
   grey teal *A. gibberifrons*
   hottentot teal *A. punctata*
   mountain teal *Austral.* = chestnut teal
     blue-winged teal *A. discors*
      green-winged teal *A. carolinensis*
   whistling teal *Austral.* = pink-eared duck
   summer-teal *Brit. A. querquedula* (= garganey) *U.S. Aix sponsa* (= wood duck)
team term for a group of horses, or oxen (*cf.* yok̲e) working in concert
tear (*see also* drop) in compound names of plants
   Job's tears the large grass *Coix lachryma-jobi*
   maiden's tears the caryophyllaceous herb *Silene cucubalus* (= bladder campion)
teasel any of numerous species of dipsaceous herbs of the genus Dipsacus
teat the nipple of a mammary gland
   carbao's teats the annonaceous tree *Uvaria rufa* and its edible fruit
   nigger teats *U.S.* the compositaceous herb *Rudbeckia bicolor*
teazel = teasel
tectin a glycoprotein, often referred to as pseudochitin. It is a frequent component of the skeleton of Protozoa
-tectous pertaining to a covering
   detectous naked
tectrix the sum total of the wing coverts of birds
tectum literally a roof

tectum 1 (*see also* tectum, tectum 2) in the sense of the roof of the skull or brain
   posterior tectum a dorsal connection between the occipital elements of the chondrocranium
   synotic tectum a plate in the chondrocranial complex of vertebrates, which connects the otic capsules above the brain or a narrow cartilagenous band connecting the two parietal plates in other chondrocrania; it is often confused with, or combined with, the posterior tectum
tectum 2 (*see also* tectum, tectum 1) in the sense of other roof-like structures particularly the epistome of acarines
teff the gress *Eragrostis abyssinica*
tegmen 1 (*see also* tegmin, tegmen 2, 3) any portion of an arthropod exoskeleton which is conspicuously thicker than the surrounding portions, especially the horny or leathery forewings of orthopteran insects
tegmen 2 (*see also* tegmin, tegmen 1, 3) the inner coat of a seed
tegmen 3 (*see also* tegmin, tegmen 1, 2) the vault of crinoid echinoderms
tegmin literally a cover or lid, almost universally misspelled tegmen
tegmina *pl.* of tegmen
-tegminous pertaining to a tegmin
   ategminous used of an ovule lacking a covering
   bitegminous covered with two coats, said particularly of a seed
tegu = teju
tegula an articular sclerite of an insect wing. The term is misused for the patagium of Lepidoptera and also for various analogous structures at, or near, the base of the wing in various insects
-tegum- *see* tegmin
Tejidae = Teiidae
Teiidae a family of lacertilian reptiles with pleurodont dentition and a long bifid tongue carrying marked papillae or folds
teju any of several species of teiid lacertilian reptiles of the genus Tupinambis
-tel- *comb. form* meaning "far off"
tela any developmental tissue, but particularly plant meristem
-tele- *comb. form* meaning "perfection" and, by extension, "end" or "termination," frequently confused with -tel-. The definitions of stenotele and eurytele, which may possibly derive from this root, are entered under their initial letters
teledu the Asiatic mephitic, meline, carnivorous mammal *Mydaus meliceps*
Telegeusidae a small family of coleopteran insects superficially resembling staphylinids with which they are sometimes confused
Teleodesmacea an order of pelecypod Mollusca with the right and left lobes of the mantle more or less connected on the ventral and hinder margin
teleology the suggestion that something is shaped by a purpose or directed towards an end. It is an inexpiable heresy in the *U.S.* but it is viewed, as a literary or descriptive device somewhat more leniently by biologists of other countries
spermateleosis sometimes used as synonymous with spermatogenesis but more properly for the metamorphosis of a spermatid into a spermatozoan
Teleostei a superorder of actinopterygiian gnathostomatous chordates containing the true bony fishes

Telestacea an order of alcyonarian anthozoans distinguished by having a long central polyp with lateral polyps as side branches

-teleut- *comb. form* meaning "end"

-telic *comb. suffix* of doubtful origin meaning "pertaining to evolution"

  bradytelic said of an organism which appears to evolve more slowly than is customary in the group under discussion

  horotelic evolving at what appears to be a normal rate

  tachytelic evolving at what appears to be a rapid rate

ateliosis nannism

ammoniotelism the condition of a kidney which primarily excretes ammonia

tellin popular name of tellinid pelecypods

Tellinidae a family of pelecypod mollusks distinguished by the extremely long siphon protruding from the thin end of the pear-shaped shell

-telma- *comb. form* meaning a "pool" but used by ecologists in the sense of a "wet meadow"

telmatad a plant of wet meadows

Telogonia an order of nematode worms distinguished from the Hologonia in possessing two or more branches of the female genital tract

telome a terminal branch, or branchlet, of a vascular plant (*cf.* mesome)

Telosporidia a subclass of sporozoan Protozoa distinguished by the absence of polar capsules on the spores

Telotremata once regarded as an order of brachiopods but no longer regarded as a valid taxon

telson the primitive terminal body segment of all arthropods, found in all embryos but frequently vestigial in adults. The term is specifically applied to the stinging segment at the end of the tail of a scorpion. The telson is well developed in many crustacea as the central lobe of the tail fan

paratelum the penultimate segment of insects

-tely *subs. suffix* from both -tel- and -tele-

  hypertely 1 (*see also* hypertely 2) the condition of being very different, often pertaining, without qualification, to difference by reason of large size

  hypertely 2 (*see also* hypertely 1) pertaining to a degree of adaptation which is beyond the bounds of reason or the reason for which is not apparent to a human

  hysterotely the metameric replication of female sex organs particularly in those forms in which this does not normally occur

-temn- *comb. form* meaning "cut"

Temnocephalida a group of aberrant turbellarian Platyhelminthes furnished with adhesive disks which adapt them to an epizoic existence. They have from time to time been regarded as a separate order or a suborder of the Rhabdocoela. At one time they were classified among the Trematoda

Temnochilidae = Ostomatidae

autotemnous capable of dividing without obvious stimuli

temperate lacking extremes, used particularly by biologists of climatic conditions

  subtemperate intermediate between temperate and extreme, properly used of the colder regions between the temperate and Arctic zones, but occasionally used also of the regions between temperate and tropical zones

tem *see* [tenuischaete]

ten *see* [tenuischaete]

tenaculum 1 (*see also* tenaculum 2, 3) literally a holdfast and so used of Algae

tenaculum 2 (*see also* tenaculum 1, 3) a toothed

structure on the third abdominal segment of collembola which acts as a catch to hold the furcula (*q.v.*)

tenaculum 3 (*see also* tenaculum 1, 2) a hollow, external projection from the bony wall of some Anthozoa

tench the cyprinid fish *Tinca tinca*

Tendipedidae = Chironomidae

tendon a mass of white fibrous connective tissue attaching a muscle to a bone

tendrac = tenrec

tendril any thread-like structure with the aid of which a plant secures itself

  leaf tendril one which is produced through the metamorphosis of a leaf

-tene- *comb. form* meaning "band" in the sense of stripe

-tene as a descriptive termination of prophase meiotic stages (*see* -nema)

Tenebrionoidea a superfamily of polyphagous coleopteran insects containing not only the Tenebrionidae but also about a dozen families of allied beetles many of which live under bark

Tenebrionidae a very large family of coleopteran insects many of which are destructive pests to stored crops. They can best be distinguished by the five-five-four tarsal formula and the notched eyes. They are commonly called darkling beetles though most of the pests have separate names

teneral the condition of an insect still soft after its emergence from the pupa

tenrec without qualification the tenrecid insectivorous mammal *Centetes ecaudatus*

  hedgehog-tenrec any of several species of tenrecids of the genus Setifer

  rice-tenrec any of several species of tenrecids of the genus Oryzorictes

  striped tenrec any of several species of tenrecids of the genus Hemicentetes

  water-tenrec web-footed tenrecids of the genera Limnogale and Geogale, usually placed in the Tenrecidae since they are Madagascan, but strongly resembling the African Potamogalidae

Tenrecidae a family of insectivores confined to the island of Madagascar containing *inter alia* the tenrecs

tentacle any flexible, as distinct from articulated, appendage. Many are tactile. The term is also applied to the podia of ophiuroid echinoderms and to the sensitive trichomes such as those on insectivorous plants

  pseudotentacle a much-branched extrusion, bearing acrorrhagii at the tip, arising immediately outside the ring of tentacles in some Zoantharia

Tentaculata that class of Ctenophora that possess tentacles

tentaculum there is no such word though it is frequently used for tenaculum

Tenthredinidae a very large family of symphytan hymenopteran insects and usually the family that is intended when the term sawfly is used without qualification. These forms resemble wasps in all save the absence of a waist and many are brightly colored. Many are destructive pests

Tenthredinoidea that superfamily of hymenopteran insects which contains the sawflies. They are distinguished by the large flattened saw-like ovipositors and the absence of the constriction at the anterior end of the abdomen

tentilla tentacle-bearing zooids of a siphonophoran coelenterate

-tentor- *comb. form* meaning "tent"

tentorium 1 (*see also* tentorium 2, 3) the apodemal skeleton of the insect head

tentorium 2 (*see also* tentorium 1, 3) that transverse fold of the meninges which lies between the cerebral hemispheres and cerebellum

tentorium 3 (*see also* tentorium 1, 2) a secondary ossification, separating in some mammals the cerebral fossa from the cerebellar fossa

[tenuchaete] a mutant (gene symbol *te*) mapped at 5.6± on the X chromosome of *Drosophila melanogaster*. The phenotype has fine short bristles with dark eyes

[tenuischaete] a mutant (gene symbol *ten*) mapped at 43.9 on the X chromosome of *Drosophila melanogaster*. The phenotypic expression is a small body with sparse short bristles

teomacaztli the annonaceous shrub *Cymbopetalum penduliflourum*

[Teopod] a mutant (gene symbol *Tp*) mapped at 36 on linkage group VII of *Zea mays*. The phenotypic expression includes enlarged bracts and the plant usually has many tillers

tepal those parts of the perianth of a flower that cannot be distinguished either as petals or sepals

tepary the leguminous plant *Phaseolus actuifolius*

tephreous ash-colored

Tephritidae a large group of myodarian cycloraphous dipteran insects properly called fruit flies though this term is often misapplied to the Drosophilidae. The trypetids are commonly found on flowers or vegetation and have brilliantly colored wings which they move up and down giving them the common name of peacock-fly. Many are pests in orchards

-tephro- *comb. form* meaning "ash"

-tera- *comb. form* meaning "prodigy" though usually used in the sense of "monster"

tercel a male of many kinds of hawk. Particularly those used in falconry

tercelet = tercel

Terebellidae a family of sedentary polychaete annelids with a prominent horseshoe-shaped preoral lobe behind which is a transverse ridge bearing large numbers of long tentacles. They are both burrowing and tube-dwelling

Terebelliformia an order of phanerocephalous polychaete annelids wigh greatly reduced parapodia. The peristomium usually carries cirri, or tentacular filiments. Most are tubicolous

Terebintaceae = Anacardiaceae

terebra literally, a drill or boring tool. Specifically applied to insect ovipositors adapted for boring or piercing and to an insect mandibular sclerite equivalent to the galea of the maxilla

Terebrantia a suborder of thysanopteran insects distinguished by the possession of a saw-like ovipositor

terebrate with scattered holes, as though bored

terebratorium the protusible tip of a miracidium larva

Terebridae a family of gastropod mollusks with slender elongate many whorled shells usually bearing transverse striations on the volutes. Commonly called screw-shells

Teredidae a family of wood-boring pelecypod mollusks commonly called shipworms

terete in the form of an extremely elongate cone or tapering cylinder

-terg- *comb. form* meaning "back"

tergite the upper, or dorsal, plate of the exoskeleton of a segmented animal particularly an arthropod (*cf.* sternite)

acrotergite a narrow flange on the anterior part of the tergal plate in insects

endotergite an infolding flap or portion of the apodeme between adjacent tergites

pleurotergite the side of the metanotum in insects

tergum = tergite

mesotergum = mesonotum

-teris- *comb. form* meaning "strife"

phyteris the struggle for existence among plants

metratrem a thickened terminal portion of the uterus in some trematode platyhelminthes

termagant = ptarmigan

termen the outer margin of an insect wing

chiasma terminalization the movement of chiasmata at the end of the meiotic prophase

termite popular name of members of the insect order Isoptera

powder post termite the kalotermitid *Cryptotermes brevis*

subterranean termite usual designation of rhinotermitids

eastern subterranean termite *U.S. Reticulitermes flavipes*

Formosan subterranean termite *U.S. Coptotermes formosanus*

western subterranean termite *U.S. Reticulitermes hesperus*

wood termite popular name of kalotermitid termites

tern any of numerous charadriiform birds of the subfamily Sterninae of the family Laridae. Many are in the genus Sterna. The terns are readily distinguished from the gulls by their narrower wings and graceful flight

Aleutian tern *S. aleutica*

Antarctic tern *S. vittata*

Arctic tern *S. paradisaea*

gull-billed tern *Gelochelidon nilotica*

large-billed tern *Phaetusa simplex*

black tern *Chlidonias niger*

white-winged black tern *Chilidonias leucopterus*

bridled tern *S. anaethetus*

Cabot's tern *Thalasseus sandvichensis* (= Sandwich tern)

Caspian tern *Hydroprogne caspia*

white-cheeked tern *S. repressa*

common tern *S. hirundo*

crested tern *Thalassius bergii*

lesser crested tern *Thalasseus bengalensis*

elegant tern *Thalasseus elegans*

fairy tern *S. nereis*

Forsters tern *S. forsteri*

black-fronted tern *Chilidonias albistriata*

white-fronted tern *S. striata*

least tern *S. albifrons*

little tern *Brit.* = least tern *U.S.*

marsh tern this term from time to time has been applied to almost any coastal tern. *Afr.* = whiskered tern

black-naped tern *S. sumatrana*

roseate tern *S. dougallii*

royal tern *Thalasseus maximus*

Sandwich tern = Cabot's tern

sooty tern *S. fuscata*

whiskered tern *Chlidonias hybridus*

white tern *Gygis alba*

Ternstroemiaceae that family of parietalous dicotyledons which contains *inter alia*, the tea plant and the camellias. The family is distinguished by the very numerous stamens

-terr- *comb. form* meaning "land"

epiterranean the above ground portion of such plants as fruit both above and below ground

mediterranean (*but see* Mediterranean) dwelling far from the sea

terrapin (*see also* scale) a term that in *U.S.* has become almost synonymous with turtle

diamondback terrapin *U.S.* any of several species of testudinid chelonian reptiles of the genus Malaclemys

stinkpot terrapin the chelydrid chelonian *Sternothaerus odoratus* (= mud turtle)

terrarium a small case in which terrestrial or amphibious animals are maintained for observation, study, or decorative purposes

territorialism the custom, particularly strong in birds, of laying claim to and defending exclusive rights to a territory

Tersinidae a monotypic family of passeriform birds erected to contain the swallow-tanager of *Cent.* and *N. S.Amer.* It is a bright turquoise blue bird, with strong black barrings on the flanks, about the size of a thrush

tertial a remex (*q.v.*) which grows from the skin of the humerus

tessellate chequered

-test- *comb. form* meaning "shell" in the sense of "hard covering." The word testa was also used by the Romans for "brick" (*see* testaceous)

test a term applied to almost any hard outer covering of an organism or part of an organism

testa the *L.* form, preferred in *bot.* for the outer coat of a seed

Testacea 1 (*see also* Testacea 2) an order of sarcodinous Protozoa distinguished by possessing a shell, usually chitinous in nature, containing a single aperture through which the pseudopodia protrude

Testacea 2 (*see also* Testacea 1) an original subdivision of the Linnean group Vermes. It contained the pelecyod Mollusca and the brachiopods

testaceous brick red

Testicardines a class of the phylum Brachiopoda distinguished by the fact that the shells are hinged together

testicular in the form of a human testis and thus used in botany as descriptive of the shape of the tubers of an orchid

Testudinidae a family of cryptodiran chelonian reptiles containing the great majority of forms called tortoises in England and tortoises, turtles, and terrapins in the U.S. They are distinguished by having a completely retractile neck, four or five-clawed digits, and in lacking parietosquamosal arches in the skull

testule = frustule

tet *see* [tetraltera]

Tetanoceratidae = Schiomyzidae

teteaweka *N.Z.* the compositaceous tree *Olearia augustifolia*

Tethinidae an obscure group of seaside-dwelling myodarian cycloraphous Diptera

Tethyidae a family of gastropod Mollusca distinguished by their large size and angular tenticular folds on the head. Commonly called sea-hares

-tetra- *comb. form* meaning "four" sometimes written -tetr-

tetra 1 (*see also* tetra 2) a term loosely used by aquarists for almost any characinid teleost fish

African tetra *Nanaethiops unitaeniatus*

black tetra *Gymbocorhymbus ternetzi*

blue tetra *Mimagoniates microlepis*

cardinal tetra *Cheirodon axelrodi*

croaking tetra *Glandulocauda inequalis*

dawn tetra *Hyphessobrycon eos*

flag tetra *Hyphessobrycon heterorhabdus*

black-flag tetra = rosy tetra

glass tetra either *Roeboides microlepis* or *Moenkhausia oligolepis*

jewel tetra *Hyphessobrycon callistus*

lemon tetra *Hyphessobrycon pulchripinnis*

neon tetra *Hyphessobrycon innesi*

platinum tetra *Gephyrocharax atracaudatus*

rosy-finned tetra *Hyphessobrycon rosaceus*

silver tetra *Ctenobrycon spilurus*

red tetra *Hyphessobrycon flammeus*

yellow tetra *Hyphessobrycon bifasciatus*

black widow tetra = black tetra

tetra 2 (*see also* tetra 1) the large *Eur.* tetraonid bird *Tetrao urogallus* (= capercailzie)

Tetrabranchia that order of cephalopodous Mollusca which contains the living nautiloids and the fossil ammonites. The order is distinguished by the presence of an external, usually multilocular, shell

Tetracentraceae a monogeneric family of dicotyledenous angiosperms distinguished from the closely allied Trochodendraceae by the possession of solitary, palmate-veined leaves

tetract a four-axis sponge spicule derived from a hexactinellid by the loss of an axis

Tetractinellida a subclass of demospongious Porifera, the spicules, when present, being tetraxonic

tetrad a group of four, particularly the four cells resulting from the division of a spore mother cell or the groups of four chromosomes resulting from the pairing at meiotic prophase

Tetradynania in the Linnaean system of plant classification that class which is distinguished by possession of six stamens, four long and two short

Tetragoniaceae = Aizoaceae

Tetragonuridae a monogeneric family of stromateid acanthopterygian fishes containing the squaretails of the genus Tetragonurus. Most have tough large scales

Tetragynia those orders of the Linnaean system of plant classification distinguished by the possession of four pistils

[tetraltera] a mutant (gene symbol *tet*) mapped at 48.5 on chromosome III of *Drosophila melanogaster.* The name refers to the replacement of wings by an additional pair of halteres in the phenotype

[Tetra molting] a group of pseudo-alleles mapped at 3.0 on linkage group VI of *Bombyx mori.* The phenotypic expression is a larva which pupates after the fourth molt.

Tetrandia that class of the Linnaean classification of plants distinguished by the possession of four stamens

Tetranychidae a family of acarine arthropods, including the spider-mites or red spiders, that have long, recurved, whip-like chelicerae and distinctive female genitalia.

Tetraodontidae a family of plectognathous fishes commonly known as puffers from their habit of blowing themselves up when out of the water

Tetraodontiformes = Plectognathi

Tetraonidae the family of galliform birds which contains the grouse. They are distinguished from other galliform birds by the short bill and the hair-like feathers on the legs

Tetraonychidae a family of coleopteran insects the tarsal claws of which are so deeply cleft as to give the appearance of being duplicated. Now usually included with the Meloida

Tetraphyllidea an order of eucestode Cestoda distinguished by having a scolex with four bothridia. Found only in the intestine of elasmobranch fishes

Tetrapoda a term coined to describe those classes of gnathostomatous chordates which have limbs as distinct from fins. As so defined, the group contains the amphibia, reptiles, birds and mammals

Tetrarhyncoidea = Trypanorhyncha

Tetrasporales that order of chlorophycean Chlorophyta which contains those forms with immobile vegetative cells

Tetraxonida at one time used for an order of Porifera in contrast to Hexactinellida and Keratosa nowadays incorporated in the Demospongiae

Tetrigidae a family of small orthopteran insects commonly called pygmy locusts. The narrow head bears large globular eyes

Tettigoniidae that family of orthopteran insects which contains the longhorned grasshoppers. The long antennae are typical

intextine *see* intexine

textularid said of that type of foraminiferan shell in which the chambers are in two or three alternating rows

tf *see* [tufted]

tg *see* [tottering]

th *see* [thread] or [tilted head]

Thaisidae a family of gastropod mollusks famous in antiquity, under the name Murex (*q.v.*), as a source of dye. They are distinguished by their huge body wall, large aperture and very short spire, in addition to the secretion of colored fluids

-thal- *comb. form* meaning a "twig" or, by extension, anything young or recently produced (*cf.* -thalam-)

-thalam- *comb. form* meaning "chamber" in the sense of enclosed space (*cf.* -thal-)

-thalamic *adj. suffix* used interchangeably as a derivative of -thal- or -thalam-

   athalamic said of lichens lacking apothecia

   monothalamic 1 (*see also* monothalamic 2) said of a plant producing a single flower

   monothalamic 2 (*see also* monothalamic 1) = unilocular

   polythalamic many chambered (= multilocular)

sporothalamium a compound sporophore

-thalamous *adj. suffix* synonymous with -thalamic, the form preferred in this work

thalamus 1 (*see also* thalamus 2) that part of the axis on which parts of a flower are inserted

thalamus 2 (*see also* thalamus 1) that part of the diencephalon which lies on each side of the third ventricle

   epithalamus the dorsal portion of the thalamencephalon

   hypothalamus the ventral portion of the thalamencephalon

-thalass- *comb. form* meaning "sea"

thalassad a sea plant

Thalassidromidae = Hydrobatidae

thalassin a toxin obtained by the alcohol extraction of nematocyst bearing tentacles (*cf.* congestin and hypotoxin)

Thalassinidea a superfamily of decapod crustacea in which the abdomen is extended, the antennal scale reduced, and a suture extends the full length of the carapace on each side. The group are commonly called burrowing shrimps

thalassinous sea green

Thaliacea a class of Urochordata with the oral and atrial apertures at opposite ends of the body. All are free-swimming

-thall- *comb. form* meaning "branch" or "young shoot"

thallic pertaining to a thallus (*q.v.*) in any meaning of that word

   athallic lacking a thallus

   epithallic growing on a thallus

   heterothallic properly dioecious, but usually only applied to thallophytes

   hemiheterothallic partial dioecism

   homothallic properly monoecious but usually only applied to thallophytes

   hemihomothallic partial monoecism

   trichothallic said of a fungal or algal growth that terminates in fine hairs

thallidium a plant structure utilized for asexual reproduction

-thalline synonymous with -thallic, the form preferred in this work

thallium a derivative of thallus (*q.v.*) used by some authors

-thallo- *see* -thall-

Thallogamae = algae

thallome a thallus, or thallus-like growth, considered as a philosophical entity

thallus literally a branch or young shoot, but most commonly used in reference to solid structures of nonvascular plants

   epithallus the cortical layer of a lichen

   homothallus the medullary layer of a lichen

   merithallus = internode

   plethysmothallus a reproductive form of a heterogenerate phaeophycean alga resembling a prothallus but which reproduces asexually

   prothallus the gametophyte of a fern or the initial stages of any thallophyte

      megaprothallus that which produces archegonia

   perithallus the upper layer of calcareous algae

   protothallus = hypothallus

heteroprothally the condition of producing prothallia of different sexes

isoprothally having prothallia only of one sex

-thalmic a not infrequent, but entirely incorrect, spelling of -thalamic

thamin the Far Eastern cervine artiodactyle *Cervus eldi*

-thanat- *comb. form* meaning "death"

   thanatosis the condition of being dead or of feigning death

Thaneroconia an order of asteroid Echinodermata distinguished by the prominent marginal plates along the side of the arms

-thaum- *comb. form* meaning "a wondrous thing"

Thaumaleidae a family of minute reddish brown dipteran flies commonly called solitary midges

Thaumastocordiae a small family of hemipteran insects which feed exclusively on palm trees

thb *see* [thin bristle]

Theaceae = Ternstroemiaceae

Thebesian *epon. adj.* from Adam Christian Thebesius (1686–1732)

-thec- *see* -thek-

theca the term has been applied to almost any strucutre which contains anything, such as the sporangium of a fern or the capsule of a moss, the loculus of an anther, the calcareous exoskeleton of echinoderms, and the wall as distinct from the basal plate, of a corallite. A more specific use is for a lorica (*q.v.*) which is in direct contact with the cell membranes of some algae and thus corresponds to the cellulose cell wall of higher forms

cephalotheca a head case, particularly the portion of the pupal integument which envelopes the insect head

ceratheca = ceratotheca

ceratotheca a horn case, particularly that part of the pupal integument which envelopes an insect's antennae

derthrotheca a horny covering over the tip of a bird's bill

epitheca 1 (*see also* epitheca 2) that part of the case of anything which covers a theca, particularly the theca of the epicone of dinoflagellates, and an outer calcified wall covering the theca of a coral

epitheca 2 (*see also* epitheca 1) the larger of the two diatom frustules

gasterotheca that part of the pupal case which covers the abdomen of the contained insects

gnathotheca the covering of the lower jaw of birds

gonotheca that portion of the perisarc of a hydrozoan coelenterate which encloses the gonozoid

gonytheca the cup on the distal end of the insect femur into which the tibia articulates

hydrotheca that part of the periderm of a hydrozoan coelenterate which surrounds the hydranth

hypotheca 1 (*see also* hypotheca 2) literally the "lower case," but particularly the theca of the hypocone of dinoflagellate Protozoa

hypotheca 2 (*see also* hypotheca 1) the inner or smaller half-frustule of a diatom

nematotheca that part of the perisarc of a hydrozoan coelenterate which surrounds the nematophore

ootheca egg case

ovotheca a terminal expanded portion of an oviduct in which eggs are accumulated before laying as one mass

podotheca the integument of the foot in birds

pseudotheca a term applied to those apparent thecae in a coral which are in point of fact formed by the fusion of sclerosepta

pterotheca the wing-enclosing portion of a pupal insect case

rhamphotheca the outer surface of the bill of birds

rhinotheca the covering of the upper bill in birds

spermatheca a container of sperm

spermotheca = pericarp

-thecal *adj. suffix* meaning "pertaining to a theca"

dithecal consisting of two cells in the broadest meaning of that word

ectothecal outside the theca. Used for naked spored ascomycetes

endothecal said of those echinoderms in which the ambulacra lie between plates of the exoskeleton

epithecal said of those echinoderms in which the ambulacra lie in grooves in a plate of the exoskeleton

exothecal said of ambulacra that are directly attached on the brachiles of Crinoids

athecate lacking a theca

thecidion = achene

-thecium- *see* -thek-

amphithecium a protective layer of cells round the developing endothecium of a moss capsule

apothecium an ascocarp bearing asci on its saucer-like base

cleistothecium = cleistocarp

endothecium the inner layer of anything, particularly the inner of two layers and specifically that layer of the anther which lies immediately under the epidermis

hypothecium the tissue lying immediately below the hymenium

mesothecium the middle layer of cells in the wall of an anther

parathecium the walls of a lichen thecium

perithecium a hollow, usually flask-like, ascocarp bearing asci over its inner surface

-thecous synonymous with -thecal, the form preferred in this work

theelin = estrone

dihydrotheelin = estradiol

theelol = estriol

Theineae a suborder of parietale dicotyledons containing the families Dilleniaceae, Actinidiaceae, Euryphiaceae, Medusagynaceae, Ochnaceae, Strasburgeriaceae, Caryocaraceae, Marcgraviaceae, Quiinaceae, Theaceae, Guttiferae, Hypericaceae and Dipterocarpaceae

-thek- *comb. form* meaning "container," "case" or "cover" almost invariably transliterated -thec-. A Latin derivative of this form "apothecium" meaning "storehouse" or "warehouse" has caused a complete confusion in biological literature between -theca and -thecium a form which never existed save in the very restricted word cited

-thele- *comb. form* meaning "nipple" and by extension, sometimes used to mean "female"

polytheleous many ovaried

endothelium an epithelium of mesodermal origin that lines a cavity such as the coelom

epithelium any tissue that covers, or lines, an organ or organism. The curious use of a word which literally means "on the nipple" for this purpose derives from the fact that the word was originally used only for the covering of the lips, transferred from this to lips and nipple and thence to the whole female skin and finally to any soft skin

bracket epithelium a leaf epithelium with interlaced projections

ciliated epithelium columnar, cuboidal, or pseudostratified columnar epithelium, the free surface of which bears cilia

columnar epithelium epithelium of which each cell is in the form of a simple elongate column

pseudostratified columnar epithelium columnar epithelium in which not all the columns reach from the basement membrane to the surface

cuboidal epithelium epithelium of which each cell is more or less cuboid

myoepithelium contractile cells in the body wall of lower invertebrates including sponges

sensory epithelium any of numerous types of epithelial cells modified to receive stimuli

simple epithelium an epithelium containing only one type of cell

squamous epithelium an epithelium which consists of flattened polygonal cells spread over the surface

stratified epithelium epithelium consisting of several layers of the same type of cell

transitional epithelium an epithelium composed of many types of cells

perithelium a loose term for connective tissue cells associated with capillaries

-thely *comb. suffix* meaning "female" by derivation from -thele (*q.v.*)

monothely the term used by entomologists when polyandry is meant (*cf.* monarseny)

Thelyphonidae a family of pedipalp arthropods commonly called whip scorpions (*cf.* tarantula)

-epithem- *comb. form* meaning "covering" (*see also* remarks under eipthelium)

epithem a tissue composed of chlorophyll-less cells serving as internal hydathodes in the mesophyll of the leaves

epithema a horny excrescense on the bill of birds

Theophrastaceae a small family of primulale dicotyledonous angiosperms containing the forms commonly called joeweeds. The family is readily distinguished by the large yellow or orange seeds

theory a hypothesis which is sufficiently supported by a logical interpretation of experimental data to justify the supposition that it extends over a wide field, but which is not yet sufficiently universally verified to justify the use of the term "law" (q.v.). The words "the postulate that . . ." or "the idea that . . ." have been omitted from the definitions that follow. The inclusion of a theory, or the omissions of *obs.* after the name is no more than an objective statement of what has been said in the literature of biology and implies neither belief nor disbelief on the part of the compiler of this dictionary (*cf.* law, hypothesis)

age and area theory the older a species, the larger will be its area of distribution

alveolar theory (*obs.*) "protoplasm" consists of alveoli filled with a clear substance—the hyaloplasm

antithetic theory (*obs.*) the alternate generations of plants are distinct entities (*cf.* homologous theory)

apposition theory a plant cell wall grows solely by deposition on its interior surface

Bergh's theory *see* gonocoel theory

chromosome theory *see* particulate theory

enterocoel theory the enterocoel is the primitive, ancestral type of coelom, derived as an outpocketing of the gut as in Cnidaria

gonocoel theory the coelom is the expanded cavity of a gonad. Sometimes called Bergh's theory

nephrocoel theory the ancestral coelom is the expanded inner end of a nephridium. Sometimes called Ziegler's theory

chiasmatype theory all cross-overs occur at all chiasmata

corm theory metamerically segmented animals are in reality a chain of zooids resulting from asexual reproduction

tunica-corpus theory a plant body derives from an apical meristem divided into a peripheral tunica and a central corpus

cosmobiotic theory life evolved elsewhere in the cosmos and was transferred to the planet Earth

monoclimax theory the type of organisms in a given community are determined by climate

polyclimax theory there may be several climax communities, each influenced by some feature of the climate

metabolic differentiation theory the obsolete theory that sex determination depends on the metabolic rate of the embryo

ergodic theory different places at one time may exhibit the same sequence that one place will exhibit in successive times

field theory an organism, more particularly an embryo, is divided into or surrounded by, areas which may mutually interact without phycical contact

fission theory = corm theory

gene theory heritable characters are controlled by pairs of particulate units (genes)

gynaeocentric theory in plants, the female sex plays a more important role in evolution than the male sex

Hatschek's theory all invertebrates having a trochophore larvae are descendants of a postulated form, called Trochozoa, which was, in effect, an adult trochophore

histogen theory a plant body developed from the division of a mass of meristematic tissue itself divided into groups of cells, which gave rise to the different plant systems

homologous theory alternate generations of plants are not distinct entities (*cf.* antithetic theory), but have arisen one from the other

imbibition theory water ascends plants by a chemical process in the cell walls, rather than by transport through vessels

lock and key theory 1 (*see also* lock and key theory 2) the specificity of an enzyme for a substrate is due to an interlocking molecular configuration

lock and key theory 2 (*see also* lock and key theory 1) the exact structural interlocking of insect genitalia causes those with even minute variations to be incapable of copulation

mechanistic theory all animal activity is controlled objectively by stimuli received from the environment

cyclomeric theory metameric segmentation arose in consequence of the rearrangement of radially symmetrical parts

pseudometameric theory metameric segmentation arose in consequence of the impedance to free movement occasioned by periodic swelling of lobed gonads

micromeristic theory = particulate theory

binuclearity theory (*obs.*) there are two kinds of "chromatin." The vegetative, as in the macronucleus of Protozoa and the generative as in the micronucleus

panspermism theory spores are to be found everywhere in the atmosphere

paranotal theory the insect wing arose as a lateral expansion of the exoskeleton of the thorax

particulate theory heritable characters are controlled by particles scattered through all the cytoplasm of all the cells of the body. The chromosome theory is a refined descendant of this in which the particles or genes are confined for the most part to the chromosomes

precocity theory the difference between meiosis and mitosis results from the precocious condensation of chromosomes in meiosis

preformation theory (*obs.*) the egg or sperm contains an adult in miniature and this adult develops purely by a process of "unfolding." The antithesis is epigenesis

presence and absence theory (*obs.*) the recessive condition in inheritance indicates the absence of the gene involved

Wagner's separation theory a new race can only be formed when there is marked geographic separation from the parent species

spumoid theory = alveolar theory

telome theory a frond is a huge and flattened cluster of stems, the terminal veins of the fronds being telomes and the intercalary veins mesomes

tritubercular theory the pattern of three cusps on the molars represents the original mammalian pattern from which all later patterns are derived

vitalist theory originally the postulate that fermentation and putrefaction were due to organisms. Often nowadays transferred to the proponents of vitalism (*q.v.*)

Willis's theory the more widely distributed an organism is in a given area, the longer it has been resident in that area

Ziegler's theory = nephrocoel theory

-ther- *comb. form* confused from three *Gr.* roots and therefore variously meaning "summer," "mammal" and to "hunt," the last either in the sense of "seek" or of the activities of a Nimrod

Theraphosidae a family of large Aranaea commonly called tarantulas in the United States. Many possess a lyra

Theraponidae a family of percoid Acanthopterygii readily distinguished by their numerous horizontal stripes from which the name tiger fish is derived

-there- *see* -ther-

anoplothere an unarmed mammal, usually referring to hornless ungulates

Therevidae a small family of orthoraphous brachyceran Diptera with a hairy body and a long pointed abdomen from which the popular name stiletto fly derives

Theria a term used to describe all mammals except the Prototheria

Theridae a very large family of dipneumonomorph aranaen arthropods distinguished by possessing 8 eyes, a large abdomen, three claws on each leg, and combs on the tarsi of the 4th pair of legs. The majority of the large nest building spiders are in this family

-therm- *comb. form* meaning "heat"

-therm *comb. suffix* used to describe organisms in relation to their reactions to heat. The adjectival derivatives ending in -thermic and -thermous are equally common, but are not separately recorded but a few are entered under -therous

ectotherm = poikilotherm

endotherm = homoiotherm

eurytherm an organism capable of maintaining itself over a wide temperature range

hekistotherm an organism living above or beyond the tree line and frequently in areas of heavy snow

heliotherm an organism that warms its body in the direct rays of the sun

homoiotherm an animal that maintains a constant body temperature in spite of variations in the temperature of the environment (*cf.* poikilotherm)

megatherm an organism requiring a very high temperature

hydromegatherm an organism dwelling in tropical rain forests

mesotherm an organism of temperate zones

microtherm said of organisms requiring a minimum of heat

poikilotherm an animal whose body temperature varies according to the temperature of the environment (*cf.* homoiotherm)

stenotherm an organism capable of tolerating only a narrow temperature range

thigmotherm an animal that draws heat to its body from a warmed object in the environment

xerotherm an organism capable of withstanding both drought and heat

thermad a plant of hot springs

hypothermia the condition of a homoiothermous animal which has a lower than normal temperature

-thermic *see* under -therm, from which this *adj. suffix* is derived

-thermous *see* under -therm

-therous pertaining to -ther- in any of its meanings

brachytherous pertaining to short summers

xerotherous the condition of an organism adapted to a dry summer

thesis literally a proposition that is set down

hypothesis a hypothesis is properly a postulate which may, bolstered by facts, turn into a theory and later a law. In *biol.* hypothesis, theory and law are so confused in the literature that only the last two are used as entries. The "recapitulation hypothesis" equally common in the literature as "theory" and "law" must be sought under the latter and the "monoclimax hypothesis" is entered under "theory"

synthesis the production of substances from other, usually simpler, compounds

biosynthesis synthesis with the aid of living organisms

photosynthesis the utilization of photon energy by plants in the synthesis of carbohydrates

antithetic (*see also* theory) said of those alternations of generations in which the alternates are fundamentally different in appearance

THFA tetrahydrofolate

thi a microbial genetic marker indicating a requirement for thiamine. One gene has been mapped at 78.0 mins. for *Escherichia coli*

thi-1 - thi-5 *see* [thiamine-1] to [-5]

thi-1, -4, -8 -10 *see* [thiamine-1] - [-4], [-8] - [-10]

thi-1o *see* [Modifier of thi-1]

thiaminase I an enzyme catalyzing the synthesis of thiazole and heteropyrithiamine from thiamine and pyridine

thiaminase II an enzyme catalyzing the hydrolysis of thiamine to yield 2-methyl-4-amino-5-hydroxy-methyl-pyrimidine and 4-methyl-5-(2-hydroxy-ethyl)-thiazole

thiamine = vitamin B_1

[thiamine-1] *either* a mutant (gene symbol *thi-1*) on linkage group I of *Neurospora crassa*. The phenotypic expression is a thiamine requirement *or* a mutant (gene symbol *thi-1*) in linkage group VIII of *Chlamydomonas reinhardi*. The phenotypic expression is a requirement of thiamine and an inability to use thiamine derivatives or precursors

[thiamine-2] either a mutant (gene symbol *thi-2*) found in linkage group III of *Neurospora crassa*. The phenotypic expression is a requirement for thiamine. Thiamine derivatives cannot be used *or* a mutant (gene symbol *thi-2*) in linkage group III of *Chlamydomonas reinhardi*. The phenotypic expression is a requirement for thiamine plus pyrimidine or thiazole plus pyrimidine

[thiamine-3] *either* a mutant (gene symbol *thi-3*) found in linkage group VII of *Neurospora crassa*. The phenotypic expression is a requirement for thiamine *or* a mutant (gene symbol *thi-3*) in linkage group I of *Chlamydomonas reinhardi*. The phenotypic expression is a requirement for either thiamine or thiazole

[thiamine-4] *either* a mutant (gene symbol *thi-4*) in linkage group III of *Neurospora crassa*. The phenotypic expression is a requirement for thiamine *or* a mutant (gene symbol *thi-4*) in linkage group IV of *Chlamydomonas reinhardi*. The phenotypic expression is a requirement for either thiamine or thiazole [thiamine-5] a mutant (gene symbol *thi-5*) on linkage group IV of *Neurospora crassa*. The phenotypic expression is a requirement for thiamine

[thiamine-8] a mutant (gene symbol *thi-8*) in linkage group V of *Chlamydomonas reinhardi*. The phenotypic expression is a requirement for either thiamine or pyrimidine

[thiamine-9] a mutant (gene symbol *thi-9*) in linkage group II of *Chlamydomonas reinhardi*. The phenotypic expression is a requirement for thiamine

[thiamine-10] a mutant (gene symbol *thi-10*) in linkage group VI of *Chlamydomonas reinhardi*. The phenotypic expression is a requirement for thiamine

[thick] a mutant (gene symbol *tk*) mapped at 55.3 on chromosome II of *Drosophila melanogaster*. The name refers to the thickened legs and tarsae of the phenotype

[thick vein] a mutant (gene symbol *thv*) mapped at 49.7± on the X chromosome of *Drosophila melanogaster*. The phenotypic expression are thickened wing veins

[thick-veins] a mutant (gene symbol *tkv*) mapped at 16.0± on chromosome II of *Drosophila melanogaster*. The name is descriptive of the wing veins of the phenotype

oyster **thief** *N.Z.* the phaeophyte alga *Colpomenia sinuosa*

-thin- *comb. form* meaning a "heap of sand"

[thin bristle] a mutant (gene symbol *thb*) mapped at 47.6± on the X chromosome of *Drosophila melanogaster*. The name is descriptive of the phenotype

thinic pertaining to sand dunes and communities dwelling on them

[thin-macros] a mutant (gene symbol *thm*) mapped at 48.9 on the X chromosome of *Drosophila melanogaster*. The phenotypic expression is thin macrochaetae

Thinocoridae a small family of charadriiform birds containing the seed-snipes. They are distinguished from other charadriiforms by the very short beak with long slit nostrils having a protective operculum above them

Thinocorythidae = Thinocoridae

Thiobacteriaceae a family of pseudomonodale Schizomycetes containing the colorless, sulphur bacteria. They are mostly coccoid or rod shaped and derive their energy from the oxidation of sulphur compounds with the production of free sulphur either intra or extra-cellularly

thiolase an obsolete name for some acyltransferases

acetoacetyl-CoA thiolase = acetyl-CoA acetyltransase

3-ketoacyl-CoA thiolase = acetyl-CoA acyltransferase

thioltransacetylase A = lipoate acetyltransferase

L-cystathionine diamino dicarboxylic acid $C_7H_{14}O_4N_2S$ a transitory amino acid formed as an intermediate step in the conversion of methionine to cysteine

L-ethionine an amino acid ethyl homologue of methionine, $C_2H_5CH_2CH_2CHNH_2COOH$. In some instances it inhibits the synthesis of methionine

Thiorhodaceae a family of pseudomonodale schizomycetes commonly called the red sulphur-bacteria. They exist only in well lighted environments containing sulphides and are distinguished by the production of green, yellow, and red pigments, as well as droplets of elemental sulphur

thistle 1 (*see also* thistle 2, aphis, tail) any of very many compositaceous plants, usually with an upright structure, inverted bell-shaped head of flowers and prickly leaves.

Barnaby's thistle *Centaura solstitialis*

blessed thistle *Cnicus benedictus*

bull thistle *Cirsium pumilum*

carline thistle *Carlina vulgaris*

common thistle *U.S. Cirsium vulgare*

cotton thistle = Scotch thistle

globe thistle any of many species of the genus Echinops

golden thistle *Scolymus hispanicus*

holy thistle *Silybum marianum*

milk thistle *Brit.* any of several species of the genus Sonchus (= sow-thistle) and also *Carduus marianus*

musk thistle *Carduus nutans*

pasture thistle *Cirsium pumilum*

plume thistle *Brit.* any of several species of the genus Cnicus (= *U.S.* blessed thistle)

plumeless thistle any of several species of the genus Carduus

St. Mary's thistle = holy thistle

Scotch thistle *Onopordon acanthium*

sow-thistle *Brit.* any of several species of the genus Sonchus (= milk thistle)

star thistle *U.S.* any of several species of the genus Centaurea. *Brit.* C. calcitrapa.

swamp thistle *Cirsium muticum*

welted thistle *Carduus crispus*

yellow thistle *Cirsium horridulum*

thistle 2 (*see also* thistle 1) some plants of other families having thistle-like characteristics

blue thistle the boraginaceous herb *Echium vulgare*

hedgehog thistle any cactus of the genus Echinocactus

Kansas thistle the solanaceous herb *Solanum rostratum*

Russian thistle the chenopodiaceous herb *Salsola kali*

-thla- *comb. form* meaning "flattened"

thm *see* [thin-macros]

Thomisidae a family of dipneumonomorph araneid arthropods, distinguished by their usually short, broad flattened body, unusually wide behind, which gives them the name of crab spider, a term reinforced by their sideways gait

thoracic pertaining to the thorax

Thoracica a suborder of cirripede Crustacea containing all those forms commonly called barnacles. Distinguished from the Abdominalia by possessing a shell and from the Rhizocephala by their nonparasitic habit

Thoracostei an order of Osteichthyes containing the families Gasterosteidae and Aulorhynchidae.

thorax that portion of the body which lies between the head and abdomen

cephalothorax the anterior body of many arthropods consisting of the head fused with one or more of the thoracic segments

endothorax the apodeme of the thorax

gnathothorax either cephalothorax or, in insect larvae, the anterior region before the head and thorax have become differentiated

mesothorax the center of the three segments of the insect thorax

metathorax the posterior of the three segments of the insect thorax

microthorax the insect prothorax when it is conspicuously smaller than the mesothorax and specifically when it serves as a neck

prothorax the first of the three thoracic segments in insects

pterothorax the aliferous thoracic segments in insects

synthorax the meso- and metathorax of insects when these are fused together

thorn 1 (*see also* thorn 2, 3, back, bill, bird, tail) a sharp spine protruding from an organism, though without qualification, the word usually applies to the woody thorns of plants

thorn 2 (*see also* thorn 1, 3) any of numerous thorny plants

black thorn *U.S.* the shrubby rosaceous tree *Prunus spinosa. Brit. P. communis*

box thorn any of several species of shrubby vine of the solanaceous genus Lycium

buckthorn (*see also* aphis) any of numerous species of the rhamnaceous shrubs, many of the genus *Rhamnus*, but most particularly *E.U.S. R. cathartica* and *W.U.S. Ceanothus soriedatus*. In *S.U.S.* the term is applied to the sapotaceous shrub *Bumelia lanuginosa*

alder buckthorn *Brit. R. frangula*

false buckthorn *U.S.* the sapotaceous shrub *Bumelia lanuginosa*

buffalo thorn the annonaceous shrub *Artobotrys suaveolens*

camel's thorn the leguminous shrub *Alhagi camelorum*

Christ's thorn the apocynaceous shrub *Carissa carandas*

cock's spur thorn the rosaceous shrub *Crataegus crusgalli*

everlasting thorn = firethorn

firethorn any of several rosaceous shrubs of the genus Pyracantha but particularly *P. coccinea*

hart's thorn the rhamnaceous shrub *Rhamnus cathartica*

hawthorn (*see also* bug) *U.S.* any of many species of the rosaceous genus Crataegus. *Brit. C. oxyacantha*

Indian hawthorn the rosaceous shrub *Rathioletsis indica*

water hawthorn the aponogetonaceous aquatic herb *Aponogeton distachyus*

indigo thorn the leguminous tree *Parosela spinosa*

Jerusalem thorn either the leguminous tree *Parkinsonia aculeata* or the rhamnaceous shrubby tree *Paliurus spinachristi*

kangaroo thorn the leguminous shrub *Acacia armata*

maythorn the rosaceous shrub *Crataegus oxyacantha*

sallow-thorn *Brit.* the polygonaceous herb *Hippothae rhamnoides*

way-thorn = buckthorn

white thorn = maythorn *Brit.*

thorn 3 (*see also* thorn 1, 2) *Brit.* any of numerous species of geometrid lepidopteran insect of the genera Ennomos and Selenia

thr a yeast genetic marker indicating either a requirement for threonine or a requirement for threonine and methionine. Also a bacterial mutant indicating a requirement for threonine, mapped at 0.0 mins. for *Escherichia coli*

thr-1 -3 *see* [Threonine-1]–[-3]

Thraciidae a family of pelecypod mullusks commonly called perforated clams. They are distinguished by the very unequal valves of the gaping shell with prominent beaks, one of which is perforated

thrasher any of many mimid birds mostly of the genera Toxostoma and, occasionally, other birds

Bendire's thrasher *T. bendirei*

curve-billed thrasher *T. curvirostre*

long-billed thrasher *T. longirostre*

black thrasher *W.I.* = pearly-eyed thrasher

blue thrasher *W.I.* the turdid *Mimocichla plumbea* (= red legged thrush)

red-legged blue thrasher *W.I.* = blue thrasher *W.I.*

scaly-breasted thrasher the mimid *Allenia fusca*

brown thrasher properly *T. rufum. W.I. Mimus gundlachii* (= Bahama mockingbird)

California thrasher *T. redivivum*

Cozumel thrasher *T. gultalum*

crissal thrasher *T. dorsale*

English thrasher *W.I.* the mimid *Mimus polyglottos* (= mockingbird)

pearly-eyed thrasher *Margarops fuscatus*

grey thrasher *T. cinerum*

Le Conte's thrasher *T. lecontei*

ocellated thrasher *T. ocellatum*

sage thrasher *Oreoscoptes montanus*

Socorro thrasher *Mimodes graysoni*

Spanish thrasher *W.I. Mimus gundlachii* (= Bahama mockingbird)

Thraupidae a large family of passeriform birds containing the forms commonly called tanagers. They are distinguished by the rather heavy conical notched beak which may even be "toothsome" in some species

thread a very long, thin, object

conjunctive thread = spindle fibre

dew-thread the droseraceous insectivorous plant *Drosera filiformis*

gold thread *U.S.* any of several ranunculaceous herbs of the genus Coptis

nuclear thread = spindle fiber

[thread] a mutant (gene symbol *th*) mapped at 43.2 on chromosome III of *Drosophila melanogaster*. The aristae of the phenotype are threadlike

[thread bristle] a mutant (gene symbol *trb*) mapped at $37.0\pm$ on the X chromosome of *Drosophila melanogaster*. As [tiny-like], the phenotypic expression involves short fine bristles

threonine α-amino-β-hydroxybuteric acid. $CH^3CHOH \cdot CHNH_2COOH$. A widely distributed amino acid necessary to the nutrition of the rat

[threonine-1] a mutant (gene symbol *thr-l*) on linkage group VII of *Neurospora crassa*. The phenotypic expression is a requirement for threonine

[Threonine-2] a mutant (gene symbol *Thr-2*) in linkage group II of *Neurospora crassa*. The phenotypic expression is a requirement for threonine

[Threonine-3] a mutant (gene symbol *Thr-3* in linkage group II of *Neurospora crassa*. The phenotypic expression is the same as [Threonine-2]

thresher = thresher shark

bigeye thresher the lamnid fish *Alopias superciliosus*

extinction **threshold** the number below which a given population cannot fall without becoming extinct

Threskiornithidae that family of ciconiiform birds which contains the ibises. They are distinguished by their long neck and have legs shorter than those found in most families of the order

thrift any of numerous species of herb of the plumbaginaceous genus Armeria. Usually, without qualification *A. maritima* and its near allies

prickly thrift any of several species of the plumbaginaceous herb Acantholimon

Thripidae a family of terebrantian thysanopteran insects distinguished by a downwardly directed ovipositor and distinguished from the closely allied Heterothripidae by the 6 to 9-segmented antennae. They are commonly called thrips or common thrips

thrips a term properly applied to thysanopteran insects of the family Thripidae but usually extended to embrace the whole order. In *Brit*. thrip, rather than thrips, is sometimes used as an imagined *sing*. of the *L*. thrips

banded thrips popular name of phlaelothripid thysanoptera

red-banded thrips *U.S.* the thripid *Selenothrips rubrocinctus*

bean thrips *U.S.* the thripid *Caliothrips fasciatus*

broad-winged thrips = banded thrips

flax thrips the thripid *Thrips lini*

lily bulb thrips *U.S.* the phlaeothripid *Liothrips vaneeckei*

camphor trips *U.S.* the phlaeothripid *Liothrips floridensis*

chrysanthemum thrips *U.S.* the thripid *Thrips nigropilosus*

citrus thrips *U.S.* the thripid *Scirtothrips citri*

common thrips popular name of thripids

composite thrips *U.S.* the thripid *Microcepkalothrips abdominalis*

corn thrips the thripid *Limothrips cerealium*

flower thrips *U.S.* = wheat thrips

gladiolus thrips *U.S.* the thripid *Taeniothrips simplex*

grain thrips *U.S.* the thripid *Limothrips cerealium*

grass thrips *U.S.* the thripid *Anaphothrips obscurus*

greenhouse thrips *U.S.* the thripid *Heliothrips haemorrhoidalis*

banded greenhouse thrips *U.S.* the thripid *Hercinothrips femoralis*

Cuban-laurel thrips *U.S.* the phlaeothripid *Gynaikothrips ficorum*

iris thrips *U.S.* the thripid *Iridothrips iridis*

large legged thrips popular name of merothripids

mullein thrips *U.S.* the phlaeothripid *Haplothrips verbasci*

onion thrips *U.S.* the thripid *Thrips tabaci*

orange thrips the thripid *Euthrips citri*

pea thrips the thripid *Kakothrips robustus*

pear thrips *U.S.* the thripid *Taeniothrips inconsequens*

privet thrips *U.S.* the thripid *Dendrothrips ornatus*

sugarcane thrips *U.S.* the thripid *Thrips saccharoni*

tobacco thrips *U.S.* the thripid *Frankiniella fusca*

wheat thrips the thripid *Frankliniellla tritici*

throat 1 (*see also* throat 2, 3, fly) the posterior exit from the mouth

throat 2 (*see also* throat 1, 3) the aperture of a gamopetalous, tubular calyx or corolla

throat 3 (*see also* throat 1, 2, wort) any organism distinguished by its throat

barbthroat any of several trochilid birds of the genus Threnetes

bluethroat the turdid bird *Luscinia svecica*

cutthroat the ploceid bird *Amandina fasciata* (*cf*. finch)

firethroat the trochilid bird *Chlorostigma herrania* (cf. thornbill)

goldenthroat any of several trochilid birds of the genus Polytmus

redthroat *Austral*. the malurine *Pyrrholaemus brunneus*

ruby-throat *W.I.* the trochilid bird *Eulampis jularis* (= purple-throated carib)

spotthroat the turdid bird *Modulatrix stictigula*

starthroat any of several species of trochilid bird of the genus Heliomaster

long-billed starthroat *H. longirostris*

plain-capped starthroat *H. constantii*

whitethroat properly any of several sylviid birds of the genus Sylvia; without qualification, *S. communis W.I.* the anatine bird *Anas bahamensis* (= Bahama pintail)

lesser whitethroat *S. curruca*

yellowthroat any of several species of parulid birds of the genus Geothlypis, without qualification, *G. trichas*

Bahama yellowthroat *G. rostrata*

Belding's yellowthroat = peninsular yellowthroat

Chapala yellowthroat *G. chapalensis*

common yellowthroat *G. trichas*

grey-crowned yellowthroat *G. poliocephala*

yellow-crowned yellowthroat *G. flavovelata*

hooded yellowthroat *G. nelsoni*

peninsular yellowthroat *G. beldingi*

black-polled yellowthroat *G. speciosa*

thrombin an enzyme catalyzing the conversion of fibrinogen into fibrin

Throscidae a family of coleopteran insects closely allied to the Elateridae but distinguished from them by the fact that the prosternal spine is firmly attached to the mesosternum and they are therefore unable to jump

throstle = thrush

thrum the stipe of an anther or sometimes the anther itself

sarothrum the metatarsus of insects when it is adapted to gathering pollen

thrumb = thrum

thrush (*see also* akalat, tanager) any of numerous passerine birds, mostly of the family Turdidae to which family belong all those listed below unless another family is specifically indicated. *Brit*., without qualification, *T. philomelos* (= song thrush) *W.I.* without qualification = grand cayman thrush, but the word is also applied to several mimids particularly *Allenis fusca* (= scaly-breasted thrasher)

African thrush *Turdus pelios*

ant-thrush 1 (*see also* ant-thrush 2) any of several species of formicariids, mostly of the genus Formicarius (*cf*. antbird)

ant-thrush 2 (see also ant-thrush 1) any of several African turdids of the genus Neocossyphus

red-tailed ant-thrush *N. rufus*

white-tailed ant-thrush *N. poensis*

Aztec thrush *Ridgwayia pinicola*

babbling-thrush = babbler

olive-backed thrush = Swainson's thrush

black-billed thrush *W.I.* the mimid *Allenia fusca* (= scaly-breasted thrasher)

grey-cheeked thrush *Catharus minimus*

Chinese thrush the timaliid *Garrulax canorus*

white-chinned thrush *Turdus aurantius*

cocoa thrush *Turdus fumigatus*

golden-crowned thrush *U.S.* the parulid *Seiurus aurocapillus* (= ovenbird)

bare-eyed thrush *W.I. Turdus nudigensis Afr. Turdus tephronotus*

wall-eyed thrush *W.I.* the mimid *Margarops fuscatus* (= pearly-eyed thrasher)

white-eyed thrush *Turdus jamaicensis*

forest thrush *Cichlherminia l'herminieri*

Grand Cayman thrush *Mimocichla ravida*

greythrush the muscicapid *Colluricincla harmonica*

ground-thrush any of several species of turdid mostly of the genus Geocichla, and also the popular hame of pittids. *Austral.*, without qualification, *Zoothera dauma*

Abyssinian ground-thrush *G. piaggiae*

spotted ground-thrush *Psophocichla guttata*

Hawaiian thrush *Phaeornis obscurus* (= omao)

hermit thrush *Catharus guttatus*

North Island thrush *N.Z.* the turnagrid *Turnagra capensis* (= piopio)

South Island thrush *N.Z.* = North Island thrush

Kurrichane thrush *Turdus libonyanus*

La Selle thrush *Turdus swalesi*

laughing-thrush = babbler, particularly, but not exclusively, those of the genus Garrulax

red-legged thrush *Mimocichla plumbea*

yellow-legged thrush *W.I.* = forest thrush

missel thrush = mistle thrush

mistle thrush *Turdus viscivorus*

mistletoe thrush = mistle thrush

black-capped mockingthrush *Donacobius atricapillus*

mountain-thrush any of several species of the genus Zoothera

New Zealand thrush the turnagrid *Turnagra capensis* (= piopio)

nightingale-thrush any of several species of the turdid genus Catharus

orange-billed nightingale-thrush *C. aurantiirostris*

ruddy-capped nightingale-thrush *C. frantzii*

black-headed nightingale-thrush *C. mexicanus*

russet nightingale-thrush *C. occidentalis*

spotted nightingale-thrush *C. dryas*

nun-thrush any of several species of timaliid birds of the genus Alcippe (*cf.* babbler)

olive thrush *Turdus olivaceus*

quail-thrush any of several species of timaliid birds of the genus Cinclosoma

chestnut-breasted quail-thrush *C. castaneothorax*

chestnut quail-thrush *C. castanonotum*

cinnamon quail-thrush *C. cinnamomeum*

spotted quail-thrush *C. punctatum*

rock-thrush any of several species of the genus Monticola. *Brit.*, without qualification, *M. saxatilis*

blue rock-thrush *M. solitaria*

Cape rock-thrush *M. rupestris*

little rock-thrush *M. rufocinerea*

mottled rock-thrush *M. angolensis*

rufous thrush the muscicapid *Colluricincla megarhyncha*

ground-scraper thrush *Psophocicla litsipsirupa*

shrike-thrush *Austral.* any of several muscicipids of the genus Colluricincla

song thrush *Turdus philomelos*

Swainson's thrush *Catharus ustulatus*

trembling thrush *W.I.* the mimid *Cinclocerthia ruficauda* (= trembler)

varied thrush *Ixoreus naevius*

waterthrush *U.S.* either of two species of parulid bird of the genus Seiurus

Louisiana waterthrush *S. motacilla*

northern waterthrush *S. noveboracensis*

western thrush the muscicapid *Colluricincla rufiventris*

whistling-thrush any of several species of timaliids of the genus *Myiophoneus*

White's thrush *Zoothera dauma*

willow thrush *Catharus fuscescens* (= verry)

wood thrush *Hylocichla mustelina*

wrenthrush the zeledoniid *Zeledonia coronata*

Thujoideae a subfamily of Cupressaceae

thumb 1 (*see also* thumb 2, 3) the medial of the five terminal digits—the other four are fingers—of the primate limb

thumb 2 (*see also* thumb 1, 3) thumb-shaped animals

miller's thumb any of several cottid fishes of the genus Cottus *Brit.* *C. gobio*

thumb 3 (*see also* thumb 1, 2) in compound names of plants

hog thumb the olacaceous tree *Ximenia americana* and its edible fruit

lady's thumb *U.S.* the polygonaceous herb *Polygonum persicaria*

tear thumb any of several species of polygonaceous herbs of the genus Polygonum particularly those of the subgenus Echinocaulon

Thurniaceae a small family of farinose monocotyledonous angiosperms

tkv see [thick vein]

thy a bacterial genetic marker having a character affecting the activity of thymidylate synthetase, mapped at 61.0 mins. for *Escherichia coli*. Also a bacterial mutant indicating a requirement for thiamine and a postulated yeast genetic marker for the same

Thyatiridae a small group of drepanoid lepidopteran insects

Thylacininae a subfamily of dasyurid mursupials containing the pouched wolf (*q.v.*)

Thymallidae a monotypic family erected to contain the grayling (*q.v.*) by those who consider that its smaller mouth and larger lateral scales should cause it to be removed from the Salmonidae

thyme (*see also* mother) any of several species of labiateous herbs or low growing shrubs mostly of the genus Thymus

basil thyme either the labiateous herb *Satureia calamintha* or *Brit.* the labiateous herb *Calamintha acinos*

common thyme *T. vulgaris*

creeping thyme *T. serpyllum*

lemon thyme *Hort. var.* of *T. serpyllum*

water thyme the hydrocharitaceous aquatic herb *Elodea canadensis*

Thymelaeaceae that family of myrtiflorous dicotyledonous angiosperms which contains, *inter alia*, the leather woods and the tapir bushes. The family is distinguished by the uniseriate gamosepalous perianth

Thymelaeineae a suborder of myrtiflorous dicotyledonous angiosperms containing the families Geissolomataceae, Pennaeaceae, Oliniaceae, Thymelaeaceae and Elaeagnaceae

thymus *see* gland

Thynnidae a family of scolioidean apocritan hymenopteran insects usually included with the Tiphiidae

-thyr- *comb. form* meaning either "door" or in the shape of an "oblong shield"

Thyrididae a family of pyraloidoid lepidopteran insects commonly called window-winged or clear-wing moths by reason of the absence of scales on part or all of their wings (*cf.* Sphingidae)

nothothyrium a notch, through which the pedicel protrudes in the palintrope of the dorsal valve of a brachiopod shell

-thyro- *see* -thyr-

thyroid (*see also* gland) shield shaped

Thyronomyidae a family of hystricomorph rodents erected to contain the *Afr.* cane rats

thyroxine the principal hormone secreted by the thyroid gland and a primary controller of metabolic rate

L-thyroxine 3,5,3',5'-tetraiodothyronine. An amino

acid synthesized in the thyroid gland of higher animals but also isolated from many lower forms. Best known for its hormonal effects in growth and metamorphosis

thyrse an inflorescence resembling the conventional Bacchic staff with a racemose main axis and cymose secondary axes. The lilac is a typical example

Thysanidae a small family of chalcidoid apocritan Hymenoptera

Thysanoptera an order of insects containing the forms commonly called thrips. The wings, in those forms that possess them, are fringed with long hairs. The two jointed tarsi are swollen at the tip

Thysanura an order of primitively wingless insects containing the forms commonly called bristle-tails. They differ from the Collembola in the possession of filiform many-jointed cerci and nine abdominal segments

ti *N.Z.* the large liliaceous tree *Cordyline australis* (= cabbage tree)

ti *see* [tarsi irregular], [tiny], or [tipsy]

tibia (*see also* bone) that segment of the arthropod limb which precedes the tarsus

tichicro *W.I.* the fringillid *Ammodramus savannarum* (= grasshopper sparrow)

tick 1 (*see also* tick 2, 3, bird) popular name of ixioid acarine arachnids. All are ectoparasites and many are vectors of disease

  bird tick *U.S.* the ixodid *Haemaphysalis chordeilis*

  castor bean tick the ixodid *Ixodes scapularis* (= black-legged tick)

  cattle tick *U.S.* the ixodid *Boophilus annulatus*

  Texas cattle tick the ixodid *Margaropus annulatus*

  cayenne tick *U.S.* the oxidid *Amblyomma cajennense*

  Gulf coast tick *U.S.* the ixodid *Amblyomma maculatum*

    Pacific coast tick *U.S.* the ixodid *Dermacentor occidentalis*

  American dog-tick *U.S.* the ixodid *Dermacentor variabilis*

    brown dog-tick *U.S.* the ixodid *Rhipicephalus sanguineus*

  ear tick *U.S.* the argasid *Otobius megnini*

  relapsing-fever tick *U.S.* the argasid *Ornithodoros turicata*

    spotted-fever tick the ixodid *Dermacentor andersoni U.S.*

  fowl tick *U.S.* the argasid *Argas persicus*

  tropical horse tick *U.S.* the ixodid *Anocentor nitens*

  black-legged tick *U.S.* = castor bean tick

  rabbit tick *U.S.* the ixodid *Haemaphysalis leporispalustris*

  rotund tick *U.S.* the ixodid *Ixodes kingi*

  lone star tick *U.S.* the ixodid *Amblyomma americanum*

  gopher-tortoise tick *U.S.* the ixodid *Amblyomma tuberculatum*

  winter tick *U.S.* the ixodid *Dermacentor albipictus*

  Rocky Mountain wood tick *U.S.* the ixodid *Dermacentor andersoni*

tick 2 (*see also* tick 1, 3) some tick-like insects

  bat tick = bat fly

  sheep tick the hippoboscid dipteran insect *Melophagus ovinus* (= ked)

tick 3 (*see also* tick 1, 2, clover, trifoil, seed, weed) in popular names of birds

  beggar's tick *U.S.* any of several species of leguminous herb of the genus Desmodium and the com-

positaceous herbs *Bidens connata, B. vulgaris* and *B. frondosa* (= stick tight)

tide the cyclic lunar displacement of waters and the waters themselves

  red tide waters infested by red-pigmented dinoflagellates particularly those of the genus Gymnobinium

tiding term for a group of magpies

tieke *N.Z.* the callaeid bird *Creadion carunculatus* (= saddle back)

leaf tier any of several lepidopteran larvae that tie leaves together with their silk

  celery-leaf tier *U.S.* the larva of the pyraustid *Oeobia rubigalis*

  greenhouse leaf tier *U.S.* = celery leaf tier

  omnivorous leaf tier *U.S.* the larva of the tortricid *Cnephasia longana*

tiercel = tercel

tiger 1 (*see also* tiger 2, bittern, fish, flower, mosquito, shark) any of several species of felid carnivore of the genus Felis. Without qualification invariably *F. tigris*

  sabre-toothed tiger any extinct machaerodont carnivore

tiger 2 (*see also* tiger 1, swallowtail) *I.P.* and *Austral.* any of numerous species of danaid lepidopteran insect of the genus Danais

  blue tiger *Austral. D. melissa*

stick-tight the compositaceous herbs *Bidens connata, B. vulgaris* and *B. frondosa* (= beggar's ticks)

Tiliaceae that family of malvalous dicotyledons which contains the linden and basswood trees. The family is distinguished from the closely allied Elaeocarpaceae by the absence of hairs on the petals

tiller a supplementary plant arising from the base of the main plant, particularly in cultivated grains

[tilt] a mutant (gene symbol *tt*) mapped at 40.0 on chromosome III of *Drosophila melanogaster*. The phenotypic expression is a widely spread, warped wing

[tilted head] a mutant (gene symbol *th*) in linkage group XIII of the mouse. The name is descriptive of the phenotypic expression

Timaliidae a very large family of passeriform birds commonly called babblers. The family contains many widely divergent forms connected only by details of their bony structure

timbal the stretched membrane on which cicadas drum

timber (*see also* beetle) commercially valuable trees

generation **time** the period between the sexual maturity of one organism and the sexual maturity of its offspring. The term is also used in bacteriology to denote the length of time required to double the number of bacteria in a culture

Timeliidae = Timaliidae

timothy (*see also* bilbug) *U.S.* any of several species of grass of the genus Phleum

timucu the belonid fish *Strongylura timucu*

Tinaegeriidae = Heliodinidae

Tinamidae that family of tinamiform birds which contains the tinamous. They are neotropical birds with the tail hidden by the wings and contour feathers

Tinamiformes a group of South American birds commonly called tinamous. Their taxonomic position is uncertain since they combine a strong power of flight with a paleognathous type of skull

tinamou popular name of tinamid birds

  slaty-breasted tinamou *Crypturellus boucardi*

  great tinamou *Tinamus major*

  little tinamou *Crypturellus soui*

rufescent tinamou *Crypturellus cinnamomeus*

tinc-tinc the sylviid bird *Cisticola textrix* (= pinc-pinc)

tine 1 (*see also* tine 2) the terminal pointed sub-branches of an antler

-tine 2 (*see also* tine 1) pertaining to the wall of a pollen grain

    intine the innermost layer of a coat of a pollen grain

    exintine the middle of the three coats of a pollen grain

Tineidae a large family of tineoid lepidopteran insects containing *inter alia* the notorious clothes moths and their relatives, the depredations of which on stored clothing and furs are too well known to need description. The antennae are shorter than the wings and most have erect scales on the head

Tineoidea a superfamily of lepidopteran insects mostly small and including many case-bearers and leaf-miners as well as the clothes moths

Tingididae a family of hemipteran insects commonly called lace bugs because of the sculptured pattern on the upper surface of the body. They are distinguished from the closely allied Piesmidae by the relative broadness of the body and the absence of ocelli

tinkling the icteirid bird *Quiscalus niger* (= Greater Antillean grackle)

tin-tin = tinkling

tinsel *I.P.* any of numerous species of lycaenid lepidopteran insect of the genus Catapoecilma but usually, without qualification, *C. elegans*

[tiny] a mutant (gene symbol *ty*) mapped at 44.5 on the X chromosome of *Drosophila melanogaster*. The name is descriptive of the phenotype in populations of which the females are sterile

[tiny] a mutant (gene symbol *ti*) on linkage group I of *Neurospora crassa*. The name is descriptive of the phenotypic colonies

[tiny-bristloid] a mutant (gene symbol *tbd*) mapped at 25.0± on the X chromosome of *Drosophila melanogaster*. The phenotype is a small fly with moderate sized bristles

[tinychaete] a mutant (gene symbol *tc*) mapped at 51.6± on the X chromosome of *Drosophila melanogaster*. The name is descriptive of the condition of the chaetae of the phenotype

[tiny-like] a mutant (gene symbol *ty-1*) mapped at 36.4 on the X chromosome of *Drosophila melanogaster*. The phenotypic expression is short fine bristles

tiong the Malayan sturnid bird *Cracula religiosa*

tip 1 (*see also* tip 2, 3) in compound names of numerous Lepidoptera

    broom-tip *Brit.* the geometrid *Chesias rufata*

    buff-tip *Brit.* the notodontid *Phalera buciphala*

    chocolate-tip *Brit.* any of three species of notodontid of the genus Pygaera

    crimson tip *I.P.* the pierid *Colotis danae* (*cf.* orange tip)

    beautiful hook-tip *Brit.* the noctuid *Laspeyria flexula*

    lilac tip the pierid *Colotis celimene*

    orange-tip *U.S.* any of several pierids. *Brit.* the pierid *Euchloe cardamines*. *I.P.* any of many species of pierids of the genus Ixias. *S.Afr.* any of numerous pierids of the genus Colotis. Usually, without qualification, *C. evenina*

    purple tip *S.Afr.* any of several species of pierid of the genus Colotis

    scarlet tip *S.Afr.* the pierid *Colotis danae*

    sulphur tip *S.Afr.* the pierid *Colotis agoye*

tip 2 (*see also* tip 1, 3) in compound names of birds

    white tip the trochilid *Urosticte benjamini*

tip 3 (*see also* tip 1, 2) in compound names of plants

    tidy tip any of several species of compositaceous herb of the genus Layia

-tiph *comb. form* meaning "pool" and used by ecologists in the more specific sense of "pond"

tiphad a pond plant

tiphic pertaining to ponds and communities dwelling in them

    ombrotiphic pertaining to temporary pools

Tiphiidae a family of scolioid apocritan hymenopteran insects many hairy and having short legs

tiphium a pond community ⟩

[tipsy] a mutant (gene symbol *ti*) in linkage group VII of the mouse, It is expressed in the phenotype as a rambling uncoordinated gait

Tipulidae a very large family of slender, long-legged nematoceran dipteran insects called craneflies in the *U.S.* and daddylonglegs in *Brit.*

tirad a plant of "badlands" (*cf.* hydrotribophilous)

Tischeriidae a family of small tineioid Lepidoptera with leaf mining larvae

tissue 1 (*see also* tissue 2, 3) an aggregate of animal cells

    adipose tissue an aggregate of fat-gorged cells

    areolar tissue the loose connective tissue between the basement membrane of an epithelium and the underlying tissues

    chondroid tissue a stiffened tissue, in some ways resembling cartilage, in the proboscis and proboscis stalk of some Enteropneusta

    connective tissue those mesodermal tissues which bind and connect other tissues together. Blood, for want of a better classification, is usually regarded as a connective tissue

        periportal connective tissue the dense connective tissue of the lobules of the liver of higher vertebrates

    lymphatic tissue a tissue composed of a stroma of reticular cells supported on fibers and with lymphocytes in the mesh of the stroma

    muscular tissue the contractile elements of the animals' body

    myeloid tissue bone marrow

    nervous tissue those elements of the body which are designed to perceive or originate stimuli or conduct impulses

tissue 2 (*see also* tissue 1, 3) an aggregation of plant cells

    conducting tissue = stigmatoid tissue

    conjunctive tissue ground tissues within the stele

        medullary conjunctive tissue = pith

    connecting tissue (*see also* tissue 1) cells lacking chlorophyll lying adjacent to the veins of leaves and also that which unites the two lobes of an anther

    dot tissue = bothrenchyma

    false tissue the mycelially produced tissues of fungi

    fundamental tissue = ground tissue

    transfusion tissue the tracheidal cells in the leaves of conifers

    generating tissue = meristem

    nascent tissue = meristem

    peripheral tissue any outer tissue but particularly that which gives rise to the root hairs

    primary tissue that formed by the terminal meristems of a plant

    primordial tissue = ground tissue

    secondary tissue that formed by the lateral meristems of a plant

    stigmatoid tissue a strand of stigma-like tissue, connecting the stigma to the interior of the ovary

transmitting tissue = stigmatoid tissue

vascular tissue conducting vessels

ventilating tissue the spongy parenchyma of a leaf

glandular woody tissue the pitted vessels of conifers

tissue 3 (*see also* tissue 1, 2) *Brit.* either of two species of geometrid lepidopteran insect of the genus Triphosa. Without qualification, *T. dubitata* is usually meant

tit 1 (*see also* tit 2, babbler, flycatcher, mouse, shrike) any of numerous small birds, mostly of the family Paridae. In *Brit.*, the term titmouse is synonymous. In *N.Z.* the word used without qualification refers to the muscicapid *Petroica macrocephala*

azure tit *Parus cyanus*

green-backed tit *Parus monticolus*

bearded tit the timaliid *Panurus biarmicus*

black tit *Parus leucomelas*

    southern black tit *Parus niger*

blue tit *Parus caeruleus*

yellow-cheeked tit *Parus xanthogenys*

cinnamon-breasted tit *Parus rufiventris*

    white-breasted tit *Parus albiventris*

bushtit either of two parids of the genus Psaltriparus; without qualification, *P. minimus*

    common bushtit *P. minimus*

    black-eared bushtit *P. melanotis*

coal tit *Parus ater*

cole tit = coal tit

crested tit *Parus cristatus*

    brown-crested tit *Parus dichrous*

dusky tit *Parus funereus*

great tit *Parus major*

grey tit = coal tit

red-headed tit *Aegithalos concinnus*

marsh tit *Parus palustris*

penduline-tit any of several parids of the genus Anthoscopus; without qualification, *A. caroli*

    African penduline-tit *A. caroli*

    mouse-colored penduline-tit *A. musculus*

scrubtit the malurine *Acanthornis magnus*

shriketit any of several muscicapids of the genus Falcunculus, without qualification, *F. frontatus*

    eastern shriketit *F. frontalis*

sombre tit *Parus lugubris*

sultan tit *Melanochlora sultanea*

swamp-tit *Austral.* the sylviid *Acrocephalus australis* (= reed warbler)

long-tailed tit any of several parids of the genus Aegithalos, without qualification, *A. caudatus*

tom-tit *Brit.* = blue tit *Austral.* the malurine *Acanthiza chrysorrhoa* (= yellow-tailed thornbill)

tufted tit *U.S. Parus bicolor* (= tufted titmouse)

varied tit *Parus varius*

willow tit *Parus montanus*

wrentit the chamaeid *Chamaea fasciata*

tit 2 (*see also* tit 1) *I.P.* any of several species of lycaenid lepidopteran insects of the genera Chliaria, Hypolycaena and Zeltus

titi 1 (*see also* titi 2) *N.Z.* the procellariid bird *Pterodrama cookii* (Cook's petrel)

titi 2 (*see also* titi 1) the cyrillaceous trees *Cliftonia monophylla* and *Cyrilla racemiflora,* and the ericaceous tree *Oxydendrum arboreum*

titipounamu *N.Z.* the acanthisittid bird *Acanthisitta chloris* (= rifleman)

titis popular name of a large group of hapaloid primates frequently placed in a separate subfamily, the Callicebinae. They are sometimes placed with the

Ceboidia in spite of the fact that they have claws instead of nails

titling = titlark

titoki *N.Z.* the sapindaceous tree *Alectryon excelsum*

titterel *Brit. obs.* the scolopacid bird *Numenius phaeopus*

tiwakawaka *N.Z.* either of two species of muscicapid bird of the genus Rhipidura (= fantail)

tityra any of several cotingid birds of the genus Tityra

    black-capped tityra *T. inquisitor*

    masked tityra *T. semifasciata*

tk *see* [tail kinks] or [thick]

tkv *see* [thick-veins]

Tm *see* [Pulmonary tumors]

tmc *see* tonomacrochaetes

-tmema- *comb. form* meaning "free"

brachytmema very frequent misspelling of brachytrema

-tmes- *comb. form* meaning "cut"

diatmesis the condition of an amitotic division in which the nucleus divides into two even parts (*cf.* diaspasis)

To *see* [Tortoise] or [Tousled]

toad 1 (*see also* toad 2, bug, crab, fish, flax, lily, shell) the distinction between frog and toad can only be maintained in countries such as *Brit.* where the ranids (frogs) and bufonids (toads) are the only anuran Amphibia. In other parts of the world there is a general tendency to regard predominantly aquatic and marshland Anura as frogs and primarily terrestrial anurans as toads but there is no general agreement as to such terms as "tree frog" and "tree toad"

American toad the bufonid *Bufo americanus*

bell toad popular name of liopelmid anurans

fire-bellied toad any of several discoglossids of the genus Bombina

British toad *Bufo vulgaris*

burrowing toad the rhinophrynid *Rhinophrynus dorsalis*

Cameroon toad *Bufo superciliaris*

clawed toad properly any of several species of aglosan pipid anurans of the genus Xenopus but usually without qualification applied to *X. laevis*

Colorado River toad *Bufo alvarius*

common toad *U.S. Bufo lentiginosus. Brit. B. vulgaris*

spade-footed toad *U.S.* any of several species of pelobatid of the genus Scaphiopus. *Brit.* any of several species of pelobatid of the genus Pelobates

giant toad *U.S.* the bufonid *Bufo marinus*

green toad *Brit. B. viridis*

horned toad (*see also* toad 2) the term is sometimes also applied to cystignathid anuran Amphibia of the genus Ceratophrys in which the eyelid is developed as a horn-like protruberance

Japanese toad *Bufo formosus*

leopard toad *Bufo regularis*

midwife toad the discoglossid *Alytes obstetricans*

narrow-mouthed toad *U.S.* any of several species of microhylids of the genus Gastrophryne. The term is also applied to engystomatids

pantherine toad *Brit. Bufo mauritanica*

rococo toad *Bufo panacnemis*

running toad *Brit. B. calamita* (= natterjack)

Surinam toad the pipid *Pipa pipa*

tree toad any of several bufonids of the genus Nectophryne

variable toad *Brit.* = green toad *Brit.*

cricket-voiced toad the bufonid *Ansonia grillivoca*

toad 2 (*see also* toad 1) toad-like animals other than anuran Amphibia

    pepper toad *U.S. local* the cottid fish *Hemitriterus americanus* (= sea-raven)

    puffer toad *U.S. local* any of several species of tetraodontid fish (= puffer)

    horned toad (*see also* toad 1) U.S. any of several species of iguanid lacertilian reptile of the genus Phrynosoma

toadlet *Austral.* any of several leptodactylid anurans of the genus Pseudophryne

toanui *N.Z.* the procellariid bird *Procellaria parkinsoni* (= black petrel *N.Z.*)

toatoa *N.Z.* either of two species of podocarpaceous tree of the genus Phyllocladus. Used without qualification *P. glaucus* is usually meant

tobacco (*see also* beetle, borer, moth, seed, thrip, weed, worm) any of numerous herbs of the solanaceous genus Nicotiana, and many other herbs often of similar aspect. Without qualification *N. tabacum* is meant

    flowering tobacco any of numerous species and horticultural varieties of the solanaceous genus Nicotiana

    Indian tobacco *U.S.* the campanulaceous herb *Lobelia inflata*

    lady's tobacco any of several species of compositaceous herb of the genus Antennaria

    tree tobacco the shrub *Nicotiana glauca*

tobira the pittosporaceous shrub *Pittosporum tobira*

-toc- *see* -tok-

oxytocin a hormone produced by the neurohypophysis that stimulates uterine contraction

α-tocopherol = vitamin E

-tocous (*see also* -tok-) pertaining to the production of offspring

    monotocous fruiting only once in a life time. The plant may be an annual, biennial or perennial

    oligotocous the condition of producing few young

    polytocous bearing several young at one time or (*bot.*) continuing to fruit for many years

Todidae a small *W.I.* family of coraciiform birds commonly called todies. In general appearance they resemble the kingfishers, but have a much shorter tail, short rounded wings, and a weak habit of flight

tody (*see also* flycatcher, motmot and tyrant) any of several West Indian coraciiform todid birds of the genus Todus closely allied to the kingfishers

    broad-billed tody *T. subulatus*

    narrow-billed tody *T. angustirostris*

    Cuban tody *T. multicolor*

    Jamaican tody *T. todus*

    Puerto Rican tody *T. mexicanus*

toe 1 (*see also* toe 2) the terminal digits of the limbs of all mammals except primates, in which the terminal digits of the forelimb are called fingers

    great toe that digit of the primate foot which corresponds to the thumb on the hand

toe 2 (*see also* toe 1, fig) an organism with toe-like parts

    nigger toe the myrtaceous tree *Bertholletia excelsa* and its seeds (= Brazil nuts)

    pussy's toes any of several species of compositaceous herb of the genus Antennaria

toetoe *N.Z.* the grass *Arundo conspicua*

tofaceous the color of tufa

tohora *N.Z.* the balaenid cetacean *Balaena australis* (= southern right whale)

toitoi *N.Z.* the malurine bird *Finschia novaezeelandiae* (= brown creeper)

-tok- *comb. form* meaning "offspring" or "production of offspring" sometimes transliterated -toc-

tokay *I.P.* the saurian reptile *Gekko gecko* (*cf.* gecko)

-toke *subs. suffix* indicating a part of an organism concerned in reproduction

    atoke the anterior sexless portion of the body of some polychaete worms

    epitoke the posterior portion of the body of some polychaete worms

-toky *subs. suffix* indicating a method or condition of reproduction (*cf.* -tocous)

    arrhenotoky the parthenogenetic production of male offspring

    deuterotoky the parthenogenetic production of both male and female offspring

    epitoky the production of sexual organs by an apparently asexual form. Particularly the production of such organs in the hinder end of polychaete worms

    thelytoky the parthenogenetic production of female offspring

tollon the rosaceous shrub *Photinia arbutifolia*

-tom- *comb. form* meaning "cut"

tom (*see also* cod) a male cat, either domestic or wild. Used also for the male turkey

    madtom any of several species of ictalurid catfishes of the genus Noturus

tom *see* [tomboy]

dichotomal bifurcated

tomatillo the solanaceous herb *Physalis ixocarpa*

tomato (*see also* mite) any of very numerous horticultural varieties of the solanaceous vine *Lycopersicon esculentum*

    husk tomato the solanaceous shrub *Physalis pubescens* and its edible fruit

    strawberry tomato the solanaceous herb *Physalis alkekengi*. The term is sometimes also applied to *P. pubescens*

    tree tomato the solanaceous shrub *Cyphomandra betacea*

[tomboy] a mutant (gene symbol *tom*) mapped at 61.5± on chromosome II of *Drosophila melanogaster*. The homozygous phenotypic female has a male-like pigmentation

-tome- *comb. form* meaning "cut" or "slice" or, by inaccurate extension, "segment" for which -tomite is preferable

    dermatome *see* myotome

    geotome *ecol. jarg.* for spade or trowel

    leptome the sum total of the phloem in a vascular bundle

    mestome the sum total of those parts of a vascular bundle that do not contribute to its strength

        hadromestome the xylem portion of a vascular bundle

    myotome those dorsal mesodermal metameres which give rise to the skeletal musculature of the body in all vertebrates and to the limb musculature in lower vertebrates

    nephrotome that portion of the embryonic mesoderm which lies between the somite and the lateral plate and from which the nephric units are derived

    sclerotome a mass of cells proliferated from the ventral inner corner of the somite and from which the ribs, vertebrae, and part of the chondrocranium are derived. The outer wall of the somites is the dermatome and the inner wall is the myotome

myosclerotome a term applied to the myotome before the sclerotomic cells have become differentiated

tomentose covered with hairs. Said particularly of silvery leaves

Tomes's *epon. adj.* from Sir John Tomes (1815–1895) [fibrils] or Charles Sissmore Tomes (1846–1928) [processes]

-tomi- *comb. form* meaning a "cutting" in the sense of a part used for the reproduction of a plant

-tomite *subs. suffix* indicating "segment" (*cf.* -tome)

sclerotomite a division of the vertebral column, running from one inter-vertebral space to the next. It does not correspond to a somite which runs from the center of one vertebra to the next

caudal sclerotomite the posterior of the two sclerotomites which compose the sclerotome of a somite, and which therefore forms all the anterior portion of a vertebra except the anterior zygapophysis

cranial sclerotomite the anterior division of a sclerotomite and which, therefore forms the posterior half of one vertebra and the prezygapophysis of the vertebra behind

tomium the cutting edge of the bill of birds

paratomium the area of the beak between calmium and tomium

-tomize *subs. suffix* meaning "to divide into parts"

anatomize (*obs.*) = dissect

automize to shed a part, particularly a limb

Tom-Kelly *W.I.* the vireonid bird *Vireo altiloquus* (= black-whiskered vireo)

tommy-goff the crotalid snake *Bothrops nummifera* (= jumping viper)

-tomous *adj. suffix* from almost any -tom- *subst.*

physotomous the condition of a fish in which the swim bladder is connected to the oesophagus

tetrachotomous the condition of a plant having four lateral peduncles around the terminal flower of a cyme

tomtate the pomadasyid fish *Bathystoma aurelineatum*

-tomy *subs. suffix* indicating the state or condition of breaking or cutting

architomy the breaking of an annelid into two individuals at a position where a new head structure has been reorganized (*cf.* paratomy)

autotomy the shedding, or voluntary removal, of parts by an animal

dichotomy the condition produced by repeated, equal forking

bostrychoid dichotomy a dichotomy in which either the left or right fork is always the more vigorous

cincinnal dichotomy condition in which alternate branches develop on either side of a stem

false dichotomy = dichasium

helicoid dichotomy a condition of dichotomy in which one of the two branches, and that always on the same side, is less vigorous than the other or the effect produced when only one of two dichotomous branches continues

scorpioid dichotomy alternate dichotomy in which all branches persist

ditrichotomy the condition produced by dividing first into two and then into three

microtomy the art of cutting thin sections but sometimes extended to all phases of the preparation of materials for examination with the microscope

ultramicrotomy microtomy as applied to the electron microscope

paratomy the division of a polychaete worm when one of the parts has been completely reproduced from undifferentiated tissue (*cf.* architomy)

plasmotomy the division of a multinucleate protozoan into many parts, without mitosis of any of the nuclei

pleiotomy multiple dichotomy

ton *see* [tonochaete]

-ton- *comb. form* meaning to "stretch" or "strain" (*cf.* -tono-)

ecotone a transitional area between two associations, climaxes or seres

tongue 1 (*see also* glossa tongue 2,3) a more or less protrusible muscular organ arising from the floor of the mouth of many vertebrates (*cf.* rasping organ)

tongue 2 (*see also* tongue 1, 3) an analogous structure in other forms. In *bot.* the term is applied to the ligule (*q.v.*) and in insects any structure that projects from the mouthpart and is used in gathering food

tongue 3 (*see also* tongue 1, 2, fern) any tongue-shaped organism

adder's tongue any of several species of the liliaceous genus Erythronium Also the fern *Polypodium vulgare*, any of several ferns of the genus Ophioglossum and locally other plants of like shape

yellow adder's tongue *U.S.* the liliaceous herb *Erythronium americanum*

beard-tongue any of very numerous species of the scrophulariaceous genus Pentstemon

devil's tongue the araceous herb *Hydrosme rivieri*

flamingo tongue the ovulid gastropod *Cyphoma gibbosa*

hart's tongue the polypodiaceous fern *Phyllitis scolopendrium*. the term is sometimes spelled heart's tongue

hound's tongue any of several species of boraginaceous herb of the genus Cynoglossum

ox tongue *Brit.* the compositaceous herb *Helminthia echioides*

serpent's tongue any of many ferns of the order Ophioglossales

smooth tongue any of several species of bathylagid fishes of the genus Leuroglossus

woman's tongue *see* tree

-tonic *adj. suffix* meaning "pertaining to stress"

hypertonic having a greater osmotic concentration

hypotonic having a lower osmotic concentration

isotonic having the same osmotic concentration

allossotinic the condition of having the turgidity increased while the volume is reduced

metatonic said of a stimulus which reverses an existing course of action

paratonic said of the retardation of plant growth by light

photonic said of an increase of irritability with increase of light

thelytonic = parthenogenetic

Tonnidae a family of gastropod mollusks with a very greatly expanded body whorl so that they appear almost globular. Commonly called cask-shells

-tono- *comb. form* meaning a "chord," frequently confused with -ton-

[**tonochaete**] a mutant (gene symbol *ton*) mapped at 60.1 on the X chromosome of *Drosophila melanogaster*. The phenotypic expression is short thin bristles

[**tonomacrochaetes**] a mutant (gene symbol *tmc*) mapped

at 17.5± on the X chromosome of *Drosophila melanogaster*. The macrochaetes of the phenotype are thin

tonsil a lymphoid gland in the roof of the mouth

tonus the prolonged persistence of the contracted state in muscles

toor the leguminous herb *Cajanus indicus* and its seeds

tooth 1 (*see also* tooth 2, 3) the enamel-covered, dentine-supported, biting chewing, or grabbing, structures produced from the jaw, and some other, bones of vertebrates

canine tooth (*see also* tooth 2) a conical tooth which, when enlarged, is frequently called a fang, between the last incisor and the first pre-molar of mammals. The term is also applied to large, conical teeth in the jaws of other vertebrates

carnassial tooth the posterior, cutting premolar in carnivores

cuspid tooth = canine

biscuspid tooth = premolar

egg tooth a tooth-like structure found in some embryos for cutting the egg. In monotremes, it is borne on the prenasal processes, and in birds on the end of the beak

ethmoid tooth a single posteriorly directed bony projection from the base of the palatine commissure in Myxine

incisor tooth literally a cutting tooth, but specifically applied to the anterior teeth in a mammalian jaw

milk tooth one of the first generation of deciduous teeth in a mammal

molar tooth literally a grinding tooth but specifically applied to the posterior teeth in the jaws of mammals

premolar one which lies immediately anterior to the molar teeth in the jaw of a mammal

pseudotooth a bony projection, without the structure of a true tooth, from a vertebrate jaw

tooth 2 (*see also* tooth 1, 3) any pointed structure arising from anything, particularly when part of a series of similar structures

buccal tooth denticles on arthropod mouth parts

canine tooth (*see also* tooth 1) large conical denticles on the mandibles of some invertebrates, particularly predacious insects

cardinal tooth the grooves, ridges, etc., which serve to orientate an articulation of the shells in pelecypod mollusks

tooth 3 (*see also* tooth 1, 2) any tooth-shaped organism

bleeding tooth the neritid gastropod *Nerita peloronta*

dog's tooth (*see also* violet) *Brit.* the noctuid lepidopteran insect *Mamestra dissimilis*

saw-tooth 1 (*see also* saw tooth 2, [saw tooth]) the telinid mollusk *Tellizora cristata*

saw-tooth 2 (*see also* saw-tooth 1) any of several species of pierid lepidopteran insect of the genus Prioneris

-top- *comb. form* meaning "place" usually in *biol.* in the sense of environment

top 1 (*see also* top 2, 3) the upper end or surface of anything

top 2 (*see also* top 1, 3, shell) a pear- or turnip-shaped object terminating in a fairly sharp point

top 3 (*see also* top 1, 2) in compound names of organisms

golden top *see* grass

rattle top *U.S.* any of several species of the ranunculaceous genus Cimifuga

red top the grass *Agrostis alba*

strangle-top *U.S.* the grass *Scolochloa festucacea*

white top any of several species of compositaceous herb of the genus Erigeron but particularly *E. annuus* and *E. strigosus*

topaz either of two trochilid birds of the genus Topaza

tope (*see also* -bio-) *Brit.* the carcharhinid selachian fish *Geleus vulgaris*

topi an *E.Afr.* alcelaphine antelopid (*Damaliscus corrigum*) (*cf.* korrigum, sassaby, blesbok, bontebok)

-topic *adj. suffix* meaning "pertaining to place' usually in the sense of environment

eurytopic said of organisms which can survive a very wide range of environmental conditions

heterotopic either found in different places, or having changed its customary place

metatopic the condition of bud leaves that do not follow a phyllotactic spiral

polytopic originating in several places

stenotopic intolerant of environmental change or restricted to a narrow environmental range

tophaceous = tofaceous

-topo- *see* -top-

over**topping** the condition of the upper branches of a tree which over-shadow the lower branches

bog**torch** *U.S.* the orchid *Habenaria nivea*

torea *N.Z.* either of two haematopodid birds of the genus Haematopus *H. ostralegus* (= pied oyster-catcher *N.Z.*) *H. unicolor* (= black oystercatcher *N.Z.*)

toreta the annonaceous tree *Annona purpurea and its fruit*

torfaceous growing in bogs

torma a sclerite on the head of certain coleopteran larvae or adult Diptera

tormentil *Brit.* anglicized form of the rosaceous genus of herb Tormentilla

tornete a rhabd, lance-headed at each end

tornus the anal angle of the wings of lepidopteran insects

toro *N.Z.* the myrsinaceous tree *Suttonia salicina*

toroa *N.Z.* the diomedeid bird *Diomedea exulans* (= wandering albatross)

torose in the shape of a cylinder with lumps on it

Torpedinidae a family of rajiform Chondrichthyes commonly called electric rays

Atlantic torpedo the torpedinid ray *Torpedo nobiliana*

torpidity a condition found in some birds physiologically resembling hibernation

torque a decorative area, or color, around the neck

torsion twisting, as in the shells, and some internal organs, of gastropods

torsk *Brit.* the gadid fish *Brosme brosme* (= *U.S.* cusk)

tortoise (*see also* beetle, plant, scale, shell, tick) popular name of chelonean reptiles. *U.S.* rarely used. *Brit.* a terrestrial form as distinct from an aquatic turtle

gopher tortoise *U.S.* the testudinid *Gopherus polyphemus*

Greek tortoise *Brit.* the testudinid *Testudo graeca*

starred tortoise the testudinid *Geochelone elegans*

[Tortoise] a mutant (gene symbol *To*) found in the sex chromosome (linkage group XX of the mouse). It is probably an allele of [Mottled] from which the phenotype cannot be distinguished

Tortricidae a family of tortricoid lepidopteran insects the larvae of the great majority of which are leaf-rollers

Tortricoidea a superfamily of lepidopteran insects

containing mostly small moths of the leaf-roller and webworm type through the codling moths also belong in this super family

tortrix any of several species of tortricid lepidopteran insects

large aspen tortrix *U.S. Choristoneura conflictana*

orange tortrix *U.S. Argyrotaenia citrana*

toru *N.Z.* the proteaceous tree *Persoonia toru*

torulose a dimunitive of torose, and usually indicating more regular swellings

torulous = torous

torous knobby

torulus literally a little mound, but specifically the articulation of the insect antenna

torus a mound, specifically that part of the axis on which the parts of a flower are inserted and a thickening of the primary cell wall in a bordered pit-pair. Also the enlarged basal segment of a mosquito antenna

Torymidae a family of chalcidoid apocritan hymenopteran insects readily distinguished by their long ovipositor and large hind coxae. The great majority are a vivid metallic green in color and between 3 and 4 mm. long

totara *N.Z.* any of several species of podocarpaceous tree of the genus Podocarpus. Used without qualification *P. totara* is usually meant

totokopio the podicipedid bird *Podiceps refipectus* (= New Zealand grebe)

[tottering] a mutant (gene symbol *tg*) found in linkage group XVIII of the mouse. The name is descriptive of the phenotypic expression

toucan any of numerous large-billed tropical piciform birds of the family Ramphastidae

keel-billed toucan *Ramphastos sulfuratus*

toucanet any of several species of small ramphastid birds, mostly of the genus Aulacorhynchus

emerald toucanet *A. prasinus*

touch-me-not any of several balsaminaceous herbs of the genus Impatiens

touraco any of several species of musophagid birds, mostly of the genus Turacus

black-billed touraco *T. schuettii*

great blue touraco *Corythaeola cristata*

Cameroon touraco *T. persa*

white-cheeked touraco *T. leucotis*

violet-crested touraco *Gallirex porphyreolophus*

white-crested touraco *T. leucolophus*

Fischer's touraco *T. fischeri*

Hartlaub's touraco *T. hartlaubi*

Livingstone's touraco *T. livingstonii*

[Tousled] a mutant (gene symbol *To*) mapped at 2.3 − on the X chromosome of *Drosophila melanogaster*. The tousled appearance of the phenotype comes from the distorted placing and growth of the thoracic bristles

toutouwai *N.Z.* = robin *N.Z.*

towai *N.Z.* the cunoniaceous tree *Weinmannia sylvicola*

tower the popular name of turrid gastropod mollusks

towhee any of several species of fringillid birds, mostly of the genus Pipilo. *U.S.* without qualification, *P. erythrophthalmus*

Abert's towhee *P. aberti*

brown tohee *P. fuscus*

collared towhee *P. ocai*

red-eyed towhee rufous-sided towhee

white-eyed towhee = rufous-sided towhee

rufous-sided towhee *P. erythrophthalmus*

spotted towhee = rufous-sided towhee

green-tailed towhee *Chlorura chlorura*

white-throated towhee *P. albicollis*

toxin a poison of organic origin

hypnotoxin the material obtained from an aqueous extract of the nematocyst bearing tentacles of coelenterates (*cf.* congestin and thalassin)

toyon = tollon

tp *see* [taupe]

Tp *see* [Teopod]

TPN *obs.* = triphosphopyridene nucleotide

tpw *see* [tapered wing]

Tr *see* [Trembler]

tr a yeast genetic marker indicating a requirement for tryptophan

tra *see* [transformed]

trabant a portion of a chromosome constricted away from the rest

trabecula literally a beam, and applied in *bot.* to almost any transverse septa or transverse bars and to bars extending transversely across gymnosperm tracheids. In *zool.* the term is used for the cartilaginous rods in vertebrate embryos supporting the anterior end of the brain and and the connective tissue framework of the spleen, and as a synonym for clasp (*q.v.*) in insects

branch trace a strand of vascular tissue connecting a branch to the main vascular system of the stem

leaf trace a vascular fascicle derived directly from the leaf but lying in the stem

trach- *comb. form* meaning "wind pipe"

trachea 1 (*see also* trachea 2, 3) water-conducting vessels

metratracheal pertaining to tangential bands in wood

trachea 2 (*see also* trachea 1, 3) a tube supported by cartilaginous rings conducting air from the mouth to the lungs

trachea 3 (*see also* trachea 1, 2) a tube, usually lined with rings, or a spiral, of chitin, which conducts air from the stigmata through the body cavity of arthropods

tracheid a long lignified cell in a plant stem both conducting water and serving as a support. It differs from a trachea in having long tapering closed ends

fiber tracheid a woody fiber, too long to be called a tracheid and too short to be called a fiber

reserve tracheid one that lies in the parenchyma sheath and stores water

tracheole diminutive of trachea, particularly in the sense of trachea 2 and 3

Tracheophyta a subphylum erected to contain the Lycopodinae (under the name Lycopsida), Psilotineae (under the name Psilopsida), and the Sphenophyllales (under the name Sphenopsida) in contrast to the Pterospida (*q.v.*)

tracheron living organisms showing active movement (*cf.* nekton, plankton)

Trachinidae a small family of marine acanthopterygian fishes distinguished by the venomous dorsal spines. Commonly called weaverfishes

Trachipteridae a family of greatly elongated laterally compressed allotriognathous fishes commonly called ribbonfishes

-trachy- *comb. form* meaning "rough" or "wind pipe"

Trachylina an order of hydrozoan coelenterates often called the trachyline medusae. The hydroid generation is either lacking or reduced

Trachymedusae a suborder of hydrozoan coelenterates

the colonial or hydroid stages of which, in the few cases that they have been found, are minute. The margin of the medusa is smooth and the gonads are borne on the radial canals

tract a specific area usually delimited within a larger one. The term is also used in the sense of "system"

    ascending tract the dorsal funiculus of the spinal cord which carries impulses towards the brain

    descending tract the ventral funiculus of the spinal cord which carries impulses in both directions

    capital tract = head tract

    cervical tract = neck tract

    crural tract a pteryla pertaining to the legs in those birds in which they are feathered

    femoral tract a pteryla which usually forms an oblique band on the outer side of the thigh

    head tract the pteryla covering the head

    humeral tract = shoulder tract

    Meckel's tract that portion of the small intestine which comprises the jejunum and ileum

    neck tract a pteryla extending all over the neck but usually interrupted by apteria on the lateral margin

    optic tract that part of the optic nerve that lies between the optic chiasma and the geniculate body

    shoulder tract a pteryla overlying the humerus

    spinal tract a pteryla extending along the vertebral column from neck to tail

        corticospinal tract the voluntary motor pathway passing in mammals from the neocortex to the spinal medulla

    sternal-abdominal tract = ventral tract

    tail tract the area bearing the rectrices of birds

    uropygial tract the pteryla surrounding the uropygial gland

    ventral tract a narrow pteryla extending along the abdomen

    wing tract the pteryla bearing the reniges

Tradescantieae a tribe of the family Commelinaceae

tragopan any of several species of phasianid birds of the genus Tragopan (*cf.* pheasant)

Tragulidae a family of ruminant artiodactyle mammals commonly called chevrotains or deerlets. They are distinguished from other ruminants by the absence of horns

Tragulina an assemblage of ruminant artiodactyle mammals containing the single family Tragulidae (*q.v.*)

tragus the anterior flap which in some mammals can be closed to cover the ear hole

trailing said of that stem which lies along the ground but does not root

train (*see also* bearer) term for a group of camels

Trajian *see* column

transcendentalism in *biol.* the doctrine that form precedes function. It is, therefore, the antithesis of teleology

transduction a form of transfer of genetic information which depends on the ability of a bacteriophage to incorporate heritable characters of cells that they parasitize into their own genetic apparatus and then to maintain and transfer these characteristics to other cells that they subsequently invade

    abortive transduction is said of those cases in which the transfused genetic fragment persists without functioning

    special transduction the transfer through transduction of genes only known by other criteria

transferase a group term for those enzymes that catalyze the transfer of atoms, or groups of atoms, between molecules

dihydroxyacetonetransferase catalyzes the production of D-erythrose 4-phosphate + D-fructose 6-phosphate from D-sedoheptulose 7-phosphate and D-glyceraldehyde 3-phosphate

aminoacid acetyltransferase catalyzes the production of CoA + N-acetyl-L-[amino acid] from acetyl-CoA + amino acid

    arylamine acetyltransferase catalyzes the production of CoA and N-acetylarylamine from acetyl-CoA and arylamine

    carnitine acetyltransferase catalyzes the production of CoA + O-acetylcarnitine from acetyl-CoA and carnitine

    choline acetyltransferase catalyzes the production of CoA + O-acetylcholine from acetyl-CoA and choline

    acetyl-CoA acetyltransferase catalyzes the production of CoA + acetoacetyl-CoA from acetyl-CoA + acetyl-CoA

    aminodeoxyglucose acetyltransferase catalyzes the production of CoA + 2-acetylamino-2-deoxy-D-glucose from acetyl-CoA and 2-amino-2-deoxy-D-glucose

    H_2S acetyltransferase catalyzes the production of CoA + thioacetate from acetyl-CoA and H_2S

    imidazole acetyltransferase catalyzes the production of CoA and N-acetyl-imidazole from acetyl-CoA and imidazole

    thioethanolamine acetyltransferase catalyzes the production of CoA and S-acetyl-thioethanolamine from acetyl-CoA and thioethanolamine

    lipoate acetyltransferase catalyzes the production of CoA + S-6-acetylhydrolipoate from acetyl-CoA + dihydrolipoate

    glutaminephenylacetyltransferase catalyzes the production of CoA + α-N-phenylacetyl-L-glutamine from phenylacetyl-CoA + L-glutamine

    phosphate acetyltransferase catalyzes the production of CoA and acetylphosphate from acetyl-CoA and orthophosphate

    aminodeoxyglucosephosphate acetyltransferase catalyzes the production of CoA + 2 acetylamino-2-deoxy-D-glucose 6-phosphate from acetyl-CoA + 2-amino-2-deoxy-D-glucose 6-phosphate

acetyl-CoA acyltransferase catalyzes the production of CoA + 3-oxoacyl-CoA from acyl-CoA + acetyl-CoA

    glycine acyltransferase catalyzes the production of CoA and N-acylglycine from acyl-CoA and glycine

    glycerolphosphate acyltransferase catalyzes the production of CoA + monoglyceride phosphate from acyl-CoA + L-glycero 3-phosphate

methionine adenosyltransferase catalyzes the production of S-adenosylmethionine with ortho- and pyrophosphates from ATP and L-methionine and water

FMN adenylyltransferase catalyzes the production of FAD and pyrophosphate from ATP and FMN

NMN adenylyltransferase catalyzes the production of pyrophosphate and NAD from ATP and NMN

    pantetheinephosphate adenylyltransferase catalyzes the production of pyrophosphate and dephospho-CoA from panthetheine 4'-phosphate and ATP

    sulphate adenylyltransferase catalyzes the production of adenylylsulphate and orthophosphate from ADP and sulphate

    sulphate adenylyltransferase catalyzes the production of adenylylsulphate and pyrophosphate from ATP and sulphate

glycolaldehydetransferase catalyzes the production of

D-ribose 5-phosphate and D-xylulose 5-phosphate from D-sedoheptulose 7-phosphate and D-glyceraldehyde 3-phosphate

phosphoribosylpyrophosphate amidotransferase = amidophosphoribosyltransferase

glycine amidinotransferase catalyzes the production of L-ornithine and guanidinoacetate from L-arginine and glycine

aminotransferase a large number of enzymes which catalyze the production of L-glutamate from the indicated substrate with 2-oxoglutarate; *e.g.*, tyrosine aminotransferase catalyzes the production of L-glutamate and p-hydroxyphenylpyruvate from L-tyrosine and 2-oxoglutarate, etc. a few aminotransferases yield amino acids other than glutamate

carboxyltransferase catalyzes the production of propionyl-CoA and oxaloacetate from 2-methyl-malonyl-CoA and pyruvate

aspartate carbamoyltransferase catalyzes the production of orthophosphate and N-carbamoyl-L-aspartate from carbamoyl-phosphate and L-asparate

  ornithine carbamoyltransferase catalyzes the production of orthophosphate and L-citrulline from carbamoylphosphate and L-ornithine

3-ketoacid CoA-transferase catalyzes the transfer of CoA from succinyl-CoA to 3 oxoacid to form 3-oxo-acyl-CoA + succinate

  3-oxo-adipate CoA transferase catalyzes the transfer of CoA from succinyl-CoA to 3-oxo-adipyl-CoA

  butyrate CoA-transferase catalyzes the transfer of CoA from succinyl-CoA to butyryl-CoA

  malonate CoA-transferase catalyzes the transfer of CoA from acetyl-CoA to malonyl-CoA

  oxalate CoA-transferase catalyzes the transfer of CoA from succinyl CoA to oxalyl-CoA

  propionate CoA-transferase catalyzes the transfer of CoA from acetyl-CoA to propionyl-CoA

cholinephosphate cytidylyltransferase catalyzes the production of pyrophosphate and CDP choline from CTP and choline phosphate

  ethanolaminephosphate cytidylyltransferase catalyzes the production of CDPethanolamine and pryophosphate from CTP and ethanolamine phosphate

dimethylallytransferase catalyzes the synthesis of the production of geranylpyrophosphate and pyrophosphate from dimethalallylpyrophosphate and isopentenylpyrophosphate

glutamate formininotransferase catalyzes the formation of L-glutamate and 5-formiminotetrahydrofolate from N-formimino-L-glutamate and tetrahydrofolate

  glycine formiminotransferase catalyzes the formation of glycine and 5-formiminotetrahydrofolate from N-formiminoglycine and tetrahydrofolate

phosphoribosylaminoimidazolecarboxamide formyltransferase catalyzes the production of 5'-phosphoribosyl-5-amino-4-imidazolecarboxamide and 10-formyltetrahydrofolate from 5'-phosphoribosyl-5-formamido-4-imidazole-carboxamide and tetrahydrofolate

  phosphoribosylglycineamide formyltransferase catalyzes the production of 5'-phosphoribosyl-glycineamide and 5,10-methenyltetrahydrofolate from 5'-phosphoribosyl-N-formylglycineamide and tetrahydrofolate

sucrose 1-fructosyltransferase catalyzes the conversion of sucrose into inulin and glucose

  sucrose 6-fructosyltransferase catalyzes the conversion of sucrose into fructan and glucose

UDP glucuronyltransferase catalyzes the production of UDP from UDP glucuronate using a wide range of glucuronide acceptors

maltose 3-glucosyltransferase catalyzes the production of glucose and α-1,3-glucosyl-glucose from maltose and D-glucose

  maltose 4-glucosyltransferase catalyzes the production of 1,4-glucan and glucose from maltose

dextrin 6-glucosyltransferase catalyzes the production of 1,6-glucans from dextrins

sucrose 6-glucosyltransferase catalyzes the production of 1,6-glucan and fructose from sucrose

glucan-branching glucosyltransferase catalyzes the conversion of amylose into amylopectin

cyclodextrin glucosyltransferase forms cyclic dextrins by transferring parts of a 1,4-glucan chain

chitin UDP acetylaminodeoxyglucosyltransferase catalyzes the production of chitin and UDP from UDP-2-acetylamino-2-deoxy-D-glucose

UDPglucose-fructose glucosyltransferase catalyzes the production of UDP and sucrose from UDP-glucose and D-fructose

sucrose glucosyltransferase catalyzes the formation of 1,4-glucan and fructose from sucrose

UDPglucose-β-glucan glucosyltransferase catalyzes the conversion of UDP glucose to cellulose and UDP

UDPglucose -α-glucan glucosyltransferase catalyzes the conversion of UDP glucose to a 1,4-glucan and UDP

UDPglucose-fructosephosphate glucosyltransferase catalyzes the production of UDP and sucrose 6-phosphate from UDPglucose and fructose 6-phosphate

UDPglucose-glucosephosphate glucosyltransferase catalyzes the production of UDP and trehalose 6-phosphate from UDPglucose and D-glucose-6-phosphate

D-glutamyltransferase catalyzes the production of 5-glutamyl-D-glutamyl-R and ammonia from glutamine + D-glutamyl-R

mannose-1-phosphate guanylyltransferase catalyzes the production of pyro phosphate and GDPmannose from GTP and α-D-mannose 1-phosphate

guanidinoacetate methyltransferase catalyzes the production of S-adenosylhomocysteine and creatin from S-adenosylmethionine and guanidinoacetate

  betaine homocysteine methyltransferase catalyzes the production of dimethyl-glycine and L-methioine from betaine and L-homocysteine

  demethylthetin homocysteine methyltransferase catalyzes the production of S-methylthioglycollate and L-methionine from dimethylthetin and L-homocysteine

  serine hydroxymethyltransferase catalyzes the production of glycine and 5,10 methylenetetrahydrofolate from L-serine and tetrahydrofolate

  nicotinamide methyltransferase catalyzes the production of S-adenosylhomocysteine and N-methylinicotinamide from S-adenosylmethionine and nicotinamide

  acetylserotonin methyltransferase catalyzes the production ot S-adenosylhomocysteine and N-acetyl-5-methoxytryptamine from S-adenosyl-methionine and N-acetylserotonin

DNA nucleotidyltransferase catalyzes the transfer of 4 DNA units from 4-deoxynucleoside triphosphates

  polyribonucleotide nucleotidyltransferase catalyzes the addition of a nucleotide unit to an RNA chain from a nucleoside diphosphate

RNA nucleotidyltransferase catalyzes the addition of 4 nucleotide units to an RNA chain from 4 nucleoside triphosphates

ethanolaminephosphotransferase catalyzes the production of CMP and a phosphatidylethanolamine from CDPethanolamine and 1,2-diglyceride

cholinephosphotransferase catalyzes the production of a phosphatidylcholine and CMP from CDP-choline and 1,2-diglyceride

riboflavin phosphotransferase catalyzes the production of D-glucose and FMN from D-glucose 1-phosphate and riboflavin

phenyltransferase = dimethylallyltransferase

nucleoside ribosyltransferase catalyzes the transfer of numerous purines and pyrimidines between molecules of D-ribosyl-R molecules

nucleoside deoxyribosyltransferase catalyzes the transfer of numerous purine and pyrimidines between 2-deoxy-D-ribosyl-R molecules

adenine phosphoribosyltransferase catalyzes the production of adenine and 5-phospho-α-D-ribosyl-pyrophosphate from AMP and pyrophosphate

amidophosphoribosyltransferase catalyzes the production of glutamine and 5-phospho-α-D-ribosyl-pyrophosphate from L-glutamate and 5-phospho-β-D-ribosylamine and pyrophosphate

nicotinamide phosphoribosyltransferase catalyzes the production of nicotinamide and 5-phospho-α-D-ribosyl-pyrophosphate from nicotinamideribonucleotide and pyrophosphate

hypoxanthine phosphoribosyltransferase catalyzes the production of hypoxanthine and 5-phospho-α-D-ribosyl-pyrophosphate from IMP and pyrophosphate

orotate phosphoribosyltransferase catalyzes the production of orotate and 5-phospho-α-D-ribosyl-pyrophosphate from orotidine -5'-phosphate and pyrophosphate

uracil phosphoribosyltransferase catalyzes the production of uracil and 5-phospho-α-D-ribosyl pyrophosphate from UMP and pyrophosphate

aryl sulfotransferase catalyzes the production of an arylsulfate adenosine 3',5'-diphosphate from a phenol and 3'-phosphoadenylsulfate

3-β-hydroxysteroid sulfotransferase catalyzes the production of a steroid 3-β-sulfate and adenosine 3',5'-diphosphate from a 3-β-hydroxysteroid and 3'-phosphoadenylylsulfate

3-mercaptopyruvate sulfurtransferase catalyzes the production of pyruvate and thiocyanate from 3-mercaptopyruvate and cyanide

thiosulfate sulfurtransferase catalyzes the production of sulfite and thiocyanate from thiosulfate and cyanide

galactose-1-phosphate uridylyltransferase catalyzes the production of UDPgalactose and pyrophosphate from UTP and α-D-galactose 1-phosphate

glucose-1 phosphate uridylyltransferase catalyzes the production of pyrophosphate and UDPglucose from UTP and α-D-glucose 1-phosphate

hexose-1-phosphate uridylyltransferase catalyzes the production of UDPgalactose and α-D-glucose 1-phosphate from UDPglucose and α-D-galactose 1-phosphate

xylose-1-phosphate uridylyltransferase catalyzes the production of UDPxylose plus pyrophosphate from UTP and α-D-xylose 1-phosphate

transformation a term of bacterial genetics indicating an alteration in the characteristics of a bacterial clone

[**transformed**] a mutant (gene symbol *tra*) mapped at 45.0± on chromosome III *Drosophila melanogaster*. The gene transforms the female to a male phenotype

translocation 1 (*see also* translocation 2) the transfer of metabolites from one part of the plant to another

translocation 2 (*see also* translocation 1) the transfer of a portion of a chromosome to a non-homologous chromosome or the transfer of genes to another linkage group

aneucentric translocation one that involves the centromere

insertional translocation one that involves the transfer of materials to the body, as distinct from the end, of a non-homologous chromosome

reciprocal translocation one in which each chromosome receives a portion of the other

[**translocation**] a general term, symbol T(...;...), for chromosomal translocations in *Drosophila melanogaster*. The numbers of the chromosomes involved are given in the parenthesis

[**translucent**] a mutant (gene symbol *od*) mapped at 49.6 on linkage group I of *Bombyx mori*. The name is descriptive of the phenotypic larva

transmission communication without obvious movement of matter

humoral transmission the communication between the cells by chemical reagents

transpiration the emission of water vapor by an organism

active **transport** the passage of materials through a differentially permeable membrane against a concentration or electrical gradient

trap an inanimate device used by a predator to secure prey

Venus fly trap the droseraceous herb *Dionaea muscipula*

wentletrap any of several species of epitoniid gastropod mollusks

Trapaceae that family of myrtiflorous dicotyledonous angiosperms which contains the single genus Trapa, the fruit of which is known as the water chestnut. The peculiar fruit is typical of this aquatic

-traum- *comb. form* meaning "wound"

lemon **traveller** *S.Afr.* the pierid lepidopteran insect *Colotis subfasciatus*

trb see [thread bristle]

tre a bacterial mutant showing a utilization of trehalose or a yeast genetic marker indicating a requirement for trehalose

treader an organism which supports itself on or moves by means of its feet

marsh treader any hydrometrid hemipterous insect of the genus Limnobates

water treader any of several species of mesoveliid hemipteran insects

tread-softly *U.S.* the euphorbiaceous herb *Cnidoscolus stimulosus* (= bull-nettle)

T1 rec a bacterial genetic marker having a character affecting the resistance to phage T1, mapped at 24.2 mins. for *Escherichia coli*

T1, T5 rec a bacterial genetic marker having a character affecting the resistance to phages T1 and T5, mapped at 3.0 mins. for *Escherichia coli*

T4 rec a bacterial genetic marker affecting the resistance to phage T4, mapped at between 49.8 mins. and 54.1 mins. for *Escherichia coli*

T6, colK rec a bacterial genetic marker having a character affecting the resistance to phage T 6 and colicine K, mapped at 10.8 mins. for *Escherichia coli*

treddle *Brit. obs.* chalaza (of egg)

tree 1 (*see also* tree 2, babbler, badger, borer, creeper, cricket, hunter, mallow, oyster, runner, scale, swift, tutu) a woody plant distinguished from a shrub in part by its larger size and in part by having a single, or at least only a few, main stems

    alligator tree the hamamelidaceous tree *Liquidambar styraciflua* (= sweet gum)

    angeleen tree = cabbage tree

    angelica tree the araliaceous tree *Aralia spinosa*
        Chinese angelica tree *A. chinensis*

    annattoo tree the bixaceous tree *Bixa orellana*

    arar tree the pinaceous tree *Callitris quadrivalvis*

    balsam tree any of several leguminous trees of the genus Myroxylon
        Peru-balsam tree *M. peruiferum*
        tolu-balsam tree *M. toluiferum*
        balsam-of-Peru tree *M. pereirae*

    banyan tree the moraceous tree *Ficus benghalensis*

    baobab tree the bombacaceous tree *Adansonia digitata*

    bark tree *S.Amer.* any of several trees of the genus Cinchona. *S.Afr.* any of several trees of the genus Brachystegia
        cats ear bark tree the lauraceous tree *Cinnamomum cassia*

    bastard tree the coniferous tree *Sequoia sempervirens* (= redwood)

    bead tree any of several species of meliaceous tree of the genus Melia

    white beam tree *Brit.* the rosaceous tree *Pyrus aria*

    beaver tree the magnoliaceous tree *Magnolia glauca*

    candleberry tree the fruit of the euphorbiaceous tree *Aleurites moluccana*

    blinding tree the euphorbiaceous tree *Excoecaria bicolor*

    bo tree the moraceous tree *Ficus religiosa*

    bottle tree the sterculiaceous tree *Sterculia rupestris*

    box tree any of several trees or shrubs of the buxaceous genus Buxus

    briar tree any of several liliaceous climbing herbs of the genus Smilax

    broom tree the tree *Baccharis scoparia* of the family Compositae

    buckwheat tree the cyrillaceous tree *Cliftonia monophylla*

    cabbage tree the leguminous tree *Andira inermis* *N.Z.* the popular name of tall erect tufted liliaceous shrubs of the genus Cordyline. Without qualification *C. australis* is usually meant

    cajupet tree the myrtaceous tree *Melaleuca leucadendron*

    calaba tree the guttiferaceous tree *Calophyllus calaba*

    camphor tree the lauraceous tree *Cinnamomum camphora*

    red candlewood tree the leguminous tree *Adenanthera pavonina*

    cannonball tree the lecythidaceous tree *Couroupita guianesis*

    caoutchou tree the euphorbiaceous tree *Hevea brasiliensis* (*cf.* rubber tree)

    bird-catching tree *N.Z.* the nyctaginaceous tree *Heimerliodendron brunonianum*

    China tree the meliaceous tree *Melia azedarath*
        wild China tree the sapindaceous tree *Sapindus drummondii*

    Christmas tree *Brit.* and U.S. any coniferous tree decorated and used as a Christmas symbol *N.Z.* the myrtaceous tree *Metrosideros tomentosa* (= pohutukawa)

    cigar tree the bignoniaceous tree *Catalpa speciosa*

    cinnamon tree the lauraceous tree *Cinnamonomum zeylanicum*

    clove tree the myrtaceous tree *Eugenia aromatica*

    coffee tree the araliaceous shrubby tree *Polyscias guilfoylei*

    copal tree *U.S.* the simaroubaceous tree *Ailanthus altissima* (= tree of heaven)

    coral tree any of several trees of the leguminous genus Erythrina but particularly *E. crista-galli*

    cork tree any of several rutaceous trees of the genus Phellodendron (The "cork" of commerce comes from an oak)
        bastard cork tree *Quercus hispanica*

    silk-cotton tree the bombacaceous tree *Ceiba casearia*

    cucumber tree either the magnoliaceous tree *Magnolia acuminata* or the oxalidaceous tree *Averrhoa bilimbi*
        large-leaved cucumber tree the magnoliaceous tree *Magnolia macrophylla*

    daisy tree *N.Z.* any of numerous compositaceous trees

    dove tree the nyssaceous tree *Davidia involucrata*

    snowdrop tree any of several species of the styracaceous genus Halesia

    monkey face tree the euphorbiaceous shrub *Mallotus philippinensis*

    fever tree the rubiaceous tree *Pinckney pubens*

    flame tree the sterculiaceous tree *Sterculia acerifolia*

    cassia flower tree the lauraceous tree *Cinnamomum loureirii*
        hand flower tree the sterculiaceous tree *Chiranthodendron platanoides*

    fringe tree any of several species of woody trees of the oleaceous genus Chionanthus

    gamboge tree the guttiferous tree *Garcinia morella*

    groundsel tree any of several compositaceous trees and shrubs of the genus Baccharis. Without qualification, usually *B. halimifolia*

    gum tree any of numerous species of tree of the myrtaceous genus Eucalyptus or the pinaceous tree *Callitris quadrivalvis*
        sourgum tree the nyssaceous tree *Nyssa sylvatica*

    gutta-percha tree any of several species of the sapotaceous trees of the genus Isonandra

    heliotrope tree the boraginaceous tree *Ehretia acuminata*

    hop tree *U.S.* any of several species of rutaceous tree of the genus *Ptelea*, particularly *P. trifoliata*

    horseradish tree the moringaceous tree *Moringa oleifera*

    huanhuan tree the monimiaceous tree *Laurelia serrata*

    mountain ivy tree *N.Z.* the araliaceous tree *Nothopanax colensoi* (= orihau)

    Judas tree any of several trees of the leguminous genus Cercis, but applied in many parts of the world to other trees with fragile branches

    kamila tree = monkey face tree

    karri tree = princess tree

    kew tree the ginkgoaceous tree *Ginkgo biloba* (= ginkgo)

    lacquer tree = varnish tree

maidenhair tree the ginkgoaceous tree *Ginkgo biloba* (= ginkgo)

Peruvian mastic tree the anacardiaceous tree *Schinus molle*

milk tree *N.Z.* the moraceous tree *Paratrophis microphylla* (= turepo)

   large-leaved milk tree *N.Z. P. opaca*

necklace tree the leguminous tree *Ormosia monosperma*

nettle tree *U.S.* any of several trees of the ulmaceous genus Celtis

nosegay tree the oleaceous shrub *Plumeria rubra* (= red jasmine)

marking-nut tree the anacardiaceous tree *Semecarpus anacardium*

croton oil tree the euphorbiaceous tree *Croton tiglium*

   China wood oil tree the euphoriaceous tree *Aleurites fordii*

   Japan wood oil tree the euphorbiaceous tree *Aleurites cordata*

orchid tree any of numerous leguminous trees of the genus Bauhinia

Japanese pagoda tree the leguminous tree *Sophora japonica*

Chinese parasol tree the sterculiaceous tree *Sterculia platanifolia*

pea tree any of several trees of the leguminous genus Caranga

peepul tree the moraceous tree *Ficus religiosa*

pepper tree *N.Z.* the winteraceous tree *Wintera colorata*

   California pepper tree the anacardiaceous tree *Schinus molle*

pudding pipe tree the leguminous tree *Cassia fistula*

plane tree any of several species of tree of the platanaceous genus Platanus

   American plane tree *P. occidentalis*

   London plane tree *P. acerifolia*

   oriental plane tree *P. orientalis*

planertree *U.S.* the ulmaceous tree *Planera aquatica* (= water elm)

fish poison tree the leguminous tree *Piscidia erythrina* (= Jamaica dogwood)

pot tree any of several lecythidaceous trees of the genus Lecythis, having a gourd-shaped fruit

monkey pot tree any of several lecythidaceous trees of the genus Lecythis, having a gourd-shaped fruit

sweet potato tree the euphorbiaceous herb *Manihot utilissima* (*see* tapioca plant)

princess tree the scrophulariaceous tree *Paulownia tomentosa*

punk tree = cajuput tree

radish tree the phytolaccaceous tree *Codonocarpus cotinifolius*

rain tree the leguminous spreading tree *Samanea saman*

   golden-rain tree the sapindaceous tree *Koelreuteria paniculata*

Japanese raisin tree the rhamnaceous tree *Hovena dulcis*

rowan tree the rosaceous tree *Sorbus aucuparia* (= European mountain ash)

rubber tree any of several species and numerous agricultural varieties of the euphorbiaceous genus Hevea

   cerara rubber tree the euphorbiaceous tree *Manihot glaziovii*

sacred tree (of India) = bo tree

   St. Thomas tree the leguminous tree *Bauhinia tomentosa*

salamander tree the euphorbiaceous tree *Antidesma bunius*

salt tree the leguminous shrub *Halimondendron halodendron*

wild service tree *Brit.* the rosaceous tree *Pyrus torminalis*

silk tree *U.S.* the leguminous tree *Albizzia julibrissin*

silver tree the proteaceous tree *Leucadendron argenteum*

siris tree the leguminous tree *Albizzia lebbek*

smoke tree the anacardiaceous tree *Cotinus coggygria*

snag tree the nyssaceous tree *Nyssa sylvatica* (= black gum)

sorrel tree *U.S.* the leguminous tree *Oxydendrum arboreum. Austral.* the malvaceous tree *Hibiscus heterophyllus*

spindle tree any of many species of celastraceous trees of the genus Euonymus. *Brit.*, without qualification *E. europaeus*

staff tree the celastraceous climbing shrub *Celastrus scandens*

strawberry tree the ericaceous tree *Arbutus unedo*

Chinese tallow tree the euphorbiaceous tree *Sapium sebiferum*

tea tree *N.Z.* either of two species of myrtaceous tree of the genus Leptospermum

   red-tea tree *N.Z. L. scoparium* (= manuka)

   swamp-tea tree = cajuput tree

   white-tea tree *L. ericoides*

woman's tongue tree = siris tree

tooart tree the myrataceous tree *Eucalyptus gomphocephala*

toothache tree the rutaceous tree *Xanthoxylum americanum*

traveler's tree the musaceous tree *Ravenala madagascariensis*

tulip tree the magnoliaceous tree *Liriodendron tulipifera*

tung-ching tree the flacourtiaceous tree *Xylosoma racemosa*

turpentine tree the myrtaceous tree *Syncarpia glomulifera*

umbrella tree *U.S.* the magnoliaceous tree *Magnolia tripetala Afr.* any of several species of moraceous tree of the genus Musanga

   Texas umbrella tree the meliaceous tree *Melia azedarach* (= China tree)

varnish tree the anacardiaceous tree *Rhus verniciflua.* The term is also applied to the candleberry tree (*q.v.*)

   Japanese varnish tree the sterculiaceous tree *Sterculia plantanifolia*

velvet tree the compositaceous shrub *Gynura aurantiaca*

wax tree the anacardiaceous shrubby tree *Rhus succedanea*

wayfaring tree the caprifoliaceous shrubby tree *Viburnum lantana*

   American wayfaring tree *V. alnifolium*

yate tree the myrtaceous tree *Eucalyptus cornuta*

tree 2 (*see also* tree 1, 3) in the sense of a structural, rather than a taxonomic, definition

canopy tree one having its largest branches at the crown

elfin tree a tree dwarfed by climatic conditions

tuft tree one having an unbranched stem with a tuft of leaves or branches at the top

tree 3 (*see also* tree 1, 2) a tree-like structure

respiratory tree large, dendritic extensions that run from the anterior part of the cloaca throughout most of the length of Holothuria. They are the main respiratory system of these animals

tree-of-heaven the simarubaceous tree *Ailanthus altissima*

trefoil any clover but more particularly those of the leguminous genus Trifolium

bastard trefoil the leguminous herb *Trifolium hybridum* (= Alsike clover)

bird's foot trefoil the leguminous herb *Lotus corniculatus* (= sheep foot)

hare's foot trefoil *Brit. T. arvense*

herb's foot trefoil *Brit.* any of several species of the leguminous genus Lotus

teazel-headed trefoil *Brit. T. maritimum*

hop trefoil *Brit. T. procumbens*

marsh trefoil *Brit.* the gentianaceous herb *Menyanthes trifoliata* (= buck-bean) (*cf.* marsh clover)

moon trefoil the leguminous herb *Medicago arborea* (= tree alfalfa)

prairie trefoil the leguminous herb *Lotus americanus*

rigid trefoil *Brit.* T. scabrum

shrubby trefoil *U.S.* any of several species of rutaceous tree of the genus Ptelea

tick trefoil *U.S.* any of several species of leguminous herb of the genus Desmodium

yellow trefoil the leguminous *Medicago lupulina*

trehalase any enzyme catalyzing the hydrolysis of trehalose into glucose

Treitz's *epon. adj.* from Wenzel Treitz (1819–1872)

-trem- *comb. form* meaning "section" or "hole"

brachytrema a disc-shaped shell which ruptures to liberate moss gemmae

gonotrema the opening of the genital bursa in insects

peritrema the sclerotic plate immediately surrounding the spiracle of an insect or a horny prominence associated with the stigmata of some arachnids

phallotrema the distal opening of the intromittent organ particularly in arthropods

pharyngotrema one of the holes through which the pharynx communicates with the exterior in the Hemichordata

Tremandraceae a small family of geranialous dicotyledonous shrubs distinguished from the closely allied Polygalaceae by the peculiarly bearded flowers

tremata *pl.* of trema

peritrematal said of that portion of the arthropod trachea which leaves directly from the stigma

Trematoda a class of Platyhelminthes containing the parastic forms commonly called flukes. They are distinguished by the possession of one or two suckers at either or both ends

ditrematous having two openings particularly in a system where one opening is more common

trembler the mimid bird *Cinclocerthia ruficauda*

[Trembler] a mutant (gene symbol *Tr*) in linkage group VII of the mouse. The phenotypic expression is well described by the name

-treme anglicized form of -trema

Tremellales an order of heterobasidiomycete Basidiomycetes distinguished by the fact that the hypobasidium becomes divided into two, three, or four cells

Trentepohliaceae that family of ulotrichale Chlorophyceae distinguished by the fact that the gametes are formed in special cells differing in shape and structure from vegetative cells

trepang dried holothurian Echinodermata forming an article of commerce in the Far East (= bêche-de-mer)

Trepidariidae a family of brachycerous dipteran insects now usually included in the Micropezidae

Treponemataceae a family of spirochaetale Schizomycetes mostly of small size, though occasionally forming elongate chains in consequence of incomplete division. All are parasitic and several are dangerous human pathogens causing, *inter alia* relapsing fever, hemorrhagic syphilis, and jaundice

Treronidae a family of columbiform birds now usually included in the Columbidae

ladies' tresses *U.S.* any of several orchids of the genus Spiranthes. *Brit.* the orchid *Neottia spiralis*

trey-reach a schilbeid asiatic catfish (*Pangasianodon gigas*) which reaches a weight of over 300 pounds pounds

-tri- *comb. form* meaning "three"

tri *see* [trident]

Triacanthodidae a family of plectognathous Osteichthyes commonly called spikefishes by reason of the conversion of the pelvic fins into a strong spine with an inner basal knob which locks when inverted

triaene a sponge spicule with one long ray (the rhabdome) and three short rays (cladome)

anatriaene a triaene with the cladi curved downwards

dichotriaene triaene in which the cladi are forked

orthotriaene = plagiotriaene

plagiotriaene a triaene of which the cladi are horizontal

protriaene a triaene of which the cladi are pointed upward

Triandria that class of the Linnaean classification of plants distinguished by the possession of three stamens

triangle *Brit.* the limacodid lepidopteran insect *Heterogena asella Austral.* any of numerous species of papilionid lepidopteran insects of the genus Papilio distinguished by the absence of a "tail" on the wing

blue triangle *Austral.* (= blue fanny)

green triangle *P. macfarlanei*

pale green triangle *Austral. P. eurypylus*

green-spotted triangle *P. agamemnon*

Triaxonida = Hexactinellida a class of the phylum Porifera in which the skeleton is composed of silica

pleurotribal said of flowers which are adapted to smearing pollen along the sides of insect visitors

tribe 1 (*see also* tribe 2) term for a group of goats, though flock or herd is more common

tribe 2 (*see also* tribe 1) a primary division, in botanical taxonomy, of a subfamily, usually distinguished by a name terminating in -oidea. Tribes may be divided in subtribes, the names of which terminate in -inae, a termination reserved in *zool.* for subfamilies. In zoological taxonomy tribe ranks between a subfamily and a genus

-trich- *comb. form* meaning "hair" or by extension, "cilium"

Trichedidae = Odobaenidae

trichium a hair in any sense of that term

actinotrichium one of many slender elastic rods forming, usually when combined in bundles, the soft, as distinct from bony, fin rays of teleost fishes

bothriotrichium in some insects and arachnoids a long sensory seta that arises from a cup-like indentation

ceratotrichium a horny fin-supporting, rod overlapping the radial in the fin of an elasmobranch fish

dermotrichium a group term for ceratotrichia and lepidotrichia

epitrichium the fetal epidermis of some mammals

lepidotrichium a fin-supporting rod composed of ossified scales

macrotrichium a large hair, particularly those on the wings of insects

microtrichium a small hair particularly one on the wing of an insect

prototrichium a sensory, terminal bristle on the scales of some reptiles

trichite hardened, rod-like structures in the bodies of some Protozoa

trichidium = sterigma

Trichinellidae a family of minute trichuroid nematode worms. *Trichinella spiralis* is a notorious parasite of those who ingest undercooked pork

Trichiuridae a small family of acanthopterygian fishes with laterally compressed bodies and snake-like heads with huge teeth commonly called cutlass fish

Trichiuroidea a suborder of acanthopterygian fishes containing the families Trichiuridae, Gempylidae, and Lepidopidae. They are distinguished from the Percoidea by the beak-like premaxillaries and from the Scombroidea by the structure of the tail

Trichoceratidae = Trichoceridae

Trichoceridae a family of small slender, long-legged, nematoceran dipteran insects commonly called winter craneflies

trichode one of the scopulae through which the guests of ant colonies produce a presumably agreeable pheromone

Trichodectidae a family of mallophagan insects which are parasitic on mammals. They are distinguished from the Gyropidae by possessing filiform antennae

Trichodontidae a small family of marine scaleless acanthopterygian fishes with typical fringed lips. They are commonly called sand fish

Trichogrammatidae a family of tiny chalcidoid apocritan hymenopteran insects closely allied to the mymarids from which they differ, as well as from all other Chalcidoid families, by having three segmented tarsae

Tricholomataceae a family of agaricale basidiomycete Fungi distinguished by having very light spores and the fact that the gill tissue is composed of interwoven hyphae which gives them a dull, as contrasted to the usual waxy, appearance

trichoma a filamentous thallus

trichome 1 (*see also* trichome 2, 3) an epidermal appendage which may be either unicellular or multicellular, frequently called a plant hair. Many types of trichome are defined under hair 2

trichome 2 (*see also* trichome 1, 3) a many-celled, frequently branched, filament of bacteria or, less usually, algae

trichome 3 (*see also* trichome 1, 2) = trichode

Trichomycteridae a family of *S. Amer.* naked catfishes many of which have retrorse spines on the operculum which allow them to become parasitic on the gills of other fishes. The notorious candiru is also a member of this family

Trichonotidae a small family of acanthopterygian fishes many of which have scaly protections for the eyes. Commonly called sanddivers

Trichosomoididae a family of hair-like trichuroid nematode worms readily distinguished from the Trichuridae by the fact that the entire body is slender

Trichoptera that order of insects which contains the forms commonly called caddis flies. They are distinguished by two pairs of membranous wings that are usually densely hairy or scaly and with primitive venation

Trichopterygidae = Ptiliidae

Trichosceidae a family of brachycerous dipteran insects usually included in the Chyromyiidae

Trichotropidae a small family of gastropod Mollusca commonly called the hairy-keeled shelled snails. The name is descriptive

-trichous *adj. suffix* meaning "pertaining to hair" in any meaning of that word

atrichous lacking hairs, but used by extension in the sense of non-motile or lacking cilia

amphitrichous said of a cell having a flagellum at each end

basitrichous hairy or spiny at the base only

heterotrichous covered with different sizes of hairs, spines or cilia

holotrichous uniformly hairy, spiny or ciliated

homotrichous covered with hair, spines or cilia of the same size throughout

lophotrichous used of microorganisms having cilia in tufts like the Hypotricha

merotrichous spines, hairs or cilia in the middle portion, but lacking at the end

microtrichous clothed with microscopic hairs

telotrichous with hair, spines or cilia at the distal end only

Trichuridae a family of trichuroid nematode worms called whipworms by reason of the narrowness of the anterior portion

Trichuridea a class of filiform parasitic nematodes without lips and with a slender pharynx. The males lack a copulatory apparatus

Trichuroidea a suborder of hologonian Nematoda distinguished by their small size and absence of an esophagus

Tricladida an order of turbellarian platyhelminths distinguished by having two posterior and one anterior branch of the intestine

Tricocceae a suborder of geraniale dicotyledonous angiosperms containing the families Euphorbiaceae and Bathniphyllaceae

tricussate as decussate but with three leaves in each bunch

Tridactylidae that family of orthopterous insects which contains the forms commonly called pigmy mole crickets. They resemble small gryllotalpids save that the hind legs are adapted for jumping and the expanded tarsae of the forelegs have two segments

[trident] a mutant (gene symbol *tri*) mapped at 55.0± on chromosome II of *Drosophila melanogaster*. The phenotypic expression is a darkened thorax

triduus lasting for three days

[trifoliate] a mutant (gene symbol *tf*) mapped at 15 on chromosome 7 of the tomato. The terminal leaflet of each leaf is divided into three

Triglidae a family of scleroparous fishes. The bright colors cause them to be called sea-robins, though the most distinctive characters are the helmet like head and large pectoral fins, the lower rays of which are free and are used for a "walking action" on the sea bottom

trigon the primary triangle in which the cusps are arranged

Trigonalidae a family of apocritan hymenopteran insects variously placed in the Ichneumonoidea and other superfamilies. They are medium sized, stout-bodied, brightly colored wasps but with extremely long antennae

Trigoniaceae a small family of geraniale dicotyledons

trigonid the basic triangle of cusps on a lower molar (*cf.* trigon)

Trigynia those orders of the Linnaean system of plant classification distinguished by the possession of three pistils

trill a call involving the rapid repitition of the same note

triller any of several campephagid birds of the genus Lalage

   varied triller *L. leucomela*

   white-winged triller *L. sueurii*

trimous = triennial

Trionychoidea an assemblage of chelonian reptiles in which the shell is flat and oval and covered with a soft leathery skin. The limbs are broadly webbed. Many are called soft-shelled turtles

Triopidae = Apodidae

trip term for a group of goats, though flock or herd is more common, or for hares (*cf.* husk, down, drove)

-tripl- *comb. form* meaning "triple"

Triploblastica a term used to distinguish those phyla of the animal kingdom (*i.e.* all phyla except Protozoa, Porifera, Coelenterata and Ctenophora) which have a layer of tissue (mesoderm) between the ectoderm and the endoderm

tripton the sum total of the suspended dead matter in water. Differs from leptopel in that the term is not restricted to minute particles (= abioseton)

triquetrous three edged

-trit- *comb. form* meaning "third"

Tritomidae = Mycetophagidae

triton any of numerous cymatid gastropod mollusks particularly those of the genus Cymatium

triungulin *see* larva

trivium literally, the junction of three roads, but in *biol.* specifically applied to those three arms which do not form the bivium (*q.v.*) of an asteroid echinoderm

Trixagidae = Throscidae

Trixoscelidae = Trichoscelidae

-troch- (*see also* larva) *comb. form* meaning "wheel"

   metatroch that band of cilia on a trochophore larva which lies between the anus and the mouth

   paratroch = metatroch

   prototroch any of one to three bands of cilia on a trochophore larva (*q.v.*) which lie anterior to the mouth

   pseudotroch areas of cilia or stiff compound cilia in the supraoral region of the buccal field of the corona of some Rotifera

   telotroch that band of cilia on a trochophore larva which immediately surrounds the anus

trochal pertaining to -troch

   postrochal that part of a trochophore larva which lies posterior to the mouth

   pretrochal that part of a trochophore larva which lies anterior to the mouth

trochanter that segment of the arthropod limb, particularly in Insecta and Arachnida, which lies between the coxa and the femur

trochantin a sclerite between the mandible and gena in Orthoptera or the proximal division of the trochanter when the latter is divided in two

   entrochantin = coxopleurite

Trochidae a family of gastropod mollusks commonly called the pearly snails. Their outstanding characteristic is the large quantity of nacre which composes almost the whole of the shell

Trochilidae an enormous family of apodiform birds commonly called hummingbirds. The extremely rapid beating of the wings and hovering flight is typical

trochlea (*see also* trochlea 1, 2, 3) literally a pulley

trochlea 1 (*see also* trochlea, trochlea 2, 3) an articular surface on a large bone such as the humerus

trochlea 2 (*see also* trochlea, trochlea 1, 3) any pulley-shaped structure in a plant

trochlea 3 (*see also* trochlea, trochlea 1, 2) in insects, a thickened area at the base of the wing

Trochodendraceae a monotypic family of ranale dicotyledons. The family is distinguished by the persistent pinnate-veined leaves with bisexual bracteolate flowers

Trochodendrineae a suborder of ranale dicotyledonous angiosperms containing the families Trochodendraceae and Cercidiphyllaceae

Trochozoon a hypothetical ancestor, in the form of a trochophore larva, from which groups having such a larva might have been derived (*see* Hatschek's theory)

trochus 1 (*see also* trochus 2) a sclerite intercalated between the usual segments of the appendage of an arthropod

trochus 2 (*see also* trochus 1) the inner of two membranelles of compound cilia, replacing the single apical band in the corona of some Rotifera (*cf.* cingulum)

-trog- *comb. form* meaning "to gnaw"

Trogidae a family of coleopteran insects closely allied to the Scarabaeidae with which they are often fused. They are frequently called skin beetles from the fact that they are often found in partially dried carcasses

-troglo- *comb. form* properly meaning a "gnawed hole" but mostly used in *biol.* in the sense of "cave"

Troglodytidae that family of passeriform birds that contains the wrens. They are mostly small birds with slender curved beaks and very active and inquisitive habits

trogon any of many species of trogonid birds mostly of the genus Trogon without qualification, usually *T. elegans*

   citreoline trogon *T. citreolus*

   collared trogon *T. collaris*

   Cuban trogon *Priotelus temnurus*

   elegant trogon *T. elegans*

   Hispaniolan trogon *Temnotrogon roseigaster*

   mountain trogon *T. mexicanus*

   coppery-tailed trogon = elegant trogon

   slaty-tailed trogon *T. massena*

   violaceous trogon *T. violaceus*

Trogonidae a family of trogoniform birds commonly called trogons. They are brilliantly metallic-green birds with conspicuous stripes on the tail with very small and weak feet and legs

Trogoniformes that order of birds which contains the trogons. They are distinguished by the short stout beak with imperforate feathered nostrils and their soft metallic colored easily detached plumage

Trogositidae = Ostomatidae

Trombiculidae a family of trombidiform acarine Arthropoda containing the chiggers or red-bugs. They can usually be distinguished by the velvety appearance of their 8-shaped bodies

Trombidiformes an order of acarine Arachnida distinguished by the usual presence of a pair of stigmata on, or near, the gnathosoma; the palpi are usually free, the chelicerae seldom chelate, and there are no anal suckers. This group contains the free-living water mites, as well as numerous plant and animal parasites, including the chiggers

Trombidiidae a family of trombidiform acarine Arthropoda containing the harvest mites. They can usually be distinguished by their soft bodies and velvety appearance and the division of the body into three, somewhat distinct, regions, in contrast to the 8-shaped body of the Trombiculidae

phare**tronid** said of a type of sponge skeleton in which the calcareous spicules are united into a solid framework of square meshes

troolie the leaf of the palm *Manicaria saccifera*

troop term for a group of kangaroos, though mob is equally common. It is also used for monkeys, for buffaloes (cf. gang), and for foxes

-trop- *comb. form* meaning "turn" or, more rarely "keel." -trop- meaning "turn" is usually indicated in *biol.* usage by the *subs. suffix* -tropism (*q.v.*) and the *adj. suffices* -tropic and -tropous. Corresponding forms for -trop meaning "keel" are usually -trope and -tropal (*cf.* -stroph-)

Tropaeolaceae that family of geranialeous dicotyledons which contains, *inter alia*, the nasturtiums. The family is principally distinguished from the closely allied Geraniaceae by its separate stamens and its peculiar indehiscent fruitlets

leco**tropal** horseshoe-shaped

palin**trope** a recurved portion of the ventral valve of a brachiopod shell which protrudes into the gap between the beak and the hinge line of the dorsal valve

-troph- *comb. form* meaning "nutrition" very frequently confused with -trop

-troph *subs. suffix* indicating an organism classified by its mode of nutrition (*cf.* -trophe, -trophy) The forms -troph and -trophe are frequently interchanged

   ectotroph an external parasite

   endotroph an internal parasite

   paratroph = parasite though sometimes used for obligate parasite

   prototroph a mutant microorganism which can grow in a medium lacking a factor necessary for the growth of the wild type (*cf.* prototrophic)

-trophe *subs. suffix* indicating nutritional materials (*cf.* -troph, -trophy) The forms -trophe and -troph are frequently interchanged

   embryotrophe either embryonic nutritional material derived by mammalian embryos from the uterine wall or the sum total of the nutritional materials secured by an embryo from any source

trophi any of numerous cuticularized pieces within the mastax of a rotifer. Sometimes used as synonomous with mastax

-trophic *adj. suffix* referring to food or nutrition

   atrophic aplastic

   acrotrophic the condition of an ovary or ovariole of which the yolk is produced by terminal cells

   allotrophic said of an organism dependent on other organisms for its nutrition. Synonymous with parasitic only if all animals are regarded as parasitic on plants

   amphitrophic said of an ovule that is recurved into a sickle shape

   autotrophic said of an organism capable of utilizing inorganic materials in the synthesis of living matter or more rarely of any organism capable of securing its own food as opposed to a parasite

   auxo-autotrophic an autotroph requiring certain specific nutrients (*cf.* auxotrophic)

   auxotrophic said of microorganism requiring certain specific nutrients (*cf.* prototrophic)

   ectotrophic 1 (*see also* ectotrophic 2) said of those invertebrates, particularly polychaete annelids, in which the mouth parts project from, and are not sunk into the head (*cf.* entotrophic)

   ectotrophic 2 (*see also* ectotrophic 1) deriving nourishment from the outside, without penetration

   endotrophic said of a parasitic fungus which penetrates the cells of its host

      ectendotrophic said of a parasite which is both external and internal

      exendotrophic said of a flower fertilized from another flower either of the same, or of a different plant

   entotrophic said of those invertebrates in which the mouthparts are withdrawn into the head (many polychaete annelids) or sunk in a pit in the head (many insects) (*cf.* ectotrophic 1)

   eutrophic rich in nutrients, said particularly of bodies of water and of swamplands (*cf.* eutrophy)

   exotrophic the condition of having a lateral shoot more developed than the shoot from which it springs

   heterotrophic said of an organism capable only of utilizing organic compounds for nourishment or of one having a varied diet (*cf.* heterotrophy)

   holotrophic said of a plant the nutrition of which is based on the synthesis of carbohydrates (= holophytic nutrition) or of a predator that only preys on a single species

   hypotrophic said of an organism, such as a virus, which utilizes the substances of the host cell

   chemolithotrophic said of microorganisms that use only inorganic substances

      photolithotrophic a type of microbial nutrition that requires inorganic hydrogen donors in the medium (*cf.* photoorganotrophic)

   mesotrophic moderately nutrient, or moderately productive when used of waters or wet lands (*cf.* mesotrophy)

   metatrophic existing on organic nutrients only

   mixotrophic said of a plant capable of both holophytic and saprophytic nutrition or of any organism living on a mixed diet

   monotrophic said of a parasite that is host-specific, a predator that feeds only on one species, or a bee that visits only one species of flower

   mycotrophic a plant having a mycorrhizal association

   myrmecotrophic furnishing food for ants

   myxtrophic being nourished through the ingestion of particles

   oligotrophic 1 (*see also* oligotrophic 2) said of plants that will grow on poor soil, of any organism that requires little food and of organisms that are restricted to a narrow range of nutrients

   oligotrophic 2 (*see also* oligotrophic 1) said of environments poor in nutrition, particularly deep waters typified by low productivity

   chemoorganotrophic said of microorganisms dependent on organic substances

      photoorganotrophic a type of microbial nutrition which requires organic hydrogen donors in the medium (*cf.* photolithotrophic)

   polytrophic said of organisms deriving food from a wide area, of bees that range over a large number of flowers, or of organisms having an extremely varied diet

prototrophic said of microorganism capable of growth on simple sugars and salts without specific nutrient requirements (*cf.* auxotrophic, prototroph) and of any plant not requiring organic nutrients

syntrophic = epiphytic (cf. synthrophy)

systrophic causing, or resulting from, inadequate nutrition

trophime the condition of a developing seed plant in which the central nucleus of the embryo sac has fused with the second antherozoid

gonotrophium = soredium

trophon popular name applied to many species of muricid mollusks

-trophy *subs. suffix* indicating the condition or result of a method or type of nutrition. Most forms given under -trophic can be converted to this form but are not repeated here unless there is a marked difference in meaning (*e.g.* atrophic, atrophy)

atrophy withering away. The term is used of a withering from any cause not only from nutritional defects

embryotrophy the process of nutrition of a mammalian egg before it becomes implanted and thus haemotrophic (*cf.* embryotroph)

epitrophy the condition of growing faster or larger on the upper side

eutrophy the condition of flowers so modified that only certain insects can pollinate them (*cf.* eutrophic)

geotrophy = geoheterauxesis

haemotrophy the nourishment of an embryo through a connection with the maternal blood supply

heterotrophy literally, feeding on mixtures. Used specifically for Protozoa which ingest a variety of food materials or for symbiotic forms such as lichens which may themselves become dependent upon a symbiotic relationship with another form (*cf.* heterotrophic)

hypertrophy overgrowth. The term is not confined to overgrowth from excessive nutrition

compensatory hypertrophy the rapid overgrowth of a smaller existing organ to take the place of one which is lost. Thus in many arthropods with a large and small chela the result of the loss of the large chela is an enlargement of the small chela and a regeneration of a new small chela on the limb from which the large chela was lost

mesotrophy incomplete autotrophy (*cf.* mesotrophic)

paratrophy obligate parasitism

phototrophy unequal lateral growth due to the incidence of light

syntrophy the condition of two organisms mutually dependent for food (*cf.* commensalism and syntrophic)

tropic either of two imaginary lines running parallel to the equator at a distance of 23° 27' on each side of it. The "tropics" is the area contained between these lines

-tropic *adj. suffix* from -trop- in the sense of turning. All entries of these compound words are under the *subs. form* -tropism. (*cf.* -tropical)

tropical in the contemporary *biol.* this *adj.* is confined to the meaning "pertaining to the tropics" in the *ecol.* sense of that word in distinction from "-tropic" (*q.v.*)

neotropical pertaining to that portion of the tropics that lies in the New World

paleotropical pertaining to that portion of the tropics that lies in the Old World

Tropidopkinae a small family of South American boid snakes distinguished by the complete absence of the left lung and development of a tracheal "lung"

adrenotropin = adrenocorticotropin

corticotropin = adrenocorticotropin

adrenocorticotropin a hormone produced by the adenohypophysis that stimulates secretion by the adrenal cortex

adrenoglomerulotropin a hormone secreted by the pineal body which stimulates the secretion of aldosterone by the adrenal cortex

chorionic gonadotropin a hormone, necessary for the course of normal pregnancy, produced by the placenta

luteotropin = lactogenic hormone

melanotropin = melanocyte-stimulating hormones

somatotropin = growth hormone

thyrotropin = thyrotropic hormone

tropism a growth response of plants or sessile animals, thus corresponding to taxis in free-living forms. A positive tropism means a response in the direction from which the stimulus is coming; negative tropism means the reverse. The prefix phobo- is sometimes used to negate the operative prefix; thus phobophototropism is negative phototropism

acrotropism continued growth in the direction in which it originally started. Applied particularly to roots

aerotropism the production of movement by a gas acting chemically (*cf.* rheotropism)

aitiotropism any exogenous tropism

allotropism 1 (*see also* allotropism 2) a condition of having an abnormal mitotic cycle

allotropism 2 (*see also* allotropism 1) said of a flower having a plentiful supply of readily available nectar

anatropism the condition of an ovule, or seed derived from an ovule, which is inverted so that the micropyle is bent downwards

anthotropism the movement of a flower in response to stimuli

antitropism said of a plant that twines against the direction of the sum

aphercotropism changing direction away from an obstruction

autotropism continued growth in a straight line under endogenous control

acarpotropism the condition of a plant, the fruit of which is not shed

postcarpotropism the curving of the peduncle of a ripe fruit

chronotropism a movement due to age, applied particularly to the movement of leaves in plants

clinotropism the condition of a lateral organ of a plant that lacks a vertical plane of symmetry

anaclinotropism a tropism resulting in growth or movement in an horizontal plane

diatropism the tendency to become placed at right angles to an operative force

geodiatropism the orientation of an organ at right angles to the force field of gravity

dromotropism the spiral growth of plants occasioned by a response to touch stimulus

edaphotropism a response to soil water

electrotropism = galvanotropism

endotropism literally, the state of being "curved inward" but also used for the condition of being fertilized by pollen from another flower on the same plant

epitropism 1 (*see also* epitropism 2) the condition of an anatropic ovule with its raphe reversed in regard to the normal position

epitropism 2 (*see also* epitropism 1) = geotropism

eutropism the condition of a plant which twines in a direction following the passage of the sun

eurytropism capable of adapting to a wide range of environments

exotropism 1 (*see also* exotropism 2) said of a flower fertilized with pollen from the same plant

exotropism 2 (*see also* exotropism 1) the movement of lateral organs away from the main axis

galvanotropism movement in response to electrical stimulation

apogalvanotropism = negative galvanotropism

gamotropism the orientation of an expanded flower

agamotropism the condition of flowers which, once opened, do not again close

hemigamotropism used of flowers which alternately open, and then partially shut

geotropism movement in response to gravity

apogeotropism = negative geotropism

diageotropism (= zenogeotropism) a response which results in the growth parallel to and usually at or immediately under, the surface of the soil

progeotropism the tendency of a root tip to bend downwards

geoparallotropism a movement of an organ to become parallel to the surface of the soil (= geodiatropism)

haptotropism movement induced by contact stimulus in a plant

heliotropism the condition of turning in relation to the sun

apheliotropism = negative heliotropism

diaheliotropism the tendency of a plant, particularly a leaf, to orient itself at right angles to the rays of the sun

paraheliotropism the movement of leaves to avoid direct sunlight

hemitropism a practically meaningless word, that has been variously used to describe flowers which are specifically adapted to pollination by certain insects, the insects which take part in this process, and for ovules to which it is difficult to apply either the term amphitrophic or anatrophic

heterotropism *either* the condition of an ovule lying parallel to the hilum, *or* one which is amphitropous

homotropism 1 (*see also* homotropism 2, 3) the condition of being curved in one direction only

homotropism 2 (*see also* homotropism 1, 3) the condition of a flower fertilized by its own pollen

exhomotropism the condition of a flower which is fertilized with pollen from the same or a different plant

homotropism 3 (*see also* homotropism 1, 2) said of an organ that turns in the same direction as the body to which it belongs or in which it is placed

homolotropism growing in a horizontal direction

hydrotropism differential growth due to differential water supply

prohydrotropism turning towards moisture

magnetotropism the turning of a living object under the influence of a magnetic flux

mesotropism the condition resulting from a succession or change from xerotropic to hydrotropic

prometatropism the fertilization of a flower by pollen from another flower of the same species

nyctitropism the condition of a leaf which changes its position at night (*cf.* tuitant)

autonyctitropism automatically assuming a night posture

geonyctitropism night movements of a plant induced by gravity

orthotropism 1 (*see als* orthotropism 2) the state of standing upright

orthotropism 2 (*see also* orthotropism 1) the condition of a straignt ovule with the foramen at the opposite to the hilum

autoorthotropism the tendency of a plant organ to continue growing in the same direction

oxygenotropism movements induced by oxygen

oxytropism movements occasioned by acid

peritropism the condition of a seed that lies horizontal to the pericarp

phototropism movement stimulated by light. Without qualification, usually means positive phototropism

aphototropism = negative phototropism

diaphototropism a movement to orient an organism or part of an organism at right angles to incident light

dysphototropism requiring a medium amount of light

plagiotropism a response which results in growth away from a vertical axis

geoplagiotropism the movement of an organ into a position obliquely across the field of gravity

photoplagiotropism the arrangement of organs obliquely towards incident light

polotropism the tendency of two growing points to come together as in the conjugation of some algal cells

polytropism the condition of leaves that have the blade vertically with the two surfaces facing east and west

rheotropism the production of movement in response to flow of fluids or gases (*cf.* aerotropism)

selenotropism turning under the influence of the moon

siotropism a movement in response to shaking

skoliotropism the state of being curved

autoskoliotropism the condition of a plant structure that tends to grow in a curve

skototropism turning towards a dark place

stenotropism capable of adapting within narrow limits

stereotropism turning towards the substrate

thermotropism the condition of responding to heat

diathermotropism movement at right angles to a heat gradient

thigmotropism movement or, in plants, curvature stimulated by contact

zenotropism negative tropism

-tropo- -trop- 1 in the sense of change and used to denote climate that alternates between hot dry sun, and torrential rains

-tropous *adj. term* synonymous with -tropic. Compound forms will be found under -tropism

campylotrophous *see* ovule

-trorse *adj. suffix* indicating direction of movement (*cf.* -ad2)

antrorse directed frontwards or upwards

detrorse downwards

dextrorse towards the right hand

extrorse directed outwards

introrse turned inwards

retrorse directed backwards or downwards

trot a method of locomotion more rapid than a walk and in which the legs move in diagonal paths

lily-trotter applied to any jacanid bird, but most properly to *Actophilornis africana*

tramp's trouble *U.S.* the liliaceous herb *Smilax bonanox*

troupial any of several species of ictarid birds mostly of the genus Icterus (*cf.* oriole). Usually, without qualification, *I. icterus*

trout (*see also* perch) any of several species of salmonid fishes of the genera Salmo and Salvelinus *Brit.* without qualification, *Salmo trutta.* (= *U.S.* brown trout) (*cf.* char, splake)

    green-back trout a geographic race of the cutthroat trout

    bastard trout the marine teleost *Cynoscion nothus* (= bastard weakfish)

    brook trout *Salvelinus fontinalis*

    brown trout *Salmo trutta*

    bull trout = dolly varden

    cutthroat trout *Salmo clarki*

    Gila trout *Salmo gilae*

    golden trout *Salmo aguabonita*

    steelhead trout a sea-run rainbow trout

    lake trout *U.S. Salvelinus namaycush*

    mackinaw trout *U.S.* = lake trout *U.S.*

    native trout that trout which is native to the area in which it is so named. In the *U.S.* the term is beginning to be synonymous with cut-throat

    rainbow trout *Salmo gairdneri*

    rock trout = greenling

    sea trout *U.S.* any of several marine sciaenid fish of the genus Cynocion (= weakfish)

        sand sea trout *C. arenarius*

        silver sea trout *C. nothus*

        spotted sea trout *C. nebulosus Brit.* a sea-run *Salmo trutta*

    sleepy trout the eleotrid *Mogurnda mogurnda*

    Sunapee trout *Salvelinus aureolus*, considered by many to be identical with the arctic char (*q.v.*)

true-love-knot *Brit.* the liliaceous herb *Paris quadrifolia* (= herb-Paris)

truffle the edible, orange or black, tuber-like ascocarps of tuberale ascomycete fungi

-trull- *comb. form* meaning either a bracklayer's trowel or a small basin

trumpet (*see also* weed) any of several species of sarraceniaceous insectivorous herbs of the genus Sarracenia

    angel's trumpet either of two species of the solanaceous genus Datura, *D. suaveolens* and *D. arborea*

    humming bird's trumpet the onagraceous herb *Zauschneria californica*

trumpeter 1 (*see also* trumpeter 2, swan) any of three species of gruiform bird of the family Psophiidae. There is much confusion occasioned by the use of this single word to describe the unrelated trumpeter swan

trumpeter 2 (*see also* trumpeter 1) *Austral.* and *N.Z.* any of numerous acanthopterygean teleost fishes of the family Latridae

    bastard trumpeter *Austral.* and *N.Z.* the marine teleost *Latridopsis forsteri*

trumpet-of-death the cantharellaceous basidiomycete fungus *Craterellus cornucopioides* (= horn-of-plenty)

[truncate] a mutant (gene symbol *tc*) in linkage group XI of the mouse. The phenotypic expression is a shortened tail

[truncated] a mutant (gene symbol *td*) in linkage group

IV of *Habrobracon juglandis*. The wings of the phenotype are greatly reduced

trunk 1 (*see also* trunk 2, 3, 4) any portion of the body of an organism posterior to the neck

    alitrunk those segments of an insect thorax which bear wings

trunk 2 (*see also* trunk 1, 3, 4) a principal vessel or nerve to which other, smaller vessels or nerves are joined

    arterial trunk the main arterial vessel leaving the ventricle

    sympathetic trunk the longitudinal commissure connecting the vertebral ganglia of each side

trunk 3 (*see also* trunk 1, 2) the proboscis of an elephant

trunk 4 (*see also* trunk 1–3) the main unbranched woody "stem" of a tree

Trupaneidae = Tephritidae

protrusible said of a portion of an organism, particularly the pharynx of some invertebrates, which may be protruded without being turned inside out (*cf.* eversible)

try a bacterial mutant indicating a requirement for tryptophan

Trygonidae = Dasyatidae

tryma a drupaceous nut with a dehiscent exocarp

tryp-1–tryp-4 *see* [tryptophan -1] to [-4]

Trypaneidae = Tephritidae

Trypanorhyncha an order of eucestod cestods characterized by a scolex provided with four bothria and four protrusible spiny proboscides

Trypetidae = Tephritidae

Trypoxyloninae a subfamily of sphecid hymenopterans commonly called muddaubers because they make elongate cells of mud

trypsin an enzyme catalyzing the hydrolysis of numerous compounds having bonds with carboxyl groups of L-arginine or L-lysine

    chymotrypsin acts as trypsin, but on the carboxyl group of aromatic L-aminoacids

L-tryptophan 1-α-amino-3-indolepropionic acid. An amino acid essential to the nutrition of rats

[tryptophan-1] a mutant (gene symbol *tryp-1*) found in linkage group III of *Neurospora crassa*. The phenotypic expression is a requirement for either tryptophan or indole

[tryptophan-2] a mutant (gene symbol *tryp-2*) on linkage group VI of *Neurospora crassa*. The phenotypic expression is a requirement for either tryptophan or indole or o-aminobenzoic acid

[tryptophan-3] a mutant (gene symbol *tryp-3*) [the symbol *td* has also been used for this mutant] on linkage group II of *Neurospora crassa*. The phenotypic expression is a requirement for tryptophan and an inability to use indole

[tryptophan-4] a mutant (gene symbol *tryp-4*) in linkage group IV of *Neurospora crassa*. The phenotypic expression is a requirement for either indole or tryptophan

ts₁–ts₆ *see* [tassel seed 1] to [-6]

TSH = thyrotropic hormone

tt *see* [tilt] or [trifoliate]

Tu *see* [Tunicate]

tu *see* [tuft]

Stanley's wash tub the araceous herb *Amorphophallus campanulatus*

-tuba- *comb. form* meaning "trumpet" frequently confused with -tubi-

tube (*see also* tube 2) a hollow cylinder (*cf.* tubule)

    buccal tube that portion of the alimentary canal, particularly in invertebrates and hemichordates, which lies between the mouth and the pharynx

conjugation tube processes uniting a fertilized trichophore with an auxilary cell

Eustachian tube a duct connecting the tympanic cavity with the pharynx

Fallopian tube the female oviduct of the human (*see* Müllerian duct)

fertilization tube = pollen tube

germ-tube the initial beginnings of a hypha coming from a spore

Malpighian tube = Malpighian tubule

neural tube a term applied to the vertebrate central nervous system at a stage when the neural groove is completely invaginated, but the neural folds have not yet fused

ovule tube a thread-like extension of the amnios in plants

perichordal tube a thin layer of fusiform cells derived from the scelerotome, lying immediately around the notochord

pollen tube the tube that develops from the pollen in the course of fertilization

receptacle tube = calyx tube

sieve tube a long series of sieve tube elements placed end to end

pharyngotympanic tube = Eustachian tube

worm tube (*see also* tube 2) properly a tube surrounding an annelid worm, particularly those of the family Serpulidae

tube 2 (see also tube 1) in compound names of organisms

worm tube (*see also* tube 1) the vermetid gastropod mollusk *Vermicularia spirata*

tuber 1 (*see also* tuber 2, beetle) a swelling, usually for food storage, on the roots of a plant (*cf.* rhizome)

bulbotuber = corm

tuber 2 (*see also* tuber 1) a swelling on an animal or animal organ (*cf.* tubercle)

tuber cinereum one of a pair of nerve centers in the medulla anterior to the nucleus cuneatus

Tuberales an order of discomycete Ascomycetae distinguished by their tuber-like ascocarps which are the truffles of commerce

tubercle diminutive of tuber. Without qualification applied specifically to a prominence from the test of an echinoderm on which spines fit with a ball-and-socket joint and to any small hard prominence on an insect cuticle for which no better name can be found

scrobicular tubercle one of those echinoderm tubercles that carry secondary spines

frontal tubercle any lump rising from the dorsal surface of the head of an arthropod, but particularly the horn-like structure through which the frontal gland opens in some termites

hypothenar tubercle that which lies below the base of the external digit in anuran Amphibia

lateral tubercle a prominence from the sides of the segments of caterpillars

Morgagni's tubercle = Montgomery's gland

tuberculate covered in knobs

tuberculum the lower of the two articular processes of the vertebrate rib (*cf.* capitulum)

tuberose the amaryllidaceous herb *Polianthes tuberosa*

-tubi- *comb. form* meaning "pipe" frequently confused with -tuba-

Tubificidae a family of red, tubicolous, oligochaete annelids

Tubiflorae that order of dicotyledonous angiosperms which contains, *inter alia* the sweet potato, the convolvulus, the borages, the verbenas, the mints and lavenders, the nightshades, the foxgloves, the begonias, the cape primroses, and the Australian sandal tree. The order is distinguished by the gamopetalous corolla with the floral parts in four isomerous whorls

Tubulariae an order of hydrozoan coelenterates distinguished by the lack of a hydrotheca on the hydranth. The medusoid forms were at one time placed in a separate order the Anthomedusae

tubule diminutive of tube

collecting tubule a functional unit of a compound kidney into which several convoluted tubules open and which itself opens into the ureter

convoluted tubule that portion of the kidney unit which connects Henley's loop to the main collecting duct

Cuvierian tubule one of a mass of tubules found at the base of the respiratory tree in holothurian echinoderms. These tubules emit a sticky, thread-like secretion, in which predators, or prey, become entangled

Malpighian tubule *either* blindly ending, presumably excretory, tubules opening into the midgut of many arthropods *or* a chordate nephron

seminiferous tubule one of the tubules secreting sperm in the testes

-tubuli- diminutive of -tubi-

Tubulidentata a small order of placental mammals containing the forms commonly called aardvarks. They are distinguished, as the name indicates, by the possession of tube-like teeth

Tubulifera a suborder of thysanopteran insects distinguished from the Terebrantia by the lack of an ovipositor

tuckahoe *U.S.* the araceous herb *Peltandra virginica*

tucotuco fossorial hystricomorph rodents of the family Ctenomydae. In habit and appearance they greatly resembly pocket-gophers and, in the Argentine, occupy the same ecological niche as do these forms in *W.U.S.*

tuft 1 (*see also* tuft 2) a number of straight, or more or less straight, objects arising from a restricted area. In *bot.* the term is sometimes, though rarely, used as synonymous for cyme

apical tuft a tuft of relatively long cilia at the animal pole of a young echinoderm, and occasionally other invertebrate, larva

comal tuft one consisting of leaves at the apex of a branch (*cf.* coma)

tuft 2 (*see also* tuft 1) in compound names of organisms

candy tuft any of numerous species and very numerous horticultural varieties of the cruciferous genus Iberis. Without qualification usually refers to *I. amara*

annual candy tuft *I. umbellata*

fragrant candy tuft *I. odorata*

golden tuft the cruciferous herb *Alyssum saxatile*

purple tuft any of several cotingid birds of the genus Iodopleura

[tuft] a mutant (gene symbol *tu*) found in linkage group II of *Neurospora crassa*. The term refers to the large clusters of conidia on the phenotypic culture

[tufted] a mutant (gene symbol *tf*) in linkage group IX of the mouse. The phenotypic expression is the loss and regrowth of hair from front to back

tui the meliphagid bird *Prosthemadera novaeseelandiae*

tuitant the condition of a plant in which the leaves have assumed a "sleep position" alongside the stem (*cf.* nyctitropic)

tulip (*see also* shell) anglicized form of Tulipa a genus of liliaceous herbs. The majority of garden forms derive from *T. gesneriana* and the term tulip is also extended to many other liliaceous flowers

butterfly tulip any of numerous species of the liliaceous genus Calochortus

globe tulip any of numerous species of the liliaceous genus Calochortus

meadow tulip any of numerous species of the liliaceous genus Calochortus

star tulip any of numerous species of the liliaceous genus Calochortus

tumid swollen

tumpat-kurundu the rutaceous tree *Pleiospermum alatum* and its fruit

tun any of several species of tonnid gastropod of the genus Malea

tuna any of several species of scombrid fishes of the genera Thunnus and Euthynnus

bigeye tuna *T. obesus*

blackfin tuna *T. atlanticus*

blue-fin tuna *U.S.*, without qualification, *T. thynnus*

yellow-fin tuna *U.S.*, without qualification, usually *T. albacares*

little tuna *E. alletteratus*

skipjack tuna *E. pelamis*

tundra an area of relatively luxuriant growth of scrub and heathland lying between the southern limit of the ice and the northern limit of the tree zone

alpine tundra a tundra-like habitat occuring in mountainous regions at or around the limit of tree growth

arctic tundra that area of northern tundra which is north of the tree zone but on which there still occur a few stunted shrubs

antarctic tundra resembles the arctic tundra save that there are many fewer shrubs

forest-tundra intermittent timbered tracts on the northern verge of the limit of the growth of trees

lichen tundra an arctic area in which the predominant, or only, vegetation is lichens

shrub tundra that area of tundra in which shrubby growths are very noticeable or even predominant

sita **tunga** the *Afr.* antelope *Linnotragus spekei*

Tungidae a family of siphonapteran insects which are permanent parasites in the adult condition and therefore differ from all other fleas

tunic any covering. In *zool.* without qualification, usually applies to the more or less thickened external covering of urochordates and in *bot.* to the skin of a seed

Tunicata = Urochordata

[Tunicate] a mutant (gene symbol *Tu*) mapped at 100 on linkage group IV of *Zea mays*. The glumes of the phenotype are greatly enlarged

tunicin a polysaccharide, closely resembling cellulose, of which the tunic of urochordates is composed

tunny *Brit.*, without qualification, the scombrid fish *Thunnus thynnus* (= *U.S.* bluefin tuna)

Tupaiidae a family of oriental tupaioid primates closely resembling squirrels in general appearance

Tupaioidia a suborder of primate mammals containing the tree shrews. Though squirrel-like in general appearance they are intermediate in structure between primates and insectivores and are sometimes placed in the latter group

tupakahi *N.Z.* the coriariaceous shrub *Coriaria arborea* (= tree tutu)

tupelo the nyssaceous tree *Nyssa sylvatica*

tur a caprine caprid (*Capra caucasica*) of the Caucasus Mountains with huge recurved and incurved horns

tur a bacterial mutant showing a utilization of turanose

turacin a red pigment peculiar to feathers of birds of the family Musophagidae

turaco = touraco

turban any of several species of turbinid mollusks of the genus Turbo (*cf.* top shell)

turk's turban the verbenaceous herb *Clerodendron siphonanthus*

turbarian pertaining to a young peat bog still supporting dwarf willows

Turbellaria that class of the phylum Platyhelminthes which contains the free-living flatworms. They are distinguished from the Cestoda and Trematoda by the presence of external cilia and by the absence of the adaptations to parasitic life shown by the latter groups

-turbin- *comb. form* meaning "whirling around" and hence "top"

turbinate comb or peg-top shaped

turbine popular name of turbinid gastropod mollusks

Turbinidae a family of gastropod mollusks distinguished by their brightly colored turbinate shells

turbot *Brit.* any of several large flatfishes. Without qualification, the bothid fish *Scophthalmus maximus*

diamond turbot *U.S.* the pleuronectid *Hypsopsetta guttulata*

hornyhead turbot *U.S.* the pleuronectid *Pleuronichthys verticalis*

spotted turbot *U.S.* the pleuronectid *Pleuronichthys ritteri*

moustached **turca** the rhynocryptid bird *Pleroptochos megapodius*

Turdidae that family of passeriform birds to which the thrushes belong. They are mostly medium sized birds with short wings and short truncated tails with large strong legs and feet

turepo *N.Z.* the moraceous tree *Paratrophis nicrophylla* (= milk tree *N.Z.*)

turfaceus = torfaceus

turgescence swelling under the influence of water

turgid swollen with fluid

turgor the condition of being bloated, swollen or inflated

turion a scaley shoot arising from the ground

turkey (*see also* beard, louse, buzzard, vulture) either of two large galliform birds of the family Meleagrididae. *Brit.* and *U.S.* without qualification, *Meleagris gallopavo*. *Austral.* the otidid *Eupodotis australis* (= Australian bustard)

brush-turkey *Austral.* megapodiid bird *Alectura lathami*

common turkey *Meleagris gallopavo*

ocellated turkey *Agriocharis ocellata*

water turkey *U.S.* the anhingid bird *Anhinga anhinga*

turmeric the zingiberaceous herb *Curcuma longa* and particularly the condiment extracted from its root. The term is sometimes applied in *U.S.* to the ranunculaceous herb *Hydrastis canadensis* (= golden seal)

Turnagridae a family of passeriform birds, sometimes included with the Muscicapidae, that contains the New Zealand thrush (piopio), *Turnagra capensis*

Turneraceae a small family of parietale dicotyledonous angiosperms distinguished by the presence of a hypanthium and a tri-carpelate ovary

Turnicidae that family of gruiform birds which contains the button-quails. They are distinguished from the closely allied Pedionomidae by the possession of only three toes

turnip (*see also* aphis, flea, maggot) numerous horticultural varieties and hybrids of the cruciferous plants *Brassica rapa* and *B. campestris*. The name is extended to numerous plants having swollen edible roots or a habit of growth resembling the turnip

Indian turnip any of several species of the araceous genus Arisaema

marsh turnip the araceous herb *Arisaema triphyllum* (= jack-in-the-pulpit)

Swedish turnip any of several horticultural varieties of *B. campestris var. napobrassica* (= rutabaga)

Turnoididae a family of passeriform birds usually included in the Timaliidae

turnsole the euphorbiaceous herb *Chrocophora tinctoria*

turpentine (*see also* beetle, tree) the essential oil distilled from the balsam of several pinaceous trees but pariculary those of the genus Pinus

Turritellidae a family of gastropod Mollusca distinguished by the extremely elongate many-whorled shells

turtle (*see also* bug, head) *U.S.* any chelonian reptile. *Brit.* an aquatic chelonian as distinct from a terrestrial chelonian more usually called a tortoise

Alabama turtle the testudinid *Pseudomys alabamensis*

alligator turtle the chelydrid *Macroclemys temmincki*

bastard turtle the marine chelonid *Lepidochelys kempii* (= Kemp's loggerhead)

bog turtle *U.S.* the emydine testudinid *Clemmys muhlenbergi*

box turtle *U.S.* any of many species of emydine testudinid of the genus Terrapene

chicken turtle *U.S.* the testudinid emydine turtle *Deirochelys reticularia*

edible turtle *Brit.* = green turtle *Brit.*

green turtle the chelonid *Chelonia mydas*

hawksbill turtle the chelonid *Eretmochelys imbricata*

leathery turtle the dermochelydid *Dermochelys coriacea*

loggerhead turtle *Brit.* without qualification the cheloniid *Caretta caretta*

big-headed turtle popular name of platysternids

map turtle *U.S.* any of several species of emydine testudinid of the genus Graptemys

mud turtle popular name of kinosternine chelydrids though usually, *U.S.*, without qualification, *Kinosternum subrubrum*

musk turtle *U.S.* any of several species of kinosternine chelydrids of the genus Sternothaerus (*cf.* stink pot)

long-necked turtle *Austral.* any of several species of chelidid of the genus Chelodina

snake-necked turtle the cheliid *Emydura macquari*

painted turtle *U.S.* any of several subspecies of emydine testudinids of the monotypic genus Chrysemys

pond turtle *Brit.* the emydine testudinid *Emys orbicularis*

black pond turtle *I.P.* the emydine *Geoclemys hamiltoni*

soft-shelled turtle general name applied to trionychoids

Asiatic soft-shelled turtle any of several species of trionychid of the genus Lissemys

snapping turtle *U.S.*, without qualification, the chelydrid *Chelydra serpentina*

alligator snapping turtle *U.S.* the chelydrid *Macroclemys temmincki* (= alligator turtle)

wood turtle *U.S.* the emydine testudinid *Clemmys insculpta*

Turturidae a family of columbiform birds now usually included in the Columbidae

turumti *I.P.* the falconid bird *Falco chicquera*

tushe a term commonly applied to the tusk of a boar

tusk a protuberant mammalian tooth, usually a canine. The tusk of the elephant and the "horn" of the narwhale are, however, incisors

tussock a compact tuft of dried grass, or the like, protruding from the soil

tutsan *Brit.* the hypericaceous herb *Hypericum androsaemum*

tree tutu *N.Z.* coriariaceous shrub *Coriaria arborea* (= tupakahi)

tuturiwatu *N.Z.* the charadriid bird *Pluviorhynchus obscurus* (= New Zealand dotterel)

tw *see* [twisted]

Tw *see* [Twirler]

twablade *Brit.* the orchid *Listera ovata*

twig (*see also* beetle, borer, girdler, pruner) a terminal, slender, subdivision of a woody branch

twin one of two young produced of one birth by a viviparous animal

parabiotic twin a laboratory produced equivalent of a Siamese twin in a non-human animal

Siamese twin two individuals joined and usually sharing major blood vessels

[Twirler] a mutation (gene symbol *Tw*) found in linkage group XV of the mouse. The phenotypic expression is the waltzing syndrome

pearl twist *U.S.* popular name of the orchidaceous genus Spiranthes (= lady's tresses)

[twisted] a mutant (gene symbol *tw*) mapped 0.4± on the X chromosome of *Drosophila melanogaster*. The phenotypic expression is a counterclockwise twisting of the abdomen

twite the fringillid bird *Acanthis flavirostris* (= mountain linnet)

herb-twopence *Brit.* the primulaceous herb *Lysimachia nummularia* (= money wort)

tx *see* [taxi]

ty *see* [tiny]. Also used for a yeast genetic marker indicating a requirement for tyrosine, or a multiple requirement for tyrosine and ϕ alanine and tryptophan

ty-1 *see* [tiny-like]

clover leaf tyer the tortricoid olethreutid lepidopteran *Ancylis angulifasciana* (*cf.* strawberry leaf-roller)

-tyl- *comb. form* meaning a "knot" in the sense of lump

tylarus the thickened pad on the undersurface of the toe of a bird

Tylenchida an order of phasmid nematodes usually with several lips and always with a protrusible mouth stylet

tylenchoid *see* pharynx

Tylidae a family of passeriform birds commonly included with the Oriolidae

Tylidae a family of brachycerous dipteran insects now usually included in the Micropezidae

tyloides a ventral tubercle on the antenna of some male hymenopteran insects

Tylopoda an assemblage of ruminant artiodactyle mammals containing the single family Camelidae and their extinct allies. The characters of the family are those of the assemblage

tylose a protrusion of the membrane of an adjacent cell into the cavity of a vessel member

tylosoid an enlarged, epithelial cell closing a resin duct

tylote a rhabd knobbed at each end

tylus the end of the clypeus in Hemiptera, separated from the juga (*q.v.*) by deep grooves

-tympan- *comb. form* meaning a "drum" or "sounding board"

tympanum a cavity between the outer and inner ear, or the middle ear. The term is often applied by abbreviation to the tympanic membrane which separates it from the outer ear

type 1 (*see also* type 2) specifically, in *biol.*, the individual or group of individuals from which the species is described, or a species which serves as the type of a genus. The terms placed in parentheses are no longer recognized by the International Code

allotype a supplementary type described from the specimen of the opposite sex to the original

neallotype a type of the opposite sex from the holotype or any of the series of paratypes and of which a description is published after the description of the original type

(**androtype**) a male type

(**apotype**) a specimen used to augment information available about a type

(**autotype**) a specimen designated by the describer of a species as, in his opinion, identical with the holotype or paratypes

(**chirotype**) a specimen serving as the origin of a chironym

(**clastotype**) a fragment of a type

(**clonotype**) a specimen obtained by asexual reproduction from the original type

(**cotype**) any one of the type series (holotype and paratype)

ecotype locally adapted variants of a given plant species

generitype the type of species of a genus (*cf.* genotype in type 2)

isogenotype a type species to which two genetic names have been applied

(**gynetype**) a female type

(**haplotype**) a single species in its original place of publication

(**heautotype**) a type of specimen selected by the author of the species, but not the original specimen, on which the species was based

(**heterotype**) a so-called type which has been derived by combining the characters of two different species

holotype one specific, recorded and preserved specimen from which a species is named

genoholotype an individual species forming the holotype of a genus

isoholotype a plant specimen taken at a later period from the same bush or tree from which the holotype was originally taken

(**homotype**) a specimen which is considered, after direct comparison, to be identical with a type in those cases in which the comparer is not the original author

(**hypotype**) an individual specimen with the aid of which supplementary characteristics are ascribed to a type

(**icotype**) a type which is not described in the literature

(**idiotype**) a type identified by the original author of the species, but not taken by him in the same locality as the original type (*but see* idiotypic)

(**isotype**) (*see also* type 2) a type described from two species of the same genus

(**lectotype**) an individual selected from a series of cotypes to serve as the holotype of a revised description

(**genolectotype**) the one species of a series designated as the type of a genus in which the describer placed it subsequent to the description

(**isolectotype**) a lectotype chosen after a longer interval of time

(**paralectotype**) those specimens which remain in a series of cotypes after a lectotype has been selected from them

(**logotype**) a type determined from a written description in the absence of either a specimen or an illustration

merotype a type derived by vegetative reproduction from the original

(**metatype**) an individual, which may or may not be a topotype, but which is considered by the author of a species, after due comparison with the holotype or cotypes, to be of the same species

(**monotype**) (*see also* type 2) a holotype which is also the only known specimen

(**morphotype**) the equivalent of a type determined on another than the type form of a polymorphic species

neotype a type selected to represent the original type, usually collected from the original locality, when the original is no longer available

(**nepionotype**) the first description of an organism from a larval stage

(**orthotype**) a genus provided with a type by original designation

paratype 1 (*see also* paratype 2) specimens other than holotypes or allotypes included in the original type series by the author

(**tectoparatype**) a paratype specially prepared for microscopic examination

(**paratype 2**) (*see also* paratype 1) a specimen which is considered, after due comparison, to be identical with a type by an individual not the original author of the type

(**parallelotype**) = paratype

(**phototype**) a photograph of a type specimen or a type specimen or a type based on a photograph

(**plastotype**) a physical, three-dimensional, reproduction of a type made by a method presumed to reproduce accurately all relevant features

plesiotype any specimen which is thought to be identical with a holotype or paratype by an individual other than the original describer

(**tectoplesiotype**) a type, neither a cotype nor a paratype, which has been prepared for microscopic examination

(**proterotype 1**) see also proterotype 2) any and every specimen which has ever served to designate a species

(**proterotype 2**) (*see also* proterotype 1) all the original specimens collected at one time from which types have been selected

(**pseudotype**) a type designated in error

(**syntype**) = cotype

(**genosyntype**) a number of species forming a genus, no one of which can be selected, either as agenolectotype or a gonoholotype

topotype a specimen collected from the type locality. Usage varies as to whether this collection must be made by the original describer

type 2 (*see also* type 1) in the more generalized sense of a kind of something

    agrotype an agricultural race or variety

    archetype a postulated ancestor of a group, or a generalized structural plan of a group of organisms

    biotype a group of individuals having the same genotypic, but not the same phenotypic, constitution

    ecotype a phyletic unit adapted to a particular evironment but capable of producing fertile hybrids with other ecotypes of the same ecospecies

        climatic ecotype an ecotype affected by climatic as distinct from soil conditions

        edaphic ecotype an ecotype affected by soil as distinct from climatic conditions

    genotype a term used to designate the genetic composition (*cf.* phenotype). Frequently incorrectly used for generitype (*see* type 1)

    isotype (*see also* type 1) a form occuring in many different localities

    karyotype two individuals are said to be karyotypic when the physical appearance of their chromosomes appears to be identical

    mimotype a form which resembles another in general shape, but not in genetic makeup, and which occupies a similar ecological niche in another part of the world

    monotype (*see also* type 1) a genus containing a single species

    phaenotype = phenotype

    phenotype used to designate the physical appearance of an organism as distinct from its genetic constitution (*cf.* genotype, phenocopy)

        ecophenotype a phenotype showing non-heritable variations from the norm in consequence of an adaptation to an environment

    prototype a hypothetical ancestor

    reaction type = phenotype

    serotype an antigenic character of an organism

    wild type the stock or population having the phenotype normally found in nature or the normally encountered allele at any locus

Typhaceae the cattail family of pandanale monocotyledons. The family is characterized by the simple pistil, the hairy pedicels and the flowers borne in spikes without a perianth

Typhlopidae a family of burrowing snakes covered with uniform cycloid scales

typhlosole a longitudinal fold projecting into the intestine in many invertebrates

-typic *adj. suffix* from type in either meaning assigned above. Direct derivations from the substantive forms are not given.

    atypic not tyypical

    bitypic said of a genus containing only two species, each presenting opposite extremes of the genetic type

    dichotypic = dimorphic but, particularly applied to leaves and flowers

    hermatypic said of those reef forming corals the polyps of which contain symbiotic green algae in contrast to the ahermatypic corals which do not

    ahermatypic said of a coral which lacks symbiotic algae

    idiotypic sexual (*but see* idiotype)

    monotypic a taxon containing a single subordinate unit

    polytypic a taxon containing many subordinate units

    zelotypic asexual

-typlo- *comb. form* meaning "blind" and hence "caecum" in the sense of a blindly ending tube. Typhlosole (*q.v.*) is possibly intended to be derived from this root

tyr a bacterial genetic marker indicating a requirement for tyrosine, mapped at 48.25 mins. for *Escherichia coli*

tryA a bacterial genetic marker affecting the activity of trypotophan synthetase, A protein, mapped at 24.0 mins. for *Escherichia coli*

tyrB a bacterial genetic marker affecting the activity of tryptophan synthetase, B protein, mapped at 24.0 mins. for *Escherichia coli*

tyrC a bacterial genetic marker affecting the activity of indole 3-glycerolphosphate synthetase, mapped at 24.0 mins. for *Escherichia coli*

tyrD a bacterial genetic marker having a character affecting the activity of 3-enolpyruvylshikimate 5-phosphate to anthranilic acid mapped at 24.0 mins. for *Escherichia coli*

tyrE a bacterial genetic marker having a character affecting the activity of anthranilic acid to anthranilic acid to anthranilic deoxyribulotide, mapped at 24.0 mins. for *Escherichia coli*

tyr-1 *see* [tyrosine-1]

Tyrannidae a family of passeriform birds containing the tyrant-flycatchers. They have broad flattened and hooked beaks with large rictal bristles and a short broad tail. The feet and legs are small

tyrannulet any of numerous small species of tyrannid birds

tyrian purple a 6-6'-dibromo indigo derived from several gastropod mollusks of the family Thaisidae

Tyroglyphidae a family of free-living sarcoptiform acarine arthropods of minute size with a thin integument bearing long setae, the hysterosoma unsegmented and with the tarsi bearing sessile caruncles. The chelicerae are chelate and all species lack eyes and trachea

[tyrosinase thermo-stability] a group of pseudoalleles in linkage group I of *Neurospora crassa*. The gene controls tyrosinase thermostability

L-tyrosine β-(p-hydroxyphenyl)alanine $HOC_6H_4CH_2 \cdot CH(NH_2)COOH$. An amino acid necessary to the nutrition of rats

    L-3, 5-dibromotyrosine an amino acid analog of tyrosine found principally in anthozoan coelenterates

    diiodotyrosine an amino acid analog of tyrosine found principally in anthozoan coelenterates

    djenkotyrosine an amino acid analog $D\beta'$-methylenedithiodialanine, deriving its name from the djenkol bean from which it was first isolated

[tyrosine-1] a mutant (gene symbol *tyr-1*) found in linkage group III of *Neurospora crassa*. The phenotypic expression is a requirement for tyrosine

Tytonidae a family of strigiform birds commonly called barn owls and most readily distinguished from the true owls by the relatively small eyes and pectinate middle claws

u *see* [uniform-ripening fruit]

U *see* [Uropygial bifurcation], [Ursa] or [Upturned]

uacaris pithecid primates of the genus Cacajao. They are distinguished by their reduced tails

udder a large, pendulous, multi-nippled mammary gland such as that of the domestic cow

UDP = uridine diphosphate

UDPG usually means UDP glucose but has also been used for UDP glucuronate

ula-ai-hawane the drepaniid bird *Ciridops anna*

Ulidiidae a family of brachyceran dipteran insects usually included with the Otitidae

uliginose growing in or pertaining to swamps

ulili the scolapacid bird *Heteroscelus incanus* (= wandering tattler)

Ulmaceae that family of urticalous dicotyledons which contains, *inter alia*, the elms. The family is easily distinguished by the more lastic stamens and the suspended anatropous seeds

Ulmoideae a subfamily of Ulmaceae distinguished by the fact that flowers are borne on twigs of the previous season (*cf*. Celtidoidea)

ulna *see* bone

ulona the fleshy mouth parts of orthopteran insects

Ulotrichaceae that family of ulotrichale Chlorophyceae which are distinguished by the unbranched filaments of the uninucleate cells

Ulotrichales that order of chlorophycean Chlorophyta distinguished by their uninucleate cells united end to end in simple or branched filaments

ulua the carangid fish *Caranx melampygus*

-ulus *adj. suffix* meaning "somewhat" or "slightly"

Ulvaceae that family of ulvale Chlorophyceae in which the thallus is an expanded sheet, hollow cylinder, or ribbon two or more cells broad

Ulvales that order of chlorophycean Chlorophyta containing those forms with uninucleate cells arranged in a thallus two or three cells thick

umbel a type of inflorescence in which the pedicels appear to spring from a common point, usually resulting in a flat flower cluster

    compound umbel one which itself consists of other umbels

    cyme umbel an umbel with a centrifugal inflorescence

    partial umbel = simple umbel

    radiant umbel one in which the outer flowers are conspicuously the largest

    similiflorous umbel one in which all the flowers are identical

Umbellales = Umbelliflorae

Umbelliferae a very large family of umbelliflorous dicotyledons including the sea hollies and numerous edible plants such as parsley, cumin, fennel, the carrots, etc. The umbelliferous flower is unmistakeably distinctive

Umbelliflorae a large order of dicotyledonous angiosperms containing, in addition to the well-known Umbellifereae, the dogwoods and ginsengs. The arrangement of the simple flowers in umbels is characteristic of the order

umbellule the terminal umbel of a compound umbel

umbilicate with a depression like the navel

umbilicus literally, the navel and applied to so many depressions of this shape in various organisms as to have no specific meaning

umbles = numbles

umbo literally, the boss in the center of a round shield, and therefore applied to any prominence in the center of a round mass such as the beak of a brachiopod shell or the protuberance on the surface of many pelecypod mollusc shells

umbones plural of umbo

umbracul *comb. form* meaning "umbrella" or "parasol"

obumbrant overhanging

umbrella (*see also* bird) the main mass of a medusoid coelenterate

    exumbrella the convex oral surface of the umbrella of a medusa

    subumbrella the concave oral surface of the umbrella of a medusa

Umbridae a small family of haplomous fishes closely allied to the Esocidae but differing from them in the rounded snout. Commonly called mud minnows

umbrosous = umbraticolous

umbu the anacardiaceous tree *Spondias tuberosa* and its edible fruit

UMP uridine 5 -phosphate

un- *comb. prefix* meaning "not". Words beginning with this form are not given in the present dictionary, unless the meaning is not self-evident

un *see* [undulated] or [uneven]

un² *see* [undulating-2]

[uneven] a mutant (gene symbol *un*) mapped at 54.4 on X chromosome of *Drosophila melanogaster*. The unevenness refers to the condition of the small rough eyes of the phenotype

unau the bradypodid xenarthran mammal *Choloepus didactylus* (= two-toed sloth)

-unci- *comb. form* meaning "hook"

uncinate hooked, or having the form of a hook, or possessing an unusually prominent uncus in those forms to which this term is applied

unctuous greasy

uncus literally, a hook. Specifically, one of the hook-shaped pair of trophi in the mastax of a rotifer, a hooked process on the intromittent organ of some insects and the anterior end of the hippocampus

subuncus a small piece occuring between the uncus and the ramus of the rotiferan mastax

undulate wavy

triundulate used of a diatom having three undulations on the frustule

[**undulated**] a mutant (gene symbol *un*) in linkage group V of the mouse. The phenotypic expression involves a kinked tail and kinked vertebral column

[**undulating-2**] a mutant (gene symbol *un²*) in linkage group VI of *Habrobracon juglandis*. The name is descriptive of the wing surface of the phenotype

[**unequal wings**] a mutant (gene symbol *uq*) mapped at 0.5± on the X chromosome of *Drosophila melanogaster*. The wings of the phenotype are always short and frequently unequal

[**uneven**] a mutant (gene symbol *un*) mapped at 54.4 on the X chromosome of *Drosophila melanogaster*. The unevenness refers to the condition of the small, rough eyes of the phenotype

-unqui- *comb. form* meaning "claw," "talon," or "nail"

exunguiculate lacking a claw

unguiculus 1 (*see also* unguiculus 2) literally a small hook. Specifically the ventral of the two claws which terminate the leg of a collembolan insect (*cf.* unguis)

unguiculus 2 (*see also* unguiculus 1) an obsolete measure of length derived from a fingernail. In general, about half an inch

unguis one of the post-tarsal claws in insects and specifically the dorsal of the two claws which terminate the leg of a collembolan (*cf.* unguiculus 1)

-ungulat- *comb. form* sometimes used for "claw" but more properly meaning "hoof"

Ungulata an order of placental mammals commonly referred to as the hoofed animals. Now obsolete and divided into two separate orders the Perissodactyla and the Artiodactyla

ungulate *zool.* hoofed *bot.* clawed

uni- *comb. prefix* meaning "one" or "single". Compound words beginning with uni- are not given in this dictionary, unless the meaning is not self-evident

unicorn (*see also* caterpillar, plant, shell) a mythical, forest-dwelling, ungulate variously supposed to be the product of bad eyesight, imagination, or a fraudulent trade in narwhal tusks

uniflagellatae a class of phycomycete fungi distinguished by the possession of chitinous cell walls. The name is, however, derived from the presence of a single flagellum on the mobile gametes

[**uniform-ripening fruit**] a mutant (gene symbol *u*) mapped at 43 on chromosome 10 of the tomato. The ripening phenotypic fruit is uniform green without dark patches

Unionidae a family of pelecypod Mollusca commonly called freshwater clams. They are distinguished by the heavy equivalved shells with a very dark colored periostracum. The ligament is external and very prominent

climax **unit** a group of associated species contributing to a climax

unka-puti the Malayan gibbon *Hylobates agilis*

up *see* [upheld]

[**upheld**] a mutant (gene symbol *up*) mapped at 41.0± on the X chromosome of *Drosophila melanogaster*.

The name refers to the upright wings of the phenotype

upokohue *N.Z.* the delphinid cetacean *Cephalorhynchus hectori* (= porpoise *N.Z.*)

[**upright scutellars**] a mutant (gene symbol *ups*) mapped at 40.8 on the X chromosome of *Drosophila melanogaster*. The posterior scutella bristles of the phenotype are upright

[**Upturned**] a mutant (gene symbol *U*) mapped at 70± on chromosome II of *Drosophila melanogaster*. Wings of the phenotype are curled like those of [Curly]. Original mutant associated with a pericentric inversion of the second chromosome

[**Uropygial bifurcation**] a mutant (gene symbol *U*) in linkage group II of the domestic fowl. The name is descriptive of the phenotypic expression

ups *see* [upright scutellars]

Upupidae a small family of coraciiform birds erected to contain the hoopoes. Readily recognized by its conspicuous transverse stripes of black, white and buff and the huge black-tipped, frequently elevated crest

upw *see* [upward]

[**upward**] a mutant (gene symbol *upw*) mapped at 62.0± on chromosome II of *Drosophila melanogaster*. The wings of the phenotype are turned up

uq *see* [unequal wings]

UQ = ubiquinone

ur a yeast genetic marker indicating a requirement for uracil

-ur- *comb. form* confused from three roots and therefore variously meaning "urine," "tail" and "mountain"

ura a bacterial mutant indicating a requirement for uracil

urachus the connection between the embryonic bladder and the allantois which later becomes a ligament supporting the bladder

Uranioidea a superfamily of lepidopteran insects containing some little known moths

Uranoscopidae a family of acanthopterygian fishes with mouth and eyes directed dorsally, pelvic fins located under the throat, and possessing electric organs. Commonly called stargazers

-urceol- *comb. form* meaning urn, but used so loosely in *biol.* that it may be applied to the shape of anything between a barrel and a Florence flask

urceolate urn shaped

urchin 1 (*see also* urchin 2, cactus) any echinoid echinoderm. The word is widely supposed to be derived from the *Fr. oursin* now meaning urchin and its carrying the connotation "little bear." However, the term urchin was used for hedgehog in *Brit.* in the 12th. century and would seem a far more probable origin

cake urchin a general name for those echinoids which are intermediate between the disc-shaped and ovoid forms

green urchin *U.S. Strongylocentrocus drobaschiensis*

keyhole urchin the scutellid *Mellita quinquiesperforata*

sea urchin more commonly used in *Brit.* than urchin alone

urchin 2 = *Brit. obs.* hedgehog

urd the bean *Phaseolus mungo*

urease an enzyme catalyzing the hydrolysis of urea to carbon dioxide and ammonia

ureaceous of the color of charred wood

-ured- *comb. form* properly meaning "burn" but applied to blights which apparently "burn" leaves

Uredinales that order of heterobasidiomycete Basidiomycetae which contains those forms commonly called rusts. They are all obligate parasites of vascular plants technically distinguished by the production of basidiospores that develop directly into a hypha

urent stinging

ureter the entire kidney duct of lower vertebrates, and in higher forms, that portion of the kidney duct which transports the products of excretion from the kidney to the bladder

urethra that portion of the kidney duct which connects the bladder to the exterior

urhur the leguminous herb *Cajanus indicus* and its seeds

urial the ovine caprid mammal *Ovis vignei*. It is the common wild sheep of Central Asia and Asiatic Siberia

Uriidae = Alcidae

Urinatoridae = Gaviidae

urite an abdominal segment of an insect

urn *see* -urceol- for comments on usage

ciliated urn internal coelomic tunnels found on the mesenteries of some holothurnians

-uro- *see* -ur-

urobilin the product to which bilirubin is converted in the large intestine

Uroceridae = Siricidae

Urochordata a subphylum of acraniate chordates containing the sea-squirts and their allies. Distinguished by the presence of an outer covering, or test, a greatly reduced central nervous system and a greatly enlarged branchial region

my**urous** mouse tailed

Uropeltidae a small family of Asiatic burrowing snakes with a short tail ending in a shield

Uropygi a small order of arachnid arthropods at one time united with the Amblypygi into the Pedipalpi. The Uropygi are characterized by the possession of a multi-articulate tail giving them the name whip-scorpion

[Ursa] a mutant (gene symbol *U*) mapped at 2.7 on linkage group XIV of *Bombyx mori*. It resembles [Dirty] save that the pigmentation is brown and not black

Ursidae a family of carnivorous mammals containing the bears. They differ from the dogs and their allies (Canidae) principally in lacking a tail and in the generally larger size

urson any of several species of the Canadian cercolabid hystricomorph rodents of the genus Erethizon (= Canadian porcupines)

Urticaceae that family of urticalous dicotyledons which contains the nettles and their allies. Distinctive characteristics, apart from the many species covered with stinging hairs, are the anemophilous flowers and one-seeded ovary

Urticales that order of dicotyledonous angiosperms which contains *inter alia* the elms, the mulberries and the nettles. The order is distinguished by the bicarpellate unilocular superior ovary with a single ovule

spring **usher** *Brit.* the geometrid lepidopteran insect *Hybernia leucophaearia*

Ustilaginales that order of heterobasidiomycete Basidiomycetae which contains the forms commonly known as smuts. They are technically distinguished by the fact that the teleutospores are formed from intercalary cells of a dikariotic mycelium

ustulate = ureaceous

ustalous = ureaceous

uterus 1 (*see also* uterus 2, capsule) that part of the invertebrate female reproductive system, in which eggs or embryos are stored

uterus 2 (*see also* uterus 1) that portion of the female mammalian reproductive system in which the embryo develops

bicornuate uterus the type of uterus in which both uteri are fused but have short lateral extensions

bipartite uterus a type where paired, tubular uteri fuse at the point of junction with the vagina

duplex uterus a condition in which paired, tubular uteri open into the vagina

simplex uterus the type in which there is no trace left of the paired uteri

utick *Brit. dial.* the turdid bird *Saxicola rubetra* (= whinchat)

UTP uridine triphosphate

utricle literally, a small wine skin, but used for any small bladder, particularly a pericarp in that form

parietal utricle that layer of protoplasm of a plant cell which abuts on the cell wall

primordial utricle the protoplasm bounding the large vacuoles in a plant cell

-utricul- *comb. form* meaning a "small bladder"

utriculate possessing bladders or having the appearance of a bladder

utriculus the dorsal of the two divisions into which the auditory vesicle is divided (*cf.* sacculus)

-uv- *comb. form* meaning "grape"

uvea the vascular tunic of the eye

uvrA a bacterial genetic marker having a character affecting the reactivation of UV-induced lesions in DNA, mapped at 82.0 mins. for *Escherichia coli*

v *see* [vermilion] or [waltzer]

V_1—V_{16} *see* [virescent]

Va *see* [Varitint-waddler] or [Venae abnormeis]

va a yeast genetic marker indicating a requirement for valine

vaagmar *Brit*. the trachypterid fish *Trachypterus arcticus* (= deal-fish)

Vaccinioideae a subfamily of Ericaceae distinguished from the Arbutoideae by having an inferior ovary

vaccinous cow-colored. It appears from the context in which it is frequently used that some people believe all cows to be brown

[vacillans] a mutant (gene symbol *vc*) in linkage group VIII of the mouse. The phenotypic expression is a wobbly gait

vacuole a small cavity, usually one within a cell

contractile vacuole a vacuole found in many Protozoa which rhythimically collects, and then discharges, its contents

gas vacuole a floating organ of some algae

protid vacuole the nucleus of the tapetal layer in gymnosperms

vacuome a systm of vacuoles in a plant cell, thought by some to be homologous with the golgi apparatus

supervacuous redundant

-vag- *comb. form* meaning "to wander"

vagil wandering

-vagin- *comb. form* meaning "sheath"

vagina 1 (*see also* vagina 2) that part of the female reproductive system designed for the reception of the intromittent organ

vagina 2 (*see also* vagina 1) any part which sheaths another

vaginate sheathed

invaginate enclosed in a sheath

invagination the pushing in, from the outside of a sphere, of a pouch or tube which thus becomes invaginated in the sense of the last entry

vagrant *S.Afr.* any of numerous pierid lepidopteran insects of the genera Nepheronia, Eronia and Catopsilia. *I.P.* the nymphalid lepidopteran insect *Issoria sinha Austral. I. egista*

-val- *comb. form* meaning "strong." The word "equivalent" (equally strong) had led to the use of -valent in the sens of "value"

val *see* [valine]. This symbol is also used for a bacterial mutant indicating a valine requirement or resistance or sensitivity

-valent *see* -val-

bivalent a pair of homologous chromosomes in the prophase of meiosis

diaschistic bivalent one which appears to divide

transversely owing to the fact that the only chiasma is at the end

univalent a single chomosome which does not pair in the prophase of meiosis

valerian any of several species of herbs or shrubs of the valerianceous genus Valeriana and a few other plants

cat valerian = common valerian

common valerian *V. officinalis*

Greek valerian the polemoniaceous herb *Polemonium reptans*

marsh valerian *V. dioica*

red valerian the valerianaceous herb *Centranthus ruber*

spur valerian *Brit.* any of several species of the valerianaceous genus Centranthus

red spur valerian *Brit. C. ruder* (= *U.S.* jupiter's beard)

Valerianaceae that family of rubialous dicotyledons which contains both the valerians and the corn salads. The family is mostly herbal and thus distinguished from the Rubiaceae which are mostly shrubby

valine α-aminoisovaleric acid. $(CH_3)_2CHCHCHNH_2 \cdot COOH$. An amino acid necessary for the nutrition of rats

norvaline α-aminovaleric acid. $CH_5CH_2CH(NH_2) \cdot COOH$. An amino acid, not essential to the nutrition of rats, isomeric with valine

[valine] a mutant (gene symbol *val*) in linkage group V of *Neurospora crassa*. The phenotypic expression is a requirement for valine but not isoleucine

-vall- *comb. form* meaning "rampart" or "valley"

vallate in the form of a cylinder with a central identation, or raised rim, the whole set in a pit

circumvallation the capture of prey by a protozoan, which throws a large cup of protoplasm around the object and the adjacent water (*cf.* circumfluence)

Vallisneriaceae a family split off from Hydrocharitaceae to contain the *N.Amer.* forms

Vallisnerioideae a subfamily of Vallisneriaceae

-valv- *comb. form* properly meaning "the part of a door which swings open and shut" and thus, by extension, to any structure which opens or closes an aperture

hypovalva the valve of the inner, smaller, frustule of a diatom

valvaceous furnished with valves

valvate pertaining to a dehiscent fruit which opens along a hinge

valve 1 (*see also* valve 2, clasper) in the original sense of a folding shutter, mostly applied to the shells of pelecypod mollusks, brachiopods and diatoms and to that portion of a capsule which opens in dehiscence. The term is also specifically applied to that portion of an anther which is not fused to the stipe and to a component of male insect genitalia (*see* clasper 3)

brachial valve that valve of a brachiopod shell which is opposite to the pedicle valve

epivalve the epitheca of a diatom

cystigenic valve that portion of the shell of an ectoproct statoblast which derives from the epidermal disc

pedicle valve that valve of a brachiopod shell through which the pedicle passes

deutoplasmic valve that part of the shell of an ectoproct statoblast which derives from the yolk cell area

spathe valve a bract like envelope under some monocotyledonous flowers

Vieussens' valve the transverse velum that separates the medulla from the cerebellum

valve 2 (*see also* valve 1) in the derivative sense of a device permitting one-way flow

eustachian valve one of the valves in the base of the postcava vein

Kerckring's valves circular folds on the inner surface of the large intestine of mammals

mitral valve one of the valves in the left hand side of the atrio-ventricular canal of a four-chambered heart

pyloric valve that which closes the pyloric end of the stomach

semilunar valve a flat valve in the aorta

spiral valve 1 (*see also* spiral valve 2) a valve in the aorta of certain fishes, assisting in the separation of blood into the aortic arches

spiral valve 2 (*see also* spiral valve 1) a spiral structure in the intestines of some fish, formed by the central helicoid twisting of a detached portion of the wall

Thebesian valves one of those in the coronary vein

tricuspid valves the valves in the right hand atrioventricular canal of a four-chambered heart

valvule diminutive of valve

vampire *see* bat

vane the broad, flat portion of the feather formed of bards as opposed to the shaft or rachis

Vangidae that family of passeriform birds which contains the vanga-shrikes of Madagascar. They are metallic blue or black birds with long, robust, hooked and toothed, bills

vanilla (*see also* grass, plant) any of several orchidaceous plants of the genus Vanilla, without qualification, usually *V. planifolia* or its aromatic seed pod

Carolina vanilla the compositaceous herb *Trilisa odoratissima*

vannal *see* vein 3

vansire either of two viverrid carnivores of the genus Galidia

vapourer *Brit.* either of two species of liparid hymenopteran insect of the genus Orgyia

vaquero the rhinodermatid anuran amphibian *Rhinoderma darwinii* (= mouth breeding frog)

-var- *comb. form* meaning "ingrown" frequently confused with -vari-

Varanidae a family of lacertilian reptiles having pleurodont dentition and a very long, protractile, bifid tongue

varex literally a ridge but specifically applied to a prominent raised transverse rib on the volute of a gastropod mollusk shell

-vari- *comb. form* meaning "diverse"

gram **variable** an organism which is gram positive in one stage of its existence and gram negative in another

convariant an individual organism which has altered in respect to its contemporaries

variation 1 (*see also* variation 2) an unacceptable taxon used by some in the sense of subspecies

variation 2 (*see also* variation 1) lack of uniformity, particularly applied to the differences between individual organisms

bud variation a variation usually in the color of the flower, arising from a single bud

discontinuous variation = mutation

genovariation a mutation at a specific locus

epharmonic variation = epharmonic adaption

heterogenic variation = mutation

polymorphic variation polymorphic animals are said to show polymorphic variation

-varic- *comb. form* meaning "ridge"

divaricate widely spread apart

varices *pl.* of varex

varicose abnormally swollen (*cf.* ventricose)

variolate pitted

Variolus's *epon. adj.* from Constantio Variolus (1543–1575)

[Varitint-waddler] a mutation (gene symbol *Va*) in linkage group XVI of the mouse. The phenotypic expression involves the waltzing syndrome, spotted coat and deafness

-vas- *comb. form* meaning "vessel" in the sense of container

vas deferens literally, a duct that "carries down" or "away" and specifically one of the ducts that conveys sperm out of the body

vas efferens literally a duct that "carries to" or "brings out" and specifically one of the ducts that carries sperm from the testis to the vas deferens

vasa *pl.* of vas

vascular pertaining to, or containing, vessels

Vater's *epon. adj.* from Abraham Vater (1684–1751)

vault the roof over the calyx of a crinoid echinoderm, of which it therefore forms the aboral wall

vavisa the saururaceous aquatic herb *Anemopsis californica*

vb *see* [vibrissae]

vc *see* [vacillans]

ve *see* [veinlet]

ve- *comb. prefix* which either negates, or reverses, the meaning of the word to which it is prefixed. Used by ecologists in the sense of "micro"

vector an organism which carries and transmits a parasite

vedalia *U.S.* the coccinellid coleopteran insect *Rodolia cardinalis*

veery the turdid bird *Catharus fuscescens* (*cf.* thrush)

-veget- *comb. form* meaning "lively," "active" or "vigorous." Most modern derivative words mean the reverse of these terms

vegetable 1 (*see also* vegetable 2,3, sheep) pertaining to plants particularly those of culinary value

vegetable 2 (*see also* vegetable 1,3) pertaining to that pole of an egg or early embryo which grows more slowly than the other

vegetable 3 (*see also* vegetable 1,2) used in the sense of vegetative 1

vegetate *bot.* to sprout *zool.* to pass into a resting phase

vegetation the sum total of plant growth in a given area

vegetative 1 (*see also* vegetative 2) used in the sense of asexual in contrast to sexual, reproduction

vegetative 2 (*see also* vegetative 1) in the sense of vegetable 1 and 2

margin **veil** a membrane enclosing the hymenium of some basidiomycetes

vein 1 (*see also* vein 2–6) a vessel carrying blood to the heart. Names of veins that are self-explanatory (*e.g.* epigastric, caudal) are not given in this dictionary

anterior abdominal vein a vein formed in some amphibia, reptiles and birds from the fusion in the midline of the two epigastric veins

lateral abdominal vein *see* iliac vein

branchial vein a term sometimes, inaccurately, applied to the efferent branchial artery

cardinal vein the primary anterolateral, posteriolateral veins of the embryo and early foetus

posterior cardinal vein the principal vein draining the posterior region of the body of vertebrate embryo. It unites with the jugular at the Cuverian duct or sinus

cutaneous vein a large respiratory vein running from the lateral skin of amphibia to the brachial vein

iliac vein the vein drawing blood from the hindlimb. It continues in the lateral body wall as the lateral abdominal vein

jugular vein one of a pair of major veins draining blood from the head region to the heart. In fishes, there is an inferior jugular vein serving the lower jaws and the lower side of the gill arches

omphalo-mesenteric vein a continuation of the subintestinal vein on both sides of the liver, draining directly into the sinus venosus

portal vein a vein which carries blood to an organ other than the heart

stellate veins veins corresponding in position to the arcuate arteries in the metanephros

subclavian vein brings blood from the forelimbs and enters the jugular vein just anterior to the junction of the posterior cardinal with the cuverian

advehent veins veins derived from the posterior cardinals bringing blood to the mesonephoros

revehent vein a vein carrying blood from the mesonephrus to the posterior vena cava

vitelline veins extraembryonic veins collecting blood from the surface of the yolk sac and carrying it back to the embryo

vein 2 (*see also* vein 1,3–6, extension) the vascular strands in a leaf. Confused with nerve 4 but properly applying to those smaller vascular strands which do not cause an elevation of the leaf surface. Types of leaf venation are entered under that word

basivein = basal vein

bast vein libriform cells in the leaves of water plants

convergent vein those which first curve out and then curve in from base to apex of the leaf

intromarginal vein one which parallels the outer margin of a leaf

vein 3 the cuticular tubes that stiffen the wings of insects, circulate blood from the hemocoel and contain nerve strands and tracheae. The names given below, and shown in the figure, are those of the Comstock-Needham system, widely used in U.S. and commonly abbreviated to the letters shown. Cross or connecting veins are named from the longitudinal veins they connect, as medio-cubital cross vein (m-cu) and are always abbreviated in lower case letters. Wing veins are numbered in other systems, from the anterior to posterior in the British systems, from posterior to anterior in certain Continental systems, counting only those which reach the margin, without regard to whether the vein counted is a main vein or a branch

anal veins (A) series of longitudinal, unbranched veins in the anal (or vannal) area of the wing (1A, 2A, etc), sometimes called "vannal veins"

costal vein (C) along the costal or anterior margin of wing

subcostal vein (Sc) immediately behind the costal

cubital vein (Cu) with two principal branches (Cu_1 and Cu_2), each sometimes with additional branches (as Cu_1a, Cu_1b). Also called the cubitus

medial vein (M) basically with three branches (M_1 to M_3), also called the medius

radial vein (R) basically with five branches (R_1 to R_5), also called the radius

vannal vein = anal vein

vein 4 (*see also* vein 1-3, 5,6) U.S. popularly used for the intestine of edible crustacea

vein 5 (*see also* vein 1-4, 6) that part of the plume of a feather which is barbed

vein 6 (*see also* vein 1-5) in compound names of organisms

blackvein *I.P.* any of numerous species of pierid lepidopteran insects of the genus Aporia

blood-vein *Brit.* either of two geometrid lepidopteran insects *Acidalia imitaria* and *Timandra amata*. The latter is usually intended when there is no qualification

[**veinless**] a mutant (gene symbol *vl*) in linkage group I of *Habrobracon juglandis*. The name refers to the condition of the wing of the phenotype

[**veinlet**] a mutant (gene symbol *ve*) mapped at 0.2 on chromosome III of *Drosophila melanogaster*. The phenotypic wing veins are broken into short lengths

vel *see* [velvet]

velamen = velumen

velarium a structure analogous to the velum, found in some Schyphozoa from which the velum is absent

veld = veldt

veldt a type of grassland habitat found in *S.Afr.* in which a luxuriant growth of grass is augmented by many patches of low shrub and occasional trees

acacia veldt a type of veldt dominated by numerous trees of the genus Acacia

bush veldt a type of veldt in which shrubs, particularly of the genus Protea, are more frequent than on most veldt

grass veldt a type of veldt lacking, or almost lacking, shrubs and without trees

Veliidae a family of hemipteran insects greatly resembling the Gerridae but with shorter legs and thicker bodies

Velloziaceae a small family of liliflorous monocotyledonous angiosperms

velum literally an "awning" but applied in *biol.* to many membranous structures which do not have a supporting function. Specifically in *bot.* to the envelope within which the whole fruiting body of some Basidiomycetes are produced. In *zool.* the term is specifically applied to a delicate annular membrane projecting inwards from the edge of the bell of medusae, the partition separating the cavity of the oral funnel from the pharynx of Amphioxus, a skirt-like development from the posterior end of the proglottid of some tapeworms which envelopes the anterior region of the pro-glottid next behind, the swimming organ of marine molluscan larva which is developed from the preoral ciliated rings of the

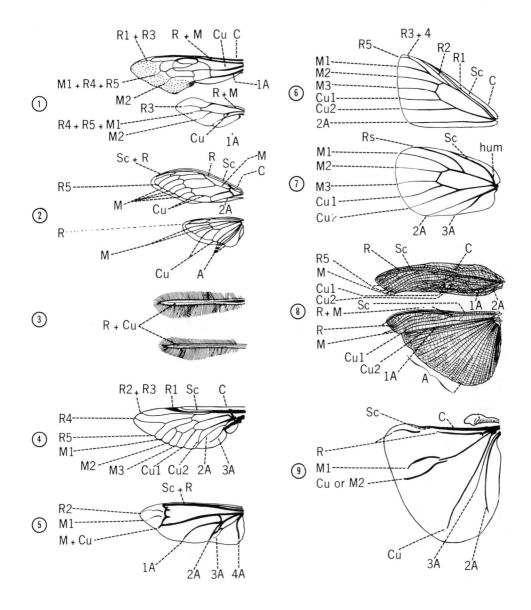

WING VENATION OF SOME INSECTS

1. Fore- and hindwings of a carpenter bee, *Xylocopa virginica* (Hymenoptera)
2. Fore- and hindwings of a cicada, *Tibicen lyricen* (Hemiptera)
3. Fore- and hindwings of a Brazilian *Thrips.* (Thysanoptera)
4. Forewing of a horsefly, *Tabanus* (Diptera)
5. Hindwing of a prinonid beetle, *Orthosoma brunneum* (Coleoptera)
6. Fore- and hindwings of the cabbage butterfly, *Pieris rapae* (Lepidoptera)
7. Fore- and hindwings of a locust, *Dissoteira carolina* (Orthoptera)
8. Fore- and hindwings of a stylopid, *Triozocera maxi* (Strepsiptera)

Abbreviations. *1A, 2A, 3A, 4A,* first, second, third and fourth anal veins. C, costal vein. Cu_1, Cu_2, first and second cubital branches (the veins commonly designated thus are, strictly speaking, branches of the first costal branch). M_1, M_2, M_3, first second and third medial branches. R_1, R_2, R_3, R_4, R_5, R_S, branches of the radial vein. *Sc,* subcostal vein

trochophore, and the membranes surrounding the pseudostome of many ciliate protozoa. It is, rarely, also used for the soft palate

transverse velum the folding in of the dorsal floor of the primitive brain which divides the diencephalon from the telencephalon

velumen literally a fleece and specifically the epidermis on the roots of aerophytes

velutinous velvety

velvet (*see also* breast) the skin covering the horns of a stag

[Velvet] a mutant (gene symbol *vel*) found in linkage group III of *Neurospora crassa*. The name is descriptive of the appearance of the phenotypic colony

[Venae abnormeis] a mutant (gene symbol *Va*) on chromosome II of *Drosophila melanogaster*. The phenotypic expression is irregularly branched or interrupted wing veins

venation the arrangement of veins, specifically those in the leaf of a plant

basal venation with veins arising from the base of the leaf

combinate venation with the lateral veins uniting before reaching the margin

compital venation that in which the veinlets intersect each other at an acute angle

costal venation with veins arising from the midrib

costate venation one in which there is more than one primary longitudinal vein

dictyodesmic venation a network venation found in some ferns

direct venation with the veins running directly out from the midrib to the margin

indirect venation that in which the veins are multiplied and reticulated between the midrib and the margin

acrodromous venation having the veins uniting at the apex of the leaf

actinodromous venation with veins arranged radially

anadromous venation the type of venation common in ferns, in which the first veins are produced towards the apex of the frond or leaf

anomodromous venation that type of venation which cannot be described by any accepted term

brachydodromous venation that in which the veins are looped

camptodromous venation that in which the secondary veins curve towards the margin but do not form loops

campylodromous venation that in which the veins curve out towards the margin and then towards the tip of the leaf

craspedodromous venation one in which the lateral veins do not branch

catadromous venation a type of venation, common in ferns, in which the first veins develop from the basal portion of the midrib

cheilodromous venation = craspedodromous venation

diadromous venation fan-shaped venation

dictyodromous venation that in which the veins form a reticular pattern

hyphodromous venation the condition in which the veins are sunk in the substance of the leaf

hypodromous venation the condition in which both anadromous and catadromous venation occur on the same side of the fern frond

metadromous venation a venation of a fern-frond in which the first set of nerves are given off either from the upper or from the lower side of the midrib

anametadromous venation a venation found in ferns in which some fronds are anadromous and some are catadromous

parallelodromous venation having parallel veins

plagiodromous venation when the tertiary veins are at right angles to the secondary

dictyogenous venation = dictydromous venation when it occurs in a monocotyledonous leaf

hinoideous venation the condition in which the veins diverge from the midrib, but remain parallel and undivided

palmatiform venation the veins palmately arranged

penniform venation as in pinnate venation, but with the tips of the lateral branches confluent

fan-nerved venation when the veins radiate from the base

flabellinerved venation = radiate venation

peltinerved venation with veins radiating from the insertion of the petiole in a peltate leaf

cryptonervous venation that which is concealed by hairs on the leaf

curvinervous venation a curved venation, occurring, where, as in the monocotyledons, one would expect it to be straight

digitinervous venation that in which the nerves of a leaf radiate from the base

falsinervous venatiin when the apparent veins do not contain vascular tissue

laterinervous venation that which consists of straight parallel veins

pedatinervous venation with three basic nerves diverging from the base, of which the outer two divide to reach the margin

polynervous venation with very numerous secondary veins

quinquinervous venation that which consists of a main rib with a pair of veins on each side

rectinervous venation having parallel veins

retinervous venation that in the form of a network

ruptinervous venation one in which the veins, usually straight, are interrupted at intervals by swelling

stellinervous venation having the veins running out from a central point like a star

vaginervous venation indeterminate venation

inophyllous venation one in which the veins are thread-like

pinnate venation the condition in which the veins leave the midrib opposite each other at a forwardly directed angle

radiate venation = palmate venation

recurrent venation that in which the veinlets recurve toward the main rib

feather-veined venation = penniform venation, though the term is sometimes also applied to direct venation

evaniscentivenousus venation a type of venation in which the lateral veisn do not reach the margin of the leaf

vendace *Brit.* either of two species of salmonid fishes of the genus Coregonus

venenatous venomous

Veneridae a family of pelecypod mollusks commonly called venus shells. They are distinguished by the equal-sized, oblong-oval, valves with a porcelain-like texture

venery the sum total of the activities of human predators particularly when acting in concert

avenous lacking veins, in the sense of vein 2

venose (Bot.) having prominent veins in the sense of vein 2

venous having veins or pertaining to veins in any sense

venter literally "belly" but specifically the belly of the archegonium in which the oosphere is formed, the swollen portion of the shell of a gastropod mollusc, the central swollen portion of a muscle, and the lower portion of the body of an acarine. Ventriculus (*q.v*) is the diminutive form of venter

ventilatorious fan shaped

ventral in *zool.* pertaining to the underside or belly. In *bot.* the ventral face of a leaf is the upper surface which was originally the inner face in the bud (*cf.* dorsal)

    dorsiventral *bot.* laminate, or flattened, as some carpellary surfaces

ventricle 1 (*see also* ventricle 2) that muscular division of the heart which pumps blood from the heart, and the cavity of this division

ventricle 2 (*see also* ventricle 1) the cavities of the brain

    common ventricle the cavity of the diencephalon

    fourth ventricle the cavity of the medulla oblonga

    lateral ventricle the cavities of the cerebral hemisphere which develop from the common ventricle, and which are referred to as ventricles one and two of the brain

    third ventricle = common ventricle

ventricose swollen on one side (*cf.* varicose)

ventricular the term is specially applied to that portion of the invertebrate pharynx, particularly that of Nematodes, which is devoid of muscles

Ventriculus liberally a little belly (*cf.* venter)

    proventriculus 1 (*see also* proventriculus 2,3) any expanded area of the alimentary canal lying immediately in front of the stomach

    proventriculus 2 (*see also* proventriculus 1,3) in birds, a glandular sac between the crop and gizzard

    proventriculus 3 (*see also* proventriculus 1,2) a pouch, frequently with internal teeth, at the junction of the foregut and midgut in insects. Analogous to the gizzard of birds

rock **venus** any of many species of venerid pelecypod Mollusca but particularly those of the genera Humilaria and Protothaca

Venus's-looking-glass any of several species of the campanulaceous genus Specularia, particularly *S. speculum-veneris* and *S. perfoliata*

verbena properly the name of a genus of verbenaceous plants but extended to other forms

    lemon verbena the verbenaceous herb *Lippia citriodora* (*cf.* orégano)

    sand verbena any of several species of the nyctaginaceous genus Abronia

Verbenaceae that family of tubiflorous dicotyledons which contains, as well as the verbenas, many semi-hardy trees or shurbs. The family is very closely related to the Labiatae from which it is distinguished by the fact that the ovary is not deeply lobed

Verbenales an order erected to contain the family Verbenaceae by those who do not hold this family to belong in the Tubiflora

Verbenineae a suborder of tubifloral dicotylendonous angiosperms containing the families Verbenaceae and Labiatae

verdin the parid bird *Auriparus flaviceps*

biliverdin the primary breakdown product of hemoglobin. It is itself converted to bilirubin

-verg- *comb. form* meaning "twig"

vergate twiggy or made of twigs

verge a term applied to the intromittent organs of some invertebrates

-verm- *comb. form* meaning "worm"

Vermes a Linnean group erected to contain all invertebrates except Insecta. It was divided into the Intestina, Mollusca, Testacea, Lithophyta and Zoophyta

Vermetidae a family of aberrant gastropod mollusks with worm-like calcareous shells frequently confused with those of serpulid annelids

-vermicul- *comb. form* meaning "worm"

vermicular worm-shaped

[vermilion] a pseudoallelic locus (gene symbol *v*) mapped at 33.0 on the X chromosome of *Drosophila melanogaster*. The name refers to eye color

Vermipsyllidae a small family of siphonapterous insects in general parasitic on large carnivores. They are distinguished by having two rows of setae on the abdominal tergites

vermis the unpaired, dorsal central portion of the cerebellum

vernal pertaining to spring

vernalization the treatment of seeds in such a manner as to shorten the maturation period of the plant developed from them

vernicose shiny

Vernonieae a tribe of tubiflorous Compositae

-verruc- *comb. form* meaning "wart"

-vers- *comb form* meaning "turn"

quaquaversal running or bending every which way

versatile capable of free movement on a support

inverse upside down or back to front

perversion used in *bot.*, in the sense of upside down and turned sideways

introvert the anterior region of the sipunclid body

vertebra one of the bones which together make the vertebral column

    caudal vertebra one of the vertebrae of the tail posterior to the pelvic symphysis

    opisthocoelous vertebra one in which the centra are convex in front and concave behind

    procoelous vertebra one in which the centra are concave in front and convex behind

    amphicoelous vertebra one in which the centra are biconcave

    lumbar vertebra one in the region that lies between the thorax and the sacrum

    notocentrous vertebra one in which the centrum is derived entirely from the neural arches

    notochordal vertebra one in which the centrum is hollow

    sacral vertebra one which is directly connected with or forms part of the sacrum

    pseudosacral vertebra one which is attached to the pelvis by its transverse processes rather than by its ribs

    urosacral vertebra one of those anterior caudal vertebrae which are fused with the sacrum

    amphyplatyan vertebra one in which both posterior and anterior surfaces of the centra are flat

    adelospondylous vertebra a vertebra with a detached neural arch

    aspidospondylous vertebra one in which the centra

are derived from the ossification of cartilaginous arches

holospondylous vertebra a vertebra consisting of a single piece

lepospondylous vertebra one which is shell-like or husk-like

phyllospondylous vertebra a shell like vertebra not articulated with other vertebra

stereospondylous vertebra one consisting of a single body or, essentially, of the intercentrum

temnospondylous vertebra one consisting of several parts or at least having the arch separated from the body

thoracic vertebrae the vertebrae of the thoracic region

vertebrate 1 (*see also* vertebrate 2) a loose term applying to those chordates that have vertebral columns

vertebrate 2 (*see also* vertebrate 1) in *bot.* means contracted at intervals, but presenting a more angular appearance than is conveyed by the term moniliform

vertex the top or anterior aspect of the insect head between the eyes, frons and occiput

verticil = whorl

Verticillales that order of dicotyledonous plants which contains the family Casuarinaceae. Characteristics of the family apply to the order

Verticillatae an order of dicotyledonous angiosperms containing the single genus Casuarina in the single family Casuarinaceae. They are much-jointed trees, the twigs strongly resembling Equisetum in their general appearance

Verticordiidae a small family of pelecypod mollusks distinguished by the heavy, solid, pearly shell

verucose warty

verculatious in the form of a cylinder with an obtuse point at each end

vervain anglicized form of the verbenaceous genus Verbena

bastard vervain any *sp.* of the verbenaceous herb genus Stachytarpheta

vervet a common guenon (*q.v.*) which is variously known as the green-monkey or the blue-monkey in spite of the fact that it is yellowish brown in color

vesicatorious blistering

vesicle a small bladder. The term was at one time used for cell, but is applied to almost any cavity particularly in *zool.* those filled with fluid but in *bot.* also those filled with gas. It is used in the sense of air sac 3

auditory vesicle the embryonic vesicle that gives rise to the inner ear of the adult

chlorophyll vesicle = chloroplast

embryo vescicle (*bot.*) = oosphere

germinal vesicle 1 (*see also* germinal vesicle 2) = oosphere

germinal vesicle 2 (*see also* germinal vesicle 1) the resting nucleus of an oocyte

Kupffer's vesicle the first rudiment of the hindgut in the early development of fish

polian vesicle an elongated sac hanging from the ring canal of the water-vascular system of holothurians and projecting into the coelom. In other echinoderms, they are smaller, blindly ending sacs, apparently serving as additional water reservoirs

madreporic vesicle an aboral extension of the axial sinus in Echinoderms placed close to the madreporite, but having no connection with it

propulsion vesicle the sperm ejaculatory portion of the sperm duct of cestodes

seminal vesicle a vessel for the storage of sperm, variously situated on the side of or at the end of the vas deferens (*cf.* receptaculum seminalis)

solenocytic vesicle a collecting vessel into which numerous solenocytes open

spermiducal vesicle a seminal vesicle consisting of an extension of the sperm duct, used by some as distinct from seminal vesicle which is, in this sense, used only of an outgrowth or appendage to the sperm duct

[vesiculated] a mutant (gene symbol *vx*) mapped at 16.3 on the X chromosome of *Drosophila melanogaster*. The name refers to the blistered and warped wings of the phenotype

vesiculous having the appearance of little bladders

-visp- *comb. form* meaning "wasp"

-vesper- *comb. form* meaning "evening"

vesperal = vespertine

Vespertilionidae a very large family of bats which lack leaves or other excresences around the nose

vespertine pertaining to the evening (*cf.* crepuscular)

Vespidae an enormous family of vespoid wasps divided into very numerous subfamilies the more important of which are recorded in this dictionary

Vespinae a subfamily of vespid hymenoptera including the forms commonly called yellowjackets and hornets

Vespoidea a superfamily of apocritan hymenopteran insects containing those forms referred to as hornets, wasps (in part) and velvet ants. They may be distinguished as possessing, like the Sphecoidea, an anal lobe on the hindwings but they differ from the Sphecoidea in that the lateral extensions of the pronotum, while not well differentiated around the lobes, do extend back to the tegulae. They differ from the bees (Apoidea) in lacking plumose hairs

vessel (*see also* member) anything which conducts or carries anything, in *biol.* usually applying to tubular ducts

lymph vessel a vessel analogous to a blood vessel, but carrying lymph

reticulated vessel one with netted thickenings in the wall

seed vessel = pericarp

vesicular vessels *see* bundle

vestal *Brit.* the geometrid lepidopteran insect *Sterrha sacraria*

vestibule 1 (*see also* vestibule 2) the chamber between the orifice and the diaphragm in Ectoprocta

vestibule 2 (*see also* vestibule 1) the inner of the two chambers of the coronal funnel in Rotifera

vestige that which remains of a previously better developed organ

coccygeal vestige termination of the neural tube in the integument

[vestigial] *either* a mutant (gene symbol *vg*) mapped at 67.0 on chromosome II of *Drosophila melanogaster*. The phenotypic expression is reduced or absent wings and balancers *or* a mutant (gene symbol *vt*) in linkage group VII of the mouse. The term applies to the nature of the tail of the phenotype

[Vestigial] a mutant (gene symbool *Vg*) mapped at 38.7 on linkage group I of *Bombyx mori*. The wings of the phenotype are vestigial

[Vestigial glume] a mutant (gene symbol *Vg*) mapped at 84 on linkage group I of *Zea mays*. The name is descriptive of the phenotypic expression

vetch (*see also* weevil) properly any of numerous leguminous herbs of the genus Vicia, though the term is often applied to other low-growing leguminous herbs

American vetch *v. americana*

bitter vetch *Brit.* any of several species of the leguminous genus Lathyrus

bitter vetch *U.S.* either *L. montanus, L. lanszwertir* or *Vicia erdilia*

> **black bitter vetch** the leguminous vine *Lathyrus niger*
> **spring bitter vetch** *L. vernus*
> **wood-bitter vetch** *Brit. V. orodus*

bush vetch *Brit. V. setium*

Carolina vetch *V. caroliniana*

common vetch *V. sativa*

crown vetch the leguminous herb *Coronilla varia*

French vetch *V. narbononsis*

hairy vetch *V. villosa*

horseshoe vetch *Brit.* any of several species of leguminous herb of the genus Hippocretis

kidney vetch any of several leguminous herbs of the genus Anthyllis, particularly *A. vulneraria*

milk vetch *Brit.* any of numerous species of herbs of the leguminous genus Astragalus

purple vetch *V. atropurpurea*

rattle vetch *Astragalus carolinianus*

Russian vetch = hairy vetch

sand vetch = hairy vetch

scarlet vetch *V. fulgens*

sensitive joint vetch *Aeschynomene virginica*

smooth vetch *Brit. V. tetrasperma*

spring vetch *Brit. V. lathyroides*

tufted vetch *Brit. V. cracca*

vetchling any of several species of leguminous herb of the genus Lathyrus

bastard vetchling *L. pratensis* (= meadow pea)

prairie vetchling *L. polymorphus*

vexillary having the vexillum folded over the other petals

-vexill- *comb. form* meaning "flag"

vexillum 1 (*see also* vexillum 2, 3) the large posterior petal of a papilionaceous flower

vexillum 2 (*see also* vexillum 1,3) one of the scoops on the front tarsi of digging Hymenoptera

vexillum 3 (*see also* vexillum 1,2) that part of a feather which borders the rachis

vg *see* [vestigial]

Vg *see* [Vestigial] and [Vestigial glumes]

viable that which is capable of exhibiting life

inviable short lived

viatical pertaining to roadsides

vibraculum an Ectoproct heterozooid in the form of a long movable bristle or seta

vibrissa 1 (*see also* vibrissa 2) a sensitive trichome

vibrissa 2 (*see also* vibrissa 1) one of the large, laterally projecting bristles on the snouts of some mammals

[vibrissae] a mutant (gene symbol *vb*) mapped at 49.3 on the X chromosome of *Drosophila melanogaster*. The phenotype has vibrissae in tufts

vicariant one species which, in any given ecological situation, represents a species found in a similar environment in another geographical location

viceroy *U.S.* the nymphalid lepidopteran insect *Limenitis archippus* which so closely mimics the danaid monarch as to be frequently mistaken for it

vicine neighboring

vicuna the Andean camellid mammal *Vicugna vicugna*.

It was at one time, and sometimes still is, included in the genus Lama

Vieussen's *epon. adj.* from Raymond de Vieussens (1641–1715)

hybrid vigor = heterosis

-vill- *comb. form* meaning "shaggy"

villus 1 (*see also* villus 2) in *bot.* an elongate, flabby trichome

villus 2 (*see also* villus 1) in *zool.* a small finger-like process, particularly those arising from the intestinal epithelium

microvillus a villus-like protuberance of a cell surface visible only by electron microscopy

vimen a long flexible shoot of a tree, like that of a willow

-vin- *comb. form* meaning "wine"

vinaceous wine colored

vinculum *pl.* vinculi. A chain, band or strap and used for so many structures of this general designation as to have no valid specific meaning

vine (*see also* looper, cactus, peach, weevil) any flowering plant having long trailing or climbing stems. Used without qualification generally means grape. The term bine (*q.v.*) is a variant spelling

Allegheny vine the fumariaceous vine *Adlumia fungosa*

balloon vine the sapindaceous vine *Cardiospermum halicacabum*

bay's-vine *U.S.* the ranunculaceous climbing plant *Clematis viorna*

blaspheme vine the liliaceous herb *Smilax laurifolia*

cinnamon vine the dioscoreaceous herb *Dioscorea batatas* and its edible tubers (= yam)

crab's-eye vine the leguminous vine *Abrus precatorius*

cross vine the bignoniaceous woody vine *Bignonia capreolata*

cypress vine any of several species of the convolvulaceous genus Quamoclit regarded by some as a subgenus of Ipomoea, particularly *Q. pinnata*

cyprus vine = cypress vine

glory vine any of several species of the leguminous genus Clianthus

grape vine = grape

honey vine U.S. the asclepiadaceous herb *Ampelamus albidus*

kudzu vine the leguminous vine *Pueraria hirsuta*

lemon vine the cactaceous scrambling shrub *Pereskia aculeata*

love vine any of several convolvulaceous parasitic herbs of the genus Cuscuta (= dodder)

marine vine *U.S.* the vitaceous vine *Cissus incisa*

matrimony vine any of several species of shrubby vine of the solanaceous genus Lycium, particularly *L. halimifolium*

moonseed vine the menispermaceous vine *Menispermum canadense* (= yellow parilla)

passion vine = passion flower

pepper vine *U.S.* the vitaceous vine *Ampelopis arborea*

pipe vine *U.S.* the aristolochiaceous shrubby climber *Aristolochia durior* (= Dutchman's pipe)

potato vine the solanaceous climbing shrub *Solanum jasminoides*

wild potato vine the convolvulaceous climbing vine *Ipomoea pandurata*

rattan vine *U.S.* the rhamnaceous vine *Berchemia scandens*

sand vine any of several species of asclepiadaceous herb of the genus Ampelamus

silk vine the asclepiadaceous shrub *Periploca graeca*

trumpet vine either the bignoniaceous woody vine *Campsis radicans* or any of several species of woody climbing vines of the bignoniaceous genus Bignonia

wonga-wonga vine the bignoniaceous woody climber *Pandorea australis*

vinegaroon the thelyphonid pedipalp *Mastigoproctus giganteus*

vio a bacterial mutant showing viomycin resistance or sensitivity

Violaceae that family of pariatelous dicotyledons which contains, *inter alia* the violets and pansies. The shape of the flower is typical *N.Z.* the whitey-wood tree (mahoe) is also a member of this family

violaceous violet colored

violet (*see also* aphis, ear, sawfly) properly, any of numerous species of violaceous herbs of the genus Viola, but extended to many other plants of similar color or habit of growth (*cf.* pansy)

    Adder's violet any of several orchids of the genus Peramium

    African violet the gesneriaceous herb *Saintpaulia ionantha*

    Australian violet *V. hederacea*

    common blue violet *V. papilionacea*

    cream violet *U.S. V. striata*

    damask violet = dame's violet

    dame's violet the cruciferous herb *Hesperis matronalis* (= rocket)

    dog's tooth violet any of several species of the liliaceous genus Erythronium

    false violet *U.S.* any of several species of rosaceous herb of the genus Dalibarda

    bird-foot violet *V. pedata*

        herb-foot violet = pansy violet

    green violet any of several species of herb of the violaceous genus Hybanthus

    horned violet *V. cornuta*

    Labrador violet = marsh violet

    marsh violet properly *V. palustris*. But also applied to the lentibulariaceous herb *Pinguicula vulgaris* (= Labrador violet)

    pansy violet *U.S. V. pedata*

    star violet *U.S.* the rubiaceous herb *Houstonia patens* and *H. minima*

    striped violet *V. striata*

    sweet violet *V. odorata*

    Usambara violet = African violet

    early yellow violet *U.S. V. rotundifolia*

    water violet *U.S.* any of several species of primulaceous aquatic herb of the genus Hottonia (= feather foil). *Brit. H. palustris*

viosterol = vitamin D$_2$

viper (*see also* bugloss, fish, grass) any of numerous poisonous snakes, mostly viperine *Brit. Vipera berus*

    asp viper *Vipera aspis*

    blow viper *U.S.* any of several species of colubrid of the genus Heterodon (= hognose snake)

    common viper *Brit.*, without qualification, *V. berus* (= adder)

    gaboon viper the viperid *Bitis gabonica*

    horned viper the viperine *Cerastes cerastes*. This is supposed by some to be the "horned asp" made famous by Cleopatra (*see* asp)

    jumping viper the crotalid *Bothrops nummifera* (= tommygoff)

    mock viper any of several species of Asiatic colubrid of the genus Psammadynastes

    mole viper any of several viperid snakes of the genus Atractaspis

    pit viper general term applying to all crotaline viperid snakes

    rhinoceros viper *Bitis nasicornis*

    sand viper *Vipera ammodytes*

    water viper *U.S.* the crotaline *Agkistrodon piscivorus* (= cottonmouth)

Viperidae a family of snakes almost all venomous, but readily distinguished from other venomous snakes by the fact that the poison fangs are movable and erectile. The family is divided into the Viperinae, commonly called vipers and which lack a sensory pit, and the Crotalinae, called pit-vipers, in which a deep heat-sensitive pit lies between the eye and the nose

Virales an order of Microtatobiotes containing the forms commonly called viruses. They are distinguished by their small size, inability to exhibit metabolic activity outside a host cell, and their simple structure consisting almost invariably of a core of nucleic acid in a protein sheath

-viren- *comb.* form meaning "green"

virent either green, or evergreen, according to the context evergreen

vireo properly, any of numerous vireonid birds, mostly of the genus Vireo, occasionally used for other vireo-like birds

    antvireo any of several species of formicariid birds of the genus Dysithanmus

        plain antvireo *D. mentalis*

    Bell's vireo *V. bellii*

    flat-billed vireo *V. nanus*

        thick-billed vireo *V. crassirostris*

    Blue Mountain vireo *V. osburni*

    black-capped vireo *V. atricapillus*

    Cozumel vireo *V. bairdi*

    Cuban vireo *V. gundlachii*

    Dwarf vireo *V. nelsoni*

    red-eyed vireo *V. olivaceus*

        white-eyed vireo *V. griseus*

    golden vireo *V. hypochryseus*

    grey vireo *V. vicinior*

    blue-headed vireo = solitary vireo

    Hutton's vireo *V. huttoni*

    Jamaican vireo *V. modestus*

    mangrove vireo *V. pallens*

    Philadelphia vireo *V. philadelphicus*

    Puerto Rican vireo *V. latimeri*

    St. Andrew vireo *V. caribaeus*

    shrike-vireo any of three species of vireolaniid bird

        green shrike-vireo *Smaragdolanius pulchellus*

        chestnut-sided shrike-vireo *Vireolanius melitophrys*

    slaty vireo *Neochloe brevipennis*

    solitary vireo *V. solitarius*

    yellow-throated vireo *V. flavifrons*

    warbling vireo *V. gilvus*

    black-whiskered vireo *V. altiloquus*

    yellow-green vireo *V. flavoviridis*

    Yucatan vireo *V. magister*

Vireolaniidae a small family of passeriform birds containing the three species of shrike-vireo of Central and South America. They may be distinguished from the closely allied pepper-shrike by the lack of lateral compression in the bill

Vireonidae that family of passeriform birds which contains the vireos. Small birds of the Americas, they are distinguished by the fact that the anterior toes are basally adherent

virescent greenish

[**virescent**] *either* a mutant (gene symbol v_2) mapped at 73 on linkage group V of *Zea mays*. The phenotypic expression is a deficiency of chlorophyll in young seedlings *or* a mutant (gene symbol v_5) mapped at 8 in linkage group VII of *Zea mays*. The young phenotypic seedlings are deficient in chlorophyll *or* a mutant (gene symbol v_3) mapped at 11 in linkage group V of *Zea mays*. The phenotypic seedlings are deficient in chlorophyll *or* a mutant (gene symbol v_4) mapped at 83 on linkage group II of *Zea mays*. The phenotypic seedlings are deficient in chlorophyll *or* a mutant (gene symbol v_{16}) mapped at 0 on linkage group VIII of *Zea mays*. The phenotypic expression is a deficiency of chlorophyll in young seedlings *or* a mutant (gene symbol v_1) mapped at 71 on linkage group IX of *Zea mays*. The phenotypic expression is a deficiency of chlorophyll in the young seedlings

virga literally a rod, but commonly applied to a spine terminating the aedeagus

virgate shaped like a birch rod, used particularly for a rotiferan mastax in which the fulcrum and the manubria are elongated

virgin 1 (*see also* virgin 2) in humans, an individual who has never engaged in sexual intercourse; in other animals, particularly insects, an uninseminated female

virgin 2 (*see also* virgin 1) anything untouched by the hand of man (*e.g.* virgin forest) or in a natural state (*e.g.* virgin wool) or produced without artifice (*e.g.* virgin oil)

virgineous a white color of the utmost purity

virosous venomous, usually of plants

virus the smallest known living entity. Many have a strand or core of ribonucleic acid or deoxyribonucleic acid. All are obligate parasites

vis(3717) *see* [Visible (3717)]

viscachas a large chinchillid hystricomorph rodent (*Lagostomus trichodactylus*) living in enormous colonies and apparently possessing a considerable degree of social organization

viscera the organs contained in the abdominal cavity

Viscoideae a subfamily of Loranthaceae differing from the Loranthoideae by the absence of a calyculus

[**Visible (3717)**] a mutant (gene symbol *vis(3717)*) in linkage group I of *Neurospora crassa*. The phenotypic expression is a semicolonial type of growth

visual pertaining to sight

Vitaceae that family of rhamnalous dicotyledons which contains the grapes. The climbing habit and the fact that the stamens are opposite the petals distinguishes it from the Rhamnaceae

vitalism the doctrine that development, or indeed all metabolic processes, are controlled by a force (entelechy) which has not yet been detected and quite possibly cannot be detected by physical equipment

vitamin an ill-defined term applied in a rather haphazard manner to nutrient factors. (*cf.* biotin, choline, inositol, nicotinic acid, pantothenic acid, folic acid)

vitamin A 3,7-dimethyl-9-(2,-6,6-trimethyl-1-cyclohexen-1-yl)-2,4,6,8-nonatetraen-1-ol. An oil soluble vitamin required by most species of animals

vitamin B₁ thiamine. 3-(4-amino-2-methylpyrimidyl-5-methyl)-4-methyl-5-β-hydroxyethylthiazolium chloride hydrochloride. A water soluble vitamin required in the nutrition of almost all organisms

vitamin B₂ riboflavin. 6,7-dimethyl-9-(D-1'-ribityl)-isoalloxazine. A water soluble vitamin forming part of the electron transfer system in the cellular metabolism of most living forms

vitamin B₆ pyridoxine. 5-hydroxy-6-methyl-3,4-pyridine dimethanol hydrochloride. A water soluble nutrient thought by many to be essential though no human requirements have been established

vitamin B₁₂ cobalamin. 5,6-dimethylbenzimidazolyl cyanocobamide. A water soluble vitamin acting as a hemopoetic agent in many animals though its specific human need has not been determined

vitamin B_C = folic acid

vitamin B_T = carnitine

vitamin C ascorbic acid L-threo-2,3,4,5,6-pentahydroxy-2-hexeno-γ-lactone. A water soluble vitamin synthesized by most mammals except man and the guinea pig

vitamin D₂ activated ergosterol 9-10-secoergosta-5,-7,10(19),22-tetra-en-3-ol. A fat soluble vitamin required by most animals for calcium metabolism

vitamin D₃ activated 7-dehydrocholesterol. 22,23-dihydro-24-demethylcalciferol. An analog of vitamin D₂ about 50 times as active in birds as in D₂ but without apparent difference in activity in most other forms

vitamin E α-tocopherol. 5,7,8-trimethyltocol. An oil soluble vitamin playing an ill defined role in fertility

vitamin F a no longer acceptable term once applied to some fatty acids

vitamin G = vitamin B₂

vitamin H = biotin. *cis*-hexahydro-2-oxo-1 H-thieno [3,4] imidazole 4-valeric acid. A water soluble vitamin the necessity for which in human nutrition has not been illustrated. It protects many animals against egg white injury

vitamin K₁ 2-methyl-3-phytyl-1,4-naphthoquinone. An oil soluble vitamin controlling production of prothrombin and therefore essential for blood clotting in all vertebrates

vitamin K₂ 2 - methyl - 3 - difarnesyl - 1,4 - naphthoquinone. Essentially similar in activity to vitamin K₁

vitamin L a so-called lactation factor of doubtful existence

vitamin M = folic acid

vitamin P a substance of unknown composition derived from citrus fruits and supposed to affect the permeability and fragility of capillary blood vessels

-vitell- *comb. form* meaning "yolk"

germ **vitellarium** an ovary producing both eggs and yolk

-viti- *comb. form* meaning "vine"

vitrella a group of clear cells around the crystalline cone in an ommatidium

Vitreoscillaceae a family of beggiatoale schizomycetes strongly resembling the Beggiatoaceae, but differing from them in never producing sulphur granules even in the presence of hydrogen sulphide

vitreous glassy, either in regard to the transparency, surface texture, or greenish yellow color

vitta a tube containing essential oil in the pericarp of an umbelliferous seed

Viverridae a large family of small, usually elongate, fissipede carnivorous mammals sometimes broadly known as the civet cats though many other forms are included

vixen a female fox

vizcacha = viscachas

vl *see* [veinless]

vo-3 *see* [vortex in 3]

Vochysiaceae a small family of geraniale dicotyledons

vogue *Brit.* the sparid fish *Box vulgaris*

vole in *Brit.* the term vole is used to distinguish the rough-tailed myomorph rodents (Cricetidae) from the smooth-tailed myomorph rodents (Muridae). In U.S., where there are no indigenous murids, the terms rat and mouse (*q.v.*) are used indiscriminately both for the indigenous Cricetidae and for the few imported murids, the term vole being generally confined to the genus Microtus and its immediate allies

    Alaska vole *Microtus miurus*

    redback vole *U.S.* any of several species of the cricetid genus Clethrionomys

    red-backed vole *Zoethrionomys rutilus*

    Alaska vole *Microtus miurus*

    redback vole *U.S.* any of several species of the cricetid genus Clethrionomys

    red-backed vole *Zoethrionomys rutilus*

    bank vole *M. glareolus*

    field vole *Brit. M. agrestis*

    meadow vole *U.S. M. pennsylvanicus*

    prairie vole *M. ochrogaster*

    sagebrush vole *Lagurus curtatus*

    narrow-skulled vole *M. gregalis*

    tundra vole *M. oeconomus*

    water vole *Brit.* the cricetid *Arvicola amphibius* (= water rat *Brit.*)

volitant capable of flight

Volkmann's (often misspelled Volkman's) *epon. adj.* for Alfred Wilhelm Volkmann (1800–1877)

-voltine *adj. suffix* meaning "pertaining to the seasons of the year"

    bivoltine having two generations per year

    multivoltine reproducing several times each season

    univoltine reproducing only once a season

voluble (*Bot.*) twining around a support

involucel an incomplete secondary involucre

involucellate provided with a secondary involucre

involucre anything wrapped around anything else, but most widely used in *bot.* for a ring of bracts surrounding several flowers, but also, rarely, for the indusium of ferns and other similarly shaped structures

    general involucre one which is at the base of a compound umbel

    partial involucre that which surrounds a partial umbel

    secondary involucre = partial involucre

volute 1 (*see also* volute 2, 3) rolled up

    involute rolled inwards

    convolute rolled around, either itself or something else

    supervolute rolled over and over

volute 2 (*see also* volute 1, 3) one turn of a spiral shell

volute 3 (*see also* volute 1, 2) popular name of volutid gastropods

volutin protein granules with a high phosphate content found in the cells of many lower animals

volution one turn of a spiral shell

    involution 1 (*see also* involution 2) of the many meanings of this word, biologists use principally the condition of a regressive change or diminution in size

    involution 2 (*see also* involution 1) the ingrowth of one or more layers of cells over the lip of the blastopore in the later stages of gastrula production in the frog

volva a membrane totally enclosing a growing basidiomycete fungus, such as the amanitaceous forms, and which remains after the cap has emerged as a veil around the bottom of the stalk

volvent = desmoneme nematocyst

Volvocaceae that family of volvocale Chlorophyceae distinguished by the fact that all are motile colonies in which the cells form a hollow sphere and are not in superimposed layers

Volvocales that order of chlorophycean Chlorophytae distinguished by the fact that the vegetative cells are flagellated and actively motile

vomer *see* bone

von Ebner's *epon. adj.* from Victor Ebner-Ritter von Rosenstein (1842–1888)

-vor- *comb. form* meaning to "devour"

-vore *subs. term* indicating "an eater of." Compounds are given under -vorous

-vorous *adj. suffix* meaning "eating." The *subs. form* -vore is more common than its equivalent -phage (*cf.* -phagous)

    calcivorous chalk-eating. Applied to forms such as some lichens and Mollusca which erode or borrow in limestone

    carnivorous flesh-eating

    erucivorous caterpillar-eating

    grammnivorous grass-eating

    granivorous seed-eating

    insectivorous insect-eating

    limivorous mud-eating

    nucivorous nut-eating

    omnivorous eating all kinds of things

    plurivorous eating many things, and specifically applied to a parasite capable of infecting numerous hosts

    univorous said of a host-specific parasite

    phytosuccivorous eating by sucking the sap of plants

    zoosuccivorous an organism that sucks the blood or other fluid from animal bodies

[**vortex in 3**] a mutant (gene symbol *vo-3*) mapped at 40.4± on chromosome III of *Drosophila melanogaster*. It is an intensifier of some of the [clumpy] alleles

vs *see* [vesiculated]

vt *see* [vestigial]

vulture any of several species of New World birds of the family Cathartidae, or Old World birds of the family Accipitridae. All are large carrion eaters.

    white-backed vulture either of two accipitrids of the genus *Pseudogyps*

        African white-backed vulture *P. africanus*

        Indian white-backed vulture *P. bengalensis*

    bearded vulture the peculiar accipitrid *Gypaetus barbatus*

    long-billed vulture the accipitrid *Gyps indicus*

    black vulture properly, the cathartid *Coragyps atratus*; *I.P.* the accipitrid *Sarcogyps calvus* (= Pondicherry vulture)

    cinereus vulture the accipitrid *Aegypius monachus*

    Egyptian vulture the accipitrid *Neophren percnopterus*

    lappet-faced vulture the accipitrid *Torgos tracheliotus*

    griffon vulture = griffon

    white-headed vulture the accipitrid *Trigonoceps occipitalis*

        yellow-headed vulture the cathartid *Cathartes burrovianus*

    hooded vulture the accipitrid *Necrosyrtes monachus*

    king vulture properly, the New World cathartid *Sarcoramphus papa* (*cf.* condor) *I.P.* the accipitrid *Sarcogyps calvus* (= Pondicherry vulture)

    palm-nut vulture the accipitrid *Gypohierax angolensis*

    Pondicherry vulture the accipitrid *Sarcogyps calvus*

    turkey vulture the cathartid *Cathartes aura*

Vulturidae an obsolete taxon of birds which at one time contained many large Accipitridae and the Cathartidae

-vul- *see* -vol-

vulva the female external genitalia but especially the female gonopore of nematode worms

w *see* [waltzing], [white] or [wide band agouti]

W *see* [Dominant spotting] or [wrinkled]

w₁ *see* [wiry foliage]

w 2 *see* [white egg 2], [-3]

wa *see* [wavy] or [washed]

wa-1 *see* [warty-like], [waved 1], [2]

[**wabbler-lethal**] a mutant (gene symbol *wl*) in linkage group III of the mouse. The phenotypic expression is with difficulty distinguished from [agitans]

[**waddler**] a mutant (gene symbol *wd*) in linkage group VIII of the mouse. The name is descriptive of the gait of the phenotype

wahoo 1 (*see also* wahoo 2, elm) the scombrid fish *Acanthocygium solanderi*

wahoo 2 (*see also* wahoo 1) the celestraceous shrub *Euonymus atropurpureus*

wainscot any of very numerous species of noctuid lepidopteran insects

    powdered wainscot *Arsilonche albovenosa*

wakaka *N.Z.* = sea lion

waldrap the threskiornithid bird *Geronticus eremita*

wood **wale** = woodpecker

walk a leisurely method of progression in which the feet are advanced alternately

night-**walker** = kinkajou

wall 1 (*see also* wall 2) in the sense of limiting margin

    adventitial wall the outer connective tissue wall of a blood vessel

    cell wall a non-living protective coat around a cell, particularly a plant cell

    primary cell wall the original cellulose wall formed round a plant cell

    secondary cell wall a wall of one or more layers, sometimes of varied composition, formed within the primary cell wall

    fruit wall a term applied to the wall of a fruit in which the pericarp cannot be clearly distinguished from accessory parts

wall 2 (*see also* wall 1) *I.P.* any of numerous satyrid lepidopteran insects of the genus Pararge

wallaby a term applied to those jumping marsupial mammals of the family Macropodidae which are too small to be called kangaroos

    Aru Island wallaby *Macropus brunii*

    banded wallaby any of several species of the genus Lagostrophus

    bridled wallaby *Onychogale frenata*

    dama wallaby *Macropodus eugenii*

    Cape York wallaby *M. coxeni*

    gazelle-faced wallaby any of several New Guinea species of the genus Dorcopsis

    black-gloved wallaby *Maur irma*

    hare-wallaby any of several species of the genus Lagorchestes (= hare-kangaroo)

    red-legged wallaby *Maeropon wilcoxi*

    red-necked wallaby *Macropodus ruficollis*

    rock wallaby any of several species of macropodid marsupial of the genus Petrogale

    black-striped wallaby *Macropodus dorsalis*

    black-tailed wallaby *Macropodus ualabatus*

    nail-tailed wallaby any of several species of macropodid of the genus Onychogale though usually, without qualification, *O. unguifera*

Wallace's *epon. adj.* from Alfred Russel Wallace (1823–1913)

wallaroo *Austral.* popular name of macropodine marsupials of the genus Osphranter. They differ from other macropodines in having hairless muzzles

walrus (*see also* bird) the odobaenid pinipede carnivore *Odobaenus rosmarus*. The walrus of the Bering Sea is thought by some to be a separate species. The generic name Trichechus, formerly used for the genus and derivative family, belongs properly to the manatee

waltzing *see* syndrome

[**waltzer**] a mutant (gene symbol *v*) in linkage group X of the mouse. The phenotypic expression is the waltzing syndrome accompanied by deafness

[**waltzing**] a mutant (gene symbol *w*) mapped at 66.3 on linkage group I of the rat. The phenotypic expression differs from the waltzing syndrome of the mouse in that the head is not shaken

wampi the rutaceous tree *Claucena lansium*

wanderer *U.S.* lycaenid papilionoid lepidopteran insect *Feniseca tarquinius* (= harvester). *I.P.* any of several species of pierid of the genus Parenonia. *S.Afr.* the acraeid *Bematistes aganice*

    false wanderer *S.Afr.* the nymphalid *Pseudacraea eurytus*

    plain wanderer *Austral.* the pedionomid gruiform bird *Pedionomus torquatus* (= collared hemipode)

wanderoo this name is variously applied to two monkeys, the cynopithecid *Macaca albibarbata* and the colobid *Presbytis cephalopterus* (*cf.* langur)

wapato either of two species of alismataceous herb of the genus Sagittaria *S. latifolia* and *S. cuneata*

wapiti either the *U.S.* cervine *Cervus canadensis* or a Siberian race of *C. elaphas* (= red deer)

waratah the proteaceous shrub *Telopea speciosissima*

warble a swelling under the skin of large vertebrate animals caused by the subdermal larvae of hypodermatid dipterans

warbler any of very many birds of the families Sylviidae and Parulidae, occasionally applied to other small birds. The former are properly the Old World warblers. In *U.S.* many sylviids are called gnat-catchers and kinglets

Adelaide's warbler the parulid *Dendroica adelaidae*

aquatic warbler the sylviid *Acrocephalus paludicola*

Arctic warbler the sylviid *Phylloscopus borealis*

Audubon's warbler the parulid *Dendroica auduboni*

Australian warbler popular name of Malurinae, more usually called wren in *Austral.*

olive-backed warbler the parulid *Parula pitiayumi*

Bachman's warbler the parulid *Vermivora bachmanii*

barred warbler the sylviid *Sylvia nisoria*

buff-bellied warbler the sylviid *Phyllolais pulchella*

black-and-white warbler the parulid *Mniotilta varia*

blackburnian warbler the parulid *Dendroica fusca*

blackpoll warbler the parulid *Dendroica striata*

black-throated blue warbler the parulid *Dendroica coerulescens*

Bonelli's warbler the sylviid *Phylloscopus bonelli*

booted warbler the sylviid *Hippolais caligate*

bracken-warbler any of several sylviids of the genus Sathrocercus

    cinnamon bracken-warbler *S. cinnamomeus*

bay-breasted warbler the parulid *Dendroica castanea*

Brewster's warbler the hybrid parulid *Vermivora leucobronchialis*

golden-browed warbler the parulid *Basileuterus belli*

brown warbler the malurine *Gerygone richmondi*

bush-warbler any of several sylviids, mostly of the genus Cettia; without qualification, *C. diphone*

    masked bush-warbler the sylviid *Apalis binotata* (*cf.* apalis)

Canada warbler the parulid *Wilsonia canadensis*

cane warbler *Afr.* the sylviid *Calamocichla rufescens* (= rufous swamp-warbler)

Cape May Warbler the parulid *Dendroica tigrina*

grey-capped warbler the sylviid *Eminia lepida*

    olive-capped warbler the parulid *Dendroica pityophila*

    rufous-capped warbler the parulid *Basileuterus rufifrons*

Cerulean warbler the parulid *Dendroica cerulea*

golden-cheeked warbler the parulid *Dendroica chrysoparia*

crescent-chested warbler the parulid *Vermivora superciliosa*

Colima warbler the parulid *Vermivora crissalis*

Connecticut warbler the parulid *Oporornis agilia*

golden-crowned warbler the parulid *Basileuterus culicivorus*

    orange-crowned warbler the parulid *Vermivora celata*

Dartford warbler the sylviid *Sylvia undata*

desert warbler the sylviid *Sylvia nana*

worm-eating warbler the parulid *Hemitheros vermivorus*

flycatcher-warbler any of several species of sylviid birds of the genus Seicercus

forest-warbler any of several sylviids of the genus Artisornis

scaly-fronted warbler the sylviid *Spiloptila clamans*

garden warbler the sylviid *Sylvia borin*

golden warbler = yellow warbler

Grace's warbler the parulid *Dendroica graciae*

black-throated green warbler the parulid *Dendroica verens*

greenish warbler the sylviid *Phylloscopus trochiloides*

grey warbler N.Z. the malurine *Gerygone igata*

black-throated grey warbler the parulid *Dendroica nigrescens*

ground-warbler either of two parulids of the genus Microlegia

    green-tailed ground-warbler *M. palustris*

    white-winged ground-warbler *M. montana*

arrow-headed warbler the parulid *Dendroica pharetra*

    pink-headed warbler the parulid *Ergaticus versicolor*

    yellow-headed warbler the parulid *Teretistris fernandinae*

hermit warbler the parulid *Dendroica occidentalis*

hill-warbler any of several species of sylviids of the genus Prinia (= prinia) *Malay*, without qualification, *P. atrogularis*

hooded warbler the parulid *Wilsonia citrina*

grasshopper-warbler any of several sylviids of the genus Locustella. *Brit.*, without qualification, usually *L. neevia*

icterine warbler the sylviid *Hippolais icterina*

Kentucky warbler the parulid *Oporornis formosus*

Kirtland's warbler the parulid *Dendroica kirtlandii*

Lawrence's warbler the hybrid parulid *Vermivora lawrencei*

Lucy's warbler the parulid *Vermivora luciae*

MacGillivray's warbler the parulid *Oporornis tolmiei*

mangrove warbler = yellow warbler

magnolia warbler the parulid *Dendroica magnolia*

marsh warbler the sylviid *Acrocephalus palustris*

melodious warbler the sylviid *Hippolais polyglotta*

morning warbler either of two species of turdids of the genus Cichladusa; usually, without qualification, *C. arguata*

    spotted morning warbler *G. guttata*

mourning warbler the parulid *Oporornis philadelphia*

moustache-warbler the sylviid *Melocichla mentalis*

myrtle warbler the parulid *Dendroica coronata*

Nashville warbler the parulid *Vermivora ruficapilla*

nuthatch-warbler any of several African sylviids of the genus Sylvietta (= crambec)

olivaceous warbler the sylviid *Hippolais pallida*

olive warbler the parulid *Peucedramus taeniatus*

Oriente warbler the parulid *Teretistris fornsi*

orphean warbler the sylviid *Sylvia hortensis*

palm warbler the parulid *Dendroica palmarum*

parula warbler the parulid *Parula americana*

    tropical parula warbler the parulid *Parula pitiayumi* (= olive-backed warbler)

pileolated warbler the parulid *Wilsonia pusilla* (= Wilson's warbler)

pine warbler the parulid *Dendroica pinus*

plumbeous warbler the parulid *Dendroica plumbea*

prairie warbler the parulid *Dendroica discolor*

prothonotary warbler the parulid *Protonotaria citrea*

reed-warbler any of several sylviids of the genus Acrocephalus. *Brit.*, *A. scirpaceus*

    African reed-warbler *A. baeticatus*

    Blyth's reed-warbler the sylviid *Acrocephalus dumetorum*

    great reed-warbler *A. arundinaceus*

    Indian great reed-warbler *A. stentoreus*

river warbler the sylviid *Locustella fluviatilis*

rock warbler the sylviid *Origma solitaria*

rufous warbler the sylviid *Agrobates galactotes*

Ruppell's warbler the sylviid *Sylvia rueppellii*

rush-warbler any of several sylviids of the genus Bradypterus

    little rush-warbler *B. baboecola*

Sardinian warbler the sylviid *Sylvia melanocephala*

Savi's warbler the sylviid *Locustella luscinoioides*

scrub-warbler any of several sylviids of the genus Scotocerca

sedge warbler variously applied to almost any bird frequenting reedy marshes. In particular, *Brit.*, the sylviid *Acrocephalus schoenobaenus*

Semper's warbler the parulid *Leucopeza semperi*

chestnut-sided warbler the parulid *Dendroica pensylvanica*

speckled warbler the sylviid *Chthonicola sagittata*

spectacled warbler the sylviid *Sylvia conspicillata*

subalpine warbler the sylviid *Sylvia cantillans*

swamp-warbler any of several sylviids of the genus Calamocichla

   greater swamp-warbler *C. gracilirostris*

   lesser swamp-warbler *C. leptorhyncha*

   rufous swamp-warbler *C. rufescens*

red warbler the parulid *Ergaticus ruber*

Sutton's warbler the probably hybrid parulid *Dendroica potomac*

Swainson's warbler the parulid *Limnothlypis swainsonii*

fan-tailed warbler U.S. the parulid *Euthluypis lachrymosa Brit.*, the sylviid *Cisticola juncidis Afr.* the sylviid *Schoenicola brevirostris*

Tennessee warbler the parulid *Vermivora peregrina*

yellow-throated warbler the parulid *Dendroica dominica*

   white-throated warbler the malurine *Gerygone olivacea*

   Indian lesser white-throated warbler the sylviid *Sylvia althaea* (*cf.* white throat)

tit-warbler any of several sylviids of the genus Leptopoecile

olive tree warbler the sylviid *Hippolais olivetorum*

Townsend's warbler the parulid *Dendroica townsendi*

Upcher's warbler the sylviid *Hippolais languida*

Virginia's warbler the parulid *Vermivora virginiae*

vitelline warbler the parulid *Dendroica vitellina*

weatern warbler the malurine *Gerygone fusca*

whistling warbler the parulid *Catharopeza bishopi*

willow-warbler any of several sylviids of the genus Phylloscopus; without qualification, *P. trochilus*

Wilson's warbler the parulid *Wilsonia pusilla*

blue-winged warbler the parulid *Vermivora pinus*

   golden-winged warbler the parulid *Vermivora chrysoptera*

   red-winged warbler the sylviid *Heliolais erythroptera*

woodland-warbler any of several sylviids of the genus Seicercus

   brown woodland-warbler *S. umbrovirens*

   yellow-throated woodland-warbler *S. ruficapillus*

wren-warbler any of several species of sylviids of the genus Calamonastes

   barred wren-warbler *C. fasciolatus*

   grey wren-warbler *C. simplex*

yellow warbler the parulid *Dendroica petechia*

[warped] a mutant (gene symbol *wp*) mapped at 47.5 on chromosome III of *Drosophila melanogaster*. The phenotypic spread wings are warped

warratau = waratah

wart a knobbly prominence (*cf.* papilla)

   copulatory wart one of one or more large papillae in a central cavity found immediately anterior to the anus in some Nematodes

[warty-like] a mutant (gene symbol *wa-1*) mapped at 64.4 on the X chromosome of *Drosophila melano-*

gaster. The phenotypic expression is a disarrangement of the ommatidia

warwinckle = motacillid bird *Motacilla lugubris*

[Washed] a mutant (gene symbol *wa*) on linkage group V of *Neurospora crassa*. The name refers to the sparse thin growth of the phenotype

[Washed eye] a mutant (gene symbol *Wi*) mapped at 43.0 on chromosome III of *Drosophila melanogaster*. The phenotype is one of the modified w pseudoalleles

dishwasher *Brit.* the motacillid bird *Motacilla alba Austral.* the muscicapid *Sisura inquieta*

wasp 1 (*see also* wasp 2) popular name of vespid and sphecid hymenopteran insects, but also applied to almost any hymenopteran insect not obviously a bee or an ant. *Birt.* without qualification *Vespa vespa*

   aphid wasp popular name of pemphredonine sphecids

   bee-killer wasp those philanthine sphecids which provision their nests with bees

   beetle wasp those philanthine sphecids which provision their nests with beetles

   cuckoo wasp popular name of chrysidids and sometimes applied to nyssonine sphecids of the genus Stizoides which lay their eggs on grasshoppers already killed by other wasps

   digger wasp any of numerous wasps which dig holes which they provision for their larvae. The term is most commonly applied to philanthine and crabronine sphecids

      spiny digger wasp the term sometimes applied to those crabonine sphecids which carry their prey impaled on the protruded sting

   ensign wasp popular name of evaniids

   fig wasp any of several species of agaontid calcidoid hymenopterans (= fig insect). *U.S.*, without qualification, *Blastophaga psenes*

   gall wasp popular name of cynipids

      oak-apple-gall wasp *Amphibolips confluentus*

      rose-root-gall wasp *U.S. Diplolepis radicum*

      mossy-rose-gall wasp *U.S. Diplolepis rosae*

   square-headed wasp popular name of crabronine sphecids

   mud wasp *Brit.* the eumenine *Eumenes coarctata*

      blue mud wasp *U.S.* the sphecid *Chalybion caeruleum*

   ruby wasp *Brit.* popular name of chrysidids (= cuckoo fly *Brit.*)

   sand wasp *Brit.* any of numerous sphecids of the genus Ammophila. Usually, without qualification, *A. sabulosa*

   sand wasp *U.S.* any of numerous sand dwelling sphecids of the subfamily Nyssoninae. They are distinguished by an elongate and triangular labrum

   thread-waisted wasp the popular name of sphecine sphecids

   weevil wasp the philanthine sphecid *Cerceris clypeata* (*cf.* beetle wasp)

wood wasp *U.S.* popular name of xiphydriid symphytans. This is the only family, except for the Orussidae, of symphytan hymenopterans usually referred to as wasps *Brit.* popular name of siricid hymenopterans (= horntail *U.S.*)

   blue wood wasp *Brit. Sirex noctilio*

   giant wood wasp *Brit. Sirex gigas*

   parasitic wood wasp popular name of orussid symphytan Hymenoptera

wasp 2 (*see also* wasp 1) other stinging organisms

sea wasp popular name of cubomedusoid schyphozoan coelenterates

watch term for a group of nightingales

watches any of several species of sarraceniaceous insectivorous herbs of the genus Sarracenia

water 1 (*see also* water 2) the fluid of that name (*see* compound names of organisms beginning with "water" under the word so compounded)

brackish water the waters of salt marshes and estuaries containing more dissolved salts than fresh water but less than sea water (*cf.* haline)

freshwater from the *biol.* point of view, water containing less than 0.5 parts per thousands dissolved salts (*cf.* haline)

sea water the water of seas and oceans usually containing between 30 and 40 parts per thousand of dissolved salts (*cf.* haline)

water 2 (*see also* water 1) in compound names of organisms

shearwater any of numerous procellariid birds, mostly of the genus Puffinus. *N.Z.* without qualification, *P. gavia*

Audubon's shearwater *P. lherminieri*
grey-backed shearwater = New Zealand shearwater
slender-billed shearwater *P. tenuirostris*
Cory's shearwater *P. diomedea*
fluttering shearwater *P. gavia*
pale-footed shearwater *P. carneipes*
pink-footed shearwater *P. creatopus*
greater shearwater *P. gravis*
little shearwater *P. assimilis*
Manx shearwater *P. puffinus*
Newell's shearwater *P. newelli*
New Zealand shearwater *P. bulleri*
sooty shearwater *P. griesus*
black-tailed shearwater *Adamastor cinereus*
short-tailed shearwater = slender-billed shearwater
wedge-tailed shearwater *P. pacificus*
Townsend's shearwater *P. auricularis*
black-vented shearwater *P. opisthomalas*

wattle 1 (*see also* wattle 2, 3) thickened brightly colored flesh on the head of a bird

wattle 2 (*see also* wattle 1, 3) *Austral.* any of numerous shrubs and trees of the leguminous genus Acacia

blue-leaved wattle *A. cyanophylla*
broom wattle *A. calamifolia*
Sydney golden wattle *A. longifolia*
green wattle *A. decurrens*

wattle 3 (*see also* wattle 1, 2) *Brit.* (*obs.*) twig

wave *Brit.* any of very many species of geometrid lepidopteran insects of the genus Acidalia

weaver's wave *A. contiguaria*

[waved-1] a mutation (gene symbol *wa-1*) in linkage group XI of the mouse. The phenotypic expression is wavy hair and whiskers

[waved-2] a mutant (symbol *wa-2*) in linkage group VII of the mouse. The phenotypic expression is identical with [waved-1]

[wavy] *either* a mutant (gene symbol *wa*) in linkage group V of *Habrobracon juglandis*. The phenotypic wings are very short and with wavy margin. *or* a mutant (gene symbol *wy*) mapped at 41.9 on the X chromosome of *Drosophila melanogaster*. The name refers to the appearance of the wings of the phenotype

wax 1 (*see also* wax 2, bill, moth, palm, plant, scale, wing work) any waxy substance

wax 2 (*see also* wax 1) in compound names of organisms

thorough wax the umbelliferous herb *Bupleurum rotundifolium*

woad waxen the leguminous shrub *Genista tinctoria* (= whin)

[waxy] a mutant (gene symbol *wx*) mapped at 69.7 on chromosome II of *Drosophila melanogaster*. The term refers to the appearance of the wings of the phenotype

[waxy endosperm] a mutant (gene symbol *wx*) mapped at 59 on linkage group IX of *Zea mays*. The term is descriptive

wc *see* [White collar]

Wc *see* [White cap of endosperm]

wd *see* [waddler] or [wilty dwarf]

we *see* [wellhaarig]

[weak] a mutant (gene symbol *wk*) mapped at 42.0± on chromosome III of *Drosophila melanogaster*. The name applies to the small and irregular bristles of the phenotype

weasel (*see also* snout) any of numerous species of several genera of small musteline Carnivora *Brit.* without qualification, *Mustela vulgaris*

least weasel *U.S. M. rixosa*
mouse weasel *M. nivalis*
snake weasel = striped weasel
striped weasel any of several species of very long, very short legged, African musteline carnivores of the genera *Poecilogale* and *Poecilictis*
long-tailed veasel *U.S. M. frenata*
short-tailed veasel *U.S. M. erminea* (= Brit. stoat)

weaver (*see also* bird) any of numerous ploceid birds

grey-backed weaver = forest weaver
grosbeak weaver *Amblyospiza albifrons*
black weaver any of several black species of the genus Ploceus; without qualification, *P. nigerrimus*
buffalo-weaver any of several species of the genera Bubalornis and Dinemellia at one time regarded as forming a separate family or subfamily; without qualification, *B. albirostris*
red-billed buffalo-weaver *B. niger*
white-headed buffalo-weaver *D. dinemelli*
brown-capped weaver *Ploceus insignis*
golden-capped weaver *Ploceus preussi*
compact weaver *Ploceus pachyrhynchus*
cutthroat weaver *Afr. Amadina fasciata* (= cutthroat finch)
forest weaver *Ploceus bicolor*
speckle-fronted weaver *Sporopipes frontalis*
golden weaver any of several yellow species of the genus Ploceus; without qualification, *P. subaureus*
black-headed weaver *Ploceus cucullatus*
hooded weaver *Spermestes cucullatus* (*cf.* mannikin)
little weaver *Ploceus luteolus*
masked weaver any of several species of the genus Ploceus; without qualification, P. intermedius
Napoleon weaver *Euplectes afra.* (= yellow-crowned bishop)
black-necked weaver *Ploceus nigricollis*
parasitic weaver *Anomalospiza imberbis*
scaly weaver *Sporpipes squamifrons*
sociable weaver *Philetarius socius*
social-weaver any of several species of the genus Pseudonigrita
black-capped social-weaver *P. cabanisi*
grey-headed social-weaver *P. arnaudi*
sparrow-weaver any of several species of the genus Plocepasser
chestnut-crowned sparrow-weaver *P. superciliosus*

stripe-breasted sparrow-weaver *P. mahali*

spectacled weaver *Ploceus ocularis*

village weaver *Ploceus cucullatus* (= black-headed weaver)

web 1 (*see also* web 2, 3, leek, worm) a net or trap of spun silk

web 2 (*see also* web 1, 3) the fold of skin connecting the toes of many water birds

web 3 (*see also* web 1, 2) that part of a feather which borders the rachis

food web an interconnected series of food chains

Weber's *epon. adj.* from Ernst Heinrich Weber (1795–1878)

wedge 1 (*see also* wedge 2) term for a group of swans in flight

wedge 2 (*see also* wedge 1) a rectangular object thinner at one end than at the other

bast wedge one of the phloem bundles in which the abaxial section is widest

weed 1 (*see also* weed 2, 3) used by gardeners in the sense of a plant growing without human encouragement. It is frequent in compound names of flowering plants but is used rather variably so that names not found in the list that follows should also be sought under plant or wort

ague weed *U.S.* the gentianaceous herb *Gentiana quinquefolia*

ashweed the umbelliferous herb *Aegopodium podagraria*

automobile weed *U.S.* the zygophyllaceous herb *Tribulus terrestris*

beetle weed the diapensiaceous herb *Galax aphylla*

bindweed any of several convolvulaceous herbs, mostly of the genus Convolvulus, and Ipomoea

hedge bindweed *C. sepium* (= rutland beauty)

bitterweed the compositaceous herb *Ambrosia artemisiifolia*

mock bishop's weed any of several species of umbelliferous herbs of the genus Ptilimnium

brookweed *Brit.* any of several species of primulaceous herbs of the genus Samolus

buffaloweed *U.S.* the compositaceous herb *Ambrosia trifida*

bugle weed any of several species of herbs of the labiateous genus Ajuga, particularly *A. virginicus*

butter weed the compositaceous herb *Senecio glabellus*. This name is also applied to the horse weed

butterfly weed the asclepiadaceous herb *Asclepias tuberosa*

buttonweed the rubiaceous herb *Spermacoce glabra* and several species of the rubiaceous genus Diodia

camphorweed either of two compositaceous herbs *Heterotheca subaxillaris*, or *Pluchea camphorata*

cancerweed the labiateous herb *Salvia lyrata*

carpenterweed the labiateous herb *Prunella vulgaris* (= heal all)

carpetweed U.S. the aizoaceous herb *Mollugo verticillata*

chaffweed any of several species of primulaceous herbs of the genus Centunculus. *Brit.*, without qualification, *C. minimus*

chickweed (*see also* wintergreen) the caryophyllaceous herb *Holostea media*

bastard chickweed the scrophulariaceous herb *Sibthorpia europaea*

blinking chickweed = water chickweed

forked chickweed *U.S.* the caryophyllaceous herbs *Paronychia canadensis* or *P. fastigiata*

giant chickweed *U.S.* the caryophyllaceous herb *Myosoton aquaticum*

Indian chickweed *U.S.* the aizoaceous herb *Mollugo verticillata*

jagged chickweed the caryophyllaceous herb *Holosteum umbellatum*

low chickweed *Stellaria humifusa*

mouseear chickweed any of several species of herb of the caryophyllaceous genera Cerastium and Malachium

sea chickweed *U.S.* the caryophyllaceous herb *Arenaria peploides*

water chickweed the portulacaceous herb *Montia fontana*

clammy weed *U.S.* any of several species of capparidaceous herbs of the genus Polanisia

clearweed *U.S.* any of several species of urticaceous herb of the genus Pilea

consumption weed the compositaceous shrub *Baccharis halimifolia*

cottonweed *U.S.* any of several species of amaranthaceous herb of the genus Froelichia

crazy weed *U.S.* the leguminous herb *Oxytropis lambertii* (= locoweed)

crossweed the cistaceous herb *Helianthemum canadense*

crunchweed the cruciferous herb *Brassica kaber*

cudweed (*see also* weed 3) any of several species of compositaceous herb of the genera Antennaria and Gnaphalium

duckweed any of several species of minute floating aquatic plants of the lemnaceous genus Lemna

tropical duckweed the araceous aquatic herb *Pistia stratiotes*. (= water lettuce)

dyer's weed the resedaceous herb *Reseda luteola* (*cf.* rocket, mignonette)

fanweed *U.S.* the cruciferous herb *Thlaspi arvense* (= penny cress)

fireweed *U.S.* any of several species of the compositaceous genus Erechtites or the onagraceous herb *Epilobium angustifolium*

flaxweed *Brit.* the cruciferous herb *Sisymbrium sophia*

horsefly weed the leguminous herb *Baptisia tinctoria* (= wild indigo)

French weed *U.S.* the cruciferous herb *Thlaspi arvense* (= penny cress)

frostweed *U.S.* the cistaceous herbs *Helianthemum canadense* or *H. bicknellii*, the compositaceous herb *Verbesina virginica*

goutweed the umbelliferous herb *Aegopodium podograria*

green *Brit.* any of several leguminous shrubs and herbs of the genus Genista

gumweed the compositaceous herb *Grindelia squarrosa*

gypsy weed the scrophulariaceous herb *Veronica officinalis*

hawkweed any of numerous species of compositaceous herbs of the genus Hieracium

bastard hawkweed any of several species of the compositaceous genus Crepis

mouseear hawkweed *H. pilosella*

climbing hempweed *U.S.* the compositaceous climbing vine *Mikania scandens*

hogweed *Brit.* the umbelliferous herb *Heracleum sphondylium* (= cow parsnip), for which the term horseweed is also used, and the compositaceous herb *Ambrosia artemisiifolia*

horseweed any of several species of the labiateous genus *Collinsonia* and the compositaceous herb *Erigeron canadensis* (*cf.* hogweed)

indigo weed the leguminous herb *Baptisia tinctoria*

ironweed any of several species of the compositaceous genus Vernonia

itchweed *U.S.* any of several liliaceous herbs of the genus Veratrum

Jamestown weed = jimson weed

jewelweed *either* of two balsaminaceous herbs *Impatiens biflora or I. pallida*

jimson weed the solanaceous herb *Datura stramonium*

Joe-pye weed the compositaceous herb *Eupatorium purpureum*

jointweed any of several species of the polygonaceous genus Polygonum

knapweed *Brit.* any of several species of compositaceous herbs of the genus Centaurea, most frequently *C. nigra*

knotweed any of several species of the polygonaceous genus Polygonum

locoweed (*Span.* loco = mad) any of several plants of the *U.S.* which stupefy or over excite grazing cattle. Usually, without qualification, the leguminous herb *Oxytropis lambertii*

mayweed the compositaceous herb *Anthemis cotula*
  scentless mayweed *Brit.* the compositaceous herb *Matricaria inodora*. (= corn feverfew)

mermaid weed any of several species of haloragaceous herb of the genus Proserpinaca

milkweed (*see also* beetle, bug, moth) any of several species of asclepiadaceous herb mostly of the genus Asclepias but also occasionally used for the lettuce
  giant milkweed the asclepiadaceous herb *Calotropis gigantea*
  poke milkweed *Asclepias exaltata*
  spider milkweed any of several species of the asclepiadaceous herb of the genus Asclepiodora

mudweed any member of the scrophulariaceous genus *Limosella* (= mudwort)

muskrat weed *U.S.* the ranunculaceous herb *Thalictrum polygamum*

neckweed *U.S.* the scrophulariaceous herb *Veronica peregrina*

necklace weed (*see also* weed 2) *U.S.* any of several species of ranunculaceous herbs of the genus Actaea (= bane berry)

pickerel weed the pontederiaceous aquatic herb *Pontederia cordata*

pigweed any of several species of chenopodiaceous herb of the genus Chenopodium particularly *C. lanceolatum, C. album,* and *C. paganum* and also several species of the amaranthaceous genus Amaranthus
  winged pigweed *U.S.* the chenopodiaceous herb *Cycloloma atriplicifolium*

pigmyweed the crassulaceous herb *Tillaea aquatica*

pinweed *U.S.* any of several species of cistaceous herb of the genus Lechea

pineapple weed the compositaceous herb *Matricaria matricarioides*

pinkweed *U.S.* the polygonaceous herb *Polygonum pennsylvanicum*

pokeweed *U.S.* the phytolaccaceous herb *Phytolacca americana*

pondweed any of numerous aquatic herbs of the potamogetonaceous genus Potamogeton, though the term is loosely applied to many other aquatics

Cape pondweed the aponogetonaceous aquatic herb *Aponogeton distachyum*

horned pondweed *Brit.* any of several species of potamogetonaceous aquatic herbs of the genus Zannichellia, particularly *Z. palustris*

povertyweed *U.S.* the chenopodiaceous herb *Monolepis nuttalliana*

prickleweed *U.S.* the leguminous shrub *Desmanthus illinoensis*

quicksilver weed the ranunculaceous herb *Thalictrum dioicum*

ragweed (*see also* bug, borer) any of several species of compositaceous herb of the genus Ambrosia, but particularly *A. artemisiifolia* (*cf.* ragwort)

rattleweed the leguminous herb *Baptisia tinctoria* (= wild indigo)

richweed the urticaceous herb *Pilea pumila* and also numerous gold or silvery flowers such as those of the ranunculaceous genus Cimicifuga or the compositaceous Ambrosia. The term is also applied to the labiateous herb *Collinsonia canadensis*

riverweed moss or lichen-like stream plants belonging to the family Podostemaceae

rosinweed any of several compositaceous herbs of the genus Silphium

scorpionweed any of numerous species of the hydrophyllaceous family Phacelia

sheepweed the malvaceous herb *Abutilon theophrasti*

shoreweed *Brit.* the plantaginaceous herb *Littorella lacustris*

shovelweed any of several species of cruciferous herbs of the genus Capsella (= shepherd's purse)

sickleweed *U.S.* any of several species of umbelliferous herbs of the genus Falcaria

silkweed = milkweed

silverweed the convolvulaceous genus of climbing plants Argyreia and the rosaceous herb *Potentilla anserina*
  Pacific silverweed *Potentilla egedei*

skeletonweed the compositaceous herb *Chondrilla juncea*

skunkweed the euphorbiaceous herb *Croton texensis*

smartweed (*see also* borer) any of several species of the polygonaceous genus Polygonum

snakeweed the polygonaceous herb *Polygonum bistorta*
  rattlesnake weed the compositaceous herb *Hieracium venosum*

snapweed *U.S.* any of several species of the balsaminaceous genus Impatiens

sneezeweed any of several species of compositaceous herb of the genus Helenium but particularly *H. autumnale*. Also the compositaceous herb *Achillea ptarmica*

soapweed *U.S.* the liliaceous herb *Yucca glauca*

squaw weed the compositaceous herb *Senecio aureus* (= golden rag wort)

staggerweed *U.S.* the papaveraceous herb *Dicentra eximia*

stinkweed *U.S.* the compositaceous herb *Pluchea camphorata* and the cruciferous herb *Thlaspi arvense* (= penny cress)

strangleweed any of several convolvulaceous parasitic herbs of the genus Cuscuta (= dodder)

styptic weed the leguminous shrub *Cassia occidentalis*

tarweed the compositaceous herb *Grindelia squarrosa*, any of several species of herbs of the compositaceous genus Madia and any of several species of boraginaceous herbs of the genus Amsinckia

thimbleweed *U.S.* either of two species of ranunculaceous herbs of the genus Anemone. *A. riparia* and *A. virginiana*

tickweed the compositaceous herb *Verbesina virginica*

tinker's weed the caprifoliaceous herb *Triosteum perfoliatum*

tobaccoweed the compositaceous herb *Elephantopus tomentosus*

trumpetweed the compositaceous herb *Eupatorium fistulosum*

tumbleweed any of numerous dried weeds, forming a more or less spherical or ovoid mass, which are blown about by the wind particularly the amaranthaceous herbs *Amaranthus graecicans* and *A. albus*. The term is also applied to the chenopodiaceous herb *Salsola kali*, and to the ranunculaceous herb *Anemone cylindrica*

wartweed *U.S.* the euphorbiaceous herb *Euphorbia helioscopia*

waterweed the hydrocharitaceous aquatic herb *Elodea canadensis*

white weed the compositaceous herb *Chrysanthemum leucanthemum*

yellow weed *Brit.* the resedaceous herb *Reseda luteola*

weed 2 (*see also* weed 1, 3) in compound names of lower plants

bootlace weed *N.Z.* any of several species of phaeophyte alga of the genus *Myriogloia*

gummy weed *N.Z.* the phaeophyte alga *Splanchnidium rugosum* (= sea cactus)

necklace weed *N.Z.* (*see also* weed 1) the phaeophyte alga *Hormosira banksii*

seaweed popular name for marine algae in general

black wireweed *N.Z.* the rhodophyte alga *Melanthalia abscissa*

weed 3 (*see also* weed 1, 2) in compound names of animals

cudweed (*see also* weed 1) *Brit.* the noctuid lepidopteran insect *Cucullia gnaphalii*

peaseweep *N. Brit.* = lapwing

weero *Austral.* the psittacid bird *Nymphicus hollandicus* (= cockatiel)

weever (*see also* weaver) any marine teleost of the genus Trachinus

weevil (*see also* curculio) this term was originally, and in Europe usually still is, confined to the long-snouted coleopteran insects once included in the group Rhynchophora but now commonly divided into the super-families, Brentoidea, Anthribooidea, Beloidea, and Cuculionoidea. In *U.S.* the term is extended to include the bruchid beetles and, in popular parlance, to any coleopteran which infests stored crops or, occasionally, to such pests as the carpet beetle

acorn weevil *U.S.* = nut weevil. *Brit.* the curculionid *Balaninus glandium*

alfalfa weevil *U.S.* curculionid *Hypera postica*

arborvitae weevil *U.S.* the curculionid *Phyllobius intrusus*

black elm bark-weevil *U.S.* the curculionid *Magdalis barbita*

red elm bark-weevil *M. armicollis*

Yosemite bark-weevil *U.S.* the curculionid *Pissodes yosemite*

bean weevil *U.S.* the bruchid *Acanthoscelides obtectus*

broadbean weevil *U.S.* the bruchid *Bruchus rufimanus*

coffee bean weevil the anthribrid *Araecerus fasciculatus*

Mexican bean weevil *U.S.* the bruchid *Zabrotes pectoralis*

cranberry weevil *U.S.* the curculionid *Anthonemus musculus*

strawberry weevil *U.S.* the curculionid *Anthonomus signatus*

birch weevil *Brit.* the curculionid *Rhynchites betulae*

biscuit weevil *Brit.* the anobiid *Anobium paniceum*

apple blossom weevil *Brit.* curculionid *Anthonomus pomorum*

boll weevil the curculionid *Anthonomus grandis*

white bud weevil *U.S.* the curculionid *Phyxelis rigidus*

silky cane weevil *U.S.* the curculionid *Metamasius hemipterus sericus*

New Guinea sugarcane weevil *U.S.* the curculionid *Rhabdoscelus obscurus*

West Indian cane weevil *U.S.* the curculionid *Metamasius hemipterus hemipterus*

sand-cherry weevil *U.S.* the curculionid *Anthonomus hirsutus*

sweetclover weevil *U.S.* the curculionid *Sitona cylindricollis*

cocklebur weevil *U.S.* the curculionid *Rhodobaenus tredecimpunctatus*

pine root collar weevil *U.S.* the curculionid *Hylobius radicis*

deodar weevil *U.S.* the curculionid *Pissodes nemorensis*

filbert weevil *U.S.* the curculionid *Curculio uniformis*

Douglas fir twig weevil *U.S.* the curculionid *Cylindrocopturus furnissi*

apple flea weevil *U.S.* the curculionid *Rhynchaenus pallicornis*

willow flea weevil *R. rufipes*

current fruit weevil *U.S.* the curculionid *Pseudanthonomus validus*

fungus weevil *U.S.* the popular name of platystomids and anthribids

dodder gall weevil *U.S.* the curculionid *Smicronyx sculpticollis*

pine gall weevil *U.S.* the curculionid *Podapion gallicola*

grain weevil any of numerous curculionids that infest stored grain (*cf.* granary weevil).

broad-nosed grain weevil *Caulophilus oryzae*

granary weevil *U.S.* Sitophilus granarius

clover head weevil *U.S.* the curculionid *Hypera meles*

hollyhock weevil *U.S.* the curculionid *Apion longirostre*

iris weevil *U.S.* the curculionid *Mononychus vulpeculus*

cloverleaf weevil the curculionid *Hypera punctata* (= clover leaf beetle)

lesser clover leaf weevil *H. nigrirostris*

pealeaf weevil *U.S.* the curculionid *Sitona lineata*

leaf-rolling weevil any of several species of curculionids, many of the genus Attelabus, that roll leaves to provide protection for their eggs

lily weevil *U.S.* the curculionid *Agasphaerops nigra*

mango weevil *U.S.* the curculionid *Sternochetus mangiferae*

New York weevil *U.S.* the belid *Ithycerus noveboracensis*

short-nosed weevils any of numerous curculionids with short and stout beaks

tooth-nosed weevil any of several species of

curculionid in which there are teeth on the inner and outer edges of the mandibles. Many are of the genus Rhynchites

nut weevil any of numerous species of curculionids, mostly of the genus Curculio, distinguished by the extremely long and slender snout with which they bore through nuts

chestnut weevil *U.S.* the curculionid *Curculio auriger*

large chestnut weevil *C. proboscideus*

small chestnut weevil *C. auriger*

Tahitian coconut weevil the curculionid *Diacalandra taitensis*

hazelnut weevil *U.S.* the curculionid *Curculio obtusus*

Asiatic oak weevil *U.S.* the curculionid *Cyrtepistomus castaneus*

gray-sided oak weevil *U.S.* the curculionid *Pandeleteius hilaris*

palm weevil *W.I.* the large curculionid *Rhynchophorus palmarum*

pea weevil *U.S.* the bruchid *Bruchus pisorum*

cowpea weevil *U.S.* the bruchid *Callosobruchus maculatus*

pecan weevil *U.S.* the curculionid *Curculio caryae*

pepper weevil *U.S.* the curculionid *Anthonomus eugenii*

pine weevil the curculionid *Hylobius abietis*

Monterey-pine weevil *U.S.* the curculionid *Pissodes radiatae*

ribbed pine weevil = white pine weevil

whitepine weevil *U.S.* the curculionid *Pissodes strobi*

pineapple weevil *U.S.* the curculionid *Metamasius ritchiei*

plum weevil the curculionid *Conotrachelus nenuphar* (= plum curculio)

cabbage seedpod weevil *U.S.* the curculionid *Ceutorhynchus assimilis*

sweet potato weevil *U.S.* the curculionid *Cylas formicarius*

West Indian sweet potato weevil *U.S.* the curculionid *Euscepes postfasciatus*

primitive weevil popular name of brentids

radish weevil *U.S.* the curculionid *Cleonus sparsus*

rice weevil the curculionid *Sitophilus oryzae*

citrus root weevil *U.S.* the curculionid *Pachnaeus litus*

strawberry root weevil *U.S.* the curculionid *Brachyrhinus ovatus*

Fuller rose weevil the curculionid *Pantomorus godmani*

scarred weevil = short-nosed weevils

clover seed weevil *U.S.* the curculionid *Miccotrogus picirostris*

Engelmann spruce weevil *U.S.* the curculionid *Pissodes engelmanni*

Sitka-spruce weevil *U.S. P. sitchensis*

bean stalk weevil *U.S.* the curculionid *Sternechus paludatus*

lodgepole terminal weevil *U.S.* the curculionid *Pissodes terminalis*

toad weevil a term applied to many short, broad, curculionids

bronze apple tree weevil *U.S.* the curculionid *Magdalis aenescens*

vegetable weevil *U.S.* the curculionid *Listroderes costirostris*

vetch weevil *U.S.* the bruchid *Bruchus brachialis* (= vetch bruchid)

black vine weevil *U.S.* the curculionid *Brachyrhinus sulcatus*

rice water weevil *U.S.* the curculionid *Lissorhoptrus oryzophilus*

short-winged weevil broad oval curculionids with short and apically rounded elytra. Many are of the genus Pterocolus

Yucca weevil *U.S.* the curculionid *Scyphophorus yuccae.*

weka *N.Z.* the rallid bird *Gallirallus australis*

weld *Brit.* the resedaceous herb *Reseda luteola*

whitwell = woodpecker

[wellhaaring] a mutation occurring in linkage group V of the mouse. The phenotypic expression is a wavy coat and curly whiskers

wels a giant silurid catfish (*Silurus glanis*) said to reach a weight of six hundred and fifty pounds

[welt] a mutant (gene symbol *wt*) mapped at 82.0– on chromosome II of *Drosophila melangaster*. The phenotypic expression is small seamed eyes

wenchman the lutjanid fish *Pristipomoides andersoni*

[wet] a mutant (gene symbol *wc*) in linkage group VIII of *Habrobracon juglandis*. The phenotypic expression is very long, and irregular, wing microchaetae

wet-my-feet = wet-my-lips

wet-my-foot = quail

wet-my-lips the phasianid bird *Goturnix communis* (= quail).

wf *see* [white flower]

wh *see* [white]

whale (*see also* bird, fish, rorqual, shark) any large marine mammal of the order Cetacea

Australian whale the balaenid *Neobalaena marginata*

baleen whale any of several mystacocetids of the genus Balaena

goose-beak whale *N.Z.* the ziphiid *Ziphius cavirostris*

beaked whale any of several physeterids with a prolonged muzzle

Beluga whale the odontocete delphinid *Delphinapterus kingii*

Biscayan whale the balaenid *Balaena biscayensis*

blue whale the balaenopterid *Megaptera novaeangliae*

ca'ing whale the delphinid *Globicephalus melas*

California grey whale the balaenopterid *Rhachianectes glaucus*

finner whale the mystacocetous *Balaenoptera musculus*

gray whale the balaenid *Eschrichiius glaucus*

Greenland whale the balaenid *Balaena mysticetus*

humpback whale any of several balaenopterids of the genus Megaptera

killer whale any of several species of the delphinid genus Orca

false killer whale the delphinid *Psuedorca crassidens*

nare whale = narwhal

narwhale (or, more commonly, narwhal) the delphinid whale *Monodon monoceros*

pigmy whale *N.Z.* the physeterid *Cogia breviceps*

pilot whale either of two delphinids of the genus Globicephala (*cf.* blackfish)

porpoise whale *N.Z.* the ziphiid *Berardius arnuxi*

right whale any whale of the family Balaenidae particularly those of the genus Balaena

southern right whale *B. australis*

Scamperdown whale *N.Z.* any of several species of ziphiid of the genus Mesoplodon

sperm whale any of several physeterid whales having sperm oil in the cavity of the skull. Most belong to the genus Physeter

pygmy sperm whale any of several physeterid whales of the genus Cogia

whalebone whale (= baleen whale) any whale of the suborder Mystacoceti

white whale the delphinid whale *Delphinapterus leucas*. This name is also sometimes extended to include the beluga whale

wharangi *N.Z.* the rutaceous bushy tree *Melicope ternata*

Wharton's *epon. adj.* from Thomas Wharton 1657–1673

whau *N.Z.* the tiliaceous tree *Entelea arborescens*

whaup the charadriid bird *Numenius arquata* (*cf.* curlew)

whauwhaupaku *N.Z.* the araliaceous tree *Nothopanax arboreum*

whd *see* [withered]

wheat (*see also* ear, jointworm, maggot, midge, mite, sawfly, strawworm, worm) any of several species and very numerous agricultural hybrids of the grass Triticum

buckwheat (*see also* tree) the polygonaceous herb *Fagopyrum esculentum* and its seeds

climbing buckwheat any of several species of polygonaceous herbs of the genus *Polygonum* particularly those of the subgenus *Tiniaria*

cow wheat *U.S.* any of several species of scrophulariaceous herb of the genus Melampyrum

duckwheat the polygonaceous herb *Fagopyrum tataricum* and its seeds

India wheat = duckwheat

wheel a type of circular ossicle with spokes found in some holothurians. The term is also applied to the corona of rotifers

whekau *N.Z.* the strigid bird *Sceloglaux albifacies* (= laughing owl)

whelk (*see also* cracker) any of numerous buccinid gastropod mollusks *Brit.*, without qualification, *Buccinum undatum*. (= *U.S.* waved whelk). *U.S.*, without qualification, usually the neptuneid genus Busycon (*cf.* conch). Some other neptuneid gastropods are locally called whelks

dog whelk popular name of nassariid gastropod Mollusca

whelp to bear young, usually applied to canines. The term is also sometimes applied to the unweaned off-spring

yarwhelp *Brit.*, the scolapaeid bird *Limosa belgiea* (*cf.* godwit)

saw whet *U.S.* the strigid bird *Aegolius acadicus*

whew *Brit. dial.* = widgeon

whiff *U.S.* either of two species of bothid fishes of the genus Citharichthys

whimbrel the scolpacid bird *Numenius phaeopus* (= *U.S.* Hudsonian curlew)

whin (*see also* chat) the leguminous shrub *Ulex europaeus*

petty whin *Brit.* any of several leguminous shrubs and

whio *N.Z.* the mergine anatid bird *Hymenolaimus malacorhynchos*. (= blue duck)

whip (*see also* bird) any long flexible organism

coachwhip 1 (*see also* coachwhip 2) *U.S.* the colubrine snake *Masticophis flagellum*

coachwhip 2 (*see also* coachwhip 1) the fouquieriaceous shrub *Fouquieria splendens*

seawhip a term commonly applied to unbranched gorgonacean alcyonarian coelenterates

whip-Tom-Kelly *W.I.* the vireonid bird *Vireo altiloquus* (= black-whiskered vireo)

whirlabout *U.S.* the hesperiid lepidopteran insect *Polites vibex*

[whirler] a mutant (gene symbol *wi*) in linkage group VIII of the mouse. The phenotypic expression is a waltzing syndrome

whiroia *N.Z.* the procellariid bird *Prion desolata*

whistler 1 (*see also* whistler 2) properly any of several birds, of the muscicapid subfamily Pachycephalinae; also used for other birds such as *Elaenia martinica* or the cairinine bird *Dendrocygna arborea*. (= West Indian tree duck)

buff-breasted whistler *Pachycephala rufogularis*

Gilbert whistler *Pachycephala inornata*

golden whistler *Pachycephala pectoralis*

mountain whistler the turdid *Myadestes genibarbis* (= rufous-throated solitaire)

olive whistler *Pachycephala olivacea*

rufous whistler *Pachycephala rufiventris*

whistler 2 (*see also* whistler 1) the sciurid rodent *Marmota caligata* (= hoary marmot)

white 1 (*see also* white 2, face, tip) any of many lepidopteran insects, particularly in *Brit.* the pierids, the term butterfly rarely being appended

broad-barred white *Brit.* the noctuid *Hecatera serena*

bath white *Pieris daplidice*

bordered white *Brit.* the geometrid *Bupalus piniaria*

cabbage white *Brit.* and *U.S.* the pierid *Pieris brassica I.P. P. canidia*

caper white *Austral.* the pierid *Anaphaeis java*

common white *S.Afr.* the pierid *Belenois creona*

forest white *S.Afr.* the pierid *Belenois zochalia*

imperial white *Austral.* the pierid *Delias harpalyce* (cf. jezebel)

large white (= cabbage white)

little white *I.P.* any of several species of pierids of the genus Euchloe (*cf.* orange tip)

marble white *Brit.* the satyrid insect *Melanargia galatea*

meadow white *S.Afr.* the pierid *Pontia helice*

small white *Brit. Pieris rapae*. *S.Afr. Dixeia doxo*

veined white *S.Afr.* any of several species of pierid lepidopteran insect of the genus Belenois

green-veined white *Brit. Pieris napi*

wood white *Brit.* the pierid *Leucophasia sinapis Austral. Delias aganippe*. *S.Afr. Leptosia alcesta*

zebra white *S.Afr.* the pierid *Pinacopteryx eriphia*

white 2 (*see also* white 1) in compound names of other organisms

bob white any of several phasianid birds of the genus Colinus. Usually, without qualification, *C. virginianus*

tassel white the saxifragaceous shrub *Itea virginica*

[white] either a mutant (gene symbol *wh*) in linkage group III of *Habrobracon juglandis*. The name is descriptive of the phenotypic eye or a pseudo-allelic locus mapped at 1.5 on the X chromosome of *Drosophila melanogaster*

whitebait *see* -bait

[White cap of endosperm] a mutant (gene symbol *Wc*) mapped at 106 on linkage group IX of *Zea mays*. The term is descriptive of the phenotypic endosperm

[white collar] a mutant (gene symbol *wc*) on linkage group VII of *Neurospora crassa*. The phenotypic expression is an absence of carotenoids

[white egg] a mutant (gene symbol *wl*) in linkage group X of *Bombyx mori*. There is no pigment in the phenotypic egg serosa

[white egg 2] a mutant (gene symbol *w2*) mapped at

3.4 on linkage group X of *Bombyx mori*. The egg changes from white to light red

[**white egg 3**] a mutant (gene symbol *w3*) mapped at 6.9 in linkage group X of *Bombyx mori*. The phenotypic egg is light purplish-brown

[**white flower**] a mutant (gene symbol *wf*) mapped at 15 on chromosome 3 of the tomato. The name is expressive of the phenotypic expression

[**white ocelli**] a mutant (gene symbol *wo*) mapped at 76.2 on chromosome III of *Drosophila melanogaster*. The name is descriptive of the phenotypic expression

[**white sheath**] a mutant (gene symbol *ws₃*) mapped at 0 on linkage group II of *Zea mays*. The phenotype lacks chlorophyll in the leaf sheaths and in neighboring parts of the stalk

[**white side egg**] a mutant (gene symbol *se*) mapped at 0 on linkage group XIV of *Bombyx mori*. The phenotypic expression is a furrowed and irregular egg

[**white virescent**] a mutant (gene symbol *wv*) mapped at 42 on chromosome 2 of the tomato. The phenotypic plant is very pale green

whiting the gadid fish *Gadus merlangus*

whydah properly any of several species of long tailed ploceid birds of the genera Vidna and Steganura, This name is also sometimes used for ploceids of the genus Coliuspasser, better known as window birds

  steel-blue whydah *V. hypocherina*
  Fischer's whydah *V. fischeri*
  paradise whydah either of two species of the genus Steganura; without qualification, *S. paradisaea*
  pin-tailed whydah *V. macroura*

wi *see* [whirler]

Wi *see* [Washed eye]

cotton**wick** the pomadasyid fish *Haemulon melanurum*

wicky *U.S.* the ericaceous shrub *Kalmia angustifolia*

wicopy the thymelaeaceous shrub *Dirca palustris*

widdy *U.S.* the rosaceous shrub *Potentilla fruiticosa*

wide-awake = tern, particularly the sooty tern

[**wide-banded agouti**] a mutant (gene symbol *w*) at locus 29.9 on linkage group IV of the rabbit. The name is descriptive of the phenotypic expression

[**wider-wing**] a mutant (gene type *ww*) at 32.9+ on the X chromosome of *Drosophila melanogaster*. As the name indicates the wings of the phenotype are short and broad

widgeon any of several species of anatine duck of the genus Mareca. *Brit.*, qithout qualification, *M. penelope. Austral. Malacorhynchus membranaceus* = pink-eared duck

  American widgeon *M. americana*
  European widgeon *M. penelope*

widow 1 (*see also* widow 2, 3, bird) a female whose mate has died

  merry widow the poeciliid teleost fish *Phallichthys amates*

widow 2 (*see also* widow 1, 3) any plant or animal likely to contribute to the production of widows

  black widow (*see also* tetra) *U.S.* the theridiid spider *Lathrodectus mactans*

widow 3 (*see also* widow 1, 2) any organism exhibiting any of the colors (black, white, yellow) associated with mourning in various parts of the world

  widow *S.Afr.* any of numerous species of satyrid lepidopteran insect of the genus Dira

wife any of several organisms, some so called by reason of their beauty

alewife the clupeid fish *Alose pseudoharengus*

  little wife *U.S.* the noctuid lepidopteran insect *Cotocala muliercola* (*cf.* underwing)
  old wife *Brit.* the sparid fish *Cantharus lineatus*
  pudding wife the labrid fish *Hilichoeres radiatus*

earwig (*see also* scorpion fly) popular name of dermapterous insects. *Brit.* without qualification *Forficula auricularia*

  European earwig the forficulid *Forficula auricularia*
  ring-legged earwig *U.S.* the labidurid *Euborellia annulipes*
  little earwig *U.S.* the labidurid *Labia minor*
  seaside earwig *U.S.* the labidurid *Anisolabis maritima*
  striped earwig *U.S.* the labidurid *Labidura bidens*
  spine-tailed earwig *U.S.* the forficulid *Doru aculeatum*
  toothed earwig *U.S.* the labiid *Spongovostox apicendentatus*

wigeon = widgeon

wiggler a mosquito larva

wight *I.P.* the hesperiid lepidopteran insect *Iton semamora*

wild (*see also* type) the antithesis of cultivated (*cf.* feral)

willet *U.S.* the scolopacid bird *Catoptrophorus semipalmatus*

royal **william** *Brit. obs.* the papilionid lepidopteran insect *Papilio machaon* (= swallow tail)

  sweet william (*see also* catch fly) the caryophyllaceous herb *Dianthus barbatus*
    wild sweet william *U.S.* the polemoniaceous herb *Philox divaricata*

stinking **willie** the compositaceous herb *Senecio jacobaea*

Willis's *epon. adj.* from Thomas Willis (1621–1675)

willow 1 (*see also* willow 2, beetle, biter, herb, midge, sawfly) any of numerous shrubs and trees of the family Salicaceae and particularly of the genus Salix; a few other willow-like plants are also called willow

  almond willow *S. amygdalina*
  balsa willow *U.S. S. pyrifolia*
  long beaked willow *U.S. S. bebbiana*
  black willow *S. nigra*
  brittle willow *S. fragilis*
  sandbur willow *U.S. S. interior*
  Colorado willow *S. irrorata*
  crack willow *U.S. S. fragilis*
  dwarf willow *S. tristis*
  Egyptian willow *S. sassas*
  goat willow *S. caprea*
  herb-like willow *S. herbacea*
  hoary willow *S. candida*
  felt-leaf willow *S. alaxensis*
  laurel-leaf willow *S. pentandra*
  peachleaf willow *S. amygdaloides*
  heart-leaved willow *S. cordata*
  Napoleon's willow *S. babylonica*
  osier willow *S. viminalis*
  prairie willow *S. humilis*
  primrose willow any of several onagraceous herbs of the genus Jussiaea
    floating primrose willow *J. diffusa*
  pussy willow *S. discolor*
  silky willow *S. sericea*
  Sitka willow *S. sitchensis*
  Virginia willow any of several speices of saxifragaceous shrubs of the genus Itea
  ward's willow *S. caroliniana*
  water willow *U.S.* any of several species of acanth-

aceous herbs of the genus Justicia, the lythraceous herb *Decodon verticillatus*

weeping willow any of several species having pendant branches, particularly *S. babylonica* (= Napoleon's willow)

   Thurlow's weeping willow *S. elegantissima*

white willow *S. alba*

yellow willow *S. vitelcina*

willow 2 (*see also* willow 1) any of several species of noctuid lepidopteran insects of the genus Caradrina

willowstrife *Brit.* the lythyraceous herb *Lythrum salicari* (= purple loosestrife)

[wilty dwarf] a mutant (gene symbol *wd*) mapped at 0 on chromosome 9 of the tomato. The phenotypic plant is small and droopy

[wilty foliage] a mutant (gene symbol *wt*) mapped at 0 on chromosome 7 of the tomato. The name is descriptive of the phenotypic expression

wing 1 (*see also* wing 2–5, louse) an organ modified to support and propel an organism in flight

   amplexiform wing said of those wings of lepidopteran insects that are locked (*i.e.* articulated) by the large humeral lobe of the hindwing projecting under the forewing rather than by a frenulum

   bastard wing a modified digit, representing a thumb, on the wing of a bird

   spurious wing = alula

wing 2 (*see also* wing 1, 3–5, stem) wing-like structures not connected with flight

   longitudinal wing one of from one to four lateral folds from the body of some nematode worms

wing 3 (*see also* wing 1, 2, 4, 5) in compound names of birds

   barwing any of several timaliid birds of the genus Actinodura

   brass-wing *W.I.* the anatine bird *Anas bahamensis* (= Bahama pintail)

   bronzewing any of several species of Australian columbid birds of the genus Phaps and its near allies

      brush bronzewing *P. elegans*

      forest bronzewing *P. chalocoptera*

   crimsonwing any of several ploceids of the genus Cryptospiza

      red-faced crimsonwing *C. reichenovii*

   greywing the phasianid bird *Francolinus africanua* (*cf.* francolin)

   lapwing any of several charadriids; without qualification, *Brit. Vanellus vanellus*

      crowned lapwing *Stephanibyx coronatus*

      grey-headed lapwing *Microsarcips cinereus*

      white-tailed lapwing *Chettusia stenura*

      long-toed lapwing *Hemiparra crassirostris*

      red-wattled lapwing *Lobivanellus indicus*

      yellow-wattled lapwing *Lobipluvia malabarica*

   merry-wing = goldeneye

   rafflewing the mergine anatid *Bucephala clangula* (= common goldeneye)

   redwing *U.S.* the icterid *Agelaius phoeniceus* (= red-winged blackbird) *Brit.* the turdid *Turdus iliacus S.Afr.* the phasianid *Francolinus levaillantii* (*cf.* francolin)

   rough-wing any of several species of hirundinids of the genus Psalidoprocne (= saw-wing swallow)

   sabre-wing any of several trochilids of the genus Campylopterus

      rufous sabre-wing *C. rufus*

      wedge-tailed sabre-wing *C. curvipennis*

      violet sabre-wing *C. hemileucurus*

shortwing any of several turdids of the genus Brachypteyx

shuffle-wing the prunellid *Prunella modularis* (= hedge-sparrow)

great sapphire-wing the trochilid *Pterophanes cyanopterus*

standard-wing the paradisaeid *Semioptera wallacei*

swallow-wing the bucconid *Chelidoptera tenebrosa*

waxwing popular name of bombycillids of the genus Bombycilla

   Bohemian waxwing *B. garnulus*

   cedar waxwing *B. cedrorum*

wing 4 (*see also* wing 1–3, 5) in compound names of insects

   anglewing *U.S.* any of numerous species of nymphalid lepidopterans of the genus Polygonia. Many have deeply notched wings and a C-shaped silvery spot on the underside

   batwing *U.S.* the geometrid lepidopteran *Dyspteris abortivaria*. *I.P.* either of two species of papilionid lepidopteran of the genus Tros. (*cf.* clubtail, rose windmill)

   birdwing any of several species of papilionid lepidopteran of the genus Troides. The common birdwing *T. helena* is one of the largest known butterflies sometimes reaching eight inches in wingspread. *Austral. P. priamus.*

      tailed birdwing *Austral. Papilio paradisea*

   bird's wing *Brit.* the noctuid *Dipterygia scabriuscula*

   clear wing a term commonly applied to any butterfly or moth lacking scales over large areas of the wing. In *U.S.* particularly sphingids of the genus Haemorrhagia (*cf.* sphinx moth)

   cloudy wing popular name of hespiriid lepidopterans of the genus Thorybes

   closewing popular name of crambid lepidopterans

   daggerwing *U.S.* any of several species of nymphalid lepidopteran of the genus Marpesia

   dusky wing popular name of hesperiids of the genus Erynnis

   glasswing *S.Afr.* the lycaenid lepidopteran *Pentila peucetia*. *Austral.* the nymphalid lepidopteran *Acreae andromacha*. *U.S.* any of several species of arctiid moth of the genus Hemihyalea and the hesperiid *Polites verna*

lacewing 1 (*see also* lacewing 2) popular name of neuropterans of the families Chrysopidae, Hemerobiidae, and Polystoechotidae

   brown lacewing popular name of hemerobiids

   common lacewing the name common lacewing, or lacewing without qualification, is usually applied to the Chrysopidae

   golden-eye lacewing *U.S. Chrysopa oculata*

lacewing 2 (*see also* lacewing 1) *I.P* and *Austral.* any of numerous nymphalid lepidopterans of the genus Cethosia

   leafwing any of numerous species of Asiatic nymphalid lepidopterans of the genus Doleschallia which, when at rest with folded wings, strongly resemble leaves. *U.S.* any of several species of nymphalid lepidopteran of the genus Anaea

   marble wing *U.S.* the geometrid lepidopteran *Marmopteryx marmorata*

   mulberry wing *U.S.* the hesperiid lepidopteran *Poanes massasoit*

   notch-wing *U.S.* the geometrid lepidopteran *Ennomos magnarius*

   purple wing *U.S.* popular name of nymphalid lepidopterans of the genus Eunica

sooty wing popular name of hesperiid lepidopterans of the genus Pholisora

underwing popular name of lepidopterans having notable underwings

clear underwing *Brit*. popular name of aegerids

crimson underwing *Brit*. either of two species of noctuids of the genus Catocala

orange underwing *Brit*. the geometrid *Brephos parthenias*

pearly underwing *Brit*. the noctuid *Agrotis saucia*

red underwing *Brit*. the noctuid *Catocala nupta*

white underwing *Brit*. the noctuid *Anarta melanopa*

yellow underwing *Brit*. any of several species of noctuids of the genera Anarta and Triphaena

wing 5 (*see also* wing 1–4, shell) in compound names of other organisms

angel's wing the pholadid mollusk *Barnea costata*

false angel's wing the petricolid mollusk *Petricola pholadiformis*

corkwing *Brit*. the labrid fish *Crenilabrus melops*

gaywings the polygalaceous herb *Polygala paucifolia*

hawkwing the strombid mollusk *Strombus raninus*

thin wing popular name of pholadomyiid pelecypods

[wingless] a mutant (gene symbol *fl*) mapped at 0+ on linkage group X of *Bombyx mori*. The term is descriptive of the phenotype

winkle *Brit*. any of several species of littorinid gastropod of the genus Littorina but particularly, without qualification, *L. littorea* (*cf*. periwinkle)

periwinkle 1 (*see also* perwinkle 2) the littorinid gastropod *Littorina littorea*

periwinkle 2 (*see also* periwinkle 1) any of several species of apocynaceous herb of the genus Vinca but particularly *V. minor*

Madagascar periwinkle *V. rosea*

winter that one of the four seasons in which the noon day sun is lowest in the sky. Anything associated with winter or coldness

Winteraceae a small family of ranale dicotyledonous angiosperms closely allied to the Illiciaceae

Winteranaceae = Canellaceae

Wirsung's *epon adj*. from Tohaun Georg Wirsung (1629–1643)

[wiry foliage] a mutant (gene symbol w_1) mapped at 28 on chromosome 4 of the tomato. The dwarfed phenotype has slender leaflets

wisdom-of-surgeons *Brit*. the cruciferous herb *Sisymbrium sophia*

wisent = European bison

wisp term for a group of snipe

witch any of many organisms of peculiar habit, dark color or low repute. *Brit*. the pleuronectid fish *Glyptocephalus cynoglossus*. *I.P.* the lycaenid lepidopteran insect *Araotes lapithis*

black witch *U.S*. the noctuid moth *Erebus odora*. *W.I*. the cuculid bird *Crotophaga ani* (= smoothbilled ani)

mountain witch *W.I*. the columbid *Geotrygon versicolor* (= crested quail dove)

shadow witch *U.S*. the orchid *Ponthieza racemose*

steppe-witch = tumble weed

[withered] a mutant (gene symbol *whd*) mapped at 61.0+ on chromosome II of *Drosophila melanogaster*. The phenotypic wings are warped and shrunken

witherod the caprifoliaceous shrub *Viburnum cassinoides*

withers the area of a horse separating the neck from the back

[without anthocyanin] a mutant (gene symbol *aw*) mapped at 30 on chromosome 2 of the tomato. The term is descriptive of the phenotype

withy a willow twig

wizard *I.P*. the nymphalid lepidopteran insect *Rhinopalpa polynice*

wk *see* [weak]

wl *see* [wabbler-lethal]

w1 *see* [white egg 1]

wo *see* [white ocelli] or [wobbly]

Wo *see* [Woolly plant]

woad (*see also* waxen) the cruciferous herb *Isatis tinctoria* and a blue dyestuff prepared from it

wobble *U.S*. *obs*. any of numerous alcid birds

[wobbly] a mutant (gene symbol *wo*) mapped at 75 on linkage group III of the rat. The name refers to the gait of the phenotype

wold = wood

forwold shrubs bordering a wood

wolf 1 (*see also* wolf 2, baneberry, eel, herring, milk) any of several large canid carnivores. The distinction between a wolf and a fox is largely one of size

Abyssinian wolf a large jackal *Thos simensis*

European wolf *Canis lupus*

gray wolf *U.S*. = timber wolf *U.S*.

Japanese wolf *Canis hodophylax*

manned wolf the *S.Amer*. canine *Chrysocyon jubatus*

pouched wolf the thylacinine dasyurid *Thylacinus cynocephalus*; the extraordinary resemblance to a wolf is frequently used to illustrate parallel evolution

prairie wolf the canid carnivore *Canis latrans* (= coyote)

red wolf *Canis niger* or *C. jubatus*

rush wolf *Canis latrans* (= coyote)

strand wolf = brown hyaena

Tasmanian wolf *Thylacinus cynocephalus*. (*cf*. Tasmanian devil)

timber wolf *U.S*. *Canis lupus*

wolf 2 non-mammalian predators

aphis wolf = aphis lion (*q.v*.) frequently with an interchange between *U.S*. and *Brit*. usage

Wolffian *epon. adj*. from Kaspar Friedrich Wolff (1733–1794)

wolverine the large nearctic musteline carnivore *Gulo luscus* (*cf*. glutton)

woma *Aust*. any of several species of pythonid snake of the genus Aspidites

old woman the compositaceous herb *Artemisia stelleriana*

wombat popular name of wombatid marsupials, often of the genus Phascolomys; they somewhat resemble very large rodents

Wombatidae a family of marsupial mammals containing the wombat (*q.v*.)

wonga-wonga (*see also* vine) the large Australian columbid bird *Leucosarcia melanoleuca*

wonkapin *U.S*. the nymphaeaceous aquatic *Nelumbo lutea*

wood 1 (*see also* wood 2, 3, apple, chat, cock, creeper, gnat, haunter, hewer, louse, nymph, pigeon, pecker, sage, shrike, snail, stainer, star, swallow, tree, wasp, wren) the hard part of the stem of a tree, consisting principally of xylem. Wood in the standing tree is called timber and, when sawn or split for use, lumber

heartwood the innermost wood, that next to the pith

partridge-wood oak-wood blotched like a partridge

breast through the action of the parasitic fungus *Stereum frustulosum*

diffuse porous wood dicotyledonous wood in which the vessels are of equal diameter and universally distributed through a growth ring

    ring-porous wood wood with the distinct rings of large and small vessels

    primary wood that developed from procambium

sapwood new wood of an exogenous tree

wood 2 (*see also* wood 1, 3) in compound names of plants

almond wood the meliaceous tree *Chukrasia tabularis*

arrow wood any of several species of the caprifoliaceous genus Viburnum, particularly *V. dentatum*

barwood = camwood

basswood (*see also* miner, roller) any of several species of tiliaceous tree of the genus Tilia. (*See also* linden, lime, and whitewood)

beef wood any of several species of tree of the casuarinaceous genus Casuarina

bitter wood the simaroubaceous tree *Pierasma excelsum*

    smaller bitter wood the annonaceous tree *Xylopia muricata*

bloodwood the myrtaceous tree *Eucalyptus corymbosa*

buttonwood the platanaceous tree *Platanus occidentalis* (= American plane tree)

boxwood *see* box 1

camwood the leguminous shrub *Bathia racemosa*

candle wood any of several species of fouquieriaceous shrubs of the genus Fouquieria

chittam wood the anacardiaceous shrub *Cotinus ovatus*

corkwood *either* the leitneriaceous shrub *Leitneria floridana or* the annonaceous tree *Annona glabra*

cottonwood any of several species of the salicaceous genus Populus, usually without qualification, *P. deltoides*. (*see also* aspen, and poplar)

    smooth-barked cottonwood *P. acuminata*

    black cottonwood *P. trichocarpa*

    great plains cottonwood *P. sargentii*

    narrow-leaved cottonwood *P. fortissima*

    southern cottonwood *P. deltoides*

    swamp cottonwood *P. heterophylla*

    valley cottonwood *P. wislizenii*

    western cottonwood *P. fremontii*

devil wood the oleaceous tree *Osmanthus americanus*

dogwood (*see also* borer, bug, midge, scale) any of numerous shrubs or trees of the cornaceous genus Cornus and extended to several plants resembling the true dogwood in habit. *Brit.*, without qualification, *C. sanguinea*

    alder dogwood *Rhamnus frangula*

    bastard dogwood *Austral.* the rhamnacean tree *Pomaderris apetala*

    flowering dogwood *C. florida*

    Jamaica dogwood the leguminous tree *Piscidia erythrina*

    alder-leaved dogwood *C. rugosa*

    poison dogwood the anacardiaceous shrub *Rhus vernix* (= poison sumac)

    red-osier dogwood *C. stolonifera*

dovewood the euphorbiaceous tree *Alchornea icifolia*

gopherwood any of numerous trees yielding a yellow lumber, including several *U.S.* leguminous trees of the genus Cladrastis

ironwood any tree yielding a hard timber. The term is almost without taxonomic value. Some of the

most common usages are the corylaceous trees *Ostrya virginiana* (= hop-hornbeam) and *Carpinus caroliniana*, the rhamnaceous tree *Reynosia latifolia*, any of several species of the sapotaceous genera Bumelia and Sideroxylon, and the leguminous tree *Millettia caffra*

    bastard ironwood *U.S.* the rutaceous tree *Zanthoxylem fagara*

    black ironwood the rhamnaceous tree *Krugiodendron ferreum*

joewood any of several species of two genera of the family Theophrastaceae

lancewood *N.Z.* the araliaceous tree *Pseudopanax crassifolium* (= horoeka)

leatherwood any of several thymelaeaceous shrubs of the genus Dirca. Usually, without qualification, *D. palustris*. A'so, the cyrillaceous shrubby tree *Cyrilla racemiflora*

lever wood *U.S.* the corylaceous tree *Ostrya virginiana* (= hop-hornbeam)

moose wood the aceraceous tree *Acer pennsylvanicum* (= striped maple), the thymelaeaceous shrub *Dirca palustris* and the caprifoliaceous shrub *Viburnum alnifolium*

opossum wood the styracaceous shrub *Halesia carolina*

pigeonwood *N.Z.* the monimiaceous tree *Hedycarya arborea* (= porokaiwhiri)

redwood any of numerous trees of which the lumber is red, purple, or brown. Amongst others *U.S.* the pinaceous tree *Sequoia sempervirens*. Also, the leguminous *Haematoxylon campechianum* and closely allied trees of the genus Caesalpinia which yield the hematoxylin of biological laboratories. The term is also used for any of several sandalwoods

    California redwood *S. sempervirens*

ribbonwood *N.Z.* either of two species of malvaceous trees of the genus Hoheria (*cf.* lacebark)

    lowland ribbonwood *N.Z.* the malvaceous tree *Plagianthus betulinus* (= manatu)

    salt-marsh ribbon wood *N.Z.* the malvaceous shrub *Plaginathus divaricatus*

rose wood the leguminous tree *Dalbergia latifolia*

round wood *U.S.* the rosaceous tree *Pyrus americana*

sandalwood any of several trees of the santalaceous genus Santalum, particularly *S. album N.Z.* the santalaceous tree *Mida myrtifolia* (= maire)

    bastard sandalwood *N.Z.* the compositaceous tree *Olearia fraversii* (= akeake). *Austral.* the myoporaceous tree *Myoporum sandwicense Brit.* the ulmaceous tree *Zelkova abelicea*

    false sandalwood the olacaceous tree *Ximenia americana* and its edible fruit

savan wood the compositaceous *Artemisia abrotanum*

shittim wood the styracaceous shrub *Halesia carolina*

sneezewood the miliaceous tree *Ptaeroxylon obliquum*

sourwood the ericaceous tree *Oxydendrum arboreum*

southern wood the compositaceous herb *Artemisia abrotanum*

spearwood the myrtaceous tree *Eucalyptus doratoxylon*

spiderwood *N.Z.* the epacridaceous tree *Dracophyllum latifolium* (= neinei)

spoonwood *U.S.* the ericaceous shrub *Kalmia latifolia* (= mountain laruel)

sweetwood the euphorbiaceous tree *Croton eluteria*

torchwood any of numerous species of trees of the rutaceous genus Amyris

twist wood the caprifoliaceous shrub *Viburnum lantana*

vermilion wood any of several leguminous trees of the genus Pterocarpus

whistlewood the aceraceous tree *Acer pennsylvanicum* (= striped maple)

whitewood any of several species of tiliaceous tree of the genus Tilia. (*see also* linden, lime, and basswood) Also the magnoliaceous tree *Liriodendron tulipifera* (= tulip tree) and the euphorbiaceous shrub *Drypetes lateriflora*

whiteywood *N.Z.* the violaceous tree *Melicytus ramiflorus* (= mahoe)

wormwood a general term for all members of the compositaceous genus Artemisia but specifically *A. absinthium*

    bastard wormwood *U.S.* any compositaceous herb of the genus Ambrosia (= ragwood)

    beech wormwood *Artemisia stelleriana*

    Roman wormwood the compositaceous herb *Ambrosia artemisiifolia*

yellowwood any of numerous trees yielding a yellow lumber, including *U.S.* several leguminous trees of the genus Cladrastis, the rutaceous tree *Acronychia laevis*. *Afr.* the taxaceous tree *Podocarpus elongatus*. *N.Z.* the rubiaceous tree *Coprosma linariifolia*. The term is also applied to the symplocaceous herb *Symplocos tinctoria*

    bastard yellowwood the taxaceous tree *Podocarpus elongatus* (= S.Afr. yellowwood)

    South African yellowwood = bastard yellowwood

zebra wood any of several leguminous trees of the genus Cynometra

zither wood any of several species of tree of the verbenaceous genus Citharexylum

wood 3 (*see also* wood 1, 2) in compound names of other organisms

    speckled wood *Brit.* the nymphalid lepidopteran *Pararge egeria*

woodling *U.S.* any of several species of noctuid lepidopteran insect of the genus Xylomiges

woodpecker *see under*-pecker

[Woolly plant] a mutant (gene symbol *Wo*) mapped at 48 on chromosome 2 of the tomato. The leaves and stem are of the phenoytpe are thickly hairy

wax work the celastraceous climbing shrub *Celestrus scandans*

worker (*see also* ergate) a female ant, which, being unable to reproduce, busies herself with other things

worm (*see also* worm 1–6, wood) in the original sense of "Vermes," any soft bodied invertebrate not obviously a "Mollusk." Now practically confined to compound words. Worm 1 covers miscellaneous invertebrates, worm 2 covers lepidopteran larvae, worm 3 covers coleopteran larva, worm 4 covers other insect larvae, worm 5 includes worm-like vertebrates and worm 6 lists the worm-like plants

worm 1 (*see also* worm, worm 2–5) almost any invertebrate lacking obvious appendages and one, the churr worm, with appendages

    acorn worm popular name of the Enteropneusta

    angleworm *U.S.* any large oligochaete annelid

    arrow worm popular name for members of the phylum Chaetognatha

    bamboo worm any of several polychaete annelid worms of the genus Axiothella

    bladder worm = cysticercus larva of cestode platyhelminths

    blood worm any of several species of red terebellid polychaetes particularly of the genera Polycirrus and Enoplobranchus

    bootlace worm popular name of members of the phylum Nemertea (*cf.* proboscis worm)

    bristle worm popular name for polychaete annelids particularly of the family Nereidae

    churr worm = mole cricket

    clam worm a general term for nereid Polychaetae particularly those of the genus Nereis

    crested worm popular name for polychaete annelids of the family Terebellidae

    earthworm any of numerous soil-dwelling oligochaete annelids, particularly of the families Megascolecidae and Lumbricidae

    eelworm the nematode *Ascaris lumbricoides*

    fanworm general term for polychaetes which have anterior feathery gills

    flatworm popular name for members of the phylum Platyhelminthes, particularly the Turbellaria

    fringe worm any of several polychaete annelids of the genus Cirrapulus

    gapeworm the syngamid nematode, *Synagamus trachealis* parasitizing the bronchi of birds and causing the disease known as "gapes"

    glass worm popular name of members of phylum Chaetognatha

    guinea worm the filariid nematode *Dracunculus medinensis*. It may reach a length of several feet and burrows through the tissues of infected humans

    hair worm popular name of elongate, attenuated nematode worms of the family Mermithidae

    horsehair worm popular name of Nematomorpha

    spiny-headed worm popular name of members of the Acanthocephala

    hook worm popular name of nematode worms of the family Ancylostomidae

    American hook worm *Necator americanus*

    dog hookworm *Ancylostoma caninum*

    Old world hookworm *Ancylostoma duodenale*

    swine kidney worm the nematode *Stephanurus dentatus*

    loa-loa worm the filariid nematode *Filaria loa*, which frequently parasitizes the human eye

    lug worm popular name of any of several polychaete annelids of the genus Arenicola, and *Brit. obs.* a large earthworm

    lurgworm any polychaete annelid of the genus Nephthys

    maw worm the ascarid nematode *Ascaris megalocephala*

    Medina worm = guinea worm

    pig nodular worm the nematode *Oesophagostemum dentatum*

    palisade worm popular name of nematode worms of the family Strongylidae

    palolo worm properly the eunicid worm *Eunice viridis* of the Pacific but sometimes applied to *E. furcata* of the Atlantic. The free swimming epitoke of the Pacific form is widely used for food (*see under* -toke)

    peanut worm (*see also* worm 2) popular name of sipunculid worms

    pin worm (*see also* worm 2) popular name of many parasitic nematodes but particularly those of the family Oxyuriformidae

    proboscis worm popular name of polychaete annelids of the genus Glycera and often applied to Nemertea

    rag worm *Brit.* nereid polychaetes used as fish bait. *U.S.* any marine worm used for the same purpose.

ribbon worm popular name of members of the phylum Nemertea

scale worm popular name of polychaete worms, such as those of the genus Lepidonotus, in which the notopodial cirrus is modified into a scale-like form

shield worm popular name of polychaete worms of the genus Spermastis

ship worm popular name of teredid pelecypods. *W.U.S.* Bankia setacea

tapeworm popular name for cestode platyhelminthes

  beef tapeworm the taeniid *Taenia saginata*

  dog tapeworm *Diphylidium caninum*

  dwarf tapeworm *U.S.Hymenolepis nana*

  fringed tapeworm *Thysanosoma actinioides*

  gid tapeworm the taeniid *Multiceps multiceps*

  double-pored tapeworm *Dipylidium caninum* (= dog tapeworm)

  pork tapeworm the taeniid *Taenia solium*

thread worm popular name for members of the phylum Nematoda occassionally applied to Nematomorpha

tongue worm popular name of members of the Echiurida

whip worm popular name of trichurid nematodes

  dog whipworm *Trichuris vulpis*

  human whipworm *Trichuris trichiura*

  winged worm popular name of polychaete annelids of the genus Chaetopterus

worm 2 (*see also* worm 1, 3–5) the larvae of lepidopteran insects. The words "larva of" are omitted from entries in this section

agave worm = maguey worm

lesser appleworm the olethreutid *Grapholitha prunivora*

armyworm *U.S.* the noctuid *Pseudaletia unipuncta*

  beet armyworm *U.S.* the noctuid *Spodoptera exigua*

  bertha armyworm *U.S.* the noctuid *Mamestra configurata*

  fall armyworm *U.S.* the noctuid *Spodoptera frugiperda*

  nutgrass armyworm *U.S.* the noctuid *Spodoptera exempta*

  southern armyworm *U.S.* the noctuid *Prodenia eridania*

  wheat head armyworm *U.S.* the noctuid *Faronta diffusa*

  yellow-striped armyworm *U.S.* the noctuid *Prodenia ornithogalli*

  western yellow-striped armyworm, *U.S.* the noctuid *Prodenia praefica*

bagworm *U.S.* popular name of psychids, without qualification, usually *Thyridopteryx ephemeraeformis*

boll-worm *U.S.* the noctuid *Heliothis armiger* (= *Brit.* scarce bordered straw)

  flax bollworm *U.S.* the noctuid *Heliothis ononis*

  pink bollworm *U.S.* the gelechiid *Pectinophora gossypiella*

green budworm *U.S.* the olethreutid *Hedia variegana*

  black-headed budworm *U.S.* the tortricid *Acleris variana*

  jack-pine budworm *U.S.* the tortricid *Choristoneura pinus*

  spruce budworm *U.S.* the tortricid *Choristoneura fumiferana*

  tobacco budworm *U.S.* the noctuid *Heliothis virescens*

purple-backed cabbageworm *U.S.* the pyraustid *Evergestis pallidata*

imported cabbageworm *U.S.* the pierid *Pieris rapae*

southern cabbageworm *U.S.* the pierid *Pieris protodice*

cross-striped cabbageworm *U.S.* the pyraustid *Evergestis rimosalis*

cankerworm *U.S.* those geometrid moths (*cf.* measuring worm) which are pests of fruit trees

  fall cankerworm *U.S.* *Alsophila pometaria*

  spring cankerworm *U.S.* *Paleacrita vernata*

carpenter worm *U.S.* cossid *Prionoxystus robiniae*

  little carpenterworm *U.S.* *P. macmurtrei*

  pecan carpenterworm *U.S.* the cossid *Cossula magnifica*

celeryworm *U.S.* the papilionid *Papilio polyxenus*

green cloverworm *U.S* the noctuid *Plathypena scabra*

spruce coneworm *U.S.* the phycitid *Dioryctria reniculella*

cutworm the larve of numerous noctuid moths. These larvae derive their name from their habit of cutting young plants off at the ground level

  army cutworm *U.S.* *Chorizagrotis auxiliaris*

  black army cutworn *U.S.* *Actebia fennica*

  red-backed cutworm *U.S.* *Euxoa ochrogaster*

  western bean cutworm *U.S.* *Loxagrotis albicosta*

  black cutworm *U.S.* *Agrotis ipsilon*

  bristly cutworm *U.S.* *Lacinipolia renigera*

  bronzed cutworm *U.S.* *Nephelodes emmedonius*

  clover cutworm *U.S.* *Scotogramma trifolii*

  dingy cutworm *U.S.* *Feltia subgothica*

  glassy cutworm *U.S.* *Crymodes devastator*

  granulate cutworm *U.S.* *Feltia subterranea*

  yellow-headed cutworm *U.S.* *Apamea amputatrix*

  w-marked cutworm *U.S.* *Spaelotis clandestina*

  western w-marked cutworm *U.S.* *Spaelotis havilae*

  dark-sided cutworm *U.S.* *Euxoa messoria*

  pale-sided cutworm *U.S.* *Agrotis malefida*

  spotted cutworm *U.S.* *Amathes c-nigrum*

  striped cutworm *U.S.* *Euxoa tessellata*

  variegated cutworm *U.S.* *Peridroma saucia*

  pale western cutworm *U.S.* *Agrotis orthogonia*

  white cutworm *U.S.* *Euxoa scandens*

corn earworm *U.S. the noctuid Heliothis zea*

filbertworm *U.S.* the elethreutid *Melissopus latiferreanus*

black-headed fireworm *U.S.* the olethreutid *Rhopobota naevana*

  cranberry black-headed fireworm the olethreutid *Rhopobota naevana*

  yellow-headed fireworm *U.S.* the tortricid *Acleris minuta*

fruit worm (*see also* worm 3) any larval insect that damages fruit trees

  cherry fruitworm *U.S.* the olethreutid *Grapholitha packardi*

  cranberry fruitworm *U.S.* the phycitid *Acrobasis vaccinii*

  currant fruitworm the carposinid *Carposina fernaldana*

  gooseberry fruitworm *U.S.* the phycitid *Zophodia convolutella*

  green fruitworm *U.S.* the noctuid *Lithophane antennata*

  tomato fruitworm *U.S.* the noctuid *Heliothis zea*

clover hayworm *U.S.* the pyralid *Hypsopygia costalis*

hornworm *U.S.* those larvae of sphingid moths which

bear a recurved horn at the posterior end. The term is sometimes extended to the adult moth

sweetpotato hornworm *U.S. Agrius cingulatus*

tobacco hornworm *U.S. Manduca sexta*

tomato hornworm *Manduca quinquemaculata*

inchworm the larva of geometrids (= spanworm, looper)

cotton leafworm *U.S.* the noctuid *Alabama argillacea*

maguey worm the larva, widely eaten in *Mex.* and *S.W.U.S.*, of several megathymids

orange-humped mapleworm *U.S.* the notodontid *Symmerista leucitys*

green-striped mapleworm *U.S.* the citheroniid *Anisota rubicunda*

measuring worm *U.S.* larva of geometrid moths. The term is also extended to some adult moths (= geometer)

melonworm *U.S.* the pyraustid *Diaphania hyalinata*

california oakworm *U.S.* the dioptid *Phryganidia californica*

red-humped oakworm *U.S.* the notodontid *Symmerista albifrons*

spiny oakworm *U.S.* the citheroniid *Anisota stigma*

orange-striped oakworm *U.S.* the citheroniid *Anisota senatoria*

pink-striped oakworm *U.S.* the citheroniid *Anisota virginiensis*

navel orangeworm *U.S.* the phycitid *Paramyelois transitella*

palmerworm *U.S.* the gelechiid *Dichomeris ligulella*

parsleyworm *U.S.* the papilionid *Papilio polyxenes asterius*

red-necked peanutworm (*see also* worm 1) *U.S.* the gelechiid *Stegasta bosqueella*

pickleworm *U.S.* the pyraustid *Diaphania nitidalis*

tomato pinworm (*see also* worm 1) *U.S.* the gelechiid *Keiferia lycopersicella*

roller worm the larva of the hesperids *Urbanus proteus*

koa seedworm *U.S.* the olethreutid *Cryptophlebia illepida*

hickory shuckworm *U.S.* the olethreutid *Laspeyresia caryana*

silkworm the bombycid *Bombyx mori*

spanworm used in the sense of looper or inchworm

Bruce spanworm *U.S.* the geometrid *Operophtera bruceata*

cranberry spanworm *U.S.* the geometrid *Anavitrinella pampinaria*

currant spanworm *U.S.* the geometrid *Itame ribearia*

elm spanworm *U.S.* the geometrid *Ennomos subsignarius*

gooseberry span-worm *U.S.* = cranberry spanworm

potato tuberworm the gelechiid *Gnorimoschema operculella*

webworm *U.S.* any of several caterpillars building web-like nests

ailanthus webworm the yponomeutid *Atteva aurea*

alfalfa webworm *U.S.* the pyraustid *Loxostege commixtalis*

beet webworm *U.S.* the pyraustid *Loxostege sticticalis*

Hawaiian beet webworm *U.S.* the pyraustid *Hymenia recurvalis*

southern beet webworm *U.S.* the pyraustid *Pachyzancla bipunctalis*

spotted beet webworm *U.S.* the pyraustid *Hymenia perspectalis*

burrowing webworm any of several acrolophids

cabbage webworm *U.S.* the pyraustid *Hellula rogatalis*

fall webworm the arctiid *Hyphantria cunea*

garden webworm *U.S.* the pyraustid *Loxostege similalis*

bluegrass webworm *U.S.* the crambid *Crambus teterrellus*

juniper webworm *U.S.* the gelechiid *Dichomeris marginella*

pale juniper webworm *U.S.* the phaloniid *Aethes rutilana*

lespediza webworm *U.S.* the epipaschiid *Tetralopha scortealis*

mimosa webworm *U.S.* the glyphipterygid *Homadaula albizziae*

oak webworm *U.S.* the tortriid *Archips fervidanus*

parsnip webworm *U.S.* the oecophorid *Depressaria heracliana*

pine webworm (*see also* worm 4) *U.S.* the epipaschiid *Tetralopha robustella*

corn root webworm *U.S.* the crambid *Crambus caliginosellus*

sod web worm *U.S.* those pyralid moths commonly called grass moths

sorghum webworm *U.S.* the noctuid *Celama sorghiella*

worm 3 (*see also* worm, worm, 2, 4–6) the larvae of coleopteran insects. The words "larva of" are omitted from entries in this section

fruitworm (*see also* worm 2) the larva of any insect that damages fruit trees

raspberry fruitworm the byturid *Byturus unicolor*

eastern raspberry fruitworm *U.S.* the byturid *Byturus rubi*

western raspberry fruitworm *U.S.* the byturid *Byturus bakeri*

glow worm (*see also* worm 4) *Brit.* the lampyrid *Lampyris noctiluca* (*cf.* firefly)

grugru worm = grugru grub

barley jointworm the eurytomid hymenopteran *Harmolita hordii*

wheat jointworm *U.S.* the eurytomid hymenopteran *Harmolita tritici*

meal worm *U.S.* properly the larva, though the term is also applied to the adult, of several species of tenebrionids

dark mealworm *U.S. Tenebrio obscurus*

lesser mealworm *U.S. Alphitobius diaperinus*

yellow mealworm *U.S. Tenebrio molitor*

corn rootworm *U.S.* the chrysomelid *Diabrotica longicornis*

northern corn rootworm *U.S. D. unidecipunctata howardi* (= spotted cucumber beetle)

western corn rootworm *U.S. D. virgifera*

cranberry rootworm *U.S.* the chrysomelid *Rhabdopterus picipes*

grape rootworm *U.S.* the chrysomelid *Fidia viticida*

western grape rootworm *U.S.* the chrysomelid *Adoxus obscurus*

strawberry rootworm *U.S.* the chrysomelid *Parifragariae*

chestnut timberworm *U.S.* lymexylid *Melittomma sericeum*

oak timberworm *U.S.* the brentid *Arrhenodes minutus*

sapwood timberworm *U.S.* lymexylid *Hylecoetus lugubris*

wireworm larvae of elaterids

abbreviated wireworm *U.S. Hypolithus abbreviatus*

Columbia Basin wireworm *U.S. Limonius subauratus*

Great Basin wireworm *U.S. Ctenicera pruinina*

dry-land wireworm *U.S. Ctenicera glauca*

eastern field wireworm *U.S. Limonius agonus*

western field wireworm *U.S. Limonius infuscatus*

Gulf wireworm *U.S. Conoderus amplicollis*

Oregon wireworm *U.S. Melanotus oregonensis*

Pacific Coast wireworm *U.S. Limonius canus*

plains false wireworm *U.S.* the tenebrionid *Eleodes opaca*

prararie grain wireworm *U.S. Ctenicera aeripennis destructor*

Puget sound wireworm *U.S. Ctenicera aeripennis aeripennis*

sand wireworm *U.S. Horistonotus uhlerii*

sugar-beet wireworm *U.S. Limonius californicus*

tobacco wireworm *U.S. Conoderus vespertinus*

wheat wireworm *U.S. Agriotes mancus*

worm 4 (*see also* worm, worm 1–3, 5,6) the larvae of insects other than Lepidoptera and Coleoptera. The words "larva of" are omitted from entries in this section

case worm = caddis worm

currant worm the tenthredinid hymenopteran *Nematus ribesii*

imported currantworm *U.S.* = currantworm

faggot worm = caddis worm

glow worm (*see also* worm 3) *N.Z.* mycetophilid dipteran *Boletophila luminosa*

screw worm properly the larva of the calliphorid dipteran *Cochliomya hominivorex*, but sometimes extended to other parasitic calliphorids

secondary screw-worm *U.S. C. macellaria*

pine false webworm (*see also* worm 2) *U.S.* the pamphiliid hymenopteran *Acantholyda erythrocephala*

wheat straw-worm *U.S.* the eurytomid hymenopteran *Harmolita grandis*

worm 5 (*see also* worm, worm 1–4, 6, eel, fish) worm-shaped vertebrates. Without qualification the cobitid fish *Acanthophthalmus kuhlii*

blind-worm *Brit.* = slow-worm *Brit.*

slow-worm *Brit.* the anguinid lacertan reptile *Anguis fragilis*

worm 6 (*see also* worm, worm 1–5) in compound names of plants

bitter worm = buckbean

wort 1 (*see also* wort 2) *Brit. obs.* word meaning plant (*cf.* plant, weed) *obs.* except in compounds

adderwort *Brit.* the polygonacean herb *Polygonum bistorta U.S. P. bistortoides*

alpine pearlwort *Brit. S. saginoides*

ashwort the compositaceous herb *Senecio tomentosus*

asterwort = aster

awlwort the cruciferous herb *Subularia aquatica*

barrenwort any of several species of the berberidaceous genus Aceranthus

bellwort any of several species of liliaceous herb of the genus Uvularia

birthwort any of numerous species of the aristolochiaceous genus Aristolochia

bladderwort any of numerous species and several genera of insectivorous aquatic plants of the family Lentibulariaceae, particularly those of the genus Utricularia

butterwort any of numerous herbs of the lentibulariaceous genus Pinguicula

coolwort *U.S.* any of several species of urticaceous herbs of the genus Pilea

crosswort any of several species of the rubiaceous genus Crucianella

damewort the caprifoliaceous shrub *Sambucus ebulus*

dropwort *Brit.* the rosaceous herb *Spiraea filipendula*

hemlock dropwort the umbelliferous herb *Oenanthe crocata* or *Tiedemannia rigida* (= cow-bane) of the same family

water dropwort *Brit.* any of several umbelliferous plants of the genus Oenanthe

fanwort any of several species of the nymphaeaceous genus of water plants Cabomba

felwort the gentianaceous herb *Gentiana amarella*

feverwort any of several species of the caprifoliaceous genus Triosteum, particularly *T. perfoliatum*

yellow fumewort *U.S.* the fumariaceous herb *Corydalis flavula*

gipsy-wort *Brit.* the labiateous herb *Lycopus europaeus*

glasswort any of several species of chenopodiaceous herb of the genus Salicornia and a few similar plants

New Zealand glasswort the chenopodiaceous herb *Salicornia australis*

starry glasswort the caryophyllaceous herb *Cerastium arvense*

woody glasswort *U.S.* the chenopodiaceous herb *Salicornia virginica*

hogwort *U.S.* the euphorbiaceous herb *Croton capitatus*

honewort any of several species of umbelliferous herb of the genus Cryptotaenia (= wild chervil)

hornwort 1 (*see also* hornwort 2) any of numerous aquatic plants of the family Ceratophyllaceae, particularly *Ceratophyllum demersum*

hornwort 2 (*see also* hornwort 1) any bryospid plant of the subclass Anthocerotidae

leadwort any of several plumbaginaceous herbs and shrubs of the genus Plumbago

liverwort any bryospid plant of the subclass Hepaticidae

leafy liverwort any of many genera of the Jungermanniales

lousewort any of several species of the scrophulariaceous genus Pedicularis

swamp lousewort *P. lanceolata*

lungwort the lichen *Lobaria pulmonaria. U.S.* the boraginaceous herb *Mertensia maritima. Brit.* any of several herbs of the boraginaceous genus Pulmonaria

golden lungwort the compositaceous herb *Hieracium murorum*

madwort *Brit.* any of several boraginaceous herbs of the genus Asperugo, *U.S. A. procumberns*

German madwort = madwort

marsh wort *U.S.* = cranberry. *Brit.* any of several species of the umbelliferous herbs of the genus Helosciadium

masterwort any of several species of the umbelliferous genus Astrantia

milkwort any of numerous species of the polygalaceous genus Polygala *Brit.* without qualification, the polygalaceous herb *Polygala vulgaris*

box leaved milkwort *P. chamaebuxus*

fringed milkwort *P. paucifolia*

sea milkwort any of several species of primulaceous herb of the genus Glaux

miterwort any of several species of loganiaceous herb of the genus Cynoctonum, and also any of several species of saxifragaceous herbs of the genus Mitella

false miter wort any of several species of the saxifragaceous genus Tiarella

moneywort *U.S.* the primulaceous herb *Lysimachia nummularia. Brit.* the scrophulariaceous herb *Sidthorpia europaea*

moonwort *U.S.* any of several species of ophioglossaceous fern of the genus Botrychium (= grape fern), but particularly *B. lunaria,* and also the cruciferous herb *Lunaria annua*

motherwort any of several species of labiateous herb of the genus Leonurus. *Brit. L. cardiaca*

mudwort any of several species of herb of the scrophulariaceous genus Limosella

mugwort the compositaceous herb *Artemisa vulgaris*

western mugwort *A. ludoviciama*

navelwort any of several species of herbs of the boraginaceous genus Omphalodes and also the crassulaceous herb *Cotyledon umbilicus*

nipplewort any of several species of compositaceous herb of the genus Lapsana, particularly *L. communis*

nosewort any of several species of tropaeolaceous herbs of the genus Tropaeolum (= "Nasturtium" of gardeners), also the ranunculaceous herb *Helleborus niger*

pearlwort any of many species of caryophyllaceous herb of the genus Sigina but particularly *S. subulata*

pennywort *U.S.* any of several species of gentianaceous herb of the genus Obolaria. Also the crassulaceous herb *Cotyledon umbilicus*

marsh pennywort any of numerous species of the umbelliferous genera Hydrocotyle and Centella. *Brit.* marsh pennywort = *U.S.* water pennywort

wall pennywort *Brit. C. umbilicus*

water pennywort any of several species of herbs of the umbelliferous genus Hydrocotyle

pepperwort any of several species of cruciferous herbs of the genus Lepidium

pile wort the compositaceous herb *Erechtites hieracifolia,* and the ranunculaceous herb *Ranunculus ficaria*

pipewort any of several species of eriocaulaceous herb of the genus Eriocaulon

hairy pipewort *U.S.* the eriocaulaceous herb *Lachnocaulon anceps*

ragwort any of several species of compositaceous herb of the genus Senecio

golden ragwort *S. aureus*

prairie ragwort the compositaceous herb *Artemisia frigeda*

purple ragwort *S. elegans*

tansy ragwort *S. jacobaea*

ribwort any of numerous species of the plantaginaceous genus Plantago (= plantain)

rupture wort any of several species of the caryophyllaceous genus Herniaria

St. John's wort any of numerous herbs of the family Hypericaceae, particularly those of the genus Hypericum

marsh St. John's wort *H. virginicum*

St. Peter's wort the hypericaceous herb *Ascyrum stans*

saltwort the chenopodiaceous herb *Salsola sali*

perennial saltwort the chenopodiaceous herb *Salicornia virginica*

sandwort any of several species of the caryophyllaceous genus Arenaria

pine barren sandwort *U.S.* the caryophyllaceous herb *A. caroliniana* (= long root)

sawwort *Brit.* the compositaceous herb *Serratula tinctoria*

soapwort *see* soap plant

spearwort *Brit.* any of several species of ranunculaceous herb of the genus Ranunculus. *U.S. R. flammula*

great spearwort *Brit. R. lingua*

lesser spearwort *Brit. R. flammula*

water plantain spearwort *R. ambigens*

spiderwort any of several species of commelinaceous herb of the genus Tradescantia, but particularly *T. virginiana*

spleenwort any of several polypodiaceous ferns of the genus Asplenium

squinancy wort = squinancy

starwort any of several species of the compositaceous genus Aster

water starwort any of several species of the callitrichaceous genus Callitriche

stichwort *Brit.* any of numerous caryophyllaceous herbs of the genus Stellaria

marsh stichwort the caryophyllaceous herb *Alsine uliginosa*

cliff stonewort the crassulaceous herb *Sedum glaucophyllum*

strapwort *Brit.* the caryophyllaceous herb *Corrigiola littoralis*

swallowwort any of several species of papaveraceous herb of the genus Chelidonium

thoroughwort any of several species of compositaceous herbs of the genus Eupatorium

throatwort the campanulaceous herb *Campanula trachelium*

toothwort any of several orobranchaceous parasitic flowering plants of the genus Lathraea. Also any of several species of herb of the cruciferous genus Dentaria

waterwort any of several species and two genera of the parietale dicotyledonous family Elatinaceae

whitlow-wort any of numerous caryophyllaceous herbs of the genus Paronychia

woundwort the leguminous herb *Anthyllis vulleraria* and any of several species of labiateous herb of the genus Stachys

woolly wound *S. lanata*

yellowwort *Brit.* the gentianaceous herb *Chlora perfoliata*

wort 2 (*see also* wort 1) *Brit.* the noctuid lepidopteran insect *Cucullia asteris* (= starwort)

wou-wou = wow-wow

wow-wow the gibbon *Hylobates leuciscus*

wp *see* [warped]

Wr *see* [Wrinkled]

grass **wrack** *Brit.* any of several species of potamogetonaceous marine monocotyledons of the genus Zostera

wrannock *Brit. obs.* = wren

wrass = wrasse

wrasse *Brit.* popular name of labrid fishes

Blackear wrasse *Halichoeres poeyi*

clown wrase *Halichoeres maculipinna*

creole wrasse *Clepticus parrai*

dwarf wrasse *Doratonotus megalipis*

Greenland wrasse *Halichoeres bathyphilus*

painted wrasse *Halichoeres caudalis*
rock wrasse the labrid fish *Halichoeres semicinctus*
yellowhead wrasse *Halichoeres garnoti*
bridal **wreath** any of several species of rosaceous shrubs of the genus Spiraea but particularly *S. vanhouttei*
purple wreath the verbenaceous woody climber *Petrea volubilis*
snow wreath the rosaceous shrub *Neviusia alabamensis*
wren (*see also* thrush, tit, warbler) *U.S.* and *Brit.*, properly, any of numerous passeriform birds of the family Troglodytidae; without qualification, *Brit.*, *Troglodytes troglodytes* (= winter wren). *N.Z.* any of several passeriform acanthisittid birds of the genus Xenicus. *Austral.* any of numerous passeriform malurine birds. Occasionally used elsewhere for other small passeriform birds
antwren any of numerous small formicariids, mostly of the genus Myrmotherula
slaty antwren *M. schisticolor*
dot-winged antwren *Microrhopias quixensis*
band-backed wren the troglodytid *Campylorhynchus zonatus*
black-backed wren the malurine *Malurus melanotus*
purple-backed wren the malurine *Malurus assimillis*
red-backed wren the malurine *Malurus melanocephalus*
spotted bamboowren the rhinocryptid *Psilorhampus guttatus*
banded wren the troglodytid *Thryothorus pleurostictus*
grey-barred wren the troglodytid *Campylorhynchus megalopterus*
white-bellied wren the troglodytid *Uropsila leucogastra*
Bewick's wren the troglodytid *Thryomanes bewickii*
wedge-billed wren the timaliid *Sphenocichal humei*
blue wren the malurine *Malurus cyaneus*
blue-and-white wren the malurine *Malurus cyanotus*
scaly-breasted wren the timaliid *Pnoepyga albiventer*
spot-breasted wren the troglodytid *Thryothorus maculipectus*
brown wren the timaliid *Pnoepyga pusilla*
bush wren *N.Z.* the acanthisittid *Xenicus longipes*
cactus wren the troglodytid *Campylorhynchus brunneicapillus*
canyon wren the troglodytid *Catherpes mexicanus*
Carolina wren the troglodytid *Thryothorus ludovicianus*
golden-creasted wren *Brit.* the sylviid *Regulus regulus* (= goldcrest)
emu-wren any of several malurines of the genus Stipiturus; without qualification, *S. malachurus*
rufous-crowned emu-wren *S. ruficeps*
mallee emu-wren *S. mallee*
field-wren any of several malurines of the genus Calamanthus; without qualification, *C. fuliginosus*
rufous field-wren *C. campestris*
striated field-wren *C. fuliginosus*
gnatwren any of several sylviids of the genera Microbates and Ramphocaenus

long-billed gnatwren *R. melanurus*
grass-wren any of several malurines of the genus Amytornis
thick-billed grass-wren *A. modestus*
stripped grass-wren *A. striatus*
western grass-wren *A. textilis*
happy wren the troglodytid *Thryothorus felix*
heath-wren any of several malurines of the genus Hylacola
mallee heath-wren *H. cauta*
chestnut-tailed heath-wren *H. pyrrhopygia*
house wren the troglodytid *Troglodytes aedon*
Kashmir wren = winter wren
marsh-wren either of two species of troglodytids of the genera Cistothorus and Telmatodytes; without qualification, *T. palustris*
long-billed marsh-wren *T. palustris*
short-billed marsh-wren *C. platensis*
rufous-napped wren the troglodytid *Campylorhynchus rufinucha*
nightingale wren the troglodytid *Microcerculus marginatus*
plain wren the troglodytid *Thryothorus modestus*
rock wren *U.S.* the troglodytid *Salpinctes obsoletus;* *N.Z.* the acanthisittid *Xenicus gilviventris*
scrub-wren any of several malurines of the genus Sericornis; without qualification, *Austral. S. maculatus*
large-billed scrub-wren *S. magnirostris*
buff-breasted scrub-wren *S. laevigaster*
white-browed scrub-wren *S. frontalis*
sedge wren = short-billed marsh-wren
spotted wren *I.P.* the timaliid *Spelaeornis formosus;* Mexico, the troglodytid *Campylorhynchus jocosus*
brown-throated wren the troglodytid *Troglodytes brunneicollis*
bar-vented wren the troglodytid *Thryothorus sinaloa*
willow-wren *Brit.* any of several sylviids of the genus Phylloscopus, particularly *P. trochilus* (= willow warbler)
winter wren the troglodytid *Troglodytes troglodytes*
wood-wren either of two troglodytids of the genus Henicorhina
grey-breasted wood-wren *H. leucophrys*
white-breasted wood-wren *H. leucosticta*
Zapata wren the troglodytid *Ferminia cerverai*
Wrights' *ep. adj.* from Sewall Wright (1889–   )
[Wrinkled] *either* a mutant (gene symbol *W*) mapped at 46.0 on chromosome III of *Drosophila melanogaster*. The phenotypic wings are not completely unfolded, *or* a mutant (gene symbol *Wr*) mapped at 76.0± on chromosome II of *Drosophila melanogaster*. The term refers to the wings of the phenotype
Wrisberg's *epon. adj.* from Heinrich August Wrisberg (1739–1808)
ws3 *see* [white sheath]
wt *see* [wet], [well] and [wilty foliage]
wutu the crotalid snake *Brothrops alternatus*
wv *see* [white virescent]
ww *see* [wider-wing]
wx *see* [waxy]
wy *see* [wavy]

X *see* sex

Xancidae a family of large gastropod mollusks distinguished by the clawlike operculum and distinct reticulation on the columella. Commonly called lamp shells, though this term is sometimes confined to the genus Xancus and should properly be reserved for the Brachiopoda

Xanclidae *see* Zanclidae

-xanth- *comb. form* meaning yellow

xanthellous yellowish

hapa**xanthic** used of annual plants having a short, single flowering period

Xanthidae a family of brachyuran crustacea distinguished from the closely allied Cancridae by the slanted or transversely folding first antenna and the nearly round cephalothorax. They differ from the Portunidae in the absence of well developed swimming legs.

xanthin any of numerous yellow carotenoid pigments

astax**anthin** green or blue oxygenated carotenoid biochrome commonly found in marine crustacea. Boiling liberates the red carotenoids

hypox**anthine** a purine (6-hydroxypurine) found in living tissue and usually produced by the hydrolysis of adenine

Xanthophyceae that class of chrysophyte algae in which the chromatophores are yellowish-green

Xantusiidae a family of lacertilian reptiles having pleurodont dentition, no osteoderms and a short scaly tongue

Xenarthra that order of placental mammals which contains the anteaters and sloths. They are distinguished by the lack of true teeth and by the large recurved claws on the feet. The armadillos are sometimes included in this group and sometimes placed in a separate order Loricata

mono**xenic** said of a parasite which is host specific

xenica *Austral.* popular name of satyrid lepidopteran insects of the genera Oreixenica and Geitoneura

Xenicidae = Acanthisittidae

-xeno- *comb. form* meaning "a host" in the opposite sense to "guest" or "foreigner"

Xenoberyces a small order of abyssal osteichthyes

xenodochae the invasion of a plant succession by foreign or anomalous plants

Xenopeltidae a monotypic family of snakes distinguished by the fact that the dentary bone is movably attached to the articular

Xenophoridae a small family of gastropod mollusca distinguished by their habit of cementing pebbles and broken shell fragments to their own shells

Xenopsyllidae = Pulicidae

Xenopterygii an order of osteichthyes containing the single family Gobiesocidae. The characteristics of the family are those of the order

Xenosauridae a family of lacertilian reptiles with pleurodont dentition, osteoderms on the body but not head, and in which the anterior portion of the tongue can be retracted

-xenous *comb. suffix* meaning pertaining to a "host" or "guest"

di**xenous** said of a parasite which may infect two species, but can pass its entire life cycle on either

hetero**xenous** = heteroecious

lipo**xenous** said of a parasite which leaves its host after it has secured adequate nourishment

meto**xenous** = metoecious

myremeco**xenous** furnishing both food and shelter for ants

pleio**xenous** said of a parasite which is not host specific

tri**xenous** said of a parasite which requires three successive hosts

-xer- *comb. form* meaning "dry"

xerad = xerophyte

xerampelinous dull purple

-xiph- *comb. form* meaning "sword"

Xiphiidae a monospecific family of scombroid acanthopterygian fishes containing the swordfish *Xiphias gladius* possessing an extremely elongate flattened bill

xiphioid sword like

-xipho- *see* -xiph-

Xiphosura an order of meristomatous arthropods commonly called horseshoe crabs or kingcrabs. They are distinguished by the horseshoe-shaped body, long pointed telson, and the presence of 5 pairs of gill books, on the opisthoma (*cf.* Merostomata)

Xiphydriidae a small family of symphytan hymenopteran insects best described as horntails lacking a horn. They are commonly called wood wasps

xl *see* [lethal]

xochinacaztli the annonaceous shrub *Cymbopetalum penduliflorum*

xuxu the cucurbitaceous vine *Sechium edule* and its edible squash-like fruit

Xyelidae a small family of symphytan hymenopteran insects distinguished from all other sawflies by the fact that the third antennal segment is longer than the remaining segments combined

-xyl- *comb. form* meaning "wood"

xyl a bacterial mutant showing a utilization of xylose

and a bacterial genetic marker having a character affecting the utilization of D-xylose, mapped at 69.8 mins. for *Escherichia coli*

-xyl *comb. suffix* used to denote dwarf woody plants

aeroxyl woody plants with a single trunk rising from the ground

microxyl dwarf woody plants having one main axis rising from the soil

geoxyl a plant having woody stems arising from an underground woody rhizome

microgeoxyl those low-lying woody plants that send up numerous stems

xylanase an enzyme hydrolyzing β-1, 4-xylan links in carbohydrates

xylem woody tissue or cells

endarch xylem that in which the most mature elements are located nearest to the center of the axis

exarch xylem that in which the most mature elements are located furthest from the center of the axis

mesarch xylem xylem in which maturation occurs in two directions outward

leptoxylem the water-conducting tissue of mosses

metaxylem 1 (*see* metaxylem 2) the central, compacted wood of a tree

metaxylem 2 (*see* metaxylem 1) xylem maturing after the completion of elongation

protoxylem xylem maturing before elongation commences or during elongation

primary xylem xylem developed in the primary part of the plant body, directly from procambium

secondary xylem xylem produced from vascular cambium

perixylematic said of the concentric bundles in the roots of grasses

dixylic having the xylem in two distinct parts

haploxylic having only one vascular bundle

-xylo- *see* -xyl-

Xylocopidae a family of apoid apocritan hymenopteran insects usually included with the Apidae

Xylocopinae a subfamily of apocritan hymenopteran insects commonly called carpenter bees because they burrow in wood. The term without qualification usually applies to the large bumblebee-like insects of the genus Xylocopa but there are many small wasp-like blue and green forms in this family

Xylomyiidae a family of brachyceran dipteran insects now usually included with the Xylophagidae

Xylophagidae a family of large orthoraphous brachyceran dipteran insects some of which resemble ichneumon flies

Xylophilidae = Euglenidae

epixylous growing on wood

Xyridaceae that family of farinose monocotyledonous angiosperms which contains the forms commonly called yellow-eyed grasses. They are distinguished as small grass-like herbs with zygomorphic flowers, subtended by an involucre of imbricated bracts

Xyridales a little-used taxon of monocotyledonous angiosperms erected to contain the single family Xyridaceae

y *see* [yellow] or [yellow fat] or [colorless fruit skin]

Y *see* [Yellow blood] or [Yellow endosperm]

golden Y *Brit.* either of two species of noctuid lepidopteran insect of the genus Plusia

silver Y *Brit.* either of two species of lepidopteran insect of the genus Plusia but usually, without qualification, *P. gamma*

yaffin = yaffingale

yaffingale *Brit.* the picid bird *Picus viridis* (= green woodpecker)

yak a Tibetan, domesticated bovine bovid (*Bos grunniens*); the animal is distinguished from other living cattle by the very long hair

yam (*see also* thrush) the dioscoreaceous herb *Dioscorea batatas* and its edible tubers. In parts of the *U.S.* the term is also applied to the sweet potato (*q.v.*)

yapock 1 (*see also* yapock 2) *Austral.* the didelphyd marsupial *Chironectes minimus.* (= water-opossum)

yapock 2 (*see also* yapock 1) a Central American webbed footed aquatic opossum of the genus Chironectes

yapok 1 (*see also* yapok 2) = yapock

yapok 2 (*see also* yapok 1) any of several species of callicebine hapaloid primates

yappingale = yaffingale

yarrow the composite herb *Achillea millefolium*

yaupon the aquifoliaceous shrub *Ilex vomitoria*

yearling an animal one year old, usually applied to those which have not yet reached sexual maturity

yeast those endomycetale ascomycete fungi that do not form hyphae

yellow (*see also* bill, legs) *Brit.* name applied to many pierid lepidopteran insects, the word butterfly rarely being appended. *S.Afr.* any of several pierids of the genus Eurema

clouded yellow any of several species of the genus Colias. *S.Afr.*, without qualification, *C. electo*

frosted yellow *Brit.* the geometrid lepidopteran insect *Fidonia limbaria*

grass yellow *I.P.* any of numerous species of yellow pierid lepidopteran insects of the genus Eurema. (= *U.S.* little sulphur) . . . *Austral.*, any of several species of the genus Terias

[**yellow**] a mutant (gene symbol *y*) mapped at 0 on the X chromosome of the *Drosophila melanogaster.* The name is descriptive of the body color

[**Yellow**] a mutant (gene symbol *ylo*) on linkage group VI of *Neurospora crassa.* The name refers to the color of the phenotypic conidia

[**Yellow blood**] a mutant (gene symbol *Y*) mapped at 25.6 in linkage group II of *Bombyx mori.* The name is descriptive of the hemolymph of the phenotypic larva

[**Yellow endosperm**] a mutant (gene symbol *Y*) mapped at 13 on linkage group VI of *Zea mays.* The term is descriptive of the phenotypic seed

[**yellow fat**] a mutant (gene symbol *y*) at locus 14.4 of the linkage group I of the rabbit. The name is descriptive of the phenotypic expression

[**yellow fruit flesh**] a mutant (gene symbol *r*) mapped at 0 on chromosome 3 of the tomato. The name is descriptive of the phenotypic expression

[**yellow-green**] a mutant (gene symbol *yg₂*) mapped at 7 on linkage group IX of *Zea mays.* The name is descriptive of the phenotypic seedling and plant

[**Yellow inhibitor**] a mutant (gene symbol I) in linkage group IX of *Bombyx mori.* The phenotypic expression is a suppression of yellow blood and rusty

[**yellow stripe**] a mutant (gene symbol *ys*) mapped at 41 on linkage group V of *Zea mays.* The phenotypic leaves are longitudinally striped in green and yellow

[**yellow virescent**] a mutant (gene symbol *yv*) mapped at 60 on chromosome 6 of the tomato. The phenotypic leaves are yellowish-green

yew any of numerous species and horticultural varieties of the coniferous genus Taxus

plum yew popular name of trees and shrubs of the family Cephalotaxaceae

yg₂ *see* [yellow-green]

yield a portion of the productivity of a population removed per unit time, for example by a predator

ylang-ylang the annonaceous tree *Canangium odoratum*

climbing ylang-ylang the annonaceous woody climbing shrub *Artobotrys odoratissima*

ylo *see* [Yellow]

yoke term for a pair of oxen working in concert (*cf.* team)

yoldia popular name of nuculanid pelecypods

yolk the intracellular food reserve of an egg

youth-and-old-age any of several species of compositaceous herb of the genus Zinnia but particularly *Z. elegans*

-ypo- *comb. form* meaning "beneath" (*cf.* -hypo-)

ypomnema an inferior calyx

Yponomeutidae a family of yponomeutoid lepidopteran insects commonly called ermine moths because many of the better known species have white, black or yellow markings on the forewings

Yponomeutoidea a superfamily of lepidopteran insects containing *inter alia* the clear-winged moths

ys *see* [yellow stripe]

yucca (*see also* bug, weevil) a genus of liliaceous plants

Yucceae a subfamily of amaryllidaceous monocotyledons. Considered by many properly to belong in the Liliaceae

Yungidae = Jyngidae

yv *see* [yellow virescent]

Z

z *see* [zeste]

Zamioideae a subfamily of Cycadaceae

Zanclidae a monospecific family of acanthopterygian osteichthyes containing the Moorish idol *Zanclus canescens* distinguished by its yellow, black and white vertically striped laterally compressed body

Zanichellieae a tribe of the family Najadaceae when this family includes the Potamogetonaceae

Zanthoxylaceae = Rutaceae

Zapodidae a family of myomorph rodents containing the so-called jumping-mice; in addition to their method of locomotion, they are distinguished by their extremely long tails

Zaproridae a family of blennioid fishes, commonly called highbrows

zb_4, zb_6 *see* [zebra striping]

Ze *see* [Zebra]

zebra 1 (*see also* zebra 2, butterfly, caterpillar, perch, plant, shell, swallowtail, white) any of several species, and several races of striped wild horses of the genus Equus

    Grevy's zebra the striped, equid mammal *Doliochohippus grevii*; it is the only living equid not in the genus Equus

zebra 2 (*see also* zebra 1) *U.S.* the nymphalid lepidopteran *Heliconius charitonius*. *I.P.* any of several species of papilionid lepidopteran insect of the genus Paranticopsis

[Zebra] a mutant (gene symbol *Ze*) on linkage group III of *Bombyx mori*. The phenotypic larva has a black band on the anterior end of each segment

[zebra striping] *either*, a mutant (gene symbol zb_4) mapped at 21 on linkage group I of *Zea mays*. The name refers to alternate dark and light stripes on the phenotypic relief *or*, a mutant (gene symbol zb_6) mapped at 84 in linkage group IV of *Zea mays*. The name refers to transverse striping in the leaves

Zeidae a family of benthonic zeomorphous fishes containing the john dory. The large lateral black spot of this fish is typical of the family

Zeiformes = Zeomorphi

Zelodoniidae a monotypic family of passeriform birds erected to contain the wrenthrush of the high mountain peaks of Costa Rica and Panama. The name is admirably descriptive of the general appearance

Zeomorphi an order of mostly deep water osteichthyes containing, *inter alia*, the Zeidae, Grammicolephidae and Caproidae. Many have extensible jaws

zeran a Central Asian "goat-gazelle" *Procapra gutturosa* (*cf.* goa)

[zeste] a mutant (gene symbol *z*) mapped at 1.0 on the X chromosome of *Drosophila melanogaster*. The phenotypic expression is yellow-eyed females

zeugite a spore, such as a teleutospore, containing two fused nuclei

Zeuzeridae a family of lepidopteran insects now usually included in the Cossidae

solomon's zigzag *U.S.* the liliaceous herb *Smilacina racemosa*

Zingiberaceae that family of scitaminous monocotyledons which contains, *inter alia*, a variety of spices including ginger, cardamon and turmeric. The outstanding characterisics of the family are the aromatic oils they contain and the single stamen

Zingiberales = Scitaminales

-zo- *comb. form* meaning animal

-zoa *comb. suffix* indicating "animal" in the plural. Taxa ending in -zoa are entered alphabetically. Compounds not taxa are entered under the singular suffix -zoon

Zoantharia a subclass of anthozoan Coelenterata distinguished by the fact that the tentacles and septa are usually in multiples of six but never eight as in the Alcyonaria. This subclass contains, *inter alia*, the sea anemones and the true corals

Zoanthidea an order of zoantharian coelenterates lacking both skeleton and pedal disk

Zoarcidae a family of blennioid fishes having very long dorsal and anal fins continuous with the caudal fin. Commonly called eel pouts

epizoarious growing on dead animals (*cf.* epizoic)

zoarium a colony of animals. Most commonly applied to Ectoprocta

-zoic *adj. suffix* meaning pertaining to an animal, or animals. The alternative form -zoous, though occasionally seen in the literature, is ignored in this work.

    coprozoic pertaining to organisms living on the excreta of other forms

    cryptozoic pertaining to animals concealing themselves in crevices, under stones, and the like

    entozoic = endozoic

    epizoic 1 (*see also* epizoic 2) said of an animal ectocommensal

    epizoic 2 (*see also.* epizoic 1) said of organisms that grow on live animals

    endozoic said of an animal living within another animal

    holendozoic said of a parasite which passes its entire life within an animal host. Used almost exclusively of fungi

    monozoic a term used to distinguish those tapeworms the body of which is not divided into proglottids

polyzoic a term used to describe those tapeworms, in distinction from monozoic forms, in which the body is typically divided into proglottids

saprozoic *see* nutrition

-zoid *see* -zooid

-zoite *adj. suffix* indicating a part of an animal produced by fission (*cf.* -zooid)

merozoite agametes produced by multiple fission of a trophozoite

sporozoite the motile product of multiple fission, either of zygotes or spores

trophozoite the growing, vegetative stage of a sporozoan protozoan

-zon- *comb. form* meaning "girdle," which has come in contemporary writing to be extended into the meaning "zone"

-zonal *comb. suffix* pertaining to a zone, particularly in *biol.* in the sense of zone 1

euryzonal pertaining to an organism occuring in many zones either aquatic or aerial

stenozonal pertaining to an organism living at a restricted altitude or depth

zonaric relating to intermediate depths

zone 1 (*see also* zone 2–4) a horizontal layer either of water or air (*cf.* horizon)

aphotic zone that zone of water which lies at such a depth that it receives no significant amount of light

euphotic zone that zone of water within which photosynthesis is possible

dysphotic zone that layer of water with enough light for animal response, but not for photosynthesis

aphytal zone those waters in which the penetration of light is too little to support photosynthesis

zone 2 (*see also* zone 1, 3, 4) a specific limited area of a land mass, particularly one in which the population is limited by physical conditions

climax zone local minor changes in the composition of a climax due to local changes in environment

eulittoral zone the true littoral zone which is variously defined in oceanography as the zone never covered by more than ten fathoms of water and in limnology as the zone of rooted aquatic plants or the zone which is subject to submerged wave action

life zone a geographic area having a characteristic fauna and flora

spray zone that area of an ocean shore which is close to high tide mark but wetted occasionally by salt spray

zone 3 (*see also* zone 1, 2, 4) in the original sense of girdle

abscission zone the area at which a deciduous leaf separates from the stem (*cf.* layer)

connecting zone the girdle connecting the valves of a diatom frustule

interzone that portion of a diatom frustule which lies between the girdle and the valves

adoral zone a row of membranelles round the peristome of some Protozoa

perimedullary zone the layer immediately abaxial to the protoxylem

zone 4 (*see also* zone 1, 2, 3) in the sense of a portion of a solid

alary zone that part of the embryonic central nervous system which lies above Monro's sulcus

basal zone that part of the embryonic central nervous system that lies below Monro's sulcus

cambial zone that region of the cambium containing the initials and their immediate derivatives

-zonic *adj. suffix* meaning pertaining to zone, particularly in the sense of zone 2 (*cf.* zonal)

hormozonic said of an organism in which the larval stages are restricted to the same environment occupied by the adult

heterozonic pertaining to organisms that can, but need not, occupy different environments in the course of their life history (*cf.* eurytope)

zonite a body division of a kinorhynch, used to avoid the word segment

Zonuridae a family of lacertilian reptiles with pleurodont dentition, a non-retractile tongue, and osteoderms on the head but not the body

-zoo- *comb. form* derived from -zo-, usually meaning animal, but used in *bot.* to denote motile in contrast to immobile

Zoochlorella a general term for green algae living in or among animal cells

zooid 1 (*see also* zooid) a motile plant gamet

antherozooid a male motile plant gamete

megazooid an algal female gamete

microzooid a male algal gamete

phytozooid = antherozooid

spermatozooid a ciliated gamete within an antheridium

sporozooid = zoospore

zooid 2 (*see also* zooid 1) an ill-defined term applied for convenience sake to individual members or distinct functional portions of aggregates of animals like Bryozoa, or colonial or compound animals like Hydrozoa. Frequently, but unfortunately, used in place of -zoite (*q.v.*)

acanthozooid an ectoproct heterozooid in the form of a thick walled spine

allozooid one which differs from the zooid from which it was derived

autozooid 1 (*see also* antozooid 2, 3) the feeding member of a hydrozoan polymorphic colony

autozooid 2 (*see also* antozooid 1, 3) the typical fully formed zooid of an Ectoproct

autozooid 3 (*see also* autozooid 1, 2) the normal polyp of an alcyonarian coelenterate (*cf.* siphonozoid)

blastozooid an individual resulting from asexual reproduction

dactylozooid a hydrozoan zooid, having tentacles furnished with nematocysts but no mouth

gastrozooid a zooid designed to digest, and frequently to capture, prey

gennylozooid = spermatozoon

gonozooid 1 (*see also* gonozooid 1) an ectoproct heterozooid with a bulbous brood chamber

gonozooid 2 (*see also* gonozooid 1) a polyp of a hydrozoan coelenterate colony that is modified to produce medusae

heliozooid an animal with rays protruding from it

heterozooid any ectoproct zooids other than autozooids are grouped under this term

isozooid one which is similar to its parent

kenozooid an ectoproct heterozooid consisting of a body wall with encased strands of tissue

machozooid = dactylozooid

mesozooid an ectoproct heterozooid consisting of a tube connecting diverging zoocial tubes

nannozooid literally, a dwarf zooid. Applied particularly to such forms in Ectoprocta

oozooid a zooid which is derived from an egg

spiral zooid a type of tentaculozooid looped into a loose spiral of one or more coils

siphonozooid a modified alcyonarian polyp lacking tentacles and serving as an inhalant water aperture

tentaculozooid a dactylozooid in the form of a single tentacle

Zoomastigophora a subclass of mastigophorous Protozoa distinguished from the Phytomastigophora by their lack of chlorophyll

-zoon *adj. suffix* meaning "an animal" The distinction from -zooid 1 is not always clear

antherozoon male motile plant gamete

coprozoon an animal growing on faecal matter either within or without the form producing the faeces

diplozoon a general term for those mastigophoran Protozoa in which all organelles are duplicated (*cf.* Diplozoa)

metazoon a many celled animal (*cf.* metabiont, metaphyte)

phytozoon = antherozooid

spermatozoon a motile male gamete

Zoophyta a somewhat indeterminate division of the Linnean group Vermes, which contained all the invertebrates except the insects

zorilla = zorille

zorille any of several species of mephitic musteline carnivores of the genus Zorilla; they occupy the same ecological niche in Africa as the skunks in America and are similar in habit and color

Zosteraceae a family of monocotyledonous angiosperms erected to contain the genera Zostera, Potamogeton, Ruppia, and Zannichellia by those who do hold them to belong in the Potamogetonaceae or Najadaceae

Zostereae a tribe of the family Najadaceae when this family includes the Potamogetonaceae

Zosteropidae that family of passeriform birds of Africa, Asia, and Australasia, commonly called white eyes. The name derived from the conspicuous white ring round the eye of almost all species

zulu *S.Afr.* any of several species of lycaenid lepidopteran insects of the genus Alaena

-zyg- *comb. form* meaning "yoke"

Zygaenidae a family of zygaenoid lepidopteran insects far better known in Europe than *U.S.* In *U.S.* they are called smoky moths but many British zygaenids commonly called burnetts are brilliantly colored in red and green. The dull colored European zygaenids are called foresters

Zygaenoidea a superfamily of lepidopteran insects containing many, little-known small moths

-zygal *comb. suffix* meaning pertaining to a syzygy (*cf* zygous)

epizygal the distal member of an echinoderm syzygy

hypozygal the proximal member of an echinoderm syzygy

zygantra an articular process, analogous to a zygapophysis but rising from the neural arch, in snakes

Zygnemataceae that family of zygematale Chlorophyceae which possess cylindrical cells permanently united in unbranched filaments

Zygnematales that order of chlorophycean Chlorophyta distinguished in possessing amoeboid, non-flagellated isogametes

zygoid = diploid

hemizygoid = haploid

Zygomycetae = Aflagellatae

Zygoneura 1 (*see also* Zygoneura 2) a zoological taxon erected by Hatschek to contain all of his Heteraxonia except the Ambulacralia and the Chordonia

Zygoneura 2 (*see also* Zygoneura 1) = Protostomia

Zygophyllaceae that family of geranialous dicotyledons which contains the lignum vitae tree. The family is readily distinguished by its prickly fruits

Zygoptera a suborder of odonatan insects distinguished by having the front and hind wings similar in shape. Commonly called damselflies

zygosis the process by which a zygote is produced

polyzygosis the fusion of more than two gametes

zygote 1 (*see also* zygote 2) the cell that results from the fusion of two gametes

heterozygote (*see also* zygoet 2) the cell which results from the fusion of two morphologically distinguishable gametes

homozygote (*see also* zygote 2) the cell which results from the fusion of two morphologically indistinguishable gametes

hypnozogote a dormant zygote

zygote 2 (*see also* zygote 1) a zygote in the sense of zygote 1, but defined by its genetic make up and the organism that developes from it

allozygote a homozygote with only recessive characters

heterozygote (*see also* zygote 1) one having different alleles at the same locus and therefore derived from the union of two genetically dissimilar chromosomes

inversion heterozygote one in which the homologous chromosomes have a different linear order as the result of inversions

homozygote (*see also* zygote 1) one having identical alleles at the same locus and therefore derived from the union of gametes of identical genetic composition

-zygous 1 *comb. suffix* meaning "pertaining to zygote" in the sense of zygote 1 (*cf.* -zygal)

azygous unpaired

-zygous 2 *comb. suffix* meaning "pertaining to zygote" in the sense of zygote 2 (*cf.* -zygal)

heterozygous the condition of an individual having two different alleles at the same locus on a pair of homologous chromosomes

homozygous pertaining to an individual having identical alleles at the same locus on a pair of homologous chromosomes

isozygous having identical genetic composition in two zygotes

microzyma an early name for those plastids thought by observers of the period to be living

-zym- *comb. form* meaning "yeast"

cozymase *obs.* = NAD

-zyme *adj. suffix* meaning enzyme

lysozyme an enzyme catalyzing the hydrolysis of the wall substances of certain bacteria (= muramidase)

pancreozymin a hormone secreted in the upper intestinal mucosa. It is active in controlling enzyme secretion by the pancreas

PRINCIPAL WORKS CONSULTED

The following list does not comprise all of the books consulted in the compilation of this dictionary and, particularly, omits journal papers. The works here cited are those which were continuously available to the compiler, or in rare cases to a reviewer, and are particularly those in which he checked and cross checked contemporary, and in many cases early, biological usage.

Abercrombie, M. *et al.* "A Dictionary of Biology," rev. ed., Baltimore, Penguin, 1960.

Allee, W. C. *et al.* "Principles of Animal Ecology," Philadelphia, Saunders, 1949.

Altman, P. L. and D. S. Dittmer, eds., "Biology Data Book," Washington, Federation American Society Experimental Biology, 1964.

American Ornithologists Union, "Check-List of North American Birds," 5th ed., Baltimore, Lord Baltimore Press, 1957.

Anon., "Nomina Anatomica," 3rd. ed., Amsterdam, Excerpta Medica Foundation, 1966.

Anon., "The Merck Index of Chemicals and Drugs," 6th ed., Rahway, N.J., Merck, 1952.

Anon., "Report of the Commission on Enzymes of the International Union of Biochemistry" London, Pergamon, 1961.

Arnet, R. H., "The Beetles of the United States," Washington, D.C., Catholic Univ. Amer. Press, 1960.

Bailey, A. M., "Birds of New Zealand," Denver, Museum of Natural History, 1955.

Bailey, R. M. *et al.*, "A List of the Common and Scientific Names of Fishes from the United States and Canada," 2nd ed., Ann Arbor, Mich., American Fisheries Society, 1960.

Bailey, L. H., "The Standard Cyclopedia of Horticulture," 3 vols., New York, Macmillan, 1944.

Baker, E. W. and G. W. Wharton, "An Introduction to Acarology," New York, Macmillan, 1952.

Ball, A. M., "The Compounding and Hyphenation of English Words," New York, Funk and Wagnalls, 1951.

Barnes, R. D., "Invertebrate Zoology," Philadelphia, Saunders, 1963.

Barnhart, L., ed., "The American College Dictionary," New York, Random House, 1951.

Barrett, C. and A. N. Burns, "Butterflies of Australia and New Guinea," Melbourne, Seward, 1951.

Bates, G. L., "Handbook of the Birds of West Africa," London, John Bale, Sons and Danielsson, 1930.

Bather, F. A. *et al.* "The Echinoderma," *in* Lankester, E. R., "A Treatise on Zoology," London, Black, 1900.

Beddard, F., "Mammalia" in Harmer, S. F., and A. E. Shipley, eds., "The Cambridge Natural History," London, Macmillan, 1902.

Benham, W. B., "The Platyhelminthes, Mesozoa, and Nemertini" *in* Lankester, E. R., "A Treatise on Zoology," London, Black, 1901.

Blair, A. P., *et al.*, "Vertebrates of the United States," New York, McGraw-Hill, 1957.

Blake, E. R., "Birds of Mexico," Chicago, The University Press, 1953.

Bond, J., "Birds of Mexico," Boston, Houghton Mifflin, 1961.

Borradaile, L. A., and F. A. Potts, "The Invertebrata," 3rd ed., [revised by G. A. Kerkut], Cambridge, The University Press, 1959.

Borror, D. J., and D. M. Delong, "An Introduction to the Study of Insects," New York, Rinehart, 1954.

Breed, R. S., *et al.* eds., "Bergey's Manual of Determinative Bacteriology," 7th ed., Baltimore, Williams and Wilkins, 1957.

Bridges, C. B., and K. S. Brehme, "The Mutants of *Drosophila melanogaster*," Washington, Carnegie Institution, 1944.

Brown, R. W., "Composition of Scientific Works," rev. ed., Baltimore, Brown, 1956.

Buchsbaum, R., "Animals Without Backbones," 2nd ed., Chicago, University of Chicago Press, 1948.

Buchsbaum, R., and L. J. Milne, "The Lower Animals," Garden City, N.Y., Doubleday, n.d.

Bullough, W. W., "Practical Invertebrate Anatomy," London, Macmillan, 1950.

Burt, W. H., and R. P. Grossenheider, "A Field Guide to the Mammals," 2nd ed., Boston, Houghton Mifflin, 1964.

Calman, W. T., "Appendiculata-Crustacea," *in* Lankester, R., "A Treatise on Zoology," London, Black, 1909.

Carpenter, J. R., ed., "An Ecological Glossary," New York, Hafner, 1962.

Chrystal, R. N., "Insects of the British Woodlands," London, Warne, 1937.

Clapper, R. B., "Glossary of Genetics and Other Biological Terms," New York, Vantage, 1960.

Clark, G. L., and G. G. Hawley, eds., "Encyclopedia of Chemistry," New York, Reinhold, 1957.

Cobb, B., "A Field Guide to the Ferns," Boston, Houghton Mifflin, 1956.

Cochran, D. M., "Living Amphibians of the World," Garden City, N.Y., Doubleday, 1961.

Cockayne, L. and E. P. Turner, "The Trees of New Zealand," 4th ed., Wellington, N.Z., Owen (Government Printer) 1958.

Collins, H. H., "Complete Field Guide to American Wildlife," New York, Harper, 1959.

Comstock, J. H., "An Introduction to Entomology," 9th ed., Ithaca, N.Y., Comstock, 1950.

Conant, R., "A Field Guide to Reptiles and Amphibians," Boston, Houghton Mifflin, 1958.

Cooke, A. H., et al., "Molluscs, Brachiopods (Recent), Brachiopods (Fossil)" in Harmer, S. F. and A. E. Shipley, eds., "The Cambridge Natural History," London, Macmillan, 1895.

Crookes, M., "New Zealand Ferns," 6th ed., New Zealand, Whitcombe and Tombs, 1963.

Crosby, E. C., et. al., "Correlative Anatomy of the Nervous System," New York, Macmillan, 1962.

Dewar, D., "Birds of the Indian Hills," London, The Bodley Head, 1915.

Dewar, D., "Birds of the Plains," London, The Bodley Head, 1909.

Dewar, D., "Bombay Ducks," London, The Bodley Head, 1906.

Dewar, D., "Himalayan and Kashmiri Birds," London, The Bodley Head, 1923.

Dixon, M. and E. C. Webb, "Enzymes," London, Longmans Green, 1957.

Ealand, C. A., "Insect Life," London, Black, 1921.

Emerton, J. H., "The Common Spiders of the United States," New York, Dover, 1961.

Esau, K., "Plant Anatomy," New York, Wiley, 1953.

Evans, A. H., "Birds," in Harmer, S. F. and A. E. Shipley, "The Cambridge Natural History," London, Macmillan, 1899.

Fernald, M. L., "Gray's Manual of Botany," 8th ed., New York, American Book Co., 1950.

Fox, R. and J. W. Fox, "Comparative Entomology," New York, Reinhold, 1964.

Fraser, J. and A. Hemsley, eds., "Johnson's Gardeners' Dictionary and Cultural Instructor," 2nd ed., London, Routledge, circa 1917.

Friedman, H., "The Birds of North and Middle America," Washington, U.S. Government Printing Office, 1901.

Gadow, H., "Amphibia and Reptiles," in Harmer, S. F. and A. E. Shipley, "The Cambridge Natural History," London, Macmillan, 1901.

Gamble, F. W. et al., "Flatworms and Mesozoa, Nemertines, Rotifers, Thread-worms, Sagitta, Gephyrea and Phoronis, Earthworms, and Leeches, Polychaete Worms, Polyzoa," in Harmer, S. F. and A. E. Shipley, eds., "The Cambridge Natural History [reprint], New York, Hafner, 1959.

Gibson-Hill, C. A., "An Annotated Checklist of the Birds of Malaya," Singapore, Government Printing Office, 1949.

Giese, A. C., "Cell Physiology," Philadelphia, Saunders, 1957.

Goodrich, E. S., "Studies on the Structure and Development of Vertebrates," 2 vols. [reprint] New York, Dover, 1958.

Gove, P. B., "Webster's Third New International Dictionary," Springfield, Mass., Merriam, 1964.

Gray, P. ed., "Encyclopedia of the Biological Sciences," New York, Reinhold, 1961.

Gruneberg, H., "The Genetics of the Mouse," 2nd ed. The Hague, Martinus Nijhoff, 1952.

Ham, A. W., "Histology," 3rd ed., Philadelphia, Lippincott, 1957.

Hanson, H. C., "Dictionary of Ecology," New York, Philosophical Library, 1962.

Harmer, S. F., et al., "Hemichordata, Ascidians and Amphioxus, Fishes," in Harmer, S. F. and A. E. Shipley, eds., "The Cambridge Natural History," [reprint], New York, Hafner, 1958.

Hart, H. C., "Scriptural Natural History. II. Animals Mentioned in the Bible," London, Religious Tract Society, 1888.

Hartog, M., et al., "Protozoa, Porifera (Sponges), Coelenterata, Ctenophora, Echinodermata" in Harmer, S. F. and A. E. Shipley, eds., "The Cambridge Natural History," London, Macmillan, 1906.

Herald, E. S., "Living Fishes of the World," New York, Doubleday, 1961.

Hervey, G. F. and J. Hems, "Freshwater Tropical Aquarium Fish," London, Spring Books, 1963.

Hilgendorf, F. W., "Weeds of New Zealand," 6th ed. (revised by J. W. Calder), Wellington, N.Z., Whitcombe and Tombs, 1960.

Hingston, R. W. G., "A Naturalist in Malaya," London, Witherby, 1920.

Hoerr, N. L. and A. Osol, eds., "Blakiston's New Gould Medical Dictionary," 2nd ed., New York, McGraw-Hill, 1956.

Holland, W. J., "The Moth Book," New York, Doubleday, Doran, 1941.

Hough, J. N., "Scientific Terminology," New York, Rinehart, 1958.

Hutton, F. W. and J. Drummond, "The Animals of New Zealand," Christchurch, N.Z., Whitcombe and Tombs, 1904.

Hunt, L. M. and D. G. Groves, eds., "A Glossary of Ocean Science and Undersea Technology Terms," Arlington, Compass, 1965.

Hyman, L. H., "The Invertebrates," Acanthocephala, Aschelminthes and Entoprocta," New York, McGraw-Hill, 1951.

Hyman, L. H., "The Invertebrates: Echinodermata," New York, McGraw-Hill, 1955.

Hyman, L. H., "The Invertebrates: Platyhelminthes and Rhynchocoela," New York, McGraw-Hill, 1951.

Hyman, L. H., "The Invertebrates: Protozoa through Ctenophora," New York, McGraw-Hill, 1950.

Hyman, L. H., "The Invertebrates: Smaller Coelomate Groups," New York, McGraw-Hill, 1959.

Innes, W. T., "Exotic Aquarium Fishes," Philadelphia, Innes, 1935.

Isemonger, R. M., "Snakes of Africa," Cape Town, Nelson, 1962.

Jackson, B. D., "A Glossary of Botanic Terms," 4th ed., New York, Hafner, 1953.

Jacobs, M. B., *et al.*, "Dictionary of Microbiology," New York, van Nostrand, 1957.

Jaques, H. E., "How to Know the Beetles," Dubuque, Iowa, Brown, 1951.

Jenkins, J. T., "The Fishes of the British Isles," London, Warne, 1925.

Johannsen, O. A. and F. H. Butt, "Embryology of Insects and Myriapods," New York, McGraw-Hill, 1941.

Johns, C. A., "Flowers of the Field," 26th ed., London, Society for Promotion of Christian Knowledge, 1889.

Jollie, M., "Chordate Morphology," New York, Reinhold, 1962.

Kelsey, H. P. and W. A. Dayton, "Standardized Plant Names," Harrisburg, Pa., McFarland, 1942.

Kenneth, J. H., ed., "A Dictionary of Scientific Terms," 7th ed., New York, Van Nostrand, 1960.

Kingsley, J. S., "Outlines of Comparative Anatomy of Vertebrates," 3rd. ed., Philadelphia, Blakiston, 1926.

Klots, A. B., "A Field Guide to the Butterflies," Boston, Houghton Mifflin, 1951.

Knight, R. L., ed. "Dictionary of Genetics," Waltham, Chronica Botanica, 1948.

Langley, L. L. and E. Cheraskin, "A Physiology of Man," 3rd., ed., New York, Reinhold, 1965.

Lawrence, G. H. M., "Taxonomy of Vascular Plants," New York, MacMillan, 1951.

Leftwich, A. W., "A Dictionary of Zoology," Princeton, N.J., Van Nostrand, 1963.

Lutz, F. E., "Field Book of Insects," New York, Putnam, 1918.

Lydekker, R., "A Handbook to the Marsupialia and Monotremata," London, Allen, 1894.

Lyons, A. B., "Plant Names Scientific and Popular," Detroit, Nelson, Baker, 1900.

MacBride, E. W., "Invertebrata," *in* Heape, W., "Textbook of Embryology" London, Macmillan, 1914.

Macfadyen, A., "Animal Ecology," London, Pitman, 1957.

MacKinnon, D. L. and R. S. J. Hawes, "An Introduction to the Study of Protozoa," Oxford, Clarendon, 1961.

McNicoll, D. H., "Dictionary of Natural History Terms," London, Lovell Reeve, 1863.

Mackworth-Praed, C. W. and C. H. B. Grant, "Birds of Eastern and North Eastern Africa," 2 vols. London, Longmans Green, 1952 and 1955.

Mackworth-Praed, C. W. and C. H. B. Grant, "Birds of the Southern Third of Africa," 2 vols., London, Longmans Green, 1962 and 1963.

Maximow, A. A. and W. Bloom, "A Textbook of Histology," 5th. ed., Philadelphia, Saunders, 1958.

Melander, A. L., "Source Book of Biological Terms," New York, The City College, 1937.

Miall, L. C., "The Natural History of Aquatic Insects," London, Macmillan, 1895.

Milne, C., "A Botanical Dictionary or, Elements of Systematic and Philosophical Botany," 3rd ed., London, Paternoster-Row, 1805.

Montagu, G., "Ornithological Dictionary of British Birds," 2nd ed. by J. Rennie, London, Hurst, Chance, 1831.

Moore, L. B. and N. M. Adams, "Plants of the New Zealand Coast," Aukland, Paul's Book Arcade, 1963.

Morris, P. A., "A Field Guild to the Shells," Boston, Houghton-Mifflin, 1960.

Morris, P. A., "A Field Guide to Shells of the Pacific Coast and Hawaii," Boston, Houghton-Mifflin, 1960.

Murray, J. A. H., "New English Dictionary," 10 vols. and suppl., Oxford, Clarendon, 1888–1933.

Needham, J., "A History of Embryology," 2nd ed., New York, Abelard Schumann, 1959.

Neilson, W. A., ed., "Websters New International Dictionary," 2nd ed., Springfield, Mass., Merriam, 1961.

Newton, A., "A Dictionary of Birds," London, Black, 1893–1896.

Nordenskiold, E., "The History of Biology," [trans. L. B. Eyre] New York, Tudor, 1949.

Nybakken, O. E., "Greek and Latin in Scientific Terminology," Iowa, State College Press, 1959.

Odum, E. P., "Fundamentals of Ecology," 2nd ed., Philadelphia, Saunders, 1959.

Ornithological Society of New Zealand, "Check List of New Zealand Birds," Wellington, N.Z., Reed, 1953.

Palseneer, P., "Mollusca" *in* Lankester, E. R., ed., "A Treatise on Zoology," London, Black, 1906.

Pennak, R. W., "Collegiate Dictionary of Zoology," New York, Ronald, 1964.

Pennak, R. W., "Fresh-Water Invertebrates of the United States," New York, Ronald, 1953.

Peters, J. A., "Dictionary of Herpetology," New York, Hafner, 1964.

Peterson, R. T., "Field Guide to the Birds," 2nd ed., Boston, Houghton-Mifflin, 1958.

Peterson, R. T., "A Field Guide to Western Birds," Cambridge, Mass., Houghton-Mifflin, 1961.

Petrides, G. A., "A Field Guide to Trees and Shrubs," Boston, Houghton-Mifflin, 1958.

Pratt, H. S., "A Manual of the Common Invertebrate Animals," Philadelphia, Blakiston, 1948.

Reid, G. K., "Ecology of Inland Waters and Estuaries," New York, Reinhold, 1961.

Roberts, A., ed., "Our South African Birds," Cape Town, United Tobacco Co., 1941.

Rose, A. and E. Rose, eds., "The Condensed Chemical Dictionary," 6th ed., New York, Reinhold, 1961.

Ross, H. H., "A Textbook of Entomology," New York, Wiley, 1961.

Sanderson, I. T., "Living Mammals of the World," Garden City, N.Y., Doubleday, 1957.

deSchauenses, R. M., "The Species of Birds of South

America, with their Distribution," Naberth, Pa., Livingston, 1966.

Schmidt, K. P. and R. F. Inger, "Living Reptiles of the World," Garden City, N.Y., Doubleday, 1957.

Scott, P., "A Colored Key to the Wildfowl of the World," London, The Wildfowl Trust, 1957.

Sedgwick, A., "A Student's Textbook of Zoology," 2 vols., London, Sonnenschein, 1905.

Sedgwick, A. *et al.*, "Peripatus, Myriapods, Insects, Part I," *in* Harmer, S. F. and A. E. Shipley, eds., "The Cambridge Natural History," London, Macmillan, 1922.

Serventy, D. L. and H. M. Whittell, "Birds of Western Australia," 3rd ed., Perth, Western Australia, Paterson Brokensha, 1962.

Sharp, D., "Insects, Part II," *in* Harmer, S. F. and A. E. Shipley, eds., "The Cambridge Natural History," London, Macmillan, 1901.

Slijper, E. J., "Whales," [*trans.* A. J. Pomerans], New York, Basic Books, 1962.

Smith, A. H., "The Mushroom Hunter's Field Guide," Ann Arbor, University of Michigan Press, 1963.

Smith, G. *et al.*, "Crustacea, Trilobites, Introduction to Arachnida, and King-Crabs, Eurypterida, Scorpions, Spiders, Mites, Ticks, Etc., Tardigrada (Water-Bears), Pentastomida, Pycnogonida," *in* Harmer, S. F. and A. E. Shipley, eds., "The Cambridge Natural History," Macmillan, 1923.

Smith, G. M., "Cryptogamic Botany, vol 1, Algae and Fungi," 2nd ed., New York, McGraw-Hill, 1955.

Snow, C. H., "The Principal Species of Wood; their Characteristic Properties," New York, Wiley, 1903.

South, R., "Butterflies of the British Isles," London, Warne, 1906.

Southern, H. N., "Handbook of British Mammals," Oxford, Blackwell, 1964.

Srb, A. M. and R. D. Owen, "General Genetics," San Francisco, Freeman, 1953.

Steen, E. B., "Medical Abbreviations," Philadelphia, Davis, 1960.

Stein, J., "The Random House Dictionary of the English Language," New York, Random House, 1966.

Step, E., "Animal Life of the British Isles," London, Warne, 1921.

Stern, C., "Principles of Human Genetics," 2nd ed., San Francisco, Freeman, 1949.

Swain, R. B., "The Insect Guide," Garden City, N.Y., Doubleday, 1948.

Swanepoel, D. A., "Butterflies of South Africa," Capetown, Maskew, Miller, 1953.

Swanson, C. P., "Cytology and Cytogenetics," Englewood Cliffs, N.J., Prentice-Hall, 1957.

Titmuss, F. H., "A Concise Encyclopedia of World Timbers," New York, The Philosophical Library, 1959.

de la Torre-Bueno, J. R., "A Glossary of Entomology," Brooklyn, N.Y., Brooklyn Entomological Society, 1937.

Tuxen, S. L., *ed.*, "Taxonomists Glossary of Genitalia in Insects," Copenhagen, Munksgaard, 1956.

Tweny, C. F. and L. E. C. Huges, eds., "Chamber's Technical Dictionary," 3rd ed., New York, Macmillan, 1961.

Van Tyne, J. and A. J. Berger, "Fundamentals of Ornithology," New York, Wiley, 1961.

Vaurie, C., "The Birds of the Palaearctic Fauna," 2 vols., London, Witherby, 1959 and 1965.

Ward, H. B. and G. C. Whipple, "Fresh-Water Biology," New York, Wiley, 1918.

Weaver, J. E. and F. E. Clements, "Plant Ecology," 2nd ed., New York, McGraw-Hill, 1938.

Webster, N., *see either* Cove, P. B. *or* Neilson, W. A.

Wheeler, W. M., "Ants," New York, Columbia University Press, 1910.

Wilson, G. B. and J. H. Morrison, "Cytology," New York, Reinhold, 1961.

Winter-Blythe, M. A., "Butterflies of the Indian Region," Bombay, The Natural History Society, 1957.

Woods, R. S., ed., "The Naturalist's Lexicon," Pasadena, Abbey Garden Press, 1944.

Woods, R. S., "Addenda to the Naturalist's Lexicon," Pasadena, Abbey Garden Press, 1947.

Yancey, P. H., "Origins from Mythology of Biological Names and Terms," rev. ed., reprinted from B.I.O.S.: vol, XVI, 1952.

Yarrell, W., "A History of British Birds," 3 vols., London, Van Voorst, 1843.

Young, J. Z., "The Life of Vertebrates," 2nd ed., New York, Oxford University Press, 1962.